ANNUAL REVIEW OF BIOCHEMISTRY

ANNUAL REVIEW OF BIOCHEMISTRY

ESMOND E. SNELL, *Editor*
University of California, Berkeley

PAUL D. BOYER, *Associate Editor*
University of California, Los Angeles

ALTON MEISTER, *Associate Editor*
Cornell University Medical College

CHARLES C. RICHARDSON, *Associate Editor*
Harvard Medical School

VOLUME 43

1974

ANNUAL REVIEWS INC.   4139 EL CAMINO WAY   PALO ALTO, CALIFORNIA 94306

ANNUAL REVIEWS INC.
Palo Alto, California, USA

International Standard Book Number 0-8243-0843-3
Library of Congress Catalog Card Number 32-25093

Assistant Editors	Carol Reitz
	Sharon Hawkes
Indexers	Mary Glass
	Susan Tinker
Subject Indexer	Beryl Glitz

FILMSET BY TYPESETTING SERVICES LTD, GLASGOW, SCOTLAND
PRINTED AND BOUND IN THE UNITED STATES OF AMERICA

CONTENTS

vi

RECOLLECTIONS

George W. Beadle
Professor of Biology, University of Chicago, Chicago, Illinois

In these exciting times when elementary and high schools teach modern biology, including many of the intricacies of biochemical genetics, the long slow process by which our present knowledge in this area was gained is not often fully appreciated. A third of a century elapsed before Mendel's work was "rediscovered" and properly appreciated. Archibald E. Garrod's (1) prophetic appreciation of the relation of genetics and biochemistry, beginning soon after the so-called rediscovery of Mendel, lay fallow for more than forty years despite the fact that he published widely and relatively voluminously. As late as a quarter of a century after the Mendel work came to light, Harvard's distinguished professor of biology, William Morton Wheeler (2), ridiculed genetics as a small bud on the great tree of biology, a bud so constricted at the base as to suggest its eventual abortion. Wheeler's colleague in paleobotany, Jeffrey (3), also expressed his disbelief in the work of the then flourishing school of *Drosophila* genetics. Fortunately, neither succeeded in significantly retarding the rapid advances then being made, many of them by two Harvard contemporaries, Edward M. East and W. E. Castle. It is of interest to note that Thomas Hunt Morgan (4) remained a skeptic about Mendelian interpretations for the first ten years after the rediscovery, that is until he established the sex-linked nature of the white eye trait in *Drosophila*.

Now that genetics is widely accepted as one of the most basic aspects of all biology, it is perhaps of interest that some of us old enough to have participated in or otherwise to know something of the history of present day genetics now record our recollections. I attempt to do so in the limited area of biochemical genetics of which I have had a small part.

Of the myriads of environmental influences large and small that have to do with the course of one's life, few are likely to be long remembered with any degree of clarity or confidence. Yet behavioral scientists are increasingly aware that what happens early in life can be of the greatest significance in later years. Unfortunately when one attempts to recall such thoughts and events as have influenced later attitudes and behavior, the uncertainties are many. Thus it is with a good deal of doubt and temerity that I attempt to record events influential in that part of my life that has had to do with biochemical genetics.

I was born in 1903 of parents who owned and operated a 40-acre farm near the small town of Wahoo, Nebraska. Both had grown up in similarly small

1

communities: father in Kendallville, Indiana and mother in Galva, Illinois. Both were inherently intelligent but limited to high school in formal education.

Because of its small size our farm was highly diversified, with field crops such as alfalfa, potatoes, and corn; truck crops including asparagus and strawberries for market; plus cattle, horses, hogs, and chickens. All of these were supplemented by retail selling of produce, including out-of-state apples and potatoes purchased in carload lots. In this and other ways I was intimately involved in matters of biological significance. We kept rabbits, ferrets, bees, cats, dogs, and for a time, a pet coyote. Hunting, fishing, and trapping were enjoyable pastimes. With these plus routine chores and farm work, life was never dull.

Mother died when I was four and a half. My older brother, a younger sister, and I were in part raised by a series of housekeepers, some very good, some poor, and one or two terrible.

My earliest years of formal school were in a genuine little red, one-teacher, wooden schoolhouse in town, which was a mile and a half from home. During my twelve years in this and other local schools I was exposed to perhaps a dozen teachers. Like our housekeepers, they were a thoroughly mixed lot.

With the accidental death of my older brother it was tacitly assumed I would eventually take over the family farm, a prospect I looked forward to with a certain amount of confidence and pleasure. But neither father nor I had reckoned with a young high school teacher of physics and chemistry, Bess MacDonald. She did not pretend to be, nor was she, a profound authority in either of the subjects she taught. But she did have a remarkable knack of interesting us, for example, in chemistry by challenging us with unknowns to identify by classical qualitative methods. But more than that, she took a personal interest in our aspirations and hopes.

I spent many nonschool hours with her at her home, during which she convinced me I should go on to college, even though I might eventually return to the farm. My psychological insight is not sufficient to describe our rather unusual relationship. Perhaps for me she was a kind of mother-substitute.

Father was not keen on the college idea, being convinced that a farmer did not need all that education. But determination won and I enrolled at the University of Nebraska College of Agriculture, fully intending to return to the farm. Had it not been tuition-free with an opportunity to work for living expenses, I doubt if I could have managed.

Again my plans were modified by teachers. In my first year I was so impressed by a required course in English that I thought to follow it up. In fact I was offered part-time employment reading student papers during my second year. Fortunately for English, the professor went off to Palestine to study the literature of the Bible. His successor did not pick up the commitment.

In rapid succession thereafter I became enamored of entomology, ecology, and genetics. I was happy with general and organic chemistry and did well in both, but was not carried away to the point of proposing to major in that general area.

After my second year I was given a summer job classifying genetic traits in a wheat hybrid population, this for Professor Keim of the Agronomy Department. In my spare time I read about genetics and found my interest increasing markedly.

I was given other assignments including laboratory instruction in an agricultural high school program given by the College at that time. I read student papers and examinations in the elementary genetics course. I had charge of a laboratory supply department set up to provide samples of crop plants and other materials for instruction in high schools that gave courses in agriculture. During summer periods I grew various exotic crop plants for this purpose, collected and mounted representative weed seeds, made up orders, mailed them out, and kept records. In my senior year I worked on a special problem on root development and survival of fall-seeded grasses of economic importance. I also devised a key for the identification by vegetative characters of local native grasses.

Keim was a remarkable person in many ways. He did not profess to be a great scholar. But he had an uncanny ability to size students up and encourage them, which he did with a kind of understanding I have never been able fully to fathom. Some he sent back to the farm, some to be county agricultural agents, others to teach high school, and a few to go on to graduate school. I've known half a dozen or more of the latter and have never known one to be a misfit.

The nearest he ever came to an error of judgment that I know of was in getting me a teaching assistantship at Cornell and admission to graduate school to work on the ecology of the pasture grasses of New York State. It might not have been a mistake if I had seen eye-to-eye with the professor who was to sponsor my thesis research. But I did not, and soon resigned my teaching assistantship to work in genetics and cytology with Professor R. A. Emerson. That was 1926. Shortly thereafter he gave me a part-time research assistantship.

This surely was one of the best things that ever happened to me. Emerson was the perfect employer, graduate advisor, and friend. He turned problems over to me. One of my special assignments was to complete a summary of all genetic linkage studies in maize up to that time (5). I had half time for course work and for my own thesis research.

These were indeed exciting times for all of us working with Emerson. He was the outstanding plant geneticist of his time and was a tremendously stimulating person to work with and under, and his group of graduate students at the time were outstanding. They included George F. Sprague, Marcus Rhoades, Barbara McClintock, H. W. Li, and perhaps a half dozen others.

Emerson's contributions to genetics came at a time when support for the new science was minimal and the doubters many. He moved from the University of Nebraska College of Agriculture to Cornell University in 1914, in part because he felt his work was judged by the Nebraska authorities to be too theoretical ever to be useful agriculturally. Thus it is of considerable interest to note that in addition to his remarkable work in basic genetics, which of course indirectly but significantly furthered the art and science of plant breeding, Emerson conscientiously assumed direct responsibility for more than his fair share of plant breeding. By genetically transferring resistance to the disease anthracnose to commercially desirable dry beans, he saved the important bean industry of New York State from utter collapse. He also succeeded in transferring disease resistance to commercially grown cantaloupes.

Emerson was one of the first of the early American workers fully to appreciate the work of Mendel, this at a time when even T. H. Morgan was still a skeptic. As I pointed out in 1960 (6), Emerson never published until he had extracted the truth from his experimental material and verified it not once but many times in many ways. Predecessors had studied the inheritance of plant and aleurone colors in corn and had been distracted by incidental modifying factors and apparent inconsistencies with Mendelian principles to the point that some of them actually renounced those principles. It was Emerson's persistence, clear thinking, and hard-headed checking of facts that established the truth and showed beyond doubt that these apparently complex systems of inheritance in reality have an understandable genetic basis. His papers on kernel and plant color inheritance in maize are outstanding as solid experimental work, sound reasoning, and clear presentation. His early studies of quantitative characters, carried on in part through collaboration with E. M. East of Harvard, importantly influenced genetic thinking. His work on variegated pericarp led to the concept of unstable genes, another significant milestone in the history of genetics.

Important as were his own scientific contributions, in many ways it is Emerson the man most vividly remembered by those privileged to know him well. He was cordial in his relations with his friends and colleagues. The contagious enthusiasm and zest, so clearly displayed in his scientific work, were extended to other activities, bowling and hunting for example. During corn season he was first in the experimental garden and among the last to leave, an example that no doubt increased the productiveness of all who worked with him. Bag lunches eaten in the shade of the garden shed during these periods of intense field activity were of special interest to students and other associates. It was there that the unpublished lore of corn genetics and geneticists was most likely to be recalled. It was also a setting in which Emerson became best known to his students. It was also in such informal ways that he did much of his teaching. He was freely available to students but it was his policy that they come at their own instigation. At all times he was willing to be helpful but he did not direct student research in any formal manner.

Emerson's research materials were freely available, not only to his own colleagues and students but as well to investigators elsewhere. This generosity played an important part in making corn the best known of all higher plants from a genetic point of view and had the effect of interesting investigators throughout the world as well as significantly increasing general genetic understanding.

With the growth of the corn group the system of communicating unpublished information through conversation became inadequate. During 1932 at the International Genetics Congress at Ithaca a "corn meeting" was held where it was decided that a central clearinghouse of information and seed stocks would be established at Cornell. Out of this there evolved a series of mimeographed "corn news letter" edited by Marcus Rhoades and sent to all interested corn geneticists. Later this became the *Maize Genetics Cooperation News Letter,* a somewhat more formal organization for the dissemination of information not published in formal journals and for recording seed lines available for research.

One of the groups of mutant types being worked on from the earliest days of

Emerson's research were those affecting chlorophyll synthesis and function. I vividly recollect Emerson's attempts to interest plant physiologists and biochemists working on photosynthesis in making use of such mutants as tools in fathoming the physiology and biochemistry of chlorophyll structure and function. None responded, otherwise biochemical genetics might have moved forward more rapidly.

In my own attempts to improve my understanding of chemistry in relation to genetics, I audited courses in physical chemistry and biochemistry. The latter was given by James B. Sumner and it was during this period that he first crystallized the enzyme urease from the jack bean. Biochemists will recall the long lag between this accomplishment, and acceptance and confirmation of it as authentic and thus a significant forward step.

The time was clearly ripe for the new discipline of biochemical genetics. But few biochemists or geneticists were then intellectually or psychologically prepared, despite the fact that Archibald E. Garrod had a quarter of a century earlier clearly suggested a one-to-one relation between gene action and enzyme activity, and had published both repeatedly and voluminously (7).

My own graduate research in cytogenetics was both rewarding and significant. In part I worked on the genetic control of meiosis using corn lines in which chromosome behavior was markedly modified genetically. The asynaptic mutant was the first, polymitotic a second, and sticky chromosomes a third. I also worked closely with Emerson on the relation of corn to its nearest wild relative, a Mexican plant known as teosinte, this, incidentally, a relationship still not fully resolved and which I am now again actively investigating.

A significant turning point in my career came in 1931 with the completion of my graduate work. I had hoped to be awarded a National Research Council Fellowship to continue my corn cytogenetics work at Cornell, by far the best place to continue in terms of facilities and associates. But the wise chairman of the Fellowship Board, Charles E. Allen of the University of Wisconsin, intervened, pointing out that remaining for postdoctoral work in the same institution in which one took his PhD degree was in principle less desirable than moving to another institution where, other things being equal, new experiences and insights were more likely to be acquired. He said he would approve the award if I would accept my second choice as a place to continue. That was the California Institute of Technology where Thomas Hunt Morgan had recently moved from Columbia to establish a new Division of the Biological Sciences. Emerson approved and I concurred, little realizing at the time that this would be another best thing that ever happened to me.

Caltech biology was indeed tremendously stimulating. Among those who were there in genetics and related areas when I arrived as a research fellow were Morgan, Sturtevant, Bridges, Dobzhansky, Schultz, Anderson, Emerson (son of R. A. Emerson), Belar, and the Lindegrens. Darlington, Haldane, and Karpechenko spent time there as visiting scholars.

General enthusiasm was at a high level and persons in other fields were caught up in it. Linus Pauling took a personal interest in genetic crossing over. R. A. Millikan delighted in escorting visitors to Biology where he could give a masterly account of *Drosophila* investigations. Charles Lauritsen and associates were building a million volt X-ray tube which became available for medical and biological use.

At first I concentrated on my corn cytogenetics program but soon became actively interested in *Drosophila,* then by far the most favorable organism for genetic study. I worked with Dobzhansky, Emerson, and Sturtevant at various times on genetic recombination in the hope that this would tell us significantly more about the nature of the gene. It didn't do as much as we had hoped, though decades later it became clear that if we had really learned enough about recombination a good deal more about the nature of the gene could have been revealed.

An additional and significant turning point in my career began with the arrival in 1933–1934 of Boris Ephrussi from Paris as a Rockefeller Foundation Fellow. He was actively interested in tissue culture and tissue transplantation as a means of learning more about gene action.

We spent long hours discussing the curious situation that the two great bodies of biological knowledge, genetics and embryology, which were obviously intimately interrelated in development, had never been brought together in any revealing way. An obvious difficulty was that the most favorable organisms for genetics, *Drosophila* as a prime example, were not well suited for embryological study, and the classical objects of embryological study, sea urchins and frogs as examples, were not easily investigated genetically.

What might we do about it? There were two obvious approaches: one to learn more about the genetics of an embryologically favorable organism, the other to better understand the development of *Drosophila.* We resolved to gamble up to a year of our lives on the latter approach, this in Ephrussi's laboratory in Paris which was admirably equipped for tissue culture, tissue or organ transplantation, and related techniques.

Morgan arranged to continue my Caltech salary, then $1500 annually, which was 33% less than the previous year because of the great depression. Only years later did I find that this stipend was almost surely provided by Morgan personally. Caltech was in dire financial straits at that time and though Morgan was extremely frugal with Institute funds, he remained always generous in personally supporting causes he thought worthy. Leaving a wife and small son in Pasadena where living costs were unbelievably low at that time, I went to Paris to work with Ephrussi. Fortunately living costs were also very modest there, provided one could do with bare necessities. My daily subsistence expenses, room and food, were approximately two dollars.

In Ephrussi's laboratory we tried tissue culture without remarkable success or promise. We switched to *Drosophila* larval embryonic bud transplantation which turned out to be successful despite assurances from the Sorbonne's great authority on the metamorphosis of the blow fly that we could not succeed.

We knew from Sturtevant's work on naturally occurring mosaic flies that the character vermilion eye (absence of brown component of the two normal eye pigments) was nonautonomous in the sense that if one eye and a small part of the adjacent tissue were vermilion and the remainder wild type, the genetically vermilion eye would produce both pigment components. Obviously an essential part of the brown pigment system was produced outside the eye and could move to it during development. We confirmed this by transplanting genetically vermilion embryonic eye buds in the larval stage to wild-type host larvae. Although it was thought a priori by some to be extremely difficult if not impossible, a technique for doing this was devised. It involved two people working cooperatively through paired binocular dissecting microscopes focussed on one recipient larvae.

We confirmed the existence of a diffusable substance which we called vermilion-plus substance (8). A second mutant lacking brown eye pigment was found to behave similarly—the so-called cinnabar character. Reciprocal transplants between the two mutants lacking brown pigment showed that there were two substances involved, one a precursor of the second. We postulated that one gene was immediately concerned with the final chemical reaction in the formation of substance 1 and the second with its conversion to substance 2.

We investigated the twenty some other eye-color mutants then known in *Drosophila* and found just these two in direct control of the two postulated chemical reactions (9).

Since most biologically significant reactions are enzymatically catalyzed, we assumed the two eye-color genes, cinnabar and vermilion, directly controlled the two postulated enzymes. This was the origin in our minds of the one gene/one enzyme concept, although at that time we did not so designate it.

In formulating this interpretation we were much encouraged by the previous related work of Caspari and others (10) on related pigmentation in the meal moth *Ephestia* and also the work of Scott-Moncrieff (11) and earlier workers on the genetic control of anthocyanin pigments in higher plants.

It is of interest and I believe of some significance that Jaques Monod, then an instructor at the Sorbonne, took a keen interest in our work and spent a good share of his spare time in Ephrussi's laboratory following progress and discussing results with us. Later when Ephrussi returned to Caltech for a year where we continued our collaboration, Monod also came as a visiting investigator.

An obviously important next step was the identification of the two brown pigment precursors. Ephrussi and Khouvine worked on this aspect of the problem in Paris and I at Harvard with Kenneth Thimann and later at Stanford University with Tatum and Clarence Clancy. Tatum demonstrated a functional relation of one of the precursors to tryptophane and he and Haagen-Smit at Caltech came close to identifying it (12).

Butenandt, Weidel & Becker (13) in Germany took up the search and were able to identify the so-called vermilion-plus substance by trying then known relatives of tryptophane; it was kynurenine.

As an interesting sidelight, kynurenine had been isolated and identified years before by Clarence Berg of the University of Iowa, son of a Wahoo harness-maker who had lived only a few miles from the Beadle farm. Had we only known, we could have got kynurenine from him.

At about this stage in our work Tatum's father, then a pharmacologist at the University of Wisconsin, came to Stanford on a family visit. One day as he was visiting our laboratory he called me aside to tell me that he was concerned about the professional future of his son. "Here you have him in a position in which he is neither a pure biochemist nor a bona fide geneticist. I'm very much afraid he will find no appropriate opportunity in either area." I recall my response very clearly: "Professor Tatum, do not worry, it is going to be all right."

I recall another episode illustrating the doubts held as to the future of the new hybrid approach. We had tried earlier to interest in joining our group a young biochemist at Columbia University who had been recommended by Professor Hans Clarke. He declined because the three part-time positions he then held were financially somewhat more rewarding than the one for which we were responsible. On again meeting him more than two decades later he told me he had many times regretted not seeing more clearly the opportunities in biochemical genetics.

At about this time, 1940–1941, Tatum gave a course at Stanford on comparative biochemistry. Auditing his lecture one day it suddenly occurred to me that there was a much easier approach than we had been following for identifying genes with known chemical reactions. If, as we believed, all enzymatically catalyzed reactions were gene controlled in a one-to-one relation, it would obviously be much less time consuming to discover additional such relations by finding mutant organisms which had lost the ability to carry out specific chemical reactions already known or postulated. For two reasons the obvious organism to use for such an approach was the red bread mold *Neurospora*. First, its cytogenetics had already been worked out by the mycologist B. O. Dodge (14), whom I had earlier met at Cornell University, and the Carl Lindegrens whom I knew from my early years at Caltech. Second, we knew from the work of Nils Fries (15) in Sweden that many filamentous fungi not too distantly related to *Neurospora* could grow on chemically defined media containing a proper balance of inorganic salts, a source of carbon and energy such as a sugar, plus one or more known vitamins.

So why not determine the minimal nutritional requirements of *Neurospora,* produce mutant types by X or ultraviolet irradiation and then test these for loss of ability to synthesize one or more components of the minimal medium? We soon found the minimal medium to consist of simple inorganic compounds, a suitable carbon and energy source such as sucrose, plus the one vitamin biotin. That was 1941 and fortunately biotin had just become commercially available as a concentrate sufficiently free of amino acids and other vitamins to serve our purpose.

The 299th culture from a single ascospore, whose parent culture had been X rayed, proved not to grow on minimal medium but did so with added Vitamin B_6. It was then a simple matter to determine that a genetic unit, presumably a single gene, had been mutated by crossing the mutant strain grown on a supplemented culture medium with the original strain of the appropriate mating type and then testing cultures from the eight single spores derived from a single meiotic event. Our test showed that four such cultures required Vitamin B_6 while four did not, indicating change in a single genetic unit (16).

Could we produce more such mutant types with other requirements? The answer was yes, for other vitamins and for various essential amino acids. In sequences of biosynthetic reactions leading to a given endproduct we could identify genes for individual steps, in general one gene and one only for a specific biosynthetic step.

In addition to biochemical mutants, which for the most part are normal morphologically when grown on properly supplemented media, a variety of morphologically altered types were found, some quite bizarre in appearance. During the time we were accumulating these along with scores of nutritionally altered mutants Doctor Charles Thom, a widely recognized authority on fungi, especially of the genus *Penicillium* and related genera, paid us a visit. As we toured the laboratories he was obviously keenly interested but made few comments. After we had demonstrated a fair sample of the work under way and a number of morphologically diverse mutant types, Doctor Thom called me aside and said "You know what you need here?"

"What," I asked.

"A good mycologist," was the answer. "Those cultures you call mutants are not mutants at all. They are contaminants."

To the question of how, when crossed with the original type, they could segregate according to established Mendelian principles he had no answer. I'm sure he left convinced we were the most inept mycologists he had ever seen. He had never been an ardent admirer of genetics and we obviously failed to influence him in that regard.

At this stage of our investigations it was obvious that we could increase our rate of progress significantly by supplementing our research personnel. We were fortunate in obtaining the additional financial help needed for this from the Rockefeller Foundation which, through grants to the Stanford Biology group, had made the initial work possible. Herschel K. Mitchell, Norman H. Horowitz, David M. Bonner, Francis Ryan, Mary Houlahan, and others joined the team. Through C. Glen King of the Nutrition Foundation we received support for graduate students including Adrian M. Srb, August Doermann, David Regnery, Frank C. Hungate, Taine T. Bell, and Verna Coonradt.

Although the Research Corporation did not support our work financially, its officers gave us much appreciated encouragement in the following way: The Rockefeller Foundation had earlier made a $200,000 grant to the C. V. Taylor group of biologists at Stanford, of which Tatum and I were members. Knowing Taylor's persistence, persuasiveness, and ambition for his group, the officers of the Foundation had placed a condition on the grant, namely that he not apply for additional funds from the Rockefeller Foundation during a following ten year period. I of course knew of this, and thus inquired of Frank Blair Hanson of the Rockefeller Foundation if there was any objection to our applying to the Research Corporation for supplemental support of our special project. There was not, so on that same day I approached the Research Corporation and was told they would provide the needed $10,000. Just as the details of how formally to apply were being discussed a telephone call came to me from Hanson of the Rockefeller Foundation, saying that they had reconsidered our special situation and felt that since they had provided initial support they thought it appropriate to provide the requested supplement. On reporting this to the Research Corporation officers, I was immediately told it was right and proper that the Rockefeller Foundation should continue the support, but that if we would send them a carbon copy of our formal request they would agree to provide the $10,000 if the Rockefeller Foundation for any reason did not do so. That is the kind of confidence that really inspires a research team. The Rockefeller Foundation did make the grant, but it was only years later that I learned it had been Warren Weaver who had recommended that the exception be made. His record of judging projects that paid off, scientifically speaking, was one of remarkable success and I have always been grateful that we did it no serious damage.

During this visit to the Research Corporation R. E. Waterman, who had worked earlier with R. R. Williams in isolating and characterizing thiamine, pointed out to me that G. W. Kidder of Amherst was working with a protozoan and had obtained results very much like ours. He added that Doctor Williams knew the details, and that I would have a good chance of seeing him if I were to hurry over to the 42nd Street Airlines Terminal where he was waiting for an airport limousine. I did find him and was led to believe Kidder indeed had results very much like ours in *Neurospora*. We were of course anxious to learn more about it. On doing so we found that the work that had so understandably impressed Williams had to do with special cultural conditions under which *Tetrahymena vorax* could synthesize thiamin (17) and was not at all designed to answer the types of questions we were asking.

By 1942 we had gone a fair way in the process of identifying genes with specific chemical reactions. Then the classical work of Garrod (7) was rediscovered, or perhaps more correctly, properly appreciated, by J. B. S. Haldane and Sewall Wright (18, 19). Back in the early part of the century, very soon after the rediscovery of Mendel's paper and the confirmation of his principles, Garrod had demonstrated that the human disease alcaptonuria was a simple Mendelian recessive trait characterized by an inability to further degrade 2,5-dihydroxyphenyl acetic acid (alcapton or homogentisic acid), a metabolic derivative of phenylalanine. Unlike their normal counterparts who further degrade alcapton, alcaptonurics excrete it in the urine where, upon exposure to air, it oxidizes to a blackish compound. Not only did Garrod correctly deduce the relation of gene to enzyme and to chemical reaction, he also used alcaptonurics to identify intermediate compounds in the sequence of reactions between phenylalanine and alcapton. In a like manner he characterized several other genetically controlled metabolic reactions in man.

On learning of this long-neglected work it was immediately clear to us that in principle we had merely rediscovered what Garrod had so clearly shown forty years before. There were three differences of significance: First, we could produce many examples. Second, our experimental organism was far better suited to both chemical and genetic investigation. Third, ours was a time far more favorable for acceptance of the obvious conclusions.

Like Mendel, Garrod was far ahead of his time, but unlike Mendel, his work was not buried in a relatively obscure journal: Garrod published in standard journals and wrote a widely distributed book, *Inborn Errors of Metabolism*, first published in 1909 with a second edition in 1923 (7). His work was well known to Bateson, the early British enthusiastic advocate of Mendelism. Bateson and his associate Punnett advised Garrod on the genetic aspects of his studies of biochemical defects in man, and Bateson's classical 1909 book, *Mendel's Principles of Heredity* (20), referred to it in some detail. For reasons most difficult to understand it then dropped out of the genetic literature until revived in 1942.

In giving a seminar on biochemical genetics at the University of California at Berkeley, in the late 1940s I pointed out that among others, Goldschmidt's 1938 book *Physiological Genetics* (21) failed to mention Garrod's contribution. Professor Goldschmidt, who was in the audience, came up after the seminar and explained that he had known of Garrod's work and could not understand how he had omitted mention of it. Clearly, like many others, he failed to appreciate its full significance, else he could not have forgotten it.

In retrospect one wonders how such important findings could be so thoroughly unappreciated and disregarded for so many years. Obviously the time was not ready for their proper appreciation. Even in 1941 when Tatum and I first reported our induced genetic-biochemical lesions in *Neurospora* few people were ready to accept what seemed to us to be a compelling conclusion, namely that in general one gene specifies the sequence of one enzyme (or polypeptide chain). In 1945 I gave a series of some two dozen Sigma Xi lectures in as many colleges and universities of the country. The skeptics were many, the converts few. Even at the time of 1951 Cold Spring Harbor Symposium on Quantitative Biology the skeptics were still many. In fact the believers I knew at the time could be counted on the fingers of one hand, despite the eloquent and persuasive additional evidence presented at that meeting by Horowitz & Leupold (22).

In speculating on the long-continued reluctance of geneticists and others to accept the simple gene/enzyme concept so clearly implied in Garrod's early work, the anthocyanin studies, and the more recent microorganism studies, Horowitz (personal communication) tells me A. H. Sturtevant had once pointed out to him that this was because of a widespread belief in the so-called pleiotropic (many effects) action of genes. In the sense that the terminal results of a single gene mutation may appear multiple, this can be said to be correct. But in terms of the primary effect of such a mutation in replacing a single amino acid in a polypeptide chain for example, it is clearly not.

With the working out of the Watson-Crick double helix structure of DNA, its method of replication, and its role in protein synthesis, the difficulty in accepting the concept of one gene/one enzyme largely disappears, for it can now be stated as one functional DNA sequence/one primary polypeptide chain.

The work I have discussed was but a small part of a prelude to the magnificent new era of biology ushered in through the elucidation of the structure of DNA two decades ago. Our knowledge of living things at the molecular level has continued to increase exponentially. In a real sense genetics has come to be recognized as an integral and basic part of all biology, of biochemistry, biophysics, immunology, virology, physiology, the behavioral sciences, plant and animal breeding, and all the rest.

Largely as a result of its advances, the opportunities and challenges have never been greater in the areas of biology. Nor have the intellectual rewards to those adequately prepared and sufficiently motivated.

In my own situation, I tried a quarter of a century ago what I thought of as an experiment in combining research in biochemical genetics with a substantial commitment to academic administration. I soon found that, unlike a number of my more versatile colleagues, I could not do justice to both. Finding it increasingly difficult to reverse the decision I had made, I saw the commitment to administration through as best I could, often wondering if I could have come near keeping up with the ever increasing demands of research had I taken the other route. My doubts increased with time.

As one bit of evidence that occasional satisfactions do accrue to academic administrators, I cite an example involving James D. Watson. On his return from the Cambridge Medical Research Council Unit shortly after he and Francis Crick had worked out the double helix structure of DNA, Watson continued research as a Senior Research Fellow at the California Institute of Technology, Division of Biology. His draft board address, however, remained Chicago, and at this time the board members concluded his deferment from military service had been sufficiently long and thereupon reclassified him 1A.

Being convinced his potential contributions to science would far outweigh anything he might do to promote the mission of the military, we set out to convince the authorities that his deferment should be continued. Successive appeals to higher and higher levels were consistently denied and the Watson file grew correspondingly thicker. Finally, through the help of the National Research Council, the appeal was carried to the highest level, the Presidential Review Board. At this level previous decisions were reversed and Watson assigned to what in Washington was facetiously referred to as "the rare bird category," a designation that seemed especially appropriate to Watson, a dedicated bird watcher.

Those of us involved were of course much pleased that our efforts had been successful. Personally I was never quite able to decide the appropriate sentiment to express to the military, condolences or congratulations.

Now, on retirement from administrative duties, I have returned to a relatively simple research project of four decades ago with Emerson, namely the origin of *Zea mays,* Indian corn. It involves a combination of genetics, ecology, archeology, biochemistry, and other related disciplines in ways I am glad to say I find intellectually and emotionally satisfying.

Literature Cited

1. Garrod, A. E. 1902. *Lancet* 2:1617
2. Wheeler, W. M. 1923. *Science* 57:61–71
3. Jeffrey, E. C. 1925. *Science* 62:3–5
4. Morgan, T. H. 1909. *Am. Breeders Assoc.* 5:365–68
5. Emerson, R. A., Beadle, G. W., Fraser, A. C. 1950. *Cornell Univ. Mem.* 180:3–83
6. Beadle, G. W. 1950. *Genetics* 35:1–3
7. Garrod, A. E. 1923. *Inborn Errors of Metabolism.* London: Hodder & Stoughton. 2nd ed. 216 pp.
8. Ephrussi, B., Beadle, G. W. 1935. *C. R. Acad. Sci.* 201:98–101
9. Beadle, G. W., Ephrussi, B. 1936. *Genetics* 21:225–47
10. Caspari, E. 1933. *Arch. Entwicklungsmech. Organ.* 130:353–81
11. Scott-Moncrieff, R. 1936. *J. Genet.* 32:117–70
12. Tatum, E. L., Haagen-Smit, A. J. 1941. *J. Biol. Chem.* 140:575–80
13. Butenandt, A., Weidel, W., Becker, E. 1942. *Naturwissenschaften* 28:63–64

14. Dodge, B. O. 1927. *J. Agr. Res.* 35: 289–305
15. Fries, N. 1938. *Symb. Bot. Upsal.* 3:1–188
16. Beadle, G. W., Tatum, E. L. 1941. *Proc. Nat. Acad. Sci. USA* 27:499–506
17. Kidder, G. W., Dewey, V. C. 1942. *Growth* 6:405–18
18. Haldane, J. B. S. 1942. *New Paths in Genetics.* New York/London: Harper. 206 pp.
19. Wright, S. 1941. *Physiol. Rev.* 21:487–527
20. Bateson, W. 1909. *Mendel's Principles of Heredity.* Cambridge. 369 pp.
21. Goldschmidt, R. 1938. *Physiological Genetics.* New York: McGraw. 325 pp.
22. Horowitz, N. H., Leupold, U. 1951. *Cold Spring Harbor Symp. Quant. Biol.* 16: 65–72

BIOCHEMISTRY OF DRUG DEPENDENCE[1]

×838

A. E. Takemori

Department of Pharmacology, University of Minnesota, Health Sciences Center
Medical School, Minneapolis, Minnesota

CONTENTS

Research in the field of drug abuse has mushroomed in recent years as a result of the concern of the government and the medical scientists over the use and abuse of stimulants, tranquilizers, hallucinogens, hypnotics, and narcotics in epidemic proportions by the public. The abuse of alcohol, of course, is an old problem. Since it would be impossible to discuss all the drugs of abuse in this brief review, I have chosen to restrict the discussion to the narcotic analgesics.

The organization and preparation of this review was simplified due to the excellent review on the same topic by Dole (1) in the 1970 volume of this series. The reader is encouraged to read that review since the present one represents a sequel. Additionally, a number of books and reviews on this subject have been published since Dole's review. Comprehensive treatises on the biochemistry of drug dependence are found in two recent multiauthored books (2, 3). All current theories on the biochemical mechanism of tolerance, physical dependence, and abstinence syndrome are capably discussed in these volumes. Several symposia (4–8) and reviews (9–15) which

---

[1] Literature coverage is limited by necessity from 1969 to June 1973. Older references are cited when pertinent.

touch on this subject are also available. Due to these excellent monographs, this review concentrates on newer developments in two areas of narcotics research, namely the narcotic analgesic receptors and substances which alter the abstinence syndrome and the development of tolerance and/or physical dependence.

NARCOTIC ANALGESIC RECEPTORS

Narcotic analgesics are generally believed to exert their pharmacologic effects by interacting with specific receptors located in the CNS. The concept of specific analgesic receptors is strengthened by the fact that narcotic drugs have common structural features, enantiomers of narcotic drugs exhibit large potency differences, and structurally related competitive antagonists can counteract the action of all the narcotic analgesics. However, the exact location, isolation, and constitution of the receptors remain elusive. The subsequent discussion will record the recent progress in the characterization of analgesic receptors in CNS. Other narcotic receptors such as those in the guinea pig ileum will not be covered due to lack of space but are amply discussed elsewhere (3, 7).

Localization of Receptors

DETERMINATION BY AUTORADIOGRAPHY   The localization of ^{14}C-morphine in the CNS by autoradiography was initially attempted by Miller & Elliott (16). Autoradiographs of the brain after an analgesic parenteral dose of radioactive morphine in the rat showed activity was concentrated in the choroid plexuses and the ventricles. Subsequently it was shown that the choroid plexus tissue actively accumulated morphine and other narcotic drugs in vitro (17, 18).

Attempts to prepare autoradiographs of brain after high intravenous doses of ^{14}C-levorphanol, a potent synthetic narcotic drug, were unsuccessful (19). However, recently Teschemacher, Schubert & Herz (20), after establishing the site of morphine action around the fourth ventricle in rabbits (*vide infra*), employed intraventricular injections of ^{14}C-morphine to localize the site of action more precisely. When the autoradiographs were correlated with the antinociceptive action, structures lining the rostral parts of the fourth ventricle were implicated as the site of action of morphine. From the comparison of concentrations of morphine in structures at different distances from the ventricular wall after intraventricular administration to those in the brain after systemic administration of equieffective doses the authors concluded that the analgesic receptors were located around 1.2 mm away from the fourth ventricular wall.

DETERMINATION OF SITES OF ACTION   Tsou & Jang (21) were the first to use the microinjection technic to delineate the site of action of morphine in rabbits. From the observation of analgesic effects after microinjections of morphine into various parts of the brain, they concluded that the main site of action of morphine resided in the periventricular grey matter of the third ventricle at a distance of 1–2 mm from the ventricular space. The authors also showed that intravenous injection of the narcotic antagonist, nalorphine, antagonized the analgesic action of morphine given by periventricular microinjection. Also, microinjections of nalorphine into the periventricular grey matter antagonized the analgesia resulting from intravenously administered morphine. When nalorphine was injected into other parts of the brain, it was ineffective in antagonizing the effects of morphine.

Foster, Jenden & Lomax (22) studied the analgesic effects of morphine and N-methylmorphine by microinjections into the brain of rats. Injections of either drug into the periventricular grey matter of the rostral hypothalamus caused a marked analgesia in a majority of the animals. Buxbaum, Yarbrough & Carter (23) also used microinjections of morphine in rats and showed that dose-response relationship for analgesia could be obtained if injections were made into the anterior thalamic nuclei. They also observed analgesic effects when morphine was injected into other thalamic and hypothalamic areas.

Herz et al (24) developed a method which consisted of plugging certain parts of the ventricular system so that they were able to inject drugs into various specific and limited portions of the ventricular system in rabbits. Using this technic, the authors concluded that the relevant sites of action of morphine were located in the periventricular grey matter surrounding the aqueduct and structures on the floor of the fourth ventricle. They also used microinjections of morphine and observed a marked antinociceptive effect when injections were made in the medial parts of the hypothalamus, subthalamus, and mesencephalon. Continuation of these studies revealed that the narcotic antagonists, nalorphine and levallorphan, were also most effective in antagonizing the analgesic effects of systematically administered morphine when injected into a restricted part of the ventricular system which represented part of the aqueduct and the fourth ventricle (25).

DETERMINATION OF SITES FOR PRECIPITATED WITHDRAWAL   It is relevant to note from the above discussion that intraventricular injections of morphine have been shown to produce tolerance and physical dependence in rats (26), rabbits (27, 28), and monkeys (29). It was also shown in these studies that an intraventricular injection of a narcotic antagonist precipitated withdrawal signs.

Herz et al (30) further localized the fourth ventricle as the most sensitive site where nalorphine will produce abstinence signs in morphine-dependent rabbits. Wei, Loh & Way (31, 32) used the method of applying naloxone crystals, a potent narcotic antagonist, to various parts of the brain of morphine-dependent rats through stereotaxically placed cannulae. Abstinence signs were most frequently observed when the naloxone was placed in the medial thalamus and medial areas of the diencephalic-mesencephalic junctures. They postulated that these were primary sites for the development of narcotic dependence.

ISOLATION OF RECEPTORS

May et al (33) first used the affinity labeling method (34–36) to see whether or not compounds could be designed which would interact irreversibly with analgesic receptors. Among the series of benzomorphan derivatives they synthesized, the hydrobromide salts of N-(2-bromoethyl)-α-5,9-dimethyl-6,7-benzomorphan and

N-(2-bromopropyl)-α-5,9-dimethyl-6,7-benzomorphan produced prolonged CNS depression with a low degree of analgesic activity. The authors could not conclude from their experiments whether this effect was due to the alkylation of certain sites in the CNS.

Portoghese et al (37), also using the affinity labeling approach, have synthesized a number of N-acylanileridines having alkylating capacity. Ethyl p-(4-ethoxy-carbonyl-4-phenyl-1-piperidinoethyl) fumaranilate significantly blocked analgesic activity of morphine in mice 2 hr after intraperitoneal administration. Pretreatment with a potent narcotic antagonist, naloxone, completely blocked the analgesic activity of the alkylator. Naloxone also protected the analgesic receptors against inactivation. The authors suggested that the fumaramido ethyl ester of anileridine may initially form a reversible complex with receptors which results in analgesia and has the capacity subsequently to alkylate the receptors which results in inactivation. Further studies revealed that this inactivation lasted only for several hours and the antagonism of morphine analgesia was not a simple noncompetitive type (Takemori & Portoghese, unpublished results).

Most of the active synthetic narcotic analgesics are of the D form while their L isomers are either inactive or very much less active. Taking advantage of this fact, Goldstein, Lowney & Pal (38) used radioactive levorphanol (active l-isomer) and dextrorphan (inactive d-isomer) to devise a method whereby stereospecific binding to brain fractions could be determined in the presence of nonspecific inter-actions. They found that about half of the total levorphanol binding by brain homogenates of mice was a nonsaturable type (trapped and dissolved drug), nearly half of the binding was nonspecific, saturable type, and only about 2% of the binding was stereospecific. Studies of binding to subcellular fractions revealed that stereo-specific binding occurred mainly in membranes separated by flotation from the crude nuclear fraction and to some extent in the crude mitochondrial and microsomal membrane fractions. Raising the concentration of levorphanol from 1.95×10^{-6} M to 3.9×10^{-5} M increased the stereospecific binding by around tenfold.

Further studies by Pal, Lowney & Goldstein (39) showed that the stereospecific binding was unchanged after extraction of 70% of the protein by Triton X-100 or sodium dodecyl sulfate, largely destroyed by treatment with neuraminidase or pronase, and enhanced by treatment with trypsin. It had a sharp pH optimum at 7.4, was enhanced by EDTA, was greatly inhibited by Ca^{2+} or Mg^{2+}, and was unaffected by p-chloromercuribenzoate, mercaptoethanol, or iodoacetic acid. The effect with Ca^{2+} is consonant with pharmacologic findings since it has been shown that Ca^{2+} inhibits and EDTA enhances morphine analgesia (39a). The material responsible for the stereospecific binding was completely extractable in chloroform methanol and had the general properties of a proteolipid. The stereospecific binding increased with increasing concentrations of levorphanol but could not be saturated. At the highest concentration used, 51 nmol or 3×10^{16} molecules of levorphanol were bound per mouse brain. At the lowest concentration used in the previous study (38), about 1 nmol or 6×10^{14} molecules were bound per brain. The concentration of etorphine, a very potent narcotic, in the brain of rats after an ED_{50} dose was 3×10^{-9} M and assuming 1% of the drug in the brain is attached to analgesic receptors, the analgesic receptor density has been estimated at about 10^{-11} M (1).

Using these figures, the amount of receptor sites occupied at the ED_{50} dose is about 3×10^{10} molecules per rat brain.[2] Assuming that all of the drug in the brain was bound to receptors, Goldstein, Lowney & Pal (38) estimated that 7×10^{11} molecules could be bound by mouse brain at the ED_{50} dose. If the material responsible for stereospecific binding represents analgesic receptors, clearly much of the narcotic stereospecifically bound to this material is not involved in evoking a pharmacologic effect.

Recently, Terenius (40) used l-methadone (active isomer) and d-methadone (much less active isomer) to demonstrate a stereospecific binding of radioactive dihydromorphine to synaptic plasma membrane fraction of cerebral cortex but not to other fractions such as nuclei and mitochondria. Dihydromorphine was used generally at a concentration of 6×10^{-10} M and the binding material was saturated at a concentration of 1×10^{-8} M of the narcotic. Four days of treatment of animals with morphine did not change the stereospecific binding. Although other narcotics and narcotic antagonists were claimed to inhibit the binding of dihydromorphine, no data were shown. The author suggested that this high affinity binding material could represent the "narcotic receptor."

More recently Pert & Snyder (41) made use of ^{3}H-naloxone with a very high specific activity to perform binding studies with homogenates and subcellular fractions of brains in rats, mice, and guinea pigs. They showed the binding of 5×10^{-9} M naloxone by brain homogenates was stereospecific by employing levorphanol and dextrorphan. The affinity of drugs for the naloxone binding sites was studied by determining the concentration of drug required to decrease the specific binding of 5×10^{-9} M ^{3}H-naloxone by 50% ("ED_{50}"). The binding affinities of the various analgesics and antagonists generally paralleled their pharmacologic potencies. The "ED_{50}s" of antagonists ranged from $2–10 \times 10^{-9}$ M while those of the narcotics ranged from $2–20 \times 10^{-9}$ M, with the notable exception of codeine which had a value of 2×10^{-5} M. The authors discussed the possibility of codeine being "activated" to morphine in the intact animal. Other classes of drugs and neurotransmitters did not bind in this system. The binding of naloxone was confined to nervous tissue. In brain, the greatest amount of binding took place in the corpus striatum and in subcellular fractions, and most of the binding material was concentrated in the microsomal fraction which is rich in membrane fragments, while half the total binding activity was recovered in the crude mitochondrial-synaptosomal fraction. The authors believe that their results represent a direct demonstration of the "opiate receptor."

Additional studies revealed that this stereospecific binding was inhibited by Ca^{2+} and enhanced by EDTA (42). Furthermore, when binding was performed with brain homogenates of mice tolerant to and dependent on morphine, the author observed an increase in the amount of binding which he interpreted as a quantitative increase of the number of opiate receptors.

The major difference between the stereospecific binding material of Terenius (40) and Pert & Snyder (41) and that of Goldstein, Lowney & Pal (38) is that the concentration of drug used for binding was four to five orders of magnitude larger

[2] An average whole brain of an adult rat was assumed to be 1.75 g.

in Goldstein et al's study. Whereas binding saturation was not achieved by Goldstein et al, the concentrations used by these workers fully saturated the binding materials of both Terenius and Pert & Snyder. The extent of stereospecificity increased with increasing concentrations of levorphanol in the system of Goldstein et al while Pert and Snyder showed that stereospecific binding decreased with increasing concentrations in their system.

While it is true that the binding functions of various narcotic analgesics to the binding material of Pert & Snyder (41) are generally correlated with the pharmacologic potencies of the drugs, the binding should rightly be compared with the brain concentrations of the narcotics after an equieffective dose ED_{50}s of analgesics which were determined after parenteral injections are influenced by several processes which determine the bioavailability of the drug at the receptor site in the CNS. A ranking of various narcotic analgesics according to their brain concentrations has been made by Herz & Teschemacher (11). This ranking of receptor activity includes the affinity to the receptor. In this ranking, morphine becomes very much more potent than other drugs. For example, levorphanol, which is about five times more potent than morphine when administered parenterally, is only about one eighth as potent. Similarly, methadone, which is equipotent or slightly more potent than morphine parenterally, ranks only one thirtieth as potent. Using this type of ranking, the binding functions of various narcotics (41) do not correlate very well with the analgesic potencies in vivo.

Simon (43) also reported the stereospecific binding of the potent narcotic, etorphine, in rat brain. The "ED_{50}" of several analgesics and antagonists for inhibition of binding appear to be within an order of magnitude reported by Pert & Snyder (41). The K_m for binding of etorphine was 5×10^{-10} M. The binding was not inhibited by metabolic inhibitors or by prostaglandin E_1 or E_2. It was inhibited by p-chloromercuribenzoate and N-ethylmaleimide, which implicates sulfhydryl groups in the binding. This contrasts from the binding material of Goldstein et al (38) who concluded that sulfhydryl or disulfide bonds were not essential for binding of levorphanol.

Two recent methods should be mentioned which may aid in the eventual isolation of receptor material. Winter & Goldstein (44) have synthesized a photochemical affinity labeling reagent, a radioactive analog of levorphanol [^{3}H-N-β-(p-azidophenyl) ethylnorlevorphanol]. This compound has potent analgesic effects and has, upon irradiation in vitro, the capacity to form covalent binding of the radioactive label to those macromolecules with which it is in contact. Unfortunately, the binding of this compound to macromolecules has been completely nonspecific. The other method devised by Simon, Dole & Hiller (45) involved the synthesis of a pharmacologically active morphine derivative, 6-succinylmorphine, and the coupling of this congener to sepharose for affinity chromatography. As yet, the use of this method has not yielded any stereospecific binding material.

PHARMACOLOGIC CHARACTERIZATION OF RECEPTORS

There have been numerous papers on the structure-action relationship of narcotic analgesics and their antagonists. From this type of study, the shape of the analgesic receptor and the possible types of interaction between the analgesic and receptor have been postulated (7, 10, 12, 46). Another approach to define the nature of receptors is to use the concept of pAx (47) to quantitatively measure drug antagonism and to identify and classify agonists. Usually, the concept of pAx is used with isolated tissue preparations to identify agonists which act on similar receptors. However, Cox & Weinstock (48) were the first to successfully use this concept in vivo.

Our group has in recent years applied the concept of pAx in intact animals for the characterization of analgesic receptors. The definition of the apparent pA_2 in vivo then becomes the negative logarithm of the molar dose of the injected antagonist which reduces the effect of a double dose of an agonist to that of a single dose. Although pA_2 is equal to the log of the affinity constant (K_B) of the antagonist for the receptor, this is not entirely true in vivo. However, the K_B in vivo should be proportional to the real K_B if it is assumed that the concentration of the antagonist at the receptor site is proportional to the injected dose. The consistently reproducible apparent pA_2 value for morphine-naloxone under a variety of conditions attests to the utility of the apparent pA_2 value as a pharmacologic constant (49).

With the use of an antagonist with no analgesic properties, naloxone, apparent pA_2 values for a series of narcotic-type and narcotic-antagonist-type analgesics have been determined to see whether or not the two types of analgesics interacted with similar receptors (50). The apparent pA_2 values with the narcotic agents morphine, levorphanol, and methadone were similar and those with the narcotic-antagonist analgesics, pentazocine, nalorphine, and cyclazocine were also the same. However, the two groups of analgesics had significantly different apparent pA_2 values. It was concluded that the narcotic and narcotic-antagonist analgesics interact either with two different receptor populations or with the same receptor in a different manner.

In another application of apparent pA_2s, morphine was shown to induce some type of qualitative change in analgesic receptors (51). Mice were pretreated with a $>$ ED_{99} dose of morphine and after 2 hr when the analgesic effect of morphine was no longer evident, dose-response curves for morphine with and without naloxone were determined. Naloxone shifted the dose-response curves much more to the right than when nonpretreated or control mice were used while the dose-response curve of morphine alone remained unchanged. The apparent pA_2 value changed significantly from 6.96 to 7.30 and further changed to 7.80 when mice tolerant to and dependent on morphine were used (52). This meant that the apparent affinity constant of naloxone for the analgesic receptor more than doubled when mice were exposed to one acute injection of morphine and increased by over sevenfold when mice were made tolerant to morphine. This indicated that morphine caused a structural change in the analgesic receptors such that the affinity of the receptors for narcotic antagonists was increased. Subsequently it was shown that the efficacy of not only naloxone but those of the antagonists, nalorphine and diprenorphine (M5050) also increased after morphine pretreatment (52a). Earlier Orahovats, Winter & Lehman (53) showed that nalorphine was more effective in antagonizing morphine analgesia in partially tolerant rats than in nontolerant rats. These findings are not in accord with theories of tolerance which infer decreases in number of pharmacologic receptors (54) or increases in number of silent receptors (54, 55) since the above data indicate a qualitative change in analgesic receptors with exposure to morphine rather than a quantitative one. Snyder (42) showed a quantitative change in his opiate receptor when the binding material from morphine-dependent mice was examined. If isolated binding material has anything to do with pharmacologic receptors, it should show the same qualitative-type changes observed pharmacologically, i.e. morphine exposure should alter the binding constant of the naloxone rather than increase the amount of binding.

Further studies showed that the increased efficacy of naloxone induced by morphine could also be induced by pretreatment with other narcotics such as levorphanol and methadone (52a). On the other hand, pretreatment with the narcotic-antagonist analgesic, pentazocine (51) or naloxone (52a), did not alter the efficacy of naloxone. The increase in naloxone efficacy due to narcotic pretreatment appeared to be a sensitive indicator of the development of tolerance and was observed much before tolerance could be detected (56). Cycloheximide is known to inhibit the development of tolerance (57) and also inhibited the increased efficacy of naloxone due to morphine treatment (52).

In view of the above pharmacologic findings, the following hypothetical analgesic receptor is proposed (Figure 1). Different analgesics are viewed as binding with a single species of receptors but having different positions of molecular binding. This interaction has also been suggested by Portoghese (46) as one of several possibilities. Since narcotic and narcotic-antagonist analgesics and narcotic antagonists are structurally similar, it is reasonable to assume that the protonated nitrogen is a common site of attachment to the receptor. Narcotic drugs and narcotic-antagonist analgesics have differing interactions since they have different apparent pA_2s with naloxone. The antagonists interact at a site different from the above two but in close association with the narcotic site and display an apparent competitive antagonism at the narcotic site. This separate site enables changes in antagonistic activity to be observed without changes in narcotic activity. When narcotic-antagonist analgesics act as antagonists of narcotic drugs, they are also presumed to interact

HYPOTHETICAL RECEPTOR MODEL

Figure 1   Hypothetical interactions between narcotic analgesics or antagonists and the analgesic receptor. N = narcotic analgesic site, A = antagonist site, NA = narcotic–antagonist analgesic site.

The small circle represents the protonated nitrogen, the rectangles and square are the substituents on the nitrogen, and the large circles are the ring structure portion of the molecule. The positions of the rings illustrate different positions of molecular binding to the receptor.

at the antagonist site. Evidence for this is that when narcotics induce a "better fit" at the antagonist site (increased naloxone efficacy), the antagonistic efficacy of nalorphine, a narcotic-antagonist analgesic, is also increased. Chronic interaction at the narcotic site desensitizes the site, resulting in tolerance, and concomitantly increases the binding constant of the antagonists for their site, resulting in increased sensitivity to antagonist, a fact consonant with pharmacologic findings. The "pure" antagonists such as naloxone are thought to antagonize the agonistic action of narcotic-antagonist analgesics at the narcotic-antagonist site in the traditional competitive manner. The present concept of receptors is capable of explaining the current data but it probably would and should be modified as additional information emerges. Finally, this hypothetical receptor is viewed as a working model for analgesia and the development of tolerance to analgesia. Physical dependence and withdrawal on the other hand may involve other receptors. Receptors involving neurotransmitters have been proposed (54). Also there is evidence that naloxone acts noncompetitively to precipitate withdrawal in levorphanol-dependent mice (58).

COMPOUNDS WHICH ALTER NARCOTIC ABSTINENCE SYNDROME, TOLERANCE, AND/OR PHYSICAL DEPENDENCE

Compounds reported to alter the development of narcotic tolerance or dependence and withdrawal signs are summarized in Tables 1–4. The list of compounds is not meant to be exhaustive; rather it is meant to acquaint the reader with the wide variety of compounds that have been studied.

Tables 1 and 2 list compounds which inhibit or enhance the development of tolerance and/or dependence. The compounds were administered to the experimental animals usually before exposure to the narcotic drug and throughout the course of chronic narcotic treatment. Tables 3 and 4 list compounds which attenuate or exacerbate withdrawal signs. These compounds were administered just before either abrupt withdrawal from the narcotic or injection of a narcotic antagonist to acutely precipitate the withdrawal signs.

Substances Which Inhibit The Development of Tolerance and/or Dependence

Inspection of Table 1 reveals that a variety of compounds inhibit the development of tolerance and dependence in several species. One might expect the narcotic antagonists to inhibit tolerance and dependence if competition with narcotics for similar receptors is assumed. However, since naloxone appears to precipitate withdrawal noncompetitively (58), the role of antagonists to inhibit dependence may be more complicated than a simple competition for receptors.

Among the inhibitors of protein synthesis, agents affecting the process of transcription, nucleic acid synthesis, and translation appear to inhibit the development of tolerance. To date, only cycloheximide (57, 64) and actinomyin D (64) have been shown to inhibit the development of dependence. The implications are that narcotics stimulate protein synthesis to form a unique protein (tolerance factor?) (78), specific enzymes (79, 80), or silent or neurotransmitter receptors (54). It must be pointed out, however, that other inhibitors such as ethionine (81) or chloramphenicol

Table 1 Compounds which inhibit the development of tolerance and/or dependence

Compound	Adaptation[a] Studied	Animal[b] Used	Method of Producing Adaptation	Assay Used	Ref.
Narcotic Antagonists					
Nalorphine	T	R	morphine injections	tail-flick	53
	D	M	morphine pellets	withdrawal signs	59
Levallorphan	D	M	morphine pellets	withdrawal signs	59
	D	Mo	morphine injections	withdrawal signs	60
Naloxone	T&D	M	morphine injections	tail-flick & withdrawal signs	61
Inhibitors of Protein Synthesis					
Actinomycin D	T	M	morphine injections	tail pressure	62
	T	R	morphine injections	tail-flick	62
	AT	R	continuous morphine infusion	tail pressure	63
	D	M	morphine pellets	withdrawal jumps	64
8-Azaguanine	T	M	morphine injections	hot plate & tail pressure	65
	T	M	morphine injections	hot plate	66
6-Mercaptopurine	AT	R	continuous morphine infusion	tail pressure	67
5-Flurouracil	AT	R	continuous morphine infusion	tail pressure	67
Puromycin	AT	R	continuous morphine infusion	tail pressure	67
	AT	M	single narcotic injection	lenticular opacity	68
Cycloheximide	AT	R	continuous morphine infusion	tail pressure	67
	T&D	M	morphine injections	tail flick & withdrawal jumps	57
	T	R	morphine injections	hot plate	69
	T	M	morphine pellets	tail flick	52
	D	M	morphine pellets	withdrawal jumps	64

Alkylators	T	R	morphine injections	hot plate	70
Cyclophosphamide	T	R	morphine injections	hot plate	70
Depletors of Biogenic Amines					
Tetrabenazine (+DOPA reverses effect)	T	M	morphine injections	tail pressure	71
p-Chlorophenylalanine (PCPA)	T&D	M&R	morphine pellets	tail flick; withdrawal jumps & body weight loss	61 72
5,6-Dihydroxytryptamine	T&D	M	morphine pellets	same as above	73
1-Phenyl-3-(2-thiazolyl)-2-thiourea	T&D	M	morphine pellets	same as above	74
Miscellaneous Compounds					
3-Methylsulfonyl-10,2-(1-methyl-2-piperidyl)-ethyl phenothiazine	T / D	R / Mo	morphine injections / morphine injections	tooth pulp stimulation / withdrawal signs	75 75
Caffeine	T / D	R / Mo	morphine injections / morphine injections	tooth pulp stimulation / withdrawal signs	75 75
Dichloroisoproterenol	T&D	M	morphine pellets	tail flick; withdrawal jumps & body weight loss	76
Calcium ion	T	M	morphine injections	electrical stimulation	77

[a] T = tolerance; D = dependence; AT = acute tolerance.
[b] R = rat; M = mouse; Mo = monkey.

(67) have been shown not to inhibit the development of tolerance. Efforts to detect increased amounts or rate of synthesis of brain proteins using radioactive amino acid precursors and acrylamide gel electrophoresis have failed (64, 82). Our attempts to see whether there was an altered incorporation of intracerebrally administered radioactive orotic acid or parenterally administered radioactive glycine into total brain RNA, RNA of various subcellular fractions of brain, or rapidly labeled RNA in brains of morphine-tolerant rats have also been unsuccessful (83). Perhaps present methods are not sensitive enough to detect small changes which may occur in a specific protein in the brain of tolerant-dependent animals. If changes are occurring in specific parts of the brain and in areas represented by the estimated analgesic receptor density (1), the task of detecting these changes will be formidable indeed. An alternate possibility is that chronic narcotic administration may induce a miscoding so that a change in quality rather than quantity of certain brain proteins occurs (84). However, there is no experimental evidence for such a postulate. Another relevant point is that the doses of inhibitors used to inhibit tolerance and dependence are very toxic. Some of the side effects which are unrelated to inhibition of protein synthesis have been discussed (84).

It appears that depletors of norepinephrine (NE), dopamine (DA), and 5-hydroxytryptamine (5-HT) all inhibit the development of tolerance and dependence. In their review, Way & Shen (85) concluded that 5-HT may be more intimately involved in the adaptation processes than the other two biogenic amines. Way and his co-workers (86) first showed that the brain turnover of 5-HT was increased in mice tolerant to and dependent on morphine. p-Chlorophenylalanine, a depletor of 5-HT, inhibited the development of tolerance and dependence (61, 72), but 6-hydroxydopamine, a depletor of NE and DA, had no effect on these processes (87). Further evidence for the involvement of 5-HT is the fact that tryptophan, a precursor of 5-HT,

Table 2   Compounds which enhance the development of tolerance and/or dependence

Compound	Adaptation[a] Studied	Animal[b] Used	Method of Producing Adaptation	Assay Used	Ref.
Brain homogenates of morphine-tolerant rats & dogs	Transfer of T	M	morphine injections	tail pressure	78
Cyclic 3′,5′-adenosine monophosphate (cAMP)	T&D	M	morphine pellets	tail flick; withdrawal jumps & body weight loss	100
Tryptophan	T&D	M	morphine pellets	tail flick; withdrawal jumps & body weight loss	15
p-Chlorophenylalanine	AT	C	morphine injections	respiratory min-vol.	102

[a] T = tolerance; D = dependence; AT = acute tolerance.
[b] M = mouse; C = decerebrate cat.

Table 3   Compounds which attenuate withdrawal signs

Compound	Animal[a] Used	Withdrawal signs studied	Ref.
Modifiers of Biogenic Amines			
α-Methyltyrosine (αMT)	R	ptosis, diarrhea, shakes, hypothermia	96
	M	jumping	97
	Mo	vomiting, hyperactivity, foaming at mouth, cramps, tremors, vocalization	110
αMT + Dopa	R	reversal of αMT effect	96
	M	reversal of αMT effect	97
Diethyldithiocarbamate	R	See ref. 96 (*vide supra*)	96
Disulfiram	M	See ref. 97 (*vide supra*)	97
α-Methyldopa	M	jumping, running, diarrhea, tremors, writhing, convulsions	111
p-Chlorophenylalanine	D	rating of 21 withdrawal signs	112
Nialamide	D	same as above	112
Haloperidol	R	withdrawal aggression, hypothermia	113
	H	human withdrawal signs	113
Imipramine	R	shakes	96
Amino Acids & Derivatives			
Alanine	M	See ref. 111 (*vide supra*)	111
Serine	M	See ref. 111 (*vide supra*)	111
Tryptophan	M	See ref. 111 (*vide supra*)	111
5-Hydroxytryptophan	M	See ref. 111 (*vide supra*)	111
5-Hydroxytryptamine	M	See ref. 111 (*vide supra*)	111
Tyrosine	M	See ref. 111 (*vide supra*)	111
Phenylalanine	M	See ref. 111 (*vide supra*)	111
Dopa	M	See ref. 111 (*vide supra*)	111
	D	rating of 21 withdrawal signs	112
Adrenergic Blockers			
Propanolol	D	See ref. 112 (*vide supra*)	112
Phenoxybenzamine	D	See ref. 112 (*vide supra*)	112
Miscellaneous Compounds			
KCN	M	See ref. 111 (*vide supra*)	114
Nicotinamide	M	See ref. 111 (*vide supra*)	114
Deoxypyridoxine	M	See ref. 111 (*vide supra*)	114
Choline	R	scoring of 9 withdrawal signs	115
Whole mouse extracts of morphine-pelleted mice and extracts of their urine	M	See ref. 111 (*vide supra*)	116

[a] M = mouse; D = dog; R = rat; Mo = monkey; H = human.

accelerated the development of tolerance and dependence (15). Way's group suggests that enzymes for 5-HT synthesis may be involved since cycloheximide, which inhibits the development of tolerance and physical dependence, also inhibited the increase in brain 5-HT turnover associated with the adaptive processes (15).

Increased turnover of brain 5-HT in tolerant-dependent animals which was observed by Way's group has been challenged by four laboratories (88–91) and confirmed by two others (92, 93). Increased brain 5-HT turnover in rats injected with a single dose of morphine but not in chronically treated rats has been reported (94). Inhibition of the development of dependence by *p*-chlorophenylalanine has also been challenged (88, 90, 92, 95, 96). These authors concluded that there was no causal relationship between the synthesis of 5-HT and the phenomenon of dependence. The main difficulty in repeating the observations of Way's group appears to be in the methodology. Ample discussion of this and other points of differences can be found elsewhere (15, 72, 98, 99).

Table 4   Compounds which exacerbate withdrawal signs

Compound	Animal[a] Used	Withdrawal signs studied	Ref.
Modifiers of Biogenic Amines			
Reserpine	M	jumping, running, diarrhea, tremors, writhing, convulsion	117
	D	rating of 21 withdrawal signs	112
	D	restlessness, vomiting, tremors, convulsion	118
Dopa	D	same as above	118
Nialamide	D	same as above	118
α-Methyldopa	D	same as above	118
α-Methyl-*m*-tyrosine	D	same as above	118
Imipramine	M	used procedure of Ref. 117 (*vide supra*)	119
5-Hydroxytryptophan	D	See ref. 112 (*vide supra*)	112
Pargyline	M	jumping	92
	M	jumping	120
6-Hydroxydopamine	M	jumping, body weight loss	87
Miscellaneous Compounds	castrated		
Testosterone	M	See ref. 117 (*vide supra*)	121
Sodium Glutamate	M	same as above	114
DPN	M	same as above	114

[a] M = mouse; D = dog.

The significance of the inhibition of tolerance and dependence by the miscellaneous agents listed in Table 1 is difficult to assess without additional data. Bhargava et al (76) have suggested that the inhibition by the β-adrenergic blocker, dichloroisoproterenol, is due to a decrease in the production of cyclic adenosine monophosphate (cAMP). They have later shown that cAMP actually enhanced the development of tolerance and dependence (100).

It should be mentioned to complete the picture that cholinergic mechanisms do not appear to play a primary part in the development of tolerance and dependence. The acetylcholinesterase inhibitors, physostigmine and diisopropylfluorophosphate, did not materially affect the development of tolerance and dependence (101). However, cholinergic mechanisms may play a modulating role in tolerance and dependence and may also play a prominent role in the expression of abstinence signs.

Compounds Which Enhance the Development of Tolerance and/or Dependence

The finding that inhibitors of protein synthesis block the development of tolerance prompted Ungar & Cohen (78) to postulate a polypeptide "tolerance factor." They proceeded to demonstrate that tolerance can be transferred by injecting naive animals with brain homogenates from morphine-tolerant animals. These results could have proved to be very significant in explaining the phenomenon of tolerance but unfortunately six different laboratories have failed to confirm these findings (103–108).

Ho, Loh & Way (100) suggest that the accelerating effect of both cAMP and tryptophan on tolerance and dependence is related to the enhancement of 5-HT turnover. p-Chlorophenylalanine, a compound which inhibits 5-HT synthesis, not only failed to block tolerance and dependence in several laboratories (vide supra) but one group of investigators reported an increase in the development of acute tolerance to respiratory depression (102). The latter report is concerned with acute tolerance, which may be different from long term tolerance, and with respiratory effects, which may be a different type receptor population from the analgesic ones. Nevertheless not all compounds which increase brain 5-HT turnover are associated with the development of tolerance and physical dependence. For example, lithium salts have been reported to increase brain 5-HT turnover by 90% (109) yet they hardly qualify as agents that produce tolerance or dependence of the morphine-type.

Compounds Which Attenuate or Exacerbate Withdrawal Signs

Among the compounds influencing withdrawal signs, those that modify the amount or actions of central biogenic amines appear to be the most relevant (Tables 3 and 4). However, the effect of these compounds on withdrawal signs of various species are not without inter- and intraspecies differences. For example, α-methyldopa attenuates withdrawal signs in mice (111) but exacerbates them in dogs (118). Also imipramine inhibits withdrawal signs in rats (96) yet enhances them in mice (119). Nialamide, a monoamine oxidase inhibiter, has been reported to attenuate (112) and exacerbate (118) withdrawal signs in dogs.

Although all the biogenic amines appear to be involved in the expression of withdrawal signs, the inhibition of withdrawal signs with α-methyltyrosine, a central depletor of NE and DA (Table 3) suggests that the full expression of abstinence syndrome in morphine-dependent mice, rats, and monkeys requires the integrity of the central stores of these catecholamines. That DA is more important in the display of withdrawal signs is supported by the fact that repletion of DA levels after α-MT treatment restored withdrawal signs (96, 97) while repletion of NE levels did not (97). Recently, Iwamoto, Ho & Way (122) have shown that withdrawal jumping response in morphine-dependent mice is accompanied by a sudden rise in central DA level while the levels of NE and 5-HT remained unchanged. Additionally, Lal et al (113) have reported that haloperidol, a central blocker of catecholamine effects, can markedly reduce abstinence signs in rats and humans. If this report that haloperidol can substantially relieve abstinence syndrome in human narcotic addicts can be substantiated, continued studies of morphine dependence and withdrawal in the rodent-model should be useful and fruitful.

The effects of other compounds such as amino acid derivatives and the miscellaneous compounds listed in Tables 3 and 4 are very difficult to assess. Most of these compounds were studied in one laboratory and unfortunately the findings were reported descriptively without any attempt to formulate any hypotheses.

To complicate matters, Collier, Francis & Schneider (123) have reported that some compounds can modify certain signs of morphine abstinence in mice but not others. They also observed that one of the factors determining the effect of modifying compounds on withdrawal signs is the time when the compound is administered in the course of dependence induction and withdrawal. From results using atropine, p-chlorophenylalanine, and indomethacin, they suggested that morphine dependence is multipartite and that acetylcholine, 5-HT, and prostaglandins play intimate roles in the physical dependence process.

CONCLUDING REMARKS

Although there are some disagreements as to the exact location of analgesic receptors, there is general agreement that they are very near the ventricular system. In view of this, transport of narcotics into the cerebrospinal fluid such as that shown for morphine (124) may become important in the bioavailability of narcotics at the receptor site. Isolation of the receptors has yet to be accomplished. Even if the receptors were isolated, it would be extremely difficult to prove that the binding of narcotic drugs to the isolated receptor material initiates a pharmacological response. There is ample proof that pharmacologic effects of narcotics and their antagonists are changed during the development of tolerance. These changes represent a change in the binding characteristics of the drug to the analgesic receptors. Thus, for any isolated binding material purported to be the analgesic receptor it must be shown that the same type of qualitative change in the binding material occurs with the development of tolerance. Such attempts should be encouraged in the hope that the crucial step of isolation can finally be accomplished.

With the variety of compounds modifying the development of tolerance or dependence and withdrawal signs, no one unifying hypothesis can be postulated.

The evidence at present implicates protein synthesis and biogenic amines in these phenomena. It is important to point out that the analgesic effects of narcotics have been dissociated from the processes of tolerance and dependence (101). One can modify the acute effects of narcotics without altering the development of tolerance and dependence and vice versa. It has also been shown that the development of tolerance and dependence can actually be accelerated with certain agents. These findings should promote the design of new experiments to aid in the elucidation of the mechanism of tolerance and dependence.

Literature Cited

1. Dole, V. P. 1970. *Ann. Rev. Biochem.* 39:821–40
2. Clouet, D. H., Ed. 1971. *Narcotic Drugs: Biochemical Pharmacology.* New York: Plenum. 506 pp.
3. Mule, S. J., Brill, H., Ed. 1972. *Chemical and Biological Aspects of Drug Dependence.* Cleveland: CRC. 561 pp.
4. Steinberg, H., Ed. 1969. *Scientific Basis of Drug Dependence.* London: Churchill. 429 pp.
5. Pharmacology Society Symposium. 1970. *New Concepts and Approaches to the Study of Drug Dependence and Tolerance.* In *Fed. Proc.* 29:2–32
6. Harris, R. T., McIsaac, W. M., Schuster, C. R. Jr., Ed. 1970. *Drug Dependence.* Austin: Univ. Texas Press. 342 pp.
7. Kosterlitz, H. W., Collier, H. O. J., Villarreal, J., Ed. 1973. *Agonist and Antagonist Actions of Narcotic Analgesic Drugs.* London: MacMillan 290 pp.
8. The Biochemical Psychopharmacology Series. *First International Conference on Narcotic Antagonists,* Vol. 8. New York: Raven. In press
9. Cochin, J. 1970. *Univ. Mich. Med. Cent. J.* 36:225–34
10. Portoghese, P. S. 1970. *Ann. Rev. Pharmacol.* 10:51–76
11. Herz, A., Teschemacher, Hj. 1971. *Advan. Drug Res.* 6:79–119
12. Lewis, J. W., Bentley, K. W., Cowan, A. 1971. *Ann. Rev. Pharmacol.* 11:241–70
13. McIlwain, H. 1971. *Essays Biochem.* 7:127–58
14. Kuschinsky, K. 1972. *Klin. Wochenschr.* 50:401–9
15. Way, E. L., Ho, I. K., Loh, H. H. 1973. *New Concepts in Neurotransmitter Regulation,* ed. A. J. Mandell, 279–95. New York: Plenum
16. Miller, J. W., Elliott, H. W. 1955. *J. Pharmacol. Exp. Ther.* 113:283–91
17. Takemori, A. E., Stenwick, M. W. 1966. *J. Pharmacol. Exp. Ther.* 154:586–94
18. Hug, C. C. Jr. 1967. *Biochem. Pharmacol.* 16:345–59
19. Mellett, L. B., Woods, L. A. 1959. *J. Pharmacol. Exp. Ther.* 125:97–104
20. Teschemacher, Hj., Schubert, P., Herz, A. 1973. *Neuropharmacology* 12:123–31
21. Tsou, K., Jang, C. S. 1964. *Sci. Sinica* 13:1099–1109
22. Foster, R. S., Jenden, D. J., Lomax, P. 1967. *J. Pharmacol. Exp. Ther.* 157:185–95
23. Buxbaum, D. M., Yarbrough, G. G., Carter, M. E. 1970. *Pharmacologist* 12:211
24. Herz, A., Albus, K., Metys, J., Schubert, P., Teschemacher, Hj. 1970. *Neuropharmacology* 9:539–51
25. Albus, K., Schott, M., Herz, A. 1970. *Eur. J. Pharmacol.* 12:53–64
26. Watanabe, H. 1971. *Jap. J. Pharmacol.* 21:383–91
27. Herz, A., Teschemacher, Hj. 1973. *Experientia* 29:64–65
28. Herz, A., Teschemacher, Hj., Albus, K., Zieglgänsberger, S. See Ref. 7, 104–5
29. Eidelberg, E., Barstow, C. A. 1971. *Science* 174:74–76
30. Herz, A., Teschemacher, Hj., Albus, K., Zieglgänsberger, S. 1972. *Psychopharmacologia* 26:219–35
31. Wei, E., Loh, H. H., Way, E. L. 1972. *Science* 177:616–17
32. Wei, E., Loh, H. H., Way, E. L. 1973. *J. Pharmacol. Exp. Ther.* 185:108–15
33. May, M., Czoncha, L., Garrison, D. R., Triggle, D. J. 1968. *J. Pharm. Sci.* 57:884–87
34. Baker, B. R. 1967. *Design of Active-Site-Directed Irreversible Enzyme Inhibitors.* New York: Wiley. 325 pp.
35. Cohen, L. A. 1968. *Ann. Rev. Biochem.* 37:695–726
36. Singer, S. J. 1970. *Ciba Foundation Symposium on Molecular Properties of Drug Receptors.* ed. R. Porter, M. O'Conner, 229–46. London: Churchill

37. Portoghese, P. S., Telang, V. G., Takemori, A. E., Hayashi, G. 1971. *J. Med. Chem.* 14:144–48
38. Goldstein, A., Lowney, L. I., Pal, B. K. 1971. *Proc. Nat. Acad. Sci. USA* 68: 1742–47
39. Pal, B. K., Lowney, L. I., Goldstein, A. See Ref. 7, 62–69
39a. Kakunaga, T., Kaneto, H., Hano, K. 1966. *J. Pharmacol. Exp. Ther.* 153 : 134–41
40. Terenius, L. 1973. *Acta Pharmacol. Toxicol.* 32:317–19
41. Pert, C. B., Snyder, S. H. 1973. *Science* 179:1011–14
42. Snyder, S. H. 1973. Presented at *Current Status of Pharmacological Receptors,* Pharmacology Society Symposium. In *Fed. Proc.* In press
43. Simon, E. See Ref. 42. In press
44. Winter, B. A. Goldstein, A. 1972. *Mol. Pharmacol.* 6:601–11
45. Simon, E. J., Dole, W. P., Hiller, J. M. 1972. *Proc. Nat. Acad. Sci. USA* 69: 1835–37
46. Portoghese, P. S. 1966. *J. Pharm. Sci.* 55:865–87
47. Schild, H. O. 1957. *Pharmacol. Rev.* 9: 242–46
48. Cox, B. M., Weinstock, M. 1964. *Brit. J. Pharmacol.* 22:289–300
49. Takemori, A. E. *Determination of Pharmacological Constants.* See Ref. 8. In press
50. Smits, S. E., Takemori, A. E. 1970. *Brit. J. Pharmacol.* 39:627–38
51. Takemori, A. E., Oka, T., Nishiyama, N. *J. Pharmacol. Exp. Ther.* In press
52. Takemori, A. E., Tulunay, F. C. 1973. *Pharmacologist* 15:242
52a. Tulunay, F. C., Takemori, A. E. Unpublished data
53. Orahovats, P. D., Winter, C. A., Lehman, E. G. 1953. *J. Pharmacol. Exp. Ther.* 109:413–16
54. Collier, H. O. J. 1966. *Advan. Drug Res.* 3:171–88
55. Castles, T. R., Campbell, S., Gouge, R., Lee, C. C. 1972. *J. Pharmacol. Exp. Ther.* 181:399–406
56. Tulunay, F. C., Takemori, A. E. 1973. *Pharmacologist* 15:242
57. Loh, H. H., Shen, F. H., Way, E. L. 1969. *Biochem. Pharmacol.* 18:2711–21
58. Cheney, D. L., Judson, B. A., Goldstein, A. 1972. *J. Pharmacol. Exp. Ther.* 182: 189–94
59. Huidobro, F., Maggiolo, C., Contreras, E. 1963. *Arch. Int. Pharmacodyn.* 144: 196–205
60. Seevers, M. H., Deneau, G. 1968. In *Addictive States,* ed. A. Wikler, 199–205. Baltimore: Williams & Wilkins
61. Shen, F. H., Loh, H. H., Way, E. L. 1970. *J. Pharmacol. Exp. Ther.* 175:427–34
62. Cohen, M., Keats, A. S., Krivoy, W., Ungar, G. 1965. *Proc. Soc. Exp. Biol. Med.* 119:381–84
63. Cox, B. M., Ginsburg, M., Osman, O. H. 1968. *Brit. J. Pharmacol.* 33:245–56
64. Cox, B. M. See Ref. 7, 219–31
65. Yamamoto, I., Inoki, R., Tamari, Y., Iwatsubo, K. 1967. *Jap. J. Pharmacol.* 17:140–42
66. Spoerlein, M. T., Scrafani, J. 1967. *Life Sci.* 6:1549–64
67. Cox, B. M., Osman, O. H. 1970. *Brit. J. Pharmacol.* 38:157–70
68. Smith, A. A., Karmin, M., Gavitt, J. 1967. *J. Pharmacol. Exp. Ther.* 156:85–91
69. Feinberg, M. P., Cochin, J. 1972. *Biochem. Pharmacol.* 21:3082–85
70. Feinberg, M. P., Cochin, J. 1968. *Pharmacologist* 10:188
71. Takagi, H., Kuriki, H. 1969. *Int. J. Neuropharmacol.* 8:195–96
72. Ho, I. K., Lu, S. E., Stolman, S., Loh, H. H., Way, E. L. 1972. *J. Pharmacol. Exp. Ther.* 182:155–65
73. Ho, I. K., Loh, H. H., Way, E. L. 1973. *Eur. J. Pharmacol.* 21:331–36
74. Bhargava, H. N., Ho, I. K., Way, E. L. 1972. *Abstr. Int. Congr. Pharmacol., 5th, San Francisco,* p. 21
75. Matsuda, L. 1970. *Arzneim. Forsch.* 20: 1596–604
76. Bhargava, H. N., Chan, S. L., Way, E. L. 1972. *Proc. West. Pharmacol. Soc.* 15:4–7
77. Kaneto, H. See Ref. 2, 300–9
78. Ungar, G., Cohen, M. 1966. *Int. J. Neuropharmacol.* 5:183–92
79. Schuster, L. 1961. *Nature.* 189:314–15
80. Goldstein, D. B., Goldstein, A. 1961. *Biochem. Pharmacol.* 8:48
81. Kato, R. 1967. *Jap. J. Pharmacol.* 17:499–508
82. Hahn, D. L., Goldstein, A. 1971. *J. Neurochem.* 18:1887–93
83. Dodge, P. W., Smits, S. E., Takemori, A. E. Unpublished data
84. Shuster, L. See Ref. 2, 408–23
85. Way, E. L., Shen, F. H. See Ref. 2, 229–53
86. Way, E. L., Loh, H. H., Shen, F. H. 1968. *Science* 162:1290–92
87. Friedler, G., Bhargava, H. N., Quock, R., Way, E. L. 1972. *J. Pharmacol. Exp. Ther.* 183:49–55
88. Algeri, S., Costa, E. 1971. *Biochem. Pharmacol.* 20:877–84
89. Cheney, D. L., Goldstein, A., Algeri, S., Costa, E. 1971. *Science* 171:1169–71

90. Marshall, I., Grahame-Smith, D. G. 1971. *J. Pharmacol. Exp. Ther.* 179:634–41
91. Schechter, P. J., Lovenberg, W., Sjoerdsma, A. 1972. *Biochem. Pharmacol.* 21:751–53
92. Maruyama, Y., Hayashi, G., Smits, S. E., Takemori, A. E. 1971. *J. Pharmacol. Exp. Ther.* 178:20–29
93. Haubrich, D. R., Blake, D. E. 1969. *Fed. Proc.* 28:793
94. Yarbrough, G. G., Buxbaum, D. M., Sanders-Bush, E. 1973. *J. Pharmacol. Exp. Ther.* 185:328–35
95. Cheney, D. L., Goldstein, A. 1971. *J. Pharmacol. Exp. Ther.* 177:1169–71
96. Schwartz, A. S., Eidelberg, E. 1970. *Life Sci.* 9:613–24
97. Maruyama, Y., Takemori, A. E. 1973. *J. Pharmacol. Exp. Ther.* 185:602–8
98. Hitzemann, R. J., Ho, I. K., Loh, H. H. 1972. *Science* 178:645–47
99. Cheney, D. L., Costa, E. 1972. *Science* 178:647
100. Ho, I. K., Loh, H. H., Way, E. L. 1973. *J. Pharmacol. Exp. Ther.* 185:347–57
101. Bhargava, H. N., Way, E. L. 1972. *J. Pharamacol. Exp. Ther.* 183:31–40
102. Florez, J., Delgado, G., Armijo, J. A. 1973. *Neuropharmacology* 12:355–62
103. Tirri, R. 1967. *Experientia* 23:278
104. Smits, S. E., Takemori, A. E. 1968. *Proc. Soc. Exp. Biol. Med.* 127:1167–71
105. Cox, B. M., Ginsburg, M. See Ref. 4, 77–86
106. Tilson, H. A., Stolman, S., Rech, R. H. 1972. *Res. Commun. Chem. Path. Pharmacol.* 4:581–86
107. Goldstein, A., Sheehan, P., Goldstein, J. 1971. *Nature* 233:126–29
108. Way, E. L. Personal communication
109. Perez-Cruet, J., Tagliomonte, A., Tagliomonte, P., Gessa, G. 1970. *Pharmacologist* 12:257
110. Pozuelo, J., Kerr, F. W. L. 1972. *Mayo Clinic Proc.* 47:621–28
111. Huidobro, F., Contreras, E., Croxatto, R. 1963. *Arch. Int. Pharmacodyn.* 146:444–54
112. Kaymaçalan, S., Tulunay, F. C., Ayhan, I. H., Kiran, B. K. 1972. *Turkish Scientific and Technical Research Council, Project No. TAG/164.* Ankara, Turkey: Turkish Sci. Tech. Res. Counc. 46 pp.
113. Lal, H., Puri, S. K., Karkalas, Y. 1971. *Pharmacologist* 13:263
114. Huidobro, F., Contreras, E., Croxatto, R. 1963. *Arch. Int. Pharmacodyn.* 146:455–62
115. Pinsky, C., Frederickson, R. C. A., Vasquez, A. J. 1973. *Nature* 242:59–60
116. Huidobro, F., Miranda, H. 1968. *Biochem. Pharmacol.* 17:1099–1105
117. Maggiolo, C., Huidobro, F. 1962. *Arch. Int. Pharmacodyn.* 138:157–68
118. Gunne, L. M. 1965. *Arch. Int. Pharmacodyn.* 157:293–98
119. Chiosa, L., Dumitrescu, S., Banaru, A. 1968. *Int. J. Neuropharmacol.* 7:161–64
120. Iwamoto, E. T., Shen, F. H., Loh, H., Way, E. L. 1971. *Fed. Proc.* 30:278
121. Huidobro, F., Larrain, G. 1965. *Arch. Int. Pharmacodyn.* 155:205–15
122. Iwamoto, E. T., Ho, I. K., Way, E. L. *J. Pharmacol. Exp. Ther.* In press
123. Collier, H. O. J., Francis, D. L., Schneider, C. 1972. *Nature* 237:220–23
124. Wang, J. H., Takemori, A. E. 1972. *J. Pharmacol. Exp. Ther.* 183:41–48

FUNGAL SEX HORMONES

×839

Graham W. Gooday

Department of Biochemistry, University of Aberdeen
Aberdeen AB9 1AS, Scotland

CONTENTS

INTRODUCTION

The fungus is a cornucopia of a seemingly infinite variety of strange and wonderful metabolites. The first volume of the *Annual Review of Biochemistry* and many succeeding volumes have charted their discovery. Many of these metabolites are of great interest and importance to man, but most have little or no apparent role to play in the lives of the fungi. However, the following pages are concerned with metabolites that we can be sure are of great interest to the cells producing them—the fungal sex hormones.

These hormones are produced by cells as very specific molecules that have profound morphogenetic effects on other cells of the same or closely related species. They regulate the temporal and spatial coordination of the sequence of events leading to the sexual pairing and fusion of nuclei from two cells. The hormones are released from their site of synthesis, and by their diffusion they define the cells that will be involved in the sexual interactions. With one notable exception they are produced in very small amounts and are all active in very small amounts. Fungal sex hormones tend to be inherently unstable and to be destroyed by recipient cells. The use of the word "hormone" for these substances is clearly fully justified (1, 2).

There is good evidence for the action of hormones over the entire taxonomic

35

spectrum of fungi, but this review concentrates on our understanding of the biochemistry of the few hormones that are completely or partially characterized. The reader is directed to a number of recent sources for accounts of the uncharacterized hormones and of the biological and historical backgrounds to those discussed here (1–10).

STEROIDS

Antheridiol

The only completely characterized steroid sex hormone outside the animal kingdom is antheridiol, the female hormone of the water mold *Achlya*. The involvement of diffusible sex hormones controlling the sexual reproduction of *Achlya* species has been beautifully demonstrated by Raper (1, 3, 11, 12). Female cells of *A. bisexualis* and *A. ambisexualis* were shown to secrete a biologically active metabolite, hormone A, now called antheridiol, to which male cells respond by producing many branched sinuous filaments, the antheridial hyphae. The number of branches produced is dependent on the amount of antheridiol and this forms the basis for the bioassay. The responding male cells then secrete hormone B, which has a reciprocal effect, causing female cells to produce the oogonia. The antheridial hyphae grow towards a developing oogonium, become appressed to it, and delimit the multinuclear antheridia, from which nuclei migrate to fertilize the female gametes.

Antheridiol was extracted and purified by Raper & Haagen-Smit in 1942 (13) by laborious techniques involving selective precipitations, partitionings, and adsorptions, and the product (2 mg from 1440 liters of female culture filtrate) was biologically active at a few pg/ml. Although it was impossible to characterize this product then, comparison with the biological activity of authentic antheridiol shows that it must have been substantially pure, and this has recently been directly confirmed by showing that antheridiol and 23-deoxyantheridiol were present in a

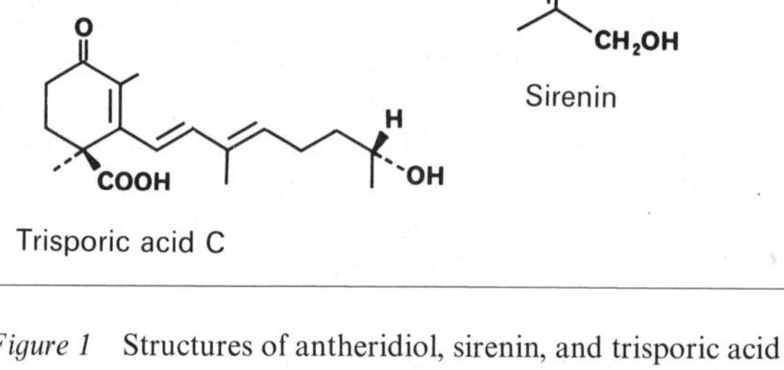

Figure 1   Structures of antheridiol, sirenin, and trisporic acid C.

surviving sample (14). Antheridiol was re-extracted by McMorris & Barksdale in 1967, who reported its spectral properties and molecular formula (15). High resolution mass spectrometry indicated a steroid nucleus, and further spectral studies enabled two plausible structures to be suggested (16), the more favored one proving to be correct (Figure 1). The first synthesis of antheridiol quickly followed (17) and gave the two C-22,23 trans-isomers, one of which was biologically active. The stereochemistry of 22S,23R was suggested by comparing a cometabolite of antheridiol, 23-deoxyantheridiol, with its synthetic C-22 epimer (14), and was confirmed by stereospecific synthesis of the 7-deoxysterols with the four possible corresponding isomeric dihydroxybutenolides. These were identified by their circular dichroism, and oxygenation of the suspected isomer gave a product identical with natural antheridiol (18). Two higher yielding syntheses have recently been reported: one giving all four C-22,23 epimers (19) and one giving stereochemically pure antheridiol (20). A synthesis designed without prior knowledge of the absolute stereochemistry has given what is probably the C-22 epimer of antheridiol admixed with about 5% antheridiol (21). An isomer, with the γ-lactone ring at C-24 instead of C-23, has also been synthesized (22).

The structural and stereochemical requirements for antheridiol hormone activity are highly specific. The three C-22,23 stereoisomers of antheridiol all show less than 0.1% the activity of antheridiol. 7-Deoxy-7-dihydroantheridiol and its 3-acetate each have about 5% the activity of antheridiol. A wide range of related sterols are either totally inactive or have such low activity as could be explained by contaminating antheridiol. The synthetic isomer (22) of antheridiol shows some antagonism, as it inhibits the formation of antheridia when added with antheridiol in the ratio 200:1 (T. C. McMorris, personal communication).

The structure of hormone B, produced by male cells responding to antheridiol, has not yet been elucidated. It is produced in much smaller quantities, only in the presence of antheridiol. It appears to be a closely related sterol, and it is very difficult to separate the two. An attractive possibility is that hormone B is a transformation product of antheridiol, but this seems unlikely considering the diffusion losses that would be involved. Hormone B is produced by some hermaphroditic strains without requiring the stimulus of added antheridiol (23). It is currently being characterized and appears to be a 7-keto-C_{29} sterol, esterified at C-3, and lacking the lactone ring of antheridiol (A. W. Barksdale, T. C. McMorris, and G. P. Arsenault, personal communication).

Cholesterol, 24-methylene cholesterol, 7-dehydrofucosterol, and fucosterol have been identified as major sterols of A. bisexualis and other members of the Saprolegniales (24, 25). Plausible biosynthetic routes to antheridiol have been suggested, and C. R. Popplestone and A. M. Unrau (personal communication) have synthesized possible precursors to investigate these routes. By demonstrating radioactive incorporation of successive precursors and products these researchers have shown that A. bisexualis can produce antheridiol from 7-dehydrofucosterol via fucosterol and its 22-dehydro and 22-dehydro-29-carboxyl derivatives.

Antheridiol is active in very low concentrations. The most receptive male strains will respond to a 10 pM solution by branching. Strains of Achlya species can be arranged in a sexual series, ranging from strong female to strong male, with some intermediate strains (12, 26). Those that can act as males respond to added antheridiol by branching and by secretion of hormone B (6, 23, 27, 28). Barksdale has shown that antheridiol can elicit the complete series of male responses. Thus antheridial branches will grow towards plastic particles with adsorbed antheridiol and will wrap around them and delimit antheridia in which meiosis then occurs (6, 26, 29). Male *Achlya* cultures efficiently remove added antheridiol from the medium, and it cannot be extracted back from the cells (27). The male response to antheridiol can be modulated by a number of factors (6, 30): (a) the concentration of antheridiol determines the rate and intensity of response, successively higher concentrations being required for branching, chemotropism, and delimitation of antheridia; (b) different strains of *Achlya* species show wide differences of response; (c) the nutrients available to the culture directly control the response.

Addition of antheridiol-containing extracts to male cultures can result in increases in protein synthesis, RNA synthesis, cellulase activity, and oxygen uptake that precede and accompany the resulting branching (31-37). These responses can be inhibited by actinomycin D, p-fluorophenylalanine, puromycin, and cyclo-heximide.

The concomitant increases in cellulase and branching in response to antheridiol have been interpreted as an example of the well-documented involvement of localized wall-softening in apical growth. Cellulose is a major component of the wall of *Achlya,* and it is envisaged that the cellulase weakens the wall to allow "blow-outs." Evidence in favor of this hypothesis includes the observations that the cellulase had no clear nutritional role when cellulose was provided as sole carbon source (33) and that cellulase activity and branching always showed a positive correlation in a wide range of experimental conditions (38). However, this interpretation must be treated with caution, as the cellulase is released as an extracellular enzyme and could well have a nutritional role in a natural environment. The release of extracellular catabolic enzymes, not always with clear nutritional or morphogenetic functions, commonly accompanies microbial differentiation, and the control of the formation of these enzymes in general is poorly understood.

Other Sterols

Sterols are implicated in the regulation of sexual reproduction of the important plant parasitic fungi, *Phytophthora* and *Pythium,* close relatives of *Achlya.* These fungi are apparently unable to synthesize sterols and do not require sterols for vegetative growth. However, they require added sterols for the formation of the sexual oospores (39-41). 3-β-Hydroxysterols with a Δ five double bond and a ten-carbon side chain, such as fucosterol, the precursor of antheridiol, are the most active (40-42). An exciting possibility is that the wide range of active sterols are metabolized to give a specific hormone, possible similar to antheridiol, just as insects make ecdysone from cholesterol in their diet. Preliminary work indicates that cholesterol and sitosterol are rapidly metabolized by *Phytophthora* and *Pythium* (43; C. G. Elliott, personal communication).

There is evidence that sterols are involved in regulating sexual reproduction in yeasts. In *Saccharomyces cerevisiae* two hormones have been described: one produced by each mating type and active on the opposite mating type. The *a* and *α* hormones act on vegetative cells of the *α* and *a* mating types respectively, causing an increase in cell volume, a prerequisite for mating (44). Their chromatographic and spectral properties suggest that they are steroids, possibly related to ergosterol (45). The expansion of *α* cells in response to *a* hormone is inhibited by actinomycin D, chloramphenicol, and cycloheximide (44), and by gluconolactone, an inhibitor of *β*-glucanase that thus may be enzymically active in cell expansion (46). Similar expansions can also be obtained by addition of testosterone and estradiol respectively and by the addition of auxins (46), and so their effects appear to be less specific than that of the peptide "*α* factor" (discussed in detail later) that inhibits cell division and causes marked elongation of *a* cells (47).

Other reports of sterols being implicated in sexual reproduction in fungi have recently been reviewed (39). In no case is there any direct evidence that sterol hormones are involved.

OTHER TERPENOIDS

Trisporic Acid

Trisporic acid is a sex hormone for the Mucorales, an order of fungi including many common molds. Sexual reproduction involves fusion of two cells of opposite mating type, designated $(+)$ and $(-)$ (8, 10). In the absence of a compatible partner, vegetative cells of *Mucor* differentiate to give large numbers of the asexual sporangiospores. In the presence of a compatible partner, this developmental sequence is supplanted by the formation of sexual hyphae, the zygophores. The zygophores fuse in mated pairs and the end result is the formation of a large resistant spore, the sexual zygospore. Trisporic acid regulates the formation of the zygophores. In the absence of trisporic acid, unmated cells of both mating types form sporangiospores; in the presence of trisporic acid they form zygophores. The number of zygophores formed by *Mucor mucedo* can readily be counted and this is the basis of the bioassay (48–51). The hormonal control of zygophores had been demonstrated by Burgeff in 1924 (52), but it was over thirty years later that he obtained the first cell-free activity (53), and although Plempel investigated this in great detail, the active principle was not fully characterized (50, 54).

Identification of trisporic acid as a sex hormone had a quality of serendipity, as it had been characterized as a fungal metabolite several years previously. It had been observed that mated fermentations of some members of the Mucorales were bright orange, having a much higher carotene content than unmated fermentations, and the trisporic acids were characterized as metabolites of mated *Blakeslea trispora* that caused a marked increase in carotene content when added to unmated cultures (55–61). This work used submerged fermentations, and so the morphogenetic action of trisporic acid was not suspected until a sex hormone extracted from *B. trispora* by Ende (62) and from *M. mucedo* by Gooday (48, 63), which caused zygophore formation in unmated cultures of *M. mucedo,* was shown to have very similar chemical characteristics. Austin, Bu'Lock & Gooday (64) directly compared the Mucor hormone and authentic trisporic acid C, and showed them to be identical.

The major component from *M. mucedo* and *B. trispora* is the C-13 alcohol, trisporic acid C (ca 80%) (Figure 1), with the C-13 ketone, trisporic acid B (ca 15%), and very much smaller amounts of the C-13 deoxy derivative, trisporic acid A (49, 64, 65). A C-13 dehydro derivative has also been described (61). The natural products are mixtures of 9-*cis*- and 9-*trans*-isomers (49, 55, 61). The stereochemistry of trisporic acid as 13-R has been determined from degradation studies (61, 66) and as 1-S from circular dichroism of the hydrogenation product. It was also shown that $[2\text{-}^{14}C]$ mevalonate selectively labeled the 1-methyl group (67). This determination of the stereochemistry of the ring by Bu'Lock and co-workers is of particular interest, as trisporic acid is a metabolite of β-carotene, and this has enabled them to suggest the stereochemistry of cyclization during β-carotene biosynthesis.

Trisporic acid C (racemic 7-*trans*-9-*trans*-methyl ester) has been synthesized by Edwards and co-workers and shown to be biologically active (68). Other syntheses include a mixture of the 9-*cis*- and 9-*trans*-isomers (69), their anhydro derivatives, and trisporone (S. Isoe, personal communication).

Mated fermentations of *B. trispora* produce a family of related neutral metabolites, the major component being trisporol C (49, 66). Other cometabolites include trisporone and anhydrotrisporone (70), and the C_{18}-ketone (13-apocarotenone) (71, 72).

To date, trisporic acid has only been directly identified as a metabolite from *M. mucedo, B. trispora,* and *Phycomyces blakesleeanus* (57, 62, 64, 65, 73). There is little doubt that it is a regulator of sexual reproduction throughout the Mucorales, because interspecific sexual interactions involving the mutual stimulation of zygophore formation are widespread in these fungi (8).

Until recently, trisporic acid was only detected in mated cultures, but careful work by Sutter and co-workers has shown that very small amounts are present in unmated (+) *B. trispora* cultures (74). Retinal, almost certainly a precursor of trisporic acid, is a very minor metabolite of *P. blakesleeanus* (75), and unmated cultures of *B. trispora* produce small amounts of very complex families of neutral metabolites related to, or precursors of, trisporic acids (74, 76–78; J. D. Bu'Lock, personal communication).

Biosynthesis of the trimethyl cyclohexyl ring of the trisporic acids had only been demonstrated at the C_{40} level in the cyclic carotenoids, and this led to the idea that β-carotene, the major carotenoid of the Mucorales, is the biosynthetic precursor of trisporic acids (64, 79). There is now considerable evidence to support this suggestion. Circumstantial evidence includes the coinhibition of carotene and trisporate biosyntheses by diphenylamine (54, 79) and the impairment of sexuality in carotene-deficient mutants of *P. blakesleeanus* (80; R. P. Sutter, personal communication). More directly, Bu'Lock and co-workers have shown that mevalonate, β-carotene, retinal, the β-C_{18}-ketone (13-apocarotenone), and its 4-hydroxy derivative are incorporated into trisporic acids with increasing efficiency by mated cultures of *B. trispora* (64, 66, 71, 72, 78). There is no direct evidence of the nature of the enzymes involved, but a series of mixed function oxygenases could be involved, because barbiturate, which can increase their activity, does lead to increased trisporate biosynthesis (65).

The overriding biochemical interest in trisporic acid is in the control of its de novo biosynthesis. Unmated cultures, in effect, do not produce it; when brought together in mated pairs, they institute a collaborative biosynthesis, and in B. *trispora* it becomes a major metabolite. This control, in which the sex hormone is only biosynthesized in the presence of a compatible partner, must be of great importance to the fungi, as the resultant sexual zygophores are formed at the expense of the asexual spores that would be the everyday means of dispersal.

A series of models has been suggested and rejected in turn to explain the mechanism of the collaborative biosynthesis. The following observations have to be taken into account: (a) The mutual switch-on of trisporate biosynthesis and consequent zygophore formation can occur by diffusion through the medium; cell contact is not necessary (48, 52, 81). (b) Both mating types of B. *trispora* contribute to the final yield of trisporic acids, as labeling cells of either mating type with ^{14}C results in ^{14}C-trisporic acid after mixing (51, 82). (c) Inhibitors of protein and RNA biosynthesis will inhibit the de novo trisporate biosynthesis that results a few hours after ($+$) and ($-$) cells are mixed (7, 51, 82–84).

Control of trisporate biosynthesis thus requires at least two attributes from each mating type: the production of an inducer with the ability to derepress the necessary enzyme biosynthesis in the opposite mating type, and the ability to respond to the corresponding inducer from the opposite mating type. The stumbling block has been the demonstration of these inducers. A simple batch exchange of media between unmated cultures does not lead to normal trisporate synthesis, and so either continual supplies of inducers or more cross-diffusion steps are required. Sutter and co-workers (74, 76) and Ende and co-workers (77, 82) have shown that both ($+$) and ($-$) unmated cultures will synthesize very small amounts ($<1\%$ normal mated yield) of trisporic acids from neutral precursors from the opposite mating type. These conversions apparently do not require new protein synthesis, because they are little affected by cycloheximide (74, 76). These authors have now presented evidence for mechanisms whereby the rate of production of these precursors can be increased in mated cultures, apparently both by factors from the opposite mating type (74) and by trisporic acid itself (77).

Bu'Lock and co-workers have investigated related cometabolites and precursors of trisporic acids in mated and unmated cultures of B. *trispora,* and the respective abilities of these cultures to transform or react to these metabolites. The zygophore bioassay, with M. *mucedo,* shows that ($+$) and ($-$) both respond to a few nanograms of the 9-*cis*- and 9-*trans*-trisporic acids B and C (49). The ($-$) responds much more than the ($+$) to chemically methylated trisporates (49, 64), and B. *trispora* ($-$) readily hydrolyzes these esters whereas ($+$) scarcely does. To date, no other characterized metabolites show significant selective activity towards ($-$) M. *mucedo,* but the trisporate precursors, trisporol C and β-C_{18}-ketone, show some activity towards ($+$) M. *mucedo.* These activities are mirrored by B. *trispora,* as when fed with β-C_{18}-ketone, ($+$) can convert it to trisporic acid, but ($-$) accumulates the intermediate, trisporol C (72).

The $(+)$ and $(-)$ *B. trispora* show quite different patterns of transformation products from methyl trisporates, the $(+)$ favoring oxidations, and the $(-)$ favoring hydrolysis and reductions (49). The $(+)$ and $(-)$ *B. trispora* also have marked differences in some of their neutral metabolites related to trisporate, those from $(+)$ being predominantly methyl esters with a 4-hydroxyl groups, and those from the $(-)$ being predominantly 4-keto-alcohols (J. D. Bu'Lock, personal communication). Bu'Lock suggests that there is more than one possible sequence to trisporic acids (i.e. a "metabolic grid") with correspondingly different products from side reactions, and that $(+)$ and $(-)$ differ in the patterns of these reactions (85). Cross-diffusion of the products could then derepress the main route to trisporic acids in mated cultures. In this view, the gene products from the mating type locus would be two forms of a regulator, differing in their patterns of affinities for gene sites and for derepression.

Continual cross-diffusion of inducers would then be required if those produced by one mating type acted as substrates for the opposite mating type, but, because of their different sequence, were not present in the opposite mating type.

Two recent unconfirmed reports may also have to be taken into account in any final scheme: that trisporate biosynthesis can be induced by agents unable to cross an ultrafilter and therefore possibly macromolecules (86) and that it can be induced by volatile metabolites from the opposite mating type (87). Unfortunately no work has been done with isogenic strains and it is difficult to decide which observations are due to random strain differences and which to true sexual differences (8). For example, Bu'Lock and co-workers (88) have shown that different strains have considerable differences in the relationship between trisporate synthesis and the ratio of $(+):(-)$ that was formerly considered a significant factor in any proposed mechanism. Genetic evidence is little help, as we know only that mating type appears to be controlled by a pair of allelomorphs at one locus, but it is not clear whether a single pair of genes or two groups of genes are involved (89).

The most important action of trisporic acid is to cause the formation of zygophores, but more work has centered on its more accessible effect of increasing isoprenoid biosynthesis. Its action as a carotenogenic stimulator in *B. trispora, M. mucedo,* and *P. blakesleeanus* is well documented (8, 48, 54, 59, 81, 90–92). Inhibition of this action by cylcoheximide suggests that it involves enzyme derepression (60). β-Carotene is the precursor of trisporic acid, and so the system appears self-amplifying. Another role for the increased carotene content has recently been described in *M. mucedo* as the resistant protective material of pollen grains, sporopollenin, has been discovered in the outer wall of the zygospore. It is a polymer of oxygenated cross-linked carotenoids, and ^{3}H-β-carotene was incorporated into the sporopollenin in *Mucor* zygospores (93). Bu'Lock & Osagie (94) have confirmed the report (60) that sterol biosynthesis is increased concomitantly with carotenogenesis in *B. trispora* treated with trisporic acid, and have further shown increases in biosynthesis of the polyprenols and ubiquinones. Clearly trisporic acid acts at an early stage to stimulate all isoprenoid biosynthesis. Bu'Lock has recently drawn an analogy between well-documented cases of the triggering of microbial differentiation by nutrient starvation and this stimulation of isoprenoid biosynthesis with its consequent drain on acetyl-CoA, NADPH, and ATP (72, 95).

Mention must be made of the way that $(+)$ and $(-)$ zygophores grow towards one another in pairs over considerable distances. This phenomenon, zygotropism, must be controlled by volatile chemicals but had remained intractable despite considerable efforts (8). However, a suggestion that the inducers of trisporate biosynthesis might also serve as zygotropic hormones (8) is supported by work by Ende and co-workers, showing that volatile extracts from $(+)$ and $(-)$ cultures can give rise to trisporic acid and also attract the resultant zygophores in the opposite mating type (87).

Sirenin

Haploid plants of the water mold *Allomyces* produce male and female gametangia, usually in pairs. When mature, the motile male and female gametes are released from the gametangia. The smaller male cells swim actively and fuse with the more sluggish females to form diploid motile zygotes. Machlis showed that the female gametes synthesize a metabolite, sirenin, that attracts the male gametes by chemotaxis (96–98). Sirenin can be bioassayed by counting the number of male gametes attracted to the surface of a cellophane membrane separating them from the test solution (99, 100).

Production of sufficient sirenin for characterization was elegantly accomplished by Machlis and co-workers (101). The growth of large volumes of the fungus had to be carefully designed to obtain the maximum yield. Presence of significant numbers of male gametes had to be avoided, as they would give rise to zygotes and a culture overrun with the resulting diploid plants, and so the fungus used was a predominantly female hybrid between two *Allomyces* species.

Nutting, Rapoport & Machlis (102, 103) determined the structure of sirenin as a bicyclic sesquiterpenediol (Figure 1) and suggested its stereochemistry by analogy with the corresponding hydrocarbon, sesquicarene, that had just been described. Biologically active sirenin has now been synthesized by several routes: first as the racemate (104–110), and more recently as the separate *d*- and *l*-sirenins (111). Plattner & Rapoport have confirmed the stereochemistry, both by direct conversion of $(-)$-monodeoxysirenin into $(-)$-sesquicarene and by circular dichroism of the two related bicyclic ketones. Syntheses have also given isomers and deoxy isomers (107, 108). The biosynthesis of sirenin is almost certainly via farnesyl pyrophosphate, and several plausible schemes can be suggested.

Machlis (112), using an improved bioassay, has tested the available isomers and analogs of sirenin. Only *l*-sirenin has biological activity, and other isomers show no antagonism. Chelated trace elements and 3 mM Ca^{2+} are required for maximal response (100). Five different motile cells are formed during the life cycle of *Allomyces*: haploid and diploid zoospores, male and female gametes, and zygotes. Only the male gametes respond to sirenin. Only the zoospores and zygotes are chemotactic to amino acids (113, 114). At fertilization, the male gametes lose their response to sirenin and the resulting zygote very quickly gains the response to amino acids (115). Male gametes of different species of *Allomyces* show considerable differences in the sensitivity of their response to sirenin (99), but the most sensitive respond to a solution of 22 pg/ml compared with the 400 μg/ml casein hydrolysate required for the chemotaxis of zygotes (113).

The male gametes specifically inactivate sirenin, and Carlile & Machlis (116) suggest that this could be a mechanism whereby they maintain their sensitivity over a very wide concentration gradient (10^{-10} to 10^{-5} M). In a more detailed analysis, Machlis (112) has shown that this uptake of sirenin follows first order kinetics from between 5 and 400 nM solutions. The sirenin cannot be extracted back from the spores and so must be metabolized to an inactive product. The gametes lose their ability to respond to sirenin after taking it up (e.g. a period of 45 min was required to regain the full response from gametes that had previously been exposed for 30 min to 5 nM sirenin). It is possible that the metabolism of sirenin is part of the chemoreception mechanism (116). As the cells can respond to a concentration gradient, the primary site of reception must be at the cell surface, and the ultimate site of action must be the flagellum (112).

HYDROCARBONS

Although the subject of their investigation is not fungal, mention must be made of the work by Müller, Jaenicke & co-workers (117, 118), showing that the volatile algal sex attractants of the seaweeds *Ectocarpus* and *Fucus* are two partly unsaturated hydrocarbons: ectocarpen and fucoserraten. These are released by the nonmotile eggs and specifically attract the motile male gametes.

There has been a preliminary characterization of hydrocarbons as sex hormones in *Neurospora crassa* (119–122). These materials can be extracted both from mating types and from mated cultures. They increase the fertility of mated (*A/a*) crosses of *Neurospora* and increase the number of *A* protoperithecia. The number of active compounds and their natural regulatory roles are not clear. Spectral and analytical data indicate that they are unsaturated hydrocarbons (ca 350 mol wt).

PEPTIDES

Yeast α Factor

Sexual reproduction in the yeast *Saccharomyces cerevisiae* is controlled by the two alternative mating-type alleles, *a* and α, and haploid cells of opposite mating type conjugate to give diploid zygotes. In 1956 Levi showed that a diffusible product from α cells causes the formation of elongate copulatory processes by *a* cells without need for cell contact (123). Duntze, MacKay & Manney (47) have extracted this α factor from culture filtrates of α cells. It can be bioassayed by observing the elongation and concomitant cessation of normal growth by budding of test *a* cells. Genetic analysis of sterile mutants of α cells strongly suggests that secretion of α factor is controlled by the mating type genes, as more than half of these nonmating mutants have lost the ability to produce α factor (47, 124).

Duntze and co-workers (125) have purified and partly characterized α factor as a peptide with molecular weight about 1400. Leucine, glycine, proline, glutamic acid, tyrosine, tryptophan, and possibly histidine are hydrolysis products. The extracted α factor is complexed with copper ions, which remain associated with the peptide during all purification procedures including electrophoresis at pH 6.5, but can be removed by electrophoresis at pH 3.6. Although the resulting copper-free factor retains its biological activity, it is probable that during the bioassay it recomplexes with traces of Cu^{2+} in the medium or even within the cells.

The effect of α factor on a cells is very specific. Cell division is inhibited; the cells form elongate processes and eventually become long, irregularly shaped giant cells up to 10 times the length and 30 times the dry weight of vegetative haploid cells. However, α cells show no response to α factor (47, 126).

DNA replication is inhibited in the responding a cells, but rates of protein and RNA synthesis are unchanged for several hours (47, 126, 127). Bücking-Throm and co-workers (127), using asynchronous and synchronous cultures and a range of temperature-sensitive cell cycle mutants, have shown that α factor specifically and reversibly arrests a cells at, or just before, the point of initiation of DNA synthesis.

Thus addition of α factor to an asynchronous culture leads to the accumulation of unbudded, mononucleate G1 phase cells. Subsequent washing away of α factor is quickly followed by a pulse of DNA synthesis, which continues after readdition of α factor. Careful observations of mating cells suggest that there is a reciprocal effect on α cells, as both α and a cells accumulate as the characteristic unbudded cells at the early stage of mating (127, 128), but as yet no corresponding a factor has been described. Hartwell (128) has shown that asynchronous haploid cultures require an interval of courtship before cells of opposite mating type can fuse. The cell cycles of both cells are synchronized during this courtship as they are un-budded and mononucleate at the time of fusion. However, Sena et al (129) suggest a more cautious interpretation of the role of α factor, as they have observed inhibition of mating associated with morphological alterations characteristic of α factor action.

COMPARATIVE BIOCHEMISTRY

Although there are very few characterized fungal sex hormones, it is worth discussing them in relation to other regulatory molecules. Table 1 illustrates their essential hormonal properties: the extreme specificity and sensitivity of their production and action; their instability and relatively small size, which would allow a concentration gradient to be maintained; and the rigid control of their bio-synthesis by different cells of the same species.

A clear comparison may be made between the sterol, antheridiol, and animal steroid hormones. Animals have made great use of the capacity of the steroid nucleus for elaboration to give specific regulatory molecules. Antheridiol has a larger side chain at C-17 than animal hormones but does share with them the oxygen functions at either end of the nucleus and is probably also derived from cholesterol. Perhaps the closest comparison may be made with the C_{27} insect hormones, such as ecdysone. Similarly, sirenin is like the "paper factors" with insect juvenile hormone activity, such as juvabione. Ecdysone and juvabione are metabolites of some plants, where a protective role has been ascribed to them, but

Table 1   Properties of fungal sex hormones

	Antheridiol	Hormone B	Trisporic Acid	Sirenin	α Factor
Formula (mol wt)	$C_{29}H_{42}O_5$ (470)	Sterol (ca 500)	$C_{18}H_{26}O_4$ (306)	$C_{15}H_{24}O_2$ (236)	Peptide (ca 1400)
Probable precursor	Cholesterol	—	Retinal	Farnesyl pyrophosphate	—
Specificity of production and activity	*Achlya* spp.	*Achlya* spp.	Mucorales	*Allomyces* spp.	*Saccharomyces* sp.
Site and control of synthesis	♀ Cells	♂ Cells in response to antheridiol	(+) and (−) cells in collaboration	♀ Gametes	α Cells
Optimal yield (M)	6×10^{-9}	< antheridiol	5×10^{-4} (*B. trispora*) 5×10^{-7} (*M. mucedo*)	10^{-6} (Hybrid)	4×10^{-8}
Sensitivity of bioassay	10^{-11} M	—	10^{-8} M	10^{-10} M	10^{-8} M
Morphogenetic action	Antheridia formed by ♂	Oogonia formed by ♀	Zygophores form by (+) and (−)	Chemotaxis of ♂	Elongation of *a* cells
Biochemical action	Increases in cellulase, respiration, proteins, RNA	—	Increases in isoprenoids	—	Inhibition of DNA synthesis
Stability	Destroyed by ♂	—	Photolabile	Destroyed by ♂	—
References	6, 16, 26, 37	6, 23	8, 51, 64, 72	100, 101, 112	47, 125, 127

with the precedents of antheridiol and sirenin such metabolites as these may have hormonal roles in higher plants (130).

Antheridiol is clearly more closely related to the C_{29} phytosterols, such as stigmasterol, with a C_2 side group at C-24, than to the typical fungal sterols, such as ergosterol, with a C_1 side group. This is consistent with current phylogenetic views, as *Achlya* is a member of the Oomycetes, a small primitive group of fungi with cellulosic walls and other attributes more akin to green plants than to the majority of fungi.

Trisporic acid has similarities to a number of biologically active molecules, such as its precursor, retinal, and the plant hormone, abscisic acid. There is an unconfirmed report that trisporic acid has abscisin-like activity (131), but this would be the only case of nonspecific activity of a fungal sex hormone and could well be due to a decomposition product.

The peptide yeast α factor, specifically inhibiting DNA synthesis, has parallels with the peptide antibiotics such as those formed during spore formation in *Bacillus* spp., some of which bind cations and at least one of which (edeine) inhibits DNA synthesis, and also with the fungal α-amanitin, which inhibits RNA synthesis.

The special attribute that distinguishes fungal sex hormones from other hormones is that they are all concerned with the intercellular coordination of events leading to the sexual fusion between two cells. Sexual reproduction is a highly desirable but expensive luxury for a microorganism. Asexual sporulation occurs readily when triggered by starvation. But the chances of meeting a compatible partner are so small that any opportunity must be seized for sexual fusion, even at the expense of vegetative growth and asexual reproduction. The specificity and instability of sex hormones help to ensure that these expensive sexual excursions occur only when there is a good chance of a compatible mating.

ACKNOWLEDGMENTS

I thank the many authors who have sent me reprints and preprints of their publications, and those authors cited in the text who have sent me details of unpublished work. In particular I thank Dr. J. D. Bu'Lock for many stimulating discussions.

Literature Cited

1. Raper, J. R. 1967. *Handbuch der Pflanzenphysiologie,* ed. W. Ruhland, 18 : 214–34. Berlin : Springer
2. Machlis, L. 1972. *Mycologia* 64 : 235–47
3. Raper, J. R. 1971. *Plant Physiology,* ed. F. G. Steward, 6A : 167–222. New York : Academic
4. Raper, J. R. 1973. In *Biology Data Book,* ed. P. L. Altman, D. S. Dittmer, 2 : 665–68. Bethesda : FASEB
5. Machlis, L. 1966. In *The Fungi,* ed. G. C. Ainsworth, A. S. Sussman, 2 : 415–33. New York : Academic
6. Barksdale, A. W. 1969. *Science* 166 : 831–37
7. Gooday, G. W. 1972. *Biochem. J.* 127 : 2–3P
8. Gooday, G. W. 1973. *Symp. Soc. Gen. Microbiol.* 23 : 269–94
9. Köhler, K. See Ref. 1, 282–320
10. Burnett, J. H. 1968. *Fundamentals of Mycology.* London : Arnold. 546 pp.
11. Raper, J. R. 1939. *Science* 89 : 321–22
12. Raper, J. R. 1940. *Mycologia* 32 : 710–27
13. Raper, J. R., Haagen-Smit, A. J. 1942. *J. Biol. Chem.* 143 : 311–20
14. Green, D. M., Edwards, J. A., Barksdale, A. W., McMorris, T. C. 1971. *Tetrahedron* 27 : 1199–1203
15. McMorris, T. C., Barksdale, A. W. 1967. *Nature* 215 : 320–21
16. Arsenault, G. P., Biemann, K., Barksdale, A. W., McMorris, T. C. 1968. *J. Am. Chem. Soc.* 90 : 5635–36
17. Edwards, J. A., Mills, J. S., Sundeen, J., Fried, J. H. 1969. *J. Am. Chem. Soc.* 91 : 1248–49
18. Edwards, J. A., Sundeen, J., Salmond, W., Iwadare, T., Fried, J. H. 1972. *Tetrahedron Lett.* 9 : 791–94
19. McMorris, T. C., Seshadri, R. 1971. *Chem. Commun.* 1646
20. McMorris, T. C., Arunachalam, T., Seshadri, R. 1972. *Tetrahedron Lett.* 26 : 2673–76
21. Smith, G. A., Williams, D. H. 1972. *J.*

Chem. Soc. Perkin Trans. I 2811–16
22. McMorris, T. C. 1970. *J. Org. Chem.* 35 : 458–60
23. Barksdale, A. W., Lasure, L. L. 1974. *Bull. Torrey Bot. Club.* In press
24. McCorkindale, N. J., Hutchinson, S. A., Pursey, B. A., Scott, W. T., Wheeler, R. 1969. *Phytochemistry* 8 : 861–67
25. Popplestone, C. R., Unrau, A. M. 1973. *Phytochemistry* 12 : 1131–33
26. Barksdale, A. W. 1967. *Ann. NY Acad. Sci.* 144 : 313–19
27. Barksdale, A. W. 1963. *Mycologia* 55 : 164–71
28. Ibid 1965. 57 : 493–501
29. Ibid 1963. 55 : 627–32
30. Ibid 1970. 62 : 411–20
31. Thomas, D. S., Mullins, J. T. 1967. *Science* 156 : 84–85
32. Mullins, J. T. 1968. *J. Elisha Mitchell Sci. Soc.* 84 : 195–98
33. Thomas, D. S., Mullins, J. T. 1969. *Physiol. Plant.* 22 : 347–53
34. Warren, C. O., Mullins, J. T. 1969. *Am. J. Bot.* 56 : 1135–42
35. Thomas, D. S. 1970. *Can. J. Bot.* 48 : 977–79
36. Warren, C. O., Sells, B. H. 1971. *J. Gen. Microbiol.* 67 : 367–69
37. Kane, B. E., Reiskind, J. B., Mullins, J. T. 1973. *Science* 180 : 1192–93
38. Mullins, J. T. 1973. *Mycologia* 65 : 1007–14
39. Hendrix, J. W. 1970. *Ann. Rev. Phytopathol.* 8 : 111–30
40. Elliott, C. G. 1972. *J. Gen. Microbiol.* 72 : 321–27
41. Elliott, C. G., Hendrie, M. R., Knights, B. A. 1966. *J. Gen. Microbiol.* 42 : 425–35
42. Child, J. J., Haskins, R. H. 1971. *Can. J. Bot.* 49 : 329–32
43. Hendrix, J. W., Bennett, R. D., Heftmann, E. 1970. *Microbios* 5 : 11–15
44. Yanagishima, N. 1969. *Planta* 87 : 110–18
45. Takao, N., Shimoda, C., Yanagishima,

N. 1970. *Develop. Growth Differentiation* 12:199–205
46. Yanagishima, N., Shimoda, C., Takahashi, T., Takao, N. 1970. *Develop. Growth Differentiation* 11:277–86
47. Duntze, W., MacKay, V., Manney, T. R. 1970. *Science* 168:1472–73
48. Gooday, G. W. 1968. *New Phytol.* 67:815–21
49. Bu'Lock, J. D., Drake, D., Winstanley, D. J. 1972. *Phytochemistry* 11:2011–18
50. Plempel, M. 1963. *Planta* 59:492–508
51. van den Ende, H., Stegwee, D. 1971. *Bot. Rev.* 37:22–36
52. Burgeff, H. 1924. *Bot. Abh.* 4:1–135
53. Burgeff, H., Plempel, M. 1956. *Naturwissenschaften* 43:473
54. Plempel, M. 1965. *Planta* 65:225–31
55. Sebek, O., Jäger, H. 1964. *Abstr. 148th Meet. Am. Chem. Soc.* 9Q
56. Caglioti, L. et al. 1964. *Chim. Ind. Milan* 46:961–66
57. Caglioti, L. et al. 1966. *Tetrahedron Suppl.* 7:175–87
58. Cainelli, G., Grasselli, P., Selva, A. 1967. *Chim. Ind. Milan* 49:628–29
59. Thomas, D. M., Goodwin, T. W. 1967. *Phytochemistry* 6:355–60
60. Thomas, D. M., Harris, R. C., Kirk, J. T. O., Goodwin, T. W. 1967. *Phytochemistry* 6:361–66
61. Reschke, T. 1969. *Tetrahedron Lett.* 39:3435–39
62. van den Ende, H. 1967. *Nature* 215:211–12
63. Gooday, G. W. 1968. *Phytochemistry* 7:2103–5
64. Austin, D. J., Bu'Lock, J. D., Gooday, G. W. 1969. *Nature* 223:1178–79
65. Bu'Lock, J. D., Winstanley, D. J. 1971. *J. Gen. Microbiol.* 69:391–94
66. Austin, D. J., Bu'Lock, J. D., Drake, D. 1970. *Experientia* 26:348–49
67. Bu'Lock, J. D., Austin, D. J., Snatzke, G., Hruban, L. 1970. *Chem. Commun.* 255–56
68. Edwards, J. A., Schwarz, V., Fajkos, J., Maddox, M. L., Fried, J. H. 1971. *Chem. Commun.* 292–93
69. Isoe, S., Hayase, Y., Sakan, T. 1971. *Tetrahedron Lett.* 40:3691–94
70. Cainelli, G. P. et al 1967. *Chim. Ind. Milan* 49:748–51
71. Bu'Lock, J. D., Quarrie, S. A., Taylor, D. A., Winskill, N. 1973. *J. Gen. Microbiol.* 77:iv–v
72. Bu'Lock, J. D. 1973. *Pure Appl. Chem.* 34:435–61
73. Sutter, R. P. et al 1972. *Abstr. Ann. Meet. Am. Soc. Microbiol.* 68
74. Sutter, R. P., Capage, D. A., Harrison, T. L., Keen, W. A. 1973. *J. Bacteriol.* 114:1074–82
75. Meissner, G., Delbrück, M. 1968. *Plant Physiol.* 43:1279–83
76. Sutter, R. P. 1970. *Science* 168:1590–92
77. Werkman, B. A., van den Ende, H. 1973. *Arch. Mikrobiol.* 90:365–74
78. Bu'Lock, J. D., Jones, B. E., Taylor, D., Winskill, N., Quarrie, S. A. 1974. *J. Gen. Microbiol.* 80:301–6
79. Austin, D. J., Bu'Lock, J. D., Winstanley, D. J. 1969. *Biochem. J.* 113:34P
80. Heisenberg, M., Cerdá-Olmedo, E. 1968. *Mol. Gen. Genet.* 102:187–95
81. van den Ende, H. 1968. *J. Bacteriol.* 96:1298–1303
82. van den Ende, H., Werkman, B. A., van den Briel, M. L. 1972. *Arch. Mikrobiol.* 86:175–84
83. Bu'Lock, J. D., Winstanley, D. J. 1971. *J. Gen. Microbiol.* 68:xvi–xvii
84. van den Ende, H., Wiechmann, A. H. C. A., Reyngoud, D. J., Hendriks, T. 1970. *J. Bacteriol.* 101:423–28
85. Bu'Lock, J. D., Jones, B. E., Quarrie, S. A., Winskill, N. 1973. *Naturwissenschaften* 60:550–51
86. Reschke, T., Plempel, M. 1972. *Z. Pflanzenphysiol.* 67:343–49
87. Mesland, D. A. M., Huisman, J. G., van den Ende, H. 1974. *J. Gen. Microbiol.* 80:111–17
88. Bu'Lock, J. D., Jones, B. E., Quarrie, S. A., Winstanley, D. J. 1974. *Arch. Mikrobiol.* In press
89. Bergman, K. et al. 1969. *Bacteriol. Rev.* 33:99–157
90. Sutter, R. P., Rafelson, M. E. 1968. *J. Bacteriol.* 95:426–32
91. Feofilova, E. P., Pakhlavuni, I. K. 1972. *Microbiology* 41:223–30
92. Yuldasheva, L. S., Samokhvalov, G. I., Bekhtereva, M. N. 1972. *Microbiology* 41:29–32
93. Gooday, G. W., Fawcett, P., Green, D., Shaw, G. 1973. *J. Gen. Microbiol.* 74:233–39
94. Bu'Lock, J. D., Osagie, A. U. 1973. *J. Gen. Microbiol.* 76:77–83
95. Bu'Lock, J. D. 1973. *J. Gen. Microbiol.* 77:v
96. Machlis, L. 1958. *Nature* 181:1790–91
97. Machlis, L. 1958. *Physiol. Plant.* 11:181–92
98. Ibid 845–54
99. Machlis, L. 1968. *Plant Physiol.* 43:1319–20
100. Ibid. 1973. 52:524–26
101. Machlis, L., Nutting, W. H., Williams, M. W., Rapoport, H. 1966. *Biochemistry* 5:2147–52

102. Machlis, L., Nutting, W. H., Rapoport, H. 1968. *J. Am. Chem. Soc.* 90:1674–76
103. Nutting, W. H., Rapoport, H., Machlis, L. 1968. *J. Am. Chem. Soc.* 90:6434–38
104. Plattner, J. J., Bhalerao, U. T., Rapoport, H. 1969. *J. Am. Chem. Soc.* 91:4933
105. Grieco, P. A. 1969. *J. Am. Chem. Soc.* 91:5660–61
106. Mori, K., Matsui, M. 1969. *Tetrahedron Lett.* 51:4435–38
107. Bhalerao, U. T., Plattner, J. J., Rapoport, H. 1970. *J. Am. Chem. Soc.* 92:3429–33
108. Mori, K., Matsui, M. 1970. *Tetrahedron* 26:2801–14
109. Corey, E. J., Achiwa, K., Katzenellenbogen, J. A. 1969. *J. Am. Chem. Soc.* 91:4318–20
110. Corey, E. J., Achiwa, K. *Tetrahedron Lett.* 26:2245–46
111. Plattner, J. J., Rapoport, H. 1971. *J. Am. Chem. Soc.* 93:1758–61
112. Machlis, L. 1973. *Plant Physiol.* 52:527–30
113. Carlile, M. J., Machlis, L. 1965. *Am. J. Bot.* 52:484–86
114. Machlis, L. 1969. *Physiol. Plant.* 22:126–39
115. Ibid 1319–20
116. Carlile, M. J., Machlis, L. 1965. *Am. J. Bot.* 52:478–83
117. Müller, D. G., Jaenicke, L., Donike, M., Akintobi, T. 1971. *Science* 171:815–17
118. Müller, D. G., Jaenicke, L. 1973. *FEBS Lett.* 30:137–39
119. Vigfusson, N. V., Walker, D. G., Islam, M. S., Weiger, J. 1971. *Folia Microbiol. Prague* 16:166–96
120. Islam, M. S., Weiger, J. 1972. *Folia Microbiol. Prague* 17:316–19
121. Islam, M. S. 1973. *Mycopath. Mycol. Appl.* 51:87–97
122. Islam, M. S., Weiger, J. 1972. *Ind. J. Biochem. Biophys.* 9:345–49
123. Levi, J. D. 1956. *Nature* 177:753–54
124. MacKay, V., Manney, T. R. 1970. *Genetics* 64:s40
125. Duntze, W., Stötzler, D., Bücking-Throm, E., Kalbitzer, S. 1973. *Eur. J. Biochem.* 35:357–65
126. Throm, E., Duntze, W. 1970. *J. Bacteriol.* 104:1388–90
127. Bücking-Throm, E., Duntze, W., Hartwell, L. H., Manney, T. R. 1973. *Exp. Cell Res.* 76:99–110
128. Hartwell, L. H. 1973. *Exp. Cell Res.* 76:111–17
129. Sena, E. P., Radin, D. N., Fogel, S. 1973. *Proc. Nat. Acad. Sci. USA* 70:1373–77
130. Berkoff, C. E. 1969. *Quart. Rev.* 23:372–91
131. Spalla, C., Biffi, G. 1971. *Experientia* 27:1387

THE BIOSYNTHESIS OF MITOCHONDRIAL PROTEINS[1,2]

✣ 840

Gottfried Schatz and Thomas L. Mason[3]
Cornell University, Ithaca, New York

CONTENTS

INTRODUCTION

In this article we discuss how the proteins of a mitochondrion are synthesized and assembled into a functional organelle. These questions have received only limited

[1] With few exceptions, this review is based on information published before July 1, 1973.

[2] The following abbreviations are used: mtDNA, mitochondrial DNA; SDS, sodium dodecyl sulfate.

[3] Present address: Biochemistry Department, University of Massachusetts, Amherst, Mass. 01002, USA

51

attention in earlier reviews on mitochondrial biogenesis (1–21), since there was relatively little of substance to report. Whereas research on mitochondrial nucleic acids has, for the past 10 years, generated one exciting result after another, the study of mitochondrial proteins progressed slowly and appeared sometimes clouded by *ex cathedra* statements that were less than infallible. This has now changed (22, 23) mainly because of several new techniques which greatly facilitate the separation and characterization of insoluble membrane proteins. These techniques include SDS²-polyacrylamide gel electrophoresis (24, 25), immunoprecipitation of detergent-solubilized proteins (26, 27), covalent labeling of surface polypeptides in membranes (28–30), the use of mutants with impaired mitochondrial assembly (31–33), and pulse labeling of membrane proteins in the presence of inhibitors of either cytoplasmic or mitochondrial protein synthesis (cf below).

In attempting to review these recent developments, we have extensively discussed the various experimental methods, since their limitations are frequently ignored. We have also speculated why, in our view, certain lines of investigation have stagnated whereas others have proved successful. Finally, when results or interpretations (including our own) struck us as questionable, we have clearly said so. In other words, we have not only compiled the facts, but also presented the field as we see it. We hope that our vision will not be judged unduly myopic and that we have succeeded in giving the published data some critical perspective.

THE DUAL ORIGIN OF MITOCHONDRIAL PROTEINS

Mitochondria are formed by a close cooperation of two genetic systems. One of these, the conventional "nuclear" system, involves nuclear DNA and the protein synthesizing apparatus of the extramitochondrial cytoplasm. The other system is located in the mitochondria and consists of a relatively small DNA and mitochondrial transcription and translation machinery.

In many respects, the two systems are quite distinct from each other. First, they are physically separated. Second, they do not seem to share even a single component. Third, their DNA genomes have no base sequences in common (19). Fourth, there is no evidence to suggest that mRNA transcribed in one system is translated in the other (19). It is therefore very likely (although by no means certain) that polypeptides translated in the cytoplasm are coded for by nuclear DNA whereas polypeptides translated in mitochondria are coded for by mtDNA.

Yet there is no doubt that these two genetic systems are closely coordinated in vivo. We still know very little about how this is accomplished, but recent evidence suggests that the two systems may influence each other through their protein products. These protein products are discussed in the following sections.

THE ROLE OF CYTOPLASMIC PROTEIN SYNTHESIS

It is now well established that the great majority of the mitochondrial proteins is coded by nuclear genes, synthesized on cytoplasmic ribosomes, and subsequently imported into mitochondria. This mode of synthesis appears to hold for the proteins of the mitochondrial matrix (35–37) and the outer membrane (38–40), most proteins of the inner membrane (41–44), and most proteins functioning in the replication, transcription, and translation of mitochondrial DNA (cf below). However, there is surprisingly little solid information on the site of synthesis of individual mitochondrial proteins. The following discussion shows that many of these uncertainties reflect the limitations of present experimental techniques.

Mitochondria of Cytoplasmic Petite Mutants of Yeast

Cytoplasmic petite mutants of *Saccharomyces cerevisiae* possess respiration-deficient mitochondria which lack cytochromes aa_3, b, and c_1 (45) as well as an energy transfer system (46). These mutants have lost large segments, or in some cases, all of their mtDNA (47–50). While independently isolated cytoplasmic petite mutants may thus differ from each other with respect to the amount of genetic information retained in their mtDNA, they all seem to have lost a functional mitochondrial protein synthesizing system (51–54). All proteins still present in these defective mitochondria are therefore made by the cytoplasmic protein synthesizing system. These proteins include the enzymes of the citric acid cycle (55–57a), ferro-chelatase (58), cytochrome c (45–55), cytochrome b_2 (59), the mitochondrial F_1-ATPase (60, 61), mitochondrial polypeptide elongation factors (62–64), and three subunits of cytochrome oxidase (32, 43, 65). Petite mitochondria also possess an apparently normal outer membrane (66, 67) and a distinct inner membrane which does not differ greatly from that in wild-type mitochondria (66). Indeed, freeze-fracture studies have failed to reveal any morphological differences between the membranes of wild-type and petite mitochondria (66a). These observations show that the cytoplasmic system produces not only most of the mitochondrial enzymes, but also the bulk of both mitochondrial membranes.

Although the presence of a given protein in mitochondria of cytoplasmic petite mutants proves firmly that this protein is synthesized in the cytoplasm, the absence of a protein is difficult to interpret. On one hand, the protein could be missing, because it is normally synthesized on mitochondrial ribosomes. Alternately, the protein could be synthesized in the cytoplasm but might require a mito-chondrially synthesized component for proper assembly into the mitochondria. Such indirect effects are indeed observed. As discussed later, one of the cyto-plasmically synthesized subunits of cytochrome oxidase is missing from petite mitochondria, apparently because mitochondrially formed "binding proteins" are absent. Since cytoplasmic petite mutants often lack mitochondrial ribosomal RNA (68, 69) or even mitochondrial DNA (49, 50, 70), many of the proteins that are normally associated with these nucleic acids, such as ribosomal proteins, initiating factors, or a rifampicin-sensitive RNA polymerase (71), may also be undetectable. Finally, it cannot be assumed that the loss of an enzymic activity or a cytochrome absorption band reflects the complete loss of the corresponding protein. There is evidence that mitochondrial organelles may contain incomplete forms of cyto-chrome oxidase (32, 43, 65), cytochrome c peroxidase (72) and cytochrome c_1 (73) which lack the catalytic activity and the absorption bands of the corresponding holocytochromes. Measurements of enzymic functions or absorption spectra are thus unreliable tools for establishing the absence of enzyme proteins in mitochondrial membranes. This is often ignored.

Despite these limitations, the detection of a protein in mitochondria from cytoplasmic petite mutants (especially those lacking mtDNA) is one of the most convenient and clear-cut methods for proving cytoplasmic synthesis of this protein.

Mitochondria of Cells Grown in the Presence of Inhibitors of Mitochondrial Protein Synthesis

When eukaryotic cells are grown in the presence of inhibitors of mitochondrial protein synthesis such as chloramphenicol, erythromycin, or ethidium bromide, they accumulate respiration-deficient mitochondria. If inhibition of mitochondrial protein synthesis in the growing cells is complete, any protein retained by mitochondria is automatically identified as a product of cytoplasmic protein synthesis. This approach is conceptually quite similar to the study of cytoplasmic petite mutants of yeast, but it is more flexible. For example, it is not limited to a single species but has been applied to cells as diverse as yeast (60, 74–81), other fungi (82–87), protozoa (88–95), algae (96, 96a), mammalian cells (97–120), and plants (121). The morphology and the enzymic composition of the abnormal mitochondria formed by the drug-inhibited cells generally approach those of mitochondria from cytoplasmic petite mutants of yeast. In all cells examined thus far, the cytoplasmic system synthesizes the bulk of the mitochondrial protein.

Another advantage of this approach is that it sometimes allows one to distinguish between direct and indirect effects of blocking mitochondrial protein synthesis (cf above). For example, if the inhibitor is added to exponentially growing cells or to yeast cells undergoing derepression from glucose or anaerobiosis, the formation of some mitochondrial components is blocked immediately whereas that of others stops only after a lag (80, 122, 123). An immediate inhibition suggests that the affected component is synthesized by mitochondria, whereas a lag points to the fact that the component is synthesized in the cytoplasm and lost only because of the indirect effects mentioned earlier. This paradigm has been extensively employed by Mahler et al (80, 122) and Vary et al (123), who found that formation of cytochrome oxidase in yeast cells is stopped immediately whereas that of cytochrome c_1 and succinate dehydrogenase is affected only incompletely or with considerable delay. These latter two components are thus probably made in the cytoplasm but require mitochondrially synthesized components for their normal integration into the mitochondria. Recent experiments on the biosynthesis of cytochrome c_1 (to be discussed later) have verified this prediction (124).

In contrast to the cytoplasmic petite mutation in yeast, growth of eukaryotic cells in the presence of chloramphenicol usually does not lead to a loss of mitochondrial nucleic acids (68, 119; cf below for exceptions), thus opening a way for studying the biosynthesis of the proteins of the mitochondrial genetic system. According to Davey et al, mitochondria from yeast cells grown in the presence of chloramphenicol retain a functional protein synthesizing system (125). This observation implies that cytoplasmic ribosomes synthesize all the proteins of the mitochondrial genetic system, including all proteins of mitochondrial ribosomes.

In our opinion, however, experiments of this type are easily overinterpreted.

First, the loss of a mitochondrial component does not exclude that the component is, in fact, of cytoplasmic origin (cf above). We suspect, for example, that the inhibitory action of chloramphenicol on the formation of some enzymes of the mevalonate pathway in yeast may reflect such indirect effects (126–128). Second, the effects of inhibitors on higher cells are often not clear-cut. These obligate aerobic cells may die upon prolonged exposure to inhibitors of mitochondrial protein synthesis and experiments are thus usually limited to a few cell generations. As a result, the mitochondrial lesions are not as extensive as those observed with facultative anaerobes such as yeast, which can be grown indefinitely in the absence of mitochondrial protein synthesis. Third, inhibitors of mitochondrial protein synthesis may also affect other cellular functions. For example, chloramphenicol inhibits mitochondrial electron transport at site 1 (129–133) and may uncouple oxidative phosphorylation (134, 135) whereas erythromycin (and to some extent chloramphenicol) inhibits the uptake of amino acids into cells (44). It is often assumed that any side effects of chloramphenicol can be identified by control experiments with L(+)-*threo*-chloramphenicol, an isomer which affects most processes similar to the acute D(−)-*threo*-chloramphenicol but does not inhibit mitochondrial protein synthesis (136–138). However, Gordon et al (134) showed that this assumption is not always valid. Finally, inhibition of mitochondrial protein synthesis in intact cells by chloramphenicol or erythromycin is usually incomplete even at very high concentrations of these antibiotics (cf e.g. 138a). Therefore, we are not yet convinced that the experiments of Davey et al (125) rigorously prove that all proteins of the mitochondrial genetic system are formed in the cytoplasm. Some of these proteins could be made in mitochondria by a mechanism which is relatively resistant to chloramphenicol (cf 139). Alternately, a mitochondrially synthesized protein could normally be present in great excess and might become limiting only after extensive dilution. Indeed, three different research groups have recently reported that prolonged growth of *S. cerevisiae* cells in the presence of inhibitors of mitochondrial protein synthesis such as erythromycin, chloramphenicol, or tetracycline converts the cells to cytoplasmic petite mutants which lack a mitochondrial protein synthesizing system or sometimes even mtDNA (70, 140, 141). In the experiments of Williamson et al (70) and Carnevali et al (140), significant conversion of wild-type cells to petite mutants was observed only if the cells were grown in the presence of the drug for 10 to 20 generations. Weislogel & Butow (141) worked with yeast mutants in which a mitochondrial translation product may have been unstable or present in low amounts so that it was more quickly depleted than in wild-type cells. These special conditions may explain why the petite-inducing effect of inhibitors of mitochondrial protein synthesis had escaped detection in earlier studies. In view of these discrepancies one should still keep an open mind as to whether or not normal replication of mtDNA requires one or more proteins synthesized in mitochondria.

In summary, we feel that experiments with cells grown in the presence of inhibitors of mitochondrial protein synthesis are often difficult to interpret and yield tentative results at best. It is therefore somewhat disappointing that such studies account for a major fraction of all papers published on mitochondrial biogenesis.

Pulse-Labeling Experiments

Pulse-chase techniques should be one of the most direct means of studying the origin of mitochondrial proteins. Several laboratories have used this approach in attempting to show that cytochrome c is synthesized on cytoplasmic ribosomes (i.e. "microsomes") and only subsequently transferred to the mitochondria (142–155). These experiments were prompted by the observation that in rat liver cells 4 to 10% of the total cytochrome c was associated with the "microsomal" fraction (144, 153, 154). When rats were injected with radioactive cytochrome c precursors (such as ^{14}C-lysine or ^{3}H-δ-aminolevulinic acid) and killed 10 to 20 min later, the microsomal cytochrome c proved to be at least twice as radioactive as the mitochondrial one. With increasing time after the radioactive pulse, however, the specific radioactivity of the microsomal cytochrome c dropped whereas that of the mitochondrial cytochrome c rose (145, 150, 154). These data were widely quoted as evidence that the polypeptide chain of cytochrome c is synthesized and combined with the heme group in the cytoplasm and only then transferred into the mitochondria.

However, a closer look at the published data reveals several ambiguities that have not been fully resolved. One of the major problems is the redistribution of mitochondrial cytochrome c during homogenization of the liver cells (145, 146, 150, 153, 156). Several attempts have been made to correct for this artifact (144, 146, 147, 150) but the corrections are not always entirely convincing or internally consistent (cf 144 vs 147). Another formidable problem is the necessity of separating the small amounts of microsomal cytochrome c from other labeled polypeptides. For this reason, it has not yet been excluded that the radioactivity of cytochrome c extracted from microsomes merely reflects the presence of highly labeled contaminants. This applies especially to experiments in which the synthesis of cytochrome c by isolated rat liver microsomes was studied (147; cf also 150a). The contamination problem can be lessened by following the incorporation of a heme precursor (such as δ-aminolevulinic acid) into the hematoporphyrin moiety of cytochrome c (152, 154); however, this approach does not yield direct information on the assembly of the apoprotein moiety.

Finally, these labeling studies have not yet resolved the question as to whether or not cytochrome c is synthesized and transferred to the mitochondria as the finished cytochrome or as a precursor. This problem was studied by Kadenbach (151) and Davidian & Penniall (155). Kadenbach claimed that cytochrome c was initially assembled on microsomes as a heme-free "precytochrome c" which was converted to the holocytochrome inside the mitochondria (151). This view rests mainly on the observation that much of the radioactivity associated with pulse-labeled microsomal cytochrome c chromatographed slightly differently from cytochrome c. However, precytochrome c could also be a contaminant unrelated to cytochrome c. Indeed, Kadenbach reported that precytochrome c from different cell fractions exhibited different chromatographic behavior (151).

Davidian & Penniall (155) injected ^{3}H-δ-aminolevulinic acid into rats and followed the incorporation of the isotope into the four electrophoretically distinct variants of cytochrome c which can be isolated from rat liver (157). While variant I (which accounts for 90% of the total cytochrome c) contained most of the label, variant IV (which constitutes less than 2% of the total cytochrome c) exhibited the highest specific radioactivity. During a subsequent chase, the specific radio-activity of variant IV fell whereas that of variant I increased. Since variant IV differs from variant I by the lack of several amide groups (157), Davidian & Penniall suggested that cytochrome c is initially synthesized as variant IV and amidated to variant I only after transfer to the mitochondria. Unfortunately, however, the change of the specific radioactivities was apparently deduced from two separate experiments, one of which was plagued by a low recovery of cytochrome c. Also, no evidence was presented that the peaks of radioactivity obtained upon electrophoresis exactly matched those of the different variant forms of cytochrome c. It is also not very reassuring that Flatmark & Sletten obtained exactly the opposite result (157). They injected ^{59}Fe into rats and found that the specific radioactivity of variant I of cytochrome c rose more rapidly than the other variants. Again, the radiochemical purity of the different cytochrome c variants was not rigorously established. This could easily explain the divergent results obtained in different laboratories.

It appears, therefore, that pulse-labeling experiments have not yet conclusively established the site of synthesis of cytochrome c. Fortunately, the cytoplasmic origin of apocytochrome c is beyond question since cytoplasmic petite mutants of yeast contain unaltered cytochrome c (45, 158, 158a).

However, pulse-labeling techniques have been successfully used to demonstrate the microsomal synthesis of glutamate dehydrogenase, an enzyme of the mito-chondrial matrix (159). In these experiments, pulse-labeled glutamate dehydro-genase was extracted from liver mitochondria and liver microsomes of the rat and rigorously purified by immunoprecipitation and subsequent SDS-acrylamide gel electrophoresis. In separate experiments, purified radioactive glutamate dehydro-genase was added to the liver homogenates to assess redistribution of the endogenous enzyme during tissue fractionation. It was found that the enzyme extractable from microsomes was labeled several times more rapidly than that extractable from mitochondria. After a 6 hr chase, however, the specific radio-activities of the enzymes from the two subcellular fractions were essentially equal. These careful experiments clearly illustrate the great potential of pulse-labeling techniques for the study of mitochondrial assembly.

Pulse Labeling in the Presence of Inhibitors

It is now well established that protein synthesis on cytoplasmic ribosomes is specifically inhibited by cycloheximide (159), whereas protein synthesis on mito-chondrial ribosomes is specifically inhibited by antibacterial antibiotics such as chloramphenicol (160, 161), erythromycin (74), and lincomycin (74, 162). Pulse-labeling experiments in the presence of these inhibitors are therefore especially powerful tools for identifying the biosynthetic origin of mitochondrial proteins.

The first labeling experiments with cycloheximide-inhibited rat liver slices (129, 163), intact yeast (164), and Krebs II ascites cells (165) led to the important discovery that the antibiotic inhibited not only the labeling of cell sap protein, but also that of mitochondrial protein. Most mitochondrial proteins are therefore made in the cytoplasm and only subsequently incorporated into the mitochondria. Unfortunately, these early experiments cannot be interpreted quantitatively since they were carried out at rather low cycloheximide levels which did not completely block cytoplasmic protein synthesis. (In one of these studies—it is still widely quoted—inhibition of cytoplasmic protein synthesis was only 63%.) Since mitochondria synthesize so little of their own protein (cf below), even a slight residual activity of the cytoplasmic system may cause grossly erroneous results. In subsequent studies, this problem was overcome by resorting to much higher cycloheximide concentrations (41, 166–172) or to other inhibitors of cytoplasmic protein synthesis such as anisomycin (173), emetine (174), or pederine (175).

Since pulse labeling in the presence of inhibitors is now so widely used, one tends to overlook the pitfalls inherent in this approach. Like any inhibitor, cycloheximide is not absolutely specific but may affect a variety of other cellular functions (176–183), including mitochondrial RNA polymerase (184). The specificity of emetine is even more suspect since, in some species at least, it is also a powerful inhibitor of mitochondrial protein synthesis (102, 184a-c) as well as of nuclear RNA synthesis (184d). In addition to these side effects, one must consider possible indirect effects of these inhibitors. For example, cycloheximide could block labeling of a mitochondrial translation product by abolishing the synthesis of a cytoplasmically made "partner" protein (cf below). Whenever possible, one should therefore attempt to show that the labeling of a putative cytoplasmic product is not only sensitive to cycloheximide, but also insensitive to inhibitors of mitochondrial protein synthesis. In the latter case it is essential to show that the label resides within the protein in question and not in some tightly bound contaminant.

This approach was first fully exploited by Küntzel (168). In pioneering experiments, he showed that the incorporation of ^{3}H-lysine into mitochondrial ribosomes from Neurospora crassa was abolished by cycloheximide and not affected by chloramphenicol. He concluded that most, if not all, proteins associated with mitochondrial ribosomes are synthesized on cytoplasmic ribosomes. This important result was later confirmed (169, 185) and extended to mitochondrial ribosomes from yeast (186) and HeLa cells (175). More recently, it was impressively corroborated by Lizardi & Luck (173) who separated the ribosomal proteins from N. crassa mitochondria into at least 53 species by a combination of isoelectric focusing and SDS-acrylamide electrophoresis. They demonstrated that none of these proteins were labeled by isolated Neurospora mitochondria and that labeling of every single mitochondrial ribosomal protein in intact cells was inhibited by anisomycin but not by chloramphenicol.

In contrast, Schmitt (187) and Millis & Suyama (188) reported that the labeling of mitochondrial ribosomal proteins from yeast (187) and Tetrahymena (188) was not only sensitive to cycloheximide, but also partially sensitive to chloramphenicol. Whereas Schmitt apparently dismissed the inhibitory effect of chloramphenicol as a side effect of the antibiotic, Millis & Suyama concluded that mitochondrial ribosomes also contain several proteins made by themselves. It seems equally probable, however, that the chloramphenicol-sensitive proteins noted by Millis and Suyama were not true ribosomal components but contaminating membrane proteins. Indeed, Groot has recently shown that even very "pure" preparations of mitochondrial ribosomes may still contain membrane components which can only be removed by dissociating the ribosomes into subunits (186). Although it is perhaps still premature to reach a final verdict, the weight of evidence suggests that all proteins of mitochondrial ribosomes are made by the cytoplasmic protein synthesizing system.

Labeling studies with cycloheximide-poisoned cells were also used to investigate the biosynthesis of the mitochondrial outer membrane. It had already been shown that the proteins of this membrane are not labeled if isolated rat liver mitochondria are incubated with radioactive amino acids (cf below); essentially the same result was found when cycloheximide-inhibited *Neurospora* (189), yeast (190), and HeLa cells (175) were pulse labeled with radioactive amino acids. It is therefore generally assumed that all proteins of the mitochondrial outer membrane are synthesized in the cytoplasm.

Cytochrome c_1 from baker's yeast has recently been purified and characterized as a single heme polypeptide of apparent molecular weight 28,500. Pulse labeling of the cytochrome by intact yeast cells with ^3H-leucine was unaffected by acriflavin and completely blocked by cycloheximide (124). The apoprotein is therefore made on cytoplasmic ribosomes. Why then does formation of holocytochrome c_1 require mitochondrial protein synthesis (45, 75)? Perhaps the apoprotein must first be combined with at least one mitochondrially synthesized polypeptide before it can accept its heme group.

Other mitochondrial proteins whose cytoplasmic origin was established by labeling in the presence of antibiotics include the mitochondrial F_1-ATPase (191, 192), cytochrome c (193), a mitochondrial DNA polymerase (194), and the four small subunits of cytochrome oxidase (27, 195, 196).

Other Techniques

The cytoplasmic origin of some mitochondrial proteins has also been inferred from the fact that they are not labeled by isolated mitochondria. Proteins studied in this manner include cytochrome c (197, 198), malate dehydrogenase (197), the mitochondrial F_1-ATPase (199), the oligomycin sensitivity-conferring protein (171), a mitochondrial DNA polymerase (194), as well as all the proteins of the outer membrane (38–40) and of the mitochondrial ribosomes (173, 185). However, these results are not very convincing since they are based on negative data and since there is as yet no clear evidence that isolated mitochondria synthesize functional proteins (cf later). This uncertainty is underscored by reports (200, 201) that isolated rat liver mitochondria do not incorporate labeled amino acids into cytochromes aa_3 or b or into oligomycin-sensitive ATPase even though these components are now known to be partly made on mitochondrial ribosomes.

Since it is very likely (although not yet certain) that polypeptides coded in the nucleus are translated on cytoplasmic ribosomes, the cytoplasmic origin of some mitochondrial enzymes can also be inferred from genetic data. Polymorphic variants of mitochondrial matrix enzymes such as malate dehydrogenase (202–204), glutamate-oxaloacetate transaminase (205), and malic enzyme (204) have been uncovered in maize (202), mice (204), and humans (203, 205). These variants are inherited according to Mendelian laws and are thus apparently coded by nuclear genes. It should not be overlooked, however, that some of the genetic data collected with human patients were by necessity incomplete and therefore not as conclusive as genetic data normally obtained in simpler systems.

Particularly elegant genetic experiments have been carried out with *Paramecium aurelia* strains containing polymorphic variants of mitochondrial β–hydroxybutyrate dehydrogenase (206). When two *Paramecium* cells (each containing two identical haploid nuclei) conjugate, each partner exchanges one haploid nucleus with the other partner. If the duration of cell fusion is kept short, each conjugant retains its own pure cytoplasm (including mitochondria) but ends up with nuclear genes from both partners. The distribution of the two β–hydroxybutyrate variants in the exconjugants is consistent with a coding by nuclear genes (206). Similar although less clear-cut results were obtained for the coding of NADP-isocitrate dehydrogenase (207). In yet another approach, mitochondria from an erythromycin-resistant *Paramecium* strain possessing a variant form of mitochondrial fumarase were injected into an erythromycin-sensitive host strain containing another variant of the enzyme (208). The injected cells were then propagated in the presence of erythromycin to eliminate all of the original host mitochondria. The heterologous, erythromycin-resistant mitochondria multiplied and acquired the fumarase variant typical of the host (208). These results strongly suggest that mitochondrial fumarase in *Paramecium* is coded for by nuclear genes and probably made on cytoplasmic ribosomes.

Classical genetic techniques were used to show that nuclear genes code for leucyl-tRNA synthetase (208a) and malate dehydrogenase (209, 210) of *N. crassa* mitochondria and iso-1-cytochrome *c* of *Saccharomyces cerevisiae* (211). The elegant studies on the inheritance of iso-1-cytochrome *c* in yeast (reviewed in 212) are to date the most conclusive evidence for the nuclear coding of a mitochondrial protein.

Present genetic evidence is thus fully consistent with the concept that at least the easily solubilized mitochondrial proteins are synthesized on cytoplasmic ribosomes.

THE ROLE OF MITOCHONDRIAL PROTEIN SYNTHESIS

The mitochondrial protein synthesizing system was discovered as early as 1958 (213), but the products of this system are only now being identified. Present evidence suggests that mitochondria synthesize probably no more than a dozen hydrophobic polypeptides (43, 44, 175, 184b, 191, 213–218); three of these are associated with cytochrome oxidase (27, 196, 219–221), four with oligomycin-sensitive ATPase (32, 191), and one with cytochrome *b* (222, 223). The others are still unidentified. The combined mitochondrial translation products account for only 5–15% of the mitochondrial protein mass (e.g. 32, 172, 217, 224, 225), yet are indispensable for the assembly of a functional mitochondrion (75).

Progress in this area was hampered by two major problems. One was the lack of good methods for separating and characterizing insoluble membrane polypeptides. Initial attempts to resolve such polypeptides by electrophoresis in gels containing phenol, acetic acid, and/or urea led to conflicting results since disaggregation of the polypeptides was almost never complete (226, 227). The situation improved markedly with the introduction of SDS-acrylamide gel electrophoresis (24, 25, 228, 229) which revealed for the first time the size and the approximate number of mitochondrial translation products. Electrofocussing in the presence of urea or nonionic detergents is another promising, although as yet little explored, method (204, 230, 231). The second major stumbling block was the lack of a good system for studying mitochondrial protein synthesis. For many years, researchers incubated isolated mitochondria with radioactive amino acids and searched for the labeled product(s) (see 9, 19, 232 for reviews). However, the following discussion shows that this system was cumbersome, and at times, outright malicious. Once it was abandoned in favor of the better in vivo system, progress was relatively swift.

Incorporation of Amino Acids by Isolated Mitochondria

As first shown by McLean et al (213), isolated mitochondria can incorporate externally added amino acids into their own protein. The number of publications dealing with this phenomenon is awe inspiring (19) and still increasing (233–240). Yet these studies produced relatively limited information since they had to contend with serious experimental drawbacks of the system. First, the incorporation is very slow and responds rather unpredictably to seemingly trivial variations in the composition of the incubation medium (19, 237, 239). This necessitates careful controls to exclude an incorporation of label via simple adsorption (241) or other processes (cf below). It also makes it difficult to study the energy dependence or the mechanism of the reaction and makes it virtually impossible to render the mitochondrial translation products sufficiently "hot" for subsequent isolation. Second, isolated mitochondria usually contain bacteria and microsomes, each of which can also incorporate amino acids. Although bacteria may sometimes pose problems (242–244), their contribution is usually insignificant (40, 243–251). Nevertheless, it is advisable to work with sterile or nearly sterile mitochondrial preparations, which is not always an easy task. A contribution by microsomes is generally negligible (19) except in experiments with mitochondria from brain (252–259) or HeLa cells (260). These mitochondrial preparations are heavily contaminated with membrane-bound cytoplasmic ribosomes and have to be extensively purified (258) or, preferably, labeled in the presence of cycloheximide (258–260). Third, the polypeptides synthesized by isolated mitochondria are tightly associated with the mitochondrial inner membrane (35, 38–40) and effectively resist purification (19); in rat liver mitochondria, they are also continuously degraded to acid-soluble products (261).

These problems have led several researchers to question whether isolated mitochondria synthesize any complete polypeptide chains at all. Some have even suggested that the "incorporating" activity of isolated mitochondria may simply reflect an artifactual adsorption of radioactive amino acids to the mitochondrial membranes (241). However, these pessimistic views are probably unwarranted.

There is ample evidence that the incorporation of amino acids by isolated mito-chondria is sensitive to inhibitors of prokaryotic ribosomes such as chloramphenicol (34, 160, 161), to puromycin (100, 262–265), to ethidium bromide (101), and, in some cases, to rifampicin (100, 266, 267); it is also influenced by the physiological status of the organisms from which the mitochondria had been isolated (240, 268–276). These facts leave little doubt that the incorporating activity reflects a physiological process. Furthermore, the successful resolution of the incorporating activity from the mitochondrial membranes (168, 277–284) has quickly led to a detailed knowledge of the cofactors and enzymes involved in mitochondrial protein synthesis and has shown that this process is similar to, yet distinct from, bacterial protein synthesis (232). Finally, many of the difficulties mentioned above may simply reflect the tight coupling between mitochondrial and cytoplasmic protein synthesis which will be discussed later. Recent studies on the assembly of cytochrome oxidase (196, 219) and oligomycin-sensitive ATPase (191) suggest that the mitochondrially made components of these complexes are only synthesized in significant amounts if they can be combined with cytoplasmically made "partner" proteins. Since isolated mitochondria probably contain only limited excess pools of these "partner" proteins, mitochondrial protein synthesis should be low. The instability of the mitochondrial products (261) is probably explained by the fact that uncombined polypeptides are easy prey for intramitochondrial proteases. There is also little reason to expect that the uncombined polypeptides should purify together with any functionally defined mitochondrial component. Indeed, they are largely recovered in a rather nondescript insoluble protein fraction (1, 35, 285, 286, 287) which was once thought to contain "structural protein" (287, 288) or "miniproteins" (289). According to more recent studies, however, mitochondrial structural protein is a mixture of denatured proteins (298a–c), whereas miniproteins are probably membrane lipids (298d).

Upon electrophoresis in gels containing SDS or other dissociating agents, the polypeptides synthesized by isolated mitochondria are resolved into at least six peaks, most of which exhibit distinct shoulders (40, 41, 142, 184b, 217, 290); the apparent molecular weights of the peaks range from about 45,000 to less than 10,000. As mentioned later, the same banding pattern is obtained if mitochondria are labeled in vivo in the presence of cycloheximide (41, 184b, 217, 290). It is therefore probable that isolated mitochondria synthesize many and perhaps all of their normal protein products, albeit at a low rate. In spite of the earlier problems encountered with this in vitro system, it may yet prove to be very useful for studying the transcriptional origin of mitochondrial proteins and the interplay between mitochondrial and cytoplasmic protein synthesis.

Costantino and Attardi have recently reported that HeLa cell mitochondria incorporate the polar amino acids Ala, Arg, Asp, Cys, Glu, Gln, Gly, and Lys at less than 10% of the rate expected from the average amino acid composition of whole cells (291). They correlated these findings with earlier hybridization experiments which had suggested that mitochondrial DNA from HeLa (292) and *Xenopus* cells (293) codes for only about 11–15 tRNAs, which is substantially less than a full degenerate complement. Costantino and Attardi speculated, therefore, that the inability of animal mitochondria to incorporate several amino acids might reflect an absence of several tRNAs from these organelles. This, in turn, could explain why mitochondrially synthesized polypeptides are hydrophobic.

It could be argued, however, that the incorporation pattern reported by Costantino and Attardi merely reflected complications by internal metabolite pools or an unequal transport of different amino acids into isolated mitochondria (294). They also conflict with the earlier observations by Neupert (294a) that isolated rat liver mitochondria incorporate Arg at a significant rate. Finally, one need not invoke such a hypothesis to explain the hydrophobicity of mitochondrial translation products; those of lower eukaryotes are extremely hydrophobic (124, 220, 295), yet contain significant amounts of the polar amino acids mentioned above (23, 220). A final decision on the lack of incorporation of certain amino acids by animal mitochondria should perhaps be deferred until the amino acid composition of a distinct polypeptide made in animal mitochondria is known.

Labeling Experiments with Cycloheximide-Inhibited Cells

In 1969, two laboratories showed independently that mitochondrial protein synthesis could be studied in vivo by labeling intact cells in the presence of cycloheximide (52, 167). The following observations suggest that this is an acceptable procedure: 1. Upon fractionation of the labeled cells, virtually all of the incorporated radioactivity is associated with mitochondria (52, 167). 2. Labeling is almost completely inhibited by specific inhibitors of mitochondrial protein synthesis (44, 52, 167, 215). 3. No significant labeling is observed with cytoplasmic petite mutants which lack a functional mitochondrial genetic system (52, 225, 296, 297). 4. Upon electrophoresis in the presence of SDS or other denaturing agents, the polypeptides labeled by mitochondria in vivo exhibit almost the same banding pattern as polypeptides synthesized by isolated mitochondria (41, 184b, 217, 290). 5. The danger of side effects or indirect effects of cycloheximide is minimized since mitochondrial labeling is inferred from a lack of inhibition by cycloheximide.

This method avoids the uncertainties which always accompany experiments with isolated organelles. It also allows newly formed mitochondrial translation products to combine with cytoplasmically synthesized partner proteins which appear to be present in excess in the cycloheximide-inhibited cells (cf below). This fact may explain why the in vivo system is several orders of magnitude more active than isolated mitochondria. Since mitochondria can now easily be labeled to a very high specific radioactivity, the purification of mitochondrially synthesized polypeptides is greatly facilitated.

Although the mitochondrial translation system of cycloheximide-inhibited cells is "uncoupled" from the cytoplasmic system, it still responds to some physiological signals. Thus, the labeling of at least two polypeptide products is dependent on oxygen and sensitive to the repressive effect of glucose (44, 231, 298). The labeling of several products is also grossly altered, or even completely suppressed, in some nuclear yeast mutants with impaired mitochondrial function (31, 215, 299, 300).

This method is now widely used, yet one should always retain some skepticism toward experiments with inhibited cells. Since many cell types are quite resistant to cycloheximide, it is essential to show that the antibiotic does indeed stop the labeling of soluble cell sap proteins. Unfortunately, this control is often omitted. Moreover, an inhibition of cytoplasmic protein synthesis may indirectly prevent the accumulation of a mitochondrially formed polypeptide. Nevertheless, this labeling technique has been the method of choice for identifying mitochondrially synthesized proteins.

Other Techniques

Several attempts have been made to selectively label mitochondrial translation products in vivo in the absence of inhibitors. Schweyen & Kaudewitz (301) have used a temperature-sensitive yeast mutant (302) which specifically stops cytoplasmic protein synthesis at the nonpermissive temperature. However, this approach still suffers from the fact that it uncouples the mitochondrial system from the cytoplasmic one. To overcome this problem, Mahler et al attempted to label the mitochondrial translation products with radioactive formate (303, 304). Mitochondria, like prokaryotes (305, 306) and chloroplasts (307), initiate polypeptides with N-formyl methionine (307) whereas initiation in the eukaryotic cytoplasm proceeds via non-formylated methionine (308). In principle, formate should therefore be specifically incorporated into the N termini of mitochondrial translation products. However, the actual situation is probably more complex. Since the formyl groups may eventually be cleaved from the finished polypeptide chain, it is not yet clear whether formate permits the specific labeling of a finished polypeptide product. Formate is also quickly converted to amino acids such as serine which are then incorporated by both protein synthesizing systems (53). It is therefore still too early to assess the potential of this innovative approach.

A third (as yet untried) method is suggested by experiments of Wilkie with yeast cells (309). The arginine analog canavanine inhibits both cytoplasmic and mitochondrial protein synthesis, but mitochondrial protein synthesis is up to 100 times more sensitive. This difference can be enhanced even further by isolating mutants whose cytoplasmic system has become resistant to canavanine. These mutants probably possess an altered cytoplasmic arginine-tRNA synthetase which can effectively discriminate between arginine and the toxic analog. In such mutants, radioactive canavanine would therefore only "fool" the mitochondrial arginine-tRNA synthetase and find its way specifically into mitochondrially synthesized polypeptides. This technique clearly merits further investigation.

The Biosynthesis of Cytochrome Oxidase

It has long been known that the formation of cytochrome oxidase is dependent on both mitochondrial and cytoplasmic protein synthesis. The enzyme is missing in cytoplasmic petite mutants of yeast (45) and in chloramphenicol-grown cells (75), and its synthesis during respiratory adaptation of yeast is blocked by either chloramphenicol or cycloheximide (123, 310). Recent experiments in three different laboratories have shown that cytochrome oxidase from *N. crassa* (219, 220) and *Saccharomyces cerevisiae* (27, 196, 221) can be resolved into seven polypeptides whose apparent molecular weight ranges from 42,000 (I) to less than 5000 (VII). Previous disagreements about the total number of polypeptides (219a, 222) appear to be resolved; the remaining small discrepancies between the molecular weights reported by different laboratories are probably largely the result of different electrophoretic procedures. The seven polypeptides are present in roughly equimolar amounts regardless of whether cytochrome oxidase is isolated by conventional fraction methods (211, 311), by chromatography on oleyl methacrylate columns (220), or by immunoprecipitation (311).

The three largest polypeptides (I–III) are apparently made on mitochondrial ribosomes since they are labeled by intact cells in the presence of cycloheximide but not in the presence of chloramphenicol or erythromycin (196, 220, 221). [Initial experiments with *Neurospora* had suggested that only one of the three large polypeptides (III) was labeled in the presence of cycloheximide (219a); later experiments (219, 220) revealed, however, that this was apparently caused by unequal pools of the three large polypeptides.] The four small polypeptides (IV–VII) are made on cytoplasmic ribosomes since they are labeled in the presence of erythromycin (196) but not of cycloheximide (196, 219a, 221).

Taken at face value, these data suggest that cytochrome oxidase contains three subunits that are made in mitochondria and four subunits that are made in the cytoplasm. However, there is still some doubt as to whether all of these seven polypeptides are genuine subunits of the enzyme. As with any membrane enzyme, this question is difficult to answer since one must decide where the enzyme ends and the membrane begins. The most conservative view would be to define cytochrome oxidase as the smallest unit that can transfer electrons from ferro-cytochrome c to oxygen in vitro. Any polypeptide not required for this activity would thus be a contaminant. This definition may be too narrow, however, if one considers the function of cytochrome oxidase in vivo. For example, some subunits may be necessary for the proper attachment of the enzyme to the mitochondrial inner membrane (cf below) or for its role in oxidative phosphorylation. These subunits may then be simply contaminants as far as electron transfer by the solubilized enzyme is concerned but may well prove to be essential for achieving energy coupling by cytochrome oxidase which has been incorporated into artificial liposomes (312).

The following observations suggest that the seven polypeptides form a specific complex that may function in the oxidation of cytochrome c in vivo:

1. Cytochrome oxidase preparations isolated from different organisms by widely different procedures exhibit similar polypeptide compositions (196, 219, 221, 313). However, one of the three large polypeptides appears to be missing in the enzyme from beef heart (314, 315).

2. All seven polypeptides are specifically precipitated from a crude mitochondrial lysate by antisera against the holoenzyme or against the two smallest polypeptides (196, 316).

3. The mitochondrial synthesis of the two largest polypeptides, like that of functional cytochrome oxidase, is dependent on oxygen (27, 196).

4. One or more of the three large polypeptides (I–III) are missing in nuclear yeast mutants which have specifically lost functional cytochrome oxidase (cf below).

5. Antisera against polypeptides I and II, against polypeptide IV, or against polypeptide V inhibit the enzymic activity of purified yeast cytochrome oxidase (316).

None of these arguments is conclusive proof that all subunits are necessary for the in vitro activity of the enzyme. For example, neither conventional purification nor immunoprecipitation distinguishes between an essential subunit and a tightly bound contaminant. Even argument 5 is suspect since the activity of a "catalytic" subunit may well be affected by the binding of an antibody to an adjacent polypeptide. For these reasons, the role of the individual polypeptides (especially that of the large ones) is still under debate. Komai & Capaldi (317) have claimed that the large polypeptides are contaminants since they could not be detected in solubilized beef heart cytochrome oxidase prepared by a new procedure. However, the activity reported for this preparation is rather low and no controls were presented to exclude that the large polypeptides were still present but merely did not enter the polyacrylamide gels during electrophoretic analysis. This can easily happen (cf 23, 219).

In contrast, Tzagoloff has recently reported that the largest polypeptide I carries the heme a in the native enzyme (318), suggesting that the polypeptide is essential for the in vitro activity of solubilized cytochrome oxidase. However, the heme-carrying polypeptide was identified mainly by its inability to penetrate into SDS–acrylamide gels. Definitive information on the catalytic role of the individual polypeptides will probably require reconstitution of the active enzyme from its purified subunits.

The role of the mitochondrially synthesized polypeptides has also been studied with the aid of cytoplasmic petite mutants of yeast which lack mitochondrial protein synthesis as well as several cytochromes, including cytochrome oxidase (cf above). It has been noted earlier that mitochondria from petite mutants retain some material which cross-reacts with antisera against active yeast cytochrome oxidase (43, 319), but the identity of this material remained obscure. Ebner et al (32) have recently identified the cross-reacting material as subunits IV, V, and VI of cytochrome oxidase. While it was not surprising that the mitochondrially synthesized polypeptides I–III were absent, loss of the cytoplasmically made component VII was unexpected. The remaining cytochrome oxidase polypeptides IV-VI were only loosely bound to the mutant mitochondria as they could be detached by mild sonic irradiation. This was in marked contrast to the tight binding of all seven subunits to wild-type yeast mitochondria (196). It thus appears that the mitochondrially made cytochrome oxidase polypeptides are necessary for the tight binding of the cytoplasmically made cytochrome oxidase polypeptides to the mitochondrial inner membrane.

These data are difficult to reconcile with an earlier claim (320) that addition of heme a to isolated petite mitochondria was sufficient to reconstitute cytochrome oxidase activity. It is not clear, however, whether this reconstitution of activity indeed reflected formation of cytochrome oxidase holoenzyme. Attempts by others to repeat these observations were unsuccessful (32).

In sum, it appears that the mitochondrially made polypeptides associated with isolated cytochrome oxidase are intimately related to the biogenesis of cytochrome oxidase and probably also to the function of the enzyme in the mitochondrial membrane. Whether they are essential for the activity of the solubilized enzyme is still unanswered.

The Biosynthesis of Oligomycin-Sensitive ATPase

The elegant studies by Tzagoloff and his colleagues have shown that preparations of oligomycin-sensitive ATPase from yeast mitochondria contain at least ten distinct polypeptides (23, 26). Five of these are associated with the cold-labile F_1-ATPase, one is the oligomycin sensitivity-conferring protein (321), and four are hydrophobic polypeptides tightly associated with the mitochondrial inner membrane (191). All ten polypeptides can be precipitated from detergent-solubilized mitochondria with an antiserum against F_1 (26, 33); they are thus tightly complexed to each other even after solubilization of the mitochondrial inner membrane.

The F_1-ATPase is synthesized on cytoplasmic ribosomes. It is present in cytoplasmic petite mutants (57a, 60, 61) as well as in cells grown in the presence of chloramphenicol (60, 114) and is labeled by intact cells in the presence of chloramphenicol but not of cycloheximide (191). The oligomycin sensitivity-conferring protein is also a cytoplasmic product since it is formed in chloramphenicol-inhibited yeast cells (321) and is not labeled by isolated beef heart mitochondria (171). The four hydrophobic proteins, in contrast, are made in mitochondria, since they are labeled in the presence of cycloheximide but not in the presence of chloramphenicol (23, 32, 191). In this context it may be significant that extrachromosomally inherited, oligomycin-resistant yeast mutants have been isolated (322–326) which may be defective in one or more of the mitochondrial products associated with the oligomycin-sensitive ATPase complex.

The smallest of these hydrophobic polypeptides (molecular weight about 7500) is soluble in chloroform-methanol and has been obtained in apparently pure form (23). Its amino acid composition, the first to be published for a mitochondrial translation product, reveals an unusually high percentage of apolar amino acids but only traces of histidine and tyrosine. A very similar protein has been obtained from chloroform-methanol extracts of beef heart mitochondria and identified as the component binding the ATPase inhibitor dicyclohexylcarbodiimide (327, 328), which mimics the activity of oligomycin. It is therefore an intriguing possibility that this protein is identical with the smallest mitochondrially made subunit of oligomycin-sensitive ATPase. In any case, it would be of great interest to analyze the 7500 dalton ATPase subunits from extrachromosomally inherited yeast mutants which are resistant to oligomycin (322–326, 329, 330), since this is probably one of the most promising approaches for investigating the coding of a mitochondrial translation product (cf below). If synthesis of the four mitochondrially made ATPase subunits is prevented by the petite mutation or by growth of cells in chloramphenicol, F_1 is still made (57a, 60, 61). However, it becomes oligomycin resistant (57a, 60, 61) and can easily be detached from the mitochondrial membranes by mild sonic irradiation (60). It seems, therefore, that the mitochondrially synthesized subunits ensure the proper binding of F_1 to the mitochondrial inner membrane. As noted earlier, a similar conclusion has been reached for cytochrome oxidase.

The Biosynthesis of Cytochrome b

Yet another mitochondrially synthesized polypeptide seems to be associated with cytochrome b, although the evidence on this point is less clear-cut. Weiss (222) reported that purified *N. crassa* cytochrome b could be dissociated into one large polypeptide (mol wt 32,000) and two small ones (mol wt 10,000 and 11,000). Pulse-labeling experiments with intact cells in the presence of either cycloheximide or chloramphenicol revealed that the large polypeptide was synthesized in mitochondria whereas the two small polypeptides were synthesized in the cytoplasm. Upon gel filtration in the presence of SDS, all of the heme migrated together with the small "subunits."

More recently, however, Weiss et al have reported that extensive gel filtration of cytochrome b in the presence of bile salts completely removed the 11,000 dalton polypeptide and partially removed the 10,000 dalton polypeptide (223). Both small polypeptides were thus probably contaminants. Their apparent migration with heme in the presence of SDS (cf above) was apparently a coincidence since free heme tends to form micellar aggregates with various detergents (316, 331). Although none of the cytochrome b fractions obtained by Weiss and co-workers was completely free of 10,000 dalton polypeptide, the specific heme content correlated well with the relative amount of the large polypeptide. This led the authors to conclude that the large polypeptide, rather than one of the small ones, was the cytochrome b apoprotein. While this may be correct, it is also possible that prolonged gel filtration of cytochrome b in the presence of bile salts led to a relocation of heme from one of the small polypeptides to the large one. This could perhaps be checked by purifying cytochrome b in the presence of added radioactive heme. Antibodies against each of the three polypeptides might also help in identifying the subunit(s) of cytochrome b.

General Properties of Mitochondrially Synthesized Proteins

Several workers have noted that 10 to 40% of the total mitochondrial translation products can be extracted from mitochondria with chloroform-methanol (54, 332–335). This observation has spawned several hypotheses, even though solubility of pure proteins in organic solvents is not uncommon (336). For example, some authors suggested that mitochondria contain two genetic systems, only one of which produces the polypeptides soluble in chloroform-methanol (335). Others termed the extracted mitochondrial translation products "proteolipids" without offering any evidence for the presence of fatty acids (333, 334). [Proteolipids were originally defined by Folch & Lees (337) as lipid-protein complexes that are soluble in organic solvents but insoluble in water.] It was even suggested that all mitochondrial translation products are low molecular weight proteolipids containing esterified fatty acids (338). According to this view, the mitochondrial translation products in the molecular weight range of 20,000 to 40,000 are mainly aggregates of 7000–8000 dalton proteolipid monomers (54, 338).

It seems to us that most of these hypotheses are either needlessly extravagant or not supported by experimental facts. For example, the mitochondrially synthesized cytochrome oxidase polypeptides are not soluble in chloroform-methanol and attempts to dissociate them into smaller units have been consistently unsuccessful (295). It also appears that, in one instance at least, a mitochondrially made proteolipid (332) was actually a mixture of fatty acids (339). This does not exclude the possibility that mitochondria synthesize small polypeptides which may be precursors to larger units (23, 340–342). However, in this context we might consider the advice which the American writer George S. Kaufman once gave to his chauffeur who was speeding up the ramp to the George Washington Bridge: "Please, mister, don't cross that bridge until you get to it!"

"Virus-like Particles" in Neurospora

Several fungi such as *Aspergillus* and *Penicillum* (343, 344) or extrachromosomal mutants of *Neurospora* (345) harbor intracellular particles whose morphology resembles that of polyhedral (343, 344) or polymorphic (345) viruses. Although infectivity of these particles has not been demonstrated, they are generally termed "virus-like particles." Küntzel et al (342, 346) have recently isolated these particles from the *abnormal-1* mutant of *N. crassa*. *Abnormal-1* is an extrachromosomal, respiration-deficient mutant (347) whose traits can be transmitted to normal cells by hyphal injection of *abnormal-1* mitochondrial fractions (348).

The virus-like particles were present both in the cytoplasm and a crude mitochondrial fraction of *abnormal-1* cells, but were undetectable in wild-type cells. Thin sections of the OsO_4-fixed particles revealed virus-like polymorphic vesicles, 100–400 nm in diameter, which contained an electron-dense nucleoid and one or two surrounding membranes. The particles were composed of 8–10% RNA, 83–85% protein, and 7% phospholipid. The RNA was single stranded, sedimented with an *s* value of 33, and could be dissociated into 7–9S fragments by heat treatment in the presence of SDS (346). It hybridized to approximately 4.5% of mitochondrial DNA from wild-type *N. crassa* or the *abnormal-1* mutant (342). Total mitochondrial RNA, but not purified mitochondrial or cytoplasmic ribosomal RNA, competed with 33S RNA for binding to mitochondrial DNA (342). Most of the protein consisted of only two polypeptides, a 95,000 dalton glycoprotein and a 15,000 dalton lipoprotein (346). The lipoprotein could be cleaved into an 11,000 dalton protein and 5 phospholipid molecules. In vivo labeling experiments in the presence of cycloheximide or chloramphenicol suggested that the glycoprotein was made in the cytoplasm whereas the lipoprotein was made in mitochondria (342).

These findings raise the possibility that the transfer of *abnormal-1* characteristics by microinjection of mitochondrial fractions (348) may reflect intracellular infection of the normal host mitochondria by the virus-like particles. Even more important, the studies by Küntzel et al may open up new possibilities for exploring the mitochondrial genetic system and its interaction with the cytoplasm. It will be of interest to learn whether or not some of the protein components of the virus-like particles are also detectable in wild-type mitochondria even though the particles themselves are absent.

INTERDEPENDENCE BETWEEN CYTOPLASMIC AND MITOCHONDRIAL PROTEIN SYNTHESIS

It is now well established that the mitochondrial and the cytoplasmic protein synthesizing systems are tightly coordinated. This coordination is still little understood and probably occurs at many different levels. There is obviously some long term coupling since most, if not all, proteins of the mitochondrial system are made in the cytoplasm. Mitochondrial protein synthesis may perhaps also be influenced by the availability of lipids or heme (275, 349–351), both of which are indirect products of cytoplasmic protein synthesis. In the present context, however, we only discuss how polypeptides made by one system can modulate the activity of the other.

The earliest indications for such a coupling came from the following experiments.

1. Even though cycloheximide-inhibited cells remain viable, mitochondrial protein synthesis in these cells usually stops after 30 to 60 min (167, 172); in some cases, it can be greatly stimulated by prior incubation of the cells in chloramphenicol (220, 352).

2. The formation of functional mitochondria during adaptation of anaerobically grown yeast cells to oxygen is inhibited by either cycloheximide or chloramphenicol (353, 354). However, if the anaerobically grown cells are first aerated in the presence of cycloheximide and, after removal of the cycloheximide, aerated in the presence of chloramphenicol, a delayed adaptation occurs (44, 354). This effect is abolished if the first incubation is carried out in the presence of both cycloheximide and chloramphenicol. In the first incubation, the mitochondrial system appears to synthesize some oxygen-induced intermediates of respiratory adaptation. These intermediates can accumulate but cannot, by themselves, cause the emergence of respiration unless complemented by oxygen-induced products of cytoplasmic protein synthesis. Similar delayed responses were observed by following the formation of cytochrome oxidase during adaptation to oxygen (310) or the formation of oligomycin-sensitive ATPase during release from glucose repression (352). Since the delayed effects were rather small, each system appeared to be dependent on the simultaneous activity of the other.

These rather preliminary data suggest that mitochondrially synthesized proteins are formed in significant amounts only if they are continuously combined with cytoplasmically made partner proteins. According to this view, mitochondrial protein synthesis in cycloheximide-inhibited cells reflects the presence of excess pools of such partner proteins and stops when these pools are exhausted. In other words, cytoplasmically made proteins control mitochondrial protein synthesis. More direct evidence for this model was provided by the following observations.

1. If yeast cells are labeled with ^{3}H-leucine in the presence of cycloheximide, the mitochondrially made cytochrome oxidase subunits can be precipitated from the lysed mitochondria with an antiserum against the cytoplasmically made cytochrome oxidase subunits (196). Since the latter could not have been synthesized after the addition of cycloheximide, they must have been present in excess.

2. If yeast cells are grown in the presence of ^{3}H-leucine, all cytochrome oxidase polypeptides become equally labeled. However, if the period of labeling is decreased to 60 min, the cytoplasmically made subunits are much less radioactive than the mitochondrially made ones (27, 196). Carefully controlled pulse-labeling experiments with *N. crassa* have shown that the excess pools of some cytoplasmically made cytochrome oxidase subunits may approach 25% of the corresponding subunits in assembled cytochrome oxidase molecules (355, 356).

Attempts to detect uncombined cytoplasmically formed cytochrome oxidase subunits by immunoprecipitation have not yet been successful (357).

At present we do not know how the formation of mitochondrial translation products is stimulated by the availability of cytoplasmically made partner proteins. It could be that the hydrophobic mitochondrial products cannot leave the mitochondrial ribosomes unless they are pulled off by the corresponding cytoplasmically synthesized subunits.

It appears that cytoplasmically synthesized proteins may also affect mitochondrial protein synthesis by more indirect mechanisms. Butow et al (215, 299) and Ebner et al (31, 300) have used SDS–gel electrophoresis to show that some respiration-deficient nuclear yeast mutants synthesize an abnormal spectrum of mitochondrial translation products. This implies that nuclear mutations can either affect the synthesis of some mitochondrial products or their integration into the mitochondrial membrane. Although it is still too early to decide between these alternatives, the second one is favored by recent experiments on cytochrome oxidaseless yeast mutants. Ebner et al (32) isolated several nuclear yeast mutants specifically lacking cytochrome oxidase and analyzed three of these mutants immunochemically for residual cytochrome oxidase subunits. Each of them lacked at least one polypeptide which, in the wild type, is synthesized on mitochondrial ribosomes. One mutant lacked subunit III (cf above) whereas the two other mutants were almost completely deficient in all three mitochondrially made subunits. All mutants possessed near normal amounts of the cytoplasmically synthesized cytochrome oxidase components. At least one of these mutations can be suppressed by a typical *amber* suppressor (124). Since mitochondrial and cytoplasmic protein synthesis very probably do not share a common suppressor activity, this mutation may primarily affect an as yet unknown organizer protein which is coded by nuclear genes and presumably translated in the cytoplasm. The mutant may thus be a regulatory or assembly mutant. However, this concept will remain speculative until the proposed cytoplasmically made organizer protein has been identified.

While cytoplasmic protein synthesis can thus affect mitochondrial protein synthesis, the converse is also true. As discussed earlier, the integration of several cytoplasmic products (such as F_1, holocytochrome c_1, or the small cytochrome oxidase subunits) into the mitochondrial inner membrane requires proteins made in mitochondria. However, mitochondrial translation products may not only affect the integration of cytoplasmically made proteins, but also the rate of synthesis of these proteins. Barath and Küntzel have observed that growth of *N. crassa* in chloramphenicol or ethidium bromide stimulated the synthesis of several components of the mitochondrial genetic system such as elongation factors, met-tRNA transformylase, ribosomes, and RNA polymerase (358, 359). The mitochondrial RNA polymerase even appeared to accumulate in the soluble cytoplasm; at the same time, the cells acquired paracrystalline cytoplasmic inclusions that were tentatively identified as aggregated mitochondrial RNA polymerase. These para-crystals had been discovered earlier in two *N. crassa* mutants (*abnormal 1* and *2*) which probably have an impaired mitochondrial genetic system (83). Similarly, growth of *Tetrahymena* in ethidium bromide stimulates the synthesis of mito-chondrial DNA polymerase (360–362). Barath & Küntzel proposed, therefore, that mitochondria synthesize a repressor for the cytoplasmically synthesized proteins of the mitochondrial genetic system (359). According to this hypothesis, the repressor is only elaborated during a specific phase of mitochondrial division. When its synthesis stops, synthesis of the cytoplasmically made mitochondrial proteins begins. The availability of cytoplasmically made partner proteins in turn triggers mitochondrial protein synthesis (cf above) and renewed production of repressor. This interesting model is perhaps an oversimplification but it can be experimentally tested. Since techniques for the large scale synchronization of microbial cultures are now available (e.g. 363), it would be feasible to follow the production of specific mitochondrial proteins as a function of the cell cycle. Some encouraging steps in this direction have already been taken (364).

HOW ARE CYTOPLASMICALLY MADE PROTEINS TRANSPORTED INTO THE MITOCHONDRIA?

This question, one of the central problems of mitochondrial biogenesis, is still unanswered because we do not have a good system with which to study it. For a while, there were hopes that such a system had been found (365, 366). It was reported that rat liver microsomes that had been labeled with radioactive amino acids in vivo could donate labeled proteins (including cytochrome *c*) to mitochondria admixed in vitro. Microsomes labeled in a cell-free system were incapable of transferring cytochrome *c* into mitochondria although they still transferred some unidentified radioactive polypeptides. The transfer reaction was time dependent and appeared to require energy as well as GTP. However, it was rather slow and has apparently resisted further analysis. Since others could not reproduce it (192), its significance is still open.

It was also considered that proteins may enter the mitochondria as phospho-lipid complexes (159, 367). This is still a good possibility since certain membrane proteins are known to bind phospholipids, sometimes with considerable specificity. The phospholipid may also allow the complexed protein to penetrate through the mitochondrial membranes. Unfortunately, this hypothesis is not easy to test experimentally and has therefore not yet received the attention it deserves.

Kellems & Butow (368, 369) have recently shown that yeast mitochondria isolated in the presence of Mg^{2+} contained bound cytoplasmic ribosomes that were more resistant to dissociation by 0.4 M KCl than free cytoplasmic ribosomes or ribosomes bound to the endoplasmic reticulum. The isolated mitochondria appeared to possess a finite number of ribosome binding sites; additional binding sites could be exposed by washing the mitochondria with 2 mM EDTA. These observations raise the possibility that cytoplasmically made mitochondrial proteins are synthesized by a special class of cytoplasmic ribosomes on the mitochondrial surface. The newly synthesized polypeptides could then perhaps be discharged across the mitochondrial membrane(s) analogous to the vectorial translation processes which occur in certain regions of the endoplasmic reticulum (370, 371). If this appealing hypothesis is correct, the nascent chains of mitochondrial polypeptides should be selectively associated with mitochondria-bound cytoplasmic ribosomes. This can be tested immunochemically. A further prediction of this hypothesis is that cytoplasmically made mitochondrial proteins should not be detectable in the extramitochondrial cytoplasm (cf however, 358, 371a). Still, it would be difficult to explain how some proteins (e.g. F_1–ATPase) are carried across both the outer and the inner mitochondrial membrane and why certain mRNAs are translated only on some cytoplasmic ribosomes but not on others.

THE TRANSCRIPTIONAL ORIGIN OF MITOCHONDRIALLY SYNTHESIZED PROTEINS

Are proteins which are made on mitochondrial ribosomes coded for by mtDNA? Some researchers seem to have taken this for granted (e.g. 217) but there is really no solid evidence on this point. To stress this uncertainty, Dawid has played the devil's advocate by suggesting that none of the proteins made in animal mitochondria is coded by mtDNA (12). While this hypothesis is very probably incorrect, it has not yet been convincingly refuted. As discussed in the following sections, however, there is a good deal of tentative evidence to suggest that mtDNA does, in fact, code for mitochondrial translation products.

Effect of Transcription Inhibitors on Mitochondrial Protein Synthesis

Many authors have noted that amino acid incorporation by isolated mitochondria or by mitochondria in cycloheximide-inhibited cells is sensitive to inhibitors of transcription such as actinomycin D (68, 264, 372, 373), acriflavin (44, 101, 215, 265), cordycepin (374, 375), ethidium bromide (101, 215, 217, 260, 374, 376), or rifampicin (100, 266, 267). For various reasons, none of these results are convincing. First, actinomycin usually inhibits amino acid incorporation by isolated mitochondria only partially, probably because mitochondria are quite impermeable to the drug. Second, actinomycin as well as cordycepin may also inhibit nuclear transcription which makes it almost impossible to interpret their effect on mitochondrial protein synthesis in vivo. Third, it has long been suspected that acriflavin and ethidium bromide may also affect reactions of translation (377, 378). Grivell & Metz have now shown that ethidium bromide may block the translation of externally added poly(U) in isolated *Xenopus* mitochondria (379). Even the effect of rifampicin on mitochondrial protein synthesis should be interpreted with caution since the specificity of this drug in mitochondrial systems is still largely unknown.

  To circumvent some of these ambiguities, Mahler et al (380) made use of a yeast mutant which was thermosensitive for nuclear RNA synthesis (381, 382). Thirty minutes after a shift to the nonpermissive temperature, the ability of the mutant to incorporate ^{3}H-leucine into nascent chains on cell sap ribosomes was greatly reduced whereas the corresponding activity of the wild type was little affected. In contrast, the nonpermissive temperature did not impair the mitochondrial system of either strain as measured by 1. incorporation of ^{3}H-leucine or ^{14}C-formate into nascent chains on mitochondrial polysomes, 2. incorporation of ^{14}C-formate into F-met-puromycin, and 3. the intramitochondrial level of polysomes. All three parameters were, however, drastically lowered by adding ethidium bromide. Since the mitochondrial system responded only to ethidium bromide and the cytoplasmic system only to a temperature shift, Mahler et al concluded that transcription of mtDNA was sufficient to sustain most, and probably all, of mitochondrial protein synthesis. While this conclusion is fairly convincing, it still involves some untested assumptions about the specificity of ethidium bromide.

Transcription and Translation of Mitochondrial DNA in vitro

If mtDNA codes for proteins, it should be possible to synthesize those proteins in vitro by first transcribing mtDNA with RNA polymerase and then translating the resulting mRNAs in a reconstituted translation system. The first success along these lines was achieved in 1972 by Blossey & Küntzel (383). They reported that E. coli RNA polymerase produced transcripts of mtDNA from N. crassa which were translated by a submitochondrial Neurospora system. Upon SDS–acrylamide gel electrophoresis, the labeled protein products were resolved into at least four species which ranged in size from 11,000 to 180,000 daltons; however, the molecular weight distribution of these products was strikingly different from that of poly-peptides synthesized on mitochondrial ribosomes in vivo.

Shortly thereafter, Chuang & Weissbach (384) reported that rat liver mtDNA could act as a template in the completely heterologous transcription–translation system prepared from E. coli extracts according to Zubay et al (385). Electrophoresis of the labeled product in the presence of SDS produced a poorly defined pattern in which most of the radioactivity migrated with apparent molecular weights of 20,000–30,000. Again, the pattern was significantly different from that obtained for rat liver mitochondria labeled in vitro.

Scragg (386) has taken a somewhat different approach. He first transcribed yeast mtDNA with RNA polymerase from yeast mitochondria, then isolated the RNA product, and finally added it to a cell-free translation system from E. coli. Some of the labeled polypeptides formed in this two-step system could be precipitated with an antiserum against an insoluble subfraction derived from yeast mitochondria.

Although these studies are only in their early stages, they have already shown that mtDNA can code in vitro for RNA with messenger-like activity. As the next step, the in vitro translation products must now be identified as specific mitochondrial proteins. This might conceivably be achieved by comparing the immunological properties and the peptide maps of the in vitro products with those of well established mitochondrial translation products (cf above). However, one could think of a myriad of reasons why these coupled systems might fail to yield an identifiable product, especially when heterologous components are used. Any verification of the labeled product(s) will therefore have to be quite convincing.

Identification of Mitochondrial mRNA

One of the most serious problems that can be anticipated for the in vitro transcription–translation system is the failure to obtain correct transcripts. This problem can in theory be averted by synthesizing mitochondrial mRNA in vivo, isolating it from mitochondrial polysomes, and hybridizing it to mtDNA. If successful, such experiments could prove that mtDNA codes for some mito-chondrial proteins. Additionally, if unique species of mRNA can be obtained, perhaps by specific immunoprecipitation of polysomes (387) with antibodies against purified mitochondrial proteins, the genes for individual proteins could be located by hybridization techniques.

In practice, however, this approach is extremely difficult since the isolation and characterization of mitochondrial polysomes has proven to be a formidable task (17, 332). Several laboratories have observed fast sedimenting structures in mito-chondrial lysates from yeast (304, 388), *Euglena* (284, 389, 390), and Hela (374, 391) cells. These structures carried nascent polypeptide chains and were dissociated into monosomes by relatively mild treatment with ribonuclease. On the other hand, Ojala & Attardi (392), also working with Hela mitochondria, isolated polysome-like structures which were sensitive to ribonuclease only after treatment with pronase. They concluded that nascent chains were stabilizing the complex. Similarily, Michel & Neupert (340, 340a) described polymeric ribosomes from *N. crassa* mitochondria which were resistant to ribonuclease; they suggested that these structures were merely aggregates which were stabilized by the interaction of hydrophobic nascent chains rather than by mRNA. Although it was not demonstrated that the polymeric ribosomes could be disaggregated by proteolytic digestion alone, these structures were clearly different from well-behaved polysomes. It seems, therefore, that it will be very difficult to isolate substantial amounts of mitochondrial mRNA from mitochondrial polysomes by conventional methods.

Fortunately, an alternate approach is suggested by the recent finding that mitochondrial RNA contains stretches of poly(A). Perlman et al (393) described a heterogeneous RNA fraction from Hela cells with the following properties:
1. Its in vivo synthesis is blocked by ethidium bromide, but not by camptothecin, an inhibitor of high molecular weight RNA synthesis in the nucleus (394, 395).
2. It is selectively retained by glass fiber filters containing immobilized poly(U).
3. In acrylamide gels, it migrates heterogeneously with apparent mobilities in the range of 15 to 30S.
4. It contains covalently bound poly(A) segments of 50 to 80 nucleotides.
5. It is apparently associated with the structures resembling mitochondrial polysomes since it is released from rapidly sedimenting particles by puromycin under conditions where cytoplasmic message is not released.

Similarly, Attardi (396) has observed an RNA from Hela cell mitochondria which contains covalently linked poly(A) tracts of about the same size (4S) as reported by Perlman et al (393). While the biosynthesis of this RNA was blocked by ethidium bromide, the biosynthesis of the poly(A) segments was insensitive as indicated by the accumulation of "free" 4S poly(A). Isolation of the poly(A)-containing RNA and subsequent analysis on sucrose gradients under denaturing conditions revealed two size classes corresponding to sedimentation coefficients of 7 and 12S. Furthermore, 100% of the 7S and about 50% of the 12S RNA hybridized to the H strand of mtDNA. Approximately 30% of the 12S material hybridized with the L strand.

Cooper & Avers (388) found that RNA synthesized in isolated yeast mitochondria during an 8 min pulse with ^{3}H-uracil became associated with mitochondrial polysomes. This rapidly labeled RNA, upon extraction from the polysomes, bound to poly(U)-Sepharose, suggesting the presence of poly(A) segments. Further analysis of this material revealed poly(A) tracts containing about 70 nucleotides.

Taken together, these observations indicate the following:

1. Both H and L strands of mtDNA code for RNA molecules containing a stretch of covalently attached poly(A).
2. Some of this RNA apparently has messenger activity as suggested by its association with mitochondrial ribosomes active in protein synthesis. (However, no hybridization experiments have yet been carried out with poly(A)-containing RNA isolated directly from mitochondrial polysomes.)
3. The length of poly(A) segment in the mitochondrial RNA of Hela cells and yeast is in the range of 50–80 nucleotides and in this respect more closely resembles the poly(A) found associated with the RNA of animal viruses than the mRNA of the nucleocytoplasmic system of animal cells.

The protein-synthetic machinery of the mitochondrion has so often been compared to that of prokaryotic organisms that it is surprising to find that it synthesizes RNA containing poly(A), a class of RNA previously found only in nuclear mRNA and viral RNA of eukaryotic cells (cf 395). In this regard it is fortunate that three laboratories are in close agreement on such a potentially controversial subject.

Mitochondrial "mRNA" in "Virus-Like Particles" of Neurospora

As mentioned earlier, the virus-like particles isolated from the abnormal-1 mutant of N. crassa contain single-stranded 33S RNA which hybridizes to wild-type or abnormal-1 mtDNA (342, 346). At least part of this RNA can program the synthesis of a 10,000 dalton polypeptide in an in vitro translation system (342). This protein product corresponds in molecular weight to the 11,000 mol wt protein component found in the virus-like particles. It is thus a distinct possibility that the 33S RNA is a fairly homogeneous preparation of mitochondrial mRNA which is overproduced as a result of the abnormal-1 mutation. A closer investigation of this RNA should be of considerable interest.

Genetic Identification of Mitochondrially Coded Proteins

If mtDNA codes for mitochondrial proteins, it should be possible to isolate extrachromosomal mutants in which the primary structure of a mitochondrial protein is altered. The following discussion will show that such mutants are probably already available but that none of them has yet been shown to possess an altered mitochondrial protein.

A complete presentation of these mutants is beyond the scope of this article but can be found in several recent reviews (18, 19, 21, 397, 398). Here we confine ourselves to those mutants which appear to have fairly clear-cut biochemical lesions. Virtually all of these are *S. cerevisiae* mutants, even though mutations in mtDNA have also been found in other organisms (399–401), including perhaps human cells in culture (402).

In general, the mitochondrial mutations isolated so far confer resistance either to inhibitors of mitochondrial protein synthesis (such as erythromycin, chloramphenicol, and spiramycin) or to oligomycin, the inhibitor of oxidative phosphorylation. Resistance can be demonstrated both with growing cells and isolated mitochondria. In *S. cerevisiae,* the resistance markers are located on mtDNA since they are inherited in a non-Mendelian manner and eliminated by converting the cells to petite mutants without mtDNA. The genetic loci for resistance to inhibitors of mitochondrial protein synthesis show linkage to each other but are unlinked to the two oligomycin-resistance loci (403).

Grivell et al have convincingly shown that many of the mutations conferring resistance to inhibitors of mitochondrial protein synthesis alter the mitochondrial ribosome (404). They isolated ribosomes from erythromycin- and chloramphenicol-resistant mitochondria and tested their antibiotic sensitivity in the following two artificial systems: 1. RNA-independent synthesis of acetyl-leucyl-puromycin; 2. amino acid incorporation programmed either by poly(U) (404) or MS2 RNA (405). In each case tested, the ribosomes isolated from resistant mitochondria were also resistant in these artificial systems. These very clear-cut experiments raised the hope that a closer study of the resistant ribosomes would reveal an altered protein that was coded for by mitochondrial DNA.

To the general chagrin, however, attempts to detect an altered protein in these ribosomes by one-dimensional acrylamide gel electrophoresis were unsuccessful (405). Because ribosomes have so many proteins, it seems almost mandatory to corroborate these data by two-dimensional separation techniques (173, 406). Nevertheless, the failure to identify an altered mitochondrial gene product among the ribosomal proteins is in line with the earlier mentioned conclusion that all essential proteins of mitochondrial ribosomes are made in the cytoplasm.

By a process of exclusion, we are therefore forced to consider that the above-mentioned resistance mutants have suffered a change in their mitochondrial rRNAs. For example, one of these RNAs could exhibit an abnormal methylation (407–409) or possibly even an altered base sequence (cf 19). While any decision on this point is premature, recent experiments with *Neurospora* mutants favor the former alternative. Three independent lines of evidence suggest that the pleiotropic effects of the cytoplasmic *mi-1* (*poky*) mutation of *N. crassa* may reflect altered methylase activities.

First, the specific methylation of Lys-72 in cytochrome *c* to ε-trimethyl lysine apparently occurs at a lower than normal rate in young cultures of *mi-1* cells (410). Second, the mutation alters the chromatographic properties of some mitochondrial tRNAs perhaps by changing the extent of their methylation (411). Third, earlier work by Rifkin & Luck (412) revealed a significant excess of the large ribosomal subunit in *poky* mitochondria, possibly suggesting defective synthesis or assembly of the small subunits. Recently, Kuriyama & Luck (413) demonstrated abnormal processing of the 32S ribosomal RNA precursor molecule (414) in mutant mitochondria and significant undermethylation of mitochondrial rRNA regardless of whether it was isolated from a total mitochondrial RNA fraction or from functional mitochondrial ribosomes. No difference was detected between mutant and wild-type cytoplasmic rRNA. These findings offer fuel for the speculation that the mitochondrial genetic system controls directly or indirectly the synthesis or the activity of a mitochondrial methylating system. Although some authors are not yet convinced that mitochondrial rRNAs are methylated at all (cf 19), it is a distinct possibility that the drug-resistant mitochondrial ribosomes of S. cerevisiae contain abnormally methylated RNA. In any case, the pursuit of drug-resistant mito-chondrial ribosomes has not yet led to one of the elusive protein products of mtDNA.

The outlook is brighter for finding at least one of these products among the four hydrophobic polypeptides associated with the oligomycin-sensitive ATPase complex. As mentioned earlier, these four polypeptides are made on mitochondrial ribosomes (23, 32, 191). Moreover, three laboratories (415–417) have isolated and defined mutations in mtDNA which confer oligomycin resistance to the ATPase complex of yeast.

The ATPase complex of these mutants exhibits a resistance to oligomycin in vivo and in vitro (398, 417). Mixed reconstitution experiments with the three main components of the ATPase complex ("stripped" membranes, oligomycin sensitivity-conferring protein, and F_1-ATPase) from mutant and wild-type cells showed that the resistance factor(s) resided in the membrane fraction (326, 417). However, oligomycin-sensitive ATPase activity also depends on membrane lipids (418, 419) and it is therefore still possible (although admittedly unlikely) that oligomycin resistance is caused by an altered membrane lipid. To settle this point, it will be necessary to compare the hydrophobic ATPase subunits from wild-type and resistant strains.

Can mtDNA Code for all Mitochondrially Synthesized Proteins?

It has sometimes been argued that mitochondrial DNA may not be large enough to accommodate cistrons for all of the mitochondrial translation products (e.g. 420). This problem appeared to be particularly acute with the small animal mtDNAs (mol wt 10^7) in which the various RNA genes occupy almost one third of the molecule (19).

Although this question is not definitively settled, it need not cause us sleepless nights. While we know very little about the translation products of animal mitochondria, they seem to be quite similar to those of yeast mitochondria with respect to approximate number and molecular weight distribution (421). For a rough approximation, we may therefore correlate the total molecular weight of the well-characterized protein products of yeast and Neurospora mitochondria with the coding capacity of animal mtDNA. Such a calculation leaves almost one third of the coding capacity still unaccounted for (Table 1). If anything, this extra capacity is probably underestimated for the following reasons: 1. The calculation of Table 1

Table 1   Can animal mtDNA code for all mitochondrially synthesized proteins?

Mitochondrial product	Molecular weight	Reference	Daltons of double-stranded DNA required for coding
Large rRNA	0.56×10^6	422, 423	1.12×10^6
Small rRNA	0.33×10^6	422, 423	0.66×10^6
20 tRNAs	0.56×10^6	422	1.12×10^6
Cytochrome oxidase subunits	4×10^4	196, 219, 314	0.80×10^6
	3×10^4	196, 219, 314	0.60×10^6
	2×10^4	196, 219, 314	0.40×10^6
ATPase-complex subunits	3×10^4	23, 26	0.60×10^6
	2×10^4	23, 26	0.40×10^6
	2×10^4	23, 26	0.40×10^6
	7×10^3	23, 26	0.14×10^6
Cytochrome b subunit	3×10^4	222, 223	0.60×10^6
			Total: 6.84×10^6

reserves space for 20 tRNAs, but animal mtDNA may code for only 11 to 15 tRNAs (292, 293). 2. Animal mitochondria may not synthesize one of the large subunits of cytochrome oxidase (314, 315). The residual space on mtDNA (about 3×10^6 daltons) could probably accommodate the cistrons for as yet unidentified mitochondrial translation products or for rRNA segments removed during maturation of rRNA. It is, therefore, not unreasonable to assume that even the small mtDNAs of animal tissues are large enough to code for all mitochondrially synthesized proteins.

CONCLUDING REMARKS

Our discussion has shown that the few proteins synthesized by mitochondria are necessary for the correct assembly of a much greater number of cytoplasmically made mitochondrial proteins. This supports the view that complex biological membranes are hierarchical structures in which some molecules are "more equal than others." How is the blueprint for this three-dimensional organization laid down in a one-dimensional DNA molecule? We do not know, but a second dimension of information could perhaps be the temporal sequence in which the various membrane proteins are synthesized. This possibility can be tested with present experimental techniques.

We have also seen that definitive information on the formation of a mitochondrial protein is usually obtained only when this protein has been obtained in a pure state. We suspect that time will not change this rate-limiting step.

Literature Cited

1. Roodyn, D. B., Wilkie, D. 1968. *The Biogenesis of Mitochondria.* London: Methuen
2. Slater, E. C., Tager, J. M., Papa, S., Quagliariello, E. 1968. *Biochemical Aspects of the Biogenesis of Mitochondria.* Bari: Adriatica Editrice
3. Nass, S. 1969. *Int. Rev. Cytol.* 25:55
4. Borst, P., Kroon, A. M. 1969. *Int. Rev. Cytol.* 26:107
5. Swift, H., Wolstenholme, D. R. 1969. *Handbook of Molecular Cytology,* ed. A. Lima-De-Faria, 972. Amsterdam: North-Holland
6. Kroon, A. M. 1969. See Ref. 5, p. 943
7. Nass, M. M. K. 1969. *Science* 165:25
8. Wagner, R. P. 1969. *Science* 163:1026
9. Schatz, G. 1970. *Membranes of Mitochondria and Chloroplasts,* ed. E. Racker, 251. New York: Van Nostrand
10. Rabinowitz, M., Swift, H. 1970. *Physiol. Rev.* 50:376
11. Linnane, A. W., Haslam, J. M. 1970. *Curr. Top. Cell. Regul.* 2:101
12. Miller, P. L., Ed. 1970. *Control of Organelle Development.* New York: Academic
13. Ashwell, M., Work, T. S. 1970. *Ann. Rev. Biochem.* 39:251
14. Boardman, N. K., Linnane, A. W., Smillie, R. M. 1971. *Autonomy and Biogenesis of Mitochondria and Chloroplasts.* Amsterdam: North-Holland
15. Küntzel, H. 1971. *Curr. Top. Microbiol. Immunol.* 54:94
16. Beattie, D. S. 1971. *Sub-Cell. Biochem.* 1:1
17. Borst, P., Grivell, L. A. 1971. *FEBS Lett.* 13:73
18. Linnane, A. W., Haslam, J. M., Lukins, H. B., Nagley, P. 1972. *Ann. Rev. Microbiol.* 26:163
19. Borst, P. 1972. *Ann. Rev. Biochem.* 41:333
20. Van Den Bergh, S. G. et al 1972. *Mitochondria/Biomembranes.* Amsterdam: North-Holland
21. Sager, R. 1972. *Cytoplasmic Genes and Organelles.* New York: Academic
22. Kagawa, Y. 1972. *Biochim. Biophys. Acta* 265:297
23. Tzagoloff, A., Rubin, M. S., Sierra, M. F. 1973. *Biochim. Biophys. Acta* 301:71
24. Shapiro, A. L., Viñuela, E., Maizel, J. V. 1967. *Biochem. Biophys. Res. Commun.* 28:815
25. Weber, K., Osborn, M. 1969. *J. Biol. Chem.* 244:4406
26. Tzagoloff, A., Meagher, P. 1971. *J. Biol. Chem.* 246:7328
27. Mason, T. et al. See Ref. 20, p. 53
28. Phillips, D. R., Morrison, M. 1971. *Biochemistry* 10:1766
29. Hubbard, A. L., Cohn, Z. A. 1972. *J. Cell Biol.* 55:390
30. Schneider, D. L., Kagawa, Y., Racker, E. 1972. *J. Biol. Chem.* 247:4074
31. Ebner, E., Mennucci, L., Schatz, G. 1973. *J. Biol. Chem.* 248:5360
32. Ebner, E., Mason, T. L., Schatz, G. 1973. *J. Biol. Chem.* 248:5369
33. Ebner, E., Schatz, G. 1973. *J. Biol. Chem.* 248:5379
34. Flavell, R. 1972. *Biochem. Genet.* 6:275
35. Roodyn, D. B., Reis, P. J., Work, T. S. 1961. *Biochem. J.* 80:9
36. Truman, D. E. S., Korner, A. 1962. *Biochem. J.* 83:588
37. Beattie, D. S., Basford, R. E., Koritz, S. B. 1966. *Biochemistry* 5:926
38. Neupert, W., Brdiczka, D., Bücher, T. 1967. *Biochem. Biophys. Res. Commun.* 27:488
39. Beattie, D. S., Basford, R. E., Koritz, S. B. 1967. *Biochemistry* 6:3099
40. Work, T. S. 1968. See Ref. 2, p. 367
41. Sebald, W., Hofstötter, T., Hacker, D., Bücher, T. 1969. *FEBS Lett.* 2:177
42. Henson, C. P., Perlman, P., Weber, C. N., Mahler, H. R. 1968. *Biochemistry* 7:4445
43. Schatz, G. et al 1972. *Fed. Proc.* 31:21
44. Groot, G. S. P., Rouslin, W., Schatz, G. 1972. *J. Biol. Chem.* 247:1735
45. Slonimski, P. P. 1953. *La formation des enzymes respiratoires chez la levure.* Paris: Masson
46. Groot, G. S. P., Kováč, L., Schatz, G. 1971. *Proc. Nat. Acad. Sci. USA* 68:308
47. Mounolou, J. C., Jakob, H., Slonimski, P. P. 1966. *Biochem. Biophys. Res. Commun.* 24:218
48. Carnevali, F., Piperno, G., Tecce, G. 1966. *Atti Accad. Naz. Lincei, Rend. Cl. Sci. Fis. Mat. Natur.* 41:194
49. Goldring, E. S., Grossman, L. I., Krupnick, D., Cryer, D. R., Marmur, J. 1970. *J. Mol. Biol.* 52:323
50. Nagley, P., Linnane, A. W. 1970. *Biochem. Biophys. Res. Commun.* 39:989
51. Kužela, Š., Grečná, E. 1969. *Experientia* 25:776
52. Schatz, G., Saltzgaber, J. 1969. *Biochem. Biophys. Res. Commun.* 37:996
53. Mahler, H. R., Feldman, F., Phan, S. H., Hamill, P., Dawidowicz, K. 1974. *International Conference on the Biogenesis of Mitochondria, Bari, Italy, June 25–28, 1973,* ed. A. M. Kroon, C.

Saccone. New York: Academic. In press
54. Tzagoloff, A., Akai, A. 1972. *J. Biol. Chem.* 247:6517
55. Linnane, A. W., Still, J. 1956. *Aust. J. Sci.* 18:165
56. Schatz, G., Tuppy, H., Klima, J. 1963. *Z. Naturforsch. B* 18:145
57. Mackler, B., Douglas, H. C., Will, S., Hawthorne, D. C., Mahler, H. R. 1965. *Biochemistry* 4:2016
57a. Perlman, P. S., Mahler, H. R. 1970. *Bioenergetics* 1:113
58. Riethmüller, G., Tuppy, H. 1964. *Biochem. Z.* 340:413
59. Gregolin, C., Ghiretti-Magaldi, A. 1961. *Biochim. Biophys. Acta* 54:62
60. Schatz, G. 1968. *J. Biol. Chem.* 243:2192
61. Kováč, L., Weissová, K. 1968. *Biochim. Biophys. Acta* 153:55
62. Scragg, A. H. 1971. *FEBS Lett.* 17:111
63. Richter, D. 1971. *Biochemistry* 10:4422
64. Parisi, B., Cella, R. 1971. *FEBS Lett.* 14:209
65. Kraml, J., Mahler, H. R. 1967. *Immunochemistry* 4:213
66. Yotsuyanagi, Y. 1962. *J. Ultrastruct. Res.* 7:141
66a. Packer, L., Williams, M. A., Criddle, R. S. 1973. *Biochim. Biophys. Acta* 292:92
67. McClary, D. O., Bowers, W. D., Jr. 1968. *J. Ultrastruct. Res.* 25:37
68. Wintersberger, E. 1967. *Z. Physiol. Chemie* 348:1701
69. Wintersberger, E., Viehhauser, G. 1968. *Nature* 220:699
70. Williamson, D. H., Maroudas, N. G., Wilkie, D. 1971. *Mol. Gen. Genet.* 111:209
71. Scragg, A. H. 1971. *Biochem. Biophys. Res. Commun.* 45:701
72. Sels, A. A., Cocriamont, C. 1968. *Biochem. Biophys. Res. Commun.* 32:192
73. Ross, E., Okamoto, H., Schatz, G. In preparation
74. Clark-Walker, G. D., Linnane, A. W. 1966. *Biochem. Biophys. Res. Commun.* 25:8
75. Clark-Walker, G. D., Linnane, A. W. 1967. *J. Cell. Biol.* 34:1
76. Linnane, A. W., Biggs, D. R., Huang, M., Clark-Walker, G. D. 1968. In *Aspects of Yeast Metabolism*, ed. R. K. Mills, 217. Oxford: Blackwell
77. Kellerman, G. M., Biggs, D. R., Linnane, A. W. 1969. *J. Cell Biol.* 42:378
78. Rubin, M. S., Tzagoloff, A. 1973. *Fed. Proc.* 32:2403
79. Ball, A. J. S., Tustanoff, E. R. 1971. See Ref. 14, p. 466
80. Mahler, H. R., Perlman, P. S. 1971. *Biochemistry* 10:2979

81. Luha, A. A., Sarcoe, L. E., Whittaker, P. A. 1971. *Biochem. Biophys. Res. Commun.* 44:396
82. Howell, N., Zuiches, C., Munkres, K. 1971. *J. Cell Biol.* 50:721
83. Wood, D. D., Luck, D. J. L. 1971. *J. Cell Biol.* 51:249
84. Lambowitz, A. M., Smith, E. W., Slayman, C. W. 1972. *J. Biol. Chem.* 247:4859
85. Birkmayer, G. D., Bücher, T. 1969. *FEBS Lett.* 5:28
86. Von Jagow, G., Klingenberg, M. 1972. *FEBS Lett.* 24:278
87. Wakiyama, S., Ogura, Y. 1972. *J. Biochem.* 71:295
88. Evans, D. A., Lloyd, D. 1967. *Biochem. J.* 103:22P
89. Marchant, R., Smith, D. G. 1968. *J. Gen. Microbiol.* 50:391
90. Turner, G., Lloyd, D. 1970. *Biochem. J.* 116:41P
91. Lloyd, D., Evans, D. A., Venables, S. E. 1970. *J. Gen. Microbiol.* 61:33
92. Adoutte, A., Balmefrézol, M., Beisson, J., André, J. 1972. *J. Cell Biol.* 54:8
93. Turner, G., Lloyd, D. 1971. *J. Gen. Microbiol.* 67:175
94. Kusel, J. P., Moore, K. E., Weber, M. M. 1967. *J. Protozool.* 14:283
95. Meyer, R. R., Boyd, C. R., Rein, D. C., Keller, S. J. 1972. *Exp. Cell Res.* 70:233
96. Smith-Johannsen, H., Gibbs, S. P. 1972. *J. Cell Biol.* 52:598
96a. Heilporn, V., Limbosch, S. 1971. *Biochim. Biophys. Acta* 240:94
97. Firkin, F. C., Linnane, A. W. 1968. *Biochem. Biophys. Res. Commun.* 32:398
98. Kroon, A. M., Jansen, R. J. 1968. *Biochim. Biophys. Acta* 155:629
99. Firkin, F. C., Linnane, A. W. 1969. *Exp. Cell Res.* 55:68
100. Kroon, A. M., De Vries, H. 1970. See Ref. 12, p. 181
101. Kroon, A. M., De Vries, H. 1971. See Ref. 14, p. 318
102. Kroon, A. M., Arendzen, A. J. 1972. See Ref. 20, p. 71
103. Storrie, B., Attardi, G. 1973. *J. Cell Biol.* 56:819
104. Lenk, R., Penman, S. 1971. *J. Cell Biol.* 49:541
105. King, M. E., Godman, G. C., King, D. W. 1972. *J. Cell Biol.* 53:127
106. Smith, U., Smith, D. S., Yunis, A. A. 1970. *J. Cell Sci.* 7:501
107. De Vries, H., Kroon, A. M. 1970. *Biochim. Biophys. Acta* 204:531
108. Naum, Y., Pious, D. A. 1971. *Exp. Cell Res.* 65:335
109. Dixon, H., Kellerman, G. M., Linnane,

A. W. 1972. *Arch. Biochem. Biophys.* 152:869
110. Hamberger, A., Isaksson, O. 1973. *Exp. Cell Res.* 76:207
111. Kahri, A. I., Milner, A. J. 1969. *Ann. Meet. Am. Soc. Cell Biol., 9th, Detroit.* Abstr. 148
112. Fettes, I. M., Haldar, D., Freeman, K. B. 1972. *Can. J. Biochem.* 50:200
113. Radsak, K., Kato, K. Sato, N., Koprowski, H. 1971. *Exp. Cell Res.* 66:410
114. Gijzel, W. P., Strating, M., Kroon, A. M. 1972. *Cell Differentiation* 1:191
115. González-Cadavid, N. F., Avila-Bello, E. M. A., Ramirez, J. L. 1970. *Biochem. J.* 118:577
116. Spolsky, C. M., Eisenstadt, J. M. 1972. *FEBS Lett.* 25:319
117. Soslau, G., Nass, M. M. K. 1971. *J. Cell Biol.* 51:514
118. Sato, N., Chance, B., Kato, K., Klietmann, W. 1973. *FEBS Lett.* 29:222
119. Storrie, B., Attardi, G. 1972. *J. Mol. Biol.* 71:177
120. Ringrose, P. S., Lambert, R. W. 1973. *Biochim. Biophys. Acta* 299:374
121. Asahi, T., Majima, R. 1969. *Plant Cell Physiol.* 10:317
122. Mahler, H. R., Perlman, P. S., Mehrotra, B. D. 1971. See Ref. 14, p. 492
123. Vary, M. J., Edwards, C. L., Stewart, P. R. 1969. *Arch. Biochem. Biophys.* 130:235
124. Ross, E. et al 1974. See Ref. 53. In press
125. Davey, P. J., Yu, R., Linnane, A. W. 1969. *Biochem. Biophys. Res. Commun.* 36:30
126. Kornblatt, J. A., Rudney, H. 1971. *J. Biol. Chem.* 246:4424
127. Adams, B. G., Parks, L. W. 1969. *J. Bacteriol.* 100:370
128. Shimizu, I., Nagai, J., Hatanaka, H., Katsuki, H. 1973. *Biochim. Biophys. Acta* 296:310
129. Beattie, D. S. 1968. *J. Biol. Chem.* 243:4027
130. Freeman, K. B., Haldar, D. 1967. *Biochem. Biophys. Res. Commun.* 28:3
131. Freeman, K. B., Haldar, D. 1968. *Can. J. Biochem.* 46:1003
132. Godchaux, W., Herbert, E. 1966. *J. Mol. Biol.* 21:537
133. Mackler, B., Haynes, B. 1970. *Biochim. Biophys. Acta* 197:317
134. Gordon, P. A., Lowdon, M. J., Stewart, P. R. 1972. *J. Bacteriol.* 110:504
135. Hanson, J. B., Hodges, T. K. 1963. *Nature* 200:1009
136. Freeman, K. B. 1970. *Can. J. Biochem.* 48:469
137. Freeman, K. B. 1970. *Can. J. Biochem.* 48:479
138. Ball, A. J. S., Tustanoff, E. R. 1970. *Biochim. Biophys. Acta* 199:476
138a. Linnane, A. W. 1968. See Ref. 2, p. 333
139. Lark, K. G. 1969. *Ann. Rev. Biochem.* 38:569
140. Carnevali, F., Leoni, L., Morpurgo, G., Conti, G. 1971. *Mutat. Res.* 12:357
141. Weislogel, P. O., Butow, R. A. 1970. *Proc. Nat. Acad. Sci. USA* 67:52
142. Haldar, D., Freeman, K., Work, T. S. 1966. *Nature* 211:9
143. Freeman, K. B., Haldar, D., Work, T. S. 1967. *Biochem. J.* 105:947
144. González-Cadavid, N. F., Campbell, P. N. 1967. *Biochem. J.* 105:427
145. González-Cadavid, N. F., Campbell, P. N. 1967. *Biochem. J.* 105:443
146. González-Cadavid, N. F., Bravo, M., Campbell, P. N. 1968. *Biochem. J.* 107:523
147. González-Cadavid, N. F., Ortega, J., González, M. 1971. *Biochem. J.* 124:685
148. Kadenbach, B. 1967. *Biochim. Biophys. Acta* 138:651
149. Kadenbach, B. 1968. See Ref. 2, p. 423
150. Kadenbach, B. 1969. *Eur. J. Biochem.* 10:312
150a. Bingham, R. W., Campbell, P. N. 1972. *Biochem. J.* 126:211
151. Kadenbach, B. 1970. *Eur. J. Biochem.* 12:392
152. Penniall, R., Davidian, N. 1968. *FEBS Lett.* 1:38
153. Davidian, N., Penniall, R., Elliott, W. B. 1968. *FEBS Lett.* 2:105
154. Davidian, N., Penniall, R., Elliott, W. B. 1969. *Arch. Biochem. Biophys.* 133:345
155. Davidian, N., Penniall, R. 1971. *Biochem. Biophys. Res. Commun.* 44:15
156. Beinert, H. 1951. *J. Biol. Chem.* 190:287
157. Flatmark, T., Sletten, K. 1968. *J. Biol. Chem.* 243:1623
158. Slonimski, P. P. 1949. *Ann. Inst. Pasteur Paris* 77:774
158a. Yaoi, Y. 1967. *J. Biochem.* 61:54
159. Siegel, M. R., Sisler, H. D. 1965. *Biochim. Biophys. Acta* 103:558
160. Rendi, R. 1959. *Exp. Cell Res.* 18:187
161. Mager, J. 1960. *Biochim. Biophys. Acta* 38:150
162. Lamb, A. J., Clark-Walker, G. D., Linnane, A. W. 1968. *Biochim. Biophys. Acta* 161:415
163. Kadenbach, B. 1968. See Ref. 2, p. 355
164. Henson, C. P., Weber, C. N., Mahler,

H. R. 1968. *Biochemistry* 7:4431,
165. Ashwell, M. A., Work, T. S. 1968. *Biochem. Biophys. Res. Commun.* 32: 1006
166. Sebald, W., Birkmayer, G. D., Schwab, A. J., Weiss, H. 1971. See Ref. 14, p. 339
167. Sebald, W., Schwab, A. J., Bücher, T. 1969. *FEBS Lett.* 4:243
168. Küntzel, H. 1969. *Nature* 222:142
169. Neupert, W., Sebald, W., Schwab, A. J., Massinger, P., Bücher, T. 1969. *Eur. J. Biochem.* 10:589
170. Michel, R., Schweyen, R. J., Kaudewitz, F. 1971. *Mol. Gen. Genet.* 111:235
171. Schatz, G., Saltzgaber, J., Rouslin, W. 1970. In *Electron Transport and Energy Conversion,* ed. J. M. Tager, S. Papa, E.Quagliariello, E. Slater, 329. Bari, Italy: Adriatica Editrice
172. Hawley, E. S., Greenawalt, J. W. 1970. *J. Biol. Chem.* 245:3574
173. Lizardi, P. M., Luck, D. J. L. 1972. *J. Cell Biol.* 54:56
174. Perlman, S., Penman, S. 1970. *Biochem. Biophys. Res. Commun.* 40:941
175. Brega, A., Baglioni, C. 1971. *Eur. J. Biochem.* 22:415
176. McMahon, D. 1971. *Fed. Proc.* 30:345
177. Čihák, A., Černá, J. 1972. *FEBS Lett.* 23:271
178. Higashinakagawa, T., Muramatsu, M. 1972. *Biochem. Biophys. Res. Commun.* 47:1
179. Timberlake, W. E., Hagen, G., Griffin, D. H. 1972. *Biochem. Biophys. Res. Commun.* 48:823
180. Garber, A. J., Jomain-Baum, M., Salganicoff, L., Farber, E., Hanson, R. W. 1973. *J. Biol. Chem.* 248:1530
181. Cowtan, E. R., Yoda, B., Israels, L. G. 1973. *Arch. Biochem. Biophys.* 155:194
182. Farber, J. L., Farmar, R. 1973. *Biochem. Biophys. Res. Commun.* 51:626
183. Frizzell, R. A., Nellans, H. N., Acheson, L. S., Schultz, S. G. 1973. *Biomembranes* 291:302
184. Gallerani, R., Saccone, C. 1974. See Ref. 53. In press
184a. Chakrabarti, S., Dube, D. K., Roy, S. C. 1972. *Biochem. J.* 128:461
184b. Ibrahim, N. G., Burke, J. P., Beattie, D. S. 1973. *FEBS Lett.* 29:73
184c. Lietman, P. S. 1971. *Mol. Pharmacol.* 7:122
184d. Gilead, Z., Becker, Y. 1971. *Eur. J. Biochem.* 23:143
185. Neupert, W., Sebald, W., Schwab, A. J., Pfaller, A., Bücher, T. 1969. *Eur. J. Biochem.* 10:585
186. Groot, G. S. P. 1974. See Ref. 53. In press

187. Schmitt, H. 1972. *FEBS Lett.* 26:215
188. Millis, A. J. T., Suyama, Y. 1972. *J. Biol. Chem.* 247:4063
189. Neupert, W., Ludwig, G. D. 1971. *Eur. J. Biochem.* 19:523
190. Bandlow, W. 1972. *Biochim. Biophys. Acta* 282:105
191. Tzagoloff, A., Meagher, P. 1972. *J. Biol. Chem.* 247:594
192. Stratman, F. W., Zahlten, R. N., Hochberg, A. A., Lardy, H. A. 1972. *Biochemistry* 11:3154
193. Kadenbach, B. 1971. See Ref. 14, p. 360
194. Ch'ih, J. J., Kalf, G. F. 1969. *Arch. Biochem. Biophys.* 133:38
195. Mason, T. L., Poyton, R. O. 1972. *Fed. Proc.* 31:1401
196. Mason, T. L., Schatz, G. 1973. *J. Biol. Chem.* 248:1355
197. Roodyn, D. B., Suttie, J. W., Work, T. S. 1962. *Biochem. J.* 83:29
198. Simpson, M. V., Skinner, D. M., Lucas, J. M. 1961. *J. Biol. Chem.* 236:PC81
199. Arion, W., Racker, E. Unpublished
200. Beattie, D. S., Patton, G. M., Stuchell, R. N. 1970. *J. Biol. Chem.* 245:2177
201. Kellerman, G. M., Griffiths, D. E., Hansby, J. E., Lamb, A. J., Linnane, A. W. 1971. See Ref. 14, p. 346
202. Longo, G. P., Scandalios, J. G. 1969. *Proc. Nat. Acad. Sci. USA* 62:104
203. Davidson, R. G., Cortner, J. A. 1967. *Science* 157:1569
204. Shows, T. B., Chapman, V. M., Ruddle, F. H. 1970. *Biochem. Genet.* 4:707
205. Davidson, R. G., Cortner, J. A., Rattazzi, M. C., Ruddle, F. H., Lubbs, H. A. 1970. *Science* 169:391
206. Tait, A. 1968. *Nature* 219:941
207. Tait, A. 1970. *Nature* 225:181
208. Knowles, J. K. C., Tait, A. 1972. *Mol. Gen. Genet.* 117:53
208a. Gross, S. R., McCoy, M. T., Gilmore, E. B. 1968. *Proc. Nat. Acad. Sci. USA* 61:253
209. Munkres, K. D., Richards, F. M. 1965. *Arch. Biochem. Biophys.* 109:457
210. Munkres, K. D. 1968. *Ann. NY Acad. Sci.* 151:294
211. Sherman, F., Stewart, J. W., Margoliash, E., Parker, J., Campbell, W. 1966. *Proc. Nat. Acad. Sci. USA* 55:1498
212. Sherman, F., Stewart, J. W. 1971. *Ann. Rev. Genet.* 5:257
213. McLean, J. R., Cohn, G. L., Brandt, I. K., Simpson, M. V. 1958. *J. Biol. Chem.* 233:657
214. Thomas, D. Y., Williamson, D. H. 1971. *Nature New Biol.* 233:196
215. Weislogel, P. O., Butow, R. A. 1971.

J. Biol. Chem. 246:5113
216. Galper, J. B., Darnell, J. E. 1971. *J. Mol. Biol.* 57:363
217. Coote, J. L., Work, T. S. 1971. *Eur. J. Biochem.* 23:564
218. Swank, R. T., Sheir, G. I., Munkres, K. D. 1971. *Biochemistry* 10:3924
219. Sebald, W., Weiss, H., Jackl, G. 1972. *Eur. J. Biochem.* 30:413
219a. Weiss, H., Sebald, W., Bücher, T. 1971. *Eur. J. Biochem.* 22:19
220. Sebald, W., Machleidt, W., Otto, J. 1973. *Eur. J. Biochem.* 38:311
221. Rubin, M. S., Tzagoloff, A. 1973. *J. Biol. Chem.* 248:4275
222. Weiss, H. 1972. *Eur. J. Biochem.* 30:469
223. Weiss, H. 1974. See Ref. 53. In press
224. Sebald, W., Bücher, T., Olbrich, B., Kaudewitz, F. 1968. *FEBS Lett.* 1:235
225. Schweyen, R., Kaudewitz, F. 1970. *Biochem. Biophys. Res. Commun.* 38:728
226. Takayama, K., MacLennan, D. H., Tzagoloff, A., Stoner, D. C. 1966. *Arch. Biochem. Biophys.* 114:223
227. Braunitzer, G., Bauer, G. 1967. *Naturwissenschaften* 54:70
228. Shapiro, A. L., Scharff, M. D., Maizel, J. V., Uhr, J. W. 1966. *Nature* 211:243
229. Shapiro, A. L., Maizel, J. V. 1969. *Anal. Biochem.* 29:505
230. Fiechter, A., Mian, F. A., Ris, H., Halvorson, H. O. 1972. *J. Bacteriol.* 109:855
231. Mian, F. A., Kuenzi, M. T., Halvorson, H. O. 1973. *J. Bacteriol.* In press
232. Dawid, I. B. 1972. See Ref. 20, p. 35
233. Williams, K. L., Birt, L. M. 1971. *Eur. J. Biochem.* 22:87
234. Williams, K. L., Birt, L. M. 1971. *Eur. J. Biochem.* 22:96
235. Devlin, T. M., Ch'ih, J. J. 1972. *Arch. Biochem. Biophys.* 152:521
236. Kislev, N., Eisenstadt, J. M. 1972. *Eur. J. Biochem.* 31:226
237. Mockel, J. 1972. *Biochim. Biophys. Acta* 277:628
238. Hernandez, A., Work, T. S. 1972. *Biochem. J.* 127:6P
239. Beattie, D. S., Ibrahim, N. G. 1973. *Biochemistry* 12:176
240. Maddaiah, V. T. et al. 1973. *J. Biol. Chem.* 248:4263
241. Hochberg, A. A., Stratman, F. W., Zahlten, R. N., Lardy, H. A. 1972. *FEBS Lett.* 25:1
242. Sandell, S., Löw, H., von der Decken, A. 1967. *Biochem. J.* 104:575
243. Beattie, D. S., Basford, R. E., Koritz, S. B. 1967. *J. Biol. Chem.* 242:3366
244. Kroon, A. M., Botman, M. J., Saccone, C. 1967. *Meet. Fed. Eur. Biochem. Soc., 4th, Oslo, 1967.* Abstr. 19
245. Lado, P., Schwendimann, M. 1967. *Ital. J. Biochem.* 15:279
246. Kroon, A. M., Saccone, C., Botman, M. J. 1967. *Biochim. Biophys. Acta* 142:552
247. Grivell, L. A. 1967. *Biochem. J.* 105:44C
248. Simpson, M. V., Fournier, M. J. Jr., Skinner, D. M. 1967. *Methods Enzymol.* 10:755
249. Davies, J. W., Cocking, E. C. 1967. *Biochem. J.* 104:23
250. Yellin, T. O., Butler, B. J., Stein, H. H. 1967. *Fed. Proc.* 26:833
251. Roodyn, D. B. 1968. *FEBS Lett.* 1:203
252. Gordon, M. W., Deanin, G. G. 1968. *J. Biol. Chem.* 243:4222
253. Cunningham, R. D., Bridgers, W. F. 1970. *Biochem. Biophys. Res. Commun.* 38:99
254. Bosmann, H. B., Hemsworth, B. A. 1970. *J. Biol. Chem.* 245:363
255. Deanin, G. G., Gordon, M. W. 1970. *Fed. Proc.* 29:3545
256. Haldar, D. 1970. *Biochem. Biophys. Res. Commun.* 40:129
257. Morgan, I. G. 1970. *FEBS Lett.* 10:273
258. Haldar, D. 1971. *Biochem. Biophys. Res. Commun.* 42:899
259. Bridgers, W. F., Cunningham, R. D., Gressett, G. 1971. *Biochem. Biophys. Res. Commun.* 45:351
260. Lederman, M., Attardi, G. 1970. *Biochem. Biophys. Res. Commun.* 40:1492
261. Wheeldon, L. W., Lehninger, A. L. 1966. *Biochemistry* 5:3533
262. Truman, D. E. S., Korner, A. 1962. *Biochem. J.* 83:588
263. Kalf, G. F. 1963. *Arch. Biochem. Biophys.* 101:350
264. Kroon, A. M. 1963. *Biochim. Biophys. Acta* 76:165
265. Wintersberger, E. 1965. *Biochem. Z.* 341:409
266. Gamble, J. G., McCluer, R. H. 1970. *J. Mol. Biol.* 53:557
267. Grant, W. D., Poulter, R. T. M. 1973. *J. Mol. Biol.* 73:439
268. Giudice, G. 1960. *Exp. Cell Res.* 21:222
269. Tata, J. R. 1966. In *Regulation of Metabolic Processes in Mitochondria,* ed. J. M. Tager, S. Papa, E. Quagliariello, E. C. Slater, 489. Amsterdam: Elsevier
270. Roodyn, D. B. 1966. See Ref. 269, p. 383

271. Roodyn, D. B., Freeman, K. B., Tata, J. R. 1965. *Biochem. J.* 94:628
272. Bronsert, U., Neupert, W. 1966. See Ref. 269, p. 426
273. Hamberger, A., Gregson, N., Lehninger, A. L. 1969. *Biochim. Biophys. Acta* 186: 373
274. Malkin, L. I. 1970. *Proc. Nat. Acad. Sci. USA* 67:1695
275. Beattie, D. S. 1971. *Arch. Biochem. Biophys.* 147:136
276. Bosmann, H. B. 1971. *J. Biol. Chem.* 246:3817
277. Sala, F., Küntzel, H. 1970. *Eur. J. Biochem.* 15:280
278. Swanson, R. F., Dawid, I. B. 1970. *Proc. Nat. Acad. Sci. USA* 66:117
279. Dawid, I. B. 1970. See Ref. 12, p. 227
280. Scragg, A. H., Morimoto, H., Villa, V., Nekhorocheff, J., Halvorson, H. O. 1971. *Science* 171:908
281. Morimoto, H., Scragg, A. H., Nekhorocheff, J., Villa, V., Halvorson, H. O. 1971. See Ref. 14, p. 282
282. Richter, D., Herrlich, P., Schweiger, M. 1972. *Nature New Biol.* 238:74
283. Allen, N. E., Suyama, Y. 1972. *Biochim. Biophys. Acta* 259:369
284. Avadhani, N. G., Buetow, D. E. 1972. *Biochem. Biophys. Res. Commun.* 46: 773
285. Yang, S., Criddle, R. S. 1969. *Biochem. Biophys. Res. Commun.* 35:429
286. Yang, S., Criddle, R. S. 1970. *Biochemistry* 9:3063
287. Criddle, R. S., Bock, R. M., Green, D. E., Tisdale, H. 1962. *Biochemistry* 1:827
288. Woodward, D. O., Munkres, K. D. 1966. *Proc. Nat. Acad. Sci. USA* 55: 872
289. Dreyer, W. J., Laico, M. T., Ruoslahti, E. I., Papermaster, D. S. 1970. *Fed. Proc.* 29:2016
290. Lederman, M., Attardi, G. 1973. *J. Mol. Biol.* 78:275
291. Costantino, P., Attardi, G. 1973. *Proc. Nat. Acad. Sci. USA* 70:1490
292. Aloni, Y., Attardi, G. 1971. *J. Mol. Biol.* 55:271
293. Dawid, I. B. 1972. *J. Mol. Biol.* 63:201
294. Buchanan, J., Popovitch, J. R., Tapley, D. F. 1969. *Biochim. Biophys. Acta* 173:532
294a. Neupert, W. 1967. Thesis. Univ. Munich, Germany
295. Poyton, R. O., Schatz, G. In preparation
296. Plattner, H., Salpeter, M., Saltzgaber, J., Rouslin, W., Schatz, G. 1971. See Ref. 14, p. 175
297. Plattner, H., Salpeter, M. M., Saltzgaber, J., Schatz, G. 1970. *Proc. Nat.*

298. *Acad. Sci. USA* 66:1252
298. Ibrahim, N. G., Stuchell, R. N., Beattie, D. S. 1973. *Eur. J. Biochem.* In press
298a. Schatz, G., Saltzgaber, J. 1969. *Biochim. Biophys. Acta* 180:186
298b. Senior, A. E., MacLennan, D. H. 1970. *J. Biol. Chem.* 245:5086
298c. Zollinger, W. D., Woodward, D. O. 1972. *J. Bacteriol.* 109:1001
298d. Dreyer, W. J., Papermaster, D. S., Kühn, H. 1972. *Ann. NY Acad. Sci.* 195:61
299. Butow, R. A., Ferguson, M. J., Cederbaum, A. 1973. *Biochemistry* 12:158
300. Ebner, E., Mason, T. 1972. *Fed. Proc.* 31:1402
301. Schweyen, R. J., Kaudewitz, F. 1971. *Biochem. Biophys. Res. Commun.* 44: 1351
302. Hartwell, L. H. 1967. *J. Bacteriol.* 93: 1662
303. Feldman, F., Mahler, H. R., Dawidowicz, K. 1972. *Fed. Proc.* 31:1399
304. Mahler, H. R., Dawidowicz, K., Feldman, F. 1972. *J. Biol. Chem.* 247:7439
305. Clark, B. F. C., Marcker, K. A. 1966. *J. Mol. Biol.* 17:394
306. Adams, J. M., Capecchi, M. R. 1966. *Proc. Nat. Acad. Sci. USA* 55:147
307. Bianchetti, R., Lucchini, G., Sartirana, M. L. 1971. *Biochem. Biophys. Res. Commun.* 42:97
308. Lucas-Lenard, J., Lipmann, F. 1971. *Ann. Rev. Biochem.* 40:409
309. Wilkie, D. 1970. *J. Mol. Biol.* 47:107
310. Chen, W. L., Charalampous, F. C. 1969. *J. Biol. Chem.* 244:2767
311. Mason, T. L., Poyton, R. O., Wharton, D. C., Schatz, G. 1973. *J. Biol. Chem.* 248:1346
312. Hinkle, P. C., Kim, J. J., Racker, E. 1972. *J. Biol. Chem.* 247:1338
313. Weiss, H., Lorenz, B., Kleinow, W. 1972. *FEBS Lett.* 25:49
314. Rubin, M. S., Tzagoloff, A. 1973. *J. Biol. Chem.* 248:4269
315. Capaldi, R. A., Hayashi, H. 1972. *FEBS Lett.* 26:261
316. Poyton, R. O. In preparation
317. Komai, H., Capaldi, R. A. 1973. *FEBS Lett.* 30:273
318. Tzagoloff, A. 1973. See Ref. 53
319. Kraml, J., Mahler, H. R. 1967. *Meet. Fed. Eur. Biochem. Soc., 4th, Oslo, 1967.* Abstr. 101
320. Tuppy, H., Birkmayer, G. D. 1969. *Eur. J. Biochem.* 8:237
321. Tzagoloff, A. 1970. *J. Biol. Chem.* 245: 1545
322. Stuart, K. D. 1970. *Biochem. Biophys. Res. Commun.* 39:1045
323. Avner, P. R., Griffiths, D. E. 1970.

FEBS Lett. 10:202

324. Avner, P. R., Griffiths, D. E. 1973. *Eur. J. Biochem.* 32:301
325. Avner, P. R., Griffiths, D. E. 1973. *Eur. J. Biochem.* 32:312
326. Shannon, C., Enns, R., Wheelis, L., Burchiel, K., Criddle, R. S. 1973. *J. Biol. Chem.* 248:3004
327. Cattell, K. J., Lindop, C. R., Knight, I. G., Beechey, R. B. 1971. *Biochem. J.* 125:169
328. Stekhoven, F. S., Waitkus, R. F., Van Moerkerk, H. T. B. 1972. *Biochemistry* 11:1144
329. Mitchell, C. H., Bunn, C. L., Lukins, H. B., Linnane, A. W. 1972. *Bioenergetics* 4:363
330. Howell, N., Trembath, M. K., Linnane, A. W., Lukins, H. B. 1973. *Mol. Gen. Genet.* 122:37
331. Simplicio, J., Schwenzer, K. 1973. *Biochemistry* 12:1923
332. Kadenbach, B. 1971. *Biochem. Biophys. Res. Commun.* 44:724
333. Murray, D. R., Linnane, A. W. 1972. *Biochem. Biophys. Res. Commun.* 49:855
334. Burke, J. P., Beattie, D. S. 1973. *Biochem. Biophys. Res. Commun.* 51:349
335. Kadenbach, B., Hadváry, P. 1973. *Eur. J. Biochem.* 32:343
336. Singer, S. J. 1962. *Advan. Protein Chem.* 17:1
337. Folch, J., Lees, M. 1951. *J. Biol. Chem.* 191:807
338. Tzagoloff, A. 1972. *Bioenergetics* 3:39
339. Kadenbach, B. 1972. Presented at Meet. Eur. Biochem. Soc., 8th, Amsterdam, 1972
340. Michel, R., Neupert, W. 1974. See Ref. 53
340a. Michel, R., Neupert, W. 1973. *Eur. J. Biochem.* 36:53
341. Rowe, M. J., Lansman, R., Woodward, D. O. 1973. *Fed. Proc.* 32:2407
342. Küntzel, H., Ali, I., Barath, Z., Kind, J., Althaus, H.-H. 1974. See Ref. 53
343. Ellis, L. F., Kleinschmidt, W. J. 1967. *Nature* 215:649
344. Wood, H. A., Bozarth, R. F., Mislivec, P. B. 1971. *Virology* 44:592
345. Tuveson, R. W., Peterson, J. F. 1972. *Virology* 47:527
346. Küntzel, H., Barath, Z., Ali, I., Kind, J., Althaus, H.-H. 1973. *Proc. Nat. Acad. Sci. USA* 70:1574
347. Garnjobst, L., Wilson, J. F., Tatum, E. L. 1965. *J. Cell Biol.* 26:413
348. Diacumakos, E. G., Garnjobst, L., Tatum, E. L. 1965. *J. Cell Biol.* 26:427

349. Watson, K., Haslam, J. M., Veitch, B., Linnane, A. W. 1971. See Ref. 14, p. 162
350. Gordon, P. A., Lowdon, M. J., Stewart, P. R. 1972. *J. Bacteriol.* 110:511
351. Gordon, P. A. 1971. Thesis. Aust. Nat. Univ., Canberra, Australia
352. Tzagoloff, A. 1971. *J. Biol. Chem.* 246:3050
353. Kováč, L., Šubík, J., Russ, G., Kollár, K. 1967. *Biochim. Biophys. Acta* 144:94
354. Rouslin, W., Schatz, G. 1969. *Biochem. Biophys. Res. Commun.* 37:1002
355. Jackl, G., Sebald, W. 1972. *Meet. Fed. Eur. Biochem. Soc., 8th, Amsterdam, 1972.* Abstr. 658
356. Schwab, A. J., Sebald, W., Weiss, H. 1972. *Eur. J. Biochem.* 30:511
357. Ebner, E., Schatz, G. Unpublished
358. Barath, Z., Küntzel, H. 1972. *Nature New Biol.* 240:195
359. Barath, Z., Küntzel, H. 1972. *Proc. Nat. Acad. Sci. USA* 69:1371
360. Westergaard, O., Pearlman, R. E. 1969. *Exp. Cell Res.* 54:309
361. Westergaard, O., Marcker, K. A., Keiding, J. 1970. *Nature* 227:708
362. Westergaard, O., Lindberg, B. 1972. *Eur. J. Biochem.* 28:422
363. Shulman, R. W., Hartwell, L. H., Warner, J. R. 1973. *J. Mol. Biol.* 73:513
364. Cottrell, S. F., Avers, C. J. 1970. *Biochem. Biophys. Res. Commun.* 38:973
365. Kadenbach, B. 1966. *Biochim. Biophys. Acta* 134:430
366. Kadenbach, B. 1967. *Biochim. Biophys. Acta* 138:651
367. Kadenbach, B. 1967. *Meet. Fed. Eur. Biochem. Soc., 4th, Oslo, 1967.* Abstr. 75
368. Kellems, R. E., Butow, R. A. 1972. *J. Biol. Chem.* 247:8043
369. Kellems, R. E., Butow, R. A. 1973. See Ref. 53
370. Redman, C. M., Sabatini, D. D. 1966. *Proc. Nat. Acad. Sci. USA* 56:608
371. Andrews, T. M., Tata, J. R. 1971. *Biochem. J.* 121:683
371a. Tzagoloff, A. 1969. *J. Biol. Chem.* 244:5027
372. Kalf, G. F. 1964. *Biochemistry* 3:1702
373. Neubert, D., Helge, H. 1965. *Meet. Fed. Eur. Biochem. Soc., 2nd, Vienna, 1965.* Abstr. 84
374. Penman, S. et al. 1970. *Cold Spring Harbor Symp. Quant. Biol.* 35:561
375. Zylber, E. A., Perlman, S., Penman, S. 1971. *Biochim. Biophys. Acta* 240:588
376. Thomas, D. Y., Williamson, D. H. 1971. *Nature New Biol.* 233:196
377. Surovaya, A., Trubitsyn, S. 1972. *FEBS*

Lett. 25:349
378. Urbanke, C., Römer, R., Maass, G. 1973. *Eur. J. Biochem.* 33:511
379. Grivell, L. A., Metz, V. 1973. *Biochem. Biophys. Res. Commun.* 55:125
380. Mahler, H. R., Dawidowicz, K. 1973. *Proc. Nat. Acad. Sci. USA* 70:111
381. Hutchison, H. T., Hartwell, L. H., McLaughlin, C. S. 1969. *J. Bacteriol.* 99:807
382. Hartwell, L. H., Hutchison, H. T., Holland, T. M., McLaughlin, C. S. 1970. *Mol. Gen. Genet.* 106:347
383. Blossey, H. Ch., Küntzel, H. 1972. *FEBS Lett.* 24:335
384. Chuang, D., Weissbach, H. 1973. *Arch. Biochem. Biophys.* 157:28
385. Zubay, G., Chambers, D. A., Cheong, L. C. 1970. *The Lactose Operon*, ed. J. R. Beckwith, D. Zipser, 375. Cold Spring Harbor, NY: Cold Spring Harbor Lab.
386. Scragg, A. H. 1974. See Ref. 53. In press
387. Palacios, R., Palmiter, R. D., Schimke, R. T. 1972. *J. Biol. Chem.* 247:2316
388. Cooper, C. S., Avers, C. J. 1974. See Ref. 53. In press
389. Avadhani, N. G., Lynch, M. J., Buetow, D. E. 1970. *Exp. Cell Res.* 69:226
390. Avadhani, N. G., Buetow, D. E. 1972. *Biochem. J.* 128:353
391. Perlman, S., Penman, S. 1970. *Nature* 227:133
392. Ojala, D., Attardi, G. 1972. *J. Mol. Biol.* 65:273
393. Perlman, S., Abelson, H. T., Penman, S. 1973. *Proc. Nat. Acad. Sci. USA* 70:350
394. Bosmann, H. B. 1970. *Biochem. Biophys. Res. Commun.* 41:1412
395. Abelson, H. T., Penman, S. 1972. *Nature New Biol.* 237:144
396. Attardi, G. 1974. See Ref. 53. In press
397. Wilkie, D. 1972. See Ref. 20, p. 85
398. Griffiths, D. E. 1972. See Ref. 20, p. 95
399. Mitchell, M. B., Mitchell, H. K., Tissieres, A. 1953. *Proc. Nat. Acad. Sci. USA* 39:606
400. Beale, G. H., Knowles, J. K., Tait, A.

1972. *Nature* 235:396
401. Adoutte, A., Sainsard, A., Rossignol, M., Beisson, J. *Biochimie.* In press
402. Kislev, N., Spolsky, C. M., Eisenstadt, J. M. 1973. *J. Cell Biol.* 57:571
403. Avner, P. R., Coen, D., Dujon, B., Slonimski, P. P. 1973. *Mol. Gen. Genet.* 123:9
404. Grivell, L. A., Reijnders, L., De Vries, H. 1971. *FEBS Lett.* 16:159
405. Grivell, L. A. 1974. See Ref. 53. In press
406. Kaltschmidt, E., Wittmann, H. G. 1970. *Anal. Biochem.* 36:401
407. Lai, C. J., Weisblum, B. 1971. *Proc. Nat. Acad. Sci. USA* 68:856
408. Helser, T. L., Davies, J. E., Dahlberg, J. E. 1971. *Nature New Biol.* 233:12
409. Helser, T. L., Davies, J. E., Dahlberg, J. E. 1971. *Nature New Biol.* 235:6
410. Scott, W. A., Mitchell, H. K. 1969. *Biochemistry* 8:4282
411. Brambl, R. M., Woodward, D. O. 1972. *Nature New Biol.* 238:198
412. Rifkin, M. R., Luck, D. J. L. 1971. *Proc. Nat. Acad. Sci. USA* 68:287
413. Kuriyama, Y., Luck, D. J. L. 1974. See Ref. 53. In press
414. Kuriyama, Y., Luck, D. J. L. 1973. *J. Mol. Biol.* 73:425
415. Stuart, K. D. 1970. *Biochem. Biophys. Res. Commun.* 39:1045
416. Avner, P. R., Griffiths, D. E. 1970. *FEBS Lett.* 10:202
417. Criddle, R. S., Shannon, C., Short, L., Enns, R., Burchiel, K. 1973. *Fed. Proc.* 32:2405
418. Bulos, B., Racker, E. 1968. *J. Biol. Chem.* 243:3901
419. Toson, G., Contessa, A. R., Bruni, A. 1972. *Biochem. Biophys. Res. Commun.* 48:341
420. Busbee, D. L., Hershberger, W. K., Tidwell, T., Shaw, C. R., Wagner, R. P. 1972. *Fed. Proc.* 31:1396
421. Perlman, S., Schatz, G. Unpublished
422. Dawid, I. B., Chase, J. W. 1972. *J. Mol. Biol.* 63:217
423. Robberson, D., Aloni, Y., Attardi, G., Davidson, N. 1971. *J. Mol. Biol.* 60:473

BIOCHEMISTRY OF BACTERIAL CELL ENVELOPES[1]

✕ 841

Volkmar Braun[2] and Klaus Hantke

Max-Planck-Institut für Molekulare Genetik, Berlin-Dahlem
Ihnestrasse 63–73, Germany

CONTENTS

The study of bacterial membrane mutants and membrane alterations during virus infection is important not only to an understanding of prokaryotic cells, but also provides model conceptions concerning corresponding membrane mediated alterations in the social life of transformed eukaryotic cells (1). Since bacteria are largely subject to experimental manipulations, more definitive answers concerning the connection between an observed biochemical defect and the altered physiological

[1] Literature survey has been completed with few exceptions to July 1973.
[2] New address: Mikrobiologie II Universität, 74 Tübingen, Auf der Morgenstelle.

behavior of the cell can be obtained in these organisms. Before we discuss these topics we shall first cover recent findings on the structure and biosynthesis of membrane constituents with emphasis on the outer membrane (cell wall).

Since several recent reviews have appeared on the following subjects they will not be covered again : 1. *Membrane assembly* with emphasis on in vitro reconstitution of enzyme systems (2, 3); morphogenesis of transport systems in relation to lipid synthesis and the fluidity of the lipid bilayer (3–6); murein (7), teichoic acid (8), and lipopolysaccharide synthesis (9); 2. *growth* and *division* (10, 11) with regard to auto-lysins (7, 10, 12) and determination of shape (13); 3. structure and function of bacterial *membranes* and *mesosomes* (14–20); 4. *transport* (21–24); 5. *periplasmic proteins* (25, 26); 6. structure, function, and biosynthesis of *phospholipids* (27–29); 7. metabolism of *fatty acids* (30); 8. *membrane polysaccharides* (31, 32); 9. *pili* (33, 34); and 10. *flagella* (35, 36).

PROTEINS

Membrane proteins have mostly been studied by sodium dodecyl sulfate (SDS)–polyacrylamide-gel electrophoresis without any chemical characterization. Therefore it is largely unknown whether the protein bands observed are comprised of a single protein species or many species. For example a protein of molecular weight 40,000 which was claimed to be the major protein of the *Escherichia coli* outer membrane (37, 38) could be further resolved into several components by using an alkaline SDS-buffer system (39) or by ion exchange chromatography (40). Depending on the conditions of pretreating the membrane, of solubilization with SDS, and of gel electrophoresis rather different patterns have been observed within *E. coli* strains (37–43). Some proteins within the membrane are rather resistant to proteases (39, 42, 44). Some of these proteins are also resistant in reconstituted membranes, pointing to a restoration of the protein arrangement similar to that in vivo (39, 42). Experimental conditions have been devised to remove murein and all the proteins of the *E. coli* membrane except three without losing the rod shape (44). This is a first attempt to ascribe the information to build a rod in a self-assembly process to one or a few proteins in the outer membrane. Most bacterial membrane proteins so far investigated are listed in (3). Therefore we describe only selected systems in which important recent progress on the biochemical level has been achieved. Many functions of membrane proteins are treated in the appropriate chapters which follow.

Adenosine Triphosphatase (ATPase)

ATPase from *Streptococcus faecalis* has been isolated in pure form and characterized in considerable molecular detail (45). The ATPase is linked to the membrane by a protein called Nectin (46) and its activity is influenced by an additional membrane component which is irreversibly inactivated by dicyclohexylcarbodiimide (47). The enzyme is phosphorylated but the physiological consequence of this is so far unknown (48). Energized uptake of K^+ and cycloleucine is coupled to the activity of the ATPase (47). This system is until now the best-studied bacterial ATPase.

Considerable progress has also been made with the elucidation of the structure and function of the *E. coli* ATPase, which has been purified to near homogeneity (49) and some of its structural and functional characteristics studied. Antibodies to the purified enzyme inhibit ATP hydrolysis and ATP stimulated pyridine-nucleotide transhydrogenase by membranes from cells grown aerobically or anaerobically. Mutants with greatly reduced levels of ATPase are deficient in oxidative phosphorylation (50–52). Such mutants are also devoid of active transport of thiomethylgalactoside when respiration is poisoned by cyanide but not when respiration is functioning (53). Uncouplers of oxidative phosphorylation also inhibit anaerobically but not aerobically driven active transport (54, 55). The ATPase of *Micrococcus lysodeicticus* has also been studied extensively (15, 16, 56).

Cytochromes

Cytochromes from 18 different species of bacteria were investigated by low temperature spectroscopy (57, 58). New alpha absorption bands tentatively assigned to new *b*- and *c*-type cytochromes resulted from fourth-order finite difference analysis. Gene dose effects in wild-type *E. coli* strains diploid for segments of the chromosome between the *lac* and *gal* operons were found (59). In an *E. coli* mutant deficient in heme biosynthesis due to a defective 5-aminolaevulinic acid synthetase, cytochrome activity (respiration) could be restored by addition of 5-aminolaevulinic acid (60). Full activity was recovered after 3 hr at 37° in the absence of protein synthesis but a fermentable carbon source was required. In vitro reconstitution with hematin required ATP, which explains the need for fermentation in vivo. Thus cytochrome apoproteins are apparently synthesized and integrated into the membrane in the absence of heme and can be converted in its presence to functional enzymes in the respiration chain.

Nitrate Reductase

Under anaerobic growth conditions *E. coli* and many other strains (61) form, in the presence of nitrate, a membrane-bound formate-nitrate reductase complex which oxidizes formate to CO_2 and reduces nitrate to nitrite. Since loss of this system by mutation is not harmful to the cell it offers a suitable means to study synthesis, assembly, and regulation of a membrane-bound enzyme complex. Eight genetic loci have been identified (62, 63) and many of these mutants have been characterized in terms of physiological and biochemical properties (64). Alterations in the cytoplasmic proteins have been related to enzyme mutants (65). The enzyme complex has been resolved into active components and reconstituted (66–68).

Murein-Lipoprotein

STRUCTURE A major protein and the most extensively studied protein of the outer membrane of *E. coli* and related enterobacteriaceae is the lipoprotein covalently linked to murein (69–71). This lipoprotein also occurs free in the envelope in an amount exceeding the bound form by a factor of two (72). In cells growing logarithmically in rich medium, about 250,000 molecules are bound to the murein (73) so that a total number of 750,000 results. For comparison it is interesting that under similar

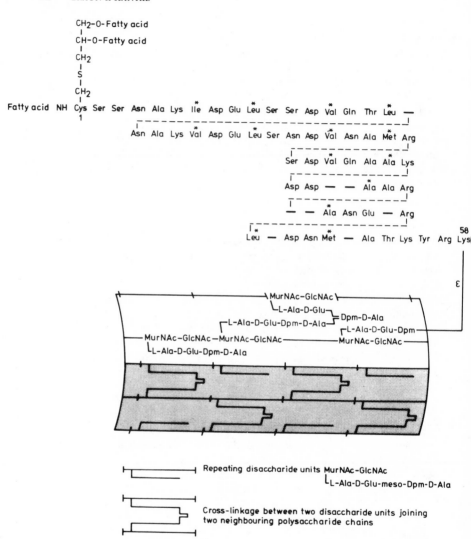

Figure 1 Structure of the murein-lipoprotein complex (rigid layer) of the outer membrane
of E. coli. The cylindrical section indicates a part of the murein with the shape of the
rod-like E. coli cell. It is not known in which direction the glycan chains span the cell
relative to the long axis of the cylinder. In this model they are arbitrarily drawn parallel.
The murein is composed of roughly 10^6 repeating units (111) to which about 10^5 lipoprotein
molecules are covalently bound (69, 73). The lipoprotein replaces D-alanine on the
diaminopimelate residue (70, 71). The amino acid sequence is represented in a way to
demonstrate the possible evolution of this molecule from a gene which coded originally for

growth conditions less than 100,000 copies of derepressed alkaline phosphatase (74), induced β-galactosidase, or individual ribosomal proteins (75) are synthesized. The primary structure of the lipoprotein (76, 77) and essential aspects of its conformation (78) have been elucidated. The polypeptide chain contains at the N-terminal end a covalently linked lipid (79) and is fixed by the ε-amino group of the C-terminal lysine to the carboxyl group at the optical L-center of statistically every 10th–12th diaminopimelate residue of the murein (Figure 1). The diglyceride residue on the N-terminal cysteine probably comes from the phospholipid pathway since the fatty acid composition is similar to that of the phospholipids. The amide-linked fatty acids in contrast consist of 65% palmitate, the rest being mainly monounsaturated fatty acids; there are virtually no cyclopropane fatty acids present at this position. It is tempting to speculate that the biosynthesis of the lipid part involves a nucleotide-activated diglyceride which is transferred to the mercapto group of the N-terminal cysteine and that in this case a nucleoside diphosphate is released. A repetitive design (see Figure 1) of the amino acid sequence is very suggestive. The first 15 amino acids are repeated nearly identically; the following stretch of 7 amino acids is essentially quadruplicated if one considers that those amino acid exchanges found are conservative and allow single amino acid deletions—usually introduced when homology parallels between similar polypeptide chains are drawn. Tripeptides extend at each end from the repetitive middle portion, and the lipid and the murein are attached there. The attachment sites are also marked by the unique occurrence of two amino acids, cysteine and tyrosine, and by the accumulation of serine (Cys-Ser-Ser-) and basic amino acids (-Lys-Tyr-Arg-Lys). The conformation of the polypeptide chain is α-helical (78). In the renatured lipoprotein an α-helical content of at least 70% was found by measurement of the circular dichroism. Starting with the repetitive part of the peptide chain every fourth or third amino acid residue is hydrophobic, and this occurs in a consistently alternating pattern. Since 3.6 amino acids make up one helical turn, the hydrophobic amino acids all are localized on one side of the helix.

LOCATION OF MUREIN-LIPOPROTEIN IN THE ENVELOPE   Murein forms the innermost layer of the outer membrane (80, 81) with the lipoprotein extending into the outer portion of the membrane. The latter concept arose from two types of experiments. In a double mutant of E. coli (W7), which can neither synthesize diaminopimelate nor convert it to lysine (Dpm-decarboxylase⁻), murein was specifically labeled with radioactive diaminopimelate. Murein was degraded with lysozyme, giving rise to

15 amino acids, was duplicated, and then only the C-terminal half was added four times (76, 77). The dashes represent hypothetical deletions of amino acids which may have occurred during evolution. The fatty acids bound as esters to the glycerylcysteine at the N-terminal end (79, 90) are mainly palmitic acid (45%), palmitoleic acid (11%), cis-vaccenic acid (24%), cyclopropylene hexadecanoic acid (12%), and cyclopropylene octadecanoic acid (8%). The fatty acids bound as amide to the N-terminal α-amino group are mainly palmitic acid (65%), palmitoleic acid (11%), and cis-vaccenic acid (11%). The stars indicate the hydrophobic amino acids at every 3.5th position.

muropeptides and a lipoprotein-linked muropeptide dimer. Lipoprotein bound to murein is thus also specifically labeled with diaminopimelate. In the separated membranes 90–95% of the lipoprotein-bound label was found in the outer membrane. A similar result was reported by Osborn et al (82). The localization of the lipoprotein including the free form was in addition proven with immunological methods (78, 83, 84). In membranes separated after a short treatment with trypsin to cleave the lipoprotein from the murein, at least 10 times more lipoprotein was found in the outer membrane (84). In smooth strains of *E. coli* and *Salmonella* the lipoprotein can hardly be reached by lipoprotein-specific antibodies (78, 83). The lipoprotein becomes increasingly reactive as antigen in mutant cells with less and less complete lipopolysaccharide (78, 83). Thus in wild-type cells it is shielded by surface structures and becomes exposed in mutant cells with defective cell surface structures.

BIOSYNTHESIS OF MUREIN-LIPOPROTEIN   Pulse-chase experiments with radioactive diaminopimelate argue against the possibility that muropeptides are already fixed to the lipoprotein in the cytoplasmic membrane (85). Small amounts would have escaped detection due to the unavoidable cross-contamination of the separated inner and outer membranes. Lipoprotein is probably fixed to the already polymerized murein of the outer membrane. This is especially likely in the light of the finding that the free form exchanges with the murein-bound form (72). Conceivably, lipoprotein is attached to murein by a mechanism similar to that forming the peptide cross-linkage (86), in that the energy of the peptide bond of D-Ala which is replaced by the lipoprotein is used for lipoprotein binding (transpeptidation). The lipoprotein is synthesized on an unusually stable messenger RNA (87) and the synthesis is surprisingly rather unaffected by puromycin (88, 89).

MUREIN

Structure

Murein (91) [also called peptidoglycan (92), mucopeptide (10), or glycopeptide] is the shape-maintaining structure of all bacteria. The exceptions are some marine bacteria where no indication for the existence of murein has been found (92–94). The chemical structures of mureins of many species from different taxonomic groups are known (95). The basic structure of the glycan strands is invariably built up from repeating units of β-1,4-N-acetylglucosamine-β-1,4-N-acetyl muramic acid. The only other sugar component detected so far is mannomuramic acid in the glycan of *Micrococcus lysodeicticus*, which replaced 2% of the usually occurring glucomuramic acid (96).

Peptide side chains cross-link the glycan chains (see e.g. Figure 1) and their composition and the mode of cross-linkage vary between different bacteria (7, 95). Since the peptide side chains are characteristic for related groups of bacteria, they are useful taxonomic markers (95).

Some substituents of the N-acetylmuramic acid residue such as glycolyl or acetyl groups (7) have been found. Several different polymeric structures are bound to the glycan chains. Wax D from mycobacteria contains an arabinogalactan mycolate bound by muramic acid-6-phosphate to the murein (97, 98). Possibly similar structures will be found in the taxonomically related *Nocardia* species (99). Wall teichoic acids are bound through phosphodiester links to muramic acid (100) as are some polysaccharides (101, 102). Several proteins are also reported to be covalently bound to the murein (103, 104, 112) of which the murein-lipoprotein of *E. coli* is the best characterized (Figure 1) (79).

The shape of the murein is apparently not determined by its chemical structure, as spherical murein derived from conditional spherical *E. coli* cells is, within the limits of detection, not different from the rod-like murein of normal *E. coli* cells (105, 106). The required knowledge on the three-dimensional organization of murein is generally lacking. Ingenious models have been proposed (12, 107, 108) without physical data (110). For *E. coli* the number of repeating units (about 2.7×10^6) and the surface area per cell have been determined from which it followed that 129 $Å^2$ are available for each repeating unit (111). From these data it was concluded that at most three murein layers could be present but only one layer was compatible with the need to construct a regular covalently fixed murein in two dimensions (111). For the Gram-positive *Bacillus subtilis* with the same repeating unit as *E. coli* a murein with 40 layers would follow if one takes the one-layered *E. coli* murein as the basic structure (111). This deduction is in accordance with the measured thickness of 300–400 Å for the *B. subtilis* wall (107) since one layer is around 10 Å thick.

Biosynthesis

The different steps of murein biosynthesis were extensively reviewed by Strominger (92) and by Ghuysen & Shockman (7). The reactions which lead to UDP-muramylpentapeptide take place in the cytoplasm. UDP-muramylpentapeptide is then transferred to the membrane-bound undecaprenol phosphate whereby UMP is released, and after this step UDP-activated N-acetylglucosamine is added (92). The way in which the undecaprenol-bound precursors are polymerized into the preexisting murein is still largely unknown. No evidence has been found in *E. coli* for polymerization of murein precursors in the cytoplasmic membrane. Only a single uncross-linked repeating unit bound to lipid has been identified in the cytoplasmic membrane (85). So one has to believe that polymerized murein in the outer membrane is the acceptor for the lipid-linked monomers.

Mirelman et al (113) succeeded in incorporating UDP-N-acetyl-D-glucosamine, UDP-N-acetylmuramic acid, and glycine into cell wall peptidoglycan using an in vitro system consisting of a crude cell wall preparation from *Micrococcus luteus* (*lysodeicticus*). This system apparently is suitable for the study of late stages of murein synthesis. Low doses of penicillin (1 µg/ml) not only inhibited the cross-linking reaction but also lowered incorporation of N-acetylglucosamine and N-acetylmuramic acid into the wall peptidoglycan by 65–70%. To explain this, two possibilities have been discussed: (*a*) that the first step in precursor polymerization is transpeptidation and the second step is transglycosylation, or (*b*) that further transglycosylation would depend on the presence of a crosslinked peptidoglycan in a kind of feed back inhibition. Previous in vivo studies have shown that about this same low penicillin concentration inhibits incorporation of precursors into the peptidoglycan of *Staphylococcus aureus* (114) but higher concentrations are required with *E. coli* (115–117). Strominger et al showed that penicillin inhibits the transpeptidase which crosslinks the peptide side chains (92, 114). The hypothesis that penicillin acts as a structural analog of D-alanyl-D-alanine is now doubted to be generally the case (7, 118). Although covalent as well as reversible binding of penicillin to various membrane proteins and enzymes of a single species (119, 120) and in different organisms (121) has been demonstrated elegantly, it remains to be shown which is the (major) killing site of penicillin. The D-Ala-D-Ala carboxy-peptidase from *Streptomyces R39* was shown to be the physiological transpeptidase responsible for peptide crosslinking during wall synthesis (122). This enzyme is noncompetitively inhibited by physiological concentrations of penicillin (118) which suggests that penicillin binds to a site of the enzyme different from the substrate binding site (-D-Ala-D-Ala). Besides penicillin, many other antibiotics interfere with the biosynthesis of murein. A good survey of research activities in this field can be obtained from papers based on a recent conference (see 89).

Although there is suggestive evidence for the role of autolysins in the final stage of murein synthesis (incorporation of precursors, murein enlargement), in addition to their better-established functions in cell division, cell separation, and remodeling of cell shape (7, 10), no unequivocal evidence for an indispensable function in murein synthesis has been reported. The cuts seen in the murein of *E. coli* at the equatorial division site of the cell after treatment with a low dose of penicillin may be due to the uncoupling of an autolytic system from the correlated biosynthetic system (117). According to this hypothesis, controlled hydrolysis of murein creates acceptor sites for newly incoming precursors. But it seems equally likely that the enzyme system responsible for local scission and rearrangement of murein during cell division has been visualized. Since the glycan chains in all species studied are short (10–50 disaccharide units) hydrolysis per se does not seem to be necessary. Hydrolases observed in vitro or in vivo under the influence of inhibitors like penicillin may normally function in vivo as transglycosidases or transpeptidases. According to the present scheme of murein synthesis the undecaprenol phosphate-linked precursors are transferred to nonreducing ends of N-acetylglucosamine residues of the glycan chains. Only membrane-bound N-acetylmuraminidases, but not the most abundant peptidases or N-acetylglucosaminidases, could create new acceptors. In *Streptococcus faecalis* such an active N-acetylmuraminidase is associated with nascent growing areas of the cell wall (109, 123). A mutant of *B. subtilis* under restrictive growth conditions stops growing and forms short, bent cells with irregularly thickened murein (124). This mutant is deficient in autolysin activity. Normal growth and conversion to rod-shaped cells could be restored by addition of lysozyme or autolysin.

In *S. aureus* and *B. subtilis* N-acetylglucosamine was identified as reducing endgroup of the glycan chains (125–127) and for *B. subtilis* it was shown that this was not the result of a secondary modification (128). In *E. coli* no free reducing endgroups have been found; instead a trehalose-like endgroup (1,1' linkage) seems to exist (129). The enzyme has been isolated which degrades murein in vitro and forms this peculiar disaccharide (129). The last examples in particular show that the mechanism of incorporation of the precursors and of murein enlargement in general are still far from being understood.

In *E. coli* a murein growth zone was detected recently (130). ^{3}H-diaminopimelate pulsed into growing cells (dap$^-$, lys$^-$) appeared first in the central area of the murein sacculus after which a rapid distribution over the whole murein surface was observed. However, in *E. coli* no substantial turnover of murein (TCA-soluble products!) has been observed (131). The murein units which presumably have to be cut out of the central area must remain bound to the enzyme or the membrane and must rapidly be reinserted. The aforementioned transglycosidases and transpeptidases may function in the rapid randomization. For the likewise rod-shaped but gram-positive *B. subtilis* random insertion into cell wall was proposed for prepolymerized peptidoglycan chains to which teichoic acids have already been attached (132). This is in agreement with the electronmicroscopically observed random intercalation of "thin pieces" into the cell wall (133). A single equatorial growth zone exists in *Streptococcus faecalis* (109, 134). Peptidoglycan is deposited in an outward direction starting from the equatorial ring. The cross wall peels apart into two layers of peripheral wall. During amino acid starvation, wall thickening occurs over the entire cell surface. Cross wall and peripheral wall synthesis in *S. aureus* starts at a single particular point and is therefore asymmetric (135, 136). The primary cell wall is cut and the new wall is built by sequential growth. The new thin cell wall underlies the old thick cell wall so that apparently no insertion of newly formed murein into old murein takes place. Thickening of the newly made wall occurs by a special apposition-growth.

TEICHOIC ACIDS

Several reviews of teichoic acids by Baddiley et al have appeared with emphasis on structural (137), biosynthetic, and functional (138, 139) aspects. Ellwood & Tempest cover in their review (140) the effects of the growth medium in chemostat cultures on the content and the composition of bacterial walls especially with regard to teichoic and teichuronic acids. We confine our discussion to some new developments.

Structure

Teichoic acids are polyolphosphate polymers which occur in the outer wall and the cytoplasmic membrane of many gram-positive bacteria. In several cases it has been shown that the wall teichoic acids are covalently bound to the 6-hydroxyl group of muramic acid residues of the peptidoglycan by a phosphodiester linkage (cited in 100). The analysis of the lysozyme degradation products of the peptidoglycan of *Bacillus licheniformis* and *B. subtilis* revealed that the teichoic acid was combined with 50% of the total peptidoglycan (141). In *Staphylococcus lactis I3* 42% of the total peptidoglycan was found to be covalently bound to teichoic acid. A detailed analysis of the peptidoglycan-teichoic acid complex showed that one teichoic acid polymer of 22–24 repeating units is bound to a peptidoglycan chain of an average length of 7.4 disaccharide units (100). This is also the length of the peptidoglycan chains which are not combined with teichoic acids. These results, together with the immunological localization of the teichoic acids at the outside of the wall and the electronmicroscopically measured thickness of the wall, are used for the construction of possible wall models without reaching a final conclusion on the arrangement of the polymers (100).

Membrane teichoic acids were found to be associated with glycolipids on the outer surface of the cytoplasmic membrane (139, 142). The only case studied in detail is the lipoteichoic acid of *Streptococcus faecalis* N.C.J.B.8191 (143). It consists of a poly-(glycerolphosphate) (28–35 units) which is bound to a glycolipid (Figure 2). Similar glycolipids free of teichoic acid are found in this and several other species (144). The lipid probably anchors the hydrophilic teichoic acid in the membrane as is the case with the lipopolysaccharide of gram-negative bacteria and the glycolipids in membranes of higher organisms.

Biosynthesis

The teichoic acids and the peptidoglycan chains to which they are attached are synthesized simultaneously and then incorporated at random into the peptidoglycan (132). The polyolphosphate chains are formed by successive transfer of the polyol-phosphate units, e.g. from CDP-ribitol or CDP-glycerol, to the endogenous acceptor in several in vitro membrane preparations (138). Using a polyglycerol-phosphate polymerase free of acceptor Mauck & Glaser succeeded in identifying a heat-stable acceptor (145). A preliminary characterization revealed glycerol-phosphate, glucosamine, and most notably fatty acids as constituents of the acceptor. The resemblance to the composition of membrane teichoic acids is obvious (143). This lipid receptor is unlike undecaprenol phosphate which, on addition to

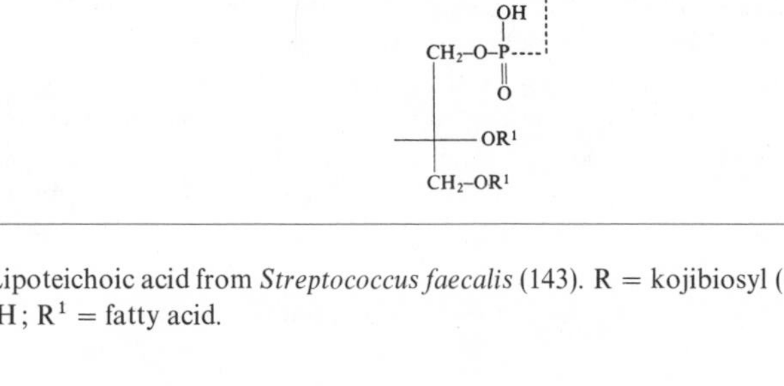

Figure 2   Lipoteichoic acid from *Streptococcus faecalis* (143). R = kojibiosyl (= diglucosyl-residue) or H; R^1 = fatty acid.

in vitro systems derived from *B. subtilis* and *S. aureus,* did not stimulate formation of polyglycerolphosphate or polyribitolphosphate (145, 146). From these experiments it was deduced that undecaprenol phosphate plays no role in teichoic acid synthesis. The opposite opinion was put forward by Baddiley and co-workers (138, 139, 147–149). N-Acetylglucosamine, bound by a rather labile ester linkage through a pyrophosphate bridge to a lipid, was isolated from *Staphylococcus lactis* N.C.T.C. 2102 which contains an N-acetylglucosamine phosphate polymer as teichoic acid. Identification of the lipid as undecaprenol phosphate was based on competition between both teichoic acid and peptidoglycan synthesis for this common intermediary substrate. In the synthesis of peptidoglycan, undecaprenol pyrophosphate is released after addition of the repeating unit to the growing peptidoglycan (150). A pyrophosphorylase creates the undecaprenol monophosphate for reentry into the biosynthetic cycle. This step is inhibited by bacitracin, which binds to the lipid (151). No inhibition occurred upon adding bacitracin to a teichoic acid-synthesizing particulate enzyme system, as undecaprenol monophosphate is released after polymerization. However, after addition of UDP-muramylpentapeptide to the same membrane system to allow peptidoglycan synthesis, part of the undecaprenol was trapped by bacitracin in the pathway of peptidoglycan synthesis, and teichoic acid synthesis was inhibited by 50% (152). Based on this and similar competition experiments in systems from three other species with different teichoic acids, undecaprenol phosphate was postulated as being an intermediate carrier in teichoic acid synthesis (147–149). The low amounts and the difficulties in isolating the lipid bound to a precursor of teichoic acid synthesis has so far prevented direct proof of the lipid structure by mass spectrometry. Nevertheless, as suggested by Baddiley et al (147) and by Strominger et al (153, 154), it is feasible that wall synthesis is controlled by the undecaprenol phosphate, since many wall polymers—peptidoglycan (155), lipopolysaccharide (156), mannan (157), capsular polysaccharide (158), glucosyl side chains of lipopolysaccharide (159, 160), and perhaps teichoic acids—may share undecaprenol phosphate as a common carrier. In this connection it will be of considerable interest to see whether the first experimental hints that undecaprenol phosphate is returned to a common pool not after each turn of each biosynthetic cycle but rather after the various polymers have been finished (147), can be substantiated by further results. Undecaprenol phosphate mediates the solubility of the hydrophilic polyol precursors in the membrane. It is yet to be seen whether it functions also in the vectorial transfer of the polymer subunits from the inner side through the lipid bilayer to the outer side of the cytoplasmic membrane and from there into the outer membrane. In the more extensively studied peptidoglycan and lipopolysaccharide biosynthesis the polymer precursors remain lipid bound up to the final incorporation into the polymer. Undecaprenol apparently serves as carrier between the individual enzymes and their precise location in the lipid bilayer will answer the mechanism of vectorial transportation.

Function

Teichoic acids seem to be of little importance for the rigidity of the cell wall (137) but are thought to contribute to magnesium binding and uptake by the cell. A first observation was that in magnesium-limited cultures, the amount of teichoic acids was increased. Under phosphate limiting conditions the teichoic acid was substituted by an uronic acid-containing polysaccharide (161) which may fulfil a similar function. When the isolated cell walls of S. *aureus* H were freed of teichoic acids by periodate oxidation of the ribitol residues, the magnesium-binding capacity of the walls was drastically lowered (162). Studies with particulate enzyme systems showed no change in teichoic acid synthesis upon addition of magnesium unless the enzyme system was purified so that the wall teichoic acid was removed. The synthesis had a sharp optimum at 10 mM magnesium. These observations show that teichoic acids can maintain the appropriate ionic conditions for cation-dependent enzyme systems in the cytoplasmic membrane (163). According to this view the teichoic acids in the wall bind especially magnesium from the environment and the teichoic acids in the cytoplasmic membrane transmit the magnesium ions for the biosynthetic enzymes. The notion that the positively charged amino group of D-alanine lowers the capacity of teichoic acids to bind magnesium (164) has been discounted because of results which showed no such effect (165).

LIPOPOLYSACCHARIDE

The development in this field is very well covered by a series of reviews (9, 166–170), the most complete survey being a book on microbial toxins (171) in which the chemical and physical structure (172–175), biosynthesis (176, 177), and genetics (178) of lipopolysaccharides (LPS) have been summarized. Various facets of the action of LPS as endotoxin have been comprehensively reviewed (179). As basis for the following discussion the complete structure of the lipopolysaccharide of *Salmonella typhimurium* is given in Figure 3.

Structure

Because the structures of a number of O-specific side chains of various *Salmonella* species are known, other gram-negative bacteria are now gaining more interest. Examples are the O-antigens of *E. coli* (180), *Serratia* (181, 182), *Citrobacter* (183), and *Shigella* (184). A group of mutants derived from different *Salmonella* species have lost their O-specificity and gained a distinct T-antigen. The T1-specific polysaccharide contains polyribose and polygalactose regions, the latter probably being composed of galactose-trisaccharide units (185) linked to glucose II of the core polysaccharide (186, 187).

Although there is a large variety in the O-specific side chains, the evidence at present suggests that the core polysaccharide and the lipid A portion are common to all *Salmonella* species except in core mutants. Core mutants have been classified from Ra-Re (Figure 3). Such core mutants have been used to elucidate (see Figure 3) the linkage of pyrophosphorylethanolamine to heptose (188), the structure of the KDO trisaccharide (189) and the lipid A portion (190). The KDO-trisaccharide is bound to a β-1,6-linked diglucosamine residue (191) to which lauric, myristic, palmitic, and 3-$D(-)$ hydroxymyristic acid in a ratio of about 1:1:1:3 are bound.

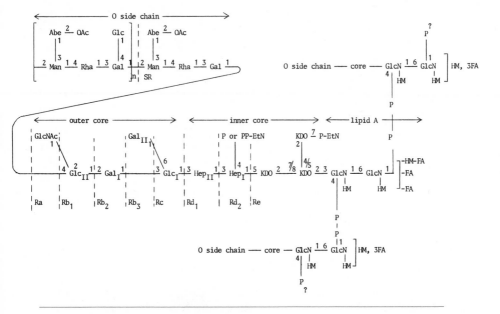

Figure 3 Tentative structure of the lipopolysaccharide of *S. typhimurium,* mainly adopted from (172). The number of chains crosslinked by pyrophosphate bridges between the diglucosamine residues was assumed to be three to suggest the polymeric structure (see also 2). The fatty acids (HM, FA) are bound to the free hydroxyl groups of the diglucosamine residues (192). The vertical dotted lines indicate the core structures of the different mutants SR, Ra–Re. The sugars in the core have roman indices for their distinction in the text. Abbreviations: Abe = abequose; Ac = acetic acid; EtN = ethanolamine; FA = fatty acid ($C_{12:0}$, $C_{14:0}$, $C_{16:0}$); Gal = galactose; Glc = glucose; GlcN = glucosamine; Hep = L-glycero-D-mannoheptose; HM = β-hydroxymyristic acid; KDO = 3-deoxy-D-mannooctulosonic acid; Man = mannose; Rha = rhamnose.

Most noticeable is the 3-myristoxymyristic acid configuration (192). About three neighboring lipopolysaccharide polymers are possibly interlinked by pyrophosphate bridges between positions 1 and 4′ of the diglucosamine portions (192). In contrast to *Salmonella* there are at least 3 different core structures known for *E. coli* (193–195) and the lipid A contains a 1,4-linked glucosamine which has also been found in *Shigella flexneri* (196).

Evidence for some protein covalently linked to a structure similar to the lipid A portion of the lipopolysaccharide has been recently reported by two groups (197–199). From a glucosamine-requiring *E. coli* mutant (J-5-G), an apparently homogeneous protein was isolated which contained radioactively labeled glucosamine, KDO, and organic phosphate in a molar ratio of 1:1:2 (199). A protein fraction isolated from *E. coli* O141:K85 and *Serratia marcescens O8* contained glucosamine, phosphorus, and fatty acids, among them lipid A-specific β-hydroxymyristic acid (197, 198). It is certain that only a small fraction of the lipopolysaccharide is bound to protein but this portion could possibly be very interesting with regard to the biosynthesis of lipid A.

Biosynthesis

The biosynthesis of the O-specific side chains is essentially known (177) so we shall refer only to the attachment of some sugars which occur as branches of the O-side chain. Polymerization of the repeating unit (Figure 3) occurs only when the abequose residue is present but is independent of the presence or absence of the glucosyl residue (159, 200, 207). The latter "modification reaction" proceeds by transfer of a UDP glucose to the undecaprenol phosphate carrier lipid before its attachment to the already polymerized O-antigen which is most likely bound to the carrier lipid (159, 201). The same mechanism accounts for the modification of *Salmonella anatum* after lysogenization by phage ε 15 (160). In *S. flexneri,* where different glucosylations produce several different antigenic structures, a lipid-linked glucose intermediate was also found (202).

The biosynthesis of the core and the lipid A part are now mainly of interest. The reconstitution of the galactosyltransferase system (Rc mutants, Figure 3) in aqueous suspension required the intimate association of phospholipids with the galactose-deficient lipopolysaccharide before addition of the transferase to obtain substantial transfer of added galactose (2, 203). In a functional monomolecular film consisting of LPS, phospholipids, and enzyme there was transfer of galactose but only two moles per mole of enzyme (204, 205). The enzyme is probably fixed in position within the film and is able to transfer galactose only to neighboring LPS molecules. This system apparently can be further developed by inserting additional enzymes of the reaction sequence [glucosyltransferase (206)] and thus offers the possibility to study the sequential growth of the linear LPS macromolecule and the movement of the substrates and/or the enzymes within the membrane. Such studies may provide clues to the enzyme-directed translocation processes in membranes.

The definitive sequence of addition of core substituents is emphasized by the finding that phosphorylation of heptose I (Figure 3) takes place after glucose I is transferred to the inner core region (207). Phosphorylation is necessary for subsequent addition of both galactose residues (207). Mutagen-induced heptose defective strains of *Salmonella typhimurium* have been obtained (208). They are leaky and contain smooth and rough form (Re) LPS. The few heptose substituents in the Re LPS are mainly D-heptose (D-glycero-D-mannoheptose) instead of the L-glycero-D-mannoheptose typical of *Salmonella.* No further substituents can be added to core stubs containing D-heptose (208). Two heptose-deficient mutants were isolated as transketolase-negative mutants on the assumption that sedoheptulose-7-phosphate may be a precursor of heptose (209). The selection was based on the fact that transketolase is also essential for synthesis of erythrose-4-phosphate, an obligatory intermediate in the aromatic pathway. A phage selection step with C21, which is known to adsorb to core mutants with heptose, was also used. Cells with heptose-deficient LPS become sensitive to bile salts. KDO mutants have been isolated (210) which were unknown until recently but eagerly awaited because of the physiological consequences. These mutants are dependent on added D-arabinose-5-phosphate as a result of an altered KDO-8-phosphate synthetase. The mutant cells cease to grow after one generation of starvation but remain viable. It seems that the ultimate defect in the LPS structure for the viability of cells is reached with KDO mutants and that the complete KDO-lipid A structure is essential for growth. Mutants defective in the biosynthesis of glucosamine, a constituent of the lipid A, lose viability very rapidly even under osmotic protection, which balances the concurrent inhibition of murein synthesis (211, 212). The lethal effect of glucosamine deficiency is most likely due to the inability of cells to synthesize lipid A as other blocks in murein biosynthesis (e.g. dap⁻ mutants) can, at least for some time, be overcome by osmotic protection.

Decisive progress regarding topological aspects of LPS biosynthesis has been achieved by use of a new technique to separate the inner and outer membrane of *Salmonella* and *E. coli* (213). It could be shown that the complete O-antigen and the core region are synthesized and attached to each other in the cytoplasmic membrane (214). The LPS is then rapidly translocated into the outer membrane. This process is irreversible because incomplete core synthesized under nonpermissive conditions and transferred to the outer membrane is not completed during subsequent growth under permissive conditions (214, 215). The mechanism of translocation of LPS into the outer membrane is unknown and is especially puzzling since incomplete LPS molecules with very different physical properties are apparently exported as well as complete LPS molecules. It was shown that newly synthesized LPS may enter the outer membrane via discrete export sites (216). A UDP-galactose-4-epimeraseless mutant of *Salmonella typhimurium* synthesized a complete LPS when galactose was added to the growth medium. After 30 sec, 220 patches per cell of newly made complete LPS were observed with ferritin-conjugated antibodies directed against complete LPS (215). After 2–3 min growth LPS was randomized. In ultra-thin sections of plasmolyzed cells 86% of the ferritin label was located over adhesion sites between cytoplasmic and outer membrane (215). Adhesion sites between cytoplasmic and outer membrane have previously been shown in plasmolyzed cells of *E. coli* (217) and it appears that these adhesion areas are also attachment sites for a number of phages [see later, (270, 271)].

CELL ENVELOPE MUTANTS

The frequency of occurrence of "membrane mutants" among temperature-sensitive mutants is very high. From 40 temperature-sensitive *E. coli* mutants selected for inability to grow on tryptone or nutrient agar at 42°, 32 were able to grow on a medium with increased osmolality (218). One mutant studied in more detail showed an increased sensitivity towards certain dyes and certain antibiotics. This is generally taken as evidence that the targets of these substances are more readily available due to cell envelope alterations. Such mutants become sensitive towards a variety of structurally and functionally unrelated antibiotics (e.g. penicillin, rifamycin, actinomycin); dyes (gentian violet, methylene blue, acridine orange, acriflavine, dinitrophenol, triphenyl tetrazolium chloride, eosin y); and detergents (neutral and charged), pointing to a general breakdown of the membrane's permeability barrier. However, the altered membranes still show some specificity. Sensitivity towards erythromycin, rifamycin, actinomycin, bacitracin, and vanco-mycin increases in *Salmonella* mutants with increasingly defective lipopolysaccharide (Ra to Re, see Figure 3) but Rb mutants are the most sensitive to penicillins and cephalosporins and Re mutants are in contrast the most resistant towards tetracyclin (219).

A very detailed study on the relationship between structural defects in the lipopolysaccharide of *E. coli* K12 and the resistance towards antibiotics and bacteriophages revealed the very complex behavior of the outer membrane and pointed to differences between *E. coli* and *Salmonella* (220). *E. coli* mutants permissive to T4 phage mutants lacking the e gene function (lysozyme) or both e and t gene function (degradation of the host cytoplasmic membrane) are mostly temperature sensitive in growth and cell division, show an increased sensitivity towards unrelated antibiotics and colicins, and are especially sensitive towards anionic detergents (221). However, some of these mutants do not leak cytoplasmic enzymes, do not take up lactose in the absence of the *lac*-permease system, and do not permit entry of penta- and hexapeptides at a rate sufficient for nutritional requirements. Although one strain (E372) became more sensitive to colicins with different receptors it remained resistant to colicin K for which it has no receptor.

Because aerobic $^{32}P_i$ uptake can be inhibited by colicin K in cells with apparently no functional receptor (222), provided that damage of the envelope is such that it allows access of colicin K to the target, it again becomes obvious that mutant membranes can show functional integrity for some membrane effectors and not for others. Colicin-tolerant mutants were one of the first types of membrane mutants where pleiotropic effects became apparent (222–225) and later it was shown that these can be the consequence of a single mutation (226). Because most membrane mutants show pleiotropic effects they are very difficult to classify. Eight out of twelve mutants temperature sensitive in DNA synthesis would have been found in a search for conditional lysis mutants (106). Thermosensitive mutants in DNA synthesis are frequently affected in cell division and form filaments with or without septation (227, 228). A protein of the cytoplasmic membrane with a molecular weight of approximately 40,000 [protein Y (229)] is deficient in mutant cells unable to initiate (*dnaA*) or elongate (*dnaB*) DNA (230). This protein is directly related to the ability of the cell to synthesize DNA (229, 230) since it is again present at the nonpermissive temperature when the mutation for DNA synthesis is suppressed by other means, e.g., with 2% NaCl (*dnaB*) (229) or by integration of the genome of phage P2 into the bacterial chromosome (*dnaA*) (230). Other deficiencies of membrane proteins (mol wt 30,000 and 60,000) are related to the temperature-sensitive mutation but these proteins are also absent when DNA synthesis proceeds at the nonpermissive temperature by the means indicated above (230). DNA synthesis mutants are usually supersensitive towards detergents and dyes (227, 228, 231, 232) and a *dnaA* mutant shows an enhanced fluorescence in the presence of the

fluorescence probe 8-anilino-1-naphthalene sulfonate at the nonpermissive temperature (231).

A whole set of chain-forming cells which in some way are impaired in cell
division have been isolated after mutagenization and most have been shown to be
sensitive to certain drugs and dyes (233–237). A septum formed in one mutant
[*envA* (233)] consists of the opposing cytoplasmic membranes and probably only
the rigid layer between them. During normal cell division of *E. coli* one usually
observes a gradual conjoint invagination of all the layers of the outer membrane.
A membrane protein (protein X) accumulates when cell division is inhibited (238).
In a mutant strain with a defective *recA* gene, cell division is uncoupled from
DNA replication and, upon inhibition of DNA synthesis, protein X is not overproduced (239). This protein change is unrelated to the absence of an envelope
protein (mol wt 15,000) in cells that are stationary or that do not divide because
of the presence of low doses of penicillin (240). Due to the pleiotropic effects it
is generally very difficult to relate any protein change to the altered physiology
of a membrane mutation even when this is firmly established as a single-step
mutation. Overproduction of a protein, such as in a mutant refractory to colicin
E2, (241) or loss of a protein in a colicin E1-tolerant mutant cell (242) is not
necessarily the cause but may be the visible consequence of a still unobserved
sequence of membrane changes.

A number of morphological mutants from *E. coli, Bacillus subtilis,* and *Bacillus
licheniformis,* which grow under certain conditions as spheres instead of rods, have
been isolated after treatment with mutagens. The composition of the membranes
of one *E. coli* mutant was studied very extensively biochemically (106). The
mutant grows at 30° as a rod; at 42° it grows and divides as a sphere provided
at least 2 mM magnesium ions are present; otherwise, the cells lyse. The rod–sphere
transition can be reversed by lowering the temperature. No change was found in
the components of the outer membrane, or in the chemical structure of the murein,
although it is spherical at 42°. The defect most likely resides in the cytoplasmic
membrane as T4-infected cells show no inhibition of lysis in liquid culture and the
plaque morphology on agar plates resembles that of rapid lysis mutants (r phenotype,
see section below). The magnesium dependency and the fact that intracellular
vesicles surrounded by a unit membrane are frequently observed at 42° also argue
for an altered cytoplasmic membrane. The rod mutants from *B. subtilis* have been
divided into three classes (243): 1. spherical cells which can be corrected by high
salt concentration, 2. spherical cells which revert to rods when grown in very rich
medium but not in high salt, and 3. various temperature-sensitive rod mutants
(243, 244). A temperature-sensitive rod mutant (*B. subtilis* 168 ts-200 B), for
example, contained less than 20% of the normal amount of teichoic acid and also
poorly glycosylated teichoic acids after growth at the nonpermissive temperature
(244). In cases where structural or enzymatic alterations (e.g. autolysis) are noted,
one must always be aware of the possibility that one is looking at secondary
consequences of unknown primary defects.

In this context of cell envelope mutants, L-forms should be mentioned (245) from which stable variants of *E. coli* (246) and *B. subtilis* (247) which grow and divide in liquid medium have now been obtained. They completely lack cell wall structure, as has been shown by electron microscopy and by biochemical analyses. The protein composition of the cytoplasmic membrane of *B. subtilis* is different from that of the wild type. The generation time of the *B. subtilis* L-form is 1 hr, for the *E. coli* L-form apparently 2-4 hr. L-forms could be useful for studying the function of the cell wall, and the genetic and biochemical analysis of reversions could yield information about the requirements for the recovery of surface polymers and their assembly into a membrane, provided sufficient amounts of L-form cells can be grown.

MEMBRANE ASPECTS OF VIRUS-CELL INTERACTION

When sensitive cells are infected with phages, several membrane phenomena are observed. Membrane alterations during T2 and T4 infection are most comprehensively studied. Upon adsorption of the phage, cells become temporally leaky for intracytoplasmic compounds of small molecular weight (perforation) (248). The membranes become sealed again within 2 min (sealing reaction). Host cells are lysed upon simultaneous addition of many phage particles (lysis from without). Lysis from without is prevented by stepwise addition of the same number of phages, starting with a low multiplicity (lysis resistance). Related to this phenomenon is the observation that cells infected by one phage suppress after some minutes the development of a superinfecting phage (superinfection exclusion). This accounts also for lysis inhibition, in which phage particles released from some cells prevent escape of phages from the other cells leading to high phage titers. The release of most types of phages is accompanied by severe membrane alterations induced by phage functions. Phages in the lysogenic state can induce changes on the cell surface (lysogenic conversion) preventing superinfection by the same phage and allowing adsorption and infection by another phage (helper phage function). The best-studied case is the *Salmonella* phage ε15 which induces the conversion of an α-glycosidic bond of the lipopolysaccharide to a β-glycosidic bond. The activity of the bacterial α-polymerase is inhibited and replaced by a virus-induced β-polymerase. In addition the synthesis of a bacterial transacetylase is suppressed. The phage ε34 uses the altered surface structure as receptor (1, 177, 249, 250).

In the following we deal with some recently published membrane aspects of adsorption, penetration, maturation, and release of phages.

Adsorption and Penetration

The known receptors in the outer membrane of bacteria have recently been comprehensively reviewed by Lindberg (251). All types of structures at the cell surface, such as lipopolysaccharide, C–carbohydrate–teichoic acid–peptidoglycan complexes, capsular polysaccharides, proteins, pili, and flagella, can serve as receptors (251). Structural requirements for phage adsorption are highly specific. Only most recently have highly purified receptor proteins been described. A virulent phage has been isolated that binds to the protein (mol wt 150,000) which forms the surface layer of *Bacillus sphaericus* strain P1 (252). The receptor protein (mol wt 85,000) for phage T5 has been isolated from *E. coli* B and found also to be the receptor for colicin M (253, 254). Glycosidases with specificities towards murein (255), lipopolysaccharide (256–258), and capsular polysaccharides (259) and a murein peptidase (260) have been found as constituents or accessories of purified phages. They are thought to be involved in penetration, which is supported by the finding that the glycosidase of the *Salmonella* phage ε15 and the *E. coli* phage Ω8 have the same structural requirement for cleavage as the phages have for adsorption (256–258).

Adsorption and penetration of the small phages are mediated by specific phage and bacterial proteins. The A-protein of the small RNA phages MS2 and R17 mediates the specific adsorption onto the sides of the F-pilus and enters the host cell together with the RNA (261, 262). During this process the A-protein is cleaved into two components with molecular weights of 24,000 and 15,000. The A-protein (gene 3-product) of the small filamentous DNA phage M13 binds to the tip of the F-pilus, enters the cell together with the DNA, and is found in the cytoplasmic membrane (263). Models for entry of the small RNA and DNA phages into the cell by the F-pilus have been proposed (34, 264, 265) but no real understanding in molecular terms has been reached so far (see review 33). The disappearance of F-pili during infection, partially due to resorption of the pili into the cell (264, 266, 267), does not necessarily mean that phage DNA is carried into the cell in such a mechanical way. Pilus resorption occurs also in noninfected cells (267). For the DNA phages fd (268) and M13 (269) it was shown that the coat protein can enter the cell and is partially reused for packaging of newly made phage DNA. This does not seem to be a necessary step for the production of viable phages since up to 90% of the M13 or fd coat protein of the infecting phage could be degraded with the protease Nagarse without preventing formation of infective centers (269).

The adsorbed phages T1–T7, λ, and φX174 seem to be preferentially located after 4–6 min over adhesions of cell wall and cytoplasmic membrane (270, 271); 200–400 such adhesion sites account for only 6% of the total cell surface. This would imply that after random collision of phage and cell the phage either desorbs again and irreversibly adsorbs only at the adhesion sites or that a lateral diffusion takes place from the first adsorption site to the preferred adhesion sites. Clusters of different receptors would have to be organized into several hundred distinct topographical areas of the surface. Also F-pili of *E. coli* HfrH were shown by electron microscopy to originate at a wall membrane fusion (271) so that adhesion areas could be general sites for the uptake of infecting nucleic acid (see competence below). The highly specific adsorption step can be bypassed as shown by the productive infection of spheroplasts of *E. coli* B4, *Salmonella anatum*, *Aerobacter aerogenes*, *Proteus vulgaris*, and *Serratia marescens* by urea-treated wild-type T4 (272). Contraction of the tail sheath, induced by the urea treatment, seems to be essential since spheroplasts can also be infected with contracted T4 phage particles, obtained by temporary adsorption of a T4 mutant lacking gene 12 product (273). The permeability barrier of the cell can be damaged by other means as shown by the productive infection of actinomycin-permeable mutants of *E. coli* (AS19, AS27) by urea-treated T4 (274).

The mutant cells are most competent for T4 DNA uptake in the early stationary phase (10- to 100-fold more plaques than with exponentially growing cells).

Penetration of Nucleic Acid

The biochemical mechanisms for transfer of nucleic acids, which are many times longer than the cell, through the membrane system are unknown. Molecular requirements on the part of the membrane have been studied in most detail with transformable *pneumococci* (275, 276). *Pneumococci*, in addition to about 15–20 other bacterial species (277), develop a transient cellular condition called competence. They absorb homologous and heterologous double-stranded but not single-stranded DNA. However, they integrate genetically only homologous DNA. Competent *pneumococci* contain a competence factor which is a protein with a molecular weight of about 10,000; 0.1 μg protein/ml activates 10^8 *pneumococci* to the competent state within 15–20 min. A receptor for the competence factor as well as an agglutinin involved in DNA binding, reside in the cytoplasmic membrane (276). A choline-containing teichoic acid at the cell surface is a prerequisite for DNA binding, since DNA does not bind to cells grown with ethanolamine in place of choline. The biosynthetic replacement of the ethanolamine by the choline causes return of the cellular response to added competence factor within less than 10% of the generation time (278). Choline is incorporated at the equatorial growth zone of the cell. This is shown by an autolytic enzyme specific for choline-teichoic acids. The enzyme cleaves the amide bond between N-acetylmuramic acid residues and L-alanine of the peptidoglycan. After 12 min reincorporation of choline into ethanolamine cells, addition of autolysin causes splitting of the cell wall at the equatorial zone. Thus DNA binding and uptake seem to be confined to the equatorial growth zone. In the case of *B. subtilis* an average number of 20–53 DNA uptake sites per cell has been calculated from the number of DNA molecules initially attached to competent cells and from the rate of DNA uptake (279).

A potentially good system for the study of DNA transfer through cell membranes is the peculiar infection mechanism of phage T5. DNA transfer can temporally and functionally be resolved into two stages. In a first step, 8% of the total DNA is transferred into the host cell. This first-step-transfer DNA must be expressed before entry of the remaining 92% of DNA can take place (280, 281), and is probably transferred in a unique direction and immediately attached to a fast sedimenting bacterial component which appears to be membrane (282). The DNA transfer can be arrested after the first step by various means [*amber* mutants (282), energy poisoning, chloramphenicol, low temperature, high cell density, low calcium ion concentration (280)]. By simple centrifugation of the first step phage-cell complex, the phage capsids can be liberated from the cells and the naked uncoiled DNA (92% piece) was beautifully shown by electronmicroscopy to be attached to the bacterial envelope at one end and sometimes to the phage head at the other end (282). Parallel to the two steps in DNA transfer the fluorescence compound ANS (8-anilino-1-naphthalenesulfonate), dissolved into the cell (283), shows a two-step fluorescence increase. Phages T2, T3, T4, and T7 induce more than 100% increase in fluorescence upon adsorption to sensitive cells (*E. coli* B) (283). Only T5 induces a second, slower, fluorescence increase after about 6 min. The latter increase is dependent on protein synthesis but occurs also under conditions in which the second DNA transfer is arrested. The fluorescence increase of cell-bound ANS is supposed to reflect membrane alterations (284, 285). Thus the second chloramphenicol-sensitive fluorescence increase probably points to metabolic membrane changes preceding the second DNA transfer. The sequential appearance of proteins induced by the first-step-transfer DNA has already been studied in considerable detail (286, 287). The receptor for phage T5 has been isolated and this can trigger the release of T5 DNA (288). The active component of the receptor is a protein which has been isolated in a highly purified state (253, 254).

T-even phages with a contractible tail sheath inject DNA into an area of the cell envelope close to the cytoplasmic membrane but apparently not through the cytoplasmic membrane (289). This conclusion, although based on a beautiful electron-microscopic study, must nevertheless be viewed with caution since it is difficult to ascertain that the phage needle extends only 120 Å into the cell, that is, just through the thickness of the outer membrane. A detailed study on the localization of the DNA of superinfecting T4 showed that it is trapped within the cell envelope (290–292). The DNA is resistant to added deoxyribonuclease but is degraded by the periplasmic deoxyribonuclease (endonuclease I). The degradation products average eight nucleotides in length like the degradation products obtained with the isolated endonuclease I. In endonuclease I$^-$ strains the superinfecting DNA is also efficiently excluded. The DNA is degraded into a segment with a molecular weight of 90×10^6 and about 30 pieces of smaller size. The localization in the cell wall was shown by high resolution autoradiography and by the fact that the DNA degradation products are released into the medium. Nonglucosylated T4 DNA in contrast is degraded in the cytoplasm into mononucleotides which remain intracellular (restriction). Adsorption of superinfecting phage is as rapid and as complete as that of primary phage, but transfer of DNA is only 50% compared to 80% of the primary phage as measured by the resistance of adsorbed ^{32}P-labeled T4 to blending. Phage mutants which affect the host envelope differently from wild-type phage [spackle (293) and rII (294)] still exhibit breakdown of superinfecting DNA. Exclusion requires protein synthesis before addition of superinfecting phage. The data can be explained most simply if the primary phage induces changes in the cell envelope which hinder DNA release from the superinfecting phage and prevent transfer of those DNA molecules into the cytoplasm which have been injected.

Membrane Alterations After Phage Infection

Inhibition of lysis in T4-infected E. coli B requires the gene products of the rIIA and rIIB cistrons of T4 (294, 295). The primary gene product of the rIIB cistron is a protein with a molecular weight of 37,000 localized in the cytoplasmic membrane of the host (294, 295). It accounts for 10% of the membrane proteins synthesized during T4 infection, which corresponds to 1% of the total protein. The gene product of the rIIA cistron is a protein with an approximate molecular weight of 74,000 and is also localized in the cytoplasmic membrane (296). The A-protein in the membrane amounts to only 10–25% of the B-protein although both cistrons are reported to be transcribed equally well (297). Synthesis of the *rIIB* protein is not influenced by mutations in the A-cistron and vice versa. Other T4-induced membrane proteins also are known (294). They seem to be less tightly integrated into the membrane since they show the same molecular weight pattern after SDS-gel electrophoresis whether dissolved at 60° or 100°. Some of the host membrane proteins of infected or noninfected cells in contrast have smaller apparent molecular weights when dissolved at 100°, indicating that dissociation of the membrane components at 100° is more complete than at 60° [(294), see also (37, 39, 42)].

DNA-Membrane Complex

Progeny DNA of T4 was found to be synthesized on the host membrane throughout infection, as revealed by the magnesium-Sarcosyl crystals or M-band technique (298, 299). Conversion of input T4 DNA to a rapidly sedimenting form requires RNA synthesis (299). After thermal derepression of a superinfected λ, lysogenic λ DNA is converted to a fast sedimenting complex which requires activity of the early regulatory *N* gene product (300). The presence of phage DNA in a rapidly sedimenting complex has also been described for the phages ϕX174, M13, P2 (301), and P22 (cited and discussed in 300, 302, 303). In most cases it was not rigorously shown that this rapidly sedimenting DNA is in fact membrane bound. The absence of phospholipids, e.g. in the *E. coli* DNA replication complex (304), points rather to a large complex containing the DNA growing point, protein, and possibly the origin of replication which may be weakly and/or transiently attached to the cytoplasmic membrane. Caution has especially to be exerted since the recent finding that newly replicated T7 DNA was first recovered in a fast sedimenting form of 40–80S which is then fragmented, this latter depending on active gene 3 (endonuclease) and gene 6 (exonuclease) products (305). Even *E. coli* DNA polymerase I (Kornberg enzyme) was found transiently associated with the membrane after infection with phages T4, T7, and λ (306). However, this phage-directed association is dispensable (306, 307).

Virus Assembly

For some phages, assembly requires host functions. *GrowE* (308) and *mop* (309) mutants of *E. coli* are nonpermissive for the growth of several wild-type phages. *GrowE* blocks T4 growth at the level of the gene 31-mediated step in head assembly. The bulk of the proteins synthesized after T4 infection sediments with the cell envelope and no head-related structures are seen in lysates. The λ- or T5-infected *growE* mutant cells produce active tails but no active heads as is the case with T4. Phage head gene mutants can overcome the bacterial *growE* mutation. *Mop* mutants are very similar to *growE*: T4 and λ head precursors aggregate on the bacterial membrane. Preliminary evidence indicates that *mop* is a gene controlling a membrane function. *Mop* mutants are insensitive to the peptides tri-L-serine and tri-L-tyrosine that inhibit growth of the wild-type strain. This points to a deficiency in uptake through the membrane in the mutant. Differences in the membrane protein pattern after SDS-gel electrophoresis have also been noted. T4 head assembly on the cytoplasmic membrane was shown by electron microscopy (310). A thermosensitive bacterial mutant that prevents proper tail fiber assembly of T4 has also been described (311).

The A protein (gene 3-product) of the small filamentous phage M13 is essential for conversion of the phage DNA to the duplex replicative form in vivo (263). In vitro, however, gene 3-protein is dispensable for replication of the viral single strands. The A-protein is found together with the replicative form DNA in the host cytoplasmic membrane. Since the newly adsorbed phage is also localized in the cytoplasmic membrane, the A-protein may play a role in uncoating the phage or in phage maturation (see below) besides its functions during the adsorption to the tip of the F-pilus (312). During production of the phages M13, fd, f1 the cell growth rate is slower but no cell lysis occurs (264, 313, 314). Newly synthesized f1 coat protein becomes associated with the inner membrane of the host shortly after it is synthesized (315). The coat protein is released from there packed in the mature phage without being intermediarily present in the cytoplasm or the outer membrane. Cells infected with the small DNA phages have altered membranes: streptomycin-resistant cells become phenotypically sensitive (316), permeable to actinomycin (317), sensitive to deoxycholate (318), and insensitive to a variety of colicins (315). Phages having *amber* mutations in all phage genes except gene 2 kill the host (319, 320). Aberrant intracytoplasmic membranes are formed, accompanied by an increase of the percentage of cardiolipin from 4 to 20% (316, 321, 322). The same phenomenon is observed when phage formation is inhibited by phage-specific antiserum (319). The small RNA phage fr lyses cells when growth occurs at 37° but not at 31° (323). The suggestive dependency of the fr-induced cell lysis on the state of the membrane at 31° and 37° is not known.

Release of Virus

In some well-studied *E. coli* phages hydrolytic enzymes are produced during infection which degrade the rigid layer (murein-lipoprotein complex) upon release of the phage. This layer is mainly responsible for the maintenance of the structural integrity of the outer membrane. Degradation of this covalently linked poly-saccharide-peptide network (91), either by cleavage of the glycan chains as is the case with the T2 (324) and T4 (325) lysozymes or by cleavage of the peptide side chain as is the case with λ (326) and T7 endolysin (327), leads to bursting of the outer membrane and release of phage particles. Wild-type T4 is unable to lyse host bacteria in the absence of phage-specific lysozyme (*e* gene product) (328). A mutant T4, called *spackle* (*s*) can lyse cells without active lysozyme present (293). Bacteria infected with *es* phage liberate about 10% as many phage particles as bacteria infected with wild-type phage. The *e* phage-infected cells liberate 1%. Bacteria infected with *s*-mutant phage show no inhibition of lysis and are far more sensitive to lysis from without than are bacteria infected with wild-type (s^+) phage. This distinguishes phenotypically *s* from *r* mutants since the latter show no lysis inhibition but are normal with respect to lysis from without. Attempts to find differences in the membrane protein composition of *s* phage-infected cells (lack of a protein since s^+ is dominant over *s*) as observed with *rIIA* and *rIIB* mutants have been unsuccessful so far (294, 295). Besides phage lysozyme (and the *s* gene product) an additional phage gene function, *t* in T4 (329, 330) and *S* in λ (331, 332), is necessary for lysis of the host. *Su* cells infected with *amber t* and *S* mutants produce more than five times as many phage particles and lysozyme as during wild-type infection. This is not the case with lysozyme-defective infections. The *t* and *S* functions seem therefore to be responsible for the cessation of phage synthesis at the end of the latent period. The *t* and *S* functions induce progressive degradation of the cytoplasmic membrane (330, 333) which apparently is prerequisite to lysozyme reaching its target, the rigid layer of the outer membrane. In the case of the T7 endolysin a number of observations suggest a role in DNA metabolism rather than lysis (334). Hydrolysis of host membrane phospholipids before the onset of lysis was shown in T2 (335), T4 (336), and T6 (335) infected cells. Some controversy (337) on the amount and the time of appearance of free fatty acids was apparently due to fatty acid degradation by β-oxidation (338). Phospholipid hydrolysis already starts 10 min after infection (336) but is apparently balanced by phospholipid biosynthesis (339, 340). It is questionable whether phospholipases are involved in release of phage λ. Host mutants devoid of detergent-resistant and detergent-sensitive phospholipase A release phage particles with the same burst size as wild-type host cells (341). No significant release of fatty acids was observed in the mutant cells, whereas in the wild-type cell, 16% of the prelabeled lipid is released as fatty acids upon infection and this is mainly due to the detergent-resistant phospholipase A. This phospholipase was isolated in pure form and studied extensively (342). However, it apparently plays also no indispensable role in T4 release (342). In a mutant of *E. coli* K12 which is temperature sensitive in membrane lipid biosynthesis (glycerol-3-phosphate acyltransferase) no T4 phage or phage lysozyme synthesis occurs at the nonpermissive temperature (343).

Infection with bacteriophage T7 leads to an interference in the macromolecule synthesis of the host in the absence of phage gene expression. *E. coli* RNA synthesis is depressed after infection in the presence of inhibitors of protein synthesis such as chloramphenicol. The same reduction in transcription takes place after infection with T7 phage with uv-inactivated genomes. This inhibition of RNA synthesis is dependent on the multiplicity of infection. Protein synthesis and replication of the host are affected very similarly (344). The most likely explanation for this depression, which is gene expression independent, is an activation via the membrane caused by adsorption and penetration. This action of T7 resembles the killing effect of DNA-free T2 and T4 ghosts (345). Beside this passive mechanism the ratio of synthesis of phosphatidylethanolamine to phosphatidylglycerol is changed, and this depends on phage gene expression. Several minor new polar lipids occur. The capacity of cell-free extracts from T7-infected cells to synthesize phospholipids is increased by a factor of 10.

CONCLUDING REMARKS

In prokaryotic and eukaryotic cells many basic cell functions which depend on the coordinated interplay between a large number of individual reaction steps are membrane bound. This is especially obvious in processes like the electron transport chain of photosynthesis and respiration and their coupling to oxidative phosphory-lation. According to the chemiosmotic hypothesis (346), the energy of the electron transport chain is conserved in an ion gradient across the membrane and this membrane potential is transformed into ATP or used in energy-requiring processes such as active transport. In his review on conservation and transformation of energy by membranes of mitochondria and bacteria (347), Harold has focused on both prokaryotic and eukaryotic cells. A simple system to study the energized state of a membrane seems to be the purple membrane of *Halobacterium halobium*. When grown at low oxygen concentration, purple membrane patches with a hexagonal lattice structure occur on the cytoplasmic membrane (348, 349). They can be separated from the rest of the cytoplasmic membrane by density gradient centri-fugation (348), and 300–500 mg purple membrane can be obtained from a 10 liter culture. The membrane contains 25% lipid and 75% protein. The protein consists of one type of polypeptide chain which contains one retinal bound through a Schiff base to a lysine residue (348), similar to the rhodopsin of the visual receptor membrane in the retina. The purple membrane protein is therefore called bacteriorhodopsin. Bacteriorhodopsin in the isolated state (350) as well as in the organism (351) is reversibly bleached by light. During illumination, protons are released from isolated purple membranes and from intact cells. This observation and the analysis of the effect of suitable uncouplers suggest that illumination generates a pH and/or electrical gradient across the membrane. In addition to the proton translocation, synthesis of ATP, reduction in oxygen consumption, phototaxis, and biosynthesis of purple membrane have been found as responses to light (351). Illumination also leads to a reduction of tryptophan fluorescence which is regenerated in the dark, pointing to structural transitions in the protein (349). The ease in obtaining this membrane and its extreme structural simplicity make it an ideal system. Synthesis of this specialized membrane can be manipulated, and thus it seems to be suitable for the study of membrane biosynthesis, which is an important problem on its own. From an evolutionary point of view the singular appearance of rhodopsin in this microorganism is very puzzling.

In addition to its function as the cell's main power plant, many other functions such as active transport (21–24), biosynthesis of membrane macromolecules (2–9), conceivably also control of growth and assembly of the cell (8–13, 146), perhaps steps in DNA replication (303), and synthesis of certain phages (268, 269, 315) take place in the cytoplasmic membrane. We devote a great deal of this review to the outer membrane, emphasizing that the outer layer, especially in gram-negative bacteria, is more than just a protective skin. Agents that interfere with cell metabolism such as bacteriocins (352–355) or phages (251) bind to specific receptors in the outer membrane. Binding triggers subsequent events such as release of the nucleic acids from the phage particles or killing of the cells by the bacteriocins. Bacteriocins and phages can share the same receptor, e.g. T5 binds to the same receptor protein as colicin M (253, 254). Even more striking is the common receptor site for Vitamin B_{12} and the E colicins in the outer membrane of *E. coli* (356). However the E colicins interfere in very different ways with cell metabolism. Binding to the receptor is also the first step in transport of Vitamin B_{12} (357). Thus very different agents use the same receptor very specifically for different purposes. It is unknown how the transmission from a common receptor to the specific target is performed. Although in the case of colicin E3, a direct action on the target, the ribosome, was shown in vitro (358–360), the target in intact cells seems unavailable without initial binding to the receptor. Cells without functional receptor are resistant, which apparently holds for all kinds of colicins (352–355).

The outer membrane also plays a role in the exchange of genetic material, a process very important for the development of organisms. Virus infection is already a quasi-sexual process. Genes are exchanged between virus and cell if genomes are integrated into the host chromosome and then not precisely excised. The cell surface selects which virus can infect. In addition special organelles, called pili, have been invented for the exchange of genetic material (33, 34). They are long proteinaceous filaments and are necessary for mating pair formation, and probably for DNA transfer in conjugation and the exchange of R and F factors. In addition, they serve as phage receptors (251).

A whole class of so-called periplasmic proteins (25, 26) are considered to be localized outside the cytoplasmic membrane. Many of them are hydrolases and their major function may be the hydrolysis of compounds which otherwise cannot enter the cell. Some may be involved in exclusion of phage by degrading the invading nucleic acid (e.g. 290). A second large group of periplasmic proteins are the binding proteins (23) involved in active transport (21–24, 361) and chemotaxis (362, 363). Binding proteins specifically bind low molecular weight substances, mainly amino acids and sugars, and link them to the respective transport system. In chemotaxis, binding proteins serve as chemoreceptors for those chemicals that attract the organism. The binding protein for transport and chemotaxis can be the same as shown for D-galactose (364, 365).

The examples listed show the many diverse ways in which the outer membrane mediates the communication of the microorganism with its environment. In a similar way cell surface structures are of great importance for the behavior of eukaryotic cells. Examples are contact inhibition of growth and movement, regulation of metabolism, for instance by peptide hormones (366) or antigens (367), or the membrane's role and alteration during virus infection (368). In microorganisms many membrane mutants have been isolated and resultingly complex pleiotropic effects became apparent (218–247). Such mutants are being studied extensively and hopefully some general features will emerge which will, at least conceptually, help to clear up questions about the altered cell surface structure of transformed eukaryotic cells, a problem which affects us much more directly.

ACKNOWLEDGMENTS

The cited papers of the authors have been supported in part by the Deutsche Forschungsgemeinschaft.

Literature Cited

1. Braun, V. 1973. *Naturwiss. Rundsch.* 26:330–36
2. Rothfield, L., Romeo, D. 1971. *Bacteriol.* *Rev.* 35:14–38
3. Machtiger, N. A., Fox, C. F. 1973. *Ann. Rev. Biochem.* 42:575–600

4. Overath, P., Schairer, H.-U., Hill, F. F., Lamnek-Hirsch, I. 1971. *The Dynamic Structure of Cell Membranes,* ed. D. F. Hölzl-Wallach, H. Fischer, 149–64. New York: Springer

5. Mindich, L. 1973. *Membranes and Walls of Bacteria,* ed. L. Leive, 1–36. New York: Dekker

6. Kepes, A., Autissier, F. 1972. *Biochim. Biophys. Acta* 265: 443–508

7. Ghuysen, J.-M., Shockman, G. D. 1973. See Ref. 5, 37–130

8. Fiedler, F., Glaser, L. 1973. *Biochim. Biophys. Acta.* In press

9. Nikaido, H. 1973. See Ref. 5, 131–208

10. Rogers, H. J. 1970. *Bacteriol. Rev.* 34: 194–214

11. Pardee, A. B., Wu, P. C., Zussman, D. R. 1973. See Ref. 5, 357–412

12. Higgins, M. L., Shockman, G. D. 1971. *Crit. Rev. Microbiol.* 1: 29–72

13. Henning, U., Schwarz, U. 1973. See Ref. 5, 413–38

14. Freer, J. H., Salton, M. R. J. 1971. *Microbial Toxins,* Vol. IV, ed. G. Weinbaum, S. Kadis, S. J. Ajl, 67–126. New York/London: Academic

15. Salton, M. R. J. 1971. *Crit. Rev. Microbiol.* 1: 161–97

16. Salton, M. R. J., Nachbar, M. S. 1971. *Anatomy and Biogenesis of Mitochondria and Chloroplasts,* ed. N. K. Boardman, A. W. Linnane, R. S. Smillie, 42–52. Amsterdam: North-Holland

17. Reusch, V. M., Burger, M. M. 1973. *Biochim. Biophys. Acta* 300: 79–104

18. Oelze, J., Drews, G. 1972. *Biochim. Biophys. Acta* 265: 209–39

19. Razin, S. 1972. *Biochim. Biophys. Acta* 265: 241–96

20. Ellar, D. J. Unpublished observations

21. Kaback, H. R. 1972. *Biochim. Biophys. Acta* 265: 367–416

22. Oxender, D. L. 1972. *Ann. Rev. Biochem.* 41: 777–814

23. Lin, E. C. C. 1971. *Structure and Function of Biological Membranes,* ed. L. Rothfield, 285–341. New York: Academic

24. Boos, W. 1974. *Ann. Rev. Biochem.* 43: In press

25. Heppel, L. A. 1971. See Ref. 23, 223–47

26. Rosen, B. P., Heppel, L. A. 1973. See Ref. 5, 209–39

27. Cronan, J. E. Jr., Vagelos, P. R. 1972. *Biochim. Biophys. Acta* 265: 25–60

28. Goldfine, H. 1972. *Advan. Microbiol. Physiol.* 8: 1–51

29. van Deenen, L. L. M., van den Bosch, H. J. 1974. *Ann. Rev. Biochem.* 43: In press

30. Volpe, J. J., Vagelos, P. R. 1973. *Ann. Rev. Biochem.* 42: 21–60

31. Sutherland, I. W. 1972. *Advan. Microbiol. Physiol.* 8: 143–213

32. Glaser, L. 1973. *Ann. Rev. Biochem.* 42: 91–112

33. Achtman, M. 1973. *Curr. Top. Microbiol. Immunol.* 60: 79–123

34. Brinton, C. C. 1971. *Crit. Rev. Microbiol.* 1: 105–60

35. Smith, R. W., Koffler, H. 1971. *Advan. Microbiol. Physiol.* 6: 219–339

36. Doetsch, R. N. 1971. *Crit. Rev. Microbiol.* 1: 73–103

37. Schnaitman, C. A. 1970. *J. Bacteriol.* 104: 882–89

38. Schnaitman, C. A. 1970. *J. Bacteriol.* 104: 890–901

39. Bragg, P. D., Hou, C. 1972. *Biochim. Biophys. Acta* 274: 478–88

40. Moldow, C., Robertson, J., Rothfield, L. 1972. *J. Membrane Biol.* 10: 137–52

41. Wu, H. C. 1972. *Biochim. Biophys. Acta* 290: 274–89

42. Inouye, M., Lee, M.-L. 1973. *J. Bacteriol.* 113: 304–12

43. Sekizawa, J., Fukui, S. 1973. *Biochim. Biophys. Acta* 307: 104–17

44. Henning, U., Rehn, K., Hoehn, B. 1973. *Proc. Nat. Acad. Sci. USA* 70: 2033–36

45. Schnebli, H. P., Vatter, A. E., Abrams, A. 1970. *J. Biol. Chem.* 245: 1122–27

46. Baron, C., Abrams, A. 1971. *J. Biol. Chem.* 246: 1542–44

47. Abrams, A., Smith, J. B., Baron, C. 1972. *J. Biol. Chem.* 247: 1484–88

48. Abrams, A., Nolan, E. A. 1972. *Biochem. Biophys. Res. Commun.* 48: 982–89

49. Hanson, R. L., Kennedy, E. P. 1973. *J. Bacteriol.* 114: 772–81

50. Butlin, J. D., Cox, G. B., Gibson, F. 1971. *Biochem. J.* 124: 75–81

51. Kanner, B. I., Gutnick, D. L. 1972. *FEBS Lett.* 22: 197–99

52. Kanner, B. I., Gutnick, D. L. 1972. *J. Bacteriol.* 111: 287–89

53. Schairer, H. U., Haddock, B. A. 1972. *Biochem. Biophys. Res. Commun.* 48: 544–51

54. Pavlasova, E., Harold, F. M. 1969. *J. Bacteriol.* 98: 198–204

55. Klein, W. L., Boyer, P. D. 1972. *J. Biol. Chem.* 247: 7257–65

56. Salton, M. R. J., Schor, M. T. 1972. *Biochem. Biophys. Res. Commun.* 49: 350–57

57. Shipp, W. S. 1972. *Arch. Biochem. Biophys.* 150: 482–88

58. Shipp, W. S. 1972. *Arch. Biochem. Biophys.* 150: 459–72

59. Shipp, W. S. 1972. *Arch. Biochem. Biophys.* 150: 473–81

60. Haddock, B. A., Schairer, H. U. 1973.

116 BRAUN & HANTKE

Eur. J. Biochem. 35:34–45
61. Pichinoty, F. et al 1969. *Ann. Inst. Pasteur* 116:27–42
62. Guest, J. R. 1969. *Mol. Gen. Genet.* 105:285–97
63. Glaser, J. H., DeMoss, J. A. 1972. *Mol. Gen. Genet.* 116:1–10
64. Glaser, J. H., DeMoss, J. A. 1971. *J. Bacteriol.* 108:854–60
65. MacGregor, C. H., Schnaitman, C. A. 1971. *J. Bacteriol.* 108:564–70
66. Azoulay, E., Puig, J., Couchoud-Beaumont, P. 1969. *Biochim. Biophys. Acta* 171:238–52
67. Villarreal-Moguel, E. I., Ibarra, V., Ruiz-Herrera, J., Gitler, C. 1973. *J. Bacteriol.* 113:1264–67
68. MacGregor, C. H., Schnaitman, C. A. 1973. *J. Bacteriol.* 114:1164–76
69. Braun, V., Rehn, K. 1969. *Eur. J. Biochem.* 10:426–38
70. Braun, V., Sieglin, U. 1970. *Eur. J. Biochem.* 13:336–46
71. Braun, V., Wolff, H. 1970. *Eur. J. Biochem.* 14:387–91
72. Inouye, M., Shaw, J., Shen, C. 1972. *J. Biol. Chem.* 247:8154–59
73. Braun, V., Rehn, K., Wolff, H. 1970. *Biochemistry* 9:5041–49
74. Maaloee, O., Kjeldgaard, N. O. 1966. *Control of Macromolecular Biosynthesis,* New York: Benjamin. 160 pp.
75. Schleif, R. 1967. *J. Mol. Biol.* 27:41–55
76. Braun, V., Bosch, V. 1972. *Proc. Nat. Acad. Sci. USA* 69:970–74
77. Braun, V., Bosch, V. 1972. *Eur. J. Biochem.* 28:51–69
78. Braun, V. 1973. *J. Infect. Dis.* Suppl. 128:9–16
79. Hantke, K., Braun, V. 1973. *Eur. J. Biochem.* 34:284–96
80. Murray, R. G. E., Steed, P., Elson, H. E. 1965. *Can. J. Microbiol.* 11:547–60
81. de Petris, S. 1967. *J. Ultrastruct. Res.* 19:45–83
82. Osborn, M. J., Gander, J. E., Parisi, E., Carson, J. 1972. *J. Biol. Chem.* 247:3962–72
83. Mayer, H., Schlecht, S., Braun, V. 1973. *Abstr. Joint Meet. Eur. Soc. Immunol.* Strassburg
84. Bosch, V., Braun, V. 1973. *FEBS Lett.* 34:307–10
85. Braun, V., Bosch, V. 1973. *FEBS Lett.* 34:302–6
86. Izaki, K., Matsuhashi, M., Strominger, J. L. 1968. *J. Biol. Chem.* 243:3180–92
87. Hirashima, A., Inouye, M. 1973. *Nature* 242:405–7
88. Hirashima, A., Childs, G., Inouye, M. 1973. *J. Mol. Biol.* In press

89. Braun, V., Bosch, V., Hantke, K., Schaller, K. 1974. *Ann. NY Acad. Sci.* In press
90. Hantke, K., Braun, V. 1973. *Z. Physiol. Chem.* 354:813–15
91. Weidel, W., Pelzer, H. 1964. *Advan. Enzymol.* 26:193–232
92. Strominger, J. L. 1969. *Inhibitors Tools in Cell Research,* ed. Th. Bücher, H. Sies, 187–207. New York: Springer
93. Brown, A. D. 1964. *Bacteriol. Rev.* 28:296–329
94. Henning, U., Braun, V., Höhn, B. Unpubl. results. See Ref. 13
95. Schleifer, K. H., Kandler, O. 1972. *Bacteriol. Rev.* 36:407–77
96. Hoshino, O., Zehavi, U., Sinay, P., Jeanloz, R. W. 1972. *J. Biol. Chem.* 247:381–90
97. Markovits, J., Vilkas, E., Lederer, E. 1971. *Eur. J. Biochem.* 18:287–91
98. Goren, M. B. 1972. *Bacteriol. Rev.* 36:33–64
99. Vacheron, M.-J., Guinand, M., Michel, G., Ghuysen, J. M. 1972. *Eur. J. Biochem.* 29:156–66
100. Archibald, A. R., Baddiley, J., Heckels, J. E. 1973. *Nature New Biol.* 241:29–31
101. Knox, K. W., Hall, E. A. 1965. *Biochem. J.* 96:302–9
102. Johnson, R. Y., White, D. 1972. *J. Bacteriol.* 112:849–55
103. Sjöquist, J., Meloun, B., Hjelm, H. 1972. *Eur. J. Biochem.* 29:572–78
104. Fox, E. N., Wittner, M. K. 1969. *Immunochemistry* 6:11–24
105. Schwarz, U., Leutgeb, W. 1971. *J. Bacteriol.* 106:588–95
106. Henning, U., Rehn, K., Braun, V., Höhn, B., Schwarz, U. 1972. *Eur. J. Biochem.* 26:570–86
107. Kelemen, M. V., Rogers, H. J. 1971. *Proc. Nat. Acad. Sci. USA* 68:992–96
108. Tipper, D. 1970. *Int. J. Syst. Bacteriol.* 20:361–77
109. Pooley, H. M., Shockman, G. D., Higgins, M. L., Porres-Juan, J. 1972. *J. Bacteriol.* 109:423–31
110. Balyuzi, H. H. M., Reaveley, D. A., Burge, R. E. 1972. *Nature New Biol.* 235:252–53
111. Braun, V., Gnirke, H., Henning, U., Rehn, K. 1973. *J. Bacteriol.* 114:1264–70
112. Martin, H. H., Heilmann, H. D., Preusser, H. J. 1972. *Arch. Microbiol.* 83:332–46
113. Mirelman, D., Bracha, R., Sharon, N. 1972. *Proc. Nat. Acad. Sci. USA* 69:3355–59
114. Tipper, D. J., Strominger, J. L. 1968.

J. Biol. Chem. 243:3169–79
115. Nathenson, S. G., Strominger, J. L. 1961. *J. Pharmacol. Exp. Ther.* 131: 1–6
116. Rogers, H. J., Mandelstam, J. 1962. *Biochem. J.* 84:299–303
117. Schwarz, U., Asmus, A., Frank, H. 1969. *J. Mol. Biol.* 41:419–29
118. Leyh-Bouille, M. et al 1972. *Biochemistry* 11:1290–98
119. Blumberg, P. M., Strominger, J. L. 1972. *J. Biol. Chem.* 247:8107–13
120. Blumberg, P. M., Strominger, J. L. 1972. *Proc. Nat. Acad. Sci. USA* 69: 3751–55
121. Suginaka, H., Blumberg, P., Strominger, J. L. 1972. *J. Biol. Chem.* 247:5279–88
122. Pollock, J. J. et al 1972. *Proc. Nat. Acad. Sci. USA* 69:662–66
123. Pooley, H. M., Shockman, G. D. 1969. *J. Bacteriol.* 100:617–24
124. Fan, D. P., Beckman, M. M., Cunningham, W. P. 1972. *J. Bacteriol.* 109:1247–57
125. Tipper, D. J. 1969. *J. Bacteriol.* 97: 837–47
126. Hughes, R. C. 1970. *Biochem. J.* 119: 849–60
127. Warth, A. D., Strominger, J. L. 1971. *Biochemistry* 10:4349–58
128. Fiedler, F., Glaser, L. In press
129. Hartmann, R., Höltje, J.-V., Schwarz, U. 1972. *Nature* 235:426–29
130. Ryter, A., Hirota, Y., Schwarz, U. 1973. *J. Mol. Biol.* 78:185–95
131. Chaloupka, J., Strnadová, M. 1972. *Folia Microbiol.* 17:446–55
132. Mauck, J., Glaser, L. 1972. *J. Biol. Chem.* 247:1180–87
133. Frehel, C., Beaufils, A. M., Ryter, A. 1971. *Ann. Inst. Pasteur* 121:139–48
134. Higgins, M. L., Pooley, H. M., Shockman, G. D. 1971. *J. Bacteriol.* 105: 1175–83
135. Giesbrecht, P., Wecke, J. 1971. *Cytobiologie* 4:349–68
136. Giesbrecht, P. 1972. *Mikroskopie* 28: 323–42
137. Archibald, A. R., Baddiley, J., Blumsom, N. L. 1968. *Advan. Enzymol.* 30:223–53
138. Baddiley, J. 1972. *Polymerization in Biological Systems,* Ciba Found. Symp., New Ser. 7, ed. G. E. W. Wolstenholme, M. O'Connor, 87–107. Amsterdam: Associated Scientific Publ. 314 pp.
139. Baddiley, J. 1972. *Essays in Biochemistry* 8:35–77
140. Ellwood, D. C., Tempest, D. W. 1972. *Advan. Microbiol. Physiol.* 7:83–117
141. Hughes, R. C., Pavlik, J. G., Rogers, H. J., Tanner, P. J. 1968. *Nature* 219:642–44
142. Coley, J., Duckworth, M., Baddiley, J. 1972. *J. Gen. Microbiol.* 73:587–91
143. Toon, P., Brown, P. E., Baddiley, J. 1972. *Biochem. J.* 127:399–409
144. Ambron, R. T., Pieringer, R. A. 1971. *J. Biol. Chem.* 246:4216–25
145. Mauck, J., Glaser, L. 1972. *Proc. Nat. Acad. Sci. USA* 69:2386–90
146. Fiedler, F., Mauck, J., Glaser, L. See Ref. 89
147. Anderson, R. G., Hussey, H., Baddiley, J. 1972. *Biochem. J.* 127:11–25
148. Hancock, I. C., Baddiley, J. 1972. *Biochem. J.* 127:27–37
149. Hussey, H., Baddiley, J. 1972. *Biochem. J.* 127:39–50
150. Siewert, G., Strominger, J. L. 1967. *Proc. Nat. Acad. Sci. USA* 57:767–73
151. Storm, D. R., Strominger, J. L. 1973. *J. Biol. Chem.* 248:3940–45
152. Watkinson, R. J., Hussey, H., Baddiley, J. 1971. *Nature New Biol.* 229:57–59
153. Sandermann, H. Jr., Strominger, J. L. 1972. *J. Biol. Chem.* 247:5123–31
154. Willoughby, E., Higashi, Y., Strominger, J. L. 1972. *J. Biol. Chem.* 247:5113–15
155. Higashi, Y., Strominger, J. L., Sweeley, C. C. 1967. *Proc. Nat. Acad. Sci. USA* 57:1878–84
156. Wright, A., Dankert, M., Fennessey, P., Robbins, P. W. 1967. *Proc. Nat. Acad. Sci. USA* 57:1798–803
157. Scher, M., Lennarz, W. J., Sweeley, C. C. 1968. *Proc. Nat. Acad. Sci. USA* 59:1313–20
158. Troy, F. A., Frerman, F. E., Heath, E. C. 1971. *J. Biol. Chem.* 246:118–33
159. Nikaido, H., Nikaido, K., Nakae, T., Mäkelä, P. H. 1971. *J. Biol. Chem.* 246:3902–11
160. Wright, A. 1971. *J. Bacteriol.* 105: 927–36
161. Tempest, D. W., Dicks, J. W., Ellwood, D. C. 1968. *Biochem. J.* 106:237–43
162. Heptinstall, S., Archibald, A. R., Baddiley, J. 1970. *Nature* 225:519–21
163. Hughes, A. H., Hancock, I. C., Baddiley, J. 1973. *Biochem. J.* 132:83–93
164. Baddiley, J., Hancock, I. C., Sherwood, P. M. A. 1973. *Nature* 243:43–45
165. Ellwood, D. C., Tempest, D. W. 1972. *J. Gen. Microbiol.* 73:395–402
166. Lüderitz, O., Jann, K., Wheat, R. 1968. *Compr. Biochem.* 26A:105–228
167. Mäkelä, P. H., Stocker, B. A. D. 1969. *Ann. Rev. Genet.* 3:291–322
168. Osborn, M. J. 1969. *Ann. Rev. Biochem.* 38:501–38
169. Wright, A., Kanegasaki, S. 1971. *Physiol. Rev.* 51:748–84

170. Osborn, M. J. 1971. *Structure and Function of Biological Membranes,* ed. L. Rothfield, 343–400. New York: Academic
171. Weinbaum, G., Kadis, S., Ajl, S. J., Eds. 1971. *Microbial Toxins,* Vol. IV: Bacterial Endotoxins. New York: Academic
172. Lüderitz, O., Westphal, O., Staub, A. M., Nikaido, H. See Ref. 171, 145–233
173. Shands, J. W. Jr. See Ref. 171, 127–44
174. Ashwell, G., Hickman, J. See Ref. 171, 235–66
175. Novotny, A. See Ref. 171, 309–29
176. Osborn, M. J., Rothfield, L. I. See Ref. 171, 331–50
177. Robbins, P. W., Wright, A. See Ref. 171, 351–68
178. Stocker, B. A. D., Mäkelä, P. H. See Ref. 171, 369–438
179. Endotoxin Conference. 1973. *J. Infect. Dis.* 128: July Suppl. Vol.
180. Reske, K., Jann, K. 1972. *Eur. J. Biochem.* 31:320–28
181. Tarcsay, L., Wang, C.-S., Li, S.-C., Alaupovic, P. 1973. *Biochemistry* 12: 1948–55
182. Wang, C.-S., Alaupovic, P. 1973. *Biochemistry* 12:309–15
183. Keleti, J., Lüderitz, O., Mlynarcik, D., Sedlak, J. 1971. *Eur. J. Biochem.* 20: 237–44
184. Lindberg, B., Lönngren, J., Ruden, U., Simmons, D. A. R. 1973. *Eur. J. Biochem.* 32:15–18
185. Berst, M., Lüderitz, O., Westphal, O. 1971. *Eur. J. Biochem.* 18:361–68
186. Nikaido, H. 1969. *J. Biol. Chem.* 244: 2835–45
187. Hämmerling, G., Lüderitz, O., Westphal, O. 1970. *Eur. J. Biochem.* 15: 48–56
188. Lehmann, V., Lüderitz, O., Westphal, O. 1971. *Eur. J. Biochem.* 21:339–47
189. Dröge, W., Lehmann, V., Lüderitz, O., Westphal, O. 1970. *Eur. J. Biochem.* 14:175–84
190. Gmeiner, J., Simon, M., Lüderitz, O. 1971. *Eur. J. Biochem.* 21:355–56
191. Gmeiner, J., Lüderitz, O., Westphal, O. 1969. *Eur. J. Biochem.* 7:370–79
192. Rietschel, E. Th., Gottert, H., Lüderitz, O., Westphal, O. 1972. *Eur. J. Biochem.* 28:166–73
193. Schmidt, G. 1972. *Zbl. Bakt. Hyg. I. Abt. Orig.* A 220:472–76
194. Schmidt, G., Jann, B., Jann, K. 1969. *Eur. J. Biochem.* 10:501–10
195. Morton, J. J., Stewart, J. C. 1972. *Eur. J. Biochem.* 29:308–18

196. Adams, G. A., Singh, P. P. 1970. *Biochim. Biophys. Acta* 202:553–55
197. Wober, W., Alaupovic, P. 1971. *Eur. J. Biochem.* 19:340–56
198. Ibid 357–67
199. Wu, M. C., Heath, E. C. 1973. *Fed. Proc.* 32:Abstr. 481
200. Yuasa, R., Levinthal, M., Nikaido, H. 1969. *J. Bacteriol.* 100:433–44
201. Takeshita, M., Mäkelä, P. H. 1971. *J. Biol. Chem.* 246:3920–27
202. Jankowski, W., Chojnacki, T., Janczura, E. 1972. *J. Bacteriol.* 112:1420–21
203. Endo, A., Rothfield, L. 1969. *Biochemistry* 8:3500–7
204. Romeo, D., Hinckley, A., Rothfield, L. 1970. *J. Mol. Biol.* 53:491–501
205. Romeo, D., Girard, A., Rothfield, L. 1970. *J. Mol. Biol.* 53:475–90
206. Hinckley, A., Müller, E., Rothfield, L. 1972. *J. Biol. Chem.* 247:2623–28
207. Mühlradt, P. F. 1971. *Eur. J. Biochem.* 18:20–27
208. Lehmann, V. et al 1973. *Eur. J. Biochem.* 32:268–75
209. Eidels, L., Osborn, M. J. 1971. *Proc. Nat. Acad. Sci. USA* 68:1673–77
210. Rick, P. D., Osborn, M. J. 1972. *Proc. Nat. Acad. Sci. USA* 69:3756–60
211. Wu, H. C., Wu, T. C. 1971. *J. Bacteriol.* 105:455–66
212. Sarvas, M. 1971. *J. Bacteriol.* 105:467–71
213. Osborn, M. J., Gander, J. E., Parisi, E., Carson, J. 1972. *J. Biol. Chem.* 247: 3962–72
214. Osborn, M. J., Gander, J. E., Parisi, E. 1972. *J. Biol. Chem.* 247:3973–86
215. Mühlradt, P. F., Menzel, J., Golecki, J. R., Speth, V. 1973. *Eur. J. Biochem.* 35:471–81
216. Kulpa, F. C., Leive, L. 1972. *Membrane Research,* ed. C. F. Fox, 155–60. New York: Academic
217. Bayer, M. E. 1968. *J. Gen. Microbiol.* 53:395–404
218. Bilsky, A. Z., Armstrong, J. B. 1973. *J. Bacteriol.* 113:76–81
219. Schlecht, S., Westphal, O. 1970. *Zbl. Bact. Parasit. Infect. Hyg.* Abt. 1, Orig. 213:356–81
220. Tamaki, S., Sato, T., Matsuhashi, M. 1971. *J. Bacteriol.* 105:968–75
221. Sundar Raj, C. V., Wu, H. C. 1973. *J. Bacteriol.* 114:656–65
222. Takagaki, Y., Kunugita, K., Matsuhashi, M. 1973. *J. Bacteriol.* 113:42–50
223. Nagel de Zwaig, R., Luria, S. E. 1967. *J. Bacteriol.* 94:1112–23
224. Holland, I. B., Threlfall, E. J., Holland, E. M., Darby, V., Samson, A. C. R.

1970. *J. Gen. Microbiol.* 62:371–82
225. Burman, L. G., Nordström, K. 1971. *J. Bacteriol.* 106:1–13
226. Bernstein, A., Rolfe, B., Onodera, K. 1972. *J. Bacteriol.* 112:74–83
227. Hirota, Y., Ryter, A., Jacob, F. 1968. *Cold Spring Harbor Symp. Quant. Biol.* 33:677–93
228. Hirota, Y., Jacob, F., Ryter, A., Buttin, G., Nakai, T. 1968. *J. Mol. Biol.* 35:175–92
229. Inouye, M. 1972. *J. Mol. Biol.* 63:597–600
230. Lazdunski, A., Shapiro, B. M. 1973. *Biochim. Biophys. Acta* 298:59–68
231. Shapiro, B. M., Siccardi, A. G., Hirota, Y., Jacob, F. 1970. *J. Mol. Biol.* 52:75–89
232. Hirota, Y. et al 1969. *C.R. Acad. Sci.* 269:1346–48
233. Normark, S., Boman, H. G., Bloom, G. D. 1971. *Acta Pathol. Microbiol. Scand. B* 79:651–64
234. Rodolakis, A., Thomas, P., Starka, J. 1973. *J. Gen. Microbiol.* 75:409–16
235. Nagai, K., Tamura, G. 1972. *J. Bacteriol.* 112:959–66
236. Adler, H. I., Fisher, W. D., Hardigree, A. A. 1969. *Trans. NY Acad. Sci.* 31:1059–70
237. Reeve, J. N., Groves, D. J., Clark, D. J. 1970. *J. Bacteriol.* 104:1052–64
238. Inouye, M., Pardee, A. B. 1970. *J. Biol. Chem.* 245:5813–19
239. Inouye, M. 1971. *J. Bacteriol.* 106:539–42
240. Starka, J. 1971. *FEBS Lett.* 16:223–25
241. Holland, I. B., Tuckett, S. 1972. *J. Supramol. Struct.* 1:77–97
242. Rolfe, B., Onodera, K. 1971. *Biochem. Biophys. Res. Commun.* 44:767–73
243. Karamata, D., McConnell, M., Rogers, H. J. 1972. *J. Bacteriol.* 111:73–79
244. Boylan, R. J., Mendelson, N. H., Brooks, D., Young, F. E. 1972. *J. Bacteriol.* 110:281–90
245. Guze, L. B., Ed. 1968. *Microbial Protoplasts, Spheroplasts and L-Forms.* Baltimore: Williams and Wilkins
246. Gumpert, J., Schuhmann, E., Taubeneck, U. 1971. *Z. Allg. Mikrobiol.* 11:19–33
247. Gilpin, R. W., Young, F. E., Chatterjee, A. N. 1973. *J. Bacteriol.* 113:486–99
248. Puck, T. T., Lee, H. E. 1955. *Exp. Med.* 101:151–75
249. Rapin, A. M. C., Kalckar, H. M. See Ref. 171, 267–307
250. Matsuyama, T., Uetake, H. 1972. *Virology* 49:359–67
251. Lindberg, A. A. 1973. *Ann. Rev. Microbiol.* 27:205–41

252. Howard, L., Tipper, D. J. 1973. *J. Bacteriol.* 113:1491–504
253. Braun, V., Wolff, H. 1973. *FEBS Lett.* 34:77–80
254. Braun, V., Schaller, K., Wolff, H. 1973. *Biochim. Biophys. Acta* 323:87–97
255. Katz, W., Weidel, W. 1961. *Z. Naturforsch. B* 16:363–68
256. Takeda, K., Uetake, H. 1973. *Virology* 52:148–59
257. Kanegasaki, S., Wright, A. 1973. *Virology* 52:160–73
258. Reske, K., Wallenfels, B., Jann, K. 1973. *Eur. J. Biochem.* 36:167–71
259. Stirm, S., Bessler, W., Fehmel, F., Freund-Mölbert, E. 1971. *J. Virol.* 8:343–46
260. Sonstein, S. A., Hammel, J. M., Bondi, A. 1971. *J. Bacteriol.* 107:499–504
261. Krahn, P. M., O'Callaghan, R. J., Paranchych, W. 1972. *Virology* 47:628–37
262. Kozak, M., Nathans, D. 1971. *Nature New Biol.* 234:209–11
263. Jazwinski, S. M., Marco, R., Kornberg, A. 1973. *Proc. Nat. Acad. Sci. USA* 70:205–9
264. Marvin, D. A., Hohn, B. 1969. *Bacteriol. Rev.* 33:172–209
265. Brinton, C. C. 1965. *Trans. NY Acad. Sci.* 27:1003–54
266. Jacobson, A. 1972. *J. Virol.* 10:835–43
267. Bradley, D. E. 1972. *J. Gen. Microbiol.* 72:303–19
268. Trenkner, E., Bonhoeffer, F., Gierer, A. 1967. *Biochem. Biophys. Res. Commun.* 28:932–39
269. Henry, T. J., Brinton, C. C. 1971. *Virology* 46:754–63
270. Bayer, M. E. 1968. *J. Virol.* 2:346–56
271. Bayer, M. E., Starkey, T. W. 1972. *Virology* 49:236–56
272. Wais, A. C., Goldberg, E. B. 1969. *Virology* 39:153–61
273. Benz, W. C., Goldberg, E. B. 1973. *Virology* 53:225–35
274. Iida, S., Sekiguchi, M. 1971. *J. Virol.* 7:121–26
275. Tomasz, A. 1971. *Informative Molecules in Biological Systems,* ed. L. G. H. Ledoux, 4–18. Amsterdam: North-Holland
276. Tomasz, A. 1972. *Membrane Research,* ed. C. F. Fox, 311–34. New York: Academic
277. Tomasz, A. 1969. *Ann. Rev. Genet.* 3:217–32
278. Tomasz, A., Zanati, E., Ziegler, R. 1971. *Proc. Nat. Acad. Sci. USA* 68:1848–52
279. Singh, R. N. 1972. *J. Bacteriol.* 110:266–72

280. Lanni, Y. T. 1968. *Bacteriol. Rev.* 32: 227–42
281. Lanni, Y. T. 1965. *Proc. Nat. Acad. Sci. USA* 53:969–73
282. Labedan, B., Crochet, M., Legault-Demare, J., Stevens, B. J. 1973. *J. Mol. Biol.* 75:213–34
283. Hantke, K., Braun, V. Unpubl. results
284. Radda, G. K., Vanderkooi, J. 1972. *Biochim. Biophys. Acta* 265:509–49
285. Phillips, S. K., Cramer, W. A. 1973. *Biochemistry* 12:1170–76
286. McCorquodale, D. J., Lanni, Y. T. 1970. *J. Mol. Biol.* 48:133–43
287. Beckman, L. D., Hoffman, M. S., McCorquodale, D. J. 1971. *J. Mol. Biol.* 62:551–64
288. Frank, H., Zarnitz, M. L., Weidel, W. 1963. *Z. Naturforsch. B* 18:281–84
289. Simon, L. D., Anderson, T. F. 1967. *Virology* 32:279–97
290. Anderson, C. W., Eigner, J. 1971. *J. Virol.* 8:869–86
291. Sauri, C. J., Earhart, C. F. 1971. *J. Virol.* 8:856–59
292. Anderson, C. W., Williamson, J. R., Eigner, J. 1971. *J. Virol.* 8:887–93
293. Emrich, J. 1968. *Virology* 35:158–65
294. Weintraub, S., Frankel, F. R. 1972. *J. Mol. Biol.* 70:589–615
295. Peterson, R. F., Kievitt, K. D., Ennis, H. L. 1972. *Virology* 50:520–27
296. Ennis, H. L., Kievitt, K. D. 1973. *Proc. Nat. Acad. Sci. USA* 70:1468–72
297. Sederoff, R., Bolle, A., Goodman, H. M., Epstein, R. H. 1971. *Virology* 46:817–29
298. Siegel, P. J., Schaechter, M. 1973. *J. Virol.* 11:359–67
299. Earhart, C. F., Sauri, C. J., Fletcher, G., Wulff, J. L. 1973. *J. Virol.* 11:527–34
300. Hallick, L. M., Echols, H. 1973. *Virology* 52:105–19
301. Ljungquist, E. 1973. *Virology* 52:120–29
302. Goulian, M. 1971. *Ann. Rev. Biochem.* 40:855–98
303. Klein, A., Bonhoeffer, F. 1972. *Ann. Rev. Biochem.* 41:301–32
304. Fuchs, E., Hanawalt, P. 1970. *J. Mol. Biol.* 52:301–22
305. Strätling, W., Krause, E., Knippers, R. 1973. *Virology* 51:109–19
306. Majumdar, C., Dewey, M., Frankel, F. R. 1972. *Virology* 49:134–44
307. Parson, K. A., Warner, H. R., Anderson, D. L., Snustad, D. P. 1973. *J. Virol.* 11:806–9
308. Georgopoulos, C. P., Hendrix, R. W., Kaiser, A. D., Wood, W. B. 1972. *Nature New Biol.* 239:38–41

309. Takano, T., Kakefuda, T. 1972. *Nature New Biol.* 239:34–37
310. Simon, L. D. 1972. *Proc. Nat. Acad. Sci. USA* 69:907–11
311. Pulitzer, J. F., Yanagida, M. 1971. *Virology* 45:539–54
312. Tzagoloff, H., Pratt, D. 1964. *Virology* 24:372–80
313. Hoffmann-Berling, H., Mazé, R. 1964. *Virology* 22:305–13
314. Hofschneider, P. H., Preuss, A. 1963. *J. Mol. Biol.* 7:450–51
315. Smilowitz, H., Carson, J., Robbins, P. W. 1972. *J. Supramol. Struct.* 1:8–18
316. Ohnishi, Y., Kuwano, M. 1971. *J. Virol.* 7:673–78
317. Roy, A., Mitra, S. 1970. *Nature* 228:365–66
318. Roy, A., Mitra, S. 1970. *J. Virol.* 6:333–39
319. Pratt, D., Tzagoloff, H., Erdahl, W. S. 1966. *Virology* 30:397–410
320. Hohn, B., von Schütz, H., Marvin, D. A. 1971. *J. Mol. Biol.* 56:155–65
321. Ohnishi, Y. 1971. *J. Bacteriol.* 107:918–25
322. Bradley, D. E., Dewar, C. A. 1967. *J. Gen. Virol.* 1:179–88
323. Hoffmann-Berling, H., Mazé, R. 1964. *Virology* 22:305–13
324. Maas, D., Weidel, W. 1963. *Biochim. Biophys. Acta* 78:369–70
325. Tsugita, A., Inouye, M., Terzaghi, E., Streisinger, G. 1967. *J. Biol. Chem.* 243:391–97
326. Taylor, A. 1971. *Nature New Biol.* 234:144–45
327. Inouye, M., Arnheim, N., Sternglanz, R. *J. Biol. Chem.* In press
328. Mukai, F., Streisinger, G., Miller, B. 1967. *Virology* 33:398–402
329. Josslin, R. 1970. *Virology* 40:719–26
330. Josslin, R. 1971. *Virology* 44:101–7
331. Harris, A. W., Mount, D. W. A., Fuerst, C. R., Siminovitch, L. 1967. *Virology* 32:553–69
332. Reader, R. W., Siminovitch, L. 1971. *Virology* 43:607–22
333. Reader, R. W., Siminovitch, L. 1971. *Virology* 43:623–37
334. Studier, F. W. 1972. *Science* 176:367–76
335. Rampini, C. 1969. *C.R. Acad. Sci. D* 269:2040–43
336. Cronan, J. E., Wulff, D. L. 1969. *Virology* 38:241–46
337. Josslin, R. 1971. *Virology* 44:94–100
338. Cronan, J. E., Wulff, D. L. 1971. *Virology* 46:977–78
339. Furrow, M. H., Pizer, L. I. 1968. *J. Virol.* 2:594–605

340. Petersson, R. H. F., Buller, C. S. 1969. *J. Virol.* 3:463–68
341. Sakakibara, Y., Doi, O., Nojima, S. 1972. *Biochem. Biophys. Res. Commun.* 46:1434–40
342. Scandella, C. J., Kornberg, A. 1971. *Biochemistry* 10:4447–56
343. Cronan, J. E., Vagelos, P. R. 1971. *Virology* 43:412–21
344. Herrlich, P., Rahmsdorf, H. J., Pai, S. H., Schweiger, M. *Mol. Gen. Genet.* In press
345. Duckworth, D. H. 1970. *Bacteriol. Rev.* 34:344–63
346. Mitchell, P. 1972. *J. Bioenerg.* 3:5–24
347. Harold, F. M. 1972. *Bacteriol. Rev.* 36:172–230
348. Oesterhelt, D., Stoeckenius, W. 1971. *Nature New Biol.* 333:149–52
349. Blaurock, A., Stoeckenius, W. 1971. *Nature New Biol.* 233:152–55
350. Oesterhelt, D., Hess, B. 1973. *Eur. J. Biochem.* 37:316–26
351. Oesterhelt, D., Stoeckenius, W. 1973. *Proc. Nat. Acad. Sci. USA* 70:2853–57
352. Nomura, M. 1967. *Ann. Rev. Microbiol.* 21:257–84
353. Brandis, H., Smarda, J. 1971. *Bacteriocine und Bacteriocinähnliche Substanzen.* Jena: Fischer
354. Reeves, P. 1972. *The Bacteriocins.* New York: Springer
355. Luria, S. 1973. See Ref. 5, 293–320
356. Di Masi, D. R., White, J. C., Schnaitman, C. A., Bradbeer, C. 1973. *J. Bacteriol.* 115:506–13
357. White, J. C., Di Girolamo, P. M., Fu, M. L., Preston, Y. A., Bradbeer, C. 1973. *J. Biol. Chem.* 248:3978–86
358. Boon, T. 1971. *Proc. Nat. Acad. Sci. USA* 68:2421–25
359. Senior, B. W., Holland, I. B. 1971. *Proc. Nat. Acad. Sci. USA* 68:959–63
360. Bowman, C. M., Sidikaro, J., Nomura, M. 1971. *Nature New Biol.* 234:133–37
361. Boos, W. 1972. *J. Biol. Chem.* 247:5414–24
362. Mesibov, R., Adler, J. 1972. *J. Bacteriol.* 112:315–26
363. Adler, J., Hazelbauer, G. L., Dahl, M. M. 1973. *J. Bacteriol.* 115:824–47
364. Hazelbauer, G. L., Adler, J. 1971. *Nature New Biol.* 230:101–4
365. Kalckar, H. M. 1971. *Science* 174:557–65
366. Bitensky, M. W., Gorman, R. E. 1973. *Progr. Biophys.* 26:409–61
367. FASEB Conference: Membranes in Growth, Differentiation, and Neoplasma. 1973. *Fed. Proc.* 32:19–108
368. Tooze, J., Ed. 1973. *The Molecular Biology of Tumour Viruses.* Cold Spring Harbor Lab.

BACTERIAL TRANSPORT

Winfried Boos[1]

Department of Biological Chemistry, Harvard Medical School and the
Biochemical Research Laboratory, Massachusetts General Hospital
Boston, Massachusetts

CONTENTS

INTRODUCTION

It has become apparent that bacteria have developed different kinds of transport systems which can be distinguished by biochemical and genetic methods as well as by their mode of action. At the present time it might be justified to classify some of these systems in different groups.

(*a*) The phosphoenolpyruvate-dependent sugar phosphotransferase system[2]

[1] Recipient of the Solomon A. Berson Research and Development Award of the American Diabetes Association. The work performed in the author's laboratory was supported by a grant from the National Institutes of Health (GM-18498).

[2] The abbreviations used are: PTS, phosphoenolpyruvate-dependent phosphotransferase system; HPr, histidine-containing phosphocarrier protein of the PTS; PTP, phosphotransfer particle; LDH, D-lactate dehydrogenase; SDS, sodium dodecylsulfate; ANS, anilino-naphthalene sulfonate; DDA$^+$, N,N-dibenzyl,N,N-dimethylammonium cation; DNP, dinitrophenol; FCCP, *p*-trifluoromethoxycarbonyl cyanide phenylhydrazone; CCCP, carbonyl cyanide *m*-chlorophenylhydrazone; PCMB, *p*-chloromercuribenzoate; NEM, N-ethylmaleimid; PMS, phenazin methosulfate; TMG, methyl-1-thio-β-D-galacto-pyranoside.

and other systems which transport solute across the cell membrane via group translocation.[3]

(*b*)   Active transport systems similar to the *E. coli* lactose system. These systems can be observed in isolated membrane vesicles and accumulation of solute can be stimulated by respiratory chain-dependent oxidation of D-lactate and other electron donors. This type of transport system has thus far been observed only in aerobic or facultative aerobic organisms.

(*c*)   Active transport systems sensitive to a cold osmotic shock. These systems are connected with periplasmic substrate binding proteins which are released by the osmotic shock procedure and have thus far been observed only in gram-negative organisms.

One of the surprising generalizations emerging from studies with different transport systems is that a particular substrate can be transported by more than one system (1). Typical examples are transport of galactose (2, 3), arabinose (4), leucine (5, 6), arginine (7, 8), phosphate (9, 10), or K^+ (11) in *Escherichia coli* or histidine in *Salmonella typhimurium* (12). In most of these cases it has been demonstrated by kinetic analysis and isolation of mutants that these systems function independently from each other (1). Yet it has also been discussed for some cases (13, 14) that multiphasic transport kinetics might be indicative of a substrate-dependent alteration of only one system, much in analogy with negative cooperativity of allosteric enzymes (15). Mutants have been isolated which show pleiotropic defects in apparently independent transport systems for histidine in *S. typhimurium* (16) and arginine (17) and leucine (5) in *E. coli*. This indicates that some transport systems share a common component.

Faced with the task of reviewing the recent progress in bacterial transport within a limited space, I have purposely restricted myself to a more detailed and hopefully critical discussion of selected aspects of bacterial transport. Rather than attempting a comprehensive review of published data, I concentrate on certain key questions that have been posed and the types of approaches being used to answer them, with some reference to related observations in cells of higher organisms. For a more comprehensive view the reader must be referred to recent reviews (1, 18–22). Studies of bacterial transport of ions (23), peptides (24–26), or di- and tricarboxylic acids (27–32), and transport in yeast (33–35) are not covered. Also, the role of lipids in transport activity (36–39) or in biosynthesis and in vivo assembly of transport systems (36, 40–42), the relationship of transport systems or transport components to chemotaxis (43, 44), and the possible role of certain components of transport systems in enzyme induction or catabolite repression (45, 46) are not discussed.

THE RECOGNITION OF SOLUTE

Very much in analogy to the Michaelis–Menton treatment of enzymatic reactions, binding of solute on the outside of the osmotic barrier is regarded as the first step

---

[3] The terms group translocation, active transport, and facilitated diffusion are used as previously defined (50, 76, 87).

in the transport process. Thus the determined K_m of uptake is generally assumed to reflect the K_{diss} of binding to the recognition site. Difficulties in interpreting such a correlation arise from the mostly unknown complexities in the translocation step, so that K_m values for uptake do not necessarily represent the K_{diss} of solute binding (47). In the following, emphasis is placed mainly on the biochemical characterization of proteins likely to function in the solute recognition of transport systems. The importance of structural requirements of recognition sites of transport systems has recently been discussed (48).

The Phosphotransferase System

In 1964 (49) a phosphotransferase system (PTS) was discovered which catalyzes the phosphoenolpyruvate-dependent phosphorylation of certain sugars. Several reviews on PTS have appeared (19, 50–52) and the genetic and biochemical evidence for its function in sugar transport has been discussed (53, 54). The system consists of a soluble, cytoplasmic part (Enzyme I, HPr) and a membrane-bound part (Enzyme II complex). The genetic analysis of the unspecific cytoplasmic factors in *E. coli* (55) and *S. typhimurium* (56) has been reported. The composition of the Enzyme II complex differs with the organism. In gram-negative organisms such as *E. coli* and *S. typhimurium* it consists preferentially of two membrane-bound proteins IIA and IIB; while both components are required for in vitro phosphorylation, only one (IIA) carries the sugar specificity. Different, apparently constitutive IIA proteins have been isolated and purified from *E. coli* for glucose, mannose, and fructose (57). The IIB protein apparently can function in combination with a variety of the sugar-specific IIA proteins. IIB from *E. coli* was isolated to a single band on SDS containing polyacrylamide gels. With the same method its molecular weight was determined to be 36,000, and it was estimated that, if pure, this protein would account for 10% of the total *E. coli* membrane proteins (57). Thus in addition to its role in PTS, IIB might also serve a structural role in the membrane. A different type of Enzyme II complex also can be found in *E. coli* (57) and is present exclusively in the gram-positive organism *Staphylococcus aureus*. This complex consists of a membrane-bound EII (apparently different from the IIB protein in the IIA/IIB complex) which interacts with a soluble protein, Factor III. The EII for lactose has been studied extensively in *S. aureus* (58–60). Here the sugar-specific part consists of a membrane-bound EII[lac] which phosphorylates lactose via the soluble Factor III[lac] as phosphoryl donor (61). The membrane-bound EII[lac] has been purified to homogeneity (59). It has a molecular weight of 36,000, is induced by growth on galactose (59), and is inhibited by –SH reagents (60). Its function as recognition site for lactose has been demonstrated by direct binding assay (K_{diss} 0.25 μM) in the absence of phosphorylation (61). Factor III[lac] has also been purified and characterized as a protein of 33,000 with 4 identical subunits (62). A molecular weight of 35,700 and 3 identical subunits were reported subsequently (63). Both EII[lac] and Factor III[lac] are induced by growth on lactose and specifically involved in its phosphorylation (64, 65), even though only EII[lac] functions as the recognition site of the system. Similarly inducible EII complexes have been found in *S. aureus* for mannitol (64) and are indicated to exist for fructose and sorbitol as well (60).

EII activities for mannitol (66), sorbitol (67), and fructose (68) have been studied in *E. coli* mainly from the genetic aspect, and at present it is not clear whether these systems belong to the IIA/IIB, or to the EII/Factor III type. The PTS has been found in several organisms including photosynthetic bacteria (69) and mycoplasma (70) while it appears to be absent in strictly aerobic organisms (71, 72). In *Aerobacter aerogenes* a new PTS component has been found which is involved in a high affinity fructose transport system. This phosphotransfer particle (PTP) is inducible by growth on fructose and replaces HPr in the overall PTS reaction. PTP functions with a membrane-bound EII which is different from an HPr-dependent EII for fructose, also present (73). Despite the differences of PTS in different organisms it is clear that the substrate recognition site is always a membrane-bound protein and the same protein catalyzes the final phosphoryl transfer to the respective sugar.

The M Protein and other Membrane-Bound Recognition Sites of Transport Systems

The M protein is so far the only known component of the lactose transport system of *E. coli* [for reviews on this system see (74–77)]. It was first isolated by Fox & Kennedy (78) and has since been shown by a combination of biochemical and genetic methods to be the recognition site of the lactose transport system (79). This highly hydrophobic protein has a minimal molecular weight of 30,000 (80). Due to the particular isolation procedure, which involved the inactivation of an essential –SH group, the M protein has not yet been isolated in native form. However, the specificity of substrate binding to the M protein has been demonstrated by differential binding of substrate and inorganic phosphate to sonicated *E. coli* pellet (75) and also by ammonium sulfate precipitation of whole cells in the presence of ligand (81). The M protein contains at least one –SH group which is essential for the inhibition of transport by –SH reagents. From the observation that only certain substrates of the lactose system are able to protect against this inhibition, two different binding sites were postulated (82). In this connection it is interesting to note that thiodigalactoside, a substrate with good protecting abilities, is apparently not very well translocated (83) while thiomethylgalactoside or lactose, which have little or no protective abilities, are transported very well (83). Thus the ability of substrates to protect the sensitive –SH group might be inversely related to their rate of transport. Possibly the sensitive –SH group is not even part of the binding site, but rather part of either the translocation or energy coupling mechanism. This possibility is corroborated by the isolation of energy uncoupled mutants for the lactose system, whose mutations reside in the structural gene of the M protein. These mutants exhibit an increased sensitivity against –SH reagents in their transport activity, together with an increased V_{max} of transport, but appear to be essentially unaltered in their K_m of transport (84, 85).

The introduction of the use of membrane vesicles in the study of bacterial transport systems (21, 86, 87) has led to the description of a variety of transport activities of different gram-positive and -negative organisms in an in vitro system of closed membranes freed of cell wall and cytoplasmic constituents (88–94).

Characterization of the recognition site of these systems is largely restricted to the description of apparent K_ms of transport. Only in the case of amino acid transport in *E. coli,* binding components have been solubilized from the membrane vesicles with the non-ionic detergent Brij 36-T. The in vitro binding activity of these proteins has been shown to be sensitive toward –SH reagents (95). Membrane-bound binding components likely to be involved in transport have also been solubilized and purified from yeast and characterized as glyco- or lipoproteins (96–99).

Periplasmic Binding Proteins as Recognition Sites of Transport Systems in Gram-Negative Bacteria

In recent years a variety of periplasmic (100) proteins have been isolated (5, 7, 8, 101–124) from gram-negative bacteria by a method developed by Heppel and associates (125, 126). The evidence for their in vivo location outside the cytoplasmic membrane has been discussed (125, 126). No enzymatic activity has yet been associated with these binding proteins, but the general belief is that they are in some unknown fashion involved in the active transport of substrates they can bind [for earlier discussion of the relationship of these proteins to transport see (1, 20, 54, 126, 127)].

EVIDENCE FOR THEIR INVOLVEMENT IN TRANSPORT   For all systems it has been found that the release of the binding protein is accompanied by a simultaneous reduction of transport activity and, further, that the binding specificity of the protein generally reflects the transport specificity of the system. In several systems it has been shown that binding protein synthesis and transport activity are coregulated (102, 114, 128–130). Perhaps the most convincing argument comes from the combination of biochemical and genetic methods: from a wild-type strain a mutant was isolated with a defect in the β-methylgalactoside transport system and containing a structurally defective galactose-binding protein. Reversion of this mutant by selection for transport-positive phenotype yielded a galactose-binding protein with nearly wild-type properties (131). Similarly, the reversion of a strain of *S. typhimurium* defective in a high affinity histidine transport system yielded a temperature-sensitive transport revertant with a histidine-binding protein (J protein) which exhibited a decreased heat stability in comparison with the wild-type binding protein (132).

Restoration of transport activity by the addition of purified binding protein to shocked cells has been reported (103, 106, 108, 133). However, these reports could not be reproduced[4] in some cases and have since been questioned (8, 126, 134). Efforts to demonstrate binding of purified and radioactively labeled galactose-binding protein to shocked cells or isolated membrane vesicles have failed so far.[5]

Restoration of nonbacterial transport systems or transport related functions has recently been reported: Storelli et al (135) found that the addition of purified sucrase (136) from small intestinal brush ·borders to black lipid membranes

---

[4] H. Rosenberg, personal communication.

[5] W. Boos, A. S. Gordon, and H. R. Kaback, unpublished observations.

greatly increased the transport of sucrose with the simultaneous hydrolysis to glucose and fructose, a mechanism proposed to occur in vivo (137). Addition of a crude arginine-binding protein preparation solubilized from the cell wall of *Neurospora crassa* to an artificial membrane of lipids isolated from the same organism increased transmembranal transport of arginine (138). Consistent with Mitchell's chemiosmotic model (139) of energy coupling for active transport (to be discussed later), the addition to mitochondrial lipids of either cytochrome c plus cytochrome oxidase or ATPase resulted in the generation of a membrane potential (140). The successful reconstitution of Ca^{2+} uptake by the addition of purified Ca^{2+} ATPase from sarcoplasmic reticulum to vesicles of soybean phospholipids was reported (141, 142). Reversal of this reaction, i.e. the generation of ATP from ADP during Ca^{2+} efflux from these vesicles, was demonstrated earlier (143).

PROPERTIES   Molecular weights of the periplasmic binding proteins range from 22,000 (122) to 42,000 (106) and all of the proteins occur in the native form as monomers. Crystallographic data of the sulfate-binding protein indicate an asymmetric shape with an axial ratio of 4 : 1 (144). No lipophilicity can be deduced from the amino acid composition (8, 101–103, 109, 111, 112, 114, 116) though this does not exclude a function as membrane proteins (145). Cysteine or half cystine is present in some (8, 102, 109, 114, 146) but not all binding proteins (101, 103, 111, 112, 116). In contrast to the M protein and the highly hydrophobic binding components isolated from membrane vesicles (95) the binding of substrate to the periplasmic binding proteins is not inhibited by –SH reagents (8, 101–103, 112, 147). Fluorescence studies have shown that a tryptophan residue in the glutamine- (111), galactose- (148), and arabinose-binding proteins (146) undergoes a change in its chemical environment upon binding of substrate. Recent studies of the galactose-binding protein utilizing ultraviolet difference spectroscopy show that substrate interacts directly with a tryptophan residue in the active site (149). Tryptophan is absent in the ribose-binding protein (116). Equilibrium dialysis studies with pure binding proteins generally indicated one binding site per monomer with dissociation constants ranging from 0.01 μM (114) to 50 μM (104). In the case of the glutamine-binding protein the kinetics of binding have been studied by fluorescence measurements in a stopped flow apparatus. The rate constants were determined to be 9.8×10^7 $[M^{-1} \ sec^{-1}]$ for association and 16 $[sec^{-1}]$ for dissociation (111). Reexamination of the binding activity of the galactose-binding protein revealed 2 binding sites per 36,000 molecular weight. Moreover it could be shown that binding is heterogeneous and that Lineweaver–Burk plots to extrapolate to an apparent K_{diss} of 0.1 μM at free galactose concentrations below 0.3 μM and to 10 μM at free galactose concentrations above 0.3 μM (148). Similar observations have been made for mitochondrial Ca^{2+}-binding proteins (150, 151), presumed to be involved in Ca^{2+} transport.

The search for transport-related conformational change in periplasmic binding proteins has yielded some results. The ease in renaturation of the leucine-binding protein after denaturation with different agents has been used as an argument for the protein's ability to easily undergo conformational changes, even though alterations in the native protein could never be demonstrated (152, 153). Fluorescence changes of binding proteins upon binding with substrate have been observed and discussed as indications of conformational changes (111, 144, 148). The electrophoretic mobility of the galactose-binding protein increases upon binding of galactose (148) and fluorescence quenching by potassium iodide is counteracted by substrate (149). These and other observations (154) suggest that the galactose-binding protein exists as an equilibrium mixture of two different forms whose ratio is determined by the substrate concentration (22). Observations which have not been interpreted as conformational changes and which might indicate a similar situation have been made on other binding proteins: a phenylalanine-binding protein alters its elution profile from carboxymethyl Sephadex in the presence of substrate (112) and the cystine-binding protein elutes at two different positions from a salt gradient during DEAE-cellulose chromatography (114). Cooperative binding behavior has been claimed for a phenylalanine-binding protein from yeast (96). At this point reference should be made to the isolation from higher organisms of binding components (150, 151, 155–158) which have been implicated in transport. Also, mutarotases from various species have been isolated and their possible role in glucose transport has been discussed (159, 160).

THE TRANSLOCATION STEP

Still very little is known about the actual translocation mechanism of any one system. In the bacterial PTS all essential components have been isolated and purified (57, 59, 60, 62, 63, 161). Moreover, sugar phosphorylation, which occurs during or subsequent to sugar translocation, has been reconstituted with the purified components (61). Using membrane vesicles and flux-dependent differential labeling, it was demonstrated that transport-dependent sugar phosphorylation occurs unidirectionally (54). Yet the key question remains: is phosphorylation required for sugar translocation, or are translocation and phosphorylation two separable processes? Early studies with mutants of S. typhimurium defective in Enzyme I (unable to phosphorylate) were used to demonstrate the ability of the EII complex to promote facilitated diffusion of glucose and galactose, but only after induction by fucose (50), an observation made previously with glucose transport-negative strains (162). This phenomenon was later explained by the induction by fucose of two PTS-independent transport systems, both of which transport glucose and galactose (130, 163). The ability of the E. coli EII for mannitol to catalyze facilitated diffusion is suggested by the increased exit of endogenously produced mannitol, arguing for a possible separation of translocation and phosphorylation (164). In contrast EII[lac] of S. aureus seems unable to exert facilitated diffusion or translocation independent of phosphorylation (61) even though endogenously produced glucose can be phosphorylated on the inside of the membrane (165).

Other bacterial transport systems functioning via group translocation as PTS does, have been reported for purines (166) and fatty acids (167). Recently a transport mechanism for amino acids based on group translocation has been proposed to occur in kidney. The postulated mechanism involves a translocation-dependent peptidation via a membrane-bound γ-glutamyltranspeptidase with glutathione as the glutamyl donor. The glutamylamino acid is then split intracellularly and glutathione resynthesized in a series of enzymatic reactions (168, 169). Interestingly enough, one of the enzymes involved in the "γ-glutamyl cycle" has been found in bacteria (170, 171); no physiological role could be assigned to this enzyme at the time.

Our knowledge concerning facilitated and active transport systems is essentially still restricted to the formulation of transport models derived from the analysis of kinetic data (21, 75, 77, 139, 154, 172–176). All these models propose in some form a conformational or rotational change of the substrate carrier during the translocation step. Evidence for two states of the M protein in vivo has been obtained by demonstrating differences in the accessibility of its essential –SH group (177). Alterations in the energy-dependent availability of external binding sites have been studied (81). Using membrane vesicles of *E. coli* it has been shown that dansylgalactoside, a fluorogenic substrate of the lactose transport system, drastically changes its fluorescence when the membranes are activated with D-lactate (178). This demonstrates directly the alteration in the chemical environment of the substrate during the translocation step. The differential inactivation of the glucose carrier in erythrocyte membranes with fluorodinitrobenzene in dependence on unilateral preloading with substrate has been interpreted as evidence for the functional role of two conformational states of a transport carrier (179).

The kinetic analysis of binding protein-mediated systems is complicated by the large amount of bound substrate in the periplasmic space in the absence of intracellular accumulation (147). The number of binding protein molecules per cell is on the order of $1–6 \times 10^4$ (102, 106, 109, 147, 180). However, kinetic studies have led to the conclusion that the glutamate- and galactose-binding proteins of *E. coli* are involved only in entry and not in exit[6] (147, 181). The majority of transportologists do not picture periplasmic binding proteins as participating in the actual translocation step.[7] The easy release of these proteins from the cell surface by osmotic shock is the main argument against their carrying out a "membrane-bound" function. However, the definition of membrane proteins depends to some extent on the conditions of isolation (145, 184), and direct interactions of periplasmic binding proteins with the cytoplasmic membrane cannot be excluded. One often-heard[7] interpretation of the function of these proteins is that they might facilitate transport of substrate through the periplasmic space, much in analogy to oxygen transport through hemoglobin solutions (185), delivering it to the "permease proper" located within the cytoplasmic membrane. However, membrane vesicles of *E. coli* which are stripped of periplasmic binding proteins (87) do not transport glutamine, arginine, or diaminopimelic acid (186), amino acids for which only a single (binding protein-dependent) system is present in whole cells; nor do these vesicles bind the

[6] Unidirectional transport activity appears to be a general feature of active transport systems in yeast (182, 183).

[7] In a recent poll of 73 researchers in the transport field, 65 thought periplasmic binding proteins are involved in the overall transport reaction but only 12 believed that they are participating in the actual translocation step through the membrane.

respective amino acid (L. A. Heppel, unpublished observation). Moreover, uptake of these amino acids in whole cells is reduced more than 90% by osmotic shock (8, 111, 114). In contrast, amino acids whose uptake is not reduced by osmotic shock (proline, glycine, alanine) (126) are well transported in membrane vesicles (186) and no periplasmic binding proteins have been isolated for these amino acids. Other amino acids, such as leucine-isoleucine-valine, glutamic acid, and lysine, for which binding proteins have been isolated (7, 102, 113) and which are still transported in membrane vesicles, show only a partial reduction of transport in shocked cells (126). These findings indicate that binding proteins are essential for translocation and are not merely auxiliary systems or K_m factors of a membrane-bound permease. From the analysis of transport mutants it is clear that most binding protein-mediated transport systems are composed of more than one component (5, 105, 119, 129, 132, 187, 188). This is reflected by the finding that active transport is inhibited by –SH reagents even though binding activity of the purified binding proteins is generally not (8, 101–103, 112, 147). It is not clear whether this inhibition affects energy coupling or translocation. A special situation exists in the case of Vitamin B_{12} transport in *E. coli* where two surface-located binding proteins of different nature were isolated, both of which appear necessary for transport (123). One of these binding components has recently been identified as the colicin E receptor (189).

Noncompetitive interaction between transport systems has been observed when sugars are transported simultaneously via PTS and active transport systems, such as the lactose system in *E. coli* (46, 51, 190, 191). Also the PTS-independent galactose and β-methylgalactoside transport systems exhibit this mutual noncompetitive inhibition (192). The nature of this interaction is not understood but might reflect energy requirements or spatial mobility of the transport carriers involved. Transport activity (193, 194) and synthesis of transport components (195) have been related to cell division.

ENERGY COUPLING OF ACTIVE TRANSPORT

Despite many recent studies the molecular mechanism of energy coupling still remains an intriguing problem. Presently two main models are being discussed.

The Respiratory Chain Model

The respiratory chain model (21, 87, 172) pictures the transport carriers as intermediates of the respiratory chain participating in electron flow by the reversible oxidation and reduction of –SH groups. The carrier proteins thus exist in an equilibrium of two different conformations. In the oxidized state the carrier has a high affinity site for the substrate located on the outside of the membrane. With the conformational change due to reduction, the affinity of the carrier is decreased and ligand is released on the interior face of the membrane. This scheme has been derived from transport studies of sugars (172, 196, 197), amino acids (186), succinate (198), and cations (199, 200), and in membrane vesicles of both gram-positive and gram-negative organisms (87). The primary respiratory substrate in *E. coli* was found to be D-lactate (201–203) even though succinate is active to some extent (186, 201) and α-glycerophosphate can serve after induction (126, 201, 204). In addition, artificial electron donors, such as ascorbate-PMS, can drive transport in vesicles of *E. coli* as well as other organisms (89, 93, 205). From the effect of respiratory chain inhibitors on entry and exit of substrate (172, 186) and the spectroscopic analysis of the respiratory components (201), the site of energy coupling by D-lactate in *E. coli* has been placed between the dehydrogenase and cytochrome b_1 (87). The primary electron donor is replaced in LDH-negative mutants by succinate (206). Moreover, different organisms appear to have different primary electron donors: α-glycerophosphate in *S. aureus* (88, 207); NADH in *Bacillus subtilis* (89), *Bacillus licheniformis* (90), and marine Pseudomonads (208); malate in *Azotobacter vinelandii* (91); glucose as well as malate in *Pseudomonas aeruginosa* (92); and NADH and succinate in *Mycobacterium phlei* (93). Two of the transport-linked dehydrogenases for D-lactate and α-glycerophosphate have recently been isolated and purified from *E. coli* (209, 210). Several points which have been presented to prove or disprove the respiratory model, particularly in comparison to Mitchell's model of the proton-motive force, are worthy of detailed discussion.

1. From the historical and biochemical aspect the respiratory model is very attractive. The M protein of lactose transport has been shown to contain a substrate binding site sensitive to –SH reagents (75) and the respiratory model provides a means of relating this alteration in binding to energy coupling. The generality of the model is seen in that most bacterial-active transport systems are sensitive to –SH reagents and the basic scheme of respiration-driven transport can be observed in membranes of a variety of different bacteria.

2. The inhibition patterns of various electron chain inhibitors which, according to the model, should leave the carrier either in the oxidized or reduced state, have been analyzed and their differential effects on solutes fluxes determined (87). From these studies it was concluded that the reduced carrier can catalyze exit as well as facilitated diffusion while the oxidized carrier can only catalyze entry. However, vesicles preloaded with Rb^+ (in the presence of valinomycin) and inhibited with oxamate (a specific inhibitor of LDH, keeping the carrier in the oxidized state) show no net exit of Rb^+, but exchange with external substrate (199). This is not predicted by the model. Also uptake of succinate (198) and Rb^+ (199) (in the presence of valinomycin) is strongly inhibited by NEM when D-lactate, but not when ascorbate-PMS, is used as energy source. In membranes of *S. aureus* –SH reagents cause efflux of preloaded amino acid when α-glycerophosphate, but not when D-lactate, is used as energy donor (211). These findings indicate that the carriers of transport systems in membrane vesicles need not necessarily be –SH-containing proteins, but that D-lactate-driven energy coupling must involve a component sensitive to –SH reagents and not residing in the LDH (209). Indeed, recent studies by Kaback with the –SH reagent Diamine gave evidence for two independent –SH groups being involved in the D-lactate-dependent stimulation of lactose transport (212).

3. One of the weak points of the model is its inability to explain active transport of solute under anaerobic conditions. Yet recent studies show that the transport of

lactose can be driven in anaerobic membrane vesicles by either α-glycerophosphate plus fumarate or formate plus nitrate, after appropriate induction under anaerobic growth conditions (213). Isolation of the enzymatically active dehydrogenase complex and its particulate nature has recently been demonstrated (214). However, the evidence that transport under these conditions is directly coupled to electron flow rather than via the formation of ATP (215) rests solely on the failure to demonstrate stimulation of transport by exogenous ATP in membrane vesicles or shocked cells. In addition, anaerobically grown bacteria still transport lactose even without induction by α-glycerophosphate plus fumarate, demonstrating that ATP in fact can drive transport. ATP-driven reversed electron flow as advocated by Konings & Kaback (213) remains a somewhat unsatisfactory explanation for ATP-driven transport.

4. The strongest argument against a direct coupling of electron flow to transport comes from the observation that uncouplers of oxidative phosphorylation, such as DNP, azide, FCCP, and others, strongly interfere with respiratory-driven transport without affecting respiration (201). Despite the different opinions as to the mechanism of uncoupling (216–218) the obvious interpretation is that respiration cannot be sufficient for energy coupling. The isolation of *etc* (electron transfer coupling) mutants deficient in transport but normal in respiration (206) corroborates this interpretation. To explain the behavior of these mutants Hong and Kaback claim that the transport carriers are linked to shunts off the main portion of the respiratory chain and that these shunts are sensitive to uncoupling agents (206). The observation that the *etc* mutation is apparently closely linked to the gene locus of the membrane-bound Mg^{2+}, Ca^{2+} ATPase suggests another interpretation: transport might be coupled to an intermediate of oxidative phosphorylation or "high energy state of the membrane" (219, 220) at a level prior to ATP formation (221–223). The proton-motive force as an alternative for the high energy intermediate will be discussed later.

5. One of the arguments used for direct coupling to electron flow and against the participation of a transmembranal pH gradient is the unique role of D-lactate as an energy source in *E. coli* membranes. Indeed the ability of respiratory substrates to stimulate transport (201) or to decrease the ANS-dependent fluorescence of membrane vesicles (203) is not paralleled by their rate of oxidation, and both oxamate and 2-hydroxy-3-butynoate, which are specific inhibitors of the membrane-bound D-lactate dehydrogenase, affect D-lactate-driven transport but not NADH oxidation (202). However, this argument is not valid, since it is not the rate of oxidation but the ability to form a transmembranal pH gradient or electrical potential which should be compared with the ability to stimulate transport. Indeed, acidification of the medium upon D-lactate oxidation has been observed with membrane vesicles. With NADH as substrate the results were obscured by alkalization of the medium, probably caused by external oxidation of NADH ($NADH + H^+ + 1/2\,O_2 \rightarrow NAD^+ + H_2O$) (224). Furthermore, the parallel decrease of ANS fluorescence and stimulation of transport (203) may well be interpreted as the effect of a transmembranal potential (225). Thus, the question arises: can a respiratory substrate activate transport while being oxidized on the outside of the

membrane vesicle, or must it first be transported and oxidized on the inside? Membranes of mutants defective in α-glycerophosphate transport show a strongly reduced ability to be stimulated by α-glycerophosphate for proline uptake, even though they are normal with D-lactate as energy source (126). Respiration-dependent uptake of D-lactate and succinate in *E. coli* membranes has been described (198, 226) and it may be significant that the capacity for transport of succinate is five times less than that for D-lactate (226), in accordance with their respective abilities to stimulate transport of other solutes. From the differential influence of succinate on proline fluxes in electron transport particles (inside out) and ghosts (right side out) of *Mycobacterium phlei,* it has been concluded that the location of succinic dehydrogenase on the inside of the membrane is required for accumulation of proline (93, 94). Thus, it may be safe to assume that respiratory substrates stimulate transport only if they are oxidized on the inside of the membrane and that external oxidation cannot be used for energy coupling of transport. Careful studies of the sidedness of the *E. coli* membrane have shown that these vesicles are right side out (87). Nonetheless, significant external oxidation of α-glycerophosphate and succinate can be observed in membrane vesicles but not in spheroplasts of *E. coli,* indicating rearrangement of these enzymes in the membrane during preparation of the vesicles (126). Similarly it has been concluded that the preparation of membrane vesicles does in fact result in the partial transposition of the Mg^{2+}, Ca^{2+} ATPase normally found on the inside of the membrane (227). Such a rearrangement must also be the explanation for restoration of D-lactate-driven transport of proline by simply mixing membranes of two types of mutants: one defective in LDH and the other in electron transport coupling (206). Also, it has been shown that LDH preparations can restore transport in mutants defective in this enzyme (228).

6. Despite the presence of several different transport systems for K^+ in whole cells of *E. coli* (11) membrane vesicles take up hardly any K^+ even in the presence of D-lactate unless valinomycin, the well-known ionophore for K^+ (229, 230), is present (199, 200). This has been used as evidence for the generation by D-lactate oxidation of a membrane potential, interior negative, in response to which K^+ or Rb^+ is accumulated after being made permeable with valinomycin (23, 231). Such a mechanism is corroborated by the observation that valinomycin strongly inhibits transport of D-lactate-driven accumulation of sugars and amino acids but only in the presence of K^+ (21, 199, 200, 232). Yet, even though this inhibition would indicate the collapse of a transport-driving electrical potential, it is difficult to explain why quite different concentrations of valinomycin are required to inhibit the accumulation of different solutes (positively as well as negatively charged amino acids) and why uptake of glucuronate is not inhibited at all (199). Moreover, exit of Rb^+ (in the presence of valinomycin) (199) initiated by uncouplers of the proton conducting type (23, 216) is prevented by the presence of the –SH reagent PCMB, which should not be the case if the valinomycin-Rb^+ complex merely responds to an electrical potential. PCMB would have to interfere with the membrane permeability of the proton conductors, a rather unlikely supposition regarding the effects of –SH reagents on membranes (233). Mutants defective in K^+ uptake are

also defective in D-lactate-driven K^+ transport in membrane vesicles, even in the presence of valinomycin (200). This is incompatible with the model of valinomycin mediating accumulation of K^+ simply in response to a membrane potential and suggests that K^+ uptake, even in the presence of valinomycin, requires additional component(s). A similar situation might be indicated in the case of enterocholine mediated Fe^{3+} transport in E. coli, where apparently additional components are necessary to accomplish transport via the Fe^{3+}-enterocholine complex (234). D-Lactate-stimulated uptake of Rb^+ in the presence of valinomycin is accompanied by the extrusion of Na^+, a phenomenon which does not occur in the case of lactose or amino acid uptake. This and other arguments support the idea that uptake of Rb^+ in the presence of valinomycin is an electrogenic process (interior positive) coupled to the oxidation of D-lactate rather than the response to a transmembranal potential (interior negative) (199).

The original observation by Kaback & Milner (232) that oxidation of D-lactate can drive active transport of proline in the in vitro system of isolated bacterial membrane vesicles has since stimulated an extremely fruitful development in the efforts to understand energy coupling on a molecular level (87). Results obtained since have overstressed the original model of direct coupling to electron flow (172) and made the introduction of some question marks necessary (21). Even though the respiratory model is still largely consistent with the experimental findings, other models favoring a secondary coupling to a high energy intermediate of whatever nature can explain the data as well (219, 227, 235).

The Proton-Motive Force

In recent years it has become fashionable to link energy coupling of active transport in bacteria to Mitchell's chemiosmotic theory, originally developed as an explanation of oxidative phosphorylation and ion transport in mitochondria. It is outside the scope of this review and outside the competence of the author to discuss the chemiosmotic theory in detail; and the reader is referred to comprehensive reviews (139, 216, 236–241). Briefly, the respiratory chain is proposed to consist of hydrogen and electron carriers in alternating sequence organized across the membrane in loops, such that oxidation of substrate brings about the transport of protons outward across the membrane, resulting in the generation of an electrochemical potential, interior negative. Moreover, the bacterial membrane-bound Mg^{2+}, Ca^{2+} ATPase is thought to translocate protons outward, concomitant with the hydrolysis of ATP. Thus, oxidation of substrate via the respiratory chain and ATP hydrolysis both funnel into the same form of energy, i.e. an electrochemical potential, designated the proton-motive force: $\Delta p = \Psi + Z\Delta pH$ (Ψ, electrical potential; $Z\Delta pH$, chemical potential expressed in electrical dimensions $(RT/F)\ln([H^+]_{out}/[H^+]_{in})$. Transport of solute against the concentration gradient is then thought to occur via a H^+ symport or an OH^- antiport. Moreover, by the additional postulation of a transmembranal H^+/Na^+ exchange (139) transport of solute could be linked to Na^+ gradients and solute carriers would play the role of Na^+ symporters. The Na^+-dependent transport of amino acids and other solutes presumably plays a fundamental role in cells of higher organisms (242, 243) where the necessary Na^+

gradient is maintained by the membrane bound Na^+, K^+-dependent ATPase (244, 245). [For opposing views see (176, 246).]

1. One of the attractive features of the model of the proton-motive force is the provision of an elegant explanation for the action of uncouplers, which are thought to function as lipid-soluble proton conductors, inhibiting active transport by collapsing a pH gradient [for review see (139, 216, 247)]. Indeed, there is a good correlation between the ability of these compounds to uncouple and their effectiveness in increasing the conductivity of H^+ in artificial membranes (248), although other correlations fit less well (249). Also, one would predict that the transport inhibiting action of uncouplers would be pH dependent (reflecting the pK of the uncoupler), a phenomenon which indeed has been reported for the uptake of amino acids in yeast (250). The uncoupler-like effect of n-fatty acids on amino acid but not on β-methylglucoside transport in whole cells or membrane vesicles of *Bacillus subtilis* supports this supposition (251). However, even though uncouplers do act as proton conductors it has been debated whether this property is at all related to their uncoupling function (217). Alternatively, a direct interaction with energy coupling proteins has been proposed (252, 253). Moreover, CCCP, one of the classical uncouplers of the proton conductor type, has now been shown to affect D-lactate driven transport in membrane vesicles simply as an –SH reagent, fully reversible by sulfhydryl compounds (212).[8] In addition, dinitrophenol has been shown to decrease the transition temperature of bacterial membranes and is ineffective in inhibition of transport below that temperature even though it still strongly interferes with ATP production (218). These observations demonstrate that the action of uncouplers might be more complex than simple proton conduction.

2. Recently several transport systems in bacteria have been shown to be dependent on cations (255–261). In the case of TMG uptake in *Salmonella typhimurium* it has been shown that Na^+ and TMG uptake are mutually dependent (262). Some transport systems in membrane vesicles also show ion dependence (89, 186, 198). However, in none of these cases could it be demonstrated that an ion gradient was indeed the driving force for accumulation and not merely the expression of an enzymatic ion requirement.

3. Extensive studies on the anaerobic organism *Streptococcus faecalis* by Harold and co-workers [for review see (23, 216)] have demonstrated that glycolyzing cells form a transmembranal pH gradient (interior alkaline) as measured by the distribution of the lipid-soluble weak acid, dimethyloxazolidinedione (263). A measure of the electrical potential resulting from the pH gradient was obtained from the uptake of lipid soluble cations such as dimethyldibenzylammonium (DDA^+) (264). This method for determining electrical potentials had been used previously in mitochondria (239). The argument that K^+ is then taken up in response to this electrical potential is somewhat weakened by the observation that uncouplers of the proton-conducting type discharge accumulated DDA^+ but not K^+. Thus, K^+

---

[8] However, CCCP also loses its ability to function as a proton conductor in the presence of SH reagents (254).

in the absence of valinomycin cannot be in equilibrium with the electrical potential (263, 264). However, a close correlation between H^+ extrusion and K^+ uptake is indicated by the isolation of mutants in the membrane-bound Mg^{2+},Ca^{2+} ATPase showing pleiotropic defects in ATPase activity, H^+ extrusion, and K^+ uptake (264, 265). Transport of amino acids, phosphate, and certain sugars is sensitive to uncouplers of the proton conducting type (266). Valinomycin in the presence of K^+ does not inhibit uptake of these solutes [but compare here to (256)]. This would indicate that uptake of amino acids and sugars, at least in *S. faecalis,* might be driven only by a pH gradient but not by a membrane potential (216). However, the observation that glutamate uptake in glycolyzing cells is not inhibited by the proton conductor tetrachlorosalicylanilide casts some doubt on this conclusion. In addition, transport of solute is still observed in cells depending for their energy supply on arginine metabolism, and these cells do not establish a detectable pH gradient (263).

The internal pH of *E. coli* has been determined with different methods to be dependent on the external pH: at a pH_{ex} below 7.3 the pH_{in} was higher than pH_{ex}, while at pH_{ex} above 7.3 the pH_{in} was lower (267). Similar conclusions, though with a lower iso pH point (pH 6.0) were drawn by measuring buffer capacities of whole and sonicated cells (268). Consistent with these results it was found that oxidizing vesicles of *E. coli* at pH 6.6 did not take up dimethyloxazolidinedione (199), indicating that membrane vesicles might not establish a pH gradient under these conditions. Reports on the determination of a transmembranal potential in the same vesicles by following the accumulation of DDA^+ has led to conflicting results. Vesicles prepared in the presence of Na^+ and absence of K^+ take up large amounts of DDA^+, accounting for an electrical potential of 100 mV interior negative. However, the capacity to take up DDA^+ is ten times less (231) or not observable (199) in vesicles prepared in the presence of K^+ and absence of Na^+, even though both kinds of vesicle preparations are equal in the D-lactate-dependent transport of solute. An explanation offered by Hirata et al (231), that extrusion of Na^+ but not of K^+ might facilitate the accumulation of DDA^+, is not convincing. The K^+ prepared vesicles would then be expected to have a very low uptake of basic amino acids, such as lysine, and this has not been observed (186). Also if D-lactate driven transport occurs via an electrical potential, DDA^+ accumulated in response to this potential should inhibit solute uptake. Such an inhibition has not been observed (199).

Even though it is difficult to attribute energy coupling of D-lactate driven transport in *E. coli* vesicles in an obligatory way to a transmembranal potential it has clearly been shown that an electrical potential can in fact drive transport. Uptake of proline in *E. coli* vesicles and TMG in whole cells of *Streptococcus lactis* can be observed by generating a diffusion potential (interior negative) in the presence of K^+ and valinomycin and in the absence of metabolic coupling (231, 269). In the latter case it has also been shown that the imposition of an artificial proton gradient can temporarily drive TMG accumulation. Moreover, by independently measuring the proton gradient and the electrical potential it could be shown that the calculated proton-motive force under different conditions was indeed proportional to the internal concentration of TMG (270). The role of proton gradients and membrane potentials for amino acid transport has also been discussed in *Staphylococcus aureus* (256, 257) and in yeast (271).

4. Uncoupler sensitive proton extrusion and its relation to solute transport from respiring bacteria have been discussed in *Micrococcus denitrificans* (272) and in *E. coli* (273, 274). As mentioned previously the oxidation of D-lactate by *E. coli* vesicles results in the acidification of the external medium (224). However, it has been debated whether in this case acidification is in fact caused by proton extrusion. Conformational changes of proteins involved in energy coupling might result in the dissociation of previously buried acidic amino acid residues (199). This view is corroborated by the observation that membrane vesicles treated with the detergent Tween 80 and which have lost their osmotic barrier still acidify the medium upon D-lactate oxidation, and this acidification remains sensitive to "proton conductors" (199). Also, the amount and rate of proton extrusion do not account for the amount of K^+ taken up (in the presence of valinomycin) in D-lactate oxidizing vesicles (199). On the other hand, studies with the lactose transport system in intact cells of *E. coli* have clearly established that translocation of substrate occurs concomitantly with proton translocation (273–275). Thus, it was observed that nonmetabolizing, anaerobic cell suspensions took up protons in exchange for K^+ when TMG was added. The alkalization of the medium was increased in cells made permeable to K^+ by EDTA-valinomycin, or in the presence of the permeable counterion SCN^-. Temporary acidification of the external medium was observed by addition of limited amounts of air-saturated solution. The decline of the proton pulse was shown to be dependent on a functioning M protein. Moreover, anaerobic cells incubated with SCN^- took up protons upon the addition of the proton conductor FCCP, presumably due to the diffusion potential created by SCN^-. Similarly protons were taken up when TMG was added instead of FCCP. Thus the M protein acted as proton translocator (274). The stoichiometry of the lactose-proton symport was determined as 1:1 (275). In contrast, mutants uncoupled in energy-linked transport of lactose but normal in its translocation show no cotransport of protons (276). Thus, it appears that proton symport during lactose uptake in *E. coli* is indeed part of an energy coupling mechanism.

As attractive as the idea of the proton-motive force as a unifying mode of energy coupling for transport appears, it seems premature at the present time to postulate its general importance. Several observations, mainly derived from transport studies in membrane vesicles of *E. coli,* cannot be explained by the chemiosmotic hypothesis. The diversity of different kinds of transport systems in their genetic and mechanistic appearance may indicate that the chemiosmotic model may be one, but by no means the only, form of energy coupling for active transport in bacteria.

Mutants of E. coli Defective in the Mg^{2+}, Ca^{2+} ATPase

The membrane-bound Mg^{2+}, Ca^{2+} ATPase is ubiquitous in bacterial membranes. The enzyme is involved in the final steps of oxidative phosphorylation (277) and energy-dependent transhydrogenation (278, 279), and several observations link it to active transport as well. When discussing the function of this enzyme in transport it is important to specify the organism. It is obvious that the ATPase of strictly anaerobic bacteria such as *S. faecalis* will be more important for energy coupling than the ATPase of a strictly aerobic or facultative aerobic organism where the enzyme might be dispensable for energy coupling of transport. Indeed, membrane vesicles of *E. coli* exhibit respiratory driven transport without significant oxidative phosphorylation, even though they still have ATPase activity. The *E. coli* enzyme has been purified and studied extensively (279–284). Its complex nature is reflected by the multiple components of the purified enzyme (284). Evidence for the involvement of the enzyme in energy coupling is mainly derived from mutants defective in oxidative phosphorylation or in ATPase activity. The conclusions from these studies are at present somewhat controversial: Schairer & Haddock (221) found that transport of TMG via the lactose transport system was not reduced in their ATPase mutant, but in contrast to the wild type, accumulation had become dependent on respiration. In agreement with this finding is the observation that anaerobic but not aerobic cells of *E. coli* are sensitive in proline uptake to arsenate (219). Prezioso et al (285), studying lactose uptake in whole cells and membrane vesicles of the ATPase mutant AN120 (277), are in general agreement: transport of lactose in membrane vesicles is stimulated by D-lactate to a similar degree as in the wild type. On the other hand Simoni and Shallenberger isolated an ATPase mutant, DL-54, which exhibits a strongly reduced transport activity for proline in whole cells as well as in membrane vesicles (286). Using the same mutant Berger reported that proline transport is normal and is even stimulated in whole cells by D-lactate (222). Yet another ATPase mutant, NR70, was recently isolated by Rosen (287). Uptake of TMG and several amino acids was reduced to varying degrees and no transport activity of proline and lysine could be found in vesicles even in the presence of D-lactate. Moreover, while DCCD inhibited transport of proline in membrane vesicles of the wild type to some extent, it stimulated the uptake in the mutant. Similar mutations affecting the ATPase either in its activity or sensitivity towards DCCD were isolated by van Thienen & Postma (227). Here again, DCCD could restore transport of serine in vesicles of mutants insensitive to DCCD in their ATPase activity. These observations are reminiscent of the stimulation by DCCD of transhydrogenase activity (288), in which case an additional function of ATPase in stabilizing a "high energy state" of the membrane was discussed. The complexity of the ATPase can also be seen in the properties of two mutants defective in ATP-driven transhydrogenase (289). One such mutant was found to be defective in ATPase while the other was altered in its sensitivity to DCCD. Both of these mutants map in the vicinity of the *ilv* locus of the *E. coli* chromosome, as do AN120 (277), NR70 (287), the mutants isolated by Schairer & Haddock (221) as well as the ones isolated by van Thienen & Postma (227).

Studies on the influence of defects in ATPase activity on transport systems mediated by periplasmic binding proteins have been done on the glutamine (222) and the β-methylgalactoside transport systems (223). Here too the results do not reveal a general pattern. In AN120 (277) and NR70 (287) transport mediated by the galactose-binding protein was strongly reduced but could be partially restored by the addition of D-lactate (223; and Boos, unpublished). Moreover a mutant defective in ubiquinone biosynthesis and unable to oxidize D-lactate (290) still exhibited transport activity (223). Thus at least for the galactose-binding protein mediated system the Mg^{2+},Ca^{2+} ATPase is not required for energy coupling but can drive accumulation. In Berger's studies (222) with mutant DL54 (286) D-lactate did stimulate the nonbinding protein-mediated uptake of proline but did not stimulate the binding protein-mediated uptake of glutamine. Thus it was concluded that transport systems observable in membrane vesicles could be driven by respiration as well as by ATP hydrolysis while binding protein-mediated systems are driven only by ATP directly (222). The somewhat contradictory results might be caused by the use of ATPase mutants defective in different components of this multifunctional enzyme. Yet despite these uncertainties it is clear that the *E. coli* ATPase can in fact accomplish energy coupling via ATP hydrolysis even though the enzyme might not be obligatory. Indeed, direct stimulation of serine transport in membrane vesicles by ATP has been reported (227). Thus in conclusion it appears that energy coupling, at least in some systems, can be accomplished by either of two modes: ATP hydrolysis or respiration without the formation of ATP. This scheme bears close analogy to the energy-dependent reduction of NADP by NADH in *E. coli* membranes. The direct energy source is most likely a high energy intermediate which can be dissipated by uncouplers and which can be generated by either of two pathways, respiratory chain oxidation of metabolites or ATP hydrolysis in the presence of the Mg^{2+},Ca^{2+} ATPase (279, 289). It is the "high energy intermediate" which is the crux of the matter. Is it an electrochemical potential across the membrane (139, 216) or a high energy conformation of energy transducing proteins (219, 220, 235)? Obviously, the present state of knowledge still precludes the answer of this question. The problems involved in studying such a multicomponent, membrane-localized system are precisely the ones that have caused elucidation of the mechanism of oxidative phosphorylation to be one of the most persistent and elusive problems in biochemistry.

Literature Cited

1. Oxender, D. L. 1972. *Ann. Rev. Biochem.* 41:777–814
2. Rotman, B., Ganesan, A. K., Guzman, R. 1968. *J. Mol. Biol.* 36:247–60
3. Kalckar, H. M. 1971. *Science* 174:557–65
4. Brown, C. E., Hogg, R. W. 1972. *J. Bacteriol.* 111:606–13
5. Rahmanian, M., Claus, D. R., Oxender, D. L. 1973. *J. Bacteriol.* 116:1258–66
6. Guardiola, J., DeFelice, M., Klopotowski, T., Jaccarino, M. Manuscript submitted
7. Rosen, B. P. 1971. *J. Biol. Chem.* 246:3653–62
8. Rosen, B. P. 1973. *J. Biol. Chem.* 248:1211–18
9. Medveczky, N., Rosenberg, H. 1971.

Biochim. Biophys. Acta 241:494–506
10. Willsky, G. R., Bennett, R. L., Malamy, M. H. 1973. *J. Bacteriol.* 113:529–39
11. Epstein, W., Kim, B. S. 1971. *J. Bacteriol.* 108:639–44
12. Ferro-Luzzi Ames, G. 1972. *Membrane Research,* ed. C. F. Fox, 409–26 New York: Academic
13. Marcus, M., Halpern, Y. S. 1969. *J. Bacteriol.* 97:1118–28
14. Reid, K. G., Utech, N. M., Holden, J. T. 1970. *J. Biol. Chem.* 245:5261–72
15. Conway, A., Koshland, D. E. 1968. *Biochemistry* 7:4011–23
16. Küstü, S. G., Ames, G. F. 1973. *J. Bacteriol.* 116:107–13
17. Rosen, B. P. 1973. *J. Bacteriol.* 116:627–35

18. Oxender, D. L. 1972. *Metabolic Pathways*, ed. L. E. Hokin, VI:133–85. New York: Academic
19. Roseman, S. 1972. *Metabolic Pathways*, ed. L. E. Hokin, VI:41–89. New York: Academic
20. Heppel, L. A., Rosen, B. P. 1973. *Bacterial Membranes and Walls*, ed. L. Leive, 209–39. New York: Dekker
21. Hong, J. S., Kaback, H. R. 1973. *Chemical Rubber Company Critical Reviews in Microbiology*, ed. A. L. Laskin, H. Leehevalier, 2:333–76. Cleveland: Chem. Rubber Co.
22. Boos, W. 1974. *Curr. Top. Membranes Transp.* 5: In press
23. Harold, F. M., Altendorf, K. 1974. *Curr. Top. Membranes Transp.* 5: In press
24. Ames, B. N., Ferro-Luzzi Ames, G., Young, J. D., Tsuchiya, D., Lecocq, J. 1973. *Proc. Nat. Acad. Sci. USA* 70: 456–58
25. Fickel, T. E., Gilvarg, C. 1973. *Nature New Biol.* 241:161–63
26. Payne, J. W., Gilvarg, C. 1971. *Advan. Enzymol.* 35:187–244
27. Willecke, K., Gries, E. M., Oehr, P. 1973. *J. Biol. Chem.* 248:807–14
28. McMillen, M. N., Willecke, K., Pardee, A. B. 1972. *The Molecular Basis of Biological Transport*, ed. J. F. Woessner, F. Huijing, 3:249–66. New York: Academic
29. Reuser, A. J. J., Postma, P. W. 1973. *Eur. J. Biochem.* 33:584–92
30. Imai, K., Iijima, T., Hasegawa, T. 1973. *J. Bacteriol.* 114:961–65
31. Ghei, O. K., Kay, W. W. 1973. *J. Bacteriol.* 114:65–79
32. Murakawa, S., Izaki, K., Takahashi, H. 1972. *Agr. Biol. Chem.* 36:2487–93
33. Scarborough, G. A. 1973. *Int. Rev. Cytol.* 34:103–22
34. Grenson, M. 1973. *Genetics of Industrial Microorganisms*, ed. Z. Vaněk, Z. Hoštálek, J. Cudlin, 2:179–93. Prague: Academia
35. Kotyk, A. 1973. *Biochim. Biophys. Acta* 300:183–210
36. Fox, C. F. 1972. *Membrane Molecular Biology*, ed. C. F. Fox, A. Keith, 345–85. Stamford, Conn: Sinauer
37. Holden, J. T., Utech, N. M., Hegeman, G. D., Kenyon, C. N. 1973. *Biochem. Biophys. Res. Commun.* 50:266–72
38. Rosen, B. P., Hackette, S. L. 1972. *J. Bacteriol.* 110:1181–89
39. Kito, M., Aibara, S., Kato, M., Ishinaga, M., Hata, T. 1973. *Biochim. Biophys. Acta* 298:69–74
40. Willecke, K., Mindich, L. 1971. *J. Bacteriol.* 106:514–18
41. Robbins, A. R., Rotman, B. 1972. *Proc. Nat. Acad. Sci. USA* 69:2125–29
42. Glaser, M., Bayer, W. H., Bell, R. M., Vagelos, P. R. 1973. *Proc. Nat. Acad. Sci. USA* 70:385–89
43. Hazelbauer, G. L., Adler, J. 1971. *Nature New Biol.* 230:101–4
44. Adler, J., Hazelbauer, G. L., Dahl, M. 1973. *J. Bacteriol.* 115:824–47
45. Saier, M. H., Simoni, R. D., Roseman, S. 1970. *J. Biol. Chem.* 245:5870–73
46. Saier, M. H., Roseman, S. 1972. *J. Biol. Chem.* 247:972–75
47. Schachter, H. 1973. *J. Biol. Chem.* 248: 974–76
48. Christensen, H. N. 1973. *J. Bioenerg.* 4:31–61
49. Kundig, W., Ghosh, S., Roseman, S. 1964. *Proc. Nat. Acad. Sci. USA* 52: 1067–74
50. Roseman, S. 1969. *J. Gen. Physiol.* 54: 138s–84s
51. Roseman, S. 1972. *The Molecular Basis of Biological Transport*, ed. J. F. Woessner, F. Huijing, 181–218. New York: Academic
52. Simoni, R. D. 1972. *Membrane Molecular Biology*, ed. C. F. Fox, A. D. Keith, 289–322. Stamford, Conn: Sinauer
53. Lin, E. C. C. 1970. *Ann. Rev. Genet.* 4:225–62
54. Kaback, H. R. 1970. *Ann. Rev. Biochem.* 39:561–98
55. Epstein, W., Jewett, S., Fox, C. F. 1970. *J. Bacteriol.* 104:793–97
56. Cordaro, J. C., Roseman, S. 1972. *J. Bacteriol.* 112:17–29
57. Kundig, W., Roseman, S. 1971. *J. Biol. Chem.* 246:1407–18
58. Hengstenberg, W. 1970. *FEBS Lett.* 8:277–80
59. Korte, T., Hengstenberg, W. 1971. *Eur. J. Biochem.* 23:295–302
60. Simoni, R. D., Nakazawa, T., Hays, J. B., Roseman, S. 1973. *J. Biol. Chem.* 248: 932–40
61. Simoni, R. D., Hays, J. B., Nakazawa, T., Roseman, S. 1973. *J. Biol. Chem.* 248:957–65
62. Schrecker, O., Hengstenberg, W. 1971. *FEBS Lett.* 13:209–12
63. Hays, J. B., Simoni, R. D., Roseman, S. 1973. *J. Biol. Chem.* 248:941–56
64. Simoni, R. D., Smith, M. F., Roseman, S. 1968. *Biochem. Biophys. Res. Commun.* 31:804–11
65. Hengstenberg, W., Penberthy, W. K., Hill, K. L., Morse, M. L. 1969. *J. Bacteriol.* 99:383–88

66. Solomon, E., Lin, E. C. C. 1972. *J. Bacteriol.* 111:566–74
67. Lengeler, J., Lin, E. C. C. 1973. *J. Bacteriol.* 112:840–48
68. Kornberg, H. L. 1972. *Molecular Basis of Biological Transport,* ed. J. F. Woessner, F. Huijing, 157–80. New York: Academic
69. Saier, M. H., Feucht, B. U., Roseman, S. 1971. *J. Biol. Chem.* 246:7819–21
70. Cirillo, V. P., Razin, S. 1973. *J. Bacteriol.* 113:212–17
71. Romano, A. H., Eberhard, S. J., Dingle, S. L., McDowell, T. D. 1970. *J. Bacteriol.* 104:808–13
72. Phibbs, P. V., Eagon, R. G. 1970. *Arch. Biochem. Biophys.* 138:470–82
73. Walter, R. W., Anderson, R. L. 1973. *Biochem. Biophys. Res. Commun.* 52:93–97
74. Kepes, A. 1970. *Curr. Top. Membranes Transp.* 1:101–34
75. Kennedy, E. P. 1970. *The Lactose Operon,* ed. J. R. Beckwith, D. Zipser, 49–92. Cold Spring, New York: Cold Spring Harbor Lab.
76. Lin, E. C. C. 1971. *Structure and Function of Biological Membranes,* ed. L. I. Rothfield, 285–341. New York: Academic
77. Kepes, A. 1971. *J. Membrane Biol.* 4:87–112
78. Fox, C. F., Kennedy, E. P. 1965. *Proc. Nat. Acad. Sci. USA* 54:891–99
79. Fox, C. F., Carter, J. R., Kennedy, E. P. 1967. *Proc. Nat. Acad. Sci. USA* 57:698–705
80. Jones, T. H. D., Kennedy, E. P. 1969. *J. Biol. Chem.* 244:5981–87
81. Bernard-Bentaboulet, M., Kepes, A. 1973. *Biochim. Biophys. Acta* 307:197–211
82. Carter, J. R., Fox, C. F., Kennedy, E. P. 1968. *Proc. Nat. Acad. Sci. USA* 60:725–32
83. Kepes, A., Cohen, G. N. 1962. *The Bacteria,* ed. I. C. Gunsalus, R. Y. Stanier, IV:179–221. New York: Academic
84. Wong, P. T. S., Kashket, E. R., Wilson, T. H. 1970. *Proc. Nat. Acad. Sci. USA* 65:63–69
85. Wilson, T. H., Kusch, M. 1972. *Biochim. Biophys. Acta* 255:786–97
86. Kaback, H. R. 1971. *Methods Enzymol.* XXII:99–120
87. Kaback, H. R. 1972. *Biochim. Biophys. Acta* 265:367–416
88. Short, S. A., White, D. C., Kaback, H. R. 1972. *J. Biol. Chem.* 247:298–304
89. Konings, W. N., Freese, E. 1972. *J. Biol. Chem.* 247:2408–18
90. MacLeod, R. A., Thurman, P., Rogers, H. J. 1973. *J. Bacteriol.* 113:329–40
91. Barnes, E. M. 1972. *Arch. Biochem. Biophys.* 152:795–99
92. Stinnett, J. D., Guymon, L. F., Eagon, R. G. 1973. *Biochem. Biophys. Res. Commun.* 52:284–90
93. Hirata, H., Brodie, A. F. 1972. *Biochem. Biophys. Res. Commun.* 47:633–38
94. Asano, A., Cohen, N. S., Baker, R. F., Brodie, A. F. 1973. *J. Biol. Chem.* 248:3386–97
95. Gordon, A. S., Lombardi, F. J., Kaback, H. R. 1972. *Proc. Nat. Acad. Sci. USA* 69:358–62
96. Vořišek, J. 1972. *Biochim. Biophys. Acta* 290:256–66
97. Hořák, J., Kotyk, A. 1973. *Eur. J. Biochem.* 32:36–41
98. Stuart, W. D., DeBusk, A. G. 1971. *Arch. Biochem. Biophys.* 144:512–18
99. Vořišek, J. 1973. *Folia Microbiol.* 18:17–21
100. Mitchell, P. 1961. *Biological Structure and Function,* ed. T. W. Goodwin, O. Lindberg, II:581–603. London: Academic
101. Pardee, A. B. 1966. *J. Biol. Chem.* 241:5886–92
102. Penrose, W. R., Nichoalds, G. E., Piperno, J. R., Oxender, D. L. 1968. *J. Biol. Chem.* 243:5921–28
103. Anraku, Y. 1968. *J. Biol. Chem.* 243:3116–35
104. Hogg, R. W., Englesberg, E. 1969. *J. Bacteriol.* 100:423–32
105. Schleif, R. 1969. *J. Mol. Biol.* 46:185–96
106. Medveczky, N., Rosenberg, H. 1970. *Biochim. Biophys. Acta* 211:158–68
107. Nishimune, T., Hayashi, R. 1971. *Biochim. Biophys. Acta* 244:573–83
108. Wilson, O. H., Holden, J. T. 1969. *J. Biol. Chem.* 244:2743–49
109. Lever, J. E. 1972. *J. Biol. Chem.* 247:4317–26
110. Rosen, B. P., Vasington, F. D. 1971. *J. Biol. Chem.* 246:5351–60
111. Weiner, J. H., Heppel, L. A. 1971. *J. Biol. Chem.* 246:6933–41
112. Kuzuya, H., Bromwell, K., Guroff, G. 1971. *J. Biol. Chem.* 246:6371–80
113. Barash, H., Halpern, Y. S. 1971. *Biochem. Biophys. Res. Commun.* 45:681–88
114. Berger, E. A., Heppel, L. A. 1972. *J. Biol. Chem.* 247:7684–94
115. Tsay, S. S., Brown, K. K., Gaudy, E. T. 1971. *J. Bacteriol.* 108:82–88
116. Aksamit, R., Koshland, D. E. 1972. *Biochem. Biophys. Res. Commun.* 48:1348–53

117. Anraku, Y. 1971. *J. Biochem. Tokyo* 69:243–45
118. Fukui, S., Miyairi, S. 1970. *J. Bacteriol.* 101:685–91
119. Kellerman, O., Szmeloman, S., Schwartz, M. In preparation
120. Lo, T. C. Y., Rayman, M. K., Sanwal, B. D. 1972. *J. Biol. Chem.* 247:6323–31
121. DiGirolamo, P. M., Kadner, R. J., Bradbeer, C. 1971. *J. Bacteriol.* 106:751–57
122. Taylor, R. T., Norrell, S. A., Hanna, M. L. 1972. *Arch. Biochem. Biophys.* 148:366–81
123. White, J. C., DiGirolamo, P. M., Fu, M. L., Preston, Y. A., Bradbeer, C. 1973. *J. Biol. Chem.* 248:3978–86
124. Matsuura, A., Iwashima, A., Nose, Y. 1973. *Biochem. Biophys. Res. Commun.* 51:241–46
125. Heppel, L. A. 1971. *Structure and Function of Biological Membranes,* ed. L. I. Rothfield, 223–47. New York: Academic
126. Heppel, L. A., Rosen, B. P., Friedberg, I., Berger, E. A., Weiner, J. H. 1972. *Molecular Basis of Biological Transport,* ed. J. F. Woessner, F. Huijing, 133–49. New York: Academic
127. Heppel, L. A. 1969. *J. Gen. Physiol.* 54:95s–113s
128. Ferro-Luzzi Ames, G., Lever, J. E. 1970. *Proc. Nat. Acad. Sci. USA* 66:1096–1103
129. Ohta, N., Galsworthy, P. R., Pardee, A. B. 1971. *J. Bacteriol.* 105:1053–62
130. Lengeler, J., Hermann, K. O., Unsöld, H. J., Boos, W. 1971. *Eur. J. Biochem.* 19:457–70
131. Boos, W. 1972. *J. Biol. Chem.* 247:5414–24
132. Ferro-Luzzi Ames, G., Lever, J. E. 1972. *J. Biol. Chem.* 247:4309–16
133. Anraku, A., Kobayashi, H., Amanuma, H., Yamaguchi, A. 1974. *J. Biochem. Tokyo.* In press
134. Pardee, A. B. 1970. *Permeability and Function of Biological Membranes,* ed. L. Bolis, A. Katchalsky, R. D. Keynes, W. R. Loewenstein, B. A. Pethica, 86–93. Amsterdam: North-Holland
135. Storelli, C., Vögeli, H., Semenza, G. 1972. *FEBS Lett.* 24:287–92
136. Cogoli, A., Mosimann, H., Vock, C., Von Balthazar, A. K., Semenza, G. 1972. *Eur. J. Biochem.* 30:7–14
137. Semenza, G. 1972. *Transport Across the Intestine,* ed. W. L. Burland, P. D. Samuel, 78–92. Baltimore: Williams & Wilkins
138. Stuart, W. D., DeBusk, A. G. 1973. *Biochem. Biophys. Res. Commun.* 52:1046–50
139. Mitchell, P. 1970. *Symp. Soc. Gen. Microbiol., 20th, 1970* 20:121–66
140. Jasaitis, A. A., Nemeček, J. B., Severina, J. J., Skulachev, V. P., Smirnova, S. M. 1972. *Biochim. Biophys. Acta* 275:485–90
141. Racker, E. 1972. *J. Biol Chem.* 247:8198–8200
142. Meissner, G., Fleischer, S. 1973. *Biochem. Biophys. Res. Commun.* 52:913–20
143. Makinose, M., Hasselbach, W. 1971. *FEBS Lett.* 12:271–72
144. Langridge, R., Shinagawa, H., Pardee, A. B. 1970. *Science* 169:59–61
145. Guidotti, G. 1972. *Ann. Rev. Biochem.* 41:731–52
146. Parson, R. G., Hogg, R. W. Manuscript in preparation
147. Parnes, J. R., Boos, W. 1973. *J. Biol. Chem.* 248:4436–45
148. Boos, W., Gordon, A. S., Hall, R. E., Price, H. D. 1972. *J. Biol. Chem.* 247:917–24
149. McGowan, E. B., Silhavy, T. J. 1973. *Fed. Proc.* 32:457 (Abstr.)
150. Gazzotti, P., Vasington, F. D., Carafoli, E. 1972. *Biochem. Biophys. Res. Commun.* 47:808–13
151. Gomez-Puyou, A., Tueda de Gomez-Puyou, M., Becker, G., Lehninger, A. 1972. *Biochem. Biophys. Res. Commun.* 47:814–19
152. Penrose, W. R., Zand, R., Oxender, D. L. 1970. *J. Biol. Chem.* 245:1432–37
153. Berman, K., Boyer, P. D. 1972. *Biochemistry* 11:4650–57
154. Rotman, B., Ellis, J. H. 1972. *J. Bacteriol.* 111:791–96
155. Ingersoll, R. J., Wasserman, R. H. 1971. *J. Biol. Chem.* 246:2808–14
156. Wolff, D. J., Siegel, F. L. 1972. *Arch. Biochem. Biophys.* 150:578–84
157. Thomas, L., 1973. *Biochim. Biophys. Acta* 291:454–64
158. Shamoo, A. E., Albers, R. W. 1973. *Proc. Nat. Acad. Sci. USA* 70:1191–94
159. Mulhern, S. A., Fishman, P. H., Kusiak, J. W., Bailey, J. M. 1973. *J. Biol. Chem.* 248:4163–73
160. Bailey, J. M., Pentohev, P. G., Fishman, P. H., Mulhern, S. A., Kusiak, J. W. 1973. *Carbohydrates in Solution,* ed. R. F. Gould, 264–308. Washington: Am. Chem. Soc.
161. Anderson, B., Weigel, N., Kundig, W., Roseman, S. 1971. *J. Biol. Chem.* 246:7023–33
162. Kamogawa, A., Kurahashi, K. 1967. *J. Biochem. Tokyo* 61:220–30

163. Saier, M. H., Bromberg, F. G., Roseman, S. 1973. *J. Bacteriol.* 113:512–14
164. Solomon, E., Miyai, K., Lin, E. C. C. 1973. *J. Bacteriol.* 114:723–28
165. Button, D. K., Egan, J. B., Hengstenberg, W., Morse, M. L. 1973. *Biochem. Biophys. Res. Commun.* 52:850–55
166. Hochstadt-Ozer, J. 1972. *J. Biol. Chem.* 247:2419–26
167. Klein, K., Steinberg, R., Fiethen, B., Overath, P. 1971. *Eur. J. Biochem.* 19:442–50
168. Meister, A. 1973. *Science* 180:33–39
169. Tate, S. S., Ross, L. L., Meister, A. 1973. *Proc. Nat. Acad. Sci. USA* 70:1447–49
170. Doolittle, R. F., Armentrout, R. W. 1968. *Biochemistry* 7:516–21
171. Szewczuk, A., Mulczyk, M. 1969. *Eur. J. Biochem.* 8:63–67
172. Kaback, H. R., Barnes, E. M. 1971. *J. Biol. Chem.* 246:5523–31
173. Wilson, T. H., Kashket, E. R., Kusch, M. 1972. *The Molecular Basis of Biological Transport,* ed. J. F. Woessner, F. Huijing, 219–47. New York: Academic
174. Heinz, E. H., Geck, P., Wilbrandt, W. 1972. *Biochim. Biophys. Acta* 255:442–61
175. Lieb, W. R., Stein, W. D. 1972. *Biochim. Biophys. Acta* 265:187–207
176. Kimmich, G. 1973. *Biochim. Biophys. Acta* 300:31–78
177. Yariv, J., Kalb, A. J., Katchalski, E., Goldman, R., Thomas, E. W. 1969. *FEBS Lett.* 5:173–76
178. Reeves, J. P., Shechter, E., Weil, R., Kaback, H. R. 1973. *Proc. Nat. Acad. Sci. USA* 70:2722–26
179. Edwards, P. A. W. 1973. *Biochim. Biophys. Acta* 307:415–18
180. Pardee, A. B., Prestidge, L. S., Whipple, M. B., Dreyfuss, J. 1966. *J. Biol. Chem.* 241:3962–69
181. Halpern, Y. S., Barash, H., Druck, K. 1973. *J. Bacteriol.* 113:51–57
182. Crabeel, M., Grenson, M. 1970. *Eur. J. Biochem.* 14:197–204
183. Kotyk, A., Říhová, L. 1972. *Biochim. Biophys. Acta* 288:380–89
184. Coleman, R. 1973. *Biochim. Biophys. Acta* 300:1–30
185. Scholander, P. F. 1960. *Science* 131:585–90
186. Lombardi, F. J., Kaback, H. R. 1972. *J. Biol. Chem.* 247:7844–57
187. Hogg, R. W. 1971. *J. Bacteriol.* 105:604–8
188. Boos, W. 1969. *Eur. J. Biochem.* 10:66–73
189. Di Masi, D. R., White, J. C., Schnaitman, C. A., Bradbeer, C. 1973. *J. Bacteriol.* 115:506–13
190. Koch, A. L. 1971. *J. Mol. Biol.* 59:447–59
191. Kornberg, H. L. 1973. *Proc. Roy. Soc. London B* 183:105–23
192. Wilson, D. 1974. *J. Biol. Chem.* 249: In press
193. Kubitschek, H. E., Freedman, M. L., Silver, S. 1971. *Biophys. J.* 11:787–97
194. Ohki, M. 1972. *J. Mol. Biol.* 68:249–64
195. Shen, B. H. P., Boos, W. 1973. *Proc. Nat. Acad. Sci. USA* 70:1481–85
196. Kerwar, G. K., Gordon, A. S., Kaback, H. R. 1972. *J. Biol. Chem.* 247:291–97
197. Dietz, G. W. 1972. *J. Biol. Chem.* 247:4561–65
198. Rayman, M. K., Lo, T. C. Y., Sanwal, B. D. 1972. *J. Biol. Chem.* 247:6332–39
199. Lombardi, F. J., Reeves, J. P., Kaback, H. R. 1973. *J. Biol. Chem.* 248:3551–65
200. Bhattacharyya, P., Epstein, W., Silver, S. 1971. *Proc. Nat. Acad. Sci. USA* 68:1488–92
201. Barnes, E. M., Kaback, H. R. 1971. *J. Biol Chem.* 246:5518–22
202. Walsh, C. T., Abeles, R. H., Kaback, H. R. 1972. *J. Biol. Chem.* 247:7858–63
203. Reeves, J. P., Lombardi, F. J., Kaback, H. R. 1972. *J. Biol. Chem.* 247:6204–11
204. Dietz, G. W. 1971. *Fed. Proc.* 30:1062 (Abstr.)
205. Konings, W. N., Barnes, E. M., Kaback, H. R. 1971. *J. Biol. Chem.* 246:5857–61
206. Hong, J., Kaback, H. R. 1972. *Proc. Nat. Acad. Sci. USA* 69:3336–40
207. Short, S. A., White, D. C., Kaback, H. R. 1972. *J. Biol. Chem.* 247:7452–58
208. Sprott, G. D., MacLeod, R. A. 1972. *Biochem. Biophys. Res. Commun.* 47:838–45
209. Futai, M., 1973. *Biochemistry* 12:2468–74
210. Weiner, J. H., Heppel, L. A. 1972. *Biochem. Biophys. Res. Commun.* 47:1360–65
211. Short, S. A., Kaback, H. R. 1973. *Abstr. Ann. Meeting Am. Soc. Microbiol.* 73:191
212. Kaback, H. R. et al 1974. *Arch. Biochem. Biophys.* 160: In press
213. Konings, W. N., Kaback, H. R. 1973. *Proc. Nat. Acad. Sci. USA* 70:3376–81
214. Miki, K., Lin, E. C. C. 1973. *J. Bacteriol.* 114:767–71
215. Miki, K., Lin, E. C. C. 1973. *Fed. Proc.* 32:632 (Abstr.)
216. Harold, F. M. 1972. *Bacteriol. Rev.* 36:172–230

217. Wilson, D. F., Ting, H. P., Koppelman, M. S. 1971. *Biochemistry* 10:2897–902
218. Smith, P. B., Montie, T. C. 1974. Manuscript in preparation
219. Klein, W. L., Boyer, P. D. 1972. *J. Biol. Chem.* 247:7257–65
220. Boyer, P. D., Klein, W. L. 1972. *Membrane Molecular Biology,* ed. C. F. Fox, A. D. Keith, 323–44. Stamford, Conn: Sinauer
221. Schairer, H. U., Haddock, B. A. 1972. *Biochem. Biophys. Res. Commun.* 48:544–51
222. Berger, E. A. 1973. *Proc. Nat. Acad. Sci. USA* 70:1514–18
223. Parnes, J. R., Boos, W. 1973. *J. Biol. Chem.* 248:4429–35
224. Reeves, J. P. 1971. *Biochem. Biophys. Res. Commun.* 45:931–36
225. Jasaitis, A. A., Kuliene, V. V., Skulachev, V. P. 1971. *Biochim. Biophys. Acta* 234:177–81
226. Matin, A., Konings, W. N. 1973. *Eur. J. Biochem.* 34:58–67
227. Van Thienen, G., Postma, P. W. 1973. *Biochim. Biophys. Acta* 323:429–40
228. Reeves, J. P., Hong, J. S., Kaback, H. R. 1973. *Proc. Nat. Acad. Sci. USA* 70:1917–21
229. Pressman, B. C., Harris, E. J., Jagger, W. S., Johnson, J. H. 1967. *Proc. Nat. Acad. Sci. USA* 58:1949–56
230. Pinkerton, M., Steinrauf, L. K., Dawkins, P. 1969. *Biochem. Biophys. Res. Commun.* 35:512–18
231. Hirata, H., Altendorf, K., Harold, F. M. 1973. *Proc. Nat. Acad. Sci. USA* 70:1804–08
232. Kaback, H. R., Milner, L. S. 1970. *Proc. Nat. Acad. Sci. USA* 66:1008–15
233. Rothstein, A. 1970. *Curr. Top. Membranes Transp.* 1:135–76
234. Cox, G. B., et al. 1970. *J. Bacteriol.* 104:219–26
235. Boyer, P. D. 1974. *Biochim. Biophys. Acta.* In press
236. Mitchell, P. 1963. *Biochem. Soc. Symp.* 22:142–69
237. Mitchell, P. 1972. *J. Bioenerg.* 3:5–24
238. Greville, G. D. 1969. *Curr. Top. Bioenerg.* 3:1–78
239. Skulachev, V. P. 1971. *Curr. Top. Bioenerg.* 4:127–90
240. Skulachev, V. P. 1972. *J. Bioenerg.* 3:25–38
241. Henderson, P. J. F. 1971. *Ann. Rev. Microbiol.* 25:393–428
242. Schultz, S. G., Curran, P. F. 1970. *Physiol. Rev.* 50:637–718
243. Morville, M., Reid, M., Eddy, A. A. 1973. *Biochem. J.* 134:11–26
244. Skou, J. C. 1971. *Curr. Top. Bioenerg.*

4:357–98
245. Skou, J. C. 1973. *J. Bioenerg.* 4:1–30
246. Heinz, E. 1972. *Na⁺ Linked Transport of Organic Solute.* Berlin: Springer. 201 p.
247. Harold, F. M. 1970. *Advan. Microbial Physiol.* 4:45–104
248. Hopfer, U., Lehninger, A. L., Thompson, T. E. 1968. *Proc. Nat. Acad. Sci. USA* 59:484–90
249. Bakker, E. P., van den Heuvel, E. J., Wiechmann, A. H. C. A., van Dam, K. 1973. *Biochim. Biophys. Acta* 292:78–87
250. Hunter, D. R., Segel, I. H. 1973. *J. Bacteriol.* 113:1184–92
251. Sheu, C. W., Konings, W. N., Freese, E. 1972. *J. Bacteriol.* 111:525–30
252. Weinbach, E. C., Garbus, J. 1969. *Nature* 221:1016–18
253. Hanstein, W. G. 1973. *Fed. Proc.* 32:515 (Abstr.)
254. Hopfer, U. 1970. *The influence of the polar moiety of lipids on the ion permeability of bilayers.* PhD thesis. Johns Hopkins Univ., Baltimore, Md.
255. Halpern, Y. S., Barash, H., Dover, S., Druck, K. 1973. *J. Bacteriol.* 114:53–58
256. Gale, E. F., Llewellin, J. M. 1972. *Biochim. Biophys. Acta* 266:182–205
257. Niven, D. F., Jeacocke, R. E., Hamilton, W. A. 1973. *FEBS Lett.* 29:248–52
258. Frank, L., Hopkins, I. 1969. *J. Bacteriol.* 100:329–36
259. Shiio, J., Miyajima, R. 1972. *J. Biochem. Tokyo* 72:773–75
260. Eagon, R. G., Wilkerson, L. S. 1972. *Biochem. Biophys. Res. Commun.* 46:1944–50
261. Willecke, K., Gries, E. M., Oehr, P. 1973. *J. Biol. Chem.* 248:807–14
262. Stock, J., Roseman, S. 1971. *Biochem. Biophys. Res. Commun.* 44:132–38
263. Harold, F. M., Pavlasova, E., Baarda, J. R. 1970. *Biochim. Biophys. Acta* 196:235–44
264. Harold, F. M., Papineau, D. 1972. *J. Membrane Biol.* 8:27–62
265. Abrams, A., Smith, J. B., Baron, C. 1972. *J. Biol. Chem.* 247:1484–88
266. Harold, F. M., Baarda, J. R. 1968. *J. Bacteriol.* 96:2025–34
267. Kashket, E., Wong, P. T. S. 1969. *Biochim. Biophys. Acta* 193:212–14
268. White, S. H., O'Brien, W. M. 1972. *Biochim. Biophys. Acta* 255:780–85
269. Kashket, E. R., Wilson, T. H. 1972. *Biochem. Biophys. Res. Commun.* 49:615–20
270. Kashket, E. R., Wilson, T. H. 1973. *Proc. Nat. Acad. Sci. USA* 70:2866–69
271. Seaston, A., Inkson, C., Eddy, A. A. 1973. *Biochem. J.* 134:1031–43

272. Scholes, P., Mitchell, P. 1970. *J. Bioenerg.* 1:309–23
273. West, I. C. 1970. *Biochem. Biophys. Res. Commun.* 41:655–61
274. West, I. C., Mitchell, P. 1972. *J. Bioenerg.* 3:445–62
275. West, I. C., Mitchell, P. 1973. *Biochem. J.* 132:587–92
276. West, I. C., Wilson, T. H. 1973. *Biochem. Biophys. Res. Commun.* 50:551–58
277. Butlin, J. D., Cox, G. B., Gibson, F. 1971. *Biochem. J.* 124:75–81
278. Cox, G. B., Newton, N. A., Butlin, J. D., Gibson, F. 1971. *Biochem. J.* 125:489–93
279. Cox, G. B., Gibson, F., McCann, L. M., Butlin, J. D., Crane, F. L. 1973. *Biochem. J.* 132:689–95
280. Evans, D. J. 1970. *J. Bacteriol.* 104:1203–12
281. Carreira, J., Leal, J. A., Rojas, M., Muñoz, E. 1973. *Biochim. Biophys. Acta* 307:541–56
282. Kobayashi, H., Anraku, Y. 1972. *J. Biochem. Tokyo* 71:387–99
283. Roisin, M. P., Kepes, A. 1972. *Biochim. Biophys. Acta* 275:333–46
284. Hanson, R. L., Kennedy, E. P. 1973. *J. Bacteriol.* 114:772–81
285. Prezioso, G., Hong, J. S., Kerwar, G. K., Kaback, H. R. 1973. *Arch. Biochem. Biophys.* 154:575–82
286. Simoni, R. D., Shallenberger, M. K. 1972. *Proc. Nat. Acad. Sci. USA* 69:2663–67
287. Rosen, B. P. 1973. *Biochem. Biophys. Res. Commun.* 53:1289–96
288. Bragg, P. D., Hou, C. 1973. *Biochem. Biophys. Res. Commun.* 50:729–36
289. Gutnick, D. L., Kanner, B. I., Postma, P. W. 1972. *Biochim. Biophys. Acta* 283:217–22
290. Cox, G. B., Young, I. G., McCann, L. M., Gibson, F. 1968. *J. Bacteriol.* 99:450–58

SYNAPTIC MACROMOLECULES: ✕ 843
IDENTIFICATION AND METABOLISM[1]

Samuel H. Barondes

Department of Psychiatry, University of California at San Diego
La Jolla, California

CONTENTS

INTRODUCTION

The nervous system functions by evaluating environmental stimuli and generating responses. This is made possible by the convergence of signals from many neurons so that they may be integrated, and divergence to many effectors so that a coordinated response can result. Communication between the neurons occurs at specialized structures, the synapses. Each synapse is formed by a specific association between a nerve terminal of the presynaptic neuron and a receptive patch on some part of the postsynaptic neuron. A given neuron may extend thousands of nerve terminals and thousands of receptive patches, each of which forms a synapse. Integration of the state of activity of all the active synapses at a given moment determines the behavior of the organism at that time.

Due to the importance of synapses, they have been studied extensively from several points of view. Anatomists have stained them in a variety of ways; neurophysiologists have devised techniques for measuring their activity by increasingly direct means;

[1] Preparation of this review was facilitated through the use of MEDLINE bibliographies on Synaptosomes and Aspects of Synapses other than Synaptosomes.

pharmacologists have discovered a large series of drugs which interact with them; and, in the past decade, there has been growing interest in determining the biochemical basis of synaptic function. Many reviews and books have been written about synapses. A recent book (1) offers a good introduction.

Until recently most biochemical studies of synapses have been concerned primarily with the macromolecules involved in the synthesis and storage of neurotransmitters in nerve endings and with the receptors on the membranes of postsynaptic neurons. Emphasis on these types of macromolecules is due not only to their intrinsic importance but also to the availability of specific reagents (neurotransmitters, drugs, toxins) which interact specifically with them. Some of this work is summarized by Hall (2) in Volume 41 of this series and is not considered explicitly here. The work which is reviewed here is concerned with: 1. other types of synaptic macromolecules that can now be identified only by demonstration of their presence in subcellular fractions enriched in a specific synaptic component; 2. the site of synthesis of synaptic macromolecules.

Before considering specific topics, alternative definitions of a synapse should be considered. In one view a synapse is the anatomically identified junction between the presynaptic and postsynaptic cells. It includes the site of release of neurotransmitters and the accumulation of receptors on the postsynaptic membrane. The densely staining material present in the junction and its presynaptic and postsynaptic components are also generally included. In a broader view the synapse also includes the remainder of the nerve terminal that contains enzymes which synthesize its neurotransmitter, synaptic vesicles which may store this chemical, and other components involved in signaling and mediating release. In the broadest sense the synapse may be viewed as a *specialized part of two cells*. This takes heed of the obvious fact that synaptic molecules arise in other parts of their respective cells and that their synthesis and degradation may be regulated primarily by distant events in the nucleus, ribosomes, and endoplasmic reticulum of the nerve cell body. Although the synapse may be considered to be an anatomically confined entity, it must also be considered as a dynamic portion of two cells whose general processes support and control it.

SYNAPTOSOME FRACTIONS

A major advance in biochemical studies of synapses was made when Gray & Whittaker (3) and De Robertis et al (4) independently demonstrated that particles which looked like intact nerve endings were abundant in a crude mitochondrial fraction obtained from brain. The nerve endings had been sheared off axons by homogenization in isotonic sucrose and had resealed so that they could be isolated and fractionated as discrete subcellular particles. Whittaker et al (5) named these particles "synaptosomes," and fractions enriched in these particles are commonly referred to as "synaptosome fractions" or "nerve ending fractions."

The fundamental procedures used in early work on fractionation of synaptosomes from other subcellular particles have been summarized (6–8). It should be emphasized that a major goal of the early purification procedures was to obtain intact synaptosomes and their constituent synaptic vesicles in relatively high purity and fairly high yield so that aspects of neurotransmitter metabolism could be studied. For this reason some contamination of the nerve ending fractions with other plasma membranes or with membranes from endoplasmic reticulum was not as serious a problem as it is for work concerning macromolecular metabolism at nerve endings or isolation of specific components of synaptic membranes. Many of the modifications of the early procedures have indeed been concerned primarily with questions other than those involving neurotransmission. For the latter the early preparations are often acceptable.

In both early procedures the "crude mitochondrial fraction," which was found to be enriched with nerve ending particles, was fractionated on a discontinuous sucrose gradient. The procedure described by Whittaker et al (5) has found particular favor, perhaps due to the fact that the gradient used is relatively simple and consists of equal volumes of 0.8 M and 1.2 M sucrose. In this procedure cerebral cortex is homogenized in 9 volumes of 0.32 M sucrose and centrifuged at 1,000 × g for 10 min; the supernatant fluid is then centrifuged under one of a number of conditions (e.g. 10,000 × g for 30 min) to give the "crude mitochondrial fraction." The latter is layered over the gradient and centrifuged in a swinging bucket rotor. Upon centrifugation myelin is concentrated above the 0.8 M layer; synaptosomes are concentrated at the 0.8–1.2 M interface, and mitochondria are concentrated in the pellet at the bottom of the tube. The synaptosome fractions are contaminated with mitochondria, occasional myelin fragments, and many membranes of diverse origin. Whittaker (6) estimates that about half the material in this fraction is in the synaptosomes. The mitochondrial fraction produced in this way is contaminated by membranes and by some synaptosomes.

Since this early procedure is still frequently applied to a variety of problems, one important aspect of the procedure is worth mentioning: the degree of centrifugation used in collecting the crude mitochondrial fraction and the degree of washing. With higher speeds of centrifugation and reduced washing, the yield of synaptosomes is increased, as is the degree of contamination with other membranes. In the original procedure of Gray & Whittaker (3), which is still used often, the crude mitochondrial fraction is obtained by centrifugation at 17,000 × g for 55 min and is not washed before application to the sucrose gradient. In the scheme described by Whittaker et al (5) the crude mitochondrial fraction is collected at 10,000 × g for 30 min and washed once under these conditions. In light of subsequent work it has been shown that contamination with membranes of endoplasmic reticulum can be substantially reduced by decreasing the sedimentation velocity or time (e.g. 11,000 × g for 20 min) and by repeated washing. For example, Morgan et al (9) have shown that contamination with RNA, believed to be bound to endoplasmic reticulum, can be successively reduced by up to 6 washes of the crude mitochondrial fraction. Despite this, three washes are used in the procedure they developed and less than three (often none) are used in several alternative procedures proposed by others. Since all the modifications begin with the crude mitochondrial fraction, their evaluation should take proper cognizance of the degree of washing.

Modifications of several types have been proposed. Fractionation on continuous sucrose gradients improves purity but reduces yield (10, 11). When the crude mitochondrial fraction is centrifuged for 2.5 hr in a continuous 0.8–1.6 M sucrose density gradient, the zone between 1.17 and 1.24 M, which contains about 10% of the protein of the crude mitochondrial fraction, constitutes a relatively pure synaptosome fraction (11). There is relatively little contamination by myelin and mitochondria, but contamination with small membrane fragments is not eliminated. Zonal centrifugation using continuous sucrose gradients (12) and Ficoll-sucrose gradients (13) have also been used to produce fractions of relatively high purity. However, the poor yield of the fraction with the highest purity has generally prompted a reversion to discontinuous gradients in conventional swinging bucket rotors.

The major modification which has been made is the use of discontinuous gradients containing mixtures of Ficoll and sucrose (14–20). By using this high molecular weight polymer it is possible to impart increasing densities to an isotonic sucrose solution without significant increases in its osmolarity. The particles are therefore not subjected to hyperosmotic conditions that may impair some of their metabolic activities. Since Ficoll does not enter the particles or dehydrate them, separations can be made on the basis of the true density of the particles. Other advantages of Ficoll-sucrose gradients which have been proposed are: their superiority in separating synaptosomes from glial membrane fragments (17); the shorter centrifugation times required compared with sucrose gradients; and the greater sensitivity of the resultant synaptosomes to the osmotic shock employed for preparation of subsynaptosomal components (18). Morgan et al (18) state that contamination with axonal fragments which appears to be prominent if sucrose gradients are used (21) is not prominent for Ficoll-sucrose. Nevertheless the overall purity of fractions obtained with Ficoll-sucrose gradients is not strikingly different from those obtained from sucrose gradients, although the nature of the contaminants may be different. For example, Cotman & Matthews (20) find that 25–40% of the particles in the synaptosome fractions prepared on Ficoll-sucrose gradients are not synaptosomes, in agreement with the results of others using a similar technique (16). Depending on the Ficoll concentration used, the predominant contaminants are either membrane fragments of unknown origin or mitochondria (19). The synaptosome fraction prepared by Morgan et al (18) is said to be more contaminated with free mitochondria than one purified on a sucrose gradient.

Most work on synaptosomes has been done with cerebral cortex from adult rat or guinea pig. Results may be somewhat different with other tissues. For example, the mossy fiber endings of the cerebellar cortex behave anomalously because of their large size (22). Small nerve ending particles containing histamine were not sedimented efficiently into a conventional crude mitochondrial fraction (23) but could be harvested by centrifugation at $20,000 \times g$ for 30 min. Attempts to prepare synaptosomes from peripheral tissues such as vas deferens (24) or electric organ (25), which are very abundantly innervated, have been disappointing because of very small yields. A microprocedure for fractionation of superior cervical ganglion has been reported (26), but the concentration of synaptosomes in the synaptosome fraction was relatively poor. Fractions containing intact synaptosomes derived from octopus (27, 28) or squid (29) central nervous system have been described when homogenization is conducted in a high concentration of sucrose approximately isotonic with the tissue fluids of these marine organisms. Fractionation of synaptosomes derived from immature brain has been reported but the density at which these particles are isolated differs from that found with adult tissues (30). Attempts have also been made to separate the synaptosomes containing different neurotransmitters using multiple step gradients (4), continuous gradients (10), and brief centrifugation (31). Although some separation of synaptosomes containing different transmitters is apparently achieved, there is considerable overlap.

SUBFRACTIONS OF THE SYNAPTOSOME FRACTION

When synaptosomes are suspended in water or other hypo-osmotic media they burst and release much of their contents including soluble materials, mitochondria, and synaptic vesicles. The residual ghost consists of the synaptic plasma membrane, attached bits of postsynaptic membrane included in the "synaptic complex," and some of the internal contents of synaptosomes which may be clumped together and associated with the inner surface of the membrane.

The components released by lysis with water can be fractionated by application to a discontinuous sucrose gradient, like that described by Whittaker et al (5). Upon centrifugation the clear top layer contains soluble cytoplasmic constituents: the hazy layer at the 0.4 M interface is highly enriched in synaptic vesicles; the 0.4–0.6 M interface contains microsomes and some synaptic vesicles; 0.6–0.8 and 0.8–1.0 M interfaces are enriched in membranes including many "synaptic membranes;" the 1.0–1.2 M interface contains partially disrupted synaptosomes; and the pellet is enriched in mitochondria although other components are also found.

Each of these subfractions may be contaminated not only with other particles found in the original synaptosomal fraction but also with the other components of the synaptosomes themselves. Of the fractions isolated, the two that may be most highly enriched in material actually derived from synaptosomes are the soluble fraction and the synaptic vesicle fraction perhaps because both of these fractions have much different sedimentation properties than the contaminants of the synaptosomal fraction and the other constituents of synaptosomes themselves.

Soluble proteins obtained after lysis of synaptosome fraction with water and centrifugation at $100,000 \times g$ for 1 hr may be highly enriched in soluble proteins actually released from synaptosomes for several reasons: 1. Soluble proteins may be as abundant in synaptosomes as in homogenates of whole cells. Whittaker (6) estimates that 64% of the volume of synaptosomes is devoid of organelles and contains soluble protein. 2. If the synaptosomal fraction is appropriately washed before lysis, little absorbed soluble protein would be expected to be found and this would probably not be released by treatment with water [although some may be released under alkaline conditions (32)]. 3. Contaminants of the synaptosome fraction do not release much soluble protein upon treatment with water (33).

The relative contribution of contaminants to the soluble fraction obtained by lysis of the synaptosome fraction with water has been studied directly (33). Upon lysis of a washed synaptosome fraction with water, about 15% of the total protein of the fraction was recovered in the soluble fraction which remains in the supernatant fluid after centrifugation at $100,000 \times g$ for 1 hr. This result contrasts with the results after lysis of other subcellular fractions in the brain prepared under identical conditions. Of their total protein, the mitochondrial fraction released 6%, the myelin fraction 4%, and the microsome fraction 3% when treated with water under these conditions. Since all these fractions, particularly the mitochondrial fraction, are contaminated with synaptosomes, it is likely that pure mitochondria, myelin, and microsomes would release an even smaller percentage of their protein upon lysis with water. If half the protein in the synaptosome fraction is in contaminants and if they release, on the average, about 5% of their protein into the soluble fraction, then the synaptosomes must be releasing 25% of their protein to give an overall release of 15%. It may therefore be calculated that more than 80% of the soluble protein released from the fraction is derived from synaptosomes. Since other fractions, particularly the mitochondrial fraction, are heavily contaminated with synaptosomes (so that less than 5% of the mitochondrial protein is actually solubilized) it is likely that more than 90% of the soluble protein recovered by lysis of a synaptosome fraction is actually derived from synaptosomes. It is also notable that 1 hr after intracerebral injection of radioactive leucine, the specific activity of soluble proteins isolated from the synaptosome fraction is less than 5% of that of the soluble protein from whole brain homogenate (34, 35); and the radioactive proteins found in these two fractions differ strikingly, as determined by electrophoresis (36). Therefore the soluble fraction derived from synaptosomes is not heavily contaminated by adsorbed soluble proteins from the homogenate.

Because of their small size and low density, synaptic vesicles are also fairly readily purified by this procedure. If prepared properly, synaptic vesicles of considerable purity can be obtained as determined by electronmicroscopic examination (5) although this has been disputed by results obtained from enzymatic studies (19). They can be separated from contaminating soluble proteins by washing and centrifugation at $100,000 \times g$ for 1 hr. Whittaker et al (5) estimate that in a variant of the procedure described above (in which the crude mitochondrial fraction rather than the synaptosomal fraction is lysed with water and fractionated) about 12–15% of the synaptic vesicles present in the original tissue are recovered and that these are highly purified. When synaptosomes are first purified before lysis and subsequent fractionation, the yield is reduced (37). Passage of synaptic vesicles through Millipore filters of appropriate pore size removes contaminating membranes (69). High yields of purified synaptic vesicles have been obtained from electric organ of *Torpedo* by crushing the frozen organ, extraction, and zonal centrifugation (144).

Another component of the synaptosomes that should be fairly easy to purify is the mitochondria present in abundance within synaptosomes. Conventional procedures probably do not purify these very well because synaptosomal fractions like those prepared by Whittaker et al (5) are contaminated with free mitochondria which originate from other portions of the tissue. Therefore the mitochondria obtained after lysis of the synaptosome fraction may be a mixture of mitochondria released from synaptosomes and contaminating free mitochondria. In Ficoll-sucrose gradients in which two synaptosomal fractions are usually obtained, the lighter one is relatively devoid of free mitochondria (19).

In addition to contamination with mitochondria from other regions of the brain, the mitochondria obtained after lysis of a synaptosome fraction are also contaminated with membranes and undisrupted synaptosomes. Davis & Bloom (38) have found that the buoyant density of mitochondria can be increased by enzymatically depositing a reaction product in mitochondria by reacting the fraction with sodium succinate and iodonitrotetrazolium. Whereas this procedure has been used by them and others (39) to remove mitochondria from membrane fractions, it might also be useful for removing membranes from mitochondrial fractions derived from the synaptosome fraction.

Synaptic Membranes

In the past few years considerable effort has been expended to isolate the membranes surrounding synaptosomes in high purity. This has proved quite difficult because: 1. the buoyant density of contaminants (some components of endoplasmic reticulum, Golgi apparatus, lysosomes, outer mitochondrial membranes, dendritic and axonal membrane, and some glial membranes) is quite similar to that of "synaptic membranes"; 2. the positive morphological and biochemical markers for synaptic membranes are not specific. For example morphological examination of a synaptic membrane fraction, after the internal contents have been removed, shows only large membrane-lined sacs that cannot be distinguished from those derived from other sources. In cases where the synaptic complex is attached it is possible to make a definite identification, but this requires serial sectioning (40) and is complicated by the fact that many of the synaptic complexes in postsynaptic membranes may be lost during conventional purification (6). Attempts to find specific biochemical markers for synaptosomal membranes have also not been very successful. Gangliosides are enriched in neuronal plasma membranes (41, 42) but are probably not specific for the synaptosome plasma membrane. Nor is the ouabain-sensitive Na^+-K^+-ATPase. In the absence of specific biochemical markers evaluation of the purification of "synaptosomal membranes" is extremely difficult.

The synaptosomal membrane fraction whose purity has been most extensively studied is that investigated by Morgan et al (9, 18). The synaptosome fraction used as a starting material is prepared on Ficoll-sucrose gradients. Lysis is performed in alkaline media rather than in water, which is usually slightly acidic. At alkaline pH, lysis is apparently more efficient and Cotman & Matthews (20) found that the resultant membrane fraction is more readily separable from mitochondria. They found that when a synaptosome fraction obtained from a Ficoll-sucrose gradient was lysed at pH 7.1 and applied to a discontinuous sucrose gradient, there was a great deal of overlap between mitochondria measured by cytochrome oxidase and plasma membranes measured by Na^+-K^+-ATPase. In this case about half the mitochondria overlapped 89% of the plasma membranes. At pH 8.5, however, 95% of the mitochondria separated from about 85% of the membranes. At pH 7.6 intermediate results were obtained.

The procedure of Morgan et al (9, 18) emphasizes removal of contaminating membranes in two ways. First, the crude mitochondrial fraction is washed three times to reduce contamination with endoplasmic recticulum. Second, after lysis of the synaptosome fraction it is centrifuged at $11,500 \times g$ for 15 min to ensure removal of particles whose sedimentation properties are not significantly changed by the osmotic shock. This procedure results in the loss of many synaptosomal membranes so that the yield is substantially smaller than that described by others.

The products of the procedure of Morgan et al (9, 18) appear to be highly purified plasma membranes. Their estimates of the percentage of each of the known contaminants, based on chemical or enzymatic markers, are as follows: soluble proteins 0% (marker, lactic dehydrogenase); lysosomes less than 1% (markers include β-galactosidase); endoplasmic recticulum less than 1% (marker, NADPH-cytochrome c reductase); myelin less than 1% (marker, cerebrosides and sulfatides); inner mitochondrial membranes less than 1% (markers, cytochrome c oxidase and succinate dehydrogenase); outer mitochondrial membranes 5–10% (marker, mono-amine oxidase); Golgi apparatus less than 15% (marker, galactosyltransferase).

The authors (9, 18) have taken great pains to further evaluate the possible con-tamination with Golgi membranes and glial membranes. For the former, thiamine pyrophosphatase has been studied. They find that this enzyme is abundant in their plasma membrane fractions just as it has been shown to be present by others in synaptosome fractions. They conclude, however, that this enzyme cannot be used as a marker for Golgi membranes from brain just as it cannot be used as a biochemical marker in liver. The use of galactosyltransferase as a marker for Golgi membranes may be misleading, as glycosyltransferases may be associated with synaptosomal membranes as well, as discussed below.

The possibility of contamination of this preparation by glial membranes is also explicitly considered using 2',3'-cyclic AMP,3'-phosphohydrolase as a glial membrane marker. They conclude that the synaptic plasma membrane fractions are contaminated with 5–10% glial membranes. Other potential markers for glial membranes also suggest contamination in this range. However, glial membranes in brain are very heterogeneous and it is not clear that any of the markers used is representative of all glial cell types.

Despite evidence that the synaptic plasma membrane fraction is not heavily contaminated by any of the other classes of membranes mentioned above, the major difficulty in evaluating this preparation is that membranes derived from axons, dendrites, or nerve cell bodies may be identical not only in sedimentation properties but also in markers with those of "synaptosomal membranes." For these reasons it seems best to consider that the isolated membranes are contaminated with an unknown proportion of neuronal plasma membranes, although they are highly enriched in synaptosomal plasma membranes.

Synaptic Complexes

The synaptic complex may be defined to include the bits of presynaptic and post-synaptic membrane which are joined at the synapse as well as the densely staining material found in the synaptic cleft and on the inner surfaces of the presynaptic and postsynaptic membranes. Since synaptic complexes represent a very small percentage of nervous system tissue, their isolation could be very difficult. However, three factors have facilitated their isolation: 1. they have a unique morphology that can be used as a reliable positive marker; 2. they are relatively insoluble (at least in part) in the non-ionic detergent Triton X-100 whereas other components of the synaptic membrane and other membranes are far more soluble (43–45); 3. they have a higher isopycnic density than plasma membranes (38, 45).

Although the unusual staining properties of the synaptic complex are detectable by conventional electronmicroscopic procedures, they may be augmented by staining with ethanolic phosphotungstic acid (46) or bismuth iodide (47), both of which preferentially stain the entire synaptic complex. Other stains that react strongly with material in the synaptic cleft are believed to be specific for sugar-rich compounds. These include phosphotungstic acid without ethanol (48) and silver methenamine staining of tissue oxidized with periodate (49). Synaptic complex stains are presently being used in an attempt to facilitate isolation of fractions enriched in these structures (45).

The relative insolubility of the synaptic complex in Triton X-100, first reported by De Robertis et al (43), has been employed by others. Cotman & Taylor (45) recently studied solubilization of membranes at different Triton concentrations. They found that when the Triton to protein ratio was 2:1, about 70% of the membrane protein was solubilized but synaptic complexes retained their distinctive morphology. The synaptic complexes produced in this manner had densities greater than 1.1 M sucrose. Under the specific conditions they used, 50% of the total protein which remained particulate after Triton treatment sedimented through 1.5 M sucrose. Electron-micrographs demonstrate that these fractions are highly enriched in synaptic complexes.

Kornguth and colleagues (40, 50, 51) have described an alternative approach to the fractionation of synaptic complexes. It is important to point out, however, that the subfraction which they refer to as synaptic complexes contains most of the other components of the synaptosome including mitochondria, vesicles, soluble proteins, and membranes. It is referred to as synaptic complexes because it is prepared under conditions which preserve the synaptic complex better than in conventional (6, 52) procedures. In this procedure homogenization is carried out in 0.32 M sucrose buffered with potassium phosphate containing 1 mM MgCl$_2$. The crude mito-chondrial fraction is not washed and the "synaptosome fraction" is obtained by centrifugation in a sucrose gradient containing 10 mM MgCl$_2$. The magnesium apparently preserves the synaptic complex. However, it has long been known that addition of salt, particularly divalent cations, leads to coacervation of particles which markedly interferes with conventional purification of synaptosomes (6). After isolation of the synaptosomal fraction at a 1.0–1.4 M sucrose interface, the fraction is centrifuged on a cesium-chloride gradient which is formed in the presence of 0.14 M sucrose (51). Electronmicrographs of the fraction referred to as synaptic complex show that it resembles synaptosomal fractions (46, 51). The authors state that about 85% of the particles identified contain a synapse (46). Levitan et al (53) have modified this procedure and have further subfractionated a fraction from the cesium chloride gradient by lysis and application to a discontinuous sucrose gradient. The material sedimenting at the 1.0–1.2 M interface of this gradient is referred to as the synaptosomal membrane preparation. Based on enzyme markers, they estimate that this final material is contaminated with only 3% mitochondria and 3% endoplasmic reticulum. Electronmicrographs of the final preparations are not shown.

Synaptic Proteins

Work with synaptosome fractions has helped to identify many of the enzymes used in neurotransmitter synthesis and specific proteins believed to participate in neurotransmitter storage in vesicles. It has also provided a preparation which might prove useful for isolation of receptors, although this has proceeded more quickly in other tissues that have higher concentrations of a single receptor. Some of the work on these diverse subjects is reviewed by Hall (2) in Volume 41.

Synaptosome fractions have also proved useful in identifying other macromolecules in the synaptic region whose roles in neurotransmission are less clear. These include proteins which interact with cyclic AMP, and fibrous proteins. Adenyl cyclase and cyclic AMP phosphodiesterase have been shown to be concentrated in subcellular fractions enriched in synaptic membranes or microsomes (54, 55). The phosphodiesterase has been shown to be localized in the postsynaptic membrane by histochemical staining visualized with the electronmicroscope (56). A cyclic AMP-dependent protein kinase assayed after treatment with Triton X-100 is concentrated in subcellular fractions enriched in synaptic membranes or microsomes (57). Phosphoprotein phosphatases were also abundant in a fraction enriched in synaptic membranes (58). Phosphorylation of a specific protein in a fraction enriched in synaptic membranes is stimulated by cyclic AMP (59). Greengard et al (60) have speculated that phosphorylation of a postsynaptic membrane protein may play a role in neurotransmission.

Fibrous proteins have also been identified in synaptosome preparations. A protein believed to be tubulin has been identified in the soluble (36) and particulate (36, 61) components of synaptosome fractions and is the major soluble protein transported to nerve endings from the nerve cell body (36). Puszkin et al (62) have identified an actomyosin-like protein called neurostenin in the synaptosome fraction of brain. The actin-like moiety (neurin) was localized in the synaptic membrane subfraction whereas the myosin-like moiety (stenin) appeared to be localized in the synaptic vesicles (63). Berl et al (63) have speculated that exocytosis of neurotransmitter contained in synaptic vesicles is mediated by interaction of neurin of the synaptic membrane with stenin of the synaptic vesicles.

Aside from the identification of proteins with known biochemical properties in the synaptic region, attempts have been made to define the specific proteins in synaptic membranes to determine if a unique structural protein, possibly the densely staining protein of the synaptic complex, could be identified. Synaptic membrane fractions obtained by several methods have been solubilized and electrophoresed on polyacrylamide gels in systems containing sodium dodecyl sulfate. In some of these studies the membrane fractions are first treated with Triton X-100, and the Triton-soluble and Triton-insoluble constituents are electrophoresed. As discussed above, the synaptic complex is relatively insoluble in Triton; therefore the proteins associated with it should be present in the Triton-insoluble fraction.

There is considerable agreement that synaptic plasma membrane fractions prepared by a variety of procedures contain a large number of proteins that can be resolved by electrophoresis in sodium dodecyl sulfate. The major protein bands found in preparations made in a number of different ways are quite similar. Banker, Crain & Cotman (64) report that the major polypeptides in their preparations have molecular weights of 99,000, 53,000, and 42,000. These account for 40–50% of the stained protein on the gels. Electrophoresis of a mitochondrial fraction yields a different pattern of proteins. However, Karlsson et al (65) showed that the predominant proteins derived from mitochondrial fractions and microsomal fractions had molecular weights of 44,000 and 52,000, and that proteins with these molecular weights were also prominent in membrane fractions derived from preparations enriched in neuronal cell bodies and synaptosomes. They speculate that the 52,000 dalton protein may be tubulin, which has been reported to be composed of subunits with molecular weights of 53,000 and 56,000 (66) and is apparently present in membranes derived from microsome and synaptosome fractions (36, 61). A protein with a molecular weight of 95,000 was also fairly prominent in all fractions. These investigators found that the membrane fractions derived from synaptosomes had other prominent bands (65).

Morgan et al (69) find that their synaptic plasma membrane has major bands with molecular weights of 93,000, 52,000, and 39,000. They suggest that the 93,000 and the 52,000 dalton proteins may be subunits of Na^+-K^+-ATPase. Wannamaker & Kornguth (67) find that their preparation of "synaptic complexes," like a microsome fraction, have predominant proteins with molecular weights of 92,000, 53,000, 43,000, and 36,000. They conclude that there is no major specific protein in either fraction and that the major protein constituents are common to these subcellular fractions. In the synaptosomal membrane fraction purified by a modification of the procedure used by Kornguth et al, the major proteins have molecular weights of 53,000 and 42,000 (68).

From these studies it seems clear that a variety of techniques for preparing fractions enriched in synaptic membranes give rise to products whose major proteins have identical molecular weights; but these appear to be identical with those of the major proteins found in other subcellular fractions, particularly microsomes. However, interpretation of these findings should take account of the fact that microsome fractions from brain may consist not only of membranes derived from endoplasmic reticulum but also of synaptosomes and neuronal plasma membranes. Therefore the possibility that the major components of synaptic membranes may be specific for these membranes, although strongly challenged by available evidence, cannot be excluded. It is notable in this regard that one fraction of the microsome fraction, which sediments on sucrose gradients like synaptosomes, can be separated from other components with different sedimentation characteristics (61).

In some studies the polyacrylamide gels have also been stained for glyco-proteins. Bosmann et al (70) found that electrophoresis of a relatively impure synaptosome membrane fraction produced six bands which reacted with a glyco-protein stain. Triton X-100 solubilized most of the glycoproteins of a synaptic membrane fraction (71) and the glycoproteins of such fractions differ in their relative extractability in this detergent (72, 73).

Another approach to identification of proteins required for synaptic function has been to prepare antibodies to specific proteins of a synaptic complex fraction and to determine if they react with the synaptic region or influence synaptic function. Antibodies have been raised to synaptosome fractions (74–76) but the antigens were not identified. Application of the approach to specific protein components has not yet been explored.

METABOLISM OF MACROMOLECULES OF THE SYNAPTIC REGION

To communicate directly with many widely separated cells, neurons have evolved long axonal and dendritic processes. Since the terminals of these processes are often distant from the nerve cell body, their regulation by events in the cell nucleus requires the transport of macromolecules. Dendrites contain ribosomes and can therefore synthesize proteins from the mRNAs provided by the nucleus. In contrast, axons (beyond the initial segment) and axonal nerve endings contain no morphologically identified ribosomes. They are therefore dependent on the abundant protein synthesizing capacity of the nerve cell body, but the possibility of some protein synthesis at nerve endings has been proposed, as considered below.

Studies of metabolism of macromolecules in the nerve ending region have been concerned with the following questions: 1. rates of transport of specific proteins to nerve endings and their rates of turnover; 2. the possibility that RNA is transported in axons since this would have important implications for protein synthesis at nerve terminals; 3. the possibility that mitochondrial and nonmitochondrial proteins are synthesized in the synaptic region; 4. biochemical reactions that structurally modify proteins in the synaptic region such as glycosylation or other reactions which require generation or elimination of a covalent bond. Studies concerning each of these specific problems will be briefly considered.

Axoplasmic Transport of Proteins

A number of reviews of axoplasmic transport of proteins have been published (77–80). A periodically updated bibliography may be obtained from UCLA Brain Information Service. Only a brief summary will be given here.

It has been clearly established that proteins destined for the nerve terminal are synthesized in the nerve cell body and are continuously transported in the axon. Two types of studies have been done. In one a nerve is ligated and the accumulation of a specific enzyme is determined in the region of the nerve immediately proximal to the ligation. As the enzyme is transported from the nerve cell body it accumulates proximal to the constriction and the rate of accumulation is a measure of the normal rate of transport. In the other type of study the transport of labeled proteins is determined after injecting radioactive amino acid into a region enriched in nerve cell bodies. Observations are made of the appearance of labeled protein in segments of a nerve containing the axons of the labeled neurons. The eye (81) is a favorite site of injection of radioactive precursor but ganglia (80), concentrations of neurons in the brain (82), and single neurons (83) have been used. Intracerebral injection of radioactive amino acid labels many cell bodies in brain and the appearance of proteins in nerve endings can be observed by isolating synaptosome fractions (34, 35).

These studies have shown that proteins are transported in axons at two major rates: a rapid rate in the range of hundreds of millimeters per day and a slow rate in the range of several millimeters per day (77–80). Other proteins may be transported at intermediate rates (84). Three enzymes involved in amine metabolism are transported at three different rates (85). Transport of proteins in dendrites has also been observed (83).

Most of the material transported with the fast component is particulate (81). Most glycoproteins are transported at a rapid rate (86–89). The slowly moving component which contains about five times as much labeled protein has been reported to consist of 60% particulate protein and 40% soluble protein (81). A major constituent of these soluble proteins is tubulin (36), the subunit of microtubules which are abundant in axons. Mitochondria are also transported with the slow component (90).

Although the mechanism of axoplasmic transport has aroused considerable speculation, it has not yet been clarified. The nerve cell body is the source of proteins which are transported, but inhibition of protein synthesis does not prevent transport of proteins already made (34, 81, 91). Application of inhibitors of energy metabolism to the axon blocks axoplasmic transport (80), which indicates that the apparatus for transport is present in the axon. This hypothesis is supported by evidence that isolated segments of axons separated from the nerve cell body continue to show transport of proteins (77–80). The observed direction of transport remains the same, that is, it proceeds from the cut end that was closest to the nerve cell body in a distal direction. Retrograde transport of proteins has also been shown. Application of fluorescent albumin derivative or horseradish peroxidase to the region of nerve terminals leads to their uptake and transport to the cell body (92, 92a). Intact microtubule functions may be required for both rates of transport since slow and rapid axoplasmic transport are blocked by colchicine or vinblastine (93–96). Microtubule structure may remain intact despite blockade of axoplasmic transport (96). Proteins of different types are transported into the central and peripheral axons of dorsal root ganglion cells (97), which suggests a mechanism for controlling selective outflow.

The fate of proteins transported to nerve terminals has not yet been clearly established. Polypeptide hormones synthesized in the hypothalamus have long been known to be secreted from nerve terminals, but this class of molecules may be a special case. Evidence for transport of radioactive materials trans-synaptically has been presented (98), but it was not clear whether intact proteins are transported across synapses or whether this represents degradation of protein at nerve terminals and uptake of the resultant amino acids which are then incorporated into protein in the postsynaptic cell. The overall turnover of proteins at nerve endings is relatively slow—in the range of several weeks (99, 100).

Although proteins synthesized in the nerve cell body must first be transported to the synapses before directly influencing their function, this may occur very rapidly. For example, given rates of transport of hundreds of millimeters a day and the short axons (e.g. 0.1 mm) which are abundant in the central nervous system, proteins synthesized in the nerve cell body may reach nerve terminals within minutes. Abundant radioactive protein has been directly demonstrated in nerve endings in mouse brain by electronmicroscope radioautography of synaptosome fractions obtained 15 min after intracerebral injection of radioactive amino acids (101).

Presumably this protein was synthesized in the nerve cell body and transported in short axons to nerve terminals. The possibility of some local synthesis of protein at nerve endings will be considered below.

Axoplasmic Transport of RNA

Due to the great interest in the possibility of protein synthesis in axons and nerve terminals, many investigators (102–109) have attempted to determine if RNA is transported to these sites from the nerve cell body. They mostly agree on the following: 1. when radioactive precursors, usually nucleosides, are applied to a region rich in nerve cell bodies (such as the retina) radioactive RNA can subsequently be isolated from emergent nerves (such as the optic nerve); 2. radioactive nucleosides and nucleotides have been shown to migrate in the nerve from the site of application of the precursor; 3. the labeled RNA does not migrate along the nerve as a discrete wave, which contrasts strikingly with the behavior of labeled proteins transported in axons; 4. when examined, the labeled RNA was found to have molecular weights characteristic of both ribosomal RNA and transfer RNA, although ribosomes have not been detected beyond the initial axonal segment, and 4S RNA is the only type which has been clearly shown to be present in axons (110).

The major problem in evaluating these findings is to distinguish RNA transport in axons from transport of precursors incorporated into RNA in the periaxonal (glial) cells whose cell bodies are abundant in nerves. Autoradiographic studies suggest that glial RNA synthesis may explain most (if not all) of the labeled RNA found in nerves. Peterson et al (103) found that the periaxonal cells are heavily labeled but some grains were found over axons. Autilio-Gambetti et al (109) conclude from a detailed radioautographic study that only a small percent of silver grains representing RNA are found over axons, whereas abundant grains were found over the periaxonal glial cells. They also report that intracranial injection of actinomycin D reduced the labeled RNA in the optic nerve by 70–80% without affecting incorporation into the retina; but intraocular injection of actinomycin D had no effect on appearance of labeled RNA in optic nerve despite marked inhibition of RNA synthesis in the retina.

On the basis of available evidence, axoplasmic transport of RNA may be confined to transport of RNA within mitochondria which are known to be transported in axons. Mitochondrial RNA transport is supported by the finding that application of ethidium bromide to the retina along with labeled precursors reduces the amount of labeled RNA found in the tectum (107). It is also possible that 4S RNA is transported in axons since this RNA species is the only one detectable in axoplasm (110). However, transport of ribosomal RNA, of primary interest because of its potential role in protein synthesis at nerve endings, is without compelling support.

Protein Synthesis by Synaptosome Fractions

Synaptosome fractions prepared in a number of ways have been shown to incorporate amino acids into protein (16, 68, 111–123). The major difficulty in evaluating these reports is the fact that the synaptosome fractions are contaminated with some ribosome-containing particles (21, 30, 122) and with free mitochondria, both of which can incorporate amino acid into protein in vitro. In most studies about 80% of the protein synthesis by synaptosome fractions is inhibited by cycloheximide, indicating that 80S ribosomes are involved. About 20% is inhibited by chloramphenicol, which suggests some mitochondrial protein synthesis. However, one protein synthesizing system derived from a disrupted synaptosome fraction was almost totally inhibited by chloramphenicol (68).

In an attempt to distinguish protein synthesis by synaptosomes from protein synthesis by contaminants, effects of RNase and cations have been studied. It is generally agreed that protein synthesis by synaptosome fractions is resistant to RNase and is dependent on the Na^+ and K^+ concentrations in the medium (113, 114). Both findings have been interpreted as indicating that the protein synthesizing systems being studied are separated from the medium by an intact membrane. However, protein synthesis in ribosome-containing particles surrounded by a membrane might have these characteristics. Therefore these properties do not distinguish protein synthesis by synaptosomes from protein synthesis by such known contaminants. Mitochondrial protein synthesis is also insensitive to RNase; and ionic conditions might influence the uptake of labeled amino acids by synaptosomes and therefore limit their incorporation by mitochondria within synaptosomes.

Two strategies have been employed in an attempt to definitively distinguish synaptosome protein synthesis from that by contaminants: 1. electronmicroscope radioautography; 2. comparison of the labeled products synthesized by synaptosome fractions with those synthesized by mitochondria or microsomes.

About half the radioactive protein found in synaptosome fractions after in vitro labeling has been localized by electronmicroscope radioautography within rare membrane-bound sacs containing ribosomes (120, 123). Of the residual protein, about 20% has been shown to be associated with mitochondria contained within synaptosomes (120), thus suggesting that the mitochondria contained within nerve terminals retain the capacity to synthesize protein. Therefore a true but restricted synaptosome protein synthesizing capacity seems to be fairly well established. What is unclear is whether or not there is another protein synthesizing system in the nerve terminals outside the mitochondria.

Electronmicroscope radioautographic analysis of the site of incorporation of radioactive leucine in a synaptosome fraction has also shown that a fraction of the radioactivity incorporated is associated with the plasma membrane surrounding synaptosomes (120). The evidence that radioactive grains are indeed associated with this membrane is quite good. Gambetti et al (120) also showed that the washing procedure that they used before radioautography was highly efficient in removing radioactive precursor. However, it was not conclusively shown that all the radio-activity incorporated represented radioactive leucine incorporated into protein. Considering the finding that only a tiny fraction of the radioactivity in the incubation mixture was actually incorporated, it is possible that the radioactivity associated with the plasma membrane is a product of a contaminant in the radioactive leucine or a product derived from radioactive leucine after it was converted to something else. Were it shown that grains associated with the synaptosome plasma membrane are no longer found if the incubation is done in the presence of both chloramphenicol and cycloheximide, the result would have been more convincing. In the absence of this control the study does not definitively show extramitochondrial incorporation of amino acid into the proteins of synaptic plasma membrane. It does, however, show that about half the incorporation studied by a number of other groups is due to a contaminant and that mitochondria in synaptosomes can incorporate amino acids.

Other possible sources for the rare radioactive grains found over nerve endings shortly after in vivo labeling have been considered (124).

Another approach to distinguishing synaptosome protein synthesis from that due to contaminants is to compare the proteins synthesized by the synaptosome fraction with those enriched in potential contaminants. The two studies (68, 121) which used this approach both suggest that there is true synaptosome protein synthesis besides that found in synaptosomal mitochondria. However, the results differ in many ways.

The synaptic membrane fraction prepared by the procedure of Levitan et al (53) has been shown to incorporate radioactive amino acids into protein (68). About 90% of this incorporation is blocked by 0.1 mg/ml of chloramphenicol; but 0.2 mg/ml of cycloheximide had no effect. The effects of these inhibitors indicate that the protein synthesis does not occur on 80S cytoplasmic ribosomes but rather on ribosomes like these found in mitochondria and bacteria. In vivo incorporation of labeled amino acid into a synaptic membrane fraction prepared by this procedure has also been shown to be partially sensitive to chloramphenicol (124a). This was detectable if the remainder of brain protein synthesis was blocked with large doses of cycloheximide.

To evaluate the possibility that the observed protein synthesis was due to mitochondrial contamination, the authors compared the products of in vitro protein synthesis by synaptosomal membranes with that of mitochondria derived from the synaptosome fraction. Upon electrophoresis of the radioactive products in a system containing sodium dodecyl sulfate, the synaptosome membrane products resolved into three discrete peaks, only one of which overlapped with the five discrete peaks which were the products of incorporation of the synaptosomal mitochondria. The finding that the synaptosomal membrane products are so discrete is strong evidence against the possibility that they may be attributed to contamination of the preparation with bacteria. The authors conclude that synaptic plasma membranes contain a protein synthesizing system characteristic of mitochondria, although the products are different. They propose that nerve ending mitochondria may discharge a protein synthesizing system that is incorporated into synaptic membranes.

In view of other studies on mitochondrial protein synthesis and on protein synthesis by synaptosome fractions, it is difficult to accept this conclusion. Evidence of transport of RNA from mitochondria has been presented (125) but there is no evidence that mitochondrial ribosomes are ever exported. It is also noteworthy that the mitochondria active in protein synthesis are relatively poor in cytochrome oxidase (126), a marker enzyme used to evaluate mitochondrial contamination in the synaptosome membrane preparation under discussion. This raises the possibility that the authors (68) are studying protein synthesis by a population of brain mitochondria poor in this enzyme. The autoradiographic study of Gambetti et al (120) showed that chloramphenicol ($\sim$0.1 mg/ml) blocked only 20% of the incorporation of radioactivity into synaptosomes, which contrasts strikingly with the 90% inhibition by chloramphenicol observed in the study under discussion.

Using a synaptosome fraction prepared differently, Gilbert (121) has identified a cycloheximide-sensitive protein synthesizing system which he believes is truly associated with synaptosomes rather than contaminating endoplasmic reticulum. This contention is based on the finding that products of these two fractions are somewhat different when electrophoresed in sodium dodecyl sulfate. The possibility that the proteins synthesized by the synaptosome fraction are a mixture of proteins synthesized by both contaminating endoplasmic reticulum and mitochondria was not excluded.

In light of these recent studies the possibility that there is nonmitochondrial protein synthesis at nerve endings remains viable. In all these studies attention has been paid to the possibility that the observed amino acid incorporation might be due to contaminants, and arguments are presented against this. However, the results are neither completely consistent nor conclusive. Failure to visualize ribosomes at nerve endings could mean that they are obscured in some way; but conclusive evidence of nonmitochondrial synaptosome protein synthesis must be demanded in light of the absence of detectable ribosomes and the contamination of synaptosome fractions with particles of a demonstrated capacity to incorporate amino acids into proteins.

Glycosylation of Proteins by Synaptosome Fractions

Even if the capacity for nonmitochondrial protein synthesis is ultimately determined to be absent from the nerve ending, covalent addition of other residues to proteins might provide a mechanism for modulating nerve ending function. Sugars are one type of residue that can be added to proteins after they have left ribosomes. There is abundant evidence that most of the glycosyl residues of glycoproteins are added while proteins are in the smooth endoplasmic reticulum or in the Golgi apparatus (127). There is also evidence that glycosylation of proteins may occur on the plasma membranes of cells (128). A number of studies have been directed towards determining if glycosylation of proteins occurs at nerve endings (33, 115, 129–141). Both in vitro and in vivo studies have been reported.

In the in vivo studies (33, 86, 129, 136, 139, 140) a radioactive sugar is injected intracerebrally so that both the nerve endings and the nerve cell body are exposed to the precursor. Since it is known that glycoproteins are synthesized in nerve cell bodies and transported to nerve endings (86–89), incorporation of sugars into proteins associated with the synaptosome fraction was determined primarily at early times after injection of the precursor. This reduces the possiblity that transport of labeled glycoproteins from the nerve cell body, prominent within 3 hr of injection of precursor (86), might obscure local incorporation of sugars into glycoproteins at nerve endings.

Using radioactive glucosamine as precursor, evidence has been presented which suggests that this sugar is incorporated into proteins of the nerve ending in vivo (33, 86, 129, 140). Since the soluble proteins obtained by lysis of a synaptosome fraction are relatively free of contamination by soluble proteins released from mitochondria or from microsomes (33), emphasis has been placed on incorporation of glucosamine into this fraction. Within 1 hr after intracerebral injection of radioactive glucosamine, the specific activity of soluble protein derived by water lysis of a synaptosome fraction equals or exceeds that of soluble protein obtained from whole brain homogenate (33, 86, 129). In contrast, after intracerebral injection of radioactive leucine the specific activity of the soluble proteins of nerve endings was only about 5% of that of the soluble proteins of whole brain 1 hr after injection of precursor

(33–35). The relative specific activity of leucine in the soluble component of the nerve ending fraction (as compared with that of soluble protein from whole brain) rose about twentyfold during the days after injection, as labeled protein was transported to nerve endings (34, 35). In contrast the relative increase in specific activity of glucosamine labeled soluble protein of nerve endings over this period was not great when compared with the specific activity of glucosamine labeled soluble protein from whole brain (33, 86, 129). The high relative specific activity of glucosamine labeled soluble protein from nerve endings within 1 hr of injection of precursor suggests that some of this radioactive glucosamine was directly incorporated in the nerve endings themselves.

Interpretation of these in vivo studies is complicated by the possibility that the nerve ending fraction may contain a small percentage of smooth endoplasmic reticulum or Golgi apparatus which might be extremely active in glycosylation of proteins. This is highly unlikely because in vitro studies indicate that the nerve ending fraction (prepared exactly as for the above in vivo studies) contains only 4% as much fucosyltransferase as the microsome fraction (130). This was the case both with particulate fucosyltransferase assayed with endogenous acceptor and with solubilized fucosyltransferase assayed with exogenous acceptor (130). Contamination with other glycosylating membranes is also unlikely because of results of in vivo studies with radioactive fucose. In striking contrast to the results with glucosamine, which suggest local incorporation of this sugar into proteins at nerve endings, studies with fucose indicate that there is little or no local incorporation (86). Thus 1 hr after intracerebral injection of labeled fucose and glucosamine the relative specific activity of soluble proteins from the nerve ending fraction (as compared with the soluble proteins from whole brain) was high in labeled glucosamine, whereas it was low in labeled fucose (86). If the observed incorporation of glucosamine was due to contamination with smooth endoplasmic reticulum or Golgi membranes, incorporation of fucose, which occurs at these sites, should be as prominent as incorporation of glucosamine. The case for sialic acid incorporation into glycoproteins at nerve endings is also far less compelling than that for glucosamine (136).

Additional support for the incorporation of glucosamine into proteins at nerve endings has come from analysis of the radioactive glycopeptides obtained by pronase digestion of the soluble and particulate components of the synaptosome fraction and of other subcellular fractions. After administration of radioactive glucosamine in vivo the major glycopeptide associated with both the soluble and particulate components of the nerve ending fraction had a molecular weight of approximately 1250 (140). Approximately 50% of the radioactive glycopeptides and 46% of those derived by digestion of the particulate component of the synaptosome fraction were in this molecular weight range. In contrast only 22% of the labeled glycopeptides of the mitochondrial fraction, 32% of the soluble glycopeptides, and 25% of the microsomal glycopeptides were in this molecular weight range. Therefore contamination with these other subcellular fractions cannot account for the results. It is notable that 4 hr after injection of the radioactive precursor, the molecular weight distribution of labeled glycopeptides released by pronase digestion of the synaptosomal fraction was closer to that found with the other fractions than it had been 1 hr after injection. This observation supports the finding (86–89) of axoplasmic transport of labeled glycoproteins. Marinari et al (139) present data which indicate prominent labeling of glycoproteins of synaptosomal plasma membranes by axoplasmic transport, but labeling of other subfractions of the synaptosome fraction is consistent with substantial local incorporation of glucosamine at nerve endings.

A number of investigators have examined the in vitro glycosylation of proteins by synaptosome fractions (130–135, 137, 138, 141). As shown by Raghupathy et al (137) several reports of multiple glycosyltransferases in synaptosomes (131, 138) must be viewed with great caution because washing of the crude mitochondrial fraction was not sufficient to remove substantial contamination with endoplasmic reticulum. Ko, Raghupathy & Peterson (133, 134, 137) found relatively low levels of galactosyltransferase, galactosaminyltransferase, and sialyltransferase in synaptosome fractions derived from washed crude mitochondrial fractions. Reith et al (141) find that a highly purified synaptosomal plasma membrane fraction contains about half as much galactosyltransferase as microsomes and about 12% as much as a fraction presumed to be enriched in Golgi apparatus. Fucosyltransferase found in a synaptosome fraction prepared from a well-washed crude mitochondrial fraction has been shown to be only 4% as active as that found in microsomes (130). Whether or not these activities are due to contamination has not been determined.

The case for a glucosaminyltransferase in synaptosomes is stronger. Festoff et al (135) found that glucosamine is incorporated into glycoproteins by synaptosome fractions, but the possibility of contamination was not explicitly considered. In a subsequent study by Dutton et al (140) the findings of Festoff et al (135) were confirmed. They further showed that the synaptosome fraction was more active than a mitochondrial fraction and far more active than a microsomal fraction in incorporating glucosamine into glycoproteins under the conditions of incubation used by Festoff et al. Therefore contamination with these other components cannot account for the observed incorporation, and glycosylation by the nerve ending particles is highly likely.

The major glycopeptide synthesized in this in vitro study was identical to that found in synaptosome fractions after in vivo incorporation for 1 hr (140). When the synaptosomal fraction was lysed with water and subfractionated on a gradient after incorporation of glucosamine, about half the product was associated with sub-fractions enriched in mitochondria, suggesting that glucosamine incorporation by synaptosomes occurs in synaptosomal mitochondria. Glycoproteins (142) and glyco-syltransferases (143) have been found associated with a brain mitochondrial fraction.

CONCLUSION

Whereas synaptic macromolecules that participate in the synthesis of neuro-transmitters or bind neurotransmitters have been relatively easy to study because of available assays, identification of other synaptic macromolecules that may play a special role in synaptic function has been dependent on purification procedures still being refined. Improved fractionation of components of synaptosome fractions is increasing the potential for identification of unknown synaptic macromolecules, but fulfillment of the goals of this work remains to be achieved.

The origin of presynaptic macromolecules in nerve cell bodies and their transport to nerve endings has been established. Regulation of synaptic function by macro-molecular synthesis can be achieved by axoplasmic transport of proteins synthesized in the nerve cell body, or by synthesis of proteins on the ribosomes in post-synaptic dendrites. The possibility of macromolecular synthesis at nerve endings has attracted considerable attention since, if demonstrated, it might be implicated in presynaptic regulation of synaptic function. On the basis of present evidence, mito-chondria may be the only nerve ending component capable of protein synthesis. Glycosylation of proteins in nerve endings also occurs in mitochondria but does not appear to be confined to this site.

ACKNOWLEDGMENT

Preparation of this review was supported in part by grants from the National Institute of Mental Health, USPHS, and the Alfred P. Sloan Foundation.

Literature Cited

1. Pappas, G. D., Purpura, D. P., Eds. 1972. *Structure and Function of Synapses.* New York: Raven
2. Hall, Z. W. 1972. *Ann. Rev. Biochem.* 41:925
3. Gray, E. G., Whittaker, V. P. 1962. *J. Anat. London* 96:79
4. De Robertis, E., Pellegrino De Iraldi, A., Rodriguez De Lores Arnaiz, G., Salganicoff, L. 1962. *J. Neurochem.* 9:23
5. Whittaker, V. P., Michaelson, I. A., Kirkland, R. J. A. 1964. *Biochem. J.* 90:293
6. Whittaker, V. P. 1969. *Handbook of Neurochemistry,* ed. A. Lajtha, 2:327. New York: Plenum.
7. Whittaker, V. P., Barker, L. A. 1972. *Methods in Neurochemistry,* ed. R. Fried, 2:1. New York: Dekker
8. De Robertis, E., Rodriguez De Lores Arnaiz, G. See Ref. 6, p. 365
9. Morgan, I. G., Wolfe, L. S., Mandel, P., Gombos, G. 1971. *Biochim. Biophys. Acta* 241:737
10. Fonnum, F. 1968. *Biochem. J.* 106:401
11. Whittaker, V. P. 1968. *Biochem. J.* 106:412
12. Cotman, C., Mahler, H. R., Anderson, N. G. 1968. *Biochim. Biophys. Acta* 163:272
13. Day, E. D., McMillan, P. N., Mickey, D. D., Appel, S. H. 1971. *Anal. Biochem.* 39:29
14. Kurokawa, M., Sakamoto, T., Kato, M. 1965. *Biochem. J.* 97:833
15. Abdel-Latif, A. A. 1966. *Biochim. Biophys. Acta* 121:403
16. Autilio, L. A., Appel, S. H., Pettis, P., Gambetti, P. L. 1968. *Biochemistry* 7:2615
17. Cotman, C., Herschman, H., Taylor, D. 1971. *J. Neurobiol.* 2:169
18. Morgan, I. G., Reith, M., Marinari, U., Breckenridge, W. C., Gombos, G. 1972. *Glycolipids, Glycoproteins and Muco-polysaccharides of the Nervous System,* ed. V. Zambotti, G. Tettamanti, 209. New York: Plenum
19. Cotman, C. *Methods Enzymol.* In press
20. Cotman, C. W., Matthews, D. A. 1971. *Biochim. Biophys. Acta* 249:380
21. Lemkey-Johnston, N., Larramendi, L. M. H. 1968. *Exp. Brain Res.* 5:326
22. Israel, M., Whittaker, V. P. 1965. *Experientia* 21:325
23. Kataoka, K., De Robertis, E. 1967. *J. Pharmacol. Exp. Ther.* 156:114
24. Whittaker, V. P. 1966. *Pharmacol. Rev.* 18:401
25. Sheridan, M. N., Whittaker, V. P., Israel, M. 1966. *Z. Zellforsch.* 74:291
26. Giacobini, E. et al 1971. *J. Neurochem.* 18:223
27. Jones, D. G. 1967. *J. Cell Sci.* 2:573
28. Florey, E., Winesdorfer, F. 1968. *J. Neurochem.* 15:169
29. Dowdall, M. J., Whittaker, V. P. 1973. *J. Neurochem.* 20:921
30. Gonatas, N. K., Autilio-Gambetti, L., Gambetti, P., Shafer, B. 1971. *J. Cell Biol.* 51:484
31. Wofsey, A. R., Kuhar, M. J., Snyder, S. H. 1971. *Proc. Nat. Acad. Sci. USA* 68:1102
32. Fonnum, F. 1968. *Biochem. J.* 109:389
33. Barondes, S. H. 1968. *J. Neurochem.* 15:699
34. Barondes, S. H. 1964. *Science* 146:779
35. Barondes, S. H. 1968. *J. Neurochem.* 15:343

36. Feit, H., Dutton, G. R., Barondes, S. H., Shelanski, M. L. 1971. *J. Cell Biol.* 51:138
37. Eichberg, J., Whittaker, V. P., Dawson, R. M. C. 1964. *Biochem. J.* 92:91
38. Davis, G. A., Bloom, F. E. 1972. *Anal. Biochem.* 51:429
39. Cotman, C. W., Taylor, D. 1972. *J. Cell Biol.* 55:696
40. Kornguth, S. E., Flangas, L., Geison, R. L., Scott, G. 1972. *Brain Res.* 37:53
41. Spence, M. W., Wolfe, L. S. 1967. *Can. J. Biochem.* 45:671
42. Derry, P. M., Wolfe, L. S. 1967. *Science* 158:1450
43. De Robertis, E., Azcorra, J. M., Fiszer, S. 1967. *Brain Res.* 5:45
44. Cotman, C. W., Levy, W., Banker, G., Taylor, D. 1971. *Biochim. Biophys. Acta* 249:406
45. Cotman, C. W., Taylor, D. 1972. *J. Cell Biol.* 55:696
46. Bloom, F. E., Aghajanian, G. 1968. *J. Ultrastruct. Res.* 22:361
47. Pfenninger, K. H. 1971. *J. Ultrastruct. Res.* 34:103
48. Pease, D. C. 1966. *J. Ultrastruct. Res.* 15:355
49. Rambourg, A., Leblond, C. P. 1967. *J. Cell Biol.* 32:27
50. Kornguth, S. E., Anderson, J. W., Scott, G. 1969. *J. Neurochem.* 16:1017
51. Kornguth, S. E. et al 1971. *J. Biol. Chem.* 246:1177
52. Garey, R., Harper, J., Best, J. B., Goodman, A. B. 1972. *J. Neurobiol.* 3:163
53. Levitan, I. B., Mushynski, W. E., Ramirez, G. 1972. *J. Biol. Chem.* 247:5376
54. De Robertis, E., De Lores Arnaiz, G. R., Alberici, M., Butcher, R. W., Sutherland, E. W. 1967. *J. Biol. Chem.* 242:3487
55. Cheung, W. Y., Salganicoff, L. 1967. *Nature* 214:90
56. Florendo, N. T., Barrnett, R. J., Greengard, P. 1971. *Science* 173:745
57. Maeno, H., Johnson, E. M., Greengard, P. 1971. *J. Biol. Chem.* 246:134
58. Maeno, H., Greengard, P. 1972. *J. Biol. Chem.* 247:3269
59. Johnson, E. M., Veda, T., Maeno, H., Greengard, P. 1972. *J. Biol. Chem.* 247:5650
60. Greengard, P., McAfee, D. A., Kebabian, J. W. 1972. *Advances in Cyclic Nucleotide Research,* 1:327. New York: Raven
61. Feit, H., Barondes, S. H. 1970. *J. Neurochem.* 14:1355
62. Puszkin, S., Nicklas, W. J., Berl, S. 1972. *J. Neurochem.* 19:1319
63. Berl, S., Puszkin, S., Nicklas, W. J. 1973.

Science 179:441
64. Banker, G., Crain, B., Cotman, C. W. 1972. *Brain Res.* 42:508
65. Karlsson, J. O., Hamberger, A., Henn, F. A. 1973. *Biochim. Biophys. Acta* 298:219
66. Feit, H., Slusarick, L., Shelanski, M. L. 1971. *Proc. Nat. Acad. Sci. USA* 68:2028
67. Wannamaker, B. B., Kornguth, S. E. 1973. *Biochim. Biophys. Acta* 303:333
68. Ramirez, G., Levitan, I. B., Mushynski, W. E. 1972. *J. Biol. Chem.* 247:5382
69. Morgan, I. G., Breckenridge, W. C., Vincendon, G., Gombos, G. 1973. *Proteins of the Nervous System,* ed. D. Scheider, 35. New York: Raven
70. Bosmann, H. B., Case, K. R., Shea, M. B. 1970. *FEBS Lett.* 11:261
71. Waehneldt, T. V., Morgan, I. G., Gombos, G. 1971. *Brain Res.* 34:403
72. Breckenridge, W. C., Morgan, I. G. 1972. *FEBS Lett.* 22:235
73. McBride, W. J., Van Tassel, J. 1972. *Brain Res.* 44:177
74. De Robertis, E., Lapetina, E., Wald, F. 1968. *Exp. Neurol.* 21:322
75. Mickey, D. D., McMillan, P. N., Appel, S. H., Day, E. D. 1971. *J. Immunol.* 107:1599
76. Herschman, H. R., Cotman, C., Matthews, D. A. 1972. *J. Immunol.* 108:1362
77. Barondes, S. H. 1969. *Neurosciences Research Symposium Summaries,* 3:191. Cambridge, Mass: MIT Press
78. Grafstein, B. 1969. *Advan. Biochem. Psychopharmacol.* 1:11
79. Lasek, R. 1970. *Int. Rev. Neurobiol.* 13:289
80. Ochs, S. 1972. *Science* 176:252
81. Mc Ewen, B., Grafstein, B. 1968. *J. Cell Biol.* 38:494
82. Fibiger, H. C., Pudritz, R. C., McGeer, P. L., McGeer, E. G. 1972. *J. Neurochem.* 19:1697
83. Schubert, P., Lux, H. D., Kreutzberg, G. W. 1971. *Acta Neuropath. Suppl.* V:179
84. Karlsson, J. O., Sjostrand, J. 1971. *J. Neurochem.* 18:749
85. Wooten, G. F., Coyle, J. T. 1973. *J. Neurochem.* 20:1361
86. Zatz, M., Barondes, S. H. 1971. *J. Neurochem.* 18:1125
87. Karlsson, J. O., Sjostrand, J. 1971. *J. Neurochem.* 18:2209
88. McEwen, B. S., Forman, D. S., Grafstein, B. 1971. *J. Neurobiol.* 2:361
89. Elam, J. S., Agranoff, B. W. 1971. *J. Neurobiol.* 2:379
90. Jeffrey, P. L., James, K. A. C., Kidman, A. D., Richards, A. M., Austin, L. 1972. *J. Neurobiol.* 3:199

91. Peterson, R. P., Jurwitz, R. M., Lindsay, R. 1967. *Exp. Brain Res.* 4:138
92. Kristensson, K., Olsson, Y., Sjostrand, J. 1971. *Brain Res.* 32:399
92a. LaVail, J. H., LaVail, H. M. 1972. *Science* 176:1416
93. Dahlstrom, A. 1968. *Eur. J. Pharmacol.* 5:11
94. Kreutzberg, G. W. 1969. *Proc. Nat. Acad. Sci. USA* 62:722
95. Fernandez, H. L., Huneeus, F. C., Davison, P. F. 1970. *J. Neurobiol.* 1:395
96. Fernandez, H. L., Burton, P. R., Samson, F. E. 1971. *J. Cell Biol.* 51:176
97. Anderson, L. E., McLure, W. O. 1973. *Proc. Nat. Acad. Sci. USA* 70:1521
98. Grafstein, B. 1971. *Science* 172:177
99. Von Hungen, K., Mahler, H. R., Moore, W. J. 1968. *J. Biol. Chem.* 243:1415
100. Morris, S. J., Ralston, H. J., Shooter, E. M. 1971. *J. Neurochem.* 18:2278
101. Droz, B., Barondes, S. H. 1969. *Science* 165:1131
102. Austin, L., Bray, J. J., Young, R. J. 1966. *J. Neurochem.* 13:1267
103. Peterson, J. A., Bray, J. J., Austin, L. 1968. *J. Neurochem.* 15:741
104. Casola, L., Davis, G. A., Davis, R. E. 1969. *J. Neurochem.* 16:1037
105. Rahmann, H., Wolburg, H. 1971. *Experientia* 27:903
106. Bondy, S. 1971. *Exp. Brain Res.* 13:135
107. Wolburg, H. 1972. *Exp. Brain Res.* 15:348
108. Ingoglia, N. A., Grafstein, B., McEwen, B. S., McQuarrie, I. G. 1973. *J. Neurochem.* 20:1605
109. Autilio-Gambetti, L., Gambetti, P., Shafer, B. 1973. *Brain Res.* 53:387
110. Lasek, R. J. 1972. *Trans. Am. Soc. Neurochem.* 3:98 (Abstr.)
111. Morgan, I. G., Austin, L. 1968. *J. Neurochem.* 15:41
112. Gordon, M. W., Deanin, G. G. 1968. *J. Biol. Chem.* 243:4222
113. Appel, S. H., Autilio, L., Festoff, B. W., Escueta, A. V. 1969. *J. Biol. Chem.* 244:3166
114. Morgan, I. G., Austin, L. 1969. *J. Neurobiol.* 1:155
115. Bosmann, H. B., Hemsworth, B. A. 1970. *J. Biol. Chem.* 245:363
116. Austin, L., Morgan, I. G., Bray, J. J. 1970. *Protein Metabolism of the Nervous System,* ed. A. Lajtha, 271. New York: Plenum
117. Bridgers, W. F., Cunningham, R. D., Gressett, G. 1971. *Biochem. Biophys. Res. Commun.* 45:351

118. Hernandez, A., Burdett, I., Work, T. S. 1971. *Biochem. J.* 124:327
119. Goldberg, M. A. 1971. *Brain Res.* 27:319
120. Gambetti, P., Autilio-Gambetti, L. A., Gonatas, N. K., Shafer, B. 1972. *J. Cell Biol.* 52:526
121. Gilbert, J. M. 1972. *J. Biol. Chem.* 247:6451
122. Morgan, I. G. 1970. *FEBS Lett.* 10:273
123. Cotman, C. W., Taylor, D. A. 1971. *Brain Res.* 29:366
124. Droz, B., Koenig, H. L. 1969. *Cellular Dynamics of the Neuron,* ed. S. H. Barondes, 35. New York: Academic
124a. Ramirez, G. 1973. *Biochem. Biophys. Res. Commun.* 50:452
125. Attardi, G., Attardi, B. 1968. *Proc. Nat. Acad. Sci. USA* 61:261
126. Storrie, B., Attardi, G. 1973. *J. Biol. Chem.* 248:5826
127. Winzler, R. J. 1970. *Int. Rev. Cytol.* 29:77
128. Roth, S., McGuire, E. J., Roseman, S. 1971. *J. Cell Biol.* 51:536
129. Barondes, S. H., Dutton, G. R. 1969. *J. Neurobiol.* 1:99
130. Zatz, M., Barondes, S. H. 1971. *J. Neurochem.* 18:1625
131. Den, H., Kaufman, B., Roseman, S. 1970. *J. Biol. Chem.* 45:6607
132. Broquet, P., Louisot, P. 1971. *Biochimie* 53:921
133. Ko, G. K. W., Raghupathy, E. 1971. *Biochim. Biophys. Acta* 244:396
134. Ko, G. K. W., Raghupathy, E. 1972. *Biochim. Biophys. Acta* 264:1972
135. Festoff, B. W., Appel, S. H., Day, E. 1971. *J. Neurochem.* 18:1871
136. DeVries, G. H., Barondes, S. H. 1971. *J. Neurochem.* 18:101
137. Raghupathy, E., Ko, G. K. W., Peterson, N. A. 1972. *Biochim. Biophys. Acta* 286:339
138. Bosmann, H. B. 1972. *J. Neurochem.* 19:763
139. Marinari, U. M., Morgan, I. G., Mack, G., Gombos, G. 1972. *Neurobiology* 2:176
140. Dutton, G. R., Haywood, P., Barondes, S. H. 1973. *Brain Res.* 57:397
141. Reith, M., Morgan, I. G., Gombos, G., Breckenridge, W. C., Vincendon, G. 1972. *Neurobiology* 2:169
142. Bosmann, H. B., Myers, M. W., Dehond, D., Ball, R., Case, K. R. 1972. *J. Cell Biol.* 55:147
143. Bosmann, H. B. 1971. *Nature New Biol.* 234:54
144. Whittaker, V. P., Essman, W. B., Dowe, G. H. C. 1972. *Biochem. J.* 128:833

MEMBRANE RECEPTORS[1]

Pedro Cuatrecasas[2]

Department of Pharmacology and Experimental Therapeutics and
Department of Medicine, The Johns Hopkins University School of Medicine
Baltimore, Maryland

CONTENTS

INTRODUCTION

During the past few years considerable progress has been made in the study, identification, isolation, and purification of a variety of membrane-localized receptors. The general approach has been to study in detail the binding interaction between a labeled (highly radioactive in the case of hormone receptors) ligand, selected by the probability that it will be highly specific, and a tissue or organelle that contains the putative receptor structures. The binding must show absolute specificity, the affinity must be consistent with the biological activity of the ligand, and the number of binding sites must be consistent with the physiological mechanisms operative in the intact system (1). Since binding data (specificity,

[1] The survey of literature pertaining to this review was concluded in August 1973.

[2] The preparation of this review and the work described from the author's laboratory were supported by grants from the National Institute of Arthritis and Metabolic Diseases (AM14956), the National Science Foundation, American Cancer Society, and The Kroc Foundation.

affinity, number of sites) must be evaluated by careful and detailed comparisons with the biological activity of the ligand, it is desirable that the initial system used for studying the binding interaction be an intact system which is as simple as possible (such as intact, viable cells), so that the binding and biological properties can be ascertained in the same system before disruptive procedures are instituted. It is important that this system be simple, since the presence of heterogeneous cell populations, the presence of connective tissue or basement membranes, problems related to diffusion through thick multicellular matrices (as in intact tissues or slices), etc, may lead to serious complications which make interpretation of the binding data (such as rates of association, dissociation, saturability, role of non-specific binding) very difficult. When the binding activity relationships are sufficiently well established, a proper setting is established to evaluate the results of studies obtained after disrupting the cell and studying the binding properties in isolated particulate (membrane) preparations. The binding properties (specificity, affinity, number of binding sites) should in principle not be altered. Since the constancy of the number of binding sites and the detailed binding properties in these broken cell preparations are quite important, it is frequently wise to initially perform these binding studies in relatively crude membrane preparations. Long and tedious membrane purifications done prematurely may give misleading results because of damage or loss (unrecognized) of receptor structures during these procedures. In many cases, the biological activity (e.g. enzyme activity) can also be determined in the isolated membrane preparation. However, some caution must be exercised in interpreting the activity of enzymes in membrane preparations, especially the modulation of activity by hormones or regulators, since important differences may exist compared to the intact cell. It is well known, for example, that very gross changes (e.g. susceptibility to chemical and enzymatic modifications) occur in erythrocyte membranes by preparing ghosts by very simple and presumably mild procedures. Since subsequent steps to isolate receptors involve dissolution of membranes, which certainly will alter or destroy residual biological activity, an ultimate goal in such studies is the reconstitution of a biologically responsive system utilizing the isolated and purified components.

As more information is gained concerning the nature of membrane receptors and their interaction with the natural regulators (e.g. hormones), greater focus will be placed on the nature of the molecular events which occur in the membrane following formation of the initial complex. How does a hormone-receptor interaction convey information to membrane-localized enzymes and transport structures, or perhaps grossly modify the organization of membrane proteins? The answers to these questions will very likely be influenced by and be dependent on our understanding of the dynamics and structure of cell membranes.

INSULIN RECEPTORS

General Considerations

The proposition that insulin receptors are localized in the cell membrane was expressed by Levine and colleagues (1a, 2) on the basis that the hormone facilitates the transmembrane transport of sugars in a variety of extrahepatic tissues (3, 4), including striated (5–7) and cardiac (8, 9) muscle and adipose tissue (10). The first evidence for insulin binding to hormone-sensitive tissues arose from the indirect studies of Stadie and co-workers (11) who observed that rat diaphragms incubated briefly with insulin and washed showed a persistent increase in glycogen synthesis. Similar observations were made with lactating mammary gland (12) and epididymal fat pads (13). Subsequently these and other investigators attempted to directly demonstrate binding of the hormone to striated muscle and adipose tissue using labeled radiosulfur or radioiodine (14–19). These studies, however, were hindered by the low specific activity of the labeled insulin, and the specificity of binding in these studies was questioned (20–23). It was not possible to discriminate between specific binding to receptors and nonspecific, physical adsorption (24) to other tissue structures (reviewed in 20). Wolthmann & Narahara (19) demonstrated that in intact striated muscle only a very small fraction of the bound insulin was bound specifically, and this was detectable only at very low concentrations of the hormone. Furthermore, the original concept of persistent binding of labeled insulin to intact tissues was questioned by demonstrations that the biological activity of insulin in perfused isolated rat hearts disappeared very quickly upon discontinuing infusion of insulin (25). Similar studies were reported upon removal of insulin in the medium of adipose tissue (26, 27).

Perhaps the principal problem of all these early studies relates to the use of intact tissues rather than isolated cells or relatively homogeneous subcellular organelles. Because of the known propensity of insulin to adsorb to a variety of organic and inert surfaces, it is not surprising that in intact tissues insulin will bind to surfaces other than cytoplasmic membranes of the sensitive cell. It is now known that even in the absence of connective tissue or extraneous tissue, insulin can bind nonspecifically and in substantial quantity to nonreceptor regions of the cytoplasmic plasma membrane. Another serious problem with binding studies that utilize intact tissues in vitro is that the hormone is not presented directly to the sensitive cells, nor is it presented by way of the microvasculature as occurs in the normal in vivo state. The hormone must diffuse throughout the tissue, probably unevenly. Upon removal of the medium, diffusion of the hormone from the extracellular space is probably quite different from the original pathways for the inward diffusion, since the substantial quantity of nonspecifically adsorbed hormone in the interstices of the tissue may again adsorb (or bind to receptors) immediately upon desorption. Thus, a significant pool of nonreceptor bound hormone may be maintained in the extracellular compartments without significant release into the medium external to the tissue.

An early attempt to overcome this problem was described by Crofford (27), who measured the disappearance of immunoreactive insulin from the incubation medium of isolated fat cells. Although the complexities of utilizing intact tissues were avoided, it was difficult by this approach to distinguish between hormone binding and hormone degradation (the cell concentration was very high), and the contribution of nonspecific adsorption of insulin to cells was difficult to assess. Furthermore, since recent studies have shown that with physiological concentrations of insulin (or other peptide hormones) only a very small fraction (less than 5%) of the total hormone in the medium is specifically bound to tissue receptors, calculations of binding by measurements of the amount of hormone disappearing from the medium may be quite inaccurate. Further confirmation of the ineffectiveness of isolated fat cells in extracting insulin in the medium comes from studies in which the biological activity of medium containing submaximal concentrations (10 μunits per ml) of insulin which was used to stimulate cells for 30 min (37°) was only slightly decreased (5 to 10%) when subsequently tested with fresh cells.[3]

Considerable evidence has been accumulated during the past few years which indicates that insulin receptors are located on the cell surface. Trypsin (28–32) and insoluble trypsin and chymotrypsin derivatives (32) destroy the insulin responsiveness of fat cells, and very low concentrations of this protease (28, 33, 34) as well as of phospholipases (35–37) exert insulin-like effects on these cells; these effects are dependent on the catalytic integrity of these enzymes. Perhaps more direct evidence for the superficial localization of receptors is the demonstration that insulin attached covalently to insoluble polymers, such as agarose, retain biological activity (38–42) and can physically interact with fat cells by forming strong complexes (43). It is equally pertinent that large soluble dextran polymers containing covalently bound insulin are also biologically active (44–46).

In the last few years direct evidence for the localization of insulin receptors to the plasma membrane of the cell has been obtained from extensive binding studies that utilize highly radioactive ^{125}I-labeled insulin and isolated cells or microsomal membrane preparations. Since many of these studies have been reviewed recently (47–51a), only the salient features and the more recent studies are summarized in this article.

Direct Measurements of Receptor Interactions

Several recent studies have demonstrated that insulin labeled with ^{125}I can bind to insulin-sensitive cells or subcellular fractions with properties that are consistent with a specific receptor binding process. Such interactions have been demonstrated in isolated fat cells (52–54), membrane fractions from isolated fat cells (55, 56) or liver (57–62), circulating blood cells and lymphocytes (63–65), thymocytes (66), and murine mammary cells (66a). These studies have depended on the use of biologically active insulin labeled with radioiodine at a specific activity of 500 to 2000 Ci per mmole. Insulin labeled with chloramine-T (67) at an average of less than one atom of iodine per molecule and purified by adsorption methods (68) is biologically active (52). Furthermore, monoiodoinsulin separated from native insulin by ion exchange chromatography is fully active biologically (60). The importance of using properly labeled and purified insulin has been stressed (32, 52) since biologically inactive, damaged components exhibit a high degree of nonspecific binding to cells or membrane fractions.

The basic properties of the insulin-receptor interaction in the various tissues and membrane preparations studied appear to be very similar. The binding is time-dependent and a high affinity (K_i about 10^{-10} M) site is detectable (52, 55, 56, 59,

[3] P. Cuatrecasas, unpublished.

61, 65), which is experimentally saturable with respect to insulin. The existence of an additional low affinity (K_i about 10^{-9} M) site has been surmised from extrapolations of Scatchard plots and competition studies (56, 65), but direct demonstration of saturability has not been possible.

Since nonspecific binding is generally related linearly to concentration, increasing nonspecific binding occurs with increasing concentrations of ^{125}I-labeled insulin; data obtained at these higher concentrations is more difficult to interpret. Under these conditions the differences between specific and nonspecific binding may be very small. The possibility that insulin at higher concentrations may bind to receptors for other noninsulin-related processes must be considered (1, 51a), especially in view of the demonstration that insulin and somatomedin may share common receptors in at least certain tissues (69). Simply demonstrating that the native hormone at high concentrations (e.g. 10^{-5} or 10^{-6} M) can prevent the binding of low concentrations of the labeled hormone is not sufficient to establish specificity of binding. The absence of saturability of nonspecific adsorption (to glass, for example) is valid only within reasonable limits. Small but reproducible displacement of binding of ^{125}I-insulin by high concentrations of native insulin to certain types of filters has been observed frequently under certain conditions, and very large quantities of such "specific" binding can be observed in the absence of tissue when incubation mixtures in at least certain types of glass test tubes are shaken (24°, 30 min) relatively vigorously. In the latter case, it may be that minute quantities of the glass surface are released, that their adsorptive capacity is limited by the very small quantity of material, and that these inert particles may not pass through the filters used in binding assays (or may sediment upon centrifugation). In such cases, however, the binding is not readily displaced by low concentrations of the native hormone, indicating that the affinity of these latter processes is not high (K_i 10^{-9} M or less). Stereospecific binding of opiates to glass filters has been observed,[4] and the stereospecificity of trytophan binding to albumin is well known. In the study of very high affinity systems, which in cases may approach energies of covalent interactions, some caution must be exercised in the use of Scatchard plots or conventional kinetic analyses (1, 51a, 69a). Since these are designed primarily for evaluation of low affinity, reversible interactions, some aberrant behavior may be observed with interactions which, for example, demonstrate direct proportionality (linearity) of binding dependent on the concentration of ligand over a given range of ligand concentration. It is not infrequent to observe such a direct linear relationship of binding over the very low concentration range of a labeled hormone if the receptor is in great excess. Such binding will result in a straight line, parallel to the horizontal axis, in a Scatchard plot. Furthermore, the simultaneous existence of heterogeneous classes of low affinity nonspecific adsorptive sites can result in Scatchard plots which deviate from linearity and erroneously suggest the existence of a "second" specific binding site.

The high affinity (K_i about 10^{-10} M) of the site for insulin binding in the various tissues studied is very similar to the physiological concentration of this hormone. In certain tissues, such as thymocytes obtained from suckling rats (66), the estimated

[4] S. H. Snyder, C. B. Pert, and G. Pasternak, personal communication.

affinity constants for insulin binding and for biological activity (increased influx of α-aminoisobutyric acid) are about 2 to 4×10^{-8} M, and saturation (or the maximal response) is not observed until a concentration of about 10^{-6} M is reached. Since it is difficult to reconcile these data with the physiological concentrations of the hormone, the possibility must be considered (51a) that the interaction being measured may result from a small (less than 1% could be sufficient if its affinity were near 10^{-10} M) contaminant in the insulin, or it may measure a receptor for another, unrecognized hormone or growth factor which may share some cross-reactivity with insulin in a way similar to that described for insulin and somatomedin (69) as well as other hormone systems (1, 51a). It is also likely that very high concentrations of insulin may, like other unrelated proteins, cause insulin-like effects by nonreceptor interactions with the cell surface (66a). The very high concentrations of insulin required to observe biological effects on immature thymocytes (66), and the prolonged incubation periods required to observe the effects (in contrast to effects in fat tissue or muscle which are immediate), resemble the well known ability of insulin to alter the growth or metabolic properties of cells in tissue culture (70–73). Such effects are seen only when the hormone is used in very high concentrations. It is pertinent, furthermore, that Minemura, Lacy & Crofford (74) demonstrated that in isolated fat cells, which are extremely sensitive to insulin, the hormone does not change the initial rate of α-aminoisobutyric acid uptake, the initial rate of efflux, or the equilibrium concentrations; in this tissue equilibrium is achieved within 5 min at 24° (74). Great caution must therefore be exercised in "receptor" interpretations of biological effects or binding interactions which occur only at extremely high and unphysiologic concentrations of a hormone. The possible relations between insulin and other growth-promoting peptides are considered in a later section.

The maximal binding capacity of isolated fat cells is about 10,000 molecules of insulin per intact fat cell (52) and 0.1 pmole of hormone per mg of liver membrane protein (59). The rates of association and dissociation of the insulin-receptor complex, which can be determined independently, obey the kinetics of simple dissociable processes (52, 55, 59). The dissociation constant calculated from these rate constants is about 10^{-10} M; a similar value is obtained independently from equilibrium data. The binding is chemically specific with respect to insulin and insulin analogs (55, 61, 62). Native insulin competes for binding in a way expected from the biological identity of iodo- and native insulin; desalanine insulin is indistinguishable from native insulin; proinsulin binds to the receptor with an affinity about 20 times less powerful than that of native insulin; desalanine-desasparagine and desoctapeptide insulins show a further decrease in affinity; reduced insulin, and S-sulfonated and carboxymethylated chains of insulin do not compete for binding.

The insulin-receptor complex can be dissociated with acid, and the insulin released from the complex is similar to unused insulin in several physical properties, its ability to interact with membrane receptors and antibody, and its ability to stimulate glucose transport in isolated fat cells (52, 55, 62). Virtually complete dissocation of the cell-insulin complex can also be accomplished by a large excess of insulin antiserum, and the cells' ability to bind insulin after such dissociation is not impaired (52). Although virtually any tissue or subcellular fraction can be shown to inactivate insulin under appropriate conditions, it is possible to demonstrate that equilibrium binding at physiological concentrations of insulin can occur without significant inactivation in the incubation medium using intact fat cells (52), fat cell membranes (55), and crude liver cell microsomal membranes (59). Marked inactivation of the hormone, accompanied by impaired immunoreactivity and altered physical properties, occurs with liver cell membranes prepared according to Neville's method (62). The latter inactivation process affects native insulin and biologically inactive insulin derivatives such as desalanine-desasparagine insulin to the same extent; it shows lower affinity for insulin than does the process of binding, indicating that binding and inactivation are not related (62).

Other properties of the insulin-receptor interaction have been studied in fat cell membranes (55). Measurement of the rate constants at various temperatures indicate that binding is much tighter at lower temperatures because the decrease in dissociation rate is disproportionately greater than the decrease in association rate. Thermodynamic calculations suggest a ΔF of -14 kcal/mol, ΔH of -28 kcal/mol, and a ΔS of -45 kcal/mol per deg (55). Thus the heat energy for binding of insulin to membrane receptors is quite favorable and is probably the major driving force for the reaction. Binding is virtually unaffected by the nature of the buffer used or by a number of different ionic species, heavy metals, or metal-complexing agents. The optimum H^+ concentration for binding occurs sharply at about pH 7.5. An abrupt thermal transition with irreversible inactivation occurs with a midpoint at 53°C. High ionic strength causes a dramatic (up to sixfold) increase in insulin binding; this probably results from the appearance (unmasking) on the membrane of new binding sites for insulin. These are kinetically indistinguishable from those normally exposed. The effects of high ionic strength appear qualitatively similar to those of digesting membranes with phospholipases (75). Treatment of the membranes with several protein-modifying reagents suggests that tyrosyl and possibly histidyl residues may be important in the binding interaction.

The relationships between insulin binding and biological activation have been examined in intact fat cells (32, 52, 75). Evidence that the high affinity binding observed is indeed a reflection of interaction with a biologically significant structure includes: 1. the similarity in the concentration dependence of insulin binding and hormone enhancement of glucose oxidation; 2. the exact correspondence (measured by affinity) between the relative ability of insulin derivatives (e.g. proinsulin) to enhance glucose oxidation and to compete with iodoinsulin for binding; 3. the parallelism between binding and biological effectiveness of insulins from various species (61); and 4. the similarity of the consequences on binding and on biological activation of certain modifications of the cell surface which specifically alter the affinity of the receptor for insulin, such as digestions by proteolytic enzymes (32).

Subcellular Localization of Receptors

The ability of insulin-agarose derivatives to simulate the metabolic effects of insulin suggests the presence of insulin receptors on the cell surface (38). The insulin binding activity of intact fat cells is quantitatively recovered in the particulate fraction of the cell homogenate (52). The virtually complete loss of insulin binding after digestion of intact cells by trypsin-agarose is not altered by homogenization, indicating that no intracellular structures exist which are capable of specifically binding insulin to a significant extent (76). Differential centrifugation reveals that the insulin binding activity of fat and liver cell homogenates is preferentially localized in the microsomal fraction of the cell (52).

The asymmetric positioning of the insulin receptor has been examined by comparing the binding of insulin to everted and inverted membrane vesicles prepared from fat cell ghosts by procedures similar to those used with erythrocyte ghosts (77). Inverted vesicles do not bind insulin unless they are disrupted by sonication or phospholipase digestion. Insulin bound to receptors and trapped inside such vesicles is not easily released into the medium, indicating that insulin binding sites are restricted to the external surface of the membrane and that physical inversion of the normal membrane orientation is not accompanied by a major rapid reorientation of these specific membrane proteins.

Enzymatic Modification of Receptors

Early studies (28–32) demonstrated that the sensitivity of fat cells to insulin was markedly reduced by digestion of the cells with trypsin and that the response of these cells to epinephrine and to cyclic AMP (78) was unimpaired. These studies suggested the possibility that enzymatic modification of the intact cell or membranes might directly alter the insulin receptor, perhaps relatively specifically. Very mild digestion of fat cells by trypsin causes a selective fall in the affinity of the insulin-receptor interaction and a corresponding fall in the apparent affinity of insulin-stimulated glucose transport (32). This occurs under conditions where the total amount of receptor and the maximal insulin response are unaffected. More drastic digestion of fat cells with trypsin causes severe loss of insulin responsiveness and insulin binding (28, 32, 53, 54). Such digestion decreases the total amount of receptor available for binding. When adipose tissue cells are first treated with phospholipases, trypsin causes severe destruction of the receptor without effects on affinity, and this occurs with much lower concentrations of the enzyme than are required to modify the native cells or membranes (32, 75).

Although digestion with low concentrations of neuraminidase has insulin-like effects on isolated fat cells, drastic digestion of the cells causes a fall in the basal rate of glucose transport and abolishes the effects of insulin on glucose transport and lipolysis (79, 80). Such enzymatic digestion, however, has no effect on the total quantity of receptors for insulin or on the affinity of insulin-receptor interaction (79). The dissociation of the biological and receptor binding properties caused by neuraminidase digestion suggests that membrane sialic acid is not involved in the recognition function of the receptor, but may be involved in the receptor's ability to convey the binding interaction to adjacent structures.

Digestion of adipose tissue cells by neuraminidase followed by digestion by β-galactosidase, or simultaneous digestion by these two enzymes, in addition to abolishing the biological effects of insulin, causes a profound fall in hormone binding (50). These findings suggest that galactose groups are involved in the recognition function of the receptor and may be chemical constituents of this molecule. Further evidence for the glycoprotein nature of the receptor is presented in a later section.

Digestions of fat and liver cell membranes with phospholipases C or A cause a dramatic (three- to sixfold) increase in insulin binding (75). These changes appear to reflect an increase in the total binding capacity of the membranes. The new exposed binding sites are kinetically indistinguishable from those normally present in the membrane. If the cells are first digested with trypsin, subsequent phospholipase digestion still has major effects, indicating that these buried binding sites are inaccessible to macromolecules in the medium. Addition of exogenous phospholipids can partially reverse the effects of phospholipases, and various chemical compounds that interact with membrane phospholipids mimic the effects of these enzymes. Extraction of membrane lipids with various organic solvents causes effects similar to those of phospholipase digestion. Since high concentrations (2 M) of NaCl have effects which are similar to those observed after removal or hydrolysis of phospholipids, the possibility was considered that masking was occurring through the polar head groups of the phospholipids (75).

Binding of Insulin to Lymphocytes and the Relation to Transformation

Recently it has been observed (63) that during concanavalin A-mediated blast transformation in human lymphocytes, there is a dramatic appearance of cell surface receptors for insulin. It has not been possible to detect significant specific binding of insulin to untransformed human circulating lymphocytes, whether this is measured in whole lymphocytes, broken cell preparations, or detergent-solubilized membrane preparations (63).[5] These results contrast with those obtained (63) using human lymphocytes (RPMI6237) maintained in long term culture, or in cells obtained from patients with acute lymphocytic leukemia, where specific insulin binding is readily observed in both whole and broken cell preparations. The insulin binding which appears in transformed lymphocytes is specific, in that it is of high affinity $K_D \simeq 10^{-9}M$, is saturable with respect to insulin, and is not affected by unrelated peptides such as glucagon and ACTH. In addition, as with the fat cell (75), the number of available insulin binding sites is augmented about threefold by digesting transformed lymphocytes with phospholipase C (63). This effect on insulin binding is not observed when untransformed lymphocytes are treated with phospholipase C. It is likely that the insulin binding sites which appear in association with transformation are either synthesized de novo or are intially present in a form incapable of binding insulin.

The inability to detect a significant number of insulin receptors in nylon column-purified resting peripheral lymphocytes (63) contrasts with other studies reported recently (64, 65). In the studies reported by Krug et al (63), it was demonstrated that insulin binding can be detected in unfractionated peripheral lymphocytes, but not in lymphocytes purified by passage through nylon columns. These columns remove circulating macrophages, which bind very large amounts of insulin,[5] and probably a proportion of those lymphocytes bearing surface immunoglobulins (B cells). According to Gavin et al (65), the amount of insulin binding detected in unstimulated cell populations is 1/300 the amount bound by cells in permanent

[5] M. Hollenberg and P. Cuatrecasas, unpublished.

culture (Figure 1, Ref. 65). This small amount of insulin binding (an average of about three receptors per cell) could well be attributed to the presence of small numbers (0.3% or less) of macrophages, platelets, or spontaneously transformed lymphocytes in the unstimulated peripheral lymphocyte population. Archer et al (81) report that the procedures used for isolating lymphocytes yield 80–85% lymphocytes, 15–20% erythrocytes and granulocytes, and a variable number of platelets. It has not been possible in this laboratory to elicit biological responses in unstimulated, purified human circulating lymphocytes with physiological concentrations of insulin. Hadden et al (82) have described an inhibition of lymphocyte membrane Na^+-K^+-ATPase by insulin, but these studies were performed in cultured lymphocytes (RPMI $\#$ 1788), which are known to possess insulin receptors.

Whether both T and B cells bind insulin when stimulated remains to be determined since the nylon column-purified preparation, while enriched in T cells, is not entirely free from B cells. It may be significant in this respect that acute lymphocytic lymphoblasts possess insulin receptors equivalent in number to those found in normal cells transformed in vitro with concanavalin A (63). In contrast, it was not possible to detect insulin receptors in chronic lymphocytic leukemia cells.[5] These cells are also refractory to stimulation by plant lectins (phytohemagglutinin and concanavalin A) (83) and they bind considerably less lectin than do normal cells (84).[5]

More recent studies (85) have shown that the appearance of insulin receptors does not depend specifically on concanavalin A as the mitogenic stimulus since another lectin, phytohemagglutinin, as well as the chemical reagent periodate (86), also cause the appearance of receptors during blastogenesis. It will be important to determine whether such receptors also appear when lymphocytes are subjected to a more physiological stimulus, namely a specific antigen. The emergence of receptors clearly follows activation of cellular RNA and protein synthesis, and it is not a consequence of cell division (83, 85). The change in receptors precedes the morphologic changes of transformation. These emerging receptors may thus be related to the continuation of some process occurring after the initial transformation stimulus but before cell division.

There is now evidence that growth hormone receptors are present in both stimulated[6] and cultured human (87) lymphocytes. Receptors for nerve growth factor appear during transformation in a way similar to insulin (85). Receptors for other polypeptide hormones, such as glucagon and epidermal growth factor, could not be detected in transformed or untransformed cells (85).[5] The specificity of the process of receptor emergence is further suggested by studies of changes in the binding properties of lymphocytes for plant lectins during transformation (88). The appearance of new hormone receptors may in part reflect the development of new cellular capabilities or the requirement for further hormonal action to complete the process of transformation once initiated by a variety of stimuli. The possible relation of the lymphocyte insulin receptors to processes of growth and to other growth-promoting substances or hormones is considered in another section.

[6] R. S. Bockman and M. Sonenberg, personal communication.

Receptors in Insulin-Resistant States

The possibility that metabolic states characterized by insulin resistance may be associated with derangements of the insulin receptor has been examined in several systems. The number of insulin binding sites per cell as well as the affinity of the receptor sites appear to be normal in rats that are starved, treated with prednisone, or made diabetic by administration of streptozotocin, and in other species which show decreased responsiveness to insulin (89). Large adipocytes from adult obese rats also have normal binding capacities and affinities for insulin (90). The possibility that insulin resistance in obese rats might be explained by a decrease in the cell-surface receptor density has been discussed (47). A decrease in insulin binding to liver and fat cell membranes has been observed in the obese-hyperglycemic syndrome of the mouse, a recessively inherited trait characterized by marked obesity and insulin resistance (91–93). Such studies have inherent difficulties which relate to the need for comparative analyses with tissues from animals (lean litter mates) that are metabolically and nutritionally different from the obese animals. For example, the lipid or protein composition (the usual reference point) of the membranes may be different in the two states. A special problem is that since in the obese cells the cell volume is increased to a greater degree than the cell surface (membrane), and since the membranes used for study include intracellular as well as cytoplasmic membranes, there may be a relative dilution of surface receptor sites in the final preparation. The studies (91–93) of the obese-hyperglycemic mice have considered these and other complications and have included several kinds of controls before concluding that the alteration in insulin binding is explained by a decrease in the total number of receptor sites available rather than in the affinity of these sites for insulin. It is not yet known whether these observations are causally related to metabolic defects observed in this syndrome (1). It is of interest that Kaplan et al (94) have described kinetic and electrophoretic abnormalities of cyclic AMP phosphodiesterase in fat cells from the genetically obese mouse.

Archer et al (81) have recently studied insulin receptors in lymphocytes from insulin-resistant obese and acromegalic patients and from insulin-sensitive hypopituitary subjects. Small but significant differences were observed in the concentration of insulin required to inhibit 50% of maximal binding between the cells from the normals and the obese, but not the acromegalic or hypopituitary patients. As discussed earlier, the extremely low amount of insulin binding to circulating cells required the use in these studies of very large quantities of cells $(0.5–1.5 \times 10^8$ cells/ml), and the total amount of ^{125}I-labeled insulin bound varied between 2 to 4% of the total present in the medium. Since the amount of insulin binding in unstimulated cells, if present at all, is at least 300 times lower (Figure 1, Ref. 65) than in cultured or transformed cells, or in macrophages (63),[5] the differences in insulin binding in these studies could be readily explained by very small differences in the composition of cells in the samples. If the diabetic cells contained one or two more macrophages, transformed lymphocytes, or perhaps even B cells per 100 cells, the differences in insulin binding observed could be explained on the basis of differences in cell populations rather than insulin receptors. Because of these difficulties, it is likely that answers to the question of whether human diabetes or other insulin-resistant states are characterized by defective insulin receptors will probably require the use of other techniques or of other tissues besides circulating cells.

Solubilization and Purification of Receptors

The insulin binding macromolecules can be quantitatively extracted in soluble form from liver and fat cell membranes with non-ionic detergents such as Triton X-100 (76, 95) and from lymphocytes with the detergent NP-40 (96). Gavin et al (97) have described the appearance of insulin receptors in the medium of lymphocytes incubated in the absence of detergents. This material may prove useful in studying the properties of receptors in the absence of detergents. In these studies, however, the quantity of receptor solubilized was not reported, and the residual binding in the lymphocytes was not determined. In view of the fact that in virtually all tissues, including lymphocytes (96), insulin receptors appear to be integral components (98) of cell membranes, which require virtual dissolution (with detergents) of the membrane to achieve reasonable extraction, it is unlikely that reasonably useful quantities of receptor can be obtained without the use of detergents. The insolubility of the receptor, or its localization in the membrane, must be considered as a relative state, and some fraction, however minute, may be released by simple in vitro manipulations. It can be readily shown (76, 95),[5] for example, that small amounts (less than 10% of the total present) of insulin receptor can be solubilized by simply shaking (in neutral buffers or distilled water) liver or fat cell membranes in the absence of detergents. In the absence of detergents, however, these structures precipitate with time after they are concentrated to characterize their properties.

The kinetics and specificity of the interaction of insulin with the detergent extracted receptor of liver and fat cell membranes are similar to those observed in the original particulate system (76, 95). The binding interaction is not affected by NaCl, neuraminidase, or phospholipase digestion (76). It is totally suppressed by concentrations of trypsin too low to affect binding in particulate fractions. Some physico-chemical parameters of the binding macromolecule have been determined (76). Gel filtration and density gradient centrifugation experiments indicate a Stokes radius of 70 Å and a sucrose sedimentation coefficient of 11S. These parameters are unaltered by phospholipase digestions and delipidation of the membranes before solubilization. It has been calculated that the receptor has a molecular weight of about 300,000, a frictional ratio of about 1.5, and an axial ratio of about 9. Since this data does not correct for the amount of detergent bound to the protein, it is expected that the true molecular weight may be lower (perhaps by 20 to 30%). A recent study has described some properties of the interaction of non-ionic detergents and membrane proteins (99). The molecular parameters of the solubilized insulin receptor are very similar to those described for the acetylcholine receptor solubilized from *Electrophorus electricus* (100).

Purification of the soluble insulin binding macromolecules is complicated by the continual requirement for detergent for maintenance of solubility and the extraordinary scarcity (10^{-4}% of the protein of a rat liver homogenate) of the binding protein in biological tissues (101). About sixtyfold purification can be achieved by ammonium sulfate precipitation followed by ion exchange chromatography (101). Several insulin-agarose derivatives have been synthesized that can selectively adsorb, in affinity chromatography experiments, the insulin binding protein from the detergent extracts of the membranes (101). Combination of these various procedures affords a purification of the receptor of about 250,000, which approaches theoretical purity (101). Because of the ability of certain plant lectins to interact with detergent-solubilized insulin receptors (102, 103), agarose affinity columns containing con-canavalin A and especially wheat germ agglutinin can effectively adsorb the receptor molecules (102). Elution can be achieved with buffers containing sugar molecules specific for the lectins. Substantial purification (5000-fold) can be achieved with these columns. These procedures have the special advantage of avoiding possible contact with insulin during purification. Despite the availability of methodology that permits very substantial purification of the receptor in small-scale experiments, considerable difficulties are encountered in attempts to purify large quantities of the receptor. The principal problems arise from the very large quantities of starting material, which require protracted time for chromatography on affinity columns.

The formidable nature of the purification of insulin receptors from mammalian membranes can perhaps be appreciated by considering that whereas a nearly 500,000-fold purification is needed to purify these receptors, acetylcholine receptors from *Torpedo* or electric tissue only require a purification of 100- to 500-fold (104, 105).

Plant Lectins, Cell Growth, and Mechanism of Action of Insulin

It has recently been demonstrated that the plant lectins, concanavalin A (102, 106) and wheat germ agglutinin (102), have marked insulin-like properties on isolated fat cells. Very low concentrations of these proteins are as effective as insulin in enhancing the rate of glucose oxidation and in inhibiting epinephrine-stimulated lipolysis in fat cells. In addition, these plant lectins can interact directly with the insulin receptor of fat cells in a way which markedly perturbs the insulin-receptor interaction (103). Low concentrations of wheat germ agglutinin enhance the apparent affinity of the insulin-receptor interaction in liver and fat cell membranes. Higher concentrations of this plant lectin, as well as concanavalin A, competitively displace the binding of insulin to receptors in these tissues. Such effects are equally apparent with insulin binding macromolecules solubilized from membranes, indicating that the plant lectins directly interact with insulin receptors. The fact that these lectins can bind directly to insulin receptors, and even compete with insulin for binding, does not necessarily mean that the insulin-like effect of these compounds is mediated through this direct interaction with insulin receptors (102, 103). The fact that digestion of cells with trypsin markedly alters the insulin response while causing only a slight or no change in the response to concanavalin A (103, 106) may suggest that the concanavalin A effect could be distal to the insulin receptor. This is not conclusive, however, since the sensitivity of the receptor molecules to tryptic digestion may be different with respect to interactions with concanavalin A and insulin if, as is likely, the exact same portion of the molecule is not involved in binding both ligands (103). The nature of the interaction of concanavalin A and wheat germ agglutinin with fat cells has been studied in detail (107).

Physiological concentrations (10^{-10} M) of insulin can inhibit adenylate cyclase activity in isolated liver and fat cell membranes, fat cell ghosts (108–112), fibroblasts (113), and isolated membranes of *Neurospora crassa* (114, 115). This effect of insulin is also mimicked by low concentrations of the lectins, concanavalin A and wheat germ agglutinin (102). It is possible that the mitogenicity of concanavalin A might be related to its insulin-like actions, and more specifically to its effects on adenylate cyclase activity or a closely related biochemical process in the cell membrane. Those concentrations of concanavalin A that lower adenylate cyclase activity in fat cell membranes correspond to those concentrations which stimulate lymphocytes (85). On the other hand, those concentrations of concanavalin A that stimulate adenylate cyclase activity correlate well with those concentrations of lectin that inhibit lymphocyte transformation and lead to cell death. Compounds that elevate lymphocyte cyclic AMP concentrations inhibit lymphocyte transformation and lymphocytes exposed for 24 hr to phytohemagglutinin have decreased levels of cyclic AMP (116, 117). These considerations are important in view of recent demonstrations of the relations between cyclic AMP and cell growth (118–120).

Mitogenic stimuli have recently been demonstrated to cause immediate, very large elevations of cyclic GMP in lymphocytes (121). Insulin has similarly been shown (122) to cause rapid and significant elevations of this cyclic nucleotide in sensitive tissues (fat cells, liver) but not in resting lymphocytes that lack insulin receptors and are apparently insensitive to this hormone. Thus in this activity at least, certain lectins can also be considered to be insulin-like. Recent evidence suggests that there may be important reciprocal relationships between cyclic AMP and cyclic GMP (121–124). Although it is not yet known by what mechanisms these cyclic nucleotides are regulated in a concerted fashion, it is quite possible that such changes may be intimately related to the processes of mitogenesis. The possibility must be considered that mitogenic stimuli are artificially eliciting fundamentally insulin-like biochemical responses (e.g. inhibition of adenylate cyclase, elevation of cyclic GMP, suppression of cyclic AMP), which are capable of activating a cell that lacks receptors for insulin or for other normally occurring stimuli that have insulin-like activity. In this hypothesis it is anticipated that antigenic activation of the lymphocyte would similarly occur by selectively triggering an insulin-like response via highly specific receptors for that antigen on the surface of the lymphocyte.

The possibility exists that the emergent lymphocyte surface structures detected by insulin binding (83, 85) may not be true receptors for insulin but for another, yet unrecognized but closely related hormone or growth factor in serum which may be important in regulating the terminal process of mitogenesis (85, 51a). It has recently been demonstrated that somatomedin (sulfation factor, thymidine factor) has potent insulin-like properties in fat cells and can effectively compete with insulin for binding to receptors in liver, fat, and chondrocyte membranes (69). Somatomedin has also been shown to inhibit adenylate cyclase activity in broken cell preparations (125). Another peptide with insulin-like and growth-promoting activities, which has properties in common with somatomedin, has recently been isolated from calf serum and from Buffalo rat hepatoma cells in culture (126–128). It is unlikely that the emergent structures for insulin in lymphocytes (83) are receptors for the growth-promoting or trophic peptide epidermal growth factor, since this peptide does not demonstrate cross reactivity with insulin for binding to receptors (129).

Under certain circumstances insulin can decrease the intracellular concentrations of cyclic AMP in adipose tissue cells (130–133) and in liver (134, 135), and these changes appear to correlate with at least some of the biological effects of insulin in these tissues. Although insulin can decrease the activity of adenylate cyclase in subcellular preparations (107–115), an effect on cyclic AMP phosphodiesterase (136–139) may also be involved. However, certain effects of insulin, like its effect on transport processes, have not yet been adequately explained on the basis of cyclic AMP, and in many situations (in the absence of substances that elevate cyclic AMP levels) the action of insulin cannot be correlated with a gross fall in the intracellular levels of cyclic AMP. The recent demonstration that insulin (and cholinergic drugs) causes a marked rise in the levels of cyclic GMP (122) may eventually help to explain the mechanisms by which insulin modulates metabolic events. The possible relationships between cyclic AMP and cyclic GMP have been discussed in this context (122). The hypothesis has been advanced (122) that a unique insulin-receptor (or cholinergic hormone-receptor) interaction at the cell membrane may act by transforming the substrate specificity of the membrane-localized adenylate cyclase into a form which now expresses preference for GTP rather than ATP. The same membrane-localized enzyme system would be responsible for the synthesis of both nucleotides, and the balance of ATP- or GTP-utilizing forms of the enzyme would depend on the relative occupancy of receptors that favor the ATP form (adrenergic hormones, glucagon, etc) compared to the occupancy of those receptors that favor the GTP form (insulin, cholinergic hormones). This could explain the simultaneous inhibition of adenylate cyclase activity and elevation of cyclic GMP levels (by enhanced synthesis) through the action of a single, unique hormone-receptor interaction. This hypothesis is attractive because it could explain the effects of insulin (and cholinergic hormones) on a unitary basis and it would place both cyclic nucleotides as initial chemical mediators. It has recently been demonstrated (140) that a substantial quantity of guanylate cyclase activity exists in isolated membranes from liver and adipose tissue cells, and that insulin and certain insulin-like substances (e.g. concanavalin A) as well as cholinergic hormones can directly stimulate the activity of this enzyme. Although this may explain the basis of the increased cyclic GMP levels observed after exposure to such agents, it remains to be demonstrated whether this enzyme is structurally related to adenylate cyclase.

It has recently been demonstrated (141) that the addition of exogenous ATP to fat cells leads to a selective suppression of the insulin-stimulated component of the glucose transport mechanism by specifically decreasing the affinity of the insulin-mediated process for glucose. This selective inactivation appears to be related to the phosphorylation of a specific protein component of the cell membrane (141, 142). These studies suggest that a chemical approach to the isolation and study of specific hormone-sensitive glucose carrier structures of membranes may be possible. It is of

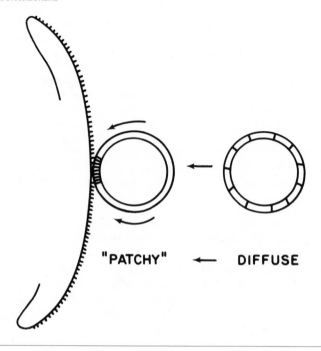

"PATCHY"   ⟵   DIFFUSE

Figure 1   Possible mechanism of action of insulin-agarose on isolated fat cell. Immobilized insulin molecules may act as a nidus for the aggregation of insulin receptors in a clustered form on the surface of the bead. Insulin receptors, normally presumed to be diffusely distributed throughout the cell surface, may be capable of undergoing spontaneous lateral diffusion within the plane of the membrane, as postulated in current models of membrane structure (98). This kind of mechanism may be operative in the binding of lymphocytes to concanavalin A-nylon fibers (148); in this case the surface receptors of the cell are known to redistribute and localize to the region of cell-fiber contact.[7]

interest that this phosphorylation reaction also leads to a block in the enhancement of glucose transport observed with concanavalin A, wheat germ agglutinin, poly-amines, and thiol reagents (141). Besides plant lectins, a number of substances are known which appear to activate the same glucose transport mechanisms as insulin, such as polyamines (143), organomercurials (144), polyene antibiotics (145), and sulfhydryl compounds (146, 147).

There is evidence that insoluble insulin-agarose derivatives (38–43) may be acting through mechanisms different from those occurring with native insulin (discussed in 1 and 47), and study of these interactions may shed light on the normal processes of receptor interaction and activation. Among several possible mechanisms considered (47) is one in which insulin-agarose may elicit biological responses by

[7] G. M. Edelman, personal communication.

acting as a polyfunctional insulin carrier which could, by its ability to crosslink surface receptors, cause a major redistribution of surface structures (Figure 1). This might lead to a shift from a diffuse receptor pattern to a clustering, aggregation, or patchy distribution of receptors. This could in effect be a model, in an extreme manner, of the normal way in which insulin initiates biological effects after its interaction with receptors. These considerations are consistent with the fluid mosaic model of cell membranes (98). It is interesting that agarose derivatives of concanavalin A or phytohemagglutinin can induce mitogenesis in B-type lymphocytes while the native, soluble plant lectins are unable to do this (149–151). There has been considerable speculation that mitogenesis by plant lectins may involve redistribution of surface receptors of the type suggested above for insulin. In a later section a more comprehensive thesis is suggested to explain in a broad sense the possible interrelationship of the insulin as well as other hormone receptors to each other and to other structures in the cell membrane.

GLUCAGON RECEPTORS

Tomasi et al (152) studied the binding of ^{131}I-labeled glucagon to liver membranes and described inhibition of this process by p-chloromercuribenzoate and 5',5'-dithiobis-2-nitrobenzoic acid. Rodbell and co-workers (153–162) have performed extensive studies on the manner by which glucagon binds to isolated liver membranes and activates adenylate cyclase in this tissue. Glucagon labeled with ^{125}I has been found to retain biological activity (155, 163) and to bind to liver membranes with properties expected for a specific receptor interaction (155, 156). The binding is saturable, indicating a finite number of membrane sites, and is reduced as expected by native glucagon but not by other polypeptide hormones or by biologically inactive peptide fragments (Fragments 1–21 and 22–29) of glucagon. The apparent affinity for binding (about 4×10^{-9} M) is approximately equal to the apparent affinity for activation of adenylate cyclase, and certain treatments of the membranes (phospholipase A, digitonin, urea) depress binding and biological responsiveness in parallel (155). The spontaneous rate of hormone-membrane dissociation is markedly accelerated by EDTA, suggesting the participation of a divalent cation in binding.

Because liver membranes inactivate glucagon very rapidly (160, 164), it has not been possible to measure accurately the rate constants of association and dissociation unless inactivation is blocked by manipulation of the experimental conditions. However, the fact that the hormone, which is bound to the membranes and is removed by acid (155), protein denaturants (165), or spontaneously, can bind again to fresh membranes while inactivated glucagon cannot, suggests that receptor binding and inactivation are separate and independent processes. Furthermore, certain peptides, reagents, and treatments of liver membranes can preferentially suppress inactivation without modifying binding (164, 165). It is interesting that when inactivation is blocked in this way the dose-response curve for activation as well as for binding are both shifted to lower concentrations, indicating that in the absence of inactivation the apparent affinity of the complex is substantially greater than judged when both processes occur simultaneously. Furthermore, such data support the view that binding is occurring with receptors. It is not yet known whether the very brisk glucagon inactivating system in these membrane preparations is physiologically relevant in the normal processes by which this hormone is inactivated. Bromer and colleagues (166) have recently reported that iodination of glucagon markedly enhances the biological activity of glucagon. Substitution of a single iodine atom increases the activity fivefold, whereas four iodines increase it by tenfold. Since this effect is evident as a shift in the glucagon dose-response curves of adenylate cyclase to lower concentrations (166), the data suggest that there has been an improvement in the apparent affinity of the receptors for glucagon. Whether this occurs by an enhancement in the intrinsic affinity of the modified molecule for the receptor or by resistance to inactivation is not yet known. These studies (166), however, indicate that extra caution will be needed in evaluating binding data for comparisons with biological properties, and that study of the products and mechanisms of the inactivating processes may be influenced by the iodinated state of the hormone.

An interesting chemical derivative of glucagon lacking the N-terminal histidine residue, Des-His-glucagon, has been described (161, 167). This derivative cannot stimulate adenylate cyclase activity in liver membranes. Grande et al (168) reported that carbamylation of the histidyl residue destroys biological activity, and Lande et al (167) found that selective blockade of the single ε-amino lysyl residue (N^{ε}-t-butyloxycarbonylglucagon) or of the α-amino group of histidine (N^{α}-trifluoroacetyl-glucagon) destroys the ability to activate adenylate cyclase activity, suggesting the requirement for free amino groups at both positions. Lande et al (167) point out that since the procedures used to prepare (161) Des-His-glucagon should have resulted in Des-histidine-N^{ε}-phenylthiocarbamylglucagon, and since simple phenyl-thiocarbamylation of the ε-lysyl group destroys biological activity despite the presence of an intact histidine, caution must be exercised in reaching conclusions regarding the regions in the glucagon molecule that are required for binding and activity (161). It has been reported (169, 170) that glucagon bound to agarose derivatives by azo linkage to the imidazole ring of histidine retains biological activity; such derivatives still possess a free α-amino group on the N-terminal histidine. These agarose derivatives have been shown in affinity chromatography experiments to selectively adsorb membrane fragments containing glucagon-sensitive adenylate cyclase activity (169).

Des-His-glucagon, although itself inactive, can block the biological activity and the binding of ^{125}I-labeled glucagon to liver membranes (159, 161). It is thus a competitive antagonist which has an affinity approximately ten times lower than native glucagon. It is especially interesting that Des-His-glucagon can cause a marked dissociation of ^{125}I-glucagon bound to membranes (159, 161).

Rodbell and colleagues (157, 159) have demonstrated that the response of the rat liver adenylate cyclase system to glucagon exhibits an absolute requirement for either GTP or ATP. It is believed that the guanyl nucleotides play a specific and obligatory role in the enzyme activation process. This process probably does not involve phosphorylation (156) since 5'-methylene guanylyl diphosphonate, which cannot donate phosphate groups, is also effective (although at 100 times higher concentrations). Similar requirements for GTP have been described (171–174) for other hormone-sensitive adenylate cyclase reactions. It has also been shown (156, 159) that GTP stimulates the rate and degree of dissociation of the ^{125}I-glucagon-membrane complex, that it decreases the total amount of hormone bound, and that it decreases the apparent affinity of the binding sites for glucagon. Since GTP does not alter the binding of Des-His-glucagon to membranes, and since digestion of membranes with phospholipase C from *Bacillus cereus* (which cleaves acidic phospholipids) abolishes the effect of GTP on glucagon dissociation from membranes, it has been postulated (162) that GTP and the histidine residue bind to a common site involved both in activation of adenylate cyclase and in the dissociation of glucagon from its receptor. Acidic phospholipids are presumably involved in the concerted effects of GTP and glucagon at this site.

Swislocki, Scheinberg & Sonenberg (175) have reported that GTP grossly alters the physical state of purified liver membranes. Gross aggregation of the membranes can be observed and GTP alters the sedimentation properties of such membranes. GTP is degraded very quickly by liver membranes and substantial binding of ^{14}C-labeled material (originating from GTP) to the membranes can be observed with very low concentrations of the nucleotide. These investigators have also reported (176) that very low concentrations of GTP, in addition to activating adenylate cyclase, can modify liver plasma membrane ATPase and NADase activities. In addition, gross changes in general membrane conformation due to small quantities of GTP can be detected by special physical measurements (176, 177). The apparent modification of overall membrane structure by GTP can thus complicate precise interpretations concerning the exact role that this nucleotide may have in the complicated processes of glucagon binding and enzyme activation.

The possible role of membrane lipids in glucagon binding and enzyme activation has been studied in detail (158, 162). Treatment of liver membranes with digitonin or phospholipase A does not alter the basal (unstimulated) activity of adenylate cyclase, but it does decrease or abolish the glucagon-stimulated activity as well as the binding of ^{125}I-glucagon to the membranes (158). Although a partial restoration of binding and enzyme sensitivity was obtained by the addition of exogenous phospholipids, it is not entirely clear whether at least some of the effects observed resulted from a solubilization of the receptors rather than from displacement of phospholipids essential to the binding process. Detergents could, of course, decrease binding by directly perturbing the receptors independently of their effects on phospholipids. The products of phospholipase A hydrolysis (lysophospholipids and fatty acids) are themselves potent surfactants that could unfavorably modify the binding interaction (162). The partial restoration of binding was not dependent on the nature of the phospholipid added, although phosphatidylserine seemed to be the most effective (158). It is not clear whether the slight restoration of activity by exogenous phospholipids reflects a true reversal of the original membrane modification (reinsertion of the cleaved phospholipids into the membrane) or whether they act nonspecifically by positive perturbations or by acting to remove residual detergent or inhibitory products of the hydrolytic processes. In this connection it is pertinent that very low concentrations of digitonin, certain detergents, and lipid suspensions can actually cause an enhancement in glucagon binding and enzyme activation (158, 165). It will be very important to determine conclusively whether phospholipids are absolutely essential to the binding of glucagon to the receptor, since the loss of this property would greatly complicate and perhaps prohibit the solubilization and purification of these binding structures.

If the binding of glucagon is indeed absolutely dependent on phospholipids, this would be in striking contrast to the results observed with insulin. Simple considerations of freedom of assessibility of interaction and reversibility make it likely that the hormone binding site of these receptors is probably exposed rather freely to the aqueous environment (where the free hormone is to be found); it thus should not be surprising (and perhaps may be expected) if phospholipids do not play a direct role in the binding interaction between hormone and receptor. Such phospholipids may, however, help to maintain the overall conformation of the receptor, and they are very likely involved in the processes by which the hormone-receptor complex conveys information to other structures (e.g. enzymes) in the membrane. Thus, loss of enzyme activation by perturbation of membrane lipids should not be surprising. In this respect the reports of Levey (178, 179) describing restoration of hormone sensitivity of detergent-solubilized heart adenylate cyclase by the addition of phosphatidylserine (for glucagon) and of phosphatidylinositol (for epinephrine) are quite interesting.

More recent studies by Rubalcava & Rodbell (162) on the role of phospholipids on the action of glucagon indicate that phospholipids do not participate directly in the binding interaction. Extensive digestion with purified phospholipase C preparations from two different sources does not diminish the number of binding sites but does decrease the affinity of the complex by tenfold. The products of hydrolysis (diglycerides and phosphorylated amines) produced by this enzyme are not potent surfactants. Phospholipase C digestion, however, can abolish the ability of glucagon to stimulate adenylate cyclase although it does not modify the basal or fluoride-stimulated activities of the enzyme. Thus, there is a rather selective loss in the ability of the hormone-receptor complex to activate the enzyme. This has led to the hypothesis that acidic phospholipids are involved in the processes by which hormone-stimulated activity is elicited (162). Since phospholipase C digestion does not alter the binding of Des-His-glucagon, and since the effects of GTP are abolished, it has been suggested that the acidic phospholipids play additional, specific functions in the glucagon-GTP-receptor interaction involving the histidine residue of the hormone (162).

Despite the specificity and affinity of binding of glucagon to liver membranes, and despite the various correlations which have been made between binding and activation of adenylate cyclase (153–162), including the rapid and reversible nature of both processes and the direct dependence of activity on the concentration of free hormone in the medium (159), Birnbaumer & Pohl (180) have recently challenged the view that the glucagon-specific binding demonstrated in previous studies actually represents receptor interactions. On the basis of time studies, they point out that hormonal stimulation is complete when only 10 to 20% of the binding sites are occupied, suggesting that 80 to 90% of the binding sites are not related to biological activation. Furthermore, Des-His-glucagon completely inhibited stimulation of adenylate cyclase but only displaced 10 to 20% of the bound ^{125}I-labeled glucagon. In addition, under certain conditions GTP caused a marked dissociation of membrane-bound glucagon without altering the glucagon-stimulated adenylate cyclase activity (180). Although the explanations for these apparently anomalous effects are unknown, several possibilities exist which must be considered. A large proportion of the binding sites may indeed be unrelated to receptors, as suggested by Birnbaumer & Pohl (180). They may, for example, be related to inactivation mechanisms, or possibly to receptors for other hormones which can cross react with glucagon. One such possibility may be receptors for glucagon-like immunoreactivity (GLI), which has been shown (181) to displace the binding of glucagon to liver membranes. Evidence was presented that specific binding sites may exist for GLI since the binding of ^{125}I-glucagon was not completely displaced (about 50%) by GLI. Since the liver is heterogeneous in its cell composition, it is even possible that membranes from different cells possess different receptors (182–184). However, it is quite possible (or probable) that the major binding of glucagon (155) is related to specific receptors despite the apparent discrepancies described above. To expect complete correspondence between the kinetics of binding and activation assumes that in the preparation of the liver membranes, a relatively long procedure, if there is inactivation of receptors or of enzyme activity this occurs equally for both processes. It should not be surprising if some enzyme activity is lost even by simple homogenization procedures which certainly must alter, perhaps in major ways, the normal structure of the membrane. Since the binding process is much more stable than the activity of the enzyme, and since uncoupling is the process perhaps most susceptible to disruption, it should not be surprising to find in an isolated membrane system that there may be an apparent excess of binding sites. Another possibility is that during the enzyme reaction (at 30 to 37°) there may be a concomitant process of enzyme inactivation which is being compensated by a truly increasing rate of activation (resulting from increasing receptor occupancy) whose balance is an apparently constant activity achieved early and without a lag period. Unfortunately it has not yet been possible to perform binding and activity studies in intact, living cells. Another possibility, discussed in a later section of this review, suggests that in a fluid membrane hypothesis in which hormone-receptor complexes must diffuse and subsequently interact with physically separate enzyme molecules, exact correspondence of the binding and activating processes may not be found.

Blecher and colleagues (163) have recently reported that glucagon binding structures can be solubilized from liver membranes with the detergent Lubrol-PX. One of the major problems in this area of research is the method used for detection of receptors once they have been extracted from membranes with detergents. Procedures have to be used that can rapidly and reliably separate the free and the receptor-bound hormones under conditions that do not cause dissociation of the complex and do not result in the simultaneous separation of nonreceptor macromolecules which nonspecifically adsorb the hormone in sufficiently large quantity to overshadow a smaller population of specific receptors. Blecher et al (163) have succeeded in this by using special Hummell–Dreyer (185, 186) columns which permit rapid binding equilibrium to be achieved and detected. Some properties of the soluble receptor have been studied, and partial purification has been achieved (163). Recent studies on the interaction of glucagon with detergents (187) and with phospholipids (188, 189) by optical techniques should prove useful in predicting the behavior of this hormone in such solutions as well as in understanding the conformational changes that may occur when polypeptide hormones bind to their membrane receptors.

Manganiello & Vaughan (190) have reported that large fat cells from adult rats are less sensitive to glucagon stimulation of lipolysis than small cells from young animals, and that the adenylate cyclase activity of membrane preparations from the bigger cells are less responsive to the hormone. They suggested that with growth there may be a selective decline in the number of glucagon receptors relative to those of epinephrine or ACTH. Livingston et al (191), using an oil flotation technique (54, 192) have shown that intact large fat cells bind less glucagon than small cells, that this is also reflected in isolated membrane preparations of these cells (measured by different procedures), and that the differences in binding could not be attributed to differences in glucagon inactivating ability. However, the decrease in ^{125}I-labeled glucagon binding was not as marked as the decrease in lipolytic response (191). In other studies they (193) demonstrated that fat cells from larger cells have elevated levels of cyclic AMP phosphodiesterase. In these cells the glucagon insensitivity may thus result from a combination of reduced numbers of hormone receptors and increased activity of phosphodiesterase.

CATECHOLAMINE RECEPTORS

Although radioactively labeled catecholamines have been used successfully for many years in the elucidation of important transport, storage, and binding functions of adrenergic neurons (194–198), it has only been in the recent past that such labeled compounds have reportedly been used to detect and measure putative β-adrenergic receptors in subcellular or intact systems. Demonstration of specific interactions of ^{3}H-norepinephrine has been described for isolated liver (199–201) and heart (202–205) microsomal preparations, spleen capsule (206), cultured myocardial cells (207), and turkey erythrocyte ghosts (208–210). The binding of catecholamines in these cases has been interpreted as representing specific receptor interactions, despite the fact that no stereospecificity was demonstrable (201, 205, 209) and the binding appears to be selective only for the catechol function of the ^{3}H-catecholamine (201, 209). In most of these studies low and high affinity binding processes have been demonstrable, and the apparent values for the high affinity system (10^{-6} to 10^{-7} M) have been thought to reflect reasonably closely the apparent affinities calculated from measurements of biological activity. In most of these cases, however, biological activity has been interpreted to mean stimulation of adenylate cyclase activity; it is well known that in many situations stimulation of this particulate enzyme in vitro requires much higher concentrations of hormone than are effective in vivo. In isolated fat cells or membranes, the binding of ^{3}H-norepinephrine indicates an apparent affinity of about 10^{-6} M whereas the half-maximal concentration required to stimulate lipolysis (when oxidation of the catecholamine is prevented) is about 6×10^{-8} M (211). In turkey erythrocyte ghosts the affinity of DL-isoproterenol (from binding) has been estimated at 4×10^{-6} M and a similar value has been estimated from adenylate cyclase stimulation (209). In intact turkey erythrocytes, however, DL-isoproterenol stimulates sodium transport maximally at 10^{-7} M and the concentration required to achieve a half-maximal response is about 10^{-8} M (212). Thus there are very significant differences in the affinity constants estimated from binding studies and from biological experiments.

The belief that measurements of binding of ^{3}H-catecholamines to intact cells or microsomal preparations reflect true β-adrenergic receptor interactions has recently been seriously questioned (211) on the following grounds: 1. The binding displays absolutely no stereospecificity in a variety of tissues (fat cells, turkey erythrocytes, and microsomes from fat cells, liver, heart, spleen, capsule, pineal gland, and superior cervical ganglion) under a wide variety of conditions. 2. The specificity of binding is directed only to the catechol moiety of the catecholamine. Noncatechol compounds such as pyrocatechol and 3,4-dihydroxymandelic acid compete for binding as well as L-norepinephrine. The ethanolamine portion of the molecule does not contribute to the binding interaction. 3. The D isomers of catecholamines and certain compounds (pyrocatechol, dihydroxymandelic acid, dopamine) which compete very effectively for binding do not stimulate lipolysis in fat cells and do not stimulate adenylate cyclase activity in microsomes from liver, fat cells, or turkey erythrocyte ghosts (211). 4. The biologically inactive D isomers of catecholamines and the catechol substances do not inhibit the L-norepinephrine stimulated lipolysis in fat cells or adenylate cyclase activity (fat cells, liver, heart, turkey erythrocytes) even at concentrations 200 to 1000 times greater than that of L-norepinephrine (211). In such studies great care must be taken when concentrations of L- or D-norepinephrine higher than 10^{-4} M are used, since nonspecific (toxic) inhibition of adenylate cyclase activity may be observed; for example, very high concentrations can nonspecifically inhibit the fluoride- or glucagon-stimulated activity of adenylate cyclase (211). It is also pertinent that Gardner et al (212) have reported that DL-dopa, dopamine, vanilmandelic acid, and 3,4-dihydroxyphenylacetic acid do not inhibit the ability of DL-isoproterenol (10^{-8} M) to stimulate sodium transport in intact erythrocytes, even when tested at concentrations 1000 times greater than isoproterenol. 5. The noncatechol, m-methanesulfonamide derivative of DL-isoproterenol, Soterenol (compound 49) (213), is as potent as L-norepinephrine biologically (lipolysis) but is ineffective in competing for ^{3}H L-norepinephrine binding (211).

On the basis of these results, the binding of ^{3}H-catecholamines to subcellular particles cannot be reconciled with the view that such binding reflects β-adrenergic receptor interactions. Under conditions where binding is totally inhibited (inactive D isomers or catechols) the biological effect is unaltered; the binding measured thus cannot be related to activity. The binding process appears to be related to a totally separate process which shares with catecholamine receptors only the property of recognizing catechol moieties. Other unusual properties of the binding process which have been described (211) include 1. The dramatic increase (fifteen- to twentyfold) in binding observed when microsomes (fat, liver, heart) are stored at $-20°$, the difference in binding (about 200-fold) to intact fat cells compared to fat cell microsomes, and the difference in binding (thirty- to sixtyfold) between intact turkey erythrocytes and ghosts. 2. The essentially irreversible nature of the binding observed with microsomes from liver, heart and fat; the binding to turkey erythrocytes is reversible. It is also of interest that the maximal binding capacity of liver, fat, and heart membranes (about 1 nmole per mg of protein) is 10^3 to 10^4 times larger than the number of receptors described for insulin (52) or glucagon (59, 155). In intact turkey or human (which are totally unresponsive to catecholamines) erythrocytes (212) the number of ^{3}H-norepinephrine binding sites is (211) much lower, about 5 per μ^2 of membrane surface, which is near the value (one per μ^2) reported for insulin receptors in intact fat cells (52). Despite the scarcity of binding sites in intact erythrocytes (e.g. compared to ghosts), no stereospecific binding can be detected in these cells (211).

The very strict requirement of ^{3}H-catecholamine binding on the presence of a catechol function suggests a possible relationship with the enzyme, catechol-O-methyltransferase (COMT). This enzyme displays very similar affinities for a wide variety of catechol substrates, even for those lacking a ring side branch, and it does not distinguish between D and L forms of norepinephrine (214–217). On the basis of considerations of specificity, it has been concluded that COMT cannot serve as a model for the β-adrenergic receptor (217). Although COMT is present in the soluble fraction of most tissues, its presence in microsomal fractions of several tissues has been established (218–220). In isolated fat cells the enzyme is found predominantly in a membrane-bound form (220). Although the particulate and soluble enzymes appear to differ in certain respects (218), both demonstrate a broad substrate specificity requiring only the catechol function (218, 219). Other features of the interaction suggest (211) that catechol binding may be related to particulate COMT, including the requirement for divalent cations, and inhibition by p-chloro-mercuribenzoate and EDTA. Well known inhibitors of COMT (tropolone, pyrogallol, quercetin, and U-0521) also inhibit ^{3}H-norepinephrine binding (211). The compulsory substrate for COMT activity, S-adenosylmethionine (a potent inhibitor of COMT) profoundly inhibits binding (211). 3-Mercaptotyramine, which selectively and irreversibly inhibits COMT by presumably forming active site disulfide complexes (221), is a most potent inhibitor of binding (211). Dithiothreitol inhibits the binding process and markedly enhances microsomal (but not soluble) COMT activity (211). It has been suggested (211) that the binding of catecholamines to microsomes may reflect binding to an altered form of COMT.

Since no significant degree of stereospecificity has yet been reported for cate-cholamine binding to tissues or microsomes, and since this is requisite for a true β-adrenergic receptor interaction, it must be concluded that if such putative receptors exist in the materials examined, they must represent a negligible fraction of the actual binding observed. It has been proposed (211) that at least two major obstacles must be faced in new attempts to discover true β-adrenergic receptor interactions: one dealing with the simultaneous presence of a large excess of other catechol binding structures, the other dealing with the requirement for much more highly radioactive probes (since the actual number of receptors must be extremely small). Possible approaches to these problems have been suggested (211).

The catechol binding protein of microsomes has been solubilized from various tissues with non-ionic detergents, and some of the properties of this protein have been reported (203, 210, 211). This macromolecule appears to be the same as that studied in intact microsome systems. Venter et al (222, 223) have described very interesting studies with catecholamines bound to glass which indicate that these compounds are indeed active by interactions with superficial cell surface structures. Important studies utilizing catecholamines and histamine bound to agarose beads have been described (224, 225) for exploring surface receptors of leukocytes.

ACETYLCHOLINE RECEPTORS

Several important discussions have appeared presenting different points of view and approaches concerning the nature of membrane proteins, membrane structure, and the interpretations of electrical activity in excitable membranes (226–232). Early attempts (233–235) to identify and isolate acetylcholine receptors from electric tissues of fish by measuring the binding of α-tubocurarine or gallamine triethiodide were unsuccessful because of interference of binding by nonspecific macromolecules. Cholinergic ligands also bind to a variety of hormones, mucopolysaccharides, and enzymes (236, 237).

DeRobertis and colleagues (232, 238, 239) reported as early as 1967 that special proteolipids (i.e. hydrophobic lipoproteins) with high affinity for binding cholinergic drugs and having other properties suggestive of receptors could be isolated with organic solvents (chloroform-methanol) from grey matter, nerve ending membranes, and electric organs of fish. The proteolipid isolated from the electric organ of *Electrophorus* (232, 239), chromatographed on Sephadex LH-20 columns, demonstrated high affinity for acetylcholine, hexamethonium, and *p*-(trimethylammonium)-benzene diazonium fluoroborate (TDF). Two sets of binding sites were reported, having K_i values of 10^{-7} and 10^{-5} M, the quantity of the latter being tenfold more than the former. These studies have been criticized because of the utilization of organic solvents in the extraction procedures, which might be expected to denature proteins on the basis of information gained with conventional globular proteins and on theoretical grounds. It is possible, however, that the organic solvents used still contained sufficient residual water molecules to permit unusual hydrophobic proteins to remain in solution. It is of special interest, however, that despite the absence (or near absence) of water, the active (binding) sites of these lipoproteins presumably remain intact since the cholinergic ligands remain tightly bound in the organic solvents. A recent study (240) has challenged the procedures used by DeRobertis and colleagues since it was found that chloroform-methanol treatment eliminates the binding activity of whole microsomes (electroplax) and does not solubilize the receptor in active form. It was further described (240) that peaks (from Sephadex LH-20) capable of binding acetylcholine could be created in the presence of protein-free and receptor-free extracts. It was suggested that the binding is nonspecific and that the receptor-ligand interaction may be artifactual.

Fiszer de Plazas & DeRobertis (241) have recently studied the proteolipid isolated from *Electrophorus* electroplax and compared its properties with that which has now been isolated by many laboratories (described below) from the same tissue with detergents. In particular, they demonstrated (*a*) that this protein could bind ^{131}I-labeled α-bungarotoxin with an affinity of about 9×10^{-7} M, (*b*) that one molecule of the toxin binds per 37,000 mol wt equivalent of the receptor proteolipid, (*c*) that the bound toxin could not be displaced with acetylcholine or decamethonium, and (*d*) that although the labeled toxin could not be extracted into the organic phase in membrane extractions (presumably disrupting the toxin-membrane complex) the binding activity was lost from the protein residue and appeared in the extract. Furthermore, the quantity of material extracted from various membranes (3 to 7×10^{14} receptor molecules per gram of tissue) is very close to the number estimated by other workers using very different (α-bungarotoxin binding) procedures. They concluded that the cholinergic proteolipid extracted with organic solvents is very similar to the protein-α-bungarotoxin complex separated by other laboratories by the use of strong detergents. DeRobertis and colleagues have reported (242, 243) that incorporation of the proteolipid into artificial lipid bilayer membranes induced conductance changes upon interaction with acetylcholine which were strongly potentiated by uranyl ions.

O'Brien and colleagues (244, 245) have used the binding of [^{3}H]-muscarone to detect binding interactions with lyophilized particulate fractions of *Torpedo* electroplax which have properties highly suggestive of receptor interactions. A single binding constant (K_i, 7×10^{-7} M) has been described. All highly potent neuromuscular agents were very effective blockers of binding. Eldefrawi et al (246) measured the binding of [^{3}H]-acetylcholine (label in acetate moiety) to *Torpedo* electroplax membranes after inactivating acetylcholinesterase with organophosphates. The binding was antagonized by nonradioactive nicotinic ligands, and it was reversible. Scatchard plots indicated two binding sites having affinities (K_i) of 8×10^{-9} M and 7×10^{-8} M.

The use of snake venom toxins in the identification, isolation, and purification of acetylcholine receptors has received a great deal of attention in the last few years. Lee and associates (247, 248) have purified and characterized α-bungarotoxin, which blocks neuromuscular transmission irreversibly, and blocks the depolarizing action of acetylcholine in the neuromuscular junction of vertebrates and the electric tissue of *Torpedo* and *Electrophorus*. The action of the toxin is partly protected by *d*-tubocurarine. On the basis of physiological studies (247–255) it has been concluded that the toxin combines irreversibly with the cholinergic receptor at the motor endplate. Lester (256, 257) has more recently reported detailed electrophysiological studies with cobra toxin. Because of difficulties arising from diffusion barriers and limited access of toxin molecules to receptors, emphasized by Lester (256), it has been difficult to correlate in precise ways (kinetics, relative rates, saturation, quantitation of sites, specificity, etc) binding studies of the toxin to isolated membranes or membrane extracts with the biological effects in intact systems. However, the use of purified membrane vesicular fragments which retain excitability (permeability) properties and cholinergic sensitivity (258–261) offers opportunities for precise correlations of receptor interactions and biological effects.

In 1970 Changeux and colleagues (262) used detergents to isolate a protein from *Electrophorus* which had properties characteristic of a cholinergic receptor. This interaction was detected by measuring the binding of [^{3}H]-decamethonium. They demonstrated (249) that α-bungarotoxin blocks the in vitro binding of decamethonium. However, only 50% of the binding activity could be prevented despite the fact that total inactivation can be observed with the toxin. Miledi et al (250) labeled the toxin with ^{131}I and demonstrated virtually irreversible binding to isolated membranes from *Torpedo* electric tissue. Carbachol and curare could block this binding, but again only about 50% of the bound toxin could be blocked. Miledi et al (251) demonstrated that the toxin binding material could be solubilized with non-ionic detergents, and they presented preliminary data on the behavior of the complex by gel filtration. Raftery et al (263) were also among the first to show specific binding of ^{125}I-labeled α-bungarotoxin and the putative cholinergic receptors of *Electrophorus* electroplax membranes; the binding activity was readily solubilized with Triton X-100.

Bourgeois et al (264) utilized immunofluorescence to demonstrate that binding of *Naja nigricollis* venom was localized in the innervated face of a single electroplax from *Electrophorus*. It was not possible, however, by this technique to distinguish labeling on membrane areas that underlie the nerve terminals (subsynaptic areas) from labeling on extrasynaptic areas. In other electronmicroscopic studies, Bourgeois et al (265) demonstrated by high resolution autoradiography that the [^{3}H]-α-toxin (14.8 Ci per mmole) binds only to the innervated (caudal) side of the electroplax and that it binds both between the synapses and under the nerve terminals. However, the binding density under the synapses is about 100 times greater than outside the synapses. The density of receptors in the subsynaptic area was estimated to be 33,000 toxin molecules per μ^2. This figure is in the same range as the estimates in muscle which had been suggested by others (251–253, 255) using different techniques; the range varies from 100,000 per μ^2 for frog muscle (251) to 12,000 per μ^2 in mouse diaphragm (252). The density found by Kasai & Changeux (258) in excitable microsacs is at least 100 times lower than these, which probably means that these membrane vesicles were derived primarily from extrasynaptic areas of the innervated surface, as would be expected from the fact that the latter constitutes about 98% of the total surface of the innervated membrane. It has been estimated (265) that, assuming a molecular weight of 50,000 and a density of 1.37, the maximum number of receptor protein molecules that can fit in a single layer is about 40,000 per μ^2. The data therefore suggest that the subsynaptic membrane surface is occupied almost exclusively by receptor protein. This would leave virtually no room for acetylcholinesterase, which has been estimated to exist in stoichiometric quantities with the receptor, and very little room for other proteins or even phospholipids. It is thus possible that relatively few molecules of acetylcholinesterase are actually located in the subsynaptic membrane. It is also possible that the calculations described above are based on erroneous estimates of membrane surface (which may be highly invaginated and tortuous and thus difficult to estimate accurately). These figures of course assume that all of the bound toxin is bound specifically to receptors and that nonspecific binding does not occur in the synaptic region to a greater extent than in the extrasynaptic membrane. This may not be the case, and it cannot be excluded that there are also important differences in accessibility of the various regions of the membrane for the toxin. As indicated earlier, even in isolated membranes not more than half of the label can be blocked by d-tubocurarine. It will be important in this regard to examine controls in which native, unlabeled toxin or d-tubocurarine is used (to saturate the receptors) before addition of the labeled toxin, which should give an idea of the nonspecific labeling which may occur. It is of interest to compare the extraordinarily high receptor density described for acetylcholine receptors with that of insulin receptors in isolated fat cells, which has been estimated to be 1 to 10 sites per μ^2, depending on the size of the cell (52, 107). From other studies it is very likely that the receptor density for other polypeptide hormones and catecholamines in mammalian cell membranes is very close to that described for insulin receptors. These considerations have important implications regarding the ease with which these receptor structures can be purified compared to those for acetylcholine, especially in the electric organs of the fish.

In addition to the studies described above, others (266–273) have demonstrated and studied in detail presumably specific interactions of labeled α-bungarotoxin with membrane proteins from electric organs of fish. In the absence of denaturants (such as sodium dodecyl sulfate) there is generally good agreement that the molecular weight of detergent-solubilized native receptor protein is about 300,000. When the effect of detergents on the hydrodynamic properties of the receptor is considered (271) by performing sucrose density gradients containing varying ratios of H_2O and D_2O, a corrected mol wt of 360,000 is obtained. Meunier et al (270) estimated, on the basis of sodium dodecyl sulfate gel electrophoresis, that the subunit had an apparent mol wt of 55,000. Schmidt & Raftery (272), utilizing receptor protein purified by affinity chromatography, demonstrated major bands of mol wt 42,000 and 34,000 and a minor band of 26,000; they conclude that the complexity of their gel patterns may reflect the occurrence of several species of receptor in the tissue. It is of interest that many of the molecular parameters (Stokes radius, sedimentation coefficient, frictional ratios, axial ratios) described for the bungarotoxin binding protein are extremely similar to the values calculated for insulin receptors solubilized from membranes (76).

Several laboratories (104, 105, 272, 274–276) have purified the detergent-solubilized cholinergic receptor from electric organs of fish by using affinity chromatography. Although columns containing α-bungarotoxin are very effective in adsorbing the receptor protein, difficulty has been encountered in eluting the protein in active form. For this reason the most successful approaches have been with resins containing specific quaternary ammonium functions. The methodology for purification is now relatively well advanced, and several groups can purify the receptor proteins in milligram quantities. It can be anticipated that considerable information should be forthcoming soon concerning the physicochemical properties and nature of the ligand (acetylcholine) interaction utilizing the purified protein. Although much optimism may be generated by the successes of these purifications, it should be pointed out that purification of receptors from mammalian sources will be a much more formidable task. For example, it has been estimated that beginning with the isolated electroplax membranes, complete purification of the acetylcholine (α-

bungarotoxin) receptor requires between 100- to 500-fold purification. In contrast, purification of detergent-solubilized insulin receptors (101) from liver membranes requires about 500,000-fold purification; estimates for other polypeptide hormones are in the same range. Furthermore, great problems are encountered in obtaining sufficient starting materials for purification. Even soluble, cytosol receptors for estrogens (uterus) require at least 100,000-fold purification (277, 278).

Neurotoxins have also been shown (279, 280) to bind to membranes obtained from guinea pig and hog brain, suggesting the existence of acetylcholine receptors in this tissue. These proteins have been solubilized with detergents, and many properties have been described.

Miledi & Potter (251), studying the effect of α-bungarotoxin on neuromuscular transmission of frog sartorius and rat diaphragm, demonstrated that after denervation the toxin binding capacity increased sharply over a 20 day period, then decreased as the muscle cells atrophied. This increase correlated with that in acetylcholine sensitivity generally observed after denervation. They also studied (252) desensitization induced by acetylcholine, in which state the toxin binding ability of the muscles was decreased. It was also observed that only half of the binding could be blocked by d-tubocurarine and that, after a single exposure to ^{131}I-labeled toxin, virtually no dissociation occurred upon repeated washing over a period of eight days, indicating that for practical purposes the binding was irreversible; others have observed similar results with intact muscle. Since it is known that the toxin membrane or receptor complex is not covalent and can be dissociated with protein denaturants, this extraordinarily slow apparent rate of dissociation is not consistent with any of the measured or estimated affinity constants and suggests that problems of trapping, diffusibility, and accessibility are at least in part responsible for this apparent irreversibility of the toxin in intact tissues. This is reminiscent of the early studies (some 20 years ago) of binding of native and of ^{125}I-insulin to intact muscles, which demonstrated virtually no reversibility after washing (discussed earlier in this review).

Barnard et al (252) studied the binding of [^{3}H]-α-bungarotoxin to diaphragm endplates by radioautography. The time course of inhibition of transmission was found to differ from the time course of binding of toxin. All of the fibers still responded when about 45% of the receptors (binding sites) were blocked, but upon further binding a steep rise in inhibition occurred at a point where 70% of the sites were bound. It was deduced from this that if spare receptors exist, they do not constitute more than 50% of the total present. Since it was not determined whether nonspecific or noncurare displaceable sites are more rapidly labeled than the others, it could well be that virtually no excess receptors exist in the muscle endplate. It was estimated from these (252) studies that there are about five cholinergic receptor sites per transmitter molecule available for single impulse transmission. Fambrough & Hartzell (253) have reported that the number of receptors per motor endplate of the rat diaphragm varies directly with rat size; in single muscle fibers the binding of ^{131}I toxin was found to localize to the endplate. Berg et al (255) extended these studies on muscle receptors and demonstrated that, in contrast to normally innervated adult muscle, toxin binding in the chronically denervated muscle or in muscle of neonatal rats occurred in substantial quantities along the entire length of the muscle. Porter et al (254) have more recently reported detailed autoradiographic studies of toxin localization in muscle. It was clear that substantial quantities of receptor-like molecules existed in regions outside the end-plates, but that it was possible to distinguish these. These studies should caution against the assumption that all toxin binding molecules observed in muscle membranes or extracts of these membranes are specific cholinergic receptors. Stalc & Zupancic have reported (282) that α-bungarotoxin modifies the activity of intact, membrane-bound acetylcholinesterase irreversibly and that this effect is also prevented by d-tubocurarine. Although their results were interpreted as supporting the view that acetylcholine receptors and esterase activity are present on the same macromolecule (but different parts), most recent studies indicate that the activities are present in structurally different and separate molecules; the studies of receptor purification, for example, have claimed physical separation of the toxin binding and cholinesterase activities. These results (282) suggest, however, that in the membrane these separate molecules may be coupled in such a way as to permit coordination in function.

Several reports (251, 255, 281) describe solubilization of the muscle toxin binding proteins with detergents. Chiu et al (281) describe at least two distinct toxin binding macromolecular species. One, found in the endplates only, has an apparent mol wt of 360,000. The other, primarily extrajunctional, has a mol wt of about 200,000; it is not associated with the acetylcholine receptor and has little affinity for d-tubocurarine. The latter may account for the binding which occurs to intact muscle (about 50%) which is not displaced by curare (discussed earlier). The twentyfold increase in binding which occurs in denervated muscle was accounted for almost exclusively by the specific endplate component and not by an increase in the extrajunctional type of toxin binding protein.

Radioactive derivatives of α-bungarotoxin have been used to study acetylcholine receptors in cultured muscle cells (283–288). Although binding sites are distributed over the entire cell surface, discrete areas of high concentration are seen (283). The receptor density in these clusters has been estimated to be 9000 per μ^2, and 900 per μ^2 in other areas of the muscle (284). The appearance of receptors has been studied during differentiation of a myogenic cell line (285). In these studies cell-bound toxin had a half-life of dissociation of 7 hr, and this dissociation was not correlated with a loss of binding activity. This finite rate of dissociation is very pertinent to the earlier discussion in this section which considered the apparent irreversible nature of toxin binding to intact tissues. It is of interest that the appearance of receptors paralleled the appearance of fused fibers during differentiation of the cells in culture (285). α-Bungarotoxin has also been used to probe acetylcholine receptors of cultured sympathetic neurons (286). Cultured muscle cells stimulated intermittently over prolonged periods are less sensitive to applied acetylcholine and bind less toxin than inactive fibers (288).

Denburg et al (289) have identified a specific nicotine binding macromolecule from axon plasma membranes from the lobster. The K_i of the binding is 40 nM and it is competitively blocked by acetylcholine and procaine. The macromolecule resembles in several ways the postsynaptic acetylcholine receptors, including its concentration in the tissue, and blockade by cholinergic drugs and by α-bungarotoxin. This presumably nicotinic receptor is thought to occur on the internal surface of the axon plasma membrane, and it may be a component common to both the Na^+ and K^+ gates. Such receptors have been postulated to exist by Nachmansohn (228–230), who believes that such molecules may play a central role in axonal conduction; the possible role of such macromolecules in conduction merits serious attention. Benzer & Raftery (290) and Henderson & Wang (291) have described the solubilization and characterization of specific tetrodotoxin binding components from nerve axon membranes. These structures may form part of the sodium ion channel involved in nerve impulse propagation.

An important recent report (292) which provides considerable strength to the supposition that the α-bungarotoxin binding proteins are indeed acetylcholine receptors, describes the production in rabbits of antibodies and of a specific syndrome after the injection of acetylcholine receptor purified from *Electrophorus*. The animals acquired a flaccid paralysis and demonstrated abnormal electromyographs characteristic of neuromuscular blockade, which were dramatically reversed by the administration of anticholinesterases.

Another important technique used to study acetylcholine receptors in vivo and in isolated membranes is affinity labeling. Unfortunately, these interesting studies cannot be considered in detail in this review. The reader is referred to several reviews of this technique (293–297).

The considerable progress in the field of acetylcholine receptors, particularly that related to isolation and purification, promises to permit rapid elucidation of the mechanisms underlying the changes in ionic fluxes that result from synaptic activation by acetylcholine. Underlying such optimism, however, is the generally held view that the receptor molecules may themselves constitute the ionophores, channels, pores, or carriers that are triggered to operate by acetylcholine activation. It may be, however, that to trigger the biological response, the ligand-receptor interaction must be communicated to entirely separate molecules, either by directly contiguous interactions or through as yet unknown chemical mediators. This general mechanism, for example, appears to be operative in the action of polypeptide, adrenergic, and possibly muscarinic cholinergic hormone interactions with membranes in peripheral tissues. The receptor (or "binding") macromolecule may be only one of the essential components of the activated complex of the membrane. In this respect, it is of interest that cyclic GMP is rapidly being implicated as an important chemical mediator of acetylcholine action (muscarinic) in a variety of tissues (122–124, 298–303), and that the enzyme, guanylate cyclase, can be stimulated directly by acetylcholine in isolated membrane preparations (140).

RECEPTORS FOR OTHER POLYPEPTIDE HORMONES

Numerous reports have rapidly been appearing concerning the identification of polypeptide hormone receptors in membrane preparations from various tissues. All of these are based fundamentally on the same principles described in this report for insulin, glucagon, and acetylcholine; the nature of the interaction of a highly radioactive (^{125}I- or ^{131}I-labeled) hormone is studied to determine whether the specificity, affinity, and number of binding sites are compatible with the known biological behavior of the hormone. For practical considerations, the description of these important studies will unfortunately be very brief and uncritical.

Pastan et al (304) described that brief exposure (1°) of thyroid slices to thyrotropin (TSH), followed by washing, resulted in a persistent effect on glucose oxidation. This effect was reversed by addition of anti-TSH serum, or by digestion with trypsin, suggesting that hormone binding to superficial cell sites was the initial effect of the hormone. Winand & Kohn (305) studied the binding of ^3H-thyrotropin and ^3H-exophthalmogenic factor to plasma membranes of cells from retro-orbital tissues. They found that γ-globulin from patients with malignant exophthalmus (but not from normals) increased the binding of these hormones by a mechanism which did not involve binding of the globulin to the membranes. These studies (305) have led to an interesting hypothesis concerning the pathogenesis of human exophthalmus. The binding of ^{125}I-labeled TSH to plasma membranes of thyroid (306) and to cultured thyroid cells (307) has been described. The binding is of very high affinity, the number (307) of sites per cell is only about 500 (although the density was not calculated), and the binding can be correlated with stimulation of adenylate cyclase activity (306) in the membranes. Fayet et al (307a) have recently described similar studies.

Lin & Goodfriend (308–310) have described the preparation of iodinated angiotensin and its use in the study of specific tissue receptors. Brecher et al (310a) studied the binding of angiotensin to adrenal capsule membranes. Lefkowitz and colleagues (311, 312) have isolated monoiodo adrenocorticotropic hormone (ACTH)-^{125}I and have studied its binding to membranes from adrenal glands. In the absence of calcium the binding of ACTH is unaltered but activation of adenylate cyclase by the hormone (but not by fluoride) is inhibited (313); high concentrations of calcium suppress both processes. The binding to membrane preparations has been used (314, 315) to develop assays for measuring concentrations of ACTH in plasma. Hofman et al (316, 317) used a series of synthetic, homogeneous radioactive (^{14}C) and unlabeled analogs of ACTH to study binding to membranes from adrenal and other tissues. Major active and binding sites were shown to reside in different sections of the ACTH molecule. Corticotropin$_{11-20}$ amide binds to membranes but fails to activate adenylate cyclase (317). Certain peptides whose in vivo activity was greater than that of corticotropin$_{1-24}$ were less active in stimulating enzyme activity, which emphasizes (317) the caution that must be exercised in predicting in vivo potency of ACTH analogs on the basis of adenylate cyclase assays. The active site of ACTH was postulated to be located on the NH_2 terminal region (positions 1 to 13), and important binding sites occupy the segment comprising positions 14 to 23. The sequence Lys-Lys-Arg-Arg (positions 15 to 18) was thought (316) to be a likely binding site.

Renal plasma membranes with specific capacity to bind ^{125}I-calcitonin have been described (318–320). Specific binding to membrane preparations, suggestive of receptor interactions, has been described for FSH (321), luteinizing hormone releasing factor (321a–322), prolactin and lactogenic hormones (323), antidiuretic hormone (324), thyrotropin releasing hormone (325–328c), human chorionic gonado-tropin (329–331a), luteinizing hormone (332–336, 331a), vasoactive intestinal polypeptide (337), and human growth hormone (338). Sonenberg (339) has described an extremely sensitive method of detecting hormone-membrane interactions by measurements of intrinsic membrane fluorescence. Some interesting studies have also been described (340–343a) for the binding of ^{3}H- or ^{125}I-labeled oxytocin to bladder epithelial cells, fat cells, uterus, skin epithelium, and mammary cells. Walter et al (344) have prepared and successfully used an affinity labeling derivative of oxytocin, bromoacetyl oxytocin, which irreversibly inhibits hormone-stimulated adenylate cyclase activity, possibly by inactivating specific hormone receptors.

The specific binding of ^{125}I-labeled epidermal growth factor to epidermis and corneal epithelial cells has been studied (345, 346). Epidermal growth factor, which at very low concentrations has effects on DNA and RNA synthesis in contact-inhibited human fibroblasts (suggestive of cell growth stimulation), binds specifically to these cells and to membrane preparations (129, 347). The interaction of ^{125}I nerve growth factor with membrane preparations of superior cervical ganglia has been studied in detail (348–350). Very careful attempts to correlate the binding of ^{3}H-vasopressin and the activation of adenylate cyclase in kidney membranes were described by Bockaert et al (351a).

The interaction of ^{125}I-labeled cholera toxin with intact cells (fat cells, erythro-cytes) and membrane preparations (fat, liver, intestinal epithelium, erythrocytes) has been studied in detail (351–355). The toxin has been shown to interact specifically with monosialogangliosides (G_{M1}) of cell membranes, and there is substantial evidence (352) that these glycolipids are the specific membrane receptors for the toxin. These gangliosides (receptors) can be incorporated spontaneously into intact cells, and the biological response can be greatly enhanced by thus increasing the number of receptors. This is probably the first instance in which the specific chemical nature of a membrane receptor has been identified and in which in vitro reconstitution of a biologically responsive receptor system has been possible. Cholera toxin, a protein of mol wt about 100,000 consisting of separate ganglioside binding and active subunits, appears to exert its biological effects primarily if not entirely by stimulating the activity of membrane-bound adenylate cyclase. The basic mechanisms involved in this effect probably closely resemble the stimulation of the enzyme by polypeptide hormones, catecholamines, and prostaglandins. For this reason the cholera toxin system may be an excellent, simple model for studying the molecular mechanisms involved in hormone-receptor and receptor-enzyme interactions. Activation of enzyme, for example, is preceded by a peculiar and long lag phase which probably involves processes occurring strictly in the cell membrane (353, 356). These may be simply a gross exaggeration of the processes that normally transpire between binding of hormones to their receptors and the translation of this interaction to adenylate cyclase (or other structures) in the membrane.

OTHER MEMBRANE RECEPTORS

Ashwell (357, 358) and colleagues have demonstrated that the liver has highly specific and efficient mechanisms for recognizing and removing from the circulation a number of selectively desialidated serum glycoproteins which possess exposed galactose residues. It has been shown that the hepatic plasma membranes are the primary locus of binding (359), and the detailed binding of native, asialo, and agalacto serum glycoproteins to plasma membranes from rat liver has been studied (360). Morell & Scheinberg (361) have solubilized (with Triton X-100) and studied these binding sites. Recently Aronson et al (362) described studies that suggest that the liver plasma membrane sites for binding asialo glycoproteins may be galactosyl-transferase enzymes.

Pert & Snyder (363, 366a) and Kuhar et al (366b) utilized ^{3}H-naloxone, a powerful opiate antagonist, to detect opiate receptors in mammalian brain and guinea pig intestine. The binding observed to particulate membrane fractions is stereospecific and the specificity of binding closely parallels the pharmacological potency of the substances tested. These receptors have now been studied in great detail by Snyder and colleagues (363, 366a, 366b). Simon et al (364) utilized a similar approach, using a ^{3}H derivative of the potent narcotic analgesic, etorphine, to detect stereospecific and saturable binding to rat brain homogenates. Winter & Goldstein (365) reported the synthesis and use of an analog of levorphanol which may prove very useful in affinity labeling studies of opiate receptors.

Alivisatos et al (366) described interesting studies which suggested that the interaction of serotonin with specific receptors in the central nervous system may involve formation of a Schiff base between the ethylamine residue of serotonin and an appropriate carbonyl residue at the receptor site. Stabilization of this bond by reduction may provide (366) a means of permanently labeling receptors as a preliminary step in their isolation. Evidence has been presented by Hardin et al (367) that the plant hormones, auxins, interact specifically with membrane structures to cause the release from the membranes of transcriptional factors that stimulate RNA polymerase activity. Venis (368) has described the isolation by affinity chromatography of auxin acceptor proteins from pea and maize shoots.

Identification and isolation of sugar binding protein structures in membrane preparations from taste buds, which may represent specific taste receptors, have been described (369–372). A membrane protein from isolated fat cells that can be phosphorylated with ^{32}P-ATP may be a component of the specific D-glucose carrier that is specifically stimulated by insulin (141, 373). The important group of cell membrane receptors for plant lectins has been discussed in detail recently (374). A discussion of cell surface receptors in immunocompetent cells is beyond the scope of this review. Presumed receptors for prostaglandins have been described (375, 376) in extracts of membranes from isolated fat cells. Moore & Wolff recently (377) presented detailed studies describing the interaction of ^{3}H-prostaglandins and beef thyroid membranes.

MEMBRANE FLUIDITY AND THE INTERACTION OF HORMONE-RECEPTOR COMPLEXES WITH OTHER FUNCTIONAL COMPONENTS OF MEMBRANES

Little is known of the manner by which hormone-receptor complexes, once formed, alter or regulate various membrane-localized functions such as enzyme activities, ion transport, sugar transport, etc. A general mechanism which may be applicable and is consistent with currently evolving concepts of membrane structure will be considered briefly here. The hormone-adenylate cyclase regulatory system will be used to exemplify these concepts since it has been one of the most thoroughly studied.

Since no evidence exists to suggest that hormone-receptor (glucagon, ACTH, catecholamines, etc) complexes regulate the activity of adenylate cyclase through the production of intermediary chemical substances (mediators), the traditional concepts visualize direct interactions between these structures. Speculation has centered on whether the receptors are integral components of the enzymic complex, whether such receptors surround the enzyme as regulatory subunits, and on the manner by which subunit interactions may be understood by analogy with well-known multisubunit soluble enzymes. The conventional models thus assume the existence of prescribed and fixed arrangements of interrelated molecules whose properties may change by simple but specific perturbations of one or another of the components of the complex by hormones or other regulators (e.g. GTP, cations, etc).

This general model suffers from having unnecessary constraints, not taking into account the importance of membrane fluidity, and being difficult to reconcile with certain experimental observations. At least in isolated fat cells, the same adenylate cyclase (378–387) is stimulated by no less than seven different hormones (epinephrine, glucagon, ACTH, TSH, LH, secretin, growth hormone, and dexamethasone) and is probably inhibited by at least three other hormones (insulin, oxytocin, prosta-glandins). If each hormone receptor is to constitute a discrete macromolecular entity, as appears likely from experimental evidence and on theoretical grounds based on specificity and conformational considerations, it becomes extraordinarily difficult to physically accommodate all of these receptor macromolecules in direct contiguity with a single enzyme complex. Judging from the estimated (200,000–300,000) mol wt of insulin (75), glucagon (163), and acetylcholine (100, 271) receptors, and from the fact that the receptors must be situated on the membrane such that their hormone binding sites are exposed to the external, aqueous environment, the existence of such a supermacromolecular complex seems most unlikely.

An alternative view is that in the free state hormone receptors (R) are not physically associated with adenylate cyclase. Formation of the hormone-receptor (H·R) complex (Figure 2) endows that receptor with the new property of having special affinity and specificity for adenylate cyclase (AC). Formation of the hormone-receptor-adenylate·cyclase complex (H·R-AC) would thus be a second, independent step whose rate would be dependent on the concentration of the individual com-ponents (H·R and AC), the diffusion properties of these components in the plane of the membrane, the temperature, etc. The normal activity state of the free AC molecules would be relatively quiescent or inhibited. The association of some H·R complexes with AC would result in stimulation of the enzyme; others would cause inhibition, which, although real, may be difficult to detect experimentally (because of the already low state of activity of the free AC) unless stimulatory hormones are added simultaneously. It is very difficult experimentally to demonstrate insulin inhibition of basal (unstimulated) adenylate cyclase activity of membrane prepara-

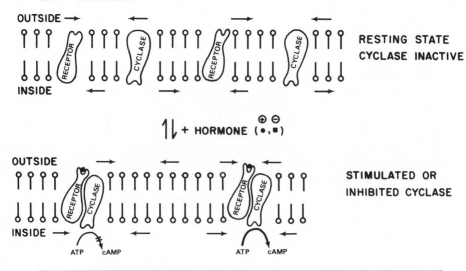

Figure 2 General two-step fluidity hypothesis for the mechanism of modulation of adenylate cyclase activity of cell membranes by hormones. The central feature is that the receptors and the enzyme are discrete and separate structures which acquire specificity and affinity for complex formation only after the receptor has been occupied by the hormone. These structures can combine after binding of the hormone because of the fluidity of cell membranes. The hormone binding sites of the receptor are on the external face, exposed to the aqueous medium, and the catalytic site of the enzyme is facing inward toward the cytoplasm of the cell.

tions (108, 110, 111). Inhibitory H · R complexes could in principle bind to the same or different regions of AC with the formation of binary or ternary complexes, respectively, if both types of hormones are present simultaneously.

A basic premise in this scheme is the assumption that biological membranes are in a relatively fluid state (98, 388–390), that H · R complexes once formed are relatively free to diffuse laterally along the plane of the membrane and interact (by random encounters) with AC (which is probably also in a relatively fluid state) in a way analogous to known interactions in solution.

In this scheme hormones would alter enzyme activity by a process involving at least two discrete, sequential, and dissociable reactions:

$$1.\ H + R \rightleftharpoons H \cdot R \quad \text{and} \quad 2.\ H \cdot R + AC \rightleftharpoons H \cdot R\text{-}AC$$

This scheme provides considerable flexibility and versatility for control mechanisms. Regulation of control by a given hormone would depend on the number of hormone receptors such that maximal effectiveness would result by having at least a stoichiometric quantity of receptors relative to the number of AC molecules. Depending on the affinity of reaction 2, a large excess of receptors would not necessarily affect the steady state degree of control achieved but only the time required to achieve this steady state. Clearly, a H·R complex present in quantities lower than AC (because of a deficiency of R or of a low concentration of H) would not achieve a maximal response. In addition to the number of R molecules (and their affinity for H), important factors in determining the relative effectiveness of various hormones would be the specific affinity of that H·R for AC, the diffusion properties of H·R (which would likely affect only the rate of formation of the steady state concentration of H·R-AC), and the relative ability of the particular H·R to modify AC once the complex has formed.

This scheme has some important consequences on the interpretation of experimental data and on the design of new experiments to further elucidate basic mechanisms. The theory predicts that finite discrepancies must exist in the kinetics of hormone binding and of activation of biological responses. Whether these are large enough to be experimentally detectable cannot be answered yet. If the rates of reaction 2 (formation of H·R-AC) are sufficiently slow, "lag" periods may be present in the onset of a biological response even when large, supersaturating concentrations of the hormone are used to immediately saturate all R. Such lag periods, which may depend on the fluid state of the membrane, should be temperature sensitive, and they may well depend on the specific phospholipid and sterol composition of the particular membrane under study. Recent observations in this laboratory (391) indicate that such lag periods can indeed be detected in the effects of certain hormones on adenylate cyclase, and that these are temperature dependent. Furthermore, it would not be surprising if in isolated membranes difficulties are encountered in the proper coupling of H·R to AC, even if each of these is individually fully intact, since the fluidity and diffusion properties may be altered.

Since in this theory the relation of hormone binding and activation are clearly complex and subject to separate modulation, the discrepancies of glucagon binding and activation of AC (180) discussed earlier may be explained by these considerations. For example, the decay of the H·R-AC complex need not follow the same pathway of complex formation. It is therefore possible, for example, that the affinity of the H for R is lower in this complete complex than in the simple HR complex. Furthermore, the hormone may dissociate faster than the residual R-AC. This could serve as a special regulatory mechanism which could, for example, explain the ability of GTP to increase glucagon-stimulated enzyme activity while at the same time increasing the rate of dissociation of the membrane-bound glucagon. In this scheme the interaction of HR with AC may be regulated by special effectors. For example, GTP, other nucleotides, or divalent cations could favor the formation of this complex, thus enhancing the biological effectiveness of the H while decreasing the apparent affinity of the overall H-membrane interaction. This theory also predicts that isolation and purification of hormone receptors is unlikely to result in the concomitant purification of a macromolecule complexed to AC, unless perhaps the isolation is performed in the presence of the hormone.

It is also quite possible in this scheme to conceive of a specific hormone affecting two or more separate and independent membrane functions (e.g. transport, adenylate cyclase, ATPase, etc) simultaneously and still doing this by binding to a single unique receptor, provided the H · R is able to subsequently complex with and thus modify these separate macromolecular components. For example, this may explain the reported ability of insulin to inhibit membrane-bound enzymes such as adenylate cyclase (108–115) and to stimulate membrane-bound guanylate cyclase (140) and Na-K-ATPase (82) in broken cell preparations. Also, growth hormone has been reported (392) to modify several membrane-bound enzymes. Further, it is now known (393) that prostaglandins can directly modify membrane localized Mg^{2+}-dependent Na^+-K^+-activated ATPase, adenylate kinase, and phosphodiesterase in addition to adenylate cyclase in isolated membrane preparations. It is also possible in this scheme to visualize how a weak stimulatory hormone could act to inhibit a more potent hormone if the former H · R has higher affinity for AC (thus effectively competing for binding of the other H · R) but is less effective in activating the enzyme once it is in the complex. These considerations may be relevant to other membrane receptor systems.

Literature Cited

1. Cuatrecasas, P. 1974. *Biochem. Pharmacol.* In press
1a. Levine, R., Goldstein, M. S., Klein, S., Huddlestun, B. 1949. *J. Biol. Chem.* 179: 985
2. Levine, R. 1965. *Fed. Proc.* 24: 1071
3. Levine, R., Goldstein, M. S., Huddlestun, B., Klein, S. 1950. *Am. J. Physiol.* 163: 70
4. Goldstein, M. S., Henry, W. L., Huddlestun, B., Levine, R. 1953. *Am. J. Physiol.* 173: 207
5. Haft, D., Mirsky, I. A., Perisutti, G. 1953. *Proc. Soc. Exp. Biol. Med.* 82: 60
6. Kipnis, D. M. 1959. *Ann. NY Acad. Sci.* 82: 354
7. Park, C. R., Johnson, L. H., Wright, J. H. Jr., Batsel, H. 1957. *Am. J. Physiol.* 191: 13
8. Park, C. R., Reinwein, D., Henderson, M. J., Cadenas, E., Morgan, H. E. 1959. *Am. J. Med.* 26: 674
9. Fisher, R. B., Lindsay, D. B. 1956. *J. Physiol. London* 131: 526
10. Crofford, O. B., Renold, A. E. 1965. *J. Biol. Chem.* 240: 14
11. Stadie, W. C., Haugaard, N., Marsh, J. B., Hills, A. G. 1949. *Am. J. Med. Sci.* 218: 265
12. Hills, A. G., Stadie, W. C. 1952. *J. Biol. Chem.* 194: 25
13. Haugaard, N., Marsh, J. B. 1952. *J. Biol. Chem.* 194: 33
14. Stadie, W. C., Haugaard, N., Vaughan, M. 1952. *J. Biol. Chem.* 199: 729
15. Stadie, W. C., Haugaard, N., Vaughan, M. 1953. *J. Biol. Chem.* 200: 745
16. Malaisse, W., Franckson, J. R. M. 1965. *Arch. Int. Pharmacodyn. Ther.* 155: 484
17. Garratt, C. J., Jarrett, R. J., Keen, H. 1966. *Biochim. Biophys. Acta* 121: 143
18. Bewsher, P. D., Hillman, C. C., Ashmore, J. 1966. *Mol. Pharmacol.* 2: 227
19. Wohltmann, H. J., Narahara, H. T. 1966. *J. Biol. Chem.* 241: 4931
20. Narahara, H. T. 1972. In *Handbook of Physiology, Endocrinology,* Vol. I, ed. R. O. Greep, B. Astwood, 333. Baltimore: Williams and Wilkins
21. Ball, E. G., Jungas, R. L. 1964. *Recent Progr. Horm. Res.* 20: 183
22. Ferrebee, J. W., Johnson, B. B., Mithoefer, J. C., Gardella, J. W. 1951. *Endocrinology* 48: 277
23. Garratt, C. J., Cameron, J. S., Menzinger, G. 1966. *Biochim. Biophys. Acta* 115: 179
24. Newerly, K., Berson, S. A. 1957. *Proc. Soc. Exp. Biol. Med.* 94: 751
25. Bleehen, N. M., Fisher, R. B. 1954. *J. Physiol. London* 123: 260
26. Beigelman, P. M. 1962. *Metabolism* 11: 1315
27. Crofford, O. B. 1968. *J. Biol. Chem.* 243: 362
28. Kono, T. 1969. *J. Biol. Chem.* 244: 1772
29. Kono, T. 1969. *J. Biol. Chem.* 244: 5777
30. Fain, J. N., Loken, S. C. 1969. *J. Biol. Chem.* 244: 3500
31. Rodbell, M. 1964. *J. Biol. Chem.* 239: 375
32. Cuatrecasas, P. 1971. *J. Biol. Chem.* 246: 6522
33. Kuo, J. F., Holmlund, C. E., Dill, I. K. 1966. *Life Sci.* 5: 2257
34. Rieser, P., Rieser, C. H. 1964. *Proc. Soc. Exp. Biol. Med.* 116: 669
35. Rodbell, M. 1966. *J. Biol. Chem.* 214: 130

36. Blecher, M. 1965. *Biochem. Biophys. Res. Commun.* 21:202
37. Blecher, M. 1966. *Biochem. Biophys. Res. Commun.* 23:68
38. Cuatrecasas, P. 1969. *Proc. Nat. Acad. Sci. USA* 63:450
39. Turkington, R. W. 1970. *Biochem. Biophys. Res. Commun.* 41:1362
40. Oka, T., Topper, Y. J. 1971. *Proc. Nat. Acad. Sci. USA* 68:2066
41. Oka, T., Topper, Y. J. 1972. *Nature New Biol.* 239:216
42. Blatt, L. M., Kim, K. H. 1971. *J. Biol. Chem.* 246:4895
43. Soderman, D. D., Germershausen, J., Katzen, H. M. 1973. *Proc. Nat. Acad. Sci. USA* 70:792
44. Suzuki, F., Daikuhara, Y., Ono, M., Takeda, Y. 1972. *Endocrinology* 90:1220
45. Armstrong, K. J., Noall, M. W., Stouffer, J. E. 1972. *Biochem. Biophys. Res. Commun.* 47:354
46. Tarui, S., Saito, Y., Suzuki, F., Takeda, Y. *Endocrinology* 91:1442
47. Cuatrecasas, P. 1973. *Fed. Proc.* 32:1838
48. Desbuquois, B., Cuatrecasas, P. 1973. *Ann. Rev. Med.* 24:233
49. Tell, G. P., Krug, F., Cuatrecasas, P. 1974. In *Handbook of Experimental Pharmacology,* ed. A. Hasselblatt, F. von Bruchhausen, 32: Chap. 2. Berlin: Springer. In press
50. Cuatrecasas, P. 1972. In *Insulin Action,* ed. I. B. Fritz, 137. New York and London: Academic
51. Cuatrecasas, P. 1972. *Diabetes* 21, Suppl. 2:396
51a. Hollenberg, M. D., Cuatrecasas, P. 1974. *Biochemical Action of Hormones.* New York: Academic. In press
52. Cuatrecasas, P. 1971. *Proc. Nat. Acad. Sci. USA* 68:1264
53. Kono, T., Barham, F. W. 1971. *J. Biol. Chem.* 246:6210
54. El-Allawy, R. M. M., Gliemann, J. 1972. *Biochim. Biophys. Acta* 273:97
55. Cuatrecasas, P. 1971. *J. Biol. Chem.* 246:7265
56. Hammond, J. M., Jarett, L., Mariz, I. K., Daughaday, W. H. 1972. *Biochem. Biophys. Res. Commun.* 49:1122
57. House, P. D. R., Weidemann, M. J. 1970. *Biochem. Biophys. Res. Commun.* 41:541
58. House, P. D. R. 1971. *FEBS Lett.* 16:339
59. Cuatrecasas, P., Desbuquois, B., Krug, F. 1971. *Biochem. Biophys. Res. Commun.* 44:333
60. Freychet, P., Roth, J., Neville, D. M. 1971. *Biochem. Biophys. Res. Commun.* 43:400
61. Freychet, P., Roth, J., Neville, D. M. 1971. *Proc. Nat. Acad. Sci. USA* 68:1833
62. Freychet, P., Kahn, R., Roth, J., Neville, D. M. 1972. *J. Biol. Chem.* 247:3953
63. Krug, U., Krug, F., Cuatrecasas, P. 1972. *Proc. Nat. Acad. Sci. USA* 69:2604
64. Gavin, J. R. III, Roth, J., Jen, P., Freychet, P. 1972. *Proc. Nat. Acad. Sci. USA* 69:747
65. Gavin, J. R. III, Gorden, P., Roth, J., Archer, J. A., Buell, D. N. 1973. *J. Biol. Chem.* 248:2202
66. Goldfine, I. D., Gardner, J. D., Neville, D. M. Jr. 1972. *J. Biol. Chem.* 247:6919
66a. O'Keefe, E., Cuatrecasas, P. 1974. *Biochim. Biophys. Acta.* In press
67. Hunter, W. M., Greenwood, F. C. 1962. *Nature London* 194:495
68. Yalow, R. S., Berson, S. A. 1964. *Methods Biochem. Anal.* 12:69
69. Hintz, R. L., Clemmons, D. R., Underwood, L. E., Van Wyk, J. J. 1972. *Proc. Nat. Acad. Sci. USA* 69:2351
69a. Henderson, P. J. F. 1973. *Biochem. J.* 135:101
70. Temin, H. M. 1967. *J. Cell Physiol.* 69:377
71. Blaker, G. J., Birch, J. R., Pirt, S. J. 1971. *J. Cell Sci.* 9:529
72. Hershko, A., Mamont, P., Shields, R., Tomkins, G. 1971. *Nature New Biol.* 232:206
73. Sheppard, J. R. 1972. *Nature New Biol.* 236:14
74. Minemura, T., Lacy, W. W., Crofford, O. B. 1970. *J. Biol. Chem.* 245:3872
75. Cuatrecasas, P. 1971. *J. Biol. Chem.* 246:6532
76. Cuatrecasas, P. 1972. *J. Biol. Chem.* 247:1980
77. Bennett, V., Cuatrecasas, P. 1973. *Biochim. Biophys. Acta* 311:362
78. Kono, T., Barham, F. W. 1971. *J. Biol. Chem.* 246:6204
79. Cuatrecasas, P., Illiano, G. 1971. *J. Biol. Chem.* 246:4938
80. Rosenthal, J. W., Fain, J. N. 1971. *J. Biol. Chem.* 246:5888
81. Archer, J. A., Gorden, P., Gavin, J. R. III, Lesniak, M. A., Roth, J. 1973. *J. Clin. Endocrinol. Metab.* 36:627
82. Hadden, J. W., Hadden, E. M., Wilson, E. E., Good, R. A., Caffey, R. G. 1972. *Nature New Biol.* 235:174
83. Robbins, J. H. 1964. *Science* 146:1648
84. Novogrodsky, A., Biniamnov, M., Ramot, B. 1972. *Blood* 40:311
85. Hollenberg, M. D., Cuatrecasas, P. 1974. In *Control of Proliferation in Animal*

Cells, ed. B. Clarkson, R. Baserga, 214. New York: Cold Spring Harbor Laboratory. In press

86. Novogrodsky, A., Katchalski, E. 1972. *Proc. Nat. Acad. Sci. USA* 69:3207

87. Lesniak, M. A., Roth, J., Gorden, P., Gavin, J. R. III. 1973. *Nature New Biol.* 241:20

88. Krug, U., Hollenberg, M. D., Cuatrecasas, P. 1973. *Biochem. Biophys. Res. Commun.* 52:305

89. Bennett, V., Cuatrecasas, P. 1972. *Science* 176:805

90. Livingston, J. N., Cuatrecasas, P., Lockwood, D. 1972. *Science* 177:626

91. Kahn, C. R., Neville, D. M., Gorden, P., Freychet, P., Roth, J. 1972. *Biochem. Biophys. Res. Commun.* 48:135

92. Freychet, P. et al 1972. *FEBS Lett.* 25:339

93. Kahn, C. R., Neville, D. M., Roth, J. 1973. *J. Biol. Chem.* 248:244

94. Kaplan, J. C., Pichard, A. L., Laudat, M. H., Laudat, P. 1973. *Biochem. Biophys. Res. Commun.* 51:1008

95. Cuatrecasas, P. 1972. *Proc. Nat. Acad. Sci. USA* 69:318

96. Gavin, J. R. III, Buell, D. N., Roth, J. 1972. *Science* 178:168

97. Gavin, J. R. III, Mann, D. L., Buell, D. N., Roth, J. 1972. *Biochem. Biophys. Res. Commun.* 49:870

98. Singer, S. J., Nicolson, G. L. 1972. *Science* 175:720

99. Makino, S., Reynolds, J. A., Tanford, C. 1973. *J. Biol. Chem.* 248:4926

100. Meunier, J. C. et al 1972. *Biochemistry* 11:1200

101. Cuatrecasas, P. 1972. *Proc. Nat. Acad. Sci. USA* 69:1277

102. Cuatrecasas, P., Tell, G. P. E. 1973. *Proc. Nat. Acad. Sci. USA* 70:485

103. Cuatrecasas, P. 1973. *J. Biol. Chem.* 248:3528

104. Karlsson, E., Heilbronn, E., Widlund, L. 1972. *FEBS Lett.* 28:107

105. Olsen, R. W., Meunier, J. C., Changeux, J. P. 1972. *FEBS Lett.* 28:96

106. Czech, M. P., Lynn, W. S. 1973. *Biochim. Biophys. Acta* 297:368

107. Cuatrecasas, P. 1973. *Biochemistry* 12:1312

108. Illiano, G., Cuatrecasas, P. 1972. *Science* 72:906

109. Ray, T. K., Tomasi, V., Marinetti, G. V. 1970. *Biochim. Biophys. Acta* 211:20

110. Hepp, K. D. 1971. *FEBS Lett.* 12:263

111. Hepp, K. D. 1972. *FEBS Lett.* 20:191

112. Hepp, K. D. 1972. *Eur. J. Biochem.* 31:266

113. DeAsua, L. J., Surian, E. S., Flawia,

M. M., Torres, H. N. 1973. *Proc. Nat. Acad. Sci. USA* 70:1388

114. Flawia, M. M., Torres, H. N. 1973. *J. Biol. Chem.* 248:4517

115. Flawia, M. M., Torres, H. N. 1973. *FEBS Lett.* 30:74

116. Smith, J. W., Steiner, A. L., Newberry, W. M. Jr., Parker, C. W. 1971. *J. Clin. Invest.* 50:432

117. Smith, J. W., Steiner, A. L., Parker, C. W. 1971. *J. Clin. Invest.* 50:442

118. Burger, M. M., Bombik, B. M., Breckenridge, B. M., Sheppard, J. R. 1972. *Nature New Biol.* 239:161

119. Otten, J., Johnson, G. S., Pastan, I. 1971. *Biochem. Biophys. Res. Commun.* 44:1192

120. Perry, C. V., Johnson, G. S., Pastan, I. 1971. *J. Biol. Chem.* 246:5785

121. Hadden, J. W., Hadden, E. M., Haddox, M. K., Goldberg, N. D. 1972. *Proc. Nat. Acad. Sci. USA* 69:3024

122. Illiano, G., Tell, G. P. E., Siegel, M. I., Cuatrecasas, P. 1973. *Proc. Nat. Acad. Sci. USA* 70:2443

123. George, W. J., Polson, J. B., O'Toole, A. G., Goldberg, N. D. 1970. *Proc. Nat. Acad. Sci. USA* 66:398

124. Lee, T. P., Kuo, J. F., Greengard, P. 1972. *Proc. Nat. Acad. Sci. USA* 69:3287

125. Tell, G. P., Cuatrecasas, P., Van Wyk, J. J., Hintz, R. L. 1973. *Science* 180:312

126. Pierson, R. W., Temin, H. M. 1972. *J. Cell Physiol.* 79:319

127. Dulak, N. C., Temin, H. M. 1973. *J. Cell Physiol.* 81:153

128. Dulak, N. C., Temin, H. M. 1973. *J. Cell Physiol.* 81:161

129. Hollenberg, M. D., Cuatrecasas, P. 1973. *Proc. Nat. Acad. Sci. USA* 70:2964

130. Butcher, R. W., Sneyd, J. G. T., Park, C. R., Sutherland, E. W. 1956. *J. Biol. Chem.* 241:1651

131. Jungas, R. L. 1966. *Proc. Nat. Acad. Sci. USA* 56:757

132. Manganiello, V. C., Murad, F., Vaughan, M. 1971. *J. Biol. Chem.* 246:2195

133. Butcher, R. W., Baird, C. E., Sutherland, E. W. 1968. *J. Biol. Chem.* 243:1705

134. Jefferson, L. S., Exton, J. H., Butcher, R. W., Sutherland, E. W., Park, C. R. 1968. *J. Biol. Chem.* 243:1031

135. Exton, J. H., Lewis, S. B., Ho, R. J., Robison, G. A., Park, C. R. 1971. *Ann. N. Y. Acad. Sci.* 185:85

136. Loten, E. G., Sneyd, J. G. T. 1970. *Biochem. J.* 120:187

137. Vaughan, M. 1972. In *Insulin Action,* ed. I. B. Fritz, 297. New York and London: Academic
138. House, P. D., Poulis, P., Weidemann, M. J. 1972. *Eur. J. Biochem.* 24:429
139. Thompson, W. J., Little, S. A., Williams, R. H. 1973. *Biochemistry* 12:1889
140. Siegel, M., Cuatrecasas, P. In preparation
141. Chang, K. J., Cuatrecasas, P. 1974. *J. Biol. Chem.* In press
142. Chang, K. J., Marcus, N., Cuatrecasas, P. In preparation
143. Lockwood, D. H., Lipsky, J. J., Meronk, F. Jr., East, L. E. 1971. *Biochem. Biophys. Res. Commun.* 44:600
144. Minemura, T., Crofford, O. B. 1969. *J. Biol. Chem.* 244:5181
145. Kuo, J. F. 1968. *Arch. Biochem. Biophys.* 127:406
146. Lavis, V. R., Williams, R. H. 1970. *J. Biol. Chem.* 245:23
147. Czech, M. P., Fain, J. N. 1972. *J. Biol. Chem.* 247:6218
148. Edelman, G. M., Rutishauser, U., Millette, C. F. 1971. *Proc. Nat. Acad. Sci. USA* 68:2153
149. Andersson, J., Edelman, G. M., Moller, G., Sjoberg, O. 1972. *Eur. J. Immunol.* 2:233
150. Greaves, M. F., Bauminger, S. 1972. *Nature New Biol.* 235:67
151. Ono, M., Maruta, H., Mizuno, D. 1973. *J. Biochem.* 73:235
152. Tomasi, V., Koretz, S., Ray, T. K., Dunnick, J., Marinetti, G. V. 1970. *Biochim. Biophys. Acta* 211:31
153. Pohl, S. L., Birnbaumer, L., Rodbell, M. 1971. *J. Biol. Chem.* 246:1849
154. Birnbaumer, L., Pohl, S. L., Rodbell, M. 1971. *J. Biol. Chem.* 246:1857
155. Rodbell, M., Krans, H. M. J., Pohl, S. L., Birnbaumer, L. 1971. *J. Biol. Chem.* 246:1861
156. Rodbell, M., Krans, H. M. J., Pohl, S. L., Birnbaumer, L. 1971. *J. Biol. Chem.* 246:1872
157. Rodbell, M., Birnbaumer, L., Pohl, S. L., Krans, H. M. J. 1971. *J. Biol. Chem.* 246:1877
158. Pohl, S. L., Krans, H. M. J., Kozyreff, V., Birnbaumer, L., Rodbell, M. 1971. *J. Biol. Chem.* 246:4447
159. Birnbaumer, L., Pohl, S. L., Rodbell, M. 1972. *J. Biol. Chem.* 247:2038
160. Pohl, S. L., Krans, H. M. J., Birnbaumer, L., Rodbell, M. 1972. *J. Biol. Chem.* 247:2295
161. Rodbell, M., Birnbaumer, L., Pohl, S. L., Sundby, F. 1971. *Proc. Nat. Acad. Sci. USA* 68:909

162. Rubalcava, B., Rodbell, M. 1973. *J. Biol. Chem.* 248:3831
163. Blecher, M., Giorgio, N. A. Jr., Johnson, C. B. 1972. In *The Role of Membranes in Metabolic Regulation,* ed. M. A. Mehlman, R. W. Hanson. New York: Academic
164. Desbuquois, B., Cuatrecasas, P. 1972. *Nature New Biol.* 237:202
165. Desbuquois, B., Cuatrecasas, P. 1974. *Biochim. Biophys. Acta.* In press
166. Bromer, W. W., Boucher, M. E., Patterson, J. M. 1973. *Biochem. Biophys. Res. Commun.* 53:134
167. Lande, S., Gorman, R., Bitensky, M. 1972. *Endocrinology* 90:597
168. Grande, F., Grisolia, S., Diederich, D. 1972. *Proc. Soc. Exp. Biol. Med.* 139:855
169. Krug, F., Desbuquois, B., Cuatrecasas, P. 1971. *Nature New Biol.* 234:268
170. Johnson, C. B., Blecher, M., Giorgio, N. A. Jr. 1972. *Biochem. Biophys. Res. Commun.* 46:1035
171. Goldfine, I. D., Roth, J., Birnbaumer, L. 1972. *J. Biol. Chem.* 247:1211
172. Krishna, G., Harwood, J. P., Barber, A. J., Jamieson, G. A. 1972. *J. Biol. Chem.* 247:2253
173. Bockaert, J., Roy, C., Jard, S. 1972. *J. Biol. Chem.* 247:7073
174. Leray, F., Chambout, A. M., Hanoune, J. 1972. *Biochem. Biophys. Res. Commun.* 48:1385
175. Swislocki, N. I., Scheinberg, S., Sonenberg, M. 1973. *Biochem. Biophys. Res. Commun.* 52:313
176. Swislocki, N. I., Scheinberg, S. 1973. *Fed. Proc.* 32:555A
177. Aizono, Y., Roberts, J., Sonenberg, M., Swislocki, N. I. 1973 *Fed. Proc.* 32:489A
178. Levey, G. S. 1971. *Biochem. Biophys. Res. Commun.* 43:108
179. Levey, G. S. 1971. *J. Biol. Chem.* 246:7405
180. Birnbaumer, L., Pohl, S. L. 1973. *J. Biol. Chem.* 248:2056
181. Bataille, D. P., Freychet, P., Kitabgi, P. E., Rosselin, G. E. 1973. *FEBS Lett.* 30:215
182. Christoffersen, T. et al 1972. *Arch. Biochem. Biophys.* 150:807
183. Reik, L., Petzold, G. L., Higgins, J. A., Greengard, P., Barrnett, R. J. 1970. *Science* 168:382
184. Bitensky, M. W., Russell, V., Robertson, W. 1968. *Biochem. Biophys. Res. Commun.* 31:706
185. Hummel, J. P., Dreyer, W. J. 1962. *Biochim. Biophys. Acta* 63:530

186. Cuatrecasas, P., Fuchs, S., Anfinsen, C. B. 1967. *J. Biol. Chem.* 242:3063
187. Bornet, H., Edelhoch, H. 1971. *J. Biol. Chem.* 246:1785
188. Schneider, A. B., Edelhoch, H. 1972. *J. Biol. Chem.* 247:4986
189. Schneider, A. B., Edelhoch, H. 1972. *J. Biol. Chem.* 247:4992
190. Manganiello, V., Vaughan, M. 1972. *J. Lipid Res.* 13:12
191. Livingston, J. N., Cuatrecasas, P., Lockwood, D. H. 1974. *J. Lipid Res.* 15:26
192. Gliemann, J., Østerlind, K., Vinten, J., Gammeltoft, S. 1972. *Biochim. Biophys. Acta* 286:1
193. DeSantis, R. A., Gorenstein, T., Livingston, J. N., Lockwood, D. H. 1974. *J. Lipid Res.* 15:33
194. Kirshner, N. 1962. *J. Biol. Chem.* 237:2311
195. Stjarne, L. 1964. *Acta Physiol. Scand.* 62, Suppl. 228:1
196. Molinoff, P. B., Axelrod, J. 1971. *Ann. Rev. Biochem.* 40:465
197. Iversen, L. L. 1967. *The Uptake and Storage of Noradrenaline in Sympathetic Nerves.* London: Cambridge Univ. Press
198. Shore, P. A. 1972. *Ann. Rev. Pharmacol.* 12:209
199. Marinetti, G. V., Ray, T. K., Tomasi, V. 1969. *Biochem. Biophys. Res. Commun.* 36:185
200. Tomasi, V. S., Ray, T. K., Dunnick, J. K., Marinetti, G. V. 1970. *Biochim. Biophys. Acta* 211:31
201. Dunnick, J. K., Marinetti, G. V. 1971. *Biochim. Biophys. Acta* 249:122
202. Lefkowitz, R. J., Haber, E. 1971. *Proc. Nat. Acad. Sci. USA* 68:1773
203. Lefkowitz, R. J., Haber, E., O'Hara, D. 1972. *Proc. Nat. Acad. Sci. USA* 69:2828
204. Lefkowitz, R. J., Levey, G. S. 1972. *Life Sci.* 11, Pt. 2:821
205. Lefkowitz, R. J., Sharp, G., Haber, E. 1973. *J. Biol. Chem.* 248:342
206. Fiszer de Plazas, S., DeRobertis, E. 1972. *Biochim. Biophys. Acta* 266:246
207. Lefkowitz, R. J., O'Hara, D. S., Warshaw, J. 1973. *Nature New Biol.* 244:79
208. Schramm, M., Feinstein, H., Naim, E., Lang, M., Lasser, M. 1972. *Proc. Nat. Acad. Sci. USA* 69:523
209. Bilezikian, J. P., Aurbach, G. D. 1973. *J. Biol. Chem.* 248:5575
210. Bilezikian, J. P., Aurbach, G. D. 1973. *J. Biol. Chem.* 248:5584
211. Cuatrecasas, P., Tell, G. P. E., Sica, V., Parikh, I., Chang, K. J. 1974. *Nature* 247:92
212. Gardner, J. D., Klaeveman, H. L., Bilezikian, J. P., Aurbach, G. D. 1973. *J. Biol. Chem.* 248:5590
213. Uloth, R. H., Kirk, J. R., Gould, W. A., Larsen, A. A. 1966. *J. Med. Chem.* 10:88
214. Axelrod, J., Tomchick, R. J. 1958. *J. Biol. Chem.* 233:702
215. Axelrod, J., Lerner, A. B. 1963. *Biochim. Biophys. Acta* 71:650
216. Crout, J. R. 1961. *Biochem. Pharmacol.* 6:47
217. Giles, R. E., Miller, J. W. 1967. *J. Pharmacol. Exp. Ther.* 156:201
218. Inscoe, J. K., Daly, J., Axelrod, J. 1965. *Biochem. Pharmacol.* 14:1257
219. Axelrod, J., Inscoe, J. K., Daly, J. 1965. *J. Pharmacol. Exp. Ther.* 149:16
220. Traiger, G. J., Calvert, D. N. 1969. *Biochem. Pharmacol.* 18:109
221. Lutz, W. B., Creveling, C. R., Daly, J. W., Witkop, B. 1972. *J. Med. Chem.* 15:795
222. Venter, J. C., Dixon, J. E., Maroko, P. R., Kaplan, N. O. 1972. *Proc. Nat. Acad. Sci. USA* 69:1141
223. Venter, J. C., Ross, J. Jr., Dixon, J., Mayer, S. E., Kaplan, N. O. 1973. *Proc. Nat. Acad. Sci. USA* 70:1214
224. Melmon, K. L., Bourne, H. R., Weinstein, J., Sela, M. 1972. *Science* 177:707
225. Weinstein, Y., Melmon, K. L., Bourne, H. R., Sela, M. 1973. *J. Clin. Invest.* 52:1349
226. Hodgkin, A. L. 1964. *The Conduction of the Nerve Impulses.* Springfield, Ill.: C C Thomas
227. Tasaki, I. 1968. *Nerve Excitation.* Springfield, Illinois: C C Thomas
228. Nachmansohn, D. 1971. *Proc. Nat. Acad. Sci. USA* 68:3170
229. Nachmansohn, D. 1970. *Science* 168:1059
230. Neumann, E., Nachmansohn, D., Katchalsky, A. 1973. *Proc. Nat. Acad. Sci. USA* 70:727
231. Keynes, R. D. 1972. *Nature* 239:29
232. DeRobertis, E. 1971. *Science* 171:963
233. Chagas, C. 1959. *Ann. NY Acad. Sci.* 81:345
234. Trams, E. G. 1964. *Biochim. Biophys. Acta* 79:521
235. Ehrenpreis, S. 1962. *Nature* 194:586
236. Ehrenpreis, S., Fleish, J. H., Mittag, T. W. 1969. *Pharmacol. Rev.* 21:131
237. Hasson-Voloch, A. 1968. *Nature* 218:330
238. DeRobertis, E., Fiszer, S. F., Soto, E. F. 1967. *Science* 158:928

239. LaTorre, J. L., Lunt, G. S., DeRobertis, E. 1970. *Proc. Nat. Acad. Sci. USA* 65:716
240. Levinson, S. R., Keynes, R. D. 1972. *Biochim. Biophys. Acta* 288:241
241. Fiszer de Plazas, S., DeRobertis, E. 1972. *Biochim. Biophys. Acta* 274:258
242. Parisi, M., Rivas, E. F., DeRobertis, E. 1971. *Science* 172:56
243. Reader, T. A., Parisi, M., DeRobertis, E. 1973. *Biochem. Biophys. Res. Commun.* 53:10
244. O'Brien, R. D., Gilmour, L. P. 1969. *Proc. Nat. Acad. Sci. USA* 63:496
245. O'Brien, R. D., Gilmour, L. P., Eldefrawi, M. E. 1970. *Proc. Nat. Acad. Sci. USA* 65:438
246. Eldefrawi, M. E., Britten, A. G., Eldefrawi, A. T. 1971. *Science* 173:338
247. Chang, C. C., Lee, C. Y. 1963. *Arch. Int. Pharmacodyn.* 144:241
248. Lee, C. Y., Tseng, L. F. 1966. *Toxicon* 3:281
249. Changeux, J. P., Kasai, M., Lee, C. Y. 1970. *Proc. Nat. Acad. Sci. USA* 67:1241
250. Miledi, R., Molinoff, P., Potter, L. T. 1971. *Nature* 229:554
251. Miledi, R., Potter, L. T. 1971. *Nature* 233:599
252. Barnard, E. A., Wieckowski, J., Chiu, T. H. 1971. *Nature* 234:207
253. Fambrough, D. M., Hartzell, H. C. 1972. *Science* 176:189
254. Porter, C. W., Chiu, T. H., Wieckowski, J., Barnard, E. A. 1973. *Nature New Biol.* 241:3
255. Berg, D. K., Kelly, R. B., Sargent, P. B., Williamson, P., Hall, Z. W. 1972. *Proc. Nat. Acad. Sci. USA* 69:147
256. Lester, H. A. 1972. *Mol. Pharmacol.* 6:623
257. Lester, H. A. 1972. *Mol. Pharmacol.* 8:632
258. Kasai, M., Changeux, J. P. 1971. *J. Membrane Biol.* 6:1
259. Kasai, M., Changeux, J. P. 1971. *J. Membrane Biol.* 6:24
260. Kasai, M., Changeux, J. P. 1971. *J. Membrane Biol.* 6:58
261. Cartaud, J., Benedetti, E. L., Kasai, M., Changeux, J. P. 1971. *J. Membrane Biol.* 6:81
262. Changeux, J. P., Kasai, M., Huchet, M., Meunier, J. C. 1970. *C. R. Acad. Sci. Paris* 270:2864D
263. Raftery, M. A., Schmidt, J., Clark, D. G., Wolcott, R. G. 1971. *Biochem. Biophys. Res. Commun.* 45:1622
264. Bourgeois, J. P. et al 1971. *FEBS Lett.* 16:92
265. Bourgeois, J. P. et al 1972. *FEBS Lett.* 25:127
266. Franklin, G. I., Potter, L. T. 1972. *FEBS Lett.* 28:101
267. Eldefrawi, M. E., Eldefrawi, A. T., Seifert, S., O'Brien, R. D. 1972. *Arch. Biochem. Biophys.* 150:210
268. Eldefrawi, M. E., Eldefrawi, A. T. 1972. *Proc. Nat. Acad. Sci. USA* 69:1776
269. Raftery, M. A., Schmidt, J., Clark, D. G. 1972. *Arch. Biochem. Biophys.* 152:882
270. Clark, D. G., Wolcott, R. G., Raftery, M. A. 1972. *Biochem. Biophys. Res. Commun.* 48:1061
271. Meunier, J. C., Olsen, R. W., Changeux, J. P. 1972. *FEBS Lett.* 24:63
272. Schmidt, J., Raftery, M. A. 1973. *Biochemistry* 12:852
273. Fulpius, B., Cha, S., Klett, R., Reich, E. 1972. *FEBS Lett.* 24:323
274. Changeux, J. P., Meunier, J. C., Huchet, M. 1971. *Mol. Pharmacol.* 7:538
275. Schmidt, J., Raftery, M. A. 1972. *Biochem. Biophys. Res. Commun.* 49:572
276. Schwyser, R., Frank, J. 1972. *Helv. Chim. Acta* 55:2678
277. Sica, V., Nola, E., Parikh, I., Puca, G. A., Cuatrecasas, P. 1973. *Nature New Biol.* 244:36
278. Sica, V., Parikh, I., Nola, E., Puca, G. A., Cuatrecasas, P. 1973. *J. Biol. Chem.* 248:6543
279. Bosmann, H. B. 1972. *J. Biol. Chem.* 247:130
280. Moore, W. J., Loy, N. J. 1972. *Biochem. Biophys. Res. Commun.* 46:2093
281. Chiu, T. H., Dolly, J. O., Barnard, E. A. 1973. *Biochem. Biophys. Res. Commun.* 51:205
282. Stalc, A., Zupancic, A. O. 1972. *Nature New Biol.* 239:91
283. Vogel, Z., Sytkowski, A. J., Nirenberg, M. W. 1972. *Proc. Nat. Acad. Sci. USA* 69:3180
284. Sytkowski, A. J., Vogel, Z., Nirenberg, M. W. 1973. *Proc. Nat. Acad. Sci. USA* 70:270
285. Patrick, J., Heinemann, S. F., Lindstrom, J., Schubert, D., Steinbach, J. H. 1972. *Proc. Nat. Acad. Sci. USA* 69:2762
286. Greene, L. A., Sytkowski, A. J., Vogel, Z., Nirenberg, M. 1973. *Nature* 243:166
287. Giacobini, G., Filogamo, G., Weber, M., Boquet, P., Changeux, J. P. 1973. *Proc. Nat. Acad. Sci. USA* 70:1708
288. Cohen, S. A., Fishbach, G. D. 1973. *Science* 181:76
289. Denburg, J. L., Eldefrawi, M. E., O'Brien, R. D. 1972. *Proc. Nat. Acad. Sci. USA* 69:177

290. Benzer, T. I., Raftery, M. A. 1973. *Biochem. Biophys. Res. Commun.* 51: 939

291. Henderson, R., Wang, J. H. 1972. *Biochemistry* 11: 4565

292. Patrick, J., Lindstrom, J. 1973. *Science* 180: 871

293. Silman, I., Karlin, A. 1969. *Science* 164: 1420

294. Meunier, J. C., Changeux, J. P. 1969. *FEBS Lett.* 2: 224

295. Keifer, H., Lindstrom, J., Lennox, E. S., Singer, S. J. 1970. *Proc. Nat. Acad. Sci. USA* 67: 1688

296. Mautner, H. G., Bartels, E. 1970. *Proc. Nat. Acad. Sci. USA* 67: 74

297. Reiter, M. J., Cowburn, D. A., Prives, J. M., Karlin, A. 1972. *Proc. Nat. Acad. Sci. USA* 69: 1168

298. Kuo, J. et al 1972. *J. Biol. Chem.* 247: 16

299. Ferrendelli, J. A., Steiner, A. L., McDougal, D. R., Kipnis, D. M. 1970. *Biochem. Biophys. Res. Commun.* 41: 1061

300. Yamashita, K., Field, J. B. 1972. *J. Biol. Chem.* 247: 7062

301. Schultz, G., Hardman, J. G., Davis, J. W., Schultz, K., Sutherland, E. W. 1972. *Fed. Proc.* 31: 440

302. Strom, T. B., Deisseroth, A., Morganroth, J., Carpenter, C. B., Merrill, J. P. 1972. *Proc. Nat. Acad. Sci. USA* 69: 2995

303. Hadden, J. W., Hadden, E. M., Haddox, M. K., Goldberg, N. D. 1972. *Proc. Nat. Acad. Sci. USA* 69: 3024

304. Pastan, I., Roth, J., Macchia, V. 1966. *Proc. Nat. Acad. Sci. USA* 56: 1802

305. Winand, R. J., Kohn, L. D. 1972. *Proc. Nat. Acad. Sci. USA* 69: 1711

306. Amir, S. M., Carraway, T. F. Jr., Kohn, L. D. 1973. *J. Biol. Chem.* 248: 4092

307. Lissitzky, S., Fayet, G., Verrier, B. 1973. *FEBS Lett.* 29: 20

307a. Fayet, G. et al 1973. *FEBS Lett.* 32: 299–302

308. Lin, S. Y., Goodfriend, T. L. 1970. *Am. J. Physiol.* 218: 1319

309. Goodfriend, T. L., Lin, S. Y. 1970. *Circ. Res. Suppl.* I: 26–27

310. Lin, S. Y., Ellis, H., Weisblum, B., Goodfriend, T. L. 1970. *Biochem. Pharmacol.* 19: 651

310a. Brecher, P., Tabacchi, M., Pyun, H. Y., Chobanian, A. V. 1973. *Biochem. Biophys. Res. Commun.* 54: 1511–17

311. Lefkowitz, R. J., Roth, J., Pricer, W., Pastan, I. 1970. *Proc. Nat. Acad. Sci. USA* 65: 745

312. Lefkowitz, R. J., Roth, J., Pastan, I. 1971. *Ann. N Y Acad. Sci.* 185: 195

313. Lefkowitz, R. J., Roth, J., Pastan, I. 1970. *Nature* 228: 864

314. Lefkowitz, R. J., Roth, J., Pastan, I. 1970. *Science* 170: 633

315. Wolfsen, A. R., McIntyre, H. B., Odell, W. D. 1972. *J. Clin. Endocrinol. Metab.* 34: 684

316. Hofmann, K., Wingender, W., Finn, F. M. 1970. *Proc. Nat. Acad. Sci. USA* 67: 829

317. Finn, F. M., Widnell, C. C., Hofmann, K. 1972. *J. Biol. Chem.* 247: 5695

318. Marx, S. J., Woodard, C. J., Aurbach, G. D. 1972. *Science* 178: 999

319. Marx, S. J., Fedak, S. A., Aurbach, G. D. 1972. *J. Biol. Chem.* 247: 6913

320. Marx, S. J., Woodard, C., Aurbach, G. D., Glossmann, H., Keutmann, H. T. 1973. *J. Biol. Chem.* 248: 4797

321. Means, A. R., Vaitukaitis, J. 1972. *Endocrinology* 90: 39

321a. Spona, J. 1973. *FEBS Lett.* 34: 24–26

321b. Rao, C. V., Saxena, B. B. 1973. *Biochim. Biophys. Acta* 313: 372–89

321c. Dufau, M. L., Charreau, E. H., Catt, K. J. 1973. *J. Biol. Chem.* 248: 6973–82

322. Grant, G., Vale, W., Rivier, J. 1973. *Biochem. Biophys. Res. Commun.* 50: 771

323. Rapoport, S. I., Thompson, H. K. 1973. *Science* 180: 968

324. Campbell, B. J., Woodward, G., Borberg, V. 1972. *J. Biol. Chem.* 247: 6167

325. Grant, G., Vale, W., Guillemin, R. 1972. *Biochem. Biophys. Res. Commun.* 46: 28

326. Labrie, F., Barden, N., Poirier, G., DeLean, A. 1972. *Proc. Nat. Acad. Sci. USA* 69: 283

327. Poirier, G., Labrie, F., Barden, N., Lemaire, S. 1972. *FEBS Lett.* 20: 283

328. Wilber, J. F., Seibel, M. J. 1973. *Endocrinology* 92: 888

328a. Hinkle, P. M., Tashjian, A. H. Jr. 1973. *J. Biol. Chem.* 248: 6180–86

328b. Eddy, L. J., Hershman, J. M., Taylor, R. E. Jr., Barker, S. B. 1973. *Biochem. Biophys. Res. Commun.* 54: 146–49

328c. Barden, N., Labrie, F. 1973. *J. Biol. Chem.* 248: 7601–6

329. Dufau, M. L., Catt, K. J., Tsuruhara, T. 1972. *Proc. Nat. Acad. Sci. USA* 69: 2414

330. Danzo, B. J., Midgley, A. R. Jr., Kleinsmith, L. J. 1972. *Proc. Soc. Exp. Biol. Med.* 139: 88

331. Dufau, M. L., Catt, K. J. 1973. *Nature New Biol.* 242: 246

331a. Lee, C. Y., Ryan, R. J. 1973. *Biochemistry* 12: 4609–15

332. Moudgal, N. R., Moyle, W. R., Greep, R. O. 1971. *J. Biol. Chem.* 246:4983
333. Lee, C. Y., Ryan, R. J. 1972. *Proc. Nat. Acad. Sci. USA* 69:3520
334. Rajaniemi, H., Vanha-Perttula, T. 1972. *Endocrinology* 90:1
335. Gospodarowicz, D. 1973. *J. Biol. Chem.* 248:5042
336. Gospodarowicz, D. 1973. *J. Biol. Chem.* 248:5050
337. Desbuquois, B., Laudat, M. H., Laudat, P. 1973. *Biochem. Biophys. Res. Commun.* 53:1187
338. Lesniak, M. A., Roth, J., Gorden, P., Gavin, J. R. III. 1973. *Nature New Biol.* 241:20
339. Sonenberg, M. 1971. *Proc. Nat. Acad. Sci. USA* 68:1051
340. Bockaert, J., Imbert, M., Jard, S., Morel, F. 1972. *Exp. Mol. Pathol.* 8:230
341. Thompson, E. E., Freychet, P., Roth, J. 1972. *Endocrinology* 91:1199
342. Soloff, M., Swartz, T., Morrison, M., Saffran, M. 1973. *Endocrinology* 92:104
343. Roy, C., Bockaert, J., Rajerison, R., Jard, S. 1973. *FEBS Lett.* 30:329
343a. Soloff, M. S., Swartz, T. L. 1973. *J. Biol. Chem.* 248:6471–78
344. Walter, R., Schwartz, I. L., Hechter, O., Dousa, T., Hoffman, P. L. 1972. *Endocrinology* 91:39
345. Covelli, I., Rossi, R., Mozzi, R., Frati, L. 1972. *Eur. J. Biochem.* 27:225
346. Frati, L., Daniele, S., Delogu, A., Covelli, I. 1972. *Exp. Eye Res.* 14:135
347. Hollenberg, M. D., O'Keefe, E., Cuatrecasas, P. In preparation
348. Banerjee, S. P., Snyder, S. H., Cuatrecasas, P., Greene, L. A. 1973. *Proc. Nat. Acad. Sci. USA* 70:2519
349. Snyder, S. H., Banerjee, S. P., Cuatrecasas, P., Greene, L. A. 1974. In *Dynamics of Degeneration and Growth in Neurons.* Oxford: Pergamon
350. Herrup, K., Shooter, E. M. 1974. *Proc. Nat. Acad. Sci. USA* 70:3884–88
351. Cuatrecasas, P. 1973. *Biochemistry* 12:3547
351a. Bockaert, J., Roy, C., Rajerison, R., Jard, S. 1973. *J. Biol. Chem.* 248:5705–11
352. Cuatrecasas, P. 1973. *Biochemistry* 12:3558
353. Ibid 12:3567
354. Ibid 12:3577
355. Cuatrecasas, P., Parikh, I., Hollenberg, M. D. 1973. *Biochemistry* 12:4253
356. Bennett, V., Cuatrecasas, P. In preparation
357. Morell, A. G., Irvine, R. A., Sternlieb, I., Scheinberg, I. H., Ashwell, G. 1968. *J. Biol. Chem.* 243:155
358. Morell, A. G., Gregoriadis, G., Scheinberg, I. H., Hickman, J., Ashwell, G. 1971. *J. Biol. Chem.* 246:1461
359. Van Lenten, L., Ashwell, G. 1971. *J. Biol. Chem.* 246:1889
360. Van Lenten, L., Ashwell, G. 1972. *J. Biol. Chem.* 247:4633
361. Morell, A. G., Scheinberg, I. H. 1972. *Biochem. Biophys. Res. Commun.* 48:808
362. Aronson, H. N. Jr., Tan, L. Y., Peters, B. P. 1973. *Biochem. Biophys. Res. Commun.* 53:112
363. Pert, C. B., Snyder, S. H. 1973. *Science* 179:1011
364. Simon, E. J., Hiller, J. M., Edelman, I. 1973. *Proc. Nat. Acad. Sci. USA* 70:1947
365. Winter, B. A., Goldstein, A. 1972. *Mol. Pharmacol.* 8:601
366. Alivisatos, S. G. A. et al 1971. *Science* 171:809
366a. Pert, C. B., Snyder, S. H. 1973. *Proc. Nat. Acad. Sci. USA* 70:2243–47
366b. Kuhar, M. J., Pert, C. B., Snyder, S. H. 1973. *Nature New Biol.* 245:447–50
367. Hardin, J. W., Cherry, J. H., Morre, D. J., Lembi, C. A. 1972. *Proc. Nat. Acad. Sci. USA* 69:3146
368. Venis, M. A. 1972. *Biochem. J.* 127:29
369. Dastoli, F. R., Price, S. 1966. *Science* 154:905
370. Hiji, Y., Kobayashi, N., Sato, M. 1971. *Comp. Biochem. Physiol.* 39B:367
371. Lo, C.-H., Ma, T. 1973. *Biochim. Biophys. Acta* 307:343
372. Hiji, Y., Sato, M. 1973. *Nature New Biol.* 244:91
373. Chang, K.-J., Marcus, N., Cuatrecasas, P. In preparation
374. Lis, H., Sharon, N. 1973. *Ann. Rev. Biochem.* 42:541
375. Kuehl, F. A. Jr., Humes, J. L. 1972. *Proc. Nat. Acad. Sci. USA* 69:480
376. Kuehl, F. A. Jr., Humes, J. L., Ham, E. A., Cirillo, V. J. 1972. *Intra-Sci. Chem. Rep.* 6:85
377. Moore, W. V., Wolff, J. 1973. *J. Biol. Chem.* 248:5705–11
378. Hardman, J. G., Robison, G. A., Sutherland, E. W. 1971. *Ann. Rev. Physiol.* 33:311
379. Birnbaumer, L., Pohl, S. L., Rodbell, M. 1969. *J. Biol. Chem.* 244:3468
380. Rodbell, M., Birnbaumer, L., Pohl, S. L. 1970. *J. Biol. Chem.* 245:718
381. Moskowitz, J., Fain, J. N. 1970. *J. Biol. Chem.* 245:1101

382. Fain, J. N., Kovacev, V. P., Scow, R. O. 1965. *J. Biol. Chem.* 240:3522
383. Braun, T., Hechter, O. 1969. *Endocrinology* 85:1092
384. Swizlocki, N. I. 1970. *Biochim. Biophys. Acta* 201:242
385. Butcher, R. W., Baird, C. E. 1968. *J. Biol. Chem.* 243:1713
386. Fain, J. N. 1968. *Endocrinology* 83:548
387. Blecher, M., Merlino, N. S., Ro'Ane, J., Flynn, P. D. 1969. *J. Biol. Chem.* 244:3223
388. Gitler, C. 1972. *Ann. Rev. Biophys. Bioeng.* 1:51
389. McConnell, H. M., McFarland, B. G. 1970. *Quart. Rev. Biophys.* 3:91
390. Radda, G. K. 1971. *Current Top. Bioenerg.* 4:81
391. Craig, S., Cuatrecasas, P. In preparation
392. Rubin, M. S., Swislocki, N. I., Sonenberg, M. 1973. *Arch. Biochem. Biophys.* 157:243
393. Johnson, M., Ramwell, P. W. 1973. *Prostaglandins* 3:703

METABOLIC ALTERATIONS OF FATTY ACIDS

Armand J. Fulco

Department of Biological Chemistry, UCLA Medical School and Laboratory of Nuclear Medicine and Radiation Biology, University of California, Los Angeles, California

CONTENTS

INTRODUCTION

This article is a logical extension of last year's review by Volpe & Vagelos (1). These authors considered saturated fatty acid biosynthesis and its regulation, whereas this review covers the metabolic alteration of fatty acids and especially the introduction of double bonds and hydroxy groups into the fatty acyl chain. Hydrogenation and hydration of double bonds in the fatty acyl chain are also covered to the extent that these reactions are distinct from those involved in either fatty acid synthesis or β oxidation. The latter subject has been included in a number of recent reviews including those by Green & Allmann (2), Stumpf (3), and Bishop & Stumpf (4). The conversion of polyunsaturated fatty acids to prostaglandins has also been reviewed recently (5) and is not covered here. Those metabolic reactions of fatty acids that center at the carboxyl carbon are covered in part by a companion review by van den Bosch in this volume (6) and by a number of other recent reviews in this series (7–10).

UNSATURATED FATTY ACID BIOSYNTHESIS

Research on unsaturated fatty acid biosynthesis has accelerated rapidly in the past decade. This effort was stimulated by the initial studies of Bloch and his group (11) on monounsaturated fatty acid biosynthesis and by Mead and co-workers (12) on the metabolism of polyunsaturated fatty acids.

To date, many of the major metabolic pathways for the formation of unsaturated fatty acids have been mapped, a number of cell-free desaturation systems have been studied, and some initial success has been achieved in attempts to understand both the biochemical mechanisms of the desaturation reaction and those factors involved in the regulation of unsaturated fatty acid biosynthesis. In this review I hope to integrate the recent advances in these areas into the overall picture that has emerged within the past decade.

Metabolic Pathways

A large variety of naturally occurring unsaturated fatty acids can be distinguished by differences in the position, type and number of double bonds and by variations in chain length and carbon branching. Despite this diversity, however, there appear to be only two distinct biochemical mechanisms for the introduction of *cis* double bonds (13). In many bacteria, a *cis*-3 double bond is introduced into a medium chain length fatty acid (usually C_{10}) by dehydration of a fatty acid synthesis intermediate, the β-hydroxyacyl thioester derivative of acyl carrier protein (ACP), by a specific enzyme, β-hydroxydecanoyl thioester dehydrase (14–18). Chain elongation of the *cis*-3 derivative gives rise to long-chain unsaturated fatty acids (11). This so-called "anaerobic" pathway is identical (except for the formation of the β,γ-unsaturated intermediate) to the pathway for saturated fatty acid biosynthesis in *E. coli* (1) and is not considered in this review. A second mechanism for *cis* double bond formation involves the direct, oxygen-dependent desaturation of long-chain fatty acids (19, 20). Although certain bacteria and presumably all nonparasitic eucaryotic organisms utilize the O_2-dependent or aerobic mechanism to produce unsaturated fatty acids, differences in the nature and availability of the activated substrates, the cofactor requirements other than O_2, and specificities of the enzyme systems themselves probably account for the differences in metabolic pathways found among various groups of organisms. Recent work on the comparative aspects of these pathways is discussed below.

BACTERIA   The anaerobic pathway for the formation of monounsaturated fatty acids (to date found only in bacteria) leads generally to palmitoleic (9-hexadecenoic) and vaccenic (11-octadecenoic) acids as the major products, while the aerobic pathway found in certain bacteria yields primarily palmitoleic and oleic (9-octadecenoic) acids (11, 21). There are exceptions to these generalizations, however, particularly in the aerobic pathway. The bacilli, for example, produce a variety of isomeric hexa-decenoic acids by O_2-dependent desaturation. *B. megaterium* KM desaturates palmitate and stearate to form exclusively the *cis*-5 derivatives (22). Fulco (23, 24) showed that *B. megaterium, B. subtilis,* and certain strains of *B. pumilis, B.*

licheniformis, and *B. alvei* also desaturate palmitate exclusively in the 5 position, while other bacilli tested, including various strains of *B. brevis, B. stearothermophilus, B. macerans,* and *B. licheniformis* produce various mixtures of the 8-, 9-, and 10-hexadecenoic isomers. Dart & Kaneda (25) showed that *B. cereus* desaturates palmitic and stearic acids to the Δ^{10} isomers while Kaneda (26) found that three psychrophilic species of Bacillus (*B. insolitus, B. psychrophilus,* and *B. globisporus*) produce exclusively the Δ^5 isomers. Among the mycobacteria, the Δ^{10} and Δ^9 isomers of hexadecenoic and octadecenoic acids are often found together (27, 28). Apparently the positional isomers are not interconvertible but are produced independently by O_2-mediated desaturation of the saturated precursors (27). Curiously, although whole cells of *M. phlei* desaturate palmitate predominantly in the 10 position, a cell-free desaturating system obtained from the same organism produces only the Δ^9 isomer (29).

Asselineau and her co-workers (30) have isolated a series of monounsaturated fatty acids from *M. phlei* containing 20–27 carbon atoms. These included normal chain Δ^5-monoenoic acids of 22, 24, or 26 C atoms, branched-chain Δ^5 derivatives of 25 and 27 C atoms, and a series which included 4-eicosenoic, 6-docosenoic, and 8-hexacosenoic acids. They suggest that the Δ^5 acids may play a role in the synthesis of the mycolic acids of the same organism.

Leptospira canicola has been found to desaturate palmitate to a mixture of the *cis-*Δ^9- and the *cis-*Δ^{11}-hexadecenoic acids whereas stearate is desaturated at the 9 position only (31). This seems to be the first demonstration in bacteria of the formation of a *cis*-11-monoenoic acid by O_2-dependent desaturation of the corresponding saturated acid. Recently, the desaturation of (1-^{14}C)-labeled palmitate and stearate was reported to occur in *Alcaligenes faecalis* (32), an organism known to produce 11-octadecenoic acid and cyclopropane fatty acids (33), both typical products of the anaerobic pathway. However, an oxygen requirement for desaturation was not demonstrated and since there was some randomization of the (1-^{14}C) label, unambiguous interpretation of these results is not yet possible.

It was originally thought that polyunsaturated fatty acids could not be synthesized by bacteria (21), but with our present knowledge of the O_2-dependent pathway in these organisms, there seems to be no reason why this should be the case. Nevertheless, the occurrence of polyunsaturated fatty acids in bacteria is extremely rare. Apparently, the only unequivocal evidence for the synthesis of diunsaturated fatty acids in bacteria was reported by Fulco (34), who showed that 5,10-hexadecadienoic was produced by the direct desaturation of added (1-^{14}C) palmitate in a strain of *B. licheniformis.* He later showed (35) that the biosynthesis of this unique dienoic acid resulted from the cooperative action of two distinct desaturation systems. One system, present in the bacterium at both 35° and 20°, desaturated palmitate to 10-hexadecenoic acid, while a second system, active only at the lower temperature, resulted in the conversion of palmitic acid to 5-hexadecenoate. Both the Δ^5 and Δ^{10} monoenoic acids were shown to be precursors of the 5,10-dienoic acid. Recently Sklan, Budowski & Volcani (36) reported the synthesis of linoleate from stearate and oleate in a supernatant fraction from calf rumen liquor. Sklan & Budowski (37) found that rat colonic contents incubated aerobically carried out the same transformation. In both cases, the biosynthesis of linoleic acid was attributed to bacteria but proof of this assumption must await further experimentation. This reviewer considers synthesis by protozoa a more likely possibility. Asselineau and others (38) have isolated from *M. phlei* several highly unsaturated fatty acids (the so-called phleic acids) of obscure metabolic origin, whose main member has the structure $n\text{-}C_{16}H_{33}[CH{=}CHCH_2CH_2]_5COOH$.

FUNGI, PROTOZOA, ALGAE, AND PLANTS   Because of the extensive work in this area in the past few years, comprehensive coverage of recent developments is beyond the scope of this review. The reader is referred to the excellent review by Stearns (39) and to the book by Hitchcock & Nichols (20) for a more complete coverage of the pathways of unsaturated fatty acid biosynthesis in these organisms, particularly in algae and plants. The short review by Erwin & Bloch (21), although somewhat dated, is also helpful.

Recent work in blue-green algae and in higher plants has blurred the once clearcut distinction between the pathways leading to polyunsaturated fatty acids in these organisms and in the metazoa. It was originally proposed (13, 21) that, starting with oleic acid as a common precursor, algae and higher plants inserted additional double bonds between the pre-existing double bond and the terminal methyl group to produce successively linoleic acid (40) and α-linolenic (9,12,15-octadecatrienoic) acid, whereas animals could desaturate only between the original double bond and the carboxyl group to yield 6,9-octadecadienoic acid or, by chain elongation and further desaturation, 5,8,11-eicosatrienoic acid (41). The animal could also carry out the same reactions on dietary linoleic acid to obtain γ-linolenic (6,9,12-octadecatrienoic) acid and arachidonic acid (42, 43). Linoleate and α-linolenate were thus considered to be typical plant fatty acids while γ-linolenate and arachidonate were characteristic of animal pathways. If one excludes the protozoa, no one has yet demonstrated clearly de novo biosynthesis of linoleate and α-linolenate in animals. However, numerous representatives among the algae and metaphyta are now known to synthesize the typical animal polyunsaturated fatty acids. Among the algae, the unsaturated fatty acid composition and pathways have been most intensively studied in the blue-green algae. Compositional data has been used to assess certain geochemical relationships (44), as an aid in determining phylogenetic positions (45), and especially in the study of the possible evolutionary significance of polyunsaturated fatty acids within this group (46–49). Thus Kenyon, Rippka & Stanier (48) examined 32 axenic strains of filamentous blue-green algae and Kenyon (49) studied 34 strains of unicellular blue-green algae; they found that four metabolic groups could be recognized according to the major fatty acid of highest degree of unsaturation found in each strain. These groups included: 1. those in which there is little or no desaturation of oleate; 2. those in which linoleate is desaturated toward the methyl end of the molecule to give α-linolenate; 3. those in which linoleate is desaturated towards the carboxyl end to give γ-linolenate; and 4. those in which octadecatetraenoate is synthesized. Thus, within this group of prokaryotic organisms are found pathways which were considered typical of animals (i.e. pathway 3) and of higher plants (pathway 2). In addition, the blue-green algae of group 1 resemble bacteria in their inability to desaturate oleate. These authors conclude that the blue-green algae are a living record of the transition stages between the bacteria and the higher protists with respect to fatty acid biosynthesis.

Holz and his co-workers have analyzed the polyunsaturated fatty acids of marine and freshwater cryptomonads (50) and marine dinoflagellates (51) and have found a diversity similar to the blue-green algae but with certain differences (including the synthesis of C_{20} and C_{22} polyunsaturated fatty acids) that suggest further evolution in the pathways for polyunsaturated fatty acid biosynthesis. In a sense, certain protozoa represent the evolutionary apex of biosynthetic pathways leading to polyunsaturated fatty acids. A number of phytoflagellates contain within the same organism the polyunsaturated fatty acids characteristic of animals (i.e. γ-linolenic, arachidonic) and of plants (linoleic, α-linolenic). The same is true for some amoebae and ciliates (21). Indeed, Gellerman & Schlenk (52) concluded from a study of unsaturated fatty acid metabolism in *Ochromonas danica* that chain elongation and desaturation of the proximal part of already unsaturated fatty acid chains appear to be subject to the same structural requirements as in the rat and the same effects apply to desaturation in the distal part of the chain.

One can surmise that separate plant and animal pathways arose during further evolution by the loss of either the ability to desaturate between the pre-existing double bond(s) and the terminal methyl group (animals) or between the pre-existing double bond(s) and the carboxyl group (most higher plants). A number of plants, particularly those low on the evolutionary scale, have retained the ability to desaturate on both the carboxyl and methyl sides of pre-existing double bonds. Schlenk & Gellerman (53) were the first to clearly demonstrate the presence of arachidonic acid in mosses and ferns, and these workers, in collaboration with Anderson, have recently broadened and extended these findings (54, 55). Among the fungi, the phycomycetes synthesize γ-linolenic acid while the ascomycetes and basidiomycetes produce α-linolenic acid (56).

ANIMALS   The pathways for polyunsaturated fatty acid biosynthesis in animals have been elaborated in great detail in recent years. Fortunately for this reviewer, Mead (12) has summarized this work up to 1969. More recently Brenner and his co-workers (57–60), Sprecher et al (61–65), Bridges & Coniglio (66–68), and a number of other workers (69–72) have continued to study the biosynthetic pathways and inter-conversions of the polyunsaturated fatty acids in animals, particularly in the rat. Ayala et al (60), utilizing subcellular fractions from rat liver and testes, compared the metabolic fate of linoleic, arachidonic, and 7,10,13,16-docosatetraenoic acids. Conversion of linoleic to arachidonic by microsomes in both tissues proceeded through γ-linolenic and 8,11,14-eicosatrienoic acids as originally demonstrated by Mead et al in whole animals (12). Testicular mitochondria carry out the retro-conversion of 7,10,13,16-docosatetraenoate to arachidonate, but there was no evidence in vitro of the conversion of 4,7,10,13,16,19-docosahexaenoic acid to 7,10,13,16,19-docosapentaenoate, a process demonstrated in whole animals by Schlenk, Sand & Gellerman (69).

Ayala et al propose that the synthesis of acids of the linoleic family proceeds in two stages: a rapid one in which arachidonic acid is made and a second, slower stage in which the C_{22} and C_{24} acids are synthesized. They also suggest that there is a cycle between microsomes and mitochondria that acts to conserve essential polyunsaturated C_{20} and C_{22} acids by means of synthesis and partial degradation, respectively. Ullman & Sprecher (65) have concluded from both in vitro and in vivo studies that the Mead pathway for arachidonic acid synthesis from linoleate (i.e. through γ-linolenate and 8,11,14-eicosatrienoate) is the only significant pathway in the rat for synthesis of this essential fatty acid. The C_{24}-polyunsaturated fatty acids of rat testes, 9,12,15,18-C_{24}-tetraenoic and 6,9,12,15,18-C_{24}-pentaenoic acids, were shown by Bridges & Coniglio (66) to be derived from linoleic acid via C_{22}-tetraene and C_{22}-pentaene, respectively.

In insects, the pathways for unsaturated fatty acid biosynthesis are less well known than in vertebrates. The de novo biosynthesis of polyunsaturated fatty acids remains an open question, although there is little doubt that some insects produce monoenoic acids by the same pathway as higher animals. Municio and his co-workers (73) have studied in vitro the elongation and desaturation of C_{10}, C_{12}, C_{14}, and C_{16} saturated fatty acids during development in insects and have shown distinct differences between homogenates prepared from larvae or pharate adults of *Ceratitis capitata*. Larval homogenates desaturate and elongate the (^{14}C)-substrates according to their chain length, with elongation decreasing and desaturation increasing with increasing chain length of substrate. These reactions, however, are insignificant in homogenates prepared from adults. Thompson & Barlow (74) injected (1-^{14}C)-acetate in adult male *Galleria mellonella* and demonstrated significant incorporation of labeled acetate into 9-eicosenoic acid, as well as in myristic, palmitic, palmitoleic, stearic, and oleic acids. Degradative analysis indicated that synthesis of monounsaturated acids proceeded through direct desaturation of the saturated analogs. These workers also found that several insect parasites (*Itoplectis conquisitor* and *Exeristes comstockii*) could, to a limited extent, metabolize and desaturate fatty acids independently of the host (75, 76).

ALTERNATE PATHWAYS IN PLANTS AND ANIMALS   There is always the temptation, after elucidating a biosynthetic pathway in one or more members of a particular class or family of organisms, to assume that all members of the group have the same pathway. Although this may often be true, two recent examples are enough to show the danger inherent in such generalizations. More than a decade ago, Fulco & Mead (77) showed that the rat could not convert *cis*-12-octadecenoic acid to linoleic acid, even though such a transformation would be expected if this unsaturated acid served as a substrate for the Δ^9-desaturase responsible for the conversion of stearate to oleate in the same animal. Although from these results it would seem likely that *cis*-12-octadecenoate would be an ineffective precursor of linoleate in other animal systems as well, such is not always the case. Gurr et al (78) have recently shown that goat, pig, hen, and *Chlorella vulgaris* convert *cis*-12-octadecenoate into linoleate in good yield; rabbit and mouse do so in very poor yield, whereas rat, hamster, and *Candida utilis* do not desaturate this substrate.

In the plant kingdom, the biosynthetic route for the formation of α-linolenate has always been assumed to proceed as follows: oleate → linoleate → α-linolenate (i.e. the typical plant pathway). Although the generality of the oleate to linoleate conversion has been well demonstrated in a large number of systems ranging from algae, fungi, and protozoa to higher plants, the direct conversion of linoleate to α-linolenate has been clearly demonstrated only in algae and fungi (39). Recently, Jacobson, Kannangara & Stumpf (79, 80) have provided compelling evidence that the major pathway for the formation of α-linolenate in spinach is not by further desaturation of linoleate but rather by chain elongation of 7,10,13-hexadecatrienoic acid. Using a disrupted chloroplast system, these workers showed that under conditions where (^{14}C) acetate was readily converted to α-linolenate, (^{14}C) oleate was a totally ineffective precursor. Furthermore, a specific elongation system which converted 5,8,11-tetradecatrienoate to the C_{16} triene and this, in turn, to α-linolenate was demonstrated. This elongation system did not act on saturated or monounsaturated substrates. Thus the typical plant pathway leading from oleate to α-linolenate may, in higher plants, be the exception rather than the rule.

Enzymology and Substrate Specificity

The first cell-free system which carried out fatty acid desaturation was described by Bloomfield & Bloch (81) in 1960 and since that time more than a dozen distinct cell-free systems have been described. Most of these are listed in Table 1. Systems utilizing intact chloroplast preparations, those in which direct desaturation of a substrate was not clearly demonstrated or those which are similar in essentials to a system listed in Table 1, have been omitted. Several of these are described in recent books and reviews (4, 11, 12, 19, 20, 39).

It is obvious from Table 1 that the enzymology of O_2-mediated desaturation remains in a primitive state. Only two active systems have been solubilized; the soybean cotyledon preparation (90), which has not been fractionated, and the photoauxotrophic Euglena gracilis system (86), which has been separated into three protein components. The other systems are either particulate preparations or homogenates which have not been further characterized. Despite this, a great deal of information has been obtained that allows several general conclusions. First of all, both O_2 and a reduced pyridine nucleotide are required for activity in every cell-free system that carries out the direct removal of two hydrogens to form a cis double bond. Furthermore, it would seem that more or less complex electron transport chains are involved in the desaturation process in most of these systems. A number of notable differences among the various systems are also evident, including differences in the components of the electron transport chains and in the types of acyl derivatives that will serve as substrates. One may distinguish three types of acyl derivatives desaturated in one system or another. Animal microsomal systems seem to desaturate only acyl-CoA esters; when the free fatty acids are added, both CoA and ATP must generally be included for maximal desaturation. The particulate cell-free bacterial system studied (82) also utilizes the CoA esters, as do the particulate stearate desaturases of the yeast Saccharomyces cerevisiae (81), the fungus Neurospora crassa (84), and the etiolated heterotrophic alga E. gracilis (86, 87). When E. gracilis is grown photoauxotrophically, however, a soluble desaturating

Table 1 Cell-free fatty acid desaturation systems[a]

Source and type of cell-free system	Major substrate and product	Components of system and cofactors	Reference
Mycobacterium phlei (bacterium) particulate	stearyl-CoA → oleate[b]	100,000 × g particles, O_2, NADPH, Fe^{+2}, FAD (or FMN)	(29, 82)
Saccharomyces cerevisiae (yeast) particulate	palmityl-CoA → palmitoleyl-CoA	100,000 × g particles, O_2, NADPH (or NADH)	(81)
Torulopsis utilis (yeast) microsomal	oleyl-phosphatidyl choline → linoleyl-phosphatidyl choline	microsomes, 100,000 × g supernatant, O_2 NADPH	(83)
Neurospora crassa (fungus) microsomal	stearyl-CoA → (oleyl-CoA) → oleyl-phospholipid → linoleyl-phospholipid	microsomes, O_2, NADH	(84)
Euglena gracilis, photoauxotrophic (alga) soluble	stearyl-ACP → oleate[b]	desaturase protein, flavoprotein, ferredoxin, O_2, NADPH	(85, 86)
Euglena gracilis, dark grown heterotrophic (algal) particulate	stearyl-CoA → oleate[b]	particles, supernatant O_2, NADPH	(86, 87)
Carthamus tinctorius seeds (plant) membrane fragments	oleyl-CoA → linoleyl-CoA	particles, O_2, NADH	(88)

Potato tubers (plant) microsomal	oleyl-CoA → linoleyl-phospholipid	microsomes, O_2, NADPH (or NADH)	(89)
Soybean cotyledon (plant) soluble	stearate → oleate[b], oleate → linoleate[b]	$100,000 \times g$ supernatant, CoA, ATP, Mg^{+2}, O_2, NADPH	(90)
Bovine mammary tissue microsomal	stearyl-CoA → oleate, palmityl-CoA → palmitoleate	microsomes, O_2, NADH	(91)
Rat brain homogenate	stearyl CoA → oleate[b]	homogenate, O_2, NADH	(92)
Rat liver microsomal	stearyl-CoA → oleyl-CoA	microsomes, O_2, NADH (or NADPH), (or ascorbate[c])	(93–97)
Hen liver microsomal	stearyl CoA → oleate[b]	acetone-extracted microsomes, lipid, O_2, NADH (or NADPH)	(98)
Rat liver microsomal	linoleyl-CoA → γ-linolenyl-CoA	microsomes, supernatant, O_2, NADPH (or NADH)	(99, 100)
Rat liver microsomal	Systems leading to other polyunsaturated fatty acids	microsomes, O_2, NADPH (or NADH)	(55, 60, 63–65, 101, 102)

[a] Limited to those systems that carry out the direct removal of 2H from a fatty acyl chain to form a *cis* double bond.
[b] Product hydrolyzed before analysis.
[c] In high concentrations can partially replace NADPH but desaturation activity is very low.

system can be isolated which desaturates stearyl-ACP much more efficiently than stearyl-CoA (stearyl-ACP is not a substrate for the particulate desaturase obtained from the etiolated heterotrophic organism). Surprisingly, the purified reconstituted soluble system desaturates the ACP and CoA derivatives of stearate equally well. In spinach chloroplasts (87) the desaturation system is specific for stearyl-ACP. In view of these results and less direct evidence by others (19) it seems likely that as more higher plant systems are investigated, additional monodesaturases will be found which preferentially utilize the saturated acyl-ACP thioesters as substrates.

The role of acyl-ACPs as desaturation substrates in nonphotosynthetic systems, on the other hand, would seem at best limited. Unlike green plants and bacteria, animals do not utilize ACP or an ACP-like protein in acyl transfer reactions or in fatty acid synthesis (1, 103) and all the animal desaturases so far studied require the CoA esters as substrates. Although yeasts do contain an ACP-like protein, it is tightly bound to the fatty acid synthetase complex and does not participate in the enzymatic desaturation of fatty acids (86, 104). The conversion of oleate to linoleate does not seem to involve the ACP derivative but here the picture is quite complex. Vijay & Stumpf (88, 105) have shown that oleyl-CoA is converted to linoleyl-CoA by a particulate desaturase from *Carthamus tinctorius* (safflower) seeds. Ben Abdelkader et al (89), using (1-^{14}C) oleyl-CoA as a substrate for a desaturase associated with a microsomal fraction from aged slices of potato tuber, were able to obtain 40% conversion to linoleate. However, all of the labeled linoleate and the remaining (1-^{14}C) oleate were incorporated into phospholipids, particularly phosphatidylcholine. Similar results were obtained by Baker & Lynen (84) using microsomes from *N. crassa*. They presented indirect evidence that oleyl phospholipid was an intermediate in the conversion of oleyl-CoA to linoleyl phospholipid. Gurr et al (106), using whole cells and isolated chloroplasts of *Chlorella vulgaris*, found a tight coupling between incorporation of (1-^{14}C) oleate into the 2 position of phosphatidylcholine and the appearance of label in linoleate. Direct proof that oleyl phospholipid can serve as a desaturation substrate has recently been obtained by Talamo, Chang & Bloch (83). They showed that a particulate enzyme system from the yeast *Torulopsis utilis* catalyzes the conversion of oleyl-CoA or oleyl phospholipid to linoleate. Incubation of particles with (^{3}H)-oleyl-CoA in the absence of a supernatant fraction yields oleyl phospholipid. This material is then desaturated to linoleate either in situ after addition of supernatant or after isolation by the complete enzyme system. Oleyl phosphatidylcholine was the most active of the various phospholipid fractions tested for desaturation. Furthermore, synthetic 1,2-di-(^{14}C)-oleyl phosphatidylcholine is converted to linoleate with high efficiency.

To this point we have considered substrate specificity in the desaturation reaction in terms of the group attached to the carboxy carbon of the fatty acid. Obviously, however, the nature of the fatty acyl chain is also of great importance. A. T. James and his co-workers (107, 108) have studied the effects of double bond position on the desaturation of monoenoic acids and the efficiency of conversion of homologous (^{14}C) labeled fatty acids to the corresponding Δ^9-monoenoic acids by goat mammary gland microsomes, hen liver microsomes, and whole cells of the yeast *Torulopsis bombicola* and of the alga *Chlorella vulgaris*. They also studied Δ^9-desaturation by these systems of a series of isomeric methylstearic acids. Their findings can be summarized as follows: In each of the four systems tested with homologous saturated fatty acid substrates (C_{12}–C_{19}), C_{18} was the most active, followed by C_{17}. Activity with C_{19} was always low (except in hen) but activity with C_{16} varied from system to system. Thus in *T. bombicola* and with hen microsomes, C_{16} was desaturated almost as well as C_{17} and C_{18}, while in goat mammary microsomes and in *C. vulgaris* it was much less active. A second activity peak at shorter chain lengths was clearly observed in two of the systems: at C_{14} for hen microsomes and at C_{15} in chlorella. Indeed, in the former system, C_{14} was desaturated almost as well as C_{18}. The isomeric methyl stearates were tested in the hen microsomal Δ^9-desaturating system and in all cases activity was lower than with stearate itself. The two most active isomers were 2-methyl- and 18-methyl-stearate (i.e. *n*-nonadecanoate) but the 17-methyl and 4-methyl derivatives were also desaturated reasonably well. The 3 and 16 substituted isomers showed only very low activity, whereas there was no significant desaturation when the methyl group was located at or between carbon atoms 5 to 15 of the stearyl chain. On the basis of this finding, Brett et al (107) proposed that substrate (between C-5 and C-15) is so closely bound to the enzyme surface that the methyl groups cannot easily be accommodated. In a hen liver desaturating system [Johnson et al (109)] and in rat liver microsomes [Paulsrud et al (110)], maximal Δ^9 desaturation was observed with stearic acid; palmitic and margaric acids were less active.

Similar specificity patterns were observed in several bacterial systems (22, 29), although in *Micrococcus lysodeikticus* palmitate was significantly favored over stearate for Δ^9 desaturation. Quint & Fulco (111) used a variety of methods to determine in vivo the substrate specificities for six desaturases from bacilli that insert a *cis* double bond at the 5 and 10 positions of the acyl chain, respectively. Five of the six Δ^5 desaturases and the Δ^{10} desaturase showed maximal activity with palmitate. Activity for the *n*-saturated substrates decreased in the order $C_{16} > C_{17} > C_{18}$. Myristic acid (C_{14}) was not desaturated, while C_{15}, tested with two Δ^5 desaturases, had about half the activity of C_{16} in both cases. The branched chain acids, iso-C_{16}, and iso- and anteiso-C_{17} were also desaturated but iso- and anteiso-C_{15} acids were not touched. Various monounsaturated C_{16} and C_{18} acids were also desaturated at the 5 position. The positional specificity of desaturation was always the same relative to the carboxyl carbon of the substrate, regardless of substrate chain length, branching, or the presence of a double bond in the 9 or 10 position. The authors concluded that the substrate is attached to the desaturase at the carboxyl carbon of the fatty acid and that the carboxyl binding site of the enzyme must be at a fixed distance from the active (hydrogen removal) site to account for the absolute positional specificity of double bond insertion. On the other hand, the efficiency of desaturation (substrate specificity) must depend on a second binding site which anchors the substrate to the enzyme near the terminal methyl group of the fatty acid. This site is visualized as a hydrophobic pocket on the enzyme surface which may vary slightly in shape from enzyme to enzyme but appears to be located, for the desaturases studied, about 15 carbon atoms away from the carboxyl binding site (assuming that the fatty acid substrate is fully extended along the enzyme surface).

Mechanisms of Desaturation

The chemical mechanism(s) of O_2-dependent desaturation of fatty acids have not yet been elucidated, primarily because no stable desaturation system has yet been fractionated into its components, rigorously purified, and then studied with this end in view. Ideally, such a study would elucidate the stereochemistry of hydrogen removal to form the *cis* double bond, the nature of the intermediates (if any) and the primary acceptors of the hydrogens removed, the role of oxygen in this process, and the nature and function of the individual components of the electron transport chain involved in the overall reaction. The stereochemistry of hydrogen removal, first studied by Schroepfer & Bloch (112) and later by Morris and his co-workers (113), has been considered in detail in reviews by Bloch (11) and Morris (114). Hence, this work is not considered here except to note that in all Δ^9-desaturating systems studied, the 9-D and 10-D hydrogens of the fatty acyl chain are stereospecifically removed to form the *cis* double bond. In *Corynebacterium diphtheriae,* isotope effects observed in the formation of oleate suggest that hydrogen removal at carbon 9 precedes hydrogen removal at carbon 10 of stearate (112). However, from similar experiments carried out with chlorella cells and goat mammary gland microsomes, Morris (114) concludes that desaturation reactions in these systems probably involve concerted removal of a pair of hydrogen atoms from the 9 and 10 positions.

To date, no intermediate in the desaturation reaction has been detected. A wealth of data gathered chiefly from work with the systems listed in Table 1 would seem to rule out hydroxystearate as an intermediate or a precursor to oleate; whether some enzyme-bound oxygenated intermediate will eventually be implicated remains to be seen. Recent work has shed little additional light on this problem, and the reader is referred to several excellent books and reviews (4, 11, 19, 20, 114) for its further discussion.

Recent research has partially elucidated the nature and function of the individual components of the electron transport chains involved in O_2-mediated desaturation of fatty acids. Nagai & Bloch (85, 86) showed that desaturation of stearyl-ACP by a soluble system from *E. gracilis* required at least three distinct protein components, a flavoprotein which catalyzed the oxidation of NADPH, ferredoxin, and the desaturase, in addition to O_2 and NADPH. Inhibition was observed with KCN (50% at 10^{-5} M) but desaturation was not affected by high CO concentrations (an inhibitor of cytochrome P-450). NADPH could not be replaced by high ascorbate concentrations or by formamidine sulfinic acid, an electron donor for certain ferredoxin-requiring enzyme reactions. However, low concentrations of dithionite partially replaced NADPH in the ferredoxin-dependent stearyl ACP desaturation by spinach chloroplasts. Based on these results, Bloch (11) proposed the electron transport chain for oxygen activation in the desaturation reaction in these systems shown in Scheme 1. In the particulate preparation from *M. phlei* (82) the necessary cofactors included FAD and Fe^{+2} in addition to O_2 and NADPH. Thus at first glance one might propose a similar electron transport scheme for this system as well, assuming that the FAD was a component of a NADPH oxidase while the Fe^{+2} served as the active portion of a nonheme iron protein. However, detailed comparison

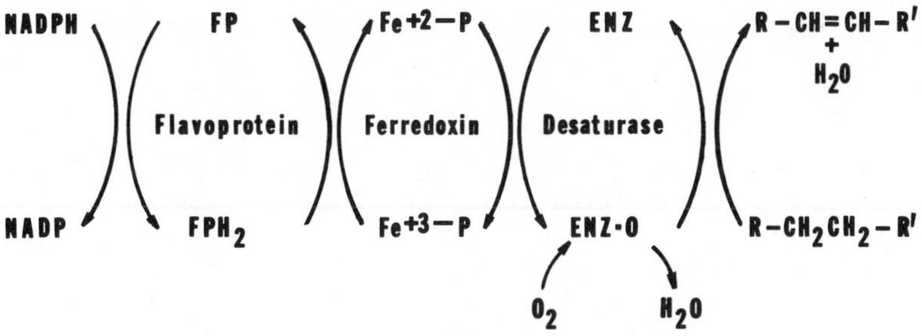

NADPH
FP
Fe+2—P
ENZ
R—CH=CH—R′
+
H2O
Flavoprotein
Ferredoxin
Desaturase
NADP
FPH2
Fe+3—P
ENZ·O
R—CH2CH2—R′
O2
H2O

Scheme 1

of the two systems leads to certain conflicts. For example, in *M. phlei,* NADPH, FAD (or FMN), and Fe^{+2} are all absolute requirements. Fe^{+3} does not replace Fe^{+2}, even in the presence of all other components. If Scheme 1 holds, it is not clear why NADPH and FAD are needed in the presence of Fe^{+2}, since they presumably function only to reduce iron to the ferrous state. Similarly, if this is their function, then Fe^{+3} should serve in place of Fe^{+2} when NADPH and FAD are present. Also, in the *M. phlei* system, neither KCN nor EDTA were good inhibitors.

Oshino and his co-workers (93–97) have presented evidence that the microsomal desaturation system from rat tissue, responsible for the conversion of stearate to oleate, consists of at least four protein components which include NADH-cytochrome b_5 reductase (F_{P_2}), NADPH-cytochrome *c* reductase (F_{P_1}), cytochrome b_5, and a terminal component called cyanide-sensitive factor (CSF) which seems to function as an oxygen-activating enzyme (93–97). Other workers, including Holloway & Wakil (115), Gaylor et al (116), and Brenner (58), have also made significant contributions to our developing concepts of the electron transport system involved in desaturation and in general their findings supplement and confirm those of Oshino's group. Oshino & Sato (96) propose the following scheme for electron transport among the various components of the rat liver microsomal desaturating system.

NADPH
Fp1
Stearyl-CoA
Oleyl-CoA
Ascorbate
Cytochrome b5
CSF
NADH
Fp2
O2
H2O

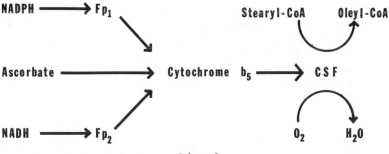

Scheme 2

This scheme is essentially the same as that proposed by Bloch (Scheme 1) for the *E. gracilis* system except that cytochrome b_5 replaces ferredoxin, and two flavoproteins, specific either for NADPH or NADH, mediate electron transport between the reduced pyridine nucleotides and the nonheme iron protein. Indeed, in the soluble system of Nagai & Bloch (86), the crude (but not the purified) preparation could utilize NADH as well as NADPH. Brenner's postulated scheme for electron transport involved in the desaturation of linoleate to γ-linolenate (58) again is in essential agreement with Schemes 1 and 2. Thus, the overall process of electron transport may well be the same, in the broad sense, for all O_2-dependent desaturations (although there are still unresolved problems with the bacterial system). However, the specifics of O_2 involvement remain completely unknown and the solution to this problem must await the purification and rigorous study of the desaturase itself.

Regulation and Control

A multitude of interlocking controls and regulatory mechanisms are known to operate in various organisms which act to maintain or adjust the level of specific unsaturated fatty acids in lipids of structural or functional importance in biological systems, particularly membranes (117–119). Of special interest to this reviewer are those regulatory mechanisms which adjust the composition of the unsaturated fatty acid component of membranes to compensate for changes in the environmental temperature. The commonly observed inverse relationship between temperature and unsaturated fatty acid composition almost certainly represents an adaptation on the part of the organism to maintain the integrity and function of cell membranes. Rose (120) has reviewed the work in this area to 1968. A number of more recent studies illustrate this adaptation process in bacteria (23, 24, 34, 35, 121–130), in yeast and fungi (131–135), in algae (136), in higher plants (137, 138), and in vertebrates (139–142). Until quite recently, however, the mechanism of these effects remained obscure. A number of possibilities exist; these include changes in 1. relative rates of β oxidation of saturated and unsaturated fatty acids, 2. relative rates of incorporation of saturated or unsaturated fatty acids into membrane lipids (including rates of transfer of lipids, particularly phospholipids, from one pool to another), and 3. relative rates of biosynthesis of saturated and unsaturated fatty acids (including both desaturation and biohydrogenation). Examples of the last two of these processes are now known. Sinensky (122) has shown that the proportion of oleate to stearate incorporated from the medium into *E. coli* phospholipids increased with decreasing temperature. In *E. coli* extracts, the relative rates of transacylation of palmityl CoA and oleyl CoA were shown to vary as a function of temperature. Temperature in turn was shown to affect the relative activity of a transacylase (as opposed to an effect on enzyme synthesis). The net result of this mechanism, then, is to increase the ratio of unsaturated to saturated fatty acids in the cell membrane as the environmental temperature is lowered.

A direct effect of temperature on unsaturated fatty acid biosynthesis seems to be the more common means of changing membrane composition. Thus, Harris & James (138, 143) have shown in several plant tissues that the temperature effect can be explained in terms of the increased solubility of O_2 (an obligate cosubstrate) as the temperature decreases. At constant temperature, desaturation increased with increasing oxygen tension, while at constant oxygen tension the rate of desaturation actually decreased somewhat with decreasing temperature. The results reported by Rinne (137) in an enzyme system obtained from developing soybean cotyledons and by Brown & Rose (132) in yeast also suggest that the increase in unsaturated fatty acid biosynthesis at lower temperatures in these systems can be ascribed to increased solubility of O_2.

This explanation, however, apparently does not apply to all systems. Fulco (24, 129, 130, 144) has intensively investigated the temperature-mediated control of desaturation in *Bacillus megaterium* and has demonstrated three control mechanisms which regulate the level of Δ^5-desaturating enzyme, and hence the rate of unsaturated fatty acid biosynthesis, in response to temperature changes in the growth or incubation medium. One control process is that of desaturase induction. A culture growing at 35° does not synthesize unsaturated fatty acids. When the culture is transferred to 20°, however, the synthesis of desaturase begins and continues at a high rate for at least one hour. This hyperinduction, so-called because levels of desaturating enzyme far exceed those found at normal 20° growth, is blocked by chloramphenicol, and the evidence suggests that both the desaturase-synthesizing system and the desaturase itself are absent in cultures growing at 35° and must be induced at 20° by a process requiring protein synthesis. Wide variations in O_2 levels do not affect hyperinduction (145). A second control process responsive to temperature is the irreversible inactivation of desaturating enzyme, which follows first order kinetics at all temperatures. Near 20°, a decrease of less than 2° in temperature of the incubation medium results in a twofold increase in the half-life of the desaturating enzyme. Temperature-mediated irreversible inactivation of the desaturating system is probably the most important mechanism for regulating the steady-state level of desaturase during the exponential phase of cell growth at temperatures near 20°. A third process, the zero-order decay of the desaturase synthesizing system, is observed when hyperinduced cultures are transferred from 20° to 35°. Present evidence (145) suggests that the rapid turn off of desaturase synthesis at 35° as well as the eventual cessation of hyperinduction at 20° involve the synthesis of a repressor which is in some way coupled to new DNA synthesis. Kepler & Tove (146) have recently studied a complementary system in bacilli, namely the induction of biohydrogenation of oleic acid in *B. cereus* by an increase in culture temperature. Oleate is reduced to stearate at 37° but not at 20°, and the appearance of the biohydrogenation system after an increase in temperature is blocked by either chloramphenicol or rifampin.

Several factors other than temperature operate in various systems to affect unsaturated fatty acid biosynthesis. In plants, these include light (147–151), aging or development (89, 152), and the levels of various ions (79, 80, 153). In higher animals, the effects of diet (and particularly the competitive effects of various fatty acids in the diet) have been intensively studied (58, 63, 64, 154–157), as have the roles of age, cofactor levels, and levels of acyl acceptors such as glycero-3-phosphate (59, 91, 158–160). Actis-Dato, Catala & Brenner (161) report the presence of a circadian rhythm of fatty acid desaturation in liver microsomes from mice exposed to light-dark cycles.

One rhythm was observed for stearate desaturation and a second for the desaturation of linoleate or γ-linolenate.

HYDROGENATION AND HYDRATION OF UNSATURATED FATTY ACIDS

The hydrogenation of an isolated double bond in the hydrocarbon chain of a fatty acid (as opposed to the reduction of *trans*-2 double bonds as in fatty acid biosynthesis) appears to be a relatively rare event in nature. Reiser (162) was probably the first to clearly demonstrate that sheep rumen contents could hydrogenate polyunsaturated fatty acids, and he attributed this action to rumen bacteria. Wright (163, 164) later showed that both rumen bacteria and rumen protozoa could hydrogenate unsaturated lipids. However, it was not until the work of Tove and colleagues (146, 165–170) that a specific organism (*Butyrivibrio fibrisolvens*) was identified and some details of the hydrogenation process elucidated. They showed that in the rumen two systems are involved in the complete hydrogenation of linoleic acid: one specific for the conversion of linoleic acid to a monoenoic acid and the other for the hydrogenation of the monoene to stearic acid (165). *B. fibrisolvens* carries out only the former reaction yielding *trans*-11-octadecenoic acid via the intermediate *cis*-9, *trans*-11-octadecadienoic acid (166). They later (167) isolated and partially purified an enzyme from the cell envelope of *B. fibrisolvens* that carried out the first step, the conversion of linoleic acid to *cis*-9,*trans*-11-octadecadienoic acid. The same enzyme could convert linolenic acid to the *cis*-9,*trans*-11,*cis*-15 isomer. Stereospecific addition of hydrogen to carbon atom 13 of linoleic acid in the D configuration was demonstrated and it was deduced that the mechanism of isomerization involves either the protonation of an enzyme-bound carbanion or a concerted reaction (169). The presence of the *cis*-9,*cis*-12-diene system and a free C-1 carboxyl group was an absolute requirement for isomerization (168). No enzyme has yet been isolated that hydrogenates the isomerized diene to *trans*-11-octadecenoic acid, but Rosenfeld & Tove (170) showed that in whole cells incubated in D_2O, deuterium was incorporated at the *cis* double bond reduced by the organism. This reduction, which takes place stereospecifically, was found to occur by *cis* addition to the D side of *cis*-9,*trans*-11-octadecadienoic acid. The distribution of deuterium at the reduced carbon atoms shows an isotope effect that can be explained if the reduction occurs by addition of a proton and hydride ion mediated by an unknown carrier. Both a rumen spirochete, *Treponèma* (*Borrellia*) sp. (171) and rat gut (172) have also been shown to hydrogenate linoleic acid (171).

Hydration of an isolated fatty acid double bond, like hydrogenation, seems to be a relatively rare phenomenon. In 1962, Wallen, Benedict & Jackson (173) reported the isolation of a pseudomonad which converted oleate to 10-hydroxy stearate. Schroepfer and his co-workers (174–179) subsequently studied this conversion both in vivo and in cell-free preparations. In a deuterium-oxide enriched medium, the organism forms 10-D-hydroxystearic acid from oleic acid with the incorporation of one atom of stably bound deuterium at carbon atom 9 in the L configuration (174, 175). The isolation of a soluble preparation which catalyzed the formation of 10-D-hydroxystearate from oleate (176) facilitated the elucidation of the reaction mechanism. The enzyme (oleate hydratase) catalyzed the addition of water across the *cis* double bond of oleic acid and also catalyzed the formation of *trans*-10-octadecenoic acid from either oleic acid or 10-D-hydroxystearic acid. Although 9-D-hydroxystearic acid was not a substrate for the enzyme, palmitoleic acid was converted to 10-hydroxypalmitic acid and linoleic acid to 10-D-hydroxy-12-*cis*-octadecenoic acid (177). In addition, the same enzyme carried out the stereospecific hydration of 9,10-*cis*-epoxystearic acid to yield one isomer of *threo*-9,10-dihydroxy-stearic acid, and of 9,10-*trans*-epoxystearic acid to yield one isomer of *erythro*-9,10-dihydroxystearic acid (178). Kisic, Miura & Schroepfer (179) prepared nine positional isomers of DL-hydroxystearic acid (ranging from 5-hydroxy- to 15-hydroxystearic acids) and found that only the D isomer of 10-hydroxystearic acid serves as a substrate for the reverse (dehydration) reaction. Mortimer & Niehaus (180) have recently shown that *trans*-10-octadecenoic acid is formed by the pseudomonad enzyme by direct isomerization of the *cis*-9 double bond of oleate and that the alternate pathway, hydration of oleate to 10-hydroxystearate followed by dehydration to *trans*-10-octadecenoic acid, does not occur. Thus, the formation of the *trans*-10-acid from 9-D-hydroxystearic acid (176) must proceed via oleic acid as an intermediate. Other work in this area includes that by Wallen et al (181), who have reported the formation of three new 10-D-hydroxy fatty acids by anaerobic microbiological hydration of linoleic, linolenic, and ricinoleic acids and by Thomas (182), who identified several enteric bacteria that convert oleic acid to hydroxystearic acid in vitro.

HYDROXYLATION OF FATTY ACIDS

Hydroxy fatty acids occur widely in nature as intermediates in synthesis or oxidation of fatty acids or as endproducts which are incorporated into various lipids. Here we are concerned only with direct hydroxylation reactions. The distribution of naturally occurring aliphatic hydroxy acids has been reviewed by Downing (183) and Pohl & Wagner (184), while recent reviews that consider hydroxylation mechanisms include those by Ulrich (185), Morris (114), and the book edited by Boyd & Smellie (186).

α Hydroxylation and α Oxidation

The hydroxylation of fatty acids at the α-carbon may be considered an intermediate step in α oxidation by which fatty acids are degraded one carbon at a time. In higher animals, several distinct α-oxidation systems have been demonstrated. Steinberg and co-workers (187–196) have intensively investigated the metabolic defect responsible for Refsum's disease in humans, a clinical entity characterized by the accumulation of a dietary constituent, phytanic acid (3,7,11,15-tetramethylhexadecanoic acid), in the tissues. They found that the disease state was caused by the genetic lack of an enzyme, phytanic acid α-oxidase, which carries out a one-carbon degradation of phytanic acid to yield the next lower homolog, pristanic acid (2,6,10,14-tetra-methylpentadecanoic acid) which, unlike its precursor, can be totally degraded by the β-oxidation pathway (191). Most of this work has been covered in a review by Stumpf (3) and is not considered here. Tsai et al (196) describe the stereospecificity of the mitochondrial α-oxidation system for phytanic acid which proceeds through the intermediate formation of α-hydroxyphytanic acid. Although the enzymatic introduction of the α-hydroxyl group is highly stereospecific (absolute configuration unknown), the oxidative decarboxylation of α-hydroxyphytanic acid is not. Both isomers are readily oxidized although one form is decarboxylated about 50% faster than the other. A change in the configuration of the methyl group at the 3 position of phytanic acid also has little effect either on hydroxylation or decarboxylation.

A second α-oxidation system found in higher animals is that associated with brain microsomes. This system, elucidated primarily by Mead and colleagues (197–205), is responsible for the oxidative decarboxylation of the very long-chain α-hydroxy acids of brain and is distinct in several ways from the system which decarboxylates α-hydroxyphytanic acid. The latter system is located in liver mitochondria, requires NADPH and O_2, is stimulated by Fe^{+3} but inhibited by Fe^{+2}, and does not require ascorbic acid (192). The brain system, on the other hand, is microsomal and requires O_2 but no pyridine nucleotide. Mead & Hare (204, 205) have recently shown that Fe^{+2} but not Fe^{+3} is a requirement for decarboxylation, as is ascorbic acid. Decarboxylation of $(1-{}^{14}C)$-2-hydroxytetracosanoic acid was greatly reduced in brain preparations from scorbutic guinea pigs. Ascorbic acid supplementation, both in vivo and in vitro, restored enzymatic activity in the preparations from scorbutic animals, but did not affect the preparations from supplemented animals in vitro.

The α-hydroxylation step responsible for the conversion of tetracosanoic to α-hydroxytetracosanoic acid in brain has recently been elucidated by Hoshi & Kishimoto (206). They showed that cell-free homogenates from rat brain carried out α hydroxylation in the presence of O_2, Mg^{+2}, pyridine nucleotides, and a heat-stable water-soluble cofactor. Both NADPH and NADH were equally effective, the oxidized forms less so. The product (α-OH-C_{24}) was detected only as a component of ceramide or cerebroside and not as the free acid or CoA ester. The presence of CO did not inhibit activity but all heavy metal ions (except Mg^{+2}) were strongly inhibitory. The apparent K_m value for tetracosanoic acid was 4.2×10^{-6} M and the enzyme seemed quite specific for the C_{24} chain length. Although the configuration of the α-hydroxy group in the product was not determined, Hammarström (207) has shown that all of the α-hydroxy acids of beef brain cerebrosides have the D configuration.

In other work in animal systems, Levis (208) separated and partially purified two α-hydroxy fatty acid oxidases from the $100,000 \times g$ supernatant fraction from a rat kidney homogenate. These enzymes formed α-keto acids and were specific for either the D or L isomers of α-hydroxystearic acid. NAD was a requirement for full activity. The α-keto acids formed by these enzymes were readily decarboxylated by a microsomal fraction from the same homogenate. Ushijima & Nakano (209) have purified 600-fold an FMN-containing enzyme present in the light mitochondrial fraction of rat liver cells which is specific for the oxidation of short-chain aliphatic L-α-hydroxy acids.

A number of α-oxidation systems have been described in plants and micro-organisms within the past few years. Most recently the sphingolipids of certain protozoa have been shown to contain large amounts of α-hydroxy fatty acids (210), and the presence of an α-hydroxy fatty acid oxidation system has been demonstrated in the protozoan, *Crithidia fasciculata* (211). In yeast, Vesonder et al (212) have shown that *Hansenula sydowiocum* excretes free α-hydroxy fatty acids (primarily α-OH-C_{26}) into the medium when grown on a glucose solution. The α hydroxylation and decarboxylation of long-chain fatty acids in a number of yeasts had been demonstrated previously by Fulco (213).

α-Hydroxy fatty acids are common in bacteria (214–219) and cell-free extracts of *Arthrobacter simplex* have been obtained by Yano, Furukawa & Kusunose (220), which catalyze the conversion of palmitic acid to α-hydroxypalmitic acid and also the conversion of the latter to pentadecanoic acid.

Several fatty acid α-oxidation systems have been demonstrated in plants and much of this work has been reviewed by Stumpf (3) and by Morris (114). Hitchcock & Morris (221) and Hitchcock & Rose (222) have studied the stereochemistry of α oxidation of fatty acids in leaves. When (U-^{14}C)-palmitate was incubated with a particulate fraction from young pea leaves, (^{14}C) pentadecanal accumulated. L-(U-^{14}C, 2-^{3}H)-Palmitate similarly yields (^{14}C) pentadecanal which retains 90% of the tritium. D-(U-^{14}C, 2-^{3}H)-Palmitate gives pentadecanal with 87% loss of tritium (221). These results suggest that D-2-hydroxypalmitic acid is an intermediate in this reaction but that L-2-hydroxypalmitic and 2-ketopalmitic acid are not. Later work (222) confirmed that D-2-hydroxypalmitic acid accumulates in small amounts during the course of α oxidation of palmitate. Earlier work by Hitchcock, Morris & James (223), however, seemed to contradict these results by showing that the L-2-hydroxy isomer is preferentially α-oxidized in a pealeaf system. Hitchcock & Morris (221) suggest that the apparent contradiction can be explained if α oxidation in leaf systems occurs by two routes: (*a*) via D-2-hydroxypalmitate and pentadecanal, both of which can accumulate as intermediates, and (*b*) via L-2-hydroxypalmitate which is rapidly oxidized with no accumulation of intermediates. Pentadecanal would not be an intermediate in pathway (*b*) although 2-ketopalmitate, excluded as a possible intermediate in pathway (*a*), could be involved in pathway (*b*).

Markovetz & Stumpf (224) have carried out a sixtyfold purification of an α-oxidation system from germinated peanut cotyledons. Two activities (formation of 2-hydroxypalmitate and CO_2 formation from palmitate) remained with a single protein fraction. An H_2O_2-generating system was required but, unlike the pealeaf system mentioned above, O_2 was ineffective. D-2-Hydroxypalmitic acid accumulates in the reaction mixture as α oxidation proceeds, whereas L-2-hydroxypalmitate does not. Competition studies with unlabeled L-2-hydroxypalmitate would suggest that it is the intermediate substrate for further breakdown. Morris (114) suggests that α-oxidation systems in general may be stereoselective rather than completely stereo-, specific. Thus, the D-2-hydroxy acid may be the favored hydroxylation product but the L isomer may be the favored substrate for decarboxylation. At present, the bulk of experimental evidence would seem to favor this hypothesis.

ω and ω-Type Hydroxylations

These hydroxylations are characteristically of the mixed function oxidase type (i.e. involving both molecular oxygen and a reduced pyridine nucleotide) and are widespread in nature. Hitchcock & Nichols (225) have reviewed most of the work in this area up to 1971.

Coon and co-workers (226, 227) have recently succeeded in solubilizing and resolving three functional components of the liver microsomal enzyme system which catalyzes the ω hydroxylation of fatty acids (as well as the hydroxylation of alkanes and various drugs). These include cytochrome P-450, NADPH-cytochrome P-450 reductase, and phosphatidylcholine (PC). The PC is necessary for electron transfer from NADPH to P-450. Hydroxylation in this soluble system is inhibited by superoxide dismutase, while a superoxide generating system can partially substitute for NADPH and the reductase. Based on these and previous observations (228), Coon et al (227) have proposed a mechanism of substrate hydroxylation catalyzed by cytochrome P-450. The initial step would involve the combination of the oxidized form of the cytochrome with the substrate, followed by reduction (by NADPH) of the cytochrome iron. PC is presumably required to facilitate reduction. Oxygen then combines with substrate-reduced P-450 to give a ternary complex. Estabrook et al (229) have provided experimental evidence for both the initial substrate-oxidized P-450 complex and the ternary complex with O_2. Intramolecular electron transfer to yield the superoxide radical (bound to oxidized P-450) followed by attack of superoxide on the substrate would then complete the reaction. The uptake of a second electron (donated either by the reductase or a second molecule of superoxide) would yield the hydroxylated product and water. Duppel, Lebeault & Coon (230) have studied a soluble system from the yeast, *Candida tropicalis,* which ω-hydroxylates lauric acid and contains cytochrome P-450 as a functional component as well as a phospholipid and NADPH-cytochrome P-450 reductase. Most recently Gallo et al (231) showed that alkane oxidation in this organism is catalyzed by the same P-450 ω-hydroxylating system described above.

A microsomal ω and ω − 1 hydroxylating system for fatty acids has been isolated from a torulopsis species of yeast (232) which may also be cytochrome P-450 dependent. Stereochemical study of this system (233) shows that the hydroxylations take place without double bond formation and with retention of configuration.

Björkhem (234, 235) and Hamberg & Björkhem (236) have studied the mechanism of ω and ω − 1 hydroxylation of fatty acids by a rat liver microsomal system. Hydroxylation in the ω or ω − 1 position of decanoic acid occurred with the loss of one hydrogen atom from the carbon hydroxylated. 10-Hydroxydecanoic acid accounted for 92% of the products formed, while L-9-hydroxydecanoic (6%) and the D-stereoisomer (2%) accounted for the remainder. Substitution of two deuteriums at carbon 9 of the substrate resulted in significant isotope effects in the formation of the 9-hydroxy derivatives, but there was no detectable isotope effect in the formation of the 10-hydroxy acid from the substrate fully deuterated at the 10 position. The two hydroxylations at carbon 9 proceeded stereospecifically with retention of the absolute configuration. These and previous results (237) suggested that ω − 1 hydroxylation of fatty acids in microsomes may involve enzymes that differ from those responsible for the cytochrome P-450 dependent ω hydroxylations. The most recent work (235) does not directly answer this point but does demonstrate that in the ω − 1 hydroxylation of laurate to 11-hydroxylauric acid, the rate-limiting step is the cleavage of the C–H bond and thus differs from the rate-limiting step in ω hydroxylation, which is probably the reduction of cytochrome P-450 (238). Ellin et al (239) have recently shown that still another system (rat kidney cortex microsomes) carries out the ω and $\omega-1$ hydroxylation of laurate (at a ratio of about 2:1) and that both hydroxylations are a function of a cytochrome P-450-like hemoprotein called P-450$_K$ by these authors.

Ichihara, Kusunose & Kusunose (240–246) have examined ω and ω-type hydroxylations in a large number of systems and have found evidence that all are of the mixed function oxidase type and that most seem to involve cytochrome P-450 or a similar hemoprotein. In several solubilized microsomal systems (porcine kidney, rat liver), the hydroxylating preparation was separated into two fractions, I and II. Fraction I contained a cytochrome P-450 type protein while fraction II contained NADPH-cytochrome c reductase. Fraction II, however, could be replaced by ferredoxin plus ferredoxin-NADP reductase. There was no evidence for a phospholipid requirement in the reconstituted (ferredoxin) systems, although ether extraction of fraction I reduced activity for laurate ω-hydroxylation. Activity was restored by the addition of various detergents (i.e. Triton X-100).

Fatty acid and alkane ω oxidation in the bacterium *Pseudomonas oleovorans* has been studied by Coon and his co-workers (247–250). Three protein components (NADH-rubredoxin reductase, rubredoxin, and the ω-hydroxylase) are required for activity and NADH is highly superior to NADPH as an electron donor. Spinach ferredoxin, adrenodoxin, and other nonheme iron proteins could not substitute for rubredoxin. Cytochrome P-450 is not a component of this system.

Most recently, Miura & Fulco (251) have described a unique soluble system from *Bacillus megaterium* which carries out the $\omega-2$ hydroxylation of fatty acids. When palmitic acid is used as a substrate, the major product is 14-hydroxy palmitate (55%), while the 15-hydroxy (27%) and the 13-hydroxy (18%) isomers are also produced. Activity is highest for the C_{15} chain length (pentadecanoic acid) and decreases in the order $C_{15} > C_{16} > C_{14} > C_{17} > C_{18} > C_{12}$. Two protein components of the system have been separated, one of which can be completely replaced by ferredoxin (from *Clostridium pasteurianum*). NADPH and O_2 are required for activity.

Although cell-free ω-type hydroxylating systems from higher plants have not been isolated, the work by Kolattukudy and his co-workers (252–256) demonstrates that such systems must exist there.

A number of fatty acid hydroxylating systems are known which are neither ω-type hydroxylases nor α-hydroxylating systems. One recent example is a soluble preparation from avocado mesocarp which carries out the β hydroxylation of medium chain length fatty acyl-CoAs in a reaction quite distinct from the usual β-oxidation pathway. In this new system, O_2 is required, whereas NAD and CO strongly inhibit (257, 258). Unfortunately, lack of space does not permit additional discussion of these systems, or of much interesting work peripheral to the subject matter of this review.

CONCLUDING REMARKS

The present state of our knowledge of fatty acid desaturation can be summarized as follows: The major pathways for the biosynthesis of the common unsaturated fatty acids in various organisms are now known and further work in this area will serve chiefly to add details to the overall picture. Also, enough cell-free desaturating systems have now been studied to conclude that the direct removal of two hydrogens from a fatty acyl chain to produce a *cis* double bond four or more carbon atoms removed from the carboxyl group always requires both O_2 and a reduced pyridine nucleotide as cofactors; a two (or more) component electron transport chain (including both a flavoprotein and an iron-containing protein) is also involved. CoA and ACP thioesters and certain phospholipids (but not free fatty acids) have been found to serve as substrates in one or another of the desaturation systems.

On the other hand, no desaturation system has been extensively purified and we know nothing of the actual desaturation mechanism or of the role of O_2 in this process. With few exceptions, the desaturases have proven to be particle bound, relatively unstable, and resistant to purification by classical techniques. Until significant purification of a desaturase can be accomplished, there seems little hope of elucidating the actual mechanism(s) of *cis* double bond formation, and thus strong future efforts should be directed toward developing methods suitable for the fractionation and purification of fatty acid desaturases.

Like the desaturases, the ω-type fatty acid hydroxylating systems are characteristically mixed function oxidases and require O_2 and a reduced pyridine nucleotide for activity. They are also generally microsomal in eucaryotes and contain both a flavoprotein and an iron protein. Unlike the desaturases, however, the free fatty acids are utilized as substrates. Furthermore, several systems have been solubilized and partially purified, and Coon and his co-workers have proposed a hydroxylation mechanism based on their work with a cytochrome P-450 system solubilized from liver. They consider the active species to be a ternary complex of P-450, O_2, and substrate which, by intramolecular electron transfer, yields P-450-bound superoxide radical. Attack of the bound superoxide radical on the fatty acid substrate followed by the uptake of a second electron would then yield the ω-hydroxy fatty acid and water. Although the evidence for this mechanism is not conclusive, it is strongly suggestive and future work might well center on efforts to determine whether superoxide may be involved in other ω-type hydroxylating systems.

There are a number of α-hydroxylating systems that act on fatty acids but they appear to differ widely among themselves and, as a group, differ from the ω-type hydroxylases. None have been significantly purified and, as with the fatty acid desaturases, future progress in determining their mechanisms of action may well depend on the successful development of methods for the purification and characterization of these intractable systems.

ACKNOWLEDGMENTS

The author wishes to thank Ellin James, Melody Horner, and Vicky Julian, all of whom gave valuable assistance in the preparation of this review.

Literature Cited

1. Volpe, J. J., Vagelos, P. R. 1973. *Ann. Rev. Biochem.* 42:21–60
2. Green, D. E., Allmann, D. W. 1968.

Metabolic Pathways, ed. D. M. Greenberg, 2:1–36. New York: Academic. 311 pp.

3. Stumpf, P. K. 1969. *Ann. Rev. Biochem.* 38:159–212
4. Bishop, D. G., Stumpf, P. K. 1971. *Biochemistry and Methodology of Lipids,* ed. A. R. Johnson, J. B. Davenport, 361–90. New York: Wiley. 578 pp.
5. Hinman, J. W. 1972. *Ann. Rev. Biochem.* 41:161–78
6. van den Bosch, H. 1974. *Ann. Rev. Biochem.* 43:243–77
7. Gatt, S., Barenholz, Y. 1973. *Ann. Rev. Biochem.* 42:61–90
8. McMurray, W. C., Magee, W. L. 1972. *Ann. Rev. Biochem.* 41:129–60
9. Stoffel, W. 1972. *Ann. Rev. Biochem.* 40:57–82
10. Lennarz, W. J. 1970. *Ann. Rev. Biochem.* 39:359–88
11. Bloch, K. 1969. *Accounts Chem. Res.* 2:193–202
12. Mead, J. F. 1971. *Progr. Chem. Fats Other Lipids* 9:159–92
13. Fulco, A. J. 1967. *The Encyclopedia of Biochemistry,* ed. R. J. Williams, E. M. Lansford, 817–20. New York: Reinhold. 876 pp.
14. Kass, L. R., Brock, D. J. H., Bloch, K. 1967. *J. Biol. Chem.* 242:4418–31
15. Brock, D. J. H., Kass, L. R., Bloch, K. 1967. *J. Biol. Chem.* 242:4432–40
16. Rando, R. R., Bloch, K. 1968. *J. Biol. Chem.* 243:5627–34
17. Helmkamp, G. M., Rando, R. R., Brock, D. J. H., Bloch, K. 1968. *J. Biol. Chem.* 243:3229–31
18. Endo, K., Helmkamp, G. M., Bloch, K. 1970. *J. Biol. Chem.* 245:4293–96
19. James, A. T. 1968. *Chem. Brit.* 4:484–88
20. Hitchcock, C., Nichols, B. W. 1971. *Plant Lipid Biochemistry,* 130. New York: Academic. 387 pp.
21. Erwin, J., Bloch, K. 1964. *Science* 143:1006–12
22. Fulco, A. J., Levy, R., Bloch, K. 1964. *J. Biol. Chem.* 239:998–1003
23. Fulco, A. J. 1967. *Biochim. Biophys. Acta* 144:701–3
24. Fulco, A. J. 1969. *J. Biol. Chem.* 244:889–95
25. Dart, R. K., Kaneda, T. 1970. *Biochim. Biophys. Acta* 218:189–94
26. Kaneda, T. 1971. *Biochem. Biophys. Res. Commun.* 43:298–302
27. Scheuerbrandt, G., Bloch, K. 1962. *J. Biol. Chem.* 237:2064–68
28. Hung, J. G. C., Walker, R. W. 1970. *Lipids* 5:720–22
29. Fulco, A. J., Bloch, K. 1964. *J. Biol. Chem.* 239:993–97
30. Asselineau, C. P., Lacave, C. S., Montrozier, H., Prome, J. C. 1970. *Eur. J. Biochem.* 14:406–10
31. Stern, N., Shenberg, E., Tietz, A. 1969. *Eur. J. Biochem.* 8:101–8
32. Ghaneker, A. S., Nair, P. M. 1973. *J. Bacteriol.* 114:618–24
33. Kates, M. 1966. *Ann. Rev. Microbiol.* 20:13–44
34. Fulco, A. J. 1969. *Biochim. Biophys. Acta* 187:169–71
35. Fulco, A. J. 1970. *J. Biol. Chem.* 245:2985–90
36. Sklan, D., Budowski, P., Volcani, R. 1972. *Brit. J. Nutr.* 28:239–48
37. Sklan, D., Budowski, P. 1972. *Brit. J. Nutr.* 28:457–62
38. Asselineau, C. P., Montrozier, H. L., Prome, J. C., Savagnac, A. M. 1972. *Eur. J. Biochem.* 28:102–9
39. Stearns, E. M. Jr. 1971. *Progr. Chem. Fats Other Lipids* 9:453–518
40. Yuan, C., Bloch, K. 1961. *J. Biol. Chem.* 236:1277–79
41. Fulco, A. J., Mead, J. F. 1959. *J. Biol. Chem.* 234:1411–16
42. Steinberg, G., Slaton, W. H. Jr., Howton, D. R., Mead, J. F. 1956. *J. Biol. Chem.* 220:257–62
43. Mead, J. F., Howton, D. R. 1957. *J. Biol. Chem.* 229:575–82
44. Parker, P. L., Van Baalen, C., Maurer, L. 1967. *Science* 155:707–8
45. Holton, R., Blecker, H. H., Stevens, T. S. 1968. *Science* 160:545–47
46. Nichols, B. W., Wood, B. J. B. 1968. *Lipids* 3:46–50
47. Kenyon, C. N., Stanier, R. Y. 1970. *Nature* 227:1164–66
48. Kenyon, C. N., Rippka, R., Stanier, R. Y. 1972. *Arch. Mikrobiol.* 83:216–36
49. Kenyon, C. N. 1972. *J. Bacteriol.* 109:827–34
50. Beach, D. H., Harrington, G. W., Holz, G. G. Jr. 1970. *J. Protozool.* 17:501–10
51. Harrington, G. W., Beach, D. H., Dunham, J. E., Holz, G. G. Jr. 1970. *J. Protozool.* 17:213–19
52. Gellerman, J. L., Schlenk, H. 1972. *Lipids* 7:51–55
53. Schlenk, H., Gellerman, J. L. 1965. *J. Am. Oil Chem. Soc.* 42:504–11
54. Anderson, W. H., Gellerman, J. L., Schlenk, H. 1972. *Lipids* 7:710–14
55. Gellerman, J. L., Anderson, W. H., Schlenk, H. 1972. *Bryologist* 75:550–57
56. Shaw, R. 1965. *Biochim. Biophys. Acta* 98:230–37
57. Peluffo, R. O., Ayala, S., Brenner, R. R. 1970. *Am. J. Physiol.* 218:669–73
58. Brenner, R. R. 1971. *Lipids* 6:567–75
59. Castuma, J. C., Catala, A., Brenner, R. R. 1972. *J. Lipid Res.* 13:783–89
60. Ayala, S., Gaspar, G., Brenner, R. R.,

238 FULCO

Peluffo, R. O., Kunau, W. 1973. *J. Lipid Res.* 14:296–305
61. Sprecher, H. S. 1967. *Biochim. Biophys. Acta* 144:296–304
62. Stoffel, W., Ecker, W., Assad, H., Sprecher, H. 1970. *Z. Physiol. Chem.* 351:1545–54
63. Ullman, D., Sprecher, H. 1971. *Biochim. Biophys. Acta* 248:61–70
64. Lee, C., Sprecher, H. 1971. *Biochim. Biophys. Acta* 248:180–85
65. Ullman, D., Sprecher, H. 1971. *Biochim. Biophys. Acta* 248:186–97
66. Bridges, R. B., Coniglio, J. G. 1970. *J. Biol. Chem.* 245:46–49
67. Bridges, R. B., Coniglio, J. G. 1970. *Biochim. Biophys. Acta* 218:29–35
68. Bridges, R. B., Coniglio, J. G. 1970. *Lipids* 5:628–35
69. Schlenk, H., Sand, D. M., Gellerman, J. L. 1969. *Biochim. Biophys. Acta* 187:201–7
70. Santomi, S. 1972. *Rinsho Shoni Igaku* 20:143–60
71. Nakamura, M., Privett, O. S. 1969. *Lipids* 4:41–49
72. Kunau, W. H., Couzens, B. 1971. *Z. Physiol. Chem.* 352:1297–305
73. Municio, A. M., Odriozola, J. M., Pineiro, A., Ribera, A. 1972. *Biochim. Biophys. Acta* 280:248–57
74. Thompson, S. N., Barlow, J. S. 1972. *Ann. Entomol. Soc. Am.* 65:1020–23
75. Thompson, S. N., Barlow, J. S. 1972. *Can. J. Zool.* 50:1105–10
76. Thompson, S. N., Barlow, J. S. 1973. *Comp. Biochem. Physiol. B.* 44:59–64
77. Fulco, A. J., Mead, J. F. 1960. *J. Biol. Chem.* 235:3379–84
78. Gurr, M. I., Robinson, M. P., James, A. T., Morris, L. J., Howling, D. 1972. *Biochim. Biophys. Acta* 280:415–21
79. Jacobson, B. S., Kannangara, C. G., Stumpf, P. K. 1973. *Biochem. Biophys. Res. Commun.* 51:487–93
80. Jacobson, B. S., Kannangara, C. G., Stumpf, P. K. 1973. *Biochem. Biophys. Res. Commun.* 52:1190–98
81. Bloomfield, D. K., Bloch, K. 1960. *J. Biol. Chem.* 235:337–45
82. Fulco, A. J., Bloch, K. 1962. *Biochim. Biophys. Acta* 63:545–46
83. Talamo, B., Chang, N., Bloch, K. 1973. *J. Biol. Chem.* 248:2738–42
84. Baker, N., Lynen, F. 1971. *Eur. J. Biochem.* 19:200–26
85. Nagai, J., Bloch, K. 1966. *J. Biol. Chem.* 241:1925–27
86. Nagai, J., Bloch, K. 1968. *J. Biol. Chem.* 243:4626–33
87. Nagai, J., Bloch, K. 1965. *J. Biol. Chem.* 240:3702–3

88. Vijay, I. K., Stumpf, P. K. 1972. *J. Biol. Chem.* 247:360–66
89. Abdelkader, A. B., Cherif, A., Demandre, C., Mazliak, P. 1973. *Eur. J. Biochem.* 32:155–65
90. Inkpen, J. A., Quackenbush, F. W. 1969. *Lipids* 4:539–43
91. McDonald, T. M., Kinsella, J. E. 1973. *Arch. Biochem. Biophys.* 156:223–31
92. Cook, H. W., Spence, M. W. 1973. *J. Biol. Chem.* 248:1786–92; 1793–96
93. Oshino, N., Imai, Y., Sato, R. 1966. *Biochim. Biophys. Acta* 128:13–28
94. Oshino, N., Imai, Y., Sato, R. 1971. *J. Biochem. Japan* 69:155–68
95. Oshino, N., Sato, R. 1971. *J. Biochem. Japan* 69:169–80
96. Oshino, N., Sato, R. 1972. *Arch. Biochem. Biophys.* 149:369–77
97. Oshino, N. 1972. *Arch. Biochem. Biophys.* 149:378–87
98. Jones, P. D., Holloway, P. W., Peluffo, R. O., Wakil, S. J. 1969. *J. Biol. Chem.* 224:744–54
99. Nugteren, D. H. 1962. *Biochim. Biophys. Acta* 60:656–57
100. Brenner, R. R., DeTomas, M. E., Peluffo, R. O. 1965. *Biochim. Biophys. Acta* 106:640–49
101. Stoffel, W. 1963. *Z. Physiol. Chem.* 333:71–88
102. Stoffel, W. 1971. *Wiss. Veroeff. Deut. Ges. Ernaehr.* 22:12–34
103. Lynen, F. 1972. *Current Trends in the Biochemistry of Lipids*, ed. J. Ganguly, R. M. S. Smellie, 5–25. London: Academic. 479 pp.
104. Schuetz, J., Lynen, F. 1971. *Eur. J. Biochem.* 21:48–54
105. Vijay, I. K., Stumpf, P. K. 1971. *J. Biol. Chem.* 246:2910–17
106. Gurr, M. I., Robinson, M. P., Sword, R. W., James, A. T. 1968. *Biochem. J.* 110:49P–50P
107. Brett, D., Howling, D., Morris, L. J., James, A. T. 1971. *Arch. Biochem. Biophys.* 143:535–47
108. Howling, D., Morris, L. J., Gurr, M. I., James, A. T. 1972. *Biochim. Biophys. Acta* 260:10–19
109. Johnson, A. R., Fogerty, A. C., Pearson, J. A., Shenstone, F. S., Bersten, A. M. 1969. *Lipids* 4:265–69
110. Paulsrud, J. R., Stewart, S. E., Holman, R. T. 1970. *Lipids* 5:611–16
111. Quint, J. F., Fulco, A. J. 1973. *J. Biol. Chem.* 248:6885–95
112. Schroepfer, G., Bloch, K. 1965. *J. Biol. Chem.* 240:54–63
113. Morris, L. J., Harris, R. V., Kelly, W., James, A. T. 1968. *Biochem. J.* 109:673–78

114. Morris, L. J. 1970. *Biochem. J.* 118: 681–93
115. Holloway, P. W., Wakil, S. J. 1970. *J. Biol. Chem.* 245:1862–65
116. Gaylor, J. L., Morr, N. J., Seifried, H. E., Jefcoate, C. R. E. 1970. *J. Biol. Chem.* 245:5511–13
117. Machtiger, N. A., Fox, C. F. 1973. *Ann. Rev. Biochem.* 42:575–600
118. Singer, S. J., Nicolson, G. L. 1972. *Science* 175:720–31
119. Cronan, J. E. Jr., Vagelos, P. R. 1972. *Biochim. Biophys. Acta* 265:25–60
120. Rose, A. H. 1968. *J. Appl. Bacteriol.* 31:1–11
121. Shen, P. Y., Coles, E., Foote, J. L., Stenesh, J. 1970. *J. Bacteriol.* 103:479–81
122. Sinensky, M. 1971. *J. Bacteriol.* 106:449–58
123. Weerkamp, A., Heinen, W. 1972. *J. Bacteriol.* 109:443–46
124. Levin, R. A. 1972. *J. Bacteriol.* 112:903–9
125. Cullen, J., Phillips, M. C., Shipley, G. G. 1971. *Biochem. J.* 125:733–42
126. Aibara, S., Kato, M., Ishinaga, M., Kito, M. 1972. *Biochim. Biophys. Acta* 270:301–6
127. Chang, N. C., Fulco, A. J. 1973. *Biochim. Biophys. Acta* 296:287–99
128. Kaneda, T. 1972. *Biochim. Biophys. Acta* 280:297–305
129. Fulco, A. J. 1972. *J. Biol. Chem.* 247:3503–10
130. Fulco, A. J. 1972. *J. Biol. Chem.* 247:3511–19
131. Gunasekaran, M., Raju, P. K., Lyda, S. D. 1970. *Phytopathology* 60:1027
132. Brown, C. M., Rose, A. H. 1969. *J. Bacteriol.* 99:371–78
133. McMurrough, J., Rose, A. H. 1971. *J. Bacteriol.* 107:753–58
134. McMurrough, J., Rose, A. H. 1973. *J. Bacteriol.* 114:451–52
135. Kates, M., Paradis, M. 1973. *Can. J. Biochem.* 51:184–97
136. Patterson, G. W. 1970. *Lipids* 5:597–600
137. Rinne, R. W. 1969. *Plant Physiol.* 44:89–94
138. Harris, P., James, A. T. 1969. *Biochem. J.* 112:325–30
139. MacGrath, W. S. Jr., VanderNoot, G. W., Gilbreath, R. L., Fisher, H. 1968. *J. Nutr.* 96:461–66
140. Knipprath, W. G., Mead, J. F. 1968. *Lipids* 3:121–28
141. Baranska, J., Wlodawer, P. 1969. *Comp. Biochem. Physiol.* 28:553–70
142. Balnave, D. 1973. *Comp. Biochem.*

143. Harris, P., James, A. T. 1969. *Biochim. Biophys. Acta* 187:13–18
144. Fulco, A. J. 1970. *Biochim. Biophys. Acta* 218:558–60
145. Fulco, A. J. Unpublished
146. Kepler, C. R., Tove, S. B. 1973. *Biochem. Biophys. Res. Commun.* 52:1434–39
147. Appleman, D., Fulco, A. J., Shugarman, P. 1966. *Plant Physiol.* 41:136–42
148. Constantopoulous, G., Bloch, K. 1967. *J. Biol. Chem.* 242:3538–42
149. Newman, D. W. 1972. *Plant Physiol.* 43:300–2
150. Tremolieres, A. 1972. *Phytochemistry* 11:3453–60
151. Bolling, H., El Baya, A. W. 1972. *Getreide, Mehl Brot* 26:207–10
152. Zilkey, B., Canvin, D. T. 1969. *Biochem. Biophys. Res. Commun.* 34:646–53
153. Fulco, A. J. 1965. *Biochim. Biophys. Acta* 106:211–12
154. Budny, J., Sprecher, H. 1971. *Biochim. Biophys. Acta* 239:190–207
155. Sprecher, H. W. 1972. *Fed. Proc.* 31(5):1451–57
156. DeGomezdumm, I. N. T., Peluffo, R. O., Brenner, R. R. 1972. *Lipids* 7:590–92
157. Cook, H. W., Spence, M. W. 1973. *J. Biol. Chem.* 248:1793–96
158. Brenner, R. R., Catala, A. 1971. *Lipids* 6:873–81
159. Raju, P. K., Reiser, R. 1972. *J. Biol. Chem.* 247:3702–3
160. Raju, P. K., Reiser, R. 1972. *Biochim. Biophys. Acta* 280:267–74
161. Actis-Dato, S. M., Catala, A., Brenner, R. R. 1973. *Lipids* 8:1–6
162. Reiser, R. 1951. *Fed. Proc.* 10:236
163. Wright, D. E. 1959. *Nature* 184:875
164. Wright, D. E. 1960. *Nature* 195:546
165. Polan, C. E., McNeill, J. J., Tove, S. B. 1964. *J. Bacteriol.* 88:1056–64
166. Kepler, C. R., Hirons, K. P., McNeill, J. J., Tove, S. B. 1965. *J. Biol. Chem.* 241:1350–54
167. Kepler, C. R., Tove, S. B. 1967. *J. Biol. Chem.* 242:5686–92
168. Kepler, C. R., Tucker, W. P., Tove, S. B. 1970. *J. Biol. Chem.* 245:3612–20
169. Kepler, C. R., Tucker, W. P., Tove, S. B. 1971. *J. Biol. Chem.* 246:2765–71
170. Rosenfeld, I. S., Tove, S. B. 1971. *J. Biol. Chem.* 246:5025–30
171. Yokoyama, M. T., Davis, C. L. 1971. *J. Bacteriol.* 107:519–27
172. Grigor, M. R., Dunckley, G. G. 1973. *Lipids* 8:53–55
173. Wallen, L. L., Benedict, R. G.,

Jackson, R. W. 1962. *Arch. Biochem. Biophys.* 99:249–53
174. Schroepfer, G. J. Jr., Bloch, K. 1963. *J. Am. Chem. Soc.* 85:3310–11
175. Schroepfer, G. J. Jr. 1966. *J. Biol. Chem.* 241:5441–47
176. Niehaus, W. G. Jr., Kisic, A., Torkelson, A., Bednarczyk, D. J., Schroepfer, G. J. Jr. 1970. *J. Biol. Chem.* 245:3790–97
177. Schroepfer, G. J., Niehaus, W. G. Jr., McCloskey, J. A. 1970. *J. Biol. Chem.* 245:3798–801
178. Niehaus, W. G. Jr., Kisic, A., Torkelson, A., Bednarczyk, D. J., Schroepfer, G. J. Jr. 1970. *J. Biol. Chem.* 245:3802–9
179. Kisic, A., Miura, Y., Schroepfer, G. J. Jr. 1971. *Lipids* 6:541–45
180. Mortimer, C. E., Niehaus, W. G. Jr. 1972. *Biochem. Biophys. Res. Commun.* 49:1650–56
181. Wallen, L. L., Davis, E. N., Wu, Y. V., Rohwedder, W. K. 1971. *Lipids* 745–50
182. Thomas, P. J. 1972. *Gastroenterology* 62:430–35
183. Downing, D. T. 1961. *Rev. Pure Appl. Chem.* 11:196–211
184. Pohl, P., Wagner, H. 1972. *Fette Seifen Anstrichm.* 74:541–50
185. Ullrich, V. 1972. *Angew. Chem. Int. Ed. Engl.* 11:701–12
186. Boyd, G. S., Smellie, R. M. S., Eds. 1972. *Biological Hydroxylation Mechanisms.* London: Academic. 250 pp.
187. Mize, C. E., Steinberg, D., Avigan, J., Fales, H. M. 1966. *Biochem. Biophys. Res. Commun.* 25:359–65
188. Steinberg, D. et al 1967. *Science* 156:1740–42
189. Tsai, S., Herndon, J. H. Jr., Uhlendorf, B. W., Fales, H. M., Mize, C. E. 1967. *Biochem. Biophys. Res. Commun.* 28:571–77
190. Herndon, J. H. Jr., Steinberg, D., Uhlendorf, B. W. 1969. *N. Engl. J. Med.* 281:1034–38
191. Mize, C. E. et al 1969. *Biochim. Biophys. Acta* 176:720–39
192. Tsai, S., Avigan, J., Steinberg, D. 1969. *J. Biol. Chem.* 244:2682–92
193. Blass, J. P., Avigan, J., Steinberg, D. 1969. *Biochim. Biophys. Acta* 187:36–41
194. Mize, C. E. et al 1969. *J. Clin. Invest.* 48:1033–40
195. Herndon, J. H. Jr., Steinberg, D., Uhlendorf, B. W., Fales, H. M. 1969. *J. Clin. Invest.* 48:1017–32
196. Tsai, S., Steinberg, D., Avigan, J., Fales, H. M. 1973. *J. Biol. Chem.* 248:1091–97

197. Fulco, A. J., Mead, J. F. 1961. *J. Biol. Chem.* 236:2416–20
198. Mead, J. F., Levis, G. M. 1962. *Biochem. Biophys. Res. Commun.* 9:231–34
199. Mead, J. F., Levis, G. M. 1963. *Biochem. Biophys. Res. Commun.* 11:319–24
200. Levis, G. M., Mead, J. F. 1964. *J. Biol. Chem.* 239:77–80
201. Davies, W. E., Hajra, A. K., Parmar, S. S., Radin, N. S., Mead, J. F. 1966. *J. Lipid Res.* 7:270–76
202. Lippel, K., Mead, J. F. 1968. *Biochim. Biophys. Acta* 152:669–80
203. MacDonald, R. C., Mead, J. F. 1968. *Lipids* 3:275–83
204. Mead, J. F., Hare, R. 1971. *Biochem. Biophys. Res. Commun.* 45:1451–56
205. Mead, J. F., Hare, R. 1973. *J. Am. Oil Chem. Soc.* 50:91A
206. Hoshi, M., Kishimoto, Y. 1973. *J. Biol. Chem.* 248:4123–30
207. Hammarström, S. 1969. *FEBS Lett.* 5:192–95
208. Levis, G. M. 1970. *Biochem. Biophys. Res. Commun.* 38:470–77
209. Ushijima, Y., Nakano, M. 1969. *Biochim. Biophys. Acta* 178:429–33
210. Ferguson, K. A., Conner, R. L., Mallory, F. B., Mallory, C. W. 1972. *Biochim. Biophys. Acta* 270:111–16
211. Vakirtzi-Lemonias, C., Karahalios, C. C., Levis, G. M. 1972. *J. Biochem.* 50:501–6
212. Vesonder, R. F., Stodela, F. H., Rohwedder, W. K., Scott, D. B. 1970. *Can. J. Chem.* 48:1985–86
213. Fulco, A. J. 1967. *J. Biol. Chem.* 242:3608–13
214. Kawanami, J., Kimura, A., Nakagawa, Y., Otsuka, H. 1969. *Chem. Phys. Lipids* 3:29–38
215. Yano, I., Furukawa, Y., Kusunose, M. 1969. *FEBS Lett.* 4:96–98
216. Fensom, A. H., Gray, G. W. 1969. *Biochem. J.* 114:185–96
217. Yano, I., Furukawa, Y., Kusunose, M. 1970. *Biochim. Biophys. Acta* 202:189–91
218. Yano, I., Kurukawa, Y., Kusunose, M. 1970. *Biochim. Biophys. Acta* 210:105–15
219. Humphreys, G. O., Hancock, I. C., Meadow, P. M. 1972. *J. Gen. Microbiol.* 17:221–30
220. Yano, I., Furukawa, Y., Kusunose, M. 1971. *Biochim. Biophys. Acta* 239:513–16
221. Hitchcock, C., Morris, L. J. 1970. *Eur. J. Biochem.* 17:39–42

222. Hitchcock, C., Rose, A. 1971. *Biochem. J.* 125:1155–56
223. Hitchcock, C., Morris, L. J., James, A. T. 1968. *Eur. J. Biochem.* 3:419–22
224. Markovetz, A. J., Stumpf, P. K. 1972. *Lipids* 7:159–64
225. Hitchcock, C., Nichols, B. W. See Ref. 20, 221–23
226. Lu, A. Y. H., Junk, K. W., Coon, M. J. 1969. *J. Biol. Chem.* 244:3714–21
227. Coon, M. J., Strobel, H. W., Autor, A. P., Heidema, J., Duppel, W. See Ref. 186, 45–54
228. Strobel, H. W., Lu, A. Y. H., Heidema, J., Coon, M. J. 1970. *J. Biol. Chem.* 245:4851–54
229. Estabrook, R. W., Hildebrandt, A. G., Baron, J., Netter, K. J., Leibman, K. 1971. *Biochem. Biophys. Res. Commun.* 42:132–39
230. Duppel, W., Lebeault, J., Coon, M. J. 1973. *Eur. J. Biochem.* 36:583–92
231. Gallo, M., Bertrand, J. C., Roche, B., Azoulay, E. 1973. *Biochim. Biophys. Acta* 296:624–38
232. Heinz, E., Tulloch, A. P., Spencer, J. F. T. 1970. *Biochim. Biophys. Acta* 202:49–55
233. Heinz, E., Tulloch, A. P., Spencer, J. F. 1969. *J. Biol. Chem.* 244:882–88
234. Björkhem, I. 1972. *Biochim. Biophys. Acta* 260:178–84
235. Björkhem, I. 1973. *Biochem. Biophys. Res. Commun.* 50:581–87
236. Hamberg, M., Björkhem, I. 1971. *J. Biol. Chem.* 246:7411–16
237. Björkhem, I., Danielsson, H. 1970. *Eur. J. Biochem.* 17:450–59
238. Holzman, J. L., Gram, T. E., Gigon, P. L., Gillette, J. R. 1968. *Biochem. J.* 110:407–12
239. Ellin, A., Jakobsson, S. V., Schenkman, J. B., Orrenius, S. 1972. *Arch. Biochem. Biophys.* 150:64–71
240. Kusunose, M., Ichihara, K., Kusunose, E. 1969. *Biochim. Biophys. Acta* 176: 679–81
241. Ichihara, K., Kusunose, E., Kusunose, M. 1969. *Biochim. Biophys. Acta* 176: 704–12
242. Ichihara, K., Kusunose, E., Kusunose, M. 1969. *Biochim. Biophys. Acta* 176: 713–19
243. Kusunose, E., Ichihara, K., Kusunose, M. 1970. *FEBS Lett.* 11:23–25
244. Ichihara, K., Kusunose, E., Kusunose, M. 1970. *Biochim. Biophys. Acta* 202: 560–62
245. Ichihara, K., Kusunose, E., Kusunose, M. 1971. *Biochim. Biophys. Acta* 239: 178–89
246. Ichihara, K., Kusunose, E., Kusunose, M. 1972. *FEBS Lett.* 20:105–7
247. Kusunose, M., Kusunose, E., Coon, M. J. 1964. *J. Biol. Chem.* 239:1374–80; 2135–39
248. Peterson, J. A., Kusunose, M., Kusunose, E., Coon, M. J. 1967. *J. Biol. Chem.* 242:4334–40
249. Peterson, J. A., Coon, M. J. 1968. *J. Biol. Chem.* 243:329–34
250. Veda, T., Coon, M. J. 1972. *J. Biol. Chem.* 247:5010–16
251. Miura, Y., Fulco, A. J. 1974. *J. Biol. Chem.* In press
252. Kolattukudy, P. E. 1969. *Plant Physiol.* 44:315–17
253. Kolattukudy, P. E., Walton, T. J., Kushwaka, R. P. 1971. *Biochem. Biophys. Res. Commun.* 42:739–44
254. Kolattukudy, P. E. 1972. *Biochem. Biophys. Res. Commun.* 49:1040–1946
255. Kolattukudy, P. E., Walton, T. J. 1972. *Biochemistry* 11:1897–907
256. Kolattukudy, P. E. 1973. *Lipids* 8:90–92
257. Harwood, J. L., Stumpf, P. K. 1972. *Biochem. J.* 126:18
258. Harwood, J. L., Sodja, A., Stumpf, P. K. 1972. *Biochem. J.* 130:1013–18

PHOSPHOGLYCERIDE METABOLISM

H. van den Bosch
Laboratory of Biochemistry, State University of Utrecht, Utrecht, The Netherlands

CONTENTS

INTRODUCTION

Since the elucidation of the general pathways of phospholipid biosynthesis about 15 years ago, largely through the efforts of Kennedy and his co-workers (1), a vast literature on the metabolism of phospholipids has appeared. The increasing awareness of the importance of biological membranes as supporting systems for many vital biological processes certainly had a great impact on phospholipid research. The dynamic character of phosphoglycerides has provoked many interesting questions on the specificity and localization of enzymes involved in the synthesis and degrada-

243

tion of these membrane constituents. Due to limitations of space it is impossible to cover the entire field of lipid metabolism. Fortunately, reviews have appeared in the preceding two volumes of this series (2, 3) and in a book on lipid metabolism (4, 5). The metabolism and function of the membrane phospholipids of *Escherichia coli* have been reviewed (6). Although this review is confined to the most recent progress, more detailed discussions had to be limited to aspects in which the reviewer has a special interest or to areas not covered in previous reviews. For these reasons and for a possibly incomplete knowledge of the literature, I wish to apologize to all investigators whose valuable contributions have not been included in this paper. For a more extensive treatment the reader is referred to a recent review on dynamics of phosphoglycerides (7).

BIOSYNTHESIS OF PHOSPHATIDIC ACID

Acylation of Glycero-3-Phosphate in Bacteria

The specific positioning of saturated and unsaturated fatty acids at the glycerol moiety is a characteristic feature of most phosphoglycerides in biological membranes. Although the physiological significance of this asymmetry is not well understood, much attention has been focused on the step in the biosynthesis in which this specific positioning is introduced. It is now clear that at least two different enzymes are involved in the stepwise acylation of glycero-3-phosphate (GP), glycerophosphate acyltransferase and monoacylglycerophosphate acyltransferase. For bacteria this conclusion was reached using mutants of *E. coli* containing a thermolabile GP acyltransferase and an unaffected monoacyl GP acyltransferase (8). Particulate preparations from the parent strain formed 1-palmitoyl GP with palmitoyl-CoA as acyl donor and 2-acyl GP with palmitoleoyl-, oleoyl-, or *cis*-vaccenyl-CoA as acyl donors. Chemical inactivation of GP acyltransferase from the parent strain as well as heat inactivation studies with the mutant resulted in identical inactivation rates for palmitate or oleate incorporation. Since genetic criteria indicated a single lesion in the mutant, it was concluded that one enzyme conferred the specific acylation of GP with palmitate at position 1 and unsaturated fatty acids at position 2 (9). Presumably the saturated 1-acyl- and unsaturated 2-acyl lyso-phosphatidate can be selectively acylated with unsaturated and saturated fatty acyl chains, respectively, to yield an asymmetric phosphatidate. However, such a selective acylation of lysophosphatidates in *E. coli* has so far only been shown for the 1-acyl isomer. Unsaturated acyl-CoA and acyl-acyl carrier protein (ACP) thioesters reacted much faster than their saturated analogs in the acylation of 1-acyl-GP (10). The formation of 2-acyl GP during the acylation of glycero-3-phosphate suggests that this compound is an intermediate in phosphatidate biosynthesis in vivo. On the other hand, recent in vitro experiments were in line with phosphatidate synthesis proceeding primarily via 1-acyl GP as intermediate regardless of the acyl donor (11). At present it remains unclear whether one or more proteins catalyze the conversion of isomeric lysophosphatidates into phosphatidate.

The in vitro results obtained with the phosphatidate synthesizing enzymes in *E. coli* using both acyl-CoA and acyl-ACP thioesters (10) are compatible with the asymmetric positioning of fatty acids found in the membrane phosphoglycerides of this bacterium. In *Clostridium butyricum* the membrane phospholipids show an unusual nonrandom distribution, e.g. more unsaturated fatty acids at position 1 (12). Goldfine, Ailhaud & Vagelos (13), using a membrane preparation of this bacterium, found that the acylation of glycero-3-phosphate is much lower with acyl-CoA esters than with acyl-ACP esters. Regarding the latter, oleoyl-ACP reacted 4 to 8 times more rapidly than palmitoyl-ACP. When mixtures of oleoyl-ACP and palmitoyl-ACP were incubated (14), carbon-1 of glycerophosphate in phosphatidic acid was preferentially acylated by the unsaturated fatty acid and carbon-2 by the saturated fatty acid, thus mimicking the unusual distribution of fatty acids in the phospholipids of *Cl. butyricum.*

Acylation of Glycero-3-Phosphate in Animal Cells

Phosphatidate biosynthesis in mammalian systems has been studied for the most part in the endoplasmic reticulum (15). Detailed studies on the exact pathway of this process have been severely hampered by the fact that the product of acylation of glycero-3-phosphate was usually found to be phosphatidic acid with no accumulation of the intermediate monoacyl GP (16–20). However, several recent lines of evidence indicate that in mammalian phosphatidate biosynthesis at least two distinct enzymes are involved (17, 21). Yamashita & Numa (22) have succeeded in partially purifying the glycero-3-phosphate acyltransferase from rat liver microsomes by techniques involving Triton X-100 treatment, Sepharose-2B filtration and sucrose density gradient centrifugation. Although only a three- to fourfold purification was achieved and the yield was low (8%), the enzyme preparation appeared to be virtually free of 1-acyl GP acyltransferase. The latter enzyme remained bound to membrane fragments, which were nicely separated from the GP acyltransferase during the gradient centrifugation step (23). The partially purified GP acyltransferase required Ca^{2+} for activity and was stimulated by neutral phosphoglycerides. The product formed by the enzyme was identified as 1-acyl GP, even when the incubation was carried out in the presence of both palmitoyl- and linoleoyl-CoA. Phosphatidate synthesis was only observed after addition of a sucrose gradient-prepared fraction containing the membrane bound 1-acyl GP acyltransferase (23), thus clearly establishing that two distinct enzymes are involved in phosphatidic acid synthesis from GP.

In intact microsomes, saturated and unsaturated fatty acids can be used for phosphatidic acid biosynthesis (20, 22, 24–27), but during purification the GP acyltransferase studied by Numa's group progressively lost its ability to use unsaturated fatty acyl-CoA thioesters. At present it is not known whether the purification procedure applied caused a change in the specificity of one microsomal GP acyltransferase or whether a second enzyme with specificity for unsaturated acyl chains was lost during the purification. The sucrose gradient fractions containing 1-acyl GP acyltransferase effectively used saturated and unsaturated fatty acids for phosphatidate formation (23) with no apparent specificity. This is in accord with several earlier studies using intact microsomes (28–30). Okuyama & Lands (31) have studied the microsomal 1-acyl GP acyltransferase in more detail and found the selectivity of this enzyme to be dependent on the incubation conditions. Thus, when the acylation was carried out at very low 1-acyl GP concentrations (below 10 μM), palmitate and arachidonate tended to be excluded from the 2-position of phosphatidate. It should be noted, however, that the experiments of Yamashita, Hosaka & Numa (23) showing no apparent specificity, were done at 7 μM concentration of lysophosphatidate. Acylation of 2-acyl GP proceeded at rates only 8–23% of that of the 1-acyl isomer for various acyl-CoA thioesters examined, but saturated acyl-CoA gave no higher rates of acyl transfer than unsaturated ones (32). Clearly the selectivities observed during in vitro acylation of glycerophosphate do not simply predict the ultimate distribution of saturated and unsaturated fatty acids in the phosphatidate endproduct. Apparently the result is dependent on environmental factors such as pH (21), concentration of micellar and monomeric forms of acyl-CoA (33), and isomeric monoacyl GP concentration (31). Moreover, various treatments of the microsomes (22, 31) and protein concentrations (34) and whether or not mixtures of acyl-CoA and/or monoacyl GP are used (31) can effect the end result. Accordingly, phosphatidate biosynthesis from GP by microsomes has been reported to occur with little or no selectivity (17, 19, 24, 25) or with an asymmetric positioning of saturated and unsaturated fatty acids similar to that found in microsomal phosphoglycerides (20, 31).

More agreement has been reached on the specificity of mitochondrial acyltransferases. In contrast to microsomes, both whole mitochondria and mitochondrial outer membranes chiefly produce monoacyl GP (35–37). Palmitate, either supplied as palmitoyl-CoA or as palmitoyl-carnitine in the presence of CoA and carnitine palmitoyltransferase, is esterified exclusively in position 1 of the glycerol moiety (38, 39). This preferential use of saturated fatty acids for acylation of glycero-3-phosphate is found for mitochondrial acyltransferases from different rat organs as well as from liver of different species.

The GP acyltransferase from rat liver mitochondria was recently obtained in a solubilized, sixfold purified form (40). This preparation was devoid of lysophosphatidate acyltransferase and produced essentially only 1-acyl GP. As in intact mitochondria, a noticeable preference for saturated fatty acids was found with the following order of relative effectiveness: palmitate, 100%; stearate and myristate, 35%; decanoate, 20%; oleate and linoleate, less than 5%. On the other hand, linoleoyl-CoA competed effectively with palmitoyl-CoA in the acylation of 1-palmitoyl glycerophosphate by rat liver mitochondria (39). Thus the specificities of the mitochondrial acyltransferases even in vitro are consistent with the asymmetrical distribution of fatty acids found in liver glycerolipids.

In a series of elegant papers Åkesson, Elovson & Arvidson (41–43) reported on the in vivo synthesis of phosphatidate after intraportal injection of ^{3}H-palmitate and ^{14}C-linoleate. Even after very short periods following the injection, 90% of the palmitate was recovered at the 1-position in phosphatidate. On the other hand, in all instances examined, 90% of the linoleate was found at position 2, clearly indicating that phosphatidate biosynthesis in vivo occurred with a high degree of specificity during the acylation reactions. The analytical data obtained on the distribution of fatty acids in phosphatidate isolated from whole liver agree very well with such a view (20, 43).

Acylation of Dihydroxyacetonephosphate

The acylation of dihydroxyacetonephosphate (DHAP) and the subsequent reduction of the product to 1-acyl lysophosphatidate as an alternate route for phosphatidic acid biosynthesis was described several years ago by Hajra & Agranoff (44–47). Recent experiments have revealed that acyl-DHAP synthesis is not confined to liver, but occurs in several other rat tissues including kidney, heart, testis, spleen, brain, and adipose tissue (48), as well as in *Saccharomyces carlsbergensis* (49). Acyl-DHAP formation by rat liver mitochondria was 6 times more rapid than in microsomes. In contrast to an earlier report (46), in which the reduction of palmitoyl-DHAP was described as an exclusively mitochondrial process, both mitochondria and microsomes of different rat organs were found to reduce alkyl and acyl derivatives of DHAP (50). Reduction of both derivatives was fairly specific, with rates at least ten to fifty times higher with NADPH than with NADH. Reduction occurred specifically by transfer of the hydrogen from the B side at position 4 of the nicotinamide ring of NADPH.

Several studies have been undertaken to evaluate the relative contributions of the GP and DHAP pathways to the synthesis of phosphatidic acid (51–55). In earlier studies in rat liver slices (51), Mycoplasma (52), and *E. coli* (55) [2-^{3}H]-glycerol and [1,3-^{14}C] glycerol were used. Incorporation of [2-^{3}H] glycerol via the DHAP pathway would result in a loss of ^{3}H label. Consequently, the decrease in the ^{3}H/^{14}C ratio in the newly synthesized glycerolipids would be a measure for the participation of the DHAP pathway. Surprisingly, when compared with the ^{3}H/^{14}C ratio in the glycerol substrate, the glycerolipid exhibited higher ^{3}H/^{14}C ratios, except in *E. coli* where the ratios were equal. Therefore, it was concluded that acylation of DHAP is of minor importance in glycerolipid biosynthesis. Manning & Brindley (53) have found that the increased ^{3}H/^{14}C ratio in glycerolipid is preceded by a severalfold higher increase in the ^{3}H/^{14}C ratio in glycero-3-phosphate. This was ascribed to the observed isotope effect of the glycero-3-phosphate dehydrogenase, which discriminated against [2-^{3}H] glycero-3-phosphate. As pointed out by these authors, the relative contribution of the GP and DHAP pathway to glycerolipid biosynthesis can only be measured from a comparison of ^{3}H/^{14}C ratios in glycero-3-phosphate (not glycerol) and glycerolipid. In doing so it was found that 50–60% of the glycerol incorporated into lipid by rat liver slices proceeded via the DHAP pathway (53). A significant contribution of the DHAP pathway had also been reported by Agranoff & Hajra (54) for mouse liver and Ehrlich ascites tumor cells. To distinguish between both pathways these investigators made use of the specific requirements for ^{3}H-NADH and ^{3}H-NADPH in DHAP and acyl-DHAP reduction, respectively. However, Wykle and others (56) have found that in Ehrlich ascites microsomes NADH is a more effective H donor than NADPH for the reduction of acyl DHAP to acyl GP. The nucleotide specificity for these reactions probably depends on the concentration of the H donors and may differ for various tissues. In view of these facts, caution should be used in evaluation of GP and DHAP pathways based on nucleotide specificity. In addition the method described by Manning & Brindley (53) can be used to establish the relative contributions of both pathways to glycerolipid biosynthesis in vivo.

BIOSYNTHESIS OF PHOSPHATIDYLCHOLINE AND PHOSPHATIDYLETHANOLAMINE

Phosphatidate is further metabolized by either of two pathways (Figure 1). It can react with CTP to give CDP-diglyceride (reaction 2) and thus serve as precursor for anionic phosphoglycerides or it can be dephosphorylated yielding 1,2-diglyceride (reaction 1) which in turn acts as precursor for the major phosphoglycerides from mammalian tissues, namely phosphatidylcholine and phosphatidylethanolamine (reactions 4 and 5, respectively). The latter pathway is not present in bacteria. The available evidence strongly suggests that phosphatidate phosphohydrolase does not display any selectivity with respect to the fatty acid composition of its substrate. This was concluded from an identical distribution of glycerol radioactivity over the molecular species of 1,2-diglycerides and those of the precursor phosphatidate during in vivo (43) as well as in vitro experiments with rat liver slices (25). Likewise, when cholinephosphotransferase and ethanolaminephosphotransferase were studied in vitro, utilization of 1,2-diglycerides proceeded with little or no selectivity (57–59), with the only notable exception of a small preferential use of hexaenoic species by ethanolaminephosphotransferase (59). On the other hand, in vivo experiments had suggested a preference of cholinephosphotransferase for dienoic and tetraenoic diglyceride species and of ethanolaminephosphotransferase for tetraenoic and hexaenoic species (43).

Kanoh & Ohno (60) have employed the reversibility of the reaction

CDP-choline + 1,2-diglyceride ⇄ phosphatidylcholine + CMP

to study the specificity of cholinephosphotransferase. Membrane-bound phosphatidylcholines contained a high percentage of label in dienoic, tetraenoic, or

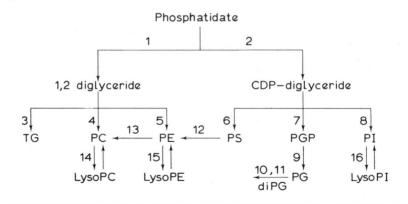

Figure 1   Major reactions in phospholipid biosynthesis. Abbreviations: TG, triglyceride; PC, phosphatidylcholine; PE, phosphatidylethanolamine; PS, phosphatidylserine; PGP, phosphatidylglycerolphosphate; PI, phosphatidylinositol; PG, phosphatidylglycerol.

hexaenoic species depending on whether the label was introduced in vivo via cholinephosphotransferase (reaction 4), acylation of 1-acyl lysophosphatidylcholine (reaction 14), or N-methylation of phosphatidylethanolamine (reaction 13), respectively. No selectivity of cholinephosphotransferase to utilize molecular species of PC for CDP-choline formation was observed. CDP-ethanolamine formation from PE was much more sensitive to inhibition by Ca^{2+} than was CDP-choline production from PC. At present it is unknown whether cholinephosphotransferase is identical to ethanolaminephosphotransferase. Although many properties of the biosynthetic system suggest such identity, the differential effect of heat (61) and acyl-CoA concentration (58) on phosphorylcholine and phosphorylethanolamine incorporation suggest that two enzymes might be involved.

The in vivo rates for PC and PE synthesis were calculated to be 20–25 and 6–8 nmol/min per g of rat liver (62, 63). This difference in synthetic activity agrees well with in vitro results on the relative activities of choline- and ethanolaminephosphotransferase (58, 60). Phosphonate analogs of phosphorylcholine (64) and of phosphorylethanolamine (65) were incorporated into rat liver phospholipids in vivo.

The secretion rate of phosphatidylcholine into bile was reported to be about 3.5 nmol/min per g liver. The secreted product contained about 70% of dier.oic species (66). Trewhella & Collins (67), using a combination of argentation chromatography and countercurrent distribution, have determined the specific radioactivities of individual molecular species of PC and PE in the livers of normal and EFA-deficient rats after injection of [^{32}P]-phosphate. In normal liver 1-stearoyl-2-arachidonoyl-GPC was found to have the slowest rate of turnover. In EFA-deficient livers 1-stearoyl-2-[5,8,11]-eicosatrienoyl-GPC turned over slowest and the arachidonoyl compound showed a faster turnover, probably as an attempt to satisfy the demand for tetraenoic species.

The relative importance of de novo synthesis, methylation of phosphatidylethanolamine, and deacylation-reacylation pathways for the synthesis of the individual molecular species of phosphatidylcholine was evaluated from experiments in which [^{32}P] phosphate, [^{3}H] choline, and [^{14}C] ethanolamine were administered simultaneously to rats (68). The data confirmed earlier observations that 1-palmitoyl-2-oleoyl-GPC and 1-palmitoyl-2-linoleoyl-GPC were mainly formed via CDP-choline. Although several investigators (69–71) concluded that N-methylation of phosphatidylethanolamine is mainly involved in the synthesis of hexaenoic lecithins, Trewhella & Collins (68) concluded from specific activity data that methylation could also be of considerable importance in the synthesis of 1-stearoyl-2-arachidonoyl-GPC. Others have ascribed the synthesis of this species mainly to a deacylation-reacylation cycle (25, 72–74). Dac LeKim, Betzing & Stoffel (75) have found that the methylation of phosphatidyl-N,N-dimethylethanolamine proceeds more rapidly as the degree of unsaturation of the substrate rises. Apparently the structure of the phosphatidyl part is very important for the methyltransferase since the related ceramide phosphoryl-N,N-dimethylethanolamine is not methylated.

The entry of free choline into phosphatidylcholine by base exchange was estimated to be about 1/20 of that incorporated via CDP-choline (62). In contrast to an earlier finding (76), evidence has recently accumulated indicating that choline and ethanolamine exchange in vivo is of minor importance (63, 64, 77–79). Similarly, exchange did not play a significant role in choline incorporation into phosphatidyl-choline by Novikoff hepatoma cells (80). On the other hand, an energy-independent, Ca^{2+}-requiring incorporation of serine, choline, and ethanolamine into their respective phosphoglycerides can easily be shown in a variety of in vitro systems (81–85). In rat liver microsomes choline was incorporated 12 times more rapidly into hexaenoic than into mono- and dienoic species (86). A similar though less pronounced preference was found for ethanolamine exchange (85). Both Bjerve (85, 87) and Kanfer (83), who have studied base exchange in detail, feel that more than one enzyme might be involved. Saito & Kanfer (88) have solubilized a membrane-bound enzyme from rat brain catalyzing a base exchange reaction.

The exchange of L-serine with the ethanolamine moiety of phosphatidyl-ethanolamine thus far appears to be the exclusive pathway of phosphatidylserine synthesis in animal tissues. The in vitro incorporation rate of L-serine can account for the rate of phosphatidylserine synthesis observed in vivo (85). The physiological function of other base exchanges remains to be assessed.

BIOSYNTHESIS OF CDP-DIGLYCERIDE AND ITS UTILIZATION IN THE SYNTHESIS OF ANIONIC PHOSPHOGLYCERIDES

Phosphatidate can be activated to form cytidine diphosphate diglyceride in a process requiring CTP (reaction 2). From in vitro studies, CDP-diglycerides have been implicated as donor of the phosphatidyl units in the synthesis of phosphatidyl-glycerol (reactions 7 and 9), cardiolipin (reactions 10 and 11), and phosphatidylino-sitol (reaction 8) in a variety of organisms as well as in the synthesis of phosphatidyl-serine (reaction 6) in bacteria. Recently, Raetz & Kennedy (89) isolated the cytosine-containing liponucleotide fraction of *E. coli* and showed it to be a mixture of almost equal amounts of CDP- and dCDP-diglycerides. Interestingly, Ter Schegget and others (90) have reported the capacity of rat liver mitochondria to incorporate dCTP into dCDP-diglyceride. The steady state level of cytosine liponucleotides in *E. coli* was estimated at about 0.04% of the total phospholipid content of this bacterium (89). The relative activity of CDP- and dCDP-diglycerides was found to vary greatly with the experimental conditions of the assay. Thus, at 0.3 mM concentrations, this ratio was about 0.6 and 5 for phosphatidylglycerophosphate and phosphatidylserine synthesis respectively. Below 0.1 mM concentrations, however, dCDP-diglyceride was more effective in both reactions. UDP-, ADP-, and GDP-diglycerides showed less than 10% of the CDP-diglyceride effectiveness in reactions 6 and 7 (89).

There is little information on the specificity of cytidyltransferase for the fatty acid composition of the phosphatidate precursors. It has been impossible to ascribe differences that have been found using various phosphatidates (91, 92) to enzyme specificity, rather than to solubility differences. This holds equally well for the different utilization rates of CDP-diglycerides for anionic phospholipid biosynthesis.

Most recent work on the synthesis of polyglycerophosphatides has centered around the mechanism of cardiolipin biosynthesis from phosphatidylglycerol. It is now clear (93–96) that cardiolipin is synthesized in bacterial systems via the reaction

2 phosphatidylglycerol → cardiolipin + glycerol

This was most convincingly demonstrated by Hirschberg & Kennedy (94), who showed that formation of cardiolipin from phosphatidylglycerol (PG) was accompanied by an equimolar release of glycerol. Experiments with cytidine-P-^{32}P-dipalmitin showed that it stimulates the reaction, presumably by a detergent effect, since the liponucleotide did not participate in the reaction as a phosphatidyl donor. Various treatments of cell cultures known to give a relative increase in cardiolipin synthesis in vivo did not give rise to an increased activity of cardiolipin synthetase in E. coli envelopes. The increased amounts of cardiolipin in these cells was ascribed to the lack of an energy requirement for the cardiolipin synthetase (96). Phosphatidyl donation by a preformed phospholipid, either PG or cardiolipin, rather than by a nucleotide intermediate has also been implicated in the synthesis of phosphatidylglucosyl diglyceride in S. faecalis (97).

Cardiolipin synthesis in mammalian mitochondria proceeds via the reaction

phosphatidylglycerol + CDP-diglyceride → cardiolipin + CMP

[2-^{3}H] Phosphatidyl [1'-^{14}C] glycerol was converted in the presence of unlabeled CDP-diglyceride to a diphosphatidylglycerol with retention of its ^{3}H/^{14}C ratio. No significant release of ^{14}C glycerol was observed (95). Stanacev and co-workers (98, 99) have demonstrated that the conversion of membrane-bound PG to cardiolipin is at least ten times higher than that of exogenously supplied PG. In addition, when CDP-diglyceride-2-^{14}C was incubated with membrane-bound phosphatidylglycerol-2-^{3}H the cardiolipin incorporated 0.7 nmol ^{14}C per nmol ^{3}H. The lack of stoichiometry was due to dilution of the ^{14}C-CDP-diglyceride with residual unlabeled liponucleotide from the preparation of membrane-bound PG. Mitochondrial cardiolipin synthetase, in contrast to PG synthesis, has a pronounced requirement for Mg^{2+}, Mn^{2+}, or Co^{2+} (100–102). Acylation of phosphatidylglycerol was catalyzed by a particular fraction of E. coli to yield acylphosphatidylglycerol (103) or bisphosphatidic acid (104). Acylphosphatidylglycerol was isolated from lipid extracts of E. coli (103) and S. typhimurium (105). A related phosphoglyceride, bis-(monoacylglycero) phosphate, was found to accumulate in cases of Niemann-Pick diseases (106) and following administration of a coronary vasodilator (107, 108). In rat liver, bis-(monoacylglycero) phosphate was found greatly enriched in lysosomes where it comprised 7% of the total lipid P (109). It would be of interest to know whether or not this compound is synthesized from phosphatidylglycerol.

Studies on the synthesis of phosphatidylinositol in yeast (110) have indicated that in this organism the lipid is synthesized in a manner similar to that found in animal cells, i.e. from CDP-diglyceride and inositol. Holub & Kuksis (111) provided evidence for an active synthesis of monoenoic and dienoic species by this pathway in rat liver. The specific activity data were consistent with the formation of tetraenoic species by means of a deacylation-reacylation retailoring (reaction 16). Similar findings were reported for brain (112). Experiments to show a preferential acylation of 1-acyl lysophosphatidylinositol with arachidonate, as was found for

1-acyl lysolecithin in liver (73), were unsuccessful due to rapid degradation of the lysocompound after intracerebral injection (112).

FORMATION AND DEGRADATION OF ALKYL- AND ALKENYL-ETHER BONDS

Largely through the efforts of Hajra, Snyder and co-workers, and of Paltauf during the last five years, a rather complete picture of the synthesis of glycerol ether bonds has emerged. This rapidly developing field was recently reviewed comprehensively by Snyder (113), and a recent book edited by Snyder (114) is devoted entirely to many aspects of the chemistry and biology of ether lipids. A new class of ether linkage-containing lipids was recently identified by Snyder's group as hydroxylalkylglycerol (115).

It is now clear (56, 116) that acyl DHAP rather than DHAP is the immediate precursor of the glycerol moiety in alkyl lipids. The replacement of the acyl chain by an alcohol, which is the crucial step in O-alkyl bond formation, is not fully understood. Both the oxygen and the hydrogens at C_1 of hexadecanol are retained in the O-alkyl moiety formed (117, 118). However, any detailed explanation of the mechanism of O-alkyl bond formation may have to account for the exchange of one tritium from C_3 in DHAP as observed in *Tetrahymena pyriformis* (119, 120). Perhaps the reaction proceeds via an enol intermediate. Such an enol form is much more favorable than in DHAP due to stabilization by intramolecular H bonding and could also explain why replacement of the ester by an alcohol does not take place in acyl GP (56, 116). Reduction of O-alkyl DHAP takes place by microsomal reductases that utilize either NADPH or NADH as H donors (56). Acylation of O-alkyl GP yields 1-alkyl-2-acyl GP, which is then dephosphorylated and converted to the alkyl analogs of glycerolipids by reactions 3, 4, and 5 (118, 121–123). An altered route for 1-alkyl-2-acyl GP synthesis, namely phosphorylation of 1-alkylglycerol followed by acylation at C_2, can account for incorporation of alkylglycerols from the diet into ether glycerolipids (124). Such incorporation was known to proceed without cleavage of the ether linkage (125–127).

Experiments in several in vivo and in vitro systems (see 113, 128, 129) using either distinctly labeled hexadecanol and DHAP or doubly labeled O-alkylglycerol consistently revealed identical isotopic ratios in O-alkyl and O-alk-1'-enyl lipids. This suggested that O-alk-1'-enyl ether bonds were formed from alkyl ether bonds without intervening separation of glycerol and alkyl moieties. Definitive proof for such a desaturation reaction came from studies using 1-[9,10-^{3}H$_2$] alkyl-2-acyl [^{14}C]GP, which in the presence of CDP-ethanolamine and NADP$^+$ was converted into alkyl-acyl GPE and alk-1'-enyl-acyl GPE, with complete retention of the isotopic ratio of the precursor (128). These experiments left unanswered whether or not the desaturation reaction took place at the neutral or phospholipid level. Recent studies by Paltauf (130, 131) and others (129, 132, 133) have clearly established 1-O-alkyl-2-acyl GPE as precursor of 1-O-alk-1'-enyl-2-acyl GPE. In intestinal mucosa (130), brain (132), and Fischer sarcomas (133) alkyl GPE is acylated to 1-alkyl-2-acyl GPE prior to conversion into ethanolamineplasmalogen.

The alkyl-acyl GPE desaturase required molecular oxygen, NADH or NADPH, and a heat-labile factor of high molecular weight from the cytosol. Several properties of the desaturase resemble fatty acyl-CoA desaturase (131, 133). In view of the strict specificity of alkyl-acyl GPE desaturase, more than one desaturase may be anticipated. When (1-^{14}C) hexadecanol was administered intracerebrally to myelinating rats, the ethanolamine phosphatides showed major amounts of radioactivity in their constituent alkyl- and alk-1'-enyl moieties, whereas the choline phosphatides did not form labeled alk-1'-enyl moieties, although their alkyl moieties were formed to similar degrees as observed for ethanolamine phosphatides (134).

The cleavage of ether bonds was reviewed in 1972 (see 7, 113). Few additional reports have appeared. The requirement for reduced pteridine as found by Tietz, Lindberg & Kennedy (135) for the oxidative cleavage of alkyl ether bonds by rat liver microsomes was confirmed by Soodsma, Piantadosi & Snyder (136). The oxidative biocleavage of glyceryl ether bonds in *T. pyriformis* did not require pteridines (137). Regardless of the exact mechanism, the results on glycerylether biocleavage were obtained with neutral lipids, and no information is available indicating whether phosphoglycerides possessing alkyl ethers are attacked by a similar enzyme system. Preliminary experiments in the *Tetrahymena* system indicated that the rate of cleavage of the glycerol ether bond in phosphoglycerides, if occurring at all, is much lower than in neutral lipids.

The alkenyl hydrolyzing enzyme reported by Ansell & Spanner (138) degraded 1-alkenyl-2-acyl GPE and 1-alkenyl GPE about equally well. By contrast, a rat liver microsomal enzyme attacked only 1-alkenyl GPC and GPE, but not the intact choline and ethanolamine plasmalogen (139, 140). Yavin & Gatt (141, 142) described an oxygen-dependent cleavage of the alkenyl ether bond in rat brain supernatant. This reaction turned out to be of nonenzymatic nature, but instead was catalyzed by a supernatant component that was identified as ascorbic acid most probably complexed with a reduced metal ion.

PHOSPHOLIPASES

Phospholipase A, hydrolyzing one acyl ester bond in phosphoglycerides, undoubtedly is the most investigated enzyme in phospholipid metabolism. Phospholipase A activity is found in almost every organism or subcellular fraction where it has been sought. Enzymes that remove acyl groups from the 2- and 1-position of the glycerol moiety, designated phospholipase A_2 and A_1, have been described. Of these, phospholipase A_2 is by far the most specific and the best characterized enzyme. In fact, phospholipase A_1, which catalyzes exclusively the hydrolysis of the 1-acyl ester bond in diacylphosphoglycerides and is inactive against any other carboxyl ester bond possessing substrates, has yet to be isolated. Some of these phospholipase A_1 preparations are even more active on 1-acyl lysophosphoglycerides than on the diacyl compounds and thus could also be designated as lysophospholipases. Although this would be consistent with the existence of positional specific lipolytic enzymes, the nomenclature is further complicated by the occurrence of enzymes, phospholipases B, that completely deacylate diacylphosphoglycerides and thus must lack positional specificity. In contrast to phospholipase A_2, the purification of phospholipase A_1, phospholipase B, or lysophospholipase has only in few instances proceeded to the point where specificity studies can be made. The area was recently reviewed (2, 3, 7) and a book was devoted to lipolytic enzymes (143).

Phospholipase A_2

The venom from snakes and bees is a rich source of phospholipase A_2. Accordingly, homogeneous enzymes have been obtained from the snakes *Crotalus adamanteus* (144, 145), *Crotalus atrox* (146, 147), *Naja naja* (148, 149), *Naja nigricollis* (150), *Agkistrodon piscivorus* (151), *Agkistrodon halys blomhoffi* (152), *Laticauda semifasciata* (153), and from the honeybee *Apis mellifica* (154, 155). Distinct enzyme species were isolated from *C. adamanteus* (144), *N. naja, Vipera russellii* (148), *N. nigricollis* (150) and *A. halys blomhoffii* (152). The apparent low activity of phospholipase A in *Bothrops* venom was attributable to the presence of a polypeptide inhibitor in the crude venom (156, 157). With the exception of the *N. naja* and *Apis mellifica* enzymes, the molecular weights of these phospholipases are all about 14,000. The relatively high number of six or seven disulfide bridges appears to be a common feature of these phospholipases (144, 150, 152, 154). In this respect, venom phospholipases resemble the pancreatic phospholipase A_2. The latter enzyme with a molecular weight of 13,800 consists of a single polypeptide chain of 123 amino acids interlinked by six disulfide bridges, whose positions have recently been established (158, 159). From a mechanistic point of view most detailed studies were done with the *C. adamanteus* enzyme by Wells and the pancreatic enzyme by de Haas and co-workers. Using dibutyryl lecithin in the monomeric state, Wells concluded from kinetic data, as well as from inhibition studies with Ba^{2+}, that Ca^{2+} and substrate binding was ordered with Ca^{2+} adding first to the enzyme (160). The kinetic data for pancreatic phospholipase A_2 obtained with micellarly dispersed diheptanoyl lecithin favored an enzymatic mechanism in which the addition of Ca^{2+} and substrate to the enzyme was random (161), although a more detailed re-examination indicated a modest heterotropic cooperativity between metal ions and substrate (162).

Several lines of evidence support the view that the venom and pancreatic enzyme operate by a different catalytic mechanism. The venom phospholipase acts optimally at pH 8.0–8.5 (160), whereas the pancreatic enzyme shows a pronounced optimum between pH 5.5 and 6.0 (161). Alkaline earth cations such as Ca^{2+}, Ba^{2+}, and Sr^{2+} induce spectral perturbations differing for both enzymes. For the venom enzyme these were interpreted as arising primarily from the removal of a positively charged group from the vicinity of a tryptophan residue (163). Ethoxyformic anhydride inactivated the enzyme by acylation of one lysine (164). This lysine is thought to donate the positively charged group responsible for the cation-induced spectral perturbation. The spectral perturbations caused by Ca^{2+} binding to the pancreatic enzyme were tentatively ascribed to effects on a tyrosine and a histidine residue (165). The metal ion binding site consists mainly of carboxyl acid groups with an additional contribution of a more neutrally ionizing residue, presumably histidine. The idea of a histidine being involved in Ca^{2+} binding is supported by inactivation experiments with *p*-bromophenacylbromide. The reaction of this reagent with the active site residue His-53 is extremely retarded in the presence of calcium or substrate analogs. Conversely, when the reagent is first introduced the modified enzyme no longer binds Ca^{2+} or substrate (166).

Another difference between the venom and pancreatic enzymes was observed in monolayer studies carried out by Verger, Mieras & de Haas (167). In contrast to the immediate action of the venom phospholipase A on a monolayer of dinonanoyl lecithin, relatively long induction times were observed with the pancreatic enzyme. The latter could be related to a slow reversible penetration of the enzyme into the monolayer. Induction times with the venom enzyme were observed only at much higher surface pressure (167). The minimal substrate requirements for the pancreatic enzyme were reinvestigated using a whole series of substrate analogs modified systematically in the stereochemical configuration, the susceptible ester bond, the phosphate moiety, the alkyl chains, the glycerol backbone, and the position of the phosphorylcholine moiety (168). Pancreatic phospholipase A occurs in zymogen form (169, 170) and in porcine pancreas is activated by a trypsin-catalyzed removal of a heptapeptide from the N-terminal end of the molecule (169). Comparative studies of the zymogen and the active enzyme indicated that both proteins are able to bind calcium and substrate monomers, but only the active enzyme is able to anchor to micellar diacyl- or monoacyllecithins (162). The working hypothesis—that the calcium and monomeric substrate binding sites pre-exist in the zymogen and that the anchoring site, which enables the active enzyme to attack organized substrates, is formed only after activation of the zymogen—is consistent with the observation that the zymogen exhibits substantial enzymatic activity towards monomeric substrates but not against micellar substrates (171). Yamaguchi, Okawa & Sakaguchi (172) have reported the presence of two thermostable phospholipases in porcine pancreas. The occurrence of phospholipase A isozymes in porcine pancreas was confirmed by van Wezel & de Haas (unpublished observations). Nieuwenhuizen, Steenbergh & de Haas (173) have isolated two precursors of the active enzyme: one containing the complete heptapeptide and one containing only the last three amino acids of the heptapeptide.

Studies by Polonovski and co-workers (174–177) revealed the presence of inactive forms of phospholipase A in erythrocytes and plasma of both humans and rats. The enzymes were activated by trypsin treatment or upon storage of plasma at $0°C$. As discussed by Paysant and others (174), the mechanism of this activation is not yet clear. Upon fractionation of whole rat plasma, the active enzyme and the unactivated form appeared in clearly distinct fractions (176). Plasma membranes isolated from perfused rat liver contained no phospholipase activity, unless treated with a lysate from platelets (178). Activation of this phospholipase occurred by treatment with lysates either before or after solubilization of the latent enzyme from plasma membranes with 1 M NaCl. Numerous papers have been concerned with the occurrence of phospholipase A_2 activities in subcellular membranes (see section on localization), but few studies on the isolation of these proteins have been made. Waite & Sisson (179) have purified the phospholipase A_2 activity of rat liver mitochondria 160-fold, and quite recently Rahman, Cerny & Peraino (198) obtained a nearly homogeneous phospholipase A_2 from rat spleen with a mol wt of 15,000. Scherphof, Scarpa & van

Toorenenbergen (180) and Waite & Sisson (181) studied the effect of various local anesthetics on phospholipases of different origins. The results were interpreted to indicate an interference of the local anesthetics with Ca^{2+} binding to the phospholipases.

Phospholipase A_1

Lipolytic activities that hydrolyze phosphoglycerides at carbon position 1 have been found in many subcellular particles (see section on localization), in postheparin plasma (189), and in bacteria (183, 184). The specificity of these enzymes, provisionally designated phospholipase A_1, has been the subject of many investigations, especially since it was recognized that lipases also exhibit phospholipase A_1 activity (185, 186). Partially purified phospholipase A_1 preparations that hydrolyzed triglycerides either not at all or at much slower rates than phosphoglycerides have been obtained from calf spleen (187), rat, calf, and human brain (188, 189), and guinea pig pancreas (190). However, determination of the relative rates of hydrolysis is complicated by the fact that optimal conditions for the hydrolysis of a given substrate may be far from optimal for other lipids. Thus, Aarsman & van den Bosch (unpublished observations) isolated a pure phospholipase A_1 from beef pancreas that required an anionic detergent such as sodium deoxycholate for activity. Under conditions giving optimal hydrolysis of diacylphosphoglycerides, the activity toward lysophosphoglycerides was completely inhibited. However, in the absence of deoxycholate the enzyme was at least 100 times more active on 1-acyl lyso-phosphoglycerides than on intact phospholipids. A highly purified phospholipase A_1 from B. megaterium spores in the presence of Triton X-100 was about 400 times more active on phosphatidylglycerol than on phosphatidylethanolamine. With sodium taurocholate this difference was only twofold. 1-Acyl lysophosphatidyl-glycerol was hydrolyzed at a rate of 0.6% of that observed with phosphatidyl-glycerol (184). By contrast a 5000-fold purified phospholipase A_1 from E. coli was slightly more active on PE than on PG and hydrolyzed 1-acyl lyso PE twice as rapidly as phosphatidylethanolamine (183). Both bacterial phospholipases A_1 were virtually without activity on triglyceride, at least under the standard incubation conditions used for phospholipids. Doi, Ohki, & Nojima (191) have reported two phospho-lipases A present in E. coli: a heat-stable, detergent-resistant enzyme of particulate nature and a heat-labile, detergent-sensitive enzyme in the soluble fraction. Although the positional specificity of these enzymes was not determined, it is very likely that the detergent-resistant enzyme may be identical to the purified phospholipase A_1 isolated by Scandella & Kornberg (183).

A phospholipase A_1 activity detected in postheparin plasma exhibited many properties similar to postheparin lipase (182, 192), and it was concluded that both activities were due to one enzyme acting on different substrates (193). Recent research using hepatectomized rats has shown that the origin of postheparin phospholipase is mainly the liver, whereas the lipase is derived entirely from extrahepatic tissues, indicating that the two activities must be related to different enzymes (194). The postheparin phospholipase A_1 was purified 7000 times and separated from lipoprotein lipase, but copurified with a monoglyceride esterase hydrolyzing monoglycerides and diglycerides but not triglycerides (195). This enzyme is most likely identical with the phospholipase A_1 of rat liver plasma membranes. Waite & Sisson showed that the latter enzyme could be solubilized by treatment of plasma membranes with heparin (196, 197). The solubilized phospholipase A_1 preparation catalyzed the hydrolysis of monoglyceride, diglyceride, and 1-acyl GPE, in addition to phosphatidylethanolamine. It is apparent that phospholipase A_1 preparations obtained so far do not exhibit the high degree of specificity observed with venom or pancreas phospholipase A_2.

Lysophospholipase

The catabolism of phosphoglycerides in mammalian tissues proceeds mainly via deacylation. Not infrequently the presence of lysophospholipases was inferred from the lack of a stoichiometric accumulation of lysophosphoglycerides during the degradation of labeled phosphoglycerides. In theory, the complete deacylation of diacylphosphoglycerides could be affected by the combined action of phospholipase A_1 and A_2, since most of these enzymes also attack 1-acyl and 2-acyl lysophospholipids respectively. Whether the catabolism of phosphoglycerides proceeds via two phospholipases A or a phospholipase A and a lysophospholipase cannot be ascertained definitively when crude enzyme preparations such as subcellular fractions are used. Further, studies to clarify these points by purification and determination of the substrate specificity of phospholipases and lysophospholipases have not always yielded unambiguous results. Instead, a further complicating factor became apparent in that some enzymes were found to catalyze the deacylation of both diacyl- and monoacylphosphoglycerides, depending on the incubation conditions.

The classical example of such an enzyme, designated phospholipase B, is that of *Penicillium notatum*. This enzyme attacked phosphatidylcholine in the presence of activators that gave a net negative charge to the lecithin micelles (199) or when ultrasonically dispersed substrate was used (200, 201). Recently, Kawasaki & Saito (201) provided evidence that one enzyme with dual activities is involved. During a 2300-fold purification yielding a homogeneous protein on disc electrophoresis, a constant ratio of about 100 was found for lysolecithin to lecithin deacylation. However, Fe^{2+} and Fe^{3+} markedly inhibited lecithin hydrolysis but were without effect on lysolecithin deacylation. A nearly homogeneous lysophospholipase from beef pancreas (202, 203) also showed phospholipase B-type activity when incubated in the presence of deoxycholate. At optimal deoxycholate concentrations for lecithin hydrolysis the lysophospholipase activity was completely inhibited and 2-acyl lysolecithin accumulated. At suboptimal concentrations the reaction proceeded until GPC formation. In accord with these findings both 1-acyl GPC and 2-acyl GPC were found to be deacylated in the absence of deoxycholate. Following heat treatment or inactivation by DFP, a parallel decrease in hydrolysis rate for both isomeric lysolecithins was observed, further supporting the view that one enzyme attacked both acyl ester positions (A. J. Aarsman and H. van den Bosch, unpublished results). Crude lysophospholipase preparations from rat liver cytosol (204) and *Culex pipiens fatigans* microsomes (205) hydrolyzed 1-acyl and 2-acyl lysophosphoglycerides.

Irregular kinetics observed for lysophospholipase activities in subcellular fractions of rat brain were discussed in light of the hypothesis that the enzyme utilized monomers rather than micelles of lysophospholipids (206). In general, studies with crude enzyme preparations have revealed that lysophospholipases do not require Ca^{2+} for activity and that most of them are inhibited by detergents (for discussion see 7, 202, and 206). A purified phospholipase A_2 from *Vipera palestinae* venom hydrolyzed 1-acyl GPC at a rate of 0.2% of that observed for lecithin (207) but required Ca^{2+} for this hydrolysis.

DEACYLATION-REACYLATION CYCLE

During the last decade acyl CoA :lysophosphoglyceride acyltransferase from various sources has been extensively studied. Largely through the efforts of Lands and his colleagues, we have gained much insight into the selectivities of this type of acyltransferases. Hill & Lands (4) have reviewed acyltransferases comprehensively and a more recent review has included paragraphs on the occurrence of acyltransferases and their specificity for the acyl donor as well as for the acyl acceptor (7). It has become clear from the studies of Lands and colleagues that the acyltransferases can distinguish minor differences in the acyl chain of acyl-CoA esters, but no single structural or physical feature of the acyl chain fits all the selectivities observed. Recent results on the selective transfers of *trans*-ethylenic (208) and acetylenic acids (209) corroborate earlier findings (210–212) that the presence of π bonds in fatty acids is an important factor in determining esterification rates at the 2-OH group of glycerol, whereas acyl transfer to the 1 position is much more sensitive to configurational differences.

Colbeau, Vignais & Piette (213) made use of acyltransferases to incorporate spin-labeled, 12-stearic acid nitroxide enzymatically into membrane phospholipids of rat liver microsomes. Acyltransferases from rat liver were able to acylate 1-acyl lyso PG (214), a finding consistent with the conclusion of Jezyk & Hughes (215) that 1-acyl lysophospholipids with a broad range of modifications in the polar headgroup can be accommodated by these enzymes. Modification of the group attached to the phosphate moiety, however, caused marked variations in the rates of acyl transfer. With the exception of short acyl chains, the fatty acyl constituents of the lysolecithin acceptor hardly influenced the rate of acyl transfer, in contrast to the tremendous influence exerted by the fatty acyl constituents in the CoA esters (216, 217). The rate of acyl transfer to the 2 position appeared to be markedly influenced by the type of bond by which the carbon chain is attached to the 1 position. Thus, great variations in acyl transfer to 1-acyl-, 1-alkyl-, and 1-alk-1'-enyl lysophosphoglycerides were found (218, 219). Whether or not acyl transfer is catalyzed by one enzyme or a group of closely related enzymes remains to be resolved.

When labeled lysophosphoglycerides were injected intravenously into rats the label disappeared rapidly from the bloodstream with a concomitant appearance of radioactivity, mainly as the corresponding diacylphosphoglyceride, in various organs such as liver, small intestine, muscle, lung, kidney, and heart (220). In using doubly labeled lysolecithin, the lecithin formed in the various organs was found to have an isotopic ratio similar to that of the injected lysolecithin, indicating a direct acylation by acyltransferases (220–222). 1-Acyl lysolecithin formed from dietary lecithin by pancreatic phospholipase A in the intestinal lumen is absorbed intact (223) and reacylated by acyltransferases in the intestinal mucosa (224–226). Interestingly, the acylation of lysophosphoglycerides by rabbit granulocytes and alveolar macrophages was increased threefold in the presence of ingestible particles (227). Similarly, stimulation of rabbit lymphocytes activated the plasma membrane acyl-transferase several times (228). Furthermore, acyltransferases constitute part of a deacylation-reacylation cycle that is thought to play an important role in the biosynthesis of some prevailing molecular species of phosphoglycerides, e.g. tetraenoic acid species (see sections on biosynthesis).

Lucy (229) has hypothesized that membrane fusion might be facilitated by factors such as lysolecithin that favor a micellar organization rather than a bimolecular arrangement for lipids in membranes. In fact lysolecithin was shown to induce fusion of different cell types (230, 231) and the Ca^{2+}-induced fusion of chicken erythrocytes was thought to be due to the observed formation of lysophospholipids (232). The energy requirement for the virus-induced fusion of Ehrlich ascites cells might be explained in this context by a conversion of micellar to bilayer regions in the membrane by acylation of lysophospholipids, thus facilitating fusion by preventing lysis of the cells (233). On the other hand, virus-induced fusion of baby hamster kidney cells was not accompanied by any gross breakdown of phospholipids (234). Much more work needs to be done before this functioning of a deacylation-reacylation cycle in membrane fusion and, in general, in the regulation of membrane permeability can be seriously considered.

LOCALIZATION OF PHOSPHOLIPID METABOLISM

Bacteria

Although phospholipids constitute essential components of all biological mem-branes, not all membranes of a given organism are equally well equipped for phospholipid biosynthesis.

In *E. coli* and *Salmonella typhimurium* the enzymes involved in the synthesis of phosphatidic acid were localized in the inner cytoplasmic membrane (235, 236). Inhibition of the acyltransferase activity of *E. coli* membranes by sulfhydryl reagents occurred only after partial denaturation of the membranes by mild sonication, heat, or toluene treatment, suggesting that the acyltransferase system is well embedded within the membrane (237). Using an unsaturated fatty acid auxotroph of *E. coli*, Mavis & Vagelos (238) investigated the dependence of glycerophosphate acyltrans-ferase and 1-acyl glycerophosphate acyltransferase activities on the phase properties of the membrane phospholipids. The shape of Arrhenius plots of the former enzymatic activity was identical for membranes containing *cis*-vaccenate, oleate, linoleate, or linolenate as the sole unsaturated fatty acid. On the other hand the effect of temperature as a function of membrane fatty acid composition was different for glycerophosphate acyltransferase and monoacyl glycerophosphate acyltrans-ferase, indicating heterogeneity in the relationship between membranous enzymes and membrane phospholipids.

With the possible exception of CDP-diglyceride: L-serine phosphatidyltransferase all the enzymes for the synthesis of phosphatidic acid, CDP-diglyceride, phosphatidylethanolamine, phosphatidylglycerol, and diphosphatidylglycerol in *E. coli* and *S. typhimurium* showed specific activities in the cytoplasmic membrane which were at least ten times higher than those observed in the cell wall (235, 236). Yet the phospholipid composition of inner and outer membranes was very similar (239). In contrast to the biosynthetic enzymes, phospholipase A_1 showed highest specific activities in the outer membrane fractions (236). A detailed study by Albright, White & Lennarz (303) on the catabolism of phospholipids in *E. coli* revealed the additional presence of a phospholipase A_2 in the cell wall and of a soluble phospholipase which was only active on phosphatidylglycerol (191, 303). The further deacylation of a 2-acyl GPE, i.e. the product of the outer membrane phospholipase A_1, was most rapidly catalyzed by the inner membrane fraction (303). These findings possibly implicate a highly dynamic translocation of phospholipids between both membranes of the cell. It will be of interest to see whether or not the membrane-bound pyrophosphatase catalyzing the hydrolysis of CDP-diglyceride into CMP and phosphatidic acid that was recently reported in *E. coli* (240) is also localized in the outer membrane.

CDP-diglyceride: L-serine phosphatidyltransferase is the only phospholipid biosynthetic enzyme found predominantly in the soluble fraction of the cell, where it appears to be associated with ribosomes (241). In growing cells accumulation of phosphatidylserine was observed only when hydroxylamine was added to the culture medium to inhibit the phosphatidylserine decarboxylase. The newly synthesized phosphatidylserine, however, was recovered in the membrane fraction rather than in the ribosomal fraction. Although other explanations can be provided (241), this again could point to the presence of phospholipid translocation proteins in *E. coli*.

Animal and Other Cells

The subcellular localization of GP acyltransferase in rat liver has been a question of considerable dispute. While many investigators reported data indicating that mitochondria have their own GP acyltransferase system (242–246), others have attributed the observed phosphatidate synthesis by mitochondrial preparations to contamination with microsomes (15, 247). Those authors who agree on the existence of mitochondrial acyltransferases (39, 242–246) localized the enzyme in the outer membrane. In some cases the specific activity of mitochondria or outer membranes was comparable to or even higher than that observed with microsomes, making the question of microsomal contamination highly unlikely. In addition several distinguishing properties of the two acylation systems were reported (39). In contrast to the microsomal acyltransferases, the mitochondrial enzymes were insensitive to sulfhydryl reagents. On the other hand, a thorough study by Davidson & Stanacev (247) indicated that the acyltransferase activities observed in the mitochondrial fractions of guinea pig and rat liver as well as of guinea pig brain, heart, and kidney could be accounted for completely by the presence of endoplasmic reticulum. The same conclusion was reached for rat brain by Possmayer, Meiners & Mudd (248). At present no single explanation can be offered to account for the controversial results. Some factors that may be of importance, such as incubation conditions and extraction procedures, have been adequately discussed in several papers (27, 39, 247, 248). Conflicting results have also been reported on the presence of GP acyltransferases in plasma membranes from rat liver (36, 249).

Glycerophosphate acyltransferase was about equally active in smooth and rough microsomal fractions (36, 250). The same is true for 1-acyl glycerophosphate acyltransferase (251). On the other hand, Yamashita and others (252) localized GP acyltransferase in smooth endoplasmic reticulum and lysophosphatidate acyltransferase in both smooth and rough subfractions of microsomes.

Phosphatidate phosphohydrolase was found primarily localized in the lysosomal fraction (253), but according to Sedgwick and Hübscher it must be regarded as a true constituent of mitochondrial and microsomal fractions as well (254). The enzyme present in the soluble fraction is active on membrane-bound phosphatidates and stimulates microsomal triglyceride and phosphatidylcholine synthesis (255–257). Presumably the lysosomal enzyme does not function in these synthetic pathways.

After some controversial results (242, 253, 258), the question of choline and ethanolamine phosphotransferase distribution in liver seemed to have been settled in favor of an exclusively microsomal localization (245, 246, 259, 260). Smooth and rough endoplasmic reticuli were about equally active (252, 259). When [^{32}P] and CDP-[^{14}C] choline were used, the incorporation of label into phospholipids of mitochondrial preparations could be explained on the basis of microsomal contamination. Less clearcut results were obtained with choline. Soto, Pasquini & Krawiec (261) have reinvestigated the subject and reached the conclusion that these cell organelles as well as nuclei can independently synthesize phosphatidylcholine. Unfortunately, only experiments with free choline as precursor were conducted. The fact that labeled mitochondria did not loose radioactivity upon reincubation with unlabeled choline was taken as evidence for the absence of base exchange. However, it should be noted that Nagley & Hallinan (79) also, for as yet unknown reasons, were unable to show release of radioactivity from choline-labeled microsomal and mitochondrial membranes, despite conditions that were known to be optimal for base exchange in vitro. Therefore, mitochondrial autonomy in the synthesis of phosphatidylcholine should be assessed using labeled precursors, such as [^{32}P] phosphate or [^{14}C] glycerol that unambiguously indicate de novo synthesis.

Cholinephosphotransferase of castor bean endosperm was exclusively recovered in a particulate fraction of microsomal nature with no activity in mitochondria or glyoxysomes (262). In spinach leaves the specific activity of chloroplasts and mitochondrial fractions was 2 and 21%, respectively, of that found for microsomes (263). By contrast, the enzyme in *Tetrahymena pyriformis* was located mainly in the mitochondria (264).

N-Methylation (265) and base exchange (82) have been regarded as microsomal processes. Quite recently ethanolamine exchange was shown to be catalyzed by mitochondrial outer membranes as well (266). The synthesis of phosphatidylserine by exchange of L-serine with the ethanolamine moiety of phosphatidylethanolamine is almost exclusively microsomal in rat liver (267). The phosphatidylserine formed in the microsomes must be readily transported to the mitochondria where phosphatidylserine decarboxylase is located (267) to account for the rapid decarboxylation of serine by intact liver slices. Such a transport may not be involved in the decarboxylation cycle of serine in *T. pyriformis,* as preliminary experiments indicated that the exchange enzyme and the decarboxylase are localized in the same particulate fraction, presumably mitochondria (268).

In the original studies on the synthesis of CDP-diglyceride in guinea pig liver, Carter & Kennedy (91) localized the cytidyltransferase in a microsomal fraction. Hostetler & van den Bosch (100) found a bimodal distribution of this enzyme in rat liver. Although the activity in microsomes was several times higher than that observed in mitochondrial inner membrane, the latter activity was about 10 times higher than could be explained by microsomal contamination. However, these relative distribution patterns were found to vary with the Mg^{2+} concentrations used in the assay. Perhaps such differences can help explain why other investigators have reported a mainly mitochondrial localization of cytidyltransferase (92, 269). A bimodal distribution pattern for CDP-diglyceride synthesis would be consistent with the involvement of this liponucleotide in mitochondrial phosphatidylglycerol and cardiolipin synthesis and in microsomal phosphatidylinositol synthesis. A bimodal distribution was also found in various plants (270, 271) and the mitochondrial site of synthesis was reported to be the inner membrane (272). In yeast, cytidyltransferase seems to be mainly mitochondrial (273).

The biosynthesis of phosphatidylglycerol occurs throughout subcellular organelles. However, the main locus appears to be the mitochondrion (100, 267, 274), especially the inner membrane (100), although significant synthetic activity is also found in plasma membranes (275), rough and smooth microsomes, and Golgi membranes (259). The multiplex occurrence of phosphatidylglycerol synthetase activity in liver is difficult to understand in view of the minute amounts of this phospholipid present in this organ. Interestingly, cardiolipin synthesis in liver was found exclusively at its major site of occurrence, the inner mitochondrial membrane (100, 276). The unavailability of cardiolipin in mitochondria for binding with anti-cardiolipin antibody suggests that the polar heads of these molecules are buried within the structure of the membrane (277).

Apart from the de novo synthesis of phosphoglycerides, an extensive reutilization of lysophosphoglycerides by acyltransferases can occur. Acyltransferases in rat liver are most abundant in the endoplasmic reticulum (15), but subsequent studies revealed the presence of these enzymes in mitochondrial outer membranes (242, 245, 278) and plasma membranes (249, 279, 280).

Numerous reports have dealt with the subcellular localization of phospholipases in mammalian tissues, especially rat liver. Because of difficulties encountered in obtaining highly purified subcellular fractions and differences in the incubation conditions used by various investigators, some controversial results have been obtained. The distribution of phospholipase A in the rat hepatocyte was recently reinvestigated and discussed by Nachbaur, Colbeau & Vignais (281). It seems justified now to conclude that rat liver mitochondria contain a Ca^{2+}-dependent phospholipase A_2 with an alkaline pH optimum (179, 282). It is localized mainly in the outer membrane (281, 283, 284), but significant activity is also found in the inner membrane (281). The soluble fraction of lysosomes contains both acidic phospholipase A_1 and A_2 activities that are inhibited rather than stimulated by Ca^{2+} (281, 284–288). Although some investigators concluded from the lack of accumulation of lysophosphoglycerides that one enzyme catalyzed the complete deacylation of phosphoglycerides (285, 286), others have concluded that two phospholipases with different positional specificities were present (281, 287, 288). Further clarification of these problems must await purification of these enzymatic activities. It is not unlikely that some activities may be associated with acidic lysosomal lipases or esterases.

There appears to be agreement on the presence of an alkaline phospholipase A_1 in rat liver microsomes (281, 282) and plasma membranes (281, 289), although the latter cell organelle contained an additional phospholipase A_2 (275, 290). The relative activities of phospholipase A_1 and A_2, both requiring Ca^{2+}, in microsomes and plasma membranes were found to be dependent on the substrate used. In both subcellular fractions phospholipase A_2 predominated with phosphatidylglycerol as substrate, whereas with phosphatidylethanolamine phospholipase A_1 was more active (290). For as yet unknown reasons the latter result does not agree with that of van Golde, Fleischer & Fleischer (259), who found the phospholipase A_2 of plasma membranes to be five times more active than phospholipase A_1 even when phosphatidylethanolamine was used as substrate. These authors showed purified Golgi membranes to contain both phospholipase A_1 and A_2, with phospholipase A_1 predominating.

With the exception of the mitochondrial phospholipase A_2 (179), none of the abovementioned phospholipases have been purified to any extent. Designating them as phospholipase A_1 or A_2, therefore, does not imply that specific enzymes are involved but indicates rather that hydrolysis of phosphoglycerides took place at the 1- or 2-acyl ester bond. At present the physiological role of these enzymes is not clear. Almost certainly the phospholipases in microsomal, mitochondrial outer and plasma membranes play a role in the dynamic turnover of fatty acids in membrane phosphoglycerides, since acyltransferases have also been reported to be present in these membranes. However, it seems appropriate to mention here that Parks & Thompson (291), in studying the in vivo labeling of phospholipids from ^{3}H-glycerol and ^{14}C-fatty acids in guinea pig liver, have recently reached the conclusion that any acyl exchange in mitochondria appears to be small when measured against the background of bulk transfer of newly synthesized phospholipid from the endoplasmic reticulum to mitochondria. The lysosomal phospholipases A function possibly in the digestion of lipids from particles taken up by the lysosomes during endocytosis or autophagy. In this respect, it is interesting to emphasize that the acidic phospholipases, unlike the other membrane-bound phospholipases, are found in the soluble fraction of lysosomes. Acidic phospholipases have been found in lysosomes of various tissues, including adrenal medulla (292), spleen (293), alveolar macrophages (294, 295), and human blood platelets (296).

Most of the lysophospholipase activity in rat liver was found in the cytosol (204, 297), but membrane-bound activities were reported for microsomes from rat liver (298), rat brain (206), and bovine adrenal medulla (299), and for particulate fractions from *Dictyostelium discoideum* (300) and *Mycoplasma laidlawii* (301).

PHOSPHOLIPID EXCHANGE

As discussed in the section on localization, several subcellular membranes lack the capacity of de novo synthesis of their main phosphoglycerides. Apparently, for their phospholipid constituents such membranes depend on the endoplasmic reticulum, most likely via an exchange of intact phospholipid. Such an exchange was demonstrated in vitro to occur between microsomes and mitochondria from rat liver (246, 302, 304), from plants (305), and from guinea pig brain (306), as well as between artificial lipid structures (307–309).

Evidence was provided indicating that the outer mitochondrial membrane participates in the exchange of phospholipids between the endoplasmic reticulum and the inner mitochondrial membrane (246, 310, 311). Phospholipid exchange and its relation to phospholipid turnover was recently discussed by Dawson (312).

Phospholipid exchange appears to be stimulated by soluble proteins. Partially purified protein factors were obtained from beef heart (308, 313) and beef liver (314). The protein from beef liver was recently obtained in homogeneous form as judged by SDS-polyacrylamide gel electrophoresis, immunoelectrophoresis, isoelectric focusing, and N-terminal amino acid analysis (315). This protein has a mol wt of 22,000, contains one disulfide bridge, and has an isoelectric point of 5.8. Although the crude pH 5.1 supernatant of beef liver stimulates the transfer of phosphatidylcholine, phosphatidylinositol, and phosphatidylethanolamine between rat liver microsomes and mitochondria, the 2680-fold purified exchange protein was specific for phosphatidylcholine (315). Results of immunological experiments suggest that specific proteins might exist for each class of phospholipid. Rabbit antiserum prepared against the purified phosphatidylcholine exchange protein completely inhibited this protein. When the antiserum was added in excess to the pH 5.1 supernatants of beef liver and brain, the exchange activities for PE and PI were unaffected, whereas PC exchange was partially inhibited (316). This indicates that, in addition to the PC-exchange protein against which the antibody is active, other activities for PC, PE, and PI exchange are present in the crude supernatants. The isolation of two protein fractions with PC-exchange activity from both beef liver (314) and beef heart (308) and of a protein fraction with a high degree of specificity for PI exchange from beef brain (317) is consistent with the multiple occurrence and specificities of exchange proteins.

The mode of action of these exchange proteins remains to be established. Activation of the $^{32}P_i$-ATP exchange in PC-deficient phosphorylating particles in the presence of PC liposomes and phospholipid exchange protein indicates that net transfer of PC can take place by the exchange protein (318). Recent experiments have shown the transfer of phospholipid molecules between two separated monolayers on one subphase containing the highly purified PC-exchange protein. Redistribution studies between a labeled phosphatidylcholine monolayer and varying amounts of protein indicated the binding of 1 mole of PC per mole of protein (319). These findings support the view that this exchange protein functions as a phospholipid carrier.

REGULATORY ASPECTS OF PHOSPHOLIPID METABOLISM

Bacteria

Although many suggestions have appeared in the literature regarding factors that may be of some importance in the control of lipid metabolism, very little is known about the mechanism by which phosphoglyceride metabolism is regulated. Glycerophosphate acylation in *E. coli* was inhibited 25% by CTP and 60% by ATP (320) while other nucleoside triphosphates had no effect. Noncompetitive inhibition of GP acyltransferases was also found with *cis*-9,10-methylenehexadecanoate (321). Although there was a tenfold increase of this fatty acid in the cells at stationary phase as compared to cells from logarithmic phase, it is difficult to prove that inhibition of GP acyltransferase by this fatty acid or by ATP is the means of regulating cellular lipid synthesis. Evidence that the synthesis of lipid is coupled to the synthesis of other macromolecules was provided by Glaser and others (322). When mutants with a temperature-sensitive GP acyltransferase were shifted to the restricted temperature, not only phospholipid synthesis but also DNA, RNA, and protein biosynthesis ceased immediately. Conversely, when wild-type or stringent cells were starved for a required amino acid, not only protein and stable RNA but also phospholipid synthesis ceased (323–325). Upon amino acid deprivation relaxed cells stop growth and protein synthesis but continue to make RNA at nearly normal rates. Such cells synthesize phospholipids at about half the normal rate observed in relaxed or stringent cells in the presence of the required amino acid and exhibit the unusual feature of turnover of phosphatidylethanolamine. Thus, synthesis and breakdown of phosphatidylethanolamine in cells deprived of a required amino acid appear to be influenced by a lesion in the ribonucleic acid control gene in the same manner as in RNA synthesis. On the other hand, Crowfoot, Esfahani & Wakil (326) have isolated an unsaturated fatty acid auxotroph which showed a stringent control of RNA accumulation but a relaxed response of lipid synthesis on amino acid deprivation. This mutant behaved in an unusual manner in that membrane phospholipids turned over continuously (327) and this turnover was stimulated by starvation for required amino acids. The precise sequence of events resulting in the stringent and relaxed control phenomena is unknown.

The phospholipid composition of *E. coli* is normally invariant under a variety of growth conditions (6). A temperature-sensitive mutant was found to have a twofold increase in the ratio of phosphatidylglycerol plus cardiolipin to phosphatidylethanolamine when grown at the restrictive temperature of 40°C. The altered phospholipid composition could be explained by an increased activity of the enzyme catalyzing the synthesis of phosphatidylglycerophosphate (328). Ohki (329) has confirmed the stability of phosphatidylethanolamine in growing *E. coli* cells. The turnover of phosphatidylglycerol during synchronous growth was found to take place in a stepwise manner at a specific cell phase. Van Golde, Schulman & Kennedy (330) have demonstrated that the loss of radioactivity from the lipid fractions of cells grown in the presence of $[2\text{-}^3H]$-glycerol-3-$[^{32}P]$ phosphate is accompanied by its appearance in oligosaccharides. This novel type of cell constituent contained equimolar amounts of glycerol and phosphate, probably derived as intact glycero-phosphate from phosphatidylglycerol or cardiolipin.

Several temperature-sensitive mutants deficient in the synthesis of *cis*-vaccenic acid (331), saturated and unsaturated fatty acids (332), phosphatidylserine (333), or phosphatidylethanolamine (334) may be valuable tools for studying aspects of regulation of bacterial phospholipid synthesis and membrane formation.

Animal Cells

Sanchez, Nicholls & Brindley (335) have concluded that in the presence of optimal substrate concentration the activity of glycero-3-phosphate acyltransferase rather than palmitoyl-CoA synthetase in isolated rat liver mitochondria is rate limiting in phosphatidate biosynthesis. However, in microsomal fractions from guinea pig intestinal mucosa and rat liver, the acyl-CoA synthetase may become rate limiting (336, 337). Tzur, Tal & Shapiro (338) have suggested a regulatory role for glycero-3-phosphate in fatty acid esterification. The level of glycero-3-phosphate in rat liver (about 0.9 μmol/g liver) was found to be not in large excess of the optimal requirements for acylation. Reduction of these levels by starvation, epinephrine injection, and ethanol ingestion could have a considerable effect on the rate of phosphatidate formation. This certainly would have to be considered at the much lower GP concentration of 0.16 μmol/g liver reported by Guynn, Veloso & Veech (339). On the other hand these authors found only minor differences in GP concentration under a variety of dietary conditions. In addition Bøhmer (340) obtained evidence indicating that tissue levels of GP change too slowly to be of importance in the regulation. Both groups found the levels of activated fatty acids to vary much more rapidly. Aas & Daae (341) found that the capacity for glycero-3-phosphate acylation in liver varied with the nutritional state of the animal. Thus, when fatty acid was the main energy source, glycerophosphate acylation (2 nmol/min/mg) was lower than upon glucose refeeding, and in controls (3 nmol/min/mg). Higgins & Barrnett (250) have reported a twofold increase in GP-acyltransferase in smooth microsomes 12 hr after phenobarbital treatment. On the other hand, Eriksson & Dallner (342) concluded that the main reason for the increase of phosphoglycerides in rat liver microsomes after phenobarbital treatment was the decreased breakdown rather than an increased rate of synthesis.

Possmayer & Mudd (248, 343) have studied the effect of various nucleotides on the acylation of GP in rat brain. Inhibition was found only for the cytidine nucleotides, particularly for CTP. Differences between the cytidine nucleotides concentrations effective in vitro and the in vivo concentrations found would seem to preclude a regulatory role for these compounds in phosphatidate formation. In addition, it has been impossible to demonstrate consistent effects of CTP on the incorporation of GP into lipids by guinea pig (344) or rat liver (248) microsomes. Mangiapane, Lloyd-Davies & Brindley (345) have measured palmitoyl-CoA synthetase, glycerophosphate acyltransferase, diglyceride acyltransferase, and phosphatidate phosphohydrolase in rat liver after subtotal hepatectomy. Only the last enzyme showed a significant increase in specific activity, which was prevented by actinomycin D injection, suggesting an increased synthesis of this enzyme. The first three enzymes seem to have sufficient reserve activity to cope with the large increase in triglyceride synthesis occurring in regenerating liver. Likewise, the increased synthesis of triglyceride upon refeeding starved rats with a high-carbohydrate diet (346) was accompanied by a two- to threefold increase in phosphatidate phosphohydrolase activity (347). A small contribution of an increased diglyceride acyltransferase activity to the augmented triglyceride synthesis in these animals cannot be excluded (348).

The decrease in phosphatidylcholine content of microsomes upon sucrose feeding has been attributed to a rapidly influenced CDP-choline pathway for lecithin synthesis (349). The activity of cholinephosphotransferase in rats fed a high-carbohydrate diet was not found changed, however, when compared with controls (348). Therefore, increased triglyceride synthesis in such animals most likely cannot be explained by a decreased demand of cholinephosphotransferase for 1,2-diglycerides, which would leave higher diglyceride concentration available for triglyceride formation. Diglyceride occupies a branch point in glycerolipid biosynthesis. Virtually nothing is known about the factors that control whether diglycerides are acylated to triglycerides or are converted to phosphoglycerides via the cytidine nucleotide pathways.

Young & Lynen (350) have found that the two biosynthetic enzymes utilizing diglycerides as substrate were altered to different degrees in animals treated with different agents. This lends support to the concept that the activity of these enzymes is important in regulating the relative amounts of diglyceride incorporated into triglyceride and lecithin. As might be expected for studies using water-insoluble substrates, specific activity data for diglyceride acyltransferase and cholinephosphotransferase showed considerable variations, ranging from the former being several times more active than the latter to the reverse situation (245, 350). In addition, such studies were done at saturating diglyceride concentrations, which are most likely far from the in vivo concentrations. However, in microsomes of hamster intestine, triglyceride synthesis was 4–5 times more active than lecithin synthesis regardless of whether aqueous dispersions of diglyceride or microsomal-bound diglycerides were used as substrate (226). The techniques using membrane-bound substrates that are currently being developed are likely to give valuable information on the ultimate fate of intermediates at the branchpoints of phosphoglyceride metabolism. Such techniques also might provide insight into the fundamental questions as to whether or not different diglyceride pools exist and if newly synthesized diglycerides freely mix with pre-existing pools before being converted to triglycerides and phosphoglycerides. Results reported by Johnston and others (371) indicate that the diglycerides synthesized from 2-monoglyceride and via the glycero-3-phosphate pathway in intestinal microsomes did not equilibrate. Only diglycerides synthesized via the latter pathway acted as precursors for phosphatidylcholine synthesis.

Bjørnstad & Bremer (351) and Treble and others (76) have provided evidence for a rapid equilibrium in rat liver between the choline moieties of CDP-choline and phosphatidylcholine. Does the reversibility of cholinephosphotransferase have some bearing on the regulation of phosphatidylcholine synthesis? Kanoh & Ohno (60) measured for the backward reaction (i.e. CDP-choline formation) a rate of 4 nmol/min/mg microsomal protein under optimal conditions. The K_m for CMP amounted to 0.19 mM indicating that the reaction in vivo proceeds far below its half-maximal rate since the CMP concentration in liver is estimated at 5–8 μM (352). Although the backward reaction showed product inhibition with a K_i for CDP-choline of 1.0 mM, the in vivo concentration of about 0.05–0.10 μmol CDP-choline/g liver (62, 352–354) would seem too low for effective inhibition. On the other hand, this concentration is close to the reported (350, 355) K_m value of 0.13 mM for CDP-choline in lecithin synthesis, which proceeds in vitro at rates of 2.5–8 nmol/min/mg (119, 166). The reversed reaction catalyzed by ethanolaminephosphotransferase (K_m for CMP $= 0.14$ mM) was likewise inhibited by CDP-ethanolamine with a K_i of 0.05 mM. This is in the range of the in vivo concentrations of 0.03–0.05 μmol/g liver for CDP-ethanolamine (63, 354, 356). These data would seem to indicate that under in vivo conditions cholinephosphotransferase is relatively more involved in the reversed reaction than is ethanolaminephosphotransferase.

In microsomal preparations of larvae of *Artemia salina* (357–358) the steady state of the reaction catalyzed by cholinephosphotransferase at low concentrations of CMP (below 1 μM) appeared to favor formation of CDP-choline. Under optimal conditions for each reaction the rate of CDP-choline synthesis from phosphatidylcholine was about twice that of phosphatidylcholine synthesis from CDP-choline. The rates were similar at low concentrations (2.2 μM) of CDP-choline and CMP. Changes in the concentrations of these nucleotides during development could influence the extent of the reversibility of phosphatidylcholine synthesis.

Weinhold (359) and Weinhold & Sanders (360) have observed a general increase in the synthesis of phosphatidylcholine, -ethanolamine, -serine, and -inositol during the late prenatal and early postnatal development of rat liver. The development patterns for each pathway, including N-methylation of phosphatidylethanolamine, and the relationships between fetal and adult activity of these pathways were different. For example, the CDP-choline pathway was near adult activity in fetal rat liver and, beginning two days before term, it increased until twice the adult value was reached in a three-day-old neonatal rat. Phosphatidylethanolamine synthesis in fetuses was only about 30% of adult activity and this began to increase 2–3 days before birth until adult values were reached. Perhaps this is an indication of two different enzymes. Similar patterns were observed for cholinephosphotransferase activity during development of rabbit liver (361) and chicken brain (362). Farrell & Zachman (363) have investigated the hypothesis that deficiency in the CDP-choline pathway in premature animals may be related to the respiratory distress syndrome. Three days after injection with corticosteroids rabbit fetuses showed a 1.4-fold increase in cholinephosphotransferase activity, whereas the N-methylation pathway was unchanged. Studies with actinomycin D and cycloheximide were consistent with the view that the corticosteroid effect was mediated on the level of RNA translation. Although the increase in cholinephosphotransferase activity would seem low at first sight, a statistically significant elevation in lung lecithin from 70 to 96 mg/g dry lung was measured. Perhaps corticosteroids may be beneficial as therapy in the respiratory distress syndrome.

The rather constant amount of phosphoglycerides in a given cellular or subcellular membrane and their continuous turnover suggest that both anabolic and catabolic processes regulate the phospholipid content and composition. Although only little is known about possible regulatory factors in anabolism, our understanding of the regulation of catabolism is even more rudimentary. The presence of masked phospholipase A activities found in certain membranes, e.g. erythrocytes and rat liver plasma membrane, that become active in vitro on certain treatments may be of some importance in this respect (174, 178). An interesting mechanism for the control of phospholipase A_1 activity in *B. subtilis* was advanced by Kent & Lennarz (364). These authors isolated an osmotically fragile mutant that yielded protoplast membranes containing only 50–70% of the phospholipid content of wild-type protoplasts as a result of an active phospholipase A_1 in the mutant. Wild-type cells were found to contain a heat-stable protein that caused a potent inhibition of the mutant phospholipase A_1. Presumably the phospholipase A_1 of the wild-type cells could be masked by this inhibitor, thereby acting as a regulator of the phospholipase activity.

The physiological role of phospholipase in bacteria is not well understood. An interesting discussion about the possible physiological role of the membrane-associated phospholipase A_1 was included in the paper of Scandella & Kornberg (183). Apparently, the enzyme is not involved in a deacylation-reacylation cycle in growing cells, since fatty acid substitution in cell cultures shifted to lower temperatures was found to result from replacement of the intact phosphoglycerides by de novo synthesis (327, 365). The enzyme could be responsible for the phospholipid breakdown observed following phage infection. However, by an ingenious technique, Ohki, Doi & Nojima (366) have isolated a mutant of *E. coli* deficient for detergent-resistant phospholipase A, which enzyme presumably is identical to the membrane-bound phospholipase A_1. Physiological studies with this mutant, including determination of growth rate, phospholipid turnover, and susceptibility to various phages, indicated that detergent-resistant phospholipase is not essential for any of these processes. Likewise, the phospholipase A_1 in sporulating *B. megaterium* was not found essential for germination because it was destroyed by heating, which does not affect the viability of the spore (184).

A particularly interesting aspect of tissue phospholipases A_2 is their possible regulatory role in the release of polyunsaturated fatty acids from phosphoglycerides for prostaglandin biosynthesis (367). In this respect it is worth mentioning that the increased synthesis of prostaglandins in the thyroid by the action of thyroid stimulating hormone (TSH) was found to be preceded by a stimulation of endogenous phospholipase A_2 which released arachidonate from phosphatidyl-inositol (368). Since TSH increases the intracellular cyclic AMP levels, the enhanced phospholipase A_2 activity may be due to stimulation by cyclic AMP as was observed for the phospholipase activities in homogenates of rat lung (369) and epididymal fat cells (370).

This section on regulatory aspects has reviewed the early papers in the research area of lipid metabolism that has just begun to develop. Hopefully, now that the pathways of phospholipid biosynthesis have been rather well established, research workers will be attracted to further explore the regulation of phosphoglyceride metabolism. Unfortunately, this task will be hampered severely by the insolubility of phosphoglycerides in aqueous media and the membrane-bound nature of most of the enzymes involved in their metabolism.

ACKNOWLEDGMENTS

I wish to thank Dr. L. L. M. van Deenen and Dr. S. A. Singal for critically reading the manuscript and for providing many useful suggestions.

Literature Cited

1. Kennedy, E. P. 1961. *Fed. Proc.* 20: 934-40
2. McMurray, W. C., Magee, W. L. 1972. *Ann. Rev. Biochem.* 41: 129-60
3. Gatt, S., Barenholz, Y. 1973. *Ann. Rev. Biochem.* 42: 61-90
4. Hill, E. E., Lands, W. E. M. 1970. *Lipid Metabolism,* ed. S. J. Wakil, 185-279. New York: Academic. 613 pp.
5. Lennarz, W. J. See Ref. 4, 155-84
6. Cronan, J. E., Vagelos, P. R. 1972. *Biochim. Biophys. Acta* 265: 25-60
7. Van den Bosch, H., Van Golde, L. M. G., Van Deenen, L. L. M. 1972. *Rev. Physiol.* 66: 13-145
8. Cronan, J. E., Ray, T. K., Vagelos, P. R. 1970. *Proc. Nat. Acad. Sci. USA* 65: 737-44
9. Ray, T. K., Cronan, J. E., Mavis, R. D., Vagelos, P. R. 1970. *J. Biol. Chem.* 245: 6442-48
10. Van den Bosch, H., Vagelos, P. R. 1970. *Biochim. Biophys. Acta* 218: 233-48
11. Okuyama, H., Wakil, S. J. 1973. *J. Biol. Chem.* 248: 5197-5205
12. Hildebrand, J. G., Law, J. H. 1964. *Biochemistry* 3: 1304-8
13. Goldfine, H., Ailhaud, G. P., Vagelos, P. R. 1967. *J. Biol. Chem.* 242: 4466-75
14. Goldfine, H., Ailhaud, G. P. 1971. *Biochem. Biophys. Res. Commun.* 45: 1127-33
15. Eibl, H., Hill, E. E., Lands, W. E. M. 1969. *Eur. J. Biochem.* 9: 250-58
16. Kornberg, A., Pricer, W. E. 1953. *J. Biol. Chem.* 204: 345-57
17. Lands, W. E. M., Hart, P. 1964. *J. Lipid Res.* 5: 81-87
18. Stoffel, W., de Tomas, M. E., Schiefer, H. G. 1967. *Z. Physiol. Chem.* 348: 882-90
19. Abou-Issa, H. M., Cleland, W. W. 1969. *Biochim. Biophys. Acta* 176: 692-98
20. Possmayer, F., Scherphof, G. L., Dubbelman, T. M. A. R., Van Golde, L. M. G., Van Deenen, L. L. M. 1969. *Biochim. Biophys. Acta* 176: 95-110
21. Lamb, R. G., Fallon, H. J. 1970. *J. Biol. Chem.* 245: 3075-83
22. Yamashita, S., Numa, S. 1972. *Eur. J. Biochem.* 31: 565-73
23. Yamashita, S., Hosaka, K., Numa, S. 1972. *Proc. Nat. Acad. Sci. USA* 69: 3490-92
24. Husbands, D. R., Lands, W. E. M. 1970. *Biochim. Biophys. Acta* 202: 129-40
25. Hill, E. E., Husbands, D. R., Lands, W. E. M. 1968. *J. Biol. Chem.* 243: 4440-51
26. Sanchez de Jimenez, E., Cleland, W. W. 1969. *Biochim. Biophys. Acta* 176: 685-91
27. Daae, L. N. W., 1973. *Biochim. Biophys. Acta* 306: 186-93
28. Stoffel, W., Schiefer, H. G., Wolf, G. D. 1966. *Z. Physiol. Chem.* 347: 102-17
29. Hill, E. E., Lands, W. E. M. 1968. *Biochim. Biophys. Acta* 152: 645-48
30. Barden, R. E., Cleland, W. W. 1969. *J. Biol. Chem.* 244: 3677-84
31. Okuyama, H., Lands, W. E. M. 1972. *J. Biol. Chem.* 247: 1414-23
32. Okuyama, H., Eibl, H., Lands, W. E. M. 1971. *Biochim. Biophys. Acta* 248: 263-73
33. Zahler, W. L., Cleland, W. W. 1969. *Biochim. Biophys. Acta* 176: 699-703
34. Lands, W. E. M., Hart, P. 1965. *J. Biol. Chem.* 240: 1905-11
35. Daae, L. N. W., Bremer, J. 1970. *Biochim. Biophys. Acta* 210: 92-104
36. Nachbaur, J., Colbeau, A., Vignais, P. M. 1971. *C. R. Acad. Sci. Paris D* 272: 1015-18
37. Daae, L. N. W. 1972. *FEBS Lett.* 27: 46-48
38. Daae, L. N. W. 1972. *Biochim. Biophys. Acta* 270: 23-31
39. Monroy, G., Rola, F. H., Pullman, M. E. 1972. *J. Biol. Chem.* 247: 6884-94

40. Monroy, G., Chroboczek Kelker, H., Pullman, M. E. 1973. *J. Biol. Chem.* 248:2845–52
41. Åkesson, B., 1970. *Biochim. Biophys. Acta* 218:57–70
42. Åkesson, B., Elovson, J., Arvidson, G. 1970. *Biochim. Biophys. Acta* 218:44–56
43. Åkesson, B., Elovson, J., Arvidson, G. 1970. *Biochim. Biophys. Acta* 210:15–27
44. Hajra, A. K. 1968. *Biochem. Biophys. Res. Commun.* 33:929–35
45. Hajra, A. K. 1968. *J. Biol. Chem.* 243:3458–65
46. Hajra, A. K., Agranoff, B. W. 1968. *J. Biol. Chem.* 243:3542–43
47. Hajra, A. K., Agranoff, B. W. 1968. *J. Biol. Chem.* 243:1617–22
48. La Belle, E. F. Jr., Hajra, A. K. 1972. *J. Biol. Chem.* 247:5835–41
49. Johnston, J. M., Paltauf, F. 1970. *Biochim. Biophys. Acta* 218:431–40
50. La Belle, E. F. Jr., Hajra, A. K. 1972. *J. Biol. Chem.* 247:5825–34
51. Okuyama, H., Lands, W. E. M. 1970. *Biochim. Biophys. Acta* 218:376–77
52. Plackett, P., Rodwell, A. W. 1970. *Biochim. Biophys. Acta* 210:230–40
53. Manning, R., Brindley, D. N. 1972. *Biochem. J.* 130:1003–12
54. Agranoff, B. W., Hajra, A. K. 1971. *Proc. Nat. Acad. Sci. USA* 68:411–15
55. Benns, G., Proulx, P. 1972. *Can. J. Biochem.* 50:16–19
56. Wykle, R. L., Piantadosi, C., Snyder, F. 1972. *J. Biol. Chem.* 247:2944–48
57. Mudd, J. B., Van Golde, L. M. G., Van Deenen, L. L. M. 1969. *Biochim. Biophys. Acta* 176:547–56
58. De Kruyff, B., Van Golde, L. M. G., Van Deenen, L. L. M. 1970. *Biochim. Biophys. Acta* 210:425–35
59. Kanoh, H. 1970. *Biochim. Biophys. Acta* 218:249–58
60. Kanoh, H., Ohno, K. 1973. *Biochim. Biophys. Acta* 306:203–17
61. Kennedy, E. P., Weiss, S. B. 1956. *J. Biol. Chem.* 222:193–214
62. Sundler, R., Arvidson, G., Åkesson, B. 1972. *Biochim. Biophys. Acta* 280:559–68
63. Sundler, R., 1973. *Biochim. Biophys. Acta* 306:218–226
64. Bjerve, K. S. 1972. *Biochim. Biophys. Acta* 270:348–363
65. Curley, J. M., Henderson, T. O. 1972. *Lipids* 7:676–79
66. Sakamoto, M., Akino, T. 1972. *Tohuku J. Exp. Med.* 106:45–59
67. Trewhella, M. A., Collins, F. D. 1973. *Biochim. Biophys. Acta* 296:34–50
68. Trewhella, M. A., Collins, F. D. 1973.

Biochim. Biophys. Acta 296:51–61
69. Arvidson, G. 1968. *Eur. J. Biochem.* 5:415–21
70. Lyman, R. L., Hopkins, S. M., Sheehan, G., Tinoco, J. 1969. *Biochim. Biophys. Acta* 176:86–94
71. Rytter, D., Miller, J. E., Cornatzer, W. E. 1968. *Biochim. Biophys. Acta* 152:418–21
72. Van Golde, L. M. G., Scherphof, G. L., Van Deenen, L. L. M. 1969. *Biochim. Biophys. Acta* 176:635–37
73. Kanoh, H. 1969. *Biochim. Biophys. Acta* 176:756–63
74. Holub, B. J., Breckenridge, W. C., Kuksis, A. 1971. *Lipids* 6:307–13
75. Lekim, D., Betzing, H., Stoffel, W. 1973. *Z. Physiol. Chem.* 354:437–44
76. Treble, D. H., Frumkin, S., Balint, J. A., Beeler, D. A. 1970. *Biochim. Biophys. Acta* 202:163–71
77. Salerno, D. M., Beeler, D. A. 1973. *Fed. Proc.* 32:601. (Abstr.)
78. Stein, O., Stein, Y. 1969. *J. Cell Biol.* 40:461–83
79. Nagley, P., Hallinan, T., 1968. *Biochim. Biophys. Acta* 163:218–25
80. Plagemann, P. G. W. 1971. *J. Lipid Res.* 12:715–24
81. Dils, R. R., Hübscher, G. 1961. *Biochim. Biophys. Acta* 46:505–13
82. Porcellati, G., Arienti, G., Pirotta, M., Giorgini, D. 1971. *J. Neurochem.* 18:1395–1402
83. Kanfer, J. N. 1972. *J. Lipid Res.* 13:468–76
84. Raghavan, S., Rhoads, D., Kanfer, J. 1972. *J. Biol. Chem.* 247:7153–56
85. Bjerve, K. S. 1973. *Biochim. Biophys. Acta* 296:549–62
86. Bjerve, K. S. 1971. *FEBS Lett.* 17:14–16
87. Bjerve, K. S. 1973. *Biochim. Biophys. Acta* 306:396–402
88. Saito, M., Kanfer, J. 1973. *Biochem. Biophys. Res. Commun.* 53:391–98
89. Raetz, C. R. H., Kennedy, E. P. 1973. *J. Biol. Chem.* 248:1098–1105
90. Ter Schegget, J., Van den Bosch, H., Van Baak, M. A., Hostetler, K. Y., Borst, P. 1971. *Biochim. Biophys. Acta* 239:234–42
91. Carter, J. R., Kennedy, E. P. 1966. *J. Lipid Res.* 7:678–83
92. Petzold, G. L., Agranoff, B. W. 1967. *J. Biol. Chem.* 242:1187–91
93. De Siervo, A. J., Salton, M. R. J. 1971. *Biochim. Biophys. Acta* 239:280–92
94. Hirschberg, C. B., Kennedy, E. P. 1972. *Proc. Nat. Acad. Sci. USA* 69:648–51
95. Hostetler, K. Y., Van den Bosch, H., Van Deenen, L. L. M. 1972. *Biochim. Biophys. Acta* 260:507–13

96. Tunaitis, E., Cronan, J. E., 1973. *Arch. Biochem. Biophys.* 155:420–27
97. Pieringer, R. A. 1972. *Biochem. Biophys. Res. Commun.* 49:502–7
98. Stanacev, N. Z., Davidson, J. B., Stuhne-Sekalec, L., Domazet, Z. 1972. *Biochem. Biophys. Res. Commun.* 47:1021–27
99. Stanacev, N. Z., Davidson, J. B., Stuhne-Sekalec, L., Domazet, Z. 1973. *Can. J. Biochem.* 51:286–304
100. Hostetler, K. Y., Van den Bosch, H. 1972. *Biochim. Biophys. Acta* 260:380–86
101. Hostetler, K. Y., Van den Bosch, H., Van Deenen, L. L. M. 1971. *Biochim. Biophys. Acta* 239:113–19
102. Domazet, Z., Stuhne-Sekalec, L., Davidson, J. B., Stanacev, N. Z. 1973. *Can. J. Biochem.* 51:274–85
103. Check, W. A., Astrachan, L. 1973. *Fed. Proc.* 32:601. (Abstr.)
104. Benns, G., Proulx, P. 1971. *Biochem. Biophys. Res. Commun.* 44:382–89
105. Olsen, R. W., Ballou, C. E. 1971. *J. Biol. Chem.* 246:3305–13
106. Rouser, R., Kritchevsky, G., Yamamoto, A., Knudson, A. G., Simon, G. 1968. *Lipids* 3:287–90
107. Yamamoto, A. et al 1971. *J. Biochem.* 70:775–84
108. Adachi, S. et al 1972. *Lipids* 7:1–7
109. Wherrett, J. R., Huterer, S. 1972. *J. Biol. Chem.* 247:4114–20
110. Steiner, M. R., Lester, R. L. 1972. *Biochim. Biophys. Acta* 260:222–243
111. Holub, B. J., Kuksis, A. 1971. *J. Lipid Res.* 12:699–705
112. Baker, R. R., Thompson, W. 1972. *Biochim. Biophys. Acta* 270:489–503
113. Snyder, F. 1972. *Advan. Lipid Res.* 10:233–59
114. Snyder, F. 1972. *Ether Lipids: Chemistry and Biology.* New York: Academic
115. Kasama, K., Rainey, W. T., Snyder, F. 1973. *Arch. Biochem. Biophys.* 154:648–58
116. Hajra, A. K. 1970. *Biochem. Biophys. Res. Commun.* 39:1037–44
117. Snyder, F., Rainey, W. T., Blank, M. L., Christie, W. H. 1970. *J. Biol. Chem.* 245:5853–56
118. Wykle, R. L., Snyder, F. 1970. *J. Biol. Chem.* 245:3047–58
119. Friedberg, S. J., Heifetz, A., Greene, R. C. 1971. *J. Biol. Chem.* 246:5822–27
120. Friedberg, S. J., Heifetz, A. 1973. *Biochemistry* 12:1100–6
121. Snyder, F., Blank, M. L., Malone, B., Wykle, R. L. 1970. *J. Biol. Chem.* 245:1800–5
122. Snyder, F., Blank, M. L., Malone, B. 1970. *J. Biol. Chem.* 245:4016–18
123. Radominska-Pyrek, A., Horrocks, L. A. 1972. *J. Lipid Res.* 13:580–87
124. Chae, K., Piantadosi, C., Snyder, F. 1973. *Biochem. Biophys. Res. Commun.* 51:119–24
125. Thompson, G. A. 1968. *Biochim. Biophys. Acta* 152:409–11
126. Blank, M. L., Wykle, R. L., Piantadosi, C., Snyder, F. 1970. *Biochim. Biophys. Acta* 210:442–47
127. Paltauf, F. 1971. *Biochim. Biophys. Acta* 239:38–46
128. Blank, M. L., Wykle, R. L., Snyder, F. 1971. *FEBS Lett.* 18:92–94
129. Horrocks, L. A., Radominska-Pyrek, A. 1972. *FEBS Lett.* 22:190–92
130. Paltauf, F. 1972. *Biochim. Biophys. Acta* 260:352–64
131. Paltauf, F., Holasek, A. 1973. *J. Biol. Chem.* 248:1609–15
132. Blank, M. L., Wykle, R. L., Snyder, F. 1972. *Biochem. Biophys. Res. Commun.* 47:1203–8
133. Wykle, R. L., Blank, M. L., Malone, B., Snyder, F. 1972. *J. Biol. Chem.* 247:5442–47
134. Schmid, H. H. O., Muramatsu, T., Su, K. L. 1972. *Biochim. Biophys. Acta* 270:317–23
135. Tietz, A., Lindberg, M., Kennedy, E. P. 1964. *J. Biol. Chem.* 239:4081–90
136. Soodsma, J. F., Piantadosi, C., Snyder, F. 1972. *J. Biol. Chem.* 247:3923–29
137. Kapoulas, V. M., Thompson, G. A., Hanahan, D. J. 1969. *Biochim. Biophys. Acta* 176:250–64
138. Ansell, G. B., Spanner, S. 1965. *Biochem. J.* 94:252–58
139. Warner, H. R., Lands, W. E. M. 1961. *J. Biol. Chem.* 236:2404–9
140. Robertson, A. F., Lands, W. E. M. 1962. *J. Clin. Invest.* 41:2160–65
141. Yavin, E., Gatt, S. 1972. *Eur. J. Biochem.* 25:431–36
142. Ibid 437–46
143. Brockerhoff, H., Jensen, R. G. 1974. *Lipolytic Enzymes.* New York: Academic. In press
144. Wells, M. A., Hanahan, D. J. 1969. *Biochemistry* 8:414–24
145. Wells, M. A. 1971. *Biochemistry* 10:4074–78
146. Wu, T. W., Tinker, D. O. 1969. *Biochemistry* 8:1558–68
147. Hachimori, Y., Wells, M. A., Hanahan, D. J. 1971. *Biochemistry* 10:4084–89
148. Salach, J. I., Turini, P., Seng, R., Hauber, J., Singer, T. P. 1971. *J. Biol. Chem.* 246:331–39

149. Currie, B. T., Oakley, D. E., Broomfield, C. A. 1968. *Nature* 220:371
150. Wahlström, A. 1971. *Toxicon* 9:45–56
151. Augustyn, J. M., Elliott, W. B. 1970. *Biochim. Biophys. Acta* 206:98–108
152. Kawauchi, S., Iwanaga, S., Samejima, Y., Suzuki, T. 1971. *Biochim. Biophys. Acta* 236:142–60
153. Tu, A. T., Passey, R. B., Toom, P. M. 1970. *Arch. Biochem. Biophys.* 140:96–106
154. Shipolini, R. A. et al 1971. *Eur. J. Biochem.* 20:459–68
155. Munjal, D., Elliott, W. B. 1971. *Toxicon* 9:403–9
156. Vidal, J. C., Stoppani, A. O. M. 1971. *Arch. Biochem. Biophys.* 145:543–56
157. Ibid 147:66–76
158. De Haas, G. H. et al 1970. *Biochim. Biophys. Acta* 221:31–53
159. De Haas, G. H. et al 1970. *Biochim. Biophys. Acta* 221:54–61
160. Wells, M. A. 1972. *Biochemistry* 11:1030–41
161. De Haas, G. H., Bonsen, P. P. M., Pieterson, W. A., Van Deenen, L. L. M. 1971. *Biochim. Biophys. Acta* 239:252–66
162. Pieterson, W. A. 1973. PhD thesis. Univ. Utrecht, Utrecht, The Netherlands. 107 pp.
163. Wells, M. A. 1973. *Biochemistry* 12:1080–85
164. Ibid 1086–93
165. Pieterson, W. A., Volwerk, J. J., De Haas, G. H. 1974. *Biochemistry* In press
166. Volwerk, J. J., Pieterson, W. A., De Haas, G. H. 1974. *Biochemistry* In press
167. Verger, R., Mieras, M. C. E., De Haas, G. H. 1973. *J. Biol. Chem.* 248:4023–34
168. Bonsen, P. P. M., De Haas, G. H., Pieterson, W. A., Van Deenen, L. L. M. 1972. *Biochim. Biophys. Acta* 270:364–82
169. De Haas, G. H., Postema, N. M., Nieuwenhuizen, W., Van Deenen, L. L. M. 1968. *Biochim. Biophys. Acta* 159:118–29
170. Arnesjö, B., Grubb, A. 1971. *Acta Chem. Scand.* 25:577–89
171. Pieterson, W. A., Vidal, J. C., Volwerk, J. J., De Haas, G. H. 1974. *Biochemistry* In press
172. Yamaguchi, T., Okawa, Y., Sakaguchi, K. 1973. *J. Biochem.* 73:187–90
173. Nieuwenhuizen, W., Steenbergh, P., De Haas, G. H. 1973. *Eur. J. Biochem.* 40:1–7
174. Paysant, M., Bitran, M., Wald, R., Polonovski, J. 1970. *Bull. Soc. Chim. Biol.* 52:1257–69
175. Paysant, M., Bitran, M., Etienne, J., Polonovski, J. 1969. *Bull. Soc. Chim. Biol.* 51:863–73
176. Paysant, M., Bitran, M., Polonovski, J. 1969. *C. R. Acad. Sci. Paris* 269:93–95
177. Etienne, J., Paysant, M., Grüber, A., Polonovski, J. 1969. *Bull. Soc. Chim. Biol.* 51:709–16
178. Torquebiau-Colard, O., Etienne, J., Polonovski, J. 1972. *C. R. Acad. Sci. Paris D* 275:2775–77
179. Waite, M., Sisson, P. 1971. *Biochemistry* 10:2377–83
180. Scherphof, G. L., Scarpa, A., Van Toorenenbergen, A. 1972. *Biochim. Biophys. Acta* 270:226–40
181. Waite, M., Sisson, P. 1972. *Biochemistry* 11:3098–3105
182. Vogel, W. C., Bierman, E. L. 1970. *Lipids* 5:385–91
183. Scandella, C. J., Kornberg, A. 1971. *Biochemistry* 10:4447–56
184. Raybin, D. M., Bertsch, L. L., Kornberg, A. 1972. *Biochemistry* 11:1754–60
185. De Haas, G. H., Sarda, L., Roger, J. 1965. *Biochim. Biophys. Acta* 106:638–40
186. Slotboom, A. J., De Haas, G. H., Bonsen, P. P. M., Burbach-Westerhuis, G. J., Van Deenen, L. L. M. 1970. *Chem. Phys. Lipids* 4:15–29
187. Lloveras, J., Douste-Blazy, L. 1968. *Bull. Soc. Chim. Biol.* 50:1493–99
188. Gatt, S., 1968. *Biochim. Biophys. Acta* 159:304–16
189. Cooper, M. F., Webster, G. R. 1970. *J. Neurochem.* 17:1543–54
190. White, D. A., Pounder, D. J., Hawthorne, J. N. 1971. *Biochim. Biophys. Acta* 242:99–107
191. Doi, O., Ohki, M., Nojima, S. 1972. *Biochim. Biophys. Acta* 260:244–58
192. Vogel, W. C., Bierman, E. L. 1967. *J. Lipid Res.* 8:46–53
193. Vogel, W. C., Brunzell, J. D., Bierman, E. L. 1971. *Lipids* 6:805–14
194. Zieve, F. J., Zieve, L. 1972. *Biochem. Biophys. Res. Commun.* 47:1480–85
195. Zieve, F. J., Freude, K. A., Zieve, L. 1973. *Fed. Proc.* 32:561. (Abstr.)
196. Waite, M. 1973. *Fed. Proc.* 32:561. (Abstr.)
197. Waite, M., Sisson, P. 1973. *J. Biol. Chem.* 248:7201–6
198. Rahman, Y. E., Cerny, E. A., Peraino, C. 1973. *Biochim. Biophys. Acta* 321:526–35
199. Dawson, R. M. C. 1958. *Biochem. J.* 70.559–70
200. Beare, J. L., Kates, M. 1967. *Can. J. Biochem.* 45:101–13

201. Kawasaki, N., Saito, K. 1973. *Biochim. Biophys. Acta* 296 : 426–30
202. Van den Bosch, H., Aarsman, A. J., De Jong, J. G. N., Van Deenen, L. L. M. 1973. *Biochim. Biophys Acta* 296 : 94–104
203. De Jong, J. G. N., Van den Bosch, H., Aarsman, A. J., Van Deenen, L. L. M. 1973. *Biochim. Biophys. Acta* 296 : 105–15
204. Van den Bosch, H., Aarsman, A. J., Slotboom, A. J., Van Deenen, L. L. M. 1968. *Biochim. Biophys. Acta* 164 : 215–25
205. Rao, R. H., Subrahmanyam, D. 1969. *J. Lipid Res.* 10 : 636–41
206. Leibovitz-Ben Gershon, Z., Kobiler, I., Gatt, S. 1972. *J. Biol. Chem.* 247 : 6840–47
207. Shiloah, J., Klibansky, C., De Vries, A., Berger, A. 1973. *J. Lipid Res.* 14 : 267–78
208. Okuyama, H., Lands, W. E. M., Gunstone, F. D., Barve, J. A. 1972. *Biochemistry* 11 : 4392–98
209. Tama, Y., Lands, W. E. M., Barve, J. A., Gunstone, F. D. 1973. *Biochim. Biophys. Acta* 296 : 563–71
210. Reitz, R. C., Lands, W. E. M., Christie, W. C., Holman, R. T. 1968. *J. Biol. Chem.* 243 : 2241–46
211. Reitz, R. C., El-Sheikh, M., Lands, W. E. M., Ismail, I. A., Gunstone, F. D. 1969. *Biochim. Biophys. Acta* 176 : 480–90
212. Okuyama, H., Lands, W. E. M., Christie, W. C., Gunstone, F. D. 1969. *J. Biol. Chem.* 6514–19
213. Colbeau, A., Vignais, P. M., Piette, L. H. 1972. *Biochem. Biophys. Res. Commun.* 48 : 1495–1502
214. Wittels, B. 1973. *J. Biol. Chem.* 248 : 2906–11
215. Jezyk, P. F., Hughes, H. N. 1973. *Biochim. Biophys. Acta* 296 : 24–33
216. Brandt, A. E., Lands, W. E. M. 1967. *Biochim. Biophys. Acta* 114 : 605–12
217. Van den Bosch, H., Van Golde, L. M. G., Eibl, H., Van Deenen, L. L. M. 1967. *Biochim. Biophys. Acta* 144 : 613–23
218. Waku, K., Nakazawa, Y. 1970. *J. Biochem.* 68 : 459–66
219. Ibid 1972. 72 : 495–99
220. Stein, Y., Stein, O. 1966. *Biochim. Biophys. Acta* 116 : 95–107
221. Akino, T., Yamazaki, I., Abe, M. 1972. *Tohoku J. Exp. Med.* 108 : 133–39
222. Illingworth, D. R., Portman, O. W. 1972. *Biochem. J.* 130 : 557–67
223. Scow, R. O., Stein, Y., Stein, O. 1967. *J. Biol. Chem.* 242 : 4919–24
224. Subbaiah, P. V., Sastry, P. S., Ganguly,

225. Mansbach, C. M. 1972. *Lipids* 7 : 593–95
226. Mansbach, C. M. 1973. *Biochim. Biophys. Acta* 296 : 386–400
227. Elsbach, P. 1968. *J. Clin. Invest.* 47 : 2217–29
228. Ferber, E., Resch, K. 1973. *Biochim. Biophys. Acta* 296 : 335–49
229. Lucy, J. A. 1970. *Nature* 227 : 815–17
230. Poole, A. R., Howell, J. I., Lucy, J. A. 1970. *Nature* 227 : 810–14
231. Gledhill, B. L., Sawacki, W., Croce, C. M., Koprowski, H. 1972. *Exp. Cell Res.* 73 : 33–40
232. Toister, Z., Loyter, A. 1973. *J. Biol. Chem.* 248 : 422–32
233. Yanovsky, A., Loyter, A. 1972. *J. Biol. Chem.* 247 : 4021–28
234. Elsbach, P., Holmes, K. V., Choppin, P. W. 1969. *Proc. Soc. Exp. Biol.* 130 : 903–8
235. White, D. A., Albright, F. R., Lennarz, W. J., Schnaitman, C. A. 1971. *Biochim. Biophys. Acta* 249 : 636–42
236. Bell, R. M., Mavis, R. D., Osborn, M. J., Vagelos, P. R. 1971. *Biochim. Biophys. Acta* 249 : 628–35
237. Négrel, R., Ailhaud, G., Mutaftschiev, S. 1973. *Biochim. Biophys. Acta* 291 : 635–49
238. Mavis, R. D., Vagelos, P. R. 1972. *J. Biol. Chem.* 247 : 652–59
239. White, D. A., Lennarz, W. J., Schnaitman, C. A. 1972. *J. Bacteriol.* 109 : 686–90
240. Raetz, C. R. H., Hirschberg, C. B., Dowhan, W., Wickner, W. T., Kennedy, E. P. 1972. *J. Biol. Chem.* 247 : 2245–47
241. Raetz, C. R. H., Kennedy, E. P. 1972. *J. Biol. Chem.* 247 : 2008–14
242. Stoffel, W., Schiefer, H. G. 1968. *Z. Physiol. Chem.* 349 : 1017–26
243. Zborowski, J., Wojtczak, L. 1969. *Biochim. Biophys. Acta* 187 : 73–84
244. Shephard, E. H., Hübscher, G. 1969. *Biochem. J.* 113 : 429–40
245. Sarzala, M. G., Van Golde, L. M. G., De Kruyff, B., Van Deenen, L. L. M. 1970. *Biochim. Biophys. Acta* 202 : 106–19
246. McMurray, W. C., Dawson, R. M. C. 1969. *Biochem. J.* 112 : 91–108
247. Davidson, J. B., Stanacev, N. Z. 1972. *Can. J. Biochem.* 50 : 936–48
248. Possmayer, F., Meiners, B., Mudd, J. B. 1973. *Biochem. J.* 132 : 381–94
249. Stein, Y., Widnell, C., Stein, O. 1968. *J. Cell Biol.* 39 : 185–92
250. Higgins, J. A., Barrnett, R. J. 1972. *J. Cell Biol.* 55 : 282–98

J. 1970. *Biochem. J.* 118 : 241–47

251. Lee, T. C., Snyder, F. 1973. *Biochim. Biophys. Acta* 291:71–82
252. Yamashita, S., Hosaka, K., Taketo, M., Numa, S. 1973. *FEBS Lett.* 29:235–38
253. Wilgram, G. F., Kennedy, E. P. 1963. *J. Biol. Chem.* 238:2615–19
254. Sedgwick, B., Hübscher, G. 1965. *Biochim. Biophys. Acta* 106:63–77
255. Johnston, J. M., Rao, G. A., Lowe, P. A., Schwartz, B. E. 1967. *Lipids* 2:14–20
256. Smith, M. E., Sedgwick, B., Brindley, D. N., Hübscher, G. 1967. *Eur. J. Biochem.* 3:70–77
257. Mitchell, M. P., Brindley, D. N., Hübscher, G. 1971. *Eur. J. Biochem.* 18:214–20
258. Bygrave, F. L. 1969. *J. Biol. Chem.* 244:4768–72
259. Van Golde, L. M. G., Fleischer, B., Fleischer, S. 1971. *Biochim. Biophys. Acta* 249:318–30
260. Ostrow, D., Getz, G. S. 1973. *Fed. Proc.* 32:601. (Abstr.)
261. Soto, E. F., Pasquini, J. M., Krawiec, L. 1972. *Arch. Biochem. Biophys.* 150:362–70
262. Lord, J. M., Kagawa, T., Beevers, H. 1972. *Proc. Nat. Acad. Sci. USA* 69:2429–32
263. Devor, K. A., Mudd, J. B. 1971. *J. Lipid Res.* 12:403–11
264. Smith, J. D., Law, J. H. 1970. *Biochem. Biophys. Acta* 202:141–52
265. Rehbinder, D., Greenberg, D. M. 1965. *Arch. Biochem. Biophys.* 109:110–15
266. Taki, T., Nishimura, K., Matsumoto, M. 1973. *Jap. J. Exp. Med.* 43:87–105
267. Dennis, E. A., Kennedy, E. P. 1972. *J. Lipid Res.* 13:263–67
268. Ibid 1970. 11:394–403
269. Vorbeck, M. L., Martin, A. P. 1970. *Biochem. Biophys. Res. Commun.* 40:901–8
270. Sumida, S., Mudd, J. B. 1970. *Plant Physiol.* 45:719–22
271. Bahl, J., Guillot-Solomon, T., Douce, R. 1970. *Physiol. Veg.* 8:55–74
272. Douce, R., Manella, C. A., Bonner, W. D. 1972. *Biochem. Biophys. Res. Commun.* 49:1504–9
273. Mangnall, D., Getz, G. S. 1971. *Fed. Proc.* 30:1226
274. Davidson, J. B., Stanacev, N. Z. 1971. *Biochem. Biophys. Res. Commun.* 42:1191–99
275. Victoria, E. J., Van Golde, L. M. G., Hostetler, K. Y. Scherphof, G. L., Van Deenen, L. L. M. 1971. *Biochim. Biophys. Acta* 239:443–57
276. Davidson, J. B., Stanacev, N. Z. 1971.

277. Guarnieri, M., Stechmiller, B., Lehninger, A. L. 1971. *J. Biol. Chem.* 246:7526–32
278. Waite, M., Sisson, P., Blackwell, E. 1970. *Biochemistry* 9:746–53
279. Stahl, W. L., Trams, E. G. 1968. *Biochim. Biophys. Acta* 163:459–71
280. Wright, J. D., Green, C. 1971. *Biochem. J.* 123:837–44
281. Nachbaur, J., Colbeau, A., Vignais, P. M. 1972. *Biochim. Biophys. Acta* 274:426–46
282. Scherphof, G. L., Waite, M., Van Deenen, L. L. M. 1966. *Biochim. Biophys. Acta* 125:406–9
283. Waite, M. 1969. *Biochemistry* 8:2536–42
284. Nachbaur, J., Vignais, P. M. 1968. *Biochem. Biophys. Res. Commun.* 33:315–20
285. Mellors, A., Tappel, A. L. 1967. *J. Lipid Res.* 8:479–85
286. Fowler, S., De Duve, C. 1969. *J. Biol. Chem.* 244:471–81
287. Stoffel, W., Trabert, U. 1969. *Z. Physiol. Chem.* 350:836–44
288. Franson, R., Waite, M., La Via, M. 1971. *Biochemistry* 10:1942–46
289. Newkirk, J. D., Waite, M. 1971. *Biochim. Biophys. Acta* 225:224–33
290. Ibid 1973. 298:562–76
291. Parks, J. G., Thompson, W. 1973. *Biochim. Biophys. Acta* 306:403–11
292. Smith, A. D., Winkler, H. 1968. *Biochem. J.* 108:867–74
293. Lloveras, J., Douste-Blazy, L. 1973. *Eur. J. Biochem.* 33:567–77
294. Franson, R. C., Waite, M., 1973. *J. Cell Biol.* 56:621–27
295. Franson, R. C., Beckerdite, S., Wang, P., Waite, M., Elsbach, P. 1973. *Biochim. Biophys. Acta* 296:365–73
296. Smith, J. B., Silver, M. J., Webster, G. R. 1973. *Biochem. J.* 131:615–18
297. Erbland, J. F., Marinetti, G. V. 1965. *Biochim. Biophys. Acta* 106:128–38
298. Bjørnstad, P. 1966. *Biochim. Biophys. Acta* 116:500–10
299. Hörtnagl, H., Winkler, H., Hörtnagl, H. 1969. *Eur. J. Biochem.* 10:243–48
300. Ferber, E., Munder, P. G., Fischer, H., Gerisch, G. 1970. *Eur. J. Biochem.* 14:253–57
301. Van Golde, L. M. G., McElhaney, R. N., Van Deenen, L. L. M. 1971. *Biochim. Biophys. Acta* 231:245–49
302. Wirtz, K. W. A., Zilversmit, D. B. 1968. *J. Biol. Chem.* 243:3596–602
303. Albright, F. R., White, D. A., Lennarz, W. J. 1973. *J. Biol. Chem.* 248:3968–77

Can. J. Biochem. 49:1117–24

276   VAN DEN BOSCH

304. Akiyama, M., Sakagami, T. 1969. *Biochim. Biophys. Acta* 187:105–12
305. Abdelkader, A. B., Mazliak, P. 1970. *Eur. J. Biochem.* 15:250–62
306. Miller, E. K., Dawson, R. M. C. 1972. *Biochem. J.* 126:823–35
307. Zilversmit, D. B. 1971. *J. Biol. Chem.* 246:2625–49
308. Ehnholm, C., Zilversmit, D. B. 1973. *J. Biol. Chem.* 248:1719–24
309. Ehnholm, C., Zilversmit, D. B. 1972. *Biochim. Biophys. Acta* 274:652–57
310. Blok, M. C., Wirtz, K. W. A., Scherphof, G. L. 1971. *Biochim. Biophys. Acta* 233:61–75
311. Wojtczak, L., Baranska, J., Zborowski, J., Drahato, Z. 1971. *Biochim. Biophys. Acta* 249:41–52
312. Dawson, R. M. C. 1973. *Sub-Cell. Biochem.* 2:69–89
313. Wirtz, K. W. A., Zilversmit, D. B. 1970. *FEBS Lett.* 7:44–46
314. Wirtz, K. W. A., Kamp, H. H., Van Deenen, L. L. M. 1972. *Biochim. Biophys. Acta* 274:606–17
315. Kamp, H. H., Wirtz, K. W. A., Van Deenen, L. L. M. 1973. *Biochim. Biophys. Acta* 318:313–25
316. Harvey, M. S., Wirtz, K. W. A., Kamp, H. H., Zegers, B. J. M., Van Deenen, L. L. M. 1973. *Biochim. Biophys. Acta* 323:234–39
317. Harvey, M. S., Helmkamp, G., Wirtz, K. W. A. Unpublished observations
318. Kagawa, Y., Johnson, L. W., Racker, E. 1973. *Biochem. Biophys. Res. Commun.* 50:245–51
319. Demel, R. A., Wirtz, K. W. A., Geurts van Kessel, W. S. M., Kamp, H. H., Van Deenen, L. L. M. 1973. *Nature New Biol.* 246:102–5
320. Kito, M., Pizer, L. I. 1969. *J. Bacteriol.* 97:1321–27
321. Kito, M., Aibara, S., Hasegawa, K., Hata, T. 1972. *J. Biochem.* 71:99–105
322. Glaser, M., Bayer, W. H., Bell, R. M., Vagelos, P. R. 1973. *Proc. Nat. Acad. Sci. USA* 70:385–89
323. Sokawa, Y., Nakao, E., Kaziro, Y. 1968. *Biochem. Biophys. Res. Commun.* 33:108–12
324. Tropp, B. E., Meade, L. C., Thomas, P. J. 1970. *J. Biol. Chem.* 245:855–58
325. Golden, N. G., Powell, G. L. 1972. *J. Biol. Chem.* 247:6651–58
326. Crowfoot, P. D., Esfahani, M., Wakil, S. J. 1972. *J. Bacteriol.* 112:1408–15
327. Crowfoot, P. D., Oka, T., Esfahani, M., Wakil, S. J. 1972. *J. Bacteriol.* 112:1396–1407
328. Bell, R. M., Davis, R. D., Vagelos, P. R. 1972. *Biochim. Biophys. Acta* 270:504–12

329. Ohki, M. 1972. *J. Mol. Biol.* 68:249–64
330. Van Golde, L. M. G., Schulman, H., Kennedy, E. P. 1973. *Proc. Nat. Acad. Sci. USA* 70:1368–72
331. Gelmann, E. P., Cronan, J. E. 1972. *J. Bacteriol.* 112:381–87
332. Harder, M. E. et al 1972. *Proc. Nat. Acad. Sci. USA* 69:3105–9
333. Cronan, J. E. 1972. *Nature* 240:21–22
334. Beebe, J. L. 1972. *J. Bacteriol.* 109:939–42
335. Sanchez, M., Nicholls, D. G., Brindley, D. N. 1973. *Biochem. J.* 132:697–706
336. Brindley, D. N. 1973. *Biochem. J.* 132:707–15
337. Lloyd-Davies, K. A., Brindley, D. N. 1973. *Biochem. Soc. Trans.* 1:436–38
338. Tzur, R., Tal, E., Shapiro, B. 1964. *Biochim. Biophys. Acta* 84:18–23
339. Guynn, R. W., Veloso, D., Veech, R. L. 1972. *J. Biol. Chem.* 247:7325–31
340. Bøhmer, T. 1967. *Biochim. Biophys. Acta* 144:259–70
341. Aas, M., Daae, L. N. W., 1971. *Biochim. Biophys. Acta* 239:208–16
342. Eriksson, L. C., Dallner, G. 1973. *FEBS Lett.* 29:351–54
343. Possmayer, F., Mudd, J. B. 1971. *Biochim. Biophys. Acta* 239:217–33
344. Davidson, J. B., Stanacev, N. Z. 1971. *Biochem. Biophys. Res. Commun.* 42:1191–99
345. Mangiapane, E. H., Lloyd-Davies, K. A., Brindley, D. N. 1973. *Biochem. J.* 134:103–12
346. Park, C. E., Marai, E., Mookerjea, S. 1972. *Biochim. Biophys. Acta* 270:50–59
347. Lamb, R. G., Fallon, H. J. 1972. *J. Clin. Invest.* 51:A53
348. Fallon, H. J., Van den Bosch, H. Unpublished observations
349. Alling, C., Cahlin, E., Scherstén, T. 1973. *Biochim. Biophys. Acta* 296:518–26
350. Young, D. L., Lynen, F. 1969. *J. Biol. Chem.* 244:377–83
351. Bjørnstad, P., Bremer, J. 1966. *J. Lipid Res.* 7:38–45
352. Domschke, W., Keppler, D., Bischoff, E., Decker, K. 1971. *Z. Physiol. Chem.* 352:275–79
353. Wilgram, G. F., Holoway, C. F., Kennedy, E. P. 1960. *J. Biol. Chem.* 235:37–39
354. Kennedy, E. P., Weiss, S. B. 1956. *J. Biol. Chem.* 222:193–214
355. Weiss, S. B., Smith, S. W., Kennedy, E. P. 1958. *J. Biol. Chem.* 231:53–64
356. Sato, R., Hasegawa, H. 1969. *Biochim. Biophys. Acta* 176:748–55

357. Ewing, R. D., Finnamore, F. J. 1970. *Biochim. Biophys. Acta* 218:463–73
358. Ibid. 474–81
359. Weinhold, P. A. 1969. *Biochim. Biophys. Acta* 187:85–93
360. Weinhold, P. A., Sanders, R. D. 1971. *Biochemistry* 10:1090–96
361. Baldwin, J., Cornatzer, W. E. 1968. *Lipids* 3:361–67
362. Freysz, L., Lastennet, A., Mandel, P. 1972. *J. Neurochem.* 19:2599–2605
363. Farrell, P. M., Zachman, R. D. 1973. *Science* 179:297–98
364. Kent, C., Lennarz, W. J. 1972. *Proc. Nat. Acad. Sci. USA* 69:2793–97
365. Bright-Gaertner, E., Proulx, P. 1972. *Biochim. Biophys. Acta* 270:40–49
366. Ohki, M., Doi, O., Nojima, S. 1972. *J. Bacteriol.* 110:864–69
367. Kunze, H., Vogt, W. 1971. *Ann. N.Y. Acad. Sci.* 180:123–29
368. Haye, B., Champion, S., Jacquemin, C. 1973. *FEBS Lett.* 30:253–60
369. Imre, S. 1972. *Acta Biochim. Biophys. Acad. Sci. Hung.* 7:247–49
370. Chiappe de Cingolani, G. E., Van den Bosch, H., Van Deenen, L. L. M. 1972. *Biochim. Biophys. Acta* 260:387–92
371. Johnston, J. M., Paltauf, F., Schiller, C. M., Schulz, L. D. 1970. *Biochim. Biophys. Acta* 218:124–33

ELECTRON MICROSCOPY OF ENZYMES

Rudy H. Haschemeyer

Department of Biochemistry, Cornell University Medical College, New York, N.Y.

Etienne de Harven

Memorial-Sloan-Kettering Cancer Center, New York, N.Y.

CONTENTS

INTRODUCTION

This review focuses on the use of electron microscopy in describing the structure and geometry of oligomeric enzymes. The value of several methods is emphasized, often with an intentionally personal point of view, and no attempt is made to completely review structural results of enzymes per se, even by reference, though a number of examples are noted. Our remarks are intended to provide the biochemical audience with a basic understanding of the power and limitations of electron microscopy for enzyme study and to offer sufficient sophistication to evaluate the quality of published results.

IMAGE CHARACTERISTICS OF NEGATIVELY STAINED PARTICLES (1–5)

Most recent studies on enzymes by electron microscopy have involved one or more of the many variations of techniques collectively called negative staining (see reviews by Haschemeyer 1, Haschemeyer & Myers 2). We present first the principles of this method and characteristics of the image obtained, so that readers unfamiliar with negative staining can appreciate the discussion to follow.

In a typical negative staining procedure referred to as the drop method, about 3 μl of a dilute enzyme solution at a concentration of approximately 2 μg/ml is placed on a thin carbon film supported by a copper mesh grid. After 30 sec most of the liquid is withdrawn by touching the edge of the droplet with filter paper, and a droplet of negative stain is immediately supplied before the residual layer of enzyme solution has dried. After about 30 sec, most of the negative stain solution is similarly removed. As water evaporates, the heavy metal salt concentrates and finally forms an amorphous glassy solid that surrounds enzyme molecules and penetrates into interstices within their macromolecular structure (Figure 1). One should keep in mind, however, that the size of the molecule of the negative stain itself might preclude its penetration into the small spaces separating enzyme subunits. For example, the phosphotungstate ions are estimated to be approximately 10 Å in size and are therefore not suitable for the highest level of resolution of today's electron microscopes which resolve 2 Å. When exposed to the electron beam, areas of the preparation containing thicker stain scatter more electrons than areas with less stain. Many of these scattered electrons do not participate in the formation of the final image because they have been stopped at various levels of the microscope column, mostly the objective aperture. Other electrons are inelastically scattered (and hence have changed wavelength) and are not focused in the image plane at positions corresponding to the object plane where scattering originated. As a result, the magnified image on the fluorescent screen, or on photographic prints, appears darker in areas corresponding to greater thicknesses of stain at the object plane. It is a significant goal of negative staining to assure that such contrast (amplitude contrast) is the only type appearing in micrographs, at least to the limit of resolution. A typical appear-

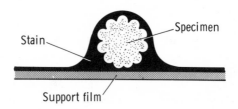

Figure 1   Schematic representation of a particle embedded in negative stain (from 2).

ance for moderately sized enzyme particles embedded in negative stain is shown in Figure 2. As anticipated by the model shown in Figure 1, many particles are surrounded by a halo darker than background, and characteristic intraparticle detail can be noted. Note that part of the contrast within the particle domain results from stain penetration between subunits on the top and bottom of the particle. The final image is a projection of both superimposed.

The purpose of negative staining is to enhance contrast in a manner that will outline morphological detail. It is critical for correct interpretation of the images that the relative darkness throughout the image reflects the relative electron scattering power of corresponding positions in the object. Properly exposed in-focus micrographs taken with a high-resolution electron microscope adequately aligned and compensated for astigmatism achieve this goal. Inadequate focus adds spurious contrast components to the image, arising primarily from interference and diffraction phenomena which are virtually impossible to relate unambiguously to object con-

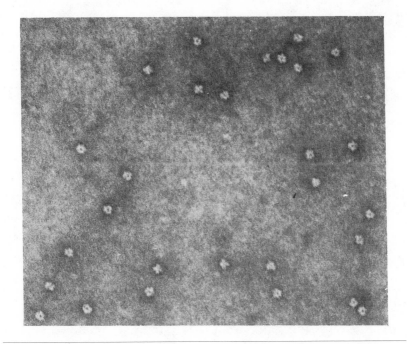

Figure 2   Electron micrograph of L-aspartate β-decarboxylase negatively stained with uranyl oxalate. Several distinctive views can be seen: a vague six-sided polygon, a cross-like presentation, and trimer-like views. These structures can be reasonably interpreted on the basis of a 12-subunit model possessing tetrahedral symmetry. The trimer views then represent particles which are predominantly stained on the bottom only. The mean molecular diameter is 150 Å.

trast. Unfortunately, such spurious contrast may look like some anticipated subunit structure and be entirely misinterpreted. Indeed, underfocus micrographs often appear somewhat sharper than in-focus electron micrographs, quite unlike the experience with light microscopy.

Johnson (3) has recently reviewed some major aspects of image formation by interference effects and Fresnel diffraction. A discussion on support films, excellently illustrated, can also be found in reviews by Koller and associates (4) and by Dubochet (5). The phenomena are briefly the following. Thin films of amorphous

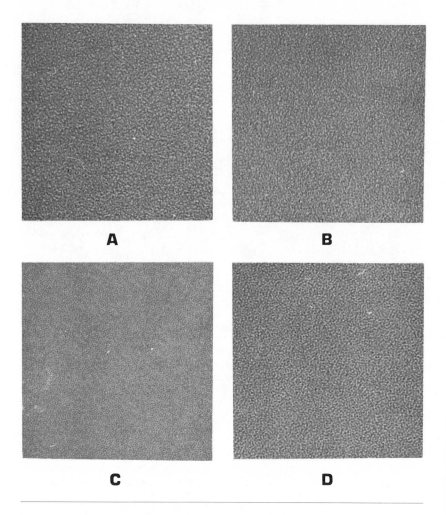

Figure 3   Bright field image of a thin carbon film as a function of focus. A: 0.1 μ underfocus. B: 0.05 μ underfocus. C: Near focus. D: 0.05 μ overfocus. Magnification × 1,000,000.

carbon commonly used as supports for bright field electron microscopy average 100 Å in thickness and show apparent granularity which has minimum size at focus. The size of this phase grain [the term phase grain or phase contrast used here for this phenomenon is common usage though not semantically precise (3)] is determined at a given accelerating potential by the degree of defocus (Figure 3). For example, at an 80 kV accelerating potential, a 0.2 μ defocus may result in a phase grain size of about 10 Å. These interference effects arise from phase shifts within structural irregularities of the carbon film. These irregularities are not contrasted in bright field illumination but are readily observed in dark field micrographs (Figure 4).

Fresnel diffraction occurs at the interface between two areas of different density. At underfocus a bright halo appears outside the area of greater density, while overfocus produces a dark fringe. It seems likely that density changes within a negatively stained enzyme are often sufficiently abrupt to produce a facsimile of Fresnel diffraction.

Slightly out of focus images of sufficiently large particles of uniform density will therefore result from the superposition of support film phase grain and of the particle image. It is not as unusual as one might expect to misinterpret this phase granularity as substructure (1, 2). For smaller particles (e.g. most oligomeric enzymes) the situation is complicated by phase changes (and possibly diffraction effects) within the particle itself. The final image will include interference contrast, which is the sum of those of the particle and of the support film plus composite nonlinear contributions resulting from particle support film interference interactions (5). The net result of defocus, then, is an unpredictably complex image subject to misinterpretation.

Sufficiently extensive empirical comparisons of defocused images with those at focus permit the following descriptive analysis. Images that show true substructure

Figure 4   Dark field image of a thin carbon film. Phase contrast does not appear in dark field imaging and thus the intensity variations observed are indicative of structure within the film. × 1,000,000.

at focus retain a recognizable impression of that structure at moderate underfocus, though defocus may superimpose additional artifactual contrast. Thus, on defocus, a uniform spherical subunit may appear to have a hole near its center or develop contrast components which give the impression of a dimer or trimer (for example) rather than a single subunit. Similarly, bar-like structures may appear to have a longitudinal groove or may appear as multiple spheres side by side in a bead-like arrangement.

Since the dimensional domain of Fresnel diffraction is the same magnitude as phase granularity within carbon support films, it was previously suggested by one of us (1, 2) that contrast variations within particles (not subunit separation) with extent greater than the apparent granularity visible in the background near the particle could be safely interpreted as being related to real structural differences within the particle itself. This conclusion is theoretically weak and clearly does not apply to support films in which phase contrast is essentially absent [e.g. crystals or amorphous aluminum oxide as described by Koller et al (4)]. Nonetheless, with thin carbon support films, the above rule of thumb is empirically observed to provide a useful guard against misinterpretation of negatively stained images.

The inescapable conclusion of the above discussion is that to interpret subunit structure of small oligomeric enzymes requires micrographs taken very near focus. Thus, any microscopist accustomed to working at high magnification using a modern electron microscope permitting visualization of phase grain near focus can follow the simple directions for negative staining (1, 2) and obtain photographs of many enzymes the first time such an effort is made.

THE PRESERVATION AND CONTRASTING OF ENZYME SUBSTRUCTURE

Enzyme molecules are subjected to unfamiliar and possible hostile environments during specimen preparation and electron microscopy. First they are diluted into low ionic strength or volatile buffer systems. Then, for negative staining, unfamiliar ions are added which concentrate past saturation, and surface tension forces of marked proportions act during final evaporation. (A general preservative effect of negative staining, however, appears to be the solidification of the stain prior to final particle dehydration.) Finally, the sample is placed in a vacuum and subjected to the radiation damage resulting from electron bombardment. Small wonder that biochemists frequently ask what relation, if any, exists between electron microscope images and true enzyme conformation in solution. This difficult question cannot be answered with certainty for any particular enzyme, but our overall positive view is based in part on the following arguments: 1. Observed electron microscopic results are often consistent with other physical measurements of shape, number of subunits, etc. 2. Similar techniques have been more extensively applied by the virologist to obtain an overall consistent viewpoint of virus structure, in some cases fortified with confirmatory X-ray data. 3. Esthetically pleasing results (e.g. consistent with some symmetry model) are in accord with current concepts of evolutionary design. 4. Even moderately distorted images may reveal information regarding enzyme structure in solution (e.g. stoichiometry).

Several suggestions can be made to enhance the credibility of electron microscope images of enzymes: 1. Particle size estimated from electron microscopy should be consistent with expectations from molecular weight estimates, thereby reducing the possibility that association or dissociation has occurred. 2. If no generalized shape accounts in its two-dimensional projections for a large proportion of particle images, disorganization is suspect. 3. Detrimental ionic effects can be tested in part by activity studies and by physical measurements such as sedimentation velocity. 4. Bifunctional cross-linking reagents such as glutaraldehyde or dimethylsuberimidate can be tried as a stabilizing fixative. 5. Different negative stains may have advantageous preservative properties for a particular enzyme. 6. Minimum exposure to negative stain is possible using the high-pressure spray gun overlap drop method. 7. Negative staining can be largely limited to contrasting particles that have adhered to the support film under favorable environmental conditions by positive interactions. These particles may be relatively stabilized by support film enzyme interactions. 8. Contamination during the electron microscope observation is minimized by using a liquid N_2-cooled anticontamination device, and beam damage is minimized by any one of several methods depending on microscope characteristics. Microscopes equipped with image intensification systems will provide a powerful tool to minimize beam damage. A detailed description of these and several other possibilities has recently been presented by one of us (1, 2). Not discussed in those reviews are several interesting possibilities recently applied to preservation of virus structure. In one of these (e.g. see Nermut 6) negative contrasting was achieved by freeze-drying a suspension of virus in negative stain solution. The excess stain appeared as a white powder on the grids and was removed by gentle blowing. In the absence of negative stain, freeze-drying may advantageously preserve structures for dark field observation. The critical point-drying method of Anderson (e.g. see review by Hayat & Zirkin 7) has been particularly useful in preserving large biological structures (Porter et al 8) and has been successfully applied to the preservation of fragile enveloped viruses (de Harven et al 9). It could in principle be used to prepare enzymes for dark field microscopy with complete elimination of surface tension artifacts. It is not clear, however, whether or not the replacement of the aqueous phase by a liquid with appropriate temperature and pressure critical points can be achieved without exposing the enzyme to intermediate solvents (e.g. ethanol) which would usually disrupt protein conformation. Critical point drying in water-soluble N_2O (Koller & Bernhard 10) deserves more attention in this respect.

In the final analysis, and independently of how many tricks have been used to preserve structure, some questions necessarily remain concerning the relationship of observed structure of the dried enzyme and true structure in solution. A good example of this is the different appearance of the enzyme L-aspartate β-decarboxylase as contrasted with different negative stains (2, 11). The appearance of the particles stained with uranyl acetate (Figure 5a), phosphotungstate (Figure 5b), and uranyl oxalate (Figure 2) is significantly different. The uranyl oxalate results are easier to interpret, after which an ad hoc explanation of the other images can be supplied. Which images (if any) nearly represent the structure in solution? We cannot say, nor is it likely that the answer will soon be forthcoming.

Some important comparisons between electron microscopic images of dried and

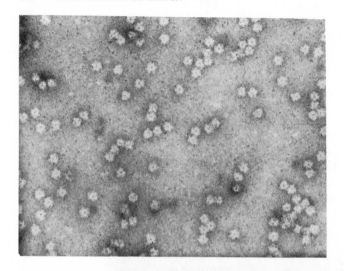

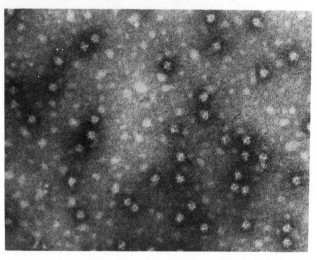

Figure 5   L-aspartate β-decarboxylase negatively stained with (*a*) uranyl acetate at pH 4.5 and (*b*) potassium phosphotungstate at pH 7.

hydrated enzyme crystals have been noted and work in this area is continuing. Using aperture-limited specimen chambers in a differentially pumped electron microscope column, and therefore permitting the study of hydrated samples, the microscopic observation on catalase by Matricardi et al (12) dramatizes the effect of dehydration on lattice spacing. Although optical transforms of wet crystal images may provide exciting molecular details in the not too distant future, contrasting individual particles in aqueous environment with hope of resolving molecular substructure does not appear feasible at this time.

THE INTERPRETATION OF ELECTRON MICROSCOPIC IMAGES OF ENZYMES

Images presented by some enzymes are nearly trivial to interpret compared to others, which may border on the impossible. The stacked hexagon arrangement of subunits in *E. coli* glutamine synthetase (Figure 6), for example, is among the most obvious.

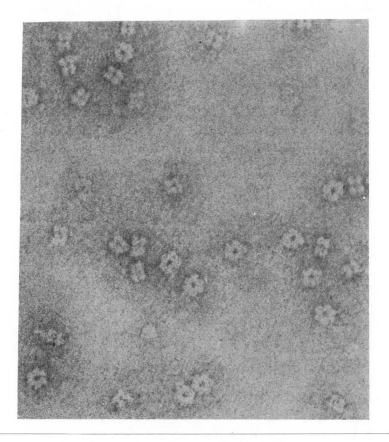

Figure 6   *E. coli* glutamine synthetase negatively contrasted with uranyl oxalate. Many clear views of the 12-subunit stacked hexagon particle are observed. Some project a hexagonal view, others a tetrameric view of a projection of the enzyme standing on edge. Further discussion is presented in the text. × 430,000.

As interpreted by Valentine et al (13), the two layers must be nearly eclipsed, or the hexagon presentation would be less obvious and clear tetramers would not be observed. The tetramer images are formed by the hexagon resting on its side with four subunits in contact with the support film. In this orientation, two subunits in each of the four positions are directly superimposed, while the pair of subunits at either end are barely projecting into thicker stain and thus not contrasted. Double bar-like images are projected by the stacked hexagon lying on its side in a manner such that stain penetrating between subunits on the top and bottom part of the particle is recorded "out of register." Further structural confirmation, if needed, was obtained from elegant photographs of small microcrystals. Few enzymes have shown such a high percentage of clear, easily interpretable projections. Note for example that if the hexameric rings had been staggered or partially rotated with respect to one another, as seems equally probable on a priori arguments, the interpretation would have been considerably complicated. The dodecameric enzyme [whose stoichiometry is well substantiated (11, 14)] L-aspartate β-decarboxylase (Figure 2) is similar in size to the *E. coli* enzyme, but a model which would, in various orientations, explain most of the observed images is much less obvious. The trimeric images were finally explained as predominantly bottom staining of a proposed tetrahedral model. If such images are present in negatively stained preparations of *E. coli* glutamine synthetase, no problem arises because in this case the top and bottom stained image is identical and directly superimposed.

It is the purpose in the remainder of this section to indicate briefly a few of the options available to find a reliable model of enzyme molecular architecture. These options are not equally applicable to all enzymes, and some require equipment that is not available to all investigators.

The Symmetry Model

The symmetry model requires that identical polypeptide chains occupy spatially equivalent positions in an enzyme. Furthermore, if the molecular domain is also uniquely defined in its extent, the molecule must possess point group symmetry. For enzymes, the only possible symmetry classes are cyclic (*n* protomers, even or odd), dihedral (2*n* protomers), and the cubic point group tetrahedral (12 protomers), octahedral (24 protomers), and icosahedral (60 protomers). (We use the term protomer as defined by Monod et al, 15.) The electron microscopist must further assume that an observable subunit is equivalent to one (the usual case) or more (e.g. unresolved dimers) protomers. It is unusual to be able to distinguish two non-equivalent subunits.

The advantage of symmetry models is obvious, namely, that by limiting choices to five symmetry classes, some hope of finding a suitable model exists. Even so, the problem is difficult if nonspherical subunits, unresolved contacts, preferential orientation, etc are present. The danger of assuming maximum symmetry even when the enzyme is thought to contain chemically identical polypeptide chains is equally obvious, for one cannot be certain that the model applies in a particular case, since several exceptions have already been noted. These exceptions and a more detailed discussion of symmetry are presented in recent reviews (1, 2, 16, 17).

Negative Staining Analog Methods

Once a proposed model has been constructed, one wishes to know whether it can account in one of its orientations for a majority of well-preserved negatively stained (or dark field) images of the enzyme. Mental visualization of these projections is often difficult and a simple analog procedure is highly recommended (1, 2, 18). The method, based on an earlier one used by Casper (19), involves positioning a model constructed of X-ray transparent material (e.g. cork, balsa, or plasticine) in a radio-opaque liquid and taking an X-ray photograph. The model can be repositioned to present a different projection and rephotographed. Quite satisfactory results are obtained provided the liquid is adjusted to the right opacity and the X-irradiation is adjusted to the right intensity. The proper adjustments for realistic contrast are readily monitored by fluoroscopy.

Investigators with mathematical inclinations may prefer to utilize computer simulation to calculate the intensity distribution of model projections and photograph the result from a cathode ray tube display. Programs of the type developed by Finch & Klug (20) and in our laboratory (11, 21) provide sufficiently realistic results (Figure 7).

Use of a Tilting Stage

Electron micrographs of the same particle at several orientations to the electron beam are obtained by use of a tilting stage that can be rotated through large angles (40–45°). This method has proved immensely valuable in the study of viruses, as

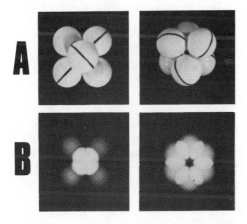

Figure 7   Computer simulation of anticipated negatively stained projections of a model composed of two staggered trimers engulfed in uniform stain with a thickness of twice that of the molecules. The model may have D_3 symmetry or tetrahedral symmetry if each sphere itself represents a dimer. The latter model projects images quite similar to those observed for L-aspartate β-decarboxylase in Figure 2.

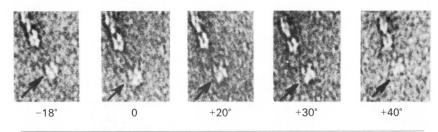

| −18° | 0 | +20° | +30° | +40° |

Figure 8   Molecules of glutamic dehydrogenase at various angles to the electron beam. The nondescript image observed without rotation is seen to change from a double hemisphere appearance at − 18° to a "cross" presentation at +40°. (From Josephs 23)

reviewed by Finch & Holmes (22). It was particularly helpful in some virus studies to rotate particles through appropriate angles so that the electron beam was first parallel to one symmetry axis and then to another. The resulting images provided elegant justification that the chosen model was correct. Such convincing results are generally not anticipated for small oligomeric enzymes, except perhaps those with cubic symmetry. For dihedral symmetry, confirmatory information might be obtained with edge-on views by rotation through angles appropriate to pass from one twofold axis to another. Rotation can be used to show that some images represent arbitrary orientations which turn into a more representative image upon rotation. This method was used by Josephs (23) (Figure 8) in studies on glutamate dehydrogenase, which is thought to be represented by a hexamer model of two staggered trimer rings. He also used the tilting stage effectively to demonstrate that images of apparent trimers were not eclipsed double trimers. He suggested that these images were the result of transverse cleavage of the hexamer, though predominantly bottom staining cannot be ruled out.

The Potential Information Content of Aggregates, Periodic Arrays, and Microcrystals

Specific association products of enzymes are useful in several ways. They may result in identical orientation of neighbors, thereby strengthening arguments that the projection observed is real. Molecular dimensions are more readily established by measurements on adjacent molecules in contact. Observations on *E. coli* glutamine synthetase aggregates provided additional evidence for the stacked hexagon model as already noted.

An elegant study for deducing symmetry and geometry from linear polymers of avidin cross-linked with a bifunctional biotin analog has been made by Green and associates (24). Their deductions are remarkable considering the small size of this tetrameric protein (mol wt ∼64,000).

Dramatic results can often be obtained from further analyses of micrographs of arrays or small crystals that exhibit periodic detail (e.g. see 25). An optical diffractometer is first used to obtain a diffraction image of the micrograph. This image itself contains valuable information regarding the arrangement and dimensions of the periodic detail. Any diffraction spots arise from periodic detail only, whereas nonperiodic detail and noise is concentrated near the center of the diffraction image. Upon construction of an appropriate optical filter (a mask at the position of the diffraction image), components of the diffraction image rich in periodic detail are recombined to provide a new image. Figure 9 shows a result of this kind obtained by Kiselev and co-workers (26) on phosphorylase B. A model employing a quite specific subunit shape is largely based on this result. The reader is cautioned to note that taking such a result too literally is somewhat hazardous, for in attempts to filter out noise from the image, very real components of enzyme structure may have been removed as well.

Labeling of Subunits

The use of antibody Fab fragments to deduce the orientation of subunits in oligomers has recently been reviewed by Green (16). Despite the limited resolution attainable (about 40 Å), further applications of this method can be anticipated, particularly if other workers find that the Fab fragment can be easily contrasted at appropriate sites in other enzymes. Green (16) also reviews the use of avidin labeling to locate the biotin-containing polypeptide chain within the complex enzyme transcarboxylase.

The use of single (or a few proximate) heavy metal derivatives to label subunits in

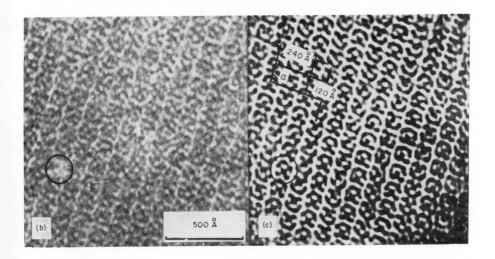

Figure 9 B: A single-layer array of phosphorylase B negative contrasted with uranyl acetate. C: Optically filtered image of B. The molecules (protein is black) in adjacent rows are seen to face in opposite directions. The averaging process results in the insertion of a molecule where none was present within the circle. A model with D_2 symmetry of banana-shaped subunits adequately accounts for these and other observed projections. (From Kiselev, Lerner & Livanova 26)

enzymes is still anticipated though not realized. Advances in high resolution dark field electron microscopy (see reviews by Dubochet 5 and Koller et al 4) indicate that single heavy atoms can be contrasted on thin films, though not as yet with an entirely favorable signal-to-noise ratio. Even more promising would seem to be the application of optical filtering described above to dark field periodic arrays of specifically labeled enzymes.

RESULTS OBTAINED FROM ELECTRON MICROSCOPY OF ENZYMES

In the first part of this section we present a short summary of a few recent applications of electron microscopy toward the analyses of enzyme structure. The examples chosen are representative of more difficult systems.

Illustrative Examples

ACETYLCHOLINESTERASE   Three native and two derived globular forms of this enzyme have been studied by Rieger and co-workers (27) using negative staining and shadowing methods. The native forms have a unique appearance consisting of a head composed of globular-appearing units connected to a tail. The smallest native form appears to have four head globules and the largest has at least ten that are visible. A proposed structure containing 4, 8, and 12 globules for the three native forms is consistent with these observations and the fact that one of the derived globular forms appears to be a stable tetramer of the globules minus the tail portion. The background of some micrographs appears to contain considerable structure, possibly unfolded protein. Nonetheless, the results seem to explain many of the unusual properties of the enzyme as isolated by different techniques. Clearly, electron microscopy is uniquely suited to provide an understanding of the unusual grape-like architecture presented by this enzyme.

ARGININOSUCCINASE   This enzyme (mol wt $\sim$200,000) was shown by Lusty & Ratner (28) to contain four identical or very similar subunits. Negatively stained micrographs of very high quality were obtained (Figure 10), and provides a good example demonstrating the difficulties frequently experienced in the interpretation of tetrameric enzyme images. Many apparent particles are poorly contrasted and some appear frayed or otherwise distorted. Note, however, that the tetrameric nature of the enzyme was not in doubt, and these authors used the micrographs to distinguish between C_4 and D_2 as the probable symmetry of the protein. They noted that the molecules often appeared both as triangles and as squares with an indication of four subunits. The results strongly favor D_2 symmetry with a tetrahedral arrangement of four nearly spherical subunits.

Argininosuccinase is also one of an increasing number of proteins that undergo association to form tubules (Dales, Schultze & Ratner 29). Electron microscopic observations revealed that the sequence of events in association was linear polymer formation, the formation of sheets one molecular layer thick, elongation and folding over of the sheets to form tubules 210 Å in width with a wall diameter of 65 Å, and

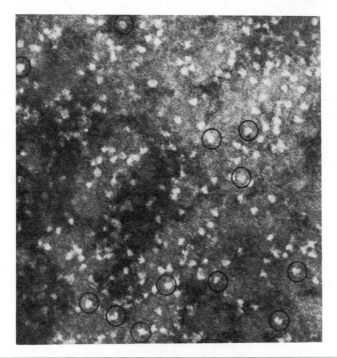

Figure 10   Micrograph of argininosuccinase negatively contrasted at pH 6.8 with potassium phosphotungstate. Magnification is × 260,000. (From Lusty & Ratner 28)

finally the grouping of tubules into paracrystals. Some of the helical parameters of these tubules were obtained by Moody (30) using optical diffraction methods.

MUSCLE PHOSPHORYLASE   Mellema, Van Breemen & Van Bruggen (31) and Hasche-meyer (1) have independently concluded that they were unable to find a unique model that could account for in-focus images of phosphorylase. Substructure was poorly contrasted although the shape of the molecule seemed well defined. Valentine & Chignell (32) had observed a rhombic projection of two-dimensional phosphory-lase crystals. Eagles & Johnson (33) reinvestigated this phenomenon and found a different subunit arrangement. They discuss several possible reasons why their observations differed from that of Valentine & Chignell (32). Optically filtered images of these crystals appear less compatible than one might hope with optically filtered images obtained by Kiselev et al (26) and the model proposed by them (see example in Figure 9). Kiselev and co-workers (26) find their banana-like subunit model compatible with images of individual molecules stained with uranyl acetate as well. These images are quite different, for example, from those obtained by one of us (1) with phosphotungstate staining. Re-examination of many of our micrographs of phosphorylase crystals show the curious phenomenon that particles lying in thicker stain near a partially dissociated crystal appear identical to profiles observed by Kiselev et al (26), while those lying nearby but in thinner stain are identical to images shown earlier (1). Perhaps the particles flatten into a more uniform mass in thin stain and are better preserved in thicker phosphotungstate and in uranyl acetate. While we do not know the reason for this rather extreme variation in observed images, it is at least comforting to know that they are real and that personal bias and field selection played no major role in the studies. For completeness with respect to the model proposed by Kiselev et al (26), we quote their own conclusions: "It is quite obvious that in the real protein molecules such a large number of gaps between subunits is absurd. The elongated subunits in the model should be regarded as an image from the sections with high density of matter, the space between them also filled, to greater or lesser extent, with protein." When the true mass distribution within this enzyme becomes known, a great deal will be learned regarding possible artifacts in electron microscopic studies of enzymes as well.

Kiselev, Lerner & Livanova (26) also showed that a variety of tubular aggregates could be formed with phosphorylase to which a little protamine was added. Single, double, and multiwalled tubes were observed. Whether or not such tendencies are physiologically significant, it is clear that the principles of quasi-equivalence outlined first by Casper & Klug (34) play important roles in some types of enzyme association, perhaps in a manner not too dissimilar from their role in virology.

GLUTAMATE DEHYDROGENASE   In a previous review (1) it was concluded that electron microscopic images observed on this enzyme by Valentine (35) showing very clear trimeric presentations in a two-dimensional array were evidence for an eclipsed double trimer structure with D_3 symmetry. This result presupposed that the number of subunits in the enzyme was six rather than three, in accord with molecular weight studies performed by other methods. Josephs (23) has presented strong evidence that the double trimer ring in the hexamer is staggered. Images consistent only with this structure were presented, although the argument is somewhat weakened by the absence of adequate large field views in which the frequency of occurrence of the nontrimeric images can be evaluated. By use of a tilting stage, Josephs (23) further demonstrated that trimer images did not arise from two layers of eclipsed stacked trimers and concluded that trimer images resulted from transverse cleavage of the hexamer.

Electron micrographs of glutamate dehydrogenase presented by Munn (36) show an extremely high fraction of trimer-like images, and he reports little success in locating images of the type Josephs (23) used to support the staggered trimer model of six subunits. At present, it seems wisest to accept the correctness of all the data and assume that glutamate dehydrogenase is a hexamer composed of staggered trimer rings which dissociate to a very high degree under most conditions for negative staining into species containing only three subunits. This "half-mer" in turn preferentially orients with its threefold axis normal to the support film.

Electron micrographs of glutamate dehydrogenase have also been presented by Fiskin, Van Bruggen & Fisher (37). Their large field micrographs appear to be of excellent quality and, although the magnification presented is insufficient to be certain, the images appear quite similar to those noted above. These authors contend that they are resolving and contrasting discrete areas of tertiary structure within a polypeptide chain. They proceed to fit models based on these areas of tertiary structure (which they term fold-ons) to their micrographs. Although evidence is presented that the intrasubunit contrast exceeds background phase contrast, it seems unlikely that this is sufficient demonstration of real structural detail since, as we noted earlier, interference effects of protein in stain superimposed on support film produces a complex image whose properties have not been treated theoretically (to our limited knowledge in this area). In addition, although the existence of such separately folded tertiary structure with more loosely connected portions of the polypeptide chain joining them with covalent linkage seems acceptable in some cases (e.g. a few are mentioned in the review by Green 16), the concept of placing such nearly equal sized fold-ons at regular positions within geometrical figures seems at least esthetically unpleasing to us in the light of current thoughts regarding evolution of oligomeric proteins. In any event, it seems doubtful that such studies can convince a majority of biochemists until supporting evidence has been obtained.

TRANSCARBOXYLASE   The "transcarboxylase story" includes one of the most elegant applications of electron microscopy to the elucidation of enzyme structure. The results, which include the famous "Mickey Mouse" profile of the enzyme and the labeling of the biotin sites with avidin, are too complex to detail here. The biochemist will find remarkable deductions and beautiful micrographs in the reviews by Green (16) and Wood (38) and references cited therein.

MULTIENZYME COMPLEXES   One of the outstanding contributions of electron microscopy to the biochemist has been to afford him the opportunity to appreciate (in part) nature's grand architectural design in the formation of multienzyme complexes. Regrettably, limitations imposed by the complexity of the systems, the space allotted to this review, as well as our limited experience in this area, all contribute to our neglect of this exciting area of research. Instead, we refer the reader to the excellent review by Reed & Cox (39) and to a selected list of recent references (40–44).

NONBACTERIAL GLUTAMINE SYNTHETASES   Electron micrographs of ovine brain glutamine synthetase stained with uranyl oxalate, phosphotungstate, and uranyl acetate were first presented by one of us (Haschemeyer 45) several years ago. Since that time, we have noted a nearly identical appearance of glutamine synthetase isolated from pea seedlings, human brain, rat liver, and chicken liver. Others have obtained very similar images with enzyme from chicken neural retina (Sarkar et al 46), hamster liver (Tiemeier & Milman 47), and pig brain (Stahl & Jaenicke 48). These images were initially described (Haschemeyer 45) as tetramers, double bars, and more complex presentations with an appearance approximating that of a capital H, N, or I. An octameric model with D_4 symmetry was proposed for the enzyme in which two nonspherical subunits formed the isologous contact between two tetramer rings. A lack of stain penetration within the dimers explained the most frequently observed image (particularly with uranyl oxalate stain), namely, the double bar. A similar conclusion was reached independently by Tiemeier & Milman (47) and by Stahl & Jaenicke (48), who were unaware of the earlier published electron microscopic study.

The techniques used for the molecular weight determinations of glutamine synthetases and their constituent polypeptide chains are subject to well-known experimental and theoretical difficulties (Haschemeyer & Haschemeyer 17). Therefore, the octameric model depends primarily on electron microscopic observations of tetrameric images with an apparent fourfold axis of symmetry. These images are relatively rare, and a number of other recurrent profiles are observed in micrographs of glutamine synthetase (Haschemeyer 45) that are inconsistent with the proposed model. Although the latter probably reflect drying artifacts, the need to rely so heavily on the tetrameric images to support the octameric model suggests that further data will be required before the structure of glutamine synthetase is securely established.

ASPARTATE TRANSCARBAMYLASE   One of the most elegant results of electron microscopy as applied to enzyme structure was obtained by Richards & Williams (49) on aspartate transcarbamylase.

In one method for preparing specimens for electron microscopy, 2% potassium phosphotungstate was first sprayed on carbon-coated collodion films which had been subjected to vacuum glow discharge to render them hydrophilic. The enzyme was then super-sprayed onto grids where the negative stain had already dried into an extremely thin layer. Electron micrographs of the enzyme prepared in this manner are shown in Figure 11. The solid triangular structure within the larger triangle has a very similar appearance to micrographs of the catalytic subunit itself, and it is reasonably inferred that the material in the arms of the larger triangle represents the three dimeric regulatory subunits of mol wt ∼34,000 each. The clearly triangular appearance of the central dense structure was used to infer that the two known catalytic subunits (containing three polypeptide chains each) must be in a nearly eclipsed position. A further description and discussion of these results is given by Cohlberg, Pigiet & Schachman (50) where references to other structural and enzymatic studies on aspartate transcarbamylase may also be found.

Richards & Williams (49) also prepared specimens for electron microscopy in more conventional ways and found that these particles displayed a much greater variability in their appearance. They conclude that this may be associated with greater variation in orientation of the molecule or with distortions in drying. A considerable number of particles appeared to have their threefold axis parallel to the support film, in contrast to the particles shown in Figure 11, where a high percentage of particles clearly assume an orientation with the threefold axis perpendicular to the film. Perhaps the super-spray method described here (or some variation of it) can be used to induce other enzymes to assume a preferred orientation on the support film.

GLUCOSE-6-PHOSPHATE DEHYDROGENASE   Glucose-6-phosphate dehydrogenase isolated from human erythrocytes can exist in monomer, dimer, or tetramer forms

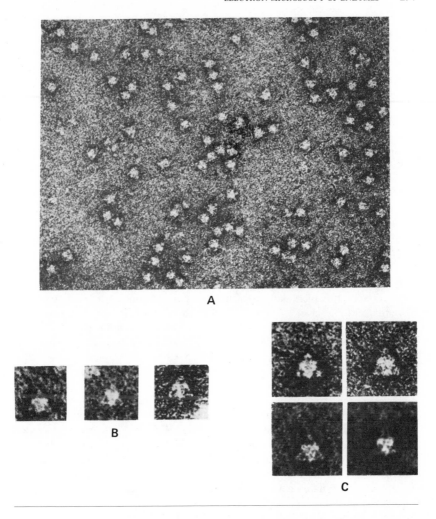

Figure 11 Electron micrograph of aspartate transcarbamylase negatively stained by the super-spray method (see text). A: General field. B: Selected particles showing the triangle within a triangle structure. C: Montage images obtained from superposition of images from 5 particles. Magnification: A: × 200,000. B and C: × 400,000 (from Richards & Williams 49)

depending on environmental conditions. This complex system was investigated by electron microscopy by Wrigley and associates (51). At first glance, their micrographs (which are of superior quality) appear to be essentially uninterpretable, in accord with the usual experience of others for such systems. However, Wrigley and co-workers (51) presented a unique statistical analysis of their micrographs in which the frequency distribution of molecules with particular dimensions for their major and minor axes are displayed as a three-dimensional histogram. These distributions were used to support models for the shape of the various molecular species. Data is also presented on the kinetics of interconversion of oligomeric species as deduced from electron micrographs. The data obtained is remarkable for its internal consistency.

The interpretations presented by Wrigley et al (51) are quite naturally subject to some uncertainties. These reservations are amply described by the authors themselves and we do not detail them here. We do wish to emphasize that this publication demonstrates that the information content of electron micrographs of enzymes can in some cases considerably exceed that which is obvious, provided an appropriate analysis is made.

Overview and General Conclusions

The identification of enzyme quaternary structure by negative staining has always been plagued by complications arising from superposition of stain penetration on the top surface and bottom of the particle. For certain particles, an interpretable image requires a view quite close to a symmetry axis. Therefore, if the particle is randomly oriented with respect to the electron beam, such views are expected to be rare events. It is thus possible that enhanced stickiness of a contaminant in a preferred orientation, or infrequent but consistent distortions of some particles, may present more frequent images of a defined type than those which truly reflect quaternary structure of the enzyme in question. It is a major frustration for the particle electron microscopist to be aware that the information he seeks may be hidden as rare events within a background of innumerable artifacts. We have little doubt that the major interpretive errors for high quality micrographs have resulted from concentrating on the appearance of a small fraction of total images and subsequently excusing the presence of other equally vivid images on the basis of drying artifacts etc.

The above problem is most frequently circumvented as the result of positive interactions between the particle and the support film to achieve a substantial fraction of images with preferred particle orientations (1, 2). If such orientation is not achieved with one negative staining method, others should be tried (e.g. variations in support film characteristics and negative stain). An outstanding example of the use of this method in structural studies on aspartate transcarbamylase has been given earlier.

A considerable effort has been made to circumvent the superposition problem by shadow-casting methods. Clearly, if only the particle topology were contrasted, interpretation would be greatly simplified. Shadowing has always been a powerful method of determining general outlines of particle shape (Hall 52), but even with the newest methods (Abermann et al 53, Henderson & Griffiths 54), well-documented and clear visualization of particle subunit structure has been rare (1). A number of laboratories, including our own, are reinvestigating shadowing methods by applying such minimal deposition of metal that the image is conveniently contrasted only for dark field observation. In this way, much of the metal granularity is limited and obscuring of subunit detail by thicker deposits of metal may be avoided (Dubochet 5).

Dark field electron microscopy of unstained biological particles has already been successfully used to study proteins, nucleic acids, and bacteriophages (Ottensmeyer 56; de Harven, Leonard & Kleinschmidt 57) and increasing applications to oligomeric enzymes can be expected. The potential resolving power of dark field electron microscopy for contrasting subunit structure has been demonstrated in images of the tail structure of bacteriophage SP50 supported on ultrathin (10–20 Å) film (de Harven et al 57). Dubochet and co-workers (58) effectively used dark field observations of lightly positively stained samples of DNA and RNA polymerase to study their interaction and to obtain some details regarding the structure of the enzyme. Further application of these methods to enzyme study can be expected.

A survey of the available literature suggests that, in general, difficulties encountered in electron microscopic studies of enzymes increase with decreasing protein molecular weight, and even for larger enzymes, with decreasing size of constituent polypeptide chains. Detail within enzymes and other small proteins composed of single polypeptide chains has not yet been unequivocally demonstrated. We refer the reader to the convincing paper by Mellema, Van Bruggen & Gruber (55) for a discussion of negative staining results on such systems. Dark field results on single polypeptide chains have been presented (Ottensmeyer 56) without the documentation of large field views required for confident evaluation. The presentation of such views has been complicated by background noise and we hope that newer support films such as aluminum oxide (Koller et al 4) will soon obviate this problem.

Oligomers with small polypeptide chains often suffer from insufficient penetration of stain between subunits. The probable reason for this is the thermodynamic requirement for a moderate sized intersubunit contact area. Thus, as the subunit size decreases, the size of the groove between two spheres (for example) decreases markedly. For larger proteins, not only is the likelihood of sufficient contrast decreased, but problems imposed by superposition of top and bottom staining are also magnified.

A majority of oligomeric enzymes are dimers and tetramers, presumably with C_2 and D_2 symmetry, respectively (17). Although there are several notable exceptions, it appears to be unexpectedly difficult to deduce the conformation of these enzymes by electron microscopy. Why this should be so is not entirely clear, but we offer the following possibilities. Both symmetry classes appear less likely to assume a preferential orientation with respect to the support film, as might be expected (2). The tetramers may assume a geometry approximating the placement of subunits at the corners of a tetrahedron, in which case there is no reason to suppose that orientation along a twofold axis will be a frequent observation. Perhaps the lack of additional subunit contacts compared to other symmetry classes, particularly for dimers, increases the probability of detrimental unfolding. The interpretation of micrographs of dimers and tetramers is often simplified by the fact that their stoichiometry is already known, but it would be very helpful if electron microscopy could be more directly interpreted. One such important case arises from the frequency with which larger oligomers are dissociated reversibly into dimers. Directly obtaining information regarding subunit shape and intersubunit staining characteristics from electron micrographs of the dimer would be an immense aid in the interpretation of micrographs of the parent oligomer. In our own laboratory, we have been unsuccessful in confidently interpreting dimer structures from three enzymes: phosphorylase, L-aspartate β-decarboxylase, and ovine brain glutamine synthetase.

Finally, we wish to emphasize that despite the above generalizations one cannot at present predict when an electron microscopic study of a particular enzyme will provide significant structural insight. Consequently, such investigations are to be encouraged whenever possible. In addition, we believe that large field, high quality micrographs of reasonably preserved enzymes should be published. even if they cannot be interpreted at present. It is expected that recurring image patterns can be used to classify enzyme structure and that data on stoichiometry and geometry on one enzyme can then be used to aid in interpreting others. It is also likely that such comparisons may permit a more positive identification of recurrent artifacts.

ACKNOWLEDGMENTS

Some results and conclusions in this chapter are based on research conducted in the authors' laboratories that was supported by grants from the Public Health Service (HL-11822 and CA-08748).

Literature Cited

1. Haschemeyer, R. H. 1970. *Advan. Enzymol.* 33:71–118
2. Haschemeyer, R. H., Myers, R. J. 1972. *Principles and Techniques of Electron Microscopy: Biological Applications,* ed. M. A. Hayat, 2:99–147. Princeton: Van Nostrand. 286 pp.
3. Johnson, H. M. 1973. See Ref. 2, 3:153–98. 321 pp.
4. Koller, T., Beer, M., Muller, M., Muhlethaler, K. 1973. See Ref. 2, 3:53–111. 321 pp.
5. Dubochet, J. 1973. See Ref. 2, 3:113–51. 321 pp.
6. Nermut, M. V. 1972. *J. Microsc.* 96:351–62
7. Hayat, M. A., Zirkin, B. R. 1973. See Ref. 2, 3:297–313. 321 pp.
8. Porter, K., Prescott, D., Frye, J. 1973. *J. Cell Biol.* 57:815–36
9. de Harven, E., Beju, D., Evenson, D. P., Basu, S., Schildlovsky, G. 1973. *Virology* 55:535–40
10. Koller, T., Bernhard, W. 1964. *J. Microsc.* 3:589–606
11. Bowers, W. F., Czubaroff, V. B., Haschemeyer, R. H. 1970. *Biochemistry* 9:2620–25
12. Matricardi, V. R., Moretz, R. C., Parsons, D. F. 1972. *Science* 177:268–70
13. Valentine, R. C., Shapiro, B. M., Stadtman, E. R. 1968. *Biochemistry* 7:2143–52
14. Tate, S. S., Meister, A. 1970. *Biochemistry* 9:2626–32
15. Monod, J., Wyman, J., Changeux, J. P. 1965. *J. Mol. Biol.* 12:88–118
16. Green, N. M. 1972. *Protein-Protein Interactions,* ed. R. Jenicke, E. Helmreich, 183–211. Berlin: Springer. 464 pp.
17. Haschemeyer, R. H., Haschemeyer, A. E. V. 1973. *Proteins: A Guide to Study by Physical and Chemical Methods.* New York: Wiley. 445 pp.
18. Nonomura, Y., Blobel, G., Sabatini, D. 1971. *J. Mol. Biol.* 60:303–32
19. Casper, D. L. D. 1966. *J. Mol. Biol.* 15:365–71
20. Finch, J. T., Klug, A. 1967. *J. Mol. Biol.* 24:289–302
21. Bowers, W. F. 1971. *The Study of Enzyme Quaternary Structure: L-Aspartate β-Decarboxylase.* PhD thesis. Cornell Univ., New York. 253 pp.
22. Finch, J. T., Holmes, E. C. 1967. *Methods Virol.* 3:351–474
23. Josephs, R. 1971. *J. Mol. Biol.* 55:147–53
24. Green, N. M., Konieczny, L., Toms, E. J., Valentine, R. C. 1971. *Biochem. J.* 125:781–91
25. Markham, R. 1968. *Methods Virol.* 4:503–29
26. Kiselev, N. A., Lerner, F. Y., Livanova, N. B. 1971. *J. Mol. Biol.* 62:537–49

27. Rieger, F., Bonn, S., Massoulie, J., Cartaud, J. 1973. *Eur. J. Biochem.* 34: 539–47
28. Lusty, C. J., Ratner, S. 1972. *J. Biol. Chem.* 247: 7010–22
29. Dales, S., Schulze, I. T., Ratner, S. 1971. *Biochim. Biophys. Acta* 229: 771–78
30. Moody, M. F. 1971. *Biochim. Biophys. Acta* 229: 779–84
31. Mellema, J. E., Van Breemen, J. F. L., Van Bruggen, E. F. J. 1968. *Proc. 4th Reg. Conf. Electron Microsc., Rome* 2: 107–8
32. Valentine, R. C., Chignell, D. A. 1968. *Nature* 218: 950–53
33. Eagles, P. A. M., Johnson, L. N. 1972. *J. Mol. Biol.* 64: 693–95
34. Casper, D. L. D., Klug, A. 1962. *Cold Spring Harbor Symp. Quant. Biol.* 27: 1–24
35. Valentine, R. C. 1968. *Proc. 4th Reg. Conf. Electron Microsc., Rome* 2: 3–6
36. Munn, E. A. 1972. *Biochim. Biophys. Acta* 285: 301–13
37. Fiskin, A. M., Van Bruggen, E. F., Fisher, H. F. 1971. *Biochemistry* 10: 2396–408
38. Wood, H. G. 1972. *Enzymes* 6: 83–115
39. Reed, L. J., Cox, D. J. 1970. *Enzymes* 1: 213–40
40. Eley, M. H., Naltihira, G., Hamilton, L., Munk, P., Reed, L. J. 1972. *Arch. Biochem. Biophys.* 152: 655–69
41. Tanaka, N. et al 1972. *J. Biol. Chem.* 247: 4043–49
42. Koike, K. 1971. *J. Biochem. Tokyo* 69: 1143–47
43. Junger, E., Reinaur, H. 1971. *Biochim. Biophys. Acta* 250: 478–90
44. DeRosier, D. J., Oliver, R. M., Reed, L. J. 1971. *Proc. Nat. Acad. Sci. USA* 68: 1135–37
45. Haschemeyer, R. H. 1968. *Trans. NY Acad. Sci.* 30: 875–91
46. Sarkar, P. K., Fischman, P. A., Goldwasser, E., Moscona, A. A. 1972. *J. Biol. Chem.* 247: 7743–49
47. Tiemeier, D. C., Milman, G. 1972. *J. Biol. Chem.* 247: 2272–77
48. Stahl, J., Jaenicke, L. 1972. *Eur. J. Biochem.* 29: 401–7
49. Richards, K. E., Williams, R. C. 1972. *Biochemistry* 11: 3393–95
50. Cohlberg, J. A., Pigiet, V. P., Schachman, H. K. 1972. *Biochemistry* 11: 3396–41
51. Wrigley, N. G., Heather, J. V., Bonsignore, A., DeFlora, A. 1972. *J. Mol. Biol.* 68: 483–99
52. Hall, C. E. 1966. *Introduction to Electron Microscopy.* New York: McGraw. 2nd ed. 397 pp.
53. Abermann, R., Salpeter, M. M., Bachmann, L. 1972. See Ref. 2, 196–217
54. Henderson, W. J., Griffiths, K. 1972. See Ref. 2, 151–193
55. Mellema, J. E., Van Bruggen, E. F. J., Gruber, M. 1968. *J. Mol. Biol.* 31: 75–82
56. Ottensmeyer, F. P. 1969. *Biophys. J.* 9: 1144–49
57. de Harven, E., Leonard, K. R., Kleinschmidt, A. K. 1971. *29th Ann. Proc. Electron Microsc. Soc. Am., Boston,* ed. C. J. Arceneaux
58. Dubochet, J., Ducommun, M., Zollinger, M., Kellenberger, E. 1971. *J. Ultrastruct. Res.* 35: 147–67

REGULATION OF AMINO ACID DECARBOXYLATION[1]

×848

David R. Morris and Robert H. Fillingame
Department of Biochemistry, University of Washington, Seattle, Washington

CONTENTS

INTRODUCTION

The amino acid decarboxylases are involved in a variety of biological processes ranging from bacterial putrefaction to neurotransmission. The enzymatic and structural properties of many of these enzymes have been studied extensively and these aspects have been reviewed in detail elsewhere (see, for example, 1–3). This review treats the amino acid decarboxylases in a regulatory context and deals with their chemical and enzymatic properties only where it is pertinent to this approach. Our intent is to emphasize regulatory questions of potential interest, even though the

[1] Abbreviations: PLP, pyridoxal 5′-phosphate; GABA, γ-aminobutyric acid; DOPA, 3,4-dihydroxyphenylalanine; dopamine, 3,4-hydroxyphenylethylamine.

answers presently available may not be firm; hence, our coverage of the literature is selective rather than exhaustive.

MICROBIAL DECARBOXYLASES

The distinction in microbial systems between biosynthetic and biodegradative (inducible) amino acid decarboxylases arose during studies of putrescine biosynthesis in *Escherichia coli*. It was found that *E. coli* produced two sets of arginine and ornithine decarboxylases (4, 5): one for the synthesis of putrescine and spermidine in cells grown in neutral minimal medium, and the other for degradation of arginine and ornithine in cells grown at low pH. Although with some decarboxylases this distinction may seem artificial, it is maintained for simplicity in the organization of this section of the review.

Biodegradative Decarboxylases

ARGININE, LYSINE, AND ORNITHINE DECARBOXYLASES   It was recognized early in this century that the decarboxylation products of arginine, lysine, and ornithine (respectively agmatine, cadaverine, and putrescine) accumulated in putrefying material together with other amines (reviewed in 6). In an investigation of the mechanism of amine production, Gale discovered the existence of six amino acid decarboxylases, including those specific for arginine, lysine, and ornithine (7). In a preliminary characterization of these enzymes, "codecarboxylase," later shown to be pyridoxal phosphate (PLP), was found to be required for enzymatic activity (8–11). More recently, the biodegradative arginine, lysine, and ornithine decarboxylases of *E. coli* have been extensively purified and characterized (12–17). These three enzymes showed extensive homologies which led to the suggestion that they may be evolutionarily related (16). In his pioneering studies, Gale also showed that these enzymes were adaptive, in that special culture conditions were required for their production. Specifically, it was found that acidic pH, low temperature, and the presence of substrate favored the induction of arginine, lysine, and ornithine decarboxylases (7). With the lysine and ornithine decarboxylases of *E. coli,* the substrate and pH requirements, but not that for low temperature, have been confirmed (16, 17). On the other hand, the amino acid requirements for induction of arginine decarboxylase have proven quite complex, as was initially suggested by Gale's studies (7). Tyrosine, methionine, and asparagine were required in addition to arginine for maximal enzyme production under anaerobic conditions (18). In aerobic cultures, maximum enzyme levels were achieved only in the presence of glutamic acid and ferrous ion in addition to the above supplements (18). It is not clear whether these required nutrients interact directly with the regulatory elements of the enzyme-forming system or if they were involved in a nonspecific sense in maintaining the proper physiological conditions for induction. It is not known whether these additional nutrients are required for maximal production of the lysine and ornithine decarboxylases.

HISTIDINE DECARBOXYLASE   Histamine was one of the first amines to be produced in defined cell culture (19) and histidine decarboxylase was among those enzymes first described by Gale in *E. coli* (7). In recent years, studies of this enzyme from *Lactobacillus* 30A by Snell and his collaborators have greatly extended our insight into the structure, function, and biological significance of the biodegradative amino acid decarboxylases. As was found with the decarboxylases described above, pH values of 5 to 6 and the presence of histidine in excess of the normal growth requirement were necessary for maximal induction of the enzyme (20). The oligomeric form of the enzyme consisted of five each of two different protein subunits of molecular weights 9000 and 29,700 (21). PLP was not required for activity, but instead the enzyme contained pyruvoyl residues attached to the amino terminus of the larger subunit (21, 22). The carbonyl group of the pyruvoyl residue appeared to function analogously to the aldehyde group of PLP in the other amino acid decarboxylases (23). The pyruvoyl residue was generated in vivo from radioactive serine without mixing with the free pyruvate pool (21). Further insight into the synthesis of the pyruvoyl group and the generation of nonidentical subunits came with the discovery of mutants of *Lactobacillus* 30A, which lacked histidine decarboxylase and accumulated a zymogen form of the enzyme (24, 25). A study of the terminal amino acid residues of the zymogen suggested that the smaller subunit of the enzyme arose from an amino-terminal fragment and the larger subunit from the carboxyl end of the zymogen. It was suggested that the pyruvoyl residue arose during cleavage of the zymogen to the enzyme form (25). Since it was possible to activate the zymogen in vitro, the nature of the cleavage reaction and the defect in the mutant should be amenable to biochemical study.

GLUTAMIC ACID DECARBOXYLASE   The biodegradative glutamic acid decarboxylase from *E. coli* has not only been rigorously characterized from a chemical and physical standpoint (26, 27), but also genetic studies have been initiated into the mechanism of its regulation. As with other biodegradative decarboxylases, maximal induction was favored by low pH. Conversely, Gale found no strong requirement for substrate (7). Later studies by Halpern (28) confirmed this conclusion and demonstrated that the glutamate requirement for maximal induction was a complex function of carbon source and temperature. For example, cells grown on glycerol require glutamate for maximal induction at 37°, but not at 30°. On the other hand, with glucose as a carbon source there was no dependence on glutamate for induction at either temperature. It is possible that these inconsistencies arose because of varying degrees of induction by high internal glutamate pools (29, 30) which could fluctuate with the growth conditions.

Halpern and his collaborators have begun a study of the genetic control of glutamic acid decarboxylase in *E. coli*. On screening various strains of *E. coli* K12, they found differences in the specific activities of glutamic acid decarboxylase in crude extracts (31). Since in all parameters tested, including immunological and kinetic behavior, the enzymes were identical (32, 33), these differences in enzyme level were attributed to different alleles of a regulatory gene, *gadR*. This gene was placed by phage transduction at approximately 72 min on the *E. coli* genetic map (31). A strain was also discovered which, by preliminary tests of immunological cross reactivity and heat stability, produced an altered glutamic acid decarboxylase (31). This mutation in the presumed structural gene for glutamic acid decarboxylase, *gadS,* mapped in the same region as *gadR* (31). Although isolation and characterization of additional mutants is clearly required, these experiments represent an encouraging beginning to our understanding of the regulation of the biodegradative amino acid decarboxylases.

A possible role for glutamic acid decarboxylase in bacterial sporulation has recently been discovered. Spores of *Bacillus* species contained high levels of glutamic acid (30) and upon germination there was a rapid decarboxylation to γ-aminobutyric acid (GABA) (34). Mutants have been isolated which require GABA for spore germination (35). Glutamic acid decarboxylase has been detected in spores of *Bacillus megaterium,* but not in the GABA-requiring mutant (36). Hence, it has been suggested that increased levels of GABA, produced via activation of glutamic decarboxylase, are involved in triggering spore germination (36).

ROLE OF BIODEGRADATIVE DECARBOXYLASES   Although one could imagine that these enzymes might be catalyzing an initial step in carbon and nitrogen utilization, this is clearly not the case in at least two instances. Putrescine, the product of ornithine decarboxylase and indirectly of arginine decarboxylase, could not be utilized by wild-type *E. coli* because of insufficient levels of putrescine-glutamate aminotransferase and γ-aminobutyraldehyde dehydrogenase (37). Although certain strains of *E. coli* could utilize glutamic acid as a carbon and nitrogen source, the ability to grow on glutamate was unrelated to the level of glutamic acid decarboxylase (38), and it was shown that glutamate was used via a transamination pathway rather than by decarboxylation (39). *E. coli* could utilize GABA only through a mutation giving rise to increased levels of GABA-glutamate aminotransferase and succinic semialdehyde dehydrogenase (40).

In an early paper on histamine formation by bacterial cultures, Koessler & Hanke suggested that the conversion of histidine to histamine could function either to neutralize the acidic pH of the culture medium or to obtain the carbon of the carboxyl group for utilization (19). After it was appreciated that bacteria required carbon dioxide for growth (41, 42), Gale extended the latter argument in proposing that amino acid decarboxylation might serve to maintain the cellular bicarbonate concentration under acidic culture conditions (6). However, it was only after a lapse of about twenty years that these proposals were rigorously tested. Becker isolated a mutant of *E. coli* lacking the biodegradative arginine and ornithine decarboxylases (43). The properties of this mutant clearly indicated that these enzymes were indispensable for growth in acid medium. Recsei & Snell have isolated and extensively characterized a mutant of *Lactobacillus* 30A lacking histidine decarboxylase (24). In a culture medium of pH 4.3, conditions under which the wild-type organism would grow, the mutant was incapable of growth even when provided with histamine and an atmosphere of 10% CO_2. On the other hand, both the mutant and the wild-type organism would grow at pH 4.3 when provided with ornithine, a substrate for which both strains produced a biodegradative decarboxylase. Recsei & Snell also demonstrated that the major defect resulting from the absence of histidine decarboxylase was an inability of the mutant to regulate the cultural pH. Hence, it was proposed that the function of the biodegradative amino acid decarboxylases was not to provide the endproducts of the reaction per se, but rather to consume protons and thus neutralize the acid products of carbohydrate fermentation (24).

Biosynthetic Decarboxylases

ORNITHINE DECARBOXYLASE   The existence of biosynthetic ornithine decarboxylase was discovered during a study of polyamine biosynthesis in *E. coli* grown in neutral minimal medium, conditions not conducive to induction of the biodegradative enzyme (4, 5). The role of this enzyme has recently been definitely established through isolation of mutants lacking the biosynthetic ornithine decarboxylase and deficient in both putrescine and spermidine (44). Biosynthetic ornithine decarboxylase was purified to homogeneity and compared with the biodegradative enzyme (17). Although both enzymes displayed very similar subunit structures and kinetic properties, the biosynthetic enzyme did not cross-react with antibodies to the biodegradative enzyme. An additional distinction between the two enzymes is their pH optima: pH 7.0 for the biodegradative enzyme as opposed to pH 8.3 for the biosynthetic. Neither enzyme displayed a metal ion requirement as was observed with the biosynthetic arginine decarboxylase (see below).

Since ornithine is an intermediate in the pathway of arginine biosynthesis, its cellular level, and hence the rate of putrescine synthesis from this source, should be regulated by the level of arginine. Thus, when growing cultures of *E. coli* were supplied with arginine, the ornithine pool dropped to undetectable levels and no putrescine was synthesized from ornithine (45, 46). Therefore, if it were not for the existence of an alternate pathway of putrescine synthesis from arginine (see below), *E. coli* would be polyamine-starved under these conditions. This would be no problem in those organisms capable of converting arginine to ornithine (see 46). For example, *Neurospora,* which has an inducible arginase, synthesized putrescine by decarboxylation of ornithine even when grown in the presence of arginine (47).

Endproduct control of putrescine and spermidine synthesis has been demonstrated in growing cultures of *E. coli* (48). Inhibition of crude (48) and partially purified (46) preparations of ornithine decarboxylase by putrescine and spermidine was observed and recently has been studied in detail with the purified enzyme (17). At saturating ornithine concentrations, 50% inhibition was produced by 5 mM spermidine or 20 mM putrescine. Although these concentrations approximate the intracellular levels of putrescine and spermidine, the interpretation of this result was complicated by the binding of the positively charged polyamines to intracellular components (discussed in 46).

Homogeneous preparations of biosynthetic ornithine decarboxylase displayed a K_m for ornithine of 5.6 mM (17). This was two orders of magnitude higher than the estimated intracellular ornithine level (45, 46). This discrepancy has been resolved in part with the discovery that certain nucleoside phosphates are positive effectors for the enzyme (49). GTP and dGTP were the most effective activators, displaying K_m values for activation of less than 1 μM. In the presence of saturating concentrations of GTP, the K_m for ornithine was lowered to a more reasonable concentration, 0.28 mM. In light of this result, it is clearly important to re-examine the endproduct inhibition of the enzyme in the presence of GTP.

The intracellular concentration of putrescine in *E. coli* was found to be a function of the osmolarity of the culture medium, high osmotic strength leading to markedly depressed putrescine levels, partially because of putrescine excretion (50). K^+-deficient cells displayed an increased rate of putrescine biosynthesis (51). Since ornithine decarboxylase was inhibited 50% by 0.1 *M* KCl, the osmotic control of putrescine synthesis might be mediated via this mechanism (51).

In addition to these controls on the activity of biosynthetic ornithine decarboxylase, it appears that the level of the enzyme is under regulation as well. When *E. coli* was made polyamine deficient, either by growth in an arginine-limited chemostat (48) or by growth of a mutant in the absence of polyamines (46, 52), the level of ornithine decarboxylase increased a maximum of fivefold. The enzyme level returned to normal upon restoration of putrescine or spermidine to the culture medium. A twofold increase in ornithine decarboxylase activity was also observed early after phage infection of *E. coli* (53). However, the physiological significance of this observation is unclear since there was apparently no change in the rate of amine synthesis during phage infection (54). The fact that the gene for biosynthetic ornithine decarboxylase, *speC,* was located very close to other genes controlling putrescine and spermidine synthesis suggests the possibility of a polyamine operon in *E. coli* (44).

ARGININE DECARBOXYLASE   Biosynthetic arginine decarboxylase catalyzes the first step in a second pathway of putrescine biosynthesis in *E. coli* (5). This pathway from arginine to putrescine functioned primarily when the cells were provided with an exogenous source of arginine (45, 46), conditions that precluded synthesis of putrescine from ornithine (see above). Biosynthetic arginine decarboxylase has been purified to homogeneity and was clearly distinct in many properties from the biodegradative enzyme (55, 56). An alkaline pH optimum and an absolute requirement for magnesium ion in addition to PLP were two distinctive catalytic properties of the biosynthetic enzyme.

Arginine decarboxylase was sensitive to feedback inhibition by putrescine and spermidine at all stages of purification (46, 48, 55). One mode of inhibition by polyamines was through competition with magnesium ion, spermidine being ten times as effective as putrescine (55). The result of this interaction between magnesium ion, as a positive effector, and polyamines, as negative effectors, would be to buffer the ratio between the cellular levels of these cations, high magnesium ion leading to increased polyamine synthesis and low magnesium ion slowing polyamine synthesis. This regulatory mechanism was suggested to be of physiological significance since the ratio of bound magnesium to polyamines was critical in maintaining intact active ribosomes (57).

Arginine decarboxylase lies at the beginning of one of two metabolic branches originating from arginine, the other leading via arginyl tRNA synthetase to protein. In *E. coli* growing in the presence of an exogenous source of arginine, approximately 80% of the arginine was converted to protein and 20% to polyamines (48). Removal of the exogenous source of arginine had no effect on the rate of protein synthesis, because of endogenous arginine synthesis, while the rate of arginine conversion to putrescine decreased between four- and fiftyfold depending on the particular strain

(45, 46). The potential deficit in polyamine synthesis was compensated for by increased conversion of ornithine to putrescine (see above). How might this partitioning of arginine between the synthesis of proteins and of polyamines be regulated? Since the K_m value for arginyl tRNA synthetase was approximately tenfold lower than that of arginine decarboxylase, a large decrease in pool level on switching from exogenous to endogenous arginine could bring about preferential inhibition of arginine decarboxylase (55). However, there are indications that the explanation may be more complex. Since the intracellular arginine level decreased only 20% on removal of exogenous arginine, it was suggested that exogenous arginine may have unique access to the biosynthetic arginine decarboxylase, perhaps through a coupling of transport and decarboxylation (45, 46). Experimental support for this concept came from a study of chemostat-grown cultures of *E. coli,* which indicated that exogenous arginine was preferred for putrescine synthesis over the endogenous supply (58). Recent studies of a mutant of *E. coli* lacking ornithine decarboxylase suggested that even under conditions of polyamine deficiency, endogenously formed arginine could not be efficiently converted to putrescine (44). The existence of this mutant clearly opens the way to additional studies of the utilization of arginine in polyamine synthesis.

As was the case with biosynthetic ornithine decarboxylase, the rate of production of biosynthetic arginine decarboxylase was also regulated. During polyamine limitation the level of arginine decarboxylase was increased as much as fourfold (46, 48, 52). The isolation and genetic mapping of mutants in biosynthetic arginine decarboxylase (*speA*), agmatine ureohydrolase (*speB)*, and biosynthetic ornithine decarboxylase (*speC,* see above) have given us new insight into possibilities for regulating the expression of these genes (44, 59). All three genes were closely linked to *metK,* a gene controlling the production of S-adenosylmethionine, a precursor of spermidine. Thus, the genetic evidence to date suggests a clustering of the genes for putrescine and spermidine synthesis into what might be a single regulatory unit.

A BIOSYNTHETIC LYSINE DECARBOXYLASE?   Wild-type *E. coli* grown under conditions for induction of the biodegradative lysine decarboxylase contained high levels of cadaverine, whereas only traces appeared in cells grown in neutral minimal medium (60). On the other hand, starvation of a polyamine auxotroph in minimal medium led to the production of significant quantities of cadaverine and the corresponding analog of spermidine (61). On investigating the mechanism of cadaverine production in this mutant, Leifer & Maas have detected two low-level activities of lysine decarboxylase in cells grown on minimal medium (62). One activity was maximal at pH 6, corresponding to the biodegradative lysine decarboxylase (16), while the other had a pH optimum of 9. The establishment of the pH 9 form as a biosynthetic lysine decarboxylase clearly awaits physical separation of the two activities. However, inhibition of both activities by putrescine and spermidine provided an explanation for the lack of cadaverine production except under conditions of putrescine and spermidine deficiency (63).

S-ADENOSYLMETHIONINE DECARBOXYLASE   S-adenosylmethionine decarboxylase functions in the generation of the *n*-propylamine moiety of spermidine and spermine

(64). This enzyme was purified to homogeneity and apparently contained covalently bound pyruvate as the carbonyl group involved in enzymatic activity (65). Since this enzyme is located at an important branch point in metabolism, S-adenosylmethionine being involved in transmethylation reactions as well as in spermidine synthesis, it seems likely that S-adenosylmethionine decarboxylase is regulated. A recent preliminary report indicated that both the activity of the enzyme and its cellular level were controlled by spermidine in *E. coli* (66). Since S-adenosylmethionine decarboxylase displayed an absolute requirement for magnesium ion (64), and in light of the interactions between magnesium ion and polyamines in the regulation of biosynthetic arginine decarboxylase (see above), it will be interesting to see if these same regulatory relationships exist with S-adenosylmethionine decarboxylase.

DIAMINOPIMELIC ACID DECARBOXYLASE   Diaminopimelic acid decarboxylase catalyzes the last step of lysine biosynthesis in many microorganisms (see 67 for review). Enzyme synthesis has been shown to be repressed by lysine in a number of organisms (68–71) and lysine is also a competitive inhibitor (72, 73). The physiological significance of the latter observation is uncertain since the enzyme was relatively insensitive to lysine inhibition.

Diaminopimelic acid decarboxylase is involved in an interesting regulatory event during the sporulation of *Bacillus* species (71, 74). Early in the sporulation process, the first enzyme of the lysine pathway became insensitive to feedback inhibition by lysine, and diaminopimelic acid decarboxylase was inactivated. The net result of this was to direct the flow of the lysine pathway to the synthesis of dipicolinic acid and diaminopimelic acid, products present in high concentration in mature spores. It will be interesting to know if the inactivation of diaminopimelic acid decarboxylase was due to the increased proteolytic activity during sporulation (75, 76) or whether another specific modification mechanism was involved.

ANIMAL DECARBOXYLASES

Synthesis of Polyamines

In animal tissues the only known mode of putrescine synthesis proceeds via L-ornithine decarboxylase, and the first step in spermidine and spermine synthesis is the enzymatic decarboxylation of S-adenosyl-L-methionine (77–79). The existence of these decarboxylases in animals was first shown five years ago. Concomitant with these discoveries were exciting observations concerning their regulation which probably triggered the flood of studies in recent years. Although this section covers most of the regulatory phenomena concerning these enzymes, aspects not covered in recent reviews (77–79) are emphasized.

ORNITHINE DECARBOXYLASE   The detection of ornithine decarboxylase activity in animal tissues was reported independently by three laboratories in 1968 (80–82). In that same year, the enzyme was shown to rise dramatically in liver after partial hepatectomy or growth hormone treatment (80, 82, 83) and in chick embryos (82). Interest in this enzyme was further stimulated by the report of Russell & Snyder (84) that ornithine decarboxylase turned over with a very short apparent half-life, a result which suggested an important regulatory role.

In contrast to the regulation of enzyme levels (see below), control through low molecular weight effectors does not seem to be of major physiological importance with mammalian ornithine decarboxylase. Putrescine and spermidine were competitive feedback inhibitors of the enzyme from prostate (K_i values being 1 and 3 mM respectively) (81). On the other hand, enzyme isolated from other sources has not been susceptible to feedback inhibition (85–87). Other compounds have been tested (79, 88, 89), e.g. nucleotides and molecules from neighboring metabolic pathways, but no low molecular weight effectors have been found.

The purification of ornithine decarboxylase has proven to be a problem, although it is apparent that a convenient source of pure enzyme is essential for detailed regulatory studies. Three laboratories have reported purification of ornithine decarboxylase to what appeared to be homogeneity. Purifications of 300-fold from rat prostate (88) and 175-fold from regenerating liver (89) were reported. However, Ono et al (85), starting with liver of thioacetamide-treated rats, which has comparable specific enzymatic activities to the above tissues, have recently reported a 5000-fold purification. The final specific activity of this preparation was 20 to 60 times those previously reported (88, 89). This procedure yielded 1 mg of protein from 7 kg of rat liver, which illustrates the potential problem that will be encountered in approaching regulation of this enzyme with immunochemical techniques.

As indicated above, ornithine decarboxylase is an adaptive enzyme. Increases in activity have now been observed in many tissues in response to a variety of stimuli. We attempt to summarize in Table 1 the systems in which changes have been observed. In addition, large changes in activity were observed during the development of a number of animals (82, 120, 135–137); high activities were found in some tumors (82, 138, 139); and changes were observed during the cell cycle (134, 140). As shown in Table 1, the magnitude of the response varied immensely depending on the tissue and stimulus. The very large increases, i.e. twenty- to several hundred-fold, have generally been associated with macromolecular synthesis and cell growth and division.

Table 1 should also emphasize the complexity of regulation of this enzyme in vivo, particularly in a tissue such as liver. Shrock et al (92) alerted workers in this area to the pitfalls in trying to interpret the physiological significance of changes in ornithine decarboxylase activity. Growth-promoting stimuli were capable of elevating activity in the liver, but so were other environmental factors such as stress, sham operations, and feeding (92, 106; Table 1). Activity was also affected by many hormones (Table 1). The pituitary gland was required for the early but not late increases observed during liver regeneration (91). It has recently been shown that ornithine decarboxylase activity increased in a biphasic manner after partial hepatectomy, the early phase being modified by age as well as by the pituitary gland (94). Furthermore, intact adrenals were required for optimal response to growth hormones (101). The complex factors operating in vivo make it likely that regulation of this enzyme can be most fruitfully approached in an in vitro system.

In all systems studied to date, protein synthesis was required for elevations of

Table 1   Response of ornithine decarboxylase to stimuli

Tissue (Stimulus)	Fold Increase (X)	Reference
Liver		
Partial hepatectomy (rat)	15–70	80, 82, 90–96
Partial hepatectomy (mouse)	3–4	97
Thioacetamide	15–90	85, 90, 95, 98–100
Growth hormone	4–35	83, 101–106
Glucocorticoids	20–60	95, 106–109
Insulin	25	107
Glucagon	15	107
Thyroxine	10	107
Theophylline or dibutryl cyclic AMP	7–10	108, 109
Epidermal growth factor	3	110
Feeding or infusion of amino acids	8–15	90, 92, 96, 103, 106
Hypertonic infusion mannitol	20–30	92
Circadian cycles	12	106
Celite injection	25–50	92
Puromycin	14	109, 111
Miscellaneous drugs	2–7	112
Combination of triiodothyronine,		
amino acids, glucagon, heparin	50	96
Kidney		
Growth hormone	5–100	105, 113
Glucocorticoids	9	113
Unilateral nephrectomy	5	113
Epidermal growth factor	8	110
Folic acid	6	99
Heart		
Growth hormone	2	105
Induced hypertrophy	3	114–116
Skeletal muscle		
Growth hormone	2	105

ornithine decarboxylase activity, as judged by inhibitor experiments (84, 90, 92, 100, 101, 103, 118, 123, 125, 127, 132). A simple interpretation of these results is that de novo synthesis of ornithine decarboxylase was required. However, other possibilities remain, e.g. inhibition of synthesis of an activator. Other approaches, such as immunochemical, need to be taken to show that the amount of enzyme is increasing. Mixing experiments appeared to rule out the presence (in excess) of diffusable inhibitors in tissues prior to stimulation (82, 118, 128).

Experiments with actinomycin D suggested that increases in ornithine decarboxylase were at least partially dependent on RNA synthesis (84, 90, 92, 98, 102, 103, 108, 111, 113, 123, 127, 131, 132). Actinomycin D, when given at the time of the stimulus, was shown to completely block increases when activity was measured a few hours later (84, 90, 92, 102, 113, 123, 127). For example, Russell & Snyder (84)

Table 1   (Continued)

Tissue (Stimulus)	Fold Increase (X)	Reference
Adrenals		
Growth hormone	5	117
Adrenocorticotropin	15	117
Spleen, thymus		
Growth hormone	5–10	105
Skin (epidermis)		
Epidermal growth factor[a]	4–30	118
Prostate		
Androgens	6	119
Testes		
Gonadotropin	3	120
Epidermal growth factor[a]	4–15	110
Mammary gland		
Pregnancy	20	121
Uterus, oviduct, ovaries		
Estrous cycle	>20	122–124
Estrogens[a]	15–50	123, 125, 126
Luteinizing hormone	10	122, 127
Gonadotropin	10	122
Cultured lymphocytes		
Mitogens	30–100	128–130
Amino acids	5	129, 131
Stationary HTC cells		
Dilution with fresh medium	250	132, 133
Serum addition	35	133
Amino acid addition	4–25	133
Don C cells		
Cell cycle	2	134

[a] Increases have been observed both in the intact animal and in vitro.

demonstrated that addition of actinomycin at the time of partial hepatectomy completely blocked the increase observed after 4 hr. However, when actinomycin was given 1 hr after the operation, the normal increase in activity was observed at 4 hr. These (and other experiments in this paper) were consistent with a new species of relatively stable RNA being synthesized in the initial hour of regeneration. However, experiments with actinomycin have also pointed to the possibility of translational control (103, 108, 111, 131, 132). For example, doses of actinomycin capable of blocking the increases in liver ornithine decarboxylase caused by dibutryl cyclic AMP were not capable of completely blocking the increases produced by theophylline or dexamethazone (108, 111). Even with dibutryl cyclic AMP and actinomycin there was a transient (twofold) increase in activity during the initial hour and then a decline in activity. Such a phenomenon could be due to increased translation of pre-existing message (due to dibutryl cyclic AMP) and message degradation due to actinomycin. Similarly, much of the early increase caused by dilution of HTC (rat hepatoma) cells, or addition of amino acids to cultures of lymphocytes (Table 1), took place in the presence of actinomycin, although at later times after addition of inhibitor the activity was greatly reduced (131, 132). Actinomycin did not block the increases in ornithine decarboxylase observed during the initial hours of lymphocyte transformation, when activity was measured within 3 hr of actinomycin addition, although activity was greatly reduced after 6 hr of actinomycin (R. H. Fillingame, unpublished results). Thus, although actinomycin may inhibit increases in ornithine decarboxylase activity, the possibility remains that the mechanism responsible for the putative increase in synthesis of ornithine decarboxylase is an increase in translation of pre-existing message (which turns over) rather than an increase in the concentration of this mRNA.

As was mentioned previously, one novel feature of ornithine decarboxylase is its apparent rapid rate of degradation. In liver the half-life, measured after inhibition of protein synthesis, was 10–20 min (84, 102, 141), and less than 30 min in a hepatoma (78). However, this method is open to criticism, e.g., the inhibitors may affect the rate of degradation of specific enzymes (142, 143). The kinetics of rapid increases and decreases in ornithine decarboxylase activity following hormone treatment (90, 101, 107–111, 122, 127) were consistent with an enzyme of short half-life [see Schimke (144) for rationale]. For example, ornithine decarboxylase activity in liver increased thirtyfold after growth hormone and then declined, the half-life being 24 min in the absence of any inhibitors (102). The apparent half-life of ornithine decarboxylase did not change during liver regeneration (84). Thus if the absolute amount of enzyme did increase (see above), it was probably due to increased synthesis of the enzyme and not decreased degradation.

Experiments in other systems have suggested that the levels of ornithine decarboxylase could be modulated by changing half-life. Dilution of stationary HTC cells with fresh medium increased the half-life by as much as eighteenfold (133). However, since the level of the enzyme increased several hundredfold, changes in degradation could not be entirely responsible for the increase in activity observed (144). Addition of individual amino acids or serum also changed the half-life somewhat (133). Addition of certain nonessential amino acids to lymphocyte cultures resulted in a threefold increase in half-life and an increase in activity which may have been due solely to the changes in enzyme degradation (129, 131). When lymphocytes were cultured in a medium containing usual levels of these amino acids, a twofold increase in the half-life of this enzyme took place during the initial day of transformation in the absence of alterations in culture conditions (130). The shorter half-life, observed at early stages of transformation, could be increased by further amino acid supplementation, whereas the stability at later times could not be increased (R. H. Fillingame, unpublished results). The shift in half-life observed under constant culture conditions might be due to the increased ability of the more transformed cells to take up amino acids.

Kay & Lindsay (87) recently found that addition of putrescine or spermidine to lymphocyte cultures led to large and rapid decreases in the level of ornithine decarboxylase activity. The exogenous polyamines had no discernible effect on the half-life of ornithine decarboxylase, which suggests that synthesis of the enzyme was decreased. The effect was apparently not directly on RNA synthesis since activity declined more rapidly after putrescine treatment than after treatment with actinomycin D. Relevant to this effect was the observation of Schrock et al (92) that putrescine administered to rats prevented the increase in ornithine decarboxylase observed in the initial phase of liver regeneration.

S-ADENOSYL-L-METHIONINE DECARBOXYLASE Under many conditions, adenosyl-methionine decarboxylase appears to be the rate-limiting enzyme in the synthesis of spermidine and spermine from putrescine (77–79). The propylamine moiety of decarboxylated adenosylmethionine is then transferred to putrescine or spermidine, forming spermidine or spermine respectively. The initial detection and characterization of adenosylmethionine decarboxylase in animals were done by Pegg & Williams-Ashman (145, 146). In contrast to the *E. coli* enzyme, this enzyme did not require magnesium ion, but rather was activated by putrescine and to a lesser extent by spermidine. Because partially purified preparations of the enzyme catalyzed stoichiometric formation of CO_2 and spermidine (or spermine) (146, 147), it was tentatively advanced that a single enzyme or complex might be responsible for both decarboxylation and the appropriate propylamine transfer. Raina & Hannonen (148) were the first to partially separate decarboxylase and propylamine transferase activities by ammonium sulfate fractionation. All three activities (adenosyl-methionine decarboxylase, and spermidine and spermine synthase) have now been separated from each other by routine purification procedures (149–153), and these activities could be recombined to yield stoichiometric formation of CO_2 and spermidine or spermine (150, 151). Although Feldman et al (154, 155) retained stoichiometric formation of spermidine during a 350-fold purification of adenosyl-methionine decarboxylase, they did not assay directly for spermidine synthase activity. Since spermidine synthase activity was present in great excess in this tissue (152), stoichiometry would be understandable if the activities were only partially separated by purification. The three activities do not appear to be complexed in crude extracts, as purified preparations of adenosylmethionine decarboxylase (lacking spermidine or spermine synthase activity) and the activity of crude homogenates have the same Stokes radius (152). However, disruption of a naturally occurring complex on the initial homogenization cannot be ruled out. The lack of a complex was substantiated by the fact that the levels of the decarboxylase and synthases were independently regulated (discussed below). The aforementioned questions have been discussed in greater detail elsewhere (78, 79, 156, 157).

Purified (149) as well as relatively crude (146) preparations of adenosylmethionine decarboxylase were activated both by putrescine and to a lesser extent by spermidine. The possibility that there are separate decarboxylases, each activated by a separate amine, has not been ruled out. However, activation by the two amines was not additive (147). In fact the maximal rate of decarboxylation observed in the presence of putrescine was reduced by spermidine (147). Since putrescine activated de-carboxylation at much lower concentrations than spermidine (78, 79, 146, 152,

155), it is probably the physiologically important activator. In this regard, Raina et al (93, 148) have shown that putrescine stimulated spermine formation at low concentrations (presumably by stimulating the decarboxylase), whereas at higher concentrations it inhibited spermine formation [by competitive inhibition of the spermine synthase (93, 146, 147)]. Thus putrescine could affect the synthesis of spermidine and spermine by activating adenosylmethionine decarboxylase, by serving as a substrate in the spermidine synthase reaction and by competitively inhibiting the spermine synthase reaction. Spermidine, on the other hand, could modulate the rates of spermidine and spermine synthesis by either inhibiting or activating decarboxylation (depending upon the concentration of putrescine, see above) and by serving as a substrate in the spermine synthase reaction.

In contrast to ornithine decarboxylase, only minor changes in the level of adenosylmethionine decarboxylase were observed in regenerating or growth hormone-treated liver (94, 104, 141, 158), the uterus after estrogen treatment (123, 126, 159), and cardiac hypertrophy (114). More substantial increases were observed during embryonic development (120, 135, 160), and large changes were found in the prostate of androgen-treated castrated rats (119), the mammary gland during lactation (121), granulomas (158), and transforming lymphocytes (129, 130, 161). The ratio of putrescine-stimulated to spermidine-stimulated decarboxylase activity did not change during lymphocyte transformation (130), despite thirtyfold changes in the level of activity, again suggesting that there was but a single decarboxylase present. Inhibitors of protein and RNA synthesis were shown to inhibit the uterine increases in adenosylmethionine decarboxylase (159). As shown by Jänne and co-workers, the levels of spermidine and spermine synthase change independently of each other and of adenosylmethionine decarboxylase during liver regeneration (94, 141, 158).

In normal liver, the putrescine concentration (controlled by ornithine decarboxylase) appeared to limit the rate of spermidine and spermine formation (93). However, shortly after partial hepatectomy the putrescine concentration was sufficiently high to maximally stimulate adenosylmethionine decarboxylation (94, 141). Thus, although ornithine decarboxylase limits spermidine and spermine formation in normal liver, adenosylmethionine decarboxylase is probably rate-limiting at the later stages of regeneration.

Adenosylmethionine decarboxylase appears to turn over rapidly, but the apparent half-life was somewhat greater than ornithine decarboxylase, i.e. 30 to 60 min (78, 126, 129, 130, 141, 159, 161, 162). The same reservations apply to these measurements, done with inhibitors of protein synthesis, that were cited with respect to the half-life of ornithine decarboxylase (see above). Hannonen et al (141) have shown that while the adenosylmethionine decarboxylase of regenerating liver turned over rapidly, the spermidine and spermine synthases did not. A potent but reversible inhibitor of adenosylmethionine decarboxylase, methylglyoxal bis(guanylhydrazone) (163), apparently interacted with the decarboxylase so as to decrease its rate of cellular degradation by at least thirtyfold (130, 161, 162). This inhibition of degradation was probably responsible for marked increases in the level of the decarboxylase observed during inhibition of polyamine synthesis (130, 161, 162, 164, 165). In contrast to the decarboxylase, the level of spermidine synthase activity was not affected by methylglyoxal bis(guanylhydrazone) (165).

Synthesis of Biogenic and Related Amines

GLUTAMIC ACID DECARBOXYLASES   The presence of L-glutamic acid decarboxylase activity in the neural tissues of animals has been known for many years (166, 167). Only in recent years has the occurrence of this enzyme been documented in other tissues (168, 169). Glutamic acid decarboxylase may play an important physiological role in the nervous system in that the substrate of this enzyme, L-glutamate, may be an excitatory transmitter, and the product, γ-aminobutyric acid (GABA), an inhibitory transmitter (170–172).

There are at least two glutamic acid decarboxylase activities in some animals. Glutamic acid decarboxylase I has recently been purified to apparent homogeneity from mouse brain (173). The enzyme was PLP-dependent and inhibited by carbonyl reagents (174). Partially purified preparations of glutamic acid decarboxylase I were strongly inhibited by anions, chloride being competitive with glutamate (175). This finding led to the suggestion that chloride levels could be important in controlling GABA release from nerve endings (172, 175). The rat brain enzyme was inhibited by ATP, the degree of inhibition depending upon the PLP concentration but not in a competitive manner (176). Glutamic acid decarboxylase I appeared to be preferentially localized in the synaptosomal fraction of homogenates from both rodent and chick brains (169, 177, 178), although a significant fraction of the activity was found in other fractions as well.

Glutamic acid decarboxylase II has been detected in kidney and other tissues of a number of species (168, 169). This enzyme was not influenced by added PLP and actually was stimulated by some carbonyl reagents, e.g. amino-oxyacetic acid and semicarbazide (169). In contrast to the brain enzyme, anions stimulated glutamic acid decarboxylase II, 0.1 M chloride activating dialyzed homogenates by as much as sixfold. This enzyme appeared to be localized in the mitochondrial fraction (169).

Haber et al (169) have recently studied the two forms of glutamic acid decarboxylase in the brain of chick embryos. In 7-day embryos the chloride-activated form(s) of the enzyme predominated, the activity of the whole brain homogenates increasing twofold on addition of chloride. During development the activity of the chloride-inhibited form increased relative to the chloride-activated form, and by two days post-hatching the activity of the whole brain homogenate was inhibited by 70% on addition of chloride. At 14 days of development, the activity isolated from the synaptosomal fraction was inhibited by chloride, whereas the activity in the mitochondrial fraction was activated by chloride.

The activity of glutamic acid decarboxylase in brain in vivo appears to be critically dependent upon maintenance of adequate PLP levels. This conclusion rests upon comparison of measurements of glutamic acid decarboxylase activity, in homogenates assayed in the absence of exogenous PLP, before and after reduction of brain PLP levels. In pyridoxine-deficient neonatal rats, the large increases in activity normally observed during development were significantly reduced (179). The decreased activity was not due to lack of apoenzyme, since glutamic acid decarboxylase activity in brains of Vitamin B_6-deficient animals was actually higher than the activity in brains of control animals when it was assayed in the presence of saturating PLP (179). Similarly, administration of pharmacological agents that reduce PLP concentrations (via inhibition of pyridoxal kinase) resulted in large reductions in glutamic acid decarboxylase activity in young animals (180). Again the reduction in activity appeared to be due to a decreased saturation of the enzyme by PLP, rather than a decline in the amount of apoenzyme. Other PLP-dependent enzymes were not sensitive to these changes in PLP levels (180, 181). These same agents, as well as other inhibitors of glutamic acid decarboxylase, induce lethal convulsions and death of the animal (180, 182). Based largely on this evidence, Scriver & Whelan (183) suggested that a rare convulsive disorder in man, termed "pyridoxine-dependency," may be due to an altered form of glutamic acid decarboxylase. Experiments to date (reviewed in 183) indicated that pyridoxine metabolism and the availability of PLP were normal in pyridoxine-dependent patients. Thus it was suggested that the dependency is due to a glutamic acid decarboxylase which required greater than normal levels of PLP for activity. To our knowledge there has not been a direct test of this interesting proposal.

As mentioned in the preceding paragraph, the total amount of glutamic acid decarboxylase activity, measured in the presence of saturating PLP, actually increased as the result of a Vitamin B_6-deficient diet (179). Although an increase in the number of enzyme molecules has not been directly demonstrated, it is possible that the reduced GABA levels (or increased glutamate levels) led to the increased accumulation of glutamic acid decarboxylase apoenzyme, e.g. by a classical derepression mechanism. Kraus demonstrated that elevation of the mouse brain L-glutamate levels led to increased glutamic acid decarboxylase activity (184). Actinomycin D partially prevented the elevation and the effect was tentatively described as "substrate induction." Conversely the amount of glutamic acid decarboxylase activity present in the brains of young mice or chick embryos was reduced when GABA levels were artificially increased (180, 185, 186). The decline in activity could not be attributed to inhibition by GABA in the extract (186). Interestingly, both glutamic acid decarboxylase I and II were reduced on elevation of the GABA levels in chick embryo optic lobes (185).

HISTIDINE DECARBOXYLASE   Mammalian L-histidine decarboxylase has been the subject of several recent reviews (3, 187–189). Two enzymes in animal tissues are capable of decarboxylating L-histidine: one specific for histidine and the other a nonspecific L-aromatic acid decarboxylase (see below). Assay methods for distinguishing between the two activities (3, 187–189) and the contrasting enzymatic properties have been reviewed (3, 188, 189). The specific histidine decarboxylase is an adaptive enzyme that changes in level in a number of tissues in response to a variety of physiological perturbations summarized elsewhere (3, 187, 188). We restrict discussion to recent data concerning the regulatory mechanism(s) responsible for these adaptive changes.

Inhibitors of protein synthesis were shown to block not only the increase in histidine decarboxylase activity observed after administration of gastrin to fasted rats (190), but also the increase observed during inflammation (191, 192) and in the liver of a normal rat cross-circulated with a rat bearing a Walker carcinosarcoma (193). Even if one assumes that these inhibitors were affecting protein synthesis only, the results do not distinguish between increased synthesis of histidine decarboxylase and other possibilities (discussed in the previous section on ornithine decarboxylase). In two of these cases, actinomycin D did not block the increases in activity (190, 191).

Histidine decarboxylase appeared to turn over rather rapidly as judged by inhibitor experiments. Administration of cycloheximide to freely feeding rats led to a decline in enzyme activity in the gastric mucosa, the half-life being roughly 2 hr (190). Rapid declines in activity have been observed without the use of inhibitors. After injection of gastrin into fasted rats, histidine decarboxylase activity reached a peak and then declined with a half-life of about 2 hr (190). The rate of degradation of histidine decarboxylase may change in the gastric mucosa depending upon the state of the animal, since the basal level in fasted rats did not decline after puromycin treatment (190). In the experiments of Schayer & Reilly (191) the basal activities of liver and lung tissue declined with a 2 hr half-life after inhibition of protein synthesis. Activity declined more slowly in the hypothalamus of rats after cycloheximide treatment (194).

Increases in histidine decarboxylase are often preceded by release of the decarboxylated product, histamine, from the tissue in question (187). Kahlson suggested that depletion of tissue histamine levels was responsible for the increase in histidine decarboxylase activity. According to this hypothesis, regulation would be on enzyme synthesis (187), as histamine was not a direct feedback inhibitor (195). Nearly all of the suggestive evidence relating to this hypothesis has come from the gastric mucosal system, which has been reviewed by Kahlson & Rosengren (187), although recent experiments in other systems support the concept (194). Recent experiments now suggest that histidine decarboxylase activity is controlled by endogenous gastrin in a manner independent from the control of gastric acid secretion (196). Antrectomy, the surgical removal of the antrum of the stomach (the main source of endogenous gastrin), abolished the increase in histidine decarboxylase activity normally observed after feeding or under other conditions that stimulated gastric secretion (196–199). The effect was due to the lack of gastrin, since pentagastrin administration to antrectomized animals led to increases in histidine decarboxylase (196, 197, 199). These results strongly argue that gastrin is the primary effector of histidine decarboxylase activity and that the effect is independent of a histamine feedback system. In this context it has been shown that gastrin does not directly affect histidine decarboxylase activity assayed in vitro (190).

L-AROMATIC AMINO ACID DECARBOXYLASE   This amino acid decarboxylase, widely distributed in animal tissues and capable of decarboxylating a variety of L-aromatic amino acids, has been discussed in several recent reviews (2, 3, 189). The large body of evidence suggesting that a single enzyme is responsible for decarboxylation of a variety of aromatic amino acids was particularly well summarized by Aures et al (189). The substrate specificity of a homogeneous preparation of the enzyme from hog kidney (200) was consistent with this view. However, the possibility remains that there are specific decarboxylases for some of these substrates (3, 189), as was shown to be the case for L-histidine (see above).

The aromatic amino acid decarboxylase functions in the biosynthesis of the catecholamines [dopamine (3,4-dihydroxyphenylethylamine), the product of DOPA (3,4-dihydroxyphenylalanine) decarboxylation; norepinephrine and epinephrine] and the biosynthesis of 5-hydroxytryptamine (serotonin) (189, 201, 202). The rate-limiting step in both pathways appears to be the initial hydroxylation of either tyrosine (with formation of DOPA) or tryptophan (with formation of 5-hydroxy-tryptophan), and regulation is thought to occur here (201–205). Thus, various physiological stimuli which caused large changes in tyrosine (or tryptophan) hydroxylase generally had little effect on aromatic amino acid decarboxylase (206). For example, administration of nerve growth factor to rats led to dramatic increases in tyrosine hydroxylase in the superior cervical ganglia, while DOPA decarboxylase activity increased only in proportion to the increase in volume of the sympathetic ganglia (207).

Some examples of adaptive changes in aromatic amino acid decarboxylase are known. The specific activity of DOPA decarboxylase activity increased during rat brain development (208); the kinetics and magnitude of the changes in tyrosine hydroxylase and dopamine-β-hydroxylase were comparable. In mice, increases in 5-hydroxytryptophan decarboxylase activity were observed during aggressive behavior (209).

In patients with Parkinson's disease, the amount of dopamine and its major metabolites appeared to be reduced in certain portions of the brain (210, 211). L-DOPA is used in treatment of patients with parkinsonism and the conversion of L-DOPA to dopamine may be responsible for the therapeutic effects (211). Lloyd & Hornykiewicz (212) have made the interesting observation that DOPA decarboxy-lase, in patients shortly after death, was drastically reduced in those portions of the brain normally rich in dopamine. However, it is not known whether this reduction occurred prior to or after the disease-related degeneration of these regions of the brain. Other possible biochemical defects in parkinsonism have been reviewed (213). Of interest to the treatment of Parkinson's disease was the finding that administra-tion of L-DOPA to rats and mice led to decreased DOPA decarboxylase activity in the liver without corresponding changes in the brain (214–216). Using antibody prepared to the purified hog kidney enzyme, it was shown that a decrease in enzyme protein was responsible for the decline (214, 215). Whether the decline was due to a decrease in enzyme synthesis was not shown. DOPA decarboxylase activity in the erythrocytes of patients with Parkinson's disease declined during DOPA therapy (216).

ACKNOWLEDGMENT

We thank Dr. E. A. Boeker for reviewing the manuscript prior to publication and for her advice concerning several of the areas covered.

Literature Cited

1. Guirard, B. M., Snell, E. E. 1964. *Compr. Biochem.* 15:138–99.
2. Meister, A. 1965. *Biochemistry of the Amino Acids,* 325–38. New York:
   Academic. 2nd ed. 1084 pp.
3. Boeker, E. A., Snell, E. E. 1972. *The Enzymes* 6:217–53
4. Morris, D. R., Pardee, A. B. 1965.

Biochem. Biophys. Res. Commun. 20: 697–702

5. Morris, D. R., Pardee, A. B. 1966. *J. Biol. Chem.* 241:3129–35

6. Gale, E. F. 1946. *Advan. Enzymol.* 6:1–32

7. Gale, E. F. 1940. *Biochem. J.* 34:392–413

8. Gale, E. F., Epps, H. M. R. 1944. *Biochem. J.* 232–42

9. Taylor, E. S., Gale, E. F. 1945. *Biochem. J.* 39:52–58

10. Gunsalus, I. C., Bellamy, W. D., Umbreit, W. W. 1944. *J. Biol. Chem.* 155:685–86

11. Baddiley, J., Gale, E. F. 1945. *Nature* 155:727–28

12. Blethen, S. L., Boeker, E. A., Snell, E. E. 1968. *J. Biol. Chem.* 243:1671–77

13. Boeker, E. A., Snell, E. E. 1968. *J. Biol. Chem.* 243:1678–84

14. Boeker, E. A., Fischer, E. H., Snell, E. E. 1969. *J. Biol. Chem.* 244:5239–45

15. Boeker, E. A., Fischer, E. H., Snell, E. E. 1971. *J. Biol. Chem.* 246:6776–81

16. Sabo, D. L. O. 1973. *Purification, properties and catalytic mechanism of inducible E. coli lysine decarboxylase.* PhD thesis. Univ. Washington, Seattle. 154 pp.

16a. Sabo, D. L., Boeker, E. A., Byers, B., Waron, H., Fischer, E. H. 1974. *Biochemistry.* In press

16b. Sabo, D. L., Fischer, E. H. 1974. *Biochemistry.* In press

17. Applebaum, D. 1972. *Purification and characterization of induced and biosynthetic ornithine decarboxylases of Escherichia coli.* PhD thesis. Univ. Washington, Seattle. 157 pp.

18. Melnykovych, G., Snell, E. E. 1958. *J. Bacteriol.* 76:518–23

19. Koessler, K. K., Hanke, M. T. 1919. *J. Biol. Chem.* 39:539–84

20. Guirard, B. M., Snell, E. E. 1964. *J. Bacteriol.* 87:370–76

21. Riley, W. D., Snell, E. E. 1970. *Biochemistry* 9:1485–91

22. Riley, W. D., Snell, E. E. 1968. *Biochemistry* 7:3520–28

23. Recsei, P. A., Snell, E. E. 1970. *Biochemistry* 9:1492–97

24. Recsei, P. A., Snell, E. E. 1972. *J. Bacteriol.* 112:624–26

25. Recsei, P. A., Snell, E. E. 1973. *Biochemistry* 12:365–71

26. Strausbauch, P. H., Fischer, E. H. 1970. *Biochemistry* 9:226–33

27. Strausbauch, P. H., Fischer, E. H. 1970. *Biochemistry* 9:233–38

28. Halpern, Y. S. 1962. *Biochim. Biophys. Acta* 61:953–62

29. Britten, R. J., McClure, F. T. 1964. *Macromolecular Biosynthesis,* ed. R. B. Roberts, 13–56. Baltimore: Garamond/ Pridemark. 702 pp.

30. Nelson, D. L., Kornberg, A. 1970. *J. Biol. Chem.* 245:1128–36

31. Lupo, M., Halpern, Y. S. 1970. *J. Bacteriol.* 103:382–86

32. Lupo, M., Halpern, Y. S. 1970. *Biochim. Biophys. Acta* 206:295–304

33. Lupo, M., Halpern, Y. S., Sulitzeanu, D. 1969. *Arch. Biochem. Biophys.* 131:621–28

34. Foerster, H. F. 1972. *J. Bacteriol.* 111:437–42

35. Foerster, H. F. 1971. *J. Bacteriol.* 108:817–23

36. Foerster, C. W., Foerster, H. F. 1973. *J. Bacteriol.* 114:1090–98

37. Kim, K. H. 1963. *J. Bacteriol.* 86:320–23

38. Marcus, M., Halpern, Y. S. 1969. *J. Bacteriol.* 97:1509–10

39. Marcus, M., Halpern, Y. S. 1969. *Biochim. Biophys. Acta* 177:314–20

40. Dover, S., Halpern, Y. S. 1972. *J. Bacteriol.* 109:835–43

41. Valley, G., Rettger, L. F. 1927. *J. Bacteriol.* 14:101–37

42. Gladstone, G. P., Fildes, P., Richardson, G. M. 1935. *Brit. J. Exp. Pathol.* 16:335–48

43. Becker, F. F. 1967. *Fed. Proc.* 26:812 (Abstr.)

44. Rundles-Dunn, S., Maas, W. K. 1973. Personal communication

45. Morris, D. R., Koffron, K. L. 1969. *J. Biol. Chem.* 244:6094–99

46. Morris, D. R., Wu, W. H., Applebaum, D., Koffron, K. L. 1970. *Ann. NY Acad. Sci.* 171:968–76

47. Davis, R. H., Lawless, M. B., Port, L. A. 1970. *J. Bacteriol.* 102:299–305

48. Tabor, H., Tabor, C. W. 1969. *J. Biol. Chem.* 244:2286–92

49. Hölttä, E., Jänne, J., Pispa, J. 1972. *Biochem. Biophys. Res. Commun.* 47:1165–71

50. Munro, G. F., Hercules, K., Morgan, J., Sauerbier, W. 1972. *J. Biol. Chem.* 247:1272–80

51. Rubenstein, K. E., Streibel, E., Massey, S., Lapi, L., Cohen, S. S. 1972. *J. Bacteriol.* 112:1213–21

52. Morris, D. R., Jorstad, C. M. 1970. *J. Bacteriol.* 101:731–37

53. Bachrach, U., Ben-Joseph, M. 1971. *FEBS Lett.* 15:75–77

54. Dion, A. S., Cohen, S. S. 1972. *J. Virol.* 9:419–22

55. Wu, W. H., Morris, D. R. 1973. *J. Biol. Chem.* 248:1687–95

56. Wu, W. H., Morris, D. R. 1973. *J. Biol. Chem.* 248:1696–99

57. Weiss, R. L., Kimes, B. W., Morris, D. R. 1973. *Biochemistry* 12:450–56
58. Tabor, H., Tabor, C. W. 1969. *J. Biol. Chem.* 244:6383–87
59. Maas, W. K. 1972. *Mol. Gen. Genet.* 119:1–9
60. Astrachan, L., Miller, J. F. 1973. *J. Virol.* 11:792–98
61. Dion, A. S., Cohen, S. S. 1972. *Proc. Nat. Acad. Sci. USA* 69:213–17
62. Leifer, Z., Maas, W. K. 1973. *Fed. Proc.* 32:659 (Abstr.)
63. Leifer, Z., Maas, W. K. 1973. Personal communication
64. Tabor, H., Rosenthal, S. M., Tabor, C. W. 1958. *J. Biol. Chem.* 233:907–14
65. Wickner, R. B., Tabor, C. W., Tabor, H. 1970. *J. Biol. Chem.* 245:2132–39
66. Su, C. H., Cohen, S. S. 1973. *Polyamines in Normal and Neoplastic Growth,* ed. D. H. Russell, 299–306. New York: Raven. 429 pp.
67. Vogel, H. J. 1965. *Evolving Genes and Proteins,* ed. V. Bryson, H. J. Vogel, 25–40. New York: Academic. 629 pp.
68. Patte, J. C., Loviny, T., Cohen, G. N. 1962. *Biochim. Biophys. Acta* 58:359–60
69. White, P. J., Kelly, B., Suffling, A., Work, E. 1964. *Biochem. J.* 91:600–10
70. Barnes, I. J., Bondi, A., Moat, A. G. 1969. *J. Bacteriol.* 99:169–74
71. Grandgenett, D. P., Stahly, D. P. 1971. *J. Bacteriol.* 106:551–60
72. White, P. J., Kelly, B. 1965. *Biochem. J.* 96:75–84
73. Grandgenett, D. P., Stahly, D. P. 1968. *J. Bacteriol.* 96:2099–2109
74. Forman, M., Aronson, A. 1972. *Biochem. J.* 126:503–13
75. Levisohn, L., Aronson, A. I. 1967. *J. Bacteriol.* 93:1023–30
76. Spudich, J. A., Kornberg, A. 1968. *J. Biol. Chem.* 243:4600–5
77. Tabor, H., Tabor, C. W. 1972. *Advan. Enzymol.* 36:203–68
78. Williams-Ashman, H. G., Jänne, J., Coppoc, G. L., Geroch, M. E., Schenone, A. 1972. *Advan. Enzyme Regul.* 10:225–45
79. Williams-Ashman, H. G. 1972. *Biochemical Regulatory Mechanisms in Eukaryotic Cells,* ed. E. Kun, S. Grisolia, 245–69. New York: Interscience. 530 pp.
80. Jänne, J., Raina, A. 1968. *Acta Chem. Scand.* 22:1349–51
81. Pegg, A. E., Williams-Ashman, H. G. 1968. *Biochem. J.* 108:533–39
82. Russell, D. H., Snyder, S. H. 1968. *Proc. Nat. Acad. Sci. USA* 60:1420–27
83. Jänne, J., Raina, A., Siimes, M. 1968. *Biochim. Biophys. Acta* 166:419–26
84. Russell, D. H., Snyder, S. H. 1969. *Mol. Pharmacol.* 5:253–62
85. Ono, M., Inoue, H., Suzuki, F., Takeda, Y. 1972. *Biochim. Biophys. Acta* 284:285–97
86. Raina, A., Jänne, J. 1968. *Acta Chem. Scand.* 22:2375–78
87. Kay, J. E., Lindsay, V. J. 1973. *Biochem. J.* 132:791–96
88. Jänne, J., Williams-Ashman, H. G. 1971. *J. Biol. Chem.* 246:1725–32
89. Friedman, S. J., Halpern, K. V., Canellakis, E. S. 1972. *Biochem. Biophys. Acta* 261:181–87
90. Fausto, N. 1969. *Biochim. Biophys. Acta* 190:193–201
91. Russell, D. H., Snyder, S. H. 1969. *Endocrinology* 84:223–28
92. Schrock, T. R., Oakman, N. J., Bucher, N. L. R. 1970. *Biochim. Biophys. Acta* 204:564–77
93. Raina, A., Jänne, J., Hannonen, P., Höltta, E. 1970. *Ann. NY Acad. Sci.* 171:697–708
94. Hölttä, E., Jänne, J. 1972. *FEBS Lett.* 23:117–21
95. Cavia, E., Webb, T. E. 1972. *Biochim. Biophys. Acta* 262:546–54
96. Gaza, D. J., Short, J., Lieberman, I. 1973. *FEBS Lett.* 32:251–53
97. Russell, D. H., McVicker, T. A. 1971. *Biochim. Biophys. Acta* 244:85–93
98. Fausto, N. 1970. *Cancer Res.* 30:1947–52
99. Raina, A., Jänne, J. 1970. *Fed. Proc.* 29:1568–74
100. Ono, M., Inoue, H., Takeda, Y. 1973. *Biochim. Biophys. Acta* 304:495–504
101. Jänne, J., Raina, A. 1969. *Biochim. Biophys. Acta* 174:769–72
102. Russell, D. H., Snyder, S. H., Medina, V. J. 1970. *Endocrinology* 86:1414–19
103. Fausto, N. 1971. *Biochim. Biophys. Acta* 238:116–28
104. Russell, D. H., Lombardini, J. B. 1971. *Biochim. Biophys. Acta* 240:273–86
105. Sogani, R. K., Matsushita, S., Mueller, J. F., Raben, M. S. 1972. *Biochim. Biophys. Acta* 279:377–86
106. Hayashi, S., Aramaki, Y., Noguchi, T. 1972. *Biochem. Biophys. Res. Commun.* 46:795–800
107. Panko, W. B., Kenney, F. T. 1971. *Biochem. Biophys. Res. Commun.* 43:346–50
108. Beck, W. T., Bellantone, R. A., Canellakis, E. S. 1972. *Biochem. Biophys. Res. Commun.* 48:1649–55
109. Beck, W. T., Canellakis, E. S. See Ref. 66, pp. 261–75

110. Stastny, M., Cohen, S. 1972. *Biochim. Biophys. Acta* 261:177–80
111. Beck, W. T., Bellantone, R. A., Canellakis, E. S. 1973. *Nature* 241:275–77
112. Russell, D. H. 1971. *Biochem. Pharmacol.* 20:3481–91
113. Brandt, J. T., Pierce, D. A., Fausto, N. 1972. *Biochim. Biophys. Acta* 279:184–93
114. Russell, D. H., Shiverick, K. T., Hamrell, B. B., Alpert, N. R. 1971. *Am. J. Physiol.* 221:1287–91
115. Feldman, M. J., Russell, D. H. 1972. *Am. J. Physiol.* 222:1199–1203
116. Matsushita, S., Sogani, R. K., Raben, M. S. 1972. *Circ. Res.* 31:699–709
117. Levine, J. H., Nicolson, W. E., Liddle, G. W., Orth, D. N. 1973. *Endocrinology* 92:1089–95
118. Stastny, M., Cohen, S. 1970. *Biochim. Biophys. Acta* 204:578–89
119. Pegg, A. E., Lockwood, D. H., Williams-Ashman, H. G. 1970. *Biochem. J.* 117:17–31
120. MacIndoe, J. H., Turkington, R. W. 1973. *Endocrinology* 92:595–605
121. Russell, D. H., McVicker, T. A. 1972. *Biochem. J.* 130:71–76
122. Kobayashi, Y., Kupelian, J., Maudsley, P. V. 1971. *Science* 172:379–80
123. Kaye, A. M., Icekson, I., Lindner, H. R. 1971. *Biochim. Biophys. Acta* 252:150–59
124. Hernandez, O., Ballesteros, L. M., Mendez, D., Rosado, A. 1973. *Endocrinology* 92:1107–12
125. Cohen, S., O'Malley, B. W., Stastny, M. 1970. *Science* 170:336–38
126. Russell, D. H., Taylor, R. L. 1971. *Endocrinology* 88:1397–1403
127. Kaye, A. M., Icekson, I., Lamprecht, S. A., Gruss, R., Tsafriri, A., Lindner, H. R. 1973. *Biochemistry* 12:3072–96
128. Kay, J. E., Cooke, A. 1971. *FEBS Lett.* 16:9–12
129. Kay, J. E., Lindsay, V. J. 1973. *Exp. Cell Res.* 77:428–36
130. Fillingame, R. H. 1973. *Accumulation of polyamines during lymphocyte transformation: An analysis of its relation to nucleic synthesis and accumulation.* PhD thesis. Univ. Washington, Seattle. 132 pp.
131. Kay, J. E., Lindsay, V. J., Cooke, A. 1972. *FEBS Lett.* 21:123–26
132. Hogan, B. L. M. 1971. *Biochem. Biophys. Res. Commun.* 45:301–7
133. Hogan, B. L. M., Murden, S., Blackledge, A. See Ref. 66, pp. 239–48
134. Friedman, S. J., Bellantone, R. A., Canellakis, E. S. 1972. *Biochim. Biophys. Acta* 261:188–93
135. Russell, D. H. 1971. *Proc. Nat. Acad. Sci. USA* 68:523–27
136. Dzodzoe, Y. G. G., Rosengren, E. 1971. *Brit. J. Pharmacol.* 41:294–301
137. Anderson, T. R., Schanberg, S. M. 1972. *J. Neurochem.* 19:1471–81
138. Russell, D. H., Levy, C. C. 1971. *Cancer Res.* 31:248–51
139. Williams-Ashman, H. G., Coppoc, G. L., Weber, G. 1972. *Cancer Res.* 32:1924–32
140. Mitchell, J. L. A., Rusch, H. P. 1973. *Biochim. Biophys. Acta* 297:503–16
141. Hannonen, P., Raina, A., Jänne, J. 1972. *Biochim. Biophys. Acta* 273:84–90
142. Kenney, F. T. 1967. *Science* 156:525–27
143. Hershko, A., Tomkins, G. M. 1971. *J. Biol. Chem.* 246:710–14
144. Schimke, R. T. 1969. *Current Topics in Cellular Regulation,* ed. B. L. Horecker, E. R. Stadtman, 1:77–124. New York: Academic. 314 pp.
145. Pegg, A. E., Williams-Ashman, H. G. 1968. *Biochim. Biophys. Res. Commun.* 30:76–82
146. Pegg, A. E., Williams-Ashman, H. G. 1969. *J. Biol. Chem.* 244:682–93
147. Pegg, A. E., Williams-Ashman, H. G. 1970. *Arch. Biochem. Biophys.* 137:156–65
148. Raina, A., Hannonen, P. 1970. *Acta Chem. Scand.* 24:3061–64
149. Jänne, J., Williams-Ashman, H. G. 1971. *Biochem. Biophys. Res. Commun.* 42:222–29
150. Jänne, J., Schenone, A., Williams-Ashman, H. G. 1971. *Biochem. Biophys. Res. Commun.* 42:758–64
151. Raina, A., Hannonen, P. 1971. *FEBS Lett.* 16:1–4
152. Hannonen, P., Jänne, J., Raina, A. 1972. *Biochem. Biophys. Res. Commun.* 46:341–48
153. Hannonen, P., Jänne, J., Raina, A. 1972. *Biochim. Biophys. Acta* 289:225–31
154. Feldman, M. J., Levy, C. C., Russell, D. H. 1971. *Biochem. Biophys. Res. Commun.* 44:675–81
155. Feldman, M. J., Levy, C. C., Russell, D. H. 1972. *Biochemistry* 11:671–77
156. Williams-Ashman, H. G., Coppoc, G. L., Schenone, A., Weber, G. See Ref. 66, pp. 181–97
157. Russell, D. H. See Ref. 66, pp. 1–13
158. Raina, A., Jänne, J., Hannonen, P., Hölttä, E., Ahonen, J. See Ref. 66, pp. 167–80

159. Russell, D. H., Potyraj, J. J. 1972. *Biochem. J.* 128:1109–15
160. Russell, D. H., McVicker, T. A. 1972. *Biochim. Biophys. Acta* 259:247–58
161. Fillingame, R. H., Morris, D. R. 1973. *Biochem. Biophys. Res. Commun.* 52:1020–25
162. Pegg, A. E., Corti, A., Williams-Ashman, H. G. 1973. *Biochem. Biophys. Res. Commun.* 52:696–701
163. Williams-Ashman, H. G., Schenone, A. 1972. *Biochem. Biophys. Res. Commun.* 46:288–95
164. Fillingame, R. H., Morris, D. R. See Ref. 66, pp. 249–60
165. Hölttä, E., Hannonen, P., Pispa, J., Jänne, J. 1973. *Biochem. J.* 136:699–76
166. Wingo, W. J., Awapara, J. 1950. *J. Biol. Chem.* 187:267–71
167. Roberts, E., Frankel, S. 1951. *J. Biol. Chem.* 188:789–95
168. Whelan, D. T., Scriver, C. R., Mohyuddin, F. 1969. *Nature* 224:916–17
169. Haber, B., Kuriyama, K., Roberts, E. 1970. *Biochem. Pharmacol.* 19:1119–36
170. Krnjevik, K. 1970. *Nature* 228:119–24
171. Johnson, J. L. 1972. *Brain Res.* 37:1–19
172. Roberts, E., Kuriyama, K. 1968. *Brain Res.* 8:1–35
173. Wu, J. Y., Matsuda, T., Roberts, E. 1973. *J. Biol. Chem.* 248:3029–34
174. Roberts, E., Simonsen, D. G. 1963. *Biochem. Pharmacol.* 12:113–34
175. Susz, J. P., Haber, B., Roberts, E. 1966. *Biochemistry* 5:2870–77
176. Tursky, T. 1970. *Eur. J. Biochem.* 12:544–49
177. Salganicoff, L., DeRobertis, E. 1965. *J. Neurochem.* 12:287–309
178. Fonnum, F. 1968. *Biochem. J.* 106:401–12
179. Bayoumi, R. A., Smith, W. R. D. 1972. *J. Neurochem.* 19:1883–97
180. Tapia, R., de la Mora, M. P., Massieu, G. H. 1969. *Ann. NY Acad. Sci.* 166:257–66
181. Tapia, R., Pasantes, H. 1971. *Brain Res.* 29:111–22
182. Baxter, C. F. 1969. *Ann. NY Acad. Sci.* 166:267–80
183. Scriver, C. R., Whelan, D. T. 1969. *Ann. NY Acad. Sci.* 166:83–96
184. Kraus, P. 1968. *Z. Physiol. Chem.* 349:1424–27
185. Haber, B., Sze, P. Y., Kuriyama, K., Roberts, E. 1970. *Brain Res.* 18:545–47
186. Sze, P. Y. 1970. *Brain Res.* 19:322–25
187. Kahlson, G., Rosengren, E. 1968. *Physiol. Rev.* 48:155–96
188. Levine, R. J., Noll, W. W. 1969. *Ann. NY Acad. Sci.* 166:246–56
189. Aures, D., Hakanson, R., Clark, W. G. 1970. *Handbook of Neurochemistry,* ed. A. Lajtha, IV:165–96. New York: Plenum. 516 pp.
190. Snyder, S. H., Epps, L. 1968. *Mol. Pharmacol.* 4:187–95
191. Schayer, R. W., Reilly, M. A. 1968. *Am. J. Physiol.* 215:472–76
192. Schayer, R. W., Reilly, M. A. 1972. *Eur. J. Pharmacol.* 20:271–80
193. Ishikawa, E., Toki, A., Moriyama, T. 1970. *J. Biochem. Tokyo* 68:347–58
194. Schwartz, J. C., Lampart, C., Rose, C. 1972. *J. Neurochem.* 19:801–10
195. Levine, R. J., Watts, D. E. 1966. *Biochem. Pharmacol.* 15:841–49
196. Johnson, L. R. 1971. *Gastroenterology* 61:106–18
197. Aures, D., Johnson, L. R., Way, L. W. 1970. *Am. J. Physiol.* 219:214–16
198. Håkanson, R., Liedberg, G. 1970. *Eur. J. Pharmacol.* 12:94–103
199. Håkanson, R., Liedberg, G. 1972. *Eur. J. Pharmacol.* 18:31–36
200. Christenson, J. G., Dairman, W., Udenfriend, S. 1970. *Arch. Biochem. Biophys.* 141:356–67
201. Iversen, L. L. 1969. *Handbook of Neurochemistry,* ed. A. Lajtha, IV:197–220. New York: Plenum. 516 pp.
202. Costa, E., Neff, N. H. 1969. *Handbook of Neurochemistry,* ed. A. Lajtha, IV:45–90. New York: Plenum. 516 pp.
203. Udenfriend, S., Zaltzman-Nirenberg, P., Gordon, R., Spector, S. 1966. *Mol. Pharmacol.* 2:95–105
204. Weiner, N., Cloutier, G., Bjur, R., Pfeffer, R. I. 1972. *Pharmacol. Rev.* 24:203–21
205. Pletscher, A. 1972. *Pharmacol. Rev.* 24:225–32
206. Thoenen, H. 1972. *Pharmacol. Rev.* 24:255–67
207. Thoenen, H. 1971. *Proc. Nat. Acad. Sci. USA* 68:1598–1602
208. Lamprecht, F., Coyle, J. T. 1972. *Brain Res.* 41:503–6
209. Eleftheriou, B., Church, R. L. 1968. *Physiol. Behav.* 3:323–25
210. Cotzias, G. C., Van Woert, M. H., Schiffer, L. M. 1967. *New Engl. J. Med.* 276:374–79
211. Calne, D. B., Sandler, M. 1970. *Nature* 226:21–24
212. Lloyd, K., Hornykiewicz, O. 1970. *Science* 170:1212–13
213. Klawans, H. L. Jr. 1968. *Dis. Nerv. Syst.* 29:805–16
214. Dairman, W., Christenson, J. G.,

Udenfriend, S. 1971. *Proc. Nat. Acad. Sci. USA* 68:2117–20

215. Dairman, W., Christenson, J. G., Udenfriend, S. 1972. *Pharmacol. Rev.* 24:269–89

216. Tate, S. S., Sweet, R., McDowell, F. H., Meister, A. 1971. *Proc. Nat. Acad. Sci. USA* 68:2121–23

THE SODIUM-POTASSIUM ADENOSINETRIPHOSPHATASE

× 849

June L. Dahl and Lowell E. Hokin
Department of Pharmacology, University of Wisconsin Medical School
Madison, Wisconsin

CONTENTS

INTRODUCTION

The active transport of sodium and potassium is a fundamental and major energy-requiring cellular process underlying the maintenance of cell volume, absorption processes in kidney and intestine, and excitability in nerve and muscle. The active transport of sugars and amino acids in many animal tissues is also dependent on proper distribution of sodium and potassium ions. Studies with resealed erythrocyte ghosts (1) and internally dialyzed squid axon (2–4) provide evidence that ATP is the energy source for this transport. In fact, much of the energy metabolism of a variety of cells is directed at supplying ATP for sodium transport (5). In the erythrocyte where the data are most accurate, three Na^+ are pumped outward and two K^+ inward for every ATP hydrolyzed (6–10).

In 1957, Skou found a Mg^{2+}-dependent adenosinetriphosphatase in the microsomal fraction of crab nerve which was markedly activated by the simultaneous presence of sodium and potassium (11). Since that time, numerous lines of evidence have supported Skou's suggestion that this enzyme system (hereafter referred to as the NaK ATPase) is involved in sodium and potassium transport. For example, tissues with high transport activity possess high NaK ATPase activity (12); both transport and the NaK ATPase require ATP, Na^+, and K^+, and are inhibited by cardiac glycosides. In addition, the NaK ATPase in erythrocytes displays the same asymmetry as the sodium pump; that is, sodium and ATP activate both transport and enzyme activity from the inside surface, and potassium and cardiac glycosides affect them from the outside (1, 13–16).

For a more detailed discussion of the physiological role of the sodium pump and the evidence linking such transport to the NaK ATPase, the reader is referred to a recent review by Baker (17).

In order to prove that the NaK ATPase constitutes the sodium pump, it will be necessary to position the purified enzyme in a chemically defined membrane system and show that it can translocate ions. In this connection, Jain et al (18, 19) claim to have induced electrogenic processes by inserting a particulate rat brain fraction containing the NaK ATPase into black lipid membranes. This observation could not be confirmed in our laboratory (20) using the same enzyme preparation as that used by Jain et al (18, 19) or the purified NaK ATPase from the rectal gland of *Squalus acanthias* (21). Recently, an ATP-dependent, ouabain-inhibitable sodium transport has been demonstrated in vesicles reconstituted from purified NaK ATPase preparations (21a, 21b).

Although it is believed that the NaK ATPase is the ion transport system or part of it, our understanding of how this enzyme works is still too limited to provide a molecular mechanism for the sodium pump. We attempt here to provide an overview of the progress that has been made with the NaK ATPase, and we touch upon the special problems that such a particulate membrane enzyme presents. Additional reviews written in the last few years include those by Skou (22), Albers (23), Heinz (24), Glynn (25), Charnock & Opit (26), Whittam & Wheeler (27), Bonting (28), Schoner (29), Schwartz, Lindenmayer & Allen (30), and Hokin & Dahl (31).

CHARACTERISTICS OF THE OVERALL REACTION

The NaK ATPase catalyzes the hydrolysis of the γ-phosphate of ATP if Mg^{2+}, Na^+, and K^+ are all present, and this hydrolysis is inhibited by cardiac glycosides, such as ouabain. The enzyme does not show absolute specificity for ATP. Other nucleoside triphosphates can serve as alternate, albeit far less effective, substrates. Several investigators have found ouabain-sensitive hydrolysis of CTP (32–34) and ITP (33, 34), marginal hydrolysis of GTP (34, 35), but no hydrolysis of UTP (32). Hegyvary & Post (36) reported relative activities of the NaK ATPase with various nucleotides of $100:49:2.3:2.4:0.6$ for ATP, dATP, CTP, ITP, and GTP and UTP. Jensen & Norby (37) have cited many references to nucleotide preference of the NaK ATPase and to inhibition of ATP hydrolysis by nucleotides. The products of ATP hydrolysis, ADP and P_i, inhibit the reaction (38–40), whereas AMP does not.

Although Na^+ and K^+ are both required to activate the enzyme, high concentrations of either ion will inhibit activation by the other. Optimal enzyme activity has been observed at Na^+/K^+ ratios between 10:1 and 5:1. The requirement for Na^+ is absolute, but numerous other ions can substitute for K^+ and do so with varying degrees of efficiency. The order of effectiveness has been found to be $K^+ > Rb^+ > NH_4^+ > Cs^+ > Li^+$ (22). Tl^+ can also substitute for K^+ and has been found to have ten times the affinity of K^+ for the enzyme (41, 42).

The presence of Mg^{2+} is essential for enzymatic activity. Many investigators have found that optimal activity is obtained when the Mg^{2+} concentration is approximately equal to that of the substrate. Discrepancies have arisen because the optimal ratio in fact varies with the ATP concentration (39, 43). When either Mg^{2+} or ATP is present in large excess over the other, inhibition occurs.

Divalent ion effects on the enzyme have been studied by several investigators. Both Mn^{2+} and Co^{2+} can substitute for Mg^{2+} but with one-tenth the efficiency (44–46); Fe^{2+}- and Ca^{2+}-dependent ATP hydrolysis have been reported (45, 46). These ions inhibit the Mg^{2+}-dependent reaction, as do Zn^{2+}, Cu^{2+}, Ba^{2+}, Sr^{2+}, and Be^{2+} (46–51). The modulation of NaK ATPase activity by Ca^{2+} is particularly interesting because of its possible physiological significance (52, 53).

There are many reports in the literature concerning inhibition of the NaK ATPase by various chemicals, drugs, and other biologically active agents. Several of these have been useful in studying the mechanism of the enzyme reaction and are discussed within that context. Inhibition by several classes of drugs has been found. The reader is referred to the recent detailed review by Glick (54) on inhibitors of the NaK ATPase and to the pertinent section in the review by Hokin & Dahl (31).

Minor activities inhibited by ouabain but not requiring the simultaneous presence of Na^+ and K^+ have been reported (22, 55–59). These may be due to intrinsic cations bound very tightly to some membrane preparations (60). A Na^+ ATPase activity inhibited by K^+ at low levels of ATP (57–59) has been interpreted to indicate that two different enzymes with different affinities for ATP contribute to the total hydrolytic activity (61). Others have concluded that these ouabain-sensitive activities arise from a single enzyme system (57, 62, 63). They suggest that spontaneous turnover of the NaK ATPase could account for the Na^+ ATPase activity (63, 64); its inhibition by K^+ can be explained by the fact that K^+ reduces the affinity of the enzyme for ATP, and thus at low ATP concentrations can effectively reduce enzyme activity (62, 63).

Crude NaK ATPase preparations contain an ATPase activity dependent only on Mg^{2+} and not inhibited by ouabain. The earlier speculation that this activity is part of the NaK ATPase has been fairly well discarded now that essentially all of the Mg^{2+} ATPase can be removed on purification of the NaK ATPase (21).

COURSE OF ATP HYDROLYSIS

There is considerable evidence that ATP hydrolysis catalyzed by the NaK ATPase proceeds through a multiple-step reaction sequence. The ultimate goal is to incorporate these reactions into a model for Na^+ and K^+ transport, presumably involving sequential conformational changes in the enzyme which effect the ion translocations.

It has been postulated that the NaK ATPase hydrolyzes ATP in a stepwise fashion involving Na^+-dependent phosphorylation of the enzyme and K^+-dependent hydrolysis of the phosphoenzyme. Several lines of investigation support the following sequence of reactions

$$E_1 + ATP \xrightleftharpoons{Mg^{2+}, Na^+} E_1 - P + ADP \qquad\qquad 1.$$

$$E_1 - P \xrightleftharpoons{Mg^{2+}} E_2 - P \qquad\qquad 2.$$

$$E_2 - P + H_2O \xrightleftharpoons{K^+} E_2 + P \qquad\qquad 3.$$

$$E_2 \rightleftharpoons E_1 \qquad\qquad 4.$$

Different conformations of the phosphorylated ($E_1 - P$ and $E_2 - P$) and non-phosphorylated (E_1 and E_2) forms of the enzyme appear to be involved as well as multiple (allosteric) interactions of cations and ATP with the enzyme. There is some evidence that E_2 has high K^+ affinity and E_1 has high Na^+ affinity. More complex sequences showing specific involvement of Na^+ and K^+ have been presented (62).

ATP Binding

An early step in the reaction sequence should involve the formation of an ATP-enzyme complex. Initially, conclusions about enzyme-substrate interactions were based on indirect (kinetic) evidence (38, 65–69). More recently, two groups have demonstrated ATP binding to the enzyme in the absence of Mg^{2+} using direct methods involving flow dialysis (36, 37, 70). Dissociation constants for the complex of $0.12 \, \mu M$ (70) and $0.22 \, \mu M$ (36) were reported. The binding of ATP was inhibited by prior treatment of the enzyme with ouabain and also by K^+ and other inorganic cations that substitute for K^+ in the stimulation of the NaK ATPase. Na^+ antagonized the inhibition by K^+, but itself had no direct effect. It was concluded that the 6-amino group of the purine ring and the γ-phosphate group are essential for tight binding. ADP was effective in displacing ATP from its binding site. In fact, the binding of ADP to the enzyme has also been studied (71). By a pulse experiment, Hegyvary & Post (36) showed that ATP bound in the absence of Mg^{2+} was a precursor of the phosphoenzyme intermediate (Step 1), thus casting doubts on conclusions reached from kinetic experiments that Mg ATP is the true substrate for the enzyme (38). A complete kinetic study of Mg^{2+} and ATP effects, such as described for other enzymes (72), has not been done for NaK ATPase. Even if detailed kinetic analyses were available, it is entirely possible that direct physical measurements using techniques such as nuclear magnetic resonance spectroscopy (73) would invalidate the conclusions of the kinetic data.

From kinetic studies, Siegel & Goodwin (62) and Post et al (63) concluded that K^+ remains bound to the enzyme after dephosphorylation (Step 3) and thus reduces the affinity for ATP. Thus K^+ must dissociate before Na^+ can activate E-P formation from ATP. The sequence presented earlier can be modified to reflect these ideas

$$E_2 - P + K^+ \rightleftharpoons K - E_2 P \qquad\qquad 5.$$

$$KE_2 P + H_2 O \rightleftharpoons KE_2 + P \qquad\qquad 6.$$

$$KE_2 \rightleftharpoons E_2 + K^+ \qquad\qquad 7.$$

$$E_2 \xrightarrow{Na^+} E_1 \qquad\qquad 8.$$

Post et al (63) suggested an activating effect of ATP in addition to its role as phosphate donor and included a complex of ATP with $K - E_2$ in order to activate the reaction. Schonfeld et al (74) also postulated a recombination of the enzyme with ATP which rejects K^+ from the cation binding sites so that a new cycle can start. Others (75, 76) have concluded that there must be an effect of ATP on the system that precedes phosphorylation.

The Phosphorylated Intermediate

As early as 1960, Skou (65) postulated that phosphorylation of the enzyme might be involved as an intermediate step in the overall reaction. This was based on his observation of an Mg^{2+}-dependent ATP-ADP exchange reaction catalyzed by a microsomal fraction from the leg nerve of the shore crab. However, at this time Na^+-stimulated exchange could not be demonstrated, and in brain microsomes Stahl et al (77) could separate by differential centrifugation most of the exchange activity. Stronger evidence for a phosphorylated intermediate was the finding of a very rapid Na^+-dependent incorporation of ^{32}P from $[\gamma$-$^{32}P]ATP$ into NaK ATPase preparations (66, 78–86). The ^{32}P remained bound to the protein after trichloroacetic acid precipitation. If K^+ was present simultaneously with Na^+, the radioactivity recovered in the acid-insoluble denatured protein was greatly reduced. If K^+ was added to the enzyme system after Na^+, the ^{32}P in the enzyme was rapidly discharged. The labeling was dependent on Mg^{2+} and specific for Na^+ (66). The same cations that substituted for K^+ in the overall hydrolysis reaction, namely, Li^+, NH_4^+, Rb^+, Cs^+, or Tl^+, activated dephosphorylation (66, 87). The nucleotide specificity for phosphorylation was essentially the same as for the overall reaction (34, 35, 88). Low concentrations of ouabain inhibited the K^+-dependent dephosphorylation and high concentrations prevented phosphorylation (89). N-ethylmaleimide (NEM) and oligomycin did not inhibit phosphorylation but did inhibit the K^+-dependent dephosphorylation (90–95). Bader et al (96) examined six tissues from eleven different species and found that although the specific activities of the NaK ATPase from these various sources varied by more than 400-fold, the range in the ratio of NaK ATPase activity to the level of phosphorylated intermediate was only twofold. This ratio, which is considered the turnover number of the enzyme, ranged from 5000 to 15,000. These findings support the view that the phosphorylated protein is a functional intermediate in the reaction sequence.

Chemical Characterization of the Phosphorylated Intermediate

The enzyme-phosphate bond has been characterized as an acylphosphate on the basis of its pH hydrolysis profile (82, 84, 97); its sensitivity to dephosphorylation by hydroxylamine (82, 84), acylphosphatase (84), or molybdate; and its phosphorylation of alcohols under acidic conditions (98, 99). Attempts to characterize the phosphorylated residue on the enzyme have been hampered by the lability of the acyl-phosphate bond. The initial concern that there might be phosphate migration during the acid-denaturation procedure has been minimized by identification of an acyl-phosphate under mild conditions, i.e., termination of the reaction with 0.1% sodium dodecyl sulfate followed by isolation of the labeled protein by gel filtration (100, 101).

Kahlenberg et al (102, 103) provided evidence that the acylphosphate was an L-glutamylphosphate residue. Post & Orcutt (104), on the other hand, concluded that the phosphorylated intermediate is an aspartyl-β-phosphate residue. Digestion of the phosphorylated enzyme to a limit with pronase yielded a tripeptide containing a dicarboxylic amino acid and lysine. That the active-site phosphate was on the amino-terminal side of lysine was deduced by high voltage electrophoresis of tryptic digests. Based on the fact that the phosphorylated peptide and synthetic Pro-Asp(P)-Lys were resistant to the action of carboxypeptidase B, whereas synthetic Pro-Glu(P)-Lys was sensitive to carboxypeptidase B, they concluded that the active site was a β-carboxyl group of an aspartic acid residue. Recently the acylphosphate has been more directly demonstrated to be β-aspartylphosphate (104a) by the method of Degani & Boyer (104b).

One interesting feature of the acylphosphate bond in the NaK ATPase is that its lability at high pH (8–10) is much influenced by its neighboring groups. Thus, at high pH values, acetylphosphate is rather stable, the phosphorylated peptide resulting from pronase digestion of the NaK ATPase is less so, and the phos-phorylated denatured NaK ATPase is even less stable (105).

Certain arguments have been raised that the acylphosphate in the enzyme is not an obligatory intermediate in the overall NaK ATPase reaction. One of the most serious has been that hydroxylamine can discharge ^{32}P from or partially prevent phosphorylation of some native mammalian preparations of NaK ATPase without inhibiting enzyme activity (106–108). If the discharge of ^{32}P involved hydroxyl-aminolysis of the acylphosphate bond, one would anticipate the enzyme to be inhibited by formation of a stable hydroxamate according to the following reactions

$$\text{Enzyme} + \text{ATP} \rightleftharpoons \text{Enzyme-O-P} + \text{ADP}$$

$$\text{Enzyme-O-P} + \text{NH}_2\text{OH} \rightleftharpoons \text{Enzyme-NHOH} + \text{P}$$

This discharging effect of hydroxylamine is apparently due, however, to a K^+-like effect of either contaminating ammonium ions in the hydroxylamine or an ammonium-like action of hydroxylamine itself, since N-methylhydroxylamine, which can discharge ^{32}P from acid-denatured phosphorylated enzyme preparations (98, 109), does not discharge ^{32}P from native phosphorylated enzyme (110) as does hydroxylamine (109). With electric organ and kidney microsomes, hydroxylamine will, in fact, partially replace K^+ in activating the NaK ATPase reaction (93, 111). It is now generally assumed that in native enzyme of mammalian origin, hydroxylamine is inaccessible to the susceptible acylphosphate residue (112).

K^+-Dependent Phosphatase Activity

Further support for the postulated sequence is the finding that NaK ATPase preparations invariably exhibit an ouabain-inhibitable, K^+-dependent phosphatase activity (113–126). There is a parallel enrichment of the two activities on purification of the NaK ATPase (126–128). p-Nitrophenylphosphate, acetylphosphate, umbelliferonephosphate, and carbamylphosphate will serve as substrates. Since these substrates are either acylphosphates or have energies of hydrolysis higher than a typical phosphate ester, the observations are consistent with the phosphorylated intermediate in the enzyme being an acylphosphate.

The close association of the two enzyme activities suggests that the phosphatase reaction may represent the final step in the reaction sequence catalyzed by the NaK ATPase and that the K^+-activated phosphatase is the expression of the ability of the transport ATPase to hydrolyze phosphate esters related to its natural substrate provided by the Na^+-dependent phosphorylation of the enzyme. Similarities and differences in the properties of the K^+-dependent phosphatase and the NaK ATPase have been noted (115, 118–121, 129, 130). In general, the differences between the two enzyme activities in sensitivities to certain ions and inhibitors become less when acetylphosphate is the substrate (116) or when the phosphatase assay conditions are made similar to the usual assay conditions for the NaK ATPase, i.e. the K^+-dependent phosphatase is assayed in the presence of Na^+ and ATP (121).

In the presence of Mg^{2+}, the p-nitrophenylphosphatase is activated by various monovalent cations but not by Na^+. The activating effect of cations is in the order: $K^+ = Rb^+ > Cs^+ > Li^+$ (123). Tl^+ can activate both the acetylphosphatase and the p-nitrophenylphosphatase of the beef brain microsomes with a K_m approximately one-tenth that of K^+ (41, 131). Thus, there is an obvious similarity between the cation sensitivity of the K^+-dependent phosphatase and the NaK ATPase. A variety of inhibitors and stimulators of the K^+-dependent phosphatase have been studied (113, 119, 121, 132–137). Simultaneous addition of ATP and Na^+ produces a large stimulation of activity at low levels of K^+ although separately both Na^+ and ATP are inhibitory (121, 123, 133). This synergistic effect of Na^+ and ATP has been interpreted by some as evidence for participation of a phosphorylated intermediate in the reaction. Robinson (135) attributes p-nitrophenylphosphatase activity to a new form of the phosphoenzyme derived from E_2-P through the action of K^+; others (123) include a modifying site for ATP but reject the participation of a phosphorylated intermediate.

Potassium-dischargeable labeling of NaK ATPase preparations by substrates for the K^+-dependent phosphatase reaction has been reported (100, 120, 138, 139). The phosphorylated intermediate formed from acetylphosphate-^{32}P appears to have similar chemical properties to that formed by $[\gamma\text{-}^{32}P]$ATP; furthermore, each substrate competes with the other for phosphorylation, suggesting that the same active site participates in both K^+-dependent acetylphosphatase and NaK ATPase activities. However, certain differences in ion effects and rates of turnover of the intermediate were observed with the two substrates (138, 139). Inturrisi & Titus (140) found phosphorylation of the NaK ATPase by p-nitrophenylphosphate only when the enzyme was inhibited by ouabain. The phosphorylation bore no clear relation to K^+-dependent phosphatase activity.

Perhaps some of the best evidence that the K^+-dependent phosphatase is a component of the NaK ATPase comes from work with intact cells. Garrahan and co-workers (141–146) found that the K^+-dependent phosphatase and the NaK ATPase in resealed red cell ghosts had the same asymmetrical requirements for K^+, Na^+, Mg^{2+}, and substrate. A good correlation has been found between the activities of the NaK ATPase and the K^+-dependent phosphatase in red cells with different pumping rates. ATP effects on K^+-dependent phosphatase have also been noted in red cells (144). When ATP is added, there is an increase in the K_m of K^+ and an increase in sensitivity to ouabain. This supports the view that, apart from providing the necessary energy for active transport, ATP also plays a role in promoting cyclical changes in selectivity for these ions, a role most active transport schemes require.

Various Forms of Phosphoenzyme

Evidence that there are two forms of the phosphorylated intermediate comes from the selective actions of Mg^{2+} and certain inhibitors on the partial reactions postulated for the NaK ATPase reaction. Certain NaK ATPase preparations have been demonstrated to catalyze a Na^+-dependent ADP-ATP exchange reaction. As mentioned earlier, Skou (65) had originally observed a Mg^{2+}-dependent ADP-ATP exchange activity in his microsomal NaK ATPase preparation from crab nerve. Subsequent investigations raised doubts about the participation of this exchange reaction in the NaK ATPase system, as no evidence was found for an ADP-ATP exchange at Na^+ and Mg^{2+} concentrations comparable to those used to phosphorylate enzyme protein and catalyze the overall hydrolysis of ATP (80, 147). Stahl et al (77) were able to separate by differential centrifugation the bulk of the ADP-ATP exchange system from the NaK ATPase in brain microsomes without affecting the activity of the latter. A residual ADP-ATP exchange activity remained firmly attached to the microsomes. Investigators working with a variety of tissues (90–93, 148–150) were able to demonstrate an exchange activity at low Mg^{2+} concentrations, which had absolute specificity for Na^+ and ATP, did not depend on K^+, and was inhibited by cardiac glycosides. Cation effects on the exchange have been investigated (151, 152). The fact that the optimum Mg^{2+} concentrations for exchange and the overall hydrolysis reaction are different is consistent with the view that there are at least two forms of phosphorylated intermediate. The conversion of E_1-P to E_2-P is postulated to be dependent on higher Mg^{2+} concentrations and to be essentially irreversible; E_2-P cannot react with ADP in an exchange reaction.

The selective action of certain inhibitors has also been useful in establishing step 2 in the proposed reaction sequence. Fahn et al (90, 91) found that whereas NEM and oligomycin inhibited the NaK ATPase reaction, there was enhancement of the exchange reaction and no reduction in the amount of phosphorylated intermediate formed from inhibited as compared to native enzyme. Thus, NEM and oligomycin are presumed to inhibit conversion of E_1-P to E_2-P, which explains why they can inhibit the overall reaction but stimulate exchange activity and not inhibit

K^+-dependent phosphatase activity under the usual assay conditions. In addition, 1,2 dithioglycerol arsenite has been found to inhibit hydrolysis while potentiating ADP-ATP exchange and enzyme phosphorylation. Further support derives from comparisons of the phosphorylated forms of native and NEM-treated NaK ATPase. The phosphorylated intermediate formed from native enzyme is cleaved by K^+ but not by ADP, where that formed after NEM treatment responds to ADP but not to K^+. No differences were noted in the electrophoretic patterns of peptide digests of these phosphorylated forms. Physiological ligands influence the sensitivity of NEM inhibition of phosphorylation and dephosphorylation (153, 154). Post et al (87) have reviewed the evidence that the transformation of E_1-P to E_2-P represents a change in conformation.

Direct identification of the transient E_1-P form has been difficult. E_2-P is the form usually observed in the presence of Na^+, Mg^{2+}, and ATP and the absence of K^+. Tobin et al (52) claim that their formation of a phosphorylated intermediate in the presence of Ca^{2+}, which is resistant to K^+ but very sensitive to ADP, constitutes identification of E_1-P in the native enzyme. They conclude that Ca^{2+}, a potent inhibitor of the NaK ATPase, may under appropriate circumstances stimulate phosphorylation and inhibit conversion of E_1-P to E_2-P. From kinetic studies, Middleton (155) and Stone (156) have concluded that the rate-limiting step in the reaction sequence is the conversion of E_1-P to E_2-P. Robinson (133), in contrast to other investigators, has suggested that the E_1-P to E_2-P conversion may not be an essential part of the translocation scheme, but that it plays a regulatory role in cation transport.

Reversal of the Reaction

In the last few years, it has been shown that the entire pump cycle is reversible. The reversal has been accomplished under special conditions using erythrocyte ghosts where defined interior and exterior compartments can be maintained experimentally (157–161). If the pump can run in either direction, it should be possible to find partial reactions of the NaK ATPase that represent steps in either direction. As previously mentioned, reversal of the transphosphorylation reaction (ADP-ATP exchange) has been demonstrated under conditions that presumably block the conversion of E_1-P to E_2-P. Reversal of a later step also appears possible since the native enzyme can be phosphorylated by P_i in the presence of Mg^{2+} and ouabain (89, 162, 163). More recently, phosphorylation by P_i and Mg^{2+} has been accomplished without ouabain or monovalent cations (100, 164–166). This latter phosphorylation is inhibited by Na^+ and by added nucleotides, whereas Li^+, NH_4^+, K^+, Rb^+, Cs^+, and Tl^+ permit formation of the phosphoenzyme. Digestion with pepsin of the phosphorylated protein prepared from P_i both in the presence and absence of ouabain and peptide mapping by electrophoresis on paper in one dimension suggested that the phosphorylation was at the same site as that phosphorylated by $[\gamma\text{-}^{32}P]$ATP (87, 167–170). Obviously, peptide mapping by electrophoresis in one dimension is a limited criterion.

The phosphorylation by ^{32}P would seem to be energetically unfeasible since a high energy phosphate bond is being formed on the enzyme. But the intermediate is not turning over under the reaction conditions and the small amount formed can be detected only by its radioactive label. The conditions that permit phosphorylation may lead to a conformational change in the enzyme—the large entropy change that would arise from the burial of hydrophobic groups exposed in the native enzyme may provide the requisite energy for acylphosphate formation.

Skvortsevich et al (171) and Dahms & Boyer (172) have recently reported a rapid Mg^{2+}- and K^+-dependent exchange of water oxygens with inorganic phosphate catalyzed by microsomal NaK ATPase preparations. This $P_i \rightleftharpoons HOH$ exchange was inhibited by Na^+ and ouabain and had no nucleotide requirement. It is felt that this exchange reaction represents a dynamic reversal of a late step in ATP hydrolysis by the NaK ATPase. Dahms & Boyer (172) speculate that a covalent phosphoenzyme, an acylphosphate, is in fact a catalytic intermediate in the $P_i \rightleftharpoons HOH$ exchange reaction. They were unable to demonstrate a $P_i \rightleftharpoons ATP$ or $ATP \rightleftharpoons HOH$ exchange indicating the apparent irreversibility of the overall reaction in vitro. In principle, it should be possible to synthesize ATP from P_i and the enzyme, but the yield of ATP would be less than the amount of enzyme in the system.

Interaction with Na^+ and K^+

Since the NaK ATPase is associated with the active transport of Na^+ and K^+, considerable attention has been focused on the role these cation activators play in the enzymatic process. Since the ions activating the NaK ATPase are the ones to be transported, their activating effects may be related to their combination with their translocation sites. It is not known whether there is one class of monovalent cation sites on the NaK ATPase which alternately prefers Na^+ then K^+ for enzyme activation (and presumably transport), or if there are distinct coexisting classes of sites for each ion. Of course there may be additional monovalent sites with only regulatory functions.

Virtually all of the investigations relating to sodium and potassium ion interactions with the enzyme have been kinetic in nature (62, 63, 67, 68, 122, 133, 136, 153, 173–179). Two specific models have emerged to account for the observed kinetics: 1. an allosteric model with multiple interacting sites; and 2. a multiple site/multiple affinity model with competitive interactions. Data from studies with intact membranes have similarly led to models in which Na^+ and K^+ compete for a common site or in which binding of each ion at its respective site influences binding of the other ion at its site.

Most models for sodium and potassium transport invoke some sequential changes in binding affinities of these ions. Various stages of the reaction sequence indicated previously have been found to respond preferentially to one ion or the other and an antagonism between them has frequently been observed. Such was the case with ATP binding as discussed previously. Robinson (51), using a nonconventional kinetic approach, reported a sufficient change in K^+ affinity after formation of a phosphorylated intermediate to permit binding of K^+ in a high Na^+ environment. Ideally, one would like to obtain direct evidence for changes in cation affinities and relate these to cation effects on the various steps in the postulated reaction sequence.

A more direct approach to Na^+ binding to the NaK ATPase has recently been used, namely, pulsed NMR of $^{23}Na^+$ (180). The K_D for Na^+ binding to the partially purified brain enzyme of Uesugi et al (126) agreed closely with that determined kinetically. With the purified dogfish rectal gland enzyme (21), Na^+ binding was absolutely dependent on ATP; with the brain enzyme, it was partially dependent. These effects of ATP were found in the absence of Mg^{2+}; this is consonant with the findings of Hegyvary & Post (36), who found ATP binding in the absence of Mg^{2+}.

Other Ion Exchange Reactions Associated with the Sodium Pump

Other ouabain-sensitive ion exchange activities which must be accounted for in any proposed model for the sodium pump have been observed in erythrocytes and squid giant axon preparations under special conditions (181–186). When erythrocytes or squid giant nerve fibers partially poisoned with alkaline dinitrophenol are incubated in K^+-free media, an ouabain-sensitive $Na^+:Na^+$ exchange is observed. An ouabain-sensitive $K^+:K^+$ exchange is also observed when erythrocytes containing P_i are incubated in K^+-containing medium. This exchange requires nucleotide, but the specificity of such a requirement is not clear.

Glynn & Hoffman (185) have compared nucleotide requirements for $Na^+:K^+$ and $Na^+:Na^+$ exchange. Although $Na^+:K^+$ exchange varied roughly in parallel with the concentration of ATP and was little affected by ADP, the $Na^+:Na^+$ exchange varied with the concentration of ADP, being relatively large when this concentration was high and almost absent when there was little ADP in the cells. These results point to a role for ADP in $Na^+:Na^+$ exchange, though not in the exchange of sodium for potassium. ADP apparently acts by reversing the reaction of which it is a product. Why should such a reversal be necessary for $Na^+:Na^+$ exchange? An attractive hypothesis is that the transphosphorylation is associated directly with the movement of sodium, so that a transfer of phosphate from ATP to E is associated with an outward movement of sodium and a transfer from E_1-P to ADP is associated with an inward movement. This would explain why the backward and forward shuttling of sodium requires ADP as well as ATP, whereas only ATP is needed for the forward running of the pump.

Evidence linking potassium entry with dephosphorylation comes from experiments showing that the reverse process, an outward movement of potassium through the pump, requires P_i. In red cells (and possibly in other cells, but nobody has looked), part of the downhill outward movement of potassium is inhibited by ouabain. When this outward movement of potassium is into a medium lacking potassium but rich in sodium, it appears to be associated with the backward running of the entire pump cycle and it is accompanied by incorporation of P_i into ATP, as discussed earlier. When the medium contains K^+, however, the whole cycle does not run backwards but, provided P_i is present inside the cells, an exchange of internal for external potassium ions takes place superimposed on the normal forward running of the pump. One may imagine that the dephosphorylation accompanying potassium entry is usually followed by rephosphorylation at the expense of P_i and is associated with an outward movement of potassium ions.

INTERACTION WITH CARDIAC GLYCOSIDES

In 1953, Schatzmann (187) observed that strophanthin k inhibited the active movements of Na^+ and K^+ in erythrocytes without affecting the energy-yielding reactions of the cell. Since that time, considerable literature has accumulated on the effects of various cardiac glycosides on Na^+ and K^+ transport in a variety of tissues (188–190).

Soon after Skou's discovery of the NaK ATPase in 1957 (11), the enzyme was found to be sensitive to cardiac glycosides (191, 192). The highly specific nature of this inhibition has led to speculation that the NaK ATPase is the pharmacologic receptor for these drugs and that their effects on the enzyme are responsible for their therapeutic and toxic effects. Evidence for and against this hypothesis has recently been presented (31, 190). We are particularly concerned here with the mechanism of the NaK ATPase reaction and the cation transport process itself. Attention has been directed to elucidate the structural features of the cardiotonic steroids essential for interaction with the enzyme, the nature of the binding between steroid and enzyme, as well as the mechanism by which inhibition of enzyme activity is produced. As will become apparent, the interaction between cardiac glycosides and the NaK ATPase is exceedingly complex. Because of this and the contradictory nature of some of the data, it is not yet possible to make definitive statements about the significance of all of the findings.

There are numerous studies on the relative inhibitory potencies of a large number of cardiac glycosides and their aglycones on Na^+ and K^+ transport and the NaK ATPase. The concentrations of compounds required for 50% inhibition generally range from 10^{-7} to $10^{-6} M$. Because there is a wide variation in the sensitivity of different tissues and species (193–196) and because the inhibition is dependent on time and temperature (163, 197–200) and on the concentration of ligands in the incubation media (89), there has been considerable variation in the reported inhibitory potency of a given compound. Nevertheless, there is uniformity of opinion about the importance to inhibitory potency of certain structural features (188, 189, 201). In general, the bufadienolides (six-membered di-unsaturated lactone ring in the β configuration at C-17) are more potent than the cardenolides (five membered mono-unsaturated lactone ring in the β configuration at C-17). An unsaturated lactone ring attached at C-17 in the β configuration to a cyclopentenophenanthrene nucleus and a β-hydroxyl at C-14 are essential structural features. Reduction in inhibitory potency results from saturation or disruption of the lactone ring, α configuration at C-17, dehydrogenation of the hydroxyl at C-3, or epimerization of this hydroxyl from β to the α position. Considerable variation in the A/B ring area can be made without destroying ability of the steroid to inhibit the enzyme.

From structure activity relationships several investigators have attempted to deduce the nature of the steroid binding site on the NaK ATPase. A characteristic feature of these compounds is the rather rigid nonpolar concavity which exists on the α surface as a result of the *cis* configuration of the C/D ring juncture. Portius & Repke

(202) postulated that the steroid interacts with a complementary surface of the enzyme by hydrophobic interactions. It is possible that this involves the nonpolar concavity of the steroid. Dissociation of the steroid from the enzyme by heat or acid treatment may be viewed as evidence for a nonpolar interaction.

Albers et al (163) found that the rate of inhibition by cardiac glycosides was inversely related to the number of hydroxyl and sugar substituents. This and the ease of removal of bound ouabain by methanol suggested to them a lipophilic interaction. Tobin & Sen (200) estimated the entropy change for ouabain interaction with the enzyme. They arrived at a high value which is consistent with a large conformational change. They believed that this ruled out a single-bond type and suggested that a large number of noncovalent bond types are involved. The presence of hydrogen bonding and stronger polar interactions has been postulated. Based on structure activity studies, Wilson et al (203) constructed a model which involves three distinct loci of interaction of the cardiotonic steroid with the enzyme. Middleton (155) proposed a model involving a mixed anhydride formed via nucleophilic attack of the phosphate of the acylphosphate residue on the cardiac glycoside lactone ring. Recently Yoda (204) has reported the dissociation rate constants of various complexes of cardiac monoglycosides with the NaK ATPase formed in the presence of Mg^{2+} and P_i. He found that the dissociation rate constant was dependent on the nature of the sugar but not the steroid moiety and presented a model for the binding site of the sugar moiety.

Other investigations have been aimed at elucidating the mechanism of inhibition of enzyme activity. It has already been mentioned that whereas high concentrations of ouabain inhibit formation of the phosphorylated intermediate from $[\gamma\text{-}^{32}P]ATP$ in the presence of Na^+ and Mg^{2+}, lower concentrations of the glycoside inhibit dephosphorylation. In the presence of Mg^{2+}, Na^+, K^+, and $[\gamma\text{-}^{32}P]ATP$, ouabain increases the steady state level of E-P. Ouabain apparently interacts with the enzyme to give an intermediate that is relatively stable even in the presence of K^+. Other partial reactions associated with NaK ATPase activity such as the ADP-ATP transphosphorylation reaction and the K^+-dependent phosphatase are also inhibited by cardiac glycosides.

The interaction between the NaK ATPase and cardiac glycosides has been studied either by following the binding of radioactive cardiac glycosides to enzyme preparations or by measuring residual inhibition of the enzyme after dilution of the system sufficiently so that inhibition due to free steroid would be negligible. While a direct relationship between binding and inhibition has been reported (163, 199, 205–207), ouabain binding has been found to be less sensitive than enzyme activity to various treatments (208). Harris (208) attributes this to the greater dependence of enzyme activity on membrane integrity. A good example of this is the fact that phospholipase C treatment destroys enzyme activity but not ouabain binding (209, 210).

Various physiological ligands influence the interaction between cardiac glycosides and the NaK ATPase (74, 75, 89, 211–218). These effects have led to hypotheses about the form of the enzyme with which these drugs interact and to speculations about the roles played by Na^+, K^+, ATP, and Mg^{2+} in the enzyme reaction and the transport process. It has only recently been appreciated that the interaction between cardiac glycosides and the enzyme is a relatively slow process and the dissociation of ouabain-enzyme complexes is even slower than the binding process. Most recent studies have been directed at measuring rates of association and dissociation and, in fact, these rates for a given complex are dependent on the nature of the ligands present (74, 75, 196, 213, 215). Species differences in sensitivity to cardiac glycosides have been found to be principally related to differences in the dissociation rates of drug-enzyme complexes (194, 195). When effects on steady state levels of ouabain-enzyme complexes are being studied, the possibility that prolonged incubation times may affect these levels should be kept in mind.

Conditions favoring formation of the phosphorylated intermediate favor binding, namely, the presence of $Mg^{2+} + Na^+ + ATP$, $Mg^{2+} + Na^+ +$ other nucleotides that phosphorylate the enzyme (216), or $Mg^{2+} + P_i$. How these ligands modify the rate and extent of interaction with the enzyme has been the subject of vigorous investigation. Some have concluded that the formation of the phosphorylated enzyme per se is not obligatory, but that some type of conformation including that induced by phosphorylation of the enzyme is required. According to this view, the glycoside does not bind directly to the phosphorylated enzyme but rather to a receptor allosterically related to this site. Schwartz feels "that the demonstration that glycosides inhibit the hydrolysis-transport cycle (a transmembrane event) via interaction at an external site constitutes the most convincing argument that inhibition is an allosteric event" (30).

Both Na^+ and K^+ affect interaction of ouabain with the enzyme. Na^+ has been found to increase inhibition by ouabain (219, 220). It had been known for a long time that partial inhibition of sodium transport and the NaK ATPase by low concentrations of cardiac glycosides could be relieved by K^+ (221). Glynn (221) had first suggested that K^+ and cardiac glycosides may compete for the same site because the antagonism showed certain features of competitive inhibition, but further analysis has shown that this is not the case (188, 189).

In fact, when one looks at rates of interaction between the enzyme and cardio-active drugs, a complex picture of Na^+ and K^+ effects emerges. There appears to be general agreement that when the steroid-enzyme complex is formed in the presence of $Na^+ + Mg^{2+} + ATP$, K^+ decreases the rates of association and dissociation; when binding occurs in the presence of $Mg^{2+} + P_i$, K^+ and Na^+ decrease the rate of association and are essentially without effect on the rate of dissociation (74, 75, 212–215). When ouabain binding is studied in the presence of Mg^{2+} alone, K^+ and Na^+ have been reported to decrease the rate of association (74, 75, 215). Hansen & Skou (218) report decreases in steady state levels by Na^+ and K^+ in both the $Na^+ + Mg^{2+} + ATP$ and the $Mg^{2+} + P_i$ systems. Others (198, 207, 212) have reported no such effects on steady state levels.

Attention has also been directed to the effects of ouabain on intact cell preparations. Perrone & Blostein (222), using inside-out and right-side-out membrane vesicles, confirmed previous findings with squid giant axons (223) and red cell ghosts (224) that ouabain inhibits transport and by inference the NaK ATPase only on the external surface of the membrane. Binding of ouabain and inhibition of transport depend on the Na^+ and K^+ concentrations in the medium. The rate of ouabain binding and the amount bound at the steady state have been reported to be increased by Na^+ and decreased by K^+ in many mammalian cells (224, 225). In contrast, ouabain binding with squid axon is increased by external Na^+, but K^+ seems to have little effect (226).

Attempts have been made to evaluate the quantitative relationship between ouabain binding, Na^+-dependent phosphorylation, and enzyme activity. All these calculations are based on the assumption that specific binding sites are saturated and there is no nonspecific binding. A stoichiometry of close to 1:1 for binding of cardiac glycosides per mole of enzyme has been reported. Albers (163) reported the stoichiometry of ouabain binding sites to phosphorylation sites in electric organ and cat brain as 2:1 and 1:1, respectively. Erdmann & Schoner (196) also reported that the stoichiometry varied with species. Kyte (227) found one cardiac glycoside binding site for each molecule of the polypeptide chain that is phosphorylated.

LIPID REQUIREMENT

Numerous studies point to a requirement for lipid in the NaK ATPase reaction. Treatment of the enzyme from a wide variety of sources with detergents, phospholipases, and solvents leads to inactivation. In those instances where the lipid content of the enzyme has been measured after such treatments, a correlation has been found between the amount of lipid removed and the loss of enzyme activity (210, 228).

While there is general agreement that lipid-depleted preparations can be reactivated by adding phospholipids, there has been some disagreement about the nature of the lipid required for restoration of activity after detergent (particularly deoxycholate) or phospholipase treatment. Early work was handicapped by the fact that the specific activities of the enzyme preparations were low and the reactivation small. A preparation widely used in these studies was first described by Tanaka & Abood (229) and Tanaka & Strickland (230). These authors used a deoxycholate solubilized beef brain microsome fraction which had been further fractionated with ammonium sulfate. The preparation was inactive and markedly depleted in phospholipids. They found that the NaK ATPase in their preparation was stimulated by several phospholipids, particularly commercial lecithin, commercial lysolecithin, and phosphatidic acid; however, no data were given on the purities of the phospholipid preparations. Sun et al (231) also obtained partial reactivation of this preparation by lecithin and further found the K_m for ATP to be the same before and after treatment with phospholipase C.

Fenster & Copenhaver (232), using essentially the Tanaka-Strickland preparation, provided evidence that contaminating phosphatidylserine in the commercial lecithin was responsible for activation of the NaK ATPase. Other phosphatide preparations which were apparently free of phosphatidylserine gave, at most, marginal stimulation of the NaK ATPase. These results supported an earlier observation of Ohnishi & Kawamura (233) that phosphatidylserine would restore the activity of the NaK ATPase in phospholipase A-treated horse erythrocyte ghost preparation. Wheeler & Whittam (234, 235) claimed that only phosphatidylserine could reconstitute the

Tanaka-Strickland preparation. They did show activating effects of their phosphatidylinositol and phosphatidic acid preparations but they attributed this to contaminating phosphatidylserine in these crude preparations. However, examination of their data suggests a true stimulation by phosphatidylinositol. Paper chromatography revealed no phosphatidylserine in their phosphatidic acid preparation. Formby & Clausen (129) found that phosphatidylserine and phosphatidylinositol gave a significant activation of the NaK ATPase in deoxycholate-treated synaptosomes from rat brain. Taniguchi & Tonomura (236) obtained similar results. Formby & Clausen (129) found that the residue of the synaptosomes after deoxycholate treatment was enriched with respect to phosphatidylserine.

Tanaka & Sakamoto (237) carried out a systematic investigation of the structural requirements of various phospholipids for activation of the NaK ATPase in the Tanaka-Strickland preparation. They found that the NaK ATPase was activated by either mono- or dialkylphosphates; both were most active when they possessed a ten-carbon chain. Activation by didecylphosphate was comparable to that obtained with several natural phospholipids, including phosphatidylserine. With respect to the natural phosphatides, Tanaka & Sakamoto (237) concluded that the minimum structural requirement for activation of the NaK ATPase was a phosphate plus one or two fatty chains. Tanaka (238) found that of the pure phosphatides tested for restoration of NaK ATPase activity, only phosphatidylserine and phosphatidic acid gave good stimulations, with weaker stimulations produced by bovine brain lecithin, lysolecithin, and soybean phosphatidylinositol. Tanaka et al (239) have studied the kinetics of binding and activation with various phospholipids, using the Tanaka-Strickland preparation. The specific activity of the Tanaka-Strickland preparation, even after activation by phospholipids, is generally two orders of magnitude lower than the purified NaK ATPase (see above). Hokin & Hexum (240) found that the partially purified beef brain enzyme of Uesugi et al (126) was rapidly inactivated by incubation with protease-free phospholipase A and was almost fully reactivated by Inosithin, a commercial preparation of phosphatidylinositol-enriched soybean phosphatides. Of the chromatographically pure phosphatides tested, phosphatidylserine, phosphatidic acid, and phosphatidylinositol were highly effective in restoring activity as was didodecyl phosphate. Inclusion of defatted serum albumin was necessary during phospholipase A treatment for maximum reconstitution by phospholipids. Hokin & Hexum (240) concluded that any one of many acidic phospholipids was capable of reconstituting activity. They further found that on purification of the NaK ATPase, the enzyme preparation became enriched with respect to phosphatidylserine. They suggested that under physiological conditions the lipid responsible for maintaining activity of the NaK ATPase may be phosphatidylserine but that the specificity for this phosphatide is not high.

Palatini et al (241) found that both phosphatidylserine and diphosphatidylglycerol were required for maximal activation of a Lubrol-solubilized, ammonium sulfate-fractionated NaK ATPase preparation. Kimelberg & Papahadjopoulos (242) reported a similar type of requirement (phosphatidylserine and phosphatidylglycerol) when they used Tanaka-Strickland enzyme preparations in which enzyme inhibition by the deoxycholate treatment was essentially complete.

Requirements for reconstitution are quite different when solvents are used to deplete lipids; it appears that cholesterol is required under these conditions with only marginal effects seen with phospholipids (243, 244).

At least three groups have reported that when phospholipids are totally depleted, reactivation is not possible (210, 245, 246). Activation by chelating agents of NaK ATPase preparations that have not been depleted of lipids has been observed (228, 247). Specht & Robinson (247) speculate that although a requirement for lipids is undoubtedly a characteristic of the NaK ATPase, at least part of the efficacy of phospholipid stimulation resides in its ability to chelate. Stahl (228) postulated that EGTA, phospholipids, and detergents might all act by displacing endogenous Ca^{2+} from a critical point on the native enzyme and thereby stimulate the system. This is not a consistent phenomenon, however. For example, Hokin & Hexum (248) observed no stimulation of the partially purified beef brain NaK ATPase of Uesugi et al (126) by phospholipids that were effective in reactivating the phospholipase A-treated enzyme.

Karlsson et al (249) found a high sulfatide content in kidney medulla, avian salt gland, and elasmobranch rectal gland, tissues specialized in sodium transport. In addition, when they fed ducks hypertonic saline for eight days, which led to hypertrophy of the salt glands, there was a 200% increase in the specific activity of the NaK ATPase. The sulfatide content of the gland increased to the same extent. The correlation between sulfatide content and the NaK ATPase is unique among the lipids examined and suggests a possible involvement of the former in the enzyme activity. Kawai et al (250) found that the relative phosphatidylserine and cholesterol ester content increased with purification of the enzyme.

In an effort to elucidate the role of lipids in NaK ATPase function, the effects of lipid depletion and restoration on the partial reactions associated with the enzyme have been studied. Among the reactions showing lipid dependence are E-P formation and decomposition, ATP-ADP exchange, the K^+-dependent phosphatase, and the steady state level of E-P (210, 228, 238, 239).

The effects of phospholipase C and phospholipase A treatment on ouabain binding appear to be conflicting. Phospholipase A treatment showed either no effect (209) or inhibition (210). Purified phospholipase C appreciably decreased enzyme activity but only slightly decreased ouabain binding capacity (208–210). Chipperfield & Whittam (251) reported a threefold stimulation of ouabain binding by phosphatidylserine of the Tanaka-Strickland preparation. Taniguchi & Iida (252, 253) have provided evidence that different conformations of the NaK ATPase that bind ouabain show different susceptibilities to removal of phospholipids.

Nonlinear Arrhenius plots of the NaK ATPase (254, 255) with an inflection point or transition temperature in the range of 15° to 20° have been reported. Recent studies have shown that enzyme preparations depleted of lipids by detergent or phospholipase treatment show linear Arrhenius plots and that reactivation of the enzyme with various phospholipids restores nonlinearity (242, 256, 257).

The role of phospholipids in the functioning of the NaK ATPase has been a subject for speculation. Among various suggestions as to the role of phospholipids is that they create a negative charge and hydrophobicity near but not at the active site (236, 251), provide ion-specific sites (242), and are components of the active site itself

(242). The fluidity of the fatty chains allows the necessary freedom for conformational and topographical changes associated with ion translocations.

Recently, Simpkins & Hokin (258) studied the purified rectal gland NaK ATPase (21) by ESR and high and low angle X-ray diffraction. In the ESR studies, the 7- and 12-nitroxide stearates were inserted into the membrane as spin labels. Strong evidence was provided by both techniques that the bulk of the phospholipid was in the form of a bilayer and in fact was hardly distinguishable from that in natural isolated membrane preparations. Phospholipase A treatment produced some change in the ESR spectrum of the enzyme labeled with the 12-nitroxide stearate, indicative of reduced mobility of the spin label. Phospholipase A treatment produced a greater change in the ESR spectrum of membranes labeled with spin-labeled NEM. Since previous work had indicated that spin-labeled NEM reacted almost exclusively with sulfhydryl groups in membranes (259), the change produced by phospholipase A in the ESR spectrum of the spin-labeled NEM-treated enzyme suggested increased exposure of buried sulfhydryl groups to a more polar environment. It was suggested that "inactivation of the NaK ATPase by phospholipase A digestion may be due to a conformational change in the protein as a result of destabilization of the protein associated with disruption of the phospholipid bilayer" (258).

PURIFICATION

The fact that the NaK ATPase is an intrinsic component of the cell membrane presents special obstacles to purification. Essentially two approaches have been used. In one of these, the enzyme remains attached to membrane fragments and is purified by removing less tightly bound proteins from these fragments. The other approach involves removing the enzyme from its lipid-protein environment, i.e. solubilizing it so that classical enzyme purification procedures can be applied. There are problems inherent in both approaches, but considerable progress in purification has been made.

The term soluble has been problematic for workers in the transport field. As operationally defined, a soluble component is one found in the supernatant fluid after centrifugation at $100,000 \times g$ for 1–2 hr. It has been argued that detergents do not solubilize the enzyme in the traditional sense, that in fact solubilization is a consequence of cleavage of the membrane into fragments containing more than one protein and considerable phospholipid. The fragment size and content of phospholipid and detergent determine its sedimentability in the ultracentrifuge. Detergents such as SDS go beyond the point of solubilization and dissociate the enzyme into its component polypeptides and lipid. Accompanying this is an irreversible inactivation of the enzyme which precludes purification of an enzymatically active species. Classical purification procedures have not proven very useful in purifying the NaK ATPase, and once this membranous enzyme is solubilized, it tends to revert to the membranous form when free detergent and part of the bound detergent are removed (126).

A word should be said about sources of the enzyme. Brain and kidney, because they are rich in NaK ATPase, are the most commonly used mammalian sources.

The enzyme is particularly abundant in the outer medulla of the kidney. Considerably richer sources of the enzyme are organs that are highly specialized for sodium and potassium pumping, such as the rectal gland of *Squalus acanthias* (21) and the electric organ of *Electrophorus electricus* (23). Membrane fractions from these sources can be obtained an order of magnitude higher in specific activity than those from the best mammalian sources. Most commonly, the membranes are isolated operationally as a microsome fraction. This fraction is a composite of many different membranous types. In the specialized organs, the plasma membrane is the major membranous component of the microsome fraction.

Extraction of extraneous protein from the microsome fraction has been accomplished with high concentrations of chaotropic agents (260), such as NaI. Nakao et al (261) reported considerable purification of the NaK ATPase from a variety of tissues by treatment of microsomes with 2 M NaI followed by diluting to 0.8 M NaI and collecting the membranes by centrifugation. This procedure extracted from two thirds to four fifths of extraneous protein from the membrane leaving the NaK ATPase membrane-bound. Yields as high as 95% (126) have been obtained by this method. The Mg^{2+} ATPase is almost completely removed.

Often treatment of the microsomes with concentrations of certain detergents that do not solubilize the NaK ATPase assists in achieving high specific activities. Deoxycholate has been most often used for this purpose. This detergent removes inert protein, reduces the level of the Mg^{2+} ATPase, and often effects a considerable activation of the NaK ATPase. As the concentration of detergent is raised, the enzyme is progressively solubilized and inactivated. Jorgensen & Skou (262) carried out a detailed study of the effect of three detergents–deoxycholate, SDS, and Lubrol 14–on the NaK ATPase in the microsomal fraction from the outer medulla of the rabbit kidney. Incubation of microsomes with deoxycholate under appropriate conditions increased the specific activity of the NaK ATPase from 45 to 270 μmol P_i/mg protein per hour. Lubrol 14 and SDS also activated the enzyme. Activation by all three detergents was optimal at the critical micelle concentration of each detergent. Higher concentrations inhibited. In the case of deoxycholate, the activated NaK ATPase remained associated with the membrane. The work of Rostgaard (263) has provided electron microscopic verification of Jorgensen and Skou's suggestion that activation of the NaK ATPase by detergents is due to exposure of latent enzyme sites resulting from opening of microsomal vesicles.

Towle & Copenhaver (33) started with a deoxycholate-solubilized (230) membrane fraction from rabbit kidney and partially purified the NaK ATPase by ammonium sulfate fractionation, chromatography on 8% agarose, and glycerol gradient centrifugation. The final specific activity was approximately 300 μmol P_i/mg protein per hour and the yield was 2.2 mg protein. The enzyme was considerably inactivated by deoxycholate but reactivation was achieved by addition of phosphatidylserine.

Jorgensen et al (128) purified deoxycholate-activated microsomes from outer medulla of the rabbit kidney to a specific activity of about 1000 μmol P_i/mg protein per hour by isopycnic zonal centrifugation with a yield of about 10–15 mg protein. The enzyme could be further purified to a specific activity of about 1500 μmol P_i/mg protein per hour by rate zonal centrifugation with a loss of about a third of the enzyme. Because of the deoxycholate activation and the fact that no polyacrylamide gel patterns were given, it is difficult to assess the purity of the preparation.

Kyte (264) purified the deoxycholate-activated microsome fraction from the outer medulla of dog kidney to a maximum specific activity of 800 μmol P_i/mg protein per hour by treatment with KI, solubilization of impurities with deoxycholate at low salt concentration, followed by solubilization of the NaK ATPase with deoxycholate at higher salt concentration. The final step was chromatography on agarose, which indicated a highly polydisperse preparation with molecular weights in the millions. This result suggests that the enzyme purification used here did not involve solubilization but was rather based on the technique of extraction of impurities from an essentially membrane-bound enzyme. The final yield was 0.46 mg of protein.

The strategy followed in our laboratory has been solubilization of the enzyme with the nonionic detergent Lubrol WX. The use of this detergent was suggested by some initial observations of Swanson et al (265) that Lubrol rendered the enzyme in rat brain microsomes unsedimentable on centrifugation at $100,000 \times g$ for 1 hr. Medzihradsky et al (266) found the enzyme in Lubrol extracts to be rather unstable but considerable stabilization could be effected by extracting in the presence of ATP and adding either Na^+ or K^+. Interestingly, the concentrations of Na^+ or K^+ for half-maximal stabilization were the same as the respective concentrations for half-maximal activation of the enzyme, suggesting that sites for stabilization by monovalent cations are the same as those for activation of the enzyme. NaI-treated beef brain microsomes could be solubilized with Lubrol on a very large scale (267, 268). In all of these studies, the Lubrol-solubilized NaK ATPase chromatographed on 6% agarose as a sharp symmetrical peak with an apparent molecular weight of 670,000, suggesting that at least a reproducible fragment had been solubilized. Correction for nonprotein components in the preparation gave a molecular weight close to 300,000, in fairly good agreement with the value of 250,000 obtained by radiation inactivation (269, 270).

Shirachi et al (271) solubilized rat brain microsomes with lower concentrations of Lubrol, which seemed to afford a more stable extract. They found that prior treatment with NaI necessitated the use of much higher concentrations of Lubrol. The enzyme was further purified by centrifugation in a sucrose density gradient. Banerjee et al (272) compared Lubrol W and Triton X-100 for solubilization of guinea pig kidney cortex NaK ATPase. They also omitted the NaI treatment stage and found that lower concentrations of detergent could be used. They claimed that Triton X-100 was superior to Lubrol W for extraction and suggested that the reverse observation reported by Medzihradsky et al (266) may have been due to interference by Lubrol in the latter's protein assays. However, that laboratory had carefully worked out a protein assay where interference by Lubrol was eliminated. Species and organ differences probably account for the discrepancy (271).

Uesugi et al (126) reported the large scale partial purification of beef brain NaK ATPase. Three grams of NaI-treated beef brain microsomal protein were solubilized with Lubrol, and the Lubrol extract was submitted to zonal centrifugation in a sucrose density gradient which had the twofold effect of removing free Lubrol from the enzyme and partially purifying the enzyme. The concentrated fraction from the zonal centrifugation was then purified by a novel ammonium sulfate fractionation procedure. The yield of enzyme was 30–50 mg of protein with a specific activity as high as 750 μmol P_i/mg protein per hour. The enzyme was purified approximately 100 times over that in the starting homogenate and was extremely stable.

Both the enzyme of Kyte (264) and that of Uesugi et al (126) showed two predominant proteins on SDS-polyacrylamide gel electrophoresis, the larger one having an apparent molecular weight of 84,000 (264) and 94,000 (126), and the smaller one a molecular weight of 55,000. These proteins are discussed in detail below.

Following the same strategy worked out by Uesugi et al (126) for purification of the beef brain enzyme, except that the NaI extraction step was omitted, Hokin et al (21) recently purified the NaK ATPase in Lubrol extracts from membranes of the rectal gland of S. acanthias. The final specific activity of the preparation was 1500 μmol P_i/mg protein per hour. Although the enzyme was activated on Lubrol treatment, this activation was eliminated on removal of free Lubrol by zonal centrifugation. From 90 to 95% of the protein was accounted for on SDS-polyacrylamide gel electrophoresis by a protein with a molecular weight of 97,000 and a protein of 55,000 mol wt (see below). Ten rectal glands yielded 30–50 mg of protein. The enzyme was highly stable at 0° in the frozen state. At the last stage of purification, electron microscopy of negatively stained preparations revealed vesicles, rods, or rings. The rods and rings were approximately 80 Å in diameter. Projections from these rods and rings of about 35–55 Å in diameter were seen at regular intervals. The vesicular form of the enzyme bore a striking resemblance to reconstituted oligomycin-sensitive mitochondrial ATPase.

Recently, Lane et al (273) reported the purification of the NaK ATPase from the outer medulla of dog kidney. The purification steps were NaI treatment of microsomes, solubilization with 0.35% deoxycholate, precipitation with glycerol, resolubilization with 3:1 cholate:deoxycholate, and ammonium sulfate fractionation. The specific activity of the enzyme was 1500 μmol P_i/mg protein per hour. SDS-polyacrylamide gel electrophoresis showed a 89,000 mol wt protein and a 56,000 mol wt protein. The yield of enzyme was 16–18 mg protein/100 g of tissue. Although the authors obtained considerable activation of the enzyme by the NaI and deoxycholate treatments, they believe the subsequent treatments removed the activation.

Dixon & Hokin (274) have purified the NaK ATPase from the electric organ of E. electricus using the same purification protocol used for the rectal gland of the dog-fish. The specific activity of the final purified enzyme was 1200. Over 90% of the protein could be accounted for by a 92,500 and a 47,500 mol wt protein. The purified enzyme was highly stable both at 0° and in the frozen state.

Nakao et al (275, 276) have purified the enzyme from pig brain microsome by subjecting them sequentially to deoxycholate treatment, NaI treatment, Lubrol extraction, and aminoethyl-cellulose chromatography. They claim a specific activity of 7000 in certain fractions. It is difficult to assess the validity of this value since it is achieved only with a modified protein determination used to detect minute amounts of protein and the polyacrylamide gel patterns are unclear and show only faint amounts of protein.

In conclusion, we wish to emphasize that there are essentially two criteria for assessing the purity of NaK ATPase preparations: specific activity and SDS-polyacrylamide gel electrophoresis. Both methods have their drawbacks. Specific activity can be misleading if the enzyme has been activated or inactivated by detergents. Different SDS-polyacrylamide gel systems vary in their resolving power, the sensitivity of staining of protein, and their ability to permit full penetration of all the protein into the gel. All of these factors must be assessed in evaluating the purity of a given preparation.

POLYPEPTIDE COMPONENTS

The Phosphorylated Peptide

As mentioned above, the purest preparations of the NaK ATPase are composed almost entirely of two proteins. The larger one has a molecular weight in the range from 84,000 to 100,000 as estimated by gel electrophoresis in SDS (21, 126, 264, 273) or 139,000 as estimated by gel filtration (281). It is probable that the slightly different molecular weights determined by polyacrylamide gel electrophoresis are real rather than entirely due to methodological differences. For example, Dixon & Hokin (274) ran the enzyme from the rectal gland and the electric organ side by side in replicate and consistently found a molecular weight of 93,000 for the subunit from the electric organ and 97,000 for the subunit from the rectal gland.

It is the larger of these two peptides that is phosphorylated during the course of ATP hydrolysis (97, 126, 277–280). Hart & Titus (282) used NEM, whose inhibition of the enzyme is influenced by physiological ligands, to further characterize the phosphorylated peptide. From their sulfhydryl protection data, they concluded (282) that the peptide

> appears to possess binding sites for sodium, ATP, potassium, and ouabain, while the data for sulfhydryl exposure accompanying phosphorylation appear to reflect conformational changes in the peptide that parallel the potential energy states of the phosphorylated intermediates. Since the binding sites for potassium and ouabain are situated on the exterior surface of the plasma membrane, while the binding sites for sodium and ATP are situated on the interior surface, the results of the current study imply that the 98,000-dalton peptide may in fact completely traverse the membrane.

Neither Uesugi et al (126) nor Hart & Titus (280) were successful in their attempts to further separate peptides within this band.

The amino acid compositions of the phosphorylated polypeptides from the NaK ATPase of dog kidney (281), rectal gland of S. acanthias, and electric organ of E. electricus (283) are quite similar, being rich in nonpolar amino acids. Only one NH_2-terminal amino acid is found in purified preparations of this polypeptide. Its identity varies with source of the enzyme being glycine, alanine, and serine for dog kidney, rectal gland, and electric organ, respectively (281, 283).

It would thus appear that although the kinetic parameters of the enzyme have remained remarkably constant throughout evolution (284), some variation in the phosphorylated subunit has been permitted.

Glycoprotein Component of Purified NaK ATPase Preparations

A glycoprotein has been reported as the second major protein component of purified NaK ATPase preparations (21, 273, 274, 281). In NaK ATPase preparations from dog kidney and rectal gland, the molecular weight of this component is close to 55,000 based on SDS-polyacrylamide gel electrophoresis. It purifies together with the catalytic subunit (21), suggesting that both polypeptide chains are components of the same protein. Cross-linking with suberimidate (281) also suggests this. The glycoprotein is rich in hydrophobic amino acids and contains glucosamine (281, 283), galactosamine (281), glucose, galactose, mannose, fucose (283), and sialic acid (281, 283). This polypeptide component purified from the NaK ATPase from dog kidney, rectal gland of the dogfish, and electric organ has NH_2-terminal alanine (281, 283).

The question arises whether the glycoprotein is a true component of the NaK ATPase or a protein that binds tenaciously to the NaK ATPase and purifies together with it. It is certainly true that the glycoprotein is always present in NaK ATPase preparations considered to be the purest, but to date there is no direct proof that it is a necessary component of the enzyme. One important criterion which should be fulfilled is that the molar ratio of catalytic subunit to glycoprotein be constant in the pure enzyme from different species. Dixon & Hokin (274) have compared this ratio for the purified enzyme from the rectal gland of *S. acanthias* and from the electric organ from *E. electricus* and have obtained values of 2.16 and 1.95. However, in the preparations from the dog kidney, this ratio is 1 (264, 273). This discrepancy is open to several interpretations: 1. The glycoprotein region on gels from the dog kidney is contaminated with other protein which lowers the ratio. 2. The gel scanning technique is not quantitative in all cases; for example, Hokin et al (21) found that plots of peak size vs quantity of protein applied were no longer linear when more than 2 μg of protein were run. 3. The glycoprotein is not related to the NaK ATPase. However, the fact that the rectal gland and electric organ enzymes gave virtually identical stoichiometric ratios would appear to be more than coincidental. It should also be emphasized that this stoichiometric ratio gives a minimum molecular weight of the enzyme of 250,000 which agrees with that obtained by radiation inactivation (269, 270).

As in the case of the phosphorylated subunit, there are chemical differences in the glycoprotein from different sources, which again could easily be due to evolutionary changes not necessarily critical to the functioning of the glycoprotein in the NaK ATPase. For example, the carbohydrate composition is different for the dog kidney enzyme, rectal gland enzyme, and electric organ enzyme. The molecular weight of the glycoprotein from electric organ is definitely lower than that from rectal gland, the former being 47,000 as compared to 55,000 for the latter (274).

One means of resolving the question of the essentiality of the glycoprotein would be reconstitution from purified components. However, the only means of isolating the protein components is dissociation with SDS followed by separation by gel filtration or by polyacrylamide gel electrophoresis in the presence of a strong dissociation agent. So far, it has not been possible to reconstitute activity from purified protein components and lipids.

TRANSPORT MODELS

In essentially all models for sodium and potassium transport, cation movement is effected by a series of conformational changes in the NaK ATPase (26, 76, 89, 163, 170, 285–287). Reference has been made to the kinetic evidence for conformational changes in the enzyme. With most of the models, the conformational changes in the NaK ATPase are geared to the cyclic phosphorylation and dephosphorylation of the enzyme. The enzyme is postulated to exist in two major conformations. One conformation is that of the dephosphorylated enzyme in which Na^+-specific sites face inward. Binding of Na^+ to these sites permits phosphorylation of the enzyme by ATP or possibly MgATP. This converts the enzyme to the second major conformational form which has K^+-specific sites facing outward. The conversion of the first conformation to the second carries Na^+ to the outside surface of the membrane with concomitant loss of Na^+ specificity and gaining of K^+ specificity. Binding of K^+ to the second conformational form of the enzyme permits dephosphorylation of the enzyme, converting the enzyme back to its first conformation. This conversion carries K^+ to the inside surface of the membrane with concomitant loss of K^+ specificity and regaining of Na^+ specificity, thus completing the cycle.

Skou (76) by contrast has proposed that ATP may produce alterations in the conformation(s) of the enzyme, which lead to translocation without E-P formation, and has suggested a simultaneous rather than sequential transfer of Na^+ and K^+.

In the novel model by Yang (288), the phosphate group itself acts as a cation carrier. "Active transport of sodium takes place along a chain of phosphomonoester groups in the membrane. Pyrophosphate linkages are synthesized at the inner boundary of the membrane by a sodium-dependent kinase and hydrolyzed by a potassium-dependent pyrophosphate at the outer boundary of the membrane" (288).

Models for sodium and potassium transport have recently been reviewed by Caldwell (289).

Literature Cited

1. Hoffman, J. F. 1960. *Fed. Proc.* 19:127
2. Caldwell, P. C., Hodgkin, A. L., Keynes, R. D., Shaw, T. I. 1960. *J. Physiol.* 152:561
3. Brinley, F. J. Jr., Mullins, L. J. 1968. *J. Gen. Physiol.* 52:181
4. deWeer, P. 1968. *Nature* 219:730
5. Whittam, R. 1964. *The Cellular Functions of Membrane Transport,* ed. J. F. Hoffman, 139. New York: Prentice-Hall
6. Glynn, I. M. 1962. *J. Physiol.* 160:18P
7. Gardos, G. 1964. *Experientia* 20:387
8. Sen, A. K., Post, R. L. 1964. *J. Biol. Chem.* 239:345
9. Whittam, R., Ager, M. E. 1965. *Biochem. J.* 97:214
10. Garrahan, P. J., Glynn, I. M. 1967. *J. Physiol.* 192:217
11. Skou, J. C. 1957. *Biochim. Biophys. Acta* 23:394
12. Bonting, S. L., Caravaggio, L. L. 1963. *Arch. Biochem. Biophys.* 101:37
13. Post, R. L., Merritt, C. R., Kinsolving, C. R., Albright, C. D. 1960. *J. Biol. Chem.* 235:1796

14. Dunham, E. T., Glynn, I. M. 1961. *J. Physiol.* 156:274
15. Whittam, R. 1962. *Biochem. J.* 84:110
16. Hoffman, J. F. 1962. *Circulation* 26:1201
17. Baker, P. F. 1972. *Metab. Pathways* VI:243
18. Jain, M. K., Strickholm, A., Cordes, E. H. 1969. *Nature* 222:871
19. Jain, M. K., White, F. P., Strickholm, A., Williams, E., Cordes, E. H. 1972. *J. Membrane Biol.* 8:363
20. Hilden, S., Ostroy, F., Hokin, L. E. Unpublished observations
21. Hokin, L. E. et al 1973. *J. Biol. Chem.* 248:2593
21a. Goldin, S., Tong, S. Personal communication
21b. Rhee, H. M., Hilden, S., Hokin, L. E. Unpublished observations
22. Skou, J. C. 1965. *Phys. Rev.* 45:596
23. Albers, R. W. 1967. *Ann. Rev. Biochem.* 36:727
24. Heinz, E. 1967. *Ann. Rev. Physiol.* 29:21
25. Glynn, I. M. 1968. *Brit. Med. Bull.* 24:165
26. Charnock, J. S., Opit, L. J. *The Biological Basis of Medicine,* ed. E. E. Bittar, N. Bittar, I:69. New York: Academic
27. Whittam, R., Wheeler, K. P. 1970. *Ann. Rev. Physiol.* 32:21
28. Bonting, S. L. 1970. *Membranes and Ion Transport,* ed. E. E. Bittar, 257. London: Wiley
29. Schoner, W. 1971. *Angew. Chem.* 10:882
30. Schwartz, A., Lindenmayer, G. E., Allen, J. C. 1972. *Current Topics in Membranes and Transport,* ed. F. Bonner, A. Kleinzeller, III:1. New York: Academic
31. Hokin, L. E., Dahl, J. L. 1972. *Metab. Pathways* VI:269
32. Hokin, L. E., Yoda, A. 1964. *Proc. Nat. Acad. Sci. USA* 52:454
33. Towle, D. W., Copenhaver, C. J. Jr. 1970. *Biochim. Biophys. Acta* 203:124
34. Matsui, H., Schwartz, A. 1966. *Biochim. Biophys. Acta* 128:380
35. Schoner, W., Beusch, R., Kramer, R. 1968. *Eur. J. Biochem.* 7:102
36. Hegyvary, C., Post, R. L. 1971. *J. Biol. Chem.* 246:5234
37. Jensen, J., Norby, J. G. 1971. *Biochim. Biophys. Acta* 233:395
38. Hexum, T., Samson, F. E. Jr., Himes, R. H. 1970. *Biochim. Biophys. Acta* 212:322
39. Moake, J. L., Ahmed, K., Bachur, N. R., Gutfreund, D. E. 1970. *Biochim. Biophys. Acta* 211:337
40. Baskin, L. S., Leslie, R. B. 1968. *Biochim. Biophys. Acta* 159:509
41. Britten, J. S., Blank, M. 1968. *Biochim. Biophys. Acta* 159:160
42. Skulskii, I. A., Manninen, V., Jarnefelt, J. 1973. *Biochim. Biophys. Acta* 298:702
43. Schoner, W., von Ilberg, C., Kramer, R., Seubert, W. 1967. *Eur. J. Biochem.* 1:334
44. Atkinson, A., Hunt, S., Lowe, A. G. 1968. *Biochim. Biophys. Acta* 167:469
45. Atkinson, A., Lowe, A. G. 1972. *Biochim. Biophys. Acta* 266:103
46. Rendi, R., Uhr, M. L. 1964. *Biochim. Biophys. Acta* 89:520
47. Ting-Beall, H. P., Clark, D. A., Sueleter, C. H., Wells, W. W. 1973. *Biochim. Biophys. Acta* 291:229
48. Donaldson, J., Minnich, J. L., Barbeau, A. 1971. *Can. J. Biochem.* 50:888
49. Bowler, K., Duncan, C. J. 1970. *Biochim. Biophys. Acta* 196:116
50. Toda, G. 1968. *J. Biochem. Tokyo* 64:457
51. Robinson, J. D. 1973. *Arch. Biochem. Biophys.* 156:232
52. Tobin, T., Akera, T., Baskin, S. I., Brody, T. M. 1973. *Mol. Pharmacol.* 9:336
53. Baker, P. F. 1972. *Progr. Biophys. Mol. Biol.* 24:177
54. Glick, N. B. 1972. *Metabolic Inhibitors: A Comprehensive Treatise,* ed. R. M. Hochster, M. Kates, J. H. Quastel, III:1. New York: Academic
55. Fujita, M., Nagano, K., Mizuno, N., Nakao, T., Nakao, M. 1967. *J. Biochem. Tokyo* 61:473
56. Fujita, M. et al 1968. *Biochem. J.* 106:113
57. Czerwinski, A., Gitelman, H. J., Welt, L. G. 1967. *Am. J. Physiol.* 213:786
58. Neufeld, A. H., Levy, H. M. 1969. *J. Biol. Chem.* 244:6493
59. Blostein, R. 1970. *J. Biol. Chem.* 245:270
60. Goldfarb, P. G. S., Rodnight, R. 1970. *Biochem. J.* 120:15
61. Neufeld, A. H., Levy, H. M. 1970. *J. Biol. Chem.* 245:4962
62. Siegel, G. J., Goodwin, B. 1972. *J. Biol. Chem.* 247:3630
63. Post, R. L., Hegyvary, C., Kume, S. 1972. *J. Biol. Chem.* 247:6530
64. Kanazawa, T., Saito, M., Tonomura, Y. 1970. *J. Biochem. Tokyo* 67:693
65. Skou, J. C. 1960. *Biochim. Biophys. Acta* 42:6
66. Post, R. L., Sen, A. K., Rosenthal, A. S. 1965. *J. Biol. Chem.* 240:1437
67. Squires, R. F. 1965. *Biochem. Biophys. Res. Commun.* 19:27

68. Robinson, J. D. 1967. *Biochemistry* 6:3250
69. Masiak, S. J., Green, J. W. 1968. *Biochim. Biophys. Acta* 159:340
70. Norby, J. G., Jensen, J. 1971. *Biochim. Biophys. Acta* 233:104
71. Kaniike, K., Erdmann, E., Schoner, W. 1973. *Biochim. Biophys. Acta* 298:901
72. Kuby, S. A., Noltner, E. A. 1962. *Enzymes* 6:515
73. Mildvan, A. S., Cohn, M. 1970. *Advan. Enzymol.* 33:1
74. Schonfeld, W., Schon, R., Menke, K. H., Repke, K. R. H. 1972. *Acta Biol. Med. Ger.* 28:935
75. Siegel, G. J., Josephson, L. 1972. *Eur. J. Biochem.* 25:323
76. Skou, J. C. 1971. *Curr. Top. Bioenerg.* 4:357
77. Stahl, W. L., Sattin, A., McIlwain, H. 1966. *Biochem. J.* 99:404
78. Rose, S. P. R. 1963. *Nature* 199:375
79. Charnock, J. S., Post, R. L. 1963. *Nature* 199:910
80. Albers, R. W., Fahn, S., Koval, G. J. 1963. *Proc. Nat. Acad. Sci. USA* 50:474
81. Charnock, J. S., Rosenthal, A. S., Post, R. L. 1963. *Aust. J. Exp. Biol. Med. Sci.* 41:675
82. Nagano, K. et al 1965. *Biochem. Biophys. Res. Commun.* 19:759
83. Gibbs, R., Roddy, P. M., Titus, E. 1965. *J. Biol. Chem.* 240:2181
84. Hokin, L. E., Sastry, P. S., Galsworthy, P. R., Yoda, A. 1965. *Proc. Nat. Acad. Sci. USA* 54:177
85. Blostein, R. 1966. *Biochem. Biophys. Res. Commun.* 24:598
86. Kanazawa, T., Saito, M., Tonomura, Y. 1967. *J. Biochem. Tokyo* 61:555
87. Post, R. L., Kume, S., Tobin, T., Orcutt, B., Sen, A. K. 1969. *J. Gen. Physiol.* 54:306s
88. Albers, R. W., Koval, G. J. 1962. *Life Sci.* 1:219
89. Sen, A. K., Tobin, T., Post, R. L. 1969. *J. Biol. Chem.* 244:6596
90. Fahn, S., Koval, G. J., Albers, R. W. 1966. *J. Biol. Chem.* 241:1882
91. Fahn, S., Hurley, J. R., Koval, G. J., Albers, R. W. 1966. *J. Biol. Chem.* 241:1890
92. Siegel, G. J., Albers, R. W. 1967. *J. Biol. Chem.* 242:4972
93. Fahn, S., Koval, G. J., Albers, R. W. 1968. *J. Biol. Chem.* 243:1993
94. Stahl, W. L. 1968. *J. Neurochem.* 15:499
95. Banerjee, S. P., Khanna, V. K., Sen, A. K. 1971. *Biochem. Pharmacol.* 20:1649
96. Bader, H., Post, R. L., Bond, G. H. 1968. *Biochim. Biophys. Acta* 150:41
97. Alexander, D. R., Rodnight, R. 1970. *Biochem. J.* 119:44P
98. Bader, H., Sen, A. K., Post, R. L. 1965. *Biochim. Biophys. Acta* 118:106
99. Bader, H., Wilkes, A. B., Jean, D. H. 1970. *Biochim. Biophys. Acta* 198:583
100. Dudding, W. F., Winter, C. G. 1971. *Biochim. Biophys. Acta* 241:650
101. Nagano, K. et al 1967. *Biochim. Biophys. Acta* 143:239
102. Kahlenberg, A., Galsworthy, P. R., Hokin, L. E. 1967. *Science* 157:434
103. Kahlenberg, A., Galsworthy, P. R., Hokin, L. E. 1968. *Arch. Biochem. Biophys.* 126:331
104. Post, R. L., Orcutt, B. 1973. *Organization of Energy Transducing Membranes*, ed. M. Nakao, L. Packer, 25. Tokyo: Tokyo Univ. Press
104a. Nishigaki, I., Chen, A., Hokin, L. J. Biol. Chem. In press
104b. Degani, C., Boyer, P. D. 1973. *J. Biol. Chem.* 248:8222
105. Ratanabanangkoon, K., Hokin, L. E. 1973. *Arch. Biochem. Biophys.* 158:695
106. Chignell, C. F., Titus, E. 1966. *Proc. Nat. Acad. Sci. USA* 56:1620
107. Schoner, W., Kramer, R., Seubert, W. 1966. *Biochem. Biophys. Res. Commun.* 23:403
108. Sachs, G., Long, M. M., Tsuji, T., Hirschowitz, B. I. 1971. *Biochim. Biophys. Acta* 223:117
109. Yoda, A., Kahlenberg, A., Galsworthy, P. R., Dulak, N. C., Hokin, L. E. 1967. *Biochemistry* 6:1886
110. Chignell, C. F., Titus, E. 1968. *Biochim. Biophys. Acta* 159:345
111. Charnock, J. S., Opit, L. J., Potter, H. A. 1967. *Biochem. J.* 104:17C
112. Formby, B. 1973. *Biochim. Biophys. Acta* 298:291
113. Judah, J. D., Ahmed, K., McLean, A. E. M. 1962. *Biochim. Biophys. Acta* 65:472
114. Rega, A., Garrahan, P. J., Pouchan, M. I. 1968. *Biochim. Biophys. Acta* 150:742
115. Emmelot, P., Bos, C. J. 1966. *Biochim. Biophys. Acta* 121:375
116. Bader, H., Sen, A. K. 1966. *Biochim. Biophys. Acta* 118:116
117. Albers, R. W., Rodriguez de Lores, G., De Robertis, E. 1965. *Proc. Nat. Acad. Sci. USA* 53:557
118. Nagai, K., Yoshida, H. 1966. *Biochim. Biophys. Acta* 128:410
119. Fujita, M. et al 1966. *Biochim. Biophys. Acta* 117:42

120. Israel, Y., Titus, E. 1967. *Biochim. Biophys. Acta* 139:450
121. Yoshida, H., Nagai, K., Ohashi, T., Nakagawa, Y. 1969. *Biochim. Biophys. Acta* 171:178
122. Robinson, J. D. 1969. *Biochemistry* 8:3348
123. Koyal, D., Rao, S. N., Askari, A. 1971. *Biochim. Biophys. Acta* 225:11
124. Albers, R. W., Koval, G. J. 1966. *J. Biol. Chem.* 241:1896
125. Brooker, G., Thomas, L. J. Jr. 1971. *Mol. Pharmacol.* 7:199
126. Uesugi, S. et al 1971. *J. Biol. Chem.* 246:531
127. Ahmed, K., Judah, J. D. 1964. *Biochim. Biophys. Acta* 93:603
128. Jorgensen, P. L., Skou, J. C., Solomonson, L. P. 1971. *Biochim. Biophys. Acta* 233:381
129. Formby, B., Clausen, J. 1968. *Z. Phys. Chem.* 349:909
130. Askari, A., Koyal, D. 1971. *Biochim. Biophys. Acta* 225:20
131. Inturrisi, C. E. 1969. *Biochim. Biophys. Acta* 173:567
132. Winter, C. G. 1972. *Biochim. Biophys. Acta* 266:135
133. Robinson, J. D. 1970. *Arch. Biochem. Biophys.* 139:164
134. Albers, R. W., Koval, G. J. 1972. *J. Biol. Chem.* 247:3088
135. Robinson, J. D. 1972. *Biochim. Biophys. Acta* 274:542
136. Albers, R. W., Koval, G. J. 1973. *J. Biol. Chem.* 248:777
137. Robinson, J. D. 1971. *Mol. Pharmacol.* 7:238
138. Sachs, G., Rose, J. D., Hirschowitz, B. I. 1967. *Arch. Biochem. Biophys.* 119:277
139. Bond, G. H., Bader, H., Post, R. L. 1971. *Biochim. Biophys. Acta* 241:57
140. Inturrisi, C. E., Titus, E. 1970. *Mol. Pharmacol.* 6:99
141. Garrahan, P. J., Pouchan, M. I., Rega, A. F. 1969. *J. Physiol.* 202:305
142. Rega, A. F., Pouchan, M. I., Garrahan, P. J. 1970. *Science* 167:55
143. Rega, A. F., Garrahan, P. J., Pouchan, M. I. 1970. *J. Membrane Biol.* 3:14
144. Garrahan, P. J., Pouchan, M. I., Rega, A. F. 1970. *J. Membrane Biol.* 3:26
145. Garrahan, P. J., Rega, A. F. 1971. *Nature New Biol.* 232:24
146. Garrahan, P. J., Rega, A. F. 1972. *J. Physiol.* 223:595
147. Swanson, P. D., Stahl, W. L. 1966. *Biochem. J.* 99:396
148. Stahl, W. L. 1967. *Arch. Biochem. Biophys.* 120:230
149. Stahl, W. L. 1968. *J. Neurochem.* 15:511
150. Blostein, R. 1968. *J. Biol. Chem.* 243:1957
151. Banerjee, S. P., Wong, S. M. E. 1972. *J. Biol. Chem.* 247:5409
152. Wildes, R. A., Evans, H. J., Chiu, J. 1973. *Biochim. Biophys. Acta* 307:162
153. Banerjee, S. P., Wong, S. M. E., Khanna, V. K., Sen, A. K. 1972. *Mol. Pharmacol.* 8:8
154. Banerjee, S. P., Wong, S. M. E., Sen, A. K. 1972. *Mol. Pharmacol.* 8:18
155. Middleton, H. W. 1970. *Arch. Biochem. Biophys.* 136:280
156. Stone, A. J. 1968. *Biochim. Biophys. Acta* 150:578
157. Garrahan, P. J., Glynn, I. M. 1967. *J. Physiol.* 192:237
158. Lant, A. F., Whittam, R. 1968. *J. Physiol.* 199:457
159. Lant, A. F., Priestland, R. N., Whittam, R. 1970. *J. Physiol.* 207:291
160. Glynn, I. M., Lew, V. L. 1970. *J. Physiol.* 207:393
161. Lew, V. L., Glynn, I. M., Ellory, J. C. 1970. *Nature* 225:865
162. Lindenmayer, G. E., Laughter, A. H., Schwartz, A. 1968. *Arch. Biochem. Biophys.* 127:187
163. Albers, R. W., Koval, G. J., Siegel, G. J. 1968. *Mol. Pharmacol.* 4:324
164. Dahms, A. S. 1971. *Fed. Proc.* 30:1170
165. Toda, G., Rogers, F. N., Post, R. L. *Abstr. Am. Chem. Soc. Meet. Washington DC, September 1971.* No. 210
166. Post, R. L., Toda, G. *IUB Meet., Buenos Aires, November 1972*
167. Chignell, C. F., Titus, E. 1969. *Proc. Nat. Acad. Sci. USA* 64:324
168. Siegel, G. J., Koval, G. J., Albers, R. W. 1969. *J. Biol. Chem.* 244:3264
169. Schwartz, A., Matsui, H., Laughter, A. H. 1968. *Science* 160:323
170. Post, R. L., Kume, S., Rogers, F. N. 1973. *Mechanisms in Bioenergetics,* ed. F. Azzone, L. Ernster, 203. New York: Academic
171. Skvortsevich, E. G., Panteleeva, N. S., Pisareva, L. N. 1972. *Dokl. Biochem.* 206:363
172. Dahms, A. S., Boyer, P. D. 1973. *J. Biol. Chem.* 248:3155
173. Hoffman, J. F., Tosteson, D. C. 1971. *J. Gen. Physiol.* 58:438
174. Lindenmayer, G. E., Schwartz, A. 1973. *J. Biol. Chem.* 248:1291
175. Ahmed, K., Judah, J. D., Scholefield, P. G. 1966. *Biochim. Biophys. Acta* 120:351

176. Priestland, R. N., Whittam, R. 1968. *Biochem. J.* 109:369
177. Garrahan, P. J., Glynn, I. M. 1967. *J. Physiol.* 192:175
178. Garrahan, P. J. 1969. *Nature* 222:1000
179. Robinson, J. D. 1968. *Nature* 220:1325
180. Ostroy, F., Hokin, L. E. Unpublished observations
181. Baker, P. F. et al 1969. *J. Physiol.* 200:459
182. deWeer, P. 1970. *Nature* 226:1251
183. deWeer, P. 1970. *J. Gen. Physiol.* 56:584
184. Glynn, I. M., Hoffman, J. F., Lew, V. L. 1971. *Phil. Trans. Roy. Soc. London* 262:91
185. Glynn, I. M., Hoffman, J. F. 1971. *J. Physiol.* 218:239
186. Sachs, J. R. 1972. *J. Clin. Invest.* 51:3244
187. Schatzmann, H. J. 1953. *Helv. Phys. Acta* 11:346
188. Glynn, I. M. 1964. *Pharmacol. Rev.* 16:381
189. Glynn, I. M. 1969. *Digitalis,* ed. C. Fisch, B. Surawicz, 30. New York: Grune & Stratton
190. Lee, K. S., Klaus, W. 1971. *Pharmacol. Rev.* 23:193
191. Post, R. L., Merritt, C. R., Kinsolving, C. R., Albright, C. D. 1960. *J. Biol. Chem.* 235:1796
192. Dunham, E. T., Glynn, I. M. 1961. *J. Physiol.* 156:274
193. Repke, K., Est, M., Portius, H. J. 1965. *Biochem. Pharmacol.* 14:1785
194. Tobin, T., Henderson, R., Sen, A. K. 1972. *Biochim. Biophys. Acta* 274:551
195. Tobin, T., Brody, T. M. 1972. *Biochem. Pharmacol.* 21:1553
196. Erdmann, E., Schoner, W. 1973. *Biochim. Biophys. Acta* 307:386
197. Allen, J. C., Schwartz, A. 1969. *J. Pharmacol. Exp. Ther.* 168:42
198. Allen, J. C., Schwartz, A. 1970. *J. Mol. Cell. Cardiol.* 1:39
199. Allen, J. C., Lindenmayer, G. E., Schwartz, A. 1970. *Arch. Biochem. Biophys.* 141:322
200. Tobin. T., Sen, A. K. 1970. *Biochim. Biophys. Acta* 198:120
201. Kupchan, S. M., Mokotoff, M., Sandhu, R. S., Hokin, L. E. 1967. *J. Med. Chem.* 10:1025
202. Portius, H. J., Repke, K. 1964. *Arzeim. Forsch.* 14:1073
203. Wilson, W. E., Sivitz, W. I., Hanna, L. T. 1970. *Mol. Pharmacol.* 6:449
204. Yoda, A. 1973. *Mol. Pharmacol.* 9:51
205. Schwartz, A., Allen, J. C., Harigaya, S. 1969. *J. Pharmacol. Exp. Ther.* 168:31
206. Hansen, O. 1971. *Biochim. Biophys. Acta* 233:122
207. Barnett, R. E. 1970. *Biochemistry* 9:4644
208. Harris, W. E., Swanson, P. D., Stahl, W. L. 1973. *Biochim. Biophys. Acta* 298:680
209. Taniguchi, K., Iida, S. 1971. *Biochim. Biophys. Acta* 233:831
210. Goldman, S. S., Albers, R. W. 1973. *J. Biol. Chem.* 248:867
211. Allen, J. C., Schwartz, A. 1970. *Mol. Pharmacol.* 12:240
212. Lindenmayer, G. E., Schwartz, A. 1970. *Arch. Biochem. Biophys.* 140:371
213. Akera, T., Brody, T. M. 1971. *J. Pharmacol. Exp. Ther.* 176:545
214. Skou, J. C., Butler, K. W., Hansen, O. 1971. *Biochem. Biophys. Acta* 241:443
215. Van Winkle, W. B., Allen, J. C., Schwartz, A. 1972. *Arch. Biochem. Biophys.* 151:85
216. Tobin, T., Baskin, S. I., Akera, T., Brody, T. M. 1972. *Mol. Pharmacol.* 8:256
217. Lishko, V. K., Malysheva, M. K., Grevizirskaya, T. I. 1972. *Biochim. Biophys. Acta* 288:103
218. Hansen, O., Skou, J. C. 1973. *Biochim. Biophys. Acta* 311:51
219. Schatzmann, H. J. 1965. *Biochim. Biophys. Acta* 94:89
220. Matsui, H., Schwartz, A. 1966. *Biochim. Biophys. Res. Commun.* 25:147
221. Glynn, I. M. 1957. *J. Physiol.* 136:148
222. Perrone, J. R., Blostein, R. 1973. *Biochim. Biophys. Acta* 291:680
223. Baker, P. F., Manil, J. 1968. *Biochim. Biophys. Acta* 150:328
224. Gardner, J. D., Conlon, T. P. 1972. *J. Gen. Physiol.* 60:609
225. Baker, P. F., Willis, J. S. 1972. *J. Physiol.* 224:441
226. Baker, P. F., Willis, J. S. 1972. *J. Physiol.* 224:463
227. Kyte, J. 1972. *J. Biol. Chem.* 247:7634
228. Stahl, W. L. 1973. *Arch. Biochem. Biophys.* 154:47
229. Tanaka, R., Abood, L. G. 1964. *Arch. Biochem. Biophys.* 108:47
230. Tanaka, R., Strickland, K. P. 1965. *Arch. Biochem. Biophys.* 111:583
231. Sun, A. Y., Sun, G. Y., Samorajski, T. 1971. *J. Neurochem.* 18:1711
232. Fenster, L. J., Copenhaver, C. J. Jr. 1967. *Biochim. Biophys. Acta* 137:406
233. Ohnishi, T., Kawamura, H. 1964. *J. Biochem.* 56:377

234. Wheeler, K. P., Whittam, R. 1970. *J. Physiol.* 207: 303
235. Wheeler, K. P., Whittam, R. 1970. *Nature* 225: 449
236. Taniguchi, K., Tonomura, Y. 1971. *J. Biochem.* 69: 543
237. Tanaka, R., Sakamoto, T. 1969. *Biochim. Biophys. Acta* 193: 384
238. Tanaka, R. 1969. *J. Neurochem.* 16: 1301
239. Tanaka, R., Sakamoto, T., Sakamoto, Y. 1971. *J. Membrane Biol.* 4: 42
240. Hokin, L. E., Hexum, T. D. 1972. *Arch. Biochem. Biophys.* 151: 453
241. Palatini, P., Dabbeni-Sala, F., Bruni, A. 1972. *Biochim. Biophys. Acta* 288: 413
242. Kimelberg, H. K., Papahadjopoulos, D. 1972. *Biochim. Biophys. Acta* 282: 277
243. Noguchi, T., Freed, S. 1971. *Nature* 230: 148
244. Jarnefelt, J. 1972. *Biochim. Biophys. Acta* 266: 91
245. Hegyvary, C., Post, R. L. 1969. *The Molecular Basis of Membrane Function*, ed. D. C. Tosteson, 519. New Jersey: Prentice-Hall
246. Roelofsen, B., Zwaal, R. F. A., Van Deenen, L. L. M. 1971. *Advan. Exp. Med. Biol.* 14: 209
247. Specht, S. C., Robinson, J. D. 1973. *Arch. Biochem. Biophys.* 154: 314
248. Hokin, L. E., Hexum, T. D. 1971. Unpublished observations
249. Karlsson, K. A., Samuelsson, B. E., Steen, G. O. 1971. *J. Membrane Biol.* 5: 169
250. Kawai, K., Nakao, M., Nakao, T., Fujita, M. 1973. *J. Biochem. Tokyo* 73: 979
251. Chipperfield, A. R., Whittam, R. 1973. *J. Physiol.* 230: 467
252. Taniguchi, K., Iida, S. 1972. *Biochim. Biophys. Acta* 288: 98
253. Taniguchi, K., Iida, S. 1973. *Mol. Pharmacol.* 9: 350
254. Bowler, K., Duncan, C. J. 1968. *Comp. Biochem. Physiol.* 24: 1043
255. Charnock, J. S., Cook, D. A., Casey, R. 1971. *Arch. Biochem. Biophys.* 147: 323
256. Taniguchi, K., Iida, S. 1972. *Biochim. Biophys. Acta* 274: 536
257. Priestland, R. N., Whittam, R. 1972. *J. Physiol.* 220: 343
258. Simpkins, H., Hokin, L. E. 1973. *Arch. Biochem. Biophys.* 159: 897
259. Simpkins, H., Panko, E., Tay, S. 1971. *J. Membrane Biol.* 5: 334
260. Hatefi, Y., Hanstein, W. G. 1969.

261. Nakao, T., Tashima, Y., Nagano, K., Nakao, M. 1965. *Biochem. Biophys. Res. Commun.* 19: 755
262. Jorgensen, P. L., Skou, J. C. 1969. *Biochim. Biophys. Acta* 233: 366
263. Rostgaard, J., Moller, O. J. 1971. *Exp. Cell Res.* 68: 356
264. Kyte, J. 1971. *J. Biol. Chem.* 246: 4157
265. Swanson, P. D., Bradford, H. F., McIlwain, H. 1964. *Biochem. J.* 92: 235
266. Medzihradsky, F., Kline, M. H., Hokin, L. E. 1967. *Arch. Biochem. Biophys.* 121: 311
267. Kahlenberg, A., Dulak, N., Dixon, J., Galsworthy, P., Hokin, L. E. 1969. *Arch. Biochem. Biophys.* 131: 253
268. Uesugi, S., Kahlenberg, A., Medzihradsky, F., Hokin, L. E. 1969. *Arch. Biochem. Biophys.* 130: 156
269. Mizuno, N. et al 1969. *Biochim. Biophys. Acta* 168: 311
270. Kepner, G. R., Macey, R. I. 1968. *Biochim. Biophys. Acta* 163: 188
271. Shirachi, D. Y., Allard, A. A., Trevor, A. J. 1970. *Biochem. Pharmacol.* 19: 2893
272. Banerjee, S. P., Dwosh, I. L., Khanna, V. K., Sen, A. K. 1970. *Biochim. Biophys. Acta* 211: 345
273. Lane, L. K., Copenhaver, J. H. Jr., Lindenmayer, G. E., Schwartz, A. Unpublished observations
274. Dixon, J. F., Hokin, L. E. Unpublished observations
275. Nakao, T., Nakao, M., Mizuno, N., Komatsu, Y., Fujita, M. 1973. *J. Biochem. Tokyo* 73: 609
276. Nakao, T. et al 1973. *J. Biochem. Tokyo* 73: 781
277. Kyte, J. 1971. *Biochem. Biophys. Res. Commun.* 43: 1259
278. Avruch, J., Fairbanks, G. 1972. *Proc. Nat. Acad. Sci. USA* 69: 1216
279. Collins, R. C., Albers, R. W. 1972. *J. Neurochem.* 19: 1209
280. Hart, W. M., Titus, E. O. 1973. *J. Biol. Chem.* 248: 1365
281. Kyte, J. 1972. *J. Biol. Chem.* 247: 7642
282. Hart, W. M., Titus, E. O. 1973. *J. Biol. Chem.* 248: 4674
283. Hokin, L. E., Hackney, J. F., Perrone, J. R., Dixon, J. F. Unpublished observations
284. Kline, M. H., Hexum, T. D., Dahl, J. L., Hokin, L. E. 1971. *Arch. Biochem. Biophys.* 147: 781
285. Jardetsky, O. 1966. *Nature* 211: 969
286. Hokin, L. E. 1969. *J. Gen. Physiol.* 54: 327s

Proc. Nat. Acad. Sci. USA 62: 1129

287. Stein, W. D., Lieb, W. R., Karlish, J. D., Eilam, Y. 1973. *Proc. Nat. Acad. Sci. USA* 70:275

288. Yang, J. H. 1970. *Proc. Nat. Acad. Sci. USA* 67:59

289. Caldwell, P. C. 1970. *Membranes and Ion Transport,* ed. E. E. Bittar, I:433. New York: Wiley-Interscience

MECHANISM OF ENZYME ACTION   ✕ 850

Albert S. Mildvan

The Institute for Cancer Research, Fox Chase Center for Cancer and Medical Sciences
Philadelphia, Pennsylvania

CONTENTS

INTRODUCTION

A review of so vast a subject must be selective and inevitably suffers from serious omissions, since advances in the mechanism of individual enzymes of interest to the author are stressed. Subsequently, some general chemical principles of enzyme action which appear to be emerging are listed. It is felt that this method of progression

357

from the particulars of individual enzymes to the general is more useful than a premature theoretical search for a unifying first principle common to all enzymes. Where available, structural data from crystallography and magnetic resonance are stressed, since it is apparent that these powerful techniques, while not providing a denouement, at least offer numerous clues in the search for enzyme mechanisms. The diverse enzymes considered catalyze the transfer of nucleotidyl and phosphoryl groups, HCO_3^- and carboxyl groups, hydrolysis, addition-elimination reactions, and oxidation-reduction reactions involving two electrons, one electron, and atom transfers.

NUCLEOTIDYL AND PHOSPHORYL TRANSFER ENZYMES

Chemical Mechanisms

Chemical mechanisms of phosphoryl (and nucleotidyl) transfer reactions have been the subject of several recent reviews (1–4), the last two stressing the role of metal ions. Briefly, phosphoryl transfer reactions are nucleophilic displacements on phosphorus and, like analogous displacements on carbon (5) or on metal ions (3), can take place by mechanisms varying between two extreme or limiting cases: 1. In the dissociative or S_N1 mechanism, the initial departure of the leaving nucleophile yields the planar triply coordinate metaphosphate anion as a reactive chemical intermediate (1, 6), which then combines with the entering ligand on either face of the metaphosphate plane.

$$1.$$

2. In the associative or S_N2 mechanism, the entering nucleophile first binds to phosphorus along the axis of the trigonal bipyramid, resulting in the formation of a pentacoordinate phosphorane either as a transition state with a lifetime $\tau = 10^{-13}$ sec or as a reactive intermediate ($\tau > 10^{-13}$ sec). The leaving nucleophile then dissociates along the opposite axis of the trigonal bipyramid.

$$2.$$

As described by Equation 2, the stereochemistry of the S_N2 displacement is in line, that is, the bond angle between entering and leaving groups is 180°.

A stereochemical possibility unique to the associative mechanism is adjacent displacement, in which the bond angle between entering and leaving groups is 90°. This possibility results from the ability of the pentacoordinate phosphorane intermediate to undergo pseudorotation (7, 8) which, with certain restrictions (9), interchanges axial and equatorial ligands.

As in carbon chemistry (4) and in coordination chemistry (3), mechanisms intermediate between the limiting dissociative and associative cases are found. Model studies (1, 2) indicate that both of the limiting mechanisms of phosphoryl transfer are substantially accelerated by electron withdrawal from the leaving group, lowering its pK_a. Such electron withdrawal may be accomplished by selective protonation, metal coordination, or appropriate covalent substituents.

Protonation, metal coordination, or further esterification of the transferable phosphoryl group itself profoundly retards the S_N1 mechanisms (1), presumably because of the decreased availability of lone pair electrons of oxygen for π bonding to phosphorus as metaphosphate is expelled, but accelerates the S_N2 mechanism by charge neutralization (4) and possibly by electron withdrawal from phosphorus. In addition, metal coordination could accelerate S_N2 displacements by five additional mechanisms (3). Chelation of the phosphoryl group could (a) induce strain in the ground state, (b) stabilize the phosphorane transition state, or (c) serve as a template for pseudorotation. Metal ions could also (d) coordinate and promote the nucleophilicity of the entering ligand or (e) simultaneously coordinate the entering ligand and the phosphoryl group undergoing nucleophilic attack, transforming a bimolecular to a unimolecular reaction (3, 4).

Since more modes of accelerating S_N2 displacements have been defined and greater stereochemical control is possible with them, it might be anticipated that more enzymes would utilize associative rather than dissociative pathways. However, the validation of this point remains speculative and must await elucidation of the various enzyme mechanisms. With these basic concepts in mind, we consider enzyme-catalyzed phosphoryl and nucleotidyl transfer reactions.

Nucleotidyl Transfer Enzymes

STAPHYLOCOCCAL NUCLEASE As a result of extensive studies of its primary, secondary, and tertiary structure by chemical degradation (10) and synthesis (11), crystallographic (12, 13), and magnetic resonance techniques (14), a great deal of information is available on the active site of this nucleotidyl transfer enzyme. Comprehensive reviews of each aspect of this work have been published (10, 15, 16). The crystal structure of the ternary complex of the enzyme, Ca^{2+}, and thymidine 3',5'-diphosphate at ~ 2.5 Å resolution reveals three of the four oxygens of the 5'-phosphate, including the C–O–P bridge oxygen, to be hydrogen bonded by the guanidinium groups of Arg 35 and 87 (13, 17). Crystallographic studies of simple complexes of guanidinium and phosphate reveal that significant P–O bond lengthening of 0.02–0.03 Å (17, 18) could occur from such hydrogen bonding. The enzyme-bound Ca^{2+}, coordinated by three or four carboxylate ligands, is at a distance from the phosphorus (4.7 ± 0.2 Å), which could permit an intervening water or hydroxyl ligand. Alternatively the value of 4.7 Å could result from a time or space average of 50% inner sphere complex (3.1 Å) and 50% second sphere complex (6.3 Å) (vide infra).

A second Ca^{2+} binding site detected by various methods in solution (14, 19) was not located in the crystal structure. Incisive NMR studies in solution using selectively deuterated enzyme preparations (14, 16) reveal the perturbation of two tyrosines by deoxythymidine 3',5'-diphosphate due to direct interaction but not by Ca^{2+} in agreement with the crystallographic data. The studies further reveal perturbation of a third tyrosine due to indirect interaction with a negatively charged phosphate group of the inhibitor. A histidine group is also perturbed in the ternary enzyme–Ca^{2+}–pdTP complex and a fourth tyrosine is perturbed as a second mole of Ca^{2+} is bound (14, 16). An elegant series of experiments involving the chemical synthesis of portions of the enzyme with specific amino acid substitutions has established the absolute requirement for Arg 35, which hydrogen bonds to the phosphate in the crystal structure, and for Asp 19, 21, and 40 and Glu 43 (11), which serve as ligands for Ca^{2+} in the crystal structure (15).

Assuming identity of the crystal and solution structures, and further assuming the polynucleotide substrate to bind in the same way found for the competitive inhibitor, thymidine 3',5'-diphosphate, the enzyme would bring about incipient protonation of the leaving sugar oxygen by hydrogen bonding to the substrate. Such an effect would promote either an S_N1 or S_N2 displacement by water. By also hydrogen bonding to two other oxygen atoms of the nucleotidyl phosphoryl group, the enzyme would inhibit an S_N1 mechanism and activate an S_N2 mechanism. Moreover, the propinquity of the bound metal, which could increase the nucleophilicity of an attacking water, would promote an in-line S_N2 mechanism and exert little catalytic effect on an S_N1 mechanism. A catalytic role for Ca^{2+} is suggested by the failure of trivalent lanthanide ions to activate this enzyme even though they occupy the Ca^{2+} binding site (20). Alternatively, Cotton & Hazen (15) have pointed out the possibility of a nucleophilic attack or stabilization of the transition state by Tyr 113. A nucleophilic attack by Tyr 113 would require adjacent displacement and pseudorotation.

Dunn et al (21) have argued that the exclusive displacement of the poorer of the two possible leaving groups in the synthetic substrate deoxythymidine-3'-phosphate-5'-p-nitrophenyl phosphate (i.e. deoxythymidine-5'-OH rather than nitrophenol) can only be rationalized with the crystal structure by an in-line displacement. If pseudorotation were at all possible and no kinetic barriers to product dissociation were operative, the better leaving group would depart. From the structure of the pdTp-enzyme complex (15), a metal-coordinated water molecule is in an ideal position to displace the observed leaving group in an in-line process, whereas Tyr 113 is not.

Hence, with staphylococcal nuclease, although the geometry of the reaction is not established, the available structural and kinetic evidence suggests an in-line associative nucleophilic displacement on phosphorus. Proof of an in-line geometry in solution would render a metal-bound water ligand a strong possibility for the attacking nucleophile.

An in-line geometry was predicted (22) for the related internal nucleotidyl transfer reaction catalyzed by ribonuclease from NMR (23) and crystallographic studies (24) of the geometry of the active site. In-line geometry has been established by an incisive stereochemical analysis of reaction products using as substrates diastereomers of uridine 2',3'-cyclic phosphorothioate (25–27). More recent crystallographic (28) and NMR (29) studies of the ribonuclease complex of a dinucleoside phosphonate more like the true substrate yield a geometry that requires an in-line mechanism for the depolymerization reaction.

DNA POLYMERASE   Despite many studies of DNA synthesis by these enzymes (30), few papers on the detailed mechanism of this nucleotidyl transfer reaction have been published (31). The reaction catalyzed is

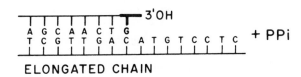

The templates that determine the base sequence of the product are either polydeoxy-ribonucleotides or polyribonucleotides, the initiator is either an oligodeoxy-ribonucleotide or an oligoribonucleotide, and the substrates are deoxyribonucleoside triphosphates. A proposed distinction between enzymes that utilized only poly-deoxyribonucleotide templates from those using only polyribonucleotide templates, and the introduction of the term "reverse transcriptase" for the latter are probably not justified on chemical grounds since DNA polymerases and reverse transcriptases can, with varying efficiencies, copy either template (32–34).

It is now clear that tightly bound, stoichiometric Zn^{2+} is present in homogeneous DNA polymerases purified from four diverse sources: a bacterial enzyme [*E. coli* polymerase I (35, 36)], two viral enzymes [T_4 bacteriophage (36) and avian myeloblastosis virus (36a)], and an animal enzyme [sea urchin polymerase (35)]. With *E. coli* polymerase I, removal of Zn^{2+} by prolonged dialysis against the chelating agent *o*-phenanthroline results in loss of activity in direct proportion to the amount of Zn^{2+} removed (36). Restoration of full activity is accomplished by the addition of Zn^{2+}, and to a less extent by Co^{2+} or Mn^{2+}. With all of the above polymerases, *o*-phenanthroline inhibits catalysis, while its nonchelating analog metaphenanthro-line fails to inhibit (35, 36). Inhibition of DNA polymerase activity by *o*-phenanthro-line has also been reported with extracts of human lymphocytes (35) and with partially purified polymerases from chicken embryos (37) and Rous Sarcoma virus (38, 39). These findings indicate that *E. coli* polymerase I meets the classical criteria for a Zn^{2+} metalloenzyme (40) and suggest that all DNA polymerases contain functional Zn^{2+}. The minimal stoichiometries found for Zn^{2+} in the various DNA polymerases analyzed are as follows: *E. coli* DNA polymerase I (36), T_4 phage DNA polymerase (36), sea urchin nuclear polymerase (35), and avian myeloblastosis polymerase (reverse transcriptase) (36a) are 1.0 ±0.15, 1.0, 4.2, and ~ 1 g atoms of zinc per mole of enzyme. Kinetic analysis of the inhibition of sea urchin polymerase and *E. coli* DNA polymerase I by *o*-phenanthroline reveals that it is wholly or in part competitive with respect to DNA (30, 35), suggesting that the enzyme-bound Zn^{2+} interacts with DNA. This suggestion is supported by the observation that DNA polymerase enhances the effect of Zn^{2+} on the nuclear quandrupolar relaxation rate of Br^- ions indicating that the enzyme-bound Zn^{2+} can coordinate Br^- ions (36). This enhanced effect is progressively reduced on titration with an oligonucleotide, $(dT)_{6-9}$, to one third of its original value with an endpoint suggesting that one molecule of oligonucleotide binds to one molecule of polymerase and displaces about two thirds of the Br^- from the enzyme-bound Zn^{2+}. The presence of Zn^{2+} at the DNA binding site is further supported by the detection of two Zn^{2+} per mole of *E. coli* RNA polymerase (41), which must also bind a DNA template but does not require an oligonucleotide initiator. However, the first nucleotide on the newly synthesized chain may serve as the initiator. Recently five RNA polymerases from animal sources have been shown to be inhibited by *o*-phenanthroline but not by *m*-phenanthroline (39). A mechanistic role for Zn^{2+} rather than simple DNA binding is suggested by the specific inhibition by *o*-phenanthroline of RNA chain initiation with RNA polymerase (41) and by the negative finding by B. M. Alberts, C. Springgate, L. A. Loeb, and A. S. Mildvan that the gene 32 protein, a noncatalytic protein that binds tightly to single-stranded DNA, does not contain significant amounts of Zn^{2+} (less than 0.1 g atoms/mole).

A reasonable mechanistic role for Zn^{2+} at the DNA site consistent with the above results is to coordinate and promote the nucleophilicity of the 3'-OH group of deoxyribose at the growing point of the initiator strand (36, 42, Figure 1). Model reactions of this type (43) and their microscopic reverse (44, 45) have been described. This role for Zn^{2+} would be assisted by a nearby general base to deprotonate the 3' hydroxyl group. In this regard it is of interest that several, but not all, DNA polymerases appear to be sulfhydryl enzymes (30, 31).

Unlike DNA analogs, deoxynucleoside triphosphate substrates do not displace Br^- ions from the enzyme-bound Zn^{2+} of *E. coli* polymerase I (36). In an important series of papers reviewed elsewhere (31), Kornberg and his colleagues established that this enzyme binds template in absence of the added divalent cation activator, Mg^{2+} (46), but requires Mg^{2+} for the tight binding of the substrates (47). A quantitative study of the binding of the divalent activator, Mn^{2+}, and nucleotides to *E. coli* polymerase I using EPR and water proton relaxation (42) revealed that the enzyme binds Mn^{2+} at three types of sites: one very tight site ($K_D \leq 10^{-6}$ M), five intermediate sites ($K_D \approx 10^{-5}$ M), and ~ 12 weak sites ($K_D = 10^{-3}$ M). From kinetic studies, the tight and intermediate sites may function catalytically but the weakest sites are probably inhibitory. The tight and intermediate sites all interact with deoxynucleoside triphosphates or monophosphates, suggesting that the enzyme pre-aligns substrate molecules prior to polymerization (42). Moreover, relaxation rates of water in solutions of ternary complexes of deoxynucleoside triphosphates and monophosphates are the same (42) and the affinities for triphosphates and monophosphates are similar (42, 46, 47), suggesting that both compounds interact with the enzyme-bound Mn^{2+} in the same way, presumably through their α-phosphoryl groups. These findings are consistent with an associative nucleophilic displacement mechanism for DNA polymerase (42, Figure 1).

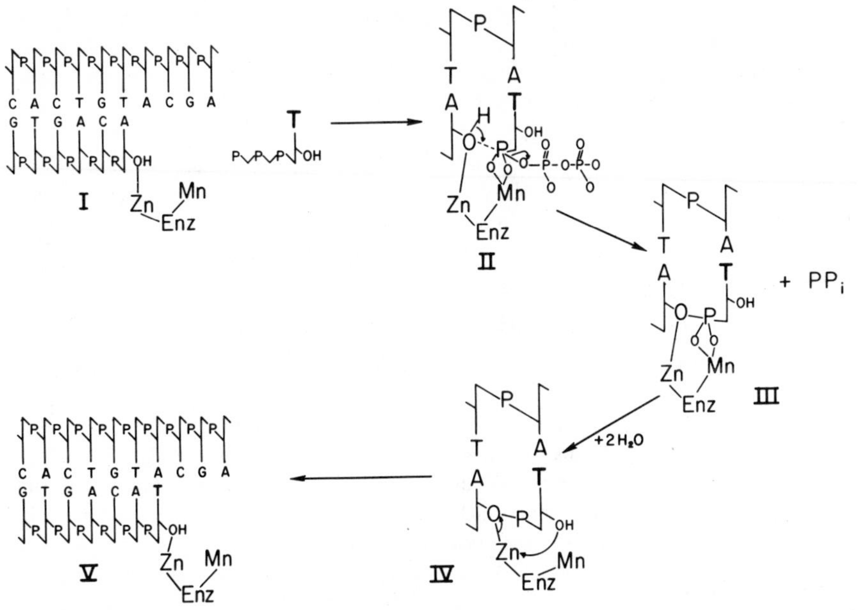

Figure 1 Proposed mechanism of DNA polymerase (42) based on nuclear relaxation studies (36, 42).

The geometry of this displacement, in-line or adjacent, is unknown. Another unknown aspect of the mechanism of DNA polymerase of profound biological importance is an understanding of the ability of this class of enzymes to catalyze the copying of a template with extremely high fidelity (48). Thus the error rate with *E. coli* polymerase I (48, 49) and T_4 phage polymerase (50) is 1 in 10^5 to 10^6 base pairs and in sea urchin polymerase (30) and the small polymerase from calf thymus (51), which apparently lack self-correcting mechanisms, it is 1 in 10^4 to 10^5 base pairs. These low error rates correspond to a free energy ΔG of discrimination between the correct and the incorrect base pair of -5.6 to -8.4 kcal/mole. Yet, the ΔG of discrimination available in the Watson–Crick base pair vs the wrong base pair is little more than the ΔG of an additional hydrogen bond, -2 kcal/mole, and possibly an additional -1 kcal/mol due to the chelate effect of a pair of hydrogen bonds, which would predict an error rate of 1 in 10^2 base pairs. In accord with this view, the additional fidelity of the GC base pair with three hydrogen bonds, over AT which has two is approximately 10^2 as found with lymphocyte polymerases (49). Hence, DNA polymerases such as sea urchin and *E. coli,* which copy with high fidelity, must amplify by at least two orders of magnitude (i.e. by -2.8 kcal/mol) the discrimination between the correct and incorrect base pair. The existence of such amplification by these polymerases has been established by detection of DNA polymerases with low fidelity such as those present in leukemia cells (49), in avian myeloblastosis virus (AMV) (52), and those induced by mutant T_4 bacteriophage (50). With AMV polymerase an error rate of 1 in 600 approximating the theoretical maximum of 1 in 10^2 has been observed (52). Elucidation of the mechanism of the selective amplification of correct base pairing and high fidelity copying would contribute to a molecular explanation of mutation rates of various species.

It has long been known that *E. coli* DNA polymerase I catalyzes several reactions in addition to polymerization (31). The recent observation of a homogeneous mini-polymerase from calf thymus that lacks reversibility and exonuclease activities indicates that these ancillary reactions are not obligatory components of polymerization (53).

Phosphoryl Transfer Enzymes

METAL-PHOSPHORUS DISTANCES   Unlike nucleotidyl transfer reactions which appear to be S_N2, the enzyme-catalyzed transfer of an unsubstituted phosphoryl group could proceed by either an S_N1 or an S_N2 mechanism. Almost all such reactions require divalent metal ions as added cofactors or in metalloenzymes. From the above mechanistic considerations, direct coordination by the metal ion of the phosphoryl group undergoing transfer or of the attacking nucleophile, or both, would strongly suggest an S_N2 mechanism and would argue against an S_N1 mechanism, while coordination of the leaving group would facilitate either an S_N1 or S_N2 mechanism. Hence, a determination of the location of the metal ion with respect to the phosphoryl group undergoing transfer in the active complex would yield insight into the mechanism. As pointed out in detail elsewhere (54–56), nuclear relaxation induced by paramagnetic metal ions such as Mn^{2+} can be used to determine distances between the paramagnetic metal ion and magnetic nuclei (^{19}F, ^1H, ^{31}P, ^{13}C) of substrates and substrate analogs in enzyme complexes. When such studies are carried out at several magnetic fields to yield an independent evaluation of the correlation time, the precision attainable in such distance calculations (± 0.1 Å) exceeds that of protein crystallography. This is not surprising since the nuclear relaxation rate obtained from the NMR data is inversely related to the sixth power of the distance, while the electron density map derived from the crystallographic data is related to the first power of the distance. Small molecule crystallography is a higher resolution technique with a precision better than ± 0.01 Å. From the results of small molecule crystallography and molecular model studies (summarized in 56, 57) the Mn^{2+} to phosphorus distance is 2.9 ± 0.1 Å for an inner sphere monodentate complex of tetrahedral phosphate; 3.6 ± 0.2 Å for an axial oxygen complex of a pentacoordinate phosphorane, i.e. a distorted inner sphere complex; 4.9 ± 0.4 Å for Mn^{2+} chelated by an adjacent pyrophosphate in a triphosphate chain; and 6.1 ± 0.5 Å for a second sphere complex in which an inner sphere water or a ligand of comparable size intervenes between the Mn^{2+} and phosphate. Corresponding distances in Ca^{2+} complexes would be 0.2 Å greater and in Mg^{2+} complexes 0.1 Å smaller.

PYRUVATE KINASE   From ^{31}P-NMR studies at one frequency of the ternary complex of muscle pyruvate kinase, Mn, and phosphoglycolate, a competitive analog of the substrate phosphoenolpyruvate, an Mn to P distance of 3.5 ± 0.3 Å was obtained, suggesting a distorted inner sphere complex (57, Figure 2). An alternative explanation of this distance, suggested by the EPR spectrum of Mn^{2+} (58), is that it represents a time average of two complexes: 32% inner sphere (2.9 Å) and 68% second sphere (6.1 Å). This possibility arises because of the inherently higher frequency or "shutter speed" of the EPR experiment (2×10^{10} sec^{-1}) as compared with that of the nuclear relaxation experiment (5×10^{6} sec^{-1}). Hence exchange between the two complexes

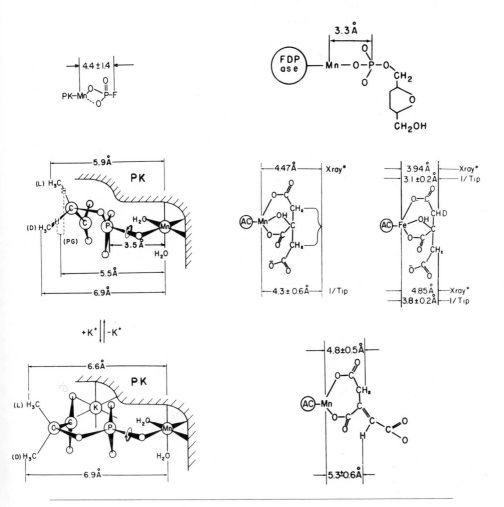

Figure 2 Dimensions of inner sphere enzyme-metal-substrate bridge complexes in solution detected by nuclear relaxation. The enzymes are PK, pyruvate kinase (56, 57, 59); FDPase, fructose diphosphatase (90); AC, aconitase (203).

with a rate constant k such that 5×10^6 sec^{-1} $< k < 2 \times 10^{10}$ sec^{-1} would result in an average structure in the NMR experiment but not in the EPR experiment. The addition of the activator K$^+$, which appears to form a metal bridge between the enzyme and the carboxyl group of phosphoenolpyruvate (PEP) and its analogs (57), causes a reorientation of the inner sphere complex as indicated by proton NMR and an increase in the amount of inner sphere complex suggested by EPR (58). A pyruvate kinase-Mn-O$_3$PF bridge complex had previously been found by ^{19}F NMR (59). A re-evaluation of the Mn to F distance in this complex (56), using better estimates of the correlation time at one frequency (3.7×10^{-9} sec) and the extent of formation of the ternary complex in the NMR experiment (4%) from computer-analyzed titration curves (57), yields an Mn to F distance of 4.3 ± 1.5 Å (Figure 2) in agreement with the Ca to F distance of 4.2 Å in the monodentate crystalline Ca-O$_3$PF complex (60). Hence the phosphoenolpyruvate analogs phosphoglycolate and FPO$_3^{2-}$ (which is also a product of the fluorokinase reaction) form inner sphere E-Mn-phosphoryl bridge complexes (Figure 2), consistent with an associative or S$_N$2 nucleophilic displacement on the phosphorus of phosphoenolpyruvate by ADP (61). Although phosphoenolpyruvate itself could not be studied by NMR because of slow exchange, it competes with these analogs in the NMR and in kinetic studies (57, 59, 62), and it yields an EPR spectrum for enzyme-bound Mn very similar to that observed in the quaternary pyruvate kinase-Mn-K-phosphoglycolate complex (58).

The general proposal of homologous mechanisms for enzymes that catalyze phosphoryl transfer from phosphoenolpyruvate based on a similarity of their metal complexes (pyruvate kinase, carboxykinase, carboxylase, and carboxytransphosphorylase) (63) has been extended by Rose (64) to include all enzymes capable of enolizing pyruvate, such as pyruvate carboxylase and transcarboxylase. Potent inhibition of these enzymes by oxalate would be expected since oxalate is isoelectric with the enolate of pyruvate and could therefore function as a transition state analog (65, 66). Such inhibition has indeed been found with pyruvate carboxylase (67), transcarboxylase (68), and pyruvate kinase (69). The lack of stereo-specificity of protonation in the formation of pyruvate from phosphoenolpyruvate in the carboxytransphosphorylase reaction has been ascribed to the release of the enolate of pyruvate from the enzyme and its subsequent ketonization in solution (70).

The geometry of a kinetically active complex of pyruvate kinase, Mn^{2+}, K$^+$, phosphate, and pyruvate, which catalyzes the enolization of pyruvate has been elucidated by ^{13}C and ^{31}P NMR (71, Figure 3). The distances from Mn to the

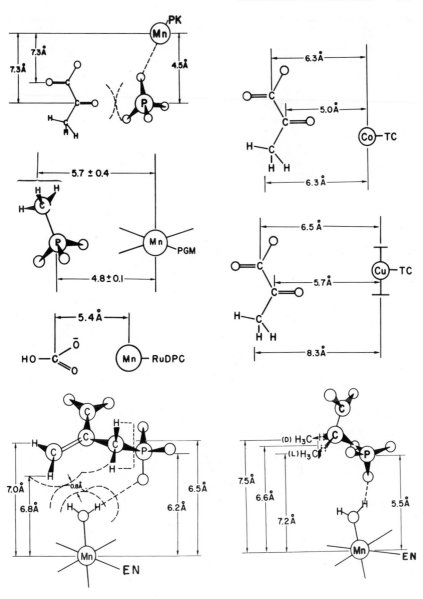

Figure 3   Dimensions of second sphere enzyme-metal-----substrate bridge complexes from crystallographic data or nuclear relaxation data in solution. The enzymes are SN, staphylococcal nuclease (12, 13, 15); PK, pyruvate kinase (71); PGM, phosphoglucomutase (114); PC, pyruvate carboxylase (71); TC, transcarboxylase (178); RuDPC, rubulose diphosphate carboxylase (190a); EN, enolase (198).

carboxyl and carbonyl carbon atoms of pyruvate (7.3 ± 0.1 Å) establish the substrate to be in the second coordination sphere, while the lower limit Mn^{2+} to phosphorus distance (4.5 ± 0.4 Å) suggests either a distorted inner sphere complex or the rapid averaging of 6% inner sphere complex with 94% second sphere complex (Figure 3). A mechanism for pyruvate kinase consistent with the geometry is given in Figure 4.

The distance between the bound monovalent cation and bound paramagnetic divalent cation, originally determined by Tl^+ NMR to be 5 and 8 Å in the presence and absence of phosphoenolpyruvate (72, 73), has been confirmed by frequency-dependent proton NMR using monomethylammonium as the activator (74). The possible widespread applicability of using monomethylammonium rather than K^+ or Tl would render such distance determinations much easier because of the general availability of proton NMR spectrometers (74).

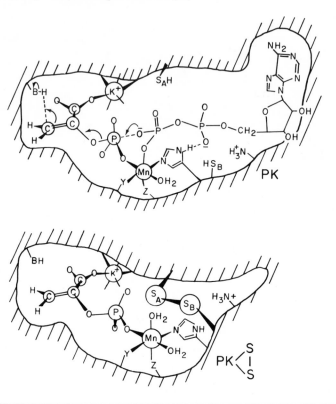

Figure 4 Upper   Mechanism of pyruvate kinase showing the mode of binding of metals and substrates to the native enzyme, from nuclear relaxation data (56, 57, 59, 61, 71). Also shown are the nonessential thiol (S_AH), the essential thiol (S_BH), and lysine ammonium groups (83–85). *Lower*   Proposed occlusion of the ADP site in the disulfide modification of pyruvate kinase in which the essential and nonessential thiols are linked (85).

The unidentified group that stereospecifically protonates C_3 of phosphoenol-pyruvate (75) or deprotonates pyruvate (76, 77) has been found by nuclear relaxation to hinder the methyl rotation of L-phospholactate but not D-phospholactate (78). The motions of other atoms of phosphoenolpyruvate analogs are also hindered by groups on the enzyme. The maximum possible immobilization, such that rotation occurs only with the protein, takes place at the reaction center phosphorus atom (78). Analysis of our data using a refinement of the relaxation equations (79) yields the same results within experimental error. These observations suggest that "freezing at reaction centers on enzymes" (FARCE) may contribute to activation of substrates on enzymes as suggested by theoretical kinetic considerations (80–82).

With respect to the ADP binding site of pyruvate kinase, a functional lysyl ammonium group and a thiol group have been detected by a combination of methods: chemical modification (83, 84) combined with binding studies using EPR and nuclear relaxation (85), and kinetic studies of the detritiation of pyruvate, a partial reaction (85). Modification of the thiol group consisted of converting it to a disulfide, a reaction mediated by 5,5'-dithiobis-2-nitrobenzoate. While the modifica-tions of both groups abolish ADP or ATP binding and preserve the binding and detritiation of pyruvate, the thiol group is closer to the Mn^{2+} and phosphoenol-pyruvate binding sites, as indicated by large decreases in the number and exchange rate of rapidly exchanging water ligand on Mn^{2+} and in the affinity for phospho-enolpyruvate in the disulfide enzyme (85, Figure 4). The finding that the modified purine ring of a fluorescent analog of ATP remains highly mobile as this analog binds to pyruvate kinase (86), along with the previously detected low base specificity of this enzyme (87), suggest that the thiol and possibly the lysine ammonium groups interact with the pyrophosphate and ribose protons of ADP.

Hopefully more detailed information will be provided by forthcoming crystallo-graphic data. The 6 Å structure of cat muscle pyruvate kinase has recently been solved (88) and reveals a tetrahedral arrangement of four elliptical subunits with no apparent clefts in the subunits. Subunit contacts are not identical across each axis of the molecule, providing a possible explanation of the tendency of the enzyme to be dissociated into active dimers by urea (89). In agreement with the NMR evidence (57, 59), the crystallographic data reveal the Mn^{2+} binding site on the enzyme to be very near the phosphoenolpyruvate binding site (H. Muirhead, private com-munication). Structural changes in the crystal have been observed on adding ADP and phosphoenolpyruvate. A mechanism consistent with these structural and kinetic observations is given in Figure 4.

From ^{31}P nuclear relaxation studies (90), an S_N2 mechanism seems likely for the reaction catalyzed by the alkaline form of fructose diphosphate, since an enzyme-Mn-fructose-1-phosphate bridge complex with a Mn to P distance of 3.3 ± 0.3 Å has been detected (Figure 2).

CREATINE KINASE As implied above, a dissociative or S_N1 mechanism for phosphoryl transfer is more difficult to establish than an S_N2 mechanism. An S_N1 mechanism may be operative in the reaction catalyzed by creatine kinase, as suggested by the observation that planar trigonal monoanions such as nitrate and formate (91), which are analogs of metaphosphate, protect the enzyme from modification of its essential thiol. However the finding that M^{2+}-ADP and creatine raise the affinity of the enzyme for nitrate as detected kinetically (91) and by proton relaxation (92, 93) and, conversely, that nitrate raises the affinity of the enzyme for creatine and ADP as detected by kinetics (91) and proton relaxation (92, 94) suggest the formation of a complex with significant ADP---NO_3^- binding and creatine---NO_3^- binding. Such a complex would mimic a transition state structure for an in-line S_N2 mechanism (92). A crucial question in this connection is the location of the bound divalent cation.

A model of the substrates at the active site, based on superposition of the geometry of several inactive and active ground state complexes determined by proton NMR, suggests that the Mn coordinates the leaving ADP group (95). Moreover, the long Mn^{2+}-creatine-CH_2 distance (9.8 Å) did not differ in the enzyme-creatine-Mn-ADP complex and in the equilibrium mixture, suggesting that Mn^{2+} remains on the α- and β-phosphorus atoms in the quaternary ATP complex. X-ray diffraction studies of the phosphoglycerate kinase-Mn-ADP system reveal an enzyme-metal-substrate bridge complex to both phosphoryl groups of ADP (96). However, more recent data on the ground state and transition state complexes of creatine kinase, substrates, and formate indicate the Mn^{2+} to creatine distance to be ≤ 8.5 Å (93). The Mn to formate proton distance (< 5.1 Å) is compatible with but does not establish direct co-ordination of formate (93, 97). EPR and proton relaxation data reveal a progressive decrease in the symmetry of the ligand field at Mn^{2+} as the ternary and quaternary complexes are formed (91–94, 97), consistent with the entrance of ligands on Mn^{2+} from the protein and possibly from the formate. Nucleotides or creatine analogs with a higher V_{max} appear to form larger fractions of the more asymmetric complex (93, 94, 97). Hence, while the Mn^{2+} appears not to coordinate the transferable phosphoryl group in the ground state complexes, it may well do so in the transition state and in the ternary enzyme-ATP-Mn complex. If this were the case, an S_N2 mechanism would be indicated, as with pyruvate kinase, and the metal would have to migrate during the course of the reaction. The failure of Cr^{III} ADP (98–100) and Cr^{III} ATP (100) to serve as substrates for various kinases may be due in part to the inability of this trivalent metal to migrate in these stable complexes.

OTHER KINASES   Low resolution X-ray studies reveal that the polypeptide chains of P-glycerate kinase (101, 102) and of the subunits of hexokinase (103) are folded into two globular domains connected by a narrow bridge of electron density. With P-glycerate kinase, the binding of Mg ADP is confined to one lobe (96, 101). The role of the second lobe is unknown. The binding sites for ATP, glycerate, or phosphoglycerate remain to be elucidated. No such bilobed structure is detected in the subunits of pyruvate kinase at a resolution of 6Å (88).

PHOSPHOENZYME INTERMEDIATES   A number of kinases, including some that do not show ping-pong kinetics such as acetate kinase (104), phosphoglycerate kinase (105), and hexokinase (106, 107), have been found to catalyze exchange reactions, suggesting a covalent E-P intermediate in the phosphoryl transfer. In these cases phosphoenzymes have been isolated which can phosphorylate the appropriate substrates (104, 105, 108). However, in all such cases, with the exception of nucleoside diphosphate kinase (109), the partial reactions and the discharge of E-P are orders of magnitude slower than the overall phosphoryl transfer, suggesting the possibility of side reactions rather than true E-P intermediates (110). The possibility of such side reactions would suggest the random capture of a reactive metaphosphate intermediate and an S_N1 mechanism. Alternatively, an accelerated discharge of E-P by the appropriate combination of substrates or substrate analogs at a rate compatible with the overall reaction would establish such E-P as intermediates. The combination of an E-P intermediate in a ternary substrate complex with an immobilized phosphoryl group and immobilized substrates would require at least one right angle displacement on phosphorus, hence an S_N2 mechanism with pseudo-rotation.

PHOSPHOGLUCOMUTASE   Phosphoglucomutase catalyzes the reversible transfer of a phosphoryl group from the 1 or 6 position of glucose 1,6-diphosphate to a serine hydroxyl group on the enzyme, yielding a phosphoenzyme intermediate (111). Direct observations of this enzyme-bound group by ^{31}P-NMR (112, 113) establish it to be a phosphoserine from its chemical shift and that of the protons to which it is coupled and shows it to be highly immobilized or "frozen to the enzyme" from its relaxation rates, suggesting direct activation of the phosphoryl group (W. J. Ray Jr. and A. S. Mildvan, to be published). The metal ion required for this reaction is bound at a site 6 Å from the phosphoserine P, suggestive of a second sphere complex, although only 20% of the phosphorus was visualized in this relaxation experiment, leaving open the possibility of an extensively broadened additional signal due to an inner sphere component. A predominantly (96%) second sphere complex with only a 4% inner sphere contribution is supported by the detection of a 4.8 ±0.1 Å Mn to P distance in the ternary complex of the dephosphoenzyme with Mn and methyl phosphonate, an analog of phosphate (114). The methyl phosphonate is very weakly bound to the enzyme at this site with a dissociation constant of 0.1 M. This site is absent on the phosphoenzyme. On both forms of the enzyme, a second binding site for methyl phosphonate is found 10.5±0.5 Å from the metal with a lower dissociation constant of 1 mM, which apparently serves to position the binding of either of the phosphates of the intermediate glucose 1,6-diphosphate (114, 115). Hence, a one-dimensional model of the active site may be constructed with the metal ion as the reference point, the weak phosphate binding site (corresponding to the phosphoserine phosphate hence to the phosphoryl transfer site) at ~5 Å from the metal and the tighter binding phosphate positioning site 10–11 Å away. The glucose 1,6-diphosphate, which interchanges its phosphoryl groups 100 times faster than it dissociates (111), must therefore alternatively shift the binding of its two phosphates between the two sites without dissociating from the enzyme. The calculated distances, together with molecular model studies, suggest two possible roles for the metal ion. By coordinating the 3-OH group of the sugar the metal-sugar bond could serve as the pivot for the phosphate shift. Alternatively, or additionally, by coordinating the serine hydroxyl, the metal could increase its nucleophilicity on phosphorus and its effectiveness as a leaving group from phosphoserine. Although the second role is suggested by the decrease in water relaxation caused by the phosphoenzyme-Mn complex as compared with the dephosphoenzyme-Mn-complex (115), at present a choice cannot be made among these suggested roles for the divalent cation. Similarly an S_N2 mechanism is suggested by the immobilization of phosphorus and by the possibility of an inner sphere contribution to the Mn-P distance.

ALKALINE PHOSPHATASE   Presence of a metal on a phosphoryl transferring enzyme provides no assurance that the metal is directly involved in phosphoryl transfer. Thus with alkaline phosphatase, no direct interactions of Cl^- with enzyme-bound Zn^{2+} (116) or water with enzyme-bound Mn^{2+} (117) were detected by nuclear relaxation. Similarly no direct interaction of phosphate with enzyme-bound Co^{2+} (118) or Mn^{2+} (118, 119) was detected by ^{31}P nuclear relaxation. An Mn^{2+} to phosphate distance of 7.3 Å was calculated from NMR data (120), indicating a second sphere complex. These results are in accord with crystallographic data on the enzyme, which at 7.7 Å resolution indicate that substrates cannot easily gain direct access to the metal site (121).

HYDROLASES

Carboxypeptidase A

A mechanism for peptide hydrolysis (122) consistent with crystallographic data (123–125), amino acid sequence (126), chemical modification (127, 128), and kinetic studies (129), involves holding of the carboxy terminus of the substrate by Arg 145, polarization of the carbonyl group by coordination to the enzyme-bound Zn, protonation of the leaving amide by Tyr 248, and activation of water by general base catalysis by Glu 270, or nucleophilic attack by this residue. Early nuclear relaxation studies of enzyme-Mn-product complexes (130) and an elegant fluorescence quenching study of active enzyme-Co^{2+}-substrate complexes (131) are consistent with this proposed role for the metal. From kinetic studies with chemically modified forms of carboxypeptidase (132) and from the observations that ester and peptide substrates have opposite effects on the water proton relaxation rate in the presence of Mn-carboxypeptidase (133), the mechanisms of peptide and ester hydrolysis appear to differ (132). Mn-carboxypeptidase appears to have either several water ligands or an unusually short Mn-H_2O distance, assuming relaxation rates at very low magnetic fields obey the same theory as at high fields (134). From model studies, Rogers & Bruice (134a) have proposed a mechanism involving a penta-coordinate metal complex in which a water ligand attacks a coordinated tetrahedral intermediate.

Mechanisms for ester hydrolysis involving the formation of an inner sphere E-M-S complex (as in the Zn-carbonyl mechanism) or the donation of a coordinated hydroxyl to the substrate by the metal (as in the Zn-OH mechanism) have been neatly ruled out by the incisive observation that Co^{III}-carboxypeptidase, which exchanges ligands very slowly, is fully active in ester hydrolysis though it has lost its peptidase activity (135). A water molecule coordinated to Co^{III}-carboxypeptidase could still, of course, donate a proton to the ester carbonyl group although no evidence on this point is available.

Although there is general agreement from crystallographic and solution data that Tyr 248 is directly involved in catalysis, the view that substrate binding induces Tyr 248 to move 12 Å to approach the active site, as indicated by the crystallographic data, is not in accord with recent experiments in solution (136). Thus in most (136, 137) but not all (138) crystalline forms of carboxypeptidase A the environment of Tyr 248, chemically modified by arsanilazo or nitro chromophores, differs from that in solution such that at pH 8.2, arsanilazo Tyr 248 and nitrotyrosine 248 are very near the Zn in solution even in absence of substrate as indicated by spectroscopic titrations. In the crystal they are more distant from the Zn as indicated by spectral changes (136–138) and by a two-unit increase in pK, respectively (139). The argument that the chemical modifications of Tyr 248, having converted it to a chelating agent for Zn, have shifted a labile equilibrium in solution to the tyrosine-on-Zn form may be only partially answered by the observation that Tyr 248 is more specifically modifiable in the crystal than in solution (136, 140), suggesting a different environment even in the native state. The crystals used for the X-ray analysis appear therefore to be atypical with respect to the mobility of Tyr 248. The availability of active chromophoric derivatives of this enzyme modified at this functional residue should provide a valuable probe of its environment in the course of enzymatic turnover.

Chymotrypsin

The previous section exemplifies the value of comparisons between crystallographic and solution data on the structure and function of enzymes. The X-ray model of chymotrypsin (141–143) has recently gained support in solution by a cogent application of the nuclear Overhauser effect (144). This NMR method indicates that the inhibitor D-tryptophan binds to chymotrypsin very near a valine methyl group, as indicated by a decrease in the intensity of the NMR spectrum of the former upon irradiating the protons of the latter. Valine 213 had been shown to be in contact with the tryptophan ring in the X-ray structure of the chymotrypsin-N-formyl-tryptophan complex (143), and interactions with valine are difficult to detect in solution by other methods.

The existence of the hydrogen-bonded charge relay system (serine-to-histidine-to-aspartate) originally inferred from the crystal structure to promote the nucleophilicity of Ser 195, and subsequently extended to the other serine proteases, trypsin, subtilisin, and elastase from analogous crystallographic data, has received powerful support from NMR studies in solution (145, 146). A downfield resonance in H_2O solutions of chymotrypsinogen and chymotrypsin has been assigned to a highly deshielded proton, hydrogen bonded between His 57 and Asp 102. The electron density at this proton as measured by its chemical shift is further decreased by protonating a group with a pK of 7.2 (presumably His 57) or by the formation of a tetrahedral adduct at serine with boronic acids. Thus the proton at or near N-1 of His 57 senses events at serine that is near N-3 of His 57, indicating that the charge relay is operative. As the tetrahedral intermediate forms, the operation of the charge relay system would transfer negative charge from Asp 102 through nine bonds to the oxyanion hole, a constellation of positive residues near the negative $\bar{0}$ of the tetrahedral intermediate (147). The enthalpic contribution to the rate acceleration thus induced may be very small, as judged by the small values of the chemical shift (G. Robillard, unpublished) and by a recent model study (146a) which indicates only a threefold rate acceleration caused by a carboxyl group placed in tandem with an imidazole acting as a general base. A larger entropic contribution to the enzymatic rate enhancement might result from the positioning and immobility of the residues that form the oxyanion hole (82). The precise location of the protons in the low pH form of this enzyme is uncertain. A recent ^{13}C-NMR experiment with enrichment at C-2 of histidine in such a system suggests an inversion of the usual pK values of the carboxyl and imidazole residues in the charge relay (147a).

BICARBONATE AND CARBON DIOXIDE UTILIZING ENZYMES

Carbonic Anhydrase

The simplest enzyme-catalyzed reaction of CO_2 is its reversible hydration catalyzed by carbonic anhydrase. Coleman (148) has reviewed the various proposed mechanisms. The crystal structure of this enzyme reviewed by Lindskog et al (149) reveals the Zn to be in a 15 Å conical cavity tetrahedrally coordinated by three imidazole ligands from the protein and by a fourth ligand, presumably water. One of the imidazole ligands (His 117) appears coordinated to Zn via its N-1 nitrogen, while the other two are more typically coordinated through their N-3 nitrogens. Additional histidines, 63 and 128, are in the cavity and a lattice of eight molecules, presumably water, may fill the cavity. In a two-dimensional X-ray analysis (150) Co^{2+}, Cu^{2+}, and Mn^{2+} bind in a manner indistinguishable from Zn^{2+}, yet the Mn^{2+} enzyme has low activity (151) and the Cu^{2+} enzyme has no activity (151). Proton relaxation data on the Co^{2+}- (152) and Mn^{2+}-substituted enzymes (153) indicate a rapidly exchanging water ligand on the metals at high pH. Strangely, the number of rapidly exchanging water molecules diminishes by one to a low value as the pH is decreased with a pK = 7.0 (Co^{2+}) or 7.8 (Mn^{2+}). This behavior is opposite in sense to the deprotonation of the Fe^{3+}-bound water ligand on metmyoglobin (154) and to the exchange rate of Cl^- on Zn carbonic anhydrase (155) or Co^{2+}-carbonic anhydrase (156), but it parallels the decrease in the enzyme-catalyzed rate of hydration of CO_2. Hence, a basic group on the enzyme appears to be necessary for rapid water exchange and rapid hydration of CO_2 and the protonated form of this base may be necessary for rapid Cl^- binding to the Zn enzyme. The simplest proposal is that this group is the Zn-bound water itself (148, 157), a suggestion in accord with the observation that the pK of water on Cu^{2+} is reduced by 1–2 units when Cu^{2+} is coordinated to imidazole ligands (158). This proposal may be rationalized with a more rapid water exchange on Co^{2+} or Mn^{2+} at high pH, since there is no assurance that the rapidly exchanging water ligand on Co^{2+} or Mn^{2+} detected in the proton relaxation experiments represents the hydroxyl ligand involved in catalysis. Indeed, a hydroxyl ligand could facilitate the expulsion of a water ligand from the same metal ion (3). Such a process would be consistent with the cyclic mechanism proposed by Kaiser & Lo (158a). Another alternative consistent with the water exchange data is that the imidazole ligand of His 117, which may be unusually coordinated to Zn via its N-1 nitrogen, is deprotonated with a pK of ~ 7.0 (159). An essential role for the nonliganding histidines near the active site has been ruled out by the observation that their modification does not abolish activity (160).

The interaction of sulfonamide inhibitors with the enzyme-bound Mn^{2+} in solution, detected by nuclear relaxation rates of their protons (161, 162), is in accord with the crystallographic data (149, 150).

In the dehydration reaction Koenig (163, 164) has re-examined the well known problem of the nature of the substrate (148), and has made the important point that H_2CO_3 rather than $H^+ + HCO_3^-$ represents a kinetically more competent substrate, since with the latter substrate pair a second order rate constant for proton binding greater by two orders of magnitude than diffusion would be required to fit the kinetic data. With H_2CO_3 as the substrate bringing in its own proton, a second order rate constant more nearly approaching the diffusional limit would suffice (163–165). The suggestion of proton transfer mediated by buffers other than the substrate itself (165, 165a) seems unlikely since kinetic measurements using ^{13}C NMR were carried out in absence of extraneous buffers (164).

Biotin Enzymes

Biotin covalently attached to an enzyme represents a mobile carboxyl carrier, as suggested by kinetic data indicating that biotin enzymes catalyze modified ping-pong mechanisms in which the biotin is first carboxylated then migrates to another site where the carboxybiotin transcarboxylates an acceptor (166, 167). A comprehensive review of this class of enzymes has been written by Moss & Lane (168).

A general scheme for most biotin enzymes is

$$E - biotin + HCO_3^- + ATP \overset{Mg^{2+}}{\rightleftharpoons} ADP + P_i + E - biotin - CO_2$$
$$E - biotin - CO_2 + acceptor \rightleftharpoons E - biotin + acceptor - CO_2$$

The first step is a ligase or synthetase reaction requiring an added divalent cation; the second step is a transcarboxylation reaction. The early studies of the mechanism of step 1 were carried out with crotonyl CoA carboxylase (169) and with propionyl CoA (170). Experiments with $H_2^{18}O$ and $HC^{18}O_3$ (170) showed that of the three oxygens introduced by the substrate HCO_3^-, two appeared in the carboxylated acceptor and one in the orthophosphate formed in step 1. On the grounds of parsimony, a concerted mechanism for step 1 was proposed (169, 170) in which the biotin N-1' nitrogen displaced an oxyanion from HCO_3^- which simultaneously attacked the terminal phosphoryl group of ATP. The need for a monatomic oxyanion transfer in a concerted mechanism renders a stepwise mechanism for step 1 with carboxyphosphate as a chemical intermediate an attractive alternative. Polakis et al (171) have obtained evidence for such an intermediate by showing that the biotin carboxylase portion of acetyl CoA carboxylase catalyzes phosphoryl transfer from carbamyl phosphate, an analog of carboxyphosphate, to ADP in a reaction that requires divalent cation and added biotin. The latter requirement can be met by N-1' substituted biotin and may therefore reflect a conformational effect rather than direct participation in a concerted process (171).

The detailed mechanisms of the second partial reactions depend on the individual enzymes and acceptors considered. Moss & Lane (172) have argued against concerted mechanisms in general for the second reaction, pointing out that acetyl CoA carboxylase, propionyl CoA carboxylase, and pyruvate carboxylase catalyze slow biotin-independent decarboxylations of their respective carboxylated acceptor substrates. Hence, these enzymes are capable of activating their carboxylated substrates presumably by enolization and protonation without concerted attack by biotin. Because of the low rate of these decarboxylation reactions and because H_2O may replace biotin in a concerted mechanism, this observation does not rule out concerted mechanisms for the second reactions. Nevertheless, with pyruvate carboxylase, the early finding that oxalate is a potent inhibitor which binds to the enzyme more than 10^3 times tighter than the substrate pyruvate (67) may be explained by assuming it to be transition state analog of the enolate of pyruvate with which it is isoelectronic. This observation suggests that pyruvate carboxylase may indeed deprotonate and enolize pyruvate prior to carboxylating this substrate with carboxy-biotin.

More cogent evidence against our previously proposed concerted mechanism for pyruvate carboxylase (67, 173) has emerged from a refinement of our magnetic resonance data (71, 174). Pyruvate carboxylase is a Mn^{2+} metallobiotin enzyme and the Mn^{2+} has clearly been shown to participate in the second partial reaction, since the substrate pyruvate and inhibitors of the second reaction interact with Mn, as indicated by water proton relaxation studies (67) and by studies of the relaxation rates of the protons of the substrates pyruvate (173), α ketobutyrate (173), and oxalacetate (175). From the early studies of the paramagnetic effect of Mn^{2+} on the relaxation rates of the methyl protons of pyruvate at the one frequency available (60 MHz), an Mn to CH_3 distance on the order of 5.0 ± 0.6 Å was calculated (173, 54) using estimated correlation times $\leq 1.7 \times 10^{-10}$ sec. This distance was consistent with an inner sphere complex ($r \leq 5.7$ Å). The advent of multiple frequency ^{13}C-NMR (71) and multiple frequency proton NMR (174) has permitted the more accurate determination of the individual correlation times ($3.7 - 5.3 \times 10^{-9}$ sec) and distances to the carbon atoms and protons of pyruvate. The revised distances indicate pyruvate to be in the second coordination sphere of Mn^{2+} rather than in the inner sphere (Figure 3). Similarly, the other substrates, α ketobutyrate and oxalacetate, form second sphere complexes. The nature of the inner sphere ligands of Mn^{2+} are not clear because the high magnetic moment of Mn^{2+} results in a relaxation rate of the second sphere ligand in excess of the exchange rate of the inner sphere ligand, obscuring the latter in the NMR experiment. To circumvent this difficulty, analogous experiments were carried out with transcarboxylase which catalyzes the same half reaction (176) but contains Co^{2+}, which is less paramagnetic than Mn^{2+}, and Zn^{2+} which is nonparamagnetic (177). However, five preparations of trans-carboxylase were also found to contain Cu^{2+} such that the sum of $Zn^{2+} + Co^{2+} + Cu^{2+}$ was ~ 12 metal ions per enzyme molecule or two metal ions per biotin (178). Correcting for the paramagnetic effects of Cu^{2+}, distances from Co^{2+} to pyruvate only slightly less than second sphere distances were obtained (Figure 3). Whether the shorter distance from the metal to the carbonyl carbon in transcarboxylase (5.0 Å) as compared with pyruvate carboxylase (7.1 Å) represents a distorted second sphere complex or the averaging of 2% inner sphere complex ($r = 2.9$ Å) and 98% second sphere complex ($r = 5.7$ Å) is unknown at present. From the lower limit distance (4.46 Å) as much as 6% inner sphere complex could be present (178, and unpublished observations).

It is not clear how the formation of second sphere complexes of pyruvate on these two enzymes that catalyze the same reaction would facilitate the deprotonation of pyruvate, its enolization, and its subsequent carboxylation by carboxybiotin at C-3. Three roles for the metal ion are possible: 1. Despite its propinquity to the carbonyl oxygen of pyruvate, the metal plays no direct role in the enolization of pyruvate but merely stabilizes the proper orientation of other catalytic groups at the active site. While this possibility cannot be excluded, it tends to ignore numerous model reactions directly involving metal-catalyzed decarboxylations of α keto acids (179, 180) and aldolization of pyruvate (181); 2. The metal ion promotes the acidity of an inner sphere water ligand which can then protonate the carbonyl oxygen of pyruvate and thereby catalyze the enolization process as suggested by the specific orientation of pyruvate with respect to the metal, with its carbonyl oxygen pointed toward the metal ion (shown in Figure 3). Model reactions involving the enolization of second sphere ligands have not been reported. Using metal complexes with stable inner coordination spheres such as Cr^{III} (H_2O)$_6$ and Co^{III} (en)$_3$ at pH 3 and 25°, we have been unable to detect significant detritiation of 3-trito pyruvate restricted to the second coordination sphere (< 1% of that induced by Zn). We have observed rapid detritiation of pyruvate in proportion to the amount of inner sphere Cr^{III}-pyruvate complex that forms as detected by decreases in the relaxation rate of water protons and by the disappearance of enzymatically active pyruvate (A. Mildvan and R. Abramson, unpublished observations). Hence no chemical evidence for specific acid catalysis of the enolization of pyruvate in a second sphere complex has been obtained; 3. The second sphere transcarboxylase-pyruvate complex detected by NMR is not the complex where pyruvate is enolized but rather represents an intermediate in the formation of an inner sphere complex in which pyruvate is enolized. As mentioned above, the presence of a small amount of inner sphere complexes where facile decarboxylation of oxalacetate or enolization of pyruvate take place is possible. The displacement of pyruvate from the enzyme by concentrations of oxalate, consistent with its K_I (173), merely argues that the complex detected by NMR is on the kinetic pathway to the enolization complex but does not establish that it is the enolization complex itself. In fact, the ping-pong nature of the reactions catalyzed by transcarboxylase (166) and pyruvate carboxylase (167) indicates that in the major kinetic pathway, pyruvate combines with enzyme-biotin-CO_2 rather than with enzyme-biotin. Because such complexes may be short-lived, higher frequency spectral techniques may be required for the unambiguous detection of inner sphere complexes if present.

Ribulose Diphosphate Carboxylase

Ribulose diphosphate carboxylase catalyzes either a carboxylase (182, 183) or an oxygenase (184–186) reaction of ribulose diphosphate, both resulting in a cleavage between C_2 and C_3 of this compound.

The early isotopic studies of the carboxylation reaction which have been

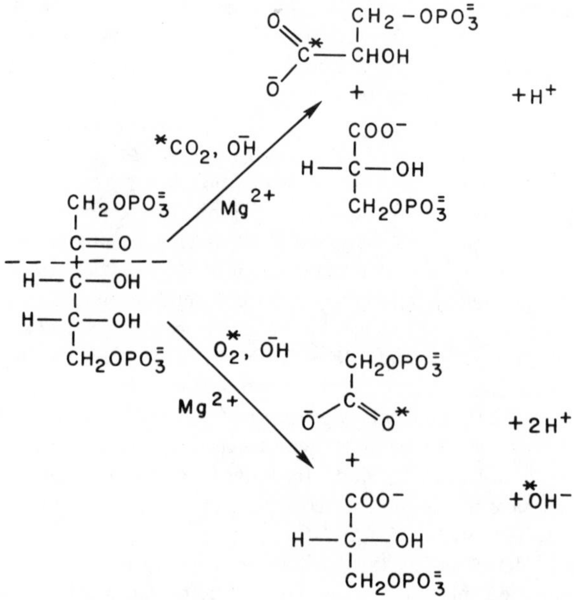

$$CH_2-OPO_3^=$$
$$\overset{O}{\underset{\bar{O}}{\diagdown}}\overset{*}{C}-CHOH$$
$$+$$
$$COO^-$$
$$H-C-OH$$
$$CH_2OPO_3^=$$

$+H^+$

$$*CO_2, \; \bar{O}H$$
$$Mg^{2+}$$

$$CH_2OPO_3^=$$
$$C=O$$
$$H-C-OH$$
$$H-C-OH$$
$$CH_2OPO_3^=$$

$$O_2^*, \; \bar{O}H$$
$$Mg^{2+}$$

$$CH_2OPO_3^=$$
$$\bar{O}\overset{C}{\diagup}\overset{*}{O}$$
$$+$$
$$COO^-$$
$$H-C-OH$$
$$CH_2OPO_3^=$$

$+2H^+$

$+\overset{*}{O}H^-$

thoroughly reviewed elsewhere (187) and recent [18]O studies of the oxidative cleavage (186) are consistent with the following mechanisms (186, 187)

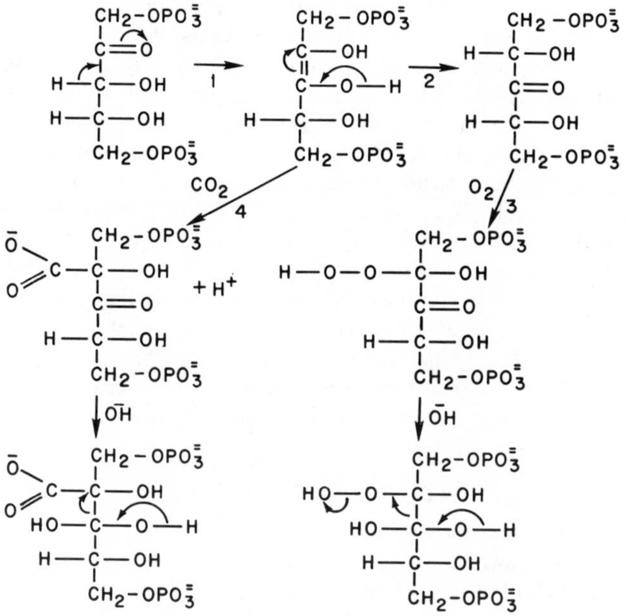

$$CH_2-OPO_3^=$$
$$C=O$$
$$H-C-OH$$
$$H-C-OH$$
$$CH_2-OPO_3^=$$

$\xrightarrow{1}$

$$CH_2-OPO_3^=$$
$$C-OH$$
$$C-O-H$$
$$H-C-OH$$
$$CH_2-OPO_3^=$$

$\xrightarrow{2}$

$$CH_2-OPO_3^=$$
$$H-C-OH$$
$$C=O$$
$$H-C-OH$$
$$CH_2-OPO_3^=$$

$\underset{4}{CO_2}$

$\underset{3}{O_2}$

$$\bar{O}\overset{C}{\diagup}\overset{O}{\diagdown}\;\;CH_2-OPO_3^=$$
$$C-OH$$
$$C=O$$
$$H-C-OH$$
$$CH_2-OPO_3^=$$

$+H^+$

$$H-O-O-C-OH \quad CH_2-OPO_3^=$$
$$C=O$$
$$H-C-OH$$
$$CH_2-OPO_3^=$$

$\downarrow \bar{O}H$

$\downarrow \bar{O}H$

$$\bar{O}\overset{C}{\diagup}\overset{O}{\diagdown}\;\;CH_2-OPO_3^=$$
$$C-OH$$
$$HO-C-O-H$$
$$H-C-OH$$
$$CH_2-OPO_3^=$$

$$CH_2-OPO_3^=$$
$$HO-O-C-OH$$
$$HO-C-O-H$$
$$H-C-OH$$
$$CH_2-OPO_3^=$$

$$\downarrow$$

$$
\begin{array}{c}
\overset{\displaystyle \quad\quad CH_2-OPO_3^{\equiv}}{\underset{\displaystyle}{\overset{O}{\underset{\bar{O}}{\diagup}}C-CHOH}}
\end{array}
$$

+

$$
\begin{array}{c}
COO^- \\
\mid \\
H-C-OH \quad +H^+ \\
\mid \\
CH_2-OPO_3^{\equiv}
\end{array}
$$

$$\downarrow$$

$$
\begin{array}{c}
CH_2-OPO_3^{\equiv} \\
\mid \\
O\!\!=\!\!C\diagdown_{O} \qquad +2H^+
\end{array}
$$

+

$$+\,O\bar{H}$$

$$
\begin{array}{c}
COO^- \\
\mid \\
H-C-OH \\
\mid \\
CH_2-OPO_3^{\equiv}
\end{array}
$$

In the mechanism, a derivative of ribulose diphosphate that has lost a proton at C_3 and is either the *cis*-enediol, the C_2 carbanion, or the C_3 keto form undergoes either electrophilic attack by CO_2 or radical attack by O_2 at C_2. The subsequent cleavage is initiated by $O\bar{H}$ attack at C_3.

Strong evidence for the transient existence of the 6-carbon carboxylated intermediate of ribulose diphosphate has recently been obtained by the demonstration that 2 carboxyribitol diphosphate, the reduced analog of this intermediate

$$
\begin{array}{c}
\bar{O}\diagdown \quad CH_2-OPO_3^{\equiv} \\
C-C-OH \\
O\diagup H-C-OH \\
H-C-OH \\
CH_2-OPO_3^{\equiv}
\end{array}
$$

in the presence of a divalent cation, is a potent but slow acting inhibitor of carboxylation, with an apparent $K_I \sim 10^{-8}\,M$ (188). A similar dissociation constant was obtained from uv difference spectra of the enzyme, suggesting that conformational changes in the protein tighten the binding of this inhibitor (188). Chemical oxidation of this reduced analog also in the presence of a divalent cation in a nonenzymatic reaction yields the presumed carboxylated intermediate of the enzymatic reaction, 2-carboxy-3-ketoribulose diphosphate (189). This unstable compound either spontaneously decarboxylates in a reaction which is a model of the reverse of step 4, or at pH 9 spontaneously cleaves to yield two molecules of 3-PGA, a model of the enzymatic cleavage. However, unlike the enzymatic reaction, the model reaction yields the L isomer of 3-PGA from C_1 and C_2.

The enzymatic reactions (184–187) as well as the tight binding of the reduced analog CRDP (188) and some or all of the model reactions involving decarboxylation and cleavage of carboxyketo ribulose diphosphate (189) require the presence of a divalent cation. Calvin had previously detected a ternary enzyme-Mg-CO_2 complex (190), which raises the possibility of direct activation of CO_2 by Mg in a metal bridge complex. However a strictly analogous role for Mg in the oxygenase reaction seems unlikely.

The divalent cation is capable of binding to the enzyme (187). However such binding is very weak ($K_D \sim 0.6$ mM) (190a). In the presence of saturating levels of HCO_3^-, one tight binding site for Mn^{2+} per subunit appears with a dissociation constant ($K_D = 10$ μM) of the same order as the K_M, suggesting a preferred order reaction in which CO_2 binding precedes Mn binding. Bicarbonate is not directly coordinated to the enzyme-bound Mn^{2+} since the water proton relaxation rate increases by 40% as bicarbonate binds. Moreover the distance from bound Mn to the carbon atom of HCO_3^- determined by its ^{13}C relaxation rate (5.4 Å) is consistent with only 2% of an inner sphere complex ($r = 3.2$ Å) and 98% second sphere complex ($r = 6.0$ Å) (190a). Hence if bicarbonate binds as an analog of the activated and distorted molecule of CO_2, this substrate is not directly activated by the metal. Another portion of the substrate may interact with the metal, since Mn^{2+} is necessary for the tight binding of the 6-carbon atom inhibitor, carboxyribitol diphosphate (188). Upon binding to the Mn enzyme, this inhibitor may displace water from the inner sphere of Mn (as indicated by proton NMR) and HCO_3^- from the second sphere of Mn as indicated by ^{13}C NMR (190a). Hence, although direct activation of the true substrate CO_2 by the enzyme-bound Mn^{2+} is not strictly ruled out, an attractive alternative is that the metal coordinates the oxygens at C_2 and/or C_3 of ribulose diphosphate facilitating carbanion formation at C_2, a necessary step in the carboxylation and oxygenation reactions.

Ribulose diphosphate carboxylase, which constitutes 16% of the soluble protein extracted from spinach leaves, consists of eight subunits (187). Phosphoenolpyruvate carboxylase, an alternative carboxylating enzyme from the same source, has the same total molecular weight (560,000) but consists of only four subunits of 130,000 mol wt (H. Miziorko, unpublished), contrary to our previous report of 12 subunits (190b). The latter error was based on preparations heavily contaminated by ribulose diphosphate carboxylase.

LYASES AND RELATED REACTIONS

Elimination Mechanisms

In an elimination reaction two ionic groups are removed from the substrate,

an electrophile, typically a proton (often assisted by a base) and a nucleophile $\overline{X}$ (often assisted by an acid or metal ion) resulting in the formation of a double bond. Depending on the sequence of events, mechanisms of elimination fall into three classes (191, 192). In the carbanion or E-1CB mechanism the proton is removed first; in the carbonium or E-1 mechanism the nucleophile $\overline{X}$ is removed first, and in the concerted or E-2 elimination both are removed within 10^{-13} sec via a single transition state. With organic model reactions determination of the rate equation often suffices to define the mechanism. With enzyme-catalyzed elimination reactions this does not suffice, since other chemical steps may be slow enough to contribute to the rate equation. Careful isotopic exchange studies and a search for primary or secondary isotope effects are generally necessary (193).

Fumarase and Enolase

Extensive isotopic studies have been carried out with fumarase (193–195) and enolase (196). With fumarase the exchange rate of the hydroxyl group of malate with water exceeded the rates of the proton exchange and the overall conversion rate of substrate to product. These findings and the detection of a secondary kinetic isotope effect at C_3 in the dehydration of malate (197) strongly suggest (193) but do not establish a carbonium ion mechanism (193, 194, 197) since deprotonation of the enzyme may be rate-limiting. With muscle enolase, the proton exchange at C_2 of 2-phosphoglycerate exceeded all other measurable exchange rates. This observation together with a sizable secondary C_3-deuterium isotope effect in the hydration of phosphoenolpyruvate establish a carbanion mechanism (196).

A carbanion intermediate would result from nucleophilic attack of water or hydroxyl ion at C_2 of phosphoenolpyruvate. The role of the required divalent ion in activating water for nucleophilic attack is suggested by NMR studies with yeast enolase (198). The enzyme-bound Mn was found by water proton relaxation studies to coordinate 2 rapidly exchangeable water ligands. Binding of the substrate α-(dihydroxyphosphenyl methyl) acrylate (Figure 3) or of substrate analogs decreased this number to 0.3 to 1.0. Yet direct coordination of the substrates or analogs by Mn^{2+} did not take place, since the distances from bound Mn to the phosphorus and protons of the substrate and analogs indicated second sphere complexes (Figure 3). Hence, the binding of the substrates or analogs may have immobilized a water ligand on the Mn^{2+} such that it exchanged slower than 10^6 sec^{-1}. This may represent a kinetic manifestation of "freezing at reaction centers on enzymes" (FARCE). A carbanion mechanism for enolase established by the isotopic exchange studies (196) and consistent with the stereochemistry (199) and the geometry (198) as elucidated by NMR is given in Figure 5. The phosphate of the substrate may function as a general base assisting the metal in deprotonation of the water (198). The product of the reaction 2-phosphoglycerate would form an E-M-S bridge complex via its 3-OH group. This complex could not be studied by NMR due to slow exchange, but kinetic studies (198) suggested that Mn^{2+} decreased the K_M of 2-phosphoglycerate consistent with a metal bridge complex.

Aconitase

An analogous role for the enzyme-bound Fe^{2+} was proposed for the elimination reaction catalyzed by aconitase (200) as part of the overall isomerization of citrate to isocitrate. In order to satisfy the stereochemical data which require a flipover of the intermediate cis-aconitate on the enzyme (201, 202), a water ligand substitution on the enzyme-bound Fe^{2+} was suggested as the driving force for a conformation change of the coordinated cis-aconitate in the ingenious ferrous wheel mechanism (200). In accord with this mechanism, an active aconitase-Fe-citrate, and inactive aconitase-Mn-citrate and aconitase-Mn-trans-aconitate complexes of the required

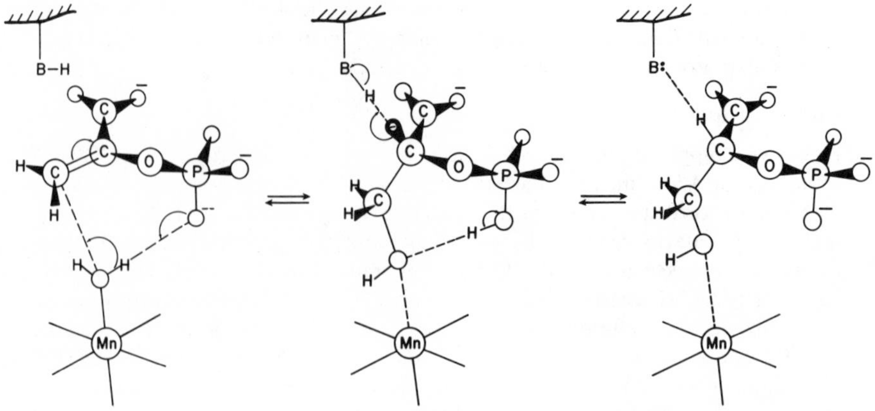

Figure 5 Mechanism of enolase consistent with nuclear relaxation data [see Figure 3 (198)], isotope exchange studies (196), and stereochemistry (199). From (198).

structure were detected by proton relaxation studies of the substrate and analog (Figure 2) (203). Moreover the required rapid water exchange rate on the enzyme-bound metal was also detected by water proton relaxation (204). However, the failure of metals other than Fe^{2+} (such as Mn^{2+}) to activate aconitase even though Mn^{2+} forms complexes of the appropriate geometry indicates the need for a mechanism unique to Fe^{2+} (203). A turnstile or Bailar twist mechanism, which would accomplish the same geometric effect as the ferrous wheel mechanism, has been proposed (203). Such processes have been shown to occur in model reactions with Fe^{2+} (205) but not with Mn^{2+}. The spin state of the added Fe^{2+} is appropriate for such a mechanism (203–205). A summary of the alternative mechanisms for aconitase is given in Figure 6 of reference 203. In addition to the added Fe^{2+}, which activates the enzyme, the homogeneous enzyme contains one or two high spin Fe^{3+} ions (206, 207) of unknown function which do not interact with the substrate (203).

Role of Metals

The role of the metal ion in the enolase (198) and aconitase (203) reactions as well as in the related enolization reactions catalyzed by yeast aldolase (208) and D-xylose isomerase (3, 192, 209) appears to be α activation (192).

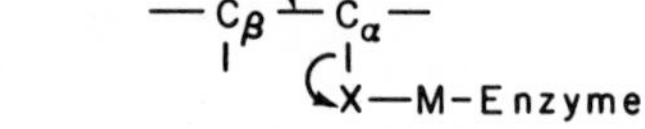

$$-B: -\overset{\frown}{\underset{\displaystyle -C_\beta}{}} \overset{\text{H}}{\underset{\displaystyle -C_\alpha-}{|}} \\ \underset{X-M-Enzyme}{\overset{|}{C}}$$

With β-methyl aspartase, activation at the γ position has been suggested (210) and with histidine deaminase, a complex consistent with δ activation has been detected (3, 192, 211). Thus, enzymes use metals to activate certain elimination or enolization reactions by placing the metal such that it coordinates a basic group one carbon removed from the carbon atom to be deprotonated (192). More distant coordination and activation is achieved as in histidine deaminase when permitted by the conjugated structure of the substrate (192).

Role of General Bases

Elimination reactions require the enzyme to remove protons from the substrate. Such proton activation by general base catalysis is most clearly demonstrated by the transfer of the same proton of one substrate to another, or to the product in an isomerization reaction (193). We suggest the term "conservative proton transfer" for such processes. Because of partial occlusion of the active site by the substrate or the protein, the nature of the basic group itself may not uniquely determine whether conservation of a proton will occur. Thus conservative proton transfers have been detected in steroid isomerase where imidazole may be the base (212) and in phosphoglucose isomerase where carboxylate may be the base (213).

A carboxylate base that does not conserve protons is involved in the mechanism of 2 keto-3-deoxy-6-phosphogluconate aldolase as indicated by trapping with 3-bromopyruvate which is both a substrate and an affinity label (214). Stereochemical arguments (214) suggest that after it deprotonates pyruvate, the conjugate acid of the carboxyl group of glutamate could by an internal rotation about its C_α-C_β bond protonate the C_4 oxygen of the adduct formed by attack of the pyruvate carbanion on 3-phosphoglyceraldehyde to yield KDP gluconate. An essential histidine detected by photo-oxidation in transaldolase (215) may function in the same way.

OXIDOREDUCTASES

Dehydrogenases

Although a direct hydride transfer has long provided the simplest explanation for the net transfer of a proton and two electrons from substrates to nicotinamide coenzymes, other more complicated possibilities have long been considered. Model reactions for atom transfer as well as hydride transfer have been found (216) and lactic dehydrogenase has been found to catalyze the oxidation of NADH by the preformed superoxide radical anion (216a). Strong evidence in favor of hydride transfer has been obtained with glutamate dehydrogenase (217) and with yeast alcohol dehydrogenase (218). Glutamate dehydrogenase has been found to catalyze the displacement of sulfite from trinitrobenzene sulfonate by NADH or NADPH (217). Nucleophilic displacement by hydride ion provides the most reasonable mechanism for this reaction. With yeast alcohol dehydrogenase an important study of substituent effects on the rate of reduction of aromatic aldehydes revealed a ρ value (2.2) indistinguishable from that for the nucleophilic attack of $C\bar{N}$ on aldehyde carbonyl groups (218). Hydride transfers from NADH to an aldehyde would be facilitated by polarization of the carbonyl group of the aldehyde on binding to the enzyme. The binding of aromatic aldehydes to yeast alcohol dehydrogenase was found to have a ρ value of -0.85 consistent with carbonyl polarization upon binding. Since alcohol dehydrogenases are Zn-metalloenzymes, such a role has long been suggested for Zn (3, 219) and highly appropriate model reactions have been discovered (220, 220a). The finding that the Zn-free apoenzyme of liver alcohol-dehydrogenase retains its ability to bind the substrate butanol and substrate analogs with unaltered affinities (221) argues against but does not rule out such a role for Zn, since the coordination of substrates by Zn may contribute little to the affinity. The combination of an NMR study of yeast alcohol dehydrogenase (222) with a 3 Å crystallographic study of liver alcohol dehydrogenase (223) indicates the substrate to be within coordinating distance (2–3 Å) of the bound Zn (Figure 6B). The structure of NADH on spin-labeled yeast alcohol dehydrogenase, using proton and ^{31}P-NMR (222), reveals an open conformation for the nucleotide and anticonformations for the adenine-ribose and pyridine-ribose interactions (Figure 6A). The presence of the aldehyde analog, isobutyramide, alters the conformation of the bound NADH as indicated by a tilt of the adenine ribose portion away from the spin label (Figure 6B). These NMR models of NADH and substrate on yeast alcohol dehydrogenase are closely superimposable onto the 3 Å crystallographic model of the liver alcohol dehydrogenase–ADP ribose–o-phenanthroline complex (223). A combination of the NMR model, which locates the pyridine ring and the substrate, with the crystallographic model, which locates the modified cysteine residue [# 43 (224)] and the Zn (Figure 6B), indicates an alignment of pyridine, substrate, and Zn such that the substrate is stacked onto the pyridine within coordination distance (2–3 Å) from the Zn. Direct coordination of the carbonyl group by Zn would facilitate polarization of this group and hydride transfer. Although the enzyme from yeast is a tetramer and that from liver is a dimer and some kinetic differences have long been known (225), a 40% homology exists in the amino acid sequences at the coenzyme binding sites of the two enzymes (224, 226). Moreover, the secondary and tertiary structures of the NAD binding sites on lactate, malate, glyceraldehyde 3-P, and alcohol dehydrogenases are very similar, consisting of six antiparallel β pleated sheets connected by several helices or chains (227). Hence, unless an unusual series of coincidences have occurred, the conformation of NADH is indistinguishable on yeast and liver alcohol dehydrogenases, and in the crystalline state and in solution. The dihedral angle of the pyridine ring (although always approximately perpendicular to the adenine ring) differs by $\sim 30°$ in lactate (228), malate (229), and alcohol dehydrogenases (222). The present results are thus consistent with earlier results which failed to detect fluorescence transfer (230) or diamagnetic chemical shifts due to ring stacking (231) on various dehydrogenases. In the reverse reaction ethanol must be deprotonated. Shore & Gutfreund have detected the rapid partial loss of a proton from liver alcohol dehydrogenase as NAD binds and the completion of this deprotonation on forming the inactive ternary complex with trifluoroethanol (232). These results have been interpreted to indicate the appearance of a proton acceptor, presumably to facilitate deprotonation of the substrate. By isolating the hydride transfer step using stopped flow kinetics, Shore has detected a large kinetic isotope effect ($k_H/k_D = 6$) with liver alcohol dehydrogenase (233).

Most dehydrogenases do not contain metals. Hence the polarization of the oxidized substrate must be accomplished by other means. In the high resolution X-ray structure of lactic dehydrogenase, the imidazolium group of His 195 appears to accomplish this by hydrogen bonding, possibly assisted by Arg 109 (234). A proton

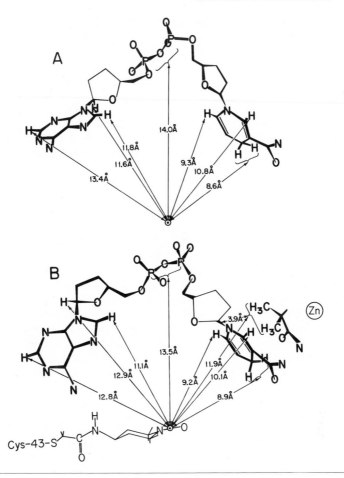

Figure 6 *A* Geometry of the binary complex of spin-labeled yeast alcohol dehydrogenase-NADH from nuclear relaxation data (222). *B* Geometry of the ternary complex of spin-labeled yeast alcohol dehydrogenase-NADH-isobutyramide from nuclear relaxation data (222). Also shown are the positions of Cys 43 and of Zn from the crystallographic model (223) and the amino acid sequence (224) of horse liver alcohol dehydrogenase. From (222).

transfer from His was predicted from steady state (235) and stopped flow kinetic studies (236).

Flavoproteins

Like nicotinamide coenzymes, flavin coenzymes may also undergo nucleophilic attack (237) when reduced by two electron donors as established by several model reactions (238–240) and in enzymatic studies with D-amino acid oxidase (241–245) and lactate oxidase (246).

The rates of the model reactions correlate with the electron affinity of the flavin when small sulfur nucleophiles are used. Nucleophilic attack occurs at the 5 (nitrogen) position unless access is sterically hindered, when it occurs at the 4a position (239, 240).

A second mechanism involving a charge transfer complex may operate in the two electron reduction of flavin by ring compounds such as NADH, since an intermediate ring-stacked complex is detected kinetically. Within this complex the electron transfer occurs by an unknown mechanism (247–249). In both the nucleophilic and the charge transfer mechanisms and also in the reoxidation of reduced flavin by O_2, the 10a position appears to be inactive (247–249).

Cogent evidence for a carbanion intermediate in a flavoenzyme reaction was first obtained in the reactions of β-chloroalanine catalyzed by D-amino acid oxidase (241). With O_2 the usual amino acid oxidation was catalyzed yielding chloropyruvate. Under N_2 the elimination of HCl was catalyzed yielding pyruvate. Mixtures of O_2 and N_2 yielded mixed products but the rate of total keto acid production was constant. This finding together with a substrate kinetic isotope effect ($k_H/k_D = 2$) at the α C-H position but no effect on the nature of the products suggest the rate-limiting formation of a common carbanion intermediate which can then undergo either oxidation or elimination (241).

A larger kinetic isotope effect of 3 was found by stopped flow kinetics (244). The initial abstraction of a proton from the substrate was further documented by the detection of an intramolecular proton transfer from C_2 to C_3 during the enzyme-catalyzed elimination of HCl from α amino β chlorobutyric acid (242). The carbanion which forms rapidly attacks the flavin ring, yielding spectroscopically distinct intermediates (242). In accord with this view the preformed nitroethane carbanion is a substrate for D-amino acid oxidase (243, 245) and forms an adduct at the 5 position which, after the elimination of nitrite, may either be trapped by CN^-, and analyzed spectroscopically, or by hydroxyl ions from the solvent from which it proceeds to the oxidized product, acetaldehyde (243, 245). This work provides the first evidence for a kinetically competent and isolable substrate-flavin adduct in a flavoenzyme reaction. With less reactive substrates, the lowering of the kinetic barrier to the formation of a carbanion intermediate or transition state may be accomplished by interaction with the flavin ring (245), as suggested by the requirement for oxidized flavin and by the observation of rapid spectral changes in the elimination reaction of α amino β chlorobutyrate catalyzed by D-amino acid oxidase (242).

However, with glucose oxidase no early intermediates between oxidized and reduced flavin were detected although such findings are rendered difficult by the rate-limiting deprotonation of the substrate (250). Thus in their ionic reactions, flavin coenzymes, like nicotinamide coenzymes, may undergo nucleophilic attack by two electron donors. The flavin ring system, because of its greater symmetry and ability to delocalize spins in a one electron adduct than nicotinamide can also stabilize radical intermediates and thereby participate in one electron processes as well.

These one electron reactions of flavoproteins have been reviewed elsewhere (251) and are exemplified by flavodoxin in the next section.

One Electron Transfer Reactions

One of the last frontiers in the field of enzyme mechanisms is a chemical explanation for the vast body of spectroscopic and kinetic data on one electron transfer reactions. Crystal structures of flavodoxin (252–255), cytochrome c (256–259), b_5 (260, 261), and c_2 (262), iron-sulfur proteins (263–266) and their appropriate models (267) are beginning to rationalize such data in terms of chemical structures and possible mechanisms.

FLAVODOXIN   One electron transfers and semiquinone intermediates are clearly involved in the function of flavodoxin, a class of flavoproteins that replace ferredoxins in various one electron transfer processes (268). The crystal structures of a Clostridial flavodoxin reveal partial exposure of the face of only one of the flavin rings to the solvent (252–255). The enhanced effect of flavodoxin semiquinone on the water proton relaxation rate suggested this to be the dimethylbenzene ring which has 15% of the spin density rather than the pyrimidine ring which has none (269). The crystal structure of a homologous flavodoxin from Desulfovibrio (254, 255) indicates the dimethylbenzene ring to be the closest to the surface of the protein. The crystal structures of the two flavodoxins reveal the stacking of an aromatic residue (probably tyrosine) over the face of the center flavin ring, the locus of the highest (85%) spin density (252–255). Two pathways for electron transfer to flavin are thus suggested: direct transfer to the exposed dimethylbenzene ring edge or face, and transfer mediated through a stacked aromatic residue. It is seen later that these are precisely the alternatives under active consideration for electron transfer reactions to cytochromes and to iron-sulfur proteins (3). Ludwig (252) has made the important suggestion that the redox potential of flavin may be controlled by the protein that constrains the FMN ring to a planar conformation, since in the absence of proteins, the oxidized species are planar while fully reduced species are bent by as much as 35.5°.

CYTOCHROMES   In a review of the role of metals in enzyme catalysis (3) it was suggested that of the two well-defined mechanisms of electron transfer to metal complexes, outer sphere and inner sphere, the former mechanism was operative in reactions of cytochromes and nonheme iron proteins. Recent crystallographic data support this suggestion. The crystal structures of oxidized and reduced cytochrome c (256–259) and oxidized cytochrome c_2 (262) indicate the exposure of the edge of one of the four pyrrole rings to the solvent. The previously observed contact shifted heme methyl resonances of cytochrome c (270, 271), when precisely assigned using a cross saturation method (272, 273), indicates maximal electron density at this pyrrole and its diagonal pyrrole, suggesting more facile electron transfer from (or to) this exposed edge. From the crystal structure of cytochrome c, Tyr 67 overlies the diagonal pyrrole to permit electron transfer from this residue to the oxidized heme (258, 259). If nothing else happened, this would leave a Tyr cation radical. However, the introduction of an electron to Try 59 via the surface Tyr 74, with which it lies parallel, would create a radical anion of Try 59. An electron transfer from the radical anion of Try 59 to the radical cation of Tyr 67 would require a conformation change of Try 59 to align the ring planes. Such a conformation change is indeed seen in the crystal structure of reduced cytochrome c (259). Stable radical intermediates have not been detected kinetically in the rapid reactions of cytochrome c, but are merely suggested by Dickerson as a formalism (258, 259). In the electron transfer process they may all be part of a single transition state. NMR studies suggest the presence of spin density on Tyr 67 and the His ligand (272a), and also indicate spin density at the methionine sulfur ligand of the iron, suggesting these sites as possible loci for one electron reduction (273). In reduced cytochrome c this pathway seems obstructed by Phe 82, an invariant residue which has moved to cover this portion of the heme pocket (259). Indeed Phe 82 itself may be involved in electron transport to the heme. Although all of the invariant aromatic residues are present in *Candida krusei* cytochrome c, the electron exchange between the oxidized and reduced forms ($10^3 \ M^{-1} \ sec^{-1}$) is an order of magnitude slower than that for horse heart cytochrome c, reflecting the role of the tertiary structure of the protein in the electron transfer process (274).

The crystal structure of cytochrome c_2 is compatible with the outer sphere transfer of an electron to the exposed heme edge, possibly coupled with a proton transfer through a network of Ser 89 to Tyr 52 to Tyr 70. The latter Tyr 70, now satisfied by a proton, releases the methionine ligand to the iron so that it can become more precisely axial in the reduced state (262).

The crystal structure of cytochrome b_5 (260, 261) suggests that the mechanism of electron transfer to and from this hemeprotein must differ significantly from that for c-type cytochromes. Thus the heme is more deeply buried in a hydrophobic crevice with decreased access to solvent and the Fe is symmetrically coordinated by two imidazole ligands, one of which (His 63) is stacked with an aromatic residue (Phe 58) which could provide a path for electron transfer. In reduced cytochrome b_5 an ion, possibly Na^+, is bound near an axial imidazole ligand and might also facilitate electron transfer to the heme (F. S. Mathews, personal communication). Crystallographic studies also reveal a hydrophobic groove on the surface of the molecule containing two phenylalanine residues which by limited rearrangement could provide a path for an electron to enter the heme. Tryptic removal of portions of this site including one of the phenylalanine residues (275) abolishes interaction of cytochrome b_5 with its reductase. The general inaccessibility of the heme to the surface renders cytochrome b_5 different from all other redox proteins studied thus far, and indicates the need for further studies of the mechanism of electron transfer in this system.

IRON-SULFUR PROTEINS   Iron-sulfur proteins constitute a well defined class of electron carriers whose unifying structural component is one or more Fe atoms each surrounded by a distorted tetrahedron of sulfur atoms. As pointed out in a valuable review of the recent literature (276), the one-Fe rubredoxins in their oxidized state are high spin Fe^{III} coordinated to four cysteines; the Fe in the plant-type ferredoxins consists of Fe_2^{III} clusters with bridging sulfur atoms, and the bacterial ferredoxins consists of Fe_4^{III} clusters with bridging sulfur atoms. The crystal structures of a rubredoxin (264), a high potential protein with one Fe_4 cluster (263), and a bacterial ferredoxin with two Fe_4 clusters have been solved (265). An excellent metallo-organic model of the Fe_4 clusters has been synthesized (267) which reproduces all of the crystallographic and spectroscopic properties of the prosthetic groups of the corresponding proteins (266, 267). Similarly, an accurate model of the Fe_2 cluster has recently been synthesized (267a). In all cases the crystal structures reveal at least one sulfur ligand exposed to the solvent. Moreover, spin density at these sulfurs in the reduced state was established by the observation of superhyperfine coupling to ^{33}S (277) and in the oxidized state by the detection of contact-shifted cysteine methylene protons (278, 279). Hence an outer sphere electron transfer via a sulfur ligand represents a reasonable pathway for reduction of these systems (3). With the Clostridial ferredoxin, tyrosine residues are found stacked onto the Fe_4 cluster (280), and shifted ^{13}C resonances, but not proton resonances of such tyrosines have been detected (281), consistent with the transfer of spin density to Tyr. Hence in these systems as in cytochromes, the participation of aromatic residues in electron transport is suggested but not established.

Coenzyme B_{12} Enzymes

Enzymes utilizing coenzyme B_{12} catalyze a variety of isomerizations exemplified by the diol dehydrase and ethanolamine deaminase reactions. Owing largely to the early isotopic studies of Abeles (282, 283), these processes have been shown to involve cleavage of the bond between cobalt and the C-5 carbon atom of the 5-deoxyadenosyl residue, and the transfer of hydrogen from the substrate to the coenzyme resulting in the formation of 5-deoxyadenosine as an intermediate hydrogen carrier (284–286).

   A major advance in our understanding of the mechanism of the carbon to cobalt bond cleavage has resulted from the detection of EPR signals by Babior and co-workers (287, 288) on mixing the ethanolamine deaminase-coenzyme complex with substrates. Similar findings have since been made with diol dehydrase (289, 290) and glycerol dehydrase (291). With diol dehydrase, addition of the substrate propandiol caused the steady state appearance of ~0.75 unpaired electrons per molecule of enzyme-coenzyme complex, while addition of the product analogs chloroacetaldehyde and hydroxyacetaldehyde yielded ~2 unpaired electrons indicating complete homolytic cleavage of the carbon-cobalt bond of the coenzyme (290). The EPR spectra consisted of a signal at $g = 2.2$ due to low spin Co^{II} and multiple signals at $g = 2.0$, presumably due to radicals derived from 5-deoxyadenosine and from the substrates (290). The latter point has been established by changes in the EPR spectrum at $g = 2.0$ with isotopic substitution in chloroacetaldehyde (291a). Thus the use of ^{13}C rather than ^{12}C chloroacetaldehyde produced a ~ 10 G broadening of the EPR signal and the use of chloroacetaldehyde deuterated rather than protonated at C_2 resulted in an ~ 5 G narrowing of the signal. These observations establish the presence of spin density on carbon atoms, and to a lesser extent on the methylene protons derived from chloroacetaldehyde. Rapid quench experiments indicate that the EPR signals with chloroacetaldehyde are half developed well within 40 msec and are therefore due to kinetically competent intermediates. Analogous and independent observations of superhyperfine interaction and rapid signal appearance have been made with ethanolamine deaminase using the active substrate propanolamine (B. Babior, W. Orme-Johnson, and H. Beinert, private communication). While these EPR studies were made at temperatures well below freezing, evidence for the formation of radicals has also been obtained at room temperature. Thus, increased water proton relaxation (290) consistent with radical formation and optical spectra typical of B_{12r} have been obtained on adding the substrate (292) or the product analog chloroacetaldehyde (293) to the diol dehydrase-coenzyme complex. The transitory disappearance of the nitroxide signal of spin-labeled coenzyme B_{12} at room temperature is also consistent with the turnover of radical intermediates (294). Radical mechanisms for B_{12} enzymes variously proposed (285, 290, 291) are summarized as follows for diol dehydrase.

$$CH_3-\underset{OH}{\underset{|}{CH}}-\underset{OH}{\underset{|}{CH_2}}$$
$$+$$
$$\underset{\overset{|}{\underset{Co}{\diagdown\diagup}}}{CH_2-R}$$

$$\left[CH_3-\underset{OH}{\underset{|}{CH}}-\underset{OH}{\underset{|}{CH_2}} \quad \cdot CH_2-R \right]$$
$$\overset{\bullet}{\underset{Co}{\diagdown\diagup}}$$

$$\rightleftharpoons$$

$$\left[CH_3-\underset{OH}{\underset{|}{CH}}-\underset{OH}{\underset{|}{\overset{\bullet}{C}H}} \quad CH_3-R \right]$$
$$\underset{Co}{\diagdown\diagup}$$

$$\left[CH_3-\underset{\underset{Co}{\diagdown\diagup}}{CH\text{=}CH} \quad CH_3-R \right]$$
$$\underset{OH}{\overset{HO}{\diagdown}}$$

$$\left[CH_3-\underset{\underset{Co}{\diagdown\diagup}}{CH-CH} \quad CH_3-R \right] \quad \overset{O-H}{} \quad \underset{OH}{}$$

$$-H_2O \updownarrow$$

$$\left[CH_3-\underset{\underset{Co}{\diagdown\diagup}}{CH}-\overset{O}{\overset{\|}{C}}\diagdown_H \quad CH_3-R \right]$$

$$\left[CH_3-\overset{\bullet}{C}H-\overset{O}{\overset{\|}{C}}\diagdown_H \quad CH_3-R \right]$$
$$\overset{\bullet}{\underset{Co}{\diagdown\diagup}}$$

$$\updownarrow$$

$$\left[CH_3-CH_2-\overset{O}{\overset{\|}{C}}\diagdown_H \quad \cdot CH_2-R \right]$$
$$\overset{\bullet}{\underset{Co}{\diagdown\diagup}}$$

$$CH_3-CH_2-\overset{O}{\overset{\|}{C}}\diagdown_H$$
$$+$$
$$\underset{\overset{|}{\underset{Co}{\diagdown\diagup}}}{CH_2-R}$$

These mechanisms may well be an oversimplification, since the unusually high primary kinetic isotope effects in the hydrogen transfer (284) and the detectable spin density on protons as well as on carbon of the substrate analog detected by EPR (291a) suggest H^0 atom transfer processes. Proton relaxation studies indicate that these radical transfers are separated from solvent by as much as 10 Å (290). The nature of the driving force for homolytic carbon-cobalt cleavage is unknown (290).

Now that the mechanism of the carbon-cobalt bond cleavage has been clarified, the appropriate models for this reaction (295) and its microscopic reverse (296) may be chosen. Following the proposed formation of the substrate-cobalt bond, a rearrangement takes place. Cogent models for such rearrangements involving intermediate π complexes on cobalt have recently been found with a cobaloxime (297) and with cobalamine itself (298).

GENERAL STATEMENTS ON ENZYME CATALYSIS

From the preceding specific considerations on individual enzymes, the following generalities on enzyme mechanisms appear to be emerging. Although all are not strictly established as principles, they are at least interesting enough to bear critical examination. The list is, of course, by no means complete.

1. All nucleotidyl transfer reactions and most phosphoryl transfer reactions appear to utilize an associative or S_N2 mechanism. The essential metals, by appropriate coordination, promote the attack of a nucleophile, stabilize the trigonal bipyramidal transition state, or promote the departure of the leaving group.

2. Nucleotidyl transfer enzymes that copy or transcribe DNA or RNA templates (DNA and RNA polymerases) contain tightly bound Zn (35, 36, 41). The role of the Zn is catalytic, possibly to activate the 3'OH group of the growing polynucleotide chain (35, 36, 42).

3. A mechanistic homology of many pyruvate- and phosphoenolpyruvate-utilizing enzymes exists (63) such that these enzymes (pyruvate kinase, phosphoenolpyruvate carboxykinase and carboxylase, pyruvate carboxylase and transcarboxylase) enolize pyruvate before adding a positively charged group (proton or carbonium ion) to C_3 (64).

4. Enzymes (pyruvate kinase, phosphoglucomutase) appear to immobilize those atoms of a substrate directly involved in the bond breaking and bond forming steps in catalysis. The catalytic effect of such "freezing at reaction centers on enzymes" (FARCE) is currently under discussion and lies between the factors 10^2 and 10^8 (78, 80–82).

5. Most but not all enzymes require a specific stereoisomeric form of their substrates (or a specific epimer in the case of sugar substrates such as phosphoglucose isomerase), indicating the need for at least moderately precise orientation of substrates with respect to catalytic groups.

6. Substrates or coenzymes on a number of enzymes (aconitase, alcohol dehydrogenase, pyruvate kinase, phosphoglucomutase, epimerases) must undergo conformational changes prior to the bond breaking and bond forming steps, presumably to achieve proper orientation with respect to other substrates or groups on the enzyme (57, 112–115, 200–203, 222).

7. The inner coordination sphere of a metal ion is used for catalysis by some enzymes (Figure 2).

8. Those lyases that require metal ions for activity (enolase, aconitase, histidine deaminase, D-xylose isomerase, aldolase, and possibly ribulose diphosphate carboxylase) place the metal such that it coordinates a basic group-one carbon

removed from the atom to be deprotonated, or on a group conjugated to this carbon atom (192, 198, 203, 208–211).

9. The second coordination sphere of a metal ion is utilized by some enzymes, often in an unknown manner, to promote catalysis (Figure 3).

10. A hydrogen-bonded charge relay system (serine → histidine → carboxylic group) facilitates the deprotonation of serine and the stabilization of the tetrahedral intermediate in serine proteases. The factor by which this accelerates the rate is unknown (141–143, 145–147).

11. The two electron reductions of NAD and of flavines on their respective enzymes appear to proceed via nucleophilic mechanisms, a hydride transfer in the former and a carbanion attack in the latter. Resonance stabilization of these full or incipient anionic species by the aromatic rings of the coenzymes may facilitate these processes (217, 218, 237–246).

12. With the possible exception of cytochrome b_5 (260, 261), all one electron transferring proteins (flavodoxins, cytochrome c, c_2, and iron-sulfur proteins) have one or more atoms of their prosthetic groups with unpaired spin density accessible to solvent, facilitating outer sphere electron transfer reactions (252–266).

13. Enzyme reactions involving coenzyme B_{12} appear to involve homolytic cleavage of the carbon to cobalt bond, subsequent one electron steps, and radical intermediates, possibly including hydrogen atoms (287–294). Solvent access to these intermediates is prevented by steric hindrance (290).

14. Finally, what is more a working hypothesis than an established generality, is the view that enzyme reactions are monomechanistic and thereby actually simpler than organic and coordination reactions which often proceed simultaneously by multiple pathways. Enzymes combine several known catalytic techniques into one "supermechanism" whose combination yields the tremendous rate accelerations observed. Only by determining with appropriate model reactions the quantitative contribution to the rate acceleration of each catalytic technique and comparing their product with the observed enzymatic rate will we understand enzymatic catalysis.

Before ascribing this combination of techniques to any undue wisdom of Mother Nature or her various male consorts, we must recall that she has had 10^9 years to evolve enzymes while man has had but 10^2 years to comprehend and duplicate them. Thus, like ordinary mortals, the forces of nature plod slowly along by trial and error (299) occasionally making an order of magnitude advance, occasionally a mistake resulting in the disappearance of a species.

ACKNOWLEDGMENTS

This review has benefited from the helpful comments of many colleagues, especially R. H. Abeles, C. B. Anfinsen, T. C. Bruice, H. J. Bright, F. A. Cotton, B. M. Dunn, B. Furie, R. K. Gupta, E. E. Hazen, W. P. Jencks, S. H. Koenig, L. A. Loeb, J. P.

Klinman, H. Miziorko, and I. A. Rose, as well as numerous others cited in the text who have sent preprints or have permitted me to quote their unpublished observations.

This work was supported by Grant GB-27739X2 from the National Science Foundation, US PHS Grants AM-13351, CA-06927 and RR-05539 from the National Institutes of Health, and an appropriation from the Commonwealth of Pennsylvania.

Literature Cited

1. Benkovic, S. J., Schray, K. J. 1973. *Enzymes* 8:201
2. Gillespie, P., Ramirez, F., Ugi, I., Marquarding, D. 1973. *Angew. Chem.* 12:91
3. Mildvan, A. S. 1970. *Enzymes* 2:445
4. Cooperman, B. 1973. In *Metal Ions in Biological Systems,* ed. H. Sigel, Vol. V. New York: Dekker. In press
5. Ingold, C. K. 1953. *Structure and Mechanism in Organic Chemistry,* 310. London: Bell
6. Bunton, C. A., Fendler, E. J., Humeres, E., Yang, K. U. 1967. *J. Org. Chem.* 32:2806
7. Berry, R. S. 1960. *J. Chem. Phys.* 32:933
8. Westheimer, F. H. 1968. *Accounts Chem. Res.* 1:70
9. Muetterties, E. L., Schunn, R. A. 1966. *Quart. Rev. Chem. Soc.* 20:245
10. Anfinsen, C. B., Cuatrecasas, P., Taniuchi, H. 1971. *Enzymes* 4:177
11. Chaiken, I. M., Anfinsen, C. B. 1971. *J. Biol. Chem.* 246:2285; Sanchez, G. R., Chaiken, I. M., Anfinsen, C. B. 1973. *J. Biol. Chem.* 248:3653
12. Arnone, A. et al 1969. *Proc. Nat. Acad. Sci. USA* 64:420
13. Arnone, A. et al 1971. *J. Biol. Chem.* 246:2302
14. Markley, J. L., Jardetzky, O. 1970. *J. Mol. Biol.* 50:223
15. Cotton, F. A., Hazen, E. E. Jr. 1971. *Enzymes* 4:153; Cotton, F. A., Bier, C. J., Day, V. W., Hazen, E. E. Jr., Larsen, S. 1971. *Cold Spring Harbor Symp. Quant. Biol.* 36:243
16. Roberts, G. C. K., Jardetzky, O. 1970. *Advan. Protein Chem.* 24:447
17. Cotton, F. A., Day, V. W., Hazen, E. E. Jr., Larsen, S. 1973. *J. Am. Chem. Soc.* 95:4834
18. Pearson, W. B., Ed. 1964. *Structure Reports* 29:360. Leipzig: Akad. Verlag
19. Cuatrecasas, P., Fuchs, S., Anfinsen, C. B. 1967. *J. Biol. Chem.* 242:3063
20. Furie, B., Eastlake, A., Schechter, A. N.,

Anfinsen, C. B. 1973. *J. Biol. Chem.* 248:5821
21. Dunn, B. M., Di Bello, C., Anfinsen, C. B. 1973. *J. Biol. Chem.* 248:4769
22. Roberts, G. C. K., Dennis, E. A., Meadows, D. H., Cohen, J. S., Jardetzky, O. 1969. *Proc. Nat. Acad. Sci. USA* 62:1151
23. Meadows, D. H., Roberts, G. C. K., Jardetzky, O. 1969. *J. Mol. Biol.* 45:491
24. Wyckoff, H. W. et al 1967. *J. Biol. Chem.* 242:3984
25. Usher, D. A. 1969. *Proc. Nat. Acad. Sci. USA* 62:661
26. Usher, D. A., Richardson, D. I., Eckstein, F. 1970. *Nature London* 228:663
27. Usher, D. A., Erenrich, E. S., Eckstein, F. 1972. *Proc. Nat. Acad. Sci. USA* 69:115
28. Richards, F. M. et al 1971. *Cold Spring Harbor Symp. Quant. Biol.* 36:35
29. Griffin, J. H., Schechter, A. N., Cohen, J. S. 1973. *Ann. NY Acad. Sci.* 222:693
30. Loeb, L. A. 1974. *Enzymes.* 10:173
31. Kornberg, A. 1969. *Science* 163:1410
32. Cavalieri, L. F., Carroll, E. 1970. *Biochem. Biophys. Res. Commun.* 41:1055
33. Wells, R. D., Flügel, R. M., Larson, J. E., Schendel, P. F., Sweet, R. W. 1972. *Biochemistry* 11:621
34. Loeb, L. A., Tartof, K. D., Travaglini, E. C. 1973. *Nature New Biol.* 242:66
35. Slater, J. P., Mildvan, A. S., Loeb, L. A. 1971. *Biochem. Biophys. Res. Commun.* 44:37
36. Springgate, C. F., Mildvan, A. S., Abramson, R., Engle, J. L., Loeb, L. A. 1973. *J. Biol. Chem.* 248:5987
36a. Poiesz, B., Battula, N., Loeb, L. A. 1974. *Biochem. Biophys. Res. Commun.* In press
37. Stavrianopoulos, J. G., Karkas, J. D., Chargaff, E. 1972. *Proc. Nat. Acad. Sci. USA* 69:1781
38. Levinson, W., Faras, A., Woodson, B., Jackson, J., Bishop, J. M. 1973. *Proc. Nat. Acad. Sci. USA* 70:164
39. Valenzuela, P., Morris, R. W., Faras, A., Levinson, W., Rutter, W. J. 1973.

Biochem. Biophys. Res. Commun. 53:
1036

40. Vallee, B. L. 1955. *Advan. Protein Chem.*
10:317

41. Scrutton, M. C., Wu, C. W., Goldthwait, D. A. 1971. *Proc. Nat. Acad. Sci.
USA* 68:2497

42. Slater, J. P., Tamir, I., Loeb, L. A.,
Mildvan, A. S. 1972. *J. Biol. Chem.* 247:
6784

43. Sigman, D. S., Wahl, G. M., Creighton,
D. J. 1972. *Biochemistry* 11:2236

44. Benkovic, S. J., Dunikoski, L. K. Jr.
1971. *J. Am. Chem. Soc.* 93:1526

45. Murakami, Y., Takagi, M. 1969. *J. Am.
Chem. Soc.* 91:5130

46. Englund, P. T., Kelly, R. B., Kornberg,
A. 1969. *J. Biol. Chem.* 244:3045

47. Englund, P. T., Huberman, J. A., Jovin,
T. M., Kornberg, A. 1969. *J. Biol. Chem.*
244:3038

48. Trautner, T. A., Swartz, M. N., Kornberg, A. 1962. *Proc. Nat. Acad. Sci. USA*
48:449

49. Springgate, C. F., Loeb, L. A. 1973.
Proc. Nat. Acad. Sci. USA 70:245

50. Hall, Z. W., Lehman, I. R. 1968. *J. Mol.
Biol.* 36:321

51. Chang, L. M. S. 1973. *J. Biol. Chem.*
248:6983

52. Springgate, C. F., Battula, N., Loeb,
L. A. 1973. *Biochem. Biophys. Res.
Commun.* 52:401

53. Chang, L. M. S., Bollum, F. J. 1973. *J.
Biol. Chem.* 248:3398

54. Mildvan, A. S., Cohn, M. 1970. *Advan.
Enzymol.* 33:1

55. Mildvan, A. S., Engle, J. L. 1972.
Methods Enzymol. 26:654

56. Mildvan, A. S., Nowak, T., Fung, C. H.
1973. *Ann. NY Acad. Sci.* 222:192

57. Nowak, T., Mildvan, A. S. 1972. *Biochemistry* 11:2819

58. Reed, G. H., Cohn, M. 1973. *J. Biol.
Chem.* 248:6436

59. Mildvan, A. S., Leigh, J. S., Cohn, M.
1967. *Biochemistry* 6:1805

60. Perloff, A. 1972. *Acta Crystallogr. B* 28:
2183

61. Mildvan, A. S., Cohn, M. 1966. *J. Biol.
Chem.* 241:1178

62. Nowak, T., Mildvan, A. S. 1970. *J. Biol.
Chem.* 245:6057

63. Miller, R. S. et al 1968. *J. Biol. Chem.*
243:6030

64. Rose, I. A. 1972. *Crit. Rev. Biochem.* 1:
33

65. Wolfenden, R. 1972. *Accounts Chem.
Res.* 5:10

66. Lienhard, G. E., Secemski, I. I., Koehler,
K. A., Lindquist, R. N. 1971. *Cold Spring
Harbor Symp. Quant. Biol.* 36:45

67. Mildvan, A. S., Scrutton, M. C., Utter,
M. F. 1966. *J. Biol. Chem.* 241:3488

68. Northrop, D. B., Wood, H. G. 1969. *J.
Biol. Chem.* 244:5820

69. Reed, G. H., Morgan, S. D. 1973. *166th
Meet. Am. Chem. Soc., Chicago* Biol.
165

70. Willard, J., Rose, I. A. 1973. *Biochemistry*
12:5241

71. Fung, C. H., Mildvan, A. S., Allerhand,
A., Komoroski, R., Scrutton, M. C. 1973.
Biochemistry 12:620

72. Reuben, J., Kayne, F. J. 1971. *J. Biol.
Chem.* 246:6227

73. Kayne, F. J., Reuben, J. 1970. *J. Am.
Chem. Soc.* 92:220

74. Nowak, T. 1973. *J. Biol. Chem.* 248:7191

75. Rose, I. A. 1970. *J. Biol. Chem.* 245:6052

76. Rose, I. A. 1960. *J. Biol. Chem.* 235:1170

77. Robinson, J., Rose, I. A. 1972. *J. Biol.
Chem.* 247:1096

78. Nowak, T., Mildvan, A. S. 1972. *Biochemistry* 11:2813

79. Werbelow, L., Marshall, A. G. 1973. *J.
Am. Chem. Soc.* 95:5132

80. Reuben, J. 1971. *Proc. Nat. Acad. Sci.
USA* 68:563

81. Bruice, T. C. 1970. *Enzymes* 2:217

82. Page, M. I., Jencks, W. P. 1971. *Proc.
Nat. Acad. Sci. USA* 68:1678; Jencks,
W. P., Page, M. I. 1972. *Enzymes,
Structure and Function,* Proc. 8th
FEBS Meeting, Amsterdam 29:45

83. Hollenberg, P. F., Flashner, M., Coon,
M. J. 1971. *J. Biol. Chem.* 246:946

84. Flashner, M., Hollenberg, P. F., Coon,
M. J. 1972. *J. Biol. Chem.* 247:8114

85. Flashner, M., Tamir, I., Mildvan, A. S.,
Meloche, H. P., Coon, M. J. 1973. *J. Biol.
Chem.* 248:3419

86. Barrio, J. R. et al 1973. *FEBS Lett.* 29:
215

87. Plowman, K. M., Krall, A. R. 1965.
Biochemistry 4:2809

88. Muirhead, H., Stammers, D. K. 1973.
Biochem. Soc. Trans. In press

89. Cottam, G. L., Mildvan, A. S. 1971. *J.
Biol. Chem.* 246:4363

90. Benkovic, S. J., Villafranca, J. J., Kleinschuster, J. J. 1973. *Arch. Biochem.
Biophys.* 155:458

91. Milner-White, E. J., Watts, D. C. 1971.
Biochem. J. 122:727

92. Reed, G., Cohn, M. 1972. *J. Biol. Chem.*
247:3073

93. Reed, G. H., McLaughlin, A. C. 1973.
Ann. NY Acad. Sci. 222:118

94. McLaughlin, A. C., Cohn, M., Kenyon,
G. L. 1972. *J. Biol. Chem.* 247:4382

95. Cohn, M., Leigh, J. S. Jr., Reed, G. H.
1971. *Cold Spring Harbor Symp. Quant.
Biol.* 36:533

96. Blake, C. C. F., Evans, P. R. 1973. *9th Int. Congr. Biochem., Stockholm,* p. 40 (Abstr.)
97. Cohn, M. 1972. *Enzymes, Structure and Function,* Proc. 8th FEBS Meet., Amsterdam 29:59
98. Hirsch, K., Mildvan, A. S., Kowalsky, A. 1969. *158th Am. Chem. Soc. Meet. New York, Biology Section* (Abstr. 52)
99. Foster, D., Mildvan, A. S. 1972. *Bioinorg. Chem.* 1:133
100. De Pamphilis, M. L., Cleland, W. W. 1973. *Biochemistry.* 12:3714
101. Blake, C. C. F., Évans, P. R., Scopes, R. K. 1972. *Nature New Biol.* 235:195
102. Wendell, P. L., Bryant, T. N., Watson, H. C. 1972. *Nature New Biol.* 240:234
103. Steitz, T. A., Fletterick, R. J., Hwang, K. J. 1973. *J. Mol. Biol.* 78:551
104. Anthony, R. S., Spector, L. B. 1970. *J. Biol. Chem.* 245:6739
105. Walsh, C. T., Spector, L. B. 1971. *J. Biol. Chem.* 246:1255
106. Kaji, A., Colowick, S. P. 1965. *J. Biol. Chem.* 240:4454
107. Walsh, C. T. Jr., Spector, L. B. 1971. *Arch. Biochem. Biophys.* 145:1
108. Cheng, L. Y., Inagami, T., Colowick, S. P. 1973. *9th Int. Congr. Biochem.,* p. 90 (Abstr.)
109. Wålinder, O., Zetterqvist, Ö., Engstrom, L. 1969. *J. Biol. Chem.* 244:1060
110. Solomon, F., Rose, I. A. 1971. *Arch. Biochem. Biophys.* 147:349
111. Ray, W. J., Peck, E. J. Jr. 1972. *Enzymes* 6:408
112. Ray, W. J. Jr. 1973. *166th Am. Chem. Soc. Meet., Chicago.* AGFD 5
113. Mildvan, A. S. 1973. *166th Am. Chem. Soc. Meet., Chicago.* INOR 95
114. Ray, W. J. Jr., Mildvan, A. S. 1973. *Biochemistry.* 12:3733
115. Ray, W. J. Jr., Mildvan, A. S., Long, J. W. 1973. *Biochemistry.* 12:3724
116. Cottam, G. L., Ward, R. L. 1970. *Arch. Biochem. Biophys.* 141:768
117. Cottam, G. L., Thompson, B. C. 1972. *J. Magn. Resonance* 6:352
118. Csopak, H., Drakenberg, T. 1973. *FEBS Lett.* 30:296
119. Zukin, R., Hollis, D. P., Gray, G. A. 1973. *Biochem. Biophys. Res. Commun.* 53:238
120. Zukin, R., Hollis, D. P., Gray, G. A. 1973. *Biochem. Biophys. Res. Commun.* 53:686
121. Knox, J. R., Wyckoff, H. W. 1973. *J. Mol. Biol.* 74:533
122. Lipscomb, W. N. 1972. *Chem. Soc. Rev.* 1:319
123. Lipscomb, W. N. 1970. *Accounts Chem.*

Res. 3:81
124. Hartsuck, J. A., Lipscomb, W. N. 1971. *Enzymes* 3:1
125. Quiocho, F. A., Lipscomb, W. N. 1971. *Advan. Prot. Chem.* 25:1
126. Bradshaw, R. A., Ericsson, L. H., Walsh, K. A., Neurath, H. 1969. *Proc. Nat. Acad. Sci. USA* 63:1389
127. Simpson, R. T., Vallee, B. L. 1966. *Biochemistry* 5:1760
128. Vallee, B. L., Riordan, J. F. 1968. *Brookhaven Symp. Biol.* 21:91
129. Auld, D. S., Vallee, B. L. 1970. *Biochemistry* 9:4352
130. Navon, G., Shulman, R. G., Wyluda, B. J., Yamane, T. 1968. *Proc. Nat. Acad. Sci. USA* 60:86; 1970. *J. Mol. Biol.* 51:15
131. Latt, S. A., Auld, D. S., Vallee, B. L. 1972. *Biochemistry* 11:3015
132. Riordan, J. F., Vallee, B. L. 1963. *Biochemistry* 2:1460
133. Quiocho, F. et al 1971. *Cold Spring Harbor Symp. Quant. Biol.* 36:561
134. Koenig, S. H., Brown, R. D., Studebaker, J. 1971. *Cold Spring Harbor Symp. Quant. Biol.* 36:551
134a. Rogers, G. A., Bruice, T. C. 1974. *J. Am. Chem. Soc.* In press
135. Kang, E. P., Storm, C. B., Carson, F. W. 1972. *Biochem. Biophys. Res. Commun.* 49:621
136. Johansen, J. T., Vallee, B. L. 1971. *Proc. Nat. Acad. Sci. USA* 68:2532
137. Johansen, J. T., Vallee, B. L. 1973. *Proc. Nat. Acad. Sci. USA* 70:2006
138. Quiocho, F. A., McMurray, C. H., Lipscomb, W. N. 1972. *Proc. Nat. Acad. Sci. USA* 69:2850
139. Riordan, J. F., Muszynska, G. 1973. *9th Int. Congr. Biochem.* p. 64 (Abstr.)
140. Johansen, J. T., Livingston, D. M., Vallee, B. L. 1972. *Biochemistry* 11:2584
141. Sigler, P. B., Blow, D. M., Matthews, B. W., Henderson, R. 1968. *J. Mol. Biol.* 35:143
142. Blow, D. M. 1971. *Enzymes* 3:185
143. Steitz, T. A., Henderson, R., Blow, D. M. 1969. *J. Mol. Biol.* 46:337
144. Bothner-By, A. A., Gassend, R. 1973. *Ann. NY Acad. Sci.* 222:668
145. Robillard, G., Shulman, R. G. 1972. *J. Mol. Biol.* 71:507
146. Robillard, G., Shulman, R. G. 1973. *Ann. NY Acad. Sci.* 222:220
146a. Rogers, G. A., Bruice, T. C. 1974. *J. Am. Chem. Soc.* In press
147. Robertus, J. D., Kraut, J., Alden, R. A., Birktoft, J. J. 1972. *Biochemistry* 11:4293
147a. Hunkapillar, M., Smallcomb, S.,

Whitaker, D., Richards, J. H. 1973. *Biochemistry.* 12:4732

148. Coleman, J. E. 1971. *Progr. Bioorg. Chem.* 1:159
149. Lindskog, S. et al 1971. *Enzymes* 5:587
150. Kannan, K. K. et al 1971. *Cold Spring Harbor Symp. Quant. Biol.* 36:221
151. Lindskog, S., Nyman, P. O. 1964. *Biochim. Biophys. Acta* 85:462
152. Fabry, M. E. R., Koenig, S. H., Schillinger, W. E. 1970. *J. Biol. Chem.* 245:4256
153. Lanir, A., Gradsztojn, S., Navon, G. I. 1973. *FEBS Lett.* 30:351
154. Mildvan, A. S., Rumen, N. M., Chance, B. 1971. *Probes Struct. Funct. Macromol. Membranes* 2:205
155. Ward, R. L. 1969. *Biochemistry* 8:1879
156. Ward, R. L., Fritz, K. J. 1970. *Biochem. Biophys. Res. Commun.* 39:707
157. Khalifah, R. G. 1971. *J. Biol. Chem.* 246:2561
158. Perrin, D. D., Sharma, V. S. 1966. *J. Inorg. Nucl. Chem.* 28:1271
158a. Kaiser, E. T., Lo, K. W. 1969. *J. Am. Chem. Soc.* 91:4912
159. Pesando, J. 1973. PhD thesis, Albert Einstein College of Medicine, New York
160. Khalifah, R. G., Edsall, J. T. 1972. *Proc. Nat. Acad. Sci. USA* 69:172
161. Lanir, A., Navon, G. 1971. *Biochemistry* 10:1024
162. Lanir, A., Navon, G. 1972. *Biochemistry* 11:3536
163. Koenig, S. H., Brown, R. D. III 1972. *Proc. Nat. Acad. Sci. USA* 69:2422
164. Koenig, S. H., Brown, R. D. III, Needham, T. E., Matwiyoff, N. A. 1973. *Biochem. Biophys. Res. Commun.* 53:624
165. Khalifah, R. G. 1973. *Proc. Nat. Acad. Sci. USA* 70:1986
165a. Lindskog, S., Coleman, J. E. 1973. *Proc. Nat. Acad. Sci. USA* 70:2505
166. Northrop, D. B., Wood, H. G. 1969. *J. Biol. Chem.* 244:5820
167. Barden, R. E., Fung, C. H., Utter, M. F., Scrutton, M. C. 1972. *J. Biol. Chem.* 247:1323
168. Moss, J., Lane, M. D. 1971. *Advan. Enzymol.* 35:321
169. Lynen, F. et al 1961. *Biochem. Z.* 335:123
170. Kaziro, Y., Hass, L. F., Boyer, P. D., Ochoa, S. 1962. *J. Biol. Chem.* 237:1460
171. Polakis, S. E., Guchhait, R. B., Lane, M. D. 1972. *J. Biol. Chem.* 247:1335
172. Moss, J., Lane, M. D. 1972. *J. Biol. Chem.* 247:4952
173. Mildvan, A. S., Scrutton, M. C. 1967.

Biochemistry 6:2978
174. Scrutton, M. C., Reed, G. H., Mildvan, A. S. 1973. *Advan. Exp. Med. Biol.* 40:79
175. Scrutton, M. C., Mildvan, A. S. 1970. *Arch. Biochem. Biophys.* 140:131
176. Wood, H. G., Lochmüller, H., Riepertinger, C., Lynen, F. 1963. *Biochem. Z.* 337:247
177. Ahmad, F., Lygre, D. G., Jacobson, B. E., Wood, H. G. 1972. *J. Biol. Chem.* 247:6299
178. Fung, C. H., Mildvan, A. S., Leigh, J. S. 1974. *Biochemistry.* In press
179. Kornberg, A., Ochoa, S., Mehler, A. H. 1948. *J. Biol. Chem.* 174:159
180. Steinberger, R., Westheimer, F. H. 1951. *J. Am. Chem. Soc.* 73:429
181. Gallo, A. A., Sable, H. Z. 1973. *Biochim. Biophys. Acta* 302:443
182. Quayle, J. R., Fuller, R. C., Benson, A. A., Calvin, M. 1954. *J. Am. Chem. Soc.* 76:3610
183. Horecker, B. L., Hurwitz, J., Weissbach, A. 1956. *J. Biol. Chem.* 218:785
184. Ogren, W. L., Bowes, G. 1971. *Nature New Biol.* 230:159
185. Andrews, T. J., Lorimer, G. H., Tolbert, N. E. 1973. *Biochemistry* 12:11
186. Lorimer, G. H., Andrews, T. J., Tolbert, N. E. 1973. *Biochemistry* 12:18
187. Siegel, M. I., Wishnick, M., Lane, M. D. *Enzymes* 6:169
188. Siegel, M. I., Lane, M. D. 1972. *Biochem. Biophys. Res. Commun.* 48:508
189. Siegel, M. I., Lane, M. D. 1973. *J. Biol. Chem.* 248:5486
190. Akoyunoglou, G., Calvin, M. 1963. *Biochem. Z.* 338:20
190a. Miziorko, H., Mildvan, A. S. 1974. *J. Biol. Chem.* In press
190b. Miziorko, H., Nowak, T., Bayer, M. E., Mildvan, A. S. 1971. *162nd Am. Chem. Soc. Meet., Washington, D.C.* (Abstr.)
191. Hine, J. 1962. *Physical Organic Chemistry*, 186. New York: McGraw. 2nd ed.
192. Mildvan, A. S. 1971. *Bioinorg. Chem.*, 100:390
193. Rose, I. A. 1970. *Enzymes* 2:281
194. Hansen, J. N., Dinovo, E. C., Boyer, P. D. 1969. *J. Biol. Chem.* 244:6270
195. Alberty, R. A., Miller, W. G., Fisher, H. F. 1957. *J. Am. Chem. Soc.* 79:3973
196. Dinovo, E. C., Boyer, P. D. 1971. *J. Biol. Chem.* 246:4586
197. Schmidt, D. E. Jr., Nigh, W. G., Tanzer, C., Richards, J. H. 1969. *J. Am. Chem. Soc.* 91:5849
198. Nowak, T., Mildvan, A. S., Kenyon, G. L. 1973. *Biochemistry* 12:1690

199. Cohn, M., Pearson, J. E., O'Connell, E. L., Rose, I. A. 1970. *J. Am. Chem. Soc.* 92:4095
200. Glusker, J. P. 1968. *J. Mol. Biol.* 38:149
201. Rose, I. A., O'Connell, E. L. 1967. *J. Biol. Chem.* 242:1870
202. Gawron, O., Glaid, A. J. III, Fondy, T. P. 1961. *J. Am. Chem. Soc.* 83:3634
203. Villafranca, J. J., Mildvan, A. S. 1972. *J. Biol. Chem.* 247:3454
204. Villafranca, J. J., Mildvan, A. S. 1971. *J. Biol. Chem.* 246:5791
205. Pignolet, L. H., Lewis, R. A., Holm, R. H. 1971. *J. Am. Chem. Soc.* 93:360
206. Villafranca, J. J., Mildvan, A. S. 1971. *J. Biol. Chem.* 246:772
207. Kennedy, C., Rauner, R., Gawron, O. 1972. *Biochem. Biophys. Res. Commun.* 47:740
208. Mildvan, A. S., Kobes, R. D., Rutter, W. J. 1971. *Biochemistry* 10:1191
209. Schray, K. J., Mildvan, A. S. 1972. *J. Biol. Chem.* 247:2034
210. Bright, H. J. 1967. *Biochemistry* 6:1191
211. Givot, I. L., Mildvan, A. S., Abeles, R. H. 1970. *Fed. Proc.* 29:531
212. Talalay, P., Wang, V. S. 1955. *Biochim. Biophys. Acta* 18:300; Talalay, P., Benson, A. M. 1972. *Enzymes* 6:591
213. Rose, I. A., O'Connell, E. L. 1961. *J. Biol. Chem.* 236:3086; 1973. *J. Biol. Chem.* 248:2225
214. Meloche, H. P., Glusker, J. P. 1973. *Science* 181:350
215. Brand, K., Tsolas, O., Horecker, B. L. 1969. *Arch. Biochem. Biophys.* 130:521
216. Westheimer, F. H. 1959. *Enzymes* 1:259
216a. Bielsky, B. H. J., Chan, P. C. 1973. *Arch. Biochem. Biophys.* 159:873
217. Bates, D. J., Goldin, B. R., Frieden, C. 1970. *Biochem. Biophys. Res. Commun.* 39:502
218. Klinman, J. P. 1972. *J. Biol. Chem.* 247:7977; 1973. *Proc. 1st Int. Symp. Alc. Aldehyde Metab. Syst.* New York: Academic. In press
219. Theorell, H., Yonetani, T. 1963. *Biochem. Z.* 338:537
220. Creighton, D. J., Sigman, D. S. 1971. *J. Am. Chem. Soc.* 93:6314
220a. Shinkai, S., Bruice, T. C. 1973. *Biochemistry* 12:1750
221. Iweibo, I., Weiner, H. 1972. *Biochemistry* 11:1003
222. Sloan, D. L. Jr., Mildvan, A. S. 1974. *Biochemistry.* In press
223. Branden, C. I. et al 1973. *Proc. Nat. Acad. Sci. USA* 70:2439
224. Jornvall, H. et al. See Ref. 222; Jornvall, H., Markovic, O. 1972. *Eur. J. Biochem.* 29:167
225. Sund, H., Theorell, H. 1963. *Enzymes* 7:25
226. Bridgen, J., Kolb, E., Harris, J. I. 1973. *FEBS Lett.* 33:1
227. Buehner, M., Ford, G. C., Moras, D., Olsen, K. W., Rossmann, M. G. 1973. *Proc. Nat. Acad. Sci. USA.* 70:3052
228. Chandrasekhar, K., McPherson, A., Adams, M. J., Rossmann, M. G. 1973. *J. Mol. Biol.* 76:503
229. Webb, L. E., Hill, E. J., Banaszak, L. J. 1973. *Biochemistry* 12:5101
230. Velick, S. F. 1961. In *Light and Life,* ed. W. D. McElroy, B. Glass, 108. Baltimore: Johns Hopkins Press
231. Lee, C. Y., Eichner, R. D., Kaplan, N. O. 1973. *Proc. Nat. Acad. Sci. USA* 70:1593
232. Shore, J. D., Gutfreund, H. 1973. *Proc. 1st Int. Symp. Alc. Aldehyde Metab. Syst.* New York: Academic. In press; Brooks, R. L., Shore, J. D., Gutfreund, H. 1973. *J. Biol. Chem.* 247:2382
233. Shore, J. D., Gutfreund, H. 1970. *Biochemistry* 9:4655
234. Adams, M. J. et al 1973. *Proc. Nat. Acad. Sci. USA* 70:1968
235. Novoa, W. B., Winer, A. D., Glaid, A. J., Schwert, G. W. 1959. *J. Biol. Chem.* 234:1143
236. Holbrook, J. J., Gutfreund, H. 1973. *FEBS Lett.* 31:157
237. Hamilton, G. A. 1971. *Progr. Bioorg. Chem.* 1:83
238. Müller, F., Massey, V. 1969. *J. Biol. Chem.* 244:4007
239. Hevesi, L., Bruice, T. C. 1972. *J. Am. Chem. Soc.* 94:8277
240. Hevesi, L., Bruice, T. C. 1973. *Biochemistry* 12:290
241. Walsh, C. T., Schonbrunn, A., Abeles, R. H. 1971. *J. Biol. Chem.* 246:6855
242. Walsh, C. T., Krodel, E., Massey, V., Abeles, R. H. 1973. *J. Biol. Chem.* 248:1946
243. Porter, D. J. T., Voet, J. G., Bright, H. J. 1972. *Z. Naturforsch. B* 27:1052
244. Voet, J. G., Porter, D. J. T., Bright, H. J. 1972. *Z. Naturforsch. B* 27:1054
245. Porter, D. J. T., Voet, J., Bright, H. J. 1973. *J. Biol. Chem.* 248:4400
246. Walsh, C. T., Schonbrunn, A., Lockridge, O., Massey, V., Abeles, R. H. 1972. *J. Biol. Chem.* 247:6004
247. Bruice, T. C., Main, L., Smith, S., Bruice, P. Y. 1971. *J. Am. Chem. Soc.* 93:7327
248. Main, L., Kasperek, G. J., Bruice, T. C. 1972. *J. Chem. Soc., Chem. Commun.* 1972:847; 1972. *Biochemistry* 11:3991
249. Porter, D. J. T., Blankenhorn, G.,

Ingraham, L. L. 1973. *Biochem. Biophys. Res. Commun.* 52:447

250. Bright, H. J., Gibson, Q. H. 1967. *J. Biol. Chem.* 242:994; Weibel, M. K., Bright, H. J. 1971. *J. Biol. Chem.* 246: 2734

251. Hemmerich, P. 1970. *Vitam. Horm.* 28:467; Hemmerich, P. et al 1972. *Z. Naturforsch. B* 27:1030–51

252. Ludwig, M. L. et al 1971. *Cold Spring Harbor Symp. Quant. Biol.* 36:369

253. Andersen, R. D. et al 1972. *Proc. Nat. Acad. Sci. USA* 69:3189

254. Watenpaugh, K. D., Sieker, L. C., Jensen, L. H., LeGall, J., Dubourdieu, M. 1972. *Z. Naturforsch. B* 27:1094

255. Watenpaugh, K. D., Sieker, L. C., Jensen, L. H., LeGall, J., Dubourdieu, M. 1972. *Proc. Nat. Acad. Sci. USA* 69:3185

256. Dickerson, R. E. 1972. *Ann. Rev. Biochem.* 41:815

257. Takano, T., Swanson, R., Kallai, O. B., Dickerson, R. E. 1971. *Cold Spring Harbor Symp. Quant. Biol.* 36:397

258. Dickerson, R. E. 1973. *Ann. NY Acad. Sci.* 227:599

259. Takano, T., Kallai, O. B., Swanson, R., Dickerson, R. E. 1973. *J. Biol. Chem.* 248:5234

260. Mathews, S. F., Argos, P., Levine, M. 1971. *Cold Spring Harbor Symp. Quant. Biol.* 36:387

261. Mathews, S. F., Levine, M., Argos, P. 1972. *J. Mol. Biol.* 64:449

262. Salemme, F. R., Freer, S. T., Xuong, N. H., Alden, R. A., Kraut, J. 1973. *J. Biol. Chem.* 248:3910

263. Carter, C. W. Jr., Freer, S. T., Xuong, N. H., Alden, R. A., Kraut, J. 1971. *Cold Spring Harbor Symp. Quant. Biol.* 36:381

264. Watenpaugh, K. D., Sieker, L. C., Herriott, J. R., Jensen, L. H. 1971. *Cold Spring Harbor Symp. Quant. Biol.* 36:359

265. Sieker, L. C., Adman, E., Jensen, L. H. 1972. *Nature* 235:40

266. Carter, C. W. Jr. et al 1972. *Proc. Nat. Acad. Sci. USA* 69:3526

267. Herskovitz, T. et al 1972. *Proc. Nat. Acad. Sci. USA* 69:2437

267a. Mayerle, J. J. et al 1973. *Proc. Nat. Acad. Sci. USA* 70:2429

268. Knight, E. Jr., Hardy, R. W. F. 1966. *J. Biol. Chem.* 241:2752; 1967. *J. Biol. Chem.* 242:1370

269. Palmer, G., Mildvan, A. S. 1972. In *Structure and Function of Oxidation Reduction Enzymes.* ed. Å. Åkeson, Å. Ehrenberg, p. 385. Oxford: Pergamon

270. Kowalsky, A. 1965. *Biochemistry* 4: 2382

271. Wüthrich, K. 1969. *Proc. Nat. Acad. Sci. USA* 63:1071

272. Gupta, R. K., Redfield, A. G. 1970. *Science* 169:1204

272a. Oldfield, E., Allerhand, A. 1974. *Proc. Nat. Acad. Sci. USA.* In press

273. Redfield, A., Gupta, R. K. 1971. *Cold Spring Harbor Symp. Quant. Biol.* 36: 405

274. Gupta, R. K. 1973. *Biochim. Biophys. Acta* 292:291

275. Strittmatter, P., Huntley, T. E. 1970. *8th Int. Congr. Biochem. Switzerland,* p. 21 (Abstr.)

276. Orme-Johnson, W. H. 1973. *Ann. Rev. Biochem.* 42:159

277. Tsibris, J. C. M. et al 1968. *Proc. Nat. Acad. Sci. USA* 59:959

278. Poe, M., Phillips, W. D., McDonald, C. C., Lovenberg, W. 1970. *Proc. Nat. Acad. Sci. USA* 65:797

279. Phillips, W. D., Poe, M., McDonald, C. C., Bartsch, R. G. 1970. *Proc. Nat. Acad. Sci. USA* 67:682

280. Adman, E. T., Sieker, L. C., Jensen, L. H. 1973. *J. Biol. Chem.* 248:3987

281. Packer, E. L., Sternlicht, H., Rabinowitz, J. C. 1972. *Proc. Nat. Acad. Sci. USA* 69:3278

282. Zagalak, B., Frey, P. A., Karabatsos, G. L., Abeles, R. H. 1967. *J. Biol. Chem.* 241:3028

283. Frey, P. A., Essenberg, M. K., Abeles, R. H. 1967. *J. Biol. Chem.* 242:5369

284. Essenberg, M. K., Frey, P. A., Abeles, R. H. 1971. *J. Am. Chem. Soc.* 93:1242

285. Babior, B. M. 1970. *J. Biol. Chem.* 245: 6125

286. Babior, B. M., Carty, T., Abeles, R. H. 1974. *J. Biol. Chem.* In press

287. Babior, B. M., Gould, D. 1969. *Biochem. Biophys. Res. Commun.* 34:441

288. Babior, B. M., Kon, H., Lecar, H. 1969. *Biochemistry* 8:2662

289. Foster, M. A., Hill, H. A. O., Williams, R. J. P. 1971. *Biochem. Soc. Symp.* 37: 187

290. Finlay, T. H., Valinsky, J., Mildvan, A. S., Abeles, R. H. 1973. *J. Biol. Chem.* 248:1285

291. Cockle, S. A., Hill, H. A. O., Williams, R. J. P., Davies, S. P., Foster, M. A. 1972. *J. Am. Chem. Soc.* 94:275

291a. Valinsky, J., Abeles, R. H., Mildvan, A. S. 1974. *J. Biol. Chem.* In press

292. Abeles, R. H., Lee, H. A. Jr. 1964. *Ann. NY Acad. Sci.* 112:695

293. Finlay, T. H., Valinsky, J., Sato, K., Abeles, R. H. 1972. *J. Biol. Chem.* 247: 4197

294. Law, P. Y., Brown, D. G., Lien, E. L.,

Babior, B. M., Wood, J. M. 1971. *Biochemistry* 10:3428

295. Hill, H. A. O., Pratt, J. M., Williams, R. J. P. 1971. *Methods Enzymol.* 18:5

296. Halpern, J., Maher, J. P. 1965. *J. Am. Chem. Soc.* 87:5361

297. Silverman, R. B., Dolphin, D., Babior, B. M. 1972. *J. Am. Chem. Soc.* 94:4028

298. Silverman, R. B., Dolphin, D. 1973. *J. Am. Chem. Soc.* 95:1686

299. Hartley, B. S. et al 1972. *8th FEBS Meet.* 29:151

UNUSUAL POLYSACCHARIDES

✖ 851

Sam Kirkwood

Department of Biochemistry, College of Biological Sciences
University of Minnesota, St. Paul, Minnesota

CONTENTS

INTRODUCTION

When I was approached by the Editorial Committee to write a chapter on the above topic, I inquired about what they intended by the adjective "unusual." It should be recorded that they had no constructive ideas on this subject and left the matter entirely to me!

We are presently in an era in which biochemistry, and to a degree biology, are dominated by the chemistry of macromolecules. One must be impressed with the successes in interpreting biological function on the basis of the structures and conformations of these substances. In this connection, however, the function of one large group of macromolecules has been almost lost to view. These are the classical polysaccharides (in contradistinction to glycoproteins, glycolipids, peptidoglycan, etc). We still know very little about the structure-function relationships of the host of polysaccharides that have been isolated and investigated by carbohydrate chemists. These substances are, in this sense, unusual macromolecules and it is

401

thought worthwhile to devote this chapter to a discussion of the light appearing at the end of this particular tunnel.

When the matter of polysaccharide function arises in biochemistry, attention is usually directed to the area of the structure-function relationships in bacterial cell walls and mammalian glycoproteins. The two most recent reviews in this series concerning polysaccharides have dealt with precisely these topics (1, 2). However the majority of the substances involved, while carbohydrate in nature, are not polysaccharide in the classical sense of the word. Polysaccharides of the classical type are involved in bacterial cell wall structure (teichuronic acid and the group A polysaccharide of the streptococci, for instance), but little is known about their function. The reason for the failure of polysaccharides to take what the carbohydrate chemists have hoped would be their rightful place in biochemistry and biology is not hard to find. Many of these substances appear to have the general function of composing the matrix in which the fibrous materials of biological support systems are embedded. It has long been realized that the matrix substance is at least as important in determining the final structure as is the nature of the fibers, but the chemical interactions resulting in the production of the matrix gel have proven difficult to unravel. Consequently, interpreting the biological function of these molecules at the molecular level has been equally difficult.

In this review I confine myself to the structural polysaccharides, but much of what is known about reserve polysaccharides is pertinent to the discussion and indeed it would appear that there was a great overlap in the evolutionary processes that developed both types of material.

ARCHITECTURAL PLANS OF CELL WALLS AND OTHER SUPPORT STRUCTURES

It would be a mistake to allow the success of the research on structure and function of bacterial cell walls to leave the impression that they represent the central architectural theme in biological support structures. In fact they are one extreme variation on the central theme. The general plan involves a fibrous substance embedded in an amorphous support matrix. The fibrous material resists tension and the matrix resists compression. Evolution has experimented with many different chemical substances for both the fibrous phase and the matrix. The fibrous material that came out on top in the plant kingdom was the polysaccharide cellulose, while the winner in the animal line of evolution was the glycoprotein, collagen. Along the way many other substances were tried: a variety of fibrous polysaccharides in algae, chitin in fungi, yeast, and arthropods (3, 4), and so on. The bacterial cell wall is an extreme variation of the use of chitin as the fibrous phase. In this case the fibers are cross-linked by covalent linkages to form one enormous molecule. Similar experimentation went on with respect to matrix materials. It is from the matrix materials of plants that many classical polysaccharides have been isolated and it is in this form that they exert their biological function. In animals the matrix is often largely mucopolysaccharide (5) although not always so. In the arthropods, for instance, it is largely protein (3).

It appears that the nature of the matrix is more important in determining the mechanical properties of various plant and animal support structures than is the fibrous phase. For this reason it has been necessary for evolution to produce a sophisticated array of matrix materials whose properties can be finely adjusted by modulation of their chemical structure, and this modification must be under close genetic control. Thus in the arthropods the chitinous exoskeleton varies in consistency from very hard and inflexible in certain applications to quite soft and pliant in others. The organism controls this by altering the protein matrix by various means, including the process of tanning (3). Similarly, higher plants must have a great degree of latitude in the flexibility of different structures and this is provided by varying the chemical nature of the polysaccharide matrix material. When greatest stiffness is required, the nonpolysaccharide material, lignin, is incorporated into the system (4, p. 45–46). These two cases are examples of convergent evolution occurring at the molecular level in two widely separated parts of the evolutionary scheme. A particularly striking example of this phenomenon is seen when one compares Wharton's Jelly, which occurs in the human umbilical cord (5), with the mucilage of the seed of the mustard plant (6): Wharton's Jelly consists of collagen fibers colloidally dispersed in a mucopolysaccharide matrix, whereas the seed mucilage consists of cellulose fibers colloidally dispersed in a polysaccharide matrix of the pectin type (6). Similar architecture, similar mechanical properties, but very different chemical bases.

Thus the role of many conventional polysaccharides is to serve as the fibrous and matrix materials in support structures of plant and animal organisms. The mode of their interaction determines the mechanical nature of the support structure and must be a process of considerable chemical sophistication. Since the interaction appears in many cases to involve noncovalent bonding between polysaccharide molecules, the interpretation of this particular biological function at the molecular level lies in this domain of chemistry. There have been significant recent advances in attempts to deal with this complex problem and we appear to be well on our way to an understanding of it.

POLYSACCHARIDE COMPOSITION OF PLANT CELL WALLS

Fibrous Phase

One can easily gain the impression that the fibrous polysaccharide in plant cell walls is invariably the β-1,4-glucan, cellulose (7, 8). However, many algae use polysaccharides other than cellulose for their fibrous component (9, 10) and even in higher plants certain specialized tissues use materials other than cellulose. Thus the cell walls of certain seeds, e.g., ivory nut, date (11), and coffee bean (12, 13) contain as their major fibrous component a mannan. The grape uses callose (β-1,3-glucan) in certain structures (14). It would appear that there was a whole range of experimental fibrous materials during the evolution of the plant kingdom, beginning with peptidoglycan in blue-green algae (15); chitin in fungi (16) and yeast (17), although a few fungi use cellulose (18); a range of polysaccharides, including cellulose, in algae (19–21); and finally cellulose, for most purposes, in higher plants (22). Chitin is particularly interesting in that it spans the plants and animal kingdoms but does not rise very high in either. There appear to be three major fibrous support systems in nature: cellulose in higher plants, collagen in animals, and chitin at lower evolutionary levels in both.

The algae are interesting because during their evolution there was major experimentation with fibrous polysaccharides as support materials. As early as 1894, Correns (23) pointed out that cell walls of the algal genus *Caulerpa* gave no tests for cellulose, callose, or chitin and he postulated that they must utilize a new and unknown substance. Events have proven him correct and sixty years later it was shown to be β-1,3-xylan (10). In 1952 Nicolai & Preston (24) reported attempts to utilize X-ray diffraction patterns of cell walls as a means of classifying algae. In 1961, while extending this work, Frei & Preston (10) observed that cell wall material from certain Siphonales (*Bryopsis, Caulerpa, Penicillus, Udotea,* and *Halimeda*) gave an X-ray diagram quite different from that of cellulose. They showed that this new diagram was due to the β-1,3-xylan that Mackie & Percival (25) and Iriki et al (26) had isolated from *Caulerpa* and that the latter authors had also isolated from *Bryopsis, Udotea,* and *Halimeda*. Thus these plants are using a β-1,3-xylan in place of a β-1,4-glucan as their fibrous support material. Frei & Preston went on to show that *Codium, Acetabularia, Dasycladus,* and *Batophora* use a mannan for their fibrous support that is identical to the β-1,4-mannan from ivory nut (10). Some algae use a mixture of two fibrous polysaccharides in their cell walls. *Halicystis* walls contain both the xylan mentioned above and cellulose (10), and the freshwater alga *Hydrodicyton* contains both cellulose and a mannan (27). Finally the red alga *Porphyra* (order bangiales) contains xylan in the cell wall proper and mannan in a "cuticle" that is situated on the outside of the cell wall (10).

There can be no doubt that the period of evolution of the algae was a time of great experimentation to determine which of the available polysaccharides was the best fibrous support material. Cellulose, xylan, and mannan (and perhaps others) were tried singly and in all possible combinations of two. Preston's group has already shown that investigations in the algal cell wall area are particularly important because the range of material available improves the chances of interpreting the functions of these fibrous polysaccharides at the molecular level.

Matrix Materials

In the higher plants the total collection of matrix polysaccharides is classified into a group of acidic polysaccharides called pectins plus a heterogeneous group of neutral polysaccharides called hemicelluloses. A high proportion of all pectin preparations is made up of chains of β-1,4-polygalacturonic acid. These chains contain varying amounts of L-rhamnose and varying degrees of branching. The carboxyl groups are esterified to different degrees with methanol [0–85% of the available carboxyl groups (28)] and the degree of branching, the amount of rhamnose, and the degree of esterification are all controlled by the plant to adjust the texture of the matrix and therefore the mechanical properties of the plant tissue. In addition, pectin preparations contain various proportions of nonuronide sugars, some in the form of separate polysaccharides. L-Arabinose, D-galactose, L-rhamnose,

D-xylose, L-fucose, D-glucose, 2-O-methyl-L-fucose, and 2-O-methyl-D-xylose all have been isolated from pectic substances (29). The proportions of the nonuronide sugars are again variable from preparation to preparation, indicating that the plant controls cell wall texture by this means. The hemicelluloses are composed of a complex mixture of polysaccharides consisting of xylans, mannans, galactans, glucomannans, galactoglucomannans, and arabinogalactans (30). The composition of cells with respect to these polymers is variable, indicating still another means of controlling cell wall texture.

Finally, the seaweeds (algae) contain an interesting group of matrix poly-saccharides, including fucoidan (sulfated L-fucose polymer), the various carrageenans (D- and L-galactose polymers with various residues masked by derivatization), porphyran (similar to the carrageenans but with a different system of masking), and finally the classic gel-forming substance, agar. Agar is composed of D- and L-galactose and D-glucuronic acid and, like the other seaweed polysaccharides, has masked residues along the chain. The seaweed matrix poly-saccharides are of particular interest because our first solid information as to the chemical basis of the organization of matrix polysaccharides in general comes from studies of their interactions (see below). Some clear ideas on how these inter-actions are controlled have also emerged.

Clearly, the matrix polysaccharides are sophisticated materials and play a sophisticated role in the organization of cell walls and other support structures. Since analogous materials with similar functions occur in the animal kingdom, information gained from the seaweed polysaccharides has direct application to animal systems.

CONFORMATIONS OF POLYSACCHARIDES IN SITU AND IN VITRO

Fibrous Materials

CELLULOSE AND CHITIN   Cellulose (β-1,4-glucan) was one of the first biological substances to be investigated by X-ray diffraction (31). The unit cell proposed by Meyer & Misch in 1937 (32) indicates that the conformation of cellulose is that of a flat ribbon. More recently it has been proposed that this ribbon is folded into higher orders of structure (33–35), the most sophisticated conformation being that of Manley (35), who proposed that the ribbon was wound into a helix. Investigations of the arrangement of the cellulose chains over the whole surface of the cell have yielded very interesting results. Preston & Astbury in 1937 traced the orientation of the cellulose chains over the whole surface of a cell of the alga *Valonia* by means of X-ray diffraction (36, 37). This is the largest known single cell and can reach the size of a pigeon egg. This investigation, which must be regarded as a masterpiece in cell topology, revealed that all cellulose chains began and terminated at definite poles of the cell. One array was a helix that wound slowly around the cell, beginning at one pole and terminating at the other. A second array was composed of a series of meridians cutting across the turns of the helix, but in every case beginning and terminating at the same poles. Thus cellulose, in this organism, clearly has a conformation with elements of secondary, tertiary, and quaternary structure. The secondary structure is the Meyer-Misch ribbon, the tertiary structure is the folding of this ribbon, and the quaternary structure is an array of these folded cellulose chains that extends over the whole surface of the cell. A large factor in determining this conformational complexity is the association between cellulose and the polysaccharides of the matrix (hemicelluloses and pectins), and this complicated structure is the result of polysaccharide-polysaccharide interaction between what are regarded as conventional polysaccharides. (Some convention!)

Chitin has a formal structural resemblance to cellulose [it is a β-1,4-poly-(N-acetylglucosamine)] and appears to function in a similar fashion. It is particularly interesting since it clearly serves as an alternative to cellulose at the lower evolutionary levels of the plant kingdom and as an alternative to collagen at the corresponding levels of the animal kingdom. It too has secondary, tertiary, and quaternary structure and is involved in sophisticated structures clearly based on protein-chitin interactions in animals and polysaccharide-chitin interactions in plants. The interpretation of these structures at the molecular level will be interesting because they involve combination between the same fibrous material and two very different types of matrix material. Figure 9 in (3) will serve to show the incredible detail that nature can incorporate into a chitinous membrane.

XYLAN IN ALGAL CELL WALLS   As mentioned above, Preston and his co-workers (see 9 for a discussion) discovered that many algae use fibrous polysaccharides other than cellulose in their cell walls. Further research into the structure of those cells using β-1,3-xylan has proven particularly interesting. In their original paper on this subject, Frei & Preston (10) came to the conclusion that the xylan chains in these walls were arranged in the form of helices. On the basis of examination with polarization microscopy, electron microscopy, and X-ray diffraction, it was concluded that the microfibrils in these walls were arrays of double helices of xylan in hexagonal packing (38). This arrangement accounts well for the difference in mechanical properties of xylan and cellulose fibers and also for several aspects of fiber structure as seen under the electron microscope. The authors postulate that the polysaccharides callose, yeast glucan, paramylon, and laminarin might also possess a helical structure. Atkins et al (39) made a more sophisticated investigation of the xylan in the wall of the alga *Penicillus dumetosus* using X-ray diffraction, infrared absorption, and model building and came to the conclusion that the fibers were in fact made up of three xylan chains intertwined to form a three-stranded helix. Each single helix had six xylose residues per turn in a pitch of 18 Å. Polarized infrared spectra are in accord with the suggested conformation and support the proposed bonding force that causes the triple helix to form and stabilize, i.e., a cyclic triad of hydrogen bonds formed between the hydroxyls at C-2 of the xylose residues, one from each individual chain. Model building, together with the X-ray pattern, indicates that the helix is probably right-handed and that water is stoichiometrically bound on its surface (39, 40).

This complex conformation is, of course, strikingly reminiscent of the triple helix of collagen, the major support material in the animal kingdom, and is another example of convergent evolution at the molecular level—similar architecture of support materials at two widely separated points on the evolutionary scale and based on an entirely different chemistry.

Matrix Materials

CARRAGEENANS AND OTHER NATURAL SULFATED POLYSACCHARIDES   The sulfated polysaccharides constitute an important group of matrix materials, but they have an oddly limited occurrence in nature. They are found in marine algae and in animals but apparently not in bacteria or higher plants. Rees has drawn attention to the fact that this whole group of materials has a common structural theme (41), a repeating disaccharide unit structure masked by derivatization of some of the residues in the sequence. Some representatives of the group definitely have sophisticated helical conformations, both in the solid state and in solution, and model building experiments indicate that this could be a common feature of the whole group. Even more interesting is the observation that their gel-forming ability can be explained on the basis of their conformations and we thus appear to be well on the way to an interpretation of the function of matrix materials at the molecular level.

The carrageenans, a family of gel-forming polysaccharides that occur in marine algae of the class rhodophycaceae, have considerable commercial importance as food additives. They are composed of D-galactose units linked alternately α-1,3- and β-1,4- and, typically, have certain units along the chain derivatized by sulfation or 3,6-anhydride formation. The various carrageenans are assigned Greek letter prefixes and the differences in their structures involve the nature of the derivatization. Their primary structures have been adequately discussed (30, 42) and are not discussed here as this would only confuse the main issue, which is the nature of their conformations. The most interesting of these molecules from the standpoint of conformation are κ- and i-carrageenans. X-ray diffraction studies on oriented fibers of these polysaccharides show that both are double helices (43) with a complete turn of the single helix having three disaccharide residues in 24.6 Å (κ) and 26.0 Å (i). In i-carrageenan the second chain is displaced exactly half a pitch from the first. This extremely important work indicates that matrix polysaccharides, too, possess sophisticated conformations. Rees and his group further show that an appreciable fraction of this conformation is retained in solution and that the gel-forming ability of the polysaccharide can be interpreted on this basis. The conformation is broken up in a controlled fashion by the nature of some of the masked residues and it is apparently by this means that the plants regulate cell wall texture. Model building studies indicate that the animal sulfated polysaccharides, chondroitin 4-sulfate, chondroitin 6-sulfate, dermatan sulfate, and keratan sulfate, may well have this same conformation (44). This, of course, means that studies on the carrageenans have a good chance of useful application in the animal kingdom. Atkins et al (45) have investigated oriented films of chondroitin 6-sulfate by X-ray diffraction and, more recently, Atkins & Laurent have investigated films of chondroitin 4-sulfate, dermatan sulfate, and heparan sulfate by the same means (46). They conclude that all these substances have helical conformations. Agar, the classical gel-forming polysaccharide, is not a sulfated polysaccharide but it has many structural features in common with the carrageenans. Although there appears to be no definitive information on its conformation, Rees (42) believes that it too is helical, and if this is so the helical structure must also be broken up by the masking groups.

GLYCURONANS   The two polyuronans whose conformations have been subjected to the most investigation are alginic acid and the pectic substances. Both are important matrix materials; they are also polyelectrolytes and bind ions in a specific manner. Thus they involve ions in the matrix structures of cell walls and other support materials.

Alginic acid is produced by the brown seaweeds (phaeophycaceae) and occurs in the cell wall as its calcium and magnesium salts. The sodium salt is of considerable industrial importance (30). Similar materials have been found in bacteria (47, 48). The alginic acids are polymers of D-mannuronic acid and L-guluronic acid joined by β-1,4- and α-1,4-linkages respectively. Current ideas suggest that they are largely co-polymers made up of blocks of mannuronic acid, guluronic acid, and mannuronic and guluronic acids in alternating sequence (49). Their sequences are quite variable, according to their source, and all indications are that the plants control cell wall texture by controlling the sequence (see below). Atkins et al (50) have shown that alginic acid preparations give one of two different X-ray diagrams and they were able to correlate one of these with stretches of β-1,4-linked polymannuronic acid and the other with stretches of α-1,4-linked polyguluronic acid. Further investigations by Atkins et al (51) have shown that the polymannuronic chain has a flat ribbon-like conformation similar to that of cellulose, while the polyguluronic chain has an entirely different rigid, rod-like conformation. Rees and his co-workers have shown, by measurements of circular dichroism, that Ca^{2+} in calcium gels of alginic acid is specifically associated with the carboxylate n orbitals of contiguous α-L-guluronate residues (52). It had been previously shown that alginate gel stiffness increased with increasing content of L-guluronate (53).

The pectic substances in cell walls of higher plants function in a manner similar to alginate in seaweeds. They too are polyelectrolytes with a D-galacturonan backbone, and their gels involve metal ions. It would appear that they are more sophisticated in their function than the alginates, which is not surprising in view of the relative points of occurrence on the evolutionary scale. Fibers of pectin and sodium pectate were investigated as early as 1945 by X-ray analysis and it was concluded that the molecules had a helical conformation (54). Rees & Wight (55) have extended these studies by model building computations and conclude that the galacturonan backbone is in the conformation of a right-handed, threefold helix; there is no possibility of it forming double helices. Residues of L-rhamnose, which occur in all pectin preparations, form kinks in the regular helical chain, thus providing one means whereby the plant controls the gel texture of pectic substances (see below). Grant et al (56) investigated the interaction between pectin gels and divalent metal ions by means of circular dichroism. They propose an interesting structural model for the cooperative binding of ions observed in both pectic substances and alginates. The polysaccharide chains associate into assemblies with interstices into which cations may fit (the "egg box" model). The conformation of the polysaccharide chains controls the size of the cavities in the egg box, which in turn controls which metal ions are complexed. The known binding strengths of various sized cations can be accounted for to a surprising degree on the basis of this model.

Finally, hyaluronic acids are polymers made up of alternating units of N-acetyl-glucosamine and glucuronic acid, the N-acetylglucosaminyl units being linked β-1,4- and the glucuronyl units being linked β-1,3-. These substances are important matrix materials in the animal kingdom and combine structural features of the polyuronans with those of the fibrous support material, chitin. Atkins and his co-workers have investigated hyaluronic acid by X-ray diffraction (57, 58) and find that it is a helix with three disaccharide units per turn. Some interesting comparisons between this structure and that of the polyuronans are discussed below.

THE ROLE OF POLYSACCHARIDE CONFORMATION IN BIOLOGY

Rees & Scott (59) have pointed out that homopolymers of the pyranose forms of glucose, galactose, mannose, xylose, and arabinose have conformations that are very restricted by steric forces alone. In fact they can all be classified into one of four characteristic shapes: Type A–extended and ribbon-like; Type B–helical and flexible; Type C–rigid and crumpled; and Type D–very flexible and extended. Most fibrous support materials fall into Type A and most matrix materials fall into Type B.

Fibrous Materials

The extended ribbon-like shape characteristic of Type A polysaccharides lends itself both to efficient packing and strong bonding between chains. Thus it is not surprising that most fibrous support polysaccharides fall in this class. Among the polysaccharides that can be constructed from the available monosaccharides, cellulose and chitin form the most rigid structures and it is therefore not surprising that they are the most important structural polysaccharides in nature. It is interesting to observe the consequences of moving away from the cellulose-chitin type structure. The hydrogen atom adjacent to the glycosidic linkage in these structures is axial. Structures in which it is equatorial [mannan, xylan, and poly(mannuronic acid)] have a considerably less rigid, more flexible chain and can adopt other conformations. The triple helix of β-1,3-xylan is a case in point (39, 40). Irregularities in sequence have the same effect: the insertion of glucuronic acid residues into the chitin structure to form hyaluronic acid and the insertion of β-1,3 linkages into the cellulose structure to form oat and barley gums are examples of this effect (14, 57).

The picture that emerges is that the options open to nature in the form of poly-saccharide material that will pack efficiently and aggregate strongly are extremely limited and amount, in fact, to cellulose and chitin.

Matrix Materials

The major characteristic of these materials is their ability to form gels and it is through this property that they function in nature. It has long been recognized that

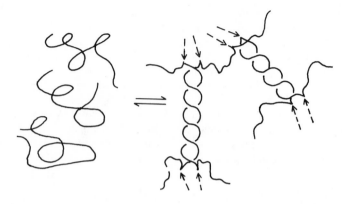

Figure 1 Rees' model for the molecular basis of gel formation by the carrageenans and other matrix polysaccharides of similar conformational type. As a warm solution of the polysaccharide is cooled, those sections of the chain that have unbroken stretches of the repeating disaccharide unit combine into double helices. This regular structure is broken up by kinks in the chain caused by derivatized monosaccharide units (indicated by the dotted arrows). The whole then forms a network that entraps solvent to form the gel structure.

a gel-forming substance must possess two structural characteristics: it must be a long molecule and the molecules must associate at certain points referred to as "junction zones." Association of the molecules by connection at the junction zones then creates a network that can entrap solvent and form a gel (Figure 1). How effective this can be is made plain from the fact that agar gels containing more than 99% water can be prepared. Thus, the simplest way to prepare a gel from a polysaccharide would be to introduce a limited number of connections between chains by means of covalent cross-linkages. This is, of course, what is done in the preparation of the gel filtration medium Sephadex, which is a cross-linked dextran. Nature has done the same thing in the case of peptidoglycan. However, it is evident both from the chemical structures of the materials involved and the reversibility of formation of the gels on cooling and heating that most naturally occurring matrix materials do not gel because of covalent cross-linking.

Some time ago, Rees and his co-workers suggested that the seaweed matrix materials, κ- and ι-carrageenan, produce gels by forming double helices (43). As mentioned earlier, these polysaccharides have a regular disaccharide repeating sequence masked by derivatization of some residues in the chain. Stretches of the regular sequence form helices that have been shown to exist both in the solid state and in solution (42). These helical segments interact with similar segments in other chains to form a double helix. These stretches of double helix, then, are the junction zones. Some of the masked monosaccharide units in the chain form kinks in the regular helical structure and these distribute the junction zones along the chain (Figure 1).

The chemical details of the kinking phenomenon, are worth pursuing. One of the members of the repeating disaccharide unit is 3,6-anhydro-D-galactose (the other is D-galactose 4-sulfate) and the regular alternation of these two residues in the chain results in a helical conformation. Replacement of one of the 3,6-anhydro units in the chain by D-galactose 6-sulfate reverses the conformation of the pyranose ring to the other chair form and the result is a discontinuity in the chain, which Rees refers to as a kink (Figure 1). The most recent presentation of Rees' persuasive views (60) should be consulted for details. He points out, on the basis of model building, that the gelling properties of an array of matrix materials may well depend on this same process. These are: κ-, ι-, and λ-carrageenans, agar, chondroitin, chondroitin sulfate, dermatan sulfate, keratan sulfate, and hyaluronic acid (44). This is an impressive array of materials distributed across a broad segment of the evolutionary scale. Since Rees' publication, Atkins' group has shown that hyaluronate, chondroitin sulfate, and dermatan sulfate exist in helical conformation in the solid state (45, 46, 57, 58).

The breadth of application of the Rees mechanism merits a brief review of the evidence supporting it. In the first instance it was based on the double helical nature of the carrageenans in the solid state as determined by X-ray diffraction (43). It was felt that these conformations, at least to a degree, were retained when the polymers passed into solution and this was supported by observations of optical rotation during the process of gelation and melting (41). One prediction from this model is that cutting the polysaccharide chain at the kink points would leave the junction zones intact to form double helices but would destroy the polysaccharide "arms" entrapping solvent in the gel (Figure 2, compare with Figure 1). Rees has succeeded in testing this prediction. The polysaccharides κ-carrageenan and agarose can be dekinked by means of the Smith degradation and in both cases the dekinked material behaves precisely as predicted. It will not gel but measurements of optical rotation indicate the formation of double helices at lower temperatures which melt as the temperature is raised (61). A further prediction is that if a nongelling

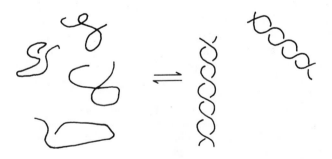

Figure 2   The result expected if a solution of dekinked polysaccharide were cooled in a fashion similar to that shown in Figure 1. The formation of double helices is expected (and in fact observed), but there will be no gel formation expected (and in fact not observed) since solvent entrapment is impossible.

Figure 3 Rees' picture of the interaction between dekinked polysaccharide and galactomannan that accounts for restoration of gel-forming capacity. The double helix of the dekinked polysaccharide (carrageenan) is shown at C and the galactomannan chains at G.

polysaccharide could be found that would bond with the dekinked helical material, then mixtures of the two should gel, even though neither gels alone under the same conditions. The added polysaccharide, when it formed the association, would simply replace the arms removed by the Smith degradation. To verify this prediction Rees investigated the interaction between dekinked κ-carrageenan and dekinked agarose and a series of naturally occurring polysaccharides known collectively as the galactomannans (60). These materials have a β-1,4-mannan backbone with single unit side branches of α-1,6-D-galactose. The proportion of galactose varies according to the source of the galactomannan and structural evidence indicates that the galactose side chains occur in blocks. A great deal of evidence indicates that the mannan chain between the blocks would have a type A ribbon-like conformation (51, 59, 60, 62). These stretches of ribbon conformation would be expected to interact with helices while the galactose-substituted stretches would be expected to replace the arms lost in the dekinking (Figure 3). These predictions are also fulfilled; a mixture of the two materials gels under conditions in which neither will gel alone. A further prediction is that this phenomenon will not occur if one uses a galactomannan with such a high proportion of galactose that the ribbon conformation stretches are negligible. Such material is available and will not gel when mixed with the dekinked material.

Rees has extended his ideas to the mode of gel formation by the pectic substances and by alginic acid. This involves participation by metal ions as well as association between the polysaccharide chains. Again the model accounts nicely for much of what is known about the properties of these materials. For instance, Hauge & Smidsroed observed that the guluronic acid chains in alginic acid are unique among the polyuronans in the strength of binding metal ions (63). This is predictable from the model, the polyguluronate chains having a conformation more favorable to binding than either polymannuronate or polygalacturonate (56). Further, polyguluronate binds Sr^{2+} preferentially to Ca^{2+} (63), whereas polygalacturonate shows no such preference. This, again, can be accounted for on the basis of the conformations of the two chains (52). Finally, as mentioned earlier, Rees' group has produced evidence that when Ca^{2+} complexes with alginate it specifically interacts with the n orbitals of the guluronic acid residues and this interaction is involved in the sol-gel transition (52).

Physiological Control of Matrix Properties by Alterations in Polysaccharide Conformation

Obviously the properties of matrix material in cell walls and other support structures must vary according to the function of the tissue in which it is located. There is reason to believe that they must also vary in the same cell at different periods of time. A rapidly growing plant cell, for instance, must have a more flexible wall than the same cell when it is mature. Thus there must be some mechanism, subject to close control, whereby the gelling properties of the matrix material can be controlled, both at the time of synthesis and when it is already in place. There is now evidence for the modulation of the conformation of matrix polysaccharides by means of alterations in the primary structure while the material is in situ.

It has been pointed out that the helical conformation of the carrageenan-type polysaccharides is due to stretches of a repeating disaccharide sequence, one member of which is 3,6-anhydrogalactose. This regular structure is kinked by replacement, at the kink points, of 3,6-anhydro sugar units by the corresponding 6-sulfate. It has been known for some time that a facile chemical reaction (treatment with base) converts galactose 6-sulfate to 3,6-anhydrogalactose. The reaction goes readily when the galactose 6-sulfate is bound in a polysaccharide. There is no reason why this reaction should not be catalyzed enzymatically; if it occurred while the polysaccharide was in situ, it would have a profound effect on the stiffness of the plant. It would remove kinks, increase the percentage of helical structure and therefore the proportion of junction zones, and thus increase the rigidity of the gel. One way, then, that a seaweed could increase the overall stiffness of its structure to adapt to increased mechanical stress in the environment would be to dekink a portion of its matrix material. Rees and his co-workers have investigated the seaweed *Porphyra umbilicalis* in an attempt to determine whether such processes go on. This seaweed uses the polysaccharide porphyran as a major matrix material. It is related in structure to the carrageenans, a major difference being that the 6-sulfate and 3,6-anhydro units are derived from L- rather than D-galactose. Rees (64) has detected an enzyme activity in *Porphyra umbilicalis* that removes 6-sulfate from the polysaccharide with concomitant formation of 3,6-anhydro residues. The enzyme attacks high molecular weight material much more readily than smaller fragments, indicating that it was designed to act specifically on intact polysaccharide. Lawson & Rees (65) have identified a similar enzyme in *Gigartina stellata* that acts on the carrageenans, the matrix material in this species. Finally, Rees & Conway (66) have produced evidence in support of these ideas by a study of the composition of porphyran isolated from *P. umbilicalis* growing under different conditions of mechanical stress. The samples were very variable in their content of both L-galactose 6-sulfate and 3,6-anhydro-L-galactose, but the sum of these materials was essentially constant within experimental error, indicating that one was formed from the other after the chains had been synthesized. In all cases, plants collected from environments where they were exposed to strong wave action gave preparations higher in 3,6-anhydro sugar content, indicating that the plants adapted to stress by converting the 6-sulfate to the 3,6-anhydride.

Similar arguments have been made in connection with the function of the matrix material alginic acid. In this case the stiffness of the gel formed is a function of the relative proportions of mannuronic and guluronic acids in the polymer. As pointed out earlier, the guluronic acid blocks in this gel have a much stiffer conformation than those of mannuronic acid and, further, the guluronic acid complexes with metal ions to stiffen the gel even further. Mackie (67) has drawn attention to the possibility that the plant could control the flexibility of structure by altering the proportions of the two monomers. Frei & Preston (68) had earlier observed by X-ray diffraction that in a single plant, alginic acid high in guluronic acid occurred in rigid tissue, whereas that high in mannuronic acid occurred in flexible tissue.

Similar arguments can be presented for the pectic substances. In this case it is necessary that the texture of the matrix be extremely variable. The pectic substances in walls having a supporting function possess a structure so regular that it can be detected by X-ray diffraction (69). The texture of the gel varies from this extreme to that characteristic of a viscous liquid (6). There is good reason to believe that the growth rate of plant cell walls is controlled by the texture of their pectic gels, whose flow rate controls the extension rate of the cell wall as new cellulose fibrils are synthesized in the matrix (70). It is commonly found that plant tissues capable of rapid growth have highly branched pectic substances with relatively high concentrations of L-rhamnose residues (42). Rees' group has pointed out that L-rhamnose residues serve as kinks in the pectic substances. It is evident that the presence of these residues and of branched chains will tend to prevent the cooperative association demonstrated by Grant et al (56), and will thus loosen the structure of the gel.

The Genetic Control of Polysaccharide Structure

It is evident from the discussion above that the functions of conventional polysaccharides are much more sophisticated than is generally realized and these functions must certainly be under genetic control. Such genetic processes are just now being unraveled. Friis & Ottolenghi (71) have demonstrated that the binding of the dye alcian blue to yeast cells is under the control of a single gene, which mapped as being linked to the character *ural*. Dye binding was shown to be to the cell wall mannan and appeared to correlate with its phosphate content. Thus the dye binding might well be indicative of a distinct aspect of the cell wall mannan structure, in which case the work indicates that the cell wall polysaccharide structure is under specific genetic control. Spencer et al (72) used nuclear magnetic resonance to follow the segregation of a terminal α-1,3-mannosyl residue in the mannan of a cross between two *S. cereviseae* strains, only one of which possessed this unit. They found that the presence of this structural feature was controlled by a single gene. More recently, (73–75), Ballou and his co-workers have used immunochemical methods to follow the genetic control of mannan biosynthesis with striking success. Two wild-type strains of *S. cereviseae* which differed in the structure of their cell wall mannans were crossed: one type was characterized by the presence of mannotetraose side chains, while the other had mannosylphosphorylmannotriose side chains in place of the mannotetraose units. Genetic analysis led to the conclusion that this difference was controlled by a single dominant gene involved in the synthesis of an α-1,3-mannosyl transferase that adds terminal α-1,3 units to the α-1,2-linked mannotriose side chains in the mannan. This gene prevents the expression of mannosylphosphate incorporation on the cell surface (see below). The authors mapped the gene precisely and it occurs on chromosome V, tightly linked to the character *ura3* and the centromere. They also succeeded in producing a series of mutants defective in cell wall mannan synthesis by treating the cells with a mutagen. The selection of mutants was based on their failure to react with an antiserum specific for the mannotetraose chain of the mannan, which has the structure αMan(1,3)αMan(1,2)αMan(1,2)Man. Two classes of mutants were obtained: one deficient in the formation of the α-1,3-mannosyl transferase that completes the side chain and the other deficient in one of two α-1,2-mannosyl transferases. One of these enzymes links mannose to the backbone of the mannan and the other links mannose to the first side-chain unit. It was concluded that these mutations involve structural genes for the various mannosyl transferases and not regulatory genes that control the expression of a particular mannan chemotype. Finally, a mutant deficient in a mannosylphosphate transferase was observed and the presence of the mannosyl-phosphate group was shown to determine the binding of alcian blue. Strains possessing both mannosylphosphate transferase and α-1,3-mannosyl transferase incorporate mannosylphosphate into the cell wall but not at the surface where it can react readily with the dye. Thus the gene controlling the production of α-1,3-mannosyl transferase also has an influence on the overall architecture of the mannan in the cell wall.

In summary, we are at last commencing to understand the relationship between chemical structure and biological function in the "classical" polysaccharides. Their primary structures are by no means as nondescript as has been commonly assumed and these dictate well-defined conformations in solution, resulting in noncovalent bonding between molecules. This, in turn, determines biological function. The principles involved seem generally applicable and will give us a much better under-standing of the function of support structures in all living things. The investigation of polysaccharide conformation, its biological implications, and its genetic control are obviously fruitful future fields of research and we owe a real debt to those workers who have opened up this area.

Literature Cited

1. Heath, E. C. 1971. *Ann. Rev. Biochem.* 40: 29–56
2. Glaser, L. 1973. *Ann. Rev. Biochem.*
3. Jeuniaux, C. 1971. *Compr. Biochem.* 26C: 595–632

42: 91–112

416 KIRKWOOD

4. Rogers, H. J., Perkins, H. R. 1968. *Cell Walls and Membranes,* 114–60. London: Spon
5. Bailey, A. J. 1971. *Compr. Biochem.* 26B: 297–423
6. Grant, G. T., McNab, C., Rees, D. A., Skerrett, R. J. 1969. *Chem. Commun.* 805–6
7. Bonner, J., Galston, A. W. 1952. *Principles of Plant Physiology.* San Francisco: Freeman
8. Bonner, J., Varner, J. E. 1965. *Plant Biochemistry.* New York: Academic
9. Rogers, H. J., Perkins, H. R. See Ref. 4, pp. 114–34
10. Frei, E., Preston, R. D. 1961. *Nature* 192: 939–43
11. Meier, H. 1958. *Biochim. Biophys. Acta* 28: 229–40
12. Wolfrom, M. L., Laver, M. L., Patin, D. L. 1961. *J. Org. Chem.* 26: 4533–35
13. Wolfrom, M. L., Patin, D. L. 1964. *J. Agr. Food Chem.* 12: 376–77
14. Stone, B. A. 1958. *Nature* 182: 687–90
15. Rogers, H. J., Perkins, H. R. See Ref. 4, p. 114
16. Rogers, H. J., Perkins, H. R. See Ref. 4, pp. 153–60
17. Moreno, R. E., Kanetsuna, F., Carbonell, L. M. 1969. *Arch. Biochem. Biophys.* 130: 212–17
18. Foster, J. W. 1949. *Chemical Activities of Fungi.* New York: Academic
19. Preston, R. D., Cranshaw, J. 1958. *Nature* 181: 248–50
20. Sponsler, O. L. 1930. *Nature* 125: 633–34
21. Roelofson, P. A. 1959. *The Plant Cell Wall.* Berlin: Borntraeger
22. Rogers, H. J., Perkins, H. R. See Ref. 4, pp. 68–89
23. Correns, C. 1894. *Ber. Deut. Bot. Ges.* 12: 355
24. Nicolai, E., Preston, R. D. 1952. *Proc. Roy. Soc. B* 140: 244–74
25. Mackie, I. M., Percival, E. E. 1959. *J. Chem. Soc.* 1151–56
26. Iriki, Y., Suzuki, T., Nishizawa, K., Miwa, T. 1960. *Nature* 187: 82–83
27. Kreger, D. R. 1960. *Proc. Kon. Ned. Akad. Wetensch. Ser. C* 63: 613 (Cited in Ref. 10)
28. Doesburg, J. J. 1973. *Phytochemistry,* ed. L. P. Miller, 1: 270–96. New York: Van Nostrand
29. Worth, H. G. J. 1967. *Chem. Rev.* 67: 465–73
30. Towle, G. A., Whistler, R. L. See Ref. 28, 1: 198–48
31. Nishikawa, S., Ono, S. 1913. *Proc. Phys. Math. Soc. Japan* 7: 131–38
32. Meyer, K. H., Misch, L. 1937. *Helv.*

33. Mühlethaler, K. 1969. *J. Polymer Sci. C* 28: 305–15
34. Sarko, A., Marchessault, R. H. 1969. *J. Polymer Sci. C* 28: 317–31
35. Manley, R. S. 1964. *Nature* 204: 1155–57
36. Preston, R. D., Astbury, W. T. 1937. *Proc. Roy. Soc. B* 122: 76–97
37. Preston, R. D. 1952. *The Molecular Architecture of Plant Cell Walls.* London: Chapman and Hall
38. Frei, E., Preston, R. D. 1964. *Proc. Roy. Soc. B* 160: 293–327
39. Atkins, E. D. T., Parker, K. D., Preston, R. D. 1969. *Proc. Roy. Soc. B* 173: 209–21
40. Atkins, E. D. T., Parker, K. D. 1969. *J. Polymer Sci. C* 28: 69–81
41. Rees, D. A., Steele, I. W., Williamson, F. B. 1969. *J. Polymer Sci. C* 28: 261–76
42. Rees, D. A. 1969. *Advan. Carbohyd. Chem.* 24: 267–332
43. Anderson, N. S., Campbell, J. W., Harding, M. M., Rees, D. A., Samuel, J. W. B. 1969. *J. Mol. Biol.* 45: 85–99
44. Rees, D. A. 1969. *J. Chem. Soc. B* 217–226
45. Atkins, E. D. T., Gaussen, R., Isaac, D. H., Nandawar, V., Sheehan, J. K. 1972. *J. Polymer Sci. B* 10: 863–65
46. Atkins, E. D. T., Laurent, T. C. 1973. *Biochem. J.* 133: 605–6
47. Lenker, A., Jones, R. S. 1964. *Nature* 204: 187–88
48. Carlson, D. M., Matthews, L. W. 1966. *Biochemistry* 5: 2817–22
49. Haug, A., Larsen, B., Smidsroed, O. 1966. *Acta Chem. Scand.* 20: 183–90
50. Atkins, E. D. T., Mackie, W., Smolko, E. E. 1970. *Nature* 225: 626–28
51. Atkins, E. D. T., Mackie, W., Parker, K. D., Smolko, E. E. 1971. *J. Polymer Sci. B* 9: 311–16
52. Morris, E. R., Rees, D. A., Thom, D. 1973. *Chem. Commun.* 245–46
53. Smidsroed, O, Haug, A. 1972. *Acta Chem. Scand.* 26: 79–88
54. Palmer, K. J., Hartzog, M. B. 1945. *J. Am. Chem. Soc.* 67: 2122–27
55. Rees, D. A., Wight, A. W. 1971. *J. Chem. Soc. B* 1366–72
56. Grant, G. T., Morris, E. R., Rees, D. A., Smith, P. J. C., Thom, D. 1973. *FEBS Lett.* 32: 195–98
57. Atkins, E. D. T., Phelps, C. F., Sheehan, J. K. 1972. *Biochem. J.* 128: 1255–63
58. Atkins, E. D. T., Sheehan, J. K. 1972. *Nature New Biol.* 235: 253–54
59. Rees, D. A., Scott, W. E. 1971. *J. Chem. Soc. B* 469–79
60. Dea, I. C. M., McKinnon, A. A., Rees,

D. A. 1972. *J. Mol. Biol.* 68 : 153–72
61. McKinnon, A. A., Rees, D. A., Williamson, F. B. 1969. *Chem. Commun.* 701–2
62. Palmer, K. J., Ballentyne, M. 1950. *J. Am. Chem. Soc.* 72 : 736–41
63. Haug, A., Smidsroed, O. 1970. *Acta Chem. Scand.* 24 : 843–54
64. Rees, D. A. 1961. *Biochem. J.* 81 : 347–52
65. Lawson, C. J., Rees, D. A. 1970. *Nature* 227 : 392–93
66. Rees, D. A., Conway, E. 1962. *Biochem. J.* 84 : 411–16
67. Mackie, W. 1971. *Biochem. J.* 125 : 89P
68. Frei, E., Preston, R. D. 1962. *Nature* 196 : 130–34
69. Roelofson, P. A., Kreger, D. R. 1951.

J. Exp. Bot. 2 : 332–43
70. Frey-Wyssling, A., Mühlethaler, K. 1965. *Ultrastructural Plant Cytology.* Amsterdam : Elsevier
71. Friis, J., Ottolenghi, P. 1970. *C. R. Trav. Lab. Carlsberg* 37 : 327–41
72. Spencer, J. F. T., Gorin, P. A. J., Rank, G. H. 1971. *Can. J. Microbiol.* 17 : 1451–54
73. Antalis, C., Fogel, S., Ballou, C. E. 1973. *J. Biol. Chem.* 248 : 4655–59
74. Raschke, W. C., Kern, K. A., Antalis, C., Ballou, C. E. 1973. *J. Biol. Chem.* 248 : 4660–66
75. Ballou, C. E., Kern, K. A., Raschke, W. C. 1973. *J. Biol. Chem.* 248 : 4667–73

PEPTIDE SYNTHESIS

×852

M. Fridkin and A. Patchornik
Department of Organic Chemistry, The Weizmann Institute of Science, Rehovot, Israel

CONTENTS

INTRODUCTION

Peptide synthesis has progressed dramatically since the synthesis of oxytocin was achieved by du Vigneaud and his colleagues (1). The obvious question which arises by looking at milestones such as ACTH (2), glucagon (3, 4), secretin (5–7), or calcitonin (8–10) is whether the methodology which yielded this impressive list of biologically active peptides is still applicable to reach the next target, namely, chemical synthesis of proteins. In this chapter we try to evaluate the different approaches currently used for the synthesis of peptides in view of the aims of this field.

This review gives information on recent improvements of classical methodology including new coupling methods, new protecting groups, detection and estimation of racemization, new purification techniques, and new synthetic strategies. Further and more detailed information on these subjects can be found in a number of books (11, 12), reviews (13–24), and the proceedings of several symposia (25–31). For extensive accounts on earlier achievements and problems in peptide synthesis, the books of Schröder & Lübke (32), Bodanzsky & Ondetti (33), and Greenstein & Winitz (34) are valuable sources.

AIMS OF PEPTIDE SYNTHESIS

Prior to the discussion of the current methodology of peptide synthesis, it is perhaps appropriate to point again to the objectives in this field. The major aims are well established:

1. Verification of structures of naturally occurring peptides as determined by degradation techniques.

2. Studies on the relationships between structure and activity of biologically active peptides and proteins, aiming to establish the molecular mechanisms by which these compounds act in their natural environment.

3. Synthesis of peptides for medical purposes.

The usefulness of peptide synthesis for proof of structure has been shown in the elucidation of the sequence of the naturally occurring nonapeptide bradykinin (35). Unambiguous synthesis is of exceptional value in proving the structure of complicated biologically active polypeptides, particularly because closely related, but wrong, sequences might have significant activity.

Structure-activity relationship studies of biologically active peptides (e.g. hormones) are directed, among other aims, toward the understanding of how such compounds initiate a chain of chemical reactions eventually leading to the specific physiological response. Techniques of affinity labeling (36, 37) and affinity chromatography (38, 39) and information about the chemistry of the active peptides may shed light on problems such as the shape or size of receptors or the nature of the hormone-receptor complex, and may even lead to the isolation of the receptor (40). Chemical studies in the field of peptide hormones may also lead to the design of antagonists which in some cases (e.g. gastrin, angiotensin II, and LRH) may be more valuable than the hormones themselves. Useful information on structure-activity relationships of peptide hormones may be found in several excellent review articles and references (41–44).

The stimulating studies on the role in binding and catalysis of individual amino acids in semisynthetic derivatives of bovine pancreatic ribonuclease (e.g. 45–47) illustrate the importance of peptide synthesis in understanding the mode of action of enzymes.

The role of synthetic peptides in medicine is rather specialized, oxytocin, vasopressin, angiotensin, and an active fragment of ACTH being the most important representatives. The releasing hormones TRH and LRH are presently being used for diagnostic purposes and, together with related peptides (e.g. GRH), may be of significant therapeutic value in the future. The increasing use of synthetic insulin will depend, of course, on the ease of its synthesis, the identity with the natural hormone, and the availability of the latter.

STRATEGIES OF PEPTIDE SYNTHESIS

In the following section we evaluate currently used strategies of peptide synthesis in view of the abovementioned targets. Categorization of the various synthetic approaches is difficult because they share many common features. Classification of strategies could be based on several logical lines: synthesis in homogeneous (solution) or heterogeneous (liquid/solid-phase) systems; maximal and minimal protection of side-chain functional groups; methods involving isolation and non-isolation of intermediate derivatives; and fragment and stepwise condensations (33). The last classification, in our opinion, best organizes the different methodological approaches and is used in this review. We feel, however, that such division is rather arbitrary and therefore, we try to discuss interrelations of the various techniques.

Fragment Condensation: This approach is represented in the synthesis of many biologically active peptides such as ACTH (2), glucagon (3, 4), secretin (7), and calcitonin (8–10), or in the progress in synthesis of ribonuclease T (48, 49) and yeast iso-1-cytochrome c (50). The synthetic strategy is based on the synthesis of several fragments of the polypeptide to be synthesized, followed by coupling of the segments to the final desired product.

Stepwise Synthesis: The strategy is based on the repetitive addition of single N-protected amino acids, usually in molar excess, to a growing amino component, starting generally from the C-terminal amino acid of the chain to be synthesized. Synthetic techniques, such as the solid-phase synthesis (51), polymeric reagents (52, 53), stepwise synthesis using active esters (5, 6, 54, 55) and symmetrical (56, 57) and mixed (58, 59) anhydrides fall into this category.

N-Carbonic Anhydrides: Most fragment syntheses in fact use stepwise procedures for the preparation of fragments. We feel, however, that the combination of the N-carbonic anhydride method (60, 61) (a stepwise-stereotype technique) with the fragment condensation approach deserves a separate discussion in view of its stimulating achievements (62–67).

Fragment Condensation

This approach is usually started by architectural segmentation of the synthetic target, P (see Scheme 1). Fragments ($F_1 ---- F_i$) are chosen so that they preferably have proline or glycine (which are not prone to racemization) at their C terminal (33), but unfortunately these two amino acids are not equally spaced in all sequences. The fragments are first synthesized by an entirely stepwise manner or by condensation of shorter fragments, followed by their linking using various techniques.

Suppression of racemization during C-terminal activation and reasonable yields are prerequisite for successful fragment condensations. Side-chain functional groups can be maximally (3, 4, 8, 9) or partially (48–50) protected during the synthetic procedures, and are removed at the end of the synthesis.

The azide method (68, 69) ($X = N_3$, see Scheme 1) has been used traditionally as the safest way of fragment condensation using N-acylpeptide hydrazides [prepared via protected hydrazides (70–72) or via esters (68) as intermediates]. The low degree of racemization when coupling is done properly and minimal side-chain protection are the most useful features of this method. The work of K. Hofmann and associates on the synthesis of ribonuclease T is a classical example of the use of the azide method throughout a complicated synthesis (48, 49). The main drawbacks

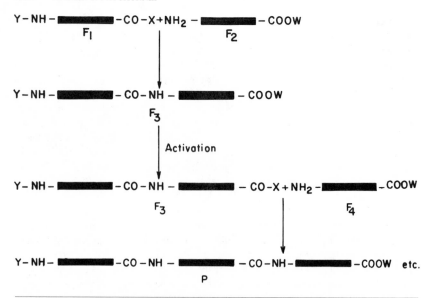

Scheme 1   General scheme of peptide synthesis by the fragment condensation approach.
$-CO-X$ = activated form of carboxylic terminal; W = carboxyl terminal protection;
Y = amino terminal protection.

associated with the azide method are the numerous side products emerging from transformation of the azide group into amide and from intramolecular Curtius rearrangements of the azides to isocyanates (73). Separation of the main product from urea-type derivatives, resulting from such isocyanates on reactions with amines, may often be a difficult or even impossible task. In addition, yields in the azide coupling may be relatively low, especially when long fragments are being coupled.

Racemization during activation of N-acylpeptides with N,N'-dicyclohexyl-carbodiimide (DCCD) prohibited the application of this useful reagent for condensation of fragments with C-terminal amino acids other than glycine or proline. Racemization could, however, be markedly decreased or totally suppressed by addition of N-hydroxysuccinimide (I, HONSu) (74, 75), 1-hydroxy-1H-benzotriazole (II, HOBt) (76, 77), or 3-hydroxy-4-oxo-3,4-dihydro-1,2,3-benzotriazine (III) (78) to the reaction mixture. Formation of highly reactive esters between the additives and the carboxylic component, which in turn react rapidly with the amino component, competes successfully with racemization. Formation of a complex (IV) between the amino component and the 1-hydroxybenzotriazole active ester was proposed to explain the properties of these useful derivatives (79). The DCCD plus HONSu method has been recently applied to the synthesis of glucagon (3, 74) and calcitonin (80). The DCCD plus HOBt method has been successfully applied to the synthesis of ACTH (81) or sequence 1–34 of human parathyroid hormone (82).

The convenient synthetic manipulations of these procedures in addition to the relatively high yields and low degree of racemization are very promising features for the synthesis of long and complicated peptides. Minor drawbacks of the methods are the few rare side reactions (78, 83, 84) and occasional difficulties in removing dicyclohexylurea from intermediate peptides.

In view of the experimental data available in the literature, one can conclude that classical fragment condensation can be applied safely, using currently available techniques, to the synthesis of peptides of the region of 30–50 amino acid residues. The main advantages of this approach derive from the fact that fragments are purified and well characterized before coupling to form larger fragments, and the relatively easy separation of the reaction mixture at the end of the reactions. It is especially true for synthesis with minimal side-chain protection when purification techniques can be used in aqueous solvents. Easy purification is due, of course, to large differences in size and composition, and hence in physicochemical properties between desired products, starting materials, and excess reagents. It must, however, be borne in mind that purification operations might be very laborious and time-consuming but purity of final products is often excellent.

Few problems are associated with the fragment condensation approach. The low solubility of large fragments, especially when maximal side-chain protection is being used, will result in a reduced concentration of the reacting species and will cause a gradual decrease in reaction rate as the synthesis proceeds (a reduced concentration will also be an obvious result from the increase in molecular weight). Additional and similar effects might result from increasing steric hindrance. It must also be remembered that under these conditions unimolecular side reactions, such as the azide-isocyanate rearrangement, might be relatively favored.

Modification of side-chain protection may, however, increase the solubility of large fragments (85). Moreover, the insoluble fragments might potentially be used as their own supports for a solid-phase synthesis, provided they swell and become highly solvated in appropriate solvent (85).

Direct coupling of a carboxylic and an amino function of simple peptide derivatives using diphenylphosphoryl azide as a reagent was described recently (86). This reaction proceeds via conversion of the free carboxyl into its azide. Evaluation of its utility for the coupling of large fragments is expected.

Stepwise Peptide Synthesis

Stepwise synthesis usually starts at the C terminal of the peptide chain to be synthesized and adds one N^α-protected amino acid at a time in a manner shown in Scheme 2.

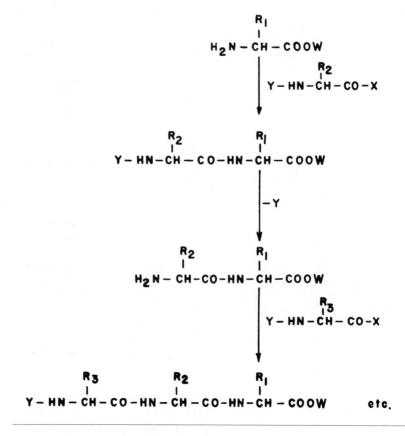

Scheme 2 Stepwise peptide synthesis. Y = α-amino protecting group; W = carboxyl protecting group; X = −OC$_6$H$_4$NO$_2$-*p*; –O–Ⓟ (see polymeric reagents listed in Scheme 7); −OCOCH(R)NHY; −OCOOCH(CH$_3$)$_2$.

This strategy was proposed by several investigators (54, 55, 68, 87) and has been used in combination with fragment condensation (e.g. 88) or in the so-called entirely stepwise synthesis (5, 6). The strategy gained strong momentum with the development of the ingenious solid-phase technique by Merrifield (51) (see Scheme 3) which led to a major breakthrough in peptide synthesis.

Using the stepwise approach for peptide synthesis one might anticipate the following advantages. First, racemization could be markedly reduced throughout synthesis if the α-amino protecting group (Y, see Scheme 2) and the activating moiety of the acylating agent (X, see Scheme 2) are well selected. Second, yields are as a rule relatively high and the reaction approaches completion when the acylating agents are used in excess (89). Easy and quantitative removal of reagents from the products are prerequisite for success of such an approach.

In a fragment-condensation synthesis use of one component in excess is generally less useful, primarily because solubility of fragments and hence concentration is rather limited, and of course because of economical considerations.

SYNTHESIS IN HOMOGENEOUS PHASE

Active esters   The benefits associated with the repetitive peptide synthesis were outlined earlier. The use of active esters, mainly p-nitrophenyl esters ($X = OC_6H_4NO_2$-p, see Scheme 2) (90), is perhaps the most prominent proof of the feasibility of such strategy in peptide synthesis, with oxytocin (54, 55) and ACTH 25–39 pentadecapeptide (88) as representatives. Synthesis of the hormone secretin (27 amino acid residues) entirely in a stepwise manner using protected amino acid active esters as reagents demonstrates a successful confrontation of the method with a challenging synthesis of a complicated polypeptide (5, 6). The acylating agents were applied initially in slight excess (10–20%) and were gradually increased with progress of synthesis up to threefold excess. Coupling reactions occurred in DMF and were practically finished within a few hours, but the reaction mixtures were kept at room temperature overnight to ensure acylation of the last traces of unreacted α-amino components. Products often precipitate directly from solution and are isolated simply by dilution of the mixture, usually with ethylacetate. Excess of active esters and the p-nitrophenol produced are soluble in this solvent and are removed when the insoluble protected peptide is washed with ethylacetate.

The synthesis proved to have the following advantages: yields were consistently high (average of 94% for the lengthening of the chain by one amino acid); racemization was very low; no significant change in reaction rates occurred with progress of synthesis; and, above all, protected intermediates obtained as solids could be analyzed, characterized, and purified. The final product was indistinguishable from natural secretin when compared by physicochemical and biological assays as well as by degradative analyses using proteolytic enzymes. The main drawback of the techniques was loss of products resulting from thorough washings with ethylacetate and minor impurities due to some residual active esters. The analogous synthesis of the same peptide on Merrifield's polymer, using p-nitrophenyl esters as reagents, unfortunately could not be completed since ammonolysis of the completed product, terminating with C-valine, could not be performed at that time.

With increasing length, the synthesized amino component, i.e. the growing peptide chain, becomes nearly insoluble, whereas the acylating agents are soluble in the reaction solvent. This is, in fact, a solid-phase synthesis without a polymeric carrier (85). The efficiency of such reactions, as mentioned before, depends largely on solvation of the peptide in the reaction solvent and will probably be reduced with increasing size or complexity of the peptide. We feel, however, that more hydrophobic protecting groups for side-chain functions might enable convenient synthesis of larger peptides ($\sim$ 50 amino acid residues) by this and related methods.

Anhydrides   Two additional repetitive procedures, using an excess of symmetrical (56, 57) [$X = OCOCH(R)NH$–Y, see Scheme 2] and mixed carbonic (58, 59) [$X = -OCOOCH(CH_3)_2$, see Scheme 2] anhydrides as acylating reagents, were reported in the literature. In the former technique, excess reagents are removed by reaction with an excess of N-(2-aminoethyl)-piperazine and the resulting basic amide removed from the neutral product with citric acid–H_2O and $NaHCO_3$–H_2O extractions. In the second method quantitative hydrolysis of excess anhydrides is effected by reaction with $KHCO_3$–H_2O at 0° for 30 min. The resulting N-protected amino acid K-salts are removed by washing the product with H_2O. The symmetrical anhydrides have also proved valuable in connection with the solid-phase synthesis (91). The advantages of the mixed carbonic anhydride method, high yields ($\sim 96\%$ for each coupling step), fast reactions, low degree of racemization, and convenient chemical manipulations, were demonstrated by Tilak (58, 59) in the repetitive stepwise synthesis in solution of a complex-protected nonapeptide glucagon fragment

$$
\begin{array}{ccccc}
Bu^t & OBu^t & Bu^t & Bu^t & \\
| & | & | & | & \\
Z\text{–Thr–Ser–Asp–Tyr–Ser–Lys–Tyr–Leu–Asp–OEt} \\
| & | & | & | & \\
Bu^t & Bu^t & BOC & OBu^t &
\end{array}
$$

Peptide intermediates could be checked and analyzed throughout synthesis and the products could thus be purified or used as such for next couplings. Beyerman (92) has further demonstrated the usefulness of the so-called REMA (Repetitive Excess Mixed Anhydride Method) synthesis by preparing human growth hormone sequence 1–10.

Two undesired side reactions might be associated with peptide synthesis (route 1, Scheme 3) by the mixed anhydride method. First, wrong-side aminolysis of the

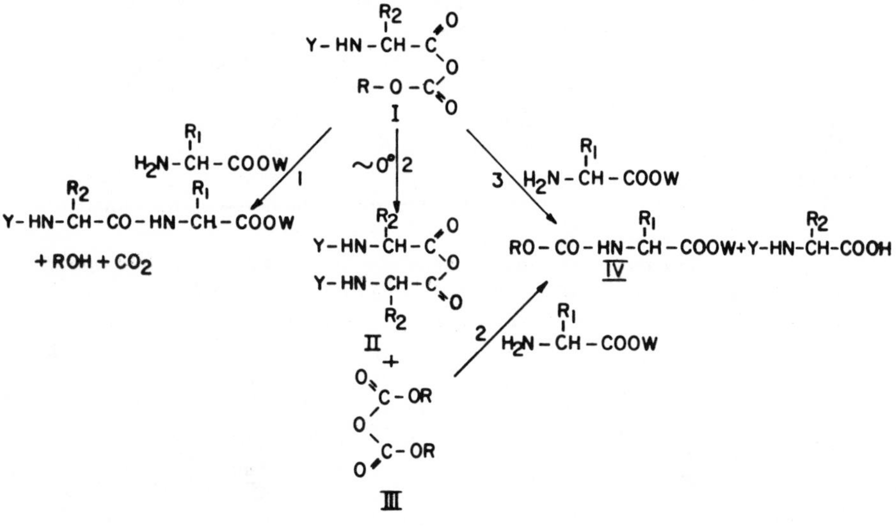

Scheme 3 Application in synthesis and side reactions of mixed carbonic anhydrides. W = carboxyl protecting groups; Y = α-amino protecting group.

anhydride (I) will block the amino component irreversibly (route 3). This reaction depends largely upon steric factors and solvents and can be avoided by using, in relevant cases, other techniques for coupling (92). Second, anhydride (I) can disproportionate (92) at $\sim 0°$ to yield symmetrical anhydride (II) and a dialkylpyrocarbonate (III). The latter will irreversibly block the amino component (route 2). This side reaction is negligible at temperatures below $-15°$.

It seems that the RENA method, provided the right precautions are taken, is reliable for the synthesis of biologically active peptides and their analogs, as well as for structural proof, in the range of 10–30 amino acid residues. Applicability of the method to the synthesis of longer polypeptides should of course be checked by confrontation with a concrete synthesis.

Handle techniques   Separation of excess acylating agents from the products could also be effected by using handle techniques. In such an approach, the C-terminal of the growing peptide chain is protected by a basic function (93, 94), (see Scheme 2)

$$W = -OCH_2 - \bigcirc - N=N - \bigcirc - N \begin{matrix} CH_3 \\ CH_3 \end{matrix} \qquad -OCH_2 - \bigcirc - N$$

allowing separation of products on cation exchange columns or by extraction into aqueous acid solutions. The 4-picolyl ester (94) method has been applied very successfully to the preparation of various peptides such as angiotensin (95) and bradykinin (96) and appears to be very promising.

SYNTHESIS IN HETEROGENEOUS PHASE

Solid-phase peptide synthesis   This revolutionary technique, launched by Merrifield (51), is characterized by the stepwise addition of soluble N^α-protected amino acids to the amino component which is attached through its carboxyl function to a suitable insoluble polymeric carrier (see Scheme 4). To approach quantitative acylation reactions, soluble reagents are used in a large excess and are easily removed from the polymer by filtration and washings at the end of the coupling reaction. The final product is cleaved by various techniques from the polymer and is subjected to conventional purification procedures.

The attractive features of the solid-phase synthesis, namely, simplicity, speed, and possible automatization or mechanization, led in a relatively short time to the preparation of an impressive list of biologically active peptides. Moreover, with the new synthetic tool in his hands, the chemist or biochemist has started to challenge problems such as the total synthesis of proteins or the relation between the primary structure and biological activities of these molecules.

Looking at the remarkable success achieved in the solid-phase synthesis of many short and medium-size peptides, e.g. bradykinin (97), angiotensin (98), oxytocin (99), valine-gramicidine A (100), B, and C (101), or α-MSH (102), one might accept the method as a rapid and efficient way for the synthesis of peptides in the range of 10–20 amino acid residues. In the same range the method may also be applicable for proving proposed structures or in structure-function relationship studies.

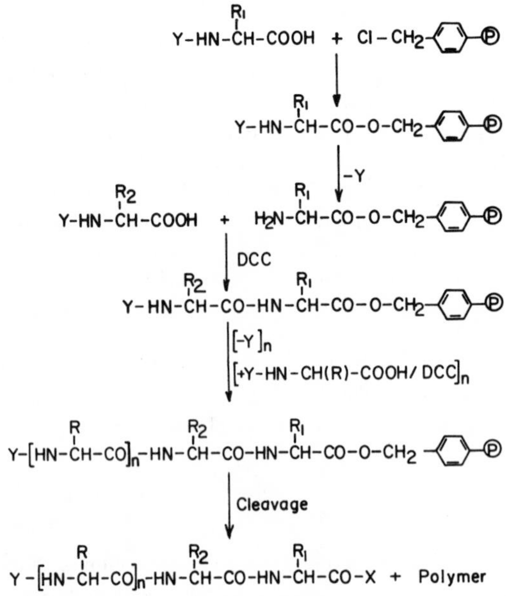

$$Y-HN-\overset{R_1}{\underset{|}{CH}}-COOH \; + \; Cl-CH_2-\langle\!\!\!\!\!\bigcirc\!\!\!\!\!\rangle\!\!-\textcircled{P}$$

$$Y-HN-\overset{R_1}{\underset{|}{CH}}-CO-O-CH_2-\langle\!\!\!\!\!\bigcirc\!\!\!\!\!\rangle\!\!-\textcircled{P}$$

$$\downarrow -Y$$

$$Y-HN-\overset{R_2}{\underset{|}{CH}}-COOH \; + \; H_2N-\overset{R_1}{\underset{|}{CH}}-CO-O-CH_2-\langle\!\!\!\!\!\bigcirc\!\!\!\!\!\rangle\!\!-\textcircled{P}$$

$$\downarrow DCC$$

$$Y-HN-\overset{R_2}{\underset{|}{CH}}-CO-HN-\overset{R_1}{\underset{|}{CH}}-CO-O-CH_2-\langle\!\!\!\!\!\bigcirc\!\!\!\!\!\rangle\!\!-\textcircled{P}$$

$$\left| [-Y]_n \right.$$
$$\left[+Y-HN-CH(R)-COOH/\,DCC \right]_n$$

$$Y-\left[HN-\overset{R}{\underset{|}{CH}}-CO\right]_n-HN-\overset{R_2}{\underset{|}{CH}}-CO-HN-\overset{R_1}{\underset{|}{CH}}-CO-O-CH_2-\langle\!\!\!\!\!\bigcirc\!\!\!\!\!\rangle\!\!-\textcircled{P}$$

$$\downarrow Cleavage$$

$$Y-\left[HN-\overset{R}{\underset{|}{CH}}-CO\right]_n-HN-\overset{R_2}{\underset{|}{CH}}-CO-HN-\overset{R_1}{\underset{|}{CH}}-CO-X \; + \; Polymer$$

Scheme 4   Merrifield's solid-phase synthesis. Y = α-amino protecting groups; P = polymer; X = OH, NH_2, $NHNH_2$, or O-alkyl, when cleavage is effected by acidolysis, ammonia, hydrazine, or alcoholysis conditions, respectively.

However, the syntheses of more complicated polypeptides such as cobrotoxin (62 amino acid residues) (103), an active portion of bacterial nuclease (104), lysozyme (129 amino acid residues) (105), or even ribonuclease A (106), reveal some of the inherent difficulties of the solid-phase technique. It seems that unambiguous synthesis of long and complicated polypeptides still cannot be attained without modification of the currently used methodology.

Some of the built-in difficulties of the solid-phase technique are as follows:

1. Incomplete couplings and deprotections will lead to an accumulation of truncated peptides (peptide chains which are unavailable for further elongation) and of deleted sequences (truncated peptides which resume elongation) on the resin (107, 108). Separation of the desired product from these incorrect but closely related sequences is often impossible. Incorrect sequences might also be a result of various side reactions (109–113).

2. Nonspecific cleavage of side-chain protection can yield branching, for example, at ε-lysine sites (114).

3. As a result of racemization, correct sequences but with different configurations will be obtained.

4. Diacylamine formation (115, 115a), possibly followed by chain shortening (116) or chain lengthening (117), might emerge from using a large excess of activated N-blocked amino acid derivatives at each coupling step.

5. Decomposition or modification of certain amino acids may occur due to the drastic conditions often needed to remove the completed peptide chain from the resin.

An additional objective difficulty, particularly prominent in the solid-phase synthesis, is the inability of the currently available analytical procedures and criteria for purity to distinguish between complicated, closely related polypeptides and the lack of suitable purification techniques to separate such compounds.

As a result of the abovementioned difficulties, the preparation of high molecular weight peptides must frequently be followed by highly sophisticated purification techniques. Often the scientist must conclude his synthesis by saying that a product or products possessing similar properties to the desired peptide X have been synthesized, rather than that a peptide X was prepared chemically.

Numerous attempts have been made toward improving the solid-phase method and solving some of the difficulties involved.

1. The problem of incorrect sequences has naturally attracted much attention and has been attacked from several directions. Since coupling and deprotection reactions may not be quantitative under standard conditions, and since every individual peptide bond might create completely different synthetic problems, it is essential to monitor directly the course of each coupling reaction and, more importantly, to determine with high precision the amount of residual α-amino groups left after each coupling. In addition, the exact amount of α-amino groups liberated after each deprotection step should be known. The ideal monitoring technique would combine the following useful features: convenience in continuous operations, speed, simplicity, high sensitivity, and directness. Several methods based on titrimetric (e.g. 118–122), colorimetric (e.g. 123, 124), fluorometric (125, 126), C^{14} (127), or Cl^{36} (128) analysis have been developed and applied with only partial success to monitor solid-phase synthesis. The method of Brunfeldt and his associates (120–122), based on potentiometric titrations of the α-amino groups with perchloric acid, is very promising particularly in combination with automatic peptide synthesis.

Use of terminating agents (highly reactive acylating agents) for masking unreacted α-amino groups, and thus stopping the growth of wrong sequences, was originally proposed by Merrifield (51). The idea has been further evaluated (129–131) but it is still uncertain whether the amino groups which failed to couple will react with the terminating agents, especially in view of the findings of Bayer et al (107) and of Marshall et al (128).

Intensive attempts are being made to understand the sources for the formation of incorrect sequences. It is quite clear that gradual alteration of the physical character of the solid matrix facilitates formation of truncated and deleted sequences. Swelling and solvation of the polymeric carrier are continuously changing with the progress of synthesis due to the increasing contribution of the peptide to physical properties of the conjugate peptide-polymer. Similar effects might also result from the frequent changes of solvents with different polarity during the synthetic procedures. Peptide chains located at different sites in the polymer would therefore be subjected to effects of exposure and burial and might have different susceptibility toward incoming reagents and thus form incorrect sequences. An interesting

discussion of the outcomes of the dynamic solvation changes of the polymer matrix
is given in the important article of Marshall et al (128).

The use of rigid, macroporous polymers (132, 133) on the one hand, and
macroreticular polymers with low degree of crosslinking (134) on the other hand,
was proposed to avoid these environmental effects. Better swelling of the polymers
with lower degrees of crosslinking can lead to more efficient reactions, as was clearly
shown in two different syntheses of encephalitogenic peptide (135, 136). In another
interesting modification, silica and glass (137–139) (V, VI) or Teflon (140)-supported
carriers were used to facilitate surface-directed peptide synthesis. Incorrect sequences
due to diffusion effects could thus be reduced and in addition some technical
advantages like column synthesis or shorter reaction times are visualized.
Sheppard (141) has suggested the use of polymeric carriers whose solvation
properties would be similar to those of the growing peptide chain. Thus, solvation
changes with progress of synthesis might be suppressed. No doubt further investigation along the various abovementioned lines would help minimize formation of
wrong sequences.

$$\underline{V} \qquad \underline{VI}$$

2. Use of highly selective protecting groups for side-chain functions is of utmost
importance in the solid-phase technique, since purification of products is performed
only after cleavage of the completed product from the resin. Good protecting
groups should of course be stable during the synthetic manipulations and be
removed under mild conditions at the end of the synthesis. The use of protecting
groups which proved to be safe in other synthetic strategies should be evaluated
here with extra care since they are subjected to multiple treatments with reagents
like acids or amines. Numerous variations in the originally synthetic scheme of
Merrifield (142) have been recommended to minimize side reactions resulting from
improper protection. Some of the important and relevant investigations are
described below.

The benzyloxycarbonyl group, traditionally used in solid-phase synthesis for the
protection of the ε-amino group of lysine, is partially cleaved under the acidic
conditions for removal of N^α-Boc protecting groups, leading to branching at the
ε-amino sites (114). Use of the more acid-stable chloro (143, 144) and bromo
(145, 146) derivatives of the benzyloxycarbonyl group that were introduced and
investigated much earlier in connection with the conventional peptide synthesis
(147–149a), and the even more stable diisopropylmethoxycarbonyl group (150), has
been suggested to overcome this side reaction. The use of the isonicotinyloxy-

carbonyl group (151), which is very stable even under drastic acidolysis conditions and can be easily removed by zinc dust in 50% aqueous acetic acid, is also very promising for the same purpose.

Several groups have been recommended for protection of the imidazole ring of histidine to avoid use of sodium in liquid NH_3 [known to cause rupture of peptide bonds(152)] needed for removal of the long-used imbenzyl group. The use of variants such as imdinitrophenyl group, removable under mild thiolysis conditions (153); imtosyl group, removable by liquid HF (154); imBoc group, removable by liquid HF (155); and other groups (156, 157) for imidazole protection is now being evaluated.

The use of many sulfur-protecting groups of cysteine (i.e. *p*-methoxybenzyl, diphenylmethyl, trityl, ethylcarbamoyl or benzyloxycarbonyl) has been tested in order to substitute the commonly used S-benzyl group for a more easily removable group (158). The most interesting suggestion for sulfhydryl self-protection by using symmetrical cystine peptides has also been tested (159, 160). We feel that the 4-picolyl group (161), removable by electrolytic reduction, might be useful in this respect.

The HF-labile *m*-bromobenzyl (146) and 2,6-dichlorobenzyl (145, 162) have been suggested as stable protecting groups for the OH function of tyrosine. The latter group was found to reduce the severe intramolecular, acid-catalyzed benzyl rearrangement occurring during acidolysis of peptides containing O-benzyl tyrosine (162).

It was found (155) that some ornithine-containing peptides are generated when the guanidino group of arginine is protected by the nitrogroup. Such side reaction was not detected when the guanidino group was protected by the tosyl group.

Finally, the usefulness of the newly recommended protecting groups should be evaluated in the preparation of complicated polypeptides.

3. The problem of racemization during solid-phase synthesis did not attract much attention since the synthetic strategy is based on the safe stepwise building of the peptide chain. Indeed a proof of the absence of racemization in such synthesis was given by Bayer and his colleagues (163). In view of the findings of Jorgensen & Windridge (164, 165) that solid state couplings with Boc-His (Bz) and DCCD gave much higher degrees of racemization ($\leq 36\%$) than the corresponding solution couplings ($\leq 2.9\%$), it seems appropriate to check this vital aspect in a more detailed manner.

4. Dicyclohexylcarbodiimide has been traditionally used as a coupling agent for the chain-lengthening step in solid-phase synthesis since it combines high efficiency with simple technical manipulations. Overactivation, however, which may lead to the formation of undesired side products is a severe disadvantage of this useful reagent (115–117). Over-reaction could probably be minimized by applying a smaller excess of the incoming N-protected amino acids or by using moderately active coupling agents. Very promising results have recently been obtained using O-nitrophenyl esters (166) and symmetrical anhydrides of protected amino acids (91) as coupling agents. We feel that studies on other potentially useful derivatives such as N-hydroxysuccinimide esters (e.g. 167, 167a), 2-pyridyl thiolesters (168), or 2-pyridyl esters (169) should be extended.

5. Removal of the completed peptide chain from the polymer is often done by drastic acidolysis (anhydrous HF or HBr in trifluoro acetic acid). Under these conditions sensitive amino acids such as tryptophan, methionine, cysteine, or even tyrosine might be modified by destructive oxidation or by the generated benzylic cations. In addition, due to steric and swelling effects, the yield of the final cleavage step might be far from quantitative. Several new polymers have been designed to ensure milder and quantitative removal of products from the carrier as peptides with free carboxyls, esters, amides, or hydrazides (see Scheme 5). Use of such polymers may, however, require revision of some of the protecting groups presently used.

There is no doubt that as a result of the many methodological improvements suggested so far and the many more to come, the valuable approach of Merrifield will lead to better yields, more convenient synthetic manipulations, and will hopefully clear up some of the present synthetic ambiguities.

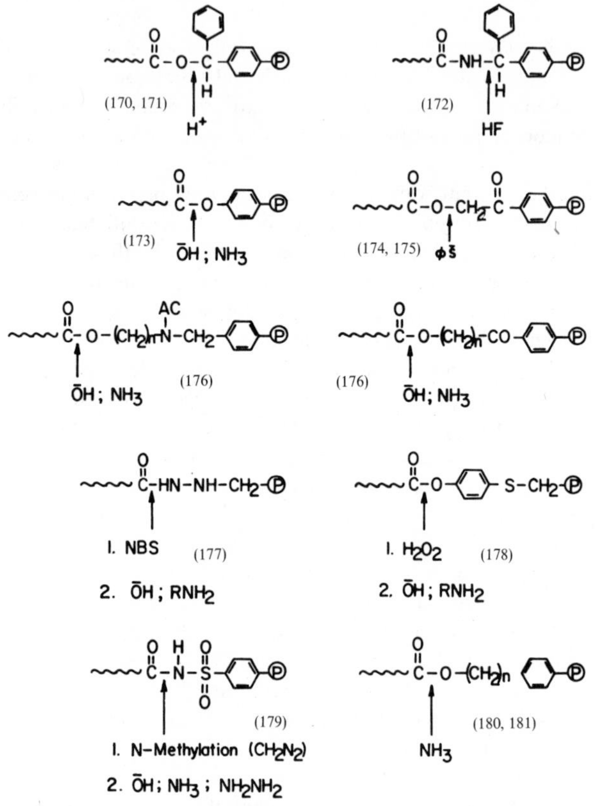

Scheme 5   New types of polymeric supports.

The soluble-polymeric-support approach of Shemyakin (182) regained interest due to the most promising work of Bayer and his associates with soluble, polyethylene glycol, polyvinyl alcohol, or polyvinylamine copolymerized with polyvinyl-pyrrolidone as supports (183, 184). Although very little detailed information is published, it seems that the syntheses carried out in homogeneous phase are highly efficient and rapid, probably due to diminishing diffusion and steric effects. Reagents can be applied in excess and easily separated from the polymer at the end of the reaction by dialysis or ultrafiltration. In addition, fragment condensations, suggested by several investigators for the solid-phase synthesis (e.g. 185–187), might be more efficient here because of reduced hindrance. Moreover, the progress of the reactions can be easily monitored, e.g. by ^{19}F-NMR technique (184, 188), and therefore be continued, repeated, or stopped at will. In the latter case, fragments might be removed from the carrier, analyzed, purified, and further condensed with other fragments in a conventional manner to form larger peptides. The major drawback of the technique is that cleavage of the final product, attached to the carrier via an ester bond from the support, is done by alkaline hydrolysis which is known to cause some racemization.

The idea of combining the useful features of the solid-phase and the classical fragment-condensation techniques was in fact proposed by Marshall & Merrifield (98). Anfinsen and his associates (189) explored a similar approach and described a procedure (190) for the preparation of N-protected peptide hydrazides, based on hydrazinolytic cleavage of the benzyl ester linkage anchoring the peptide chain to Merrifield's resin. Other investigators (174, 175, 178, 179, 191) have used modified polymers, usually containing labile polymer peptide bonds, to prepare various N-protected peptide derivatives. Wang & Merrifield have recently reported the synthesis of a t-alkyloxycarbonylhydrazide resin (VII) (192, 193) which could be used to prepare N-protected peptide hydrazides directly without the intervention of a cleavage step with hydrazine, and a t-alkyl alcohol resin (193, VIII), designed for preparations of N-protected segments with free carboxyl groups. We believe that such an approach, although not yet thoroughly explored in the synthesis of complicated peptides, is by far more promising than the entirely stepwise synthesis on a resin, since incorrect sequences and hence chemical ambiguity can be markedly reduced.

Polymeric reagents A new approach to facilitate organic syntheses was success-fully demonstrated by the use of polymers as chemical agents (22) according to the following general scheme

$$\text{P}\!-\!A_n + B \rightarrow \text{P}\!-\!A_{n-1} + AB$$

According to this scheme, the A residues, which are attached to the polymer P via chemically active bonds, are transferred during the chemical reaction to B with the formation of A–B. Using polymeric reagents in chemical reactions, one might anticipate the following advantages:

1. Increased yields and improved synthetic procedures as a result of the use of an excess of insoluble or soluble reagents which can be removed readily at the end of the reaction by filtration or by differential solubility of reagents and products.

2. Products of high purity obtained as a result of the technical advantages intrinsically involved in the novel synthetic procedures.

3. Continuous synthesis on columns and automatization of synthesis when insoluble polymeric reagents are employed.

The use of polymeric active esters in peptide synthesis (52, 53) represents a successful application of the polymeric reagents approach. Polymeric active esters can be prepared by the conversion of N-blocked amino acids or peptides into the corresponding polymeric esters according to Scheme 6.

$$\text{Y—HN—CH—COOH} + \text{HX—Ⓟ} \rightarrow \text{Y—HN—CH—CO—X—Ⓟ}$$
$$\underset{R}{\big|} \qquad\qquad\qquad\qquad \underset{R}{\big|}$$

Scheme 6   General scheme for preparation of polymeric active esters. $Y = \alpha$-amino protecting group; P = polymeric carrier; X = oxygen, sulfur, or nitrogen.

Potentially useful carriers, Ⓟ—XH, should contain functional hydroxy or thio groups to which the carboxylic terminal of the amino acid derivatives can be coupled to yield an active ester bond. Some of the polymers which were prepared and investigated are shown in Scheme 7.

The coupling between the polymer, Ⓟ–XH, and the carboxylic derivatives is carried out in appropriate organic solvent, using the conventional coupling agents for the synthesis of active esters, such as DCCD (52, 53, 198, 200) or mixed anhydrides (53, 197). At the end of the reaction, excess reagent is washed off the polymer, which is now ready for peptide synthesis. The use of polymeric active esters in peptide synthesis is illustrated in Scheme 2. The reagent, $\text{Y–HN–CH(R}_2\text{)–CO–O–}$Ⓟ, (2–3 equivalent) is coupled while suspended in an organic solvent such as CH_2Cl_2, $CHCl_3$, DMF, or CH_3CN, with an equivalent of the appropriate soluble amino acid ester containing a free α-amino group. The soluble N- and C-blocked dipeptide obtained is isolated by filtration of the polymer and evaporation of the reaction solvent. After selective removal of the N-blocking group, the coupling reaction can be repeated with an insoluble active ester of another N-blocked amino acid. Further repetition of this set of reactions leads obviously to elongation of the peptide chain and formation of a peptide with a predetermined amino acid sequence.

The experimental data available in the literature (e.g. 194, 198, 200, 205) show that yields in such syntheses are usually high (90–100% for each coupling step); products with high purity (often without purification of soluble intermediates) are obtained;

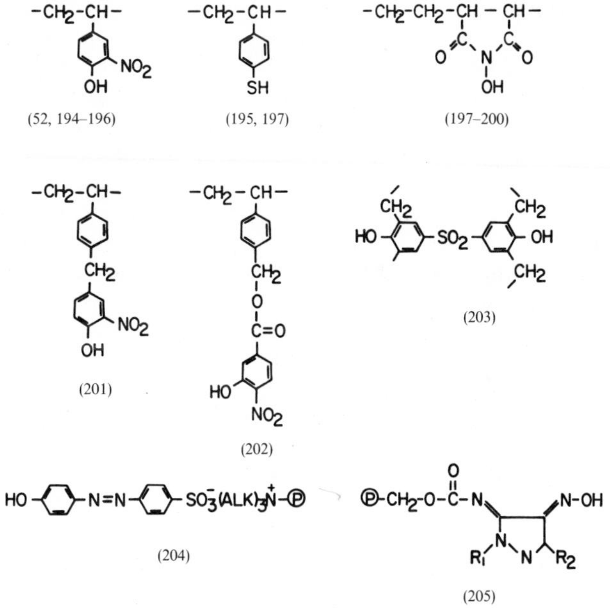

Scheme 7   Polymers used for preparation of polymeric active esters.

and a continuous synthesis on columns is possible (206). The method has been used successfully for the synthesis of many relatively simple peptides as well as for the synthesis of the naturally occurring nonapeptide bradykinin (194) and a heptapeptide corresponding to residues 159–165 of bovine carboxypeptidase A (200). The applicability of the method to the synthesis of longer peptides (15–30 amino acid residues), where difficulties might arise from low solubilities or steric effects, has not yet been evaluated. Alanylation of insulin and of poly-ε-benzyloxycarbonyl lysine (200) indicates, however, that such syntheses are feasible.

A technique related to the polymeric reagents approach is based on the use of polymers as condensing agents in peptide synthesis (22). Such reagents, when employed in excess, can also facilitate synthesis and ease purification of products. Several agents such as polymeric carbodiimides (207, 208), chlorosulfonated polystyrene (208), and polymer carrying N-ethoxycarbonyl-2-ethoxy-1,2-dihydro-quinoline functional group (poly-EEDQ) (209, 210), have been prepared and used for the synthesis of various short peptides. The method has not yet been applied to the synthesis of long peptide and cannot be evaluated at this stage.

Our feeling is that combination of the various repetitive stepwise syntheses described in this chapter with the fragment condensation approach and with improved methodology for coupling of large fragments will be the answer to the most challenging problems of peptide synthesis.

N-CARBOXYANHYDRIDES   A fruitful combination of the fragment condensation and the stepwise synthesis strategies was demonstrated by the stimulating work of the Merck group (60, 61). The use of N-carboxyanhydrides of amino acids (NCA) for controlled stepwise synthesis of peptidic fragments in aqueous media without isolation of intermediates is the attractive characteristic of this method.

NCA derivatives could be easily and inexpensively prepared (e.g. 60, 61, 211), and they react under strictly defined conditions instantaneously, often almost quantitatively, and with no racemization with the free α-amino group of amino acids or peptide derivatives to yield the corresponding elongated peptides. As shown in Scheme 8, the resulting product bears a free α-amino group that enables several repetitions of the coupling reaction using desired N-carboxyanhydrides.

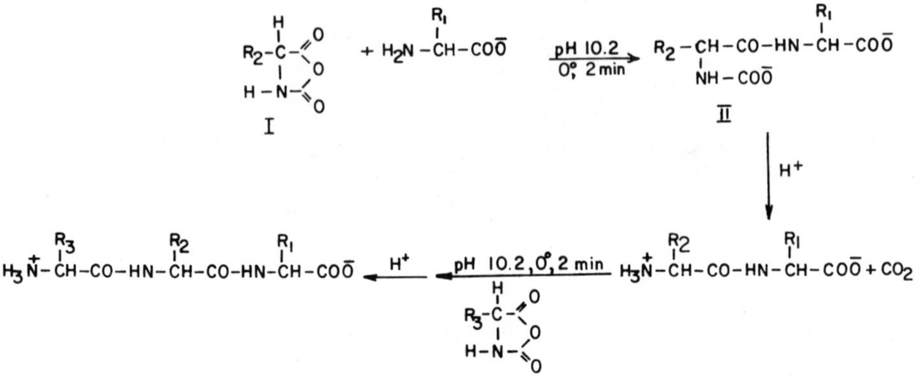

Scheme 8   Synthesis of a tripeptide by the N-carboxyanhydride method.

Polyfunctional amino acids, with the exception of lysine and cystine, are used without side-chain protection (60, 61). Due to some side reactions (see below), the repetitive synthesis must be interrupted after several steps and the product could be conveniently subjected to purification techniques in aqueous solvents.

Several difficulties, some of them technical, are associated with the NCA method. The pH must be carefully controlled during synthesis. At pH values below 10 the peptide-carbamate (II, see Scheme 8), a synthetic intermediate, tends to lose CO_2, free α-amino group is generated, and oligomerization (over reaction) takes place. Similar effects may occur at pH values higher than 10.5 due to some hydrolytic decomposition of the N-carboxyanhydride (I) or slow dissolution of the NCA in the reaction media. Additional minor complications might result from nonspecific reaction of NCA with side-chain functions and mainly with hydroxyl groups. Side reactions could, however, be minimized by using carefully developed techniques (60, 61, 212). Over-reaction could also be reduced by using the thio analogs (IX) of the N-carboxyanhydrides (2,5-thiazolidinediones) as the activated monomers (213,

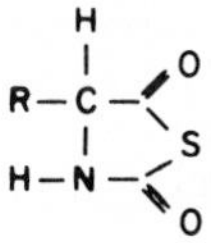

$$\begin{array}{c} \overset{H}{\underset{|}{R-C}} \\ \underset{|}{\overset{|}{H-N}} \end{array} \begin{array}{c} \diagup O \\ \diagdown S \\ \diagdown O \end{array}$$

$$\begin{array}{c} NH-CHR-CO \sim\sim \\ | \\ O=C-S^- \end{array}$$

$$\underline{IX} \qquad\qquad\qquad \underline{X}$$

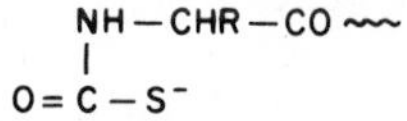

214). The stepwise synthesis proceeds in the same manner, but the thiocarbamates (X) obtained after each coupling step, as well as the thioanhydride itself (IX), are more stable toward basic hydrolysis with a consequent diminishing of byproduct formation. Unfortunately, the thioanhydrides do give some racemization.

The initial stepwise synthesis of relatively large fragments and subsequent

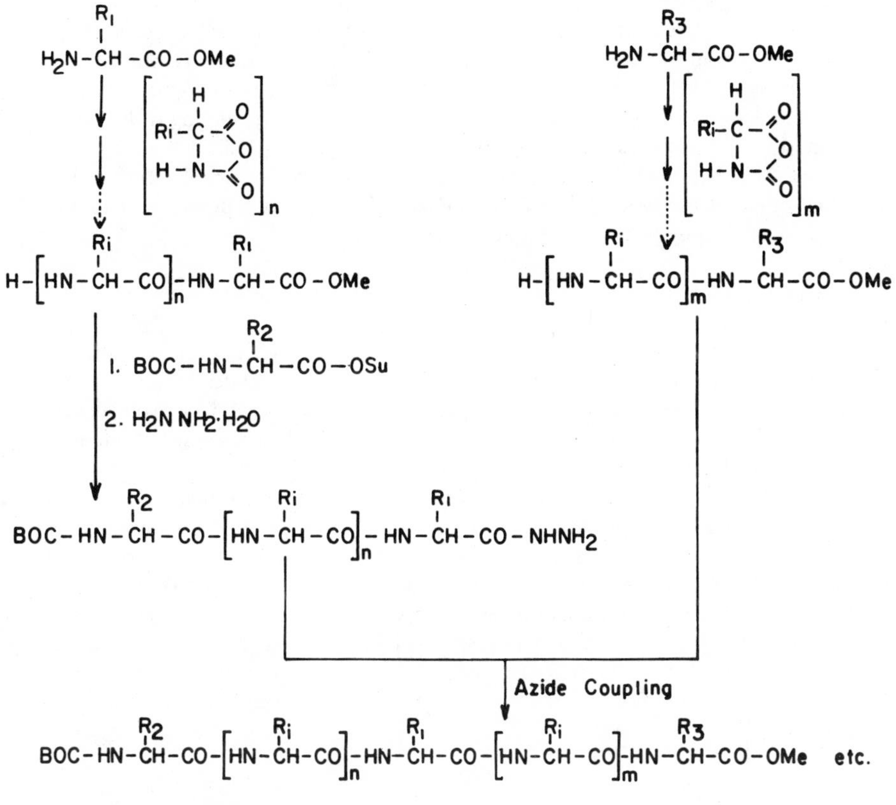

Scheme 9   Polypeptide synthesis via the N-carboxyanhydride method.

couplings of these fragments to form longer polypeptides (see Scheme 9) was the strategy used by the Merck group for the synthesis of the S protein of ribonuclease (62–67).

Fragments representing the amino acid sequence of the S protein were prepared by a combination of the N-carboxyanhydride method with the use of *t*-butyloxycarbonyl amino acid succinimide esters (167). The oligopeptide fragments (19 in all) were linked together after analysis and purification by the azide method (69), which was the only possible way to combine the stripped derivatives. After deprotection, purification, and oxidation, the final product had consistent, but rather difficult to ascertain, biological activity. It was obtained in minute amounts, which prevented full characterization. The major drawbacks of the present synthesis seem to result mainly from use of only the azide method for the linkage of the fragments. Typical side reactions (e.g. Curtius rearrangements) and characteristic diminishing of yields in the coupling of large fragments call for laborious and often inadequate purification techniques. Some modification of the present methodology—use of protected hydrazides for C-terminal residues (70–72); use of imidazole-protected NCA of histidine instead of the N-thiocarboxyanhydride analog which is sensitive to racemization (214); protection of OH-function in tyrosine, serine, and threonine (211); and above all, better understanding and hence better governing of the azide method—will undoubtedly lead to better yields and higher specific activities.

Partial protection of the fragments (especially carboxylic functions) will probably allow other methods to be used for fragments linkage, but purification of the products will consequently be more difficult.

CONCLUSION

We have tried to evaluate the different methodological approaches for the synthesis of peptides. The ability of the peptide chemist to prepare peptides of intermediate length using different synthetic approaches is now well established. Our strong belief, however, is that nonambiguous synthesis of complicated polypeptides will result only from combining the useful features of various methods presently used.

Several major aspects of peptide synthesis will probably attract much attention in future research in this field: thorough study of the coupling of large fragments will lead to the development of highly efficient techniques for this purpose; sensitive, general, and convenient methods for detection and estimation of racemization after each step of synthesis will be developed; above all, attempts will be made to develop suitable analytical techniques that will ease the struggle with complicated mixtures of synthetic high molecular weight peptides. Such techniques include, of course, sensitive and accurate tools for detection and separation of closely related peptides.

ACKNOWLEDGMENTS

The authors wish to express their deep gratitude to Prof. J. Rudinger for the very stimulating discussions and his most helpful suggestions.

Literature Cited

1. DuVigneaud, V. et al 1953. *J. Am. Chem. Soc.* 75:4879
2. Schwyzer, R., Sieber, P. 1963. *Nature* 199:172
3. Wünsch, E. 1967. *Z. Naturforsch. B* 22:1269
4. Wünsch, E., Wendlberger, G., Jaeger, E., Scharf, R. 1968. *Proc. 9th Eur. Peptide Symp. Orsay,* ed. E. Bricas, Amsterdam: North-Holland. 229 pp.
5. Bodanszky, M., Williams, N. J. 1967. *J. Am. Chem. Soc.* 89:685
6. Bodanszky, M., Ondetti, M. A., Levine, S. D., Williams, N. J. 1967. *J. Am. Chem. Soc.* 89:6753
7. Ondetti, M. A. et al 1968. *J. Am. Chem. Soc.* 90:4711
8. Sieber, P., Brugger, M., Kamber, B., Riniker, B., Rittel, W. 1968. *Helv. Chim. Acta* 51:2057
9. Sieber, P., Riniker, B., Brugger, M., Kamber, B., Rittel, W. 1970. *Helv. Chim. Acta* 53:2135
10. Guttman, S. et al 1968. *Helv. Chim. Acta* 51:1155
11. Law, H. D. 1970. *The Organic Chemistry of Peptides.* London: Wiley-Interscience
12. Jakubke, H. D., Jeschkeit, H. 1969. *Aminosäuren, Peptide, Proteine: Eine Einführung* Berlin: Akademie-Verlag
13. Jones, J. H. 1970. *Amino-acids, Peptides and Proteins,* Vol. 2, Specialist Periodical Reports. London: The Chemical Society. 143 pp.
14. Jones, J. H. 1971. See Ref. 13, Vol. 3. 219 pp.
15. Jones, J. H., Ridge, B. 1972. See Ref. 13, Vol. 4. 309 pp.
16. Schröder, E., Lübke, K. 1968. *Fortschr. Chem. Org. Naturstoffe* 26:48
17. Bodanszky, M. 1970. *Peptides: Chemistry and Biochemistry,* Proc. 1st Am. Peptide Symp. Yale, 1968, ed. S. Lande, B. Weinstein, p. 1. New York: Dekker
18. Marglin, A., Merrifield, R. B. 1970. *Ann. Rev. Biochem.* 39:841
19. Gish, D. 1970. *Protein Sequence Determination,* ed. S. B. Needham. London: Chapman & Hall. 276 pp.
20. Wünsch, E. 1971. *Angew. Chem. Int. Ed.* 10:786
21. Schwyzer, R. 1972. *Ciba Found. Symp.* Amsterdam: North-Holland. Vol. 7, 23 pp.
22. Patchornik, A., Fridkin, M., Katchalski, E. 1973. *The Chemistry of*

Polypeptides, Essays in Honor of Prof. L. Zervas, ed. P. G. Katsoyannis. New York: Plenum. In press
23. Marshall, G. R., Merrifield, R. B. 1971. *Biochemical Aspects of Reactions on Solid Supports,* ed. G. R. Stark. New York: Academic. 111 pp.
24. Klausner, Y. S., Bodanszky, M. 1972. *Synthesis,* 453
25. Scoffone, E., Ed. 1969. *Peptides 1969.* Proceedings of the 10th European Peptide Symposium, Abano, Italy. Amsterdam: North-Holland
26. Nesvadba, H., Ed. *Peptides 1971.* Proceedings of the 11th European Peptide Symposium, Vienna, Austria. Amsterdam: North-Holland
27. Hanson, H., Jakubke, H. D., Eds. 1972. *Peptides 1972.* Proceedings of the 12th European Peptide Symposium, Reinhardsbrunn Castle, East Germany. Amsterdam: North-Holland
28. Proceedings of a Symposium on the Chemistry of Peptides, Santa Monica, California, 1970. *1971 Intra-Sci. Chem. Rep.* 5
29. Lande, S., Weinstein, B. 1970. *Peptides: Chemistry and Biochemistry,* Proceedings of the 1st American Peptide Symposium, Yale, 1968. New York: Dekker
30. Lande, S., Ed. 1972. *Peptides: Chemistry and Biochemistry.* Proceedings of the 2nd American Peptide Symposium, Cleveland, 1970. London: Gordon & Breach
31. Meienhofer, J., Ed. 1972. *Chemistry and Biology of Peptides.* Proceedings of the 3rd American Peptide Symposium, Boston, 1972. Ann Arbor: Ann Arbor Sci.
32. Schröder, E., Lübke, K. 1965. *The Peptides,* Vols. I & II. New York: Academic
33. Bodanszky, M., Ondetti, A. M. 1965. *Peptide Synthesis.* New York: Interscience
34. Greenstein, J. P., Winitz, M. 1961. *Chemistry of the Amino Acids,* Vol. 2. New York: Wiley
35. Boissonnas, R. A., Guttmann, S., Jaquenoud, P. A. 1960. *Helv. Chim. Acta* 43:1349
36. Singer, S. J. 1967. *Advan. Protein Chem.* 22:1
37. Shaw, E. 1970. *Physiol. Rev.* 50:244
38. Cuatrecasas, P., Anfinsen, C. B. 1971. *Ann. Rev. Biochem.* 40:259
39. Wilchek, M., Givol, D. See Ref. 26, p. 203

40. Cuatrecasas, P. 1972. *Proc. Nat. Acad. Sci. USA* 69: 1277
41. Rudinger, J. 1971. *Drug Design,* ed. E. J. Ariens, 5: Chap 9. New York–London: Academic. 319 pp.
42. Rudinger, J. 1972. *Endocrinology 1971,* Proceedings of the 3rd Int. Symp. Endocrinology, London, 1971, ed. S. Taylor. London: Heinemann. p. 12
43. Rudinger, J., Pliška, V., Krejči, I. 1972. *Recent Progress in Hormone Research,* Vol. 28. New York–London: Academic. 131 pp.
44. Schwyzer, R., Schiller, P., Fauchère, J. L., Karlaganis, G., Pelican, G. M. 1971. *Structure-Activity Relationships of Protein and Polypeptide Hormones,* P. 2. Excerpta Med. Int. Congr. Ser., No. 241, p. 167
45. Hofmann, K. et al 1971. *Bioorg. Chem.* 1: 66
46. Gutte, B., Lin, M. C., Galdi, D. G., Merrifield, R. B. 1972. *J. Biol. Chem.* 247: 4763
47. Lin, M. C., Gutte, B., Galdi, D. G., Moore, S., Merrifield, R. B. 1972. *J. Biol. Chem.* 247: 4768
48. Beecham, J. et al 1971. *J. Am. Chem. Soc.* 93: 5526
49. Camble, R. et al 1972. *J. Am. Chem. Soc.* 94: 2091
50. Moroder, L., Borin, G., Marchiori, F., Scoffone, E. 1972. *Biopolymers* 11: 2191
51. Merrifield, R. B. 1963. *J. Am. Chem. Soc.* 85: 2149
52. Fridkin, M., Patchornik, A., Katchalski, E. 1966. *J. Am. Chem. Soc.* 88: 3164
53. Wieland, T., Birr, C. 1966. *Angew. Chem. Int. Ed.* 5: 310
54. Bodanszky, M., du Vigneaud, V. 1954. *Nature* 183: 1324
55. Bodanszky, M., du Vigneaud, V. 1959. *J. Am. Chem. Soc.* 81: 5688
56. Weygand, F., Huber, P., Weiss, K. 1967. *Z. Naturforsch. B* 22: 1084
57. Weygand, F., DiBello, C. 1969. *Z. Naturforsch. B* 24: 314
58. Tilak, M. A., Hendricks, M. L., Wedel, D. S. See Ref. 30, p. 351
59. Tilak, M. A. 1970. *Tetrahedron Lett.,* p. 849
60. Denkewalter, R. G. et al 1966. *J. Am. Chem. Soc.* 88: 3163
61. Hirschmann, R. et al 1967. *J. Org. Chem.* 32: 3415
62. Denkewalter, R. G., Veber, D. F., Holly, F. W., Hirschmann, R. 1969. *J. Am. Chem. Soc.* 91: 502
63. Strachan, R. G. et al 1969. *J. Am. Chem. Soc.* 91: 503
64. Jenkins, S. R. et al 1969. *J. Am. Chem. Soc.* 91: 505
65. Veber, D. F. et al 1969. *J. Am. Chem. Soc.* 91: 506
66. Hirschmann, R. et al 1969. *J. Am. Chem. Soc.* 91: 507
67. Hirschmann, R., Denkewalter, R. G. 1970. *Naturwissenschaften* 57: 145
68. Curtius, T. 1902. *Chem. Ber.* 35: 3226
69. Honzl, J., Rudinger, J. 1961. *Collect. Czech. Chem. Commun.* 26: 2333
70. Hofmann, K., Lindenmann, A., Magee, M. Z., Khan, N. H. 1952. *J. Am. Chem. Soc.* 74: 470
71. Schwyzer, R., Surbech-Wegmann, E., Dietrich, H. 1960. *Chimica* 14: 366
72. Boissonnas, R. A., Guttmann, S., Jaquenoud, P. A. 1960. *Helv. Chim. Acta* 43: 1349
73. Schnabel, E. 1962. *Liebigs Ann. Chem.* 659: 168
74. Wünsch, E., Drees, F. 1966. *Chem. Ber.* 99: 110
75. Weygand, F., Hoffmann, D., Wünsch, E. 1966. *Z. Naturforsch. B* 21: 426
76. König, W., Geiger, R. 1970. *Chem. Ber.* 103: 788
77. König, W., Geiger, R. 1970. *Chem. Ber.* 103: 2034
78. König, W., Geiger, R. 1970. *Chem. Ber.* 103: 2024
79. König, W., Geiger, R. 1972. See Ref. 31, p. 343
80. Riniker, B., Brugger, M., Kamber, B., Sieber, P., Rittel, W. 1969. *Helv. Chim. Acta* 52: 1058
81. Sieber, P., Rittel, W., Riniker, B. 1972. *Helv. Chim. Acta* 55: 1243
82. Andreatta, R. H. et al 1973. *Helv. Chim. Acta* 56: 470
83. Gross, H. Bilk, L. 1968. *Tetrahedron* 24: 6935
84. Weygand, F., Steglich, W., Chytil, N. 1968. *Z. Naturforsch. B* 23: 1391
85. Wünsch, E., Wendlberger, G., Thamm, P. See Ref. 20, p. 792
86. Shiori, T., Ninomiya, K., Yamada, S., 1972. *J. Am. Chem. Soc.* 94: 6203
87. Boissonnas, R. A., Guttmann, S., Jaquenoud, P. A., Walter, J. P. 1955. *Helv. Chim. Acta* 38: 1491
88. Schwyzer, R., Sieber, P. 1966. *Helv. Chim. Acta* 49: 134
89. Bodanszky, M., Bodanszky, A. 1967. *Am. Sci.* 55: 185
90. Bodanszky, M. 1955. *Nature* 175: 685
91. Wieland, T., Birr, C., Flor, F. 1971. *Angew. Chem. Int. Ed.* 10: 336
92. Beyerman, H. C., 1972. See Ref. 31, p. 351
93. Wieland, T., Racky, W. 1968. *Chimia* 22: 375

94. Camble, R., Garner, R., Young, G. T. 1968. *Nature* 217:247
95. Garner, R., Young, G. T. 1971. *J. Chem. Soc. C* 50
96. Young, G. T., Schafer, D. J., Elliott, D. F., Wade, R. 1971. *J. Chem. Soc. C* 46
97. Merrifield, R. B. 1964. *J. Am. Chem. Soc.* 86:304
98. Marshall, G. R., Merrifield, R. B. 1965. *Biochemistry* 4:2394
99. Manning, M. 1968. *J. Am. Chem. Soc.* 90:1348
100. Fontana, A., Gross, J. E., See Ref. 27, p. 229
101. Noda, K., Gross, E. 1972. See Ref. 31, p. 241
102. Blake, J., Li, C. H. 1971. *Int. J. Protein Res.* 3:185
103. Aoyagi, H. et al 1972. *Biochim. Biophys. Acta* 263:823
104. Ontjes, D. A., Anfinsen, C. B. 1969. *J. Biol. Chem.* 244:6316
105. Barstow, L. E. et al See Ref. 31, p. 231
106. Gutte, B., Merrifield, R. B. 1971. *J. Biol. Chem.* 246:1922
107. Bayer, E. et al 1970. *J. Am. Chem. Soc.* 92:1735
108. Shapira, R., Chou, F. C. H., Chawla, R. K., Kibler, R. F. 1971. *J. Am. Chem. Soc.* 93:267
109. Khosla, M. C., Smeby, R. R., Bumpus, F. M. 1972. *J. Am. Chem. Soc.* 94:4721
110. Gisin, B. F., Merrifield, R. B. 1972. *J. Am. Chem. Soc.* 94:3102
111. Beyerman, H. C., de Leer, E. W. B., van Vossen, W. 1972. *J. Chem. Soc. Chem. Commun.* 929
112. Brunfeldt, K., Christensen, T., Roepstorff, P. 1972. *FEBS Lett.* 25:184
113. Sano, S., Kawanishi, S. 1973. *Biochem. Biophys. Res. Commun.* 51:46
114. Yaron, A., Schlossmann, S. F. 1968. *Biochemistry* 7:2673
115. Brenner, M., 1967. *Peptides*. Proceedings of the 8th European Peptide Symposium, Nordwijk, p. 1. Amsterdam: North-Holland
115a. Fankhauser, P., Schilling, B., Brenner, M. 1972. See Ref. 27, p. 162
116. Weygand, F., Wünsch, E. Unpublished results cited in Ref. 20, p. 788
117. Mitchell, A. R., Roeske, R. W., 1970. *J. Org. Chem.* 35:1171
118. Dorman, L. 1969. *Tetrahedron Lett.* 2319
119. Losse, G., Ulbrich, R. 1971. *Z. Chem.* 11:346
120. Brunfeldt, K., Roepstorff, P., Thomsen, J. 1969. *Acta Chem. Scand.* 23:2906
121. Brunfeldt, K. et al. See Ref. 31, p. 183
122. Brunfeldt, K., Christensen, T., Villemoes, P. 1972. *FEBS. Lett.* 22:238
123. Esko, K., Karlsson, S., Porath, J. 1968. *Acta Chem. Scand.* 22:3342
124. Gisin, B. F. 1972. *Anal. Chim. Acta* 58:248
125. Garden, J. II, Tometsko, A. M. 1972. *Anal. Biochem.* 46:216
126. Felix, A. M., Jimenez, M. H. 1973. *Anal. Biochem.* 52:377
127. Krumdieck, C. L., Baugh, C. M. 1969. *Biochemistry* 8:1568
128. Hancock, W. S., Prescott, D. J., Vagelos, P. R., Marshall, G. R. 1973. *J. Org. Chem.* 38:774
129. Wieland, T., Birr, C., Wissenbach, H. 1969. *Angew. Chem.* 81:782
130. Wissmann, H., Geiger, R. 1970. *Angew. Chem. Int. Ed.* 9:908
131. Markley, L. D., Dorman, L. C. 1970. *Tetrahedron Lett.* 1787
132. Tilak, M. A., Hollinden, C. S. 1971. *Org. Prep. Proc. Int.* 3:183
133. Losse, A. 1971. *Tetrahedron Lett.* 4989
134. Sano, S., Tokunaga, R., Kun, K. A. 1971. *Biochim. Biophys. Acta* 244:201
135. Stewart, J. M., Matsneda, G. R. See Ref. 31, p. 221
136. Shapira, R., Chou, F. C. H., Kibler, R. F. See Ref. 31, p. 225
137. Bayer, E., Jung, G., Halasz, I., Sebestian, I. 1970. *Tetrahedron Lett.* 4503
138. Parr, W., Grohmann, J. K. 1971. *Tretrahedron Lett.* 2633
138a. Parr, W., Grohmann, J. K. 1972. *Angew. Chem. Int. Ed.* 11:314
139. Scott, R. P. W., Chou, K. K., Kucera, P., Zolty, S. 1971. *J. Chromatogr. Sci.* 9:577
140. Tregear, G. W., Niall, H. D., Potts, J. T., Leeman, S. E., Chang, M. M. 1971. *Nature New Biol.* 232:87
141. Sheppard, R. C. See Ref. 26, p. 111
142. Stewart, J. M., Young, J. D. 1969. *Solid Phase Peptide Synthesis.* San Francisco: Freeman
143. Erickson, B. W., Merrifield, R. B. See Ref. 31, p. 191
144. Erickson, B. W., Merrifield, R. B. 1973. *J. Am. Chem. Soc.* 95:3757
145. Yamashiro, D., Noble, R. L., Li, C. H. See Ref. 31, p. 197
146. Yamashiro, D., Li, C. H. 1972. *Int. J. Peptide Protein Res.* 4:181
147. Boissonnas, R. A., Preitner, G. 1953. *Helv. Chim. Acta* 36:875
148. Kisfaludy, L., Dualszky, S. 1960. *Acta Chim. Acad. Sci. Hung.* 24:301

149. Channing, D. M., Turner, P. B., Young, G. T. 1951. *Nature* 167:487
149a. Blaha, K., Rudinger, J. 1965. *Collect. Czech. Chem. Commun.* 30:585
150. Sakakibara, S., Fukuda, T., Kishida, Y., Honda, I. 1970. *Bull. Chem. Soc. Jap.* 43:3322
151. Veber, D. F., Brady, S. F., Hirschmann, R. See Ref. 31, p. 315
152. Guttmann, S., 1963. *Peptides,* Proc. 5th Eur. Peptide Symp., ed G. T. Young, New York: Pergamon. 41 pp.
153. Chillemi, F., Merrifield, R. B. 1969. *Biochemistry* 8:4344
154. Fujii, T., Sakakibara, S. 1970. *Bull. Chem. Soc. Japan* 43:3954
155. Yamashiro, D., Blake, J., Li., C. H. 1972. *J. Am. Chem. Soc.* 94:2855
156. Losse, G., Krychowski, U. 1971. *Tetrahedron Lett.* 4121
157. Losse, G., Krychowski, U. 1970. *J. Prakt. Chem.* 312:1097
158. Hammerström, K., Lunkenheimer, W., Zahn, H. 1970. *Makromol. Chem.* 133:41
159. Lunkenheimer, W., Zahn, H. 1970. *Liebigs. Ann. Chem.* 740:1
160. Lunkenheimer, W., Zahn, H. 1970. *Angew. Makromol. Chem.* 10:69
161. Gosden, A., Stevenson, D., Young, G. T. 1972. *J. Chem. Soc. Chem. Commun.* 1123
162. Erickson, B. W., Merrifield, R. B. 1973. *J. Am. Chem. Soc.* 95:3750
163. Bayer, E. et al 1970. *J. Am. Chem. Soc.* 92:1738
164. Windridge, G. C., Jorgensen, E. C. 1971. *Intra-Sci. Chem. Rep.* 5:375
165. Windridge, G. C., Jorgensen, E. C. 1971. *J. Am. Chem. Soc.* 93:6318
166. Bodanszky, M., Funk, K. W. 1973. *J. Org. Chem.* 38:1296
167. Ragnarsson, U., Lindeberg, G., Karlsson, S. 1970. *Acta. Chem. Scand.* 24:3079
167a. Gut, V., Rudinger, J. 1968. *Peptides* Proc. 9th Eur. Peptide Symp., Orsay, ed. E. Bricas, 185. Amsterdam: North-Holland
168. Lloyd, K., Young, G. T. 1971. *J. Chem. Soc. C* 2890
169. Dutta, A. S., Morley, J. S. 1971. *J. Chem. Soc. C* 2896
170. Southard, G. L., Brooke, G. S., Pettee, J. 1971. *Tetrahedron* 27:2701
171. Southard, G. L., Brooke, G. S., Pettee, J. See Ref. 25, p. 95
172. Pietta, P. G., Marshall, G. R. 1970. *J. Chem. Soc. Chem. Commun.* 650
173. Inukai, N., Nakano, K., Murakami, M. 1968. *Bull. Chem. Soc. Jap.* 41:182

174. Weygand, F. 1968. *Peptides,* Proc. 9th Eur. Peptide Symp. Orsay, ed E. Bricas, p. 183. Amsterdam: North-Holland
175. Mizoguchi, T., Shigezane, K., Takamura, N. 1969. *Chem. Pharm. Bull.* 17:411
176. Tilak, M. A., Hollinden, C. S. 1968. *Tetrahedron Lett.* 1297
177. Wieland, T., Lewalter, J., Birr, C. 1970. *Annalen* 740:31
178. Marshall, D. L., Liener, I. E. 1970. *J. Org. Chem.* 35:867
179. Kenner, G. W., McDermott, J. R., Sheppard, R. C. 1971. *J. Chem. Soc. Chem. Commun.* 636
180. Bayer, E., Bretmaier, E., Jung, G., Parr, W. 1971. *Z. Physiol. Chem.* 352:759
181. Parr, W., Yang, C., Holzer, G. 1972. *Tetrahedron Lett.* 101
182. Shemyakin, M. M., Ovchinnikov, Y. A., Kiryushkin, A. A. 1965. *Tetrahedron Lett.* 2323
183. Mutter, M., Hagenmaier, H., Bayer, E. 1971. *Angew. Chem.* 83:883
184. Bayer, E., Mutter, M. 1972. *Nature* 237:512
185. Weygand, F., Ragnarsson, U. 1966. *Z. Naturforsch. B* 21:1141
186. Visser, S., Kerling, K. E. T., 1970. *Rev. Trav. Chim.* 89:880
187. Yajima, H., Kawatani, H., Watanabe, H. 1970. *Chem. Pharm. Bull. Jap.* 18:1333
188. Bayer, E., Hunziker, P., Mutter, M., Sievers, R. E., Uhmann, R. 1972. *J. Am. Chem. Soc.* 94:265
189. Anfinsen, C. B., Ontjes, D., Ohno, M., Corley, M., Eastlake, A. 1967. *Proc. Nat. Acad. Sci. USA* 58:1806
190. Ohno, M., Anfinsen, C. B. 1967. *J. Am. Chem. Soc.* 89:5994
191. Flanigan, E., Marshall, G. R. 1970. *Tetrahedron Lett.* 2403
192. Wang, S. S., Merrifield, R. B. 1969. *J. Am. Chem. Soc.* 91:6488
193. Wang, S. S., Merrifield, R. B. 1972. *Int. J. Peptide Protein Res.* 4:309
194. Fridkin, M., Patchornik, A., Katchalski, E. 1968. *J. Am. Chem. Soc.* 90:2953
195. Sklyarov, L. Yu, Gurbunov, V. I., Shchukina, L. A. 1966. *Zh. Obshch. Khim.* 36:2220
196. Williams, R. E. 1972. *J. Polym. Sci. A-1,* 10:2123
197. Patchornik, A., Fridkin, M., Katchalski, E. 1967. *Peptides,* Proc. 8th Eur. Peptide Symp. Nordwijk, p. 91. Amsterdam: North-Holland
198. Laufer, D. A., Chapman, T. M.,

Marlborough, D. I., Vaidya, V. M., Blout, E. R. 1968. *J. Am. Chem. Soc.* 90:2696

199. Wildi, B. S., Johnson, J. H. 1968. *Abstr. 155th Nat. Meet. Am. Chem. Soc.* A-8

200. Fridkin, M., Patchornik, A., Katchalski, E. 1972. *Biochemistry* 11:466

201. Kalir, R., Fridkin, M., Patchornik, A. In press

202. Panse, G. T., Laufer, D. A. 1970. *Tetrahedron Lett.* 4181

203. Wieland, T., Birr, C. 1966. *Angew. Chem. Int. Ed.* 5:310

204. Wieland, T., Birr, C. 1967. *Chimia* 21:581

205. Guarneri, M., Ferroni, R., Giori, P., Benassi, C. A. See Ref. 31, p. 213

206. Warshavski, A. To be published

207. Wolman, Y., Kivity, S., Frankel, M. 1967. *J. Chem. Soc. Chem. Commun.* 629

208. Fridkin, M., Patchornik, A., Katchalski, E. See Ref. 25, p. 164

209. Brown, J., Williams, R. E. 1971. *Can. J. Chem.* 49:3765

210. Williams, R. E., Brown, J., Lauren, D. R. 1972. *Polym. Repr.* 13(2):823

211. Hirschmann, R. et al 1971. *J. Am. Chem. Soc.* 93:2746

212. Iwakura, Y., Uno, K., Oya, M., Kataki, R. 1970. *Biopolymers* 9:1419

213. Dewey, R. S. et al 1968. *J. Am. Chem. Soc.* 90:3254

214. Hirschmann, R. 1968. *Peptides,* Proc. 9th Eur. Peptide Symp. Orsay, ed. E. Bricas, p. 139. Amsterdam: North-Holland

BIOSYNTHESIS OF SMALL PEPTIDES

Kiyoshi Kurahashi

Institute for Protein Research, Osaka University
Yamada, Suita-shi, Osaka 565, Japan

CONTENTS

INTRODUCTION

Recent progress in studies of the biosynthetic mechanism of gramicidin S and tyrocidines has revealed a novel mechanism for the assembly of amino acids with a sequence determined by protein templates (cf 1–6). The biosynthetic mechanism of these two cyclic peptides features the involvement of thiol groups of enzyme proteins in activating the constituent amino acids and participation of 4′-phosphopantetheine in the transfer of peptides to make a head to tail condensation to the next activated amino acid residue. The parallelism between the synthesis of gramicidin S or tyrocidine and fatty acid synthesis was discussed by Lipmann (4) and Laland (6). The involvement of aminoacyl thioester in linear gramicidin synthesis was recently reported by Bauer et al (7). The synthesis of polymyxin-group antibiotics seems to follow the similar mechanism of gramicidin S and tyrocidine synthesis in that aminoacyl adenylates and aminoacyl thioesters participate in the synthesis, as will be discussed later. The presence of 4′-phosphopantetheine in these enzyme fractions

445

has not been confirmed. Shimura's group obtained preliminary evidence for the involvement of 4'-phosphopantetheine in bacitracin synthesis (8). Thus, the synthetic mechanism of gramicidin S and tyrocidine seems prevalent in the synthesis of antibiotic peptides, but well-documented studies are still limited to gramicidin S and tyrocidine; and extensions of investigation to other antibiotic peptides are urgently needed.

Since Perlman & Bodanszky (9) and Maier & Gröger (10) recently made an extensive literature survey on related subjects, this review presents only the current problems concerning each specific peptide whose biosynthetic mechanism has been under investigation. Readers are recommended to consult review articles (1–6, 9–18) and original papers cited for more complete literature surveys of structural and in vivo studies.

CURRENT PROBLEMS OF BIOSYNTHESIS OF SPECIFIC OLIGOPEPTIDES

Gramicidin S and Tyrocidines

The biosynthetic mechanism of gramicidin S and tyrocidine has been studied by various workers (cf 1–6), and the findings may be considered a breakthrough in the elucidation of the synthesis of peptide antibiotics.

Gramicidin S and tyrocidine synthetase were fractionated into two and three complementary fractions, respectively, as summarized in Table 1 (1, 3, 5, 6, 19–28). The initiation point of gramicidin S and tyrocidine synthesis is the D-phenylalanine residue adjacent to L-proline (1, 5, 29–31). The light enzyme of both synthetases activates L- and D-phenylalanine to the respective adenylate and transfers the amino acid to a thiol on the enzyme molecule, where racemization takes place in the ratio of 70% of the D isomer to 30% of the L isomer (29, 32, 33). The light enzyme of gramicidin S synthetase could replace the light enzyme of tyrocidine synthetase, indicating a close correlation between gramicidin S and tyrocidine synthesis (1, 22, and S. G. Laland, personal communication). Roskoski et al (26) failed to confirm the above results. This discrepancy seems to result from the noncomplementary mixing of the fractions of the synthetases, as shown in Table VIII of (26). Fraction I of gramicidin S synthetase is heavy enzyme and II is light enzyme, as shown in Table 1. To avoid confusion, the author suggests using the nomenclature as shown in Table 1 henceforth. It is quite interesting to compare the molecular structure of the two light enzymes and the mode of complex formation between the two light enzymes and gramicidin S heavy enzyme or tyrocidine intermediate enzyme.

Kurahashi et al isolated gramicidin S nonproducing mutants of *Bacillus brevis* ATCC 9999 which were devoid of gramicidin S heavy enzyme activity (1). However, a further examination of the crude extract of a mutant revealed that the L-proline-activating subunit of the heavy enzyme was present at the region of light enzyme in sucrose density gradient centrifugation and that D-Phe-L-Pro diketopiperazine could be synthesized by this extract (34). These results suggest that the site of interaction of gramicidin S heavy enzyme or tyrocidine intermediate enzyme with light enzyme is in the L-proline-activating subunit and that not only gramicidin S light enzyme and

Table 1   Features of GS and Ty synthetase[a]

Enzyme	Fraction Number	Nomenclature[b]	Amino Acid Activated	PPan	Molecular Weight
GS synthetase	II	GSLE	L- and D-Phe	0	100,000
	I	GSHE	Pro, Val, Orn, Leu	1	280,000
Ty synthetase	I	TyLE	L- and D-Phe	0	100,000
	II	TyIE	Pro, L- and D-Phe (Trp)	1	230,000
	III	TyHE	Asn, Gln, Phe (Trp, Tyr), Val, Orn, Leu	1	440,000

[a] According to the table of Lee et al (27) with some modifications. GS, gramicidin S; Ty, tyrocidine; PPan, 4'-phosphopantetheine.

[b] LE, light enzyme; IE, intermediate enzyme; HE, heavy enzyme.

tyrocidine light enzyme, but also the L-proline-activating subunit of gramicidin S heavy enzyme and of tyrocidine intermediate enzyme should possess a homologous region. The above findings coincide with the conclusion that the transfer of the amino acid from D-phenylalanine thioster on gramicidin S light enzyme to L-proline thioester on gramicidin S heavy enzyme does not involve 4'-phosphopantetheine (35, 36). The use of a synthetic thiol ester of D-phenylalanine to circumvent the first activation step, L-phenylalanyladenylate formation, was cleverly demonstrated by Roskoski et al (35).

The presence of 4'-phosphopantetheine on gramicidin S and tyrocidine heavy enzymes and its implication in peptidyl transfer were proven and discussed by Laland et al (6, 28, 37) and Lipmann et al (4, 31, 36, 38). More recently the presence of one mole of 4'-phosphopantetheine per mole of tyrocidine intermediate enzyme was shown (27), and the finding that the intermediate enzyme activated L-proline, L- and D-phenylalanine (or tryptophan) (23) was also confirmed by amino acid binding studies (27). By incubating tyrocidine light enzyme and tyrocidine intermediate enzyme together with L-^{14}C-proline and L-^{3}H-phenylalanine and by reisolating the intermediate enzyme by sucrose density gradient centrifugation, Lee et al (27) proved the presence of D-Phe-L-Pro dipeptide, D-Phe-L-Pro-L-Phe tripeptide, and D-Phe-L-Pro-L-Phe-D-Phe tetrapeptide bound to the intermediate enzyme through thioester bonds. Now it appears that peptidyl transfer in the protein template mechanism of peptide biosynthesis requires the 4'-phosphopantetheine arm (4, 6). Whether this generalization is true or not awaits the elucidation of the biosynthetic mechanism of other small peptides.

Lee et al (27) succeeded in disaggregating the intermediate and heavy enzymes of tyrocidine synthetase by incubating crude extracts of *B. brevis* with DNase. Both intermediate and heavy enzymes were split into 70,000–75,000 molecular weight subunits. They also showed the separation of subunits by sodium dodecylsulfate gel electrophoresis. Besides 70,000 molecular weight subunits, both intermediate and heavy enzymes yielded a band at the region of 20,000 mol wt. It is important to identify this subunit as a peptidyl carrier protein containing 4'-phosphopantetheine to conclude the analogy between fatty acid synthetase and gramicidin S or tyrocidine synthetase as was proposed (4, 6).

As far as the mechanism of tyrocidine synthesis concerns, we can conclude the following: 1. Initiation of peptide synthesis is the formation of D-Phe-L-Pro dipeptide bound to intermediate enzyme through a thioester bond. 2. Elongation of the peptide takes place towards the C-terminus by a head to tail condensation of the peptide to the successive amino acid facilitated by a 4'-phosphopantetheine arm. 3. Activation of amino acids is catalyzed by each subunit specific to each amino acid in the sequence to synthesize an aminoacyl adenylate followed by transfer to an enzyme-thiol to form an aminoacyl thioester-enzyme complex. 4. Spatial arrangement of the subunit determines sequence of the peptide (protein template). 5. Upon completion of the linear decapeptide bound to a thiol on the L-leucine-activating subunit of heavy enzyme, release and cyclization take place. 6. Conversion of L amino acids to D isomers takes place in the form of aminoacyl thioester bound to light or intermediate enzyme, although the mechanism of D-tryptophan formation by intermediate enzyme has not been thoroughly studied. Thus, the questions raised previously for the synthesis of tyrocidine and gramicidin S (1, 9) were answered rather completely by the recent progress in the field.

One further problem is the cyclization mechanism of gramicidin S. It was shown that the incubation of the four constituent amino acids other than L-leucine with gramicidin S synthetase produced enzyme-bound di-, tri-, and tetrapeptides in the sequence found in gramicidin S (5, 29, 39). Upon addition of L-leucine, most of the intermediary peptides diminished in quantity, resulting in gramicidin S synthesis, but none of the intermediary peptides larger than the pentapeptide was detected (29). These results were taken to suggest that gramicidin S was formed by an antiparallel doubling reaction between the two carboxyl activated (enzyme-thiol bound) penta-peptidyl units (4–6). Laland's group (40) carried out an elaborate experiment by making use of the property of gramicidin S synthetase to incorporate amino acid analogs into the product (41). They isolated two types of heavy enzyme differentially labeled: one was charged with ^{14}C-azetidine-2-carboxylic acid containing di-, tri-, and tetrapeptides and the other with ^{3}H-proline containing peptides. Upon mixing the two heavy enzymes together with the unlabeled constituent amino acids, ATP and Mg^{2+}, Laland's group obtained two types of cyclic decapeptides, one containing exclusively ^{14}C-azetidine-2-carboxylic acid and the other ^{3}H-proline (gramicidin S), and if any, very little hybrid molecules which contained both proline and its analog. They concluded that cyclization took place intramolecularly between the two penta-peptides on one heavy enzyme molecule and that light enzyme was not involved in the cyclization reaction (6, 37, 40). However, to make this mechanism plausible, we have to postulate that there is another thiol site on the heavy enzyme on which a pentapeptide attaches and awaits the completion of the second pentapeptide. Saxholm et al suggest that this thiol site may be the D-leucine binding site which was found on the heavy enzyme besides the L-leucine binding site; this was based on their studies on the inhibition of gramicidin S synthesis by D-leucine (42). One contradiction to the above notion was seen in the results of Roskoski et al (35). It was shown that gramicidin S could be synthesized by incubating light and heavy enzymes, both of which were previously charged with the respective amino acids and freed of aminoacyl adenylates. In this experiment no growing peptide was bound to the heavy enzyme, thence a pentapeptide formed on one heavy enzyme molecule must have interacted with another pentapeptide on a different heavy enzyme molecule intermolecularly to make a head to tail condensation. Still further work seems necessary to clarify the cyclization mechanism of gramicidin S.

The gramicidin S synthetase activity appears abruptly at the late logarithmic phase of growth of B. brevis and disappears when the cells enter into the stationary phase (1, 19, 43). Both light and heavy enzymes follow the above pattern, indicating that there is a common regulatory mechanism of the synthesis of light and heavy enzymes (1). The studies of the effect of chloroamphenicol, actinomycin D, and rifamycin on the synthesis of gramicidin S synthetase during the growth of B. brevis cells indicated that the synthesis was regulated at the transcriptional level (S. Ohara & K. Kurahashi, unpublished).

To study further the regulatory mechanism of the synthesis of gramicidin S synthetase, an attempt was made to isolate various mutants of B. brevis which could not produce gramicidin S (1, 34). Three classes of mutants were identified by complementation tests in vitro with the use of purified light and heavy enzymes: one lacked light enzyme activity; the second, both light and heavy enzyme activity; and the third, heavy enzyme activity. Similar mutants were obtained by Saito's group (44). The isolation of the second class of mutants supports the above idea that a common regulatory mechanism governing the synthesis of both light and heavy enzymes is operating. Two mutants belonging to the third class were further differentiated (34). One mutant lacked only the L-leucine-activating subunit and the molecular weight of the heavy enzyme of this mutant was shown to be smaller than that of the wild type (mol wt 280,000). The other mutant was shown to be totally devoid of heavy enzyme by sucrose density gradient centrifugation of the extract, but it was proven to possess the L-proline-activating subunit as discussed above. This finding makes it possible to isolate the L-proline-activating subunit and clarify its molecular structure and properties.

These results suggest that the mutational events which result in the failure of production of gramicidin S have a broad spectrum: mutations in the structural genes of subunits which activate each amino acid, aberrations in the regulatory mechanism, and mutations which affect the ability of complex formation between heavy and light enzymes, transfer of amino acid from an aminoacyl adenylate to an enzyme-thiol, and possibly the synthesis of the 4'-phosphopantetheine arm.

A close correlation between the antibiotic peptide synthesis and sporulation has been noted by various workers (cf 45), but no direct genetic link has been shown. Isolation of various mutants that have lesions at the various active sites of gramicidin S synthetase will serve to solve the above problem. Furthermore, the recent work by Sarkar & Paulus (46) on the inhibition of RNA polymerase of tyrocidine and linear gramicidin producing *B. brevis* by tyrocidine and linear gramicidin but not by gramicidin S may open up a new area of investigation on the correlation between sporulation and antibiotic production.

Linear Gramicidins

Gramicidins are linear pentadecapeptides with antibiotic activities produced by the same *B. brevis* strain which produces tyrocidine. This peptide possesses a formyl group at the N-terminal valine residue and ethanolamine at the C-terminal tryptophan residue. Frøyshov et al (47) showed that the synthesis of linear gramicidin was not inhibited by inhibitors of protein synthesis. At first glance at the structure, one supposes that the initiation point of peptide growth may be at the formylvaline residue (by analogy with protein synthesis), and that the termination takes place by ethanolaminolysis of the thioester bond of pentadecapeptide bound to the enzyme, if one assumes that the synthesis also involves the thiol group and 4′-phospho-pantetheine. We have searched for an enzyme which formylates L-valine or activates formylvaline without success. Recently Bauer et al (7) succeeded in isolating a pentadecapeptide with the same sequence as linear gramicidin by incubating necessary ingredients together with partially purified linear gramicidin synthetase and treating the reisolated enzyme with alkali. Contrary to expectation, the peptide contained neither the formyl group nor the ethanolamine moiety. When crude extracts were used, they could isolate formylpentadecapeptides in addition to the desformyl ones. They also showed that the enzyme-bound pentadecapeptide could be converted to linear gramicidin by chemical formylation and ethanolaminolysis. The results indicate that formylation does not take place before completion of the pentadecapeptide bound to the enzyme and precedes the attachment of the ethanolamine moiety. The involvement of tetrahydrofolic acid in this step was seen by the complete inhibition of linear gramicidin production in cultures by aminopterine (7).

One possible way in which ethanolamine may form at the C terminus of the peptide would be the synthesis of a hexadecapeptide with a serine residue at the C terminus followed by decarboxylation of the peptide after its release from the enzyme. However, neither ^{14}C-serine nor ^{14}C-ethanolamine was incorporated into linear gramicidin by whole cells (7).

Synthesis of the pentadecapeptide requires at least two enzyme fractions (7). Fractionation of linear gramicidin synthetase, quantitative determination of 4′-phosphopantetheine, elucidation of the formylation reaction, and identification of the precursor of ethanolamine are necessary to clarify the mechanism of linear gramicidin biosynthesis.

Polymyxins

That the polymyxin group of antibiotics are also synthesized by a mechanism that does not involve the ribosome-RNA system has been shown by various workers (48–52) with the use of inhibitors of protein synthesis in vivo. The interesting features of the structure of this series of antibiotics are the presence of N^{α}-6-methyloctanoyl (or isoctanoyl)-α,γ-diaminobutyric acid (MOA-DAB or IOA-DAB) at the N terminus of the peptide, the presence of six DAB residues in the molecule, and the cyclic structure with a linear side chain. One encounters the following problems in searching for an enzymatic system of biosynthesis of such peptides: 1. Will MOA (or IOA) or MOA-DAB (or IOA-DAB) initiate synthesis of the peptide or MOA (or IOA) be added after completion of the peptide as in the case of linear gramicidin synthesis? 2. Does cyclization take place after completion of the linear chain or are the linear side chain and cyclic peptide formed separately and conjugated together afterwards? 3. Are there six different subunits activating six DAB in the peptide or does one DAB activating enzyme suffice the requirement? 4. Is the di- and tri-DAB sequence found in the structure formed first and then incorporated into polymyxin or not?

Jayaraman, Monreal & Paulus (53) partially purified a DAB-activating enzyme from crude extracts of *B. polymyxa* 2459s. This enzyme activity was not detected in preparations from polymyxin negative strains derived from 2459s, suggesting that this DAB-activating enzyme played a role in the biosynthesis of polymyxin. Ito et al (51) could not detect DAB-activating and D-leucine-activating activities in *B. colistinus* Koyama, which produces colistin A and B, the polymyxin group antibiotics. However, they did isolate an intermediate from growing cultures of *B. colistinus* by pulse labeling with ^{14}C-DAB and identified it as MOA-DAB, the N-terminal residue of colistin A (54). Ito et al (55) further extended their work to obtain a cell-free system to synthesize colistin with the use of chemically synthesized MOA-DAB and other constituent amino acids as substrates. L-^{14}C-threonine was incorporated into a product which had the same Rfs as authentic colistin A on paper chromatograms. In this report, however, it was not shown whether or not MOA-DAB was essential for the synthesis. If acyl-DAB is an essential substrate, then the synthesis of colistin must be initiated by the formation and activation of acyl-DAB, contrary to the case of linear gramicidin synthesis (7). These points need further investigation.

Komura & Kurahashi have partially purified a DAB-activating enzyme from crude extracts of *Aerobacillus polyaerogenes* which also produces colistin A and B (56). The enzyme preparation possesses ATP-^{32}PP$_i$ exchange activities dependent mainly on DAB, L- and D-leucine, and L-threonine, which are the constituent amino acids. The enzyme binds these amino acids as an acid-stable form which can be released by the treatment with alkali or mercuric acetate, suggesting the presence of aminoacyl thioesters.

Bacitracins

Bacitracins are also branched cyclic peptides with antibiotic activities. The N-terminal isoleucine residue forms a thiazoline structure with the adjacent L-cysteine residue. Ishihara et al (57) obtained a cell-free system from the lysate of protoplasts of *B. licheniformis* ATCC 10716 which incorporated L-^{14}C-histidine into a compound that was identified as bacitracin by paper chromatography. The enzyme system is soluble and none of the inhibitors of protein synthesis has any effect on the synthetic activity. They recently succeeded in fractionating the system into three parts (8; K. Shimura, personal communication): Component I with a mol wt of 230,000 activated L-isoleucine, L-cysteine, L-leucine, and L-glutamic acid; Component II with a mol wt of 240,000, L-lysine, and L-ornithine; and Component III with a mol wt of 300,000, L-isoleucine, L- and D-phenylalanine, L-histidine, L-aspartic acid, and L-asparagine. They also obtained preliminary results indicating that each of the three components had one equivalent of 4'-phosphopantetheine. Binding of the constituent amino acids seems to occur through thioester bonds. All these results suggest the similarity between tyrocidine synthesis and bacitracin synthesis. However, whether conjugation of activated (thioester) amino acids alone suffices to bring about spontaneous thiazoline ring formation and cyclization with the pentapeptide side chain cannot be predicted at present. Pfaender et al (58, 59) also prepared a partially purified enzyme which synthesized bacitracin A. They obtained enzymes of various molecular weights, indicating that the enzyme tends to dissociate and reassociate under the experimental conditions.

Malformin

This peptide produced by *Aspergillus niger* is the smallest cyclic peptide whose biosynthetic mechanism has been studied in a cell-free system. Yukioka & Winnick (60) obtained a cell-free system capable of synthesizing malformin in the presence of the constituent amino acids, Mg^{2+}, and ATP. One specific requirement (KCl) is different from gramicidin S or tyrocidine synthetase. The details of the mechanism of biosynthesis are still unknown.

Edeine

Edeine is an antibiotic peptide produced by *B. brevis* strain Vm 4. The structure of this peptide is unique in that four out of the five amino acid residues are unusual amino acids that are not found in proteins and in that the C-terminal residue is spermidine (in edeine B, it is guanylspermidine). Kurylo-Borowska & Tatum (61) obtained a cell-free system which incorporated labeled glycine and tyrosine into edeine. They partially purified the enzyme system and fractionated it into two complementary fractions by Sephadex G-200 column chromatography; Fraction I with a mol wt of 280,000 and Fraction II with a mol wt of 100,000, both of which were required for edeine synthesis (62). However, DEAE-cellulose column chromatography of each fraction revealed a more complex picture when the resolved fractions were tested for the ATP-$^{32}PP_i$ exchange activity dependent on α-tyrosine and the constituent amino acids of edeine. Fraction IA activates α-tyrosine and glycine; Fraction IB, β-tyrosine and diaminopropionic acid; Fraction IIA, α-tyrosine, isoserine, 2,6-diamino-7-hydroxyazelaic acid, and glycine; and Fraction IIB, isoserine and diaminopropionic acid. These results suggest that the multienzyme complex for peptide biosynthesis may be split in different combinations of subunits. They found β-tyrosine amide in the reaction mixture during the synthesis of β-tyrosine from α-tyrosine. This finding is analogous to the formation of D-phenylalanine amide during the conversion of L-phenylalanine into D-phenylalanine in the presence of ammonium ion with phenylalanine racemase (gramicidin S light enzyme) (19, 24, 33). They also observed the formation of glycylspermidine during the synthesis of edeine. If the termination of peptide elongation takes place by the addition of glycyl-spermidine, then the mechanism would be different from that proposed for linear gramicidin synthesis where C-terminal tryptophan was conjugated before ethanol-amine incorporation. Disclosure of experimental details of these studies is awaited with interest.

Actinomycins

Actinomycin consists of a chromophore, actinocin, and two 16-membered peptide lactones. Biosynthesis of this group of chromopeptide antibiotics has been studied extensively with growing or resting cells (cf 12, 63). Phenoxazinone synthetase catalyzes the oxidative condensation of o-aminophenols to the corresponding phenoxazinones (64). Because this enzyme was found to catalyze the condensation of 4-methyl-3-hydroxyanthraniloyl-pentapeptide at a greater rate than that of 4-methyl-3-hydroxyanthranilic acid (63, 65), Katz proposed a scheme for the biosynthesis of actinomycin that involved the synthesis of anthraniloyl-pentapeptide lactone and the condensation of the two molecules (63, 65). He suggested that the amino group for 4-methyl-3-hydroxyanthranilic acid might be protected by an acyl group during the synthesis of the pentapeptide lactone, thus preventing premature condensation of partially completed anthraniloyl peptides. Orlova et al (66, 67) reported that N-acetyl-4-methyl-3-hydroxyanthranilic acid was synthesized by actinomycin-producing *Actinomyces* but not by nonproducer strains. D-Valine, which inhibits actinomycin production, brings about the accumulation of N-acetyl-4-methyl-3-hydroxyanthranilic acid. These findings support the above hypothesis and suggest that the hexapeptide with N-acetyl-4-methyl-3-hydroxyanthranilic acid as its N terminus may be synthesized by a stepwise addition of amino acids followed by deacylation and condensation catalyzed by phenoxazinone synthetase. The mechanism of activation of N-acetyl-4-methyl-3-hydroxyanthranilic acid requires study.

So far no cell-free enzyme system that can synthesize anthraniloyl peptides has been established. If the above mechanism were correct, then the synthesis of hexapeptides would follow a similar pathway as in the case of gramicidin S or tyrocidine synthesis catalyzed by a multienzyme complex. No actinomycin derivative with a single or uncompleted peptide chain has been found, despite production of numerous analogs formed by substitutions with various amino acid analogs (12, 63). This may suggest that the presumable deacylase has a specificity towards N-acetyl-4-methyl-3-hydroxyanthraniloyl-pentapeptide. Isolation of actinomycin monolactone by Perlman, Walker & Perlman (68) suggests that lactone ring formation may take place after condensation of 4-methyl-3-hydroxyanthraniloyl-pentapeptides.

Walker, Otani & Perlman (69) recently reported the presence of an L-valine-activating enzyme in *S. antibioticus* extracts that was separate from L-valine-tRNA ligase. The enzyme was present in the extract of cells which were actively synthesizing actinomycin. The fact that phenoxazinone synthetase was shown to be synthesized much earlier than actinomycin production (70) may mean that the synthesis of this enzyme and amino acid activating enzymes are subject to a different regulatory mechanism.

Quinoxaline Antibiotics

Quinoxaline antibiotics are a family of heterodetic cyclodepsipeptide antibiotics. Considering that the quinoxaline moiety is derived from tryptophan (71), the synthetic pathway may be similar to that postulated above for actinomycin synthesis. If we assume that the quinoxaline ring should be the N-terminal residue and that a stepwise synthesis of the peptide portion should be accomplished, the last step will be condensation of the two molecules of quinoxaline-2-carboxylic acid-tetrapeptide, forming lactone bonds between the hydroxyl group of D-serine and the carboxyl group of N-methylvaline, instead of the oxidative condensation of two anthraniloyl peptides to form actinomycin.

Cell-free synthesis of echinomycin was reported by Arif et al (72), but the detailed mechanism has not been clarified.

Other Depsipeptides

Labeled precursor studies in vivo have been carried out for the synthesis of serratamolide (73, 74), sporidesmolide (75), valinomycin (76, 77), pyridomycin (78), and etamycin (79), but no clue for the mechanism of the biosynthetic pathway of such peptides has been presented. Recently Kleinkauf succeeded in obtaining a cell-free system that can synthesize valinomycin using L-alanine and L-^{14}C-valine as precursors (H. Kleinkauf, personal communication).

Penicillins and Cephalosporins

The nucleus of penicillin, 6-aminopenicillanic acid, has been considered to be derived from cysteinylvaline, and δ-(α-aminoadipyl)-cysteinylvaline has been postulated as a precursor of cephalosporin (15, 16, 80). Loder & Abraham (81, 82) succeeded in synthesizing δ-(L-α-aminoadipyl)-L-cysteinylvaline in a cell-free system. They incubated a particulate fraction obtained by ultrasonic treatment of mycelium of *Cephalosporium* sp. C91 together with L-^{14}C-valine, δ-(L-α-aminoadipyl)-L-cysteine, and an energy-regenerating system. This combination of substrates only gave rise to the product. Any other combinations of the constituent amino acids and dipeptides failed to produce δ-(L-α-aminoadipyl)-L-cysteinylvaline, indicating that the synthesis of this tripeptide follows the similar mechanism as that of glutathione synthesis as was proposed by Demain (15).

Loder (82) suggests that penicillin synthesis follows a mechanism similar to that of cephalosporin, based on the results of cell-free experiments. When a Sephadex G-25 filtrate of *Penicillium chrysogenum* extracts was incubated with ^{14}C-phenylacetyl-CoA in the presence of either 6-aminopenicillanic acid or isopenicillin N, ^{14}C-label was incorporated into benzylpenicillin, indicating that isopenicillin N may be the precursor of naturally occurring various penicillins.

Peptide-Type Ergot Alkaloids

By opening up the ring structure of the peptide portion of the ergotamine or ergotoxine molecule one can visualize lysergylhydroxyalanylphenylalanylproline or lysergylhydroxyvalylphenylalanylproline, respectively, as their possible precursor peptide structures. The incorporation of labeled L-alanine, L-phenylalanine, L-valine, and L-proline into the peptide portion of the alkaloids in vivo has been reported (83–89). Erge, Wenzel & Gröger (90) reported that inhibitors of protein synthesis did not affect the synthesis of the peptide portion of ergotoxine. Since lysergic acid is activated as lysergyl-CoA (91), the first step of ergometrine synthesis may be similar to hippuric acid synthesis, that is, the condensation of lysergyl-CoA with L-alanine. If this is the case, lysergylalanine very likely may not be an intermediate of ergotamine synthesis as was concluded by Floss et al (87, 92) from their results of in vivo incorporation studies. Synthesis of the acceptor tripeptide, alanyl (or α-hydroxyalanyl) phenylalanylproline, of lysergic acid may be a prerequisite in the synthesis of ergotamine. Tripeptide synthesis may follow a mechanism similar to that of glutathione synthesis.

Ohashi et al (93) proposed a pathway for the synthesis of ergokryptine via formation of D-lysergyl-L-valine and conjugation with L-Leu-L-Pro-lactam, based on the results of their cell-free experiments, but the experimental conditions seem to require further improvements to make the proposal conclusive.

Ferrichrome and Pulcherriminic Acid

Ferrichrome and pulcherriminic acid are naturally occurring hydroxamic acids. The peptide part of ferrichrome is a cyclic hexapeptide containing three residues each of glycine and of hydroxyornithine. Glycine, L-ornithine, N^{δ}-hydroxyornithine, and N^{δ}-acetyl-N^{δ}-hydroxyornithine were shown to be incorporated in vivo (17, 94). Pulcherriminic acid is a derivative of L-Leu-L-Leu diketopiperazine. L-Leucine was found to be the precursor (95, 96). Activation of the constituent amino acids as aminoacyl adenylates and involvement of enzyme-bound aminoacyl thioesters such as occur in gramicidin S or D-Phe-L-Pro diketopiperazine synthesis seem plausible, but no cell-free system has yet been obtained.

Peptidoglycan

Biosynthesis of the peptide portions of cell wall peptidoglycan has been studied extensively by Strominger and his associates (cf 18). Stepwise synthesis of UDP-MurNAc-pentapeptide has been elucidated to follow a mechanism (97–99) similar to that of glutathione synthesis. On the contrary, synthesis of interpeptide bridges of peptidoglycans involves an activation step of amino acids as aminoacyl tRNAs and the interpeptide bridges elongate towards the amino terminus (100–103). In this connection it is interesting to see whether or not the β-lysine chain of streptothricin elongates towards the N terminus (104).

CONCLUDING REMARKS

It seems well established that the biosynthesis of gramicidin S and tyrocidine is carried out by a novel mechanism which may be called a protein template mechanism (3–6). The possible generality of this mechanism in oligopeptide biosynthesis has been discussed above. The ambiguity of the protein template mechanism in relation to amino acid specificity contrasts with the strict fidelity of the ribosome-RNA system of protein synthesis. It is notable that a given strain of bacteria may produce various analogs of peptide antibiotics in which some constituent amino acids are replaced by other structurally related ones (cf 9–14). Fujikawa et al (105 and cf 1) and Roskoski et al (31) presented evidence that the relative specificity of tyrocidine synthetase towards available amino acids determines the composition of amino acids in tyrocidine synthesized in vitro.

Although the poly(γ-D-glutamyl) capsule of *B. licheniformis* should no longer be considered as a small peptide, the biosynthetic mechanism of this peptide may be relevant to the subject under discussion. The recent work of Troy (106) established that the substrate for this synthesis was L-glutamic acid, and not D-glutamic acid. The enzyme is particulate, but the synthesis is not ribosomal. He further obtained preliminary evidence that activation of L-glutamic acid occurs via glutamyladenylate and that the growing chain might be anchored on a thiol group of the particulate enzyme. The direction of peptide elongation has not yet been determined. If the biosynthetic mechanism involved the adenylate of L-glutamate and transfer of the glutamyl moiety to a thiol group of the enzyme, followed by racemization, the initial reaction would be very similar to that of gramicidin S synthesis. The next step may be different in that γ-glutamyl polymer synthetase may require only one other fraction which would anchor the growing chain as a thioester; this might be a ligase.

One antibiotic peptide in which the biosynthetic mechanism has been reported to be different from that of gramicidin S or tyrocidine is mycobacillin, a cyclic tridecapeptide. Bose and his associates (107) postulated a mechanism that involved activation of L-proline as phophoryl L-proline and a stepwise addition of the constituent amino acids to the activated growing peptide in a similar manner to glutathione synthesis.

One precaution that is required before starting cell-free studies of peptide biosynthesis is illustrated in the studies of bacilysin. On hydrolysis, bacilysin yields L-alanine and L-tyrosine (108), but DL-^{14}C-tyrosine does not function as a precursor of the antibiotic when added to a growing culture of *B. subtilis* A14 (109). It appears that a precursor of L-tyrosine in the synthetic pathway leading from shikimic acid to aromatic amino acids is involved in the synthesis of this peptide. This finding indicates the importance of precursor-product studies in vivo before searching for a cell-free system.

Discussion of the synthesis of small peptides in this review is confined to the ones produced by microorganisms. The peptides of animal origin whose bio-synthetic mechanisms are known to be different from that of protein synthesis are limited to dipeptides and tripeptides (carnosine, glutathione, ophthalmic acid, etc). Most of the peptides with physiological activities are derived from proteins (nogens) by the action of specific proteinases. However, the discovery that many hormone-releasing hormones present in hypothalamus are small peptides, some of which have a pyroglutamyl N terminus and an amide at the C-terminal residue, raises the question of whether or not they can also be produced as split products of proteins. At present the similarities found between growth hormone-releasing hormone and the N-terminal decapeptide of β-chain of hemoglobin may favor this proposition (110). However, the possibility of de novo synthesis of these peptides in hypo-thalamus cannot be excluded because of the recent findings that the precursor amino acids can be incorporated into thyrotropin-releasing hormone, luteinizing hormone-releasing hormone, and follicle-stimulating hormone-releasing hormone in vitro (111–113).

ACKNOWLEDGMENTS

The author wishes to thank all colleagues who sent a collection of reprints related to the subject. He is especially grateful to Drs. S. G. Laland, K. Shimura, Z. Kurylo-Borowska, E. Katz, K. Yoshida, H. Kleinkauf, and P. Pfaender who made unpublished results and manuscripts available. Thanks are also due to Drs. E. Katz, T. Shiba, S. Sakakibara, M. Yamada, Y. Imae, and Miss K. Mimaki for their cooperation in preparing this manuscript.

The work cited in this review that originated in the author's laboratory was supported in part by the United States Public Health Service (AM 04600) and by an aid from the Naito Foundation and Hoansha.

Literature Cited

1. Kurahashi, K. et al 1969. *Cold Spring Harbor Symp. Quant. Biol.* 34:815–26
2. Saito, Y., Otani, S., Otani, S. 1970. *Advan. Enzymol.* 33:337–80
3. Lipmann, F., Gevers, W., Kleinkauf, H., Roskoski, R. Jr. 1971. *Advan. Enzymol.* 35:1–34
4. Lipmann, F. 1971. *Science* 173:875–84
5. Kleinkauf, H., Gevers, W. 1969. *Cold Spring Harbor Symp. Quant. Biol.* 34:805–13
6. Laland, S. G., Zimmer, T.-L. 1973. *Essays Biochem.* 9:31–57
7. Bauer, K., Roskoski, R. Jr., Kleinkauf, H., Lipmann, F. 1972. *Biochemistry* 11:3266–71
8. Ishihara, H., Endo, Y., Shimura, K. 1972. *Seikagaku* 44:488
9. Perlman, D., Bodanszky, M. 1971. *Ann. Rev. Biochem.* 40:449–64
10. Maier, W., Gröger, D. 1972. *Pharmazie* 27:491–505
11. Gottlieb, D., Shaw, P. D., Eds. 1967. *Antibiotics II: Biosynthesis.* Berlin: Springer. 466 pp.
12. Katz, E. 1970. *Progr. Antimicrob. Anticancer Chemother.* 2:1138–59
13. Katagiri, K., Yoshida, T., Sato, K. 1973. *Antibiotics: Mode of Action,* ed. J. W. Corcoran. Berlin: Springer. In press
14. Taylor, A. 1970. *Advan. Appl. Microbiol.* 12:189–276
15. Demain, A. L. 1966. *Biosynthesis of Antibiotics I,* ed. J. F. Snell, 29–94. New York: Academic. 234 pp.
16. Abraham, E. P. 1971. *Pure Appl. Chem.* 28:399–412
17. Emery, T. F. 1971. *Advan. Enzymol.* 35:135–85
18. Osborn, M. J. 1969. *Ann. Rev. Biochem.* 38:501–38
19. Tomino, S., Yamada, M., Itoh, H., Kurahashi, K. 1967. *Biochemistry* 6:2552–60
20. Itoh, H., Yamada, M., Tomino, S., Kurahashi, K. 1968. *J. Biochem.* 64:259–61
21. Fujikawa, K., Suzuki, T., Kurahashi, K. 1968. *Biochim. Biophys. Acta* 161:232–46
22. Fujikawa, K., Sakamoto, Y., Kurahashi, K. 1971. *J. Biochem.* 69:869–79
23. Kambe, M., Sakamoto, Y., Kurahashi, K. 1971. *J. Biochem.* 69:1131–33
24. Gevers, W., Kleinkauf, H., Lipmann, F. 1968. *Proc. Nat. Acad. Sci. USA* 60:269–76
25. Kleinkauf, H., Gevers, W., Lipmann, F. 1969. *Proc. Nat. Acad. Sci. USA* 62:226–33
26. Roskoski, R. Jr., Gevers, W., Kleinkauf, H., Lipmann, F. 1970. *Biochemistry* 9:4839–45
27. Lee, S. G., Roskoski, R. Jr., Bauer, K., Lipmann, F. 1973. *Biochemistry* 12:398–405
28. Gilhuus-Moe, C. C., Kristensen, T., Bredesen, J. E., Zimmer, T.-L., Laland, S. G. 1970. *FEBS Lett.* 7:287–90
29. Gevers, W., Kleinkauf, H., Lipmann, F. 1969. *Proc. Nat. Acad. Sci. USA* 63:1335–42
30. Bredesen, J. E., Zimmer, T.-L., Laland, S. G. 1969. *FEBS Lett.* 3:169–72
31. Roskoski, R. Jr., Kleinkauf, H., Gevers, W., Lipmann, F. 1970. *Biochemistry* 9:4846–51
32. Yamada, M., Kurahashi, K. 1969. *J. Biochem.* 66:529–40
33. Takahashi, H., Sato, E., Kurahashi, K. 1971. *J. Biochem.* 69:973–76

34. Kambe, M., Imae, Y., Kurahashi, K. 1974. *J. Biochem.* In press
35. Roskoski, R. Jr., Ryan, G., Kleinkauf, H., Gevers, W., Lipmann, F. 1971. *Arch. Biochem. Biophys.* 143:485–92
36. Kleinkauf, H., Roskoski, R. Jr., Lipmann, F. 1971. *Proc. Nat. Acad. Sci. USA* 68:2069–72
37. Laland, S. G., Frøyshov, Ø., Gilhuus-Moe, C., Zimmer, T.-L. 1972. *Nature New Biol.* 239:43–44
38. Kleinkauf, H., Gevers, W., Roskoski, R. Jr., Lipmann, F. 1970. *Biochem. Biophys. Res. Commun.* 41:1218–22
39. Frøyshov, Ø., Zimmer, T.-L., Laland, S. G. 1970. *FEBS Lett.* 7:68–71
40. Stoll, E., Frøyshov, Ø., Holm, H., Zimmer, T.-L., Laland, S. G. 1970. *FEBS Lett.* 11:348–52
41. Winnick, R. E., Winnick, T. 1961. *Biochim. Biophys. Acta* 53:461–68
42. Saxholm, H., Zimmer, T.-L., Laland, S. G. 1972. *Eur. J. Biochem.* 30:138–44
43. Yamada, M., Kurahashi, K. 1968. *J. Biochem.* 63:59–69
44. Iwaki, M., Shimura, K., Kanda, M., Kaji, E., Saito, Y. 1972. *Biochem. Biophys. Res. Commun.* 48:113–18
45. Schaeffer, P. 1969. *Bacteriol. Rev.* 33:48–71
46. Sarkar, N., Paulus, H. 1972. *Nature New Biol.* 239:228–30
47. Frøyshov, Ø., Zimmer, T.-L., Laland, S. G. 1972. *FEBS Lett.* 20:249–50
48. Paulus, H., Gray, E. 1964. *J. Biol. Chem.* 239:865–71
49. Monreal, J., Paulus, H. 1970. *Biochim. Biophys. Acta* 199:280–82
50. Daniels, M. J. 1968. *Biochim. Biophys. Acta* 156:119–27
51. Ito, M., Aida, K., Uemura, T. 1970. *Biochim. Biophys. Acta* 213:244–47
52. Nefelova, M. V., Pattakhov, A. A., Silaev, A. B. 1971. *Antibiotiki* 16:499–504
53. Jayaraman, K., Monreal, J., Paulus, H. 1969. *Biochim. Biophys. Acta* 185:447–57
54. Ito, M., Aida, K., Uemura, T. 1970. *Progr. Antimicrob. Anticancer Chemother.* 2:1128–37
55. Ito, M., Koyama, Y., Aida, K., Uemura, T. 1970. *Biochim. Biophys. Acta* 215:418–20
56. Komura, S., Kurahashi, K. 1973. *Seikagaku* 45:484
57. Ishihara, H., Sasaki, T., Shimura, K. 1968. *Biochim. Biophys. Acta* 166:496–504
58. Pfaender, P. et al 1973. *FEBS Lett.* 32:100–4
59. Simlot, M. M., Pfaender, P., Specht, D. 1973. *FEBS Lett.* In press
60. Yukioka, M., Winnick, T. 1966. *J. Bacteriol.* 91:2237–44
61. Kurylo-Borowska, Z., Tatum, E. L. 1966. *Biochim. Biophys. Acta* 114:206–9
62. Kurylo-Borowska, Z., Tatum, E. L. 1970. *Progr. Antimicrob. Anticancer Chemother.* 2:1123–27
63. Katz, E. See Ref. 11, 276–341
64. Katz, E., Weissbach, H. 1962. *J. Biol. Chem.* 237:882–86
65. Salzman, L., Weissbach, H., Katz, E. 1969. *Arch. Biochem. Biophys.* 130:536–46
66. Orlova, T. I., Silaev, A. B., Zheltova, A. O. 1968. *Biokhimiya* 33:162–66
67. Orlova, T. I., Nefelova, M. V., Kulida, L. I., Silaev, A. B. 1970. *Biokhimiya* 35:648–51
68. Perlman, K. L., Walker, J., Perlman, D. 1971. *J. Antibiot.* 24:135–36
69. Walker, J. E., Otani, S., Perlman, D. 1972. *FEBS Lett.* 20:162–66
70. Marshall, R., Redfield, B., Katz, E., Weissbach, H. 1968. *Arch. Biochem. Biophys.* 123:317–23
71. Yoshida, T., Katagiri, K. 1969. *Biochemistry* 8:2645–51
72. Arif, A. J. et al 1970. *Ind. J. Biochem.* 7:193–95
73. Bermingham, M. A. C., Deol, B. S., Still, J. L. 1970. *Biochem. J.* 119:861–69
74. Bermingham, M. A. C., Deol, B. S., Still, J. L. 1972. *Biochem. J.* 127:45–46p
75. Butler, G. W., Russell, D. W., Clarke, R. T. J. 1962. *Biochim. Biophys. Acta* 58:507–13
76. MacDonald, J. C. See Ref. 11, 268–70
77. MacDonald, J. C., Slater, G. P. 1968. *Can. J. Biochem.* 46:573–78
78. Ogawara, H., Maeda, K., Umezawa, H. 1968. *Biochemistry* 7:3296–3302
79. Hook, D. J., Vining, L. C. 1973. *J. Chem. Soc. Chem. Commun.* 185–86
80. Abraham, E. P., Newton, G. G. F. See Ref. 11, 1–16
81. Loder, P. B., Abraham, E. P. 1971. *Biochem. J.* 123:477–82
82. Loder, P. B. 1972. *Postepy Hig. Med. Dosw.* 26:493–500
83. Vining, L. C., Taber, W. A. 1963. *Can. J. Microbiol.* 9:291–302
84. Majer, J., Kybal, J., Komersová, I. 1967. *Folia Microbiol.* 12:489–91
85. Nelson, U., Agurell, S. 1969. *Acta Chem. Scand.* 23:3393–97
86. Minghetti, A., Arcamone, F. 1969. *Experientia* 25:926–27
87. Floss, H. G., Basmadjian, G. P., Tcheng,

M., Spalla, C., Minghetti, A. 1971. *Lloydia* 34:442–45
88. Floss, H. G., Basmadjian, G. P., Tcheng, M., Gröger, D., Erge, D. 1971. *Lloydia* 34:446–48
89. Gröger, D., Erge, D. 1970. *Z. Naturforsch.* 25:196–99
90. Erge, D., Wenzel, A., Gröger, D. 1972. *Biochem. Physiol. Pflanzen* 163:288–307
91. Maier, W., Erge, D., Gröger, D. 1972. *Biochem. Physiol. Pflanzen* 163:432–42
92. Floss, H. G., Basmadjian, G. B., Gröger, D., Erge, D. 1971. *Lloydia* 34:449–50
93. Ohashi, T., Takahashi, H., Abe, M. 1972. *J. Agr. Chem. Soc. Japan* 46:535–40
94. Emery, T. F. 1966. *Biochemistry* 5:3694–3701
95. MacDonald, J. C. 1965. *Biochem. J.* 96:533–38
96. Uffen, R. L., Canale-Parola, E. 1972. *J. Bacteriol.* 111:86–93
97. Ito, E., Strominger, J. L. 1962. *J. Biol. Chem.* 237:2689–2703
98. Ito, E., Strominger, J. L. 1964. *J. Biol. Chem.* 239:210–14
99. Nathenson, S. G., Strominger, J. L., Ito, E. 1964. *J. Biol. Chem.* 239:1773–76
100. Kamiryo, T., Matsuhashi, M. 1969. *Biochem. Biophys. Res. Commun.* 36:215–22
101. Kamiryo, T., Matsuhashi, M. 1972. *J.*

Biol. Chem. 247:6306–11
102. Thorndike, J., Park, J. T. 1969. *Biochem. Biophys. Res. Commun.* 35:642–47
103. Plapp, R., Strominger, J. L. 1970. *J. Biol. Chem.* 245:3667–82
104. Voronina, O. I., Khokhlov, A. 1972. *Postepy Hig. Med. Dosw.* 26:541–48
105. Fujikawa, K., Sakamoto, Y., Suzuki, T., Kurahashi, K. 1968. *Biochim. Biophys. Acta* 169:520–33
106. Troy, F. A. 1973. *J. Biol. Chem.* 248:305–15
107. Sengupta, S., Bose, S. K. 1972. *Biochem. J.* 128:47–52
108. Walker, J. E., Abraham, E. P. 1970. *Biochem. J.* 118:563–70
109. Roscoe, J., Abraham, E. P. 1966. *Biochem. J.* 99:793–800
110. Veber, D. F. et al 1971. *Biochem. Biophys. Res. Commun.* 45:235–39
111. Mitnick, M., Reichlin, S. 1971. *Science* 172:1241–43
112. Johansson, K. N. G., Hooper, F., Sievertsson, H., Currie, B. L., Folkers, K. 1972. *Biochem. Biophys. Res. Commun.* 49:656–60
113. Johansson, K. N. G., Currie, B. L., Folkers, K. 1973. *Biochem. Biophys. Res. Commun.* 50:8–15

X-RAY STRUCTURAL STUDIES OF FERREDOXIN AND RELATED ELECTRON CARRIERS

× 854

L. H. Jensen

Department of Biological Structure, University of Washington, Seattle, Washington

CONTENTS

INTRODUCTION

Ferredoxin is the name applied by Mortenson et al (1) to a nonheme, iron-containing protein with a low redox potential isolated from the anaerobic bacterium, *Clostridium pasteurianum*. Similar low-potential, iron-containing proteins have been isolated from a number of bacteria including *Clostridium acidi-uracii* (2), *Mircococcus lactolyticus* (3), and *Chromatium* (4).

Tagawa & Arnon (5) reported that the iron-containing protein found in spinach leaves had several properties similar to bacterial ferredoxin, including a low redox potential. They proposed the name "chloroplast ferredoxin" and noted that it appeared to be the same substance recognized earlier and reported under such names as "photosynthetic pyridine nucleotide reductase" (6, 7) and "haem protein reducing factor" (8). Tagawa & Arnon showed that it was functional in photosynthesis as an electron carrier.

In isolating ferredoxin from bacterial extracts, Lovenberg et al (9) and Mortenson (10) had observed a red protein fraction. The protein was subsequently isolated from *C. pasteurianum* in crystalline form by Lovenberg & Sobel (11), who showed that it was a nonheme iron protein and that it could substitute for ferredoxin, at least in some reactions. They proposed the name "rubredoxin." Similar proteins have been isolated from a number of bacteria including *Peptostreptococcus elsdenii* (12), *Clostridium sticklandii* (13), and *Desulfovibrio gigas* (14).

In 1963 Bartsch (15) reported that an iron-containing protein could be isolated

461

in relatively large amounts from the photosynthetic bacterium, *Chromatium*. He showed that it was probably functional as an electron carrier in some unknown role. He also noted that although there were similarities to the ferredoxin-type iron proteins, there were marked differences such as the high redox potential. This latter property gave rise to the name "high potential iron protein" (HiPIP).

These proteins all belong to the large class now designated as iron-sulfur proteins and specifically to the nonheme iron group in which the Fe is bonded to one or more sulfur ligands. Tsibris & Woody (16) have subdivided this type on the basis of the number of iron atoms per molecule and the presence of other prosthetic groups. If other prosthetic groups are absent, then they list the following types, where S* designates labile sulfur:

1. 1-Fe proteins (rubredoxin, except for that from *P. oleovorans*).
2. 2 Fe-2S* (chloroplast ferredoxin, adrenodoxin, putidaredoxin).
3. 4 Fe-4S* (HiPIP, some bacterial ferredoxins).
4. nFe-nS* ($n = 8$ for most bacterial ferredoxins).

Three-dimensional structures of proteins from each of these types except the second have been reported (17–19). This review summarizes the structural work on these proteins with emphasis on the geometry of the iron-sulfur complexes. For more general and other reviews of the iron-sulfur proteins see (20–23).

RUBREDOXIN

Rubredoxin was first isolated from *C. pasteurianum* and characterized by Lovenberg & Sobel in 1965 (11). It transfers one electron on oxidation-reduction and has E'_0 of -0.06 V. Similar proteins are found in other bacteria. All except that from *P. oleovorans* have one Fe atom and four cysteines per molecule of weight ~ 6000. No function is known for the rubredoxins, again with the exception of the one from *P. oleovorans* that functions in the ω-hydroxylation of fatty acids and hydrocarbons (24).

Rubredoxin from *C. pasteurianum* has been studied by X-ray methods and its structure was solved at 3 Å resolution with phases based on a single HgI_4^{2-} derivative (17). The electron density map showed the course of the main chain and the highest peak in the map was taken as the Fe position. It was clearly evident that the main chain was linked to the Fe atom at four points, presumably the four cysteine S atoms being coordinated to the Fe atom as had been postulated from the chemical sequence (25). The linkage of the chains to the Fe atom was tetrahedral, although in the 3 Å map it appeared to be considerably distorted from a regular array.

When the resolution was increased to 2.5 Å, three of the four cysteine S atoms were clearly resolved from the Fe atom and the position of the fourth could be estimated (17). It was clear that the S atoms were not as greatly distorted from a regular tetrahedral array as had appeared in the 3 Å resolution map. To derive an estimate of the precision of the bond angles and bond lengths in the FeS_4 complex, two sets of Fe and S coordinates were estimated independently from the 2.5 Å resolution electron density map by different observers. The mean S–Fe–S bond

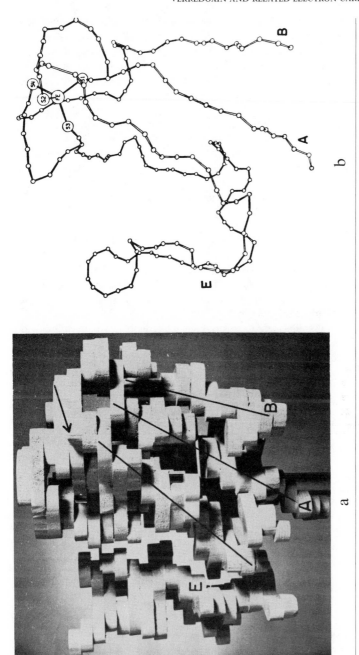

Figure 1 (*a*)   Balsa wood model of *P. aerogenese* rubredoxin based on 3 Å electron density map. A and B are N and C termini of the main chain. E is hairpin loop. Fe atom is at tip of arrow. Three sections of antiparallel chain shown by three straight lines. Thickness of balsa sections = 1 Å. (*b*) Model of chain corresponding to *a*. (From Herriott et al 17.)

angle was 109°, with a range of 24° and an estimated standard deviation of 8°; the mean Fe–S bond length was 2.30 Å, with a range of 0.2 Å in the mean values and an estimated standard deviation of 0.3 Å in the individual observations. The estimated standard deviations were derived from a comparison of the two sets of coordinates, and although they are useful quantities, they are so large that nothing at all can be said about deviations of individual bond angles or lengths from mean values. In particular, none of the bond angles differ significantly from the tetrahedral value. Note that this does not imply that the bond angles are tetrahedral.

The 2.5 Å resolution map showed much more detail than the one at 3 Å resolution. Although the increased resolution is not great, the amount of data is increased by $(3/2.5)^3$, a factor of ~ 1.7, and this can be taken as a measure of the increased information in the map (26).

When the 2.5 Å resolution map was calculated, the sequence for *C. pasteurianum* rubredoxin was unknown. An attempt was made, however, to deduce the sequence from the map, and although there were many uncertainties, 29 of the amino acids were correct by comparison with the subsequently determined chemical sequence (27).

The 2 Å data set has almost twice as many reflections as the 2.5 Å one, and the 2 Å resolution electron density based on the more extensive data shows great detail (28). Although the composition suggested 55 amino acids, it was evident that no electron density could be observed beyond residue 54 other than what could be attributed to solvent molecules. It was assumed, therefore, that the number of amino acids is 54 but the electron density tends to be diffuse toward the chain termini, presumably stemming from the fact that they are not firmly hydrogen bonded to the rest of the molecule.

In the 2 Å map, all four cysteine S atoms are resolved from the Fe, but covalently bonded C, N, and O atoms separated by distances in the range of 1.2–1.54 Å are still unresolved. To achieve greater resolution, two courses may be followed by protein crystallographers. One of these is to derive phases for higher resolution data from heavy atom isomorphous derivatives as is done for the lower resolution data. The difficulty with this approach, however, is that lack of isomorphism of the native and derivative crystals and other sources of error lead to increasing inaccuracies in experimental phases for the high resolution data. Nevertheless, experimental phases were determined for data to a resolution of 1.7 Å (29). Figure 2 shows the electron density in a plane through Phe 29 at resolutions of 3, 2.5, 2, and 1.7 Å with a skeletal model of the side chain superimposed. The improved electron density as shown by the contours is evident as the resolution is increased. For Phe 29 the resolution appears to be better at 1.7 Å than at 2 Å, but other groups showed no improvement, or at best only marginal improvement, and this approach was not pursued further.

An alternate course, and one which now appears to hold real promise, is to use calculated phases from a model derived from the highest resolution map based on experimental phases. This method is advantageous in extending resolution in that data for only native protein crystals need be collected. The chief advantage, however, is that the model can be improved by a refinement process similar to that used for small structures (29–31).

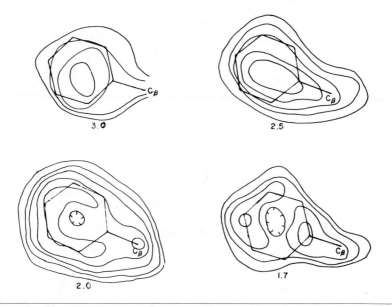

Figure 2 Electron density through Phe 29 ring at 3, 2.5, 2, and 1.7 Å resolution, experimental phases. (From Watenpaugh et al 29.)

In the course of deriving atomic coordinates for a model of rubredoxin from the 2 Å map, the sequence was redetermined (32). Comparison with the subsequently determined chemical sequence showed that 40 of the 54 amino acids were correctly identified and most of the rest were within an atom or two of the correct identification.

At this point the determination of the sequence was incidental to deriving the model for rubredoxin. The primary goal was to derive a model that could be refined in the conventional crystallographic sense to give an improved view of the molecule generally and in particular to provide more reliable bond lengths and angles for the FeS_4 complex.

The model for rubredoxin has now been subjected to eight refinement cycles and the R index, which may be taken as a measure of the error, has been reduced from 0.39 to 0.13 (31). The improvement for Phe 29 shown in Figure 3 may be compared with the same group in Figure 2.

Bond lengths and angles in the FeS_4 complex and their standard deviations are shown in Figure 4. These are now very much improved over the values derived from the 2.5 Å map. In contrast to what could be inferred from the earlier results, it is now evident that at least four of the S–Fe–S angles are significantly different from the tetrahedral value, 109.5° (any angle that differs by 3 σ from this value has a probability greater than 0.99 of actually being different).

The differences between the observed angles and the tetrahedral value are not surprising. In fact, the differences are in the same range as has been observed in a

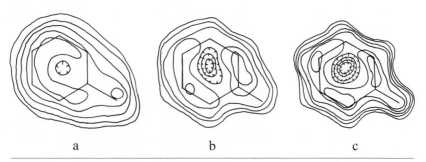

Figure 3   Electron density through Phe 29. (*a*) 2 Å resolution, experimental phases; (*b*) 2 Å resolution, calculated phases from refinement; (*c*) 1.5 Å resolution, calculated phases from refinement. (From Watenpaugh et al, *Trans. Am. Crystallogr. Assoc.* 9:95.)

model structure (33) and not much greater than angular distortions resulting from packing forces in small molecule structures. What is surprising, however, is that one of the Fe–S bond lengths is over 8 σ different from the mean value of the other three.

When assessing whether this bond length is really different from the other three, one must always consider the basis of the statistical treatment of the data. The standard deviations will be valid estimates of the precision of the results only if the errors are random, i.e. if no systematic errors are present, if the model is a proper one, and if the weights are correct, at least on a relative basis. Although none of these criteria are completely met in the refinement of the rubredoxin model based on 1.5 Å data, there is good evidence indicating that the standard deviations are realistic (31). The chief uncertainty appears to be systematic errors, but allowance has been made for the known ones. Nevertheless, unsuspected ones could remain, although the presence of any large enough to account for the observed difference in the Fe–S$^\gamma$42 bond length are not now known.

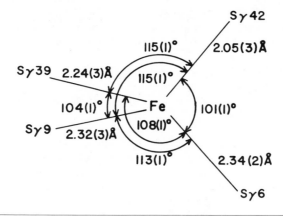

Figure 4   Bond lengths and angles in FeS$_4$ complex

HIGH POTENTIAL IRON PROTEIN (HiPIP)

HiPIP was first isolated from the photosynthetic bacterium *Chromatium* (15) and shown to have four Fe and four labile S atoms (S*) per molecule. It has an 86-residue polypeptide chain, including four cysteines, and a molecular weight of 9650 (34). On oxidation-reduction the molecule transfers one electron (15, 35). In contrast to other iron-sulfur proteins, HiPIP is distinguished by the unusually high E'_0 of $+0.35$ V. For comparison E'_0 for rubredoxin is -0.06 V (1) and for clostridial ferredoxin, -0.41 V (5).

HiPIP from *Chromatium* crystallizes in the monoclinic space group $P2_12_12_1$ with a = 42.5 Å, b = 42.0 Å, and c = 38.0 Å. The differences in the X-ray intensities for reflections h, k, l and $\bar{h}, \bar{k}, \bar{l}$, i.e. the Bijvoet differences which arise from anomalous scattering, were used by Kraut et al (36) in an attempt to locate the Fe atoms. It is these atoms that account for the major part of anomalous scattering for CuK_α radiation. The vector map based on the Bijvoet differences suggested strongly that the four Fe atoms were clustered about a single site, possibly tetrahedrally arranged with the Fe atoms too close to be resolved with 4 Å resolution data. The positions of the Fe atoms that would explain the main vector peaks approached a cubic close-packed arrangement. Kraut et al noted that a possible explanation was that the Fe atoms were located near the center of roughly spherical molecules in a cubic close-packed array.

In extending the X-ray study of HiPIP, Strahs & Kraut (37) collected 4 Å resolution data for two derivatives, $PtCl_6^{2-}$ and $UO_2F_5^{3-}$, to derive a set of phases. The electron density map based on these phases showed a region of outstanding density in the same location as deduced from the vector map of the Bijvoet differences. As a check that this was, in fact, the location of the anomalous scattering atoms,

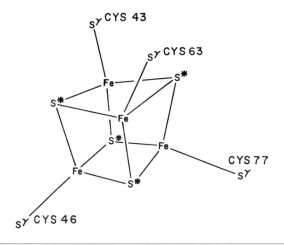

Figure 5   HiPIP complex

i.e. the Fe atoms, a Fourier map was calculated in which the coefficients were the Bijvoet differences and the phases were those based on the derivatives. Such a map approximates the distribution of the anomalous scattering atoms in the structure, and it did confirm that the main peak in the electron density map was, in fact, the location of the Fe atoms and that there were no other minor sites. Moreover, arrangements of the Fe atoms other than tetrahedral appeared to be ruled out, but unfortunately the outline of a single molecule was uncertain so that the location of the cluster within it could not be determined. Carter et al have now extended the work to higher resolution, first to 2.25 Å (18) and subsequently to 2 Å (38). The four iron atoms are indeed arranged in a tetrahedral cluster near the center of the protein molecule, confirming the earlier suggestion (36). The HiPIP molecule is roughly the shape of a prolate ellipsoid with approximate axial dimensions of 35 and 20 Å. Cysteine residues established from the sequence (34) at positions 43, 46, 63, and 77 bind to the four Fe atoms in the cluster. The positions of the labile S* atoms were inferred from the electron density map. Four lobes of density in the cluster not attributable to the Fe atoms formed a second tetrahedron concentric with the Fe one and apparently of similar dimensions. The complex is shown in Figure 5.

Refinement of the HiPIP model with 2 Å data (18, 38) has now provided much improved dimensions of the complex and a measure of their precision. In fact, in the earlier work it had been found that after exposure of reduced HiPIP crystals to the X-ray beam in collecting diffraction data, they were extensively oxidized. It was possible, however, to maintain the crystals in the reduced state, and 2 Å resolution data sets for both the reduced and oxidized native crystals have been collected and used in refining models for both forms (38). Bond lengths, bond angles, and Fe–Fe distances for the models are given in Table 1. The dimensions of the complexes in

Table 1   Mean values of bond lengths and angles in HiPIP complex (From 38)

	HiPIP$_{oxidized}$	HiPIP$_{reduced}$
Bond lengths (Å)		
Fe–Fe	2.72	2.81
rms deviation[a]	0.04	0.04
Fe–S*	2.26	2.32
rms deviation	0.08	0.09
Fe–S$^\gamma$	2.20	2.22
rms deviation	0.02	0.03
Bond angles (°)		
Fe–S*–Fe	74	76
rms deviation	1.3	2.4
S*–Fe–S*	104	104
rms deviation	2.4	2.6
S*–Fe–S$^\gamma$	115	116
rms deviation	4.9	5.3

[a] root-mean-square deviation

the oxidized and reduced forms are closely similar but probably not identical. Thus if the assumption is made that the Fe tetrahedron is a regular one, i.e. only one population of Fe–Fe distance is assumed, then the standard deviation in that distance is the root-mean-square deviation/$\sqrt{5}$, and the iron tetrahedron in reduced HiPIP appears to be significantly larger than the one in the oxidized form.

FERREDOXIN

Ferredoxin was isolated from *C. pasteurianum* and characterized by Mortenson et al in 1962 (1). Similar proteins are found in other bacteria and are characterized by low redox potentials, about -0.4 V, the presence of eight Fe and eight labile S atoms, and eight cysteines per molecule of weight ~ 6000.

Various arrangements for the Fe and S atoms in the 8Fe–8S* ferredoxins have been postulated, the earliest being a linear array (39). In the initial X-ray work, an attempt was made to confirm this for ferredoxin from *P. aerogenese* (40). From the low resolution vector map it was concluded that the Fe and S atoms could not be in a single linear array because this would have led (in general) to eight columns of high-vector density radiating from the origin although the map did not rule out several short linear arrays.

In contrast to HiPIP, the complexity of the vector map suggested that a definitive answer to the question of the arrangement of the Fe and S atoms would probably

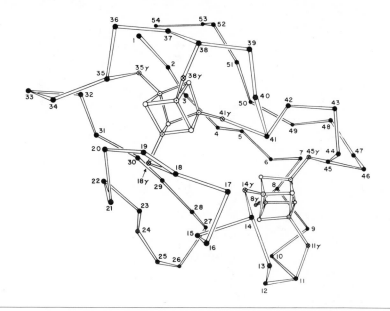

Figure 6   Plot of C^{α}, Fe, and S positions in molecule of *P. aerogenese* ferredoxin Fe, $\odot$; labile S, $\bigcirc$; cysteine S, $\otimes$; C^{α}, $\bullet$. (From Adman et al 43.)

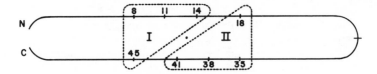

Figure 7   Schematic diagram of ferredoxin chain showing possible approximate twofold axis. (From Adman et al 43.)

require an electron density map. Numerous attempts were made to obtain iso-morphous heavy atom derivatives for *P. aerogenese* ferredoxin by a variety of methods. Success was ultimately achieved by the soaking technique in which UO_2^{2+}, Sm^{3+}, and Pr^{3+} were "driven on" by relatively high concentrations (41).

The most prominent features in the electron density map were two high-density clusters (41). Although adjacent Fe and S atoms were not resolved, it was apparent that the central, cube-like region of each cluster had four Fe atoms at opposite corners and four S atoms at the remaining four corners. The protein chain was linked to each cube-like region by coordination of the cysteine sulfurs to the Fe atoms. Both clusters appeared to be identical to the one found in HiPIP accounting for the eight Fe, eight S*, and the eight cysteine S atoms in the molecule.

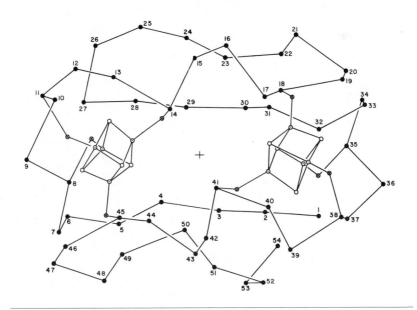

Figure 8   Plot of C^α, Fe, and S atoms showing approximate twofold axis. (From Adman et al 19.)

From the sequence (42) it is known that residues 8, 11, 14, 18, and 35, 38, 41, 45 are cysteine. It might have been expected that each of these groups of four cysteines would be coordinated in one of the complexes. Instead, cysteines 8, 11, 14, and 45 are coordinated in one complex, and cysteines 18, 35, 38, and 41 in the other as shown in Figure 6 (19). In spite of the crossover between the two halves of the molecule in bonding to the complexes, the way in which the chain is linked to them is a symmetric one as can be seen in the schematic diagram in Figure 7. In fact, the two halves of the molecule and the two complexes are related by an approximate twofold axis as shown in Figure 8.

There are two tyrosines in the *P. aerogenese* ferredoxin molecule, Tyr 2 and Tyr 28. Each is positioned next to one of the FeS_4 complexes and both have a ring edge exposed at the surface of the molecule, suggesting the possibility that the R groups may be involved in the two-electron transfer by the molecule. In the five ferredoxins for which sequence information is available, these groups, although not invariant, are all aromatic.

COMPARISON OF Fe_4S_8 GROUPS IN HiPIP, FERREDOXIN, AND SYNTHETIC ANALOG

The ferredoxin model has been refined to an R of 0.24 for 2 Å resolution data. Bond lengths and angles in the Fe_4S_8 complexes have improved in the sense that the range for each type has decreased as the refinement has progressed (43). Mean bond lengths and angles in the complexes at the present stage are shown in Figure 9. Comparison of these values with corresponding ones in Table 1 shows that the complexes in ferredoxin are not significantly different from the ones in either oxidized or reduced HiPIP.

As suggested above, the Fe–Fe tetrahedron in reduced HiPIP appears to be

Table 2 Mean values of bond lengths and angles in Fe_4S_8 core of $(Et_4N)_2[Fe_4S_4(SCH_2Ph)_4]$ (From 44)

	Number	Distance or Angle
Bond lengths (Å)		
Fe–S*	8	2.310(3)
Fe–S*	4	2.239(4)
Fe–S$^\gamma$	4	2.251(3)
Fe–Fe	2	2.776(10)
Fe–Fe	4	2.732(5)
S–S	2	3.645(3)
S–S	4	3.586(7)
Bond angles (°)		
Fe–S*–Fe	12	73.8(3)
S*–Fe–S*	12	104.1(2)
S*–Fe–S$^\gamma$	12	111.7–117.3

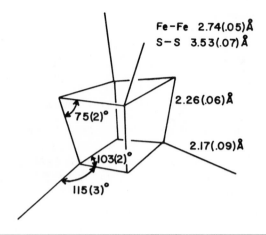

Figure 9   Average dimensions of the ferredoxin complex. Numbers in parentheses are estimated standard deviations in the mean values times 2 to allow for preliminary nature of refinement.

significantly larger than that in the oxidized form. This was confirmed by a difference Fourier map calculated with coefficients which were the differences in amplitudes between reflections from the oxidized and reduced crystals (38). Both the difference map and an analysis of the individual Fe–S* bond lengths suggest that two pairs of opposite $Fe_2S_2^*$ faces move closer by about 0.1 and 0.2 Å while the other pair moves apart by about 0.1 Å on oxidation (38).

The synthesis and structure of a compound containing an Fe_4S_8 core has been reported by Herskovitz et al (44). Table 2 lists the principal bond lengths and angles in the analog structure. The dimensions are close to those of the complexes in HiPIP and in ferredoxin. There are two sets of Fe–Fe distances with a mean value of 2.75 Å. Likewise there are two sets of Fe–$S_{sulfide}$ bond lengths: eight with a mean of 2.310(3) Å and four with a mean of 2.239(4) Å. The $Fe_4S_4^*$ is thus distorted from tetrahedral symmetry, and is actually tetragonal (point group $D_{2d} \sim \bar{4}2m$). The precision of the HiPIP and ferredoxin clusters precludes a definitive statement about deviations from tetrahedral symmetry at present, but Carter et al (38) indicate that an orientation for the reduced HiPIP cluster can be shown which displays two sets of Fe–S* bond lengths: eight with a mean of 2.35 Å and four with a mean of 2.25 Å.

Although the complexes in HiPIP and ferredoxin are similar, the large difference in the redox potentials was surprising. Carter et al cite the difficulty in accounting for a 0.75 V difference corresponding to over 17 cal/mol free energy change by differences in the protein environment if only two oxidation states for the complexes are assumed. However, a "three-state" hypothesis in which reduced HiPIP and oxidized ferredoxin contain the same state resolves the difficulty and nicely explains the available data (38).

Further work in improving the precision of the protein models will be necessary to know if or to what extent the $Fe_4S_4^*$ cores of the HiPIP and ferredoxin complexes deviate from tetrahedral symmetry and to establish with greater certainty the differences in geometry of the complexes in oxidized and reduced HiPIP. The availability of a genuine synthetic analog of the 4Fe–4S* and 8Fe–8S* proteins opens new vistas for studying and understanding the structure and electronic properties of these important structures.

Literature Cited

1. Mortenson, L. E., Valentine, R. C., Carnahan, J. E. 1962. *Biochem. Biophys. Res. Commun.* 7:448–52
2. Buchanan, B. B., Lovenberg, W., Rabinowitz, J. C. 1963. *Proc. Nat. Acad. Sci. USA* 49:345–53
3. Valentine, R. C., Jackson, R. F., Wolfe, R. S. 1963. *Biochem. Biophys. Res. Commun.* 7:453–56
4. Arnon, D. I. 1965. *Science* 149:1460–70
5. Tagawa, K., Arnon, D. I. 1962. *Nature* 195:537–43
6. Arnon, D. I., Whatley, F. R., Allen, M. B. 1957. *Nature* 180:182–85
7. SanPietro, A., Lang, H. M. 1958. *J. Biol. Chem.* 231:211–29
8. Davenport, H. E., Hill, R., Whatley, F. R. 1952. *Proc. Roy. Soc. London B* 139:346–58
9. Lovenberg, W., Buchanan, B. B., Rabinowitz, J. C. 1963. *J. Biol. Chem.* 238:3899–913
10. Mortenson, L. E. 1964. *Biochim. Biophys. Acta* 81:71–77
11. Lovenberg, W., Sobel, B. E. 1965. *Proc. Nat. Acad. Sci. USA* 54:193–99
12. Mayhew, S. G., Peel, J. L. 1966. *Biochem. J.* 100:80p
13. Stadtman, T. C. 1965. In *Non-heme Iron Proteins*, ed. A. SanPietro, 439–45. Yellow Springs, Ohio: Antioch
14. LeGall, J., Dragoni, N. 1966. *Biochem. Biophys. Res. Commun.* 23:145–49
15. Bartsch, R. G. 1963. In *Bacterial Photosynthesis*, ed. H. Gest, A. SanPietro, L. P. Vernon, 315–26. Yellow Springs, Ohio: Antioch
16. Tsibris, J. C. M., Woody, R. W. 1970. *Coord. Chem. Rev.* 5:417–58
17. Herriott, J. R., Sieker, L. C., Jensen, L. H., Lovenberg, W. 1970. *J. Mol. Biol.* 50:391–406
18. Carter, C. W. Jr., Freer, S. T., Xuong, Ng. H., Alden, R. A., Kraut, J. 1971. *Cold Spring Harbor Symp. Quant. Biol.* 36:381–85
19. Adman, E. T., Sieker, L. C., Jensen, L. H. 1973. *J. Biol. Chem.* 248:3987–96
20. Malkin, R., Rabinowitz, J. C. 1967. *Ann. Rev. Biochem.* 36:113–48
21. Orme-Johnsen, W. H. 1973. *Ann. Rev. Biochem.* 42:159–204
22. Lippard, S. J. 1973. *Accounts Chem. Res.* 6:282–88
23. Mason, R., Zubieta, J. A. 1973. *Angew. Chem. Int. Ed. Eng.* 12:390–99
24. Peterson, J. A., Coon, M. J. 1968. *J. Biol. Chem.* 243:329–34
25. Bachmayer, H., Yasunobu, K. T., Whiteley, H. R. 1967. *Biochem. Biophys. Res. Commun.* 26:435–40
26. Jensen, L. H. 1974. *Ann. Rev. Biophys. Bioeng.* 4: In press
27. McCarthy, K. 1972. PhD thesis. George Washington Univ., Washington DC
28. Watenpaugh, K. D., Herriott, J. R., Sieker, L. C., Jensen, L. H. 1970. *Am. Crystallogr. Assoc. Abstr.,* 78, Tulane Univ., New Orleans
29. Watenpaugh, K. D., Sieker, L. C., Herriott, J. R., Jensen, L. H. 1971. *Cold Spring Harbor Symp. Quant. Biol.* 36:359–67
30. Watenpaugh, K. D., Sieker, L. C., Herriott, J. R., Jensen, L. H. 1970. *Am. Crystallogr. Assoc. Abstr.,* 44, Carleton Univ., Ottawa
31. Watenpaugh, K. D., Sieker, L. C., Herriott, J. R., Jensen, L. H. 1973. *Acta Crystallogr. B* 29:943–56
32. Herriott, J. R., Watenpaugh, K. D., Sieker, L. C., Jensen, L. H. *J. Mol. Biol.* In press
33. Churchill, M. R., Wormald, J. 1971. *Inorg. Chem.* 10:1778–82
34. Dus, K., Tedro, S., Bartsch, R. G., Kamen, M. D. 1971. *Biochem. Biophys. Res. Commun.* 43:1239–45
35. Mayhew, S. G., Petering, D., Palmer, G., Foust, G. P. 1969. *J. Biol. Chem.* 244:2830–34
36. Kraut, J., Strahs, G., Freer, S. T. 1968. In *Structural Chemistry and Molecular Biology,* 55–64. San Francisco: Freeman
37. Strahs, G., Kraut, J. 1968. *J. Mol. Biol.* 35:503–72

38. Carter, C. W. Jr. et al 1972. *Proc. Nat. Acad. Sci. USA* 69:3526–29
39. Blomstrom, D. C., Knight, E. Jr., Phillips, W. D., Weiher, J. F. 1964. *Proc. Nat. Acad. Sci. USA* 51:1085–92
40. Sieker, L. C., Jensen, L. H. 1965. *Biochem. Biophys. Res. Commun.* 20:33–35
41. Sieker, L. C., Adman, E., Jensen, L. H. 1972. *Nature* 235:40–42
42. Tanaka, M. et al 1971. *J. Biol. Chem.* 246:3953
43. Adman, E. T., Sieker, L. C., Jensen, L. H. 1974. To be published
44. Herskovitz, T. et al 1972. *Proc. Nat. Acad. Sci. USA* 69:2437–41

X-RAY STUDIES OF PROTEIN INTERACTIONS

Anders Liljas[1] and Michael G. Rossmann

Department of Biological Sciences, Purdue University
Lafayette, Indiana

CONTENTS

INTRODUCTION

This review is one in a long series of summaries of X-ray diffraction studies of protein structure. In recent years the reports have appeared biannually. Blow & Steitz (1) and Dickerson (2) have covered the field up to 1971. Other recent reviews

---

[1] Present address: Department of Molecular Biology, Wallenberg Laboratory, University of Uppsala, Uppsala, Sweden.

have been written by Hess & Rupley (3), Blake (4), Vainshtein & Borisov (5), Blundell & Johnson (6), Matthews (7), and Matthews & Bernhard (8). Since it is almost impossible to cover the entire field within the limits of one of these reviews, we confine ourselves here primarily to a consideration of different types of protein interactions.

A summary of the current state of protein crystallography is given in Table 1. It includes all proteins where the polypeptide chain has been traced from beginning to end. It should be compared to the table given by Matthews (7) where most crystallized proteins that have been investigated by X-ray methods are listed.

PROTEIN-PROTEIN INTERACTIONS

The Folding Problem

Among the exciting unsolved problems of protein structure is the nature of the code that relates the primary amino acid sequence to the folded up tertiary structure. The high correlation between primary and tertiary structures has been verified in many ways. Anfinsen's work on the reduction and reoxidation of disulfide bonds in ribonuclease (96) and his more recent work on the folding of staphylococcal nuclease (97) provide evidence that a polypeptide chain knows its final conformation both in the presence and absence of guiding S–S bonds. The code is, however, degenerate. There are now more and more examples of rather different primary sequences which give rise to essentially the same tertiary structure, for instance myoglobin and the α and β chains of hemoglobin. Nor does a unique sequence give rise to a unique structure, but rather to a family of closely related conformations. The same protein may also exhibit conformational changes induced by pH, by the binding of ligands, or by the oxidation state of a metal atom (for instance the iron in hemoglobin and cytochrome c).

Secondary Structure

Astbury's (98) differentiation of protein structures into α and β types can now be seen with higher resolution. Dickerson & Geis (99) give a good review of possible helical and sheeted structures, and a summary of those found in proteins studied at high resolution is given in Table 2. The directional character of hydrogen bonds (Ramakrishnan & Nageshwar 100, Donohue 101, Richards et al 102) thus greatly influences the accurate three-dimensional character of protein molecules.

All known β-pleated sheets in globular proteins have a left-handed propeller like twist, suggesting a preference for the allowed β region in the Ramachandran diagram (Ramachandran & Sasisekharan 107, Chothia 108). This is generated by a right-handed rotation of successive peptide planes along each extended chain within the sheet. The rotation can vary from near $0°$ per residue in concanavalin A (Edelman et al 109, Hardman & Ainsworth 28) to $20°$ in carbonic anhydrase (Liljas et al 15) and more.

Many fairly successful attempts at relating primary sequence to α-helix and β-bend formation have been reported (Nagano 110, Lewis et al 111). Robson & Pain (112) give a recent review on the prediction of helical regions while Dickerson et al (33) test various prediction methods on the structure of cytochrome c.

Table 1 Stage of analysis of protein structures at high resolution[a]

Protein	Species	Symmetry of oligomeric aggregate	Resolution (Å)	Ref.	Refinement Type	Final R	Ref.	Coord. Ref.	Functional binding studies Res. (Å)	Ref.	Functional mechanism (Ref.)
Adenylate kinase	porcine muscle		3.0	9							
Alcohol dehydrogenase	horse liver	2	2.9	10					2.9	10	
Bence–Jones protein	human	(2)	3.5	11							
Carbonic anhydrase B	human		2.3	12					2.3	12	13, 14
Carbonic anhydrase C	human		2.0	15					2.0	16	17, 19
Carboxypeptidase A	bovine		2.0	17	M	0.44	18	18	2.0	18	
α-Chymotrypsin	bovine	2	2.0	20	MRE	20	20	20	2.5	21, 22	21
	bovine		2.8	23					2.8	23	
γ-Chymotrypsin	bovine		2.7	24					2.7	24	
Chymotrypsinogen	bovine		2.5	25							
Concanavalin A	jack bean	222	2.8	26					3.6	27	
			2.4	28				28	2.8	29	
			2.0	30						31	
Cytochrome b5	calf liver		1.5*	32							
Cytochrome c	horse ferri		2.8	33	M		34	34			35
	bonito ferri		2.8	33							
	bonito ferro		2.3	36							
	tuna ferri		2.8	37							
	tuna ferro		2.45	36	M		34	34			
Cytochrome c2	R. rubrum		2.0	38							
Cytochrome c550	M. denitrificans		4.0	38a							
Elastase	porcine		3.5	39					3.5	39	
Fab	human		2.8	40					4.5	41	
Ferredoxin	P. aerogenase		2.0	42	DL	0.34	43				

Table 1 (Continued)

Protein	Species	Symmetry of oligomeric aggregate	Resolution (Å)	Ref.	Refinement Type	Refinement Final R	Refinement Ref.	Coord. Ref.	Functional binding studies Res. (Å)	Functional binding studies Ref.	Functional mechanism (Ref.)
Flavodoxin	*clostridium MP*		1.9*	44							
	D. vulgaris		2.0*	45							
Glyceraldehyde-3-Phosphate dehydrogenase	lobster	222	3.0	46					3.0	46	
Hemoglobin	bloodworm		2.5	47							
	chironomus		2.5	50							
	horse oxy	2	2.8	52	M		52		2.5	51	48, 49
	horse deoxy	2	2.8	53							
	human deoxy	2	3.5	54					3.5	55	
	lamprey		2.0	56	M		56				56
High potential iron protein	chromatium		2.0	43	DL	0.17	43		2.0	43	
Insulin	porcine	32	1.9*	57	ME						
			?	Univ. of Peking 1972†							
Lactate dehydrogenase	dogfish M₄ apo	222	3.1	58							
			2.0	59	M		60	60	2.8	61, 62	
	dogfish M₄ ternary	222	3.0	59					3.0	59	
Lysozyme	hen egg white		2.0	63	MRE		63	63	3.0	63	63
			2.5	64	MRD	0.35	64				
	human		2.5*	Phillips*							

Protein	Source	Space group	Resolution	Ref	Method	R	Ref	Ref	Resolution	Ref	Ref
Malate dehydrogenase	porcine	2	2.5*	65					2.5	66	
Myogen	carp		1.85	67	M		67	67			
Myoglobin	sperm whale		1.4	68	MRE		68, 69	68	2.8	70	
	seal		2.5*	71							
Nuclease	S. aureus		2.0	72	MRD	0.23			2.0	72	
Pancreatic trypsin inhibitor	bovine		1.5	73			73				
Papain	papaya latex		2.8	74					2.8	74	
Phosphoglycerate kinase	horse		3.0	75							
Phosphoglycerate mutase	yeast	222	3.5	76							
	yeast		3.0	76							
Rhodanese	bovine liver	2	3.9	77							
Ribonuclease A	bovine		2.0	78							
			2.5	79							
Ribonuclease-S	bovine		2.0	80				81	2.0	82	83
Rubredoxin	C. pasteurianum		1.5	84	DL	0.13	84				
Subtilisin BPN'	B. amyloliquefaciens		2.0	85	M	0.44	86	86	2.5	87	
Subtilisin Novo	B. subtilis		2.8	89	M		89				88
Thermolysin	B. Termoproteolyticus		2.3	90							
Trypsin	bovine		2.7	91					2.7	91	
Trypsin-PTI	bovine		2.6	92	MR		92			92	
Trypsin-Soybean trypsin inhibitor	porcine		2.6*	93							

a Symbols used in this table are as follows: * = personal communication, † = indirect communication, M = model building refinement (34, 94), R = real space refinement (95), E = energy refinement (69), D = difference fourier refinement, L = least squares refinement.

Table 2 Secondary structure[a]

Protein	Helix %	No.	Comments	β structure %	No. β	No. Strands	Comments	β(I)	β(II)	γ	Ref.
Carbonic anhydrase	20	7	$\alpha, \alpha_{II}, 3_{10}$	37	2	13	P, A, (C)	3	1		15
Carboxypeptidase A	38	9	α, α_{II}	17	1	8	P, A	10*	3*		18
Chymotrypsin (elastase, trypsin)	14	3	$3_{10}, \alpha$	45	2	12	A, 2C	11	6		20
Concanavalin A	2	1		57	3	18	A, across twofold axis				28, 30
Cytochrome b_5	52	6	$\alpha, 3_{10}$	25	1	5	P, A	1*	1*		103
Cytochrome c	39	5						3	3		33
Ferredoxin				4	1	2	A				42
Flavodoxin	30	4		30	1	5					44, 45
Insulin	52	3	$\alpha, 3_{10}$	6	1	1	A, across twofold axis				57
Lactate dehydrogenase (malate dehydrogenase)	45	10		20	3	12	P, A	6	4		104
Lysozyme	40	6	$\alpha, \alpha_{II}, 3_{10}$	12	2	6	A	4	2		63
Myogen	57	6	$\alpha, 3_{10}$	4	1	2		5			67
Myoglobin (hemoglobin)	79	8	$\alpha, 3_{10}, \pi$					6*			68
Nuclease	24	3		14	1	3	A				72
Pancreatic trypsin inhibitor	28	2	α, PP	33	1	3	A				105
Papain	28	5		15	1	7	A				74
Ribonuclease	26	3	$\alpha, 3_{10}$	35	1	6	(P), A	1*	1*		81
Subtilisin	31	8		16	2	8	P, A	16*	1*		85, 89
Thermolysin	38	7		22	1	10	P, A, (C), CL			1	90

[a] Symbols used in this table are as follows: $\alpha = \alpha$ helix, $\alpha_{II} = \alpha_{II}$ helix, $3_{10} = 3_{10}$ helix, $\pi = \pi$ helix, PP = polyproline (helices or parts thereof in these conformations often in a distorted fashion).
A = Antiparallel β structure, P = Parallel β structure, C = Cylindrical β structure, CL = Clover leaf β structure, $+3_{10}$ conformations in helices not counted as bends.
* Data from Crawford et al (106).

Wu & Kabat (113) attempt a different approach by looking for persistence of conformations among classes of tripeptides. Unlike helix and β-bend formations, which concern themselves only with interactions between amino acids near each other along the polypeptide chain, longer range interactions have not yet been used in the prediction of structure.

Explicit information of secondary structure of a given protein is best shown in terms of its main chain hydrogen bonding (e.g. Wyckoff et al 81). However, an implicit mode of representation of secondary structure is the distance which maps interactions between residues. A cross on this plot represents the close approach between residues in the horizontal and vertical residue sequences

(Phillips 114, Ooi & Nishikawa 114a). An extension of this concept is to represent the inter-residue distance as a contour plot (Figure 1), somewhat analogous to the Patterson function in crystallographic investigations. Just as in a Patterson function it is unable to give information of the hand of the fold but does represent a unique

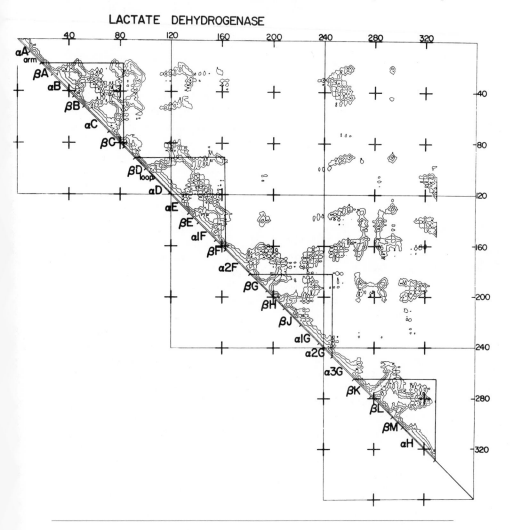

Figure 1 A distance for lactate dehydrogenase. The distance between the C_α atoms of the ith and jth residues have been plotted as a square matrix. The results are presented as a contour map. Contours are drawn at 0, 4.0, 8.0, ..., 16.0 Å. Different parts of the structure have been identified along the diagonal. The four structural domains between which there is less contact, are marked.

statement of the structure. An α helix will be represented by a series of interactions along the diagonal; two strands of a parallel pleated sheet will be represented by a series of short interactions running in a line parallel to but at a distance from the diagonal; while two antiparallel strands will produce interactions in a line perpendicular to the diagonal. The diagram is a two-dimensional statement of the three-dimensional structure and could be used to compare protein structure in a way analogous to the rotation function (Rossmann & Blow 115).

Tertiary Structure: A Tool in the Study of Evolution

Among the rapidly increasing number of known structures (Table 1), a variety of recurring patterns have been recognized. While structural and functional similarities may be suggestive of gene duplication, these can best be established on the basis of amino acid sequence. This is shown to be the case for the immunoglobulin molecules. Repetitive amino acid sequences along heavy and light chains are suggestive of a number of very similar tertiary structural domains along the polypeptide chains (Edelman et al 116, Welscher 117). This has recently been confirmed by the tertiary structures of Fab fragments (Poljak et al 40, Padlan et al 41) and Bence–Jones proteins (Schiffer et al 11, Epp et al 118).

If the divergent evolution of duplicated genes has progressed sufficiently far, tests based on amino acid homologies can be unreliable (Kretsinger 119, McLachlan 120). The only remaining evidence of a common origin may then be in a conservation of tertiary structure and function. In all cases where the structure is known it should be consulted in making sequence alignments (Dickerson 121).

An example of conservation of structure and function with only distant sequence homology has been found in carp muscle albumin by Kretsinger (119). He suggests that this protein has been produced by gene triplication where the second and third domains have retained the common function of binding calcium in equivalent positions.

An example of conservation of structure and function with apparently no clear sequence homology has been given by Rao & Rossmann (122). They have shown that a certain fold of about 60 amino acids occurs twice in lactate dehydrogenase (LDH) (see Table 3): once in flavodoxin as well as once in subtilisin. At one end of this structure there is a site whose function is to bind a mononucleotide or substrate (according to the needs of the enzyme concerned). It is of interest to note the suggestion of a common evolutionary origin for NAD and flavin binding proteins by Baltscheffsky & Baltscheffsky (124).

Recently it has been shown that the structure represented by the first two domains of LDH, corresponding to residues 22–165, occurs also in soluble malate dehydrogenase (s-MDH) (Hill et al 65), in alcohol dehydrogenase (L-ADH) (Brändén et al 10, Jörnvall 125) and in glyceraldehyde-3-phosphate dehydrogenase (GPD) (Buehner et al 46; Table 3 and Figure 2). Functionally the two domains common to all these enzymes are concerned with coenzyme binding. Comparison of the sequences of LDH (Taylor et al 126), L-ADH (Jörnvall 127), and GPD (Davidson et al 128) shows, however, no extensive relationship (K. Olsen, private communication). Thus, in spite of the lack of sequence evidence, the

Table 3   Domains in proteins and residues involved in binding of cofactors, substrates, or in catalysis

Protein	First unit	Second unit	Third unit	Fourth unit	Repetitive pattern of domains	Ref.
Bence–Jones protein (Fab)	1–110	111–220			AA	11, 40
Carbonic anhydrase	44–149					15
	His 63, Glx 66, His 93, 95, 118					
Carboxypeptidase	1–127	128–189	190–307			18
	His 69, Glu 72	Arg 145	His 196, Tyr 248 Glu 270			
Chymotrypsin (elastase, trypsin)	27–130	130–230			AA	20, 39, 91
	Phe 41, His 57, Asp 102	Glu 192, Gly 193, 216 Ser 195, 214				
Concanavalin A	1–87	88–234				28, 109
Cytochrome c	1–47	48–91				33
	Cys 14, 17, His 18	Met 80				
Ferredoxin	1–26	27–54			AA	42
	Cys 8, 11, 14, 18	Cys 35, 38, 41, 45				
Flavodoxin	1–48	49–138				44, 45
GPD	1–77	90–149	150–331		AAB	46
High potential iron protein	1–42	47–86				123
		Cys 63, 72				
Lactate dehydrogenase	22–91	92–165	166–265	266–329	AABB	104
	Asp 53	Arg 101, 109 Cys 165	Arg 171, 173 His 195			
Lysozyme	40–85					63
	Asp 52					
Myogen	1–33	34–71	72–108		AAA	119
Nuclease	1–100	101–140				72
	Arg 35, 87	Tyr 113				
Papain	10–111	112–207				74
	Cys 25	His 159, Asn 175				
Subtilisin	1–100	100–176	177–275			85, 89
	Asp 32, His 64	Ser 125, 130, Gly 127, Asp 155	Asp 218, Ser 221			
Thermolysin	1–157	158–316				90
	His 142, 146	Glu 166, His 231				

structural results favor the hypothesis that the first two domains in LDH, L-ADH, and GPD and part of the structure of flavodoxin represent divergent developments of the same gene and represent the very fundamental biological necessity of a protein fold able to bind nucleotides. The alternative concept of convergent evolution towards a nucleotide binding fragment is less attractive with regard to the available information.

The preliminary structures of adenylate kinase (AK) and phosphoglycerate kinase (PGK) have recently been reported (9, 75, 76). PGK consists of two well-separated domains (Watson et al 129, Blake et al 130), one of which is involved in ADP binding. This domain consists of six parallel strands in a β-pleated sheet with four helices antiparallel with regard to the sheet, two on each side. Similarly AK

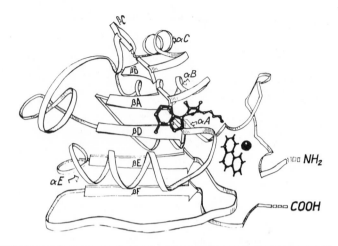

Figure 2 Diagrammatic representation of the dinucleotide binding fragment in liver alcohol dehydrogenase. The same fragment has been found in malate dehydrogenase, lactate dehydrogenase, and glyceraldehyde-3-phosphate dehydrogenase. Sequence homologies show it to be present also in yeast alcohol dehydrogenase. Comparison with Figure 1 shows this structure to consist of two domains of similar structure, one for each nucleotide. The presence of this mononucleotide binding fragment in flavodoxin is noteworthy.

consists of five strands and three helices. It is probable that their connectivity and coenzyme binding bear significant relation to that found in the dehydrogenases. The structure of hexokinase (Steitz et al 131), which like PGK consists of two domains, is also likely to be determined at high resolution in the near future.

The two structural repeats, which each bear the function of binding one mononucleotide, in the first half of LDH have already been mentioned and can be quickly seen on a distance plot (Figure 1). A similar repeat of a different structure in the second half of one LDH subunit can also be seen on a distance plot but has not yet been associated with a functional similarity.

The chymotrypsin-trypsin family of enzymes shows a repetitive structural domain (Table 3) although there is no functional similarity or sequence homology. Residues 16 through 124 (α-chymotrypsin) and 125 through 230 have a similar hydrogen bonding pattern and fold into two separated rough cylinders (Birktoft & Blow 20).

The presence of two domains in staphylococcal nuclease has been indicated by Anfinsen (97). Jardetzky et al (132) have shown that these domains show the greatest structural integrity during denaturation. Attempts have been made to recognize structural domains based only on visual inspection of known proteins (Wetlaufer 133). The implied criteria are the recognition of independently folded units, spatially separated from each other and connected by a minimum of peptide chains.

Frequently structural domains have also been recognized chemically by controlled proteolysis (Porter 134).

The distance plot (e.g. Figure 1) can be used in the structural recognition of domains as has been shown for LDH. An alternative method of recognizing a proposed structural domain could be by searching for units within which no sequence longer than n amino acids ($n = 3$, might be useful) makes contacts to structure only without the domain. Many other criteria could be considered. An analysis of various proteins into structural domains is shown in Table 3. Common domains within a single polypeptide chain are listed. Indicated also are important functional residues. Inspection shows that the active center frequently occurs between structural domains although within subunits of oligomeric enzymes.

Functional and sequence similarities between domains could be of evolutionary significance. These more highly preserved folds may aid in the prediction of structures for proteins when the function is clear or conversely predicting function where only the structure is known.

Quaternary Structure

The majority of proteins exist as aggregates of identical subunits. However, due to their larger size only a few oligomeric proteins have been studied in detail by X-ray crystallography (see Table 1). The frequency with which different types of interactions between subunits have been found in these examples is given in Table 4.

Table 4 Chemical nature of subunit interaction

Protein	Symmetry	Contact	van der Waal's contacts	H bonds	Ion pairs	Ref.
α-Chymotrypsin	2	A	443	9	1	20
		B	57	6		
Concanavalin A	222	∥ to A		2		30
		∥ to B	142	14	6	
		∥ to C	174	14		
Hemoglobin	2	$\alpha_1\beta_1$	110	5		52
oxy		$\alpha_1\beta_2$	80	1		
		$\alpha_1\alpha_2$				
		$\beta_1\beta_2$				
deoxy	2	$\alpha_1\beta_1$	98	5		54
		$\alpha_1\beta_2$	69	1	1	
		$\alpha_1\alpha_2$			2	
		$\beta_1\beta_2$			1	
Insulin	32	OP	111	8		57
		OQ	99	2	1	

Matthews & Bernhard (8) tabulate the number of subunits (and where available their symmetry) per molecule in a large number of examples. It is quite clear that

the symmetries 2, 222, and 32 are greatly preferred to the point groups 3, 4, 5 ....
The requirement for high symmetrical arrangements has been considered for viruses
from the point of view of genetic economy and for oligomeric enzymes for
functional advantages (Monod et al 135, Koshland et al 136). The important
difference between the considerations of Monod et al and Koshland et al is that
the former requires maximum symmetry at all times, whereas the latter permits
equilibria between the enzyme in its maximum symmetry state and lower
symmetrical forms. The changes of crystallographic symmetry with varying degrees
of substitution of coenzyme in the dehydrogenases (Matthews & Bernhard 8) may
have a bearing on this question. If molecules, only some of whose subunits have
bound coenzyme, had retained their high symmetry, then a randomization of the
filled coenzyme site, exhibited by low occupancy, would probably be observed in
the crystals.

Some general patterns for subunit interactions have started to emerge.
Extended main chains are mostly hidden or may be susceptible to proteolytic
attack which in γ-chymotrypsin (Segal et al 24), in subtilisin BPN' (Robertus et al
88), or in the trypsin-trypsin inhibitor complex (Rühlmann et al 92) has been
demonstrated to involve the temporary formation of a short antiparallel β
structure. Extended main chain can be shielded by side chains, by a helical segment
as in GPD (Buehner et al 46) or to some extent in LDH (Adams et al 104), or by
dimerization of the protein through the formation of a continuous β structure
between subunits as in concanavalin A (Edelman et al 109, Hardman &
Ainsworth 28), insulin (Blundell et al 57), and alcohol dehydrogenase (Brändén et
al 10).

Another type of subunit interaction involving pleated sheet structure is found in
concanavalin A, GPD, and to some extent also in LDH. When one surface of the
sheet constitutes the subunit boundary a twofold related sheet can be stacked on
top of it.

Despite the similarity of the coenzyme binding structure in the dehydrogenases,
the subunit interactions involving this part can be very different in various enzymes.
In the case of ADH the twofold axis in the dimer is located next to the βF strand
making a 12-stranded sheet structure through the dimer. In GPD and LDH the
βF strand is in the inside of one subunit. The Q axis in LDH or in s-MDH (Rossmann
et al 137) bears a strong resemblance with one of the twofold axes in GPD. In the
latter case the coenzyme binding sites across this axis are spaced further apart due to
additional pieces of polypeptide chain. The corresponding axes to P and R in LDH
are however translated by the size of a subunit along the Q axis. Thus, while in
LDH the coenzyme sites are on the outside of the molecule, they are close to the
subunit interfaces in GPD. This different association of the GPD subunits as compared to LDH might be linked with its cooperative properties (Conway &
Koshland 138). These molecules thus show, despite their similar tertiary structure,
that the aggregation of protein subunits is a more recent evolutionary event. Furthermore the absence of the Q axis in L-ADH would indicate the earlier separate
development of L-ADH as opposed to LDH, s-MDH, and GPD (Table 5).

Small deviations from high symmetry have been observed in many cases and

Table 5   Evolutionary development of the nucleotide binding protein

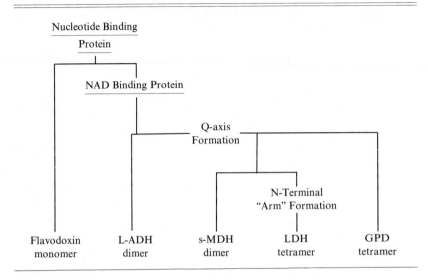

may be the result of important functional requirements. Examples of asymmetry in oligomers are given by insulin (Blundell et al 57) and chymotrypsin (Tulinsky et al 23), while the appearance of small translational elements in the quaternary structure has been observed for concanavalin A (Jack et al 27) and hexokinase (Steitz et al 131). Larger deviations from high symmetry are observed in the interactions of nonidentical subunits of similar structure. Twofold translation axes are found in hemoglobin (Muirhead et al 139), while a rotation of around 60° relates the V_L dimer and C_L dimer in a Bence–Jones protein (Schiffer et al 11) or the corresponding structures in the Fab fragment (Poljak et al 40).

The association of nonidentical subunits can result in high symmetry as is the case in aspartate transcarbamylase which is in the 32 point group. The creation of a central cavity is similar to that of a small spherical virus (Warren et al 140).

The assembly of larger protein aggregates has been studied primarily for viruses. Here the requirements are for maximum symmetry where the exact equivalence of all subunits cannot be obtained (Caspar & Klug 141). Although the structure of the tobacco mosaic virus (TMV) protein has not yet been studied in detail (Barrett et al 142), its mode of aggregation into cylindrical rods has been established (Butler & Klug 143). In TMV, as in the spherical virus cowpea chlorotic mottle virus (CCMV), the presence of nucleic acid has a controlling effect on the assembly. Bancroft (144) has shown that in the absence of RNA a variety of closed surfaces can be formed by the CCMV protein subunit, yet only one type of virus coat exists in the presence of RNA.

A useful method for describing subunit interactions is by means of a diagram first used in representing LDH subunit contacts (Figure 3) (Adams et al 104). Here each subunit is portrayed on a two-dimensional surface with as many straight edges as there are subunit contacts. The interactions are marked along each of the corresponding edges. A suitable solid figure can then be constructed of the same

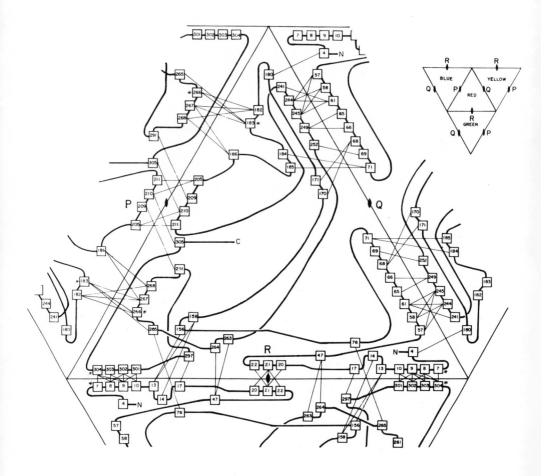

Figure 3 Dogfish M_4 LDH main chain interactions between different subunits (C_α–$C_\alpha \leqq 6$ Å). Diagrammatic representation of contacts between α-carbon atoms in neighboring subunits. Each subunit is given a color code: red, yellow, green, or blue. Thus the red subunit contacts the other three by virtue of the Q, R, and P axes, respectively.

symmetry as the oligomer, with each of the subunits convoluted to a surface and each contact area convoluted to a line.

PROTEIN-METAL INTERACTIONS

General Observations

The structural information concerning metal proteins has been rapidly increasing in the last few years (Table 6). No attempt will be made to review the binding of heme groups (see Dickerson 2) nor of the metals used in heavy atom derivatives (see Eisenberg 147).

The very close proximity in sequence number of at least two protein ligands to the metal ions has been pointed out by Waara et al (145) (see Table 2). With the exception of the zinc ion in insulin, all other metal atoms bound specifically have at least two protein ligands no more than four residues apart. This phenomenon bears a strong resemblance with chelates. Another striking feature in Table 6 is the clear preference of a particular metal for certain protein ligands. Thus iron invariably has been found to bind to cysteines (or inorganic sulfur) in the nonheme iron proteins. Calcium ions exclusively bind to oxygen atoms, and thus frequently to the acidic residues glutamate or aspartate, the binding then being of ionic character. The proteins that bind zinc ions donate two or three histidyl ligands to the metal and occasionally a glutamate. The binding has in these cases a definite amount of covalent character.

Concanavalin A and thermolysin (Table 6) are examples of double metal sites, a feature that had not previously been observed. Such metal ions have two or more ligands in common. When the calcium in thermolysin is substituted with lanthanide ions then the double site becomes only singly occupied (Matthews & Weaver 148). Double metal sites will probably become increasingly well known when more structures of multimetal proteins are determined. The copper ions in lactase (Malmström & Ryden 149) could, for instance, be expected to have double sites.

A common feature of the zinc enzymes (Table 6) is the immediate environment of the metal ligands. One of the histidines which is coordinated to the zinc is hydrogen bonded to an acidic residue: carboxypeptidase A, His 69-Asp 142 (Quiocho & Lipscomb 17); carbonic anhydrase C, His 117-Glu 115 (Liljas et al 15); thermolysin, His 142-Asp 170 (Colman et al 90). This presumably enhances the metal binding, increases the pK of the histidine, and affects the overall charge of the metal complex, and may thus be important in catalysis.

Functional Roles of Metal Ions in Proteins

Vallee & Walker (150) have summarized the roles of metal ions in proteins. In Table 6 well-established roles for metals in structurally investigated proteins are listed. Some of these roles will be discussed below, while others have been covered in other reviews of this series (Blow & Steitz 1, Dickerson 2, Jensen 151).

STABILIZATION OF STRUCTURE AND CHANGES OF CONFORMATION   The remarkable heat stability of thermolysin is highly dependent on the presence of calcium ions

Table 6 Metal sites in proteins

Protein	Metal	Ligands* 1	2	3	4	5	6	Coordination	Comments	Proposed Role of Metal†	Ref.
Carbonic anhydrase C	Zn	His 93 $N_{\varepsilon2}$	His 95 $N_{\varepsilon2}$	His 118 $N_{\delta1}$	H_2O			tetr	distorted	a, c, e, f	145
Carboxypeptidase A	Zn	His 69 $N_{\delta1}$	Glu 72	His 196 $N_{\delta1}$	H_2O			tetr	distorted	c, e, f	17
Thermolysin	Zn	His 142 $N_{\varepsilon2}$	His 146 $N_{\varepsilon2}$	Glu 166	H_2O			tetr	distorted	a, (e?)	148
	Ca 1	Asp 138	Glu 177	Asp 185	O'Glu 187	Glu 190	H_2O	oct	double site, 3.8 Å apart, distorted	a, (b?)	
	Ca 2	Glu 177	O'Asn 183	Asp 185	Glu 190	H_2O	H_2O	oct			
	Ca 3	Asp 57	Asp 59	O'Gln 61	H_2O	H_2O	H_2O	oct	distorted		
	Ca 4	O'Tyr 193	O_2Thr 194	O'Thr 194	O'Ile 197	Asp 200	$2H_2O$	oct			
Insulin	Zn	His B10 $N_{\varepsilon2}$	His B10 $N_{\varepsilon2}$	His B10 $N_{\varepsilon2}$	H_2O	H_2O	H_2O	oct	solvent not well defined	a	57
Concanavalin A	Mn	Glu 8	Asp 10	Asp 19	His 24 $N_{\varepsilon2}$	H_2O	H_2O	oct	double site, 5.3 Å apart	a, b	26, 28, 30
	Ca	Asp 10	O'Tyr 12	Asn 14	Asp 19	H_2O	H_2O	oct			
S. Nuclease	Ca	(Asp 19)	Asp 21	Asp 40	Glu 43	H_2O		oct		(e?)	72
Myogen	Ca 1	Asp 51	Asp 53	Ser 55	O'Phe 57	Glu 59	Glu 62			a	67**
	Ca 2	Asp 90	Asp 92	Asp 94	O'Lys 96	H_2O	Glu 101			a	
Rubredoxin	Fe	Cys 6	Cys 9	Cys 38	Cys 41			tetr	distorted, one short Fe-S	c, d	84
Ferredoxin	Fe_4S_4 1	Cys 8	Cys 11	Cys 14	Cys 45			tetr	Fe_4S_4 clusters closely resemble each other, distorted cube	c, d	42
	Fe_4S_4 2	Cys 18	Cys 35	Cys 38	Cys 41			tetr		c, d	
HiPIP	Fe_4S_4	Cys 43	Cys 46	Cys 63	Cys 77			tetr		c, d	43
Myoglobin	Cu	His A10	Lys A14	Asn GH4					Not in native myoglobin		146
	Zn	Lys A14	His GH1	Asn GH4							

* Protein ligands less than four residues apart are underlined.

** Private communication, R. H. Kretsinger (1973).

† a = Stabilize the protein structure; b = Bring about functionally advantageous conformational changes; c = Produce a distorted metal coordination; d = Oxidation-reduction or transfer of electrons; e = Participate in binding of substrates or cofactors; f = Activate the enzyme substrate complex once formed.

(Feder et al 152). The two calcium ions of the double site as well as the zinc ion is bound between the two domains with ligands to the metal sites from both halves of the molecule (Table 3). Not only would the absence of metal ions weaken the links between the two parts of the enzyme but the negative charges of the ligands of the double site would furthermore repel each other.

Another aspect of the stabilization of protein structure by metal ions is revealed by denaturation and renaturation studies. Carbonic anhydrase is renatured at higher guanidinium concentrations in the presence of zinc ions (Yazgan & Henkens 153). The zinc ion was found to bind at an early stage of the renaturation, thereby increasing the rate of refolding. In light of the structure it seems likely that the zinc ion initially binds to the chelate formed by His 93 and His 95.

Conformational changes have been observed in an X-ray study of binding of calcium and manganese ions to demetallized concanavalin A. The binding of manganese to the innermost part of the double site (Reeke et al 30, Hardman & Ainsworth 28) has to precede the calcium binding and might produce a small conformational change of the protein. The calcium is then able to bind to the outer part of the double site, thereby probably producing a larger change of the protein conformation resulting in a quaternary structural change. Similarly hexamers of insulin have been shown to undergo quaternary structural changes by changing the amount of bound zinc (Dodson et al 154).

DISTORTED METAL COORDINATION   One theory on enzyme mechanisms suggests that the conformation of the active site might be in an energetically poised or "entatic" state (Vallee & Williams 155). This has especially been studied for metalloenzymes where various spectral methods have been valuable tools to examine the coordination of chromophoric metals (Lindskog 156, Coleman 14). In the case of carboxypeptidase (Latt & Vallee 157) and carbonic anhydrase (Coleman & Coleman 158) it has been concluded from spectral data that the enzymatically active Co(II) enzymes have distorted tetrahedral symmetry in agreement with X-ray investigations (Quiocho & Lipscomb 17, Liljas et al 15). With the small differences in position between the metal in the native carbonic anhydrase and in the Me(II)-substituted enzyme (Waara et al 145) it is likely that the same metal ligands from the protein are being used with only minor adjustments in the binding directions. In inactive complexes spectral data indicate a tendency towards a square planar geometry around the metal (Taylor & Coleman 159). Thus it might be possible that the active complex needs a distorted tetrahedral arrangement. The solvent ligand in carbonic anhydrase is probably the most flexible but also the one closest involved in catalysis.

In the distorted tetrahedral arrangement of cysteine sulfurs around the iron in rubredoxin (Watenpaugh et al 84) one of the Fe–S bonds is significantly shorter than the other three. The Fe_4S_4 clusters of ferredoxin and high potential iron protein constitute another example of metal ions in a distorted arrangement. The pseudo-cube-shaped clusters do allow neither the iron nor the sulfur atoms to assume normal geometries. The observed structure of the clusters is rather a compromise between "conflicting architectural demands" (Carter et al 43).

PROTEIN-SOLVENT INTERACTIONS

Interactions with Water

STRUCTURAL RESULTS   Since protein crystals consist of about 50% water (Matthews 160), the X-ray diffraction methods applied to protein crystals could provide some structural information about the protein-solvent interface. Difficulties can exist in the interpretation of the solvent densities at the resolutions obtained. The errors in the intensities and the phase angle determination can easily create relatively high peaks of meaningless electron density. If, however, for various reasons a certain peak is assumed to be real, it is then necessary to assign the correct molecular or ionic species. As in the case of small molecule crystallography, neutron diffraction can aid in determining molecular species by the difference in scattering between hydrogen and deuterium (Schoenborn 161). Calculation of structure factors and subsequent refinement using difference electron density maps or least squares methods (Watson 68, Watenpaugh et al 84) can also provide some discrimination between artefacts and physical realities.

Table 7 gives a summary of the number and locations of water molecules observed in protein structures. In the case of rubredoxin, one of the most carefully refined proteins, the number approaches the average value for the hydration water of proteins 0.3–0.5 g H_2O/g proteins (Kuntz et al 163).

Where the position of water molecules have been published they are exclusively

Table 7   Bound water molecules in protein crystals

Protein	Main chain N–H	C-O	Polar Side Chains	Second or Third Layer	Buried	Active Center	Total No.	Resolution (Å)	Ref.
Carbonic anhydrase						9		2.0	15
Carboxy-peptidase A					10	several		2.0	19
α-Chymotrypsin	19	32	26	6	12	4	50	2.0	20
Lysozyme					3	15	185	2.0	63, 162
Myoglobin	22	41	30				91	1.4	68
	8	31	56	10			106	Neutron diffraction 2.0	161
Ribonuclease-S	4	11	4			3	18	2.0	80, 82
Rubredoxin							127	1.5	84
Subtilisin BPN'					3	14	150	2.5	86

hydrogen bonded to polar groups. Not surprisingly, main chain carbonyls are involved more frequently than main chain amides (Kauzmann 164). No static "icebergs" surrounding hydrophobic residues in contact with the solvent have yet been observed.

The main part of the bound water molecules are in immediate contact with the protein, but occasionally chains of water molecules are formed, sometimes bridging over to other polar groups on the same or neighboring protein molecules or just extending into the solvent (Schoenborn 161, Birktoft & Blow 20). In many instances buried water molecules without access to the solvent have been found within protein structures. These water molecules were presumably hydrogen bonded and subsequently trapped during the folding of the protein.

Almost invariably water molecules have been found in the active centers of enzymes (Table 7). The expulsion of water molecules when substrates or inhibitors bind might be an important contribution to binding constants (Phillips 162, Liljas et al 15). Some of these water molecules are often involved in the catalytic mechanism (13, 17, 165).

EFFECT ON FOLDING   It has long been recognized that the specific folding of a protein molecule depends on its solvent surrounding. The structure of water, the physiological solvent, and its interaction with the protein is thus of much interest and many excellent discussions of this matter have been published (101, 164, 166–168).

One old concept of protein folding is that hydrophobic residues in the unfolded protein are surrounded by cages or icebergs of ordered water molecules (Kauzmann 164). These water molecules are released when the hydrophobic residues become buried during the folding of the enzyme and thereby produce a favorable entropy of the folded protein. On the other hand Brandts et al (169) conclude that the hydrophobic contribution in the denaturation of ribonuclease is not at all apparent. Furthermore, Lee & Richards (170) have investigated the accessible surface area for a water molecule in the structures of myoglobin, lysozyme, and ribonuclease-S. They found that the contribution to the accessible surface area is slightly more than half for the polar atoms in all three proteins. The main chain amide nitrogen and carbonyl oxygen atoms had very low accessibility when compared to side chain atoms. Most polar groups in the interior of a protein are hydrogen bonded (see Table 2). There is good evidence from other protein structures that the picture presented by Lee & Richards (170) is of a general nature (see Table 7).

An alternative hypothesis for the folding of protein molecules is to put more emphasis on the charged and polar groups. In the denatured stage these would hydrogen bond to one or several water molecules each. If there is a possibility for interaction with other charged or polar groups within the protein, a number of water molecules would be released with a gain in entropy. The number of van der Waals interactions would simultaneously be maximized, thus strongly favoring the formation of compact protein molecules.

Interactions with Anions and Cations

Proteins exhibit both direct and indirect interactions with ions in the solvent. In an excellent review von Hippel & Schleich (167) have described these interactions. It has long been recognized that protein solubility and crystallization, subunit association, protein denaturation, enzyme activity or inhibition, etc are dependent on the type of ions present in the solvent. Anions and cations can be ordered into a lyotropic series with regard to their ability to perform such interaction with proteins (Hofmeister 171). Anions are usually ordered in the following manner:

$$PO_4^{3-}, citrate, SO_4^{2-}, F^-, Cl^-, CH_3COO^-, Br^-, NO_3^-, HCO_3^-, I^-, N_3^-,$$
$$NCS^-, NCO^-$$

The action of the ions is thus less related to their electrostatic interaction with the protein than to their ability to modify the properties of water. The order of the ions is related primarily to their structure making and structure breaking properties with regard to the solvent. The series above is ordered in increasing structure breaking ability and the phenomena mentioned are mostly the result of indirect action of the ions in solution on the protein.

Table 8 Protein–anion and protein–cation interactions (substrates, coenzymes, or heavy atom modifications are not included)

Protein	Ion in Native Structure	Ligands	Other Ions Binding to this Site	Comment	Ref.
Carbonic anhydrase	OH^- ?	Zn^{2+}, Thr 197 Lys 22, Arg 243	I^-, Br^-, Cl^-, I^-, Br^-, SCN^-, RSO_2NH^-	active site	16
α-Chymotrypsin	SO_4^{2-}	Ser 195, Tyr 146, His 57	SeO_4^{2-}	active site involved in dimer formation	23
	SO_4^{2-}	Asn 95, Lys 177	SeO_4^{2-}		
	SO_4^{2-}	Thr 224	SeO_4^{2-}		
	SO_4^{2-}	Arg 154	SeO_4^{2-}		
	SO_4^{2-}	NH_3^+, Ala 149	SeO_4^{2-}		
	SO_4^{2-}	Asn 245	SeO_4^{2-}		
	SO_4^{2-}	Asn 236	SeO_4^{2-}		
Elastase pH5	SO_4^{2-}	Ser 195, His 57, N′ 193	$2H_2O$ (pH 8.5)	active site	39
	SO_4^{2-}	Arg 145, Arg 230	empty (pH 8.5)		
Lactate dehydrogenase apoenzyme	SO_4^{2-}	Arg 171, His 195	citrate	substrate site	62
	SO_4^{2-}	Arg 173	citrate	between subunits	
Malate dehydrogenase	SO_4^{2-}			substrate site	66
Myoglobin	SO_4^{2-}	His E 7		absent in deoxy Mb	70
Ribonuclease	SO_4^{2-}	His 12, His 119, N′ 120, $2H_2O$	PO_4^{3-}, ASO_4^{3-}	substrate site	78, 81
	SO_4^{2-}	Lys 7, Lys 41			
Subtilisin novo	Na^+	Asp 197, O′ 169, O′ 174	K^+, Tl^+ (1.5 Å)		89

Most proteins have been crystallized from salt solutions of high molarity. Nevertheless there are very few specific binding positions for the solvent ions observed in the electron density maps (Table 8). The absence of solvent ions (excluding water) bridging between crystallographically related molecules re-emphasizes that the crystallization is more due to the *indirect* interaction of ions with the protein. The ions used for crystallization are always structure making ions such as sulfate, phosphate, or citrate.

Table 8 gives examples of the *direct* interaction between solvent ions and proteins. Hybridization experiments with subunits of lactate dehydrogenase from different species have been performed in the presence of various salts.(Chilson et al 172, Markert & Massaro 173, Jaenicke et al 174). It was found that sulfate, citrate, and several other compounds stabilized the tetramer whereas halides and thiocyanate promoted the hybridization in good agreement with the lyotropic series. Sulfate and citrate have been found to bind at a site between subunits (Adams et al 62) close to the molecular twofold axis P (Rossmann et al 137). An attempt to bind iodide to this site gave a negative result (A. Liljas, unpublished results). It therefore seems possible that not only the indirect effect of structure making anions in the solvent but also a specific binding between subunits of this type of anions inhibit the hybridization.

In carbonic anhydrase, structure breaking anions bind with increasing affinity to the active site close to the metal ion, thereby inhibiting the enzyme (Lindskog et al 175). The electrostatic interaction with the metal cannot be of any great importance since iodide is a much better inhibitor than fluoride. The active site of carbonic anhydrase is a 15 Å deep cavity. In the native structure a number of ordered water molecules have been observed bound in the active site (Liljas et al 15). The anions are found to bind at the bottom of the cavity close to the zinc ion displacing a few water molecules. Iodide fits snugly into a pocket surrounded by Leu 196, Thr 197, Thr 198, and two of the histidines liganded to the zinc ions (Bergstén et al 16). The inhibitor binding in carbonic anhydrase is probably essentially due to the interaction between less hydrated anions and the surrounding nonpolar atoms.

Relationship Between Protein Structure in Solution and Crystal

Little doubt remains today that the conformation and activity of a protein in solution is very similar to that in crystals. Different accounts on this problem have been given by Edsall (176), Rupley (177), Drenth et al (89), and Matthews (7), among others. Currently the conformation of carboxypeptidase in solution and crystals is being extensively discussed.

Chemical modification studies of the A_γ form of this enzyme with diazotized arsanilic acid (Johansen & Vallee 178, Johansen et al 179) raise doubts of the similarity between the protein in solution and in crystals as well as of the catalytic mechanism proposed by Lipscomb and co-workers (Quiocho & Lipscomb 17). The reagent specifically modifies Tyr 248 (Johansen & Vallee 178). This residue moves by 12 Å to become directly involved in catalysis (Quiocho & Lipscomb 17). A difference in color for the modified enzyme in crystals (yellow) and in solution (red) was observed indicating conformational differences. Furthermore it was found that the red color was due to an interaction between the modified tyrosyl residue and the zinc ion (Johansen & Vallee 178). These findings suggest that in crystals Tyr 248 is not able to approach the metal.

Recently Quiocho et al (180) and Johansen & Vallee (180a) have discussed the conformational differences between crystals and solution for various forms of the enzyme. The enzyme used in the crystallographic studies, carboxypeptidase A_α (CPA_α), exhibits red color both in crystals and in solution when modified. Those crystals are elongated along the a axis and have an activity which is one third the enzymatic activity in solution (Quiocho et al 180). Crystals of all the generally available α, β, and γ forms of the arsanilazo enzyme are, however, yellow and elongated along the b axis (Quiocho et al 180, Johansen & Vallee 180a). Furthermore, the ratio of the activity in crystals to the one in solution is 1/300 (Quiocho & Richards 181, Johansen & Vallee 180a). Tyr 248, when modified, is obviously not able to come in close contact with the zinc ion in most crystal forms, the only exception known being the one used in the X-ray studies. It is difficult to draw conclusions about the significance of these differences and to what extent they invalidate the catalytic mechanism and the conformational changes during catalysis, as proposed by Lipscomb and co-workers (Quiocho & Lipscomb 17), without further examination of carboxypeptidase behavior in solution and in crystals.

PROTEIN-NUCLEIC ACID INTERACTIONS

The central biological role played by the binding of nucleotides to protein makes a study of their interactions of great significance.

Analysis of the dihedral angles of nucleotide structures (182–185) shows that these are fairly constrained to limited ranges. Furthermore, these ranges are not significantly altered whether a mononucleotide is subject to crystal lattice forces or a polynucleotide is held in a helical form by the π electrons between base pairs. It is therefore a reasonable assumption that when a nucleotide binds to protein it will obey the same rules as found elsewhere. This is certainly the case for an actinomycin–deoxyguanosine crystalline complex (Sobell et al 186) and is also consistent with NMR results on mononucleotides (Barry et al 187).

Various nucleotides and nucleosides have been investigated when bound to alcohol dehydrogenase (ADPR), flavodoxin (FMN), glyceraldehyde-3-phosphate dehydrogenase (NAD), lactate dehydrogenase (NAD), malate dehydrogenase (NAD), staphylococcal nuclease (pdTp), phosphoglycerate kinase (ADP), and ribonuclease-S (UpcA) (see Table 1 for references). The binding of deoxyguanosine to actinomycin should be considered in this context (Sobell et al 186). Unfortunately, for none of these cases is it possible to determine accurate conformations. Conversely, the known constraints in conjunction with the electron density have been used to obtain reasonable dihedral angles as has been done for the coenzyme structure of lactate dehydrogenase (Chandrasekhar et al 61) and malate dehydrogenase (Webb et al 66). Such studies could be improved by using systematic refinement procedures. Figure 4 shows diagrammatically the interactions between the whole of the

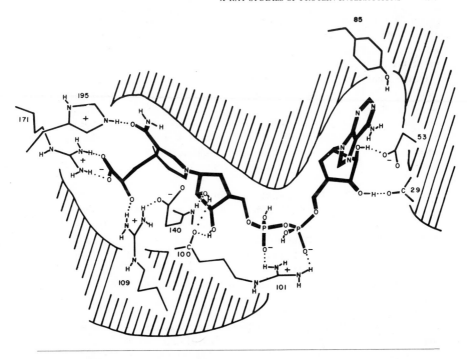

Figure 4 Diagrammatic representation of dinucleotide binding as observed in an LDH:NAD–pyruvate ternary complex.

nicotinamide adenine dinucleotide coenzyme and lactate dehydrogenase in a ternary complex. Adams et al (59) and Voet (188) suggest probable functions of the different moieties in NAD.

The binding of nucleotides, and in particular the phosphate moiety, has frequently a significant effect on the protein structure. In ribonuclease-S (Richards et al 82) the imidazole ring of His 119 has four distinct positions controlled by the phosphate binding mode. In LDH the phosphates probably act as the trigger for the conformational changes of the "loop" residues (94–114) (Adams et al 59). A movement of 12.5 Å in parts of the main chain and 23 Å for one guanadinium group is observed. The phosphates in 2′,3′-diphosphoglycerate, when bound to hemoglobin, generate a variety of conformational changes involving distances of 1–2 Å, thus locking the molecule into its deoxy quaternary structure (Arnone 55).

DYNAMIC INTERACTIONS: CATALYSIS

The function of proteins frequently involves dynamic interactions with other molecules. The exemplary work on hemoglobin by Perutz and his co-workers

(Perutz 48) has been reviewed by Dickerson (2). Suggested catalytic mechanisms of lysozyme, carboxypeptidase A, the serine proteinases, papain, and ribonuclease were reviewed by Blow & Steitz (1), while Dickerson (2) considered the dynamic properties of cytochrome c.

Of the interesting accounts which have been recorded in the last two years we would like to focus on serine proteinases and on lactate dehydrogenases. The serine proteinases provide an example of a now well-established general catalytic mechanism, whereas lactate dehydrogenase may offer a basis for future comparative work on dehydrogenases. Simultaneously, Jensen (151) is considering the nature of ferredoxin and related electron carriers in a parallel article.

Serine Proteinases

SUBSTRATE BINDING AND CATALYSIS   Two of the more exciting events in recent protein crystallography are the determination of the structure of trypsin in complex with the pancreatic trypsin inhibitor (Rühlmann et al 92) and the even more recent elucidation of the trypsin complex with soybean trypsin inhibitor (Sweet et al 93). The structures of trypsin and its pancreatic trypsin inhibitor were previously known (73, 91, 105) and several proposals of the interaction between trypsin or α-chymotrypsin and the trypsin inhibitor turned out to be essentially right (91, 105, 189, 190).

The enzyme and the pancreatic inhibitor bind together like rigid molecules, excluding water molecules from the inhibitor enzyme contact area. No significant movements of main chain atoms were observed (Rühlmann et al 92) relative to the component structures. Lys 15I (I after residue number indicates an inhibitor residue) binds via a salt bridge to Asp 189 in the specificity pocket of the enzyme. The polypeptide chain between Pro 13I and Lys 15I forms a short segment of antiparallel β structure with the enzyme in a fashion similar to the way peptide substrates bind in γ-chymotrypsin (Segal et al 24). The bond to be broken in a real substrate corresponds to Lys 15I–Ala 16I. This bond in PTI is not broken by the enzyme (Wilson & Laskowski 191), in good agreement with the X-ray results. The O_γ atom of Ser 195 is found in the position characteristic for the acylenzyme of α-chymotrypsin (Henderson 21) and covalently linked to the carbonyl carbon of Lys 15I. The complex can thus be described as a tetrahedral adduct with regard to this atom. The negatively charged oxygen at the tetrahedral carbon is hydrogen bonded to the amide nitrogens of Gly 193 and Ser 195. The structural alterations needed to form the tetrahedral adduct from the components is minimal, making the energy of the transition state low. In the complex $N_{\varepsilon2}$ of His 57 interacts with O_γ of Ser 195, accepting the seryl proton transferred upon the formation of the covalent bond as discussed by Henderson et al (192). $N_{\delta1}$ of His 57 is hydrogen bonded to Asp 102. His 57 is not able to move into a position where a proton donation to the amide nitrogen of Ala 16I is possible. Instead this nitrogen is hydrogen bonded to the carbonyl oxygen of Gly 36I. These structural features may be the reason why the peptide bond between Lys 15I and Ala 16I is not broken. The association constant $K_{assoc} > 10^{14}$ is due to the tight interaction and the many types of contacts between the enzyme and the inhibitor, although the contact area is small.

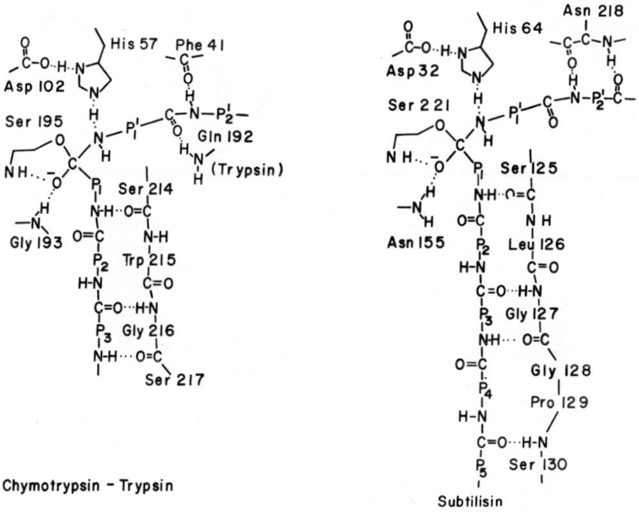

Chymotrypsin – Trypsin

Subtilisin

Figure 5   Similarity of substrate binding sites in chymotrypsin and subtilisin.

The trypsin-PTI complex is a beautiful example of how an ideal substrate interacts with the related enzyme and provides considerable evidence that the proposed enzyme mechanism (Henderson et al 192) is essentially right.

The remarkable finding that the catalytic sites of subtilisin and chymotrypsin closely resemble each other even though the overall folding is totally different (Wright et al 85, Alden et al 193) has now been extended in a number of interesting studies (87, 88, 194, 195). Figure 5 illustrates parts of the similarities. Robertus et al (87) have compared the positions of 27 active site atoms in α-chymotrypsin and subtilisin BPN'. The mean deviation was found to be no more than 0.8 Å. Furthermore the way the residues in the charge relay systems (Ser-His-Asp) are rigidly held by a number of hydrogen bonds to main chain and side chain atoms are similar (Wright 195). Due to these rigid arrangements, only very small movements of the protein within the active site region are observed.

With the close similarities of the active sites of chymotrypsin, elastase, trypsin, and subtilisin it seems very likely that the catalytic mechanisms are essentially identical. Slightly different proposals have been made for chymotrypsin (Henderson et al 192) and subtilisin (Robertus et al 88). The subtilisin mechanism (Figure 6) is probably preferable since it provides a stereochemical mechanism for stabilizing transition state with respect to both the Michaelis complex and the acylenzyme. The only observed tetrahedral intermediate, the trypsin-PTI complex (Rühlmann et al 92), fits excellently with the proposed mechanism.

SUBSTRATE SPECIFICITY   The serine proteinases also provide an interesting example of how the substrate specificity can be studied using X-ray methods. It has been

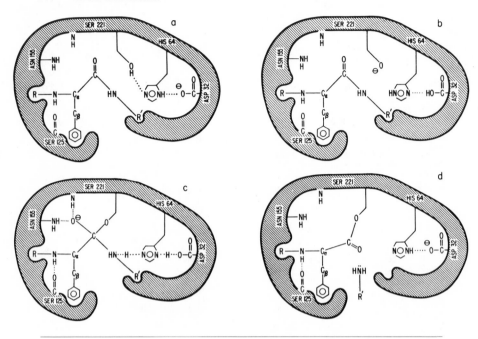

Figure 6 A schematic representation of the mechanism of action of subtilisin BPN′ for a good polypeptide substrate. Substrate residues $P_2–P_4$ are represented by R, residues P'_1 and P'_2 by R′. (*a*) Michaelis complex, (*b*) activated charge relay system, (*c*) tetrahedral addition product, (*d*) acyl intermediate, with leaving group diffusing away. (From Robertus et al 87)

well known for proteinases that not only the residue on the amino-terminal side is important for the specific binding of substrates, but a number of residues on each side of the scissile bond influence the binding of the peptide (Schechter & Berger 196).

The most striking difference between the serine proteinases is the specificity pocket next to the catalytic site. Chymotrypsin has a narrow and smooth hydrophobic pocket just big enough to bind aromatic side chains. Subtilisin has a more cleft-like site for the P1 residue which is less smooth than in chymotrypsin, thus allowing a wider range of residues close to the scissile bond. Elastase has the entrance to the specificity pocket blocked by residues Val 216 and Thr 226, both being glycyl residues in chymotrypsin. Thus elastase prefers only smaller uncharged residues at this site. Residue 189 in the specificity pocket is serine in chymotrypsin and aspartate in trypsin, thus yielding higher affinity for basic lysyl and arginyl residues in the latter. The other binding sites are in general less specific but certainly do exist. In position P2 the space for side chains is limited with a preference for small hydrophobic residues in chymotrypsin (24, 87, 88). The stereo-specificity for L-amino acids (Morihara et al 197) is easily accounted for in the study of subtilisin BPN' (Robertus et al 87). A similar consideration has also been given by Segal et al (24).

In case of the 3.5 Å investigation of peptide binding to elastase (Shotton et al 39) the binding mode differed from the one observed in γ-chymotrypsin. In light of the structure of the trypsin-PTI complex (Rühlmann et al 92) it becomes clear that the binding observed in elastase is the leaving group binding to the P'_n sites. In one of the difference maps (Shotton et al 39) a weak ridge of positive electron density was found in a place identical to the peptide binding observed in γ-chymotrypsin and corresponding to the position for a peptide in an acylenzyme complex.

Structure-Function Relationships in Dehydrogenases

A number of complexes of dogfish M_4 lactate dehydrogenase have been studied to elucidate relevant facts concerning the catalytic mechanism for the enzyme. Co-enzyme and coenzyme fragments can be soaked into crystals of the apoenzyme (Chandrasekhar et al 61). Soaking with the coenzyme causes a loss of symmetry (Adams et al 198). Only the molecular axis Q (Rossmann et al 137) is retained. 5'-AMP and 5'-ADP cause the same structural change at low pH whereas adenosine is able to bind but unable to produce the loss of symmetry. It seems likely that the phosphates are involved in the structural change mentioned, the pH difference being due to titration of a phosphate. The ternary complexes provide detailed information concerning coenzyme and substrate binding as well as concomitant changes in protein structure (Adams et al 59). The molecule in these crystals has again 222 symmetry. The coenzyme is bound in an open conformation with the adenine in a hydrophobic cleft and the rest of the coenzyme held against a hydro-phobic side of the active site cavity (Figures 2 and 4). In LDH it has been found that a few hydrogen bonds control the exact position of the coenzyme. The nicotinamide is buried most deeply in the molecule. The carboxyl group of pyruvate forms an ion pair with the guanidinium group of Arg 171. In the active complex the carbonyl group of the substrate hydrogen bonds to His 195. The protein undergoes drastic conformational changes when binding coenzyme and substrate. Arg 101 moves to form an ion pair with the pyrophosphate of the coenzyme. Most outstanding is the closing of the active site cavity by the movement of a loop of the polypeptide chains. A number of positively and negatively charged residues are thereby excluded from contact with the solvent but probably form a number of ion pairs in close proximity of the nicotinamide and the substrate. The functional importance of this is not fully understood. During the catalytic reaction the A side hydrogen of the C4 position of the nicotinamide is transferred to the C2 carbon of the pyruvate. His 195 donates a proton to the pyruvate carbonyl oxygen, yielding the product lactate (Figure 7). The change in formal charge of the histidine and the nicotinamide during the reaction is probably balanced by small rearrangements of the charged residues in the active site. Kinetic data and chemical modifications are in agreement with the X-ray observations.

These observations for LDH are generally also applicable to the other known

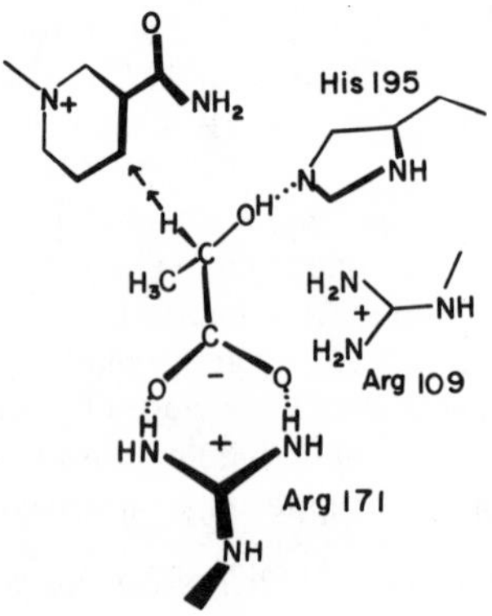

Figure 7 Diagrammatic representation of anticipated substrate binding in the active ternary intermediate of lactate dehydrogenase.

dehydrogenase structures (s-MDH, L-ADH, GPD). The coenzyme is open with a similar conformation in all cases. The nicotinamide ring is rotated by 180° in GPD to make it a B-side specific enzyme. The substrate site is in all cases equivalently positioned with regard to secondary and tertiary structure. The electron density of s-MDH suggests the presence of arginine in position 171 as in LDH. His 176 in GPD probably corresponds to His 195 in LDH. Both GPD and L-ADH have significantly shorter loops.

CONCLUSION

The numerous new structures being unveiled with ever increasing frequency (Matthews 7) have revolutionized our understanding of protein structure and function. Yet it became clear to the authors of this review that only few examples exist of deep and thorough investigations. For instance the study of hemoglobin by Perutz and others has revealed its structure and function relationship within a single polypeptide chain and has demonstrated its physiological interactions between subunits.

A challenge for future crystallographic investigations is not only the complex structures of allosteric enzymes, antibodies, and viruses, but also to attempt a deeper understanding of structure-function relationships of protein molecules. A greater emphasis on the comparison of known conformations should reveal structural domains with similar functional requirements and common genetic precursors. The current diversity of information may then become part of a unified evolutionary pattern.

ACKNOWLEDGMENTS

We are grateful to Drs. R. H. Kretsinger, M. Laskowski Jr., J. E. Johnson, M. L. Hackert, and K. W. Olsen for helpful discussions. We are indebted to the invaluable work of Sharon S. Wilder in the preparation and checking of the manuscript and to Elaine J. Hackert for typing assistance. Figures 2 and 6 are reproduced with the permission of Dr. C. I. Brändén of the Agricultural College of Sweden and Professor J. Kraut of the University of California at San Diego, respectively. We have had support from NSF Grant #GB29596x and NIH Grants #GM10704 and #AI11219 during the writing of the review.

Literature Cited

1. Blow, D. M., Steitz, T. A. 1970. *Ann. Rev. Biochem.* 39:63–100
2. Dickerson, R. E. 1972. *Ann. Rev. Biochem.* 41:815–42
3. Hess, G. P., Rupley, J. A. 1971. *Ann. Rev. Biochem.* 40:1013–44
4. Blake, C. C. F. 1972. *Progr. Biophys.* 25:83–130
5. Vainshtein, B. K., Borisov, V. V. 1974. *Advan. in Biol. Chem. USSR.* In press
6. Blundell, T. L., Johnson, L. N. 1973. *MTP Biennial Rev. Sci., Phys. Chem. II, Chem. Crystallogr.*
7. Matthews, B. W. 1974. *The Proteins,* ed. H. Neurath, R. L. Hill. New York: Academic. 3rd ed. In press
8. Matthews, B. W., Bernhard, S. A. 1973. *Ann. Rev. Biophys. Bioeng.* 2:257–312
9. Schulz, G. E., Biederman, K., Elzinga, M., Marx, F., Schirmer, R. H. 1973. *Stockholm Symp. Struct. Biol. Molecules,* 84
10. Brändén, C. I. et al. 1973. *Proc. Nat. Acad. Sci. USA* 70:2439–42
11. Schiffer, M., Girling, R. L., Ely, K. R., Edmundson, A. B. 1974. *Biochemistry.* In press
12. Kannan, K. K. et al 1973. See Ref. 9, p. 76
13. Khalifah, R. G. 1971. *J. Biol. Chem.* 246:2561–73
14. Coleman, J. E. 1971. *Progr. Bioorg. Chem.* 1:159–344
15. Liljas, A. et al 1972. *Nature New Biol.* 235:131–37
16. Bergstén, P. C. et al 1972. *Oxygen Affinity of Hemoglobin and Red Cell Acid Base Status, Alfred Benzon Symp IV,* ed. M. Rørth, P. Astrup, 363–83. New York: Academic
17. Quiocho, F. A., Lipscomb, W. N. 1971. *Advan. Protein Chem.* 25:1–78
18. Lipscomb, W. N., Reeke, G. N. Jr., Hartsuck, J. A., Quiocho, F. A., Bethge, P. H. 1970. *Phil. Trans. Roy. Soc. London B* 257:177–214
19. Hartsuck, J. A., Lipscomb, W. N. 1971. *The Enzymes,* ed. P. D. Boyer, 3:1–56. New York: Academic. 3rd ed.
20. Birktoft, J. J., Blow, D. M. 1972. *J. Mol. Biol.* 68:187–240
21. Henderson, R. 1970. *J. Mol. Biol.* 54:341–54
22. Steitz, T. A., Henderson, R., Blow, D. M. 1969. *J. Mol. Biol.* 46:337–48
23. Tulinsky, A., Vandlen, R. L., Morimoto, C. N., Marri, N. V., Wright, L. H. 1974. *Nature New Biol.* In press
24. Segal, D. M., Powers, J. C., Cohen, G. H., Davies, D. R., Wilcox, P. E. 1971. *Biochemistry* 10:3728–37
25. Freer, S. T., Kraut, J., Robertus, J. D., Wright, H. T., Xuong, Ng. H. 1970. *Biochemistry* 9:1997–2009
26. Weinzierl, J., Kalb, A. J. 1971. *FEBS Lett.* 18:268–70
27. Jack, A., Weinzierl, J., Kalb, A. J. 1971. *J. Mol. Biol.* 58:389–95
28. Hardman, K. D., Ainsworth, C. F. 1972. *Biochemistry* 11:4910–19
29. Hardman, K. D., Ainsworth, C. F. 1972. *Nature New Biol.* 237:54–55
30. Reeke, G. N. et al 1974. *Ann. NY Acad. Sci.* In press
31. Becker, J. W., Reeke, G. N. Jr., Edelman, G. M. 1971. *J. Biol. Chem.* 246:6123–25
32. Mathews, F. S., Argos, P., Levine, M. 1971. *Cold Spring Harbor Symp. Quant. Biol.* 36:387–95
33. Dickerson, R. E. et al. 1971. *J. Biol. Chem.* 246:1511–35
34. Brown, J. N., Takano, T., Dickerson, R. E. 1974. *J. Mol. Biol.* In press
35. Takano, T., Kallai, O. B., Swanson, R., Dickerson, R. E. 1973. *J. Biol. Chem.* 248:5234–55
36. Tanaka, N., Yamane, T., Tsukihara, T., Ashida, T., Kakudo, M. 1973. *Ann. Meet. Chem. Soc. Japan, 28th*
37. Takano, T., Swanson, R., Kallai, O. B., Dickerson, R. E. 1971. See Ref. 32, 397–404
38. Salemme, F. R., Freer, S. T., Xuong, Ng. H., Alden, R. A., Kraut, J. 1973. *J. Biol. Chem.* 248:3910–21
38a. Timkovich, R., Dickerson, R. E. 1974. In press

39. Shotton, D. M., White, N. J., Watson, H. C. 1971. See Ref. 32, 91–105
40. Poljak, R. J. et al. 1973. *Abstr. Int. Congr. Biochem.,* 9th, p. 301
41. Padlan, E. et al. 1973. See Ref. 9, p. 73
42. Adman, E. T., Sieker, L. C., Jensen, L. H. 1973. *J. Biol. Chem.* 248:3987–96
43. Carter, C. W. Jr. et al. 1972. *Proc. Nat. Acad. Sci. USA* 69:3526–29
44. Andersen, R. D. et al. 1972. *Proc. Nat. Acad. Sci. USA* 69:3189–91
45. Watenpaugh, K. D., Sieker, L. C., Jensen, L. H., Legall, J., Dubourdieu, M. 1972. *Proc. Nat. Acad. Sci. USA* 69: 3185–88
46. Buehner, M., Ford, G. C., Moras, D., Olsen, K. W., Rossmann, M. G. 1973. *Proc. Nat. Acad. Sci. USA.* 70:3052–54
47. Love, W. E. et al. 1971. See Ref. 32, 349–57
48. Perutz, M. F. 1970. *Nature* 228:726–39
49. Morimoto, H., Lehmann, H., Perutz, M. F. 1971. *Nature* 232:408–13
50. Huber, R., Epp, O., Steigemann, W., Formanek, H. 1971. *Eur. J. Biochem.* 19:42–50
51. Huber, R., Epp, O., Formanek, H. 1970. *J. Mol. Biol.* 52:349–54
52. Perutz, M. F., Muirhead, H., Cox, J. M., Goaman, L. C. G. 1968. *Nature* 219: 131–39
53. Bolton, W., Perutz, M. F. 1970. *Nature* 228:551–52
54. Muirhead, H., Greer, J. 1970. *Nature* 228:516–19
55. Arnone, A. 1972. *Nature* 237:146–49
56. Hendrickson, W. A., Love, W. E. 1971. *Nature New Biol.* 232:197–203
57. Blundell, T. L., Dodson, G., Hodgkin, D., Mercola, D. 1972. *Advan. Protein Chem.* 26:279–402
58. Sakabe, N., Sakabe, K., Katayama, C. 1972. *Int. Union Crystallogr., Tokyo,* Abstr. S34
59. Adams, M. J. et al. 1973. *Proc. Nat. Acad. Sci. USA* 70:1793–94
60. Adams, M. J., Ford, G. C., Liljas, A., Rossmann, M. G. 1973. *Biochem. Biophys. Res. Commun.* 53:46–51
61. Chandrasekhar, K., McPherson, A. Jr., Adams, M. J., Rossmann, M. G. 1973. *J. Mol. Biol.* 76:503–18
62. Adams, M. J., Liljas, A., Rossmann, M. G. 1973. *J. Mol. Biol.* 76:519–31
63. Imoto, T., Johnson, L. N., North, A. C. T., Phillips, D. C., Rupley, J. A. 1972. *The Enzymes,* ed. P. D. Boyer, 7:665. New York: Academic. 3rd ed.
64. Moult, J., Yonath, A., Rabinovich, D., Traub, W. 1973. See Ref. 9, p. 77
65. Hill, E., Tsernoglou, D., Webb, L., Banaszak, L. J. 1972. *J. Mol. Biol.* 72: 577–91
66. Webb, L., Hill, E., Banaszak, L. J. 1974. *J. Biol. Chem.* In press
67. Kretsinger, R. H., Nockolds, C. E. 1973. *J. Biol. Chem.* 248:3313–34
68. Watson, H. C. 1969. *Progr. Stereochem.* 4:299–333
69. Levitt, M., Lifson, S. 1969. *J. Mol. Biol.* 46:269–79
70. Nobbs, C. L., Watson, H. C., Kendrew, J. C. 1966. *Nature* 209:339–41
71. Scouloudi, H. 1969. *J. Mol. Biol.* 40: 353–77
72. Cotton, F. A., Bier, C. J., Day, V. W., Hazen, E. E. Jr., Larsen, S. 1971. See Ref. 32, 243–49
73. Deisenhofer, H., Steigemann, W. 1973. See Ref. 9, p. 98
74. Drenth, J., Jansonius, J. N., Koekoek, R., Wolthers, B. G. 1971. *Advan. Protein Chem.* 25:79–115
75. Blake, C. C. F., Evans, P. R. 1973. See Ref. 40, p. 40
76. Watson, H. C. 1973. See Ref. 40, p. 39
77. Smit, J. D. G., Ploegman, J. H., Kalk, K. H., Jansonius, J. N., Drenth, J. 1973. See Ref. 9, p. 54
78. Kartha, G., Bello, J., Harker, D. 1967. *Nature* 213:862–65
79. Carlisle, C. H., Gorinsky, B. A., Bazumdar, S. K., Palmer, R. A., Yeates, D. G. R. 1972. See Ref. 58, S36
80. Richards, F. M., Wyckoff, H. W. 1971. *The Enzymes,* ed. P. D. Boyer, 4:647–806. New York: Academic
81. Wyckoff, H. W. et al. 1970. *J. Biol. Chem.* 245:305–28
82. Richards, F. M. et al. 1971. See Ref. 32, 35–43
83. Roberts, G. C. K., Dennis, E. A., Meadows, D. H., Cohen, J. S., Jardetzky, O. 1969. *Proc. Nat. Acad. Sci. USA* 62:1151–58
84. Watenpaugh, K. D., Sieker, L. C., Herriott, J. R., Jensen, L. H. 1971. See Ref. 32, 359–67
85. Wright, C. S., Alden, R. A., Kraut, J. 1969. *Nature* 221:235–42
86. Alden, R. A., Birktoft, J. J., Kraut, J., Robertus, J. D., Wright, C. S. 1971. *Biochem. Biophys. Res. Commun.* 45: 337–44
87. Robertus, J. D. et al. 1972. *Biochemistry* 11:2439–49
88. Robertus, J. D., Kraut, J., Alden, R. A., Birktoft, J. J. 1972. *Biochemistry* 11: 4293–4303
89. Drenth, J., Hol, W. G. J., Jansonius, J. N., Koekoek, R. 1971. See Ref. 32, 107–16
90. Colman, P. M., Jansonius, J. N., Matthews, B. W. 1972. *J. Mol. Biol.* 70:701–24

91. Stroud, R. M., Kay, L. M., Dickerson, R. E. 1974. *J. Mol. Biol.* In press
92. Rühlmann, A., Kukla, D., Schwager, P., Bartels, K., Huber, R. 1973. *J. Mol. Biol.* 77:417–36
93. Sweet, R. M., Wright, H. T., Janin, J., Blow, D. M. 1973. See Ref. 9, p. 56
94. Diamond, R. 1966. *Acta Crystallogr.* 21:253–66
95. Diamond, R. 1971. *Acta Crystallogr. A* 27:436–52
96. Anfinsen, C. B., Haber, E., Sela, M., White, F. M. 1961. *Proc. Nat. Acad. Sci. USA* 47:1309–14
97. Anfinsen, C. B. 1972. *Biochem. J.* 128:737–49
98. Astbury, W. T. 1938. *Trans. Faraday Soc.* 34:378–88
99. Dickerson, R. E., Geis, I. 1969. *The Structure and Action of Proteins.* New York: Harper & Row
100. Ramakrishnan, C., Prasad, N. 1971. *Int. J. Protein Res.* 3:209–31
101. Donohue, J. 1968. *Structural Chemistry and Molecular Biology,* ed. A. Rich, N. Davidson, 443–65. San Francisco: Freeman
102. Richards, F. M., Wyckoff, H. W., Allewell, N. 1970. *The Neurosciences,* 901–12. New York: Rockefeller Univ. Press
103. Mathews, F. S., Levine, M., Argos, P. 1972. *J. Mol. Biol.* 64:449–64
104. Adams, M. J. et al. 1972. *Protein-Protein Interactions,* ed. R. Jaenicke, E. Helmreich, 139–58. New York: Springer
105. Huber, R., Kukla, D., Rühlmann, A., Steigemann, W. 1971. See Ref. 32, 141–50
106. Crawford, J. L., Lipscomb, W. N., Schellman, C. G. 1973. *Proc. Nat. Acad. Sci. USA* 70:538–42
107. Ramachandran, G. N., Sasisekharan, V. 1968. *Advan. Protein Chem.* 23:283–437
108. Chothia, C. 1973. *J. Mol. Biol.* 75:295–302
109. Edelman, G. M. et al. 1972. *Proc. Nat. Acad. Sci. USA* 69:2580–84
110. Nagano, K. 1973. *J. Mol. Biol.* 75:401–20
111. Lewis, P. N., Momany, F. A., Scheraga, H. A. 1971. *Proc. Nat. Acad. Sci. USA* 68:2293–97
112. Robson, B., Pain, R. H. 1971. *J. Mol. Biol.* 58:237–59
113. Wu, T. T., Kabat, A. 1973. *J. Mol. Biol.* 75:13–31
114. Phillips, D. C. 1970. *Brit. Biochem. Past Present,* ed. T. W. Goodwin, 11–28. London: Academic
114a. Ooi, T., Nishikawa, K. 1973. *Conformation of Biological Molecules and Polymers,* ed. E. D. Bergmann, B. Pullman, 173–87. New York: Academic
115. Rossmann, M. G., Blow, D. M. 1962. *Acta Crystallogr.* 15:24–31
116. Edelman, G. M. et al. 1969. *Proc. Nat. Acad. Sci. USA* 63:78–85
117. Welscher, H. D. 1969. *Int. J. Protein Res.* 1:267–82
118. Epp, O. et al. 1972. *J. Mol. Biol.* 69:315–18
119. Kretsinger, R. H. 1972. *Nature New Biol.* 240:85–87
120. McLachlan, A. D. 1972. *Nature New Biol.* 240:83–85
121. Dickerson, R. E. 1971. *J. Mol. Biol.* 57:1–15
122. Rao, S. T., Rossmann, M. G. 1973. *J. Mol. Biol.* 76:241–56
123. Carter, C. W. Jr., Freer, S. T., Xuong, Ng. H., Alden, R. A., Kraut, J. 1971. See Ref. 32, 381–85
124. Baltscheffsky, H., Baltscheffsky, M. 1974. *Ann. Rev. Biochem.* 43:871–95
125. Jörnvall, H. 1973. *Proc. Nat. Acad. Sci. USA* 70:2295–98
126. Taylor, S. S., Oxley, S. L., Allison, W. S., Kaplan, N. O. 1973. *Proc. Nat. Acad. Sci. USA* 70:1790–93
127. Jörnvall, H. 1970. *Eur. J. Biochem.* 16:25–40
128. Davidson, B. E., Sajgò, M., Noller, H. F., Harris, J. I. 1967. *Nature* 216:1981–85
129. Wendell, P. L., Bryant, T. N., Watson, H. C. 1972. *Nature New Biol.* 240:134–36
130. Blake, C. C. F., Evans, P. R., Scopes, R. K. 1972. *Nature New Biol.* 235:195–98
131. Steitz, T. A., Fletterick, R. J., Hwang, K. J. 1973. *J. Mol. Biol.* 78:551–61
132. Jardetzky, O., Theilmann, H., Arata, Y., Markley, J. L., Williams, M. N. 1971. See Ref. 32, 257–61
133. Wetlaufer, D. 1973. *Proc. Nat. Acad. Sci. USA* 70:697–701
134. Porter, R. R. 1959. *Biochem. J.* 73:119–27
135. Monod, J., Wyman, J., Changeux, J. P. 1965. *J. Mol. Biol.* 12:88–118
136. Koshland, D. E., Nemethy, G., Filmer, D. 1966. *Biochemistry* 5:365–85
137. Rossmann, M. G. et al. 1973. *J. Mol. Biol.* 76:533–37
138. Conway, A., Koshland, D. E. 1968. *Biochemistry* 7:4011–23
139. Muirhead, H., Cox, J. M., Mazzarella, L., Perutz, M. F. 1967. *J. Mol. Biol.* 28:117–56
140. Warren, S. G., Edwards, B. F. P., Evans, D. R., Wiley, D. C., Lipscomb,

W. N. 1973. *Proc. Nat. Acad. Sci. USA* 70:1117–21

141. Caspar, D. L. D., Klug, A. 1962. See Ref. 32, 27:1–23

142. Barrett, A. N. et al. 1971. See Ref. 32, 433–48

143. Butler, P. J. G., Klug, A. 1972. *Proc. Nat. Acad. Sci. USA* 69:2950–53

144. Bancroft, J. B., Hills, G. J., Markham, R. 1967. *Virology* 31:354–79

145. Waara, I., Lövgren, S., Liljas, A., Kannan, K. K., Bergstén, P. C. 1972. *Advan. Exp. Med. Biol.* 28:169–87

146. Banaszak, L. J., Watson, H. C., Kendrew, J. C. 1965. *J. Mol. Biol.* 12:130–37

147. Eisenberg, D. 1970. *The Enzymes,* ed. P. D. Boyer, 1:1–89. New York: Academic. 3rd ed.

148. Matthews, B. W., Weaver, L. H. 1974. *Biochemistry* In press

149. Malmström, B. G., Rydén, L. 1968. *Biological Oxidations,* ed. T. P. Singer, 415–38. New York: Interscience

150. Vallee, B. L., Walker, W. C. 1970. *The Proteins,* ed. H. Neurath, 5:1–192. New York: Academic. 2nd ed.

151. Jensen, L. H. 1974. *Ann. Rev. Biochem.* 43:461–74

152. Feder, J., Garrett, L. R., Wildi, B. S. 1971. *Biochemistry* 10:4552–56

153. Yazgan, A., Henkens, R. W. 1972. *Biochemistry* 11:1314–18

154. Dodson, E., Harding, M. M., Hodgkin, D. C., Rossmann, M. G. 1966. *J. Mol. Biol.* 16:227–41

155. Vallee, B. L., Williams, R. J. P. 1968. *Proc. Nat. Acad. Sci. USA* 59:498–505

156. Lindskog, S. 1970. *Struct. Bonding* 8:153–96

157. Latt, S. A., Vallee, B. L. 1971. *Biochemistry* 10:4263–69

158. Coleman, J. E., Coleman, R. V. 1972. *J. Biol. Chem.* 247:4718–28

159. Taylor, J. S., Coleman, J. E. 1971. *J. Biol. Chem.* 246:7058–67

160. Matthews, B. W. 1968. *J. Mol. Biol.* 33:491–97

161. Schoenborn, B. P. 1971. See Ref. 32, 569–75

162. Phillips, D. C. 1972. *Harvey Lect.* 66:135–60

163. Kuntz, I. D. Jr., Brassfield, T. S., Law, G. D., Purcell, G. V. 1969. *Science* 163:1329–31

164. Kauzmann, W. 1959. *Advan. Protein Chem.* 14:1–63

165. Coleman, J. E. 1974. *MTP Int. Rev. Sci.* In press

166. Némethy, G., Scheraga, H. A. 1962. *J. Chem. Phys.* 36:3382–3400

167. von Hippel, P. H., Schleich, T. 1969.

Structure and Stability of Biological Macromolecules, ed. S. N. Timasheff, G. D. Fasman, 2:417–574. New York: Dekker

168. Eisenberg, D., Kauzmann, W. 1969. *The Structure and Properties of Water.* Oxford, England: Oxford Univ. Press

169. Brandts, J. F., Oliveira, R. J., Westort, C. 1970. *Biochemistry* 9:1038–47

170. Lee, B., Richards, F. M. 1971. *J. Mol. Biol.* 55:379–400

171. Hofmeister, F. 1888. *Arch. Exp. Pathol. Pharmacol.* 24:247

172. Chilson, O. P., Costello, L. A., Kaplan, N. O. 1965. *Biochemistry* 4:271–81

173. Markert, C. L., Massaro, E. J. 1966. *Arch. Biochem. Biophys.* 115:417–26

174. Jaenicke, R., Koberstein, R., Teuscher, B. 1971. *Eur. J. Biochem.* 23:150–59

175. Lindskog, S. et al. 1971. *The Enzymes,* ed. P. D. Boyer, 5:587–665. New York: Academic. 3rd ed.

176. Edsall, J. T. 1968. See Ref. 101, 88–97

177. Rupley, J. A. 1969. See Ref. 167, 291–352

178. Johansen, J. T., Vallee, B. L. 1971. *Proc. Nat. Acad. Sci. USA* 68:2532–35

179. Johansen, J. T., Livingston, D. M., Vallee, B. L. 1972. *Biochemistry* 11:2585–88

180. Quiocho, F. A., McMurray, C. H., Lipscomb, W. N. 1972. *Proc. Nat. Acad. Sci. USA* 69:2850–54

180a. Johansen, J. T., Vallee, B. L. 1973. *Proc. Nat. Acad. Sci. USA* 70:2006–10

181. Quiocho, F. A., Richards, F. M. 1966. *Biochemistry* 5:4062–76

182. Sundaralingam, M. 1969. *Biopolymers* 7:821–60

183. Arnott, S., Hukins, D. W. L. 1969. *Nature* 224:886–88

184. Arnott, S., Hukins, D. W. L. 1972. *Biochem. J.* 130:453–65

185. Kim, S. H., Berman, H. M., Seeman, N. C., Newton, M. D. 1973. *Acta Crystallogr. B* 29:703–10

186. Sobell, H. M., Jain, S. C., Sabore, T. D., Ponticello, G., Nordman, C. E. See Ref. 32, 263–70

187. Barry, C. D., North, A. C. T., Glasel, J. A., Williams, R. J. P., Xavier, A. V. 1971. *Nature* 232:236–45

188. Voet, D. 1973. *J. Am. Chem. Soc.* 95:363–70

189. Blow, D. M. et al. 1972. *J. Mol. Biol.* 69:137–44

190. Laskowski, M. Jr., Sealock, R. W. 1971. *The Enzymes,* ed. P. D. Boyer, 3:375–473. New York: Academic. 3rd ed.

191. Wilson, K. A., Laskowski, M. Sr. 1971. *J. Biol. Chem.* 246:3555–61

192. Henderson, R., Wright, C. S., Hess, G. P., Blow, D. M. 1971. See Ref. 32, 63–70
193. Alden, R. A., Wright, C. S., Kraut, J. 1970. *Phil. Trans. Roy. Soc. London B* 257:119–24
194. Kraut, J. et al. See Ref. 32, 117–23
195. Wright, C. S. 1972. *J. Mol. Biol.* 67: 151–63

196. Schechter, I., Berger, A. 1967. *Biochem. Biophys. Res. Commun.* 27:157–62
197. Morihara, K., Oka, T., Tsuzuki, H. 1970. *Arch. Biochem. Biophys.* 138: 515–25
198. Adams, M. J., McPherson, A. Jr., Rossmann, M. G., Schevitz, R. W., Wonacott, A. J. 1970. *J. Mol. Biol.* 51: 31–38

PEPTIDE HORMONES

× 856

Howard S. Tager[1] *and Donald F. Steiner*

Department of Biochemistry, University of Chicago, Chicago, Illinois

CONTENTS

INTRODUCTION

The last comprehensive review in this series on the biochemistry of peptide hormones appeared in 1969 (1). However, various aspects of this subject have been reviewed frequently in the *Annual Reviews of Biochemistry, Physiology, Pharmacology,* and *Medicine,* as well as in other review series including *Recent Progress in Hormone Research, Hormonal Proteins and Peptides,* and *Vitamins and Hormones.* This volume includes a review of peptide hormone binding to cellular constituents and the possible relationship of this phenomenon to their biological effects (2). The *Atlas of Protein Sequence and Structure* by Dayhoff[2] is also a valuable source for comparisons of primary structural relationships among various peptide hormones.

In view of the breadth of this topic, our discussion is limited to structural and biosynthetic studies; peptide hormone secretion and action are not considered. This is necessitated by the abundance of new information and concepts in all these areas. Rapidly accumulating sequence information on peptide hormones has provided interesting new clues to evolutionary and functional interrelationships among many hormones. Thus, several groups of related peptide hormones appear to have evolved from a relatively small number of ancestral proteins. Likewise,

---

[1] Present Address: Department of Biochemistry, Medical College of Ohio, Toledo, Ohio 43614.

[2] Published by the National Biochemical Research Foundation, Silver Spring, Md. The last comprehensive edition appeared in 1972 (Vol. 5). Supplements are published annually.

509

recent biosynthetic studies of a variety of endocrine peptides indicate that the primary gene products differ significantly from the known peptide hormones. In at least six cases these peptides appear to be synthesized from larger precursors, and multiple molecular forms of the same hormone, representing biosynthetic intermediates as well as metabolites, may circulate in the blood. These new findings have altered earlier notions regarding the evolution, biosynthesis, and active forms of peptide hormones.

INSULIN

The last three to four years have seen considerable progress in elucidating the mechanism of insulin biosynthesis and in isolating and characterizing proinsulin (3, 4) and related peptide products from various species. It is now firmly established that proinsulin is synthesized on ribosomes in the rough endoplasmic reticulum of the β cells (5–7) and that the precursor is then transferred via an energy-dependent process to the Golgi apparatus (8–11), where proteolytic conversion to insulin begins (12–15). Conversion continues to about 95% completion over a period of hours within newly formed secretory granules after they have formed by budding from the inner lamallae of the Golgi apparatus (7, 13–15). The process may be terminated at this stage by cocrystallization of the insulin with the residual proinsulin and intermediate products (16), giving rise to the dense crystalline granule inclusion that can be seen by electron microscopy within mature granules (17, 18). This small amount of proinsulin retained within the granules is secreted subsequently with insulin and has been identified in the circulation of man and animals (19–22).

The conversion process has been studied in greater detail in secretion granule fractions isolated from islets prelabeled with radioactive amino acids in vitro (7, 12–15). Conversion in these particles exhibits a relatively sharp pH optimum at or slightly above pH 6.0 (15, 23). At least two kinds of proteolytic activity appear to be required for the conversion of proinsulin to insulin. The first is an endopeptidase with trypsin-like specificity which cleaves on the carboxyl side of the pairs of basic residues that link the connecting polypeptide chain to the termini of the insulin A and B chains (4, 24). The second is an exopeptidase having specificity similar to carboxypeptidase B, which removes the C-terminal basic residues from both insulin and the C peptide[3] (25). Both kinds of activities have been demonstrated in disrupted secretion granule preparations (7, 15), and the endopeptidase activity may be associated with the secretion granule membrane (7). Further support for this localization of the proinsulin converting enzymes within the β cells has been obtained in electron microscopic histochemical studies (26). Sufficient amounts of these enzymes for more detailed chemical characterization

---

[3] The arrangement of the mammalian proinsulin polypeptide chain is: NH_2-A chain·Arg·Arg·C-peptide·Lys·Arg·B chain-COOH. The C peptide thus becomes that portion of the connecting peptide sequence, aside from the pairs of basic residues at the ends, which is removed in the conversion to insulin (14).

have not been obtained from β granules, and studies with a variety of proteolytic inhibitors have been inconclusive in establishing their relationships to other known trypsin-like enzymes or carboxypeptidases (15). A further complication arises from the recent finding that a chymotrypsin-like cleavage occurs in rat connecting peptides during their proteolytic excision in incubated whole rat islets (27). These results suggest that the secretory granules may contain low levels of several kinds of proteolytic enzyme activities. Specificity of cleavage thus may be determined as much by tertiary structural features of the substrates as by special adaptations of the proteases. An enzyme that converts proinsulin has been isolated from whole pancreas, but it has not been fully characterized with respect to either its origin in the islet tissue of the pancreas or its cleavage specificity and mechanism of action (28).

In view of a recent report that immunoglobulin chains having extended amino-terminal regions are synthesized during in vitro translation of myeloma cell mRNA fractions (29), one might inquire whether insulin precursors larger than proinsulin also exist. However, aside from some evidence for the expected role of an N-terminal residue of methionine in the initiation of proinsulin synthesis in fetal calf pancreas (30), no convincing indications of larger precursors have been found. Islet polysomes active in proinsulin synthesis appear to be mainly trisomes, a size consistent with the expected mRNA length of about 258 nucleotides required to encode the 86-residue proinsulin polypeptide (31).

Comparative studies of insulin biosynthesis in the cod (32) and angler fish (33), as well as in such primitive vertebrates as cyclostomes (34), indicate the formation and cleavage of a proinsulin similar in size to the mammalian proteins. A requirement for trypsin-like cleavage has been demonstrated for both of the fish proinsulins, and an interesting intermediate cleavage form, having an N-terminal tripeptide A-chain extension, has been isolated from angler fish islets by Yamaji et al (35). A number of reports have appeared on the biosynthesis, isolation, and characterization of intermediate forms of mammalian proinsulins in various species (15, 24, 27, 36–40).

The proinsulin C peptide, somewhat analogous to the activation peptide in some zymogen proteins, also has become a focus of attention. Due to localization of the conversion process within secretion granules, the C peptide accumulates with insulin in equimolar amounts (41) and is secreted along with the hormone by exocytosis of the granule contents (42). C peptides from nine mammalian and one avian species have been isolated and sequenced (41, 43–50). A high rate of mutation acceptance—much higher than the rate for insulin and approaching that for the fibrinopeptides (49)—as well as the appearance of deletions in more than one region of the C-peptide sequence, suggest that structural requirements in this portion of proinsulin are less stringent than in the hormonally active portion of the molecule. Whether or not biologically important functions other than efficient peptide chain folding, sulfhydryl oxidation, and specific enzymic cleavage are also encoded in this peptide remains unanswered.

Synthesis of several mammalian C peptides has been accomplished recently by classical fragment condensation (51–56). The synthetic porcine C peptide, containing all four terminal basic residues, was tested for its ability to promote the recombination of insulin A and B chains in vitro, but it failed to influence the yield (57). Synthetic porcine and bovine C peptides cross-react well with antibodies directed against the corresponding natural proinsulins or C peptides, and fragments of these peptides have been successfully utilized to study the antigenic determinants in this region of the proinsulin molecule (53, 54, 58).

Both proinsulin and the C peptide have been detected in the circulation of man and other species by means of specific immunoassays (for a review see 59). The level of proinsulin rises slowly after a glucose load in normal subjects but does not exceed 20% of the total insulin-like immunoreactive material. Although abnormal proportions or absolute concentrations of proinsulin have been found in obesity, chronic renal failure, and patients with severe hypokalemia, the major diagnostic significance of elevated proinsulin levels has been in detecting patients with β cell tumors (60). The immunoassayable C-peptide levels have been shown by Rubenstein and co-workers to mirror changes in insulin levels and thus provide a means for evaluating endogenous insulin production in diabetic individuals in whom insulin antibodies and administered animal insulins invalidate direct insulin measurements (61, 62). The handling of proinsulin and C peptide in vivo differs significantly from that of insulin (63, 64). This factor must be taken into consideration when interpreting changes in peripheral blood levels of immunoreactive insulin and in estimating the relative biological potency of proinsulin by means of in vivo blood glucose-lowering assays. The biological activity of proinsulin on fat cells (65) and muscle tissue (66) in vitro is about 3–5% that of insulin. Higher activity is usually found in vivo, ranging from 20–30%, a result attributable to the slower turnover rate of proinsulin rather than to any proteolytic conversion of proinsulin to insulin in the circulation or tissues (59). The difference in turnover rates of proinsulin and insulin is due largely to the relatively greater uptake and degradation of insulin by the liver, a major site of insulin destruction in the intact organism (67). The precise enzymic mechanism of insulin degradation remains controversial despite a recent renewal of interest in this problem (68–71). The studies of Varandani and co-workers have brought forth new evidence implicating an initial step of reductive cleavage in this process (71–73).

The recent elucidation of the three-dimensional structure of insulin, initially at a resolution of 2.8 Å (74, 75) and with recent refinements at 1.9 Å (76) by Hodgkin and her co-workers, represents an important breakthrough in the study of peptide hormone structure. It is beyond the scope of this review to describe this structure in detail or the growing literature on the chemical and immunological properties and their structural correlations in normal, modified, or synthetic insulin molecules. Several recent reviews are cited (76–80). Modifications include the introduction of various substituent groups on the amino, histidyl, or carboxyl groups of insulin (76, 81–86), selective reduction and substitution of disulfide bonds (87), and the introduction of intramolecular crosslinks between the A1 and B1, or A1 and B29 (ε-lysine) amino groups (79, 88, 89). The latter group of derivatives, especially the series linked between A1 and B29 by dicarboxylic acids (79), are of added interest, since these bridges simulate the naturally occurring connecting polypeptide in proinsulin. Adipoylinsulin, in which adipic acid serves to crosslink the amino groups of residues A1 and B29, has been reduced and reoxidized in vitro under conditions suitable for proinsulin reoxidation with comparable yields ranging up to 75% (90). These results indicate that the role of the connecting peptide in promoting correct pairing of half-cystine residues in proinsulin can be played by a non-peptide molecular prosthesis.

None of the insulin derivatives prepared to date have exhibited higher biological activity than insulin itself. As with proinsulin, assays often indicate higher activity in vivo than in vitro (79), suggesting that many of these analogs may accumulate to a greater extent in the blood as a result of their decreased susceptibility to degradation. This cannot be the case with the identical turkey and chicken insulins which exhibit 2–4 times higher activity than bovine insulin in several in vitro bioassay systems (91). Weitzel and co-workers suggest that this heightened activity may result from enhanced receptor binding due to the substitutions of histidine for alanine and asparagine for serine at positions A8 and A9, respectively. On the other hand, substitution of the B5 histidine by alanine in a synthetic bovine insulin led to lower biological activity (92). Duck insulin may help to clarify the role of substitutions in the A8–10 region in enhancing biological activity of chicken and turkey insulins. This avian species has glutamic acid at position A8 and proline at A10, and the B chain differs only at position 30 where threonine replaces alanine (93). Bioassay results have not yet been reported.

In contrast, guinea pig insulin, which differs from porcine insulin at 17 positions (49), displays significantly lower biological activity in other mammals as well as in guinea pigs (94). In addition to altered molecular topography due to the amino acid changes, guinea pig insulin does not bind zinc and does not form dimers or higher polymers in solution (94). Associated with this rather drastic change in properties is the replacement by asparagine of the B10 histidine residue, which coordinates with zinc in two zinc porcine insulin crystals (74). By contrast, the two insulins in the mouse, an old-world rodent, closely resemble other mammalian insulins and are identical to those of the rat (95).

Another insulin of considerable interest has been isolated from the islet organs of a primitive jawless vertebrate, the Atlantic hagfish (34, 96), an animal belonging to one of two extant orders of the cyclostomes, which are believed to have diverged from the gnathostomes about 600 million years ago (97, 98). Hagfish insulin has about 10% of the activity of bovine insulin in mammalian systems (96). About 40% of its amino acid residues differ from those found in mammalian insulins, including replacement of the zinc-binding B10 histidine residue by aspartic acid (99). Nevertheless, it forms large tetragonal crystals at pH 6.0 in the absence of zinc (34). Almost all of the invariant residues in the known gnathostomian insulins (49), including the half-cystines, are conserved in hagfish insulin.

Studies with selectively degraded or synthetic insulins indicate that the absence of the A1 glycine amino group (100) or the A1 glycine residue (79, 81) results in a loss of biological activity. In the absence of the A1–4 tetrapeptide sequence (101) or the C-terminal A21 asparagine residue (102, 103), biological activity is either absent or extremely low. Although C-terminal shortening of the B chain up to residue 27 does not affect the biological activity (104, 105), further deletions toward residue 22 (102) progressively reduce biological activity (104), suggesting that the region B22–26 may play an important role in biological activity (106) as well as in dimerization (76). Replacement of the cysteine residues providing the intrachain disulfide loop, between A6–11 with two alanine residues results in a product with about 10% of normal biological activity in vivo (107). X-ray diffraction studies are required to differentiate purely local from more generalized topographical changes in these altered molecules, and thus delineate more precisely the region(s) required for cell binding and biological function.

The three-dimensional structure of proinsulin has evoked considerable interest ever since the prohormone was first isolated. Although it has been crystallized successfully in several forms by Low and co-workers (108, 109), the crystals have not been of sufficient quality to permit extensive data collection. Preliminary X-ray diffraction analyses, however, confirm the results from other physical studies indicating that proinsulin aggregates similarly to insulin to form dimers and hexamers under appropriate conditions (110, 111). The hypothesis that the insulin moiety of proinsulin has essentially the same conformation as insulin is borne out by similarities in optical rotary dispersion (ORD) and circular dichroism (CD) spectra, by the titration behavior of the tyrosine residues of both peptides (112, 113), and by the high degree of immunological cross-reactivity of proinsulin with antisera to insulin (114, 115). Little evidence for the existence of ordered structure in the connecting peptide portion of proinsulin has been obtained (112). The peptide is believed to be folded in some manner over the external surface of the insulin monomers in proinsulin hexamers, spanning the 8–10 Å gap between the C terminus of the B chain and the N terminus of the A chain (16). Proinsulin and insulin evidently can form mixed polymers, which may account for the tendency of the prohormone to crystallize with insulin during the commercial preparation of insulin (16).

In addition to the studies on proinsulin mentioned above, recent structural investigations on insulin and various derivatives of insulin have utilized a variety of physical probes, including ORD and CD (103, 112, 116), Raman spectroscopy (117), electron paramagnetic resonance spectroscopy (118), and infrared spectroscopy (119). Recent evidence on the structure of insulin fibrils produced at elevated temperatures in acid solutions indicates that these are cross-β structures having a uniform cross-section of 29×47 Å, and consisting of flattened insulin monomers packed heterologously in layers 4.7 Å thick in the direction of the fibril axis (120).

NONINSULIN INSULIN-LIKE PROTEINS

It has become increasingly clear that proteins other than insulin can share many of its biological actions and may have additional physiological functions as well. One of these is the nonsuppressible insulin-like activity (NSILA) of plasma that has been studied extensively by Froesch and co-workers (121, 122). Although it circulates normally as an inactive larger complex, a soluble active substituent of molecular weight about 7500 can be separated under acidic conditions. This single-chain peptide can compete weakly but effectively with insulin for receptor sites in membrane preparations (123, 124), and it reproduces all the known biological effects of insulin (125, 126). As suggested first by Hall, recent evidence points to the probable identity of this substance with somatomedin, the new designation (127) for the well-known serum sulfation factor believed to be causally associated with the growth-promoting actions of growth hormone in vivo (124, 128–130). Although insulin, somatomedin, and perhaps the submaxillary gland nerve growth factor, which has some structural similarities to proinsulin (131), all produce basically similar anabolic and mitogenic effects, it is clear that nonidentical receptors with sharply differing binding specificities for these peptides exist in their various target tissues (124). Whether or not lowered cyclic AMP levels (130) can account for the spectrum of responses elicited by these distinctive groups of peptides is a question that requires much further study.

GROWTH HORMONE, PROLACTIN, AND PLACENTAL LACTOGEN

The primary structures of mammalian growth hormones (GH)[4] have been actively investigated in recent years. The ovine hormone contains 189–191 residues (132, 133). The primary structure of bovine GH is practically identical to that of the ovine hormone (132–134). Partial sequences of porcine (135) and equine GH (136–138) suggest that as a group the ovine, bovine, porcine, and equine hormones vary by less than 5% of their structure. The peptide chain of human GH is about the same length as that of the ovine hormone, but less than two thirds of its residues are homologous with ovine or bovine GH (139–142). This difference in structure probably accounts for the lack of bovine GH activity in man (143). The mammalian growth hormones contain two disulfide bridges: one spans 110 residues and joins cysteine residues 53–164; the other spans only six residues and joins residues 181 and 189.

Two other peptide hormones show remarkable structural similarities with GH. The chorionic hormone, placental lactogen, contains 191 residues, 85% of which are homologous with those of human GH (140, 144, 145). This near identity in primary structure is reflected by a similar placement of disulfide bridges and some similarities in biological activity. Since the placental lactogen appears to possess a portion of the activity of human GH in promoting growth, Li and his associates have suggested that it be called chorionic somatomammotropin (146). The other hormone exhibiting structural homology with GH is pituitary prolactin. The complete primary structures of both the porcine and ovine prolactins are now known (147–149). The bovine hormone is probably nearly identical to the ovine (150). Despite earlier uncertainty as to the existence of a unique prolactin in the primate pituitary, the monkey (151) and human (152) hormones have recently been isolated, and the N-terminal sequence of the human hormone was determined by Niall et al (153).

---

[4] The abbreviations used in this article are the following: GH, growth hormone (somatotrophic hormone); LH, luteinizing hormone (interstitial cell-stimulating hormone); FSH, follicle-stimulating hormone; TSH, thyroid-stimulating hormone; HCG, human chorionic gonadotropin; ACTH, adrenocorticotrophic hormone (corticotropin); MSH, melanocyte-stimulating hormone (melanotropin); LPH, lipotrophic hormone (lipotropin); and PTH, parathyroid hormone.

Human amniotic fluid is a promising source for human prolactin (154). Ovine prolactin contains 198 amino acid residues and three disulfide bridges. Two of these appear to be homologous with the disulfide bridges of GH. The other spans a six-residue sequence near the N terminus of the hormone and is also present in human prolactin (153).

Although the amino acid sequences of ovine GH and ovine prolactin show less than 30% homology, many of the apparent amino acid interchanges appear to be relatively conservative when viewed either by side-chain functionality or by the number of nucleotide base changes necessary to alter the appropriate codons (140, 147). A common evolutionary origin of prolactin, placental lactogen, and growth hormone has been suggested (140). Even more provocative is the suggestion by Niall and co-workers that these peptide hormones contain internal structural homology (140). On this basis, they propose that the three hormones may have arisen from repeated tandem duplications of a gene coding for a relatively small peptide (140).

The partial functional similarities of GH and prolactin in mammals are paralleled by their similar actions in lower vertebrates (155). Ovine prolactin and ovine GH are both potent somatotropins in a variety of such species, including teleost fishes and reptiles. The growth hormone present in the pituitaries of modern teleosts is apparently ineffective in mammals, however (156). The pituitaries of birds, reptiles, and amphibians appear to contain separate prolactin-like and growth hormone-like hormones (155). The biological significance of these forms, especially the prolactins, is still unknown. The recently isolated growth hormones from duck and turtle have remarkably similar amino acid compositions which differ only slightly from that of human GH (157).

Prolactin, placental lactogen, and GH also share some features of secondary structure, as determined by CD, fluorescence, and titration behavior (158–160). Each of the three hormones contains about 50% α helix and all undergo similar structural transitions in acid or base. Aloj & Edelhoch have found that human GH is less easily denatured than bovine GH or ovine prolactin by treatment with acid or urea (160). The single tryptophan residue in placental lactogen appears to be more exposed to the solvent than is the corresponding residue in human GH (159). Likewise, this tryptophan residue appears to be more exposed in human than in bovine GH (161, 162). Cambiaso and co-workers interpret rates of proton exchange as indicating that the structure of human GH is more open than that of the bovine hormone (163). Thus relatively small structural or conformational changes may account for the lack of activity of bovine GH in man.

Controversy continues to surround the question of the biological activity of various fragments of growth hormone. Bornstein and co-workers claim to have isolated two peptides having opposing biological activities after limited proteolysis of human GH (164). A C-terminal, 25-residue fragment of the hormone is reported to be diabetogenic, while a smaller N-terminal fragment of the hormone appears to have an insulin-like effect (164, cf 140). However, Schwartz has reported that a cyanogen bromide fragment of human GH containing the C-terminal 21 residues lacks the inhibitory actions of Bornstein's 25-residue fragment on several glycolytic enzymes in vitro (165).

In the case of bovine GH, a 37-residue tryptic fragment derived from the large disulfide loop of the hormone appears to have some growth-promoting activity, even in man (166, 167). This peptide as isolated by Sonenberg et al is partially homologous with a similar sequence of human GH and may include the active core of the hormone. Another fragment of human or bovine GH, consisting of residues 5–117 and containing a single residue of aminoethylcysteine, exhibits some of the biological properties of the intact hormone (168), but does not stimulate skeletal growth. Synthetic sequences of human GH representing residues 87–123 and residues 124–155 appear to have growth-promoting activity in rats (169). In the course of studies of the biological activities of GH fragments, several incorrect sequences have been synthesized (164, 170, cf 140 and 169). That these peptide analogs possess some biological activity suggests that considerable latitude is allowed in the recognition of active forms.

Several recent reports suggest that circulating human GH consists of GH itself and a considerably larger form, which usually makes up 10–30% of the plasma immunoreactivity. Since this large GH is dissociated to GH by urea or by various physical treatments, it may represent a GH aggregate or GH bound to a larger protein, rather than a precursor. Such large forms of immunoreactive GH are also found in human, rat, and dog pituitary glands (171–173). The physiological significance of the large form is not known. A similar large form of placental lactogen has been detected in human pregnancy serum and in extracts of human placenta (174).

GLYCOPEPTIDE HORMONES

Since the discovery that the treatment of ovine luteinizing hormone (LH) with dilute acid results in the dissociation of the hormone into dissimilar subunits (175, 176), considerable effort has been expended in isolating this glycopeptide hormone from a variety of sources. The dissociated subunits have been separated from each other by countercurrent distribution (177) or by various chromatographic procedures (178). The α subunits of ovine (179, 180), porcine (181), and bovine (182–184) LH contain 96 amino acid residues, whereas the α subunit of human LH appears to contain only 89 residues (185). Carbohydrate is attached at asparagine residues 56 and 82 of the three nonprimate subunits and at the homologous asparagine residues 49 and 75 of the human subunit.

The primary structures of ovine LH-α as reported by Sairam, Papkoff & Li (180) and by Liu and associates (179) differ only by an inversion at positions 88 and 89. The former group proposes the sequence Cys-Ser, whereas the latter suggests Ser-Cys. The homologous sequence reported for human LH-α is Ser-Cys (185), but the sequence reported for bovine LH-α and HCG-α is Cys-Ser (182, 183, 186). The α subunit of ovine LH appears to be markedly heterogeneous at its N-terminus, due to deletion of up to eight of its N-terminal residues (179). An apparent 7-residue deletion at the N terminus of human LH-α, when compared to ovine LH-α, has been well documented (185, 187, 188). Whether this is a true deletion or an artifact due to proteolytic degradation during isolation is uncertain.

A comparison of primary structures shows the α subunits of bovine and ovine LH to be identical, while the porcine subunit differs from these by only 5, and the human by 23, residues. Many amino acid sequences within the subunits from different species remain invariant, the longest blocks of identity being 14, 11, and 8 residues. The 10 cysteine residues and their homologous positions also remain unchanged.

The carbohydrate content of LH has been difficult to assess. The total carbohydrate and the relative proportions of hexoses, fucose, hexosamines, and sialic acid appear to differ among the human (188, 189), ovine (190), bovine (191), and equine hormones (192). The equine hormone appears to be unusually high in sialic acid. Alterations in the carbohydrate content, particularly in sialic acid, have been proposed as an explanation for microheterogeneity in human LH (193).

The primary structure of the β subunit of ovine LH as determined by Liu et al (194) differs from that of Sairam et al (195) in amidation at residues 10 and 62 and also in the sequence following residue 50. The former group proposes the sequence Pro-Met-Pro whereas the latter group suggests Pro-Pro-Met-Pro. More recent evidence suggests that the latter sequence with 120 residues is correct (182). The N-terminal serine residue appears to be acylated and carbohydrate is attached at the asparagine residue at position 13. The bovine (196) and ovine (194) LH-β subunits appear to be nearly identical. The primary structure of the β subunit of porcine LH differs from that of ovine LH at about 15 of 120 positions (197). Large blocks of β subunits of the ovine, porcine, and bovine hormones have identical amino acid sequences, the longer ones containing 21, 18, and 16 residues.

Recently, two reports of the primary structure of the β subunit of human LH have appeared. Both Shome & Parlow (198) and Closset, Hennen & Lequin (199) report that the subunit contains 115 residues and that the N terminus is serine and the C terminus glycine. Aside from four other amino acid differences between the two structures, Shome & Parlow do not report the existence of methionine 43 and proline 55 in their sequence. The total length of the subunit is maintained, however, by differences in the C-terminal sequence following the last cysteine residue. Shome & Parlow report seven additional residues (198), whereas Closset et al report only five (199). Both groups suggest, however, that human LH-β contains 12 cysteine residues as do the porcine, ovine, and bovine subunits.

Another discrepancy is apparent when considering the placement of carbohydrate in human LH-β. Closset et al place carbohydrate at asparagine 13 (as in other mammalian LH-β subunits) and also at asparagine 30 (a position filled by threonine in other species) (199). On the other hand, Shome & Parlow place carbohydrate only at position 30 (198). Although this discrepancy and those in the amino acid sequences will likely be resolved soon, it does appear that the carbohydrate placement in human LH-β differs from that of other mammals.

The subunit nature of human chorionic gonadotropin (HCG) was first recognized only a few years ago by Morgan & Canfield (200, 201) and confirmed by Swaminathan & Bahl (202). Recently, Bahl and co-workers reported the amino acid sequences of the α and β subunits of HCG (186). The α subunit contains 92 amino acid residues, four residues fewer than the α subunits of ovine and porcine LH, apparently due to a deletion at the N terminus. The primary structures of

HCG-α and human LH-α are identical, except for an apparent deletion at the N terminus in human LH-α, which also may be due to the known microheterogeneity in this group of proteins.

As shown in a recent report of Bahl and co-workers, the β subunit of HCG contains 147 amino acid residues (186) compared to 119 for ovine LH-β (194, 195). HCG-β shows about 45 amino acid substitutions when compared with ovine LH-β. However, Morgan et al (203) propose a length of 145 residues for HCG-β. Although some differences remain to be resolved, both groups agree that approximately 30 residues at the C terminus of HCG-β do not appear in ovine or human LH-β (186, 203). Aside from this difference, the HCG-β and human LH-β subunits are closely homologous.

The carbohydrate placement of HCG-β differs from that of LH-β. Ovine LH-β and human LH-β have one site for carbohydrate attachment, but HCG-β has five. Three of those five attachments are at serine residues 118, 121, and 123, positions appearing beyond the regions homologous with LH-β. The other two attachments are at asparagine residues and apparently account for the bulk of the covalently bound carbohydrate (204). The HCG secreted by several clones of human choriocarcinoma cells is heterogeneous by isoelectric focusing (Hammond et al, 205), a result presumably due to variability in sialic acid content. Sialylation appears to be a significant post-transcriptional modification of HCG, since it plays a critical role in determining its biological half-life (206).

Liao & Pierce obtained the α and β subunits of bovine thyroid-stimulating hormone (TSH) in pure form (183) and subsequently determined their primary structures (207). The amino acid sequence of bovine TSH-α is identical to that of bovine or ovine LH-α and differs only slightly from that of HCG-α. Also, like the LH-α subunits, TSH-α contains carbohydrate bound at asparagine residues 56 and 82 (207). The β subunit of bovine TSH contains 114 amino acid residues and has attached carbohydrate at asparagine 23. It contains six disulfide bridges, as do the corresponding LH subunits, and if position 8 of bovine LH-β is aligned with position 1 of bovine TSH-β, the positions of the cysteine residues are identical with one minor exception. In the alignment described, only about 25% of the residues in the two hormone subunits are identical, the longest consecutive identical sequence being only four residues. Nevertheless many of these amino acid substitutions are conservative. Preliminary data of Sairam & Li suggest that the α subunit of human TSH-α is identical to that of human LH-α, and that the first eight N-terminal residues of the β subunits of bovine and human TSH are identical (208). Cornell & Pierce, on the basis of their studies on human TSH subunits (209), suggest that the α subunit, which has N-terminal valine, is shortened at its N terminus by seven residues in relation to bovine TSH-α, as is human LH-α. Unlike human TSH, however, bovine TSH contains sialic acid, nearly all in the α subunit (209). Shome & Parlow (210) have recently reported the amino acid sequence of human TSH-β, which differs from bovine TSH-β at only 12 positions.

The chemical characterization of the α and β subunits of follicle-stimulating hormone (FSH) has proceeded less rapidly. The separation of subunits from the human (211) and ovine hormones (212) has been reported, and preliminary data concerning the equine hormone also are available (213). Gonadotropin from pregnant mare serum may have characteristics similar to those of the pituitary and chorionic gonadotropins (214) and also may dissociate into subunits (215).

The chemistry of the glycopeptide hormones from nonmammalian tissues has received comparatively little attention. The evidence suggests that the fish pituitary contains a single gonadotropin rather more like LH than FSH. Donaldson and co-workers have partially purified a pituitary gonadotropin from the Pacific salmon (216) which apparently has both LH and FSH activity in the salmon. It is biologically effective in all classes of animals except mammals, although its activity is diminished in reptiles (216). Conversely, mammalian LH is only partially active or inactive in fish. A similar purification of carp gonadotropin has also been obtained by Burzawa-Gerard (217).

A partially purified turtle gonadotropin, like the salmon hormone, appears to be chemically more similar to mammalian LH than to mammalian FSH (Papkoff & Licht, 218). Nevertheless, the biological actions of the hormone in the reptile correspond to those of both LH and FSH. Licht studied the actions of ovine LH and FSH in snakes (219) and turtles (220). FSH is significantly more potent than LH in maintaining spermatogenesis and interstitial cell activity. Ovine LH is also effective in producing a variety of gonadotropic activities in lizards and at least part of its lower biological activity may be due to its shorter biological half-life (221). These results indicate that reptiles, like fish, induce gonadotropic activity with a single, subunit-containing glycopeptide hormone, which is chemically rather like LH but biologically more like FSH.

Other considerations of glycopeptide hormone specificity in lower vertebrates concern TSH. The teleost thyroid apparently responds to mammalian TSH, whereas the elasmobranch thyroid does not (222, 223). However, the elasmobranch thyroid gland apparently responds with thyroxine secretion to its own pituitary TSH (222).

Little is known about the tertiary structures or the placement of disulfide bridges in the glycopeptide hormones. Cheng & Pierce (224) and Sairam, Papkoff & Li (225) have examined the nitration behavior of the tyrosine residues in intact bovine and ovine LH, respectively. The primary structures of the two are practically identical, and the tyrosine residues of the α subunit are nitrated more readily than those of the β subunit in each case. Similar results were obtained with iodination of ovine LH (226). The intact hormones contain one or more buried tyrosine residues, although all these residues are reactive in the dissociated subunits. The correct identification of these inaccessible residues remains a matter of controversy (224–226). Semiquantitative interpretation of the CD spectrum of ovine LH also indicates the presence of tryosine residues inaccessible to solvent (227).

The optical absorbances of only about four of the seven tyrosine residues of HCG are readily perturbed by solvent (Mori, 228). Two of the others become available in the urea-denatured hormone; the characteristics of the third remain obscure. Titration of HCG also suggests the presence of inaccessible tyrosine (229). CD spectra of native HCG (230) and ovine LH (227) are not simply the sum of the spectra of their component subunits. Bewley, Sairam & Li have also indicated that native ovine LH contains little or no α helix (227). Using their criteria for the interpretation of CD spectra, it appears that the structure of HCG contains similarly low amounts of α helix (230).

The separated α and β subunits of TSH, LH, and HCG were shown by Pierce et al to have little biological activity when assayed in systems sensitive to the action of TSH or the gonadotropins (231). Catt et al found that the subunits of ovine LH and HCG have a low biological potency in vitro that parallels their in vivo potency (232). Thus the apparent activity of the subunits may be due to contamination by HCG itself, inasmuch as the biological half-lives of the HCG subunits are very much shorter than the half-life of the intact hormone (233). The apparent activity of HCG-β is neutralized by antibodies specific for HCG-α, suggesting again that the activity may be due to contamination (234). Nevertheless, significant biological activity has been attributed to the β subunit of ovine LH (235, 236).

The isolated subunits of the individual hormones TSH-α + TSH-β (183), LH-α + LH-β (183), HCG-α + HCG-β (201, 230, 231), and FSH-α + FSH-β (212) recombine under mild conditions to give products with high, if not full, biological activity. Of particular interest is the apparent interchangeability of these subunits to give hybrid products having only the hormonal character of the hormone from which the β subunit was obtained. This work, pioneered by Pierce and his associates (183, 231) and confirmed by others (193, 212, 237), has shown that 1. LH-β or HCG-β combined with LH-α, HCG-α, TSH-α, or FSH-α results in a hormone with LH-like or HCG-like activity; 2. TSH-β combined with TSH-α, LH-α, or HCG-α results in a hormone with TSH-like activity; and 3. FSH-β combined with FSH-α, LH-α, or HCG-α results in a hormone with FSH-like activity.

The altered reactivity of hybrid glycopeptide hormones with antisera directed toward the native hormones does not always parallel the altered biological activities of the hybrid hormones (238, 239). The hybrids may also be physically less stable than the natural combinations. For example, the subunits of the hybrid HCG-α + TSH-β are easily separated by gel electrophoresis, while those of TSH-α + TSH-β are not (231). Different biological assay systems may also yield different results for hybrid hormones. The ratios of biological activity for the long-term prostate weight augmentation assay to the short-term ovarian ascorbic acid depletion test for HCG hybrids are LH-α + HCG-β, 1; TSH-α + HCG-β, 5; and HCG-α + HCG-β, 16 (240). These ratios presumably reflect differences in the inherent stability of the complexes and in their biological half-lives.

As discussed above, the α subunits of LH, TSH, and HCG are nearly identical in their primary structures as well as in the placement of carbohydrate side chains. The β subunits of the corresponding hormones appear to be more variable structures, which is not surprising since these subunits confer hormonal specificity. Nevertheless, careful comparisons disclose the existence of considerable homology among the β subunits and also some similarities between α and β subunits (182). Thus, both the α and β subunits of these glycoprotein hormones may have arisen from a common ancestral molecule (182, 223).

GLUCAGON AND RELATED HORMONES

The primary structure of porcine glucagon was reported over 15 years ago (241). The relatively insoluble, 29-residue peptide contains N-terminal histidine and has an isoelectric point very near neutrality. Recently, there has been renewed interest in the structures of glucagons from other species. Bovine (242) and human (243) glucagon are identical in sequence to the porcine hormone. Furthermore, the amino acid compositions of rabbit (244) and rat glucagon (245) are identical to those of the porcine, bovine, and human hormones. Conservation of the primary structure of mammalian glucagon appears, at least at this stage, to be complete. This apparent conservatism in glucagon structure contrasts with the low but significant variability of insulin structure among mammals.

Avian glucagon does differ from mammalian glucagon in its primary structure. The duck hormone shows two amino acid substitutions when compared to the mammalian hormone (246); serine at position 16 is replaced by threonine and asparagine at position 28, by serine. The amino acid composition and partial sequence determination of turkey glucagon suggest that a single amino acid substitution has occurred at position 28 (247). The replacement of asparagine by threonine markedly alters the affinity of the hormone for a variety of antisera directed toward mammalian glucagon. Monodesamido porcine glucagon, the major contaminant in the commercially available hormone, also has altered antibody affinity (248). The deamidation apparently can occur at any one of the three glutamine residues, but not at the penultimate asparagine.

Although crystalline glucagon (249) and glucagon in concentrated solution (250, 251) appear to have small amounts of α helix, the extent of helix formation in solution is concentration dependent (252). It is generally assumed that in dilute aqueous solution, glucagon possesses little easily defined secondary structure (252–254). Nevertheless, urea and temperature changes do produce reversible changes in the CD spectrum of glucagon, suggesting that a preferred conformation does exist (255). Also, hydrogen-tritium exchange studies suggest that two classes of exchangeable hydrogen atoms occur in native glucagon (256). Viscosity as well as other studies suggest that although glucagon molecules aggregate in concentrated solution, the monomer itself exists in a compact structure (257, 258).

From ORD data, Blanchard & King have suggested that the unit of aggregation is a trimer (250); however, on the basis of sedimentation data, Swann & Hammes suggest that aggregation proceeds through a dimer to a hexamer (259). Gratzer, Creeth & Beaven have recently isolated a glucagon trimer crosslinked by suberimidate from a concentrated solution, suggesting once again a trimeric mode of aggregation (260). The pK_a of glucagon tyrosine 13 shifts on association (251). Moreover, the lack of association of the N-terminal, 27-residue cyanogen bromide fragment of glucagon (258) suggests that the C-terminal region also plays an important role in aggregation.

Various parameters of glucagon activity such as lipolysis, glycogenolysis, and insulin secretion are not equally affected in either degraded glucagon or synthetic analogs (261). The cyanogen bromide-cleaved peptide, residues 1–27 (262); the N-bromosuccinimide-cleaved peptide, residues 1–25 (262); the synthetic peptide, residues 1–23 (263), and the carboxypeptidase-digested peptide, residues 1–21 (262) all retain at least a portion of their lipolytic activity in in vitro systems. The fragment containing residues 1–27 will activate liver adenylcyclase, although the fragment containing residues 1–23 is inactive (264). A peptide larger than residues 1–23 is necessary for the stimulation of insulin secretion, but it is not known whether the fragment 1–27 is active (261).

The importance of the N-terminal histidine of glucagon in determining biological activity has been a matter of some controversy. Felts et al first showed that glucagon cycled once through the procedure of Edman degradation, (N,ε-phenylthio-carbamyl deshistidine glucagon) was not effective in activating fat cell lipolysis (262). Recently Lande, Gorman & Bitensky prepared deshistidine glucagon not otherwise modified by added groups (265). The resulting peptide had only 6% lipolytic activity in vitro although N,ε-Lys$_{12}$-butyloxycarbonyl glucagon had 10–15% activity and N,α-His$_1$-trifluoroacetyl glucagon had 30–40% activity. These data as well as data for N,α-His$_1$-carbamylglucagon (266) suggest that a free α-amino group, a free ε-amino group, and the histidyl function at position 1 are all necessary for full biological activity.

The primary structure of porcine secretin was determined in the laboratory of Mutt (267). This intestinal hormone contains 27 amino acid residues. Its C-terminal valine residue is amidated. If the amino acid sequences of porcine glucagon and secretin are aligned from their N-terminal residues, the two hormones show remark-able structural similarities. Said & Mutt and their co-workers have isolated and determined the structure of another intestinal peptide similar to glucagon, the vasoactive intestinal peptide, which appears to have a variety of physiological effects (268, 269). The peptide contains 28 amino acids and its C-terminal asparagine residue is amidated. Only about one third of its residues are identical to those of glucagon, but the homology is unmistakable. The N-terminal dipeptide of glucagon, secretin, and the vasoactive peptide is His-Ser.

Brown and his colleagues also have recently isolated (270) and sequenced (271) an intestinal hormone called gastric inhibitory polypeptide. The N-terminal 29 residues of this 43-residue peptide show about 50% identity with glucagon (269). There are few long sequences of identity, however, and the N-terminal residues are Tyr-Ala. As suggested by Bodanszky, Klausner & Said, these four hormones, glucagon, secretin, vasoactive intestinal peptide, and gastric inhibitory polypeptide, may have evolved from a primitive ancestral protein by a process of gene multiplica-tion (269).

An interesting but structurally undefined group of compounds are those called gut glucagon-like immunoreactivity or enteroglucagon. Valverde et al (272) showed that extracts of canine intestine contain two classes of compounds reactive with anti-glucagon serum, one with an apparent molecular weight approximating that of pancreatic glucagon, and another approximately 2–3 times larger. A similar hetero-geneity has been observed with extracts of human intestine (273). Attempts to purify these compounds from porcine intestine by ion-exchange chromatography (274) or by affinity chromatography (275) have been only partially successful. In fact, there appear to be more than two intestinal compounds with glucagon-like immunoreactivity. Both the high and low molecular weight forms differ from pancreatic glucagon immunologically (272). Kerins & Said have suggested that the low molecular weight gut glucagon (276) may be identical to the vasoactive intestinal peptide of Said & Mutt (268), but structural data are lacking.

Several recent studies have suggested that pancreatic glucagon, like insulin, is initially synthesized in precursor form. Thus, extracts of canine pancreas contain two components reactive with antiglucagon serum (Rigopoulou et al, 277). Glucagon itself accounted for 95 percent of the total immunoreactive components; a larger component, approximately 9000 daltons as determined by gel filtration, accounted for the remaining immunoreactivity. Furthermore, trypsin converted the large component to peptides having reactivity with antiglucagon serum indistinguishable from that of trypsin-treated glucagon. Large glucagon-like immunoreactivity has also been found in extracts of pancreas from the human, duck, ox, and rat (277).

Several other investigators have examined the immunoreactivity profiles of extracts of pancreatic islets from a variety of species after gel filtration and have correlated large components possessing glucagon-like immunoreactivity with apparently identical components biosynthetically labeled with ^{3}H-Trp. Tung & Zerega provided suggestive evidence for the biosynthesis of a 50,000–60,000-dalton precursor to glucagon by pigeon islets in vitro (278). Noe & Bauer have proposed the existence of an 11,000-dalton glucagon precursor in angler fish islets (279, 280), whereas Hellerstrom et al provide evidence for an 18,000-dalton precursor in hamster islets (281). Recently, O'Connor, Gay & Lazarus reported a 9000-dalton precursor obtained from isolated perfused rat pancreas (282).

In many cases, the high molecular weight forms are reactive with glucagon antisera (279–282) and their synthesis, as well as that of glucagon, appears to be inhibited by high concentrations of glucose (278–282). The precursor-product relationship between the early labeled, high molecular weight forms and glucagon has been difficult to assess by pulse-chase studies. Noe & Bauer suggest that conversion of the precursor to glucagon occurs during incubation in vitro (279), but Hellerstrom et al could find evidence for apparent conversion only after labeled islets were cultured for periods of several days (281).

Noe & Bauer showed that mild tryptic digestion of the angler fish glucagon precursor results in products of initially higher immunoreactivity (279). These products, like glucagon itself, are subsequently degraded to peptides with only marginal reactivity with antiglucagon sera. The physical properties of the biosynthetically labeled or the immunoreactive precursor have not been examined in detail. Observations using acrylamide gel electrophoresis indicate that the angler fish precursor may be significantly more acidic than angler fish or porcine glucagon (279, 280), but this does not appear to be the case for the precursor from the rat (282).

We have recently examined commercially prepared crystalline glucagon for the presence of high molecular weight glucagon-like components (283). A higher molecular weight fraction separated by gel filtration contains a variety of peptides, ranging in size from 3700 to 9000 daltons, which are reactive with antiglucagon serum. One highly basic peptide of about 4500 daltons was isolated from this mixture and its primary structure was determined. The N-terminal 29 residues of the 37-residue peptide correspond exactly to those of porcine or bovine glucagon. Residues 30–37 form the sequence Lys-Arg-Asn-Asn-Lys-Asn-Ile-Ala. The identification of the entire primary structure of glucagon within that of a larger peptide provides strong chemical evidence for a higher molecular weight form of the secreted hormone, and we proposed that the basic peptide is a fragment of a larger glucagon precursor (283). A trypsin-sensitive sequence at the C terminus of glucagon within the 37-residue fragment is consistent with other results concerning the tryptic sensitivity of the precursor, and suggest that, as for proinsulin, a combination of trypsin-like and carboxypeptidase B-like activities may be necessary for conversion of the precursor to the hormone product.

GASTRIN AND RELATED HORMONES

Gastrin is an unusual peptide hormone in two respects (284–299): 1. the first residue of the 17-residue hormone is pyrrolidone carboxylic acid and the last residue is phenylalanineamide; and 2. the hormone exists in two forms, one having esterified sulfate on tyrosine residue 12 and the other being unmodified. The hormone from the human antral mucosa (285) has an amino acid sequence identical to that of the hormone from the human Zollinger-Ellison pancreatic tumor (286). Human gastrin contains an unusual pentaglutamyl sequence following residue 5, and the acidic core of the peptide appears to have been conserved throughout the mammals. The bovine and ovine hormones are identical and differ from the human hormone only at residues 5 and 10 (288). The canine hormone differs from the human at residues 5 and 8 (287). The feline (289) and the porcine hormones (284) differ from the human hormone by only one residue each.

The C-terminal five residues of the 33-residue, mammalian hormone cholecystokinin-pancreozymin (290, 291), the 10-residue, amphibian hormone caerulein (292, 293), and the 9-residue, amphibian hormone phyllocaerulein (294) are identical to the C-terminal five residues of gastrin: Gly-Trp-Met-Asp-Phe. In addition, the C terminal of each of these hormones is amidated and all contain a sulfated tyrosine residue. These observations suggest that all these hormones may have arisen from a common ancestral form. The biological similarities among the gastrins, cholecystokinin and caerulein on gastric acid secretion is readily explained by the observation of Tracy & Gregory that the C-terminal tetrapeptide amide of gastrin can mimic completely the actions of the intact hormone (295).

Although much effort has been expended in defining the smallest peptide with gastrin-like activity, the question still remains open to discussion. Morley suggested that the tetrapeptide amide is the smallest active unit and that the penultimate aspartyl residue is required for activity, whereas the other three residues serve only to provide optimal binding to a recognition site (296). Lin has suggested, however, that the C-terminal tripeptide amide retains a small amount of gastrin-like activity (297). Nevertheless, Trout & Grossman, in studies of gastric acid secretion due to analogs of cholecystokinin, found that the penultimate aspartyl residue can be replaced by alanine in an octapeptide amide if this analog also contains sulfated tyrosine (298).

The question of the existence of multiple forms of natural gastrin, aside from the variations in sulfation, has received much recent study. Yalow & Berson showed that circulating immunoreactive gastrin in patients with pernicious anemia or with the Zollinger-Ellison syndrome could be fractionated into two immunoreactive components (299): the less abundant component was the heptadecapeptide gastrin of Tracy & Gregory; the more abundant component was a less acidic form with an apparent mol wt of about 7000. The ratio of the two forms varied considerably from one patient to another, but the larger component represented at least 50 percent of the total immunoreactive gastrin in the circulating plasma (300). Also, the larger component was converted by trypsin to an immunoreactive form having the charge and size characteristics of heptadecapeptide gastrin.

Berson & Yalow also identified a large, immunoreactive gastrin in extracts of human antra and duodena (301). A similar high molecular weight form was found in extracts of porcine antra by Praissman & Berkowitz (302). Gregory & Tracy isolated a high molecular weight form of gastrin from human Zollinger-Ellison tumor tissue (303). The analysis of the two forms of tumor gastrin showed that the heptadecapeptide form predominated. Amino acid analysis of the large gastrin indicated that it contains in excess of those amino acids present in human gastrin 1 aspartic acid, 1 serine, 2 glutamic acids, 4 prolines, 2 glycines, 1 alanine, 1 valine, 1 leucine, 1 histidine, and 2 lysines. The large form was converted by trypsin to native gastrin which had glutamine rather than pyrrolidone carboxylic acid at its N terminus. The conversion of the precursor by trypsin indicates that the extension is at the N terminus of heptadecapeptide gastrin.

Recently, Yalow & Berson (304) and Rehfeld (305) have suggested the existence of an immunoreactive gastrin even larger than 7000 daltons in sera from patients with Zollinger-Ellison syndrome or pernicious anemia. The component of Rehfeld is 9000–10,000 daltons, whereas that of Yalow & Berson is considerably larger. Nevertheless, both groups present evidence that trypsin can convert the large component to heptadecapeptide gastrin. Neither urea nor serum affects the elution profile of the 9000 dalton peptide from gel filtration columns and the peptide bears no immunological similarity to cholecystokinin (Rehfeld, 305). The precursor-product relationship between the large gastrins and heptadecapeptide gastrin has not yet been shown biosynthetically, but available evidence suggests that the large forms are precursors. It is not clear, however, whether or not the large gastrin isolated from the Zollinger-Ellison tumor corresponds to any of the circulating forms of large gastrin.

CORTICOTROPIN, MELANOTROPIN, AND LIPOTROPIN

The primary structures of the corticotropins (adrenocorticotrophic hormone, ACTH) from sheep (306), pig (307), ox (308), and human (309) have been known for some time. Re-evaluation of these structures by Riniker et al (310) and by Li (311) has suggested that these highly basic, 39-residue hormones differ in sequence at only three positions, namely 21, 31, and 33. The N-terminal sequence of ACTH containing residues 1–19, however, has very nearly full steroidogenic activity (312, 313). This finding prompted Li and his co-workers to suggest that the basic sequence 15–18, Lys-Lys-Arg-Arg, is necessary for biological activity. Nevertheless, recent reports indicate that the N-terminal sequences 1–14 (314) and 1–10 (315) will mimic the steroidogenic action of ACTH, albeit at concentrations several orders of magnitude greater than those necessary for the native hormone.

The α-melanotropins (α-melanocyte-stimulating hormone, α-MSH) of the pig (316), ox (317), horse (318), monkey (319), and sheep (320) are identical tridecapeptides containing N-terminal acylserine and C-terminal valineamide. The primary structure of the dogfish α-MSH is probably identical to that of the mammalian hormone (321). A very interesting aspect of α-MSH structure is that the entire sequences of the α-melanotropins are identical to the invariant sequences of residues 1–13 of mammalian ACTH. The possible precursor-product relationship between ACTH and α-MSH has not been examined biosynthetically, but Scott and his co-workers give strong evidence for such a relationship (322). They have isolated from porcine and rat pituitaries and from an MSH-producing human tumor a peptide called Corticotropin-like Intermediate Lobe Peptide. The peptide occurs in equimolar amounts with α-MSH and reacts well with antibodies directed towards the C terminus of ACTH but not with antibodies directed towards the N terminus of that hormone. The amino acid composition of the peptide corresponds exactly to that expected from the 22-residue peptide $ACTH_{18-39}$ of the appropriate species. It was suggested that α-MSH and the intermediate lobe peptide arise in vivo from proteolytic cleavage of ACTH and that ACTH may be considered a precursor form of the two peptides. Since MSH activity apparently antedates ACTH activity in evolution, Scott et al suggest that immunoreactive ACTH in the cyclostomes may serve exclusively a precursor function (322). Conversion might be accomplished by a trypsin-like cleavage of ACTH in its basic region followed by the action of carboxypeptidases to yield α-MSH and intermediate lobe peptide. Still to be approached experimentally are the mechanisms for N acylation and C amidation of α-MSH after it has been cleaved from the parent ACTH.

Yalow & Berson have suggested that human ACTH itself is initially synthesized from a larger precursor (323, 324). A large immunoreactive component is present both in plasma and in extracts of pituitary glands. Its apparent mol wt is greater than 20,000, and it is not altered by treatment with urea. Also, the precursor form, which is more acidic than human ACTH, is converted by small amounts of trypsin to an immunoreactive form having the expected characteristics of ACTH. More severe trypsin treatment, however, degrades the high molecular weight form and native ACTH at the same rate. Since trypsin alone appears to convert the suggested precursor to the usual form of the hormone, Berson & Yalow suggest that the extension is N terminal. These results taken together suggest that the sequence of reactions resulting in the biosynthesis of α-MSH proceeds in the order procorticotropin $\rightarrow$ corticotropin $\rightarrow$ α-melanotropin. Biosynthetic studies are required, however, to establish this relationship.

The lipotropins (LPH) and β-MSH appear to be structurally related to ACTH and α-MSH since all of these peptides contain the identical heptapeptide sequence Met-Glu-His-Phe-Arg-Trp-Gly. The primary structures of the 90- or 91-residue ovine (325) and porcine (326) β-lipotropins differ by only 15% of their structure, most of the changes appearing in the N-terminal portion of the hormone. The structure of the C-terminal portion of human β-LPH also has been examined (327). γ-LPH from sheep (328) and pigs (326) consists of the N-terminal 58 residues of the corresponding β-LPH and may result from proteolytic cleavage of the larger form.

The primary structures of β-MSH from the ox (329), pig (330), monkey (331), horse (332), sheep (333), and human (334) are all very similar. Within each species, the entire primary structure of β-MSH occurs within that of β-LPH. Since the region corresponding to β-MSH is circumscribed by pairs of basic amino acids, the conversion of β-LPH to β-MSH could take the same course as the conversion of proinsulin to insulin. However, biosynthetic evidence that β-LPH is a precursor of β-MSH is lacking.

PARATHYROID HORMONE

The parathyroid hormones (PTH) of ox (335, 336) and pig (337) differ by less than 10% of their 84 residues. Like a variety of simple polypeptide hormones, they contain no cysteine or disulfide bonds. They each contain two highly basic tripeptide sequences, Arg-Lys-Lys, following residues 25 and 51. The N-terminal portion of the hormone from the human has been examined by two groups, although their results differ. Brewer et al suggest that there are five amino acid substitutions in the first 34 residues of human PTH when its sequence is compared with porcine PTH (338), whereas Jacobs et al (339) suggest that there are only two differences in the amino acid sequences of the N-terminal 37 residues of the two hormones. Potts and his associates demonstrated that a synthetic peptide containing the first 34 residues of the bovine hormone retains biological activity on both kidney and bone (340). A 1–29 sequence obtained by dilute acid hydrolysis of the native hormone also possesses high biological activity (341). Shorter N-terminal peptides also appear to serve, but their activities are significantly higher in vitro than in vivo (342).

Berson & Yalow reported that circulating PTH differed immunologically from the hormone extracted from the parathyroid gland (343). Sherwood, Rodman & Lundberg (344) and Arnaud et al (345) studied the secretion of PTH from parathyroid tissue in vitro and suggested that the secreted form was smaller than the tissue form. Biosynthetic experiments using radioactive amino acids suggested a similar conversion of the glandular 9000-dalton hormone to its 7000-dalton circulating form (344). Although these investigators suggested that glandular PTH was a precursor or storage form of the hormone, Habener et al showed by venous catheterization of glandular tissue that the secreted and storage forms were the same 9000-dalton peptide (346). Nevertheless, the major portion of circulating PTH-like immunoreactivity is not identical with the glandular hormone (343–346). Canterbury & Reiss suggest that a 5000-dalton PTH-like peptide as well as the

7000-dalton peptide circulates in the plasma of hyperparathyroid humans (347). Following parathyroidectomy, the residual immunoreactivity of the various hormonal forms decreases with molecular weight in the order 9000 → 7000 → 5000.

Segre et al have examined the structure of the major form of PTH in peripheral human plasma using a variety of antisera directed toward different portions of the PTH sequence 1–84 (348). The major circulating immunoreactive form of the hormone apparently represents a C-terminal fragment of the 9000-dalton peptide. The cleavage necessary to convert the secreted form to its major circulating form apparently occurs between residues 14 and 34. Since the studies of Aurbach et al suggest that the N-terminal portion of PTH is required for biological activity (342), Segre and co-workers propose that the major immunoreactive form of PTH in the circulation may not have biological activity (348).

Fischer et al have recently identified a proteolytic activity in extracts of porcine parathyroid tissue which appears to convert intact PTH (mol wt 9000) to a smaller form (mol wt 7000) having, at least superficially, the characteristics of the major circulating form of the hormone (349). This smaller form retains a portion of the biological activity. The enzyme activity and a similar one in extracts of liver are inhibited by calcium ion and by iodoacetate, but not by soybean trypsin inhibitor. The existence of this proteolytic activity within parathyroid tissue may explain some of the discrepancies noted above concerning the size of the secreted hormone, but the exact chemical natures of the circulating and the biologically active forms of PTH require further clarification.

Early studies on the biosynthesis of parathyroid hormone proceeded from incubation of slices of bovine parathyroid glands with radioactive amino acids. The biosynthetically labeled forms appeared to have chemical properties similar to those of immunoreactive PTH (344, 350, 351). Recently, several laboratories have proposed that the 84-residue secreted hormone is synthesized in precursor form. Habener et al observed that certain human parathyroid adenomas secreted substances larger than 9000 daltons with PTH-like immunoreactivity (346). Hamilton et al first detected the biosynthesis of a larger peptide having a higher specific activity than PTH in slices of bovine parathyroid gland (352). This component, designated calcemic fraction A, was isolated and partially characterized; it had significant PTH-like biological activity and its biosynthesis also was inhibited by calcium ions (352).

Further biosynthetic studies by Cohn et al (353) showed that the labeling of the precursor began immediately, whereas label appeared in PTH only after a lag, as expected for a precursor-product relationship. Also, trypsin treatment of the immunoreactive precursor resulted in products of initially higher reactivity with antibodies directed toward PTH. The purified precursor hormone from the ox contained 109 amino acids, the most notable being seven basic residues in addition to those present in bovine PTH (353).

Kemper and co-workers also showed that the precursor form of PTH was more basic than the hormone itself and that the large peptide labeled with radioactive amino acids was converted in situ to one having the size and charge characteristics of PTH during a pulse-chase experiment (354). Small amounts of trypsin could apparently convert the labeled precursor to PTH in vitro. Molecular weight determinations suggested that the precursor contained about 20 more acid residues than PTH. Biosynthetic evidence for the existence of a PTH precursor in slices of human parathyroid adenomas also has been reported (355). The rate of conversion of proparathyroid hormone to PTH varied from one tumor to another.

Habener and co-workers recently examined the structure of the biosynthetically-labeled PTH precursor by the selective use of proteolytic cleavage and peptide analysis (356). Their data suggest that the peptide extension in the precursor is at the N terminus and that the conversion mechanism very likely includes the action of a trypsin-like enzyme. The primary structure of the N-terminal extension in bovine proparathyroid hormone has recently been reported to be Lys-Ser-Val-Lys-Lys-Arg (357, 358). It is not clear whether or not the proposed 90-residue precursor form represents a fragment of a still larger precursor form (353, 354).

CONCLUDING REMARKS

A close comparison of the primary structures of the more than 20 peptide hormones mentioned in this review suggests that most of them can be grouped into a few structural classes: 1. insulin, proinsulin, and nerve growth factor; 2. glucagon, secretin, vasoactive intestinal peptide, and gastric inhibitory polypeptide; 3. gastrin, cholecystokinin, and caerulein; 4. corticotropin, melanotropin, and lipotropin; 5. growth hormone, prolactin, and placental lactogen; and 6. luteinizing hormone, follicle-stimulating hormone, human chorionic gonadotropin, and thyroid-stimulating hormone. The structural homologies within each class may be clearer in some cases than in others, and in some instances they extend to similar modes of post-translational modification and to quaternary structure. Whether or not all the peptide hormones in each group are derived from a single ancestral peptide by gene duplication and subsequent mutational modification remains a provocative question deserving further consideration. The possible existence of homologies between the primary structures of various members of two or more of these six distinct groups also has been proposed recently by several authors (359, 360, see discussion by R. Fellows following 153), but no firm conclusions can be drawn as yet from the limited comparative data available within most of the groups.

In the interval since precursor processing was recognized as playing an important role in the biosynthesis of insulin, considerable evidence has accumulated indicating the existence of precursor forms of glucagon, gastrin, corticotropin, melanotropin, and parathyroid hormone. There appears to be no general explanation accounting for the existence of these forms, especially for those hormones lacking disulfide bonds and well-defined tertiary structure. Possible explanations include the existence of limitations on the minimum length of mRNA that can be produced or utilized in the biosynthesis of small proteins, or similar limitations on the size of the protein products that can be handled efficiently for storage or secretion. Alternatively, portions of the added nonhormonal sequences may provide signals to the secretory apparatus of the endocrine cells to indicate that the biosynthetic product is to be packaged for export.

Perhaps it is more likely that these complex biosynthetic pathways largely reflect the evolutionary history of many of these hormonal peptides. Thus post-translational processing may serve to match both the biological activity of these peptides and their metabolic characteristics (i.e. the rates of onset of their actions and rates of decay) to the precise needs of the organism. Whatever final explanation(s) may emerge, it is now clearly established that the major glandular storage form and the circulating form(s) of many of the small polypeptide hormones often represent only portions of the initial products of their structural genes.

ACKNOWLEDGMENTS

We are grateful to Drs. Hugh Niall and John G. Pierce for their helpful comments on portions of this review. We also thank Miss Roberta Erfurth for her assistance with the literature search and the preparation of the manuscript. The work of this laboratory has been supported in part by U.S. Public Health Service Grant AM-13914. H.S.T. is a recipient of a U.S. Public Health Service Postdoctoral Fellowship (No. 1-F02 AM-51612) from the Arthritis and Metabolic Disease Unit, National Institutes of Health.

Literature Cited

1. Behrens, O. K., Grinnan, E. L. 1969. *Ann. Rev. Biochem.* 38:83–112
2. Cuatrecasas, P. 1974. *Ann. Rev. Biochem.* 43:169–214
3. Steiner, D. F. et al 1969. *Recent Prog. Horm. Res.* 25:207–82
4. Chance, R. E., Ellis, R. M. 1969. *Arc. Int. Med.* 123:229–36
5. Sorensen, R. L., Steffes, M. W., Lindall, A. W. 1970. *Endocrinology* 86:88–96
6. Permutt, M. A., Kipnis, D. M. 1972. *Proc. Nat. Acad. Sci. USA* 69:506–9
7. Sun, A. M., Lin, B. J., Haist, R. E. 1973. *Can. J. Physiol. Pharmacol.* 51:175–82
8. Howell, S. L., Kostianovsky, M., Lacy, P. E. 1969. *J. Cell Biol.* 42:695–705
9. Orci, L. et al 1971. *J. Cell Biol.* 50:565–82
10. Steiner, D. F. et al 1970. *The Pathogenesis of Diabetes Mellitus. Proceedings of the Thirteenth Nobel Symposium,* ed. E. Cerasi, R. Luft, 123–32. Stockholm: Almqvist och Wiksell
11. Howell, S. L. 1972. *Nature New Biol.* 235:85–86
12. Kemmler, W., Steiner, D. F. 1970. *Biochem. Biophys. Res. Commun.* 41:1223–30
13. Grant, P. T., Coombs, T. L., Thomas, N. W., Sargent, J. R. 1971. *Subcellular Organization and Function in Endocrine Tissues. Memoirs Soc. Endocrinology,* ed. H. Heller, K. Lederis, Vol. 19, 481–495. New York: Cambridge Univ. Press
14. Steiner, D. F., Kemmler, W., Clark, J. L., Oyer, P. E., Rubenstein, A. H. 1972. *Handb. Physiol. Sect. 7: Endocrinol.* 1:175–98
15. Kemmler, W., Steiner, D. F., Borg, J. 1973. *J. Biol. Chem.* 248:4544–51
16. Steiner, D. F. 1973. *Nature* 243:528–30
17. Greider, M. H., Howell, S. L., Lacy, P. E. 1969. *J. Cell Biol.* 41:162–65
18. Lange, R. H., Boseck, S., Syed Ali, S. 1972. *Z. Zellforsch. Mikrosk. Anat.* 131:559–70
19. Sando, H., Borg, J., Steiner, D. F. 1972. *J. Clin. Invest.* 51:1476–85
20. Melani, F., Rubenstein, A. H., Steiner, D. F. 1970. *J. Clin. Invest.* 49:497–507
21. Yip, C. C., Logothetopoulos, J. 1969. *Proc. Nat. Acad. Sci. USA* 62:415–19
22. Melani, F., Rubenstein, A. H., Oyer, P. E., Steiner, D. F. 1970. *Proc. Nat. Acad. Sci. USA* 67:148–55
23. Sorenson, R. L., Shank, R. D., Lindall, A. W. 1972. *Proc. Soc. Exp. Biol. Med.* 139:652–55
24. Nolan, C., Margoliash, E., Peterson, J. D., Steiner, D. F. 1971. *J. Biol. Chem.* 246:2780–95
25. Kemmler, W., Peterson, J. D., Steiner, D. F. 1971. *J. Biol. Chem.* 246:6786–91
26. Smith, R. E. 1972. *Diabetes* 21:581–83

27. Tager, H. S., Emdin, S. O., Clark, J. L., Steiner, D. F. 1973. *J. Biol. Chem.* 248: 3476–82
28. Yip, C. C. 1971. *Proc. Nat. Acad. Sci. USA* 68: 1312–15
29. Milstein, C., Brownlee, G. G., Harrison, T. M., Mathews, M. B. 1972. *Nature New Biol.* 239: 117–20
30. Yip, C. C., Liew, C. C. 1973. *Can. J. Biochem.* 51: 783–88
31. Tijoe, T. O., Kroon, A. M. 1973. *FEBS Lett.* 33: 225–28
32. Grant, P. T., Coombs, T. L. 1971. *Essays Biochem.* 6: 69–92
33. Trakatellis, A. C., Schwartz, G. P. 1970. *Nature* 225: 548–49
34. Emdin, S. et al 1973. *Ninth Int. Congr. Biochem. Abstract Book* 452
35. Yamaji, K., Tada, K., Trakatellis, A. C. 1972. *J. Biol. Chem.* 247: 4080–88
36. Clark, J. L., Steiner, D. F. 1969. *Proc. Nat. Acad. Sci. USA* 62: 278–85
37. Tung, A. K., Yip, C. C. 1969. *Proc. Nat. Acad. Sci. USA* 63: 442–49
38. Chance, R. E. 1971. *Proceedings of the 7th Congress of the International Diabetes Federation,* ed. R. R. Rodriguez, J. J. Vallance-Owen, 292–305. Amsterdam: Excerpta Med.
39. Kitabchi, A. E., Duckworth, W. C., Stentz, F. B., Yu, S. 1972. *CRC Crit. Rev. Biochem.* 1: 59–94
40. Gutman, R. A., Lazarus, N. R., Recant, L. 1972. *Diabetologia* 8: 136–40
41. Steiner, D. F. et al 1971. *J. Biol. Chem.* 246: 1365–74
42. Rubenstein, A. H., Clark, J. L., Melani, F., Steiner, D. F. 1969. *Nature* 224: 697–99
43. Oyer, P. E., Cho, S., Peterson, J. D., Steiner, D. F. 1971. *J. Biol. Chem.* 246: 1375–86
44. Ko, A. S. C., Smyth, D. G., Markussen, J., Sundby, F. 1971. *Eur. J. Biochem.* 20: 190–99
45. Peterson, J. D., Nehrlich, S., Oyer, P. E., Steiner, D. F. 1972. *J. Biol. Chem.* 247: 4866–71
46. Markussen, J., Sundby, F. 1973. *Eur. J. Biochem.* 34: 401–8
47. Markussen, J., Sundby, F. 1972. *Eur. J. Biochem.* 25: 153–62
48. Tager, H. S., Steiner, D. F. 1972. *J. Biol. Chem.* 247: 7936–40
49. Dayhoff, M. O., Ed. 1972. *Atlas of Protein Sequence and Structure,* Vol. 5. Bethesda: Biomedical Research Foundation
50. Dayhoff, M. O., Ed. 1973. *Atlas of Protein Sequence and Structure,* Vol. 5, Suppl. 1. Bethesda: Biomedical Research Foundation
51. Geiger, R., Jäger, G., König, W., Volk, A. 1969. *Z. Naturforsch. B* 24: 999–1004
52. Yanaihara, N., Hashimoto, T., Yanaihara, C., Sakura, N. 1970. *Chem. Pharmacol. Bull.* 18: 417
53. Yanaihara, N., Hashimoto, T., Yanaihara, C., Sakagami, M., Sakura, N. 1972. *Diabetes* 21: 476
54. Yanaihara, N., Sakura, N., Yanaihara, C., Hashimoto, T. 1972. *J. Am. Chem. Soc.* 94: 8243–44
55. Naithani, V. K. 1972. *Z. Physiol. Chem.* 353: 1806–16
56. Naithani, V. K. 1973. *Z. Physiol. Chem.* 354: 659–72
57. Geiger, R., Wissmann, H., Weidenmüller, H. L., Schroeder, H. G. 1969. *Z. Naturforsch. B* 24: 1489–90
58. Naithani, V. K. 1973. *Horm. Metab. Res.* 5: 53
59. Rubenstein, A. H., Melani, F., Steiner, D. F. See Ref. 14, 515–28
60. Schein, P. S., DeLellis, R. A., Kahn, C. R., Gorden, P., Kraft, A. R. 1973. *Ann. Intern. Med.* 79: 239–57
61. Block, M. B., Mako, M. E., Steiner, D. F., Rubenstein, A. H. 1972. *Diabetes* 21: 1013–26
62. Block, M. B., Rosenfield, R. L., Mako, M. E., Steiner, D. F., Rubenstein, A. H. 1973. *N. Engl. J. Med.* 288: 1144–48
63. Stoll, R. W., Touber, J. L., Menahan, L. H., Williams, R. H. 1970. *Proc. Soc. Exp. Biol. Med.* 111: 894–96
64. Katz, A. I., Rubenstein, A. H. 1973. *J. Clin. Invest.* 52: 1113–21
65. Gliemann, J., Sørensen, H. H. 1970. *Diabetologia* 6: 499–504
66. Narahara, H. T. 1972. *Insulin Action,* ed. I. Fritz, 63. New York: Academic
67. Rubenstein, A. H., Pottenger, L. A., Mako, M., Getz, G. S., Steiner, D. F. 1972. *J. Clin. Invest.* 51: 912–21
68. Brush, J. S. 1971. *Diabetes* 20: 151–57
69. Duckworth, W. C., Heinemann, M. A., Kitabchi, A. E. 1972. *Proc. Nat. Acad. Sci. USA* 69: 3698–3702
70. Izzo, J. L., Roncone, A., Izzo, M. J., Foley, R., Bartlett, J. W. 1972. *J. Biol. Chem.* 247: 1219–26
71. Varandani, P. T., Shroyer, L. A., Nafz, M. A. 1972. *Proc. Nat. Acad. Sci. USA* 69: 1681–84
72. Varandani, P. T. 1973. *Biochim. Biophys. Acta* 295: 630–36
73. Varandani, P. T. 1973. *Biochim. Biophys. Acta* 304: 642–59
74. Adams, M. J. et al 1969. *Nature* 224: 491–95
75. Blundell, T. L. et al 1971. *Nature* 231: 506–11

76. Blundell, T. L. et al 1973. *Peptides Proc. Eur. Symp. 12th 1972,* 255–69
77. Blundell, T. L., Dodson, G. G., Dodson, E., Hodgkin, D. C., Vijayan, M. 1971. *Recent Prog. Horm. Res.* 27: 1–40
78. Humbel, R. E., Bosshard, H. R., Zahn, H. See Ref. 14, 111–32
79. Brandenburg, D. et al See Ref. 76, 270–83
80. Arquilla, E. R., Miles, P. V., Morris, J. W. See Ref. 14, 159–73
81. Africa, B., Carpenter, F. H. 1970. *Biochemistry* 9: 1962–72
82. Levy, D., Paselk, R. A. 1973. *Biochim. Biophys. Acta* 310: 398–405
83. Levy, D. 1973. *Biochim. Biophys. Acta* 310: 406–15
84. Massey, D. E., Smyth, D. G. 1972. *Eur. J. Biochem.* 31: 470–73
85. Covelli, I., Frati, L., Wolff, J. 1973. *Biochemistry* 12: 1043–47
86. Vlasov, G. P. 1972. *Zh. Obshch. Khim.* 42: 1128
87. Busse, W. D., Gattner, H. G. 1973. *Z. Physiol. Chem.* 354: 147–55
88. Lindsay, D. G. 1971. *FEBS Lett.* 21: 105–8
89. Brandenburg, D. 1972. *Z. Physiol. Chem.* 353: 869–73
90. Brandenburg, D., Wollmer, A. 1973. *Z. Physiol. Chem.* 354: 613–27
91. Weitzel, G., Renner, R., Kemmler, W., Rager, K. 1972. *Z. Physiol. Chem.* 353: 980–86
92. Weitzel, G., Weber, U., Eisele, K., Zollner, H., Martin, J. 1970. *Z. Physiol. Chem.* 351: 263–67
93. Markussen, J., Sundby, F. 1973. *Int. J. Peptide Protein Res.* 5: 37–48
94. Zimmerman, A. E., Kells, D. I. C., Yip, C. C. 1972. *Biochem. Biophys. Res. Commun.* 46: 2127–33
95. Markussen, J. 1971. *Int. J. Protein Res.* 3: 149–55
96. Weitzel, G., Strätling, W., Hahn, J., Martini, O. 1967. *Z. Physiol. Chem.* 348: 525–32
97. Falkmer, S., Patent, G. J. See Ref. 14, 1–23
98. Falkmer, S. 1972. *Gen. Comp. Endocrinol. Suppl.* 3: 184–91
99. Peterson, J. D., Steiner, D. F., Emdin, S. O., Ostberg, Y., Falkmer, S. 1973. *Fed. Proc.* 32: 577
100. Katsoyannis, P. G., Zalut, C. 1972. *Biochemistry* 11: 1128–32
101. Katsoyannis, P. G., Zalut, C. 1972. *Biochemistry* 11: 3065–69
102. Carpenter, F. H. 1966. *Am. J. Med.* 40: 750–58
103. Brugman, T. M., Arquilla, E. R. 1973. *Biochemistry* 12: 727–32
104. Weitzel, G., Eisele, K., Zollner, H., Weber, U. 1969. *Z. Physiol. Chem.* 350: 1480–83
105. Katsoyannis, P. G., Zalut, C., Harris, A., Meyer, R. J. 1971. *Biochemistry* 10: 3884–89
106. Weitzel, G., Eisele, K., Guglielmi, H., Stock, W., Renner, R. 1971. *Z. Physiol. Chem.* 352: 1735–38
107. Katsoyannis, P. G., Okada, Y., Zalut, C. 1973. *Biochemistry* 12: 2516–25
108. Fullerton, W. W., Potter, R., Low, B. W. 1970. *Proc. Nat. Acad. Sci. USA* 66: 1213–19
109. Rosen, L. S., Fullerton, W. W., Low, B. W. 1972. *Arch. Biochem. Biophys.* 152: 569–73
110. Frank, B. H., Veros, A. J. 1970. *Biochem. Biophys. Res. Commun.* 38: 284–89
111. Grant, P. T., Coombs, T. L., Frank, B. H. 1972. *Biochem. J.* 126: 433–40
112. Frank, B. H., Veros, A. J. 1968. *Biochem. Biophys. Res. Commun.* 32: 155–60
113. Frank, B. H., Veros, A. J., Pekar, A. H. 1972. *Biochemistry* 11: 4926–31
114. Rubenstein, A. H., Steiner, D. F., Cho, S., Lawrence, A. M., Kirsteins, L. 1969. *Diabetes* 18: 598–605
115. Rubenstein, A. H., Mako, M., Welbourne, W. P., Melani, F., Steiner, D. F. 1970. *Diabetes* 19: 546–53
116. Mercola, D. A., Morris, J. W. S., Arquilla, E. R. 1972. *Biochemistry* 11: 3860–74
117. Yu, N. T., Liu, C. S., Culver, J., O'Shea, D. C. 1972. *Biochim. Biophys. Acta* 263: 1–6
118. Venable, J. H. Jr. 1972. *J. Mol. Biol.* 66: 169–80
119. Capaldi, R. A., Garratt, C. J. 1971. *Eur. J. Biochem.* 23: 551–56
120. Burke, M. J., Rougvie, M. A. 1972. *Biochemistry* 11: 2435–39
121. Oelz, O., Froesch, E. R., Bünzli, H. F., Humbel, R. E., Ritschard, W. J. See Ref. 14, 685–702
122. Van Wyk, J. J. et al 1974. *Recent Prog. Horm. Res.* In press
123. Megyesi, K. et al 1974. *Proc. Nat. Acad. Sci. USA.* In preparation
124. Hintz, R. L., Clemmons, D. R., Underwood, L. E., Van Wyk, J. J. 1972. *Proc. Nat. Acad. Sci. USA* 69: 2351–53
125. Morell, B., Froesch, E. R. 1973. *Eur. J. Clin. Invest.* 3: 112–18
126. Morell, B., Froesch, E. R. 1973. *Eur. J. Clin. Invest.* 3: 119–23
127. Daughaday, W. H. et al 1972. *Nature* 235: 107

128. Hall, K., Uthne, K. 1971. *Acta Med. Scand.* 190:137–43
129. Hall, K. 1972. *Acta Endocrinol.* 70, Suppl. 163
130. Tell, G. P. E., Cuatrecasas, P., Van Wyk, J. J., Hintz, R. L. 1973. *Science* 180:312–15
131. Frazier, W. A., Angeletti, R. H., Bradshaw, R. A. 1972. *Science* 176:482–88
132. Fernandez, H. N. et al 1972. *FEBS Lett.* 25:265–70
133. Li, C. H., Gordon, D., Knorr, J. 1973. *Arch. Biochem. Biophys.* 156:493–508
134. Santome, J. A. et al 1971. *FEBS Lett.* 16:198–200
135. Mills, J. B., Howard, S. C., Scepa, S., Wilhelmi, A. E. 1970. *J. Biol. Chem.* 245:3407–15
136. Oliver, L., Hartree, A. S. 1968. *Biochem. J.* 109:19,–24
137. Zakin, M. M., Poskus, E., Dellacha, J. M., Paladini, A. C., Santome, J. A. 1972. *FEBS Lett.* 25:77–82
138. Conde, R. D., Paladini, A., Santome, J. A., Dellacha, J. M. 1973. *Eur. J. Biochem.* 32:563–68
139. Li, C. H., Dixon, J. S. 1971. *Arch. Biochem. Biophys.* 146:233–36
140. Niall, H. D., Hogan, M. L., Sauer, R., Rosenblum, I. Y., Greenwood, F. C. 1971. *Proc. Nat. Acad. Sci. USA* 68:866–69
141. Niall, H. D. 1971. *Nature New Biol.* 230:90–91
142. Bewley, T. A., Dixon, J. S., Li, C. H. 1972. *Int. J. Peptide Protein Res.* 4:281–87
143. Tanner, J. M. 1972. *Nature* 237:433–39
144. Sherwood, L. M., Handmerger, S., McLavrin, W. D., Lanner, M. 1971. *Nature New Biol.* 233:59–61
145. Li, C. H., Dixon, J. S., Chung, D. 1973. *Arch. Biochem. Biophys.* 155:95–110
146. Li, C. H. et al 1968. *Experientia* 24:1288
147. Bewley, T. A., Li, C. H. 1970. *Science* 168::1361–62
148. Li, C. H., Dixon, J. S., Lo, T. B., Schmidt, K. D., Pankov, Y. A. 1970. *Arch. Biochem. Biophys.* 141:705–37
149. Li, C. H. 1973. *Int. Res. Comm. System* (73-3) 3-2-1
150. Dayhoff, M. See Ref. 49, D-202
151. Guyda, H. J., Friesen, H. G. 1971. *Biochem. Biophys. Res. Commun.* 42:1068–75
152. Hwang, P., Guyda, H., Friesen, H. G. 1972. *J. Biol. Chem.* 247:1955–58
153. Niall, H. D. et al 1973. *Recent Prog. Horm. Res.* 29:387–416
154. Hwang, P., Murray, J. B., Jacobs, J. W., Niall, H. D., Friesen, H. G. 1974. *Biochemistry.* In press
155. Nicoll, C. S., Licht, P. 1971. *Gen. Comp. Endocrinol.* 17:490–507
156. Hayashide, T. 1970. *Gen. Comp. Endocrinol.* 15:432–52
157. Papkoff, H., Hayashide, T. 1972. *Proc. Soc. Exp. Biol. Med.* 140:251–55
158. Aloj, S. M., Edelhoch, H. 1970. *Proc. Nat. Acad. Sci. USA* 66:830–36
159. Bewley, T. A., Kawauchi, H., Li, C. H. 1972. *Biochemistry* 11:4179–87
160. Aloj, S., Edelhoch, H. 1972. *J. Biol. Chem.* 247:1146–52
161. Maddaiah, V. T., Collipp, P. J. 1972. *FEBS Lett.* 23:208–10
162. Maddaiah, V. T., Collipp, P. J., Sharma, R. K., Chen, S. Y., Thomas, J. 1972. *Biochim. Biophys. Acta* 263:133–38
163. Cambiaso, C. L., Retegni, L. A., Dellacha, J. M., Santome, J. A., Paladini, A. C. 1970. *Biochim. Biophys. Acta* 231:290–94
164. Bornstein, J., Armstrong, J. M., Ng, F., Paddle, B. M., Misconi, L. 1971. *Biochem. Biophys. Res. Commun.* 42:252–58
165. Schwartz, P. L. 1972. *Proc. Soc. Exp. Biol. Med.* 141:419–22
166. Yamasaki, N., Kikutani, M., Sonenberg, M. 1970. *Biochemistry* 9:1107–14
167. Yamasaki, N., Kangawa, K., Kobayashi, S., Kikutani, M., Sonenberg, M. 1972. *J. Biol. Chem.* 247:3874–80
168. Nutting, D. F., Kostyo, J. L., Mills, J. B., Wilhelmi, A. E. 1972. *Endocrinology* 90:1202–13
169. Chillemi, F., Aiello, A., Pecile, A. 1972. *Nature New Biol.* 238:243–45
170. Chillemi, F., Pecile, A. 1971. *Experientia* 27:385
171. Goodman, A. D., Tanenbaum, R., Rabinowitz, D. 1972. *J. Clin. Endocrinol. Metab.* 35:868–78
172. Gorden, P., Hendricks, C. M., Roth, J. 1973. *J. Clin. Endocrinol. Metab.* 36:178–84
173. Frohman, L. A., Burek, L., Stachura, M. E. 1972. *Endocrinology* 91:262–69
174. Hambley, J., Grant, D. B. 1972. *Acta Endocrinol.* 70:43–47
175. Li, C. H., Starman, B. 1964. *Nature* 202:291–92
176. Ward, D. N., Arnott, M. S. 1965. *Anal. Biochem.* 12:296–302
177. Papkoff, H., Samy, T. S. A. 1967. *Biochim. Biophys. Acta* 147:175–77

178. Lamkin, W. M., Fujino, M., Mayfield, J. D., Holcomb, G. N., Ward, D. N. 1970. *Biochim. Biophys. Acta* 214: 290–98

179. Liu, W. K. et al 1972. *J. Biol. Chem.* 247: 4351–64

180. Sairam, M. R., Papkoff, H., Li, C. H. 1972. *Arch. Biochem. Biophys.* 153: 554–71

181. Maghuin-Rogister, G., Combarnous, Y., Hennen, G. 1972. *FEBS Lett.* 25: 57–60

182. Pierce, J. G., Carlsen, R. B., Liao, T. H. 1973. *Protein and Polypeptide Hormones*, ed. C. H. Li, Vol. 1, 17–57. New York: Academic

183. Liao, T. H., Pierce, J. G. 1970. *J. Biol. Chem.* 245: 3275–81

184. Pierce, J. G., Liao, T. H., Carlsen, R. B., Reimo, T. 1971. *J. Biol. Chem.* 246: 866–72

185. Sairam, M. R., Papkoff, H., Li, C. H. 1972. *Biochem. Biophys. Res. Commun.* 48: 530–37

186. Bellisario, R., Carlsen, R. B., Bahl, O. P. 1973. *J. Biol. Chem.* 248: 6796–6809; Carlsen, R. B., Bahl, O. P., Swaminathan, N. 1973. *J. Biol. Chem.* 248: 6810–27

187. Inagami, T., Murakami, K., Puett, D., Hartree, A. S., Nureddin, A. 1972. *Biochem. J.* 126: 441–42

188. Closset, J., Hennen, G., Lequin, R. M. 1972. *FEBS Lett.* 21: 325–29

189. Stockell-Hartree, A. et al 1971. *J. Endocrinol.* 51: 169–80

190. Sherwood, O. D., Grimek, H. J., McShan, W. H. 1970. *Biochim. Biophys. Acta* 221: 87–106

191. Papkoff, H., Gan, J. 1970. *Arch. Biochem. Biophys.* 136: 522–28

192. Landefeld, T. D., Grimek, H. J., McShan, W. H. 1972. *Biochem. Biophys. Res. Commun.* 46: 436–69

193. Rathnam, P., Saxena, B. B. 1971. *J. Biol. Chem.* 246: 7087–94

194. Liu, W. K., Nahm, H. S., Sweeney, C. M., Holcomb, G. N., Ward, D. N. 1972. *J. Biol. Chem.* 247: 4365–81

195. Sairam, M. R., Samy, T. S. A., Papkoff, H., Li, C. H. 1972. *Arch. Biochem. Biophys.* 153: 572–86

196. Maghuin-Rogister, G., Dockier, A. 1971. *FEBS Lett.* 19: 209–13

197. Maghuin-Rogister, G., Hennen, G. 1972. *FEBS Lett.* 23: 225–29

198. Shome, B., Parlow, A. F. 1973. *J. Clin. Endocrinol. Metab.* 36: 618–21

199. Closset, J., Hennen, G., Lequin, R. M. 1973. *FEBS Lett.* 29: 97–100

200. Canfield, R. E., Agosto, G. M., Bell, J. J. 1970. *Gonadotropins and Ovarian Development*, ed. W. R. Butt, A. C. Crooke, M. Ryle, S. Livingstone, 161–70. London: Churchill-Livingstone

201. Morgan, F. J., Canfield, R. E. 1971. *Endocrinology* 88: 1045–53

202. Swaminathan, N., Bahl, O. P. 1971. *Biochem. Biophys. Res. Commun.* 40: 422–27

203. Morgan, F. J., Birken, S., Canfield, R. E. 1973. *Mol. Cell. Biochem.* 2: 97–99

204. Bahl, O. P. 1969. *J. Biol. Chem.* 244: 575–83

205. Hammond, J. M., Bridson, W. E., Chrambach, A., Kohler, P. O. 1972. *J. Clin. Endocrinol.* 34: 185–87

206. Van Hall, E. V., Vaitukaitis, J. L., Ross, G. T., Hickman, J. W., Ashwell, G. 1971. *Endocrinology* 89: 11–15

207. Liao, T. H., Pierce, H. G. 1971. *J. Biol. Chem.* 246: 850–65

208. Sairam, M. R., Li, C. H. 1973. *Biochem. Biophys. Res. Commun.* 51: 336–42

209. Cornell, J. S., Pierce, J. G. 1973. *J. Biol. Chem.* 248: 4327–33

210. Shome, B., Parlow, A. F. 1973. *55th Ann. Meet. Endocrine Soc. Abstr.*, A-60

211. Saxena, B. B., Rathnam, P. 1971. *J. Biol. Chem.* 246: 3549–54

212. Papkoff, H., Ekblad, M. 1970. *Biochem. Biophys. Res Commun.* 40: 614–21

213. Nuti, L. C., Grimek, H. J., Braselton, W. E., McShan, W. H. 1972. *Endocrinology* 91: 1418–22

214. Schams, D., Papkoff, H. 1972. *Biochim. Biophys. Acta* 263: 139–48

215. Gospodarowicz, D. 1972. *Endocrinology* 91: 101–6

216. Donaldson, E. M., Yamazaki, F., Dye, H. M., Philleo, W. W. 1972. *Gen. Comp. Endocrinol.* 18: 469–81

217. Burzawa-Gerard, E. 1971. *Biochimie* 53: 545–52

218. Papkoff, H., Licht, P. 1971. *Proc. Soc. Exp. Biol. Med.* 139: 372–76

219. Licht, P. 1972. *Gen. Comp. Endocrinol.* 19: 273–81

220. Licht, P. 1972. *Gen. Comp. Endocrinol.* 19: 282–89

221. Licht, P., Papkoff, H. 1973. *Gen. Comp. Endocrinol.* 20: 172–76

222. Jackson, R. G., Sage, M. 1973. *Comp. Biochem. Physiol.* 44A: 867–70

223. Fontaine, Y. A. 1968. *Progress in Endocrinology—Proceedings of the Third International Congress of Endocrinology* 184: 453–57

224. Cheng, K. W., Pierce, J. G. 1972. *J. Biol. Chem.* 247: 7163–72

225. Sairam, M. R., Papkoff, H., Li, C. H. 1972. *Biochim. Biophys. Acta* 278: 421–32

226. Yang, K. P., Ward, D. N. 1972. *Endocrinology* 91 : 317–20
227. Bewley, T. A., Sairam, M. R., Li, C. H. 1972. *Biochemistry* 11 : 932–36
228. Mori, K. F. 1972. *Biochim. Biophys. Acta* 257 : 523–26
229. Mori, K. F., Hollands, T. R. 1971. *J. Biol. Chem.* 246 : 7223–29
230. Merz, W. E., Hilgenfeldt, U., Brocker-hoff, P., Brossmer, R. 1973. *Eur. J. Biochem.* 35 : 297–306
231. Pierce, J. G., Bahl, O. P., Cornell, J. S., Swaminathan, N. 1971. *J. Biol. Chem.* 246 : 2321–24
232. Catt, K. J., Dufau, M. L., Tsuruhara, T. 1973. *J. Clin. Endocrinol. Metab.* 36 : 73–80
233. Braunstein, G. D., Vaitukaitis, J. L., Ross, G. T. 1972. *Endocrinology* 91 : 1030–36
234. Rayford, P. L., Vaitukaitis, J. L., Ross, G. T., Morgan, F. J., Canfield, R. E. 1972. *Endocrinology* 91 : 144–46
235. Yang, W. H., Sairam, M. R., Papkoff, H., Li, C. H. 1971. *Science* 175 : 637–38
236. Yang, W. H., Sairam, M. R., Li, C. H. 1973. *Acta Endocrinol.* 72 : 173–81
237. Reichert, L. E. Jr. 1972. *Endocrinology* 90 : 1119–22
238. Vaitukaitis, J. L., Ross, G. T., Reichert, L. E. Jr. 1973. *Endocrinology* 92 : 411–16
239. Donini, S., Oliveri, V., Ricci, G., Donini, P. 1973. *Acta Endocrinol.* 73 : 133–45
240. Parlow, A. F., Kovacic, N., Bhalla, R. C. 1972. *J. Clin. Endocrinol. Metab.* 35 : 319–22
241. Bromer, W. W., Sinn, L. G., Behrens, O. K. 1957. *J. Am. Chem. Soc.* 79 : 2807–10
242. Bromer, W. W., Boucher, M. E., Koffenberger, J. E. Jr. 1971. *J. Biol. Chem.* 246 : 2822–27
243. Thomsen, J., Kristiansen, K., Brunfeldt, K., Sundby, F. 1972. *FEBS Lett.* 21 : 315–19
244. Sundby, F., Markussen, J. 1972. *Horm. Metab. Res.* 4 : 56
245. Sundby, F., Markussen, J. 1971. *Horm. Metab. Res.* 3 : 184–87
246. Sundby, F., Frandsen, E. K., Thomsen, J., Kristiansen, K., Brunfeldt, K. 1972. *FEBS Lett.* 26 : 289–93
247. Markussen, J., Frandsen, E., Heding, L. G., Sundby, F. 1972. *Horm. Metab. Res.* 4 : 360–63
248. Bromer, W. W., Boucher, M. E., Patterson, J. M. Peckar, A. H., Frank, B. H. 1972. *J. Biol. Chem.* 247 : 2581–85

249. King, M. V. 1965. *J. Mol. Biol.* 11 : 549–61
250. Blanchard, M. H., King, M. V. 1966. *Biochem. Biophys. Res. Commun.* 25 : 298–303
251. Gratzer, W. B., Beaven, G. H. 1969. *J. Biol. Chem.* 244 : 6675–79
252. Srere, P. A., Brooks, G. C. 1969. *Arch. Biochem. Biophys.* 129 : 708–10
253. Edelhoch, H., Lippold, R. L. 1969. *J. Biol. Chem.* 244 : 3876–83
254. Gratzer, W. B., Beaven, G. H., Rattle, H. W. E., Bradbury, E. M. 1968. *Eur. J. Biochem.* 3 : 276–83
255. Epand, R. M. 1972. *Arch. Biochem. Biophys.* 148 : 325–26
256. McBride-Warren, P. A., Epand, R. M. 1972. *Biochemistry* 11 : 3571–75
257. Epand, R. M. 1971. *Can. J. Biochem.* 49 : 166–69
258. Epand, R. M. 1972. *J. Biol. Chem.* 247 : 2132–38
259. Swann, J. C., Hammes, G. G. 1969. *Biochemistry* 8 : 1–7
260. Gratzer, W. B., Creeth, J. M., Beaven, G. H. 1972. *Eur. J. Biochem.* 31 : 505–9
261. Assan, R., Slusher, N. 1972. *Diabetes* 21 : 843–55
262. Felts, P. W., Ferguson, M. E. C., Hagey, K. A., Stitt, E. S., Mitchell, W. M. 1970. *Diabetologia* 6 : 44–45
263. Guy-Grand, B., Assan, R. 1973. *Horm. Metab. Res.* 5 : 60
264. Spiegel, A. M., Bitensky, M. W. 1969. *Endocrinology* 85 : 638–43
265. Lande, S., Gorman, R., Bitensky, M. 1972. *Endocrinology* 90 : 597–604
266. Grande, F., Grisolia, S., Diederich, D. 1972. *Proc. Soc. Exp. Biol. Med.* 139 : 855–60
267. Mutt, V., Jorpes, J. E., Magnusson, S. 1970. *Eur. J. Biochem.* 15 : 513–19
268. Said, S. I., Mutt, V. 1972. *Eur. J. Biochem.* 28 : 199–204
269. Bodanszky, M., Klausner, Y. S., Said, S. I. 1973. *Proc. Nat. Acad. Sci. USA* 70 : 382–84
270. Brown, J. C., Mutt, V., Pederson, R. A. 1970. *J. Physiol. London* 209 : 57–64
271. Brown, J. C., Dryburgh, J. R. 1971. *Can. J. Biochem.* 49 : 867–72
272. Valverde, I., Rigopoulou, D., Marco, J., Faloona, G. R., Unger, R. H. 1970. *Diabetes* 19 : 614–23
273. Valverde, I. et al 1973. *J. Clin. Endocrinol. Metab.* 36 : 185–87
274. Moody, A. J., Markussen, J., Schaich-Fries, A., Steenstrup, C., Sundby, F. 1970. *Diabetologia* 6 : 135–40
275. Murphy, R. F., Buchanan, K. D., Elmore, D. T. 1973. *Biochim. Biophys. Acta* 303 : 118–27

276. Kerins, C., Said, S. I. 1973. *Proc. Soc. Exp. Biol. Med.* 142:1014–17
277. Rigopoulou, D., Valverde, I., Marco, J., Faloona, G., Unger, R. H. 1970. *J. Biol. Chem.* 245:496–501
278. Tung, A. K., Zerega, F. 1971. *Biochem. Biophys. Res. Commun.* 45:387–95
279. Noe, B. D., Bauer, G. E. 1971. *Endocrinology* 89:642–51
280. Noe, B. D., Bauer, G. E. 1973. *Proc. Soc. Exp. Biol. Med.* 142:210–13
281. Hellerstrom, C., Howell, S. L., Edwards, J. C., Andersson, A. 1972. *FEBS Lett.* 27:97–101
282. O'Connor, K. J., Gay, A., Lazarus, N. R. 1973. *Biochem. J.* 134:473–80
283. Tager, H. S., Steiner, D. F. 1973. *Proc. Nat. Acad. Sci. USA* 70:2321–25
284. Gregory, H., Hardy, P. M., Jones, D. S., Kenner, G. W., Sheppard, R. C. 1964. *Nature* 204:931–33
285. Bentley, P. H., Kenner, G. W., Sheppard, R. C. 1966. *Nature* 209:583–85
286. Gregory, R. A., Tracy, H. J., Agarwal, K. L., Grossman, M. I. 1969. *Gut* 10:603–8
287. Agarwal, K. L., Kenner, G. W., Sheppard, R. C. 1969. *Experientia* 25:346–48
288. Agarwal, K. L. et al 1968. *Nature* 219:614–15
289. Agarwal, K. L., Kenner, G. W., Sheppard, R. C. 1969. *J. Am. Chem. Soc.* 91:3096–97
290. Mutt, V., Jorpes, J. E. See Ref. 49, D-212
291. Mutt, V., Jorpes, J. E. 1968. *Proc. Int. Union Physiol. Sci.* 6:193
292. Anastasi, A., Erspamer, V., Endean, R. 1968. *Arch. Biochem. Biophys.* 125:57–68
293. Anastasi, A. et al 1970. *Brit. J. Pharm.* 38:221–28
294. Anastasi, A et al 1969. *Brit. J. Pharm.* 37:198–206
295. Tracy, H. J., Gregory, R. A. 1964. *Nature* 204:935–38
296. Morley, J. S. 1968. *Proc. Roy. Soc. London* 170:97–111
297. Lin, T. M. 1972. *Gastroenterology* 63:922–23
298. Trout, H. H., Grossman, M. I. 1971. *Nature New Biol.* 234:256
299. Yalow, R. S., Berson, S. A. 1970. *Gastroenterology* 58:609–15
300. Yalow, R. S., Berson, S. A. 1971. *Gastroenterology* 60:203–14
301. Yalow, R. S., Berson, S. A. 1971. *Gastroenterology* 60:215–22
302. Praissman, M., Berkowitz, J. M. 1972. *Biochem. Biophys. Res. Commun.* 48:175–80
303. Gregory, R. A., Tracy, H. J. 1972. *Lancet* 2:797–99
304. Yalow, R. S., Berson, S. A. 1972. *Biochem. Biophys. Res. Commun.* 48:391–95
305. Rehfeld, J. F. 1972. *Biochim. Biophys. Acta* 285:364–72
306. Li, C. H. et al 1955. *Nature* 176:687–89
307. Shepherd, R. G. et al 1956. *J. Am. Chem. Soc.* 78:5067–76
308. Li, C. H., Dixon, J. S., Chung, D. 1958. *J. Am. Chem. Soc.* 80:2587–88
309. Lee, T. H., Lerner, A. B., Buettner-Janusch, V. 1961. *J. Biol. Chem.* 236:2970–74
310. Riniker, B., Sieber, P., Rittel, W., Zuber, H. 1972. *Nature New Biol.* 235:114–15
311. Li, C. H. 1972. *Biochem. Biophys. Res. Commun.* 49:835–39
312. Li, C. H. et al 1970. *J. Am. Chem. Soc.* 82:5760–62
313. Blake, J., Wang, K. T., Li, C. H. 1972. *Biochemistry* 11:438–41
314. Nakamura, M. 1972. *J. Biochem.* 71:1029–41
315. Seelig, S., Sayers, G, Schwyzer, R., Schiller, P. 1971. *FEBS Lett.* 19:232–34
316. Harris, J. I., Lerner, A. B. 1957. *Nature* 179:1346–47
317. Li, C. H. 1959. *Lab. Invest.* 8:574–87
318. Dixon, J. S., Li, C. H. 1960. *J. Am. Chem. Soc.* 82:4568–72
319. Lee, T. H., Lerner, A. B., Buettner-Janusch, V. 1961. *J. Biol. Chem.* 236:1390–94
320. Lee, T. H., Lerner, A. B., Buettner-Janusch, V. 1963. *Biochim. Biophys. Acta* 71:706–9
321. Lowry, P. J., Chadwick, A. 1970. *Biochem. J.* 118:713–18
322. Scott, A. P. et al 1973. *Nature New Biol.* 244:65–67
323. Yalow, R. S., Berson, S. A. 1971. *Biochem. Biophys. Res. Commun.* 44:439–45
324. Yalow, R. S., Berson, S. A. 1973. *J. Clin. Endocrinol. Metab.* 36:415–23
325. Li, C. H., Barnafi, L., Chretien, M., Chung, D. 1965. *Nature* 208:1093–94
326. Graf, L., Barat, E., Cseh, G., Sajgo, M. 1971. *Biochim. Biophys. Acta* 229:276–78
327. Cseh, G., Barat, E., Patthy, A., Graf, L. 1972. *FEBS Lett.* 21:344–46
328. Chretien, M., Li, C. H. 1967. *Can. J. Biochem.* 45:1163–74
329. Geschwind, I. I., Li, C. H., Barnafi,

L. 1957. *J. Am. Chem. Soc.* 79:1003–4

330. Geschwind, I. I., Li, C. H., Bernafi, L. 1957. *J. Am. Chem. Soc.* 79:620–25

331. Lee, T. H., Lerner, A. B., Buettner-Janusch, V. 1961. *J. Biol. Chem.* 236:1390–94

332. Dixon, J. S., Li, C. H. 1961. *Gen. Comp. Endocrinol.* 1:161–69

333. Lee, T. H., Lerner, A. B., Buettner-Janusch, V. 1963. *Biochim. Biophys. Acta* 71:706–9

334. Harris, J. I. 1959. *Nature* 184:167–69

335. Niall, H. D. et al 1970. *Z. Physiol. Chem.* 351:1586–88

336. Brewer, H. B. Jr, Ronan, R. 1970. *Proc. Nat. Acad. Sci. USA* 67:1862–69

337. O'Riordan, J. L. H. et al 1971. *Proc. Roy. Soc. Med. London* 64:1263–65

338. Brewer, H. B. Jr, Fairwell, T., Ronan, R., Sizemore, G. W., Arnaud, C. D. 1972. *Proc. Nat. Acad. Sci. USA* 69:3585–88

339. Jacobs, J. W. et al 1973. *Fed. Proc.* 32:648

340. Potts, J. T. Jr. et al 1971. *Proc. Nat. Acad. Sci. USA* 68:63–67

341. Keutmann, H. T., Dawson, B. F., Aurbach, G. D., Potts, J. T. Jr. 1972. *Biochemistry* 11:1973–79

342. Aurbach, G. D. et al 1971. *Recent Prog. Horm. Res.* 28:353–98

343. Berson, S. A., Yalow, R. S. 1968. *J. Clin. Endocrinol. Metab.* 28:1037–47

344. Sherwood, L. M., Rodman, J. S., Lundberg, W. B. Jr. 1970. *Proc. Nat. Acad. Sci. USA* 67:1631–38

345. Arnaud, C. D. et al 1971. *Am. J. Med.* 50:630–38

346. Habener, J. F., Powell, D., Murray, T. M., Mayer, G. P., Potts, J. T. Jr.

1971. *Proc. Nat. Acad. Sci. USA* 68:2986–91

347. Canterbury, J. M., Reiss, E. 1972. *Proc. Soc. Exp. Biol. Med.* 140:1393–98

348. Segre, G. V., Habener, J. F., Powell, D., Tregear, W., Potts, J. T. Jr. 1972. *J. Clin. Invest.* 51:3163–72

349. Fischer, J. A., Oldham, S. B., Sizemore, G. W., Arnaud, C. D. 1972. *Proc. Nat. Acad. Sci. USA* 69:2341–45

350. Hamilton, J. W., Cohn, D. V. 1969. *J. Biol. Chem.* 244:5421–29

351. Hamilton, J. W., Spierto, F. W., MacGregor, R. R., Cohn, D. V. 1971. *J. Biol. Chem.* 246:3224–33

352. Hamilton, J. W., MacGregor, R. R., Chu, L. L. H., Cohn, D. V. 1971. *Endocrinology* 89:1440–47

353. Cohn, D. V., MacGregor, R. R., Chu, L. L. H., Kimmel, J. R., Hamilton, J. W. 1972. *Proc. Nat. Acad. Sci. USA* 69:1521–25

354. Kemper, B., Habener, J. F., Potts, J. T. Jr, Rich, A. 1972. *Proc. Nat. Acad. Sci. USA* 69:643–47

355. Habener, J. F., Kemper, B., Potts, J. T. Jr, Rich, A. 1972. *Science* 178:630–33

356. Habener, J. F., Kemper, B., Potts, J. T. Jr, Rich, A. 1973. *Endocrinology* 92:219–26

357. Hamilton, J. W., Niall, H. D., Keutmann, H. T., Potts, J. T. Jr, Cohn, D. V. 1973. *Fed. Proc.* 32:269

358. Potts, J. T. Jr. et al 1973. *Mount Sinai J. Med. New York* 40:448–61

359. Adelson, J. W. 1971. *Nature* 229:321–25

360. Weinstein, B. 1972. *Experientia* 28:1517–22

THE INFORMATION CONTENT OF   ✶857
PROTEIN AMINO ACID SEQUENCES[1]

T. T. Wu
Departments of Engineering Sciences and Physics, Northwestern University
Evanston, Illinois

W. M. Fitch
Department of Physiological Chemistry, University of Wisconsin Medical School
Madison, Wisconsin

E. Margoliash
Departments of Biological Sciences and Chemistry, Northwestern University
Evanston, Illinois

CONTENTS

INTRODUCTION

The rich biological information content of protein amino acid sequences stems from their being, on the one hand, simple translations of genome segments that necessarily reflect the history of genetic events to which the organism or species has been subjected, and on the other hand, effectors of the multitudinous physiological functions constituting a living creature. Therefore, as often stated (1, 2), studies of the

[1] Supported by Grant GM-19121 from the National Institutes of Health to one author (E.M.), GB-33146 from the National Science Foundation to another (T.T.W.), and GB-32274X from the National Science Foundation to the third (W.M.F.).

amino acid sequences of proteins have been used for the reconstruction of genetic events, whether resulting from experimental manipulations or occurring during the evolution of species and leading to descriptions of evolutionary history and mechanisms (3), and for the elucidation of structure-function relations in proteins (4). Moreover, the transition between the linear, relatively structureless polymer produced by the genetic-protein biosynthetic mechanisms and the precisely folded functionally competent protein molecule has long been known to be entirely spontaneous (5). The full complexity of a protein's structure in space is therefore inherent in its amino acid sequence and should be derivable from it. Recent indications of some success in this direction make it a useful subject for discussion at this time.

Structure-function relations of proteins are extensively studied subjects particular to every protein and could not possibly be covered in a single chapter; these are not discussed. Though a far smaller subject area, the molecular evolution of proteins is still in part specific to every protein, and only more general aspects are considered, particularly those relating to procedures for extracting evolutionary information from comparisons of amino acid sequences of proteins from different species. This article therefore contains mention of some of the more recent procedures developed in connection with amino acid sequence determination, an outline of the principles which may lead to the prediction of the spatial structures of proteins from their amino acid sequences, and some considerations of techniques for the reconstruction of evolutionary history from the primary structures of proteins.

THE DETERMINATION OF AMINO ACID SEQUENCES

Principles

Driven by the importance of amino acid sequences as a multifaceted source of fundamental biological information, the last several years have seen an explosive increase in the number of proteins subjected to such determinations. Thus, the first edition of the Eck & Dayhoff's *Atlas of Protein Sequence* appearing in 1965 (6) gave 65 protein primary structures, while the latest version by Dayhoff dated 1972 and the 1973 Supplement (7) list 663. This is remarkable considering how many years it took Sanger and his co-workers to establish the sequence of the 21 residues of the A chain and the 30 residues of the B chain of insulin (8–10) in the late 1940s and early 1950s, and it attests to the considerable work and ingenuity put into this enterprise. Today sequences of single polypeptide chains of over 500 residues are becoming increasingly common (see 7), and the establishment of the primary structures of smaller proteins is largely a routine procedure. Even so, except for some relatively small polypeptides such as the calcitonins (11, 12), glucagon (13), and others, for which the automated Edman procedure (14) may yield complete sequences, the principle employed is the same as that originally developed by Sanger for insulin, namely: (*a*) fragmentation of the protein chain into peptides; (*b*) isolation and purification of these smaller peptides; (*c*) determination of the amino acid sequence of each peptide by chemical and enzymic means; and (*d*) repetition of steps *b* and *c* following a second type of fragmentation of the intact protein in the hope of obtaining peptides that overlap peptides from the first fragmentation and thus make it possible to establish unambiguously the order of peptides in the whole protein chain.

The optimal size of the fragments one wishes to obtain depends on the ease with which they can be purified and their sequences established. Smaller peptides are often easier to purify and present fewer problems in structure determination, but many more of them are obtained from a given protein chain. The advent of machines (14) for carrying out the manipulations of the Edman sequential degradation from the amino-terminal residue (15–21) by extending the number of cycles through which the degradation can be usefully carried out from the common 10–15 residue limits for the different manual procedures to routine sequence yields of 30–45 residues, and in some cases far more (see, for example, 14, 22, 23), has resulted in a significant change of tactics. The difficulties of breaking the protein chain into relatively large but well defined segments and obtaining such fragments in a pure form is now outweighed by the advantage of having to deal with only a few of them when it has become possible to determine their sequences by a semi-automated procedure. Examples of smaller proteins to which this type of approach has been applied are parathyroid hormone (3, 4), placental lactogen (24), neurophysin (25, 26), a serum apolipoprotein (27, 28), and amyloid (29–31). Among the larger proteins are the immunoglobin M heavy chain (32–34), trypsin (35), and thermolysin (36).

Strategies for the determination of the complete amino acid sequences of large proteins by the semi-automatic Edman machinery have been discussed by several authors (35–38). The difficulty with this approach is primarily that as with all repetitive procedures, whether degradative or synthetic, errors are cumulative, so that if at any cycle a significant amount of unreacted amino-terminal residue remains, it will appear at the next cycle, and as the same error is repeated the amount and diversity of the background material rises until the residue at the cycle being processed can no longer be identified. Such admixture also occurs from random breaks of the peptide chain as it is subjected to not entirely innocuous reagents. Possibly the most serious cause of the drastic decrease in yield, particularly obvious at the first step in the automated procedure (down to 30% of the amount of protein chain, in some cases), is oxidative desulfuration of the phenylthiocarbamylpeptide (18) which can then no longer undergo cyclization and cleavage. Such chains are now merely alkylated and no longer take part in further cycles. The O_2 probably comes from air adsorbed on the surface of the container of the protein solution, and the effect appears far more prominently in the automated than in the manual procedures, possibly because the volume-to-surface ratio is so much smaller in the former than in the latter. Finally, some protein chains tend to have sufficient solubility in the organic solvents used for extraction to be lost from the reaction container, particularly after the progress of the sequential degradation has resulted in the removal of enough hydrophobic residues. To counteract this tendency, the peptide to be degraded can be attached to an insoluble resin matrix (39–43), the reagent may similarly be made insoluble (44), or the peptide can be reacted so as to add hydrophilic groups to some of its residue side chains (45–47). With polypeptides attached to resin beads, it is relatively simple to automate the degradation procedure since the material can be packed in a column through which reagents and solvents are percolated. Such a commercial instrument is now available and the amino acid sequence of a ribosomal protein has been determined in part through its use (48).

Ideally, sequential degradation, if it could be made to yield complete sequences of sizable protein chains routinely, would present a second major principle of structure determination on par with that originally established by Sanger. More recently, the possibility of very rapid determination of peptide structure by mass spectrometry (see 49, 50) following suitable chemical derivatization to make the material sufficiently volatile (51) is leading to what may eventually develop into a third major approach to amino acid sequence determination (52–55). A protein chain is fragmented, the peptides subjected to a preliminary crude separation, and the mixtures obtained analyzed by mass spectrometry, a technique capable of determining simultaneously the complete structure of each individual peptide in the mixture. Overlapping peptides are produced by a second fragmentation and their structures similarly determined to lead to the complete amino acid sequence of the original chain. For example, in the case of the carboxymethylated A chain of porcine insulin (55), the dipeptides produced by dipeptidylaminopeptidase I (56, 57) were transformed to N-pentafluoropropionyldipeptide methyl esters and analyzed in a gas chromatography–mass spectrometry computer system, in which the gas chromatography partially separated the peptides, the mass spectrometry determined the structures of the dipeptides, and the computer read and analyzed the data. Overlapping dipeptide sequences were produced by using the same procedure on an A-chain sample from which the amino-terminal glycyl residue had been removed by one cycle of Edman degradation. These results gave 12 possible sequences for the chain, and all but the correct sequence could be eliminated by studying the time-course of release of certain dipeptides. Though still far from being a full-fledged technique to attack long chains, this type of procedure shows obvious possibilities for further development, such as the use of other fragmentations to yield peptides larger but still within the easy operating range of the mass spectrometer, the use of other types of preliminary separations and chemical derivatizations, etc. Its major advantage is eliminating the requirement for the isolation of pure peptides, the part of the usual approach to amino acid sequence determination which commonly consumes the largest amount of labor. Its major difficulty is likely to be the rapid increase in the number of possible sequences represented by the data as the length of protein chain increases. Such ambiguities are less pronounced if longer fragments are used, but the longer the fragments the more difficult their mass spectrometric identification. Eventually, these techniques will probably be limited by such conflicting requirements, and in all cases further independent data will be required.

Some Recent Useful Procedures

Within the contexts of these major approaches, numerous procedural steps have been studied over the last few years which show promise of simplifying or improving various phases of amino acid sequence determinations. To describe them all would be beyond the space limitations of this article. In addition to those discussed above, a list of some of the more recent useful developments is provided as follows: the continued improvement of automatic amino acid analyzers, represented, for example, by the high-pressure capillary column chromatographic system and the computer control of the Durrum D-500 machine; techniques for peptide and protein hydrolysis designed to improve yields of labile residues such as tyrosine, cysteine, and tryptophan (58–65); reductive hydrolysis by hydroiodic acid of the phenylthiohydantoin derivatives of amino acids obtained from Edman sequential degradation procedures to yield free amino acids in near quantitative yields that can be analyzed with precision on automatic amino acid analyzers, thus helping to interpret more accurately the results of the degradation (23); various methods for the rapid or sensitive identification of thiohydantoin derivatives of amino acids (66–70), such as gas chromatography or with radioactively labeled reagents; gas chromatography for amino acid analysis (67, 71, 72); a new reagent, fluorescamine, which readily yields fluorescent derivatives of primary amines, making the detection of amino acids and peptides in column effluents, for example, easier and more sensitive than with other procedures (73–75); reagents other than the phenyliso-thiocyanate originally introduced by Edman (15) for the sequential degradation of peptides and presenting a variety of advantages over it, such as 4-sulfophenyl-isothiocyanate (45, 47, 76, 77) or even more highly sulfonated reagents (46), methylisothiocyanate (39, 68, 78, 79), cyanomethyldithiobenzoate (80), 4-dimethyl-amino-1-naphthylisothiocyanate (81), thioacetylthioglycollic acid (82, 83), and pentafluorophenylisothiocyanate (84); computer programs and numerical methods to increase the precision of interpretation of results from various analytical techniques, such as in the quantitation of sequential Edman degradations (23), the determination of the exact number of residues in a protein chain by amino acid analysis (85), the calculation of the amino acid composition of proteins (86), and the determination of the molecular weight of proteins by amino acid analysis (87); chemical methods for cleavage of peptide chains at specific residues, making it possible to obtain fragments different from those resulting from the common enzymic hydrolyses employed and often large enough to be useful for automated sequential Edman degradation, such as at tryptophanyl residues (88, 89), cysteinyl residues (90, 90a), asparaginyl–glycine (91, 92), and aspartyl–proline (93) bonds; chemical modifications of specific side chains to either block reaction of enzymes at these residues, such as maleylation (94) or citraconylation (95) of the ε-amino groups of lysyl residues to prevent cleavage by trypsin, or to provide a new point of enzymic cleavage such as by reaction of cysteinyl residues with ethylene imine (96), yielding aminoethylcysteinyl side chains that present sites for tryptic hydrolysis; a method for sequential degradation of peptide chains from the carboxyl terminus (97–99); and the isolation of several new proteolytic enzymes that may be used to simplify a variety of aspects of amino acid sequence determination, such as the dipeptidylaminopeptidase I (56, 57) which hydrolyzes peptide chains into dipeptides starting at the amino terminus (100–103), other dipeptidylaminopeptidases, termed II, III and IV, which can hydrolyze dipeptides with amino-terminal lysyl and arginyl residues and with a penultimate proline, respectively, situations which stop further cleavage by the dipeptidlyaminopeptidase I (104), a dipeptidylcarboxypeptidase from *Corynebacterium equi* (105), an enzyme, clostripain, which specifically cleaves at arginyl residues (106, 107), carboxypeptidases of rather generalized activity, including the ability to liberate proline from a variety of sources (108–115), some proteases with a relatively low degree of specificity such as thermolysin from a thermophilic bacterium that hydrolyzes peptide bonds on the amino-terminal side of hydrophobic residues, particularly isoleucine and valine (116–118), an enzyme that removes pyrrolidone carboxylic acid from the amino terminus of peptide chains (119), making them amenable to sequential degradation that is effective with peptides but often ineffective with entire protein chains. Reviews of methods for the determination of the amino acid sequences of proteins are found in books by Schroeder (120) and Blackburn (121), in a volume edited by Needleman (122), and in the especially valuable compendia in Volumes XI (1967) and XXV (1972) of *Methods in Enzymology* (123, 124).

Need for Automation

Notwithstanding considerable technical and conceptual advances, some of which are cited above, the establishment of the complete primary structure of a protein remains, to a diminishing but still considerable extent, a slow, manual procedure requiring highly skilled operators. It is thus an expensive undertaking and human operators are singularly ill-suited to the tedious, repetitive functions which constitute a good deal of the procedure. Moreover, as is made clear in the following sections, if we are ever to obtain a representative sample of the enormous fund of biological and structure-function information encoded in the proteins of extant species, it will be necessary to determine the primary structures of at least one or

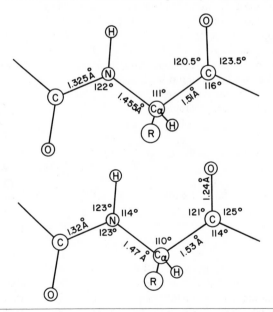

Figure 1   The two models of a peptide unit (143, 144).

possibly two orders of magnitude more proteins than those for which this information is currently available. More or less complete automation of the process of amino acid sequence determination is thus a necessary objective and should be within reach in the not too distant future. The first steps in this direction, represented for example by the automatic Edman sequential degradation machine (14), the "cascade" automated column chromatography system of Edelman & Gall (125) for the separation of peptide mixtures into their components, the recent commercial availability of an automatic machine for spotting and developing thin-layer chromatograms, etc can be taken as a basis for some optimism in this field. A second crucial objective is the adaptation of methods to the ultramicro range, since many biologically important proteins are likely to be obtainable only in much smaller amounts than currently used for complete amino acid sequence determination. Here again the first harbingers of this development, such as the newer high-sensitivity amino acid analyzers, the use of radioactive reagents to increase by severalfold the sensitivity of detection of derivatives obtained by Edman degradation (2, 70), a synthetic polypeptide carrier to prevent losses of protein permitting the use of less than 10 nmoles of protein for automatic Edman degradation (126), and the development of high-pressure liquid chromatographic systems (127, 128) yet to be employed for rapid separations of small amounts of material in amino acid sequence work all foster the prediction that the field is bound for an expansion that will dwarf the already impressive development of the last several years.

PREDICTION OF THE SPATIAL STRUCTURES OF PROTEINS

In their classical experiments with ribonuclease, Anfinsen and his collaborators (5, 129) clearly demonstrated that the three-dimensional structure of a protein in a given environment is determined by its amino acid sequence. The significance of this finding was further emphasized by the difficulty of recombining the A and B chains of insulin to recover a biologically active molecule (130–135), while proinsulin, the single-chain precursor of insulin, could be renatured in a similar fashion as ribonuclease (136–138). Thus, the amino acid sequence and the spatial-like structure of a protein are intimately related. Numerous researchers during the past twenty years have attempted to define this relationship in precise terms and predict the secondary and tertiary structures of proteins from their amino acid sequences.

Theoretical Conformation Calculations and Comparison with Experiment

The starting point of all theoretical studies is the structure of the peptide unit. From earlier studies (139–142), two slightly different models have been proposed (143, 144). They are shown in Figure 1 and probably have similar experimental errors (142). However, when a few hundred of these units are joined together, any minor differences can result in gross variations in overall spatial structure. The peptide bond in these models is assumed to be planar and rigid. Subsequent studies, however, suggested that slight rotation around the C–N bond is allowed and the N–C_α–C angle can vary somewhat (145). The rotation angle around the N–C_α bond is designated as Φ, and that around the C_α–C bond as Ψ (146–149).

The earliest predicted structures are the α helix (150) and the β sheet (151). Subsequently, a more systematic study based on the Φ, Ψ angles was undertaken. The simplest method assumed that atoms were hard spheres with fixed radii. Physically allowed conformations of the peptide unit were dictated by the minimum permitted contact distances between the atoms, and were indicated on a Φ, Ψ plot (145, 152–155). For each set of Φ, Ψ values, the interatomic distances of all pairs of nonbonding atoms were calculated and compared with the minimum permitted contact distances for these pairs. If the interatomic distances were larger, the conformation prescribed by the Φ, Ψ values was considered sterically allowable.

The next theoretical approach was based on the concept of Anfinsen and his collaborators (5, 129) that the protein molecule in a given environment attains the minimum Gibbs free energy. For peptide units, the interaction energy consisted of the Lennard-Jones or Buckingham 6-exp potential for nonbonding atoms (145), electrostatic Coulomb interaction between the amide and carbonyl dipoles (156), empirical hydrogen-bonding potentials (157–159), torsional potential (145, 160), and the free energy changes due to addition or removal of solvents (160, 161). Because of numerous parameters in these energy calculations, it becomes extremely difficult to assess the reliability of the final results. From a more fundamental point of view, one can also start with the Schrödinger equation (162) for the peptide unit. However, since it cannot be solved exactly, various approximations have been used to calculate the interaction energy. These quantum mechanical calculations (163–166) also involve many parameters and suffer from difficulties similar to those of the classical energy calculations.

For polypeptides, attempts have been made to predict the three-dimensional structures of Residues 2–20 (161) and Residues 65–72 (167) of ribonuclease by calculating the extreme energy states. For proteins, the tertiary structures of ribonuclease (168) and cytochrome c (169) have been suggested from such theoretical considerations.

The occurrence of α helices in nature was almost immediately verified experimentally (170). As more tertiary structures of proteins were established by X-ray diffraction analyses, α helices and β sheets became typical features (155). Unfortunately, the usefulness of purely theoretical conformational calculations stops here. In the case of glycine, many experimentally observed conformations in proteins are outside the theoretically predicted regions in the Φ, Ψ plot (145, 171, 172). Furthermore, many of the predicted conformations have not been observed experimentally. A similar situation exists for alanine. For polypeptides and proteins, the agreement between theory and experiment is even worse. The predicted structures of ribonuclease (161, 167, 168) and cytochrome c (169) were quite different from those subsequently obtained by X-ray crystallography (173, 174). Thus, it appears that there is no adequate purely theoretical method to predict three-dimensional structures of proteins from their amino acid sequences. The major difficulty is to try to find a global minimum of the energy function in a several hundred-dimensional space which is not only nonlinear but also contains many uncertain parameters. This problem is mathematically unsolvable.

Empirical Approaches and Estimation of Φ, Ψ Angles

The tertiary structures of many proteins have been determined by X-ray diffraction analysis (175). It was discovered that proteins with high proline contents tended to have low contents of α helices (176). This observation was subsequently developed into assignments of helix-forming and helix-breaking tendencies to individual amino acid residues (177–189) and to neighboring residues (172, 190–200) by comparing amino acid sequences with those in proteins whose tertiary structures have been determined by X-ray crystallography. Combinations of these different empirical approaches were quite successful in locating α-helical segments in proteins with known amino acid sequences (201, 202). Similar studies have been extended to β sheets (184, 188, 196, 198, 203); however, due to the relative scarcity of experimental data, the reliability of predicting the locations of β sheets is less adequate. Furthermore, the formation of β sheets requires relatively long-range interactions, so far as the protein chain is concerned (155). The empirical predictions of β bends (204, 206) are also under intensive study. To the question whether we can construct the backbone structure of a protein if all its α helices, β sheets, and β bends are precisely located, the answer is obviously negative, since other structures are experimentally observed. However, one of the ways to circumvent this difficulty would be to predict the Φ, Ψ angles for each residue of the protein by an empirical method (207, 208).

As the resolution of X-ray diffraction analyses of protein structures improved (175), it became possible to locate the atoms with reasonable precision. From these atomic coordinates, the Φ, Ψ angles of each residue have been calculated (209). A systematic scheme of estimating Φ, Ψ values for any residue in a protein with known amino acid sequence has been developed (207, 208) and consists of the following steps:

1. For example, position 4 of horse cytochrome c is occupied by a glutamyl residue. Its two nearest neighbors are valine at position 3 and lysine at position 5. To estimate the Φ, Ψ values of the glutamyl residue, the tripeptide, i.e. Val-Glu-Lys, was

Table 1 Φ, Ψ values of the middle amino acid from ten reference proteins used to estimate the choices for the glutamic acid residue in the tripeptide Val Glu Lys (208)

Val Glu Lys	Val ( ) Lys	Val Glu ( )	( ) Glu Lys	( ) Glu ( )
−75, +142	−101, −154	−69, −9	−46, −34	−104, −148
	−128, +136	−82, −33	−48, −43	−108, +94
	−133, +147	−45, −42	−114, +180	−113, +109
	−104, +125	−121, +134	−124, −30	−111, +83
	−61, −49	−65, −42	−99, −29	
	−67, −59	−29, −45	−87, −44	(and 81
	−89, +92	−68, −56	−93, +22	other sets
		−60, −39		not listed)

compared with the amino acid sequences of ten other proteins of known tertiary structures. Five different tripeptides, i.e. Val-Glu-Lys, Val ( ) Lys, Val-Glu ( ), ( ) Glu-Lys, and ( ) Glu ( ), where ( ) represents any of the twenty protein amino acids, were sought in these proteins. If they existed, the Φ, Ψ values of the middle amino acids were listed as shown in Table 1 (208).

2. Five criteria were used to select Φ, Ψ values from the tabulated data. They are, in order of decreasing reliability: (a) If there was a set of Φ, Ψ values under Val-Glu-Lys, the sets of Φ, Ψ values under Val ( ) Lys, Val-Glu ( ), and ( ) Glu-Lys were searched for those which were within $\pm 15°$ of the first set. All these sets were averaged. Thus, -75, $+142$ is under Val-Glu-Lys (Table 1). The only set under Val ( ) Lys, Val-Glu ( ), and ( ) Glu-Lys which lies within $\pm 15°$ of -75, $+142$ is -104, $+125$. They are averaged to give -90, $+134$. Any fraction of one degree is neglected. (b) If there were sets of Φ, Ψ values within $\pm 15°$ under Val ( ) Lys, Val-Glu ( ), and ( ) Glu-Lys only, they were averaged. In the present case, -61, -49 and -67, -59 under Val ( ) Lys; -45, -42; -65, -42; -68, -56; and -60, -39 under Val-Glu ( ); and -46, -34 and -48, -43 under ( ) Glu-Lys satisfy this criterion. Their average is -58, -46. (c) If there were sets under Val ( ) Lys and Val-Glu ( ) only, or under Val ( ) Lys and ( ) Glu-Lys only, they were averaged. In Table 1, -128, $+136$; -133, $+147$; and -104, $+125$ under Val ( ) Lys, and -121, $+134$ under Val-Glu ( ) are averaged to give -122, $+136$. (d) If there was a set only under Val ( ) Lys, it was checked against those sets under ( ) Glu ( ). Only when there was a similar set under the latter, was the set under the former listed. For -101, -154 under Val ( ) Lys, there is -104, -148 under ( ) Glu ( ). Similarly, for -89, $+92$ under Val ( ) Lys, there are -108, $+94$; -113, $+109$; and -111, $+83$ under ( ) Glu ( ). Because of the large amount of data under ( ) Glu ( ), a graphical plot is used instead of matching Φ, Ψ values. (e) If there were sets under Val-Glu ( ) and ( ) Glu-Lys only, they were averaged. Thus, -69, -9 and -82, -33 under Val-Glu ( ) can be averaged with -99, -29 under ( ) Glu-Lys to give -83, -24. Or, -82, -33 under Val-Glu ( ) can be averaged with -99, -29 and -87, -44 under ( ) Glu-Lys to give -89, -35. In addition, -29, -45 under Val-Glu ( ) and -46, -34 and -48, -43 under ( ) Glu-Lys are averaged to give -41, -41. In summary, the following choices of Φ, Ψ values are obtained for the glutamyl residue at position 4 of horse cytochrome c: (a) -90, $+134$; (b) -58, -46; (c) -122, $+136$; (d) -101, -154 and -89, $+92$; and (e) -83, -24; -89, -35; and -41, -41.

3. Obviously these are too many choices for one position; since for a protein of about 100 amino acids and about five choices for each position, 5^{100} different structures have to be analyzed. This is where the existence of homologous amino acid sequences becomes important. For cytochrome c, three different tripeptides, Val-Glu-Lys, Ile-Glu-Lys, and Val-Ala-Lys, have been found in these positions in different species (210). By considering the totality of these tripeptides and assuming that a common set of Φ, Ψ values should exist at that position, a first choice of -58, -47; a possible second choice of -91, $+136$; and no third choice were obtained (208). The Φ, Ψ values at position 4 of cytochrome c derived from X-ray diffraction analysis (174) are -74, -34.

4. Even though the Φ, Ψ choices at each position can be reduced to about two with the help of homologous sequences, 2^{100} is still too large a number. Model construction (207) can then be used to further restrict the choices. Starting from the N-terminal residue, a plastic model (Labquip) of the protein backbone was constructed by using the first choices. If steric hindrance occurred, namely, the chain folded back too tightly, second choices were used locally to untangle the chain. This is in essence equivalent to a theoretical calculation using the hard sphere approximation. For the variable region of the κ light chains of human immunoglobulins, the resulting model of Bence-Jones protein Ag (207) had residues 23 and 88 sufficiently close to form the disulfide bond, and the three hypervariable regions (211) could constitute a well-defined antibody-combining site.

Combination of Empirical and Theoretical Predictions

The above method can empirically provide one set (or a few sets) of Φ, Ψ values for proteins with many known homologous amino acid sequences. Other experimental findings, such as the location of S–S bonds, spectral data, metal binding sites, etc, may also be incorporated into the process of selecting the final set (or sets) of Φ, Ψ values (212). Thus, we have effectively made a zeroth order approximation to locate a point (or a few points) in the several hundred-dimensional energy function space. It may then be feasible to vary the Φ, Ψ values within $\pm 15°$, say, to obtain a local energy minimum. In fact, even this calculation is not trivial. For the neurohypophyseal hormones, two models with local minimum energy states have been proposed on the basis of this approach (212).

In summary, theoretical or empirical approaches used singly have encountered numerous difficulties in predicting the three-dimensional structures of proteins from their amino acid sequences. Nevertheless, the combination of these two approaches appears to provide useful results (207, 210, 212). Such proposed structures are yet to be subjected to the stringent test of comparison with the corresponding experimentally determined spatial structures.

RECONSTRUCTION OF EVOLUTIONARY EVENTS

Evolution, of course, had a long history prior to the appearance of the common ancestor that diverged to give the many taxa producing the proteins mentioned. For readers interested in the problems these earlier stages pose, the following are noted: Works of broad scope on the abiogenic phase of evolution (213–218), the primitive atmosphere (218–222), the origin of organic monomers (223–234), optical isomery (235–238), and the condensation of monomers (239–254); the origin of the genetic code (255–271); and thermodynamic models for the development of biological complexity (272–277). There have been recent reviews of enzyme and protein evolution by Smith (278), Fitch & Margoliash (279), Margoliash (280), and Arnheim (281); on immunoglobulin evolution by Gally & Edelman (282), and Smith, Hood & Fitch (283); and on the evolution of metabolic pathways by Brew (284), Vogel (285), and LéJohn (286).

Sequence Comparisons and Genetic Processes

Comparative studies first require that two amino acid sequences are significantly similar. There has been much activity in developing methods for finding similarity where it is not readily apparent (7, 287–304). A critical analysis of these methods has been presented by Fitch (305). Most of the methods were found wanting. If one finds significant similarity, the most likely position of any gaps can be found (306). Both problems (similarity and gap location) are attacked simultaneously in the important procedure of Needleman & Wunsch (298), which may well be the starting point for any future improvements. It would be most salutory if investigators would recognize that such methods, when they work, only prove that the sequences are similar and not necessarily homologous, which means descent from a common ancestor. Similarity could also arise by convergent evolution from unrelated ancestral genes, in which case the sequences are analogous rather than homologous. The only method so far for proving homology is that of Fitch (307), who proved that fungal cytochromes c are homologous to the plant and animal cytochromes c. The method has also been used to prove that α and β hemoglobins (308) and, more recently, protozoan and higher eukaryotic cytochromes c (309) are homologous. Although significant similarity could, in principle, arise through convergent processes (and has in fact for many morphological structures), no example of analogous proteins has yet been demonstrated, with the general result that significant similarity can so far be taken as strong presumptive evidence for homology. Examples of proteins with significant similarity include the serine proteases (310, 311), lysozyme-lactalbumin (312), prolactin-growth hormone (313), and corticotropin-melanotropin-lipotropin (314, 315). The β chain of human haptoglobin is similar to the serine proteases (316) and prothrombinase (317). Alignments of most of these sequences can be found in Dayhoff's compendium (7). Adelson (318) has recently made a wide-ranging study of the correlation between proteins secreted by various glands and the embryological origin of the glands, and Strydom (319) has observed a similarity between mammalian trypsin inhibitors and two snake venom toxins.

Testing for similarity of primary structure is fundamentally the same problem in nucleic acids as in proteins, and the papers referred to above can be consulted profitably. Of more direct interest in this connection is the paper by Sankoff & Cedergren (320). Other facets of nucleic acid sequence comparisons include the attempt at phylogenetic comparisons (321), prediction of secondary structure from primary sequence information (322–325, 416), and prediction of both the secondary structure and the primary sequence of mRNA given only the amino acid sequence encoded by the messenger (326–328). Fitch (329) has cast serious doubt upon the validity of any method purporting to accomplish the last objective, showing among other features that knowing only the amino acid sequence of typical globins, including human α hemoglobin, permits one to design an mRNA with 80% base pairing. Moreover, the resultant inferred nucleotide sequences, when they could be checked, have consistently proven to be wrong. Examining the so-called flower model of the phage coat protein gene of Min Jou et al (330), Ball

(331) concluded that the protein sequence must have been altered to accommodate the needs of secondary structure in the RNA. This conclusion was reached, however, by erroneously assuming that a null hypothesis that isn't falsified by a test is therefore proved. In fact, a more sensitive test falsified the hypothesis that the degeneracy of the genetic code was not used to optimize base pairing in the MS2 coat protein gene (329). It has also been shown that the synthesis of unwanted proteins is not minimized by either minimizing out-of-frame AUG triplets or maximizing out-of-frame terminating triplets (328).

Comparisons of protein amino acid sequences have now revealed a variety of mutational processes, all of which were expected from genetic experiments. However, the nomenclature, dating as it does from before the comparative study of proteins, can be confusing. For over a century the term homology in a biological context has meant descent from a common ancestor, while geneticists since 1936 (332) have used nonhomologous to mean a misalignment of chromosomes at synapsis. As a consequence, β and δ hemoglobin chains are homologous by descent but nonhomologous by alignment and thus the abnormal human hemoglobin Lepore, a β-δ recombinant (333), can be the result of either an homologous or a nonhomologous crossing-over, depending upon one's viewpoint. Some clarifying terminology has been proposed (305). Homology and nonhomology are preserved in the sense of the presence or absence of a common ancestor. Thus, almost all previously classified nonhomologous crossings-over can be retained with no problem. The possible confusion in the term homologous crossing-over is avoided by replacing homologous with a more precise terminology. Two alleles at the same locus are said to be orthologous, while two loci arising through gene duplication or translocation from a common ancestral gene are said to be paralogous. Orthology and paralogy are subsets of homology and their utilization eliminates the ambiguity. β and δ hemoglobin chains are therefore paralogous and hemoglobin Lepore is properly classified as the product of a paralogous crossing-over. This terminology reflects the phylogenetic nature of the genes involved. The occurrence of mutants such as hemoglobin Gun Hill (334), which is a human β hemoglobin chain missing five of its amino acids in the region of residue 95, illustrates that recombination may occur between orthologous genes improperly aligned. To avoid confusion with terminology relating to genes, an alignment where the codons paired possess a common ancestral codon is said to be conjugate in the sense of having a common derivation. Hemoglobin Gun Hill is thus the result of a nonconjugate orthologous crossing-over. All the possible forms of recombinations have been observed and, for most of them, examples are available where the product was preserved by the evolutionary process.

Nonhomologous crossing-over may have two results, depending upon the intergenic region. If the terminator of the first gene is eliminated, one of the two resultant products is a fused gene. Tryptophan synthetase is composed of two protein chains in E. coli (335) and other bacteria, but of a single gene product with two functional regions in fungi (336, 337). This is strongly suggestive of gene fusion. Other possible cases include the trpB gene product where separate parts of the enzyme have glutamine amidotransferase and anthranilate phosphoribosyltransferase activities (338) and another *E. coli* enzyme with localized aspartokinase and homoserine dehydrogenase activities (339).

On the other hand, if the nonhomologous crossing-over event pairs two regions farther apart than one gene, the two possible outcomes are complete gene duplication or deletion. As Ingram (340) recognized, myoglobin and the various hemoglobin chains, α, β, γ, and δ, probably arose in this fashion, except for α, which is not closely linked to β, γ, and δ, one cannot rule out the alternative possibility of chromosome breakage and translocation as the source of separate loci. The list of significantly similar proteins given above (serine proteases, etc) is representative of this second kind of nonhomologous crossing-over. In this second form, however, the term nonhomologous may only be an expression of ignorance. If the two intergenic regions on either side of a gene were themselves paralogous, then gene duplication could arise from a paralogous crossing-over. Moreover, once a pair of tandem genes has been produced by nonhomologous crossing-over, further gene duplications can then be produced by paralogous crossing-over.

The phrase, unequal crossing-over, lacks the evolutionary distinctions made in this nomenclature and may therefore be used as a refuge when recombination is clear but the relationship of the recombining elements is not.

The immediate result of the second form of nonhomologous crossing-over is the creation of identical copies of the gene in tandem. The existence of tandem α hemoglobin genes has been suggested (341), but the existence of homozygotes of the α hemoglobin variant Tongariki that make no normal hemoglobin whatsoever (342) makes two human α loci most unlikely. On the other hand, Hollan et al (418) have described three brothers all of whom make three α-Hb chains, a normal α, α-G, and α-J in the ratio of $2:1:1$. This strongly suggests that humans are variable in the number of their α hemoglobin loci. Multiple α hemoglobin loci have also been proposed for the mouse (343) and the Virginia white-tailed deer (344).

Hemoglobin Lepore, discussed above, can thus be described accurately as the result of a conjugate paralogous crossing-over because two paralogous genes, i.e. separate loci descended from a common ancestor, were aligned so that the codons with a common ancestral codon were paired, hence conjugate, at the time of crossing-over. Given that the crossing-over is reciprocal, one strand has a deletion and the other an insertion. Homozygotes for hemoglobin Lepore show no normal β or δ chains and hence the Lepore chain must have arisen from the strand with the deletion. The counterpart strand with the insertion of one additional gene has been called anti-Lepore and has in fact recently been observed (345–348). The amino-terminal portion of anti-Lepore chain is from the β gene and the carboxyl-terminal portion from the δ gene which, as expected, is the reverse of the situation in the Lepore chain. Hemoglobin Kenya is the product of a comparable conjugate paralogous crossing-over between human γ and β hemoglobin genes (417).

Conjugate orthologous crossing-over is recombination between two alleles in proper alignment. This is not likely to be observed in the absence of outside markers, since the two allele products must differ in at least four positions before a single recombinant (or three positions for a double recombinant) is a simpler explanation that one amino acid replacement. Galizzi (349) has postulated such an event in the β-hemoglobin chains of rabbits but the evidence is capable of simpler interpretation (305). Yanofsky (350), with appropriate markers, showed conjugate orthologous crossing-over in *Escherichia coli* tryptophan synthetase. The two recombining genes gave products differing in only one position, arginine or glutamate, but gave glycine in the recombinant so that crossing-over must have occurred between the first and second nucleotide of the codon.

Nonconjugate orthologous crossing-over leads to changes in the length of the gene. The previously mentioned hemoglobin Gun Hill is an example, as are all the gaps required to align two paralogous sequences. Thus, the gap of two amino acids in β hemoglobin between Residues 18 and 19 required for its alignment with α hemoglobin suggests that not too long after the appearance of the β locus, it was shortened by two codons. The alternative that the α locus was lengthened is discounted because the same two residues are also present in myoglobin. Given only a pair of sequences, there is no a priori way of knowing whether a gap arose from an insertion in one gene or a deletion in the other. Most nonconjugate orthologous crossings-over involve only a small number of residues and there is a suggestion (305) that deletions are expressed more often than insertions among both those protein chains selected by evolution and those naturally occurring variants whose selective fate has not yet been determined. α-2 Haptoglobin is a particularly interesting example of nonconjugate orthologous crossing-over because it involved a lengthening of the gene by 59 codons and the recombination was between distinguishable alleles (351, 352).

There is no demonstrated example of nonconjugate paralogous crossing-over, but a good candidate would appear to be clupeine Z (353). Clupeines Y_I and Y_{II} differ by a minimum of 13 genetic events, including several insertions/deletions and so are probably non-allelic, i.e. paralogous. Furthermore, a simple recombination of clupeine Y_I and Y_{II} will give rise to clupeine Z directly with all the differences between Z and Y_I lying to the right of those between Z and Y_{II}.

It is almost commonplace to assert that processes of recombination are facilitated by similar sets of nucleotide sequences in the two mispaired gene regions. Apparently the region need not be very long, as judged by a rather impressive number of cases where small deletions are consistent with mispairing of sequences only five or six nucleotides long (305). Nevertheless, it is clear that not all insertion/deletion mutants can be explained by a facilitation through mispairing of similar sequences from different regions. Moreover, for most deletions in the region coding for the amino or carboxyl terminus of proteins, the facts are not available to permit their explication. There are two cases, however, where this is not so. Baker's yeast mutants have been isolated in which the initiating AUG of the iso-1 cytochrome c gene has suffered a single nucleotide replacement, thus preventing initiation (354, 355). A revertant has been isolated which created an AUG in the triplet preceding the original initiating codon. This not only proves the original mutation but raises interesting questions. Not all AUGs are initiators, so if one can relocate the position of the initiator, then either there is flexibility of recognition or the initiating AUG is not itself part of that necessary for recognition. The latter would seem to conflict with a considerable amount of evidence. The second case is the abnormal human α-hemoglobin mutant, Constant Spring (356), where its normal terminating codon, UAA, suffered a mutation to the glutamine codon, CAA, and translation continued on for a total of 31 additional codons. There is also a variant human β-hemoglobin chain, Tak (357), that is longer than normal by 10 residues, but there is no ready explanation for it. The first extra residue, threonine, is encoded by ACX. It cannot be obtained from any terminator by a single nucleotide change regardless of whether that change is a replacement, insertion, or deletion. Moreover, the extra residues are not apparently homologous to the six carboxyl-terminal residues of myoglobin that extend beyond the β-hemoglobin chain when both are homologously aligned, nor with the extra residues of hemoglobin Constant Spring (305). And finally, if the gene is a recombinant, as would seem most likely from the foregoing discussion, there is no uninterrupted sequence of nucleotides anywhere within the β chain gene that could be the source of the added amino acids (305).

The other major process of mutational change, apart from simple nucleotide replacement, in proteins is frame-shifting through the insertion or deletion of a number of nucleotides not divisible by three. The result, however, is apparently too drastic to survive evolutionary pressures and still be detected. This could be because the product was either selectively disadvantageous or could use, and obtained, so much improvement through further change that its relationship to its progenitor was effectively erased. Until recently, the only known examples of such processes were from experimental manipulations in the case of lysozyme (358–360). However, a human α-hemoglobin chain variant [hemoglobin Wayne, (361)] has recently been found in which the last three amino acids of the normal sequence have been altered and five additional amino acids added on. By comparing the possible messengers of hemoglobin Wayne with hemoglobin Constant Spring, it was clear that the Wayne variant was produced by the loss of one adenine nucleotide in lysine codon 139 and that a terminator was reached eight codons further. The expectation is that this variant will be lost in the course of continued evolution.

The process of nucleotide replacement is too well documented to need review here. A large number of human hemoglobin variants have been cataloged (7, 357). Two particularly interesting cases are listed. Hemoglobin Hopkins-2 II (419) has three amino acid replacements, $\alpha^{112H \to D}$, $\alpha^{114P \to S}$, and $\alpha^{118T \to G}$. In every other case, the variant is either a single amino acid change from the normal or, in one case (420), a single amino acid change from the relatively common hemoglobin S (sickle). Moreover, in all those other cases, the amino acid changes are accountable by a single nucleotide replacement whereas the threonine→glycine change in position 118 requires two nucleotide replacements. The variant is sufficiently peculiar that it should not be thought of as an exception to the rule that point mutations are single nucleotide replacements, but rather as an example of some other mutational process. It is not simply explained either by recombination or frameshifting. Clustering of the three changes suggests breakage and misrepair.

The second interesting case is the simultaneous occurrence of two variants,

$\beta^{67V \to D}$ (421) and $\beta^{67V \to E}$ (422). Each of these variants can be explained by a U→A nucleotide replacement in the second position of a valine codon. However, since aspartate is encoded by GAY (Y = pyrimidine) whereas glutamate is encoded by GAR (R = purine), the preceding explanation implies the existence of two valine codons in the human population, namely GUY and GUR. The extent of such silent polymorphisms is not known for any population.

Forget et al (423) have made progress in sequencing the messenger for human α and β hemoglobin and the data for several nucleotide fragments are consistent with only one amino acid sub-sequence in the two proteins. Moreover, they are consistent with the inferences drawn from the comparison of hemoglobins Constant Spring and Wayne.

Phylogenetic Trees from Amino Sequence Comparisons and Derived Data

The construction of phylogenies for various taxa utilizing protein sequences has become commonplace following the introduction of the concept and an effective methodology in relation to eukaryotic cytochromes c in 1967 (362). Such studies have by now included vertebrate hemoglobins (363–366), primate β and δ hemoglobin chains (367), plant cytochromes c (368, 369), artiodactyl (370, 374) and hominoid (371, 374) fibrinopeptides, eukaryotic cytochromes c (309, 372, 373), snake toxins (375, 376), mammalian calcitonins (377), bacterial cytochromes (378), mistletoe toxins (379), plant and bacterial ferredoxins (380), muscular parvalbumins (381), and avian lysozymes (382). The lysozymes are an excellent example of one of the pitfalls, namely, for the phylogeny derived from the amino acid sequences to be congruent to that of the taxa from which the proteins were obtained, it is insufficient that the sequences are homologous; they must be orthologous rather than paralogous. If one tries to create a phylogeny from the lysozyme sequences of the chicken, goose, and duck, one can only conclude that the duck is more closely related to the chicken than to the goose. If one abandons the attempt to obtain the phylogeny and instead forces the data into a phylogeny acceptable to ornithologists, one can only conclude that the goose lysozyme has evolved at a prodigious rate since its common ancestor with the duck. Both conclusions are wrong because they assume that the goose lysozyme is orthologous to the other two. Arnheim & Steller (383), studying the black swan, discovered that it possesses two lysozyme loci, one like the known duck and chicken forms and a paralogous form similar to the goose's. Apparently different forms are expressed at different points in a bird's ontogenetic development. Although the original problem is thus explained, another problem is created. Why is the lysozyme gene expressed in the egg of the duck not the same locus as the one in the egg of the goose?

There are many techniques for constructing phylogenies from biological data and the best introduction to these is the discussion of Sneath & Sokal (384). There are a number of methods specifically designed for use with sequences (296, 298, 362, 385). The two methods from Gibbs' laboratory (296, 385) are presumed to have the advantage that they do not depend upon a previous alignment of the sequences.

The methods, however, give very poor results with eukaryotic cytochromes c where alignment is not a problem. The reason for this is that the measure of genetic distance depends primarily on the location of amino acid replacements and very little on the nature of those replacements.

Using orthologous sequences, one gets a putative phylogeny of the taxa from which the sequences came. One can, of course, examine a set of paralogous sequences to get a putative phylogeny of the genes where now a bifurcation implies not speciation, but gene duplication. Boyer et al (367) and Goodman & Moore (366) have combined the two types of study to investigate whether δ hemoglobin chains have arisen more than once in primate evolution. Some methods of determining genetic distance depend only indirectly on sequences. Kohne (386) has used DNA hybridization and Wilson & Sarich (387) have used immunological cross-reactivity. These two papers are interesting in part because they both are studying primates and, assuming uniformity of rates of change, they come to opposite conclusions regarding the accuracy of the paleontological dating of hominoid evolution.

Once one has a phylogeny in hand, whether produced by the preceding methods from sequence data or accepted from some outside source, one is in a position to reconstruct the evolutionary history of the gene and thereby obtain information on the number, location, and kinds of replacements of amino acids and nucleotides. The method of Dayhoff (7) deals strictly with amino acids in reconstructing ancestral sequences. It is not clear whether that procedure gives the fewest possible number of amino acid substitutions required to account for the resulting sequences, given the phylogeny, but it is clear that the most parsimonious tree in terms of amino acid replacements need not be most parsimonious for nucleotide replacements. For example, consider a position where a human protein has glycine, the corresponding monkey protein has threonine, the duck protein has proline, and the chicken protein has valine. There are six solutions for which a minimum of three amino acid replacements is required. Four of these require five nucleotide replacements and two require six nucleotide replacements. However, allowing the ancestral nodes to be alanine will, at the cost of a fourth amino acid replacement, permit the number of nucleotide replacements to be reduced from five (or six) to four.

Fitch (388) introduced a method that obtains all the most parsimonious solutions consistent with that phylogeny, given the mRNA sequences and the phylogeny. The mRNA sequences can, of course, be obtained from amino acid sequences using the genetic code. Hartigan (389) has proven the correctness of this procedure. Moore, Barnabas & Goodman (390) have given a second method of obtaining the same result except that it may, under some circumstances, give a larger minimum which would be superior since it is closer to the true number. The circumstance arises because under certain conditions the sixfold degenerate codons for arginine, leucine, and serine appear to require a fully degenerate representation not unambiguous for the amino acid encoded. Where results of the two methods can be compared, similar conclusions have been reached (366).

With an estimate of the ancestral sequences in hand, one can ask other questions.

For example, are the nucleotide substitutions fixed during evolution distributed randomly among the codons of the gene? Since they are not Poisson distributed, it is clear that they are not randomly distributed (391). This is still true even when allowance is made for a number of invariable sites, although Jukes (392) remains unconvinced. Nonrandomness has been shown for cytochrome c (373, 393), α hemoglobin (394), and β hemoglobin (395). Moreover, the number of invariable (as opposed to unvaried) sites is clearly a function of the range of species considered. This led to the concept of the covarion (COncomitantly VARIable codON), the number of codons in any one species at any one point in time that is capable of fixing an amino acid-changing nucleotide replacement. The number of covarions in mammals has been estimated for the following genes: cytochrome c, 12 (309); α hemoglobin 50 (394); β hemoglobin, 39 (395); fibrinopeptide A, 18 (396); and insulin C peptide, 18 (305). The plant cytochromes c prove to have approximately the same number of covarions as the mammalian cytochromes c (309).

The mutations fixed during evolution are only a minute fraction of the pool of variants available for selection. Thus, even though the evolutionarily fixed mutations are nonrandomly distributed, the variants occurring in a population at any one time may not be. The availability of a large collection of human hemoglobin variants gave Vogel (397) an opportunity to test this, and he concluded that both the α and β chain variants were Poisson distributed. A more careful analysis reveals, however, that the distribution of the α variants is significantly nonrandom (398). The significance of this nonrandomness is unclear, however, since there are four biases operating in data collection for hemoglobin chains: a given mutation can only be counted once although it may have arisen several times independently; mutations resulting in a change of electric charge of the protein chain are more likely to be detected than others; the number of ways of getting a given amino acid interchange is different for various pairs of amino acids even if restricted to pairs with codons only one nucleotide different; and variants leading to clinical symptoms are more likely to be discovered.

One also can ask if nucleotide replacements during evolution were distributed randomly among the three positions of the codons, making due allowance for the fact that replacements not changing the encoded amino acid are not observed. The answer is they were not randomly distributed in the evolution of eukaryotic cytochromes c ($P < 10^{-6}$), and probably not ($P = 0.01$) in the evolution of α hemoglobin (399). They do, however, appear to have been randomly distributed in the evolution of β hemoglobin ($P \sim 0.2$). In every case, the distribution shows an excess of replacements in the first position and a deficit in the second, relative to expectation.

One also can inquire whether nucleotide replacements are random with respect to the nature of the nucleotides (A, C, G, T) replaced and replacing. The preponderance of $G \rightarrow A$ replacements in both evolutionary development and human variants has been noted by many (400–405). The explanation is not finally resolved, but appears to be associated with a bias in allowable amino acid alternatives. It appears that the nucleotide composition of the cytochrome c gene is near equilibrium (309), and a more careful analysis gave strong evidence that the four nucleotides were replaced in proportion to their abundance, so that the nonrandomness was a result of a bias in the nature of the replacing nucleotide. The same, however, does not hold true for α and β hemoglobins where there is also nonrandomness in the nature of the nucleotides replaced (Fitch, unpublished).

Calculated rates of protein evolution vary considerably depending upon the assumptions. They are, however, quite uniform among the five genes for which the number of covarions has been estimated, namely about four amino acid-changing nucleotide replacements per covarion per 10^9 years (305). It is not always possible to express protein evolutionary rates in this unit but it does have the advantage that the rate of change is calculated on the basis of only those codons that can change, with no averaging in of invariable sites. The above rates vary by more than one order of magnitude if they are calculated on a per-codon or per-gene basis, namely, without distinguishing variant and invariant sites. The above rates are based on replacements required to account for the observed differences among the sequences from various taxa. Attempts have been made to estimate the total number of substitutions from those actually observed (7, 406–412), but it is clear that further work is required (305).

As pointed out previously (305, 394, 395), the equality of the above rates of evolutionary replacement is consistent with Kimura's neutral mutation hypothesis (413). The consistency proves, however, to be the result of an insensitive measure, since Langley & Fitch (308, 414) showed by a maximum likelihood estimate of the simultaneous expected nucleotide replacement rates in the α and β hemoglobin, cytochrome c, and fibrinopeptide A genes that: (a) the same gene has evolved at different rates in different lines of descent and in different intervals of the same line; and (b) the relative rates of replacement between two genes vary significantly in different intervals. The latter observation is independent of generation time since both genes have shared the same species history. It must therefore be concluded that selection has played a significant, but not necessarily the total, role in the evolution of these four genes in vertebrates. Although reluctant to reach the same conclusion, Ohta & Kimura (415) earlier provided evidence that selection played a significant role. But while the rates of replacement for each gene vary significantly over time, the total rates of all four genes average out to provide the best estimate so far of time since evolutionary divergence of species when compared to paleontological dating (305, 414).

Literature Cited

1. Nolan, C., Margoliash, E. 1968. *Ann. Rev. Biochem.* 37:727–90
2. Arnheim, N. 1973. *The Antigens,* ed. M. Sela, 377–416. New York: Academic
3. Brewer, H. B. Jr., Ronan, R. 1970. *Proc. Nat. Acad. Sci. USA* 67:1862–69
4. Niall, H. D. et al 1970. *Z. Physiol. Chem.* 351:1586–88
5. Anfinsen, C. B. 1973. *Science* 181:223–30
6. Dayhoff, M. O., Eck, R. V., Chang, M. A., Sochard, M. R. 1965. *Atlas of Protein Sequence and Structure.* Silver Spring: Nat. Biomed. Res. Found.
7. Dayhoff, M. O., Ed. 1972. See Ref. 6, 5: Suppl. I
8. Sanger, F., Tuppy, H. 1951. *Biochem. J.* 49:481–90
9. Sanger, F., Thompson, E. O. P. 1953. *Biochem. J.* 53:366–74

10. Sanger, F. 1956. *Currents in Biochemical Research,* ed. D. E. Green. New York: Interscience
11. Niall, H. D., Keutmann, H. T., Copp, D. H., Potts, J. T. Jr. 1969. *Proc. Nat. Acad. Sci. USA* 64:771–78
12. Sauer, R., Niall, H. D., Potts, J. T. Jr. 1970. *Fed. Proc.* 29:728
13. Thomsen, J., Kristiansen, K., Brunfeldt, K., Sundby, F. 1972. *FEBS Lett.* 21:315–19
14. Edman, P., Begg, G. 1967. *Eur. J. Biochem.* 1:80–91
15. Edman, P. 1950. *Acta Chem. Scand.* 4:283–93
16. Edman, P. 1956. *Acta Chem. Scand.* 10:761–68
17. Edman, P. 1963. *Thromb. Diath. Haemorrh. Suppl.* 13:17–20
18. Ilse, D., Edman, P. 1963. *Aust. J. Chem.* 16:411–16
19. Gray, W. R. 1967. *Methods Enzymol.* 11:469–75
20. Konigsberg, W. 1967. *Methods Enzymol.* 11:416–69
21. Schroeder, W. A. 1967. *Methods Enzymol.* 11:445–61
22. Barstead, P., Rudikoff, S., Potter, M., Hood, L. E. 1974. *Biochemistry.* In press
23. Smithies, O. et al 1971. *Biochemistry* 10:4912–21
24. Niall, H. D., Hogan, M. L., Sauer, R., Rosenblum, I. Y., Greenwood, F. C. 1971. *Proc. Nat. Acad. Sci. USA* 68:866–70
25. Walter, R., Schlesinger, D. H., Schwartz, I. L., Capra, J. D. 1971. *Biochem. Biophys. Res. Commun.* 44:293–98
26. Capra, J. D., Kehoe, J. M., Kotelchuck, D., Walter, R., Breslow, E. 1972. *Proc. Nat. Acad. Sci. USA* 69:431–34
27. Brewer, H. B. Jr., Lux, S. E., Ronan, R., John, K. M. 1972. *Proc. Nat. Acad. Sci. USA* 69:1304–8
28. Brewer, H. B. Jr., Lux, S. E., Ronan, R., John, K. M. 1972. *Chemistry and Biology of Peptides,* ed. J. Meienhofer, 705–12. Ann Arbor: Ann Arbor Science Publishers. 762 pp.
29. Benditt, E. P., Eriksen, N., Hermodson, M. A., Ericsson, L. H. 1971. *FEBS Lett.* 19:169–73
30. Hermodson, M. A. et al 1972. *Biochemistry* 11:2934–38
31. Franklin, E. C., Pras, M., Levin, M., Frangione, B. 1972. *FEBS Lett.* 22:121–23
32. Putnam, F. W., Shimizu, A., Paul, C., Shinoda, T., Köhler, H. 1971. *Ann. NY Acad. Sci.* 190:83–103
33. Paul, C., Shimizu, A., Köhler, H.,

34. Putnam, F. W. 1971. *Science* 172:69–72
34. Putnam, F. W., Shimizu, A., Paul, C., Shinoda, T. 1972. *Progr. Immunol. Int. Congr. Immunol. 1st 1971.* 291–307
35. Hermodson, M. A., Ericsson, L. H., Neurath, H., Walsh, K. A. 1973. *Biochemistry* 12:3146–53
36. Titani, K., Hermodson, M. A., Ericsson, L. H., Walsh, K. A., Neurath, H. 1972. *Biochemistry* 11:2427–35
37. Niall, H. D. 1971. *Agr. Food Chem.* 19:638–44
38. Hermodson, M. A., Ericsson, L. H., Titani, K., Neurath, H., Walsh, K. A. 1972. *Biochemistry* 11:4493–4502
39. Laursen, R. A. 1966. *J. Am. Chem. Soc.* 88:5344–46
40. Laursen, R. A. 1971. *Eur. J. Biochem.* 20:89–102
41. Laursen, R. A. 1972. *Methods Enzymol.* 25:344–59
42. Laursen, R. A., Horn, M. J., Bonner, A. G. 1972. *FEBS Lett.* 21:67–70
43. Previero, A., Derancourt, J., Coletti-Previero, M. A., Laursen, R. A. 1973. *FEBS Lett.* 33:135–38
44. Dowling, L. M., Stark, G. R. 1969. *Biochemistry* 8:4728–34
45. Braunitzer, G., Schrank, B., Ruhfus, A. 1970. *Z. Physiol. Chem.* 351:1589–90
46. Braunitzer, G., Schrank, B., Ruhfus, A. 1971. *Z. Physiol. Chem.* 352:1730–33
47. Inman, J. K., Hannon, J. E., Appela, E. 1972. *Biochem. Biophys. Res. Commun.* 46:2075–81
48. Terhorst, C., Möller, W., Laursen, R., Wittmann-Liebold, B. 1973. *Eur. J. Biochem.* 34:138–52
49. Beimann, K. 1972. *Biochemical Applications of Mass Spectrometry,* ed. G. R. Woller, 405–28. New York: Interscience
50. Das, B. C., Lederer, E. 1971. *New Techniques in Amino Acid, Peptide and Protein Analysis,* ed. A. Niederwieser, G. Pataki, 175–225. Ann Arbor: Ann Arbor Science Publishers
51. Kolb, B. Ibid. pp. 129–74
52. Fairwell, T., Barnes, W. T., Richards, F. F., Lovins, R. E. 1970. *Biochemistry* 9:2260–67
53. Morris, H. R., Williams, D. H., Ambler, R. P. 1972. *Biochem. J.* 125:189–201
54. Förster, H. J., Kelley, J. A., Nau, H., Biemann, K. 1972. See Ref. 28, 679–86
55. Caprioli, R. M., Seifert, W. E., Sutherland, D. E. 1973. *Biochem. Biophys. Res. Commun.* 55:67–75
56. McDonald, J. K., Zeitman, B. B., Reilly, T. J., Ellis, S. 1969. *J. Biol. Chem.* 244:2693–2709

57. McDonald, J. K., Callahan, P. X., Zeitman, B. B., Ellis, S. 1969. *J. Biol. Chem.* 244:6199–6208
58. Sanger, F., Thompson, E. O. P. 1963. *Biochim. Biophys. Acta* 71:468–71
59. Benisek, W. F., Raftery, M. A., Cole, R. D. 1967. *Biochemistry* 6:3780–90
60. Africa, B., Carpenter, F. H. 1970. *Biochemistry* 9:1962–72
61. Moore, S. 1972. See Ref. 28, 629–53
62. Inglis, A. S., Liu, T. Y. 1970. *J. Biol. Chem.* 245:112–16
63. Matsubara, H., Sasaki, R. M. 1969. *Biochem. Biophys. Res. Commun.* 35:175–81
64. Liu, T. Y., Change, Y. H. 1971. *J. Biol. Chem.* 246:2842–48
65. Hugli, T. E., Moore, S. 1972. *J. Biol. Chem.* 247:2828–34
66. Pisano, J. J., Bronzert, T. J. 1969. *J. Biol. Chem.* 244:5597–5607
67. Pisano, J. J. 1972. *Methods Enzymol.* 25:27–44
68. Waterfield, M., Haber, E. 1970. *Biochemistry* 9:832–39
69. Linenberg, A. 1970. *Mol. Biol. Biochem. Biophys.* 8:124
70. Laursen, R. A. 1969. *Biochem. Biophys. Res. Commun.* 37:663–67
71. Zumwalt, R. W., Kuo, K. C., Gehrke, C. W. 1971. *J. Chromatog.* 57:193–208, 209–17
72. Gehrke, C. W., Takeda, H. 1973. *J. Chromatog.* 76:63–75, 77–89
73. Samejima, K., Dairman, W., Udenfriend, S. 1971. *Anal. Biochem.* 42:222–36
74. Samejima, K., Dairman, W., Stone, J. Udenfriend, S. 1971. *Anal. Biochem.* 42:237–47
75. Udenfriend, S., Stein, S., Bohlen, P., Dairman, W. 1972. See Ref. 28, 655–63
76. Birr, C., Reitel, C., Wieland, T. 1970. *Angew. Chem.* 82:771–72
77. Barnett, D. R., Lee, T. H., Bowman, B. H. 1972. *Biochemistry* 11:1189–94
78. Vance, D. E., Feingold, D. S. 1970. *Anal. Biochem.* 36:30–42
79. Stepanov, V. M., Katrukha, S. P., Baratova, L. A., Belyanova, L. P., Korzhenko, V. P. 1971. *Anal. Biochem.* 43:209–16
80. Previero, A., Pechere, J. F. 1970. *Biochem. Biophys. Res. Commun.* 40:549–56
81. Ichikawa, H., Tanimura, T., Nakajima, T., Tamura, Z. 1970. *Chem. Pharm. Bull.* 18:1493–95
82. Mross, G. A., Doolittle, R. F. 1971. *Fed. Proc.* 30:1241
83. Prager, E. M., Arnheim, N., Mross, G. A., Wilson, A. C. 1972. *J. Biol. Chem.* 247:2905–16
84. Lequin, R., Niall, H. 1972. *Biochim. Biophys. Acta* 257:76–82
85. Boyer, S. H., Noyes, A. N., Boyer, M. L., Marr, K. 1973. *J. Biol. Chem.* 248:992–1003
86. Ozawa, K., Tanaka, S. 1968. *Anal. Biochem.* 24:270–80
87. Katz, E. P. 1968. *Anal. Biochem.* 25:417–31
88. Omenn, G. S., Fontana, A., Anfinsen, C. B. 1970. *J. Biol. Chem.* 245:1895–1902
89. Fontana, A. 1972. *Methods Enzymol.* 25:419–23
90. Degani, Y., Patchornik, A. 1971. *J. Org. Chem.* 36:2727–28
90a. Jacobson, G. R., Schaffer, M. H., Stark, G. R., Vanaman, T. C. 1973. *J. Biol. Chem.* 248:6583–91
91. Bornstein, P. 1969. *Biochem. Biophys. Res. Commun.* 36:957–64
92. Bornstein, P. 1970. *Biochemistry* 9:2408–21
93. Piszkiewicz, D., Landon, M., Smith, E. L. 1970. *Biochem. Biophys. Res. Commun.* 40:1173–78
94. Butler, P. J. G., Harris, J. I., Hartley, B. S., Leberman, R. 1969. *Biochem. J.* 112:679–89
95. Dixon, H. B. F., Perham, R. N. 1968. *Biochem. J.* 109:312–14
96. Cole, R. D. 1967. *Methods Enzymol.* 11:315–17
97. Stark, G. R. 1968. *Biochemistry* 7:1796–1807
98. Yamashita, S. 1971. *Biochim. Biophys. Acta* 229:301–9
99. Stark, G. R. 1972. *Methods Enzymol.* 25:369–84
100. Callahan, P. X., McDonald, J. K., Ellis, S. 1972. *Fed. Proc.* 31:1105–13
101. Lindley, H. 1972. *Biochem. J.* 126:683–88
102. Callahan, P. X., McDonald, J. K., Ellis, S. 1972. *Methods Enzymol.* 25:282–98
103. Ovchinnikov, Y. A., Kiryushkin, A. A. 1972. *FEBS Lett.* 21:300–2
104. McDonald, J. K., Callahan, P. X., Ellis, S., Smith, R. E. 1971. *Tissue Proteinases*, ed. A. J. Barrett, J. T. Dingle, 69–107. Amsterdam: North-Holland. 353 pp.
105. Lee, H. J., Larue, J. N., Wilson, I. B. 1971. *Biochim. Biophys. Acta* 250:608–13
106. Mitchell, W. M., Harrington, W. F. 1968. *J. Biol. Chem.* 243:4683–92
107. Porter, W. H., Cunningham, L. W., Mitchell, W. M. 1971. *J. Biol. Chem.* 246:7675–82

108. Hayashi, R., Aibara, S., Hata, T. 1970. *Biochim. Biophys. Acta* 212: 359–61
109. Hayashi, R., Hata, T. 1972. *Biochim. Biophys. Acta* 263: 673–79
110. Hayashi, R., Moore, S., Stein, W. H. 1973. *J. Biol. Chem.* 248: 2296–2302
111. Tschesche, H., Kupfer, S. 1972. *Eur. J. Biochem.* 26: 33–36
112. Sprössler, B., Heilmann, H. D., Grampp, E., Uhlig, H. 1971. *Z. Physiol. Chem.* 352: 1524–30
113. Carey, W. F., Wells, J. R. E. 1972. *J. Biol. Chem.* 247: 5573–79
114. Ihle, J. N., Dure, L. S. 1972. *J. Biol. Chem.* 247: 5034–40, 5041–47
115. Ichishima, E. 1972. *Biochim. Biophys. Acta* 258: 274–88
116. Matsubara, H., Sasaki, R., Singer, A., Jukes, T. H. 1966. *Arch. Biochem. Biophys.* 115: 324–31
117. Matsubara, H., Sasaki, R. 1968. *J. Biol. Chem.* 243: 1732–57
118. Ambler, R. P., Meadway, R. J. 1968. *Biochem. J.* 108: 893–95
119. Doolittle, R. F. 1972. *Methods Enzymol.* 25: 231–44
120. Schroeder, W. A. 1968. *The Primary Structure of Proteins* New York: Harper & Row
121. Blackburn, S. 1970. *Protein Sequence Determination* New York: Dekker. 292 pp.
122. Needleman, S. B., Ed. 1970. *Protein Sequence Determination* New York: Springer. 345 pp.
123. Hirs, C. H. W., Ed. 1967. *Methods Enzymol.* 11
124. Hirs, C. H. W., Timasheff, S. N., Eds. 1972. *Methods Enzymol.* 25
125. Edelman, G. M., Gall, W. E. 1971. *Proc. Nat. Acad. Sci. USA* 68: 1444–49
126. Silver, J., Hood, L. E. *Anal. Biochem.* In press
127. Hatano, H. 1973 *Res./Develop.* 24: 28–32
128. Brown, P. R. 1973. *High Pressure Liquid Chromatography* New York: Academic
129. Anfinsen, C. B., Redfield, R. R. 1956. *Advan. Protein Chem.* 11: 1–100
130. Du, Y. C., Zhang, Y. S., Lu, Z. X., Tsou, C. L. 1961. *Sci. Sinica* 10: 84–104
131. Tsou, C. L., Du, Y. C., Xu, G. J. 1961. *Sci. Sinica* 10: 332–42
132. Du, Y. C., Jiang, R. Q., Tsou, C. L. 1965. *Sci. Sinica* 14: 229–36
133. Zahn, H., Gutte, B., Pfeiffer, E. F., Ammon, J. 1966. *Ann. Chem.* 691: 225–31
134. Pruitt, K. M., Robison, B. S., Gibbs, J. H. 1966. *Biopolymers* 4: 351–64
135. Katsoyannis, P. G., Tometsko, A. 1966. *Proc. Nat. Acad. Sci. USA* 55: 1554–61
136. Steiner, D. F., Oyer, P. 1967. *Proc. Nat. Acad. Sci. USA* 57: 473–80
137. Steiner, D. F., Cunningham, D., Spigelman, L., Aten, B. 1967. *Science* 157: 697–700
138. Steiner, D. F., Clark, J. L. 1968. *Proc. Nat. Acad. Sci. USA* 60: 622–29
139. Albrecht, G., Corey, R. B. 1939. *J. Am. Chem. Soc.* 61: 1087–1103
140. Levy, H. A., Corey, R. B. 1941. *J. Am. Chem. Soc.* 63: 2095–2108
141. Wright, B. A., Cole, P. A. 1949. *Acta Crystallogr.* 2: 129–30
142. Marsh, R. E., Donohue, J. 1967. *Advan. Protein Chem.* 22: 235–56
143. Corey, R. B., Donohue, J. 1950. *J. Am. Chem. Soc.* 72: 2899–2900
144. Corey, R. B., Pauling, L. 1953. *Proc. Roy. Soc. London B* 141: 10–20
145. Ramachandran, G. N., Sasisekharan, V. 1968. *Advan. Protein Chem.* 23: 283–438
146. Schellman, J. A., Schellman, C. 1954. *Proteins* 2: 1–137
147. Ramakrishnan, C., Ramachandran, G. N. 1965. *Biophys. J.* 5: 909–33
148. Edsall, J. T. et al 1966. *J. Mol. Biol.* 15: 399–407
149. IUPAC-IUB Commission on Biochemical Nomenclature. 1970. *Biochemistry* 9: 3471–79
150. Pauling, L., Corey, R. B., Branson, H. R. 1951. *Proc. Nat. Acad. Sci. USA* 37: 205
151. Pauling, L., Corey, R. B. 1951. *Proc. Nat. Acad. Sci. USA* 37: 729
152. Ramachandran, G. N., Ramakrishnan, C., Sasisekharan, V. 1963. *J. Mol. Biol.* 7: 95–99
153. Leach, S. J., Némethy, G., Scheraga, H. A. 1966. *Biopolymers* 4: 369–407
154. Leach, S. J., Némethy, G., Scheraga, H. A. 1966. *Biopolymers* 4: 887–904
155. Dickerson, R. E., Geis, I. 1969. *The Structure and Action of Proteins.* New York: Harper & Row. 120 pp.
156. DeSantis, P., Giglio, E., Liquori, A. M., Ripamonti, A. 1965. *Nature* 206: 456–58
157. Poland, D., Scheraga, H. A. 1967. *Biochemistry* 6: 3791–3800
158. Balasubramanian, R., Chidambaram, R., Ramachandran, G. N. 1970. *Biochim. Biophys. Acta* 221: 196–206
159. McGuire, R. F., Momany, F. A., Scheraga, H. A. 1972. *J. Phys. Chem.* 76: 375–93
160. Scheraga, H. A. 1971. *Chem. Rev.* 71: 195–217
161. Gibson, K. D., Scheraga, H. A. 1967.

Proc. Nat. Acad. Sci. USA 58:420–27

162. Schiff, L. I. 1968. *Quantum Mechanics,* 3rd ed. New York: McGraw

163. Hoffmann, R., Imamura, A. 1969. *Biopolymers* 7:207–13

164. Rossi, A. R., David, C. W., Schor, R. 1970. *J. Phys. Chem.* 74:4551–55

165. Maigret, B., Pullman, B., Perahia, D. 1971. *J. Theor. Biol.* 31:269–85

166. Schor, R., Stymne, H., Wettermark, G., David, C. W. 1972. *J. Phys. Chem.* 76:670–72

167. Gibson, K. D., Scheraga, H. A. 1971. *Proc. Nat. Acad. Sci. USA* 58:1317–23

168. Scheraga, H. A. 1960. *J. Am. Chem. Soc.* 82:3847–52

169. Levinthal, C. 1966. *Sci. Am.* 214:42–52

170. Perutz, M. F. 1951. *Nature* 167:1053–54

171. Brant, D. A., Miller, W. G., Flory, P. J. 1967. *J. Mol. Biol.* 23:47–65

172. Wu, T. T., Kabat, E. A. 1971. *Proc. Nat. Acad. Sci. USA* 68:1501–6

173. Wyckoff, H. W. et al 1970. *J. Biol. Chem.* 245:305–28 (and private communication)

174. Dickerson, R. E. et al 1971. *J. Biol. Chem.* 246:1511–35 (and private communication)

175. Dickerson, R. E. 1972. *Ann. Rev. Biochem.* 41:815–42

176. Szent-Györgyi, A. G., Cohen, C. 1957 *Science* 126:697

177. Davies, D. R. 1964. *J. Mol. Biol.* 9:605–9

178. Guzzo, A. V. 1965. *Biophys. J.* 5:809–22

179. Havsteen, B. H. 1966. *J. Theor. Biol.* 10:1–10

180. Prothero, J. W. 1966. *Biophys. J.* 6:367–70

181. Cook, D. A. 1967. *J. Mol. Biol.* 29:167–71

182. Prothero, J. W. 1968. *Biophys. J.* 8:1236–55

183. Kotelchuck, D., Scheraga, H. A. 1968. *Proc. Nat. Acad. Sci. USA* 61:1163–70

184. Goldsack, D. E. 1969. *Biopolymers* 7:299–313

185. Kotelchuck, D., Scheraga, H. A. 1969. *Proc. Nat. Acad. Sci. USA* 62:14–21

186. Ptitsyn, O. B. 1969. *Mol. Biol.* 3:495–99

187. Ptitsyn, O. B. 1969. *J. Mol. Biol.* 42:501–10

188. Ptitsyn, O. B., Finkel'shtein, A. V. 1970. *Biofizika* 5:757–67

189. Ptitsyn, O. B., Finkel'shtein, A. V. 1970. *Dokl. Akad. Nauk SSSR* 195:221–24

190. Schiffer, M., Edmundson, A. B. 1967. *Biophys. J.* 7:121–35

191. Periti, P. F., Quagliarotti, G., Liquori, A. M. 1967. *J. Mol. Biol.* 24:313–22

192. Low, B. W., Lovell, F. M., Rudko, A. D. 1968. *Proc. Nat. Acad. Sci. USA* 60:1519–26

193. Kotelchuck, D., Dygert, M., Scheraga, H. A. 1969. *Proc. Nat. Acad. Sci. USA* 63:615–22

194. Pain, R. H., Robson, B. 1970. *Nature* 227:62–63

195. Robson, B., Pain, R. H. 1971. *J. Mol. Biol.* 58:237–59

196. Finkel'shtein, A. V., Ptitsyn, O. B. 1971. *J. Mol. Biol.* 62:613–24

197. Robson, O. B., Pain, R. H. 1972. *Nature New Biol.* 238:107–8

198. Kabat, E. A., Wu, T. T. 1973. *Biopolymers* 12:751–74

199. Chou, P. Y., Fasman, G. D. 1973. *J. Mol. Biol.* 74:263–81

200. Nagano, K. 1973. *J. Mol. Biol.* 75:401–20

201. Schellman, J. A., Schellman, C. G. In preparation

202. Anfinsen, C. G., Scheraga, H. A. *Advan. Protein Chem.* In preparation

203. Kabat, E. A., Wu, T. T. 1973. *Proc. Nat. Acad. Sci. USA* 70:1473–77

204. Lewis, P. N., Momany, F. A., Scheraga, H. A. 1971. *Proc. Nat. Acad. Sci. USA* 68:2293–97

205. Bunting, J. R., Athey, T. W., Cathou, R. E. 1973. *Biochim. Biophys. Acta* 285:60–71

206. Crawford, J. L., Lipscomb, W. N., Schellman, C. G. 1973. *Proc. Nat. Acad. Sci. USA* 70:538–42

207. Kabat, E. A., Wu, T. T. 1972. *Proc. Nat. Acad. Sci. USA* 69:960–64

208. Wu, T. T., Kabat, E. A. 1973. *J. Mol. Biol.* 75:13–31

209. Protein Data Bank, T. F. Koetzle, Dept. of Chem., Brookhaven Nat. Lab., Upton, NY 11973

210. Fitch, W. M., Margoliash, E. 1967. *Science* 155:279–84

211. Wu, T. T., Kabat, E. A. 1970. *J. Exp. Med.* 132:211–50

212. Honig, B., Kabat, E. A., Katz, L., Levinthal, C., Wu, T. T. 1973. *J. Mol. Biol.* 80:277–95

213. Lemmon, R. E. 1970. *Chem. Rev.* 70:95–109

214. Kenyon, D. H., Steinman, G. 1969. *Biochemical Predestination* New York: McGraw. 301 pp.

215. Oparin, A. I. 1957. *Origin of Life* New York: Dover. 495 pp.

216. Koesian, J. 1964. *Origin of Life* New York: Rheinhold. 118 pp.
217. Calvin, M. 1969. *Chemical Evolution* Oxford: Oxford Univ. Press. 278 pp.
218. Margulis, L. 1970. *Origins of Life* New York: Gordon & Breach. 376 pp.
219. Wolman, Y., Haverland, W. J., Miller, S. L. 1972. *Proc. Nat. Acad. Sci. USA* 69: 809–11
220. Lawless, J. G., Kvenvolden, K. A., Peterson, E., Ponnamperuma, C., Moore, C. 1971. *Science* 173: 626–27
221. Lawless, J. G., Kvenvolden, K. A., Peterson, E., Ponnamperuma, C., Jarosewich, E. 1972. *Nature* 236: 66–67
222. Sagan, C. 1972. *Nature* 238: 77–80
223. Cairns-Smith, A. G., Ingram, P., Walker, G. L. 1972. *J. Theor. Biol.* 35: 601–4
224. Fox, S. W., Windsor, C. R. 1970. *Science* 170: 984–86
225. Hulett, H. R., Bar-Nun, A., Bar-Nun, N., Bauer, S. H., Sagan, C. 1970. *Science* 170: 1000–2
226. Bar-Nun, A., Bar-Nun, N., Bauer, S. H., Sagan, C. 1970. *Science* 168: 470–73
227. Sagan, C., Khare, B. N. 1971. *Science* 173: 417–20
228. Hulett, H. R. et al 1971. *Science* 174: 1038–41
229. Khare, B. N., Sagan, C. 1971. *Nature* 232: 577–79
230. Wollin, G., Ericson, D. B. 1971. *Nature* 233: 615–16
231. Sanchez, R. A., Orgel, L. E. 1970. *J. Med. Biol.* 47: 531
232. Stephen-Sherwood, F., Oro, J., Kimball, A. P. 1971. *Science* 173: 446–47
233. Friedmann, N., Miller, S. L., Sanchez, R. A. 1971. *Science* 171: 1026–27
234. Dowler, M. J., Fuller, W. D., Orgel, L. E., Sanchez, R. A. 1970. *Science* 169: 1320–21
235. Frank, F. C. 1953. *Biochim. Biophys. Acta* 11: 459–63
236. Garay, A. S. 1968. *Nature* 219: 338–40
237. Segal, H. L. 1972. *FEBS Lett.* 20: 255–56
238. Mörtberg, L. 1971. *Nature* 232: 105–7
239. Paecht-Horowitz, M., Berger, J., Katchalsky, A. 1970. *Nature* 2238: 636–39
240. Degens, E. T., Matheja, J., Jackson, T. A. 1970. *Nature* 227: 492–93
241. Chang, S., Flores, J., Ponnamperuma, C. 1969. *Proc. Nat. Acad. Sci. USA* 64: 1011–15
242. Saunders, M. A., Rohlfing, D. L. 1972. *Science* 176: 172–73
243. Fuller, W. D., Sanchez, R. A., Orgel, L. E. 1972. *J. Mol. Biol.* 67: 25–33
244. Tapiero, C. M., Nagyvary, J. 1971. *Nature* 231: 42–43
245. Lohrmann, R., Orgel, L. E. 1968. *Science* 161: 64–66
246. Bishop, M. J., Lohrmann, R., Orgel, L. E. 1972. *Nature* 237: 162–64
247. Schwartz, A. W. 1972. *Biochim. Biophys. Acta* 281: 477–80
248. Ibanez, J. D., Kimball, A. P., Oró, J. 1971. *J. Mol. Evol.* 1: 112–14
249. Ibanez, J. D., Kimball, A. P., Oró, J. 1971. *Science* 173: 444–46
250. Sulston, J., Lohrmann, R., Orgel, L. E., Miles, H. T. 1968. *Proc. Nat. Acad. Sci. USA* 59: 726–33
251. Sulston, J., Lohrmann, R., Orgel, L. E., Miles, H. T. 1968. *Proc. Nat. Acad. Sci. USA* 60: 409–15
252. Sulston, J. et al 1969. *J. Mol. Biol.* 40: 227–34
253. Degani, C., Halmann, M. 1972. *Nature New Biol.* 235: 171–72
254. Usher, D. A. 1972. *Nature New Biol.* 235: 207–8
255. Woese, C. R. 1967. *The Genetic Code.* New York: Harper & Row
256. Crick, F. H C. 1968. *J. Mol. Biol.* 38: 367–79
257. Orgel, L. E. 1968. *J. Mol. Biol.* 38: 381–93
258. Woese, C. 1969. *J. Mol. Biol.* 43: 235–40
259. Sonneborn, T. M. 1965. *Evolving Genes and Proteins,* ed. V. Bryson, H. Vogel, 377–97. New York: Academic. 629 pp.
260. Fitch, W. M. 1966. *J. Mol. Biol.* 16: 1–8
261. Jukes, T. H. 1967. *Biochem. Biophys. Res. Commun.* 27: 573–78
262. Crick, F. H. C. 1967. *Nature* 213: 119
263. Woese, C. R. 1965. *Proc. Nat. Acad. Sci. USA* 54: 71–75
264. Thomas, B. R. 1970. *Biochem. Biophys. Res. Commun.* 40: 1289–96
265. Lesk, A. M. 1970. *Biochem. Biophys. Res. Commun.* 38: 855–58
266. Saxinger, C., Ponnamperuma, C. 1971. *J. Mol. Evol.* 1: 63–73
267. Rendell, M. S., Harlos, J. P., Rein, R. 1971. *Biopolymers* 10: 2083–94
268. Raszka, M., Mandel, M. 1972. *J. Mol. Evol.* 2: 38–43
269. MacKay, A. L. 1967. *Nature* 216: 159–60
270. King, J. L., Jukes, T. H. 1969. *Science* 164: 788–98
271. King, J. L. 1971. *Biochemical Evolution and the Origin of Life,* ed. E. Schoffeniels, Vol. 1. Amsterdam: North-Holland
272. Moorhead, P. S., Kaplan, M. M., Eds.

1967. *Mathematical Challenges to the Neo-Darwinian Interpretation of Evolution,* Philadelphia: Wistar Univ. Press, Monograph. 140 pp.

273. Waddington, C. H., Ed. 1968. *Towards a Theoretical Biology.* Chicago: Aldine. 3 vols.

274. Hulett, H. R. 1969. *J. Theor. Biol.* 24: 56–72

275. Eigen, M. 1971. *Naturwissenschaften* 58: 465–523

276. Black, S. 1971. *Biochem. Biophys. Res. Commun.* 43: 267–72

277. Theodoridis, G. C., Stark, L. 1971. *J. Theor. Biol.* 31: 377–88

278. Smith, E. L. 1970. *Enzymes* 1: 267–339

279. Fitch, W. M., Margoliash, E. 1970. *Evol. Biol.* 4: 67–109

280. Margoliash, E. 1972. *The Harvey Lectures* Series 66: 177–247

281. Arnheim, N. 1973. *The Antigens,* ed. M. Sela, 377–416. New York: Academic

282. Gally, J. A., Edelman, G. M. 1972. *Ann. Rev. Genet.* 6: 1–46

283. Smith, G., Hood, L., Fitch, W. M. 1971. *Ann. Rev. Biochem.* 40: 969

284. Brew, K. 1970. *Essays Biochem.* 6: 93–118

285. Vogel, H. J. 1965. *Evolving Genes and Proteins,* ed. V. Bryson, H. Vogel, 25–40. New York: Academic. 629 pp.

286. LéJohn, H. B. 1971. *Nature* 231: 164–68

287. Metzger, H., Shapiro, M. B., Mosimann, J. E., Vinton, J. E. 1968. *Nature* 219: 1166–68

288. Marchalonis, J. J., Weltman, J. K. 1971. *Comp. Biochem. Physiol.* 38: 609–25

289. Harris, C. E., Kobes, R. D., Teller, O. C., Rutter, W. J. 1969. *Biochemistry* 8: 24–42

290. Shapiro, H. M. 1971. *Biochim. Biophys. Acta* 236: 725–38

291. Fitch, W. M. 1966. *J. Mol. Biol.* 16: 9–16

292. Fitch, W. M. 1970. *J. Mol. Biol.* 49: 1–14

293. Manwell, C. 1967. *Comp. Biochem. Physiol.* 23: 383–406

294. Haber, J. E., Koshland, D. E. Jr. 1970. *J. Mol. Biol.* 50: 617–39

295. Sackin, M. J. 1971. *Biochem. Genet.* 5: 287–313

296. Gibbs, A. J., McIntyre, G. A. 1971. *Eur. J. Biochem.* 16: 1–11

297. McLachlan, A. D. 1971. *J. Mol. Biol.* 61: 409–24

298. Needleman, S. B., Wunsch, C. D. 1970. *J. Mol. Biol.* 48: 443–53

299. Sankoff, D. 1972. *Proc. Nat. Acad. Sci. USA* 69: 4–6

300. Reichert, T. A., Cohen, D. N., Wong, A. K. C. 1973. *J. Theor. Biol.* 42: 245–61

301. Bauer, K. 1970. *Int. J. Protein Res.* 3: 165–75

302. Bauer, K. 1971. *Int. J. Protein Res.* 3: 313–24

303. Bauer, K. 1971. *Naturwissenschaften* 58: 364–65

304. Bauer, K. 1972. *Biochim. Biophys. Acta* 278: 606–9

305. Fitch, W. M. 1973. *Ann. Rev. Genet.* 7: 343–80

306. Fitch, W. M. 1969. *Biochem. Genet.* 3: 99–108

307. Fitch, W. M. 1970. *Syst. Zool.* 19: 99–113

308. Langley, C. H., Fitch, W. M. 1973. *Genetic Structure of Population,* ed. N. E. Morton, 246–62. Honolulu: Univ. Press of Hawaii

309. Fitch, W. M., Margoliash, E. *Physiol. Rev.* Submitted

310. Walsh, K. A., Neurath, H. 1964. *Proc. Nat. Acad. Sci. USA* 52: 884–89

311. Hartley, B. S., Brown, J. R., Kauffman, D. L., Smillie, L. B. 1965. *Nature* 207: 1157–59

312. Brew, K., Vanaman, T. C., Hill, R. L. 1967. *J. Biol. Chem.* 242: 3747–49

313. Li, C. H., Dixon, J. S., Tung-Bin, L., Pankov, Y. A., Schmidt, K. D. 1969. *Nature* 224: 695–96

314. Harris, J. I., Ross, P. 1956. *Nature* 178: 90

315. Li, C. H., Barnafi, L., Chrétien, M., Chung, D. 1965. *Nature* 208: 1093–94

316. Barnett, D. R., Lee, T. H., Bowman, B. H. 1972. *Biochemistry* 11: 1189–94

317. Titani, K. et al 1972. *Biochemistry* 11: 4899–4903

318. Adelson, J. W. 1971. *Nature* 229: 321–25

319. Strydom, D. J. 1973. *Nature New Biol.* 243: 88–89

320. Sankoff, D., Cedergren, R. J. 1973. *J. Mol. Biol.* 77: 159–64

321. Cedergren, R. J., Cordeau, J. R., Robillard, P. 1972. *J. Theor. Biol.* 37: 209

322. Tinoco, I. Jr., Uhlenbeck, O. C., Levine, M. D. 1971. *Nature* 230: 362–67

323. Uhlenbeck, O. C., Borer, P. N., Dengler, B., Tinoco, I. Jr. 1967. *J. Mol. Biol.* 23: 483–96

324. Delisi, C., Crothers, D. M. 1971. *Proc. Nat. Acad. Sci. USA* 68: 2682–85

325. Gralla, J., Crothers, D. M. 1973. *J. Mol. Biol.* 73:497–511
326. White, H. B. III, Laux, B. E., Dennis, D. 1972. *Science* 175:1264–66
327. Mark, A. J., Petruska, J. A. 1972. *J. Mol. Biol.* 72:609
328. Fitch, W. M. 1972. *J. Mol. Evol.* 1:185–207
329. Fitch, W. M. 1974. Submitted for publication
330. Min Jou, W., Haegeman, G., Ysebaert, M., Fiers, W. 1972. *Nature* 237:82–88
331. Ball, L. A. 1973. *Nature New Biol.* 242:44–45
332. Bridges, C. B. 1936. *Science* 83:210
333. Baglioni, C. 1962. *Proc. Nat. Acad. Sci. USA* 48:1880–86
334. Bradley, T. B. Jr., Wohl, R. C., Rieder, R. F. 1967. *Science* 157:1581–83
335. Yanofsky, C. 1960. *Bacteriol. Rev.* 24:221–45
336. Bonner, D. M., DeMoss, J. A., Mills, S. E. 1965. *Evolving Genes and Proteins*, ed. H. J. Vogel, V. Bryson, 305–18. New York: Academic. 629 pp.
337. Manney, T. R., Duntze, W., Janosko, N., Salazar, J. 1969. *J. Bacteriol.* 99:590–96
338. Grieshaber, M., Bauerle, R. 1972. *Nature New Biol.* 236:232–35
339. Véron, M., Falcoz-Kelly, F., Cohen, G. N. 1972. *Eur. J. Biochem.* 28:520–27
340. Ingram, V. M. 1961. *Nature* 189:704–8
341. Ostertag, W., von Ehrenstein, G., Charache, S. 1972. *Nature New Biol.* 237:90–94
342. Abramson, R. K., Rucknagel, D. L., Shreffler, D. C., Saave, J. J. 1970. *Science* 169:194–96
343. Hilse, K., Popp, R. A. 1968. *Proc. Nat. Acad. Sci. USA* 61:930–36
344. Harris, M. J., Wilson, J. B., Huisman, T. H. J. 1972. *Arch. Biochem. Biophys.* 151:540–48
345. Yanase, T. et al 1968. *Jap. J. Hum. Genet.* 13:40
346. Lehmann, H., Charlesworth, D. 1970. *Biochem. J.* 119:43P
347. Badr, F. M., Lorkin, P. A., Lehmann, H. 1973. *Nature New Biol.* 242:107–10
348. Ohta, Y., Yamaoka, K., Sumida, I., Yanase, T. 1971. *Nature New Biol.* 234:218–20
349. Galizzi, A. 1971. *Nature New Biol.* 229:142–43
350. Yanofsky, C. 1963. *Cold Spring Harbor Symp. Quant. Biol.* 28:581
351. Connell, G. E., Dixon, G. H., Smithies, O. 1962. *Nature* 193:505–6
352. Black, J. A., Dixon, G. H. 1970. *Can. J. Biochem.* 48:133–46
353. Fitch, W. M. 1971. *Nature* 229:245–47
354. Sherman, F. et al 1970. *Symp. Soc. Exp. Biol.* 24:85–107
355. Stewart, J. W., Sherman, F., Shipman, N. A., Jackson, M. 1971. *J. Biol. Chem.* 246:7429–45
356. Weatherall, D. J., Clegg, J. B. 1972. *Synthesis Structure and Function of Hemoglobin*, ed. H. Martin, L. Nowicki, 237–39. Munich: J. F. Lehmann's Verlag
357. Lehmann, H. 1972. *Synthesis, Structure and Function of Hemoglobin*, ed. H. Martin, L. Nowicki, 359–79. Munich: H. Lehmann's Verlag
358. Inouye, M., Akaboshi, E., Tsugita, A., Streisinger, G., Okata, Y. 1967. *J. Med. Biol.* 30:39
359. Berger, H., Brammar, W. J., Yanofsky, C. 1968. *J. Mol. Biol.* 34:219–38
360. Berger, H., Brammar, W. J., Yanofsky, C. 1968. *J. Bacteriol.* 96:1672–79
361. Seid-Akhavan, M., Winter, W. P., Abramson, R. K., Rucknagel, D. L. 1972. *Blood* 40:927
362. Fitch, W. M., Margoliash, E. 1966. *Science* 155:279–84
363. Goodman, M., Barnabas, J., Matsuda, G., Moore, G. W. 1971. *Nature* 233:604–13
364. Barnabas, J., Goodman, M., Moore, G. W. 1971. *J. Comp. Biochem. Physiol.* 39:455–82
365. Barnabas, J., Goodman, M., Moore, G. W. 1972. *J. Mol. Biol.* 69:249–78
366. Goodman, M., Moore, G. W. 1973. *Syst. Zool.* 22:508–32
367. Boyer, S. H. et al 1971. *Biochem. Genet.* 5:405–48
368. Boulter, D., Ramshaw, J. A. M., Thompson, E. W., Richardson, M., Brown, R. H. 1972. *Proc. Roy. Soc. London B* 181:441–55
369. Boulter, D. 1973. *Syst. Zool.* 22:549–53
370. Mross, G. A., Doolittle, R. F. 1967. *Arch. Biochem. Biophys.* 122:674–84
371. Doolittle, R. F., Wooding, G. L., Lin, Y., Riley, M. 1971. *J. Mol. Evol.* 1:74–83
372. McLaughlin, P. J., Dayhoff, M. O. 1973. *J. Mol. Evol.* 2:99–116
373. Fitch, W. M., Markowitz, E. 1970. *Biochem. Genet.* 4:579–93
374. O'Neil, P., Doolittle, R. F. 1973. *Syst. Zool.* 22:590–95
375. Strydom, D. J. 1973. *Comp. Biochem. Physiol.* 44:269–81

376. Strydom, D. J. 1973. *Syst. Zool.* 22: 596–608
377. Staehelin, M. 1972. *J. Mol. Evol.* 1: 258–62
378. Ambler, R. P. 1973. *Syst. Zool.* 22: 554–65
379. Samuelsson, G. 1973. *Syst. Zool.* 22: 566–69
380. Yasunobu, K., Tanaka, M. 1973. *Syst. Zool.* 22: 570–89
381. Pechère,J.-F., Capony,J.-P., Demaille, J. 1973. *Syst. Zool.* 22: 533–48
382. Prager, E. M., Arnheim, N., Mross, G. A., Wilson, A. C. 1972. *J. Biol. Chem.* 247: 2905–16
383. Arnheim, N., Steller, R. 1970. *Arch. Biochem. Biophys.* 141: 656–61
384. Sneath, P. H. A., Sokal, R. R. 1973. *Numerical Taxonomy, Principles and Practice of Numerical Classification* San Francisco: Freeman
385. Gibbs, A. J., Dale, M. B., Kinns, H. R., MacKenzie, H. G. 1971. *Syst. Zool.* 20: 417–25
386. Kohne, D. E. 1970. *Quart. Rev. Biophys.* 3: 327–75
387. Wilson, A. C., Sarich, V. M. 1969. *Proc. Nat. Acad. Sci. USA* 63: 1088–93
388. Fitch, W. M. 1971. *Syst. Zool.* 20: 406–14
389. Hartigan, J. A. 1973. *Biometrics* 29: 53
390. Moore, G. W., Barnabas, J., Goodman, M. 1973. *J. Theor. Biol.* 38: 459
391. Fitch, W. M., Markowitz, E. 1970. *Biochem. Genet.* 4: 579–93
392. Jukes, T. H. 1971. *J. Mol. Evol.* 1: 46–62
393. Uzzell, T., Corbin, K. W. 1971. *Science* 172: 1089–96
394. Fitch, W. M. 1972. *Haematol. Bluttransfus.* 10: 199–215
395. Fitch, W. M. 1972. *Brookhaven Symp. Biol.* 23: 186–215
396. Margoliash, E., Fitch, W. M. 1970. *Homologies in Enzymes and Metabolic Pathways,* ed. W. J. Whelan, 33–51. Amsterdam: North-Holland
397. Vogel, F. 1969. *Humangenetik* 8: 1–26
398. Fitch, W. M. 1973. *J. Mol. Evol.* 2: 181–86
399. Fitch, W. M. 1973. *J. Mol. Evol.* 2: 123–36
400. Fitch, W. M. 1967. *J. Mol. Biol.* 26: 499–507
401. Lehmann, H., Carrell, R. W. 1969. *Brit. Med. Bull.* 25: 14–23
402. Vogel, H., Derancourt, J., Zuckerkandl, E. 1971. *Peptides,* 339–46. Amsterdam: North-Holland
403. Zuckerkandl, E., Derancourt, J., Vogel, H. 1971. *J. Mol. Biol.* 59: 473–90
404. Vogel, F. 1972. *J. Mol. Evol.* 1: 334–67
405. Vogel, F. 1972. *Humangenetik* 16: 71–76
406. Fitch, W. M., Margoliash, E. 1968. *Ann. N Y Acad. Sci.* 151: 359–81
407. Holmquist, R. 1972. *J. Mol. Evol.* 1: 115–33
408. Holmquist, R. 1972. *J. Mol. Evol.* 1: 134–49
409. Holmquist, R. 1972. *J. Mol. Evol.* 1: 211–22
410. Holmquist, R., Jukes, T. H. 1972. *J. Mol. Evol.* 2: 10–16
411. Jukes, T. H., Holmquist, R. 1972. *J. Mol. Biol.* 64: 163–79
412. Holmquist, R., Cantor, C., Jukes, T. 1972. *J. Mol. Biol.* 64: 145–61
413. Kimura, M. 1968. *Nature* 217: 624–26
414. Langley, C. H., Fitch, W. M. *J. Mol. Evol.* Submitted
415. Ohta, T., Kimura, M. 1971. *J. Mol. Evol.* 1: 18–25
416. Fink, T. R., Crothers, D. M. 1972. *J. Mol. Biol.* 66: 1–12
417. Huisman, T. H. J., Wrightstone, R. N., Wilson, J. B., Schroeder, W. A., Kendall, A. G. 1972. *Arch. Biochem. Biophys.* 153: 850–53
418. Hollán, S. R. et al 1972. *Nature* 235: 47–50
419. Ostertag, W., von Ehrenstein, G., Charache, S. 1972. *Nature New Biol.* 237: 90–94
420. Lang, A., Lehmann, H., McCurdy, P. R., Pierce, L. 1972. *Biochim. Biophys. Acta* 278: 57–61
421. Steadman, J. H., Yates, A., Huehns, E. R. 1970. *Brit. J. Haematol.* 18: 435–46
422. Gerald, P. S., Efron, M. L. 1961. *Proc. Nat. Acad. Sci. USA* 47: 1758–67
423. Forget, B. et al 1974. *Ann. N Y Acad. Sci.* In press

THE BIOSYNTHESIS OF COLLAGEN[1]

× 858

Paul Bornstein[2]

Departments of Biochemistry and Medicine, University of Washington
Seattle, Washington

CONTENTS

[1] The survey of the literature was concluded in August 1973. Abbreviations used: BAPN, β-aminopropionitrile; SLS, Segment-long spacing; SDS, sodium dodecylsulfate; NEM, N-ethylmaleimide; p-CMB, p-chloromercuribenzoate; DON, 6-diazo-5-oxonorleucine; RER, rough endoplasmic reticulum; RCM, reduced carboxymethylated.

[2] Recipient of Research Career Development Award K4-AM-42582 from the U.S. Public Health Service.

INTRODUCTION

Recent experimental evidence for a higher molecular weight biosynthetic precursor of the collagen monomer, procollagen,[3] has permitted a clearer focus on the intra- and extracellular events that lead to the biogenesis of this important structural macromolecule. It is the purpose of this review to critically evaluate the recent burgeoning literature dealing with the biosynthesis of collagen and to reinterpret relevant earlier work in light of this information. No attempt is made to provide an exhaustive survey of published work in the area of collagen biochemistry. Several comprehensive chapters treat the metabolic turnover of collagen, the molecular conformation and structure of the protein in the fibril, and the synthesis, structure, and molecular distribution of covalent crosslinks in collagen and elastin (1–7). Useful reviews dealing more specifically with the biosynthesis, structure, and cross-linking of collagen (8–12), and with the role of ascorbic acid in collagen metabolism (13) have also appeared. Three monographs summarize the proceedings of recent symposia (14–16) and place these subjects in the appropriate context of the broad field of connective tissue biology.

In the period following the demonstration that polymeric or fibrous collagen results from the aggregation of discrete subunits or monomers (roughly 1955–1970), efforts were directed to a characterization of the chain composition of the protein and the covalent structure of the individual chains. In most vertebrate tissues the triple-stranded collagen molecule was found to consist of two identical chains (α1) and a different but clearly homologous chain (α2). The fractionation of these chains in large amounts by CM-cellulose chromatography (17) permitted the application of classical techniques in protein chemistry to the elucidation of the primary structure of these chains. These studies have culminated in determining the complete amino acid sequence of an α1 chain (more than 1000 amino acids long), albeit experiments were performed partly with rat (18–24) and partly with bovine skin collagen (25–30).

As a consequence of this work, the collagen chain was defined as a linear sequence of amino acids, linked only by α-amino, α-carboxyl peptide bonds and characterized for more than 95 percent of its length by glycine in every third position. Only at the amino and carboxy terminal ends of the chain do sequences occur that do not contain the repetitive glycine structure (18, 28). The concept that short sequences (telopeptides) exist which differ in structure and possibly in func-

---

[3] The term procollagen has been used in the past to describe a soluble precursor of the collagen fiber. For this purpose, the term has been replaced by the descriptive terms soluble collagen or collagen monomer or tropocollagen. In this review, the use of pro-collagen will be reserved to describe a precursor form of the functional collagen molecule in the sense that the terms procarboxypeptidase and proinsulin are used. Soluble collagen will be referred to as collagen or collagen monomer, but not as tropocollagen. The latter term has been confused with protocollagen and suggests a misleading parallel with the nomenclature in the muscle protein field. Protocollagen will be used to refer to an experimentally produced form of procollagen deficient in hydroxyproline and hydroxylysine.

tion from the rest of the collagen molecule (31–33) is therefore valid with reference to these terminal extensions.

None of the unusual nonpeptide or isopeptide bonds and only a few of the unusual chemical functions originally postulated for collagen (34) were found to

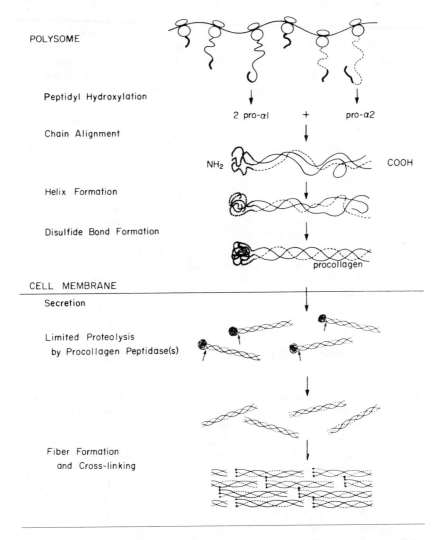

Figure 1 The biosynthesis of procollagen, conversion to collagen, and extracellular fibrogenesis. It is likely that alignment and aggregation of the three chains, helix formation, and disulfide bond formation occur concomitantly and at least in part prior to release of chains from ribosomes.

exist. Collagen is therefore a more orthodox protein than many workers in the field would have thought on the basis of information available 5–10 years ago. A distinct exception to this generalization is the mechanism by which interchain covalent bonds form in the protein. The biosynthetic scheme for crosslink formation, whose elements have so far been identified only in the related protein elastin, involves the oxidative deamination of lysyl and hydroxylysyl side chains yielding aldehydes which subsequently form a series of complex covalent bonds (3, 10). However, studies of collagen chemistry have become more complex since it is now known that genetically and structurally distinct collagens exist in different tissues of the same species. The observation, first made by Miller & Matukas (35) working with chick cartilage, has been extended to fetal human skin (36), basement membrane (37), and the collagen synthesized by embryonic chick spinal cord epithelium (38). These collagens contain three identical α chains which appear to be homologous to the $\alpha 1$ chain of the major collagen component in skin, bone, and tendon.

The primary concepts examined in this review are illustrated in Figure 1. The collagen molecule is synthesized as a biosynthetic precursor, procollagen, consisting of three pro-α chains with additional NH_2-terminal extensions. These sequences differ in structure and conformation from the rest of the molecule and may be involved in chain alignment and other functions. Hydroxylation of peptidyl proline and lysine is initiated on nascent chains and peptidyl hydroxyproline is now thought to contribute to the stability of the triple helix. Although not depicted, it is likely that at least the initial stages of chain aggregation, generation of the triple helix, and disulfide bond formation occur as a concerted process prior to release of chains from ribosomes. The mechanisms involved in the transcellular movement and secretion of procollagen are not shown and are not well understood. Finally, more than one enzymatic activity may be involved in the conversion of procollagen to collagen prior to incorporation of the protein into the extracellular fibril.

EVIDENCE AND POSTULATED FUNCTIONS FOR PROCOLLAGEN

Early evidence for chemically different forms of soluble collagen existing in the same tissue (39, 40) was difficult to evaluate in the absence of a precise characterization of the collagen monomer and because of the difficulty of purifying the protein. Schmitt (41–43) suggested that specialized regions at the ends of collagen molecules might be involved in the extracellular polymerization of collagen and subsequently removed, a concept modified by Speakman (44) to include a role for "registration peptides" in chain association and triple helix formation.

In 1971 Layman et al (45), working with normal human skin fibroblasts in culture, reported that a large fraction of the hydroxyproline-containing protein secreted into the medium differed from monomeric collagen extracted from skin in that (a) the protein was soluble under physiological conditions; and (b) under conditions in which dissociation of collagen to its constituent α chains would be expected, components with molecular weights equal to or greater than β components (α chain dimers) were observed. Since cells were incubated in the presence of

BAPN, an inhibitor of lysyloxidase (46), lysyl-derived interchain crosslinks were not likely to account for such aggregates. This medium fraction was converted to a collagen-like protein by limited cleavage with pepsin, suggesting the presence of amino acid sequences differing in conformation from the triple helix.

Fibroblasts isolated from embryonic chick tendons (47) also secreted a collagen-like protein into the culture medium which, after reduction with mercaptoethanol and chromatography on an SDS-agarose column, yielded chains with a mol wt of approximately 125,000, some 25 percent higher than that of α chains (48). In an extension of these studies Dehm et al (49) showed that the medium protein, considered to be a transport form of collagen, contained cystine and probably interchain disulfide bonds.

Evidence for the existence in tissues of a higher molecular weight biosynthetic precursor of collagen, procollagen, was obtained by Bellamy & Bornstein (50). These investigators, working with newborn rat cranial bones, demonstrated the existence of an acid-extractable collagen fraction which, on the basis of pulse-chase experiments, was shown to play an antecedent role in the biosynthesis of collagen. The constituent (pro-α) chains of procollagen were separable from $\alpha1$ and $\alpha2$ by CM-cellulose chromatography in urea-containing buffers and by acrylamide gel electrophoresis (50, 51). As in the case of the medium fraction of fibroblast cultures (45, 49, 52), cranial bone procollagen was converted to a collagen-like protein by limited proteolysis with pepsin (50).

Functions postulated for a collagen precursor include participation in the appropriate lateral aggregation and crosslinking of collagen during extracellular fibrogenesis (41–43, 53–55), initiation of triple helix formation (44, 50), inhibition of intracellular fibrogenesis (50, 56), and facilitation of transmembrane movement of the protein (45, 49). None of these functions has been demonstrated directly. However, the observation that procollagen secreted by cells in culture remains soluble under conditions expected to induce fiber formation (45, 49) supports the contention that procollagen and collagen differ in fibrogenic properties. Furthermore, in dermatosparaxis, a hereditary disorder of cattle caused by a defect in the conversion of procollagen to collagen (56, 57), the low tensile strength of skin and other tissues attests to the poor fiber-forming quality of collagen precursors. Extracted dermatosparactic collagen, in fact, polymerizes poorly under conditions that readily lead to fiber formation by normal collagen, and depolymerizes more rapidly and completely upon cooling (56). Recently Bailey & Lapière (58) have shown that the disorganized collagen fibers in dermatosparactic skin are deficient in aldimine and higher order crosslinks and that removal of the additional peptides results in a collagen preparation capable of forming a normal crosslink pattern in vitro.

Evidence is accumulating, however, that procollagen or a derivative thereof may normally play a role in the extracellular matrix. Veis and his associates (53–55) have succeeded in isolating collagen fractions from bovine and rat skin with chains higher in molecular weight than α chains and in quantities greater than would be expected if procollagen were only an evanescent biosynthetic precursor. Antibodies to bovine dermatosparactic collagen have also been shown by indirect immunofluorescence to react with constituents in the interstitial connective tissue of normal bovine and human kidney, muscle, and skin (59).

STRUCTURE OF PROCOLLAGEN

Chemical Nature and Location of Additional Sequences

The incorporation of radio-labeled cysteine or cystine into procollagen during in vitro synthesis by fibroblasts (49, 60–62), and of both cysteine and tryptophan into the pro-α1 chain of cranial bone (63, 64) indicated that the amino acid sequences unique to procollagen differed in composition from collagen. These amino acids had been considered absent from vertebrate collagens. Although cysteine was not found in the pro-α2 chain of acid-extracted bone procollagen (63), it is now known that prolonged acid extraction results in selective proteolytic release of disulfide-bonded regions from both pro-α1 and pro-α2 chains (65, 66). Cystine was also identified directly by amino acid analysis of the pro-α1 chain of chick bone collagen (63, 64) and the p-α1 and p-α2 chains of dermatosparactic collagen (56).

The amino acid composition of the pro-α1 chain of chick bone collagen (63, 64) indicated that the additional sequence unique to the precursor lacked the one third glycine and high amino acid content required to assume a triple helical conformation characteristic of collagen. Instead, the sequence was relatively rich in aspartic acid, glutamic acid, and serine, and contained a higher content of tyrosine and histidine than collagen. These data are consistent with the finding from several laboratories that the additional sequences in procollagen are selectively susceptible to proteolytic attack by pepsin (45, 49, 50, 52, 60, 65, 67) and chymotrypsin (68–70). Most, if not all, of the additional sequence in the pro-α1 chain of acid-extracted chick bone procollagen has been isolated after cyanogen bromide (CNBr) cleavage as a cystine- and tryptophan-containing fragment with a mol wt of approximately 20,000 (64). The composition of this fragment (pro-α1–CB1) was in excellent agreement with the difference in compositions of pro-α1 and α1. A portion of the sequence represented by pro-α1–CB1 is susceptible to digestion by purified bacterial collagenase, since the cystine-containing peptide isolated after collagenase digestion of pro-α1 is roughly three quarters the size of pro-α1–CB1 (71).

Furthmayr et al (72) have isolated a collagenase-resistant peptide from the p-α1 chain of dermatosparactic bovine collagen. This fragment also had a mol wt of approximately 20,000, but its amino acid composition bears little resemblance to either the CNBr- or collagenase-produced fragments isolated from chick bone procollagen (64, 71). No good explanation, other than species-specific differences, is available to account for the marked compositional differences.

Recently, Sherr et al (73) succeeded in isolating a triple-chain disulfide-bonded fragment containing tryptophan by collagenase digestion of culture medium protein from diploid human fibroblasts. Procollagen was labeled with [³H] tryptophan and the amino acid was demonstrated in both pro-α1 and pro-α2 chains after reduction. The amino acid composition of the tryptophan-containing peptide differed from that reported for the chick and bovine pro-α1 sequences, but inclusion of a segment derived from pro-α2, partial cleavage of the nonhelical region of procollagen by collagenase, and species differences may be responsible. Since the identity of this peptide as a derivative of procollagen depends very critically on the specificity of the antiserum used for immunoprecipitation, confirmation of these results will be necessary.

If the additional sequences in procollagen are to function in part to facilitate triple helix formation, their location at the NH_2-terminal ends of α chains would be expected (44, 50). This location would be consistent with the presence in all collagens thus far examined of short (12–20 amino acids) sequences, differing in primary structure from triple helical regions, at the NH_2-terminal ends of both $\alpha 1$ and $\alpha 2$ chains. However, the recent demonstration that nontriple helical regions also exist at the COOH terminus of the collagen molecule (28, 74) raises the possibility that in procollagen such COOH-terminal sequences are also extended.

Conclusive evidence for an NH_2-terminal location of at least a major part of the additional sequences in procollagen was obtained by examination of electron micrographs of SLS aggregates of procollagen. In SLS crystallites, collagen molecules are aligned laterally with NH_2- and COOH-terminal ends in register (75). SLS crystallites of dermatosparactic collagen (56, 69), procollagen synthesized by embryonic chick tendon fibroblasts (49, 76), and procollagen extracted from normal bovine skin (55) revealed an additional poorly striated region at the NH_2-terminal end of the molecule. No extension was noted at the COOH-terminal end in any of these preparations and no differences were otherwise noted in cross-striation patterns of the proteins. The fortuitous association of noncovalently linked protein with dermatosparactic collagen can be excluded, since SLS aggregates with the NH_2-terminal extension could be formed by renaturation of dissociated p-$\alpha 1$ and p-$\alpha 2$ chains (69).

SLS aggregates of dermatosparactic collagen were found to be 200–250 Å longer than segments of normal collagen; in tendon the additional region measured 130 Å. Since the additional sequences in procollagen cannot assume the triple helical conformation (presumably a more globular structure exists), there is no reason to expect a direct correlation between the degree of lateral extension and the length of these sequences. Furthermore, it seems likely that minor differences in technique may affect the reproducibility with which the less rigid NH_2-terminal ends of the crystallites can be visualized by electron microscopy.

As expected, electron microscopic examination of procollagen following limited cleavage with chymotrypsin (69) or pepsin (49) demonstrated selective removal of NH_2-terminal nontriple helical regions. Treatment of tendon procollagen with a purified tadpole collagenase produced the expected three-quarter (TC^A) and one-quarter (TC^B) fragments but did not cleave the extension at the NH_2-terminal (A) end (49).

Evidence for Disulfide Bonds

The observation that procollagen obtained both from cell culture medium and from bone could be converted to a collagen-like molecule by limited cleavage with pepsin and chymotrypsin indicated that the additional sequences in the precursor differed in conformation, and therefore in composition, from the triple helical region of procollagen. Procollagen extracted from cranial bone by acetic acid could be dissociated into its constituent pro-α chains in the absence of reduction of disulfide bonds (50, 77, 78). However, procollagen from cell culture medium was shown to contain interchain bonds, since reduction with mercaptoethanol liberated pro-α chains from higher molecular weight material (49, 60, 61, 67). In the absence of reduction, evidence for both pro-α chain dimers (61, 79) and trimers (52, 61) was obtained.

The procollagen secreted into the medium of cultured fibroblasts exists as a triple-stranded molecule with disulfide bonds linking all three chains (49, 61). Goldberg et al (61) proposed that stepwise proteolytic cleavage may yield a molecule containing a dimer of pro-α1 and a noncovalently associated pro-α2 chain; further steps then lead to excision of all disulfide-bonded sequences and eventually to a molecule with three α chains. This scheme is consistent with the observation of Smith et al (79) that pro-α dimers, present in a native human fibroblast procollagen fraction, contain predominantly pro-α1 chains.

The apparent discrepancy between the existence of interchain disulfide bonds in procollagen isolated from the medium of cultured cells and their absence in acetic acid-extracted procollagen from bone has been resolved by the recent work of Monson & Bornstein (65). It was shown that chick cranial bone procollagen, extracted at neutral pH in the presence of inhibitors of proteolytic enzymes, contained disulfide bonds linking all three chains. In addition, the molecular weights of the pro-α chains obtained by reduction and alkylation of the disulfide-bonded precursor were higher than those previously determined for pro-α1 and pro-α2 from acid-extracted procollagen, indicating that extraction of tissue at low pH resulted in a selective loss of disulfide-bonded regions from procollagen. Whether acid-extracted bone procollagen represents a physiologic intermediate in the conversion to collagen or results from inadvertent exposure of the precursor to tissue enzymes during homogenization remains to be determined.

Fessler et al (66) have also obtained a disulfide-bonded collagen precursor from chick embryo cranial bone by very rapid acetic acid extraction. Under these conditions, disulfide interchange and the possibility of artifactual interchain links are avoided. The authors suggest, on the basis of preliminary evidence, that interchain disulfide bonds are formed only after the entire procollagen molecule is synthesized and that such bonds may function to assure alignment and stabilize the structure of the precursor.

A disulfide-bonded collagen precursor was also isolated from the cell layer of confluent L-929 fibroblasts by Kerwar et al (80). Heavy polysomes obtained from these fibroblasts synthesized pro-α1 and pro-α2 chains as well as material eluting from CM-cellulose in the position of α2. In light of the work of others (60, 65), the latter fraction is likely to represent the disulfide-linked collagen precursor.

Molecular Weight

Molecular weights of procollagen, pro-α chains, and p-α chains of dermatosparactic collagen have been estimated by SDS acrylamide gel electrophoresis (50, 51, 56, 61, 81, 82), molecular sieve chromatography (48, 49, 63, 79, 81, 83), sedimentation velocity (66), and sedimentation equilibrium (56) ultracentrifugation. In addition, molecular weights of peptides derived from the additional sequences in procollagen by CNBr (64) and collagenase cleavage (71–73) were estimated by several methods.

All studies using molecular sieve chromatography or electrophoresis have been limited by a lack of appropriate standards. It is well established that in a buffer such as 1 M CaCl$_2$ the effective hydrodynamic volume of collagen chains is substantially greater than that of globular proteins of equivalent molecular weight (84). Collagen chains and fragments also migrate anomalously, relative to other proteins, during SDS acrylamide gel electrophoresis (64, 85). The slower than expected migration observed may result in part from the exceptionally high imino acid content of collagen, which may interfere with the assumption of a rod-like conformation postulated for proteins in the presence of SDS (86), and therefore reduce the amount of detergent bound. Such an interpretation is consistent with the observation (85) that collagen $\alpha2$ chains, which have a lower imino acid content than $\alpha1$ chains, migrate more rapidly than $\alpha1$ chains despite very similar molecular weights.

Very consistently, pro-α chains have been observed during electrophoresis to migrate more slowly than α chains (mol wt 95,000), but more rapidly than β components (mol wt 190,000). Published molecular weight estimates, based on collagen standards, vary from 115,000 for the pro-$\alpha1$ chain of acid-extracted chick bone procollagen (63) to 130,000 for an underhydroxylated pro-$\alpha1$ chain extracted from 3T6 fibroblasts (82). Similarly, pro-α chain dimers and trimers differ in migration from collagen β and γ marker components, both by electrophoresis, chromatography (52, 61, 65, 79), and band sedimentation analysis (66), but molecular weight estimates based on such standards must be interpreted cautiously. Pro-α chains obtained by reduction and alkylation of a disulfide-linked precursor extracted at neutral pH are clearly larger than acid-extracted pro-$\alpha1$ and pro-$\alpha2$ (65). In this regard, the low mol wts of 99,000 to 105,000 reported for p-α chains of dermatosparactic collagen (56) are not consistent with a subsequent report of the isolation of a collagenase fragment of mol wt 19,500 from the additional sequence in p-$\alpha1$ (72). This matter requires additional study. A mol wt of 75,000–80,000 was reported by Sherr et al (73) for the three-stranded peptide obtained by collagenase digestion of human fibroblast procollagen, with a value of 25,000 obtained after reduction and alkylation.

If a mol wt of approximately 130,000 is assumed for a pro-α chain, the mol wt of procollagen would be in the vicinity of 400,000. Church et al (81), however, reported evidence for a collagen precursor, isolated from 3T6 cells, with a mol wt of 500,000–600,000 (based on collagen standards). Surprisingly, this fraction was observed under conditions which would have been expected to reduce disulfide bonds. The authors suggested that this high molecular weight component may represent a single-chain collagen precursor. A number of other considerations, discussed below, make a single-chain precursor unlikely. A still higher molecular weight precursor linked by disulfide bonds to a noncollagenous moiety remains a possibility and may account for the material with a lower hydroxyproline content separated from procollagen by DEAE chromatography (79).

A number of studies have sought to determine the size of collagen synthesizing polyribosomes (87–91). If a relatively constant spacing between ribosomes on mRNA is assumed for proteins, the size of a polysome may be taken as a reflection of the length of the polypeptide produced (92). In addition to the problems commonly associated with such experiments, it is now recognized that interpretations based on observed sedimentation constants of collagen synthesizing polysomes are complicated by interactions occurring between nascent chains and possibly between polysomes and completed collagen polypeptides (89, 93, 94). Lazarides & Lukens (91) prepared polysomes from chick embryos and incubated these in an in vitro system capable of completing and releasing polypeptides. Polysomes synthesizing hydroxyproline sedimented with an average s value of 330. These data were consistent with the existence of a monocistronic collagen message coding for an α chain of mol wt 95,000. This approximation can be expanded, in view of the subsequent evidence for procollagen, to encompass the synthesis of a pro-α chain of a somewhat higher molecular weight. The data do not favor, however, the synthesis of a polycistronic mRNA required for a single-chain procollagen precursor. The kinetic synthetic data of Vuust & Piez (77, 95) are also difficult to reconcile with a single-chain model for procollagen.

Antigenicity

Dermatosparactic bovine collagen (59), procollagen in the medium of cultured human fibroblasts (96), and the pro-α1 chain of acid-extracted chick bone procollagen (71) have been used to produce antibodies in rabbits. In the case of human procollagen and the pro-α1 chain, antibodies were limited to determinants located on the additional sequences unique to the precursor chains, whereas with dermatosparactic collagen a low titer of antibodies to normal bovine collagen was also observed. The superior immunogenicity of the additional sequences in procollagen may reflect the chemical nature of this region or the greater interspecies differences in the structure of the precursor in comparison with the high degree of homology observed for the triple helical region of collagen (71).

Both p-α1 and p-α2 chains were equally effective in inhibiting hemagglutination by antibodies against dermatosparactic collagen (59), and essentially all the serological activity of the p-α1 chain was retained in a collagenase fragment of mol wt 19,500 prepared from p-α1. Similarly the collagenase fragment obtained from the sequence unique to the pro-α1 chain yielded a reaction of identity with pro-α1 when tested in double diffusion against an antiserum to pro-α1 (71).

Antibodies to dermatosparactic collagen (appropriately adsorbed to remove antibodies to normal collagen) react with collagen fibers in bovine skin and in the interstitial connective tissue of kidney and muscle, as indicated by indirect immunofluorescence (59). These findings provide support for the possibility that derivatives of procollagen persist extracellularly for relatively long periods of time and may play a role in fibrogenesis (53, 55). However, reaction of such antibodies with cleaved peptide extensions, rather than with the intact procollagen molecule, has not been excluded.

Synthesis of Protocollagen

In the presence of chelators of iron, or in the absence of oxygen or ascorbic acid, cells and tissues cultured in vitro fail to hydroxylate peptidyl proline and lysine normally (see below). Synthesis of an underhydroxylated collagen precursor, protocollagen,[4] continued at a normal rate for 1–2 hr when tendon fibroblasts were cultured in the presence of 0.3 mM α,α'-dipyridyl (76). Since secretion was markedly inhibited, the resulting protocollagen accumulated intracellularly. The mol wt of 125,000, determined by SDS-agarose chromatography for chains derived from protocollagen, was similar to values for pro-α chains obtained from procollagen secreted into the medium. SLS aggregates of protocollagen revealed NH_2-terminal extensions indistinguishable from those obtained with procollagen (76). Protocollagen and procollagen also chromatographed in a similar fashion from DEAE-cellulose (97), and the isolated chains demonstrated similar properties as indicated by elution from CM-cellulose (98) and agarose (76, 83). It is therefore likely that the unusual solubility and chromatographic properties of protocollagen, attributed to a lack of hydroxyproline (99, 100), reflect in part the fact that protocollagen as usually isolated represents an underhydroxylated collagen precursor rather than an underhydroxylated collagen. Earlier studies failed to detect the higher molecular weight of protocollagen chains (relative to collagen α chains) because sufficiently discriminating techniques were not used and perhaps because variable partial proteolysis of the protein occurred during extraction.

Tissue-Specific Procollagens

It is likely that the several genetically distinct collagens are synthesized as higher molecular weight biosynthetic precursors. Studies of the product synthesized and secreted by embryonic chick lens cells in culture suggest that it represents a higher molecular weight form of basement membrane collagen (101, 102). The presence of some 13% of the hydroxyproline as the 3-isomer and the high content of glucosyl-galactosyl hydroxylysine support the relation of these chains to basement membrane collagen. After reduction of disulfide bonds, chromatography on SDS-agarose revealed chains possessing a mol wt of approximately 140,000. However, if some 70 additional hexose units per polypeptide chain in basement membrane collagen are taken into account (103), chains synthesized by lens cells may be equivalent in length to the pro-α chains synthesized by tendon fibroblasts in culture. Some evidence was obtained for a time-dependent conversion of the secreted product to a lower molecular weight form of basement membrane collagen during in vitro incubation, although the separation achieved between the precursor and product by SDS-agarose chromatography was poor.

In analogy with their work on the procollagen synthesized by chick tendon fibroblasts, Dehm & Prockop (104) have demonstrated that cells isolated by enzymatic dissociation of embryonic chick sterna synthesize and secrete a cartilage procollagen. After reduction, cartilage procollagen consisted of a single type of

---

[4] The term protocollagen was initially applied to an underhydroxylated form of collagen. However, it is now apparent that in most instances lack of hydroxylation also results in impairment in subsequent conversion of the precursor. The protocollagen formed is therefore an underhydroxylated procollagen and this meaning can be implied by the term. Cumbersome designations such as "proprotocollagen" can therefore be avoided.

pro-α chain as shown by SDS acrylamide gel electrophoresis; these chains contained both cysteine and tryptophan. As expected, in comparison with tendon procollagen a higher proportion of lysyl residues were hydroxylated in cartilage procollagen and a higher proportion of these were glycosylated.

Recently Church et al (105) have provided evidence for synthesis of two different procollagens by cloned fibroblasts derived from the skin of a dermatosparactic calf. The procollagens were fractionated by DEAE-cellulose chromatography and were shown to have different chain compositions by acrylamide gel electrophoresis of the pepsin-treated proteins. While these results are preliminary, the synthesis by dermal fibroblasts of a procollagen with the structure (pro-α1)$_3$ might be expected in view of the evidence for a collagen with three identical chains in human skin (36).

CONVERSION OF PROCOLLAGEN TO COLLAGEN

The likelihood of involvement of a proteolytic mechanism in the conversion of a collagen precursor to collagen was anticipated by Schmitt (42, 43) and Speakman (44). Lapière et al (57) reasoned that the heritable connective tissue disorder of cattle, dermatosparaxis (56), represented a defect in conversion of procollagen, and they identified a proteolytic activity capable of converting dermatosparactic collagen to collagen in bovine tissues. Bornstein et al (68) have identified a similar activity in extracts of embryonic chick cranial bone which converted acid-extracted procollagen to collagen. The activity in bone was demonstrated in vitro despite inhibition of new protein synthesis by cycloheximide, suggesting the presence of a preformed pool of enzyme in this tissue.

Procollagen Peptidase

No published reports exist describing purification of the enzymatic activity. In studies with bovine (57) and chick bone (68) enzymes, tissues were homogenized in neutral salt solutions and the supernatant obtained after high speed centrifugation was used as the source of activity. The bovine enzyme was identified in skin, aorta, tendon, lung, and cartilage by its ability to convert dermatosparactic collagen to collagen. The assay employed was a densitometric tracing of stained acrylamide gel electrophoretograms, in which the proportions of the p-α1 and p-α2 chains of dermatosparactic collagen and the α1 and α2 chains of collagen were measured. The bovine enzyme was maximally active at pH 7.0 and was inhibited by 0.15 M EDTA and 0.1 M 2-mercaptoethanol, but not by p-CMB. Specific enzymatic activity was observed with both isolated p-α1 and p-α2 chains as well as with reconstituted fibrils formed from dermatosparactic collagen.

The chick bone enzyme is also a neutral protease which was irreversibly inhibited by pH below 4.5 and by 10 mM EDTA, but not by soybean trypsin inhibitor, NEM, p-CMB, cysteine, or diisopropyl fluorophosphate (68, 107). A preliminary size fractionation by ultrafiltration indicated a mol wt in excess of 100,000. The activity of procollagen peptidase is substantial in 1 M NaCl at 4° and probably accounts for the inability to detect procollagen in neutral salt extracts of tissues (68, 77) unless specific measures are taken to inhibit the enzyme.

Evidence for the involvement of procollagen peptidase in the conversion of procollagen to collagen has rested in part on the specificity of the enzyme for the test substrates dermatosparactic and acid-extracted procollagens. However, it is now apparent that acid-extracted procollagen represents a derivative of procollagen rather than the native protein (65). Similarly the chromatographic heterogeneity of dermatosparactic collagen chains (56), and the lower molecular weights of these chains in comparison with pro-α chains from other sources, suggest that partial proteolysis in vivo results in a truncated precursor which accumulates and is extracted from dermatosparactic tissues. It will be necessary to demonstrate that procollagen peptidase is active against native procollagen and that cleavage of X–Gln or X–Glu bonds occurs (with cyclization yielding the NH_2-terminal pyrrolidone carboxylic acid residues in the α1 and α2 chains of collagen).

The possibility exists that other enzymatic cleavages precede the action of procollagen peptidase. In pulse-chase experiments, Monson & Bornstein (65) were unable to detect chains of the type obtained from acid-extracted procollagen in substantial quantities. However, such molecules may be intermediates which were rapidly converted to collagen. Stepwise enzymatic conversion of procollagen (52, 54) could account for intermediates such as those observed by Goldberg et al (61), as well as for the presence of pro-α chain dimers (61, 79). The difficulty in these and other in vitro systems is to distinguish between a physiological scheme for the generation of collagen from procollagen and the adventitious action of ancillary enzymes.

Multiple steps in the "activation" of procollagen clearly augment the potential for cellular control of fibrogenesis, particularly if limited proteolysis occurred partially intracellularly and if inactive precursors of enzymes involved in such proteolysis also existed. The identification of the lower molecular weight products derived from procollagen would assist in defining the steps involved. Pontz et al (78) have isolated two peptide fragments of molecular weight 18,000 and 12,000 from the culture medium of chick cranial bones. They attribute the origin of these fragments to the additional sequences in the pro-α1 and pro-α2 chains of procollagen. However, if the precursor chains have a molecular weight higher than 115,000, additional peptide material will have to be accounted for.

Site of Conversion

The presence of collagen precursors in the medium of cultured fibroblasts indicates conclusively that conversion is not a requirement for secretion and strongly suggests that the processing of procollagen occurs extracellularly. Dermatosparactic tissues also secrete and utilize procollagen, or derivatives thereof, in the formation of extracellular collagen fibers (56). Culture medium procollagen, after conversion to collagen, was shown to be capable of incorporation into fibers of human fibroblast cell layers (52). A similar sequence of events was suggested for embryonic chick tendons cultured in vitro (48), but a direct relationship between the medium procollagen and tendon matrix collagen was not established.

Layman & Ross (108) have identified a procollagen peptidase activity in the medium of cultured confluent adult human fibroblasts. Little or no activity was evident in the medium during the logarithmic growth phase of the cells and no appreciable procollagen peptidase activity was found in cells at any stage of growth. Kerwar et al (80) also identified procollagen peptidase activity in the medium of 3T3 and 3T6 cells but not in the confluent cell layers. Medium from L-929 cells contained a much lower level of enzymatic activity, comparable to that found in cultures of fibroblasts from dermatosparactic tissues; again the enzyme was not detectable in the cell layer. The reason for failure to demonstrate the enzyme in cell homogenates is unclear; an inactive form may exist intracellularly.

The above observations support an extracellular site for procollagen conversion. Nevertheless, experiments to date cannot exclude some degree of intracellular modification of procollagen prior to secretion in vivo. If direct cellular participation in the conversion of procollagen to collagen occurs, the extent of this participation may vary from one tissue to another. Thus, it may be advantageous for the corneal epithelium, which elaborates a highly ordered matrix, to secrete a form of collagen that accretes rapidly onto existing fibrils. In contrast to skin, where a less ordered fiber pattern exists, procollagen may be secreted and converted extracellularly. In the extreme case of basement membranes, procollagen may escape complete conversion, and derivatives of the precursor may function as structural elements extracellularly.

HYDROXYLATION OF PEPTIDYL PROLINE

The nature and significance of the biological processes involved in the synthesis of hydroxyproline and hydroxylysine in collagen have fascinated investigators with widely different interests. Despite the fact that conflicting data, premature interpretations, and inoperative conclusions have resulted in considerable confusion in the field, important insights have been obtained.

Earlier literature established beyond serious doubt that 1. proline and lysine rather than hydroxyproline and hydroxylysine serve as precursors of the peptidyl amino acids; 2. hydroxylation occurs on polypeptide substrates rather than on intermediates such as amino acyl tRNA; and 3. molecular oxygen and not water serves as the source of the oxygen of the hydroxyl groups. Evidence for these points has been summarized in several reviews (106, 109–111) and is not discussed here. Other experiments indicated that the enzymatic processes leading to the synthesis of peptidylhydroxyproline and hydroxylysine had many features in common and led to the suggestion that a single hydroxylase was involved (112, 113). It is now known that prolylhydroxylase[5] and lysylhydroxylase are different enzymes (114–116). The purification of lysylhydroxylase and considerations specific to that enzymatic process are therefore discussed in the following section.

Subcellular Localization

Early studies of collagen biosynthesis indicated that microsomal fractions contained a high specific activity of protein-bound hydroxyproline (117, 118); incorporation

[5] These enzymes have also been referred to as protocollagen and collagen hydroxylases. The terms prolyl- and lysylhydroxylases will be used in this review since they imply a less restricted substrate specificity. It is possible, for example, that prolylhydroxylase is involved in the synthesis of both collagen and elastin.

of proline into hydroxyproline was also demonstrated by chick embryo microsomes in a cell-free system (119). Subsequently, a number of investigators were able to isolate polyribosomal complexes containing hydroxyproline (120, 121); peptide-bound hydroxyproline was released by puromycin from such complexes and therefore presumably existed in nascent chains (120, 121a).

Assays of microsomal fractions from chick embryo homogenate also revealed prolylhydroxylase activity (122, 123). It was therefore tempting to localize the enzymatic activity to the microsomal polysome-membrane complex. Several experimental observations, however, were interpreted in favor of a cytoplasmic site for the hydroxylase (111): 1. The enzyme as usually isolated appeared largely in the soluble fraction of tissues (113, 124, 125), although a substantial fraction was particulate or sedimentable (122, 123, 125); 2. when hydroxylation was inhibited by chelation of iron or by anaerobiosis in vitro, subsequent release of inhibition led to hydroxylation of substantial quantities of protocollagen which accumulated during the block (126); and 3. pulse-labeling of embryonic chick tibiae suggested a significant lag between completion of protocollagen chains and hydroxylation (127).

Adequate discussion of these questions requires a consideration of the nature of the normal substrate for prolylhydroxylase (see below). However, Diegelmann et al (128) have recently reported data supporting localization of the hydroxylase to the membranes of the RER. Ninety percent of the collagen-synthetic activity of embryonic chick bones was found in fractions sedimenting between 700 and 35,000 g. These fractions contained membrane-bound polysomes which were dissociated by detergents, as monitored by both electron microscopy and synthetic activity. Membrane-bound polysomes were capable of synthesizing peptidyl hydroxyproline in the absence of exogenous hydroxylase if cofactors were added to the system. Following detergent treatment, however, no hydroxylation was observed. Diegelmann et al (128) therefore postulated that prolylhydroxylase was attached to membranes of the polysome complex.

Preliminary morphological observations also favor a microsomal localization of the enzyme. Berg et al (129) have reported the development of antisera to highly purified prolylhydroxylase. γ-Globulin fractions of such antisera were labeled with peroxidase and used to demonstrate the intracellular localization of prolylhydroxylase. In an extension of these studies, Olsen et al (130) have used ferritin-labeled antibodies to demonstrate, at the ultrastructural level, that the enzyme is localized almost exclusively within the cisternae of the RER. These experiments employed a new and potentially useful technique whereby cells were broken after light formaldehyde fixation. Such preparations retained their ability to react specifically with antibody and could subsequently be embedded in plastic, thus retaining normal subcellular morphology to a high degree.

Properties of Prolylhydroxylase

Partially purified preparations of the enzyme were obtained from homogenates of chick embryos (112, 113, 131, 132) and newborn rat skin (133, 134) by salt precipita-tion, fractionation on calcium phosphate gels, and ion exchange and molecular sieve chromatography. By an extension of these methods, both the chick embryo (135) and rat skin (136) enzymes were purified almost to homogeneity by the criteria of analytical ultracentrifugation and acrylamide gel electrophoresis. The enzymes had a specific activity of 5–6 μmol of hydroxyproline/mg for each hour tested with either RCM *Ascaris* cuticle collagen (136) or a synthetic (Pro-Gly-Pro)$_n$ substrate (135).

Berg & Prockop (137) have purified prolylhydroxylase to homogeneity by the criterion of acrylamide gel electrophoresis using an affinity column containing RCM *Ascaris* cuticle collagen-agarose. The enzyme was eluted with (Pro-Gly-Pro)$_n$ (average mol wt 2400) and recovered in 58% yield with a 1600-fold purification from an ammonium sulfate precipitate of a crude chick embryo homogenate. This exceptionally effective purification can be attributed to the high affinity of the enzyme for large polypeptide substrates (138) and the specificity of elution achieved by (Pro-Gly-Pro)$_n$. The specific activity of the enzyme with a RCM *Ascaris* collagen substrate (32.4 μmol of hydroxyproline/mg per hour) was severalfold higher than that previously reported. The purified enzyme was devoid of lysylhydroxylase or peptidyl 3-proline hydroxylase activity.

Sedimentation equilibrium ultracentrifugation of prolylhydroxylase indicated a molecular weight of 230,000 (137). After dissociation with 0.1 M mercaptoethanol in 6 M urea, SDS-acrylamide gel electrophoresis revealed two bands of equal intensity with molecular weight of 64,000 and 60,000. The enzyme may therefore be composed of a tetramer of two dissimilar subunits. Pänkäläinen et al (139) also determined a molecular weight of 248,000 for prolylhydroxylase by sedimentation equilibrium of a somewhat less pure preparation of the chick embryo enzyme. In contrast, the molecular weight of the rat skin enzyme was judged to be in the vicinity of 130,000 with a reduction to 65,000 after mercaptoethanol treatment (136). The consistently higher molecular weight of 250,000 to 350,000, determined for prolylhydroxylase by molecular sieve chromatography (135, 140), may reflect the high frictional ratio of the native enzyme (139).

The description of the electron microscopic appearance of the enzyme performed with incompletely purified preparations (141) is now known to be incorrect (137, 142). The structure described was that of a minor contaminant particularly stable to the procedures required for microscopy. The enzyme purified by affinity chromatography was visualized by electron microscopy as a tetramer composed of monomers roughly 3.3 nm in diameter and 7nm long (142). On the basis of their data, Olsen et al (142) propose a model consisting of two V-shaped interlocking dimers.

Extensive studies have established that molecular O_2, Fe^{2+}, α-ketoglutarate, and ascorbic acid are employed as cofactors or cosubstrates by prolylhydroxylase (8, 13, 106, 110, 140, 143). Only the more recent developments in this area are discussed in this review. A role for heavy metals in prolylhydroxylation was suggested by the inhibitory effects of several chelating agents in chick embryo skin slices (144); addition of Fe^{2+} or Fe^{3+} but not other metals augmented synthesis of protein-bound hydroxyproline by embryonic chick tibiae (124). The requirement for divalent iron in the enzymatic process was subsequently firmly established (131, 143, 145, 146).

In 1966 Hutton et al (146, 147) determined that a heat-stable dialyzable factor, in addition to ascorbate and Fe^{2+}, was required for enzymatic activity of prolyl-hydroxylase. Only α-ketoglutarate (apparent K_m 3×10^{-6} M) was capable of replacing the activity of this factor. The metabolic fate of α-ketoglutarate was initially difficult to determine because α-ketoglutarate was consumed by other enzymes present in the crude preparations of prolylhydroxylase (147), and assays lacked sufficient sensitivity (148). The availability of purer enzymatic preparations and more sensitive assays for α-ketoglutarate, however, permitted the demonstration of a stoichiometric coupling of oxidative decarboxylation of α-ketoglutarate (to succinate) with hydroxylation of peptidyl proline (134). In addition to its implications for the enzymatic mechanism, the utilization of $[1-^{14}C]\alpha$-ketoglutarate permitted the development of an extremely sensitive assay for prolylhydroxylase, based on the evolution of $^{14}CO_2$, which can be used with nonradioactive substrates such as RCM *Ascaris* cuticle collagen and synthetic polytripeptides (134).

The role of ascorbic acid in peptidyl hydroxylation has been difficult to establish. A relation to collagen metabolism, and more specifically to the synthesis of collagen with a normal hydroxyproline content, was strongly suggested by studies in organ culture and of scorbutic animals (13, 149, 150). Numerous reports have also indicated that the addition of ascorbic acid to fibroblasts and related cells in culture led to an increase in the synthesis of protein-bound hydroxyproline (151–158). In the absence of ascorbic acid supplementation, cells continued to synthesize and secrete protocollagen (152, 153), but at a reduced rate (159). However, cell-free studies using partially purified prolylhydroxylase and substrates such as protocollagen, RCM *Ascaris* collagen, or synthetic polytripeptides have called into question the specificity of the ascorbic acid requirement in peptidyl hydroxylation. A number of reducing agents including tetrahydropteridine, tetrahydrofolate, and dithiothreitol were also found to be capable of stimulating hydroxylation (122, 136, 147, 155).

Recently Stassen et al (160) have proposed that ascorbic acid may play a role in the activation of a precursor of prolylhydroxylase, in addition to its possible function in the hydroxylation mechanism (see below). Gribble et al (161) had originally demonstrated that during rapid growth of L-929 fibroblasts, underhydroxylated collagen was secreted into the medium and maximal prolylhydroxylase activity did not appear until the cells approached confluency. Cell concentration by centrifugation resulted in an activation of hydroxylase activity by a process which did not require new protein synthesis (161, 162). A two- to fivefold stimulation of hydroxylase activity was also observed by addition of lactate (40 mM) to early log phase cells (163). This effect was again independent of new protein synthesis but required intact cells for its expression. It is of interest that Green & Goldberg (164) had also noted an increase in peptidyl hydroxyproline synthesis by 3T6 cells under the influence of lactate.

McGee et al (165) prepared an antibody specific to rat skin prolylhydroxylase which cross-reacted with the mouse enzyme. By using this antibody they were able to show that the marked increase in prolylhydroxylase activity associated with the progression of L-929 cells from a logarithmic to a stationary growth phase was not accompanied by an increase in enzyme protein. Similarly the activation of prolylhydroxylase performed by cell concentration or lactate administration occurred in the absence of an increase in immunologically reactive protein. These studies therefore provided direct evidence for the existence of an inactive enzyme precursor.

Although it was tempting to postulate the presence of a higher molecular weight proenzyme of prolylhydroxylase (165), the studies of Stassen et al (160) demonstrated that activation of enzyme precursor (either by ascorbate or by lactate) was associated with an increase in the molecular weight of the enzyme protein. Maximal activation of early log phase L-929 cells was achieved in 2 hr with 2.5×10^{-4} M ascorbate, whereas 6 hr in 8×10^{-2} M lactate was required to obtain a similar effect. Dithiothreitol (10 mM) produced a dissociation of active enzyme to inactive precursors. It is possible that the precursor found in log phase cells or produced by dithiothreitol is a more native form of the unfolded enzyme subunits produced by reduction in urea or SDS (136, 137).

The role of sulfhydryl groups in the activity of prolylhydroxylase was examined by Popenoe et al (132). The partially purified enzyme from chick embryos was inhibited by p-CMB and NEM but not by iodoacetamide; considerable protection against inactivation by NEM was afforded by α-ketoglutarate (0.5 mM). Dithiothreitol in concentrations of 0.05 to 0.1 mM stimulated enzymatic activity but higher concentrations produced an inhibition. The authors interpreted these findings as a requirement for a free sulfhydryl group at or near the active site of the enzyme and for disulfide linkages in the intact protein.

Prolylhydroxylase utilizes molecular O_2 in a direct displacement reaction at the C-4 position of a prolyl residue (166). The demonstration that a stoichiometric decarboxylation of α-ketoglutarate accompanied hydroxylation (134) suggested that the enzyme represented a new and unusual class of mixed function oxygenases. γ-Butyrobetaine hydroxylase from rat liver, thymine 7-hydroxylase from *Neurospora*, and several other *Neurospora* oxygenases also require α-ketoglutarate (167). Cardinale et al (168), in analogy with the work of Linblad et al (169), were able to show that one atom of $^{18}O_2$ was incorporated into hydroxyproline, whereas the second was recovered in succinate, derived from α-ketoglutarate. A mechanism involving the formation of a peroxide intermediate binding the carbonyl group of α-ketoglutarate and C-4 of the prolyl residue was proposed (168).

The precise role of divalent iron and ascorbate in this mechanism is unclear. Hurych et al (170) have suggested that Fe^{2+} is oxidized to Fe^{3+} after interaction with molecular O_2. The resulting iron-O_2 complex would then interact with α-ketoglutarate. In this scheme ascorbate would be required to reduce Fe^{3+} and should be consumed stoichiometrically. However, iron may remain in the divalent state if the mechanism of hydroxylation resembles that of γ-butyrobetaine in rat liver (171). In the latter case ascorbate may function to maintain the ferrous ion and sulfhydryl groups in the enzyme in the reduced state. The inhibition of hydroxylation observed with nitroblue tetrazolium (172) does not necessarily distinguish between the two mechanisms.

Nature of the Substrate

In the absence of a triplet code for hydroxyproline, some means must exist to specify the position of prolyl residues in amino acid sequences which will subsequently become hydroxylated. Extensive structural studies indicated that in mammalian collagens hydroxylation of peptidyl proline, yielding 4-hydroxyproline, is limited to prolyl residues that are in position Y in the repeating sequence -Gly-X-Y- (18–30). However, the single residue of 3-hydroxyproline in the α1 chain of bovine skin collagen was located in position X in the above sequence (27). Since purified prolylhydroxylase contains no peptidyl 3-proline hydroxylase activity (137), it is likely that the enzyme synthesizing 3-hydroxyproline differs in its specificity for both the pyrrolidine ring and the position of prolyl residues in the polypeptide sequence. Earthworm cuticle, which contains a substantial quantity of 3-hydroxyproline and a high content of 4-hydroxyproline, contains the latter amino acid in both positions X and Y (173). Hydroxylase preparations from earthworm cuticle were also capable of hydroxylating prolyl residues in both positions in the synthetic polytripeptide $(Pro-Gly-Pro)_n$ (174).

The specificity of mammalian prolylhydroxylase has been established by hydroxylation of protocollagen (prepared by incubation of chick embryo minces with α,α'-dipyridyl) (175) and by hydroxylation of synthetic polytripeptides (112, 174, 176). In studies of the in vitro hydroxylation of rat tendon α1-CB2, a peptide which is normally incompletely hydroxylated, Rhoads et al (177) also demonstrated that additional hydroxylation was limited to prolyl residues in position Y of the collagen triplet.

Rhoads & Udenfriend (178) found that the peptide hormone bradykinin, H-Arg-Pro-Pro-Gly-Phe-Ser-Pro-Phe-Arg-OH, served as a substrate for the chick embryo hydroxylase, and they established that only the prolyl residue in position 3 was hydroxylated. These studies were extended by McGee et al (179), who examined a series of bradykinin analogs. These workers concluded that while the minimum sequence requirement for prolylhydroxylase was X-Pro-Gly, the nature of adjacent amino acids affected both the affinity of the substrate for the enzyme and the rate of the reaction.

Synthetic polypeptides have been used extensively to examine the minimum length required for hydroxylation (180, 181), the effect of varying substrate lengths and adjacent amino acids on the kinetics of the reaction (131, 176, 179, 181–183), and the conformational requirements for hydroxylation (131, 148, 176, 179–182, 184, 185). The relative effectiveness of synthetic peptides as inhibitors of hydroxylation has also provided an indication of the substrate requirements of prolylhydroxylase (182, 186, 187). Interpretation of some of these studies has been difficult because of length heterogeneity of earlier polymer preparations (with resulting lack of homogeneity in conformation) and because of a tendency for polymers to regain a helical conformation, aggregate, and become insoluble at higher concentrations. Consequently the literature is plagued with contradictory conclusions.

No hydroxylation was observed with the tripeptide Gly-Pro-Pro (112), but Pro-Pro-Gly was capable of serving as a substrate (180, 181). On the other hand Gly-Pro-Gly-Gly was neither a substrate nor an inhibitor of prolylhydroxylase (182). In general the effectiveness of substrates of the form $(Pro-Pro-Gly)_n$ increased as n increased from 5 to 20 (182–184). In the most recent study (183) it was shown for the series $(Pro-Pro-Gly)_n$ that although the maximal velocity remained constant, the K_m, with respect to substrate concentration, decreased by about two orders of magnitude as n increased from 5 to 20. Similar conclusions were reached for less homogeneous peptides of molecular weight range 1300–8000 (182). Results indicating a plateau (131) or a reduced effectiveness of higher molecular weight peptides (181, 185) can be attributed to the difficulties mentioned above. The affinity of prolylhydroxylase for all synthetic peptides tested is, however, substantially lower than that for RCM *Ascaris* collagen or protocollagen (183).

Polyproline II with peptide bonds in the *trans* position, but not polyproline I (*cis*), is a competitive inhibitor of prolylhydroxylase (186). Prockop & Kivirikko (187) demonstrated that the inhibitory constants (K_i) of polyproline II fractions decreased with increasing molecular weights of the polymers in the range 1600 to 21,000. The results were interpreted as indicating either an extremely large binding site on the enzyme for the inhibitor or lateral movement of the enzyme along the polypeptide. These results, as well as experiments with substrates of varying size, suggest that the affinity of prolylhydroxylase for protocollagen might be exceptionally high. A very low dissociation constant (2×10^{-11} M when expressed in terms of substrate concentration) was estimated by Juva & Prockop (138). Partial hydroxylation of incompleted chains produced in the presence of puromycin markedly reduced the affinity of the enzyme for the substrate.

Juva & Prockop (138) have proposed that the reduced affinity of prolylhydroxylase for partially hydroxylated collagen chains may account for the phenomenon of incomplete hydroxylation of individual prolyl residues, first noted in early sequence studies of collagen (19). No pattern with respect to either the position of incompleted hydroxylated prolyl residues in α chains or the chemical nature of adjacent residues has yet been discerned. Additional hydroxylation of prolyl residues in collagen can, however, be achieved in vitro using large amounts of enzyme (177).

Rhoads et al (177) observed that collagens from several sources did not serve as substrates for additional hydroxylation unless the collagens had previously been heat denatured. These findings suggested that the triple helical conformation prevented hydroxylation and that a random coil or unfolded state was necessary for prolyl residues in protocollagen to be accessible to the hydroxylase. This point has been one of considerable controversy. The higher yields of hydroxyproline obtained by hydroxylation of larger synthetic polytripeptides suggested that helicity might be a requirement for hydroxylation (184). However, several laboratories using protocollagen as a substrate concluded that hydroxylation could occur in both the native and random coil forms, since heat denaturation did not increase substrate activity and sometimes decreased it (124, 131, 147, 148, 174). The possibility that the protocollagen substrate was unfolded prior to heat denaturation was not considered at the time. In contrast, Kikuchi et al (181) observed that heating the polymer $(Pro-Pro-Gly)_{15}$, but not $(Pro-Pro-Gly)_5$, at concentrations of 1 mg/ml and below, increased hydroxylation and concluded that a triple helical structure inhibited hydroxylation. Similarly *Ascaris* cuticle collagen did not serve as a substrate for prolylhydroxylase unless it was heat denatured (188).

In a more recent examination of this question, Kivirikko et al (183) concluded that the triple helical conformation did not in itself prevent hydroxylation. This conclusion was based on studies of the hydroxylation of (Pro-Pro-Gly)$_{10}$ synthesized by the Merrifield technique and reported to be homogeneous by several criteria. Significant hydroxylation was observed at 15°, a temperature assumed to be substantially below the melting temperature of the polytripeptide. However the melting temperature of (Pro-Pro-Gly)$_{10}$ varies considerably with pH, ionic strength (190, 191), and possibly with the nature of the buffer. Consequently under the conditions used for hydroxylation, the synthesis of hydroxyproline observed by Kivirikko et al (183) may have been due to the presence of partially unfolded chains in the preparation.

Berg & Prockop (70) have re-examined this issue armed with the knowledge that the denaturation temperature of protocollagen is substantially lower than that of collagen and that the presence of hydroxyproline markedly increases the stability of the triple helix (see below). Protocollagen was prepared by incubation of embryonic chick tendon fibroblasts in the presence of α,α'-dipyridyl (76) and modified by limited proteolysis with pepsin or chymotrypsin. When the substrate was tested at 15 or 20°C (temperatures at which the chick embryo prolylhydroxylase was fully active), virtually no hydroxylation occurred. However, protocollagen served as a substrate for synthesis of hydroxyproline at higher temperatures and the susceptibility to hydroxylation correlated well with the denaturation temperature of 24°C determined for enzyme-modified protocollagen (192). Berg & Prockop therefore reached the important conclusion that the triple helical conformation of protocollagen prevents its hydroxylation. Similar experiments and conclusions were reported by Murphy & Rosenbloom (193).

A final question which has generated considerable controversy relates to whether peptidyl hydroxylation occurs on nascent (ribosome-bound) polypeptides or on completed protocollagen chains. The early evidence in favor of the latter view was summarized by Prockop (111). Much of this evidence was based on observations of in vitro systems where hydroxylation was blocked by addition of metal chelators or by anaerobiosis (127, 194). Under such conditions, substantial quantities of protocollagen accumulated intracellularly (more than would be expected if all chains remained ribosome-bound), and this protocollagen could be hydroxylated after removal of the enzymatic block (126). Studies indicating a temporal lag between incorporation of proline into protocollagen and synthesis of peptidyl hydroxyproline (127, 195–197) have also been interpreted in favor of hydroxylation occurring after polypeptide chain assembly (197). However, the most carefully determined lag time of 1.5 min is not incompatible with hydroxylation of nascent chains, in view of an estimated synthesis time of 6 min for a procollagen chain (77) and the need to synthesize a precursor sequence at the NH$_2$ terminus of the chain which contains proline but not hydroxyproline.

The best evidence currently indicates that a substantial fraction, and perhaps the major part of hydroxyproline synthesis in the absence of an inhibition of prolylhydroxylase occurs prior to the release of polypeptides from ribosomes. Experimental support for this position, which avoided many of the pitfalls complicating earlier studies, derives from the demonstration of a high proportion of hydroxyproline in nascent chains released by puromycin from guinea pig granuloma and 3T6 mouse fibroblast polysomes (120, 121). Since Lazarides & Lukens (198), using the same 3T6 polysome system, confirmed that hydroxylation was not required for release of collagen chains from ribosomes, and that completed and released protocollagen chains could be hydroxylated by intact cells, studies performed with inhibitors of peptidyl hydroxylation can be reconciled with the view that hydroxylation normally occurs parri passu with chain elongation.

Stabilization of the Collagen Helix by Hydroxyproline

It appears that the solution to one of the major puzzles in collagen chemistry, the function of peptidyl hydroxyproline, may have been found. Sakakibara et al (189) synthesized (Pro-Pro-Gly)$_n$ and (Pro-Hyp-Gly)$_n$ with n equal to 5 or 10 by fragment condensation on a Merrifield resin. They demonstrated that substitution of hydroxyproline for proline resulted in a highly significant increase in the melting temperatures of both polytripeptides. For the decatripeptide the difference, monitored by optical rotation, was 35°C. The results therefore provided direct evidence for a role for hydroxyproline in the stabilization of the triple helical conformation in synthetic peptides. Berg & Prockop (192) similarly concluded that hydroxylation of protocollagen increases the thermal stability of the protein. Chymotrypsin-modified protocollagen and procollagen were obtained from chick embryo tendon cells in culture. The melting temperature for the protocollagen was 24°C, 15° lower than that observed for the procollagen. The conversion of protocollagen to procollagen by replacing an anaerobic environment with O_2 resulted in a rapid acquisition of resistance to complete digestion with pepsin at 25°C (199).

These results were confirmed and extended by Rosenbloom and co-workers (200, 201), who prepared chick tendon protocollagen samples with varying contents of hydroxyproline. Thermal stability, monitored by susceptibility to complete digestion by pepsin, varied directly with hydroxyproline content. The denaturation temperature for the protocollagen samples lacking hydroxyproline entirely was found to be 24°C. It is worthwhile to note in this regard that the increased susceptibility of protocollagen to tissue proteases at 37°C was first observed by Hurych et al (202). Their findings can now be accounted for by the absence of the protection normally afforded by the triple helix at that temperature. The conclusion of Jimenez et al (76) that intracellular protocollagen, synthesized in the presence of α,α'-dipyridyl, exists in a triple helical conformation appears to be in error in light of the above work. Resistance to pepsin digestion was probably conferred by renaturation of protocollagen occurring during lengthy extraction in acetic acid at 4°C.

The above studies therefore imply that peptidyl hydroxylation is required in vivo to enable procollagen chains to assume a triple helical conformation at temperatures prevailing in the body. Since the triple helical conformation of collagen inhibits hydroxylation, the completeness of hydroxylation of peptidyl proline and lysine, or lack thereof, may reflect the rapidity with which folding occurs in different parts of the molecule. Inhibition of hydroxylation results in the presence of protocollagen chains in random coil form and it may be the failure to assume a triple helical conformation which retards secretion of protocollagen (see below).

Evidence for the participation of hydroxyprolyl residues in the stabilization of the collagen helix requires a re-evaluation of both the atomic structures proposed for collagen and the hydrogen bonding which may contribute to its stability. The stability imparted to the collagen structure by proline and hydroxyproline had been equated and attributed to the limitation of rotation of the dihedral angle ϕ, resulting from the incorporation of the N and Cα atoms in the pyrrolidine ring (2, 203, 204). Existing models for the collagen structure therefore do not make allowances for the participation of the hydroxyl group of hydroxyproline in hydrogen bonding within the triple helix. Ramachandran et al (205) have recently re-examined a modification of their original model which proposed that water molecules form interchain bridges in the triple helix (206). They conclude that it is possible to hydrogen bond the hydrogen of the hydroxyl group of hydroxyproline in one chain and the amide NH in a second chain to the oxygen of water. In this position, one of the hydrogens of water is capable of bonding to a carbonyl oxygen in the first chain and the second hydrogen may be capable of bonding with an amino acid side chain in the second chain if suitable groups exist. This matter requires further study and other equally satisfactory structures may exist.

HYDROXYLATION OF PEPTIDYL LYSINE

Separation of lysylhydroxylase from prolylhydroxylase was achieved primarily by chromatography on DEAE-cellulose (115, 116) or DEAE Sephadex (114). Miller (114) observed that DEAE Sephadex fractions that released tritium from a [4,5-^{3}H]lysine-labeled protocollagen substrate were inactive in tritium release of protocollagen labeled with [3,4-^{3}H]proline. Kivirikko & Prockop (116) used both lysine-labeled protocollagen and a synthetic peptide, Ala-Arg-Gly-Ile-Lys-Gly-Ile-Arg-Gly-Phe-Ser-Gly (LI), as substrates to monitor enzymatic activity.

Multiple molecular weight forms of lysylhydroxylase were observed on molecular sieve chromatography of partially purified preparations (116). It is possible that this heterogeneity resulted from increased disulfide bond interchange attributable to dithiothreitol (50 μM), which was added to stabilize activity during purification. The molecular weight of 550,000 and 200,000, estimated by Kivirikko & Prockop (116) for the two predominant forms of lysylhydroxylase, do not agree with the value of 350,000 determined by Popenoe & Aronson (115). The latter value is similar, however, to the value calculated for prolylhydroxylase under similar conditions (135). In low ionic strength buffers, aggregation and insolubilization of lysylhydroxylase occurs (116).

Lysylhydroxylase, purified 300- to 600-fold, requires Fe^{2+}, O$_2$, α-ketoglutarate, and a reducing agent such as ascorbate for activity. K_m values for Fe^{2+}, α-ketoglutarate, and ascorbate were in a range similar to those determined for prolylhydroxylase (116). The activity of the enzyme is increased by bovine serum albumin, catalase, and dithiothreitol (115, 116) in a manner similar to that observed for prolylhydroxylase (131–133, 136). Glycine (0.1 M) affords some protection against inactivation (115), an effect previously noted for prolylhydroxylase (207). Lysylhydroxylase is also inhibited by p-CMB and the inhibition can be partially reversed by dithiothreitol (116). It therefore seems likely that the enzyme, like prolylhydroxylase, requires free sulfhydryl groups for activity.

Kivirikko et al (208) studied the substrate specificity of lysylhydroxylase and the nature of the enzymatic reaction using synthetic peptides. The peptides LI, LII, and LIII were designed to resemble sequences surrounding a lysyl residue normally hydroxylated in the α1 chain of a number of collagens. LII and LIII resembled LI (see above) but contained additional (Pro-Pro-Gly)$_4$ sequences. All three peptides were hydroxylated with a similar V_{max}, but the K_m of the two longer peptides was half that of LI. Occurrence of hydroxylation was established by measurement of both the hydroxylysine synthesized and the $^{14}CO_2$ released from [1-^{14}C]α-ketoglutarate. The close agreement between the two measurements and identification of succinate as a product of the reaction indicate that the enzymatic mechanisms for lysyl and prolyl hydroxylation are similar.

RCM *Ascaris* collagen could not serve either as a substrate or as a competitive inhibitor of lysylhydroxylase (208). Free lysine and the tripeptide Lys-Gly-Pro were also inactive. Ile-Lys-Gly, (Ile-Lys-Gly)$_2$, and lysine[8]-vasopressin, which contains the sequence Pro-Lys-Gly-NH$_2$, were hydroxylated but kinetic measurements indicated that these compounds were poorer substrates than either of the longer synthetic peptides. The minimum substrate requirement for lysylhydroxylase is therefore the sequence X-Lys-Gly, analogous to the sequence X-Pro-Gly required by prolylhydroxylase.

As in the case of prolyl residues, hydroxylation of lysyl residues in collagen is normally incomplete (209). When a number of collagen substrates were treated in vitro with a partially purified lysylhydroxylase preparation, substantial additional hydroxylation was observed (210). The extent of hydroxylation was monitored both by measurement of hydroxylysine synthesized and by release of CO_2 from α-ketoglutarate. Hydroxylysine synthesis did not occur unless collagen substrates were heat denatured, indicating that, as in the case of prolyl hydroxylation, the triple helical conformation of the substrate inhibited the enzymatic activity. It is of interest that the hydroxylysine content of collagen varies widely from tissue to tissue and with the age of the animal in the same tissue (211, 212). Furthermore, in Vitamin D deficiency, a substantial increase in lysyl hydroxylation has been observed in bone collagen (213, 214).

Hydroxylysyl residues in collagen serve as sites of attachment of galactosyl and galactosyl glucose moieties (215, 216) and participate in extracellular interchain crosslink formation (3, 10). The suggestion that a normal content of hydroxylysine (and by implication of the glycosylated residue) was required for secretion of collagen was based on the observation that chick tibiae, incubated in the presence of the lysine analog 4,5-dehydrolysine, failed to extrude the protein normally (217). This hypothesis is not supported by more recent data which indicate that drastic inhibition of peptidyl hydroxylation by α,α′-dipyridyl retards but does not prevent secretion of procollagen by 3T6 fibroblasts in culture (218). Similarly, a clearcut inhibition of crosslink formation by glycosylation of hydroxylysyl residues has been called into question by the isolation of a glycosylated crosslink (219). Nevertheless it is likely that the modulation of lysyl hydroxylation and hydroxylysyl glycosylation plays an important role in modifying the structural and functional characteristics of collagen fibers in tissues. As an extreme example, a deficiency in lysylhydroxylase leads to a severe connective tissue disorder, hydroxylysine-deficient collagen disease, a form of the Ehlers-Danlos syndrome, in which the mechanical properties of tissues such as skin and ligaments are impaired (see below).

INTRACELLULAR TRANSLOCATION AND SECRETION OF PROCOLLAGEN

Kinetics of Synthesis

Careful studies of the time required to synthesize pro-α chains by newborn rat cranial bones in culture were performed by Vuust & Piez (77, 95). These workers initially performed classical pulse-label experiments patterned after the studies of Dintzis (220). Plots of relative specific activities of CNBr peptides against the position of these peptides in the α1 and α2 chains (for labeling periods of 3, 7.5, and 15 min) permitted the calculation of a synthesis time of 4.8 min for an α chain. The data excluded the assembly of α chains from lower molecular weight polypeptide subunits and indicated that separate initiation points existed for α1 and α2 chains (95). In an extension of these studies, Vuust & Piez (77) designed experiments which measured translation times based on differences in specific activity between CNBr peptides in the same chain. These estimates did not depend on the measurement of total labeling time and they avoided the need to calculate the time required to equilibrate the intracellular amino acid pool or to terminate labeling rapidly. An average translation rate of 209 residues per min was calculated; a pro-α chain of 1250 amino acids would therefore require a synthesis time of about 6 min. No time lag was observed between incorporation of activity into the COOH-terminal end of the chain and the completion of the fully hydroxylated native procollagen molecule. Vuust & Piez therefore concluded that hydroxylation and helix formation occurred concomitantly with chain elongation or very rapidly after termination of polypeptide synthesis.

Role of Subcellular Structures

The manner in which procollagen is transported from its site of synthesis on the RER to the extracellular space is not well understood. A number of proposals have been made (see 221 for review). In contrast to the secretory process in exocrine glands and plasma cells, clearcut evidence for the participation of the Golgi complex in the secretion of collagen by mesenchymal cells has thus far not been obtained (221). Ross & Benditt (222), on the basis of autoradiographic and morphologic observations of healing guinea pig wounds, suggested the occurrence of intermittent communication of vesicles derived from the cisternae of the RER with the extracellular space. However, attempts to demonstrate the presence of collagen in vesicles directly by electron microscopy have not proved successful except in the case of the corneal epithelium (223, 267) and the odontoblast (268); in these cells such vesicles may be derived from the Golgi complex (267, 268). An alternative proposal, namely that the ground cytoplasm serves as an intermediate compartment during intracellular translocation (224, 225), seems less likely in view of what is known of the transport of other proteins.

Localization of procollagen to the RER has been indirect and based largely on the identification of peptide-bound hydroxyproline and prolylhydroxylase activity in this subcellular compartment (see above). Because of the tendency of collagen to aggregate, clearly interpretable results have not been obtained by subcellular fractionation. The development of antibodies to determinants specific to the procollagen molecule (59, 71, 96) raises the possibility that ferritin-coupled antibodies may be employed to localize the protein intracellularly at the ultrastructural level.

When agents which impair microtubular function, including colchicine, vinblastine, deuterium oxide, and high hydrostatic pressure, were applied to cultures of cranial bones, the rate of conversion of procollagen to collagen was retarded (221, 226). Autoradiographic studies at the light microscope level indicate that this effect is accompanied by a marked impairment of extrusion of proline-labeled material into the extracellular matrix (227). Colchicine and vinblastine have also been shown to reduce the secretion of procollagen into the medium of cultured fibroblasts (228, 229). These observations implicate microtubules in the intracellular transport of procollagen and are consistent with an extracellular site of conversion of the precursor.

Ehrlich & Bornstein (226) initially suggested that conversion of procollagen might occur intracellularly, subjacent to the cell membrane. This suggestion was based on the observation that cytochalasin B, which did not affect the proportion of procollagen to collagen in cranial bone cultures, rather specifically inhibited procollagen synthesis. However, more extensive experiments (227) now indicate that both collagen and noncollagen protein synthesis are inhibited by cytochalasin B. The effects of cytochalasin B, which are known to be highly complex, will therefore require further study before conclusions concerning the mechanism of procollagen secretion can be drawn from its use. Similarly, other metabolic effects of antimitotic agents, in addition to disaggregation or stabilization of microtubules, must be considered. If, however, microtubules are involved in procollagen transport, they may function in the intracellular translocation of either procollagen-containing vesicles or elements of the RER.

Does Secretion Require Peptidyl Hydroxylation?

Inhibition of peptidyl hydroxylation in embryonic cartilage by chelation of ferrous ions or under anaerobic conditions led to the intracellular accumulation of protocollagen as observed by autoradiography (230). Inhibition of secretion of procollagen was also shown more directly by use of α,α'-dipyridyl in cultured fibroblasts (47, 76, 83). These studies were extended by the use of a number of analogs of proline (231–234) and lysine (217) in incubation with both chick tibiae and tendon fibroblasts. Although azetidine 2-carboxylic acid, cis-fluoroproline, and dehydroproline were incorporated into collagen, these compounds also inhibited the hydroxylation of proline and lysine incorporated in their presence. Cis-hydroxyproline, however, while substituting for proline, did not alter the proportion of proline and lysine hydroxylated in the analog-containing protein (234). Since it seemed possible, at least in the case of α,α'-dipyridyl, that the effects on peptidyl hydroxylation were relatively specific and that a general impairment of the cell's secretory apparatus did not exist (235), Prockop, Rosenbloom and co-workers concluded that a relatively normal content of trans-4-hydroxyproline and hydroxylysine was required in order to permit secretion of intact procollagen at a normal rate. In the presence of α,α'-dipyridyl, or analogs of proline or lysine, fibroblasts were found to secrete largely lower molecular weight peptides, suggesting intracellular degradation of accumulated protocollagen (76, 83, 236, 237).

Modifications of the above positions will, however, be necessary as a result of experiments performed under conditions of ascorbic acid deficiency (67, 152, 153, 159). Incubation of 3T6 cells in the absence of ascorbic acid for a period of 24 hr indicates that neither the rate of collagen synthesis nor the proportion of collagen secreted into the medium is reduced. The collagen secreted into the medium was markedly underhydroxylated yet relatively normal in size, as determined by molecular sieve chromatography (67). In retrospect the observation of Gribble et al (161) that L-929 cells secrete relatively large (excluded by Sephadex G-100) underhydroxylated polypeptides into the medium during log phase of growth can be attributed to the instability of ascorbic acid at 37°C (153).

Peterkofsky (159), however, has shown that the initial rate of secretion of procollagen by 3T3 cells in culture is reduced in the absence of ascorbic acid, and that ascorbic acid deficiency reduces total collagen synthesis by chick embryo fibroblasts (but not 3T3 cells) in culture. It seems likely therefore that although a normal complement of peptidyl hydroxylation is necessary for an optimal rate of collagen synthesis and secretion (perhaps more so in normal cells and tissues than in established cell lines), intact protocollagen chains are capable of being secreted. When care was taken to prevent proteolysis, protocollagen synthesized and secreted in the presence of α,α'-dipyridyl was also shown to have high molecular weight (193, 218). The earlier suggestions that only low molecular weight degradation products are secreted when protocollagen accumulates intracellularly (76, 83, 236) can therefore be attributed to the increased susceptibility of secreted protocollagen to extracellular proteolysis. Similar considerations may apply to collagen synthesized and secreted in the presence of cis-hydroxyproline, since the collagen which accumulates intracellularly is sensitive to limited digestion with pepsin (237) and, therefore, presumably lacks a triple helical conformation. Since it has been established that peptidyl trans-4-hydroxyproline plays an important role in stabilizing the triple helix (see above), it seems possible that the native conformation of procollagen rather than peptidyl hydroxyl groups per se is required for normal secretion of the protein.

REGULATION OF PROCOLLAGEN SYNTHESIS

Very little is known of the manner in which the synthesis of procollagen is regulated. Given the extreme importance of fine control of collagen biosynthesis, both during morphogenesis and growth and for homeostasis, it is likely that controls exist to modulate cellular synthetic activity. Such controls might well supplement and act as fine tuning for other regulatory controls resulting from cellular migration and catabolism of collagen by specific collagenases. However, the means by which the large extracellular collagen pool, existing almost entirely as insoluble fibers, could be capable of feedback control at the cellular level remains a mystery. Peptide fragments released from procollagen during the extracellular conversion to collagen may possess biological properties including a feedback regulatory role, but this is conjectural.

Relatively early experiments using embryonic rabbit skin suggested that collagen mRNA was less stable than mRNA for noncollagenous proteins (238, 239). These observations were based on inhibition of nuclear RNA synthesis by actinomycin D and raised the possibility of control at the transcriptional level. Bhatnagar & Rapaka (240) observed an increase in peptidyl hydroxyproline synthesis, relative to the incorporation of proline, following addition of 6-diazo-5-oxonorleucine (DON) or of DON plus actinomycin D to embryonic chick tibiae in vitro. These workers interpreted the results in terms of post-transcriptional control resulting from a reduced degradation of a relatively stable collagen message (241). However, the stimulatory effect of DON on collagen synthesis by chick tibiae could not be confirmed in this laboratory (242).

Injection of *Xenopus* oocytes with mRNA from rabbit reticulocytes has resulted in hemoglobin synthesis (243, 244), and peptidyl hydroxyproline was demonstrated after injection of an mRNA fraction from polysomes of frog larvae (245). Collagen synthesis by injected oocytes may therefore represent a means of assaying directly for collagen mRNA activity.

Fernandez-Madrid (93, 246) suggested that the rate of collagen synthesis may depend in part on the size of collagen synthesizing polyribosomal aggregates and that a number of factors, including the availability of ascorbic acid and glycosaminoglycans, may serve to influence the degree of aggregation of polyribosomes. This hypothesis rested on the observation of a positive correlation between the sedimentation constants of hydroxyproline-containing polysomes in sucrose density gradients and the rate of collagen synthesis in a variety of systems (see 93 for references). Morphological findings also indicated that the highly organized RER of normal guinea pig fibroblasts was replaced in ascorbic acid deficiency by a disorganized pattern dominated by single ribosomes and small aggregates (247). It seems probable, however, that rapidly sedimenting polyribosomal aggregates result from interaction of nascent collagen chains with intracellular procollagen or extracellular collagen during preparation of the tissue, as well as from interpolysomal interactions due to bonding between nascent chains (89, 94). Stable aggregates are more likely to occur when collagen chains are fully hydroxylated in the presence of ascorbic acid. Rather than serving as a control mechanism, the varying size of collagen polyribosomal aggregates may therefore represent a consequence of different synthetic rates of the protein.

Initial studies of collagen synthesis by cultured fibroblasts led to the suggestion that synthesis was markedly augmented after the cells achieved confluency and ceased to proliferate (164, 248). These observations raised the possibility that elements of the control of collagen synthesis might be attributed to changes in the cell division cycle or to cell surface or extracellular changes associated with cell to cell interactions. However, Gribble et al (161) reported that during the logarithmic growth phase, L-929 cells synthesize and secrete an underhydroxylated form of collagen into the medium. Subsequently, a number of workers demonstrated that exponentially growing cultures synthesized collagen rapidly (151–153, 249–251); expressed on the basis of DNA as much or more collagen chain was synthesized by dividing cells. It is now known that the increase in collagen chain synthesis during the stationary phase, measured as peptide-bound hydroxyproline, is only apparent and results from an increase in activity of prolylhydroxylase under the particular conditions used for cell culture.

It is known that increases in hydroxylase activity occur in tissues with rapid changes in collagen synthesis in vivo (252–254). There is as yet no good evidence, however, that the enzymatic activity plays a rate-determining role in collagen production. Control of hydroxylation may represent a means of control of collagen synthesis, since it has been shown that the rate of secretion of protocollagen is reduced, leading in some cells to an inhibition of synthesis (159). Other possibilities for control include changes in the size of the intracellular proline pool (255).

The use of in vitro cell-free synthesizing systems offers considerable potential for the study of the regulation of procollagen synthesis. Lazarides & Lukens (91) reported that chick embryo polysomes were capable of completing and releasing nascent collagen chains. Both pro-α and α chains were identified. Kerwar et al (80, 256) have reported both initiation and completion of collagen polypeptides using a ribosome salt wash as a source of initiation factors. The products of polysomes from chick embryos after denaturation included α chains as well as pro-α chains, presumably due to the presence of procollagen peptidase in the S-30 supernatant fraction used as a source of soluble factors required for protein synthesis. Polysomes and supernatant fractions prepared from L-929 cells, however, synthesized only precursor chains (80) and may represent a more convenient system for study.

ALTERATIONS IN SYNTHESIS AND SUBSEQUENT PROCESSING

The procollagen chain undergoes an unusually large number of covalent modifications following peptide bond formation and prior to the expression of the normal function of the molecule within the extracellular fiber. These include: 1. peptidyl hydroxylation, 2. glycosylation, 3. limited proteolysis, and 4. interchain crosslink formation. Heritable disorders resulting in defects in any of the above steps would be expected to lead to serious disturbances in collagen structure and function. Studies of the clinical and biochemical consequences of such disorders may therefore provide useful insights into the complex process of collagen biogenesis.

A hereditary disorder of cattle, dermatosparaxis, was first described in Belgium by Hanset & Ansay (257, 258) and subsequently in the United States by O'Hara et al (259). The disorder is characterized by extreme fragility of the skin and laxity of articular ligaments. Considerable disorganization of collagen fibers exists in these tissues, as observed both by light and electron microscopy. Individual filaments display only a faint cross-striation pattern. Involvement is not limited to dermis and ligaments, but tissues such as tendon and blood vessels are affected to a much lesser extent.

Lenaers et al (56) recognized that dermatosparaxis resulted from a defect in the conversion of procollagen to collagen, and Lapière et al (57) subsequently confirmed that a neutral proteolytic activity, procollagen peptidase, was absent in several tissues from dermatosparactic animals. These and other studies of dermatosparaxis (58, 59, 69, 72) are discussed elsewhere in this review. A heritable disorder of sheep, also characterized by unusually friable skin, has been reported (260) and is likely to represent a similar defect.

Lichtenstein and co-workers (261, 262) have examined the electrophoretic pattern of collagen extracted from skin and tendon of individuals with a severe form of the Ehlers-Danlos syndrome, characterized by recurrent joint dislocations. Bands corresponding in migration to pro-α1 and pro-α2 were identified, indicating that a defect in conversion of procollagen to collagen exists in this disorder. The activity of procollagen peptidase in the medium of cultured fibroblasts from three patients was reduced. However, the synthesis of peptide-bound hydroxyproline was increased, suggesting that a product of procollagen inhibition, possibly the peptides released by procollagen peptidase, may play a role in feedback inhibition of procollagen synthesis.

In another form of the Ehlers-Danlos syndrome, characterized by severe scoliosis, recurrent joint dislocations, and hyperextensible skin and joints, Pinnell et al (263) have uncovered a different biochemical defect: a marked reduction in hydroxylysine content in several connective tissues. The dermis was normal by routine histological examination and electron microscopy but dermal collagen was abnormally soluble in denaturing agents. Fibroblasts from individuals with hydroxylysine-deficient collagen disease are markedly deficient in lysylhydroxylase activity (264). The manifestations of the disorder probably result from inadequate crosslinking of collagen, since lysyl aldehydes may not be as reactive, or lysyl-derived crosslinks as stable, as their hydroxylysyl counterparts. The abnormal crosslink patterns obtained after reduction of affected tissues with sodium borohydride (265) provide support for this suggestion.

A defect that more specifically involves the formation of covalent interchain crosslinks in elastin and collagen and bears a distinct resemblance to experimental osteolathyrism (4, 6, 7, 11) has recently been described by Rowe et al (266) in mice. Such mice are allelic at a genetic locus on the X chromosome governing coat color. Collagen from the skin of these animals is unusually soluble and both skin collagen and aortic elastin lack a normal complement of lysyl- and hydroxylysyl-derived aldehydes and crosslinks. A defect in lysyl oxidase or in copper metabolism appears likely.

SUMMARY

A number of fundamental advances have been made in our understanding of the early steps in collagen biosynthesis and of the several stages that lead from macromolecular assembly to extracellular polymerization of the protein.

1. Collagen is synthesized as a biosynthetic precursor, procollagen, with additional sequences at the NH_2-terminal ends of the three chains. The additional sequences differ in composition and conformation from the triple helical body of the molecule and are stabilized by interchain disulfide bonds. The precise structure of this additional region and the possibility of a similar domain at the COOH-terminus of the rod-like molecule remain to be established.

2. Genetically distinct procollagens exist in different tissues of the same animal.

3. The additional sequences in procollagen are likely to participate in chain alignment and to accelerate triple helix formation. However, this and other postulated functions of the precursor, such as mediation of transcellular movement and a role in extracellular fibrogenesis, will require verification.

4. The conversion of procollagen to collagen is achieved by at least one and possibly more proteolytic steps. A neutral extracellular proteolytic activity, procollagen peptidase, has been identified in normal connective tissues and a deficiency of the activity is established in a heritable disorder of cattle, dermatosparaxis. A derivative of procollagen accumulates extracellularly in dermatosparaxis but whether this represents a normal intermediate in the conversion of procollagen or is the result of haphazard cleavage by adventitious enzymes is not known. The metabolic fate and activity, if any, of the additional sequences in procollagen is unknown.

5. Antibodies have been developed to determinants unique to procollagen and these may prove useful in charting the movement of the protein through the cell and extracellularly.

6. Prolyl and lysyl hydroxylases are similar but distinct enzymes. Prolylhydroxylase is a mixed function oxygenase which utilizes O_2 with a concomitant oxidative decarboxylation of the cosubstrate α-ketoglutarate (to succinate). The enzyme requires Fe^{2+} and may employ ascorbic acid to maintain iron in the divalent state.

7. Peptidyl hydroxylation normally occurs on nascent chains and is inhibited by the triple helical conformation of the substrate. Folding of the protein may be one of the factors limiting the extent of peptidyl hydroxylation in collagen.

8. *trans*-4-Hydroxyproline contributes to the thermal stability of the triple helix, presumably by participating in interchain hydrogen bonds through the intermediary of a water molecule. The melting temperature of protocollagen is 24°C, some 15° lower than that of the normally hydroxylated protein.

9. Normally hydroxylated procollagen is secreted more efficiently than protocollagen. Whether it is the native conformation of procollagen or the hydroxyl groups directly which affect secretion has not been established. When hydroxylation is inhibited in vitro, proteolytic degradation of protocollagen frequently occurs, as at 37°C protocollagen lacks the protection afforded by the triple helical conformation. Ascorbic acid enhances peptidyl hydroxylation and secretion of procollagen and thus may serve to stimulate synthesis of the protein.

10. Genetic disorders in animals and in man have proved useful in clarifying several biochemical steps involved in the synthesis of a functional collagen molecule. Conversely, it is likely that a more adequate understanding of these processes will lead to a more rational investigation and effective therapy of disorders involving connective tissues.

ACKNOWLEDGMENTS

I am indebted to many colleagues who have contributed to the completeness of this review by making manuscripts in press available to me. I thank my co-workers Eva Marie Click, Janet Monson, and Klaus von der Mark for many stimulating discussions during the course of our work on procollagen and for helpful suggestions during preparation of the manuscript. Thanks are also due to George Cardinale, John Fessler, and Kari Kivirikko for a critical reading of parts of the manuscript. Original contributions from this laboratory were supported by NIH Grants AM11248, HD04872, and DE02600, and by a Lederle Medical Faculty Award.

Literature Cited

1. Bailey, A. J. 1968. *Comp. Biochem.* 26B: 297–423
2. Traub, W., Piez, K. A. 1971. *Advan. Protein Chem.* 25: 243–352
3. Gallop, P. M., Blumenfeld, O. O., Seifter, S. 1972. *Ann. Rev. Biochem.* 41: 617–72
4. Bornstein, P. 1974. *Duncan's Disease of Metabolism,* ed. P. K. Bondy, L. E. Rosenberg, 881–948. Philadelphia, Saunders. 7th ed.
5. Bornstein, P., Traub, W. *Proteins 3rd Ed.* Vol. 6
6. Rojkind, M. 1973. *Molecular Pathology of Connective Tissues,* ed. R. Perez-Tamayo, M. Rojkind, 1–103. New York: Dekker
7. Rojkind, M., Zeichner, M. See Ref. 6, 135–74
8. Grant, M. E., Prockop, D. J. 1972. *N. Engl. J. Med.* 286: 194–99, 242–249, 291–300
9. Kühn, K. 1969. *Essays Biochem.* 5: 59–87
10. Tanzer, M. L. 1973. *Science* 180: 561–66
11. Davison, P. F. 1973. *CRC Crit. Rev. Biochem.* 1: 201–245
12. Priest, R. E. See Ref. 6, 105–33
13. Barnes, M. J., Kodicek, E. 1972. *Vitam. Horm.* 30: 1–43
14. Balazs, E. A., Ed. 1970. *Chem. Mol. Biol. Intercell. Matrix* Vol. 1–3
15. Slavkin, H. C., Ed. 1972. *The Comparative Molecular Biology of Extracellular Matrices.* New York: Academic
16. Kulonen, E., Pikkarainen, J., Eds. 1973. *Biology of Fibroblasts.* New York: Academic
17. Piez, K. A., Eigner, E. A., Lewis, M. S. 1963. *Biochemistry* 2: 58–66
18. Kang, A. H., Bornstein, P., Piez, K. A. 1967. *Biochemistry* 6: 788–95
19. Bornstein, P. 1967. *Biochemistry* 6: 3082–93
20. Butler, W. T. 1970. *Biochemistry* 9: 44–50
21. Butler, W. T., Ponds, S. L. 1971. *Biochemistry* 10: 2076–81
22. Balian, G., Click, E. M., Bornstein, P. 1971. *Biochemistry* 10: 4470–78
23. Balian, G., Click, E. M., Hermodson, M. A., Bornstein, P. 1972. *Biochemistry* 11: 3798–806
24. Piez, K. A., Miller, E. J., Lane, J. M., Butler, W. T. 1969. *Biochem. Biophys. Res. Commun.* 37: 801–5
25. Fietzek, P. P., Wendt, P., Kell, I., Kühn, K. 1972. *FEBS Lett.* 26: 74–76
26. Wendt, P., von der Mark, K., Rexrodt, F., Kühn, K. 1972. *Eur. J. Biochem.* 30: 169–83
27. Fietzek, P. P., Rexrodt, F. W., Wendt, P., Stark, M., Kühn, K. 1972. *Eur. J. Biochem.* 30: 163–68
28. Rauterberg, J. et al 1972. *FEBS Lett.* 21: 75–79
29. Rexrodt, F. W., Hopper, K. E., Fietzek, P. P., Kühn, K. 1973. *Eur. J. Biochem.* 38: 384–95
30. Fietzek, P. P., Rexrodt, F. W., Hopper, K. E., Kühn, K. 1973. *Eur. J. Biochem.* 38: 396–400
31. Rubin, A. L. et al 1965. *Biochemistry* 4: 181–90
32. Drake, M. P., Davison, P. F., Bump, S., Schmitt, F. O. 1966. *Biochemistry* 5: 301–12
33. Davison, P. F., Schmitt, F. O. 1968. *Z. Physiol. Chem.* 349: 119–24
34. Harding, J. J. 1965. *Advan. Protein Chem.* 20: 109–90
35. Miller, E. J., Matukas, V. J. 1969. *Proc. Nat. Acad. Sci. USA* 64: 1264–68
36. Miller, E. J., Epstein, E. H. Jr, Piez, K. A. 1971. *Biochem. Biophys. Res Commun.* 42: 1024–29
37. Kefalides, N. A. 1971. *Biochem. Biophys. Res. Commun.* 45: 226–34
38. Trelstad, R. L., Kang, A. H., Cohen,

A. M., Hay, E. D. 1972. *Science* 179: 295–97
39. Fessler, J. H. 1960. *Biochem. J.* 76: 452–63
40. Veis, A. 1967. *Treatise on Collagen,* ed. G. N. Ramachandran, 1: 367–439. London: Academic
41. Hodge, A. H., Schmitt, F. O. 1958. *Proc. Nat. Acad. Sci. USA* 44: 418–24
42. Schmitt, F. O. 1960. *Bull. NY Acad. Med.* 36: 725–49
43. Schmitt, F. O. 1964. *Fed. Proc.* 23: 618–22
44. Speakman, P. T. 1971. *Nature* 229: 241–43
45. Layman, D. L., McGoodwin, E. B., Martin, G. R. 1971. *Proc. Nat. Acad. Sci. USA* 68: 454–58
46. Siegel, R. C., Pinnell, S. R., Martin, G. R. 1970. *Biochemistry* 9: 4486–92
47. Dehm, P., Prockop, D. J. 1971. *Biochim. Biophys. Acta* 240: 358–69
48. Jimenez, S. A., Dehm, P., Prockop, D. J. 1971. *FEBS Lett.* 17: 245–48
49. Dehm, P., Jimenez, S. A., Olsen, B. J., Prockop, D. J. 1972. *Proc. Nat. Acad. Sci. USA* 69: 60–64
50. Bellamy, G., Bornstein, P. 1971. *Proc. Nat. Acad. Sci. USA* 68: 1138–42
51. Ehrlich, H. P., Bornstein, P. 1972. *Biochem. Biophys. Res. Commun.* 46: 1750–56
52. Goldberg, B., Sherr, C. J. 1973. *Proc. Nat. Acad. Sci. USA* 70: 361–65
53. Clark, C. C., Veis, A. 1972. *Biochemistry* 11: 494–502
54. Veis, A., Anesey, J. R., Garvin, J. E., DiMuzio, M. T. 1972. *Biochem. Biophys. Res. Commun.* 48: 1404–11
55. Veis, A., Anesey, J., Yuan, L., Levy, S. J. 1973. *Proc. Nat. Acad. Sci. USA* 70: 1464–67
56. Lenaers, A., Ansay, M., Nusgens, B. V., Lapière, C. M. 1971. *Eur. J. Biochem.* 23: 533–43
57. Lapière, C. M., Lenaers, A., Kohn, L. D. 1971. *Proc. Nat. Acad. Sci. USA* 68: 3054–58
58. Bailey, A. J., Lapière, C. M. 1973. *Eur. J. Biochem.* 34: 91–96
59. Timpl, R., Wick, G., Furthmayr, H., Lapière, C. M., Kühn, K. 1973. *Eur. J. Biochem.* 32: 584–91
60. Burgeson, R. E., Wyke, A. W., Fessler, J. H. 1972. *Biochem. Biophys. Res. Commun.* 48: 892–97
61. Goldberg, B., Epstein, E. H. Jr, Sherr, C. J. 1972. *Proc. Nat. Acad. Sci. USA* 69: 3655–59
62. Uitto, J., Jimenez, S. A., Dehm, P., Prockop, D. J. 1972. *Biochim. Biophys. Acta* 278: 198–205

63. Bornstein, P., von der Mark, K., Wyke, A. W., Ehrlich, H. P., Monson, J. M. 1972. *J. Biol. Chem.* 247: 2808–13
64. von der Mark, K., Bornstein, P. 1973. *J. Biol. Chem.* 248: 2285–89
65. Monson, J. M., Bornstein, P. 1973. *Proc. Nat. Acad. Sci. USA.* 70: 3521–25
66. Fessler, L. I., Burgeson, R. E., Morris, N. P., Fessler, J. H. 1973. *Proc. Nat. Acad. Sci. USA* 70: 2993–96
67. Bates, C. H., Bailey, A. J., Prynne, C. J., Levene, C. I. 1972. *Biochim. Biophys. Acta* 278: 372–90
68. Bornstein, P., Ehrlich, H. P., Wyke, A. W. 1972. *Science* 175: 544–46
69. Stark, M., Lenaers, A., Lapière, C., Kühn, K. 1971. *FEBS Lett.* 18: 225–27
70. Berg, R. A., Prockop, D. J. 1973. *Biochemistry* 12: 3395–3401
71. von der Mark, K., Click, E. M., Bornstein, P. 1973. *Arch. Biochem. Biophys.* 156: 356–64
72. Furthmayr, H., Timpl, R., Stark, M., Lapière, C. M., Kühn, K. 1972. *FEBS Lett.* 28: 247–50
73. Sherr, C. J., Taubman, M. B., Goldberg, B. 1973. *J. Biol. Chem.* 248: 7033–38
74. Miller, E. J. 1972. *Biochemistry* 11: 4903–9
75. Bruns, R., Gross, J. 1973. *Biochemistry* 12: 808–15
76. Jimenez, S. A., Dehm, P., Olsen, B., Prockop, D. J. 1973. *J. Biol. Chem.* 248: 720–29
77. Vuust, J., Piez, K. A. 1972. *J. Biol. Chem.* 247: 856–62
78. Pontz, B., Müller, P., Meigel, W. 1973. *J. Biol. Chem.* 248: 7558–64
79. Smith, B. D., Byers, P. H., Martin, G. R. 1972. *Proc. Nat. Acad. Sci. USA* 69: 3260–62
80. Kerwar, S. S., Cardinale, G. J., Kohn, L. D., Spears, C. L., Stassen, F. L. H. 1973. *Proc. Nat. Acad. Sci. USA* 70: 1378–82
81. Church, R. L., Pfeiffer, S. E., Tanzer, M. L. 1971. *Proc. Nat. Acad. Sci. USA* 68: 2638–42
82. Tsai, R. L., Green, H. 1972. *Nature New Biol.* 237: 171–73
83. Ramaley, P. B., Rosenbloom, J. 1971. *FEBS Lett.* 15: 59–64
84. Piez, K. A. 1968. *Anal. Biochem.* 26: 205–312
85. Furthmayr, H., Timpl, R. 1971. *Anal. Biochem.* 41: 510–16
86. Reynolds, J. A., Tanford, C. 1970. *J. Biol. Chem.* 245: 5161–65
87. Malt, R. A., Speakman, P. T. 1964. *Life Sci.* 3: 81–84
88. Manner, G., Kretsinger, R. H., Gould,

B. S., Rich, A. 1967. *Biochim. Biophys. Acta* 134:411–39

89. Goldberg, B., Green, H. 1967. *J. Mol. Biol.* 26:1–18
90. Fernandez-Madrid, F. 1967. *J. Cell Biol.* 33:27–42
91. Lazarides, E., Lukens, L. N. 1971. *Nature New Biol.* 232:37–40
92. Warner, J. R., Knopf, P. M., Rich, A. 1963. *Proc. Nat. Acad. Sci. USA* 49:122–29
93. Fernandez-Madrid, F. 1970. *Clin. Orthop.* 68:103–81
94. Speakman, P. T. 1968. *Nature* 219:724–25
95. Vuust, J., Piez, K. A. 1970. *J. Biol. Chem.* 245:6201–7
96. Sherr, C. J., Goldberg, B. 1973. *Science* 180:1190–92
97. Müller, P. K., McGoodwin, E. B., Martin, G. R. See Ref. 16
98. Müller, P. K., McGoodwin, E. B., Martin, G. R. 1971. *Biochem. Biophys. Res. Commun.* 44:110–17
99. Kivirikko, K. I., Prockop, D. J. 1967. *Biochem. J.* 102:432–42
100. Lukens, L. N. 1970. *J. Biol. Chem.* 245:453–61
101. Grant, M. E., Kefalides, N. A., Prockop, D. J. 1972. *J. Biol. Chem.* 247:3539–44
102. Grant, M. E., Kefalides, N. A., Prockop, D. J. 1972. *J. Biol. Chem.* 247:3545–51
103. Kefalides, N. A. 1971. *Int. Rev. Exp. Pathol.* 10:1–39
104. Dehm, P., Prockop, D. J. 1973. *Eur. J. Biochem.* 35:159–66
105. Church, R. L., Tanzer, M. L., Lapière, C. M. 1973. *Nature New Biol.* 244:188–89
106. Cardinale, G., Udenfriend, S. *Advan. Enzymol.* In press
107. Bornstein, P., von der Mark, K., Ehrlich, H. P., Monson, J. M. 1973. *Miami Winter Symp.* 6:263–81
108. Layman, D. L., Ross, R. 1973. *Arch. Biochem. Biophys.* 157:451–56
109. Udenfriend, S. 1966. *Science* 152:1335–40
110. Udenfriend, S. 1970. *Chem. Mol. Biol. Intercell. Matrix* 1:371–84
111. Prockop, D. J. 1970. *Chem. Mol. Biol. Intercell. Matrix* 1:335–70
112. Kivirikko, K. I., Prockop, D. J. 1967. *Arch. Biochem. Biophys.* 118:611–18
113. Kivirikko, K. I., Prockop, D. J. 1967. *Proc. Nat. Acad. Sci. USA* 57:782–89
114. Miller, R. L. 1971. *Arch. Biochem. Biophys.* 147:339–42
115. Popenoe, E. A., Aronson, R. B. 1972. *Biochim. Biophys. Acta* 258:380–86

116. Kivirikko, K. I., Prockop, D. J. 1972. *Biochim. Biophys. Acta* 258:366–79
117. Lowther, D. A., Green, N. M., Chapman, J. A. 1961. *J. Biophys. Biochem. Cytol.* 10:373–88
118. Prockop, D. J., Peterkofsky, B., Udenfriend, S. 1962. *J. Biol. Chem.* 237:1581–84
119. Peterkofsky, B., Udenfriend, S. 1961. *Biochem. Biophys. Res. Commun.* 6:184–90
120. Miller, R. L., Udenfriend, S. 1970. *Arch. Biochem. Biophys.* 139:104–13
121. Lazarides, E., Lukens, L. N., Infante, A. A. 1971. *J. Mol. Biol.* 58:831–46
121a. Manning, S. M., Meister, A. 1966. *Biochemistry* 5:1154–65
122. Peterkofsky, B., Udenfriend, S. 1965. *Proc. Nat. Acad. Sci. USA* 53:355–42
123. Prockop, D. J., Juva, K. 1965. *Biochem. Biophys. Res. Commun.* 18:54–59
124. Prockop, D. J., Juva, K. 1965. *Proc. Nat. Acad. Sci. USA* 53:661–68
125. Hutton, J. J. Jr, Udenfriend, S. 1966. *Proc. Nat. Acad. Sci. USA* 56:198–202
126. Bhatnagar, R. S., Prockop, D. J., Rosenbloom, J. 1967. *Science* 158:492–94
127. Rosenbloom, J., Bhatnagar, R. S., Prockop, D. J. 1967. *Biochim. Biophys. Acta* 149:259–72
128. Diegelmann, R. F., Bernstein, L., Peterkofsky, B. 1973. *J. Biol. Chem.* 248:6514–21
129. Berg, R. A., Olsen, B. R., Prockop, D. J. 1972. *Biochim. Biophys. Acta* 285:167–75
130. Olsen, B. R., Berg, R. A., Kishida, Y., Prockop, D. J. 1973. *Science* 182:825–27
131. Kivirikko, K. I., Prockop, D. J. 1967. *J. Biol. Chem.* 242:4007–12
132. Popenoe, E. A., Aronson, R. B., Van Slyke, D. D. 1969. *Arch. Biochem. Biophys.* 133:286–92
133. Rhoads, R. E., Hutton, J. J. Jr, Udenfriend, S. 1967. *Arch. Biochem. Biophys.* 122:805–7
134. Rhoads, R. E., Udenfriend, S. 1968. *Proc. Nat. Acad. Sci. USA* 60:1473–78
135. Halme, J., Kivirikko, K. I., Simons, K. 1970. *Biochim. Biophys. Acta* 198:460–70
136. Rhoads, R. E., Udenfriend, S. 1970. *Arch. Biochem. Biophys.* 139:329–39
137. Berg, R. A., Prockop, D. J. 1973. *J. Biol. Chem.* 248:1175–82
138. Juva, K., Prockop, D. J. 1969. *J. Biol. Chem.* 244:6486–92
139. Pänkäläinen, M., Aro, H., Simons, K., Kivirikko, K. I. 1970. *Biochim. Bio-*

phys. Acta 221:559–65
140. McGee, J. O., Udenfriend, S. 1972. *Arch. Biochem. Biophys.* 152:216–21
141. Olsen, B. R., Jimenez, S. A., Kivirikko, K. I., Prockop, D. J. 1970. *J. Biol. Chem.* 245:2649–55
142. Olsen, B. R., Berg, R. A., Kivirikko, K. I., Prockop, D. J. 1973. *Eur. J. Biochem.* 35:135–47
143. Chvapil, M., Hurych, J. 1968. *Int. Rev. Connect. Tissue Res.* 4:68–196
144. Hurych, J., Chvapil, M. 1965. *Biochim. Biophys. Acta* 97:361–63
145. Chvapil, M., Hurych, J., Ehrlichova, E., Tichy, M. 1967. *Eur. J. Biochem.* 2:229–35
146. Hutton, J. J. Jr, Tappel, A. L., Udenfriend, S. 1966. *Biochem. Biophys. Res. Commun.* 24:179–84
147. Hutton, J. J. Jr, Tappel, A. L., Udenfriend, S. 1967. *Arch. Biochem. Biophys.* 118:231–40
148. Kivirikko, K. I., Bright, H. J., Prockop, D. J. 1968. *Biochim. Biophys. Acta* 151:558–67
149. Gould, B. S. 1968. *Treatise on Collagen,* ed. B. S. Gould, 2A:139–88. New York: Academic
150. Barnes, M. J., Constable, B. J., Morton, L. F., Kodicek, E. 1970. *Biochem. J.* 119:575–85
151. Castor, C. W. 1970. *J. Lab. Clin. Med.* 75:798–810
152. Bates, C. J., Prynne, A. J., Levene, C. I. 1972. *Biochim. Biophys. Acta* 263:397–405
153. Peterkofsky, B. 1972. *Arch. Biochem. Biophys.* 152:318–28
154. Green, H., Goldberg, B. 1964. *Proc. Soc. Exp. Biol. Med.* 117:258–61
155. Priest, R. E., Bublitz, C. 1967. *Lab. Invest.* 17:371–79
156. Peck, W. A., Birge, S. J. Jr, Brandt, J. 1967. *Biochim. Biophys. Acta* 142:512–25
157. Levene, C. I., Bates, C. J. 1970. *J. Cell Sci.* 7:671–82
158. Levene, C. I., Bates, C. J. See Ref. 16.
159. Peterkofsky, B. 1972. *Biochem. Biophys. Res. Commun.* 49:1343–50
160. Stassen, F. L. H., Cardinale, G. J., Udenfriend, S. 1973. *Proc. Nat. Acad. Sci. USA* 70:1090–93
161. Gribble, T. J., Comstock, J. P., Udenfriend, S. 1969. *Arch. Biochem. Biophys.* 129:308–16
162. Comstock, J. P., Gribble, T. J., Udenfriend, S. 1970. *Arch. Biochem. Biophys.* 137:115–21
163. Comstock, J. P., Udenfriend, S. 1970. *Proc. Nat. Acad. Sci. USA* 66:552–57
164. Green, H., Goldberg, B. 1964. *Nature*

204:347–49
165. McGee, J. O., Langness, U., Udenfriend, S. 1971. *Proc. Nat. Acad. Sci. USA* 68:1585–89
166. Fujita, Y., Gottlieb, A., Peterkofsky, B., Udenfriend, S., Witkop, B. 1964. *J. Am. Chem. Soc.* 86:4709–16
167. Abbott, M. T., Udenfriend, S. 1974. *Molecular Mechanisms of Oxygen Activation,* ed. O. Hayaishi. New York: Academic. In press
168. Cardinale, G. J., Rhoads, R. E., Udenfriend, S. 1971. *Biochem. Biophys. Res. Commun.* 43:537–43
169. Lindblad, B., Lindstedt, G., Tofft, M., Lindstedt, S. 1969. *J. Am. Chem. Soc.* 91:4604–6
170. Hurych, J. et al See Ref. 16.
171. Lindstedt, G., Lindstedt, S. 1970. *J. Biol. Chem.* 245:4178–86
172. Bhatnagar, R. S., Lie, T. Z. 1972. *FEBS Lett.* 26:32–34
173. Goldstein, A., Adams, E. 1968. *J. Biol. Chem.* 243:3550–52
174. Nordwig, A., Pfab, F. K. 1969. *Biochim. Biophys. Acta* 181:52–58
175. Hutton, J. J. Jr, Kaplan, A., Udenfriend, S. 1967. *Arch. Biochem. Biophys.* 121:384–91
176. Kivirikko, K. I., Prockop, D. J., Lorenzi, G. P., Blout, E. R. 1969. *J. Biol. Chem.* 244:2755–60
177. Rhoads, R. E., Udenfriend, S., Bornstein, P. 1971. *J. Biol. Chem.* 246:4138–42
178. Rhoads, R. E., Udenfriend, S. 1969. *Arch. Biochem. Biophys.* 133:108–11
179. McGee, J. O., Rhoads, R. E., Udenfriend, S. 1971. *Arch. Biochem. Biophys.* 144:343–51
180. Kikuchi, Y., Fujimoto, D., Tamiya, N. 1969. *FEBS Lett.* 2:221–23
181. Kikuchi, Y., Fujimoto, D., Tamiya, N. 1969. *Biochem. J.* 115:569–74
182. Hutton, J. J. Jr et al 1968. *Arch. Biochem. Biophys.* 125:779–85
183. Kivirikko, K. I., Kishida, Y., Sakakibara, S., Prockop, D. J. 1972. *Biochim. Biophys. Acta* 271:347–56
184. Prockop, D. J., Juva, K., Engel, J. 1967. *Z. Physiol. Chem.* 348:553–60
185. Susuki, F., Koyama, E. 1969. *Biochim. Biophys. Acta* 177:154–56
186. Kivirikko, K. I., Ganser, V., Engel, J., Prockop, D. J. 1967. *Z. Physiol. Chem.* 348:1341–44
187. Prockop, D. J., Kivirikko, K. I. 1969. *J. Biol. Chem.* 244:4838–42
188. Fujimoto, D., Prockop, D. J. 1969. *J. Biol. Chem.* 244:205–10
189. Sakakibara, S. et al 1973. *Biochim. Biophys. Acta* 303:198–202

602   BORNSTEIN

190. Shaw, B. 1973. PhD thesis. University of Washington, Seattle
191. Berg, R. A., Olsen, B. R., Prockop, D. J. 1970. *J. Biol. Chem.* 245: 5759–63
192. Berg, R. A., Prockop, D. J. 1973. *Biochem. Biophys. Res. Commun.* 53: 115–20
193. Murphy, L., Rosenbloom, J. 1973. *Biochem. J.* 135: 249–51
194. Blumenkrantz, N., Rosenbloom, J., Prockop, D. J. 1969. *Biochim. Biophys. Acta* 192: 81–89
195. Bhatnagar, R. S., Rosenbloom, J., Kivirikko, K. I., Prockop, D. J. 1967. *Biochim. Biophys. Acta* 149: 273–81
196. Fessler, J. H., Smith, L. A. 1970. *Chem. Mol. Biol. Intercell. Matrix* 1: 457–63
197. Lane, J., Rosenbloom, J., Prockop, D. J. 1971. *Nature* 232: 191–92
198. Lazarides, E., Lukens, L. N. 1971. *Science* 173: 723–25
199. Uitto, J., Prockop, D. J. 1973. *Abstr. Ninth Int. Congr. Biochem.* p. 425
200. Rosenbloom, J., Harsch, M., Jimenez, S. A. 1973. *Arch. Biochem. Biophys.* 158: 478–84
201. Jimenez, S., Harsch, M., Rosenbloom, J. 1973. *Biochem. Biophys. Res. Commun.* 52: 106–14
202. Hurych, J., Chvapil, M., Tichy, M., Beniac, F. 1967. *Eur. J. Biochem.* 3: 242–47
203. Piez, K. A., Gross, J. 1960. *J. Biol. Chem.* 235: 995–98
204. Rao, N. V., Harrington, W. F. 1966. *J. Mol. Biol.* 21: 577–81
205. Ramachandran, G. N., Bansal, M., Bhatnagar, R. S. 1973. *Biochim. Biophys. Acta* 322: 166–71
206. Ramachandran, G. N., Chandrasekharan, R. 1968. *Biopolymers* 6: 1649–58
207. Halme, J., Kivirikko, K. I. 1968. *FEBS Lett.* 1: 223–26
208. Kivirikko, K. I., Shudo, K., Sakakibara, S., Prockop, D. J. 1972. *Biochemistry* 11: 122–29
209. Butler, W. T. 1968. *Science* 161: 796–98
210. Kivirikko, K. I., Ryhänen, L., Anttinen, H., Bornstein, P., Prockop, D. J. 1973. *Biochemistry* 12: 4966–71
211. Miller, E. J., Martin, G. R., Piez, K. A., Powers, M. J. 1967. *J. Biol. Chem.* 242: 5481–89
212. Barnes, M. J., Constable, B. J., Morton, L. F., Kodicek, E. 1971. *Biochem. J.* 125: 925–28
213. Toole, B. P., Kang, A. H., Trelstad, R. L., Gross, J. 1972. *Biochem. J.* 127: 715–20
214. Barnes, M. J., Constable, B. J., Morton, L. F., Kodicek, E. 1973. *Biochem. J.* 132: 113–15
215. Butler, W. T., Cunningham, L. W. 1966. *J. Biol. Chem.* 241: 3882–88
216. Spiro, R. G. 1969. *J. Biol. Chem.* 244: 602–12
217. Christner, P. J., Rosenbloom, J. 1971. *J. Biol. Chem.* 246: 7551–56
218. Ramaley, P. B., Jimenez, S. A., Rosenbloom, J. 1973. *FEBS Lett.* 33: 187–91
219. Eyre, D. R., Glimcher, M. J. 1973. *Biochem. Biophys. Res. Commun.* 52: 663–71
220. Dintzis, H. M. 1961. *Proc. Nat. Acad. Sci. USA* 47: 246–61
221. Bornstein, P., Ehrlich, H. P. See Ref. 16.
222. Ross, R., Benditt, E. P. 1965. *J. Cell Biol.* 27: 83–106
223. Trelstad, R. L. 1971. *J. Cell Biol.* 48: 689–94
224. Salpeter, M. M. 1968. *J. Morphol.* 124: 387–433
225. Cooper, C. W., Prockop, D. J. 1968. *J. Cell Biol.* 38: 523–37
226. Ehrlich, H. P., Bornstein, P. 1972. *Nature New Biol.* 238: 257–60
227. Ehrlich, H. P., Ross, R., Bornstein, P. Submitted for publication
228. Dehm, P., Prockop, D. J. 1972. *Biochim. Biophys. Acta* 264: 375–82
229. Diegelmann, R. F., Peterkofsky, B. 1972. *Proc. Nat. Acad. Sci. USA* 69: 892–96
230. Juva, K., Prockop, D. J., Cooper, G. W., Lash, J. W. 1966. *Science* 152: 92–94
231. Takeuchi, T., Prockop, D. J. 1969. *Biochim. Biophys. Acta* 175: 142–55
232. Takeuchi, T., Rosenbloom, J., Prockop, D. J. 1969. *Biochim. Biophys. Acta* 175: 156–64
233. Rosenbloom, J., Prockop, D. J. 1970. *J. Biol. Chem.* 245: 3361–68
234. Rosenbloom, J., Prockop, D. J. 1971. *J. Biol. Chem.* 246: 1549–55
235. Margolis, R. L., Lukens, L. N. 1971. *Arch. Biochem. Biophys.* 147: 612–18
236. Harsch, M., Murphy, L., Rosenbloom, J. 1972. *FEBS Lett.* 26: 48–52
237. Uitto, J., Dehm, P., Prockop, D. J. 1972. *Biochim. Biophys. Acta* 278: 601–5
238. Bekhor, I. J., Bavetta, L. A. 1965. *Proc. Nat. Acad. Sci. USA* 53: 613–19
239. Bekhor, I. J., Mohseni, Z., Nimni, M. E., Bavetta, L. A. 1965. *Proc. Nat. Acad. Sci. USA* 54: 615–22

240. Bhatnagar, R. S., Rapaka, S. S. R. 1971. *Nature* 234:92–93
241. Tomkins, G. M. et al 1969. *Science* 166:1474–80
242. Jansen, H. W. B., Bornstein, P. Unpublished results
243. Gurdon, J. B., Lane, C. D., Woodland, H. R., Marbaix, G. 1971. *Nature* 233:177–82
244. Lane, C. D., Marbaix, G., Gurdon, J. B. 1971. *J. Mol. Biol.* 61:73–91
245. Rollins, J. W., Flickinger, R. A. 1972. *Science* 178:1204–5
246. Fernandez-Madrid, F., Pita, J. Jr. 1970. *Chem. Mol. Biol. Intercell. Matrix* 1:439–47
247. Ross, R., Benditt, E. P. 1964. *J. Cell Biol.* 22:365–89
248. Green, H., Goldberg, B. 1963. *Nature* 200:1097–98
249. Priest, R. E., Davies, L. M. 1969. *Lab. Invest.* 21:138–42
250. Conrad, G. W. 1970. *Develop. Biol.* 21:611–35
251. Manner, G. 1971. *Exp. Cell Res.* 65:49–60
252. Mussini, E., Hutton, J. J., Udenfriend, S. 1967. *Science* 157:927–29
253. Takeuchi, T., Kivirikko, K. I., Prockop, D. J. 1967. *Biochem. Biophys. Res. Commun.* 28:940–44
254. McGee, J. O., O'Hare, R. P., Patrick, R. S. 1973. *Nature New Biol.* 243:122–23
255. Rojkind, M., Diaz DeLeon, L. 1970. *Biochim. Biophys. Acta* 217:512–22
256. Kerwar, S. S., Kohn, L. D., Lapière, C. M., Weissbach, H. 1972. *Proc. Nat. Acad. Sci. USA* 69:2727–31
257. Hanset, R., Ansay, M. 1967. *Ann. Med. Vet.* 111:451–70
258. Ansay, M., Gillet, A., Hanset, R. 1968. *Ann. Med. Vet.* 112:449–64
259. O'Hara, P. J., Read, W. K., Romane, W. M., Bridges, C. H. 1970. *Lab. Invest.* 23:307–14
260. Prockop, D. J. et al See Ref. 16
261. Lichtenstein, J. R., Martin, G. R., Kohn, L. D., Byers, P., McKusick, V. A. 1973. *Science* 182:298–99
262. Lichtenstein, J. R., Kohn, L. D., Martin, G. R., Byers, P., McKusick, V. A. 1974. *Trans. Am. Assoc. Phys.* In press
263. Pinnell, S. R., Krane, S. M., Kenzora, J. E., Glimcher, M. J. 1972. *N. Engl. J. Med.* 286:1013–20
264. Krane, S. M., Pinnell, S. R., Erbe, R. W. 1972. *Proc. Nat. Acad. Sci. USA* 69:2899–2903
265. Eyre, D. R., Glimcher, M. J. 1972. *Proc. Nat. Acad. Sci. USA* 69:2594–98
266. Rowe, D. W. et al 1974. *J. Exp. Med.* In press
267. Hay, E. D., Dodson, J. W. 1973. *J. Cell Biol.* 57:190–213
268. Weinstock, M., Leblond, C. P. 1974. *J. Cell Biol.* 60:92–127

THE ELECTRONMICROSCOPY OF DNA

✷859

H. Banfield Younghusband and Ross B. Inman

Biophysics Laboratory and Biochemistry Department
University of Wisconsin, Madison, Wisconsin

CONTENTS

INTRODUCTION

The present status of electronmicroscopy as an important tool in the study of the biology of DNA is due almost entirely to the preparative method devised by Kleinschmidt, Zahn, and co-workers (1, 2). In this technique, DNA is mixed with a basic globular protein (usually cytochrome *c*) and the complex spread on an air-water interface. The resulting film can then be picked up, shadowed with heavy metal, and examined in the electronmicroscope. (For a recent detailed account of the method see reference 3.)

The Kleinschmidt technique provided for the first time a simple and reproducible method of preparing intact, "relaxed" DNA molecules for both quantitative and qualitative studies. The technique is used routinely in a great many laboratories today for determining the size and configuration of DNA molecules and has provided answers to a number of important questions about the structure and function of DNA.

In this short review we will attempt to deal with those areas of molecular biology in which electronmicroscopy has been particularly instrumental in elucidating biological principles. Specifically we will deal with physical mapping of genes, DNA replication, transcription, DNA-protein interactions, and the structure of genomes or portions of genomes; we have also included a section on significant modifications and alterations in the technique. The literature has been reviewed through July 1973.

MODIFICATIONS AND IMPROVEMENTS IN THE METHOD

Since the protein monolayer technique was introduced for the electronmicroscopic visualization of DNA, a number of modifications have been made. The most significant improvements have been the increased resolution of single DNA strands

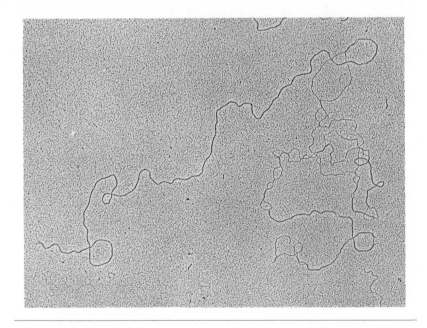

Figure 1 A DNA molecule from bacteriophage λ which has been partially denatured by alkali. The denatured sites can be identified as those regions where the duplex has become destabilized and two single strands are seen. These sites occur at fixed positions and can be mapped (25). Note the difference in appearance between single and double strands of DNA. This preparation was rotary shadowed with platinum. (Photograph courtesy of D. K. Chattoraj and S. Sachs.)

and the visualization of protein-DNA complexes. The introduction of micro-methods and techniques for producing physical reference points on DNA molecules has also been instrumental in the elevation of electronmicroscopy to its prominent position in the study of DNA.

Formamide

In the standard Kleinschmidt technique, single-stranded DNA collapses and appears as a "bush" in the electronmicroscope. Westmoreland, Szybalski & Ris (4), however, have shown that in the presence of formamide the single strands are extended and can be visualized along with double-stranded molecules. This modification allows the measurement of single-stranded DNA or single-stranded regions in double-stranded molecules and has been extremely important in the study of heteroduplex DNA molecules (see below). Formaldehyde also has the effect of extending single-stranded DNA (5, 6). (See Figure 1.)

Staining DNA

The introduction of methods (7, 8) for staining DNA with uranyl salts dissolved in organic solvents has greatly improved the contrast of the mounted DNA, especially of single-stranded regions and, in comparison to shadowing, has reduced the time involved in specimen preparation. Both double- and single-stranded DNA are stained by these methods, but the difference in appearance between the two is less distinct than in shadowed preparations. This becomes a problem in certain situations and some workers stain and shadow their preparations (3) to get maximum contrast and reasonable differentiation between single and double strands.

In addition to these general methods, there have been attempts at staining specific bases to determine polynucleotide base sequences in the electronmicro-scope (9, 10). Although these methods have failed to provide any useful base sequence information, the outlook for specifically staining a particular type of base is promising.

DNA-Protein Complexes

In the standard Kleinschmidt technique, DNA has basic protein attached to it and is adsorbed to a denatured protein monolayer at an air-water interface. Thus, the method of preparation has made it virtually impossible to study protein-DNA interactions in the electronmicroscope as the protein under study is either obscured by the cytochrome c film, denatured, or lost. These latter two problems have been solved by fixing the protein-DNA complex with either glutaraldehyde (11) or formaldehyde (D. K. Chattoraj and R. B. Inman, unpublished) prior to mounting for electronmicroscopy by the usual basic protein film technique. This method has given excellent results with the gene-32 protein of bacteriophage T4 (11), the gene-5 protein from filamentous bacterial viruses (12), a DNA-unwinding protein from Escherichia coli (13), and attachment of phage tails to DNA (D. K. Chattoraj and R. B. Inman, unpublished).

A number of attempts have been made to improve the resolution of DNA or

DNA-protein complexes in the electronmicroscope by eliminating the protein film which obscures detail. None of the methods have produced dramatic results but some appear to have good potential. Kellenberger's group (15) and Griffith et al (14) both reported a method for rendering carbon films suitable for direct attachment and visualization of nucleic acids and proteins. Dubochet et al (15) suggest that if DNA prepared by this method is viewed by dark field electronmicroscopy, then very high resolution may be attainable. Gordon & Kleinschmidt (16) have devised a method for mounting DNA directly on aluminum-mica, thereby eliminating the need for a protein film.

There have been two reports using DEAE-cellulose instead of a basic protein to mount the DNA. After picking up the DNA on carbon-collodion coated grids, the DEAE-cellulose film was dissolved in ethanol (17, 18). This method, however, has not been published in detail and cannot yet be evaluated.

The Microdiffusion Method

Lang, Kleinschmidt & Zahn (19) devised a diffusion method for preparing DNA for electronmicroscopy in which cytochrome c is spread on a dilute solution of DNA. When the DNA molecules diffuse to the cytochrome c surface film they become irreversibly attached. The amount of DNA adsorbed to the protein monolayer is controlled by the concentration of DNA in the bulk solution and by the time allowed for diffusion. As well as providing a simple means of preparing DNA for microscopy, this method has provided information about the shape of DNA molecules in solution (20) and the diffusion coefficient of DNA at zero concentration (21).

The diffusion method has been adapted to a micro-scale (22) and further modified for easier, more reproducible DNA mounting (23). The micro-method requires only 2×10^{-5} μg DNA. The standard Kleinschmidt technique has also been modified to a micro-technique (24).

Orientation of DNA Molecules

Some electronmicroscopic studies of DNA have been hampered by the absence of physical markers on DNA molecules. That is, it is impossible to orient linear molecules or identify positions on circular molecules. This problem has been solved for some types of DNA by the introduction of denaturation mapping by Inman (5) (for a review, see reference 25). This technique takes advantage of the variations in base composition along a DNA molecule and allows reproducible, unambiguous orientation of DNA molecules by locating the positions which are preferentially denatured (A + T-rich regions) at suboptimal denaturing conditions (see Figure 1). Since the technique was originally introduced (5) numerous methods have been used to achieve partial denaturation (5, 11, 13, 24–27), although high pH (24) is the most widely used. Formaldehyde is used with high temperature or high pH denaturation, and, if omitted, denaturation appears primarily at the ends of the molecules and no map can be obtained (28).

The restriction enzymes of several bacteria have been used to aid in the orientation of some circular molecules. These enzymes are sequence specific and thus make a very limited number of cuts in double-stranded DNA. The R_1 restriction endonuclease of E. coli, for example, makes only one cut in the circular Simian Virus 40 (SV40) genome (29). The scission occurs at a specific site, as demonstrated by denaturation and reannealing of the DNA (30) and by denaturation mapping (31) and has provided a reference point for the study of SV40 DNA replication (32).

Interbase Distance

The mass-to-length ratio for DNA spread by the Kleinschmidt technique is dependent on the ionic strength of the solution, pH, temperature, and other physical and chemical parameters (20, 33–35). Intercalating drugs also alter the mass-to-length ratio of DNA (36). The length of single-stranded DNA is more sensitive to the molecular environment than is double-stranded DNA (37). Because of these effects, it is difficult to make meaningful comparisons of absolute length unless one includes a standard DNA of known size in each preparation (for detailed description of errors and methods, see reference 3).

It has been suggested (38) that the base composition of DNA may also affect the mass-to-length ratio. Recent X-ray diffraction studies, however, seem to indicate that the mass-to-length ratio is independent of base composition (39).

PHYSICAL MAPPING OF GENOMES

Heteroduplex Analysis

The introduction of heteroduplex analysis of viral chromosomes by Davis & Davidson (40) and Westmoreland, Szybalski & Ris (4) has advanced cytogenetics to a level where genes and other chromosomal sites can be very accurately placed on a physical map. A heteroduplex is a double-stranded DNA molecule formed by annealing the opposite strands of related yet different DNAs. Regions of the heteroduplex that are identical in base sequence, or very nearly so, will form a double helix, while regions of nonhomology will not anneal and remain single stranded. Thus, the nonhomologous regions are easily identifiable when the heteroduplex is mounted for electronmicroscopy. Similarly, deletions or substitutions in otherwise identical DNA molecules will result in single-stranded loops or ends in heteroduplexes. The introduction of formamide (4) to extend the single strands has been extremely important in this technique (see section on Modifications and Improvements in the Method).

Heteroduplex analysis of DNA was introduced in studies on bacteriophage λ DNA (4, 40) and since that time the λ genome has been extensively and precisely mapped (41–47). These studies have utilized the vast number of genetically defined deletions, insertions, inversions, and substitutions of phage λ to precisely map most of the genes and other sites on the chromosome. These studies have also demonstrated a direct correlation between physical distance and recombination frequencies over a portion of the genome (43, 47).

A similar type of study has also been initiated with the lysogenic coliphage P2 but as yet only a limited number of genes have been mapped (48; D. K. Chattoraj, personal communication; H. B. Younghusband, J. B. Egan, and R. B. Inman, unpublished observations).

Heteroduplex studies of insertion mutants in the *E. coli lac* and *gal* genes (transducing phages carrying the mutant bacterial DNA were used) and also some phage *λ* insertions, have shown that in many cases the same piece of foreign DNA has been introduced (49–51). It was further shown that this foreign DNA can be integrated in either orientation.

Heteroduplexy has also been used to examine the tRNA genes (52) and rII region (53) of bacteriophage T4, to physically locate the integrated SV40 DNA in the SV40–adenovirus hybrids (30, 54), to study tandem duplication mutants of phage DNA (55) and inversions and hybrids of phage Mu DNA (56, 57), and to determine the orientation of foreign genetic material in some transducing phages (41, 44, 49–51, 58).

Genetic Relationships

Another important use of the heteroduplex method is the study of sequence homologies and nonhomologies in the DNA of related phages. This type of study has given some insight into the evolution of related viruses. As with the physical mapping of genes this application has been used most extensively with the lambdoid phages. The regions of sequence homology between phages *λ*, 434, 82, 80, and 21 have been mapped (44, 59). The coliphages T7 and T3 and single-stranded *φ*X174 and S13 have also been examined for regions of homology by the heteroduplex method (27, 60), as have the lysogenic *Bacillus subtilis* phages SPO2 and *φ*105 (61). In the lambdoid phages most of the regions of homology and nonhomology were at least gene-sized and were easily distinguishable. That is, the homologous regions appeared to be completely homologous and the nonhomologous regions did not base pair at all. With T3 and T7 and with SPO2 and *φ*105, however, there were extensive regions of partial homology with few gene-sized portions of complete homology. *φ*X174 and S13 had only one very short region of complete homology and the rest of the genomes were partially homologous. These observations have been discussed with respect to their possible evolutionary significance (27, 44, 59–61).

Other studies of this type have shown that *Pasteurella pestis* phage H and coliphage *φ*II are nearly identical (62), and *φ*II has been examined for regions of sequence homology with coliphages T3 and T7 (63). Heteroduplexy has also been used to study the relatedness of the genes for two tail components in bacteriophages T2 and T4 (64). This study gave a good correlation between base sequence homology and recombination frequency.

In addition to the heteroduplex analysis of various phage DNAs, Davidson, Sharp and co-workers have made elegant heteroduplex studies of sequence relationships among the F factors, F-prime factors and drug resistance, or R factors of *E. coli* (65, 66). These studies have produced physical maps for these episomes and related them to the genetic maps and have also offered a clear explanation for a number of genetic observations. Davidson and co-workers have also directly confirmed the Campbell model for the structure of prophage *λ* and mapped the *attλ* site on an episome by examination of heterotriplexes composed of DNA from an F factor, an F-prime factor carrying an integrated *λ*, and a *λ*b2b5c (67). Other studies involved the annealing of episomal DNA containing Mu-1 DNA with normal episomes, or with phage DNA, to map episomal genes and elucidate the prophage structure (68). Similarly, the structure of the *B. subtilis* prophage SPO2 is a circular permutation of the mature DNA, whereas the ϕ105 prophage is colinear with the mature phage sequence (69).

RNA-DNA Hybrids

Electronmicroscopic examination of DNA-RNA hybrids has also been employed to directly map genes. This technique has been used extensively with bacteriophage T7 for which both the in vivo early and late mRNA and the in vitro RNAs have been mapped (70–73). These studies have demonstrated that early T7 mRNA maps contiguously in the left 20% of the *r* strand of T7 DNA and that in wild-type infection virtually the entire genome, except for the terminal redundancy, is transcribed.

The direct visualization of RNA-DNA hybrids has also been used to map the relative positions of ribosomal RNA genes in mitochondrial DNA from HeLa cells (74, 75).

In an elegant modification of the RNA-DNA hybridization mapping method, Wu, Davidson and co-workers have attached the electron-opaque label ferritin to 4S RNA and were thus able to map the positions of the 4S RNA–ferritin conjugate–DNA hybrids (75, 76). 4S RNA is too short to be distinguished in the electronmicroscope as a hybrid region on a DNA molecule without the visible ferritin label. This method (76) has been applied to HeLa mitochondrial DNA and to transducing bacteriophage DNAs which carry *E. coli* tRNA genes (75–77).

DNA REPLICATION

Electronmicroscopy has proved to be an extremely useful tool in the study of DNA replication (77a). The structures of replicative intermediates have been determined by electronmicroscopy for a number of different DNAs, and the origin and direction of replication has been established for some of these. Electronmicroscopy of replicating DNA has also verified some predictions related to the structure of the growing chains in the vicinity of the replication fork.

Mode and Direction of Replication

The replication of bacteriophage λ DNA was the first to be extensively studied by electronmicroscopy. Tomizawa and co-workers (78) have shown that the early rounds of λ DNA replication proceed by a double-branched circular mechanism, that is, a theta(θ)-shaped circular structure. Schnös & Inman (79), by denaturation mapping of the first round replicative molecules, physically located the origin of replication at a point 18% from the right end of the mature λ chromosome and showed that, in the majority of the molecules, replication was bidirectional. That is, replication begins at a unique origin and both forks are growing points which progress in opposite directions around the circular molecule.

Electronmicroscopic partial denaturation studies on replicating molecules of phages P2, 186, and T7 DNA, similar to the study on λ, also identified unique origins for replication. In P2 and 186 DNA, however, the molecules were single-branched circles with unidirectional replication (80, 81). In T7 DNA the first round of replication occurs as a linear unit length molecule with an internal initiation site and replication proceeding bidirectionally to the ends of the molecule (26). A second initiation can occur before the first round of replication is complete (82). It has been suggested that this type of replicative intermediate may be unable to complete the replication of the ends of the molecule (83, 84), and this is a possible explanation for the subsequent appearance of long concatemeric T7 DNA (85, 86). The long concatemers of λ DNA which occur late in replication are head to tail tandem repeats of the mature λ genome (87, 88). The ends of these concatemers were randomly located along the genetic map. They seem to represent the product of a replicative mechanism which is different from the early replication.

The replication of mitochondrial DNA has been examined in detail by Vinograd and co-workers and is discussed in detail elsewhere (77a).

In a study of T4 DNA replication Delius, Howe & Kozinski (89) presented evidence for multiple initiation sites and bidirectional replication. They also reported the existence of "whiskers": short, single-stranded, free-ended branches extending out of the replication fork. Whiskers have since been observed in replicating 186 DNA (81), P2 DNA, and less frequently in λ replicative molecules (D. K. Chattoraj and R. B. Inman, unpublished). It seems likely that they are produced by branch migration (see below) at the replication fork (89).

The animal viruses SV40 and polyoma replicate their DNA bidirectionally via double-branched circular forms similar to λ (32, 90). In this case, however, both of the template strands remain covalently closed and appear superhelical in the electronmicroscope (91, 92). As replication proceeds around such a template, the unwinding of the parental strands will be impeded. It is obvious that some mechanism must be provided for either intermittently relaxing the superhelical template or somehow continuously removing the topological constraints in the vicinity of the replication forks. Proteins isolated from mammalian cells (93) and E. coli (93a) which can relax superhelical circular DNA are primary candidates for this "swivel" in replicating DNA.

Electronmicroscopy has also been used to study the replicative form of bacteriophage ϕX174 and PM2 DNA (94–97), the colicinogenic factor E1 (98, 99), kinetoplast DNA of Trypanosoma cruzi (100), and human adenovirus DNA (101, 102).

The Replication Fork

The structure of the replication fork has been examined in many of the above systems. Electronmicroscopic studies show the replication fork to be assymetric in that one daughter segment is always double stranded (this does not rule out the possibility of nicks) while the other is usually connected to the parental molecule by a single DNA strand and frequently contains another single-stranded region separated from the fork region by a short double-stranded segment (79–81, 103, 104). These observations give direct visual support for the involvement of Okazaki pieces in DNA replication, as described in a model (105) based on the biochemical properties of replicating DNA. These observations have also been extended into a replication unit model for DNA synthesis (104).

REPEATED SEQUENCES

It is clear from biochemical data that eukaryotic chromosomes contain many copies of some nucleotide sequences (106, 107). The arrangement of these repeated sequences in the chromosomal DNA, however, has not been clearly defined. In an examination of this problem, Thomas and co-workers (108–113) investigated the ability of short pieces (1–10 μm) of shear broken eukaryotic DNA to form circles by folding or slipping. They reached the surprising conclusion that up to 50% of some eukaryotic DNA may be arranged as a series of tandem duplications (108, 109) about 200 to 400 nucleotide pairs long which are clustered into many relatively short regions along the chromosome (110–113).

The genes coding for the 28S and 18S ribosomal RNA in *Xenopus* are known, from genetic and biochemical evidence, to be present in several hundred copies clustered together at a single locus. The rDNA has been purified from *X. laevis* and *X. mulleri* on the basis of its buoyant density, and the arrangement and homogeneity of the repeating sequences examined and compared by electron-microscopic denaturation mapping (114, 115). These studies showed that the repeats were homogeneous with the same polarity, the spacer regions were of a fixed size and composition, and amplified ribosomal DNA was indistinguishable from the normal chromosomal rDNA. They also showed that while the rRNA from two different species is virtually identical, the spacer regions between the rRNA genes are different in the two species (115).

The DNA containing the genes for 5S ribosomal RNA have been isolated from *X. laevis* and examined by denaturation mapping (116). This study also indicated a uniform polarity but a very long repeating unit was observed. It was estimated that the spacer regions between the genes were about five times as long as the 5S RNA genes.

Chow & Davidson (117) have used denaturation and renaturation of chromosomal DNA to examine the ribosomal genes of *B. subtilis*. They showed that many, or all, of the rRNA genes occur in groups with each 16S and 23S rDNA set separated from other sets by spacers. All the groups have the same polarity. The spacer segments are not homologous to each other and vary in length from 2,000 to 55,000 base pairs (117).

Electronmicroscopic denaturation mapping has been used to investigate the long concatemers which occur late during bacteriophage λ infection (see DNA replication).

BRANCH MIGRATION

Lee, Davis & Davidson (118), in a study of reannealed molecules from a circularly permuted, terminally repetitious bacteriophage DNA discovered the phenomenon of "branch migration." The single-stranded branches in a circular molecule of this type have the same base sequence as one of the adjacent strands in the duplex (118, 119). Thus, the branch can displace this strand and produce a forked branch (two single-stranded branches originating from the same point on the duplex) which can migrate along the length of the branch without changing the total number of base pairs. In a subsequent publication (120) using complex heteroduplex structures formed by annealing monomer and multimer circular DNA molecules they investigated the kinetics of this reaction.

Branch migration has been evoked as the probable origin of the whiskers [the short, single-stranded branches first reported by Delius et al (89)] seen at the replication fork of bacteriophage T4 DNA and subsequently reported for other replicating phage DNAs (81).

Although there is no evidence for in vivo branch migration it has also been suggested as a possible mechanism for driving the strand exchange which must occur in recombination intermediates of the type observed by Broker & Lehman (121) in the electronmicroscope. Recent molecular models for recombination have suggested the same type of mechanism under the term "rotary diffusion" (122, 123).

DNA-PROTEIN INTERACTIONS

Interactions between proteins and DNA control many of the functions of DNA (transcription, replication, and recombination) and thereby control the organized use and flow of genetic information. As described above, however, the visualization of protein-DNA complexes has only recently become fruitful.

These studies have been of two general types, either to examine the nature of the protein-DNA interaction (for example, DNA polymerase-DNA complexes) or to examine the topological constraints imposed on a DNA by the protein (for example, histone-DNA complexes).

A study of E. coli DNA polymerase bound to synthetic DNA (14) indicated that the binding occurred only at nicks (internal single-strand breaks) in the DNA molecules and that the attachment site on the polymerase molecule was probably a very shallow groove.

Delius and co-workers (11) have examined a number of glutaraldehyde-fixed DNA-protein complexes in the electronmicroscope. T4 bacteriophage gene-32 protein was shown to bind cooperatively to single-stranded DNA and, at saturation, produced a uniformly covered flexible rod. Gene-32 protein will induce denaturation of double-stranded DNA and it was shown, at low concentration, to preferentially interact with A + T-rich regions (11). A denaturation map produced by gene-32 protein is essentially the same as that obtained by high pH (11). The gene-5 protein from the filamentous bacterial viruses interacts with DNA in many of the ways that the T4 gene-32 protein does; however, protein-saturated, single-stranded molecules tend to coalesce into branched, rod-like structures (12). A DNA-unwinding protein has also been isolated from E. coli by Alberts and his co-workers (13) and its interaction with DNA is almost identical to that of the T4 gene-32 protein.

The attachment of the termination factor rho to DNA has also been studied (124).

There seemed to be a very limited number of rho factor attachment sites on bacteriophage fd DNA and it was suggested that the DNA molecule fits in a central channel in the rho particle.

Histone-DNA interactions have been studied in the electronmicroscope using "model systems" of bacteriophage DNA and either fractionated or nonfractionated histones (125, 126). These studies have demonstrated a cooperative binding of histones to DNA.

TRANSCRIPTION COMPLEXES

There have been a number of detailed and informative electronmicroscopic investigations of transcription complexes beginning with the classic studies of Miller and co-workers (127–131), which provided the first visual or direct demon-

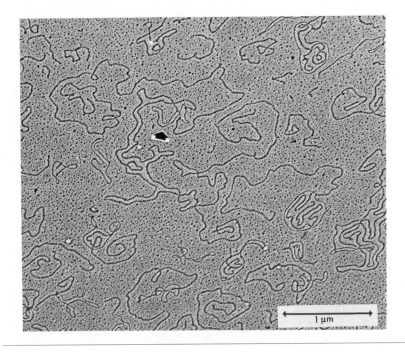

Figure 2 An in vitro SV40 transcription complex which has been incubated with T4 gene-32 protein and fixed with glutaraldehyde prior to mounting. Under the conditions used, the protein associates with the nascent RNA molecules and extends them so they are easily visualized. This is similar to the appearance of single-stranded DNA associated with gene-32 protein. The arrow indicates the SV40 circular double-stranded DNA molecule. The longest RNA chain corresponds to synthesis five times around the circular template. This preparation was stained with uranyl acetate followed by rotary shadowing with platinum. (Photograph courtesy of H. Delius and H. Westphal.)

stration of transcription in both eukaryotes (127) and prokaryotes (129). These studies also gave an indication of the size of "spacer" regions in the ribosomal RNA genes of eukaryotes, demonstrated that most if not all of these genes have the same polarity, have given an indication of the spacing of RNA polymerases on active DNA, and have shown some clear differences in these features among different species (131).

Davis & Hyman (70) have made a detailed study of the in vitro transcription of bacteriophage T7 DNA by E. coli RNA polymerase. They mapped the physical location of the initiation and termination sites and measured the size of the RNA transcripts. They also estimated the rate of chain elongation to be 45 nucleotides/sec. Subsequent studies on this system have confirmed these results (132–134).

Delius et al (134) have attached T4 gene-32 protein to the RNA of T7 transcription complexes, thus extending the RNA and making possible direct measurements of the length of the RNA attached to the template. They have also described a weak promoter near the right hand end of the T7 genome. The in vitro SV40 transcription complex was examined by the same method and found to have up to six RNA chains per DNA molecule. In this system there was no termination; rather, the RNA chains reached lengths of up to four times the length of the circular DNA. Thus, all SV40 in vitro transcription is unidirectional (134).

Bick, Lee & Thomas (135) in an examination of the T7 in vitro transcription system have described "transcription bubbles": regions of lowered duplex stability immediately behind the polymerase. The bubbles appeared as denatured regions at the point of RNA attachment when the samples are prepared by methods that are not normally denaturing.

Mitochondrial transcription complexes have up to 25 RNA molecules per DNA circle and these range in size up to one complete transcript of the mitochondrial DNA (136).

GENOME STRUCTURE

Electronmicroscopy has been instrumental in determining the structure of many viral and organal chromosomes. Early studies on the structure of superhelices were greatly assisted by visual examination of the DNA (137, 138). Electronmicroscopy has been used to examine reannealed DNA (118, 119) or partially denatured DNA (139–141) as a probe for circular permutations. This type of approach also led to the unexpected finding of single-stranded circles in annealed mixtures of adenovirus DNA, indicative of an inverted terminal repetition (142, 143).

The structure of many "complex forms" (circular dimers, trimers, and catenated circles) of viral and mitochondrial DNA have been determined by electronmicroscopy (144–146); their origin and biological significance, however, remain obscure. Other structural features of bacteriophage DNA which have been examined by DNA-DNA hybridization and electronmicroscopy include the location of the first-step transfer fragment and the single-strand interruptions in T5 (147) and T5st0 DNA (148), self-complementary sequences in the tRNA genes of the trans-ducing phage ϕ80hpsuIII$^{+,-}$ (149), and the ordering of the separated DNA fragments produced by restriction endonuclease action on DNA (150).

It seems likely that the use of restriction enzymes will become increasingly important in the electronmicroscopy of DNA due to the simplicity with which they can provide unambiguous reference points on both linear and circular molecules. Also, bacteriophage T4 gene-32 protein is likely to be used with increasing frequency to aid in the visualization and extension of single-stranded DNA.

ACKNOWLEDGMENTS

This work was supported by grants from the National Institutes of Health and the American Cancer Society.

Literature Cited

1. Kleinschmidt, A. K., Zahn, R. K. 1959. *Z. Naturforsch. B* 14:770–79
2. Kleinschmidt, A. K. 1968. *Methods Enzymol. B* 12:361–77
3. Davis, R. W., Simon, M., Davidson, N. 1971. *Methods Enzymol. D* 21:413–28
4. Westmoreland, B. C., Szybalski, W., Ris, H. 1969. *Science* 163:1343–48
5. Inman, R. B. 1966. *J. Mol. Biol.* 18:464–76
6. Inman, R. B. 1967. *J. Mol. Biol.* 28:103–16
7. Wetmur, J. G., Davidson, N., Scaletti, J. V. 1966. *Biochem. Biophys. Res. Commun.* 25:684–88
8. Gordon, C. N., Kleinschmidt, A. K. 1968. *Biochim. Biophys. Acta* 155:305–7
9. Beer, M., Moudrianakis, E. N. 1962. *Proc. Nat. Acad. Sci. USA* 48:409–16
10. Whiting, R. F., Ottensmeyer, F. P. 1972. *J. Mol. Biol.* 67:173–81
11. Delius, H., Mantell, N. J., Alberts, B. 1972. *J. Mol. Biol.* 67:341–50
12. Alberts, B., Frey, L., Delius, H. 1972. *J. Mol. Biol.* 68:139–52
13. Sigal, N., Delius, H., Kornberg, T., Gefter, M. L., Alberts, B. 1972. *Proc. Nat. Acad. Sci. USA* 69:3537–41
14. Griffith, J., Huberman, J. A., Kornberg, A. 1971. *J. Mol. Biol.* 55:209–14
15. Dubochet, J., Ducommun, M., Zollinger, M., Kellenberger, E. 1971. *J. Ultrastruct. Res.* 35:147–67
16. Gordon, C. N., Kleinschmidt, A. K. 1970. *J. Colloid Interface Sci.* 34:131–41
17. Dingman, C. W., Fisher, M. P., Kakefuda, T. 1972. *Biochemistry* 11:1242–50
18. Salzman, L. A., White, W. L., Kakefuda, T. 1971. *J. Virol.* 7:830–35
19. Lang, D., Kleinschmidt, A. K., Zahn, R. K. 1964. *Biochim. Biophys. Acta* 88:142–54
20. Lang, D., Bujard, H., Wolff, B., Russell, D. 1967. *J. Mol. Biol.* 23:163–81
21. Lang, D., Coates, P. 1968. *J. Mol. Biol.* 36:137–51
22. Mayor, H. D., Jordan, L. E. 1968. *Science* 161:1246–47
23. Lang, D., Mitani, M. 1970. *Biopolymers* 9:373–79
24. Inman, R. B., Schnös, M. 1970. *J. Mol. Biol.* 49:93–98
25. Inman, R. B. 1973. *Methods Enzymol. E* 29: In press
26. Wolfson, J., Dressler, D., Magazin, M. 1972. *Proc. Nat. Acad. Sci. USA* 69:499–504
27. Davis, R. W., Hyman, R. W. 1971. *J. Mol. Biol.* 62:287–301
28. Fuke, M., Wada, A., Tomizawa, J.-I. 1970. *J. Mol. Biol.* 51:255–66
29. Mertz, J. E., Davis, R. W. 1972. *Proc. Nat. Acad. Sci. USA* 69:3370–74
30. Morrow, J. F., Berg, P. 1972. *Proc. Nat. Acad. Sci. USA* 69:3365–69
31. Mulder, C., Delius, H. 1972. *Proc. Nat. Acad. Sci. USA* 69:3215–19
32. Fareed, G. C., Garon, C. F., Salzman, N. P. 1972. *J. Virol.* 10:484–91
33. Inman, R. B. 1967. *J. Mol. Biol.* 25:209–16
34. Lang, D. 1970. *J. Mol. Biol.* 54:557–65
35. Wellauer, P., Weber, R., Wyler, T. 1973. *J. Ultrastruct. Res.* 42:377–93
36. Freifelder, D. 1971. *J. Mol. Biol.* 60:401–3
37. Bujard, H. 1970. *J. Mol. Biol.* 49:125–37
38. Freifelder, D. 1970. *J. Mol. Biol.* 54:567–77
39. Bram, S. 1973. *Proc. Nat. Acad. Sci. USA* 70:2167–70
40. Davis, R. W., Davidson, N. 1968. *Proc. Nat. Acad. Sci. USA* 60:243–50
41. Hradecna, Z., Szybalski, W. 1969. *Virology* 38:473–77
42. Davidson, N., Szybalski, W. 1971. *The Bacteriophage Lambda*, ed. A. D. Hershey, 45–82. Cold Spring Harbor. 792 pp.

43. Parkinson, J. S., Davis, R. W. 1971. *J. Mol. Biol.* 56:425–28
44. Fiandt, M., Hradecna, Z., Lozeron, H. A., Szybalski, W. See Ref. 42, 329–54
45. Davis, R. W., Parkinson, J. S. 1971. *J. Mol. Biol.* 56:403–23
46. Blattner, F. R., Dahlberg, J. E. 1972. *Nature New Biol.* 237:227–32
47. Blattner, F. R., Dahlberg, J. E., Boettiger, J. K., Fiandt, M., Szybalski, W. 1972. *Nature New Biol.* 237:232–36
48. Chattoraj, D. K., Inman, R. B. 1972. *J. Mol. Biol.* 66:423–34
49. Hirsch, H. J., Starlinger, P., Brachet, P. 1972. *Mol. Gen. Genet.* 119:191–206
50. Malamy, M. H., Fiandt, M., Szybalski, W. 1972. *Mol. Gen. Genet.* 119:207–22
51. Fiandt, M., Szybalski, W., Malamy, M. H. 1972. *Mol. Gen. Genet.* 119:223–31
52. Wilson, J. H., Kim, J. S., Abelson, J. N. 1972. *J. Mol. Biol.* 71:547–56
53. Bujard, H., Mazaitis, A J., Bautz, E. K. F. 1970. *Virology* 42:717–23
54. Kelly, T. J. Jr., Rose, J. A. 1971. *Proc. Nat. Acad. Sci. USA* 68:1037–41
55. Busse, H. G., Baldwin, R. L. 1972. *J. Mol. Biol.* 65:401–12
56. Hsu, M.-T., Davidson, N. 1972. *Proc. Nat. Acad. Sci. USA* 69:2823–27
57. Daniell, E., Boram, W., Abelson, J. 1973. *Proc. Nat. Acad. Sci. USA* 70:2153–56
58. Miller, R. C. Jr., Besmer, P., Khorana, H. G., Fiandt, M., Szybalski, W. 1971. *J. Mol. Biol.* 56:363–68
59. Simon, M., Davis, R. W., Davidson, N. See Ref. 42, 313–28
60. Godson, G. N. 1973. *J. Mol. Biol.* 77:467–77
61. Chow, L. T., Boice, L. B., Davidson, N. 1972. *J. Mol. Biol.* 68:391–400
62. Brunovskis, I., Hyman, R. W., Summers, W. C. 1973. *J. Virol.* 11:306–13
63. Hyman, R. W., Brunovskis, I., Summers, W. C. 1973. *J. Mol. Biol.* 77:189–96
64. Beckendorf, S. K., Kim, J. S., Lielausis, I. 1973. *J. Mol. Biol.* 73:17–35
65. Sharp, P. A., Hsu, M.-T., Ohtsubo, E., Davidson, N. 1972. *J. Mol. Biol.* 71:471–97
66. Sharp, P. A., Cohen, S. N., Davidson, N. 1973. *J. Mol. Biol.* 75:235–55
67. Sharp, P. A., Hsu, M.-T., Davidson, N. 1972. *J. Mol. Biol.* 71:499–501
68. Hsu, M.-T., Davidson, N. 1972. *Proc. Nat. Acad. Sci. USA* 69:2823–27
69. Chow, L. T., Davidson, N. 1973. *J. Mol. Biol.* 75:257–64
70. Davis, R. W., Hyman, R. W. 1970. *Cold Spring Harbor Symp. Quant. Biol.* 35:269–81
71. Hyman, R. W. 1971. *J. Mol. Biol.* 61:369–76
72. Hyman, R. W., Summers, W. C. 1972. *J. Mol. Biol.* 71:573–82
73. Summers, W. C., Brunovskis, I., Hyman, R. W. 1973. *J. Mol. Biol.* 74:291–300
74. Robberson, D., Aloni, Y., Attardi, G. 1971. *J. Mol. Biol.* 55:267–70
75. Wu, M., Davidson, N., Attardi, G., Aloni, Y. 1972. *J. Mol. Biol.* 71:81–93
76. Wu, M., Davidson, N. 1973. *J. Mol. Biol.* 78:1–21
77. Wu, M., Davidson, N., Carbon, J. 1973. *J. Mol. Biol.* 78:23–34
77a. Kasamatsu, H., Vinograd, J. 1974. *Ann. Rev. Biochem.* 43:695–719
78. Tomizawa, J.-I., Ogawa, T. 1968. *Cold Spring Harbor Symp. Quant. Biol.* 33:533–51
79. Schnös, M., Inman, R. B. 1970. *J. Mol. Biol.* 51:61–73
80. Schnös, M., Inman, R. B. 1971. *J. Mol. Biol.* 55:31–38
81. Chattoraj, D. K., Inman, R. B. 1973. *Proc. Nat. Acad. Sci. USA* 70:1768–71
82. Dressler, D., Wolfson, J., Magazin, M. 1972. *Proc. Nat. Acad. Sci. USA* 69:998–1002
83. Watson, J. D. 1972. *Nature New Biol.* 239:197–201
84. Bellett, A. J. D., Younghusband, H. B. 1972. *J. Mol. Biol.* 72:691–709
85. Carlson, K. 1968. *J. Virol.* 2:1230–33
86. Schlegel, R. A., Thomas, C. A. Jr. 1972. *J. Mol. Biol.* 68:319–45
87. Wake, R. G., Kaiser, A. D., Inman, R. B. 1972. *J. Mol. Biol.* 64:519–40
88. Skalka, A., Poonian, M., Bartl, P. 1972. *J. Mol. Biol.* 64:541–50
89. Delius, H., Howe, C., Kozinski, A. W. 1971. *Proc. Nat. Acad. Sci. USA* 68:3049–53
90. Bourgaux, P., Bourgaux-Ramoisy, D. 1971. *J. Mol. Biol.* 62:513–24
91. Jaenisch, R., Mayer, A., Levine, A. 1971. *Nature New Biol.* 233:72–75
92. Bourgaux, P., Bourgaux-Ramoisy, D. 1972. *J. Mol. Biol.* 70:399–413
93. Champoux, J. J., Dulbecco, R. 1972. *Proc. Nat. Acad. Sci. USA* 69:143–46
93a. Wang, J. C. 1971. *J. Mol. Biol.* 55:523–33
94. Knippers, R., Whalley, J. M., Sinsheimer, R. L. 1969. *Proc. Nat. Acad. Sci. USA* 64:275–82
95. Knippers, R., Razin, A., Davis, R., Sinsheimer, R. L. 1969. *J. Mol. Biol.* 45:237–63
96. Dressler, D. 1970. *Proc. Nat. Acad. Sci. USA* 67:1934–42
97. Espejo, R. T., Canelo, E. S., Sinsheimer, R. L. 1971. *J. Mol. Biol.* 56:597–621
98. Inselburg, J., Fuke, M. 1971. *Proc. Nat. Acad. Sci. USA* 68:2839–42
99. Fuke, M., Inselburg, J. 1972. *Proc. Nat.*

Acad. Sci. USA 69:89–92

100. Brack, Ch., Delain, E., Riou, G. 1972. *Proc. Nat. Acad. Sci. USA* 69:1642–46
101. Sussenbach, J. S., Van der Vliet, P. C., Ellens, D. J., Jansz, H. S. 1972. *Nature New Biol.* 239:47–49
102. Van der Eb, A. J. 1973. *Virology* 51:11–23
103. Wolfson, J., Dressler, D. 1972. *Proc. Nat. Acad. Sci. USA* 69:2682–86
104. Inman, R. B., Schnös, M. 1971. *J. Mol. Biol.* 56:319–25
105. Okazaki, R. et al 1968. *Cold Spring Harbor Symp. Quant. Biol.* 33:129–43
106. Britten, R. J., Kohne, D. E. 1968. *Science* 161:529–40
107. McCarthy, B. J., Church, R. B. 1970. *Ann. Rev. Biochem.* 39:131–50
108. Thomas, C. A. Jr., Hamkalo, B. A., Misra, D. N., Lee, C. S. 1970. *J. Mol. Biol.* 51:621–32
109. Pyeritz, R. E., Lee, C. S., Thomas, C. A. Jr. 1971. *Chromosoma* 33:284–96
110. Lee, C. S., Thomas, C. A. Jr. 1973. *J. Mol. Biol.* 77:25–42
111. Pyeritz, R. E., Thomas, C. A. Jr. 1973. *J. Mol. Biol.* 77:57–73
112. Bick, M. D., Huang, H. L., Thomas, C. A. Jr. 1973. *J. Mol. Biol.* 77:75–84
113. Thomas, C. A. Jr., Zimm, B. H., Dancis, B. M. 1973. *J. Mol. Biol.* 77:85–99
114. Wensink, P. C., Brown, D. D. 1971. *J. Mol. Biol.* 60:235–47
115. Brown, D. D., Wensink, P. C., Jordan, E. 1972. *J. Mol. Biol.* 63:57–73
116. Brown, D. D., Wensink, P. C., Jordan, E. 1971. *Proc. Nat. Acad. Sci. USA* 68:3175–79
117. Chow, L. T., Davidson, N. 1973. *J. Mol. Biol.* 75:265–79
118. Lee, C. S., Davis, R. W., Davidson, N. 1970. *J. Mol. Biol.* 48:1–22
119. MacHattie, L. A., Ritchie, D. A., Thomas, C. A. Jr., Richardson, C. C. 1967. *J. Mol. Biol.* 23:355–63
120. Kim, J.-S., Sharp, P. A., Davidson, N. 1972. *Proc. Nat. Acad. Sci. USA* 69:1948–52
121. Broker, T. R., Lehman, I. R. 1971. *J. Mol. Biol.* 60:131–49
122. Sigal, N., Alberts, B. 1972. *J. Mol. Biol.* 71:789–93
123. Meselson, M. 1972. *J. Mol. Biol.* 71:795–98
124. Oda, T., Takanami, M. 1972. *J. Mol. Biol.* 71:799–802
125. Olins, D. E., Olins, A. L. 1971. *J. Mol. Biol.* 57:437–55
126. Rubin, R. L., Moudrianakis, E. N.

1972. *J. Mol. Biol.* 67:361–74
127. Miller, O. L. Jr., Beatty, B. R. 1969. *Science* 164:955–57
128. Miller, O. L. Jr., Beatty, B. R. 1969. *J. Cell. Physiol.* 74: Suppl. 1, 225–32
129. Miller, O. L. Jr., Hamkalo, B. A., Thomas, C. A. Jr. 1970. *Science* 169:392–95
130. Miller, O. L. Jr., Beatty, B. R., Hamkalo, B. A., Thomas, C. A. Jr. 1970. *Cold Spring Harbor Symp. Quant. Biol.* 35:505–12
131. Miller, O. L. Jr., Bakken, A. H. 1972. *Acta Endocrinol. Suppl.* 168:155–77
132. Harford, A. G., Beer, M. 1972. *J. Mol. Biol.* 69:179–86
133. Gómez, B., Lang, D. 1972. *J. Mol. Biol.* 70:239–51
134. Delius, H., Westphal, H., Axelrod, N. 1973. *J. Mol. Biol.* 74:677–87
135. Bick, M. D., Lee, C. S., Thomas, C. A. Jr. 1972. *J. Mol. Biol.* 71:1–9
136. Aloni, Y., Attardi, G. 1972. *J. Mol. Biol.* 70:363–73
137. Vinograd, J., Lebowitz, J., Radloff, R., Watson, R., Laipis, P. 1965. *Proc. Nat. Acad. Sci. USA* 53:1104–11
138. Follett, E. A. C., Crawford, L. V. 1967. *J. Mol. Biol.* 28:455–59
139. Doerfler, W., Kleinschmidt, A. K. 1970. *J. Mol. Biol.* 50:579–93
140. Doerfler, W., Hellmann, W., Kleinschmidt, A. K. 1972. *Virology* 47:507–12
141. Younghusband, H. B., Bellett, A. J. D. 1972. *J. Virol.* 10:855–57
142. Garon, C. F., Berry, K. W., Rose, J. A. 1972. *Proc. Nat. Acad. Sci. USA* 69:2391–95
143. Wolfson, J., Dressler, D. 1972. *Proc. Nat. Acad. Sci. USA* 69:3054–57
144. Hudson, B., Vinograd, J. 1967. *Nature* 216:647–52
145. Hudson, B., Clayton, D. A., Vinograd, J. 1968. *Cold Spring Harbor Symp. Quant. Biol.* 33:435–42
146. Rush, M. G., Eason, R., Vinograd, J. 1971. *Biochim. Biophys. Acta* 228:585–94
147. Bujard, H., Hendrickson, H. E. 1973. *Eur. J. Biochem.* 33:517–28
148. Labedan, B., Crochet, M., Legault-Demare, J., Stevens, B. J. 1973. *J. Mol. Biol.* 75:213–34
149. Wu, M., Davidson, N. 1973. *J. Mol. Biol.* 78:35–41
150. Allet, B., Jeppesen, P. G. N., Katagiri, K. J., Delius, H. 1973. *Nature* 241:120–23

EUKARYOTIC MESSENGER RNA ✘ 860

George Brawerman
Department of Biochemistry and Pharmacology, Tufts University School of Medicine
Boston, Massachusetts

CONTENTS

INTRODUCTION

Advances in the past few years have led to the coming of age of messenger RNA
(mRNA) as a well-defined biochemical entity. Since the formulation of the mRNA
theory in 1961 (1), mRNA has been shown to be a metabolically unstable species of
RNA with a base composition similar to that of DNA and a heterogeneous size

621

distribution. Knowledge of mRNA remained essentially at that level for almost a decade, owing to the lack of technology for the isolation and characterization of this material. It had been assumed that mRNA consists simply of a string of nucleotide triplets translated sequentially into the amino acids of polypeptides. Studies of the translation of bacteriophage RNAs, however, have indicated that mRNA may have a more intricate structure. It was shown that special sites exist for the initial attachment of ribosomes to the RNA (2) and that translation of different cistrons may be modulated by specific folding of the RNA molecule (3).

Since the publication of the last major review on eukaryotic mRNA in this series in 1966 (4), several developments have provided means for the isolation and assay of mRNA. Studies on rabbit reticulocyte polysomes have shown that a unique RNA component with a sedimentation coefficient of 9S and characteristics expected of the mRNA for globin can be isolated by zone centrifugation. The development of several systems for protein synthesis capable of responding to exogenous RNA has permitted the translation of a variety of mRNAs into specific proteins, thus providing a means for the assay and characterization of unique mRNA species. Studies of mRNA metabolism in animal cells have been greatly facilitated by the use of actinomycin D to selectively inhibit the labeling of ribosomal RNA (rRNA). Finally, the discovery that most eukaryotic mRNA contains a long poly(A) sequence has led to the development of precise assay methods and procedures for separation from other RNA components.

Two general reviews dealing with this subject have been published recently (5, 6).

DETECTION OF MESSENGER RNA

Rapidly Labeled Polysomal RNA and the Use of Actinomycin D

The radioactively labeled RNA present in polysomes after a short incorporation period consists primarily of a heterogeneous population of DNA-like RNA molecules with sedimentation values ranging from 10 to 30S (7–11). Similar material is present in particles not attached to polysomes (10, 11). Longer labeling periods lead to the appearance of the 18S rRNA component, first in a 45S cytoplasmic particle, then in polysomes (7, 9–11). The 28S rRNA component appears next in the cytoplasm. This pattern is not necessarily due to more rapid mRNA synthesis, but rather to the fact that newly synthesized mRNA is transferred rapidly from nucleus to cytoplasm, while the rRNA components remain for some time in the nucleus (7).

Actinomycin D can selectively inhibit rRNA synthesis when used at low concentrations (12–15), thus permitting the labeling of cytoplasmic mRNA for extended periods without interference by radioactivity in rRNA (16, 17). An actinomycin level of 0.04 μg/ml, generally effective in blocking rRNA synthesis, has little effect on mRNA synthesis in some mammalian cell types (16, 17). In other cells, however, this level can significantly affect mRNA (15,18).

The use of polysomes as a source of mRNA is not without problems. Labeled RNA from damaged nuclei or from other sources can sediment together with polysomes (10, 17). In order to distinguish between such contaminants and polysomal mRNA, agents that dissociate polysomes have been used (17). Treatment of polysomes with ethylenediamine tetraacetate (EDTA) converts the ribosomes into 50 and 30S subunits and releases the mRNA as particles sedimenting in the 20 to 120S range. Polysome disaggregation and mRNA release can be produced in intact cells by treatment with puromycin. It is also possible to identify mRNA associated with polysomes by subjecting preparations fixed with glutaraldehyde to buoyant density centrifugation in CsCl. The labeled polysomal RNA will band at the density of ribosomes, whereas RNA associated with other structures will show different buoyant densities (10, 19).

Poly(A)-Containing RNA

The occurrence of a poly(A) segment in most eukaryotic mRNA (20–24) provides a precise criterion for its identification. The poly(A) segment can be adsorbed on Millipore filters at high ionic strength (22) or annealed to poly(U) immobilized on fiberglass filters (25). In this manner, the mRNA species can be selectively adsorbed and their radioactivity measured. Actinomycin D need not be used to inhibit rRNA labeling, since the latter is not retained on the filters. Messenger RNA from unlabeled cells can be detected by annealing with labeled poly(U) (26). The poly(U) annealed to the poly(A) segments is protected from RNase and serves as a measure of the mRNA present.

ISOLATION OF MESSENGER RNA

Procedures for the isolation of mRNA have been reviewed recently (27, 28). The conventional methods for extraction with phenol can be inadequate because of the tendency of the poly(A) segment to bind to denatured proteins. Other properties of poly(A) have permitted the development of several procedures for the separation of mRNA from other RNA components.

Extraction

It had been known for some time that DNA-like RNA components are refractory to phenol extraction. Phenol treatment of mammalian nuclei in the cold at neutral pH led to effective recovery of rRNA in the aqueous phase, but left a DNA-like RNA fraction in the protein gel (29, 30). This material could be recovered by re-extraction at 60–65° (30). Re-extraction in the cold in the presence of slightly alkaline tris buffers (pH 8.3–9) led to the recovery of a similar fraction (31). It was first thought that the resistance to phenol extraction was due to the localization of this material in nuclei (30), but it was shown later that a rapidly labeled DNA-like RNA fraction in the cytoplasm behaves in a similar fashion (32). The rapidly labeled RNA of purified ascites cell polysomes had the same property (22). The behavior of mRNA during phenol extraction can be explained by the capacity of the poly(A) segment present in this RNA to bind to denatured proteins (33). This interaction is promoted by monovalent cations, and the high concentration of neutral tris buffer present during the phenol treatment provided sufficient cations to cause this binding.

On the basis of the above findings, it was possible to define optimal conditions for the extraction of mRNA (33). Monovalent cations, such as Na^+ or K^+, should be avoided when possible. While slightly alkaline conditions (pH 9) favor the extraction of mRNA, the phenol treatment at neutral pH is effective provided a low concentration of tris buffer (10 mM) is used (27). It was observed that inclusion of CHCl$_3$ favors the extraction of mRNA even when the ionic conditions are unfavorable (34). Chloroform, however, was not effective when used at 0–4°C (27). Hot phenol treatment without CHCl$_3$ is also effective for the recovery of poly(A)-containing RNA molecules (23), but the yields are not as high as with the other methods (34).

It has been reported that mRNA extracted by the phenol procedure is contaminated by a Mn^{2+}-dependent RNase that cleaves the poly(A) segment (35). It is possible that the loss of poly(A) during phenol extraction observed by Perry et al (34) is due to the action of such an enzyme.

Messenger RNA can be recovered by treatment of polysomes or other cell components with sodium dodecylsulfate (SDS), followed by zone sedimentation to separate the RNA from the SDS-protein complexes. The procedure is well suited for the isolation of mRNA species easily separable by sedimentation from the rRNA components (36).

Purification

The conventional method of zone sedimentation can be used to separate unique mRNA species from the rRNA components when the respective sedimentation coefficients are sufficiently different (36). The mRNA, however, remains contaminated with any RNA component or fragment that might sediment at a similar rate.

Several procedures are available for the separation of poly(A)-containing RNA molecules from other RNA components. Since mRNA appears to be the sole cytoplasmic RNA species that contain poly(A), these procedures should in theory be effective for its complete purification. The RNA mixture can be adsorbed either on poly(dT) coupled to cellulose (37) or poly(U) coupled to sepharose (38) under conditions that favor complementary base pairing with the poly(A) segment. The bound mRNA is recovered by elution at low ionic strength or by some other appropriate means. Optimal conditions for fractionation on poly(dT)-cellulose have been described (39). The poly(A)-containing RNA molecules can be first hybridized with poly(U), and the resulting molecules with a double- or triple-stranded segment can be separated from the unreacted single-stranded species by chromatography on hydroxylapatite (40). Some commercial cellulose preparations can adsorb specifically poly(A)-containing RNA molecules at high ionic strength (41–44) due to interaction between poly(A) and aromatic lignins that occur as impurities in the preparations (43, 44). Adsorption on cellulose nitrate membrane filters (Millipore filters) in the presence of 0.5 M KCl can also be used (33, 45). The two latter procedures appear to be effective only with poly(A) segments that contain more than 50 nucleotides (27, 43). Messenger RNA from HeLa cells has also been purified by adsorption on benzoylated DEAE-cellulose, but it is not known whether this procedure is based on a property of the poly(A) segment (46).

A one-step isolation procedure has been described, in which polysomes are treated with sarkosyl and poly(U)-sepharose is added directly to the lysate to adsorb the poly(A)-containing RNA molecules (18).

ISOLATION AND CHARACTERISTICS OF UNIQUE MESSENGER RNA SPECIES

Globin RNA

The work of Chantrenne and collaborators led to the isolation of a 9S RNA component from rabbit reticulocyte polysomes and provided circumstantial evidence for its identification as mRNA (36). Zone sedimentation of polysomes treated with SDS showed the occurrence of a minor 9S peak well resolved from the 18S rRNA. The material was labeled preferentially when the rabbits were injected with ^{32}P, 10–20 hr prior to harvesting the reticulocytes. Treatment of the polysomes with amounts of RNase sufficient to fragment polysomes resulted in the loss of the 9S component. The investigators concluded that the 9S RNA component represents the thread that holds the ribosomes together in polysomes. Lingrel and collaborators showed that the 9S component isolated in this fashion from mouse reticulocytes could promote the synthesis of mouse globin when added to a rabbit reticulocyte lysate (47, 48). It was shown later that the 9S RNA is the only reticulocyte RNA component capable of inducing globin synthesis when injected into frog oocytes (49). When polysomes and single ribosomes were used as sources of RNA, the 9S component was detected in polysomes only (50).

Zone sedimentation did not prove adequate to produce pure globin mRNA. Electrophoresis in polyacrylamide gel of the 9S RNA showed several bands. Only one of the electrophoretic components was labeled more rapidly than rRNA (50) and was active in promoting globin synthesis (51, 52). Preparations of 9S RNA purified by poly(A) selection on poly(U)-sepharose contained a single electrophoretic component that corresponded to the one active in globin synthesis (51, 52). Globin mRNA has been prepared on a large scale by fractionation of total polysomal RNA on oligo(dT)-cellulose (53). The product, however, was still contaminated by 12, 18, and 28S components.

Treatment of reticulocyte polysomes with EDTA releases the globin mRNA in a structure that sediments at approximately 15S (36, 54). The 9S RNA obtained from this particle has been reported to be free of the contaminants present in preparations obtained by zone sedimentation of total polysomal RNA (51, 52), but this is not always the case (54).

Several investigators have attempted to separate the mRNA species specific for the α- and β-globin chains. Polyacrylamide gel electrophoresis proved unsuccessful (51, 52). It has been observed that α-globin is synthesized primarily on small polysomes and β-globin on larger ones (55), probably a consequence of differences in the rates of initiation on the respective mRNA species (56). A greater difference in size distribution could be produced with the use of methylthreonine (57). Methylthreonine interferes with polypeptide chain elongation at positions where isoleucine should be incorporated, and thus inhibits progression of the ribosomes on the mRNA past the triplets for isoleucine. Such triplets are present on α-globin mRNA near the initiation site, while β-globin mRNA has only one near the terminus. Since polysomes with α-mRNA remain very small (2–3 monomers) and those with

β-mRNA become overloaded with ribosomes (7–10 monomers), it was possible to isolate RNA enriched in either α or β messenger (58). It has also been observed that rabbit reticulocytes contain some 9S RNA in a 15–20S structure not attached to ribosomes, and that this RNA is active for the synthesis of α-globin only (59, 60). This could be a consequence of the apparent poor efficiency of initiation with α-globin mRNA (55, 56).

A 10S RNA component has also been isolated from duck reticulocyte polysomes, either by zonal sedimentation of total polysomal RNA (61) or by prior release of a 20S particle from polysomes (62). This material was active in promoting duck globin synthesis in a rabbit reticulocyte lysate (62).

Ovalbumin RNA

Ovalbumin synthesis represents 50–60% of total protein synthesis in chick oviducts stimulated by estrogen (63). RNA extracted from oviducts is capable of inducing ovalbumin synthesis in a reticulocyte lysate, and the active material sediments at about 15S (63). Adsorption of oviduct RNA on Millipore filters to select for poly(A)-containing molecules yielded material enriched in activity for ovalbumin synthesis (45). Zone sedimentation of rapidly labeled RNA from heavy polysomes showed a broad 8–17S peak, and the material selected on Millipore filters had similar s values (64). Schimke and collaborators observed that antibodies to ovalbumin can bind to the nascent chains on polysomes (65), and they developed a procedure for the precipitation of polysomes specific for ovalbumin (66). The precipitate was contaminated only about 10% by nonspecific polysomes (67). Adsorption of the resulting polysomal RNA on Millipore filters produced a 25-fold enrichment in the activity to promote ovalbumin synthesis. The purified material had a sedimentation value of 15S, but migrated more slowly than the 18S rRNA when subjected to polyacrylamide gel electrophoresis (67).

Immunoglobulin RNA

Mouse myeloma tumor cells, which produce relatively large quantities of immunoglobulin (Ig), have been used for the study of Ig mRNA. As a protein normally secreted by the cells, Ig is believed to be synthesized on membrane-bound polysomes, and the latter have been commonly used as a source of mRNA. RNA extracted from microsomes of mouse plasma tumor cells contains 9–13S material active in promoting the synthesis of the light Ig chain in a reticulocyte lysate (68). The polypeptide synthesized with a 10–15S myeloma RNA fraction in an ascites cell-free system had the same size as the light Ig chain, but the material produced in a reticulocyte lysate was somewhat larger, with additional amino acids on the N-terminal end (69). Polysomes specific for Ig synthesis were selected by precipitation with antibodies to the light Ig chain (70). Zone centrifugation of the RNA extracted from the precipitated polysomes revealed a rapidly labeled 12S peak, but no such peak was seen when total polysomal RNA was used. SDS treatment of membrane-bound polysomes followed by several cycles of zone centrifugation yielded a 12S-labeled component active in promoting polypeptide synthesis, but only 38% of the synthesized material was Ig (71). Fractionation of RNA from membrane-bound polysomes on poly(dT)-cellulose led to the isolation of poly(A)-containing RNA with a major peak of activity for light chain synthesis at 13S (72) or 14S (73). The 14S component yielded a symmetrical peak after recentrifugation, and consisted mainly of Ig mRNA as judged by analysis of the polypeptides produced under its direction (73).

It was reported that Ig can bind specifically to mRNA for the heavy Ig chain and cause its precipitation (74). Polyacrylamide gel electrophoresis of the material precipitated from ^{32}P-labeled cellular RNA showed the presence of three components, two of which were active in promoting the synthesis of the heavy chain in frog oocytes.

Histone RNA

Histones are synthesized in large quantities in cultured mammalian cells during the period of DNA synthesis; a class of small polysomes engaged in histone synthesis appears in the cytoplasm (75). Inhibitors of DNA synthesis depress histone synthesis, and the small polysomes disappear concurrently. These polysomes contain a relatively homogeneous 8S RNA component (76). A similar 9–10S component was detected in sea urchin embryos and was observed to be most prominent in the early embryonic stages, when the rate of DNA synthesis is high (77). The rapidly labeled RNA component from synchronized HeLa cells in the S phase was resolved by polyacrylamide gel electrophoresis into three bands, with mol wt estimated at 1.55, 1.8, and 2.1×10^5. These sizes are roughly compatible with the sizes of histones (78). A 5–10S RNA fraction from cells engaged in DNA synthesis was active in promoting the synthesis of some histone components in a reticulocyte lysate, while the same fraction from cells not in the S phase was much less active (78). Preparative polyacrylamide gel electrophoresis has been used to obtain a 7–9S fraction active as template in an ascites cell-free system (79).

The rapidly labeled 9–10S histone RNA from HeLa cells and mouse L cells was found to lack the poly(A) sequence (80, 81).

Lens Crystallin RNA

Crystallins constitute 90% of the proteins of calf lens. They are made in epithelial and cortical fiber cells, where 60–75% of the total protein synthesis is devoted to α-crystallin production. The characterization of crystallin mRNA components has been carried out mainly by Bloemendal and collaborators (82). Zonal centrifugation of lens polysomal RNA produces peaks at 7, 10, and 14S. EDTA treatment of the polysomes releases 16 and 21S particles from which the 10 and 14S RNA components can be obtained. The latter two RNA components can promote the synthesis of crystallin proteins both in the reticulocyte lysate and in the ascites cell-free system. There is a curious anomaly in the mRNA sizes. The 14S RNA, with an estimated mol wt of 360,000, codes for A_2 α-crystallin (20,000 mol wt). RNA of this size would be sufficient for two such polypeptides. The 10S component codes for larger polypeptides.

Chick lens polysomes were shown to contain 9 and 15S RNA components, as well as 15 and 22S EDTA-released particles (83).

Myosin RNA

Heavy polysomes from embryonic chick muscle contain a rapidly labeled 26S RNA component, and material from this sedimentation zone is active in promoting myosin synthesis (84). A unique feature of this RNA is the absolute requirement of muscle ribosomal factors for its translation in heterologous cell-free systems. The 26S mRNA can be separated from rRNA owing to the capacity of the 28S rRNA component to be selectively adsorbed on Sepharose 2B at high ionic strength (85). The size of the active material, as determined by polyacrylamide gel electrophoresis, is greater than that of 28S rRNA.

Other Messenger RNAs

The silk gland of *Bombyx mori* was found to contain a 45–65S RNA component. On the basis of its peculiar base composition (40% G and 19% C), the material was considered to represent the messenger for silk fibroin, a protein in which glycine predominates (86). The oligonucleotide pattern from an RNase T_1 digest also showed a correspondence with amino acid sequences in the protein. The apparent size of the RNA (5×10^6 daltons) seems considerably greater than that expected for fibroin mRNA, since the probable mol wt of the protein is 1.7×10^5.

Avidin synthesis in hen oviduct is induced by progesterone. RNA active in promoting avidin synthesis could be obtained only from tissue stimulated by the hormone (87). Messenger activity could be detected after the RNA was purified by adsorption on Millipore filters. The active component sedimented around 8–9S.

RNA active for protamine synthesis was isolated from trout testes polysomes. The active component was soluble in 2 M NaCl and thus could be separated from the bulk of macromolecular RNA (88).

Membrane-bound polysomes from mammary glands have served as a source of 7–17S RNA active in promoting α_s-casein synthesis in a reticulocyte lysate (89).

STRUCTURE OF MESSENGER RNA

Poly(A) Segment in Messenger RNA

OCCURRENCE   Poly(A) represents the most conspicuous structural feature of eukaryotic mRNA. The poly(A) segment was originally detected in labeled messenger-like RNA by its resistance to pancreatic and T_1 RNases (21, 24), by its ability to bind to oligo(dT)-cellulose (23), to polystyrene beads (20), and to Millipore filters at high ionic strength (22), and by the resistance of the RNA to phenol extraction under specific conditions (22, 32, 90). When RNA preparations are fractionated by a poly(A) selection procedure, the template activity for protein synthesis remains associated with the poly(A)-containing molecules (51, 52, 91). This provides clear evidence for the identification of these molecules as mRNA. Studies leading to the discovery of the poly(A) segment have been reviewed recently (92).

The question has been raised whether occurrence of a poly(A) segment is a feature of all mRNA species in a given cell. Of the well-investigated species, only the mRNA for histones has been shown to lack this sequence (80, 81). When polysomal RNA preparations labeled in the presence of low concentrations of actinomycin are fractionated by poly(A) selection techniques, 60 to 85% of the radioactivity behaves like poly(A)-containing material (18, 22, 40). This proportion can vary with the labeling period, since an RNA component lacking poly(A) was found to appear in the polysome fraction more rapidly than the poly(A)-containing molecules (93). These observations suggest that the majority of mRNA contains the poly(A) sequence, but the occurrence of some species other than histone mRNA that lack poly(A) cannot be excluded.

CHARACTERISTICS   The poly(A) segment consists of a pure poly(A) sequence as indicated by its resistance to pancreatic and T_1 RNases. Also, nucleotide composition analyses of the purified segment show an AMP content of at least 99% (23, 93, 94). Various investigations indicate that the poly(A) segment is located at the 3' end of the mRNA chain. Alkaline hydrolysis of the isolated poly(A) from mouse sarcoma polysomes yielded one mole of adenosine per 170 AMP residues, a result expected if segments 170 nucleotide long were all terminated by a free 3'-OH group (93). Similar results have been obtained with the poly(A) from HeLa cells (95) and yeast (96). A segment with this configuration could have originated only from the 3' terminus of the mRNA after pancreatic RNase digestion. The location has been confirmed by the use of highly purified exonucleases that cleave from the 3' end and require the presence of a free 3'-OH (97, 98). When intact RNA was digested, the poly(A) segment was destroyed first. Prior digestion of the RNA with pancreatic RNase left the poly(A) segment completely susceptible to the exonucleases. A segment released by the RNase from an internal position would have been terminated by a 3'-OH esterified with phosphate, and would have been resistant to the exonucleases. Labeling of the 3' terminus of mRNA by periodate oxidation followed by tritiated borohydride treatment tended to confirm the location of the poly(A) at the 3' end (95).

SIZE VARIABILITY   The poly(A) segment isolated from newly synthesized mammalian mRNA behaves as a relatively homogeneous component with a sedimentation coefficient of about 4S (22, 24) and an electrophoretic mobility in polyacrylamide gel corresponding to that expected for a 7S RNA component (23, 38, 99). These values have been interpreted as indicating a size of about 200 AMP residues. More precise values have been obtained by determining the amount of adenosine released after alkaline hydrolysis, assuming that all the poly(A) segments are terminated by a free 3'-OH. This produced values of about 170 residues for the poly(A) of mouse sarcoma polysomes (93) and 100–160 for the material from HeLa cells (95). Lower eukaryotes make shorter poly(A) segments. The component from *Dictyostelium*, with an electrophoretic mobility expected for a 4–5S component, has been estimated to consist of 100 residues (100). Yeast poly(A) has a mobility expected for a 4S component and yields 1 adenosine per 50 AMP residues after alkaline hydrolysis (96).

Discrepancies in some of the values for the size of poly(A) in mammalian cells have led to the conclusion that this segment decreases in size after appearing in the cytoplasm. The poly(A) from rabbit reticulocyte polysomes, labeled in the animal for 18 hr, was reported to be 50 to 70 nucleotides long on the basis of electrophoretic mobility (20), but the newly synthesized component from duck erythroblasts was shown to have the same mobility as that of the newly synthesized mammalian poly(A) (61). In mouse L cells labeled for 1 hr, the poly(A) was relatively homogeneous with a mobility expected for a 7S component, but it was quite heterogeneous in cells labeled for 20 hr, with mobilities ranging from those expected for 7S to 4S components (99). Poly(A) segments isolated from mouse sarcoma and HeLa cell polysomes after increasing periods of labeling were observed to have gradually higher mobilities in polyacrylamide gel and to become more heterogeneous in size (101, 102). The decrease was most pronounced during the first 3–6 hr, and there was relatively little change in size distribution between 6 and 24 hr. The poly(A) size decrease also occurred in mRNA not associated with polysomes (102) and was not dependent on protein synthesis (101, 102). Values for the steady state size distribution of poly(A) were obtained by zone sedimentation of the material from unlabeled cells (103). The poly(A) peaks were detected by hybridization with radioactive poly(U). Sedimentation coefficients of 3.3S for mouse sarcoma poly(A) and 2.6S for rabbit reticulocyte poly(A) were obtained, corresponding to average sizes of about 125 and 75 residues, respectively.

Other Untranslated Segments

EVIDENCE BASED ON SIZE MEASUREMENTS   Attempts have been made to determine whether mRNA contains special sequences, such as ribosomal binding sites, that might serve as signals for the control of its function. In many instances, careful size measurements have suggested that the mRNA is substantially larger than required for coding of the polypeptide. Unfortunately, the methods used for size determination, rate of sedimentation, and mobility in polyacrylamide gel are not quite satisfactory since these parameters are highly dependent on secondary structure. Thus the size estimates for rabbit globin mRNA range from 2.2×10^5 to 1.7×10^5 daltons (50, 54, 62, 104). These values indicate that the number of nucleotides exceeding those required for coding could be as high as 210 or as low as 65. Since about 50 of these nucleotides are in the poly(A) segment, the evidence for additional untranslated sequences based on size estimates is uncertain. The size of the mRNA for the immunoglobulin light chain, estimated at 850 nucleotides (72) or 1100 nucleotides (73), appears to be substantially more than that required for coding, but the same uncertainty applies here.

RAPIDLY HYBRIDIZABLE SEQUENCES   Several investigators have obtained evidence for the occurrence in mRNA of sequences complementary to reiterated DNA segments. Rat liver polysomal RNA was shown to be capable of competing for sites on DNA complementary to rapidly hybridizable nuclear RNA sequences (105). Globin 9S RNA labeled chemically with dimethylsulfate contained material capable of rapid hybridization with DNA (106). Studies of the kinetics of hybridization of rapidly labeled polysomal RNA from *Xenopus* embryos suggested that each molecule contains a rapidly hybridizable segment (107). Using polysomal RNA from mouse L cells labeled in low actinomycin to avoid radioactive rRNA, it was concluded that 18% of the mRNA sequences are derived from highly reiterated DNA

(108). In the above studies, however, the possibility of interference by contaminating RNA components was not entirely excluded. A study of the hybridization characteristics of poly(A)-containing polysomal RNA of *Dictyostelium* indicated that 9–13% of the mRNA sequences are complementary to reiterated DNA. It was also suggested that reiterated segments might be present both at the 5′ end and near the 3′ end (109).

A SEQUENCE PAST THE TERMINATION SITE ON GLOBIN mRNA   The occurrence of two human hemoglobin variants has provided evidence for the existence of an additional polynucleotide sequence beyond the termination signal in globin mRNA. Hemoglobin Constant Spring contains α chains with 31 additional amino acids on the carboxyl terminal side. The codon for the first additional amino acid, glutamine, could have arisen through a replacement of U by C in the termination triplets UAA or UAG (110). This suggests that a complex polynucleotide sequence occurs between the termination signal and the poly(A) sequence. Hemoglobin Wayne has an α chain with five additional amino acids on the carboxyl side, and could have arisen through a frame-shift mutation caused by a base deletion in the codon for amino acid 139, thus permitting elongation past the termination codon (111).

Secondary Structure

There exists little knowledge of the extent of secondary structure in mRNA. Preparations of 9S RNA from rabbit reticulocyte polysomes have been used for melting studies. These have indicated a hyperchromicity of 18–19%, as compared to 23% for rRNA (48, 104). The preparations used for these studies, however, were not extensively purified. Comparisons of sedimentation rates and mobilities during electrophoresis in polyacrylamide gel have provided a better indication of differences in secondary structure between mRNA and rRNA. Myosin mRNA was reported to have a sedimentation coefficient of 26S, but to migrate more slowly than the 28S rRNA in polyacrylamide gel (85). Ovalbumin mRNA shows a similar behavior, with a sedimentation coefficient of about 15S and a mobility in gel lower than that of the 18S rRNA (67). Such behavior would be expected of RNA chains more extended than those of rRNA. It has been suggested that the active mouse globin mRNA may occur in a variety of configurations. It migrates as a diffuse band in polyacrylamide gel, but behaves as a sharp band in the presence of formamide (51).

BIOGENESIS OF MESSENGER RNA

The Precursor to Messenger RNA in the Nucleus

The heterogeneous DNA-like RNA synthesized in the nucleoplasm (hnRNA) is considered to be the precursor to cytoplasmic mRNA (see 92). The hnRNA is considerably larger than mRNA, with a mol wt as high as 2×10^7. Most of this RNA is rapidly destroyed in the nucleus. These characteristics have made it difficult to establish the precise relationship between hnRNA and mRNA.

The discovery of the poly(A) segment has led to the best evidence for the precursor nature of hnRNA and has permitted the formulation of a model for its conversion to mRNA. Poly(A) is present in hnRNA, where its size is similar to that of newly synthesized polysomal poly(A) (23, 93, 95, 101, 102). The poly(A) segment is also located at the 3′ end of the nuclear RNA chains (93, 95, 97, 98). It has been proposed that the polynucleotide sequences representing mRNA destined to be transferred to the cytoplasm are localized near the 3′ end of the large hnRNA chains, next to the poly(A) segment (93, 112).

Evidence for the above model was provided by hybridization studies with the 5′- and 3′-terminal portions of hnRNA from rat liver. The former could be identified by the presence of pppNp in alkaline hydrolysates of ^{32}P-labeled RNA, and the latter by chemical labeling of the 3′ terminal nucleotide in the RNA. Rapidly hybridizable sequences were present in both the 5′- and 3′-terminal regions, but only the latter were susceptible to competition by polysomal RNA (105). The tetraphosphate derived from the 5′ terminus was detected only in the very large molecules, while segments bearing the 3′-terminal label were present in smaller molecules as well as large ones. It was concluded that sequences transcribed from reiterated DNA are present at both ends of the hnRNA, that the 5′ end is removed and rapidly destroyed, and that the 3′ end is conserved in polysomal mRNA. A similar, but simpler pattern appears to occur in *Dictyostelium,* where the nuclear precursor RNA is only 25% larger than the polysomal mRNA (109). Scission of the molecules by mild KOH treatment, followed by separation of the poly(A)-containing (3′ terminal) portion, showed that most but not all the rapidly hybridizing segments are at the 5′ terminus. It was proposed that processing involves removal of a reiterated sequence about 200 nucleotides long from the 5′ end of the RNA chain.

The nature of the hnRNA sequences lost in the course of processing is of much interest. It has been estimated that about one third of the hnRNA sequences of mouse L cells are derived from reiterated DNA (108). This value does not seem to account for all the sequences lost during processing. Double-stranded RNA segments have been isolated from the large hnRNA, where they represent 3% of the total RNA (113, 114). They are destroyed during processing, as indicated by their absence from the smaller nuclear RNA molecules. A sequence rich in U, about 30 nucleotides long, has also been detected in the giant hnRNA (115). It has also been reported that hnRNA contains a sequence complementary to a portion of the polysomal RNA (116).

It is not known whether all hnRNA molecules are precursors to mRNA. Only 20–40% of the newly synthesized RNA molecules appear to contain a poly(A) segment, as judged by the capacity of these molecules to hybridize with poly(U) (40, 117). The others could be either precursor molecules still awaiting polyadenylation or RNA with a different function (40). This question is of some practical importance in the interpretation of the results described above, since in many cases the studies were done without preselecting the hnRNA molecules that contain poly(A). Measurements of the labeled poly(A) content of newly synthesized hnRNA suggest that each molecule, regardless of size, contains a single poly(A) segment (40, 117). This would be expected of molecules with poly(A) only at the 3′ terminus.

There remains some uncertainty as to the actual size of the large hnRNA, owing to its tendency to form aggregates. The hnRNA retains high *s* values after treatment with dimethylsulfoxide (DMSO), which disrupts secondary structure in nucleic acids. It has been reported, however, that the DMSO treatment can cause aggregation under certain conditions (118). Preparations of highly purified hnRNA from HeLa cells, which behave like giant RNA on gel electrophoresis, were shown to sediment like molecules the size of rRNA (119). Centrifugation of giant hnRNA from duck erythroblasts in the presence of DMSO reduces considerably its s values (120). This has been attributed either to the presence of hidden breaks produced normally during processing or to the occurrence of intermolecular networks made up of smaller molecules.

Attempts have been made to obtain direct evidence for the presence of mRNA sequences in the large hnRNA molecules. Globin mRNA was used as template for the synthesis of complementary DNA (120) or RNA (121). The labeled products served as probes for the detection of sequences homologous to those of the globin RNA. Such sequences were detected in the large hnRNA, but much of the reactive material was released as 9S molecules after DMSO treatment (120), indicating that small RNA molecules can become aggregated to the hnRNA in the course of its isolation. Messenger RNA for immunoglobulin (122) and for globin (123) was also detected in hnRNA by its ability to promote the synthesis of the specific proteins in frog oocytes. A control experiment suggested that the activity was not likely to be due to aggregated mature mRNA (123).

Poly(A) Addition

The poly(A) segment is synthesized by stepwise addition of adenylate residues to the 3' end of completed hnRNA molecules. This process is distinct from transcription and is particularly susceptible to inhibition by 3'-deoxyadenosine (92). Enzymes capable of poly(A) synthesis and requiring RNA with a free 3'-OH terminal as primer have been detected in a variety of cells (124–129). In many cases, the enzymes were shown to be localized in nuclei. They usually require Mn^{2+} for activity and are inhibited by ribonucleoside triphosphates other than ATP. GTP is particularly inhibitory (127–129). No specificity toward a particular type of RNA as primer has been observed, but preferences for either RNA or poly(A) have been reported. Yeast nuclei were found to contain two enzymes which differ with respect to this type of primer specificity (129). This suggests that different enzymes may be involved in the initiation and elongation stages of poly(A) addition.

Poly(A) addition appears to be an essential part of the processing mechanism. Cordycepin (3'-deoxyadenosine), which inhibits poly(A) synthesis, also prevents the appearance of mRNA in the cytoplasm (92). Enzyme induction is also prevented by this compound (130). The poly(A)-containing RNA molecules enter into the cytoplasm after a delay of 15–20 min, but the delay is much shorter for the histone mRNA, which lacks the poly(A) segment (80, 131). This difference has been attributed to the time interval required for polyadenylation (80). Poly(A) synthesis, however, must be very rapid, as indicated by the narrow size distribution of the newly synthesized segments (117). It also appears that poly(A) addition is a late event in processing, since mRNA molecules bearing labeled poly(A) appear in the cytoplasm with relatively little delay (93, 132). Treatment with actinomycin D to block transcription results in a gradual suppression of poly(A) synthesis (112, 117). This effect is probably due to depletion of free 3'-OH termini in the nucleoplasm. It suggests that poly(A) is not added immediately after termination of transcription. Results from kinetic experiments have suggested that most of the nuclear poly(A) is transferred to the cytoplasm (117), but other data have been interpreted as indicating a considerable turnover of poly(A) in the nucleus (132).

There are indications that poly(A) synthesis may also occur in the cytoplasm. Viral RNAs that function as messengers for protein synthesis also contain a poly(A) sequence localized at the 3' end of the molecules (133–135). In the case of Vesicular Stomatitis Virus (VSV), where the mRNA must be derived by complementary copying of virion RNA, only the viral mRNA has the poly(A) sequence (136, 137). Synthesis of the viral poly(A) is assumed to take place in the cytoplasm, since these viruses do not replicate in the nucleus. Cytoplasmic extracts of L cells infected with VSV are capable of synthesizing viral mRNA with a poly(A) segment (138). Since virion transcriptase may lack the capacity to produce poly(A), a cellular cytoplasmic enzyme could be involved. Evidence for poly(A) formation in the cytoplasm of fertilized sea urchin eggs has been obtained (139). Poly(A) synthesis has been observed in the cytoplasm of chinese hamster cells pretreated with actinomycin D to suppress nuclear poly(A) formation (140). This process, which is insensitive to cordycepin, appears to involve the addition of adenylate residues to pre-existing poly(A) in mRNA.

STABILITY OF MESSENGER RNA

The initial approach to the study of mRNA stability was to treat cells or organisms with a dose of actinomycin D sufficient to arrest all RNA synthesis, and to measure by some means the quantity of mRNA remaining in the cells at various times after administration of the drug. Rates of protein synthesis and total polysome content in the treated cells were used as a measure of mRNA. Subsequent studies, however, have indicated that high levels of actinomycin may cause an inhibition of polypeptide chain initiation, thus preventing full expression of the mRNA present in the cells (141). In HeLa cells, actinomycin caused a gradual disappearance of polysomes, with a decay curve suggesting a half-life of 3–4 hr (8), but measurements of poly(A)-containing RNA indicated little change in mRNA content after a 5 hr drug treatment (141). Moreover, the polysomes could be restored by the use of cycloheximide to slow down elongation.

Measurements of the decay of uridine-labeled mRNA in HeLa cells after a chase with unlabeled uridine gave half-life values of 2–3 days (46). In another study of the decay of poly(A)-containing RNA in these cells, which did not involve the use of a chase with cold uridine, the existence of two classes of mRNA with half-lives of 6 hr and 21 hr was suggested (142). Kinetic measurements of the labeling of poly(A)-containing RNA in exponentially growing mouse L cells indicated a mean lifetime of 14.4 hr for the mRNA, a value close to that of the generation time of these cells (143).

Some investigations have suggested that the mRNA for histone is considerably less stable than the poly(A)-containing mRNAs. These studies, however, have entailed the use of high levels of actinomycin and are subject to the uncertainty discussed above. Histone synthesis was reported to decay in actinomycin-treated mouse L cells with a half-life of 1.5 hr, as compared to a half-life of 2.5–3 hr for the bulk of cellular protein (144). Addition of the drug to synchronized HeLa cells in the S phase caused the disappearance of the rapidly labeled 7–9S polysomal RNA, with a decay curve indicating a half-life of 1 hr, while the heterogeneous polysomal RNA showed a half-life of 3 hr (76). In another study, actinomycin had a less rapid effect on histone polysomes (145). The validity of studies of histone mRNA stability in actinomycin-treated cells was questioned, since this drug inhibits DNA synthesis, and the formation of histones is tightly coupled to that of DNA (145). The existence of an unstable mRNA class with a lifetime of 2 hr, was suggested by a study of the kinetics of RNA labeling in the presence of low actinomycin in mouse 3T3 cells (15). The nature of the unstable RNA species, however, was not determined.

Studies of mRNA stability in rat liver and other tissues have also involved the use of high levels of actinomycin. Protein synthesis in rat liver remained unaffected by this drug for periods of 20 to 40 hr, an indication of high mRNA stability (146, 147). These results also indicate that protein synthesis in this tissue might be insensitive to the inhibitory effect of actinomycin observed in HeLa cells. In another study, however, involving the use of animals subjected to starvation and refeeding prior to treatment with the drug, a gradual decrease in liver polysomes was observed (148). Analysis of the decay curve indicated that 64% of the polysomes had a half-life of 3–3.5 hr, and the rest a half-life of no less than 80 hr. Albumin synthesis was particularly resistant to actinomycin (149). Crystallin synthesis was found to be sensitive to actinomycin in embryonic bovine lens, but not in the adult tissue (150). This finding was interpreted in terms of a difference in mRNA stability, but another interpretation is now possible. A comparison of the extent of labeling of free and membrane-bound polysomes in horse thyroid tissue after a 60 min incorporation suggested that the mRNA of free polysomes is labeled more rapidly (151). A class of heavy membrane-bound polysomes involved in thyroglobulin synthesis was least labeled and was also least susceptible to disaggregation induced by actinomycin. The mRNA of membrane-bound polysomes from myeloma cells also appeared to be more stable, as suggested by the kinetics of labeling of poly(A)-containing RNA from the two types of polysomes (152).

MESSENGER RIBONUCLEOPROTEINS

Cytoplasmic Particles

Messenger RNA occurs in cell extracts in association with proteins. Treatment of polysomes with EDTA causes the release of the messenger-like RNA as a heterogeneous population of particles ranging from 20–120S. A low buoyant density in CsCl, usually around 1.4 g/cc, suggests that the ratio of protein to RNA in these particles is 4/1. The cytoplasmic mRNA not associated with polysomes is found in particles with similar characteristics. In spite of numerous studies (see 153–155 for reviews), the significance of the messenger ribonucleoprotein particles (mRNP) remains obscure.

A question as yet unsettled is whether mRNPs might not represent an artifact caused by association of cellular proteins with free RNA during cell disruption. Cytoplasmic extracts have been shown to contain proteins that will bind to a variety of RNA species. The resulting complexes have a buoyant density and sedimentation characteristics similar to those of the mRNP (156–158). The binding protein is of considerable size (8–10S) and was reported to consist of two components, one acidic and the other basic, both essential for binding (158). The artificial complexes are unstable at moderate or high ionic strength (156), but their stability is greater when they are produced at physiological temperature instead of at 0–4° (154). When labeled RNA is added to cells prior to disruption in the cold, the resulting RNP complexes are less stable than the endogenous mRNPs (159). Since the native mRNPs show a high stability, it has been argued that they could not have been produced during cell disruption in the cold and that they must represent a genuine cellular form of mRNA (154).

It has also been suggested that the mRNP released from polysomes by EDTA treatment may be produced by adventitious binding of some ribosomal proteins to the mRNA. Thus, when polysomes produced by addition of poly(U) to a cell-free system for protein synthesis are treated with EDTA, the poly(U) is released as heterogeneous rapidly sedimenting material (160). It has been shown, however, that mRNA released from polysomes without the use of dissociating agents also appears as mRNP. Such a release has been produced by ribosome runoff in intact cells subjected to treatments that interfere with the initiation step in protein synthesis (19, 161, 162).

The wide range of s values of mRNPs has been attributed to the heterogeneity of the RNA component (154). It was observed, however, that RNA preparations from EDTA-released particles with widely different s values had similar size distributions (163, 164). Since the buoyant density values for these particles were uniform, the possibility that the heterogeneity is due to variations in the ratio of RNA to protein was excluded, and it was suggested instead that one or more ribosomal subunits remain attached to some of the particles (163). In contrast to the heterogeneity of the EDTA particles from mammalian cells and tissues, the mRNP from reticulocytes appears to behave on sedimentation as a homogeneous component (36, 54).

The nature of the protein components of mRNP has evoked much interest. Reliable data are scarce, however, owing to the dubious state of purity of the available mRNP preparations. Several investigators have studied the globin mRNP released from polysomes and separated from the ribosomal subunits by zone sedimentation. The 14S peak from EDTA-treated rabbit reticulocyte polysomes showed the presence of three protein bands on polyacrylamide gel electrophoresis in SDS, with mol wt estimated at 130,000, 68,000, and 45,000 (165). The smallest protein component was lost when the particles were purified by an additional cycle of zone sedimentation. The 20S peak released from polysomes by treatment with puromycin and high salt showed the presence of two bands of 78,000 and 52,000 daltons (166). A 20S EDTA-released particle from duck reticulocytes yielded one prominent band with a mol wt of 73,000 and a fainter one of 49,000 daltons (167). The protein of rat liver mRNP was studied by indirect means, since the particles cannot be separated from the ribosomal subunits. One protein, with a mol wt of 160,000, was attributed to the mRNP, and some evidence was obtained for the occurrence of a protein of similar size in globin mRNP (168). Free mRNP from rat brain, isolated by several cycles of centrifugation, showed a complex pattern of protein bands, most in the 40,000–70,000 dalton range. Similar proteins, however, were observed in the free cytoplasmic protein fraction (169).

A major question that remains to be resolved is whether the protein components of mRNP are randomly distributed along the RNA chain, or whether specific proteins are associated with unique sites on the RNA. In the first arrangement, cytoplasmic binding proteins would cover spontaneously any naked RNA segment, perhaps for protection against nucleases. Specific proteins on unique sites, on the other hand, could conceivably have regulatory functions. It was observed that the poly(A) segment in mouse sarcoma polysomes is associated with proteins that cause it to sediment at about 12–15S (25), and that a similar structure of 8S occurs in rabbit reticulocyte polysomes (103). Evidence obtained from studies with globin mRNP suggests that one of the two protein components identified in that particle is associated with the poly(A) segment (170).

The lack of functional characteristics for the mRNP presents a major obstacle to its study. The 15S globin RNP was shown to be capable of directing globin synthesis, but its efficiency as a template was no greater than that of the 9S mRNA (171, 172).

Nuclear Ribonucleoprotein Particles

Nucleoplasmic RNA occurs in nuclear extracts in the form of RNPs with buoyant densities similar to those of the cytoplasmic particles. It has been proposed that mRNA processing and transport occurs at the level of such RNA-protein complexes (see 154 and 155 for reviews). The work of Georgiev and collaborators (155) showed that treatment of rat liver nuclei with tris buffers causes the release of DNA-like RNA in the form of a 30S particle. Polydisperse particles that sediment in the 30–200S range are obtained when the rat liver RNase inhibitor is included in the extracting medium. Treatment of the larger particles with RNase causes their conversion to the 30S component but leaves their buoyant density unchanged (173). Exposure of the 30S RNP to high salt concentrations causes dissociation of the RNA and leaves a 30S protein particle that can be used to reconstitute RNPs of various sizes (174, 175). It was concluded that several 30S protein particles are normally attached to a single nucleoplasmic RNA chain and that the RNA is localized on the surface of the particles (174). Somewhat different characteristics were observed with particles obtained from rat brain nuclei (176) and from rat liver nuclei disrupted in the French press (177). In both cases particles with a wide range of s values were recovered in the presence of the RNase inhibitor, and subsequent treatment with RNase produced a smaller 40–60S component. Treatment with deoxycholate or with KCl caused the release of some proteins from the particles, thus leaving a portion of the RNA exposed (178–180). This behavior suggested that the structure of the nuclear RNP may be more complex than that postulated by Georgiev and collaborators. It was also proposed that the large particles may represent aggregates of smaller RNPs linked together through their RNA chains (180).

The nature of the protein components of the nuclear particles remains unsettled. In some of the preparations from rat liver nuclei, only a few protein bands were detected by gel electrophoresis in the presence of urea, and a single band was present after treatment with β-mercaptoethanol (175, 181–184). A similar pattern was obtained with the RNPs from rabbit liver, Ehrlich ascites cells, and human cells in culture (174, 184). Gel electrophoresis in the presence of SDS produced a single band, with a mol wt of 40,000 to 45,000 (181). More complex band patterns, however, have been observed in similar preparations, with a major 36,000 dalton component and others ranging from 39,000 to 130,000 daltons (185). Highly complex protein patterns were obtained with rat brain RNP and with liver RNP prepared from nuclei disrupted in the French press (169, 178). In these latter studies, similar protein components were observed in the soluble fraction of the nuclear extracts. This indicates either that proteins that bind to nucleoplasmic RNA are present in excess in the nucleus or that some nuclear proteins with other functions may bind to the RNA during extraction of the RNP.

The occurrence of nucleoplasmic RNA as RNP raises the possibility that the protein components may play a role in mRNA processing. This is supported to some extent by the finding that a poly(A)-addition enzyme is associated with the RNP from rat liver nuclei (186). This enzyme, which shows the usual characteristics of poly(A) polymerases, seems to prefer nucleoplasmic RNA as primer. A study of rat liver particles obtained by differential extraction of nuclei provided some indication that nuclear RNA at different stages of processing occurs as RNP (187). Particles extracted at pH 7 contained relatively small RNA chains (6–16S) particularly rich in A and with extensive sequence homology with reiterated portions of polysomal RNA. Subsequent extractions at pH 8 and at higher ionic strength yielded particles with increasingly larger RNA chains and less extensive sequence homology. A 43S particle obtained from nuclei of ascites cells appeared to represent a late stage of nuclear RNA processing, as judged by its relatively high poly(A) content and by the fact that its labeling was preferentially depressed by cordycepin (188).

Several investigators have attempted to determine whether protein components of the nuclear RNPs remain associated with the mRNA as it is transferred to the cytoplasm. The general approach has been to compare the proteins of nuclear and cytoplasmic RNPs by gel electrophoresis. The proteins of rat liver polysomes were reported to contain a component analogous to that of purified nuclear particles (189). Studies with rat liver and avian reticulocytes failed to uncover any analogy between the protein components of nuclear particles and of mRNPs released from polysomes by EDTA (167, 184). The absence of nuclear RNP protein from the polysomal particles was also demonstrated by a different approach, in which antibodies prepared against the purified protein component of rat liver nuclear particles were tested on EDTA-released polysomal RNPs. No reaction took place, as indicated by the failure of the antibodies to lower the buoyant density of the mRNPs (190).

It is clear that a definitive answer to the question of the relation between nuclear and cytoplasmic RNP will have to await the availability of well-characterized preparations of these two types of particles.

Literature Cited

1. Jacob, F., Monod, J. 1961. *J. Mol. Biol.* 3:318
2. Steitz, J. A. 1969. *Nature* 224:957
3. Lodish, H. F., Robertson, H. D. 1969. *Cold Spring Harbor Symp. Quant. Biol.* 34:655
4. Singer, M. F., Leder, P. 1966. *Ann. Rev. Biochem.* 35:195
5. Williamson, R. 1972. *J. Med. Genet.* 9:348
6. Mathews, M. B. 1974. *Essays Biochem.* In press
7. Darnell, J. E. 1968. *Bacteriol. Rev.* 32:262
8. Penman, S., Scherrer, K., Becker, Y., Darnell, J. E. 1963. *Proc. Nat. Acad. Sci. USA* 49:654
9. Girard, M., Latham, H., Penman, S., Darnell, J. E. 1965. *J. Mol. Biol.* 11:187
10. Perry, R. P., Kelley, D. E. 1968. *J. Mol. Biol.* 35:37
11. Knöchel, W., Tiedemann, H. 1972. *Biochim. Biophys. Acta* 269:104
12. Georgiev, G. P., Samarina, O. P., Lerman, M. I., Smirnov, M. N., Severtzov, A. N. 1963. *Nature* 200:1291
13. Perry, R. P. 1963. *Exp. Cell Res.* 29:400
14. Harel, L., Harel, J., Boer, A., Imbenotte, J., Carpeni, N. 1964. *Biochim. Biophys. Acta* 87:212
15. Cheevers, W. P., Sheinin, R. 1970. *Biochim. Biophys. Acta* 204:449
16. Roberts, W. K., Newman, J. F. E. 1966. *J. Mol. Biol.* 20:63
17. Penman, S., Vesco, C., Penman, M. 1968. *J. Mol. Biol.* 34:49
18. Lindberg, U., Persson, T. 1972. *Eur. J. Biochem.* 31:246
19. Schochetman, G., Perry, R. P. 1972. *J. Mol. Biol.* 63:577
20. Lim, L., Canellakis, E. S. 1970. *Nature* 227:710
21. Kates, J. 1970. *Cold Spring Harbor Symp. Quant. Biol.* 35:743
22. Lee, S. Y., Mendecki, J., Brawerman, G. 1971. *Proc. Nat. Acad. Sci. USA* 68:1331
23. Edmonds, M., Vaughan, M. H. Jr., Nakazato, H. 1971. *Proc. Nat. Acad. Sci. USA* 68:1336
24. Darnell, J. E., Wall, R., Tushinski, R. J. 1971. *Proc. Nat. Acad. Sci. USA* 68:1321
25. Sheldon, R., Jurale, C., Kates, J. 1972. *Proc. Nat. Acad. Sci. USA* 69:417
26. Kwan, S. W., Brawerman, G. 1972. *Proc. Nat. Acad. Sci. USA* 69:3247
27. Brawerman, G. 1973. *Methods Cell Biol.* 7:1
28. Brawerman, G. 1974. *Methods Enzymol.* 30:In press
29. Sibatani, A., de Kloet, S. R., Allfrey, V. G., Mirsky, A. E. 1962. *Proc. Nat. Acad. Sci. USA* 48:471
30. Georgiev, G. P., Mantieva, V. L. 1962. *Biochim. Biophys. Acta* 61:153
31. Brawerman, G., Gold, L., Eisenstadt, J. 1963. *Proc. Nat. Acad. Sci. USA* 50:630
32. Hadjivassiliou, A., Brawerman, G. 1967. *Biochemistry* 6:1934
33. Brawerman, G., Mendecki, J., Lee, S. Y. 1972. *Biochemistry* 11:637
34. Perry, R. P., LaTorre, J., Kelley, D. E., Greenberg, J. R. 1972. *Biochim. Biophys. Acta* 262:220
35. Rosenfeld, M. G., Abrass, I. B., Perkins, L. A. 1972. *Biochem. Biophys. Res. Commun.* 49:230
36. Chantrenne, H., Burny, S., Marbaix, G. 1967. *Progr. Nucl. Acid Res. Mol. Biol.* 7:173
37. Nakazato, H., Edmonds, M. 1972. *J. Biol. Chem.* 247:3365
38. Adesnik, M., Salditt, M., Thomas, W., Darnell, J. E. 1972. *J. Mol. Biol.* 71:21
39. Faust, C. H. Jr., Diggelmann, H., Mach, B. 1973. *Biochemistry* 12:925
40. Greenberg, J. R., Perry, R. P. 1972. *J. Mol. Biol.* 72:3
41. Kitos, P. A., Saxon, G., Amos, H. 1972. *Biochem. Biophys. Res. Commun.* 47:1426
42. Shutz, G., Beato, M., Feigelson, P. 1972. *Biochem. Biophys. Res. Commun.* 49:680

43. Sullivan, N., Roberts, W. K. 1973. *Biochemistry* 12:2395
44. DeLarco, J., Guroff, G. 1973. *Biochem. Biophys. Res. Commun.* 50:486
45. Rosenfeld, G. C., Comstock, J. P., Means, A. R., O'Malley, B. W. 1972. *Biochem. Biophys. Res. Commun.* 47:387
46. Murphy, W., Attardi, G. 1973. *Proc. Nat. Acad. Sci. USA* 70:115
47. Lockard, R. E., Lingrel, J. B. 1969. *Biochem. Biophys. Res. Commun.* 37:204
48. Lingrel, J. B., Lockard, R. E., Jones, R. F., Burr, H. E., Holder, J. W. 1971. *Ser. Haematol.* 4:37
49. Lane, C. D., Marbaix, G., Gurdon, J. B. 1971. *J. Mol. Biol.* 61:73
50. Gaskill, P., Kabat, D. 1971. *Proc. Nat. Acad. Sci. USA* 68:72
51. Lanyon, W. G., Paul, J., Williamson, R. 1972. *Eur. J. Biochem.* 31:38
52. Kazazian, H. H. Jr., Moore, P. A., Snyder, P. G. 1973. *Biochem. Biophys. Res. Commun.* 51:564
53. Aviv, H., Leder, P. 1972. *Proc. Nat. Acad. Sci. USA* 69:1408
54. Labrie, F. 1969. *Nature* 221:1217
55. Hunt, R. T., Hunter, A. R., Munro, A. J. 1968. *Nature* 220:481
56. Lodish, H. F. 1971. *J. Biol. Chem.* 246:7131
57. Rabinowitz, M., Freedman, M. L., Fisher, J. M., Maxwell, C. R. 1969. *Cold Spring Harbor Symp. Quant. Biol.* 34:567
58. Temple, G. F., Housman, D. E. 1972. *Proc. Nat. Acad. Sci. USA* 69:1574
59. Jacobs-Lorena, M., Baglioni, C. 1972. *Proc. Nat. Acad. Sci. USA* 69:1425
60. Gianni, A. M., Giglioni, B., Ottolenghi, S., Comi, P., Guidotti, G. G. 1972. *Nature New Biol.* 240:183
61. Pemberton, R. E., Baglioni, C. 1972. *J. Mol. Biol.* 65:531
62. Pemberton, R. E., Housman, D., Lodish, H. F., Baglioni, C. 1972. *Nature New Biol.* 235:99
63. Rhoads, R. E., McKnight, G. S., Schimke, R. T. 1971. *J. Biol. Chem.* 246:7407
64. Means, A. R., Comstock, J. P., Rosenfeld, G. C., O'Malley, B. W. 1972. *Proc. Nat. Acad. Sci. USA* 69:1146
65. Palacios, R., Palmiter, R. D., Schimke, R. T. 1972. *J. Biol. Chem.* 247:2316
66. Palmiter, R. D., Palacios, R., Schimke, R. T. 1972. *J. Biol. Chem.* 247:3296
67. Palacios, R., Sullivan, D., Summers, N. M., Kiely, M. L., Schimke, R. T. 1973. *J. Biol. Chem.* 248:540
68. Stavnezer, J., Huang, R. C. C. 1971. *Nature New Biol.* 230:172
69. Milstein, C., Brownlee, G. G., Harrison, T. M., Mathews, M. B. 1972. *Nature New Biol.* 239:117
70. Delovitch, T. L., Davis, B. K., Holmes, G., Sehon, A. H. 1972. *J. Mol. Biol.* 69:373
71. Delovitch, T. L., Baglioni, C. 1973. *Proc. Nat. Acad. Sci. USA* 70:173
72. Swan, D., Aviv, H., Leder, P. 1972. *Proc. Nat. Acad. Sci. USA* 69:1967
73. Mach, B., Faust, C., Vassalli, P. 1973. *Proc. Nat. Acad. Sci. USA* 70:451
74. Stevens, R. H., Williamson, A. R. 1973. *Proc. Nat. Acad. Sci. USA* 70:1127
75. Robbins, E., Borun, T. W. 1967. *Proc. Nat. Acad. Sci. USA* 57:409
76. Borun, T. W., Sharff, M., Robbins, E. 1967. *Proc. Nat. Acad. Sci. USA* 58:1977
77. Kedes, L. H., Gross, P. R. 1969. *Nature* 223:1335
78. Breindl, M., Gallwitz, D. 1973. *Eur. J. Biochem.* 32:381
79. Jacobs-Lorena, M., Baglioni, C., Borun, T. W. 1972. *Proc. Nat. Acad. Sci. USA* 69:2095
80. Adesnik, M., Darnell, J. E. 1972. *J. Mol. Biol.* 67:397
81. Greenberg, J. R., Perry, R. P. 1972. *J. Mol. Biol.* 72:91
82. Bloemendal, H., Berns, A., Strous, G., Mathews, M., Lane, C. D. 1973. In *RNA Viruses—Ribosomes,* 237–250. Amsterdam: North-Holland
83. Williamson, R., Clayton, R., Truman, D. E. S. 1972. *Biochem. Biophys. Res. Commun.* 46:1936
84. Heywood, S. M., Nwagwu, M. 1969. *Biochemistry* 8:3839
85. Morris, G. E. et al 1972. *Cold Spring Harbor Symp. Quant. Biol.* 37:535
86. Suzuki, Y., Brown, D. D. 1972. *J. Mol. Biol.* 63:409
87. O'Malley, B. W., Rosenfeld, G. C., Comstock, J. P., Means, A. R. 1972. *Nature New Biol.* 240:45
88. Gilmour, R. S., Dixon, G. H. 1972. *J. Biol. Chem.* 247:4621
89. Gaye, P., Houdebine, L., Denamur, R. 1973. *Biochem. Biophys. Res. Commun.* 51:637
90. Hadjivassiliou, A., Brawerman, G. 1966. *J. Mol. Biol.* 20:1
91. Morrison, M. R., Gorsky, J., Lingrel, J. B. 1972. *Biochem. Biophys. Res. Commun.* 49:775
92. Weinberg, R. A. 1973. *Ann. Rev. Biochem.* 42:329
93. Mendecki, J., Lee, S. Y., Brawerman, G. 1972. *Biochemistry* 11:792
94. Molloy, G. R., Darnell, J. E. 1973. *Biochemistry* 12:2324

95. Nakazato, H., Kopp, D. W., Edmonds, M. 1973. *J. Biol. Chem.* 248:1472
96. McLaughlin, C. S., Warner, J. R., Edmonds, M. E., Nakazato, H., Vaughan, M. H. 1973. *J. Biol. Chem.* 248:1466
97. Molloy, G. R., Sporn, M. B., Kelley, D. E., Perry, R. P. 1972. *Biochemistry* 11:3256
98. Sheldon, R., Kates, J., Kelley, D. E., Perry, R. P. 1972. *Biochemistry* 11:3829
99. Greenberg, J. R., Perry, R. P. 1972. *Biochim. Biophys. Acta* 287:361
100. Firtel, R. A., Jacobson, A., Lodish, H. F. 1972. *Nature New Biol.* 239:225
101. Sheiness, D., Darnell, J. E. 1973. *Nature New Biol.* 241:265
102. Brawerman, G. 1973. *Mol. Biol. Rep.* 1:7
103. Jeffery, W., Brawerman, G. 1974. In press
104. Williamson, R., Morrison, M., Lanyon, G., Eason, R., Paul, J. 1971. *Biochemistry* 10:3014
105. Georgiev, G. P., Ryskov, A. P., Coutelle, C., Mantieva, V. L., Avakyan, E. R. 1972. *Biochim. Biophys. Acta* 259:259
106. Williamson, R., Morrison, M., Paul, J. 1970. *Biochem. Biophys. Res. Commun.* 40:740
107. Dina, D., Crippa, M., Beccari, E. 1973. *Nature New Biol.* 242:101
108. Greenberg, J. R., Perry, R. P. 1971. *J. Cell Biol.* 50:774
109. Firtel, R. A., Lodish, H. F. 1973. *J. Mol. Biol.* 79:207
110. Clegg, J. B., Wheatherall, D. J., Milner, P. F. 1971. *Nature* 234:337
111. Seid-Akhavan, M., Winter, W. P., Abramson, R. K., Rucknagel, D. L. 1972. *Blood* 40:927
112. Darnell, J. E., Philipson, L., Wall, R., Adesnik, M. 1971. *Science* 174:507
113. Jelinek, W., Darnell, J. E. 1972. *Proc. Nat. Acad. Sci. USA* 69:2537
114. Ryskov, A. P., Farashyan, V. R., Georgiev, G. P. 1972. *Biochim. Biophys. Acta* 262:568
115. Molloy, G. R., Thomas, W. L., Darnell, J. E. 1972. *Proc. Nat. Acad. Sci. USA* 69:3684
116. Stampfer, M., Rosbash, M., Huang, A. S., Baltimore, D. 1972. *Biochem. Biophys. Res. Commun.* 49:217
117. Jelinek, W. et al 1973. *J. Mol. Biol.* 75:515
118. Birnboim, H. C. 1972. *Biochemistry* 11:4588
119. Bramwell, M. E. 1972. *Biochim. Biophys. Acta* 281:329
120. Imaizumi, T., Diggelmann, H., Scherrer, K. 1973. *Proc. Nat. Acad. Sci. USA* 70:1122
121. Melli, M., Pemberton, R. E. 1972. *Nature New Biol.* 236:172
122. Stevens, R. H., Williamson, A. R. 1972. *Nature* 239:143
123. Williamson, R., Drewienkiewicz, C. E., Paul, J. 1973. *Nature New Biol.* 241:66
124. Edmonds, M. P., Caramela, M. G. 1969. *J. Biol. Chem.* 244:1314
125. Mans, R. T., Walter, T. J. 1971. *Biochim. Biophys. Acta* 247:113
126. Giron, M. L., Huppert, J. 1972. *Biochim. Biophys. Acta* 287:438
127. Tsrapolis, C. M., Dorson, J. W., DeSante, D. M., Bollum, F. J. 1973. *Biochem. Biophys. Res. Commun.* 50:737
128. Sasaki, K., Tazawa, T. 1973. *Biochem. Biophys. Res. Commun.* 52:1440
129. Haff, L. A., Keller, E. B. 1973. *Biochem. Biophys. Res. Commun.* 51:704
130. Sarkar, P. K., Goldman, B., Moscona, A. A. 1973. *Biochem. Biophys. Res. Commun.* 50:308
131. Schochetman, G., Perry, R. P. 1972. *J. Mol. Biol.* 63:591
132. Perry, R. P., Kelley, D. E., LaTorre, J. 1974. *J. Mol. Biol.* In press
133. Yogo, Y., Wimmer, E. 1972. *Proc. Nat. Acad. Sci. USA* 69:1877
134. Eaton, B. T., Donaghue, T. P., Faulkner, P. 1972. *Nature New Biol.* 238:109
135. Armstrong, J. A., Edmonds, M., Nakazato, H., Phillips, B. A., Vaughan, M. H. 1972. *Science* 176:526
136. Mudd, J. A., Summers, D. F. 1970. *Virology* 42:958
137. Johnston, R. E., Bose, H. R. 1972. *Proc. Nat. Acad. Sci. USA* 69:1514
138. Galet, H., Prevec, L. 1973. *Nature New Biol.* 243:200
139. Slater, I., Gillespie, D., Slater, D. W. 1973. *Proc. Nat. Acad. Sci. USA* 70:406
140. Diez, J., Brawerman, G. 1974. In press
141. Singer, R. H., Penman, S. 1972. *Nature* 240:100
142. Singer, R. H., Penman, S. 1973. *J. Mol. Biol.* 78:321
143. Greenberg, J. R. 1972. *Nature New Biol.* 240:102
144. Craig, N., Perry, R. P., Kelley, D. E. 1971. *Biochim. Biophys. Acta* 246:493
145. Gallwitz, D., Mueller, G. C. 1969. *J. Biol. Chem.* 244:5947
146. Guidice, G., Novelli, G. D. 1963. *Biochim. Biophys. Res. Commun.* 12:383
147. Revel, M., Hiatt, H. H. 1964. *Proc. Nat. Acad. Sci. USA* 51:810

148. Wilson, S. H., Hoagland, M. B. 1967. *Biochem. J.* 103:556
149. Wilson, S. H., Hill, H. Z., Hoagland, M. B. 1967. *Biochem. J.* 103:567
150. Stewart, J. A., Papaconstantinou, J. 1967. *J. Mol. Biol.* 29:357
151. Vassart, G., Dumont, J. E. 1973. *Eur. J. Biochem.* 32:322
152. Storb, U. 1973. *Biochem. Biophys. Res. Commun.* 52:1483
153. Spirin, A. S. 1969. *Eur. J. Biochem.* 10:20
154. Spirin, A. S. 1972. *Front. Biol.* 27:515–38
155. Georgiev, G. P., Samarina, O. P., 1971. *Advan. Cell Biol.* 2:47
156. Baltimore, D., Huang, A. S. 1970. *J. Mol. Biol.* 47:263
157. Schweiger, A., Hannig, K. 1971. *Biochim. Biophys. Acta* 254:255
158. Stepanov, A. S., Voronina, A. S., Ovchinnikov, L. P., Spirin, A. S. 1971. *FEBS Lett.* 18:13
159. Olsnes, S. 1971. *Eur. J. Biochem.* 23:248
160. Nolan, R. D., Arnstein, H. R. V. 1969. *Eur. J. Biochem.* 9:445
161. Lee, S. Y., Krsmanovic, V., Brawerman, G. 1971. *Biochemistry* 10:895
162. Christman, J. K., Reiss, B., Kyner, D., Levin, D. H., Klett, H., Acs, G. 1973. *Biochim. Biophys. Acta* 294:153
163. Olsnes, S. 1971. *Eur. J. Biochem.* 18:242
164. Lee, S. Y., Brawerman, G. 1971. *Biochemistry* 10:510
165. Lebleu, B. et al 1971. *Eur. J. Biochem.* 19:264
166. Blobel, G. 1972. *Biochem. Biophys. Res. Commun.* 47:88
167. Morel, C., Kayibanda, B., Scherrer, K. 1971. *FEBS Lett.* 18:84
168. Olsnes, S. 1971. *Eur. J. Biochem.* 23:557
169. Matringe, H., Jacob, M. 1972. *Biochimie* 54:1169
170. Blobel, G. 1973. *Proc. Nat. Acad. Sci. USA* 70:924
171. Olsen, G. D., Gaskill, P., Kabat, D. 1972. *Biochim. Biophys. Acta* 272:297
172. Sampson, J., Mathews, M. B., Osborn, M., Borghetti, A. F. 1972. *Biochemistry* 11:3636
173. Samarina, O. P., Lukanidin, E. M., Molnar, J., Georgiev, G. P. 1968. *J. Mol. Biol.* 33:251
174. Lukanidin, E. M., Zalmanzon, E. S., Komaromi, L., Samarina, O. P., Georgiev, G. P. 1972. *Nature New Biol.* 238:193
175. McParland, R., Crooke, S. T., Busch, H. 1972. *Biochim. Biophys. Acta* 269:78
176. Stevenin, J., Mandel, P., Jacob, M. 1970. *Bull. Soc. Chim. Biol.* 52:703
177. Faiferman, I., Hamilton, M. G., Pogo, A. O. 1970. *Biochim. Biophys. Acta* 204:550
178. Faiferman, I., Hamilton, M. G., Pogo, A. O. 1971. *Biochim. Biophys. Acta* 232:685
179. Stevenin, J., Jacob, M. 1972. *Eur. J. Biochem.* 29:480
180. Stevenin, J., Zawislak, R., Jacob, M. 1973. *Eur. J. Biochem.* 33:241
181. Krichevskaya, A. A., Georgiev, G. P. 1969. *Biochim. Biophys. Acta* 194:619
182. Sarasin, A. 1969. *FEBS Lett.* 4:327
183. Niessing, J., Sekeris, C. E. 1971. *FEBS Lett.* 18:39
184. Lukanidin, E. M., Georgiev, G. P., Williamson, R. 1971. *FEBS Lett.* 19:152
185. Niessing, J., Sekeris, C. E. 1971. *Biochim. Biophys. Acta* 247:391
186. Niessing, J., Sekeris, C. E. 1972. *FEBS Lett.* 22:83
187. Drews, J. 1969. *Eur. J. Biochem.* 9:263
188. Cornudella, L., Faiferman, I., Pogo, A. O. 1973. *Biochim. Biophys. Acta* 294:541
189. Schweiger, A., Hannig, K. 1970. *Biochim. Biophys. Acta* 204:317
190. Lukanidin, E. M., Olsnes, S., Pihl, A. 1972. *Nature New Biol.* 240:90

ANIMAL RNA VIRUSES: GENOME STRUCTURE AND FUNCTION

✖ 861

Aaron J. Shatkin

Department of Cell Biology, Roche Institute of Molecular Biology, Nutley, New Jersey

CONTENTS

INTRODUCTION

Viruses that depend upon the diverse world of animal cells for their survival have evolved in various directions. Within the resulting multitude of animal RNA viruses there exist, nevertheless, unifying principles of genome structure and function shared by viruses in different groups. These similarities between viruses may not be apparent by serological or morphological criteria, but they are often reflected in the processes which constitute the virus life cycle. For example, influenza and reovirus have segmented genomes, and both yield a high frequency of recombinants by reassortment of genome segments during replication. In this review some aspects of animal RNA virus genome structure are considered in relation to the biochemistry of progeny virus formation.

CLASSIFICATION ACCORDING TO GENOME STRUCTURE AND FUNCTION

Successful multiplication of an animal virus requires that its genome be organized to direct the synthesis of viral RNA and proteins, including in some instances the enzymes involved in genome replication. Thus, the production and translation of

643

virus-specific mRNA are key steps in the viral replicative cycle. The recently proposed classification of RNA animal viruses according to their mechanism of genetic expression (1), together with a consideration of their genome structure, provides a useful basis for comparing viruses in different groups.

Class 1: Picornaviruses and Arboviruses (Genome = mRNA)

Viruses that contain single-stranded RNA as their genetic material have the potential to initiate a replicative cycle in the absence of RNA synthesis: i.e. the genome may bind to ribosomes and act directly as messenger. This relatively simple type of direct decoding of the viral genome is used by RNA bacteriophages (2). It is also found in a modified version among animal viruses of class 1 which includes members of the picornavirus and arbovirus (recently renamed togavirus) groups. Viruses in this first class contain one molecule of single-stranded RNA which is infectious under appropriate conditions when extracted free of viral protein (3). In agreement with a messenger function for the genome RNA, the RNA complementary to that present in encephalomyocarditis virus (EMC) and poliovirus is not infectious (4–6). A molecular weight of $2.6-2.8 \times 10^6$ has been determined by several independent methods for the genomes of representative picornaviruses including polio (7), a bovine enterovirus (8), and EMC virus (9). The size of the arbovirus genome is approximately 50% larger: 4×10^6 daltons for Sindbis (10), Semliki Forest (11), and Western equine encephalitis (12) viruses. Reversible conformational changes observed for arbovirus RNA (10) are consistent with an ability to form extensive base-paired regions, possibly similar to the hydrogen-bonded loops in bacteriophage RNA. The arbovirus genome inside virions is associated with protein subunits in a highly condensed nucleoid (13, 14), and an ordered secondary structure presumably is required for assembly of the ribonucleoprotein complex.

Many animal cell and virus mRNAs have polyriboadenylic acid, poly(A), covalently attached at the 3′ terminus (15–22). The function of the poly(A) is unknown, but it presumably is involved in translational or transcriptional control. Since the genome and mRNA are the same for class 1 viruses, it is not unexpected that both picornavirus [polio (23, 24), mengo (25)] and arbovirus [Eastern equine encephalitis (23), Sindbis (25, 26)] genomes contain covalently bound poly(A). In poliovirus RNA, the adenine-rich sequence is at the 3′ end of the RNA and comprises about 1% of the total genome of 7500 nucleotides. The poly(A) released by RNase A digestion of polio RNA is heterogeneous in size with an average chain length of about 90 present in the sequence -PyrpGpGp$(Ap)_{89}A_{OH}$ (24). The poly(A) may vary in length or the heterogeneity may result from nucleolytic cleavage at an occasional pyrimidine (Pyr) residue within repeating sequences of adenine. Size variability was also observed for the poly(A) in Sindbis virus RNA (26). On the basis of their greater resistance to RNase A digestion and ability to bind to nitrocellulose filters in 0.5 M KCl, 10–20% of the RNA molecules extracted from Sindbis virus were found to contain poly(A) of chain length 150–250 nucleotides. The remaining 80–90% contained a sequence of only 60–80 adenines. The two types of viral RNA molecules are of similar specific infectivity, indicating that the length of the poly(A), at least between 60 and 250 nucleotides, does not affect the ability of Sindbis RNA to initiate a productive infectious cycle (26).

Its widespread occurrence indicates that poly(A) probably has an important role in RNA function and/or replication. For example, circular double-stranded RNA has been reported in picornavirus-infected cells (27). Furthermore, an RNA ligase that converts oligo(A) to circles is present in both bacterial (28) and mammalian cells (B. Öberg, personal communication; 29). The 3' terminus of poliovirus RNA is poly(A) of variable chain length (24), and pA rather than a triphosphate is present at the 5'end (30). Termini of this type could result from cleavage of circular RNA at variable positions within a poly(A) sequence. Thus, the poly(A) could serve as a recognition sequence for an endonuclease that converts circular forms of viral RNA to linear forms that are later packaged as genomes in mature virions (31). The possibility that circular RNA is a replicative intermediate should be considered as speculative in view of the extensive evidence that linear replicative forms are involved in poliovirus RNA synthesis (3).

In addition to the structural requirements for RNA polymerase template activity and for virion packaging demanded of all viral genomes, the class 1 RNA genomes, like the RNA of bacteriophages, must also have the sequences necessary for ribosome binding and polypeptide chain initiation and termination. In contrast to the multiple initiation and termination sites in the polycistronic genomes of RNA bacteriophages (2), the genome of picornaviruses apparently contains a single poly-peptide initiation site (32, 33) and, like most animal cell mRNAs, is translated as a monocistronic mRNA (34–36). The resulting polyprotein is cleaved during synthesis to yield structural and nonstructural polypeptides by a post-translational mechanism probably involving both host and virus-specific proteases or host protease stimulated by virus infection (37). A similar mechanism operates for arbovirus protein synthesis (38–44). No extensive sequence analyses have been reported for animal virus genomes and the structural basis for a single, functional initiation site in a polycistronic RNA is not known. The largest virus-specific polyprotein observed for picornavirus-infected cells is $> 200,000$ daltons, i.e. similar to the entire coding capacity of the genome (45–48). Consequently, initiation must occur near the 5' terminus of the RNA (49–52). The 5'-terminal nucleotide of poliovirus RNA is predominantly pAp (30) in contrast to the pppGp of bacteriophage RNAs (2). Animal cell mRNA produced by enzymatic processing of larger, heterogeneous nuclear RNA also would be expected to contain a 5'-monophosphate (or hydroxyl) terminus. Thus, the very limited amount of structural information available indicates that genomes of the class 1 viruses resemble other animal cell mRNAs.

Class 2: Rhabdoviruses and Paramyxoviruses (Genome Complementary to mRNA)

A second and more complex type of genome expression occurs among the rhabdo-viruses (bullet-shaped) and paramyxoviruses (subgroup II myxoviruses). They also contain a single-stranded RNA molecule but it comprises only 1–3% of the virion mass (53, 54) in contrast to 25–30% for picornaviruses (55). The molecular weight of the rhabdovirus genome estimated for vesicular stomatitis virus (VSV) RNA by sedimentation velocity (56), electron microscopy (57), and electrophoretic mobility in polyacrylamide gels (58, 59) varies from 3.2 to 4×10^6. On the basis of sedimenta-tion constants of 50–57S determined for the RNA of Newcastle disease virus (NDV)

(60, 61), SV5 (62), and Sendai virus (63–65) a molecular weight of about 7×10^6 was calculated for paramyxovirus genomes. The same molecular weights were estimated after denaturation of the RNAs (66, 67) suggesting that paramyxovirus and rhabdovirus genomes are large, covalently linked polyribonucleotide chains. In view of a recent report (68) that NDV and Sendai virus RNA may have molecular weights of only 2.3×10^6 based on their sedimentation in dimethylsulfoxide-containing gradients, it is important to reexamine the chain length of paramyxovirus RNA by other methods including determination of the number of internal vs terminal nucleotides per molecule.

Within virus particles the genome RNA of class 2 virions is firmly associated with one species of viral structural protein (69, 70). This nucleocapsid or N protein is uniformly arranged along the RNA, one protein subunit per 10–20 nucleotides, to form a helical, RNase-resistant nucleocapsid (70, 71). The helix of SV5 and mumps virus nucleocapsids is left-handed (72). Unit lengths of about 1 μ were found for the nucleocapsids of several paramyxoviruses including NDV (71), Sendai (73), SV5 (74), and mumps and measles (75) viruses, consistent with the presence of one genome equivalent of RNA. In some instances paramyxoviruses may contain more than one nucleocapsid per virion (73), a finding predicted from earlier genetic data (76). Measurements of uncoiled nucleocapsids released from VSV also indicate that each helix contains one RNA strand of molecular weight about 3.6×10^6 (57).

The genome RNA of class 2 virions is not infectious (56, 61, 77, 78). It appears not to function directly as a messenger (65, 79–83) and contains little, if any, poly(A) (25, 84, 85). In infected cells, the viral genome is transcribed into shorter complementary sequences. The RNA-dependent RNA polymerase that transcribes the parental genome, providing the initial mRNA, is a structural component of class 2 virions (86–89). Although naked VSV RNA is not infectious, an infectious ribonucleoprotein complex has been obtained from detergent-treated virions (90, 91). The complex contains an active RNA polymerase which presumably accounts for its infectivity (91). Although absent from the genome RNA, poly(A) is present in viral genome transcripts isolated from cells infected with VSV (84, 92), NDV (93), and Sendai virus (94). In addition, RNA made in vitro by the polymerases of purified VSV (95) and NDV (93) and by extracts of VSV-infected cells (96) contains poly(A). Interestingly, uridylic acid is the predominant nucleotide in the genomes of both VSV (82) and several paramyxoviruses (78). However, long stretches of poly(U) were not detected in viral RNA genomes extracted from VSV, NDV, and representatives of class 1, 3, and 4 viruses (97). Thus poly(A) may be added to viral mRNAs by a template-independent post-transcriptional process or by an enzyme slippage mechanism at a site in the template RNA consisting of a few consecutive uracils. The latter could provide the punctuation that signals the termination of a nascent mRNA molecule (95).

Class 2 genome RNA transcripts are present in polyribosomes of infected cells and probably function as monocistronic viral mRNA (80, 81, 83, 98). Virus-specific RNA molecules large enough to be polycistronic mRNA are also present in VSV- and NDV-infected cells, and post-translational cleavage of nascent viral structural polypeptides has not been entirely ruled out. However, it seems likely that most of the class 2 virion proteins are coded for by monocistronic mRNAs (67, 69, 99, 100). No cell-free tests of mRNA activity have been reported for viral transcripts produced by class 2 virion-associated polymerases in vitro, but RNA isolated from polysomes of VSV-infected cells (101) or from Sendai-infected cells (D. Kingsbury, personal communication) directs viral protein synthesis in reticulocyte extracts. VSV RNA in the size range 12–15S codes for the smaller proteins: N, NS, M, and G, and the formation of the large (L) protein is directed by 28S VSV RNA (101). Sendai virus 50S genome RNA was inactive as an in vitro messenger, but genome transcripts ($\sim$ 18S RNA) directed the synthesis of authentic N protein and at least one other Sendai virus structural protein.

Since class 2 virion RNA is predominantly of opposite polarity to viral mRNA in infected cells (102, 103), the genome probably does not contain sites for ribosome binding or correct polypeptide chain initiation. However, the formation of transcripts from several defined regions of the genome requires multiple template sites for initiation of RNA synthesis. The structure of these internal sites and the 5′- and 3′-terminal sequences of rhabdo- and paramyxovirus genome and mRNA is currently being studied in several laboratories. It has been reported recently that VSV in vitro transcripts include products with 5′ sequences: pppApCpGp... and pppGpCp... (104).

As further evidence for the structural similarity between the rhabdo- and paramyxovirus groups, cells doubly infected with VSV and SV5 yield both parental virus types and phenotypically mixed, bullet-shaped virions containing the two surface glycoproteins of SV5 in addition to a full complement of VSV structural polypeptides (105). "Pseudotype" VSV particles containing surface structural proteins of the RNA tumor viruses have also been described (106). It would be of interest to know if internal virion proteins, for example the core particle-associated polymerases of VSV and RSV, can also be phenotypically mixed and, if so, what effect this would have on the biological activity of the virions.

Class 3: Influenza and Diplornaviruses (Segmented Genomes)

A third class of animal RNA viruses consists of virions that have discontinuous genomes, either single-stranded RNA as in influenza and other orthomyxoviruses (subgroup I myxoviruses) or double-stranded RNA as found in the diplornaviruses including reovirus, blue-tongue virus, and others (107). A recently isolated bacteriophage, $\phi 6$, also contains a segmented double-stranded RNA genome (108). On the basis of the RNA and protein composition of purified virions, the estimated total molecular weights of the influenza and reovirus genomes are $3–5 \times 10^6$ (109–112) and 15×10^6 (113), respectively. However, RNA extracted from infectious virions of either type consists of a reproducible mixture of noninfectious molecules of lower molecular weight (114–118). By polyacrylamide gel electrophoresis the influenza RNA mixture can be resolved into 5 or more distinct segments ranging in size from about 0.2 to 1.0×10^6 daltons. Each segment is associated with nucleocapsid protein subunits, and the resulting ribonucleoprotein complexes of different lengths can be separated according to the size of the influenza RNA segment that they contain (119). The double-stranded RNA mixture isolated from reovirus can be separated into ten distinct molecules: three large (L = $2.3–2.7 \times 10^6$ daltons), three medium (M = $1.3–1.6 \times 10^6$ daltons) and four small segments (S = $0.6–0.9 \times 10^6$ daltons)

(113). Infectious reovirus and empty particles have the same polypeptide composition, suggesting that reovirus RNA is not associated with an internal nucleocapsid protein (120). The arrangement of the genome segments within virions is an interesting but unsolved problem. The genetic information present in each segment is probably unique since hybridization studies revealed no base sequence homology among the different genome segments of influenza (121, 122) or the three size classes of reovirus RNA (123). Thus it is not surprising that only virus particles that contain all of the genome segments are infectious. Defective influenza and reovirus particles both are deficient in the largest genome segment (117, 124, 125), suggesting that it may be present in limiting concentrations in infected cells or is the last segment to be packaged during genome assembly and viral morphogenesis.

Analyses of the 5'- and 3'-terminal sequences also indicate that the influenza and reovirus genome segments do not arise by cleavage during extraction of the RNA from virions. Each of the influenza single-stranded segments has the 5'-terminal nucleotide, pppAp (126), and all twenty 5' ends of the reovirus double-stranded segments contain ppGp (127). The absence of a 5' triphosphate in reovirus RNA is due to the presence in purified virions of phosphohydrolase(s) that hydrolyzes the γ-phosphates from the 5' termini (128–130). The presence of unique, initiating nucleotides at the 5' termini strongly suggests that the genome segments are independently synthesized in infected cells. As further evidence for the presence of RNA segments within virions, the same number of ^{3}H-labeled 3' termini was found in reovirus RNA that had been oxidized and reduced with tritiated borohydride either in situ or after extraction from virions (131). The 3'-terminal sequence of each of the reovirus RNA segments is ... AAC_{OH} (132) and the presumably base-paired 5' ends have the complementary sequence, ppGUU .... The same oxidation-reduction technique was used to label the 3' termini of RNA isolated from purified influenza virus (133). More than seven labeled RNA segments were resolved by gel electrophoresis, and in each RNA segment the 3'-terminal base was uracil.

The genome segments of class 3 virions do not function as mRNA, and no covalently bound poly(A) has been found in the genome RNA (113, 134–137). Both influenza (138) and reovirus (139, 140) contain a virion-associated RNA polymerase that transcribes the genome RNA. The genome transcripts are present in polyribosomes (141–143) and probably function as viral mRNA during the early stages of the replicative cycle. The parental virion polymerase of influenza and reovirus may provide the initial mRNA in infected cells, but the early phase of replication of both viruses is inhibited by high levels of actinomycin (144, 145). The antibiotic does not block replication of viruses in classes 1 and 2 (146) and does not inhibit class 3 virion-associated RNA polymerase activity in vitro (138, 140). Sensitivity to an inhibitor of DNA-dependent RNA synthesis suggests that a host cell function is also required for synthesis of class 3 virion RNA.

Only one strand of each of the ten reovirus segments is copied by the virion-associated polymerase (147–149). The in vitro products correspond in size and base sequence to the viral mRNA synthesized in reovirus-infected cells (150). Correct initiation and termination of mRNA continues in vitro for long periods and the polymerase remains template-bound throughout this process (130). The prolonged reinitiation and transcription from one strand of each duplex segment by a bound polymerase activity would be facilitated if the genome segments were arranged in a circular configuration that rotated through fixed polymerase sites in the reovirus core. The molecular weights of the single-stranded reovirus RNA species coincide with the values expected for monocistronic mRNAs on the basis of size of the viral structural proteins (151). A similar relationship may hold for influenza RNA and proteins (152). Evidence in support of a monocistronic function has been obtained for reovirus mRNA. RNA synthesized in vitro was separated according to size and translated in mammalian cell-free, protein synthesizing systems (153, 154). Products similar in size to reovirus proteins were formed, and the size was a function of the molecular weight of the mRNA directing the system. In contrast to most animal mRNAs, reovirus mRNA made in vitro (155) or isolated from infected cells (156) contains no detectable poly(A). The 3'-terminal residue is cytosine and the 5' end is ppG in all reovirus mRNA species, i.e. identical to the termini of the genome segments. These findings are consistent with a precursor-product relationship between the single-stranded mRNA and double-stranded genome segments of reovirus (157, 158).

Although neither the mRNA nor genome RNA of reovirus has covalently bound poly(A), 35–50% of the total RNA present in purified virions is a mixture of low molecular weight single-stranded oligonucleotides including oligo(A) (135, 137, 159). The oligonucleotides are likely to be virus-specified since they are consistently found in the same relative proportions in all three serotypes of reovirus purified from mouse, human, or hamster tissue culture cells (136, 137). Each virion contains an average of 2500 molecules of oligomers of chain length two to nine bases with ppG or pppG at the 5' termini (160, 161). In addition, there are about 1000 molecules of oligo(A) per virus particle. The oligo(A) is in the size range 10–15 residues and has been reported to contain 5' pA (160) and 5' pppA (161). The oligomers may be abortive transcripts synthesized by the virion-associated polymerase during maturation (161). The role of the single-stranded oligonucleotides in the reovirus replicative cycle is presently unknown, and infectious virions purified free from the oligo(A) have been reported (162). Attempts to obtain highly infectious subviral particles that lack oligo(A) by growing reovirus in cordycepin-treated L cells (R. L. Ward and A. J. Shatkin, unpublished results) or by selectively digesting purified reovirus with chymotrypsin were unsuccessful (163, 164). However, the reduced infectivity of subviral particles may have been due to their failure to adsorb efficiently since a productive cycle can be initiated by oligo(A)-depleted particles added to cells in the presence of kaolin (unpublished results, C. Carter and A. J. Shatkin).

Cells infected with subviral particles produce virions that contain oligo(A). In addition, a polymerase activity that catalyzes the synthesis of oligo(A) in vitro is present in purified reovirus (C. M. Stoltzfus, A. K. Banerjee, M. Morgan, and A. J. Shatkin, unpublished results). Presumably this activity mediates the formation of virus-associated oligo(A) during viral morphogenesis.

Class 4: RNA Tumor Viruses

The genomes of the RNA tumor viruses also appear to be segmented (165). These viruses, which include various leukemia and sarcoma viruses of avian and mammalian origin, have been placed in a fourth class on the basis of their oncogenic potential (166). The single-stranded genome of RNA tumor viruses has a sedimentation constant of 60–70S, consistent with a total molecular weight of 10^7. After denaturation by heat, formamide, or dimethylsulfoxide treatment, this value is reduced to about 35S, indicating that the genome consists of 3 or 4 noncovalently associated subunits of about $2–3 \times 10^6$ daltons each. The subunits of avian RNA tumor viruses have been partially resolved into two size classes, a and b, by polyacrylamide gel electrophoresis (167). Pure clones of sarcoma virus were found to have only a subunits and nontransforming variants of the clones isolated subsequently to have only b subunits (168). It is not known if the subunits are redundant or unique, but the more slowly migrating (larger) a subunit is thought to have the gene(s) necessary for oncogenic transformation of cells (167, 169). Genetic evidence consistent with polyploidy has been obtained (170, 171). However, annealing studies with DNA transcribed from the RSV and AMV genomes in vitro are consistent with the presence in virions of 3 or 4 subunits of different base sequence (172; H. Fan & D. Baltimore, personal communication). Thus, there may be some genes necessary for virus replication that are common to a and b subunits and other genes (x) required for transformation that are present only in a subunits, i.e. $a = b + x$ (171). Fingerprint patterns of purified a and b subunits should help to resolve this question. Different-sized RNA subunits were not found in another study of transforming vs nontransforming avian RNA tumor viruses (173). In addition to the genome segments, 4S tRNA (174), 5S, 7S, and other low molecular weight RNAs (175) have been isolated from purified virions. These may include RNase degradation products of the genome RNA in some preparations (176), but the smaller RNAs present in virions have important functions in the virus life cycle. For example, the 4S RNA associated with the genome segments serves as a primer for the virion-associated reverse transcriptase. A provocative finding is the observation that this 4S RNA is predominantly a species of Met-tRNA (J. M. Bishop, personal communication). The 7S RNA from avian and murine viruses and from uninfected chick cells have similar oligonucleotide fingerprints, suggesting that the 7S RNA in virions and cells have the same, as yet unknown, functions (177). The mechanism for holding the genome segments together has not been defined, but a fraction of the low molecular weight RNA is hydrogen-bonded to the 35S subunits and could have a "linker" function.

Both the 5′- and 3′-terminal residues of the 35S RNA genome subunits have been analyzed. Heat-denatured Rous sarcoma virus (RSV) RNA was treated with polynucleotide kinase and (γ-^{32}P) ATP to phosphorylate the 5′ termini (178). Alkaline digestion released one residue of ^{32}pAp per 2.3×10^6 daltons. The same results were obtained with RNA treated with alkaline phosphatase before the kinase. These findings indicate that the subunits of RSV RNA have predominantly $_{HO}$Ap at the 5′ termini. Digestion with RNase T_1 released more than half the radioactivity as ^{32}pApGp, suggesting that the 5′-terminal sequences of the subunits may be similar. Direct determination of the terminal nucleotides requires an inordinate amount of ^{32}P-labeled purified virus since the virus contains 31% lipid, including phospholipids, and only 1% RNA. Adenosine (179) and uridine (180) have both been reported as the major 3′-terminal nucleoside of the avian myeloblastosis virus (AMV) genome subunits; 3′-terminal uridine is also present in murine sarcoma virus (MSV) 70S

RNA (181). In another study (182), the same technique of reducing oxidized 3' ends with tritiated borohydride was used to label 70S RNA from several mammalian viruses and a reptilian virus particle. Again, uridine was identified as the predominant 3'-terminal residue, but smaller amounts of adenosine, cytosine, and guanosine were also labeled with ^{3}H. On the basis of the number of ^{3}H-labeled termini observed, molecular weights of $2.2–2.9 \times 10^6$ were estimated for the genome RNA subunits. The basis for the different 3' termini found in AMV RNA may be due to the presence of more than one type of virus particle in purified virus preparations or to the low molecular weight RNA species that are associated with the 70S RNA.

The intracellular events which occur after infection with RNA tumor viruses are not completely understood. They have been summarized recently in excellent review articles (166, 183–185). The observation that DNA synthesis is required for oncogenic conversion and virus replication (186–188) was the basis for Temin's provirus theory, i.e. that DNA copied from the viral genome is integrated into the host genome and serves as a template for production of more viral RNA. The well-known discovery of RNA-dependent DNA polymerase, "reverse transcriptase," in purified RNA tumor viruses (189, 190) has stimulated many studies to test the role of DNA in cell transformation by RNA tumor viruses. It has been shown recently that DNA isolated from RSV-transformed mammalian cells that do not produce virus is capable of transforming chick cells; furthermore, the chick cells yield RSV that is indistinguishable from the original transforming virus (191). In addition, RSVα(0), a variant of RSV that lacks reverse transcriptase, is also noninfectious (192). These observations lend strong support to the idea that the intermediate involved in RNA tumor virus transformation of cells is DNA and that the virion-associated polymerase is responsible for its synthesis.

RNA complementary to the viral genome was not detected in RSV-transformed, virus-producing chick cells (193), suggesting that the viral genome and mRNA have the same polarity. In addition, it has been reported that AMV genome RNA can function as mRNA in vitro with cell-free extracts of E. coli (194). Translation of tumor virus RNA into an authentic viral protein in vitro has not been reported with a eukaryotic protein synthesizing system. Consistent with an mRNA function, poly(A) is covalently associated with the genomes of all RNA tumor viruses studied including avian and mammalian viruses (85, 134, 195–197). The position of the poly(A) within the RNA subunits is not known, but in RSV it may be present as several sequences of 10–40 consecutive adenine residues (196). Similar amounts of poly(A) were found in the low molecular weight RNAs from RSV, suggesting that at least some of the low molecular weight material may be degraded genome RNA (134). The mechanism of addition of poly(A) to, or its function in, RNA tumor virus genome subunits remains to be determined.

RNA SYNTHESIS: GENOME TRANSCRIPTION VS REPLICATION

Class 1 Virions

Class 1 virus genome transcription and replication cannot be differentiated on the basis of complementarity between the viral mRNA and genome. The genome of picornaviruses is present in polysomes of infected cells and has been shown to function as mRNA in vivo and in vitro (32, 33, 55, 198–201). The entire genome is translated and the resulting large polypeptide is cleaved during and subsequent to translation to form the viral proteins. Arbovirus RNA is infectious (12, 26) and this indicates that it can also serve as mRNA. Perhaps due to the high levels of RNase in arbovirus host cells, only recently have data been obtained on the nature of arbovirus mRNA in infected cells. Virus-specific polysomes from Sindbis-infected BHK cells contain a mixture of 16–28S RNA molecules that have covalently bound poly(A) and may represent specific parts of the viral genome (202). The predominant species of viral RNA in polysomes of Sindbis virus-infected BHK cells (203) and SFV-infected chick cells (204) is 26S RNA, and from Sindbis-infected HeLa cells (205) it is 28S RNA. The 26S RNA from Sindbis virus-infected hamster or chick cells has the same polarity as virion RNA and represents a unique one third portion of the 49S genome RNA (206). It is of sufficient molecular weight to code for the three viral structural proteins (207). In addition to the 26S RNA, lesser amounts of 15–18S RNA (205), 33S RNA (203, 204), and 42S RNA (203) were also found in arbovirus-specific polysomes. At least some of these differences may be due to the presence of defective, interfering virus in the inoculum (208). A fraction of the 28S and 15–18S RNA was converted to an RNase-resistant form after annealing with Sindbis virus-specific double-stranded RNA. This conversion was prevented by an excess of virion RNA, consistent with the smaller RNAs having the same polarity as the viral genome (205). Although the parental arbovirus genome can serve as the initial viral mRNA in infected cells, the newly formed RNAs probably code for viral proteins at the time of maximal synthesis in the virus replicative cycle. The presence of large RNAs representing portions of the viral genome in polysomes of arbovirus-infected cells is consistent with a mechanism of arbovirus protein synthesis involving both post-translational cleavage of large polypeptides as in picornaviruses and the utilization of subgenomic mRNAs as in class 2 viruses.

In both picornavirus- and arbovirus-infected cells the synthesis of viral RNA occurs in membranous "replicative complexes" that contain viral RNA, proteins, and RNA polymerase activity (209–212). Genetic studies indicate that the polymerase activities are dependent on functional viral genes (213–215). Thus initiation of virus RNA synthesis in infected cells presumably is preceded by the translation of the parental virus RNA. The resulting polymerase(s) replicates the input genome RNA by a two-step process: formation of a complementary (minus), single-stranded template that is preferentially copied to produce an excess of genome (plus) RNA. The details of this process including a description of the RNA replicative intermediates have been discussed for picornaviruses (3). A model for arbovirus RNA replication has also been proposed on the basis of RNase-resistant RNA structures obtained from Sindbis virus-infected hamster and chick cells (216). The molecular weights of nuclease-resistant RNAs support the hypothesis that they correspond to intermediates in the synthesis of genome RNA (4.0–4.6×10^6 daltons) and two abbreviated forms of the genome RNA that probably function as mRNA: 1. the 26S RNA (1.5×10^6 daltons) described above and 2. another RNA of inter-mediate size with an expected molecular weight of 2.8×10^6, representing the remaining 70% of the viral genome. In infected cells the 26S RNA is made in large quantities, the 49S in lesser amounts, and the RNA of intermediate size in very minor quantities consistent with a controlled synthesis of selected regions of the 49S genome (216). If 26S RNA represents the 5'-terminal one third of the genome RNA, the relative amounts of the 49S and 26S species could be regulated by the number of times the replicase reads through the termination sequence for the 26S RNA in the 49S genome complement, i.e. template RNA. The small amount of intermediate-size RNA as compared to 26S and 49S RNA presumably would result from a lower frequency of chain initiation at an internal site as compared to a 5'-terminal site in the template RNA (216).

The general mechanism of class 1 virus RNA synthesis is presumed to resemble that so elegantly elucidated for bacteriophage $Q\beta$ (2, 217, 218). However, in contrast to the highly purified $Q\beta$ RNA polymerase, with few exceptions the animal virus RNA polymerases that have been studied in vitro are crude cell fractions that are (a) relatively unstable, (b) of low specific activity compared to infected cells, and (c) not dependent on exogenous templates. A soluble RNA polymerase complex purified 85-fold from poliovirus-infected HeLa cells (219) or several hundredfold from EMC virus-infected BHK cells (220) remained template-bound and independent of added viral RNA. The in vitro products of the poliovirus enzyme complex included both single-stranded viral RNA and partially RNase-resistant, presumptive replicative intermediate RNA. Cell membrane fractions that synthesize predominantly single- or double-stranded RNA have been isolated from cells infected with picornaviruses (221–223) and arboviruses (224, 225). However, an answer to the question of whether the products are synthesized by separate polymerases or the same enzyme in different replicative complexes awaits purification of template-dependent RNA polymerase. A start has been made in this direction with the membrane-bound RNA polymerase isolated from EMC virus-infected BHK cells (220). The purified polymerase-RNA complex was dissociated from membranous material by detergent treatment, and the endogenous template was digested with micrococcal nuclease. The preparation was essentially inactive in the absence of added template but incorporated GMP into acid-soluble product in response to poly(C). Several polypeptides similar in molecular weight to the host- and viral-coded subunits of the $Q\beta$ polymerase (217, 218) were detected in the polymerase preparation. Unfortunately, the ability of the enzyme to utilize added EMC virus RNA as a template was not tested.

Class 2 Virions

Class 2 virions including the rhabdoviruses, VSV (86) and Kern Canyon virus (226), and the paramyxoviruses, NDV (87) and Sendai virus (88, 89), contain RNA polymerase that transcribes the viral genome. The VSV transcriptase activity is associated with the viral nucleocapsid (91). The presence of transcriptase but not replicase activity in purified virions indicates that two distinct intracellular processes, possibly mediated by two forms of an enzyme complex, are involved in viral mRNA and genome formation. Consistent with this possibility, in cells infected in the presence of cycloheximide with Sendai virus (227) or VSV (228), transcription of the parental viral RNA occurred but genome replication was prevented. Inhibition of protein synthesis at later times in the Sendai infectious cycle, when both genome and mRNA were being synthesized, resulted in the preferential cessation of genome replication (227). Under the same conditions transcription continued, again suggesting that two enzymes, or an enzyme complex consisting of stable and unstable components, are required for synthesis of mRNA and genome RNA. Additional support for a multicomponent viral RNA synthesizing activity in VSV-infected cells has been obtained using *ts* mutants (A. Huang, personal communication).

Subcellular structures containing nascent virus-specific RNA have been obtained from homogenates of Sendai- (229) and VSV-infected cells (230). The Sendai virus-specific structures were partially separated by sucrose density gradient centrifugation into several viral RNA-containing components, but no RNA polymerase activity was detectable in any of them. In another study, the structures from Sendai-infected cells were deproteinized before sedimentation analysis (231). Two classes of partially RNase-resistant RNA complexes were observed: 24S "replicative intermediate," apparently involved in genome RNA synthesis, and heterogeneous 28–60S "transcriptive intermediates" that contained genome RNA and complementary RNA of lower molecular weight. As expected for an intermediate involved in genome replication, cycloheximide treatment of infected cells inhibited [^{3}H]-uridine incorporation into the 24S RNA complexes more rapidly than into the 28–60S intermediates (231). However, interpretation of the results is subject to the reservation that deproteinization is known to result in artifactual formation of multi-stranded complexes of phage RNA in infected bacteria (232).

Replication of the paramyxovirus or rhabdovirus genome requires a template RNA molecule of the same molecular weight but complementary base sequence. In infected cells, such a molecule could be synthesized by the virion-associated RNA polymerase, but the in vitro products of these enzymes are smaller than a complete copy of the viral genome. The 57S RNA isolated from paramyxovirus virus-infected cells self-hybridizes to the extent of 25% or more indicating that both plus and minus high molecular weight RNA strands are synthesized (229, 231). The 57S RNA corresponding to transcript product is also packaged in mature virions of paramyxoviruses, but to a relatively small extent (102, 103). As noted above, paramyxoviruses sometimes have more than one nucleocapsid per virion, and it seems likely that some virions contain the two complementary strands of 57S RNA.

Rhabdovirus-specific RNA extracted from VSV-infected cells consists of a mixture of 12–15S and 28S complementary mRNA species that contain poly(A) and 23S, 28S, and 42S genome RNA molecules that do not (58, 59, 81–83, 98). The two shorter lengths of genome RNA are incorporated into defective virus particles that arise during high multiplicity passage of 42S RNA-containing, infectious virions (233, 234). On the basis of the kinetics of nucleocapsid formation and the inability to detect the 42S template for genome RNA synthesis, it was suggested that the 42S viral RNA is formed by ligation of the 23S and 28S RNA (58). However, the two smaller genome RNA species have been shown to comprise the same region of the 42S viral RNA with a specific portion of the 28S RNA deleted from the 23S RNA (59). Furthermore, RNase-resistant RNA corresponding to replicative intermediates of all three sizes of genome RNA has been reported recently, and hybridization results are consistent with the presence of both genome and complementary 42S RNA strands in VSV-infected cells (59). VSV RNA replication thus resembles paramyxovirus genome RNA synthesis and apparently involves (a) transcription to yield a 42S complementary (plus) strand followed by (b) replication in which the 42S transcript is used as a template for the production of multiple copies of genome (minus) RNA.

Class 3 Virions

Class 3 virion RNA synthesis has been discussed in detail in a comparative review (107). An important feature of both influenza and reovirus is the presence of RNA polymerase or transcriptase activity in purified virions (138–140). Ribonucleoprotein complexes isolated from influenza-infected cells contain a transcriptase activity and only a single detectable polypeptide, the nucleoprotein (NP) polypeptide (235). The reovirus polymerase activity in purified virions is associated with subviral particles that contain three detectable polypeptides similar in size to those comprising the RNA polymerase of E. coli (120, 140). Thus transcription of the influenza genome occurs on individual ribonucleoprotein complexes, each containing a distinct genome segment, while the reovirus double-stranded RNA is copied in a complex particle that contains all ten genome segments. These differences may be important reflections of the mechanisms used for assembly into infectious virions of the single-stranded influenza genome segments vs the double-stranded reovirus molecules. The influenza process may be a random assembly, as suggested by Compans (236). In contrast, the reovirus genome segments appear to remain associated with each other throughout replication, possibly by virtue of their association with viral proteins. Parental reovirions are only partially uncoated in infected cells, and the double-stranded RNA is conserved within subviral particles (237, 238). Nascent double-stranded RNA is also associated with virion-like particles (239). Thus, the mRNA synthesized by the reovirus-associated transcriptase may become associated with, and possibly linked by, nascent viral proteins during or shortly after synthesis (240). This complex, possibly assembled on the RNA synthesizing parental particles, could then function in the genome replication reaction described below. Although the RNA segments may be noncovalently linked during infection, the high yields of genetic recombinants observed for influenza and reovirus are consistent with the reassortment of genome segments, indicating that subunit association does not prevent efficient genetic interactions (241).

The RNA produced in vitro by the influenza virion polymerase, like the viral RNA in polysomes of infected chick cells (143), is complementary to the genome RNA. Similarly, only one strand of the reovirus genome segments is transcribed in vitro by a conservative mechanism (147–149). It corresponds in size and polarity to the in vivo viral mRNA (150). Thus, the virion-associated polymerases are transcriptases that probably are required for viral mRNA synthesis, especially during the early stages of the infectious cycle (242, 243). Reovirus mRNA but not genome RNA is synthesized in infected cells that have been treated with cyclo-heximide from the start of infection to block protein synthesis (242). This mRNA, presumably synthesized by the parental virion polymerase, is copied from four of the ten genome segments. The mRNA pattern corresponds to that observed during the early period of a productive infection (242). At later times in the cycle all segments are transcribed. The control of reovirus transcription appears not to be mediated by structural protein(s) present in the polymerase-containing viral core (163, 244), in contrast to another diplornavirus, blue-tongue virus (245). However, addition of an outer shell structural protein, $\sigma 3$, to active core particles blocks the reovirus RNA polymerase (246). In reovirus-infected cells, a preformed host component may be important for regulation of viral genome transcription. Infectious subviral particles similar in composition to partially uncoated virions isolated from infected cells (237, 238) were prepared by limited chymotrypsin digestion of purified virions (163). The subviral particles contained an active RNA polymerase that synthesized all ten mRNA species in vitro, but predominantly four species after infection of cycloheximide-treated cells. Extracts of cycloheximide-treated, infected cells prepared by detergent treatment transcribed all genome segments, suggesting that the preformed host component may be a membrane attachment site for parental virions (238).

Although early viral mRNA formation is catalyzed by a preformed transcriptase, influenza and reovirus genome RNA replication require protein synthesis in infected cells (142, 247–249). Thus genome replication in infected cells is differentially inhibited when protein synthesis is blocked by the presence of cycloheximide. Transcription and replication of reovirus RNA can also be separated by using conditional lethal mutants. Several mutants have been described (250, 251) that are temperature sensitive for genome RNA production but not for viral mRNA synthesis. The two synthetic processes are functionally distinct, but it remains to be determined if they are mediated by completely different proteins or if replicase is derived from transcriptase by addition or modification of one or more subunits of an enzyme complex (157, 252).

Reovirus RNA replicase activity has been observed in a subcellular large particle fraction obtained by differential centrifugation of infected cell homogenates (253). Like viral genome synthesis in infected cells (254), the in vitro formation of the complementary strands of the double-stranded RNA is asynchronous (158, 252, 254). The replicase in vitro uses viral mRNA (plus strand) preformed in vivo as template for production of the complementary (minus) strand, in agreement with the absence of free minus strands in infected cells (141, 142, 150, 239). Although the large particle fraction includes transcriptase activity, the mRNA produced in vitro does not function as a template for the formation of more double-stranded RNA (252). The replicase studied in vitro apparently does not start new RNA chains, i.e. the synthesis observed represents completion of nascent minus strands in duplexes that have been initiated in the cell before homogenization. A similar activity is present in purified bacteriophage $\phi 6$, a virus that also contains a segmented, double-stranded RNA genome (255). The initiation of minus strand synthesis would be expected to occur in a complex consisting of the viral protein(s) and template (plus) strand RNA. A complex of this type has been observed, but its ability to synthesize RNA chains is not known (240). Replicase activity in cell fractions is associated with particles that have several of the properties of virions (256) and virion cores (257). It seems likely that the synthesis of genome RNA is completed by the replicase as one of the late stages of reovirus morphogenesis (157, 256, 257). Perhaps at a late step in the maturation process the replicase subunit(s) are released and the remaining enzyme component(s) function as transcriptase and are packaged into mature virions.

Class 4 Virions

Interest in RNA tumor viruses has increased dramatically in the short time since the first reports that they contain RNA-dependent DNA polymerase (189, 190). Much effort has been made to elucidate the structure and mechanism of action of the polymerase since it apparently is required for virus replication and oncogenic transformation, but not for maintenance of the transformed state (192). The ability to produce DNA transcripts of most of the viral RNA sequences in vitro (175, 258, 259) provided a sensitive annealing probe for detecting virus-specific RNA sequences in the presence of a large excess of host RNA (260–262). This had not been possible by the usual method of selectively inhibiting host RNA synthesis with actinomycin (146) because both host cell and virus-specific RNAs are presumably made on DNA templates (166).

In murine sarcoma virus (MSV)-transformed mouse cells (261) and RSV-transformed chick cells (193, 262) 0.5–1.0% of the total RNA is virus specific, as defined by its ability to hybridize with low levels of viral DNA of high specific radioactivity. A similar value (0.3–0.5%) was observed when radioactive RNA from RSV-infected chick cells was annealed with large amounts of viral DNA transcribed in vitro by the virion reverse transcriptase (263). A higher proportion (5%) of the total RNA in MSV-transformed mouse cell nuclei was virus specific as measured by chromatography of DNA-RNA hybrids on hydroxylapatite (261), but nuclear enrichment of RSV-specific RNA in nuclei from transformed chick cells was not observed using a more stringent test for hybrid formation, resistance of viral DNA to single strand-specific nuclease, S-1, after the annealing reaction (193). Virus-specific RNA is not only present in RSV-infected chick cell nuclei but also appears to be synthesized there (263). After pulse-labeling infected cells with [^{3}H]-uridine for periods of 15 min to 4 hr nuclear and cytoplasmic fractions were separated. The RNA extracted from cell fractions and from released virions was hybridized with viral DNA prepared in vitro. The amount of virus-specific RNA was quantitated on the basis of ^{32}P-labeled virion RNA added as an internal standard. Virus-specific RNA was detected in nuclei after a short pulse of RSV-transformed chick cells, followed by a later appearance in the cytoplasm and finally in released virions (263). These experiments also emphasize the important influence of experimental conditions on the quantitation of virus-specific RNA; increasing the pulse-labeling time from 15 to 60 min resulted in a decrease in the proportion of virus-specific nuclear RNA (0.58–0.42%) and an increase in the cytoplasmic virus-specific RNA (0.0–0.6%).

The synthesis of viral RNA in nuclei and its sensitivity to actinomycin both support the postulated role of viral DNA sequences as templates for viral genome production. If these DNA sequences are integrated into the cellular genome (191), their transcription could result in the synthesis of RNA molecules that contain both viral and host sequences, as in cells transformed by DNA tumor viruses (264). The virus-specific RNA would then undergo post-transcriptional processing. It is therefore not surprising that virus-specific RNA from transformed cells is heterogeneous in size (193, 265). In addition, RNA extracted with phenol from RSV-transformed chick cells includes some virus-specific RNA that sediments faster than 70S (193), and these could be host sequence-containing precursors of viral genome RNA.

Denatured viral RNA from virus-producing MSV-transformed cells consisted predominantly of 35S and 20S molecules. The 20S RNA was absent from transformed cell lines that were not producing virus, suggesting that it is related to virus replication but not the transformed state. Some of the virus-specific RNA presumably functions as mRNA and a small fraction ($\sim 10\%$) of the 10–40S virus-specific RNA in the cytoplasm of cells transformed by murine or avian viruses is present in polyribosomes (265, 266; J. M. Bishop, personal communication). However, most of the virus-specific RNA that sediments together with polyribosomes is probably not functional mRNA since it does not have a reduced sedimentation rate after the polyribosomes are dissociated with EDTA or puromycin (266; J. M. Bishop, personal communication). The virus-specific RNA from RSV-transformed chick and rat cells (193, 262) and from MSV-transformed mouse and rat cells (261, 265) hybridizes exclusively to in vitro single-stranded DNA that is complementary to virion RNA, i.e. intracellular virus-specific RNA and virion RNA have the same sequence. Thus, as in class 1 viruses, class 4 virus mRNA and genome RNA appear to correspond to the same (plus) strand. Furthermore, the 35S genome subunits of the RNA tumor viruses are similar in molecular weight to the picornavirus genome, and it will be of interest to determine if post-translational cleavage is also of general importance in the production of class 4 viral proteins (267).

Class 4 virions undergo further changes after they are released from cells, and this maturation process is accompanied by structural rearrangements in the viral RNA (268–270). RSV harvested at 3 min intervals from the growth medium of [^{3}H]-uridine-labeled transformed cells contained radioactive RNA of genome subunit size, 30–40S, in addition to smaller amounts of 60–70S and the low molecular weight RNAs usually found in mature virions (268). Incubation of the purified immature particles at 40°C resulted in conversion of the 30–40S RNA to 60–70S RNA. More 4S RNA appears to associate with the 30–40S RNA as it is converted to faster sedimenting forms. Some of the low molecular weight RNA may function as additional primers for the reverse transcriptase or as "linkers" to hold the genome subunits together in the 60–70S RNA (268).

In another similar study, immature RSV isolated at 5 min intervals also contained very little 68S RNA, the main component of mature virions harvested at 24 hr intervals (269). Immature virions (5 min) had predominantly 55–60S RNA and lesser amounts of 36S subunit RNA and smaller RNA (269). Particles harvested at 10 min intervals contained 68S RNA as the major component, suggesting that the RNA matures during the first few minutes after the particles are released from cells. After incubation of the 5 min particles for 24 hr at 37°C, its RNA remained 55–60S but was more homogeneous, consistent with a partial maturation involving association of more low molecular weight RNA with the genome RNA (269). The properties of the reverse transcriptase and the primer-template activity of viral RNA are different for immature and mature virions (268, 269). It will be of interest to determine if these differences influence the infectivity of RSV.

MSV-RNA also undergoes structural rearrangement within virions after they are released from transformed rat cells into the culture medium (270). In this study, the sedimentation constant of MSV-RNA from 5 min or 2 hr harvests was 50S as compared to 58S for mature (24 hr) virion RNA. Genome subunits (28S RNA) were not detected in undenatured MSV-RNA from immature virions in contrast to immature RSV, but after heating at 50°C, both the 58S and 50S MSV-RNA yielded 28S subunits and small RNA species. The 58S RNA found in mature MSV was converted to the intermediate 50S RNA by heating at a lower temperature (44°C), suggesting that mature virion 58S RNA is a more helical form of the 50S RNA in immature particles. Similarly, stepwise conversion of AMV 60–70S RNA to 50–54S components and 30–40S subunits was observed in the presence of increasing concentrations of formamide (271). The physical changes that occur in the RNA and proteins (268) of avian and murine tumor viruses after the release from cells may have important biological consequences.

In virus-infected, transformed cells viral DNA sequences can be detected in the host cell DNA and virus-specific RNA and proteins are produced. RSV DNA sequences are also present in the genome of normal chickens (272–274) and MLV DNA in normal mice (275). In some normal chick cells these sequences are transcribed (276, 277) and viral group-specific (gs) antigens, but no mature infectious virus particles, are usually (but not always) synthesized (278–282). In other normal chick cells virus-specific RNA is not detected (263, 276, 277). However, the presence of similar quantities of viral DNA sequences can be demonstrated in all chick cells by nucleic acid hybridization (272–274). Furthermore, latent virus can be rescued from normal cells by infection with another avian tumor virus (283). Many normal cells including those of rat, hamster, and pheasant origin may also harbor rescuable RNA tumor viruses (284–286). Chick cells that contain viral DNA sequences but do not produce virus-specific RNA also do not yield virus recombinants after infection with RSV (171). This indicates that the genetic interactions of avian RNA tumor viruses, whether subunit reassortment or true recombination, most probably occur at the level of viral RNA (171). Thus, in normal chick cells the expression of endogenous avian virus genes is regulated at the level of transcription by a control mechanism that is host gene dependent (287).

Genome expression of transformed mammalian cells is also regulated at the transcriptional level. Virus-specific RNA from MSV-transformed rat cells that are not producing virus hybridizes to 30–40% of the sequences in viral DNA. In contrast, 60–70% of the DNA sequences hybridize with RNA from virus-producing MSV-transformed cells (288). Sarcoma virus can be rescued from nonproducer cells indicating that the entire viral genome is present, although only about one half of the viral sequences apparently are transcribed. Regulation may also occur at the level of translation since some gs-negative cells including hamster (265), rat (288), and chick cells (276) produce low but detectable levels of virus-specific RNA. It remains to be determined if this RNA contains the information for synthesis of gs-antigen or some other viral protein(s).

CONCLUSIONS

The diversity found among animal RNA viruses provides a catalog of model systems for the increasing number of investigators interested in the metabolism of eukaryotic cells. Studies of the structure and function of animal RNA viruses can contribute to a better understanding of many fundamental processes including oncogenic transformation (166), membrane biogenesis (289), cell surface modification (290–293), and control of gene expression in normal and diseased tissues (294, 295). Important in all these virus-related processes is the nature of the viral RNA genome both in virus particles and infected cells. In many animal viruses including influenza, the RNA tumor viruses and agents that cause acute [lymphocytic choriomeningitis virus (296)] and slow infections [visna (297), maedi (298), and progressive pneumonia virus (299)] the viral genome appears to consist of more than one molecule of RNA. Little is known about the mechanisms involved in packaging multiple genome RNA segments within virions or the biological significance of segmented genomes. Furthermore, defective interfering particles that contain only a portion of the viral RNA genome are present in preparations of many animal viruses including representatives of all four classes (234). Interactions between infectious and interfering particles may play an important but largely unknown role in viral pathogenesis. It can be anticipated that increased efforts to elucidate animal virus RNA structure and function will also result in valuable new information about normal cells and hopefully will provide a basis for controlling or eliminating many diseases. Finally, it should be noted that with the rapid expansion of animal virology, many findings may have been inadvertently overlooked and some described in this review superseded before it appears in print.

ACKNOWLEDGMENTS

The author is grateful to Dr. C. M. Stolzfus and many others for critically reading the manuscript and providing preprints of their work and to Mrs. K. Idselis for excellent secretarial assistance.

Literature Cited

1. Baltimore, D. 1971. *Bacteriol. Rev.* 35: 235–41
2. Stavis, R. L., August, J. T. 1970. *Ann. Rev. Biochem.* 39: 527–60
3. Bishop, J. M., Levintow, L. 1971. *Progr. Med. Virol.* 13: 1–82
4. Bechet, J. M. 1972. *Virology* 48: 855–57
5. Roy, P., Bishop, D. H. L. 1970. *J. Virol.* 6: 604–9
6. Best, M., Evans, B., Bishop, J. M. 1972. *Virology* 47: 592–603
7. Granboulan, N., Girard, M. 1969. *J. Virol.* 4: 475–79
8. Clements, J. B., Martin, J. S. 1971. *J. Gen. Virol.* 12: 221–32
9. Burness, A. T. H., Clothier, F. W. 1970. *J. Gen. Virol.* 6: 381–93
10. Arif, B. M., Faulkner, P. 1972. *J. Virol.* 9: 102–9
11. Levin, J. G., Friedman, R. M. 1971. *J. Virol.* 7: 504–14
12. Sreevalsan, T., Lockart, R. Z., Dodson,

M. L., Hartman, K. A. 1968. *J. Virol.* 2: 558–66
13. Acheson, N. H., Tamm, I. 1967. *Virology* 32:128–43
14. Strauss, J. H., Burge, B. W., Pfefferkorn, E. R., Darnell, J. E. 1968. *PNAS* 59: 533–37
15. Lim, L., Canellakis, E. S. 1971. *Nature* 227:710–12
16. Edmonds, M., Vaughan, M. H., Nakazato, H. 1971. *PNAS* 68:1336–40
17. Lee, S. Y., Mendecki, J., Brawerman, G. 1971. *PNAS* 68:1331–35
18. Darnell, J. E., Wall, R., Tushinski, R. J. 1971. *PNAS* 68:1321–25
19. Burr, H., Lingrel, J. B. 1971. *Nature New Biol.* 233:41–43
20. Philipson, L., Wall, R., Glickman, G., Darnell, J. E. 1971. *PNAS* 68:2806–9
21. Weinberg, R. A., Ben-Ishai, Z., Newbold, J. E. 1972. *Nature New Biol.* 233:111–13
22. Sheldon, R., Kates, J., Kelley, D. E., Perry, R. P. 1972. *Biochemistry* 11:3829–34
23. Armstrong, J. A., Edmonds, M., Nakazato, H., Phillips, B. A., Vaughan, M. H. 1972. *Science* 176:526–28
24. Yogo, Y., Wimmer, E. 1972. *PNAS* 69: 1877–82
25. Johnston, R. E., Bose, H. R. 1972. *PNAS* 69:1514–16
26. Eaton, B. T., Faulkner, P. 1972. *Virology* 50:865–73
27. Agol, V. I., Romanova, L. I., Cumakov, I. M., Dunaevskaya, L. D., Bogdanov, A. A. 1972. *J. Mol. Biol.* 72:77–89
28. Silber, R., Malathi, V. G., Hurwitz, J. 1972. *PNAS* 69:3009–13
29. Cranston, J., Malathi, V. G., Silber, R. 1973. *Fed. Proc.* 32:498
30. Wimmer, E. 1972. *J. Mol. Biol.* 68: 537–40
31. Yogo, Y., Wimmer, E. 1973. *Nature New Biol.* 242:171–74
32. Öberg, B. F., Shatkin, A. J. 1972. *PNAS* 69:3589–93
33. Smith, A. 1973. *Eur. J. Biochem.* 33: 301–13
34. Summers, D. F., Maizel, J. V. 1968. *PNAS* 59:966–70
35. Jacobson, M. F., Baltimore, D. 1968. *PNAS* 61:77–84
36. Holland, J. J., Kiehn, D. E. 1968. *PNAS* 60:1015–22
37. Korant, B. D. 1972. *J. Virol.* 10:751–59
38. Burrell, C. J., Martin, E. M., Cooper, P. D. 1970. *J. Gen. Virol.* 6:319–23
39. Pfefferkorn, E. R., Boyle, M. K. 1972. *J. Virol.* 9:187–88
40. Scheele, C. M., Pfefferkorn, E. R. 1970. *J. Virol.* 5:329–37

41. Strauss, J. H., Burge, B. W., Darnell, J. E. 1969. *Virology* 37:367–76
42. Schlesinger, S., Schlesinger, M. J. 1972. *J. Virol.* 10:925–32
43. Waite, M. R. F. 1973. *J. Virol.* 11: 198–206
44. Snyder, H. W., Sreevalsan, T. 1973. *BBRC* 53:24–31
45. Kiehn, E. D., Holland, J. J. 1970. *J. Virol.* 5:358–67
46. Jacobson, M. F., Asso, J., Baltimore, D. 1970. *J. Mol. Biol.* 49:657–69
47. Summers, D. F., Shaw, E. N., Stewart, M. L., Maizel, J. V. 1972. *J. Virol.* 10: 880–84
48. Roumiantzeff, M., Summers, D. F., Maizel, J. V. 1971. *Virology* 44:249–58
49. Summers, D. F., Maizel, J. V. 1971. *PNAS* 68:2852–56
50. Taber, R., Rekosh, D., Baltimore, D. 1971. *J. Virol.* 8:395–401
51. Butterworth, B. E., Rueckert, R. R. 1972. *J. Virol.* 9:823–28
52. Rekosh, D. 1972. *J. Virol.* 9:479–87
53. Blair, C. D., Duesberg, P. H. 1970. *Ann. Rev. Microbiol.* 24:539–74
54. Cartwright, B., Smale, C. J., Brown, F., Hull, R. 1972. *J. Virol.* 10:256–60
55. Baltimore, D. 1969. In *The Biochemistry of Viruses*, ed. H. B. Levy, 101–76. New York: Dekker
56. Huang, A. S., Wagner, R. R. 1966. *J. Mol. Biol.* 22:381–84
57. Nakai, T., Howatson, A. F. 1968. *Virology* 35:268–81
58. Kiley, M. P., Wagner, R. R. 1972. *J. Virol.* 10:244–55
59. Schincariol, A. L., Howatson, A. F. 1972. *Virology* 49:766–83
60. Duesberg, P. H., Robinson, W. S. 1965. *PNAS* 54:794–800
61. Kingsbury, D. W. 1966. *J. Mol. Biol.* 18:195–203
62. Compans, R. W., Choppin, P. W. 1968. *Virology* 35:289–96
63. Iwai, Y., Iwai, M., Okumoto, M., Hosokawa, Y., Asai, T. 1966. *Biken's J.* 9:241–49
64. Barry, R. D., Bukrinskaya, A. G. 1968. *J. Gen. Virol.* 2:71–79
65. Blair, C. D., Robinson, W. S. 1968. *Virology* 35:537–49
66. Duesberg, P. H. 1968. *PNAS* 60:1511–18
67. Mudd, J. A., Summers, D. F. 1970. *Virology* 42:328–40
68. Kolakofsky, D., Bruschi, A. 1973. *J. Virol.* 11:615–20
69. Mountcastle, W. E., Compans, R. W., Caliguiri, L. A., Choppin, P. W. 1970. *J. Virol.* 6:677–84
70. Wagner, R. R., Schnaitman, T. C.,

Snyder, R. M., Schnaitman, C. A. 1969. *J. Virol.* 3:611–18

71. Kingsbury, D. W., Darlington, R. W. 1968. *J. Virol.* 2:248–55
72. Compans, R. W., Mountcastle, W. E., Choppin, P. W. 1972. *J. Mol. Biol.* 65:167–69
73. Hosaka, Y., Kitano, H. 1966. *Virology* 29:205–21
74. Compans, R. W., Choppin, P. W. 1967. *PNAS* 57:949–56
75. Finch, J. T., Gibbs, A. J. 1970. *J. Gen. Virol.* 6:141–50
76. Granoff, A. 1962. *Cold Spring Harbor Symp. Exp. Biol.* 27:319–26
77. Sokol, F., Schlumberger, H. D., Wiktor, T. J., Koprowski, H., Hummeler, K. *Virology* 38:651–65
78. Compans, R. W., Choppin, P. W. 1971. In *Comparative Virology*, ed. K. Maramorosch, E. Kurstak, 407–32. New York: Academic
79. Kingsbury, D. W. 1966. *J. Mol. Biol.* 18:204–14
80. Bratt, M. A., Robinson, W. S. 1967. *J. Mol. Biol.* 23:1–21
81. Schaffer, F. L., Hackett, A. J., Soergel, M. E. 1968. *BBRC* 31:685–92
82. Mudd, J. A., Summers, D. F. 1970. *Virology* 42:958–68
83. Huang, A., Baltimore, D., Stampfer, M. 1970. *Virology* 42:946–57
84. Ehrenfeld, E., Summers, D. F. 1972. *J. Virol.* 10:683–88
85. Gillespie, D., Marshall, S., Gallo, R. C. 1972. *Nature New Biol.* 236:227–31
86. Baltimore, D., Huang, A. S., Stampfer, M. 1970. *PNAS* 66:572–76
87. Huang, A. S., Baltimore, D., Bratt, M. A. 1971. *J. Virol.* 7:389–94
88. Stone, H. O., Portner, A., Kingsbury, D. W. 1971. *J. Virol.* 8:174–80
89. Robinson, W. S. 1971. *J. Virol.* 8:81–86
90. Brown, F., Cartwright, B., Crick, J., Smale, C. J. 1967. *J. Virol.* 1:368–73
91. Szilagyi, J. F., Uryvayev, L. 1973. *J. Virol.* 11:279–86
92. Soria, M., Huang, A. S. 1973. *J. Mol. Biol.* 77:449–55
93. Weiss, S. R., Bratt, M. A. 1973. *ASM Abstr.*, 202
94. Pridgen, C., Kingsbury, D. W. 1972. *J. Virol.* 10:314–17
95. Banerjee, A. K., Rhodes, D. P. 1973. *PNAS* 70:3566–70
96. Galet, H., Prevec, L. 1973. *Nature New Biol.* 243:200–3
97. Marshall, S., Gillespie, D. 1972. *Nature New Biol.* 240:43–45
98. Wild, T. F. 1971. *J. Gen. Virol.* 13:295–310
99. Collins, B. S., Bratt, M. A. 1973. *ASM Abstr.*, 242

100. Stampfer, M., Baltimore, D. 1973. *J. Virol.* 11:520–26
101. Morrison, T., Stampfer, M., Baltimore, D., Lodish, H. 1974. *J. Virol.* 13:62–72
102. Robinson, W. S. 1970. *Nature* 225:944–45
103. Portner, A., Kingsbury, D. W. 1970. *Nature* 228:1196–97
104. Roy, P., Bishop, D. H. L. 1973. *J. Virol.* 11:487–501
105. McSharry, J. J., Compans, R. W., Choppin, P. W. 1971. *J. Virol.* 8:722–29
106. Zavada, J. 1972. *Nature New Biol.* 240:122–24
107. Shatkin, A. J. 1971. *Bacteriol. Rev.* 35:250–66
108. Semancik, J. S., Vidaver, A. K., Van Etten, J. L. 1973. *J. Mol. Biol.* 78:617–25
109. Compans, R. W., Klenk, H. D., Caliguiri, L. A. Choppin, P. W. 1970. *Virology* 42:880–89
110. Skehel, J. J. 1971. *J. Gen. Virol.* 11:103–9
111. Schulze, I. T. 1972. *Virology* 47:181–96
112. Bishop, D. H. L., Obijeski, J. F., Simpson, R. W. 1971. *J. Virol.* 8:74–80
113. Shatkin, A. J., Sipe, J. D., Loh, P. 1968. *J. Virol.* 2:986–91
114. Gomatos, P. J., Stoeckenius, W. 1964. *PNAS* 52:1449–55
115. Ada, G. L., Lind, P. E., Larkin, L., Burnett, F. M. 1959. *Nature* 180:360–63
116. Sokol, F., Szurman, J. 1959. *Acta Virol.* 3:175–80
117. Duesberg, P. H. 1968. *PNAS* 59:930–37
118. Pons, M. W., Hirst, G. K. 1968. *Virology* 34:385–88
119. Compans, R. W., Content, J., Duesberg, P. H. 1972. *J. Virol.* 10:795–800
120. Smith, R. E., Zweerink, H. J., Joklik, W. K. 1969. *Virology* 39:791–810
121. Content, J., Duesberg, P. H. 1971. *J. Mol. Biol.* 62:273–85
122. Horst, J., Content, J., Mandeles, S., Fraenkel-Conrat, H., Duesberg, P. H. 1972. *J. Mol. Biol.* 69:209–15
123. Watanabe, Y., Graham, A. F. 1967. *J. Virol.* 1:665–77
124. Nonoyama, M., Graham, A. F. 1970. *J. Virol.* 6:693–94
125. Pons, M., Hirst, G. K. 1969. *Virology* 38:68–72
126. Young, R. J., Content, J. 1971. *Nature New Biol.* 230:140–42
127. Banerjee, A. K., Shatkin, A. J. 1971. *J. Mol. Biol.* 61:643–53

128. Borsa, J., Grover, J., Chapman, J. D. 1970. *J. Virol.* 6:295–302
129. Kapuler, A. M., Mendelsohn, N., Klett, H., Acs, G. 1970. *Nature* 225:1209–13
130. Banerjee, A. K., Ward, R., Shatkin, A. J. 1971. *Nature New Biol.* 230:169–72
131. Millward, S., Graham, A. F. 1970. *PNAS* 65:422–29
132. Banerjee, A. K., Grece, M. A. 1971. *BBRC* 45:1518–25
133. Lewandowski, L. J., Content, J., Leppla, S. H. 1971. *J. Virol.* 8:701–7
134. Lai, M. C., Duesberg, P. H. 1972. *Nature* 235:383–86
135. Bellamy, A. R., Shapiro, L., August, J. T., Joklik, W. K. 1967. *J. Mol. Biol.* 29:1–17
136. Bellamy, A. R., Joklik, W. K. 1967. *PNAS* 58:1389–95
137. Shatkin, A. J., Sipe, J. D. 1968. *PNAS* 59:246–52
138. Chow, N. L., Simpson, R. W. 1971. *PNAS* 68:752–56
139. Borsa, J., Graham, A. F. 1968. *BBRC* 33:895–901
140. Shatkin, A. J., Sipe, J. D. 1968. *PNAS* 61:1462–68
141. Prevec, L., Graham, A. F. 1966. *Science* 154:522–23
142. Shatkin, A. J., Rada, B. 1967. *J. Virol.* 1:24–25
143. Pons, M. W. 1972. *Virology* 47:823–32
144. Barry, R. D., Ives, D. R., Cruickshank, F. G. 1962. *Nature* 194:1139–40
145. Kudo, H., Graham, A. F. 1965. *J. Bacteriol.* 90:936–45
146. Shatkin, A. J. 1968. In *Actinomycin—Nature, Formation and Activities,* ed. S. A. Waksman, 69–86. New York: Interscience
147. Banerjee, A. K., Shatkin, A. J. 1970. *J. Virol.* 6:1–11
148. Levin, D. H. et al 1970. *PNAS* 66:890–97
149. Skehel, J. J., Joklik, W. K. 1969. *Virology* 39:822–31
150. Hay, A. J., Joklik, W. K. 1971. *Virology* 44:450–53
151. Zweerink, H. J., McDowell, M. J., Joklik, W. K. 1971. *Virology* 45:716–23
152. Skehel, J. J. 1972. *Virology* 49:23–36
153. Graziadei, W. D., Lengyel, P. 1972. *BBRC* 46:1816–23
154. McDowell, M. J., Joklik, W. K., Villa-Komaroff, L., Lodish, H. F. 1972. *PNAS* 69:2649–53
155. Stoltzfus, C. M., Shatkin, A. J., Banerjee, A. K. 1973. *J. Biol. Chem.* 248:7993–98
156. Ward, R., Banerjee, A. K., LaFiandra, A., Shatkin, A. J. 1972. *J. Virol.* 9:61–69
157. Acs, G. et al 1971. *J. Virol.* 8:684–89
158. Sakuma, S., Watanabe, Y. 1972. *J. Virol.* 10:628–38
159. Bellamy, A. R., Hole, L. V. 1970. *Virology* 40:808–19
160. Stoltzfus, C. M., Banerjee, A. K. 1972. *Arch. Biochem. Biophys.* 152:733–43
161. Nichols, J. L., Bellamy, A. R., Joklik, W. K. 1972. *Virology* 49:562–72
162. Krug, R. M., Gomatos, P. J. 1969. *J. Virol.* 4:642–50
163. Shatkin, A. J., LaFiandra, A. J. 1972. *J. Virol.* 10:698–706
164. Joklik, W. K. 1972. *Virology* 49:700–15
165. Duesberg, P. H. 1970. *Curr. Top. Microbiol. Immunol.* 51:79–104
166. Temin, H. M., Baltimore, D. 1972. *Advan. Virol. Res.* 17:129–86
167. Duesberg, P. H., Vogt, P. K. 1970. *PNAS* 67:1673–80
168. Duesberg, P. H., Vogt, P. K. 1973. *Virology* 54:207–19
169. Martin, G. S., Duesberg, P. H. 1972. *Virology* 47:494–97
170. Vogt, P. K. 1973. In *Le Petit Symp.,* ed. L. G. Silvestri. Amsterdam: North-Holland. In press
171. Weiss, R. A., Mason, W. S., Vogt, P. K. 1973. *Virology* 52:535–52
172. Bishop, J. M. et al 1973. In *Proc. Symp. Mol. Biol. Vir. Res.,* Squaw Valley, Calif. ed. C. F. Fox, W. S. Robinson, 15–32. New York: Academic
173. Scheele, C. M., Hanafusa, H. 1972. *Virology* 50:753–64
174. Erikson, E., Erikson, R. L. 1971. *J. Virol.* 8:254–56
175. Garapin, A. C., Varmus, H. E., Faras, A. J., Levinson, W. E., Bishop, J. M. *Virology* 52:264–74
176. Rosenbergova, M., Lacour, F., Huppert, J. 1965. *C.R.Acad. Sci. Paris* 260:5145–48
177. Erikson, E., Erikson, R. L., Henry, B., Pace, N. R. 1973. *Virology* 53:40–46
178. Silber, R., Malathi, V. G., Schulman, L. H., Hurwitz, J., Duesberg, P. H. 1973. *BBRC* 50:467–72
179. Stephenson, M. L., Wirthlin, L. R. S., Scott, J. F., Zamecnik, P. C. 1972. *PNAS* 69:1176–80
180. Erikson, R. L., Erikson, E., Walker, T. A. 1971. *Virology* 45:527–28
181. Robin, J. et al 1972. *FEBS Lett.* 27:58–62
182. Maruyama, H. B., Hatanaka, M., Gilden, R. V. 1971. *PNAS* 68:1999–2001

183. Eckhart, W. 1972. *Ann. Rev. Biochem.* 41:503–16
184. Temin, H. M. 1971. *Ann. Rev. Microbiol.* 25:609–48
185. Green, M. 1970. *Ann. Rev. Biochem.* 39:701–56
186. Temin, H. M. 1964. *Virology* 23:486–94
187. Bader, J. P. 1964. *Virology* 22:462–68
188. Vigier, P., Goldé, A. 1964. *Virology* 23:511–19
189. Baltimore, D. 1970. *Nature* 226:1209–11
190. Temin, H. M., Mizutani, S. 1970. *Nature* 226:1211–13
191. Hill, M., Hillova, J. 1972. *Nature New Biol.* 237:35–39
192. Hanafusa, H., Hanafusa, T. 1971. *Virology* 43:313–16
193. Leong, J. A. et al 1972. *J. Virol.* 9:891–902
194. Siegert, W., Konings, R. N. H., Bauer, H., Hofschneider, P. H. 1972. *PNAS* 69:888–91
195. Green, M., Cartas, M. 1972. *PNAS* 69:791–94
196. Horst, J., Keith, J., Fraenkel-Conrat, H. 1972. *Nature New Biol.* 270:105–9
197. Ross, J., Tronick, S. R., Scolnick, E. M. 1972. *Virology* 49:230–35
198. Scharff, M. D., Shatkin, A. J., Levintow, L. 1963. *PNAS* 50:686–94
199. Penman, S., Scherrer, K., Becker, Y., Darnell, J. E. 1963. *PNAS* 49:654–62
200. Kerr, I. M., Brown, R. E., Tovell, D. R. 1972. *J. Virol.* 10:71–73
201. Boime, I., Aviv, H., Leder, P. 1971. *BBRC* 45:788–95
202. Eaton, B. T., Donaghue, T. P., Faulkner, P. 1972. *Nature New Biol.* 238:109–11
203. Moskowitz, D. 1973. *J. Virol.* 11:535–43
204. Kennedy, S. I. T. 1972. *BBRC* 48:1254–58
205. Rosemond, H., Sreevalsan, T. 1973. *J. Virol.* 11:399–415
206. Simmons, D. T., Strauss, J. H. 1972. *J. Mol. Biol.* 71:599–613
207. Schlesinger, M. J., Schlesinger, S., Burge, B. W. 1972. *Virology* 47:539–41
208. Shenk, T. E., Stollar, V. 1972. *BBRC* 49:60–67
209. Caliguiri, L. A., Tamm, I. 1970. *Virology* 42:112–22
210. Girard, M., Baltimore, D., Darnell, J. E. 1967. *J. Mol. Biol.* 24:59–74
211. Friedman, R. M., Levin, J. G., Grimley, P. M., Berezesky, I. K. 1972. *J. Virol.* 10:504–15
212. Grimley, P. M., Levin, J. G., Berezesky, I. K., Friedman, R. M. 1972. *J. Virol.* 10:492–503
213. Tan, K. B., Sambrook, J. F., Bellett, A. J. D. 1969. *Virology* 38:427–39
214. Burge, B. W., Pfefferkorn, E. R. 1967. *J. Virol.* 1:956–62
215. Cooper, P. D. 1969. In *The Biochemistry of Viruses,* ed. H. B. Levy, 177–218. New York: Dekker
216. Simmons, D. T., Strauss, J. H. 1972. *J. Mol. Biol.* 71:615–31
217. Blumenthal, T., Landers, T. A., Weber, K. 1972. *PNAS* 69:1313–17
218. Groner, Y., Scheps, R., Kamen, R., Kolakofsky, D., Revel, M. 1972. *Nature New Biol.* 239:19–20
219. Ehrenfeld, E., Maizel, J. V., Summers, D. F. 1970. *Virology* 40:840–46
220. Rosenberg, H., Diskin, B., Oron, L., Traub, A. 1972. *PNAS* 69:3815–19
221. Arlinghaus, R. B., Polatnick, J. 1969. *PNAS* 62:821–28
222. Girard, M. 1969. *J. Virol.* 3:376–84
223. Plagemann, P. G. W., Swim, H. E. 1968. *J. Mol. Biol.* 35:13–18
224. Martin, E. M., Sonnabend, J. A. 1967. *J. Virol.* 1:97–109
225. Sreevalsan, T., Yin, F. H. 1969. *J. Virol.* 3:599–604
226. Aaslestad, H. G., Clark, H. F., Bishop, H. L., Koprowski, H. 1971. *J. Virol.* 7:726–35
227. Robinson, W. S. 1971. *Virology* 44:494–502
228. Huang, A. S., Manders, E. K. 1972. *J. Virol.* 9:909–16
229. Robinson, W. S. 1971. *Virology* 43:90–100
230. Wagner, R. R., Kiley, M. P., Snyder, R. M., Schnaitman, C. A. 1972. *J. Virol.* 9:672–83
231. Portner, A., Kingsbury, D. W. 1972. *Virology* 47:711–25
232. Feix, G., Slor, H., Weissmann, C. 1967. *PNAS* 57:1401–8
233. Petric, M., Prevec, L. 1970. *Virology* 41:615–30
234. Huang, A. S. 1973. *Ann. Rev. Microbiol.* 27:101–17
235. Compans, R. W., Caliguiri, L. A. 1973. *J. Virol.* 11:441–48
236. Kingsbury, D. W. 1970. *Progr. Med. Virol.* 12:49–77
237. Chang, C., Zweerink, H. J. 1971. *Virology* 46:544–55
238. Silverstein, S. C., Astell, C., Levin, D. H., Schonberg, M., Acs, G. 1972. *Virology* 47:797–806
239. Gomatos, P. J. 1967. *PNAS* 58:1798–1805
240. Ward, R. L., Shatkin, A. J. 1972. *Arch. Biochem. Biophys.* 152:378–84
241. Fields, B. N. 1971. *Virology* 46:142–48
242. Watanabe, Y., Millward, S., Graham, A. F. 1968. *J. Mol. Biol.* 36:107–23

243. Krug, R. M. 1972. *Virology* 50:103–13
244. Zweerink, H. J., Joklik, W. K. 1970. *Virology* 41:501–18
245. Huismans, H., Verwoerd, D. W. 1973. *Virology* 52:81–88
246. Astell, C., Silverstein, S. C., Levin, D. H., Acs, G. 1972. *Virology* 48:648–54
247. Watanabe, Y., Kudo, H., Graham, A. F. 1967. *J. Virol.* 1:36–44
248. Scholtissek, C., Rott, R. 1970. *Virology* 40:989–96
249. Pons, M. W. 1973. *Virology* 51:120–28
250. Cross, R. K., Fields, B. N. 1972. *Virology* 50:799–809
251. Ito, Y., Joklik, W. K. 1972. *Virology* 50:189–201
252. Sakuma, S., Watanabe, Y. 1971. *J. Virol.* 8:190–96
253. Watanabe, Y., Gauntt, C. J., Graham, A. F. 1968. *J. Virol.* 2:869–77
254. Schonberg, M., Silverstein, S. C., Levin, D. H., Acs, G. 1971. *PNAS* 68:505–8
255. Van Etten, J. L., Vidaver, A. K., Koski, R. K., Semancik, J. S. 1973. *J. Virol.* 12:464–71
256. Zweerink, H. J., Ito, Y., Matsuhisa, T. 1972. *Virology* 50:349–58
257. Sakuma, S., Watanabe, Y. 1972. *J. Virol.* 10:943–50
258. Duesberg, P. H., Canaani, E. 1970. *Virology* 42:783–88
259. Spiegelman, S. et al 1970. *Nature* 227:563–67
260. Garapin, A. C., Leong, J., Fanshier, L., Levinson, W. E., Bishop, J. M. 1971. *BBRC* 42:919–25
261. Green, M., Rokutunda, H., Rokutunda, M. 1971. *Nature New Biol.* 230:229–32
262. Coffin, J. M., Temin, H. M. 1972. *J. Virol.* 9:766–75
263. Parsons, J. T., Coffin, J. M., Haroz, R. K., Bromley, P. A., Weissmann, C. 1973. *J. Virol.* 11:761–74
264. Wall, R., Darnell, J. E. 1971. *Nature New Biol.* 232:73–76
265. Tsuchida, N., Robin, M. S., Green, M. 1972. *Science* 176:1418–20
266. Fan, H., Baltimore, D. 1973. *J. Mol. Biol.* 80:93–117
267. Vogt, V. M., Eisenman, R. 1973. *PNAS* 70:1734–38
268. Canaani, E., Helm, K. V. D., Duesberg, P. H. 1973. *PNAS* 70:401–5
269. Cheung, K. S., Smith, R. E., Stone, M. P., Joklik, W. K. 1973. *Virology* 50:851–64
270. East, J. L. et al 1973. *J. Virol.* 11:709–20
271. Travnicek, M., Riman, J. 1973. *BBRC* 53:217–23
272. Rosenthal, P. N., Robinson, H. L., Robinson, W. S., Hanafusa, T., Hanafusa, H. 1971. *PNAS* 68:2336–40
273. Varmus, H. E., Weiss, R. A., Friis, R. R., Levinson, W., Bishop, J. M. 1972. *PNAS* 69:20–24
274. Baluda, M. 1972. *PNAS* 69:576–80
275. Gelb, L. D., Aaronson, S. A., Martin, M. 1971. *Science* 172:1353–55
276. Hayward, W. S., Hanafusa, H. 1973. *J. Virol.* 11:157–67
277. Bishop, J. M., Jackson, M., Quintrell, N., Varmus, H. E. 1973. In *Le Petit Symp.*, ed. L. G. Silvestri. Amsterdam: North-Holland. In press
278. Payne, L. N., Chubb, R. C. 1968. *J. Gen. Virol.* 3:379–91
279. Dougherty, R. M., DiStefano, H. S. 1966. *Virology* 29:586–95
280. Weiss, R. A. 1969. *J. Gen. Virol.* 5:511–28
281. Hanafusa, T., Hanafusa, H., Miyamoto, T. 1970. *PNAS* 67:1797–1803
282. Weiss, R. A., Friis, R. R., Katz, E., Vogt, P. K. 1971. *Virology* 46:920–38
283. Hanafusa, T., Hanafusa, H., Miyamoto, T., Fleissner, E. 1972. *Virology* 47:475–82
284. Aaronson, S. A. 1971. *Virology* 44:29–36
285. Kelloff, G., Huebner, R. J., Lee, Y., Toni, R., Gilden, R. 1970. *PNAS* 65:310–17
286. Hanafusa, T., Hanafusa, H. 1973. *Virology* 51:247–51
287. Crittenden, L. B., Smith, E., Weiss, R. A., Sarma, P. S. 1974. *Virology* 57:128–38
288. Beneviste, R. E., Scolnick, E. M. 1973. *Virology* 51:370–82
289. Choppin, P. W., Compans, R. W., Scheid, A., McSharry, J. J., Lazarowitz, S. G. 1972. In *Membrane Research,* 163–85. New York: Academic
290. Nicolson, G. L. 1971. *Nature New Biol.* 233:244–46
291. Becht, H., Rott, R., Klenk, H. D. 1972. *J. Gen. Virol.* 14:1–8
292. Poste, G., Reeve, P. 1972. *Nature New Biol.* 237:113–14
293. Birdwell, C. R., Strauss, J. H. 1973. *J. Virol.* 11:502–7
294. Huang, A. S., Baltimore, D. 1970. *Nature* 226:325–27
295. Sugiyama, T., Korant, B. D., Lonberg-Holm, K. K. 1972. *Ann. Rev. Microbiol.* 26:467–502
296. Pedersen, I. R. 1973. *J. Virol.* 11:416–23
297. Lin, F. H., Thormar, H. 1971. *J. Virol.* 7:582–87
298. Lin, F. H., Thormar, H. 1972. *J. Virol.* 10:228–33
299. Stone, L. B., Takemoto, K. K., Martin, M. A. 1971. *J. Virol.* 8:573–78

METHODS OF GENE ISOLATION[1]

Donald D. Brown and Ralph Stern[2]

Department of Embryology, Carnegie Institution of Washington, Baltimore, Maryland

CONTENTS

INTRODUCTION

Genetic analysis by gene isolation is an alternative to classical genetics, particularly in higher organisms where the collection and mapping of mutants is at best laborious and often impossible. Gene linkage can be studied by determining whether two genes are located on the same DNA molecule. It should be

[1] Abreviations used: HAP, hydroxyapatite; MAK, methylated serum albumin-Kieselguhr; rRNA, combined term for the two large RNAs in ribosomes; rDNA, structural genes for rRNA; 5S DNA, the structural genes for the small molecular weight ribosomal RNA termed 5S RNA; ssDNA, single-stranded DNA; dsDNA, double-stranded DNA.

[2] Present address: Jackson Laboratory, Bar Harbor, Maine 04609.

possible to study the initiation, termination, and promotor regions of specific genes. Their nucleotide sequences can be determined as well as their exact location with respect to the structural gene. Do related but unlinked genes share any adjacent nucleotide sequences? What are the size and function of the regions between genes termed spacer DNA? What is the rate of evolution of genes and spacers? The control mechanisms in eukaryotes might be unraveled by their reconstruction in vitro, just as has been done in bacteria. A purified gene would be mixed with the proper molecules, and exact initiation and termination by the correct RNA polymerase should take place. With such an assay it may be possible to understand control of gene action in eukaryotes. Gene isolation and characterization may well play a prominent role in the understanding of chromosome structure and chromosome pairing, and in the analysis of meiotic and mitotic recombination.

Gene isolation has become possible with the development of assay systems by which the quantitative amount of a given nucleotide sequence can be measured. Success of gene isolation depends upon three factors: the dependability of the assay system, the fraction of the genome which the gene(s) comprises, and the extent to which the gene differs from the rest of the DNA in some physical parameter.

ASSAY SYSTEMS

The first assay system used in bacteria for gene enrichment studies was transformation with purified DNA. Individual DNA fractions are assayed for the number of transformants they can produce. *Pneumococcus, Haemophilus,* and *Bacillus subtilis* are transformed readily with exogenous DNA containing a variety of drug resistance markers or wild-type alleles of auxotrophic mutations (cf 1). *Escherichia coli* spheroplasts are transformed with exogenous DNA in the presence of Ca^{2+} (2).

DNA-dependent protein synthesis via coupled transcription and translation, another potential assay system for a specific structural gene, has been carried out successfully in bacterial systems. The regulatory mechanisms of the lactose (3), tryptophan (4), and arabinose (5) operons in *E. coli* have been reproduced with a DNA-dependent system for the synthesis of their specific proteins.

Assays for gene purification have been performed most commonly by molecular hybridization with a radioactive nucleic acid probe whose sequence is complementary to the gene. The probe is usually a purified RNA labeled in vivo or in vitro. The genes most often studied with in vivo labeled RNAs are those that code for abundant, stable cellular RNAs, namely ribosomal RNA (rRNA) and transfer RNA (tRNA). Messenger RNAs are now being purified from a variety of specialized tissues (cf 6). Ordinarily, mRNAs cannot be synthesized in vivo with sufficiently high specific radioactivities to be useful for hybridization with unfractionated DNA. However, recent in vitro labeling techniques obviate this problem. Medium-level specific activities can be introduced into RNA with ^{3}H-dimethylsulfate (7). Recently, chemical iodination of cytosine residues of DNA and RNA has been described (8). If 5% of all cytosine residues in RNA were iodinated with carrier-free iodine[125] (about 1% of the total base residues), the nucleic acid would have a specific activity of about 10^8 dpm/μg. RNA iodinated at this level hybridizes with DNA (9).

Radioactive DNA complementary to purified mRNAs can be synthesized by reverse transcriptase, an RNA-dependent DNA polymerase usually purified from avian myeloblastosis virus (10, 11). Comparable activity has been described for bacterial DNA polymerase (12). These enzymes require a small oligodeoxynucleotide primer for initiation of DNA synthesis. Since most eukaryotic mRNAs have poly A residues at their 3′ termini (cf 13), oligodeoxythymidylate can be used to prime synthesis on these templates. The synthetic DNA probes are often much shorter than their RNA template and presumably are complementary to the 3′ end of the RNA template only.

The most serious technical problem with molecular hybridization as an assay system is the purity of the nucleic acid probe. This problem is accentuated when the gene to be detected is a small fraction of the total DNA. For example, when *Bombyx mori* DNA was hybridized with a radioactive fibroin mRNA preparation estimated to be more than 90% pure, a minor contaminant, ribosomal RNA, hybridized with the DNA to a greater extent than the predominant mRNA (14) since rDNA is about 80 times more abundant in the DNA than is fibroin DNA. Because of such artifacts, traditional means of assessing RNA purity, such as acrylamide gel electrophoresis, are necessary but insufficient. The most reliable method to determine specificity of the assay system is the demonstration that the material which hybridizes has, in fact, the same nucleotide sequence as the bulk of the probe itself. This method helped demonstrate the specificity of the assay for fibroin genes (14).

Assay specificity is improved if the gene has different physical properties (usually base composition) from the bulk of the DNA. The DNA is fractionated by a technique that accentuates the difference, such as density gradient centrifugation. If the gene has been separated physically from the rest of the DNA, hybridization to the gene can be distinguished from nonspecific hybridization resulting from radioactive contaminants in the probe. This method has been used to quantitate ribosomal DNA (15) and 5S DNA (16).

Impure radioactive nucleic acid probes have been used to estimate the relative abundance of a gene in a complex DNA mixture by DNA excess hybridization (17). This technique involves a kinetic analysis of probe reassociation with its complementary DNA strands in the DNA mixture.

Once a DNA component has been purified and its purity determined, a convenient source of radioactive probe for future experiments is complementary RNA synthesized from the purified DNA with *E. coli* DNA-dependent RNA polymerase.

DETERMINATION OF DNA PURITY

Transformation and DNA-dependent protein synthesis are used to detect the presence of a gene and to quantitate the extent of its enrichment. However, in principle they are not designed to estimate gene purity since they do not detect unrelated DNA which lacks biological activity. For this reason methods relying on the physical and chemical properties of a DNA component are more satisfactory for establishing its purity.

A homogeneous DNA component bands at equilibrium in neutral CsCl as a single symmetrical peak whose width is inversely proportional to its molecular weight. In alkaline CsCl gradients, duplex DNA should give a single peak if the strands have no base composition difference, or two peaks if the strands are asymmetrical. Differences in the absorbance of the two separated strands can result from differences in content of the individual bases. The two strands will reassociate to form a single DNA band; contaminants can be visualized in analytical CsCl gradients as separate peaks after reassociation. If the DNA has a reasonably high molecular weight and, if given a long time to reassociate, each homogeneous component will form a network of very high molecular weight which bands as a single hypersharp peak in CsCl (18), thus greatly aiding the visualization of minor contaminants with slightly different buoyant densities. If a pure radioactive probe labeled in vivo is available, purity can be ascertained by banding the DNA in one or more different kinds of preparative buoyant density gradients and hybridizing the probe with the DNA in many fractions across the gradients. If the DNA is a single component, the specific activity across the gradient (counts per minute of RNA hybridized per microgram of DNA) should be constant.

Another assay for purity is available if the strand of DNA complementary to the RNA product can be purified. When hybridized with its homologous RNA, the density of the DNA shifts by an amount dependent on the RNA to DNA ratio. DNA molecules that did not hybridize with RNA will not change in density and can be visualized in the ultracentrifuge as a separate lighter DNA component. This method was used to demonstrate that preparations of 5S DNA were pure (19).

Unbroken DNA molecules from viruses and mitochondria have homogeneous contour lengths which can be used to assess their purity. DNA components often have a characteristic partial denaturation pattern of single- and double-stranded regions as visualized in the electron microscope (20, 21). If the pattern is simple enough and present on every molecule it is an excellent indication of purity. Individual or repetitive genes isolated as fragments from longer DNA duplexes will be broken at random. They must be aligned by a characteristic denaturation pattern either unique to the sequence or highly repetitive (22, 23).

THE SEPARATION OF SINGLE-STRANDED DNA, DOUBLE-STRANDED DNA, AND RNA-DNA HYBRIDS

A variety of methods have been reported which separate ssDNA from dsDNA and DNA-RNA hybrids from either ssDNA, native DNA, or free RNA. Duplexes between RNA and DNA have a different buoyant density than either ssDNA and dsDNA or RNA; depending upon the ratio of RNA to DNA in the hybrid, they have other physical characteristics intermediate between completely single- or double-stranded molecules (cf 24). In general, the principles underlying the purification of one strand of a gene are usually very different from the fractionation procedures for purely dsDNA, which rely on differences in base composition or reiteration frequency of certain nucleotide sequences.

Methods to separate strands of homogeneous DNAs depend upon base composition or sequence differences. Since strand separation is important for gene characterization and in some instances for gene purification, these methods are included in this section.

Hydroxyapatite

The most widely used method for the fractionation of DNA on the basis of its strandedness is hydroxyapatite (HAP) chromatography. This method was first used for the purification of native DNA by Bernardi (25) and was subsequently shown to have a higher affinity for native DNA than for denatured DNA (26). Almost all of the studies with HAP have been carried out with low molecular weight DNA (less than 10^6).

Repetitious DNA has been separated from DNA with more complex sequences (single-copy or unique DNA) by reassociation of denatured DNA mixtures for short time periods. The repetitious DNA forms partial or complete duplexes and is separated on HAP from the ssDNA (27, 28). By varying the size of ssDNA fragments, the interspersion of repetitious and single-copy DNA was studied in the eukaryotic genome (29).

Kohne first used HAP to isolate rRNA-DNA hybrids from *E. coli* and *Proteus mirabilis* (26). The denatured DNA was hybridized with excess rRNA for short periods to minimize DNA-DNA reassociation. The hybrids and any reassociated DNA were separated from ssDNA and the excess rRNA by absorption on HAP. The duplex fraction was eluted and denatured, and the process was repeated. Purity of the complementary strand was assessed by the fraction of labeled DNA which behaved as duplex molecules after hybridization with rRNA. Besides bacterial rDNA, 5S and tRNA complementary DNA fragments have been isolated recently by a similar method (cf 24, 30, 31). The technique has been applied to the purification of those DNA sequences transcribed into RNA by a variety of animal species (32, 33). *Neurospora* tRNA-DNA (34) and rRNA-DNA (35) and yeast rRNA-DNA (36) hybrids have been purified by absorption to HAP.

Nitrocellulose

Nygaard & Hall originally demonstrated that denatured DNA binds to nitrocellulose while native DNA and RNA do not (37, cf 38). RNA-DNA hybrids that still have ssDNA regions are absorbed, and this method has been used commonly for assay of such hybrids.

ssDNA ranging in size from several hundred nucleotides to molecules of high molecular weight absorbs to nitrocellulose. However, a variety of factors affect absorption of DNA. High molecular weight native DNA is trapped under certain conditions (39), and DNA that appears to be native by other criteria can bind to nitrocellulose (40). Certain proteins are absorbed by nitrocellulose and this has been a useful method for enriching DNA fragments complexed with proteins, as discussed later, but it is also a source of artifactual binding of RNA and native DNA to nitrocellulose.

The ratio of RNA to DNA in a hybrid affects its affinity to nitrocellulose;

E. coli rRNA-DNA hybrids (41) were found to elute more easily from nitro-cellulose than do *Drosophila* rRNA-DNA hybrids (42). Nitrocellulose chromato-graphy was used to separate *B. subtilis* ssDNA from rRNA-DNA hybrids (43).

Methylated Serum Albumin

Methylated serum albumin (MAK) complexed to diatomaceous earth (Kieselguhr or Celite) fractionates DNA on the basis of size, secondary structure, and base composition (44, 45). The strands of *B. subtilis* DNA (termed H and L) are separated completely on MAK columns (46). This separation takes place even though the DNA molecules applied to the column are small fragments of the full length chromosome. This means that the sequence asymmetry which accounts for the separation is maintained along the full length of the genome. Additional fractionation of *B. subtilis* ssDNA is effected by MAK. This can be shown since rRNA and tRNA hybridize only with those fragments of the H-strand DNA that elute at the highest salt concentrations from MAK (47). Beginning with the purified H strand from MAK, Colli & Oishi hybridized the DNA with rRNA components and purified the DNA complementary to rRNA about eightyfold (48).

MAK has been used to enrich the RNA-DNA hybrids formed by the hybridiza-tion of *E. coli* (49) and mammalian (50) DNAs with their homologous 4S RNAs.

Rapidly reassociating DNA fractions from mouse have been purified from ssDNA by digestion with an endonuclease specific for ssDNA followed by chromatography of the resistant duplex fragments on MAK (51).

Other Columns

ssDNA and dsDNA are reported to be separated by column electrophoresis (52), partition chromatography (53), and chromatography on columns of silk fibroin (54) and wool cortical cell protein (55). Columns of an organomercurial-Sephadex derivative (56) or benzoylated (57) or benzoylated naphthoylated (58) DEAE-cellulose, have also been used.

Buoyant Density Centrifugation

Denatured DNA has a density in CsCl which is on an average 14 mg/cm^3 higher than native DNA. This difference is sufficient to separate a mixture of the two. Sodium iodide gradients also separate ssDNA from dsDNA (59). DNA com-ponents which reassociate rapidly and form well-matched duplexes can be separated from ssDNA by buoyant density centrifugation. Using this method, reassociated mitochondrial DNA has been separated from nuclear DNA by CsCl buoyant density centrifugation (60).

The dependence of buoyant density on the base composition of ssDNA in alkaline CsCl has been studied (61, 62). The differences in densities of the four nucleotides in alkaline CsCl make it possible to separate strands of homogeneous DNA components with asymmetric distribution of bases. Banding of ssDNA in Cs_2SO_4 has been shown to be rather complex in nature, giving rise to fractionation which does not occur in CsCl (62, 63).

RNA-DNA hybrids band at higher buoyant density than ssDNA alone. Synthetic polyribonucleotides interact to different extents with separated strands of some

DNAs effecting strand separation in CsCl (64). Summers (65) has investigated the ribopolymers most suited for this separation. Copolymers of I and G were found to be better than poly G. Slight alkali degradation of G-rich polymers also increased their binding to ssDNA. The strands of purified *X. laevis* rDNA have been separated by hybridizing the DNA with rRNA and banding in CsCl (66).

Density gradient centrifugation enriched for RNA-DNA hybrids between rRNA and DNA from *E. coli* (67) and from *B. subtilis* (43), and between tRNA and DNA from bacterial DNAs (48, 68) and animal sources (50). Other RNA-DNA hybrids separated from ssDNAs by buoyant density centrifugation are phage mRNA-DNA hybrids (69) and hybrids between total mammalian RNA and homologous DNA (33). In these kinds of experiments the molecular weight of the DNA fragments is critical, since the higher the RNA to DNA ratio the higher the buoyant density of the hybrid and the greater its separation from ssDNA. This fact has been exploited by Marks (68) who treated hybrids of *E. coli* tRNA-DNA with *Neurospora* endonuclease specific for ssDNA. The resistant fragments have an RNA to DNA ratio of about one, and they are much denser than the hybrid prior to digestion.

The difference in density in Cs_2SO_4 between partly single-stranded molecules such as RNA-DNA hybrids and completely single-stranded ones is accentuated by interaction with Hg^{2+}, which binds more tightly to ssDNA. Hybrids between rRNA and DNA from *B. subtilis* were purified by this technique (48).

Single-Strand Specific Nucleases

RNA-DNA hybrids and renatured DNA are stable to a variety of single-strand specific nucleases. *Neurospora crassa* endonuclease (70) has been used most commonly (51, 68, 71). Shark liver endonuclease digested single-stranded regions in T_4 DNA-mRNA hybrids (72). The exonuclease I of *E. coli* (43, 73) and mung bean endonuclease I (74) have also been used for their ssDNase activity. *Aspergillus* S_1 nuclease is single-strand specific but can digest either DNA or RNA (75).

The *Neurospora* enzyme can digest the earliest melting regions of native DNA (76). Partly denatured DNA was fixed with formaldehyde and digested immediately. Only the high A-T regions were digested. Resistant native DNA fragments were purified by techniques that depend upon size fractionation.

Phase Extraction

Albertsson's phase extraction techniques separate ssDNA and duplex DNA (77). Recently, a simple extraction column using the two-phase aqueous polymers has been devised by Blomquist & Albertsson (78), which should be adaptable for large amounts of DNA. DNAs with differing extents of single strandedness seem likely to be separable by these systems (79) as will RNA-DNA hybrids. The separation is sensitive to molecular weight as well as secondary structure (77).

Polynucleotides Fixed to Insoluble Matrices

ssDNA components can be purified by fixing complementary RNA or DNA molecules to some kind of insoluble support and circulating the soluble DNA mixture through the affinity column. Large amounts of sequence-specific DNA have been purified from mixtures of phage and bacterial DNAs by circulating the denatured DNAs through columns containing one of the DNAs adsorbed to nitro-cellulose (69, 80).

Poly I-coated Kieselguhr separates the strands of a variety of bacterial DNAs which do not separate either in alkaline CsCl or when complexed with poly I in solution and then banded in CsCl (81, 82).

Ribosomal RNA-cellulose columns enriched the complementary DNA strands from *B. subtilis* about twentyfold (83). *E. coli* rRNA attached to agar gave a twentyfold purification of the complementary DNA strand (84). This latter method was successful only with ssDNA molecules of about 7×10^4 daltons; higher molecular weight DNA was not fractionated. A general method for isolating ssDNA was designed which uses the same principle (85). Large amounts of RNA transcribed from SV40 DNA with *E. coli* RNA polymerase were fixed to cellulose by a water-soluble carbodiimide. Denatured DNA containing a small amount of SV40 DNA was sheared to about 3×10^5 daltons and circulated at temperature and salt conditions which promoted hybridization. The complementary SV40 DNA fragments were removed from the solution almost quantitatively.

NATIVE DNA FRACTIONATION

The major characteristic of native DNAs that allows their separation is the variation in GC content between molecules. The buoyant density, chromatographic properties, electrophoretic mobility, and interactions of DNA with a variety of ligands change as a function of GC content. These methods are also sensitive to sequence differences. An extreme example is the synthetic polynucleotides dAT and dAdT, which have vastly different physical and chemical properties. DNA molecules containing many repeats of the same simple sequence will differ invariably in one or more of these parameters from a more complex DNA of the same overall base composition.

Another important point to be made concerning native DNA fractionation is that each gene has different DNA sequences on either side. These may differ strikingly in characteristics from the sequence being purified, preventing a priori knowledge of a gene's behavior. An example is the buoyant density change of transforming markers in DNAs of different molecular weights (86). Tandem genes in eukaryotes are known to be separated by spacer DNAs which can vary in size and base composition. Table 1 presents the characteristics of native purified genes, with particular attention to these spacers. Additional evidence for the presence of spacers has been presented for the histone genes of the sea urchin (87), the ribosomal genes of *Drosophila* (42), and tRNA genes of *Xenopus laevis* (88). The function of spacers is unknown, but they evolve rapidly even between related species, and thus resemble satellite DNAs. Since the multiple spacers evolve together (19, 23), the gene and spacer behave as a more or less homogeneous and repetitious DNA component.

Table 1 Some properties of purified genes and their spacers

	Gene: Spacer Mol wt × 10⁻⁶	Base composition (% GC)[d]	Approximate of repeats[c]	Fraction of genome (%)	Ref.
rDNA					
X. laevis	4.4:4.4[a]	60:73	450	0.2	(89)
X. mulleri	4.4:4.4[a]	60:69	450	0.2	(23)
Lytechinus variegatus	~4.2:1.0	?	260	0.3	(90)
5S DNA					
X. laevis	0.08:0.5[b]	57:35	24,000	0.7	(16)
X. mulleri	0.08:1.2[b]	57:43	9,000	0.7	(19)

[a] These values refer to the relative abundance of 28S and 18S gene sequences compared to the rest of the rDNA. This includes nontranscribed plus transcribed spacer, i.e., that portion of the DNA transcribed as part of the 40S RNA precursor.

[b] These are average values. All spacers within a species may not be the same length.

[c] Number of repeats per haploid complement of DNA.

[d] The base composition of *Xenopus* bulk DNA is 39% GC.

Size

Until recently, separation of DNA by size was restricted to the purification of DNA components from isolated cell particles or viruses. Examples are mitochondrial, episomal, and plasmid DNAs. Recently, the potential of gene separation by size has been increased greatly by the discovery of DNA restriction enzymes (cf 91). These enzymes recognize specific deoxynucleotide sequences of five to six base pairs in length and hydrolyze phosphodiester bonds in both strands within such sequences. DNA fragments lacking the sequence can be isolated as high molecular weight DNA free from the smaller cleaved fragments. Mouse satellite DNA is more resistant than main band DNA to the enzyme R_1 from *E. coli* (92). *Xenopus laevis* and *Xenopus mulleri* 5S DNAs (J. Morrow and D. Carroll, personal communication) are totally resistant to the enzyme. Discrete size fragments are produced from SV40 DNA by various restriction enzymes. All of these fragments were separated by gel electrophoresis and have been arranged in order on the chromosome (93). Restriction enzymes will be invaluable for isolating exact sized fragments of DNA which are highly enriched for a given gene.

Methods for separating DNA molecules by size include sedimentation velocity (cf 94), gel electrophoresis (93), molecular sieve chromatography (95), and certain columns such as MAK (44). Varying concentrations of isopropanol precipitate DNA as a function of size (96). This method may be useful for crude fractionation of large amounts of DNA by size.

Structure

A number of methods have been devised to separate circular from linear DNA molecules. Vinograd et al introduced the use of the intercalating dyes, ethidium bromide (97) and propidium diiodide (98). Linear and open circular molecules interact with more dye than closed circular DNA, causing the former to band at a lower density in CsCl. This method has been used for purification of mitochondrial (cf 99), viral, and bacterial plasmid DNAs (cf 100). A technique has been described in which closed and open circles are separated from linear molecules by trapping the circular DNA in agar under conditions where linear molecules can be washed out of the matrix (101). The complementary strands of denatured closed circular DNA molecules reassociate faster than open circles or linear forms, since the two strands are interlocked after denaturation and therefore more likely to re-establish base pairing (102).

Sedimentation varies for molecules in different physical configurations. Closed circular, open circular (with one-strand scission), and full length linear molecules are separated by velocity gradient centrifugation (cf 100, 103). Single-stranded circles can be separated from double-stranded circles (cf 100).

Methods that separate linear molecules from circles can be exploited by generating circles. Any DNA molecule whose sequence is terminally repetitive can be digested with a 3' exonuclease to produce complementary single-stranded ends (104). These molecules will form circles. Gaps in the circles can be filled in with DNA polymerase and sealed with ligase. The technique has been applied to repetitious DNA molecules from eukaryotes (105).

Circularity and the degree of secondary structure are the best studied variations in the structure of DNA. However, a number of proposed structures have been suggested for native DNA which, if proven, might make these DNA components easy to purify. Symmetrical sequences (including palindromes) are known to occur in DNA; these might cause dsDNA to assume complex secondary structures such as cloverleaf structures (106). DNA networks have been proposed to exist in vivo to account for the complexities of the immunoglobulin genes (107). Such networks might also exist in polytene chromosomes at the junction between the highly replicated euchromatin and the under replicated heterochromatin (108, 109). Forks and rolling circles are structural changes which might be expected to change the sedimentation rates of replicating DNA to the point where they could be isolated from normal duplex molecules.

Selective Denaturation

A DNA preparation can be enriched for GC-rich DNA by selective heat denaturation of high AT DNA followed by separation of ssDNA from dsDNA.

Native DNA fragments absorbed to hydroxyapatite can be eluted with a thermal gradient (110). As DNA denatures, it is eluted from the column. There is a direct relationship between GC content and elution temperature. The technique is modified by the addition of sodium perchlorate to the elution buffer (111), which lowers the temperature at which denaturation and elution take place. The rRNA and tRNA genes in *Mycoplasma* are enriched by a partial heat denaturation of the DNA followed by fractionation of the DNA on hydroxyapatite (112). Feldmann (113) purified these genes further by denaturing the high GC fraction from HAP, hybridizing it with tRNA, and isolating the hybrid molecules from ssDNA. Similar techniques were used for *B. subtilis* rDNA (114).

The selective denaturation method probably was used first for the purification of a high GC episomal DNA from *Proteus* (115). Chromosomal *Proteus* ssDNA was removed by trapping on nitrocellulose filters. Nitrocellulose absorption of ssDNA was used to enrich for rDNA in *B. subtilis*, which is higher in GC content than the main band DNA (43, 116). Recently, native rDNA has been enriched from human DNA by partial heat denaturation and passage through a nitrocellulose column to remove the ssDNA (117).

The first step in the purification of high GC rDNA from sea urchins was denaturation of most of the DNA followed by separation of the native (including rDNA) from the denatured fraction by Albertsson's two-phase separation (117a).

The partial denaturation method conceivably could be extended in application by denaturing DNA under conditions that alter the dependence of thermal stability on GC content of DNA. Certain organic salts can reverse this dependence. Thus, dGdC melts at a lower T_m than dAdT in very high concentrations of tetramethylammonium chloride or tetracetylammonium chloride (118). Another method of reversing the T_m dependence on GC content is to complex the DNA with basic proteins such as polylysine and some histones. These proteins complex specifically with AT-rich DNA (119) and raise their T_m. By such methods it should be possible to change that fraction of DNA which has the highest T_m, then repurify the DNA and separate the native from ssDNA by one of the methods described in this section.

Density Gradients

Native DNA is fractionated most commonly by neutral CsCl buoyant density gradient centrifugation. A direct relationship between GC content and buoyant density of native DNAs in CsCl has been described (120, cf 121). Synthetic polymers (121a) and many naturally occurring minor components do not always obey this relation. Methylation of cytosine residues also changes the buoyant density of DNA (122). A variety of satellites have been purified by neutral CsCl centrifugation. Two of the first to be purified were mouse satellite DNA (123) and crab poly dAT (124). The first isolation of a DNA component of known function was accomplished by Birnstiel and co-workers (125), who isolated ribosomal DNA from *Xenopus laevis,* which has a high GC content compared to the bulk DNA (126). Although it comprises only 0.2% of the total DNA, rDNA could be isolated by repeated equilibrium centrifugation in CsCl and pooling of the densest fractions from the gradient. Neutral CsCl centrifugation is most successful for the isolation of trace DNA components if they have a large density difference from main band DNA. CsCl gradients have low capacity for DNA and serious artifacts can occur with high molecular weight DNA (15).

Useful modifications of buoyant density techniques applicable to CsCl and other density salts were the introduction of fixed angle rotors (126a, 127), which allow greater resolution than swinging bucket rotors, and zonal rotors, which fractionate larger quantities of DNA (128). Two techniques that increase the rate of attainment of equilibrium are relaxation centrifugation (129) and step gradient centrifugation (130).

Compared to CsCl, Cs_2SO_4 produces a steeper gradient with a reduced dependence of buoyant density on base composition (131). For these reasons Cs_2SO_4 is not used by itself for the fractionation of native DNAs. It is used extensively to fractionate nucleic acid-metal ion complexes that would be destroyed by CsCl (see below).

NaI (132) and KI (133) gradients have been introduced recently. The resolution of DNAs of different base compositions in these gradients seems to be comparable to CsCl. NaI gradients have been used to separate the halves of the simian adenovirus SA7 genome (134) and to purify mitochondrial DNA from yeast (135). A significant advantage of NaI gradients is their low cost. Unfortunately NaI absorbs strongly at 260 nm. Since ethidium bromide does not alter the buoyant density of DNA in NaI, it may be used to locate the DNA by fluorescence (132).

A variety of molecules have some sequence specificity in their interaction with DNA (cf 136, 137). Only a few of these bind to DNA strongly enough in high salt to be useful for equilibrium centrifugation separation. Conceivably, many more might be adapted to column or solvent extraction methods described in a later section.

Only a few metal ions interact with the bases of nucleic acid rather than with its phosphate groups (138). Of these, Ag^+ and Hg^{2+} have been used extensively because of their apparent sequence specificity. Silver combines with DNA bases, forming an extremely dense complex (139). There is evidence for three types of binding of Ag^+ to native DNAs (139, 140). The first type is found with poly dGdC but not poly dAT, and saturates at an r_b (ratio of bound silver to nucleotide residue) of 0.25. The second type of binding is found with poly dAT but not poly dGdC, and saturates at an r_b of 0.5. Binding of the first type reaches saturation at Ag^+ concentrations at least tenfold below those required to achieve saturation for the second type. This accounts for the simplified description of Ag^+ as a GC-specific reagent. However, the different plateaus mean that with higher silver concentrations, poly dAT would bind twice as much silver as poly dGdC. Daune et al (140) have proposed that the nearest neighbor frequency is an important determinant of silver binding behavior. Given these data, it should come as no surprise that the relative location of a DNA fraction in an Ag^+-Cs_2SO_4 gradient cannot always be predicted from its GC content and may change with the amount of silver added. Moreover, binding of the second type releases a proton, making the reaction pH dependent. Filipski et al (141) have shown that the proper choice of pH can improve the resolution of DNAs in Ag^+-Cs_2SO_4 gradients. In the future this parameter must receive as much attention as the silver concentration. Ag^+-Cs_2SO_4 gradients have been used to purify satellite DNAs from guinea pig (142, 143), human (144, 145), trypanosome (146), African green monkey (147), calf (141, 143), kangaroo rat (147a), dinoflagellate (147b), and crown-gall tissue culture (148). They have also been used to isolate the 5S DNA from *X. laevis* (16) and the mitochondrial DNA of yeast (149), and to study the ribosomal and fibroin genes in *Bombyx* (14) and the 5S, tRNA, and rRNA genes in yeast (150). In many of these studies the position of a component in an Ag^+-Cs_2SO_4 gradient could not be predicted from its buoyant density in CsCl. This was as true for the genes as for the satellite DNAs. In some instances Ag^+-Cs_2SO_4 gradients were capable of resolving minor components whose buoyant density in CsCl was identical to that of main band DNA. Only a few papers presented density versus silver titrations (141, 151), so the detailed behavior of most of these fractions in Ag^+-Cs_2SO_4 gradients is unknown.

The other metal ion used in Cs_2SO_4 is Hg^{2+} (152). In contrast to Ag^+, Hg^{2+} is considered to be an AT-specific reagent, but there is more than one type of binding with this metal and plateaus in titration curves are seen. At low Hg^{2+} concentrations, crab dAT binds more Hg^{2+} than does main band crab DNA (151); however, at high concentrations dAT would be expected to bind less Hg^{2+}. Similarly, two satellite DNAs of *Xenopus mulleri* that appear as light satellites in CsCl appear as heavy satellites in Hg^{2+}-Cs_2SO_4 gradients at low Hg^{2+} concentration, but as light satellites at higher Hg^{2+} concentrations (19). Hg^{2+}-Cs_2SO_4 gradients have been used to fractionate satellite DNAs from yeast (150, 153). *Drosophila* (154), human (155), African green monkey (155a), crab (151), and *Trypanosoma cruzi* (156) DNAs. They have also been used to fractionate DNA fragments of phage λ (157, 158), bacteriophage SP3 (159), adenovirus type 2 (160), CELO virus (161), and yeast mitochondrial (162), calf (163), *B. subtilis* (163a), and *E. coli* (164). The genes coding for 5S RNA, tRNA, and rRNA in yeast (150, 153) and 5S RNA and rRNA in *Xenopus mulleri* (19) have also been studied with these gradients. Hg^{2+}-Cs_2SO_4 gradients were the first step used for the isolation of these latter genes. Ag^+ and Hg^{2+} have been used together in the same gradient to produce a separation of two satellite DNAs that neither metal alone could effect (165).

Many drugs bind to DNA and some have apparent base specificity (137). From early studies it was found that DNAs from a number of different organisms bind actinomycin D in proportion to their GC content and that synthetic dAT does not bind actinomycin D. It was concluded from these and other results that actinomycin D bound to guanine residues in native DNA. However, Wells & Larson (165a) have shown that poly d(A-T-C)·poly d(G-A-T) does not bind actinomycin D, whereas poly d(T-T-G)·poly d(C-A-A), poly d(T-T-C)·poly d(G-A-A), and poly d(T-A-C)·poly d(G-T-A) do bind actinomycin D. Clearly actinomycin D is not merely a base-specific reagent; the nucleotide sequence is an important determinant of the interaction.

Kersten et al (166) first showed that the differential binding of actinomycin D to various DNAs was maintained in CsCl gradients: the more actinomycin DNA bound, the lower the density. Actinomycin interaction was used as the final step for 5S DNA purification from *X. laevis*. The change in buoyant density of five different DNA components from *Xenopus mulleri* was studied as a function of actinomycin concentration (167). These DNAs included two satellites rich in AT, 5S DNA, rDNA, and main band DNA. All five components bind different amounts of the drug and therefore have different buoyant densities. Yet in contrast to metal ion titrations, each of these components reached its respective buoyant density plateau at the same actinomycin concentration. Therefore, the best separation occurred at saturating concentrations of the drug (10 μg/ml of drug and 20 μg/ml of total DNA). The other antibiotics shown by Kersten et al (166) to bind to DNA in density gradients were compared with actinomycin for their ability to separate *Xenopus* rDNA and 5S DNA from main band DNA. None gave as good a separation as actinomycin D (19, 167). Actinomycin-CsCl gradients have been used to study the fibroin and rRNA genes of *Bombyx mori* (168) and to purify satellite DNAs of *Drosophila melanogaster* (169).

Netropsin is an antibiotic which binds to DNA with AT specificity (170). It has been used to isolate satellite DNAs of *D. melanogaster* in CsCl gradients (169).

Basic Protein—DNA Interaction

Lysine-rich histones and synthetic poly-L-lysine react cooperatively with DNA. At lower salt concentration, polylysine precipitates DNA indiscriminately and at higher salt concentration, the polylysine is dissociated from the DNA; at about 1 M NaCl, polylysine interacts specifically with AT-rich DNA and can precipitate it from a mixture with GC-rich DNA. This was shown by Leng & Felsenfeld (119, 171) for artificial mixtures of bacterial DNAs with different GC contents. The polylysine method is a practical way to purify a high GC DNA component from a large amount of starting DNA. Thus, more than 95% of *X. laevis* DNA (39% GC) was precipitated under conditions where most of the rDNA (67% GC) remained in the supernatant fluid (89). Several hundred milligrams of DNA can be processed in a few days. This is much cheaper and faster than processing the same amount of DNA by CsCl centrifugation. Poly-L-arginine interacts with high GC DNAs preferentially (119). When polylysine interacts with DNA in the presence of 2 M tetramethylammonium chloride, the base composition specificity is reversed (172).

Basic proteins retain their specificity for high AT DNA when absorbed to an insoluble material such as Kieselguhr. A variety of *B. subtilis* genes were separated partly on MAK as assayed by transformation (172a, 173, 174). Chromatography on MAK has been used to purify crab (174a, 175), calf (176), mouse (176), guinea pig (176), and human satellite (177) DNAs. Many of these satellite DNAs have an elution position unexpected from their buoyant densities. MAK chromatography also fractionates DNA by size which can obscure the base composition fractionation (174).

Poly-L-lysine Kieselguhr has been used to purify yeast mitochondrial DNA by a batch procedure (178, 179). Since nucleic acids are absorbed selectively to poly-L-lysine Kieselguhr, a crude lysate was mixed with the material, washed with an NaCl concentration that removes nuclear DNA, placed in a column, and eluted with a salt gradient. Mitochondrial DNA was recovered about 95% pure and in high yield.

Column Methods

A variety of columns fractionate DNA (cf 180). The most widely used and best characterized is HAP. Native calf, guinea pig, and mouse DNAs have been fractionated on HAP columns (177). Yeast mitochondrial DNA elutes from HAP at very high salt concentrations, even higher than synthetic and crab poly dAT. This has made it possible to isolate yeast mitochondrial DNA by a batch procedure (181). Total DNA was mixed with HAP at a phosphate molarity that elutes most of the main band DNA. After washing the HAP, it was packed into a column and the mitochondrial DNA eluted in high yield and purity with a salt gradient. This procedure has also been used for the mitochondrial and chloroplast DNAs of *Euglena gracilis* (182).

Solvent Extraction

A number of workers reported the selective loss of DNA fractions following phenol deproteinization (183–186). In most cases, the missing component was partitioned into the phenol phase. Smith et al (187) reported that variations in salt concentration and pH are factors governing the behavior of synthetic poly dAT. The dAT satellite from *C. borealis* partitions into phenol at room temperature and is recovered in the aqueous phase at $0°$ (188). Sabeur et al (189) working with *C. pagarus* found that the dAT isolated by manipulating the salt concentration of the aqueous phase is no longer native. Fractions of crab dAT differing in GC content and isolated by thermal chromatography on hydroxyapatite are decreasingly soluble in phenol as their GC content increases (190). Although these studies were not conducted on native DNA, they suggest that the phenol-water phase system might be applied to the fractionation of native DNAs differing in GC content. A different phenomenon accounts for the selective loss of kinetoplast DNA from *Trypanosoma equiperdum* DNA isolated with phenol deproteinization (186). In this case, the missing DNA is in the protein interface and may be recovered in pure form by chloroform-isoamyl alcohol extraction.

Native DNAs also differ in their partition coefficients in polyethylene glycol-dextran two-phase systems (77). These differences might allow fractionation of DNA by multiple extraction procedures utilizing countercurrent distribution methods.

Electrophoresis

Zeiger et al (191, 191a) have shown a linear relationship between electrophoretic mobility and GC content for a series of microbial DNAs. This method resolved mouse satellite DNA (191) and *Callinectes* satellite DNA (191a) from their main band DNA.

Specific Protein-DNA Interaction

In addition to the interaction of basic proteins such as polylysine and certain histones as a function of GC content and its use for the isolation of native DNA by base composition, another kind of protein-DNA interaction is the single-strand-specific gene-32 protein purified from T_4-infected bacterial cells (192). In contrast to these less specific kinds of interaction, a variety of regulatory proteins, polymerases, and enzymes have been shown to interact with specific nucleotide sequences of bacterial DNA (cf 136).

DNA complexed with protein can be purified since it is protected from DNase digestion. The *lac* operator-repressor complex was isolated by its selective absorption to nitrocellulose (193). This method resulted in the purification of the *lac* repressor binding site, which consists of 21 base pairs of DNA (193) from about 10^7 total base pairs in the *E. coli* genome. The two λ operons are similarly enriched after complexing with λ repressor (194). Another specific regulator protein, the catabolite gene activator protein (CAP), which along with cyclic AMP regulates certain bacterial operons, has been used to probe the DNA sequences with which it complexes (195). Experiments demonstrate that this method is useful for isolating promotor regions of DNA complexed with RNA polymerase (195a). The isolation of a stable RNA polymerase-DNA product by CsCl centrifugation has been reported (196). Conditions for specific binding of *E. coli* RNA polymerase to λ DNA have been described (197). Weak binding sites are competed for by the addition of tRNA, and specific binding is favored at low enzyme to DNA ratios.

Earlier studies on ribosome binding sites of various RNA phages have been successfully extended to phages ϕX174 (198) and T$_7$ (199) DNAs. DNase-resistant fragments of a ribosome-DNA complex have been sequenced. Specific protein interaction with ϕX174 DNA depended upon N-formylmethionyl tRNA, GTP, and a crude mixture of factors for the initiation of protein synthesis.

A variety of other bacterial proteins with nucleotide sequence specificity have been described but not applied to gene isolation. Bacterial DNA methylases are highly specific for certain sequences (cf 91). The use of DNA restriction enzymes for their specific cleavage site has been mentioned earlier. Both of these proteins as well as any sequence-specific protein could in principle be used to purify the DNA components with which they interact by the same method described above for regulatory proteins. Presumably they need only interact with the DNA under conditions where binding occurs without enzymatic activity, and the protected fragment could be isolated. Proteins in eukaryotic cells with the degree of specificity found in these bacterial proteins have not yet been described. The search for such proteins is handicapped by the lack of detailed genetics and mutants for specific genes and the presence of a large amount of basic proteins and DNA binding proteins with no apparent subtle nucleotide sequence specificity. Since proteins alone have been found to have the specificity to interact with long nucleotide sequences, their use for native DNA purification in the future cannot be overestimated.

CHROMOSOME AND CHROMATIN FRACTIONATION

The isolation of metaphase chromosomes might result ultimately in a homogeneous preparation of a single chromosome. Metaphase chromosomes have been isolated and fractionated according to size from Chinese hamster cells (200–203), mouse L cells (201), L-5178Y lymphoblast (201), HeLa cells (200, 204, 205), regenerating rat liver cells (206), and PHA-stimulated human lymphocytes (207). The smaller chromosomes are always obtained in higher purity due to aggregates of small chromosomes sedimenting with the larger chromosomes. Refractionation on a second gradient results in more homogeneous fractions. Wray et al (208) have reported that most of the methods used for isolating chromosomes result in loss of high molecular weight DNA. Ribosomal DNA was found to be localized to small human chromosomes (204), while the genes coding for 5S RNA (205), 4S RNA (205), mRNA (204), and heterogeneous nuclear RNA (202) were present in all size classes. One promising approach to chromosome isolation could take advantage of somatic cell hybrids between two different species whose chromosomes are very different in size and in which one set of chromosomes is lost selectively. This exact situation has been described for hybrids of mouse and human cells. The larger human chromosomes are lost selectively after a number of cell generations (209). By using appropriate mutants and conditions for selection, it might be possible to select for the maintenance of a single large human chromosome among the small mouse chromosomes.

Chromosome fractionation by microdissection is now feasible for giant polytene chromosomes. Chromosome IV of *Chironomus* has several very active genetic loci which undergo puffing and concomitant RNA synthesis. These regions can be visualized with a microscope and excised from the chromosome (210), presumably effecting a great purification of the puffed gene.

Unbroken DNA molecules which appear to be about the correct size for the duplex DNA molecule of chromosomal length have been prepared from yeast (211–213) and detected in *Drosophila* (214). The advantage of *Drosophila* is the availability of genetic deletions and different species with different sized chromosomes. Mitochondrial DNA sediments as the lowest molecular weight component in gentle DNA lysates of yeast (213). Strains of yeast that are disomic for different chromosomes have been studied. The DNA molecules containing rDNA sediment at rates different from the bulk of the high molecular weight DNA (212), and evidence is presented to show that much of the rDNA is localized on a single yeast chromosome and probably comprises most of the DNA in the chromosome (215). Such experiments should be useful to determine gene linkage.

The present state of chromatin fractionation relies on purported physical and chemical differences between euchromatin and heterochromatin or between active and inactive chromatin. Heterochromatin is more compact and sediments more rapidly than euchromatin (216). Heterochromatin enriched for satellite DNA has been extracted selectively from nuclear preparations (217, 218). The higher the protein content of chromatin the higher the T_m, and thermal elution of chromatin from hydroxyapatite has been described (219). CsCl (220) and Cs_2SO_4 (221) centrifugation of deoxyribonucleoprotein particles demonstrates heterogeneity of chromatin. Hossaing et al (222) have banded chromatin to equilibrium in chloral-hydrate gradients, a technique which obviates the necessity to use high ionic strength, a condition known to dissociate protein-DNA complexes. Aside from satellite DNA localization in heterochromatin, specific genes have not been localized unequivocally in specific chromatin fractions nor has a chromatin fraction been purified that is enriched for a known gene.

SYNCHRONIZED REPLICATION

Synchronized replication was first used as a method of gene mapping and its potential for gene isolation was noted in prokaryotes. *B. subtilis* DNA has been synchronized in its replication by germination of spores (223). *E. coli* DNA replication can also be synchronized (224). Most experiments done with these synchronized cells have focused on mapping of related genes. Thus, rDNA, 5S DNA, and tDNA are clustered near the origin of replication in *B. subtilis* (225). The principle of gene isolation in synchronized cells is to incubate the cells with 5-bromodeoxyuridine (Budr) for very short periods and then purify the substituted DNA which has a higher buoyant density. This method has been applied to *E. coli* DNA by Cutler (226).

The technique has not been used for eukaryotic cells but it holds considerable promise. Mitochondrial DNA replicates outside of the S phase of nuclear DNA (227). A preferential replication of rDNA seems to occur in Tetrahymena as a result of a change in the growth medium which initiates active metabolism (228). The acellular slime mold, *Physarum polycephalum*, is undoubtedly the best eukaryote known for the study of synchronized DNA replication. The amoeboid form is a giant syncytium with many nuclei in the same cytoplasm, all of which are known to divide synchronously (229). It has been shown (230, 231) that rDNA replication is not restricted to the S phase of *Physarum*. Because the nuclei are synchronized so accurately, Budr pulses at different parts of the S phase should give a very reproducible fractionation of the genome based on the time of replication of various sequences. A variety of methods have been developed to synchronize cultured cells but none are coordinated as accurately as the slime mold.

Heterochromatin is known to replicate late in animal cells (218). The fact that satellite DNAs are located usually in heterochromatin supports the observation that they are replicated late in the cell cycle (232). The notion is attractive that there is a relationship between replication time of a gene and its activity in the cell, and that different cell types of the same animal could have different genes that replicate early. Unfortunately there is no support for this idea beyond the correlation of lack of function and late replication of heterochromatin.

ENRICHMENT OF GENES IN VIVO

Sophisticated methods have been devised to enrich the content of certain genes in bacteria. With appropriate mutants and selective conditions, portions of the bacterial chromosome have been incorporated into transducing phages or episomes.

High frequency transducing phages such as ϕ80 and λ for *E. coli* and P22 for *Salmonella* can carry regions of the bacterial chromosomes near the prophage attachment site. The first example was λ carrying the *gal* operon of *E. coli* (233). Phage ϕ80 containing genes for tryptophan biosynthesis (234) and tyrosine tRNA (235) have been prepared. *Salmonella* phage P22 attaches to a specific site on an F' factor from *E. coli*, and a variety of *E. coli* genes have been incorporated into this phage (236).

The kinds of genes that can be integrated into these phages can be increased by several means. Deletions in the bacterial chromosome near the prophage attachment site brings new genes close enough to be incorporated into the phage. F' factors have been inserted into the *E. coli* chromosome near the ϕ80 attachment site and thus made available for incorporation into the phage (237). Many kinds of F′ factors are known which carry chromosomal genes, including the ϕ80 attachment site itself (238). These latter phages increase the variety of genes that can be brought close to the ϕ80 attachment site, since they can recombine with other F′ factors containing different genes (239). The same technique has been used successfully to generate P22 phage fused with F′ factors carrying structural genes (240). Finally, in strains lacking the normal prophage attachment site, the phage integrates with low frequency at other sites in the chromosome, in each case making the adjacent gene sequences susceptible to incorporation into the phage genome (241).

The insertion of a bacterial gene into a transducing phage represents a great enrichment of the gene, since the *E. coli* genome contains about 200 times more DNA than do the phages. Episome and phage DNAs are often circular and can be separated from bacterial chromosomal DNA by ethidium-CsCl centrifugation (described earlier). F-containing episomes of *E. coli* have been transferred into other species of bacteria whose DNA composition is sufficiently different to permit isolation of the episome by its GC content (242–245). The rDNA of *E. coli* was enriched by this method (245). An episome from *E. coli* containing rDNA was transferred to *Proteus mirabilis*. The lower GC content of *P. mirabilis* (39%) permitted its selective heat denaturation under conditions where the episomal DNA (50% GC) remained native. dsDNA and ssDNA were separated by benzoylated DEAE chromatography. Shapiro et al (246) further purified *lac* DNA from two transducing phages which contained the *lac* operon in reverse orientation. The complementary strands of the *lac* operon were both located on the same DNA strand of the two phages. When the two viral H strands were hybridized, only the *lac* operon reassociated; the phage DNA remained single-stranded. The single-stranded region was digested by *Neurospora* single-strand-specific endonuclease, leaving presumably pure reassociated duplex *lac* DNA. A similar technique resulted in the purification of the DNA for tyrosine tRNA from two strains of ϕ80 containing the genes in reverse polarity (247).

Recently, Davies & Rownd (248) have added another feature to using episomes containing drug-resistant genes. When these bacteria are grown on high levels of drugs, cells are selected which carry multiple copies of the particular determinant. The episome changes in buoyant density and abundance in the cell, presumably due to the construction of multiple similar genes in linear array. When the drug is removed from the medium, episomes with the amplified genes are diluted out and bacteria containing the normal episome reappear. The episome with multiple identical genes is clearly unstable when the selective environment is removed.

Bacterial episomes and transducing phages conceivably will be useful for the study of eukaryotic genes. Phage or episomal DNAs could be cleaved and foreign DNA fragments covalently linked to them. Methods for this in vitro integration procedure require formation of complementary single-stranded regions on the ends of molecules to be joined. Poly dT and poly dA have been added to DNA with calf thymus deoxynucleotidyl terminal transferase (249, 250). In another method certain restriction enzymes produce specific staggered cleavages at symmetrical DNA sites, leaving small single-stranded complementary ends (251, 252). These ends are long enough to hydrogen bond with each other and they can then be joined covalently with DNA ligase. The bacteria would be infected with the altered phage or episome and large amounts of the heterologous DNA could be produced. This method has now been successfully applied using an *E. coli* plasmid and *X. laevis* ribosomal DNA (252a). The restriction enzyme R_1 cleaves the plasmid once and rDNA at least twice within each repeat. Hybrid closed-circular molecules were joined with DNA ligase, and *E. coli* were transfected with the hybrid DNA molecules. Since the plasmid contains a gene for tetracycline resistance, this provided a means to select transfected bacteria. A substantial number of the tetracycline-resistant colonies contained rDNA fragments stably integrated in the plasmid DNA. If the heterologous DNA is impure to begin with, individual colonies containing integrated fragments can be isolated and tested for their content of the nucleotide sequence in question. This presumes that a specific hybridization assay is available for the DNA component. It remains to be seen whether an animal gene can function in a bacteria and thus provide a means of selecting for bacteria containing specific animal genes.

The selective replication of certain nucleotide sequences has been useful in purifying DNA of higher organisms. The relative abundance of mitochondrial DNA in yeast is affected greatly by the carbon source (253). Chloroplast synthesis can be regulated in growing plants by light. Some animals are known to undergo the selective loss of a portion of their genome during certain periods of their life cycle (cf 254). Chromosome diminution in some worms, insects, and crustaceans is known to occur. This process must involve selective regions of the genome as has been shown for *Ascaris* (255). Selective DNA replication occurs in polytenization of dipteran chromosomes. In *Drosophila* salivary glands, the euchromatic DNA containing mainly single-copy DNA is replicated about a thousandfold, while the heterochromatic portion located at the centromeres is not replicated (108, 109). In *Sciara* and *Ryncosiara,* DNA puffs are synthesized late in larval development (256). The genetic nature and function of this extra DNA are unknown but it appears to be a limited fraction of the total DNA that could be purified. Differences in the DNA between the macro- and micronuclei of the unicellular eukaryote *Stylonichia* have been reported (257). The highly replicated macronucleus does not contain all the sequences of the genome present in the micronucleus.

The production of petite mutants in yeast represents a means to enrich for regions in yeast mitochondrial DNA. The mitochondrial DNA of these respiratory-deficient cytoplasmic mutants differs from one strain to the next. Petite strains can be generated at essentially 100% efficiency by treating yeast with ethidium bromide (258). Individual strains appear to contain only restricted regions of the mito-chondrial genome replicated in many copies (259, 260). A number of mitochondrial genes have been identified in yeast and some linkage information is available. Complementation tests permit a more detailed mapping of the mitochondrial genome. Using these techniques along with molecular hybridization, it should be possible to produce strains with only a single gene region in many copies.

The best studied example of selective gene replication is the specific amplification of rDNA in oocytes (261, 262). This occurs in a wide variety of animals but has been studied in most detail in amphibians. An oocyte of *Xenopus laevis* has about 4000 haploid equivalents of ribosomal DNA, which amounts to about 70% of the total nuclear DNA. Because of this enormous enrichment, oocytes are a convenient source for rDNA purification.

CONCLUDING REMARKS

Isolation of a gene as native DNA permits any experiment that can be carried out with the purified coding strand plus a variety of others. Studies on the control of gene action require native DNA. Faithful transcription by RNA polymerase and studies dealing with specific binding of proteins involved in the control of eukaryotic genes need native DNA components. The present methods for native DNA isolation are applicable to reiterated genes such as rDNA, 5S DNA, and the genes for transfer RNA that comprise fractions of the genome larger than 0.005%. It is likely that these DNA components can be isolated from a wide variety of animal DNAs by one of the methods for native DNA fractionation described here. The histone genes appear to be present in multiple copies in some genomes (87), and these genes should be easy to purify. However, evidence is accumulating which shows that there will be few, perhaps only one, structural genes per haploid genome in eukaryotes (14, 17). One globin gene of about 300,000 daltons would comprise about $1.7 \times 10^{-5}\%$ of total human DNA. One gram of starting DNA would contain 0.17 μg of this gene. This formidable purification problem is compounded by the presence of neighboring sequences of unknown composition on either side of single genes which will vary considerably in molecules of different size containing the same gene (168). For this reason, a restriction enzyme will have to be found which specifically cleaves the DNA into homogeneous fragments, preferably pieces containing the structural gene and some sequences on either side.

Clearly, purification of important structural genes will have to be coupled with some method in which a small amount of a given gene can be increased enormously in amount. After purification has enriched the gene sequence about a thousandfold the remaining DNA would be amplified hundreds- to thousandsfold in amount. Purification could then be continued. The amplification step might be carried out in vitro by an efficient DNA polymerase, which would replicate faithfully each molecule of DNA many times. Alternatively, insertion of the DNA into a phage or bacterial episome, followed by infection and growth within a bacteria, could produce large amounts of homogeneous DNA components. This last method has the advantage of cloning individual DNA molecules from an impure mixture of DNA.

Given a pure RNA product of any gene, it should be possible to isolate the coding strand. There are no requirements for unusual base composition or nucleotide sequence of the gene which at present are necessary criteria for native DNA fractionation. Many of the best methods to process reasonably large amounts of single-stranded nucleic acids such as HAP or phase extraction do not seem to work as well for high molecular weight DNA. The larger the DNA fragments, the more the RNA-DNA hybrid will behave like ssDNA and the less successful will be the fractionation. These problems limit the technique to ssDNA molecules less than 10^6 daltons, which is still larger than most structural genes. Perhaps the use of complementary nucleic acid fixed to an insoluble support will permit the fractionation of higher molecular weight ssDNA. The enrichment of ssDNA fragments containing the sequence complementary to globin mRNA might make it possible to measure the distance between related linked globin genes (β, δ, and γ chain genes). Using ssDNA isolation methods, it should be possible to determine linkage for the genes that code for the constant and variable parts of immunoglobulin molecules.

Finally, one could isolate genes as single-stranded fragments and then replicate the complementary strand with a DNA polymerase, producing a duplex molecule suitable for gene control studies.

ACKNOWLEDGMENTS

We are grateful to Drs. D. Carroll, I. Dawid, P. Hartman, and P. Lizardi for their critical reviews of the manuscript.

Literature Cited

1. Hotchkiss, R. D., Gabor, M. 1970. *Ann. Rev. Genet.* 4:193–224
2. Mandel, M., Higa, A. 1970. *J. Mol. Biol.* 53:159–62
3. Zubay, G., Schwartz, D., Beckwith, J. 1970. *Proc. Nat. Acad. Sci. USA* 66:104–10
4. Zubay, G., Morse, D. E., Schrenk, W. J., Miller, J. H. M. 1972. *Proc. Nat. Acad. Sci. USA* 69:1100–3
5. Zubay, G., Gielow, L., Englesberg, E. 1971. *Nature New Biol.* 233:164–65
6. Brawerman, G. 1974. *Ann. Rev. Biochem.* 43
7. Smith, K., Armstrong, J. L., McCarthy, B. J. 1967. *Biochim. Biophys. Acta* 142:323–30
8. Commerford, D. L. 1971. *Biochemistry* 10:1997–2000
9. Prensky, W., Steffensen, D. M., Hughes, W. L. 1973. *Proc. Nat. Acad. Sci. USA* 70:1860–64
10. Temin, H. M., Mizutani, S. 1970. *Nature* 226:1211–13
11. Baltimore, D. 1970. *Nature* 226:1209–11
12. Loeb, L. A., Tartof, K. D., Travaglini, E. C. 1973. *Nature New Biol.* 242:66–69
13. Darnell, J. E., Jelinek, W. R., Molloy, G. R. 1973. *Science* 181:1215–21
14. Suzuki, Y., Gage, L. P., Brown, D. D. 1972. *J. Mol. Biol.* 70:637–49
15. Brown, D. D., Weber, C. S. 1968. *J. Mol. Biol.* 34:661–80
16. Brown, D. D., Wensink, P. C., Jordan, E. 1971. *Proc. Nat. Acad. Sci. USA* 68:3175–79
17. Bishop, J. O., Pemberton, R., Baglioni, C. 1972. *Nature New Biol.* 235:231–34
18. Britten, R. J., Waring, M. 1965. *Carnegie Inst. Washington Yearb.* 64:314–33
19. Brown, D. D., Sugimoto, K. 1973. *J. Mol. Biol.* 78:397–415
20. Inman, R. B. 1966. *J. Mol. Biol.* 18:464–76
21. Inman, R. B., Schnös, M. 1970. *J. Mol. Biol.* 49:93–98
22. Wensink, P. C., Brown, D. D. 1971. *J. Mol. Biol.* 60:235–47
23. Brown, D. D., Wensink, P. C., Jordan, E. 1972. *J. Mol. Biol.* 63:57–73
24. Fournier, M. J., Brenner, D. J., Doctor, B. P. 1973. *Progr. Mol. Subcell. Biol.* Vol. 3
25. Bernardi, G. 1971. *Methods Enzymol.* 21D:95–139
26. Kohne, D. E. 1968. *Biophys. J.* 8:1104–18
27. Botchan, M., Kram, R., Schmid, C. W., Hearst, J. E. 1971. *Proc. Nat. Acad. Sci. USA* 68:1125–29
28. Saunders, G. F., Shirakawa, S., Saunders, P. P., Arrighi, F. E., Hsu, T. C. 1972. *J. Mol. Biol.* 63:323–34

29. Davidson, E. H., Hough, B. R., Amenson, C. S., Britten, R. J. 1973. *J. Mol. Biol.* 77:1–23
30. Doctor, B. P., Brenner, D. J. 1972. *Biochem. Biophys. Res. Commun.* 46: 449–56
31. Brenner, D. J., Fournier, M. J., Doctor, B. P. 1970. *Nature* 227:448–51
32. Kohne, D. E., Byers, M. J. 1973. *Biochemistry* 12:2373–78
33. Becker, W. M., Hell, A., Paul, J., Williamson, R. 1970. *Biochim. Biophys. Acta* 199:348–62
34. Ray, R., Dutta, S. K. 1972. *Biochem. Biophys. Res. Commun.* 47:1458–63
35. Chattopadhyay, S. K., Kohne, D. E., Dutta, S. K. 1972. *Proc. Nat. Acad. Sci. USA* 69:3256–59
36. Lusby, E. W., DeKloet, S. R. 1970. *Biochim. Biophys. Acta* 209:263–65
37. Nygaard, A. P., Hall, B. D. 1963. *Biochem. Biophys. Res. Commun.* 12: 98–104
38. Boezi, J. A., Armstrong, R. L. 1967. *Methods Enzymol.* 7A:684–86
39. Phillips, A. P. 1969. *Biochim. Biophys. Acta* 195:186–96
40. Probst, H., Jenke, H. S. 1973. *Biochem. Biophys. Res. Commun.* 52:800–6
41. Spadari, S., Ritossa, F. 1970. *J. Mol. Biol.* 53:357–67
42. Quagliarotti, G., Ritossa, F. M. 1968. *J. Mol. Biol.* 36:57–69
43. Sgaramella, V. 1969. *Biochim. Biophys. Acta* 195:466–72
44. Mandel, J. D., Hershey, A. D. 1960. *Anal. Biochem.* 1:66–77
45. Sueoka, N., Cheng, T. Y. 1962. *J. Mol. Biol.* 4:161–72
46. Rudner, R., Karkas, J. D., Chargaff, E. 1968. *Proc. Nat. Acad. Sci. USA* 60:630–35
47. Smith, I., Colli, W., Oishi, M. 1971. *J. Mol. Biol.* 62:111–19
48. Colli, W., Oishi, M. 1970. *J. Mol. Biol.* 51:657–69
49. Marks, A., Spencer, J. H. 1970. *J. Mol. Biol.* 51:115–30
50. McFarland, E. S., Fraser, M. J. 1964. *Biochem. Biophys. Res. Commun.* 15: 351–57
51. Brahic, M., Fraser, M. J. 1971. *Biochim. Biophys. Acta* 240:23–36
52. Shack, J., Bynum, B. S. 1964. *J. Biol. Chem.* 239:2602–6
53. Kidson, C. 1969. *Biochemistry* 8:4376–82
54. Huh, T. Y., Helleiner, C. W. 1967. *Anal. Biochem.* 19:150–56
55. Freeland, G. N., Hoskinson, R. M. 1971. *J. Chromatogr.* 56:147–50
56. Cerami, A. 1969. *J. Biol. Chem.* 244:221–22
57. Udvardy, A., Venetianer, P. 1971. *Eur. J. Biochem.* 20:513–17
58. Iyer, V. N., Rupp, W. D. 1971. *Biochim. Biophys. Acta* 228:117–26
59. Birnie, G. D. 1972. *FEBS Lett.* 26:19–22
60. Chase, J. W., Dawid, I. B. 1972. *Develop. Biol.* 27:504–18
61. Riva, S., Barrai, I., Cavalli-Sforza, L., Falaschi, A. 1969. *J. Mol. Biol.* 45:367–74
62. Wells, R. D., Larson, J. E. 1972. *J. Biol. Chem.* 247:3405–9
63. Cummings, D. J., Mondale, L. 1966. *Biochim. Biophys. Acta* 120:448–53
64. Opara-Kubinski, Z., Kubinski, H., Szybalski, W. 1964. *Proc. Nat. Acad. Sci. USA* 52:923–30
65. Summers, W. C. 1969. *Biochim. Biophys. Acta* 182:269–72
66. Reeder, R. H., Brown, D. D. 1970. *J. Mol. Biol.* 51:361–77
67. Davison, P. F. 1966. *Science* 152:509–12
68. Marks, A. 1973. *J. Mol. Biol.* 76:405–14
69. Mazaitis, A. J., Bautz, E. K. F. 1967. *Proc. Nat. Acad. Sci. USA* 57:1633–37
70. Linn, S., Lehman, I. R. 1965. *J. Biol. Chem.* 240:1287–93
71. Spadari, S., Mazza, G., Falaschi, A. 1972. *Eur. J. Biochem.* 28:389–98
72. Jayaraman, R., Goldberg, E. B. 1969. *Proc. Nat. Acad. Sci. USA* 64:198–204
73. Sgaramella, V., Spadari, S., Falaschi, A. 1968. *Cold Spring Harbor Symp. Quant. Biol.* 33:839–42
74. Kedzierski, W., Laskowski, M. Sr, Mandel, M. 1973. *J. Biol. Chem.* 248: 1277–80
75. Ando, T. 1966. *Biochim. Biophys. Acta* 114:158–68
76. Landy, A., Ross, W., Foeller, C. 1973. *Biochim. Biophys. Acta* 299:264–72
77. Albertsson, P. A. 1971. *Partition of Cell Particles and Macromolecules.* New York: Wiley-Interscience. 323 pp.
78. Blomquist, G., Albertsson, P. A. 1972. *J. Chromatogr.* 73:125–33
79. Rudin, L. 1967. *Biochim. Biophys. Acta* 134:199–202
80. Riggsby, W. S. 1969. *Biochemistry* 8: 222–30
81. Lin, H. J. 1970. *Biochim. Biophys. Acta* 217:232–48
82. Lin, H. J. 1973. *Anal. Biochem.* 51:220–28
83. Smith, I., Smith, H., Pifko, S. 1972. *Anal. Biochem.* 48:27–32
84. Robberson, D. L., Davidson, N. 1972. *Biochemistry* 11:533–37
85. Shih, T. Y., Martin, M. A. 1973.

Proc. Nat. Acad. Sci. USA 70:1697–1700

86. Guild, W. R. 1963. *J. Mol. Biol.* 6: 214–29
87. Kedes, L. H., Birnstiel, M. L. 1971. *Nature New Biol.* 230:165–69
88. Clarkson, S. G., Birnstiel, M. L., Purdom, I. F. 1973. *J. Mol. Biol.* 79: 411–29
89. Dawid, I. B., вrown, D. D., Reeder, R. H. 1970. *J. Mol. Biol.* 51:341–60
90. Patterson, J. B., Stafford, D. W. 1971. *Biochemistry* 10:2775–79
91. Meselson, M., Yuan, R., Heywood, J. 1972. *Ann. Rev. Biochem.* 41:447–66
92. Botchan, M., McKenna, G., Sharp, P. A. 1974. *Cold Spring Harbor Symp. Quant. Biol.* 38:383–95
93. Danna, K. J., Sack, G., Nathans, D. 1973. *J. Mol. Biol.* 78:363–76
94. Hershey, A. D., Burgi, E. 1965. *Proc. Nat. Acad. Sci. USA* 53:325–28
95. Prunell, A., Bernardi, G. 1973. *J. Biol. Chem.* 248:3433–40
96. Cohen, R. J., Crothers, D. M. 1970. *Biochemistry* 9:2533–39
97. Radloff, R., Bauer, W., Vinograd, J. 1967. *Proc. Nat. Acad. Sci. USA* 57: 1514–21
98. Hudson, B., Upholt, W. B., Devinny, J., Vinograd, J. 1969. *Proc. Nat. Acad. Sci. USA* 62:813–20
99. Borst, P. 1972. *Ann. Rev. Biochem.* 41: 333–76
100. Helinski, D. R., Clewell, D. B. 1971. *Ann. Rev. Biochem.* 40:899–942
101. Fuke, M., Thomas, C. A. Jr. 1970. *J. Mol. Biol.* 52:395–97
102. Rush, M. G., Warner, R. C. 1970. *J. Biol. Chem.* 245:2704–8
103. Saucier, J. M., Wang, J. C. 1973. *Biochemistry* 12:2755–58
104. Thomas, C. A., Hamkalo, B. A., Misra, D. N., Lee, C. S. 1970. *J. Mol. Biol.* 51:621–32
105. Schachat, F. H., Hogness, D. S. 1974. *Cold Spring Harbor Symp. Quant. Biol.* 38:371–82
106. Sobell, H. M. 1972. *Proc. Nat. Acad. Sci. USA* 69:2483–87
107. Smithies, O. 1970. *Science* 169:882–83
108. Gall, J. G., Cohen, E. H., Polan, M. L. 1971. *Chromosoma* 33:319–44
109. Dickson, E., Boyd, J. B., Laird, C. D. 1971. *J. Mol. Biol.* 61:615–27
110. Miyazawa, Y., Thomas, C. A. Jr. 1965. *J. Mol. Biol.* 11:223–37
111. Graham, D. E. 1970. *Anal. Biochem.* 36:315–22
112. Ryan, J. L., Morowitz, H. J. 1969 *Proc. Nat. Acad. Sci. USA* 63:1282–89
113. Feldmann, H. 1973. *Z. Physiol. Chem.*

354:189–202
114. Fried, A. H., Rappaport, H. P. 1970. *Biochim. Biophys. Acta* 204:91–98
115. Wohlhieter, J. A., Falkow, S., Citarella, R. V. 1966. *Biochim. Biophys. Acta* 129: 475–81
116. Takahashi, H. 1969. *Biochim. Biophys. Acta* 190:214–16
117. Schmickel, R. D. 1973. *Pediat. Res.* 7:5–12
117a. Patterson, J. B., Stafford, D. W. 1970. *Biochemistry* 9:1278–83
118. Melchior, W. B. Jr, Von Hippel, P. H. 1973. *Proc. Nat. Acad. Sci. USA* 70: 298–302
119. Leng, M., Felsenfeld, G. 1966. *Proc. Nat. Acad. Sci. USA* 56:1325–32
120. Schildkraut, C. L., Marmur, J., Doty, P. 1962. *J. Mol. Biol.* 4:430–43
121. Szybalski, W., Szybalski, E. H. 1971. *Procedures Nucleic Acid Res.* 2:311–54
121a. Wells, R. D., Blair, J. E. 1967. *J. Mol. Biol.* 27:273–88
122. Kirk, J. T. O. 1967. *J. Mol. Biol.* 28: 171–72
123. Bond, H. E., Flamm, W. G., Burr, H. E., Bond, S. B. 1967. *J. Mol. Biol.* 27:289–302
124. Cheng, T. Y., Sueoka, N. 1964. *Science* 143:1442–43
125. Birnstiel, M. L., Wallace, H., Sirlin, J., Fischberg, M. 1966. *Nat. Cancer Inst. Monogr.* 23:431–47
126. Wallace, H., Birnstiel, M. L. 1966. *Biochim. Biophys. Acta* 114:296–310
126a. Hershey, A. D., Burgi, E., Davern, C. I. 1965. *Biochem. Biophys. Res. Commun.* 18:675–78
127. Flamm, W. G., Bond, H. E., Burr, H. E. 1966. *Biochim. Biophys. Acta* 129:310–19
128. Williamson, R. 1969. *Anal. Biochem.* 32:158–63
129. Anet, R., Strayer, D. R. 1969. *Biochem. Biophys. Res. Commun.* 34:328–34
130. Brunk, C. F., Leick, V. 1969. *Biochim. Biophys. Acta* 179:136–44
131. Szybalski, W. 1968. *Methods Enzymol.* 12B:330–60
132. Anet, R., Strayer, D. R. 1969. *Biochem. Biophys. Res. Commun.* 37:52–58
133. DeKloet, S. R., Andrean, B. A. G. 1971. *Biochim. Biophys. Acta* 247:519–27
134. Mayne, N., Burnett, J. P., Butler, L. K. 1971. *Nature New Biol.* 232: 182–83
135. DeKloet, S. R., Andrean, B. A. G., Mayo, V. S. 1971. *Arch. Biochem. Biophys.* 143:175–86
136. Von Hippel, P. H., McGhee, J. D. 1972.

Ann. Rev. Biochem. 41:231–300

137. Goldberg, I. H., Friedman, P. A. 1971. *Ann. Rev. Biochem.* 40:775–810

138. Izatt, R. M., Christensen, J. J., Rytting, J. H. 1971. *Chem. Rev.* 71:439–81

139. Jensen, R. H., Davidson, N. 1966. *Biopolymers* 4:17–32

140. Daune, M., Dekker, C. A., Schachman, H. K. 1966. *Biopolymers* 4:51–76

141. Filipski, J., Thiery, J. P., Bernardi, G. 1973. *J. Mol. Biol.* 80:177–97

142. Corneo, G., Ginelli, E., Soave, C., Bernardi, G. 1968. *Biochemistry* 7:4373–79

143. Corneo, G., Ginelli, E., Polli, E. 1970. *Biochemistry* 9:1565–71

144. Corneo, G., Ginelli, E., Polli, E. 1970. *J. Mol. Biol.* 48:319–27

145. Corneo, G., Ginelli, E., Polli, E. 1971. *Biochim. Biophys. Acta* 247:528–34

146. Riou, G., Pautrizel, R. 1967. *C.R. Acad. Sci. D* 265:61–63

147. Maio, J. J. 1971. *J. Mol. Biol.* 56:579–95

147a. Prescott, D. M., Bostock, C. J., Hatch, F. T., Mazrimas, J. A. 1973. *Chromosoma* 42:205–13

147b. Rae, P. M. M. 1973. *Proc. Nat. Acad. Sci. USA* 70:1141–45

148. Guille, E., Grisvard, J. 1971. *Biochem. Biophys. Res. Commun.* 44:1402–9

149. Bernardi, G., Piperno, G., Fonty, G. 1972. *J. Mol. Biol.* 65:173–89

150. Retèl, J., Planta, R. J. 1972. *Biochim. Biophys. Acta* 281:299–309

151. Davidson, N. et al 1965. *Proc. Nat. Acad. Sci. USA* 53:111–18

152. Nandi, U. S., Wang, J. C., Davidson, N. 1965. *Biochemistry* 4:1687–96

153. Cramer, J. H., Bhargava, M. M., Halvorson, H. O. 1972. *J. Mol. Biol.* 71:11–20

154. Blumenfeld, M., Forrest, H. S. 1971. *Proc. Nat. Acad. Sci. USA* 68:3145–49

155. Corneo, G., Ginelli, E., Polli, E. 1967. *J. Mol. Biol.* 23:619–22

155a. Kurnit, D. M., Maio, J. J. 1973. *Chromosoma* 42:23–36

156. Riou, G., Paoletti, C. 1967. *J. Mol. Biol.* 28:377–82

157. Wang, J. C., Nandi, U. S., Hogness, D. S., Davidson, N. 1965. *Biochemistry* 4:1697–1702

158. Skalka, A., Burgi, E., Hershey, A. D. 1968. *J. Mol. Biol.* 34:1–16

159. Poon, P. H., Schumaker, V. N., Romig, W. R. 1971. *Biochim. Biophys. Acta* 254:187–98

160. Doerfler, W., Kleinschmidt, A. K. 1970. *J. Mol. Biol.* 50:579–93

161. Younghusband, H. B., Bellett, A. J. D. 1971. *J. Virol.* 8:265–74

162. Piperno, G., Fonty, G., Bernardi, G. 1972. *J. Mol. Biol.* 65:191–205

163. Cohen, R. J., Crothers, D. M. 1971. *J. Mol. Biol.* 61:525–42

163a. Yamagishi, H., Takahashi, I. 1971. *J. Mol. Biol.* 57:369–71

164. Yamagishi, H. 1970. *J. Mol. Biol.* 49:603–8

165. Skinner, D. M., Beattie, W. G. 1973. *Proc. Nat. Acad. Sci. USA.* 70:3108–10

165a. Wells, R. D., Larson, J. E. 1970. *J. Mol. Biol.* 49:319–42

166. Kersten, W., Kersten, H., Szybalski, W. 1966. *Biochemistry* 5:236–44

167. Stern, R. 1973. *Carnegie Inst. Washington Yearb.* 72:15–18

168. Lizardi, P., Brown, D. D. 1974. *Cold Spring Harbor Symp. Quant. Biol.* 38:701–6

169. Peacock, J. et al 1974. *Cold Spring Harbor Symp. Quant. Biol.* 38:405–16

170. Zimmer, C. et al 1971. *J. Mol. Biol.* 58:329–48

171. Shapiro, J. T., Leng, M., Felsenfeld, G. 1969. *Biochemistry* 8:3219–32

172. Shapiro, J. T., Stannard, B. S., Felsenfeld, G. 1969. *Biochemistry* 8:3233–41

172a. Mindich, L., Hotchkiss, R. D. 1964. *Biochim. Biophys. Acta* 80:73–92

173. Saito, H., Masamune, Y. 1964. *Biochim. Biophys. Acta* 91:344–47

174. Ishida, T., Kan, J., Kano-Sueoka, T. 1971. *Procedures Nucleic Acid Res.* 2:608–17

174a. Sueoka, N., Cheng, T. -Y. 1962. *J. Mol. Biol.* 4:161–72

175. Ohki, I., Chang, H., Lohr, K., Laskowski, M. Sr. 1970. *Biochim. Biophys. Acta* 224:253–55

176. Corneo, G., Zardi, L., Polli, E. 1970. *Biochim. Biophys. Acta* 217:249–58

177. Corneo, G., Zardi, L., Polli, E. 1972. *Biochim. Biophys. Acta* 269:201–4

178. Blamire, J., Finkelstein, D. B., Marmur, J. 1972. *Biochemistry* 11:4848–53

179. Finkelstein, D. B., Blamire, J., Marmur, J. 1972. *Biochemistry* 11:4853–58

180. Kothari, R. M. 1970. *Chromatogr. Rev.* 12:127–55

181. Bernardi, G., Piperno, G., Fonty, G. 1972. *J. Mol. Biol.* 65:173–89

182. Stutz, E., Bernardi, G. 1972. *Biochimie* 54:1013–21

183. Skinner, D. M., Triplett, L. L. 1967. *Biochem. Biophys. Res. Commun.* 28:892–97

184. Morgan, A. R., Wells, R. D. 1968. *J. Mol. Biol.* 37:63–80

185. Skinner, D. M., Beattie, W. G., Kerr,

M. S., Graham, D. E. 1970. *Nature* 227:837–39
186. Riou, G., Lacome, A., Brock, C., Delain, E., Pautrizel, R. 1971. *C.R. Acad. Sci. D* 273:2150–53
187. Smith, D. A., Martinez, A. M., Ratliff, R. L. 1970. *Anal. Biochem.* 38:85–89
188. Brzezinski, A., Siemicki, R. 1970. Proc. Ann. Meeting Polish Biochem. Soc., 8th, Szczecin, p. 93 (in Polish). Cited in Laskowski, M. Sr. 1972. *Progr. Nucl. Acid Res. Mol. Biol.* 12:161–88
189. Sabeur, G., Procourt, J., Leng, M. 1973. *C.R. Acad. Sci. D* 276:2729–32
190. Bhorjee, J., Janion, C., Laskowski, M. Sr. 1972. *Biochim. Biophys. Acta* 262:11–17
191. Zeiger, R. S., Salomon, R., Peacock, A. C. 1971. *Biochemistry* 10:4219–23
191a. Zeiger, R. S., Salomon, R., Dingman, C. W., Peacock, A. C. 1972. *Nature New Biol.* 238:65–69
192. Alberts, B. M. 1971. In *Nucleic Acid-Protein Interactions and Nucleic Acid Synthesis in Viral Infections*, ed. D. W. Ribbons, J. F. Woessner, J. Schultz. Amsterdam: North-Holland
193. Gilbert, W., Maizels, N., Maxam, A. 1974. *Cold Spring Harbor Symp. Quant. Biol.* 38:845–56
194. Pirrotta, V. 1973. *Nature New Biol.* 244:13–16
195. Riggs, A. D., Reiness, G., Zubay, G. 1971. *Proc. Nat. Acad. Sci. USA* 68:1222–25
195a. Chen, C. -Y., Hutchison, C. A. III, Edgell, M. H. 1973. *Nature New Biol.* 243:233–36
196. Fukada, R., Ishihama, A. 1971. *Biochem. Biophys. Res. Commun.* 45:1255–61
197. LeTalaer, J. Y., Jeanteur, P. 1972. *FEBS Lett.* 28:305–8
198. Robertson, H. D., Barrell, B. G., Werth, H. L., Donelson, J. E. 1973. *Nature New Biol.* 241:38–40
199. Arrand, J. R., Hindley, J. 1973. *Nature New Biol.* 244:10–13
200. Mendelsohn, J., Moore, D. E., Salzman, N. P. 1968. *J. Mol. Biol.* 32:101–12
201. Maio, J. J., Schildkraut, C. L. 1969. *J. Mol. Biol.* 40:203–16
202. Pagoulatos, G. N., Darnell, J. E. 1970. *J. Mol. Biol.* 54:517–35
203. Burki, H. J., Gegimbal, T. J. Jr., Mel, H. C. 1973. *Prep. Biochem.* 3:157–82
204. Huberman, J. A., Attardi, G. 1967. *J. Mol. Biol.* 29:487–505
205. Aloni, Y., Hatlen, L. E., Attardi, G. 1971. *J. Mol. Biol.* 56:555–63
206. Hooper, D. C., Becker, F. F. 1971.

Proc. Soc. Exp. Biol. Med. 136:707–10
207. Schneider, E. L., Salzman, N. P. 1970. *Science* 167:1141–43
208. Wray, W., Stubblefield, E., Humphrey, R. 1972. *Nature New Biol.* 238:237–38
209. Weiss, M. C., Green, H. 1967. *Proc. Nat. Acad. Sci. USA* 58:1104–11
210. Daneholt, B., Hosick, H. 1973. *Proc. Nat. Acad. Sci. USA* 70:442–46
211. Petes, T. D., Fangman, W. L. 1972. *Proc. Nat. Acad. Sci. USA* 69:1188–91
212. Finkelstein, D. B., Blamire, J., Marmur, J. 1972. *Nature New Biol.* 240:279–81
213. Blamire, J., Cryer, D. R., Finkelstein, D. B., Marmur, J. 1972. *J. Mol. Biol.* 67:11–24
214. Kavenoff, R., Zimm, B. H. 1973. *Chromosoma* 41:1–27
215. Oeyen, T. B. 1973. *FEBS Lett.* 30:53–56
216. Janowski, M., Nasser, D. S., McCarthy, B. J. 1972. *Acta Endocrinol. Copenhagen Suppl.* 168:112–25
217. Schildkraut, C. L., Maio, J. J. 1968. *Biochim. Biophys. Acta* 161:76–93
218. Yunis, J. J., Yasmineh, W. G. 1970. *Science* 168:263–65
219. McConaughy, B. L., McCarthy, B. J. 1972. *Biochemistry* 11:998–1003
220. Ilyin, Y. V., Georgiev, G. P. 1969. *J. Mol. Biol.* 41:299–303
221. Wilt, F. H., Anderson, M., Ekenberg, E. 1973. *Biochemistry* 12:959–66
222. Hossainy, E., Zweidler, A., Bloch, D. P. 1973. *J. Mol. Biol.* 74:283–89
223. Yoshikawa, H., O'Sullivan, A., Sueoka, N. 1964. *Proc. Nat. Acad. Sci. USA* 52:973–80
224. Cutler, R. G., Evans, J. E. 1967. *J. Mol. Biol.* 26:81–90
225. Smith, I., Dubnau, D., Morell, P., Marmur, J. 1968. *J. Mol. Biol.* 33:123–40
226. Cutler, R. G., Evans, J. E. 1967. *J. Mol. Biol.* 26:91–105
227. Smith, D., Tauro, P., Schweizer, E., Halvorson, H. O. 1968. *Proc. Nat. Acad. Sci. USA* 60:936–42
228. Engberg, J., Mowat, D., Pearlman, R. E. 1972. *Biochim. Biophys. Acta* 272:312–20
229. Rusch, H. P. 1970. *Advan. Cell Biol.*, 1:297–327
230. Zellweger, A., Ryser, U., Braun, R. 1972. *J. Mol. Biol.* 64:681–91
231. Newlon, C. S., Sonenshein, G. E., Holt, C. E. 1973. *Biochemistry* 12:2338–45
232. Tobia, A. M., Brown, E. H., Parker,

R. J., Schildkraut, C. L., Maio, J. J. 1972. *Biochim. Biophys. Acta* 277: 256–68

233. Morse, M. L., Lederberg, E. M., Lederberg, J. 1956. *Genetics* 41: 142–56
234. Cohen, P. T., Yaniv, M., Yanofsky, C. 1973. *J. Mol. Biol.* 74: 163–77
235. Russell, R. L. et al 1970. *J. Mol. Biol.* 47: 1–13
236. Hoppe, I., Roth, J. 1974. *Genetics.* In press
237. Low, K. B. 1972. *Bacteriol. Rev.* 36: 587–607
238. Gottesman, S., Beckwith, J. R. 1969. *J. Mol. Biol.* 44: 117–27
239. Press, R. et al 1971. *Proc. Nat. Acad. Sci. USA* 68: 795–98
240. Kaye, R., Barravecchio, J., Roth, J. 1974. *Genetics.* In press
241. Shimada, K., Weisberg, R. A., Gottesman, M. E. 1972. *J. Mol. Biol.* 63: 483–503
242. Attardi, G., Naono, S., Rouviere, J., Jacob, F., Gros, F. 1963. *Cold Spring Harbor Symp. Quant. Biol.* 28: 363–72
243. Falkow, S., Marmur, J., Carey, W. F., Spilman, W. M., Baron, L. S. 1961. *Genetics* 46: 703–6
244. Falkow, S., Wohlhieter, J. A., Citarella, R. V., Baron, L. S. 1964. *J. Bacteriol.* 87: 209–19
245. Birnbaum, L. S., Kaplan, S. 1971. *Proc. Nat. Acad. Sci. USA* 68: 925–29
246. Shapiro, J. et al 1969. *Nature* 224: 768–74
247. Daniel, V. et al 1971. *Proc. Nat. Acad.*

Sci. USA 68: 2268–72
248. Davies, J. E., Rownd, R. 1972. *Science* 176: 758–68
249. Jackson, D. A., Symons, R. H., Berg, P. 1972. *Proc. Nat. Acad. Sci. USA* 69: 2904–9
250. Lobban, P. E., Kaiser, A. D. 1973. *J. Mol. Biol.* 78: 453–71
251. Mertz, J. E., Davis, R. W. 1972. *Proc. Nat. Acad. Sci. USA* 69: 3370–74
252. Boyer, H. W., Chow, L. T., Dugaiczyk, A., Hedgpeth, J., Goodman, H. M. 1973. *Nature New Biol.* 244: 40–44
252a. Morrow, J. F. et al 1974. *Proc. Nat. Acad. Sci. USA.* In press
253. Finkelstein, D. B., Blamire, J., Marmur, J. 1972. *Biochemistry* 11: 4853–58
254. Brown, D. D., Dawid, I. B. 1969. *Ann. Rev. Genet.* 3: 127–54
255. Tobler, H., Smith, K. D., Ursprung, H. 1972. *Develop. Biol.* 27: 190–203
256. Pavan, C., da Cunha, A. B. 1969. *Ann. Rev. Genet.* 3: 425–50
257. Prescott, D. M., Murti, K. G., Bostock, C. J. 1973. *Nature* 242: 576–600
258. Slonimski, P. P., Perrodin, G., Croft, J. H. 1968. *Biochem. Biophys. Res. Commun.* 30: 232–39
259. Gordon, P., Rabinowitz, M. 1973. *Biochemistry* 12: 116–23
260. Nagley, P., Linnane, A. W. 1972. *J. Mol. Biol.* 66: 181–93
261. Brown, D. D., Dawid, I. B. 1968. *Science* 160: 272–80
262. Gall, J. G. 1968. *Proc. Nat. Acad. Sci. USA* 60: 553–60

REPLICATION OF CIRCULAR DNA IN EUKARYOTIC CELLS[1]

Harumi Kasamatsu and Jerome Vinograd

California Institute of Technology, Pasadena, California

CONTENTS

INTRODUCTION

Our understanding of DNA replication in vitro and in a variety of organisms has increased substantially during the last few years. In this article we present a review of the progress made in the restricted area of replication of circular DNAs in eukaryotic cells. These DNAs include the extrachromosomal DNAs in mitochondria, chloroplasts, and kinetoplasts; the circular viral DNAs of the papova group; and the circular intermediates in the amplification of ribosomal genes in amphibian oocytes.

[1] Abbreviations used: EthBr, ethidium bromide; BUdR, 5-bromodeoxyuridine; MAK, methylated albumin kieselguhr; PDI, propidium diiodide; FUdR, 5-fluorodeoxyuridine.

In our opinion the mechanism of replication of relatively small circular DNAs will be similar to that of the larger DNAs in eukaryotic systems.

Rapid progress in the field reviewed here has occurred because of the development of new methods for isolating pulse-labeled replicating forms in buoyant EthBr-CsCl density gradients and for visualizing replicating forms by electron microscopy, a direct procedure that is often an important supplement or replacement for deductive procedures. The use of the electron microscope for visualizing nucleic acids is reviewed elsewhere in this volume (1a). The availability of restriction enzymes that cleave DNAs at specific sites and provide specific fragments has made it possible to map circular DNAs and thereby dissect certain aspects of the mechanism of replication. The reader is referred to other recent reviews (1b–7) for full coverage of the subject.

MODE OF REPLICATION OF CIRCULAR DNA

Two principal and apparently general modes of circular DNA replication have been observed in a wide variety of organisms and viruses. The essential features of the Cairns mode (8) and the rolling circle mode (9, 10) are presented diagrammatically in Figure 1.

In the rolling circle mode, initiation requires the introduction of a nick $(P \rightarrow a)$

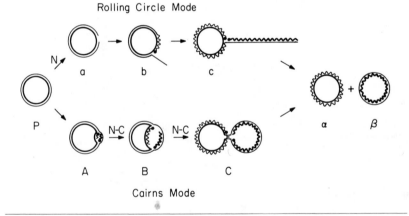

Figure 1   The two general modes of replication of closed circular DNA. The parental closed molecule, *P,* is nicked (*N*) in the rolling circle to provide a primer for nucleotide addition, which then proceeds $a \rightarrow b \rightarrow c$ by displacement synthesis followed by duplex or complement synthesis. The parental closed molecule need not be nicked in the first stage of the Cairns mode, $P \rightarrow A$, which, however, requires the introduction of a primer by transcription or by some other kind of insertion of an oligonucleotide. Displacement replication continues in the presence of a nicking-closing system (*N-C*) from $A \rightarrow B \rightarrow C$. Both modes lead to progeny circles, α and β. The progeny strands are indicated by wavy lines and the sites for nucleotide addition by black dots.

so that the covalent addition of deoxyribonucleotides to a parental strand can occur $(a \to b)$. In the Cairns mode, covalent addition to parental strands does not occur. A primer, however, appears to be required to initiate DNA synthesis. The black dot in the diagram shows the site of nucleotide addition, assuming that all polymerases incorporate deoxynucleoside 5'-triphosphates at the 3'-OH termini. The parental strands are necessarily unwound in both displacement synthesis steps, $(a \to b)$ and $(A \to B)$. This process requires the presence in the unreplicated region of a transient or permanent nick, which can act as a swivel in the circle. In the rolling circle mode, a permanent nick is present immediately ahead of the growing point; the unwinding of the parental strands occurs either by rotation of the entire duplex DNA around the swivel or by threading the displaced strand through the circle. In the Cairns mode, the necessary nick in all cases so far studied appears to be transient and is followed by a closing reaction. Since the replicating intermediates in several systems are largely closed when isolated, the relative time in the nicked state appears to be small. At the present time we do not know whether the addition of nucleotides takes place in nicked or in closed molecules, or in both. It is probable that similar nicking-closing cycles occur to shorten the unreplicated duplex that must spin during the replication of long stretches of chromosomal DNA. The two modes of replication differ in that one fork is present in the rolling circle mode and two forks are present in the Cairns mode. Fragility at one of the forks of Cairns forms, especially if the nick is close to a fork, can lead to breakage and the artifactual formation of replicating intermediates, which in the elctron microscope appear to have been formed by the rolling circle mode.

It is useful at this stage to regard replication as occurring in two conceptually separable and sequential steps. Formation of the first progeny strand with the displacement of one of the parental strands is referred to as displacement synthesis. Formation of the second progeny strand on the displaced parental strand is referred to as duplex or complement synthesis. A time lag between these two processes gives rise to asymmetry and single strandedness at the fork. The extent of this asymmetry varies widely in the Cairns mode, with relatively small asymmetries in the papova viral DNA replicative intermediates and extensive asymmetry in mitochondrial DNA replicative intermediates.

One or both forks move relative to the origin in replication via the Cairns mode. Only one fork appears to move in mitochondrial DNA replication. Displacement synthesis is unidirectional and the replication mode is said to be unidirectional. In papova viral DNA replication, both forks move in opposite directions at about the same rate. Duplex synthesis here is necessarily discontinuous. Displacement synthesis can also occur discontinuously, as first proposed by Okazaki and associates for bacterial DNA replication (11). Replication in the rolling circle mode is clearly unidirectional.

After the near completion of the displacement synthesis and varying extents of duplex synthesis, $(b \to c)$ and $(B \to C)$, α and β progeny molecules are formed in a process we call separation. In the Cairns mode, the separation of parental strands may occur by linearizing a duplex containing one of the parental strands and recircularizing it. Alternatively, separation can occur after displacement synthesis has progressed to within one complete turn of the duplex from the termination site. At this stage, the last swivel could be closed and separation would occur with the formation of a circle containing a small gap.

If extensive asymmetry exists at the time of separation, circular intermediates containing large gaps are formed. The rolling circle mode of replication requires cleavage of the intermediate shown in Figure 1C followed by circularization of the resulting linear molecule.

In the above discussion, replication is assumed and in most cases has been proven to be semiconservative.

REPLICATION OF MITOCHONDRIAL DNA

In Vivo

GENERAL MECHANISM   The mode of replication of the mtDNA in animal cells grown in culture (12), Chang rat solid hepatoma, Novikoff ascites hepatoma, regenerating rat liver, and chick embryo (13) has been established by ordering the various replicative intermediates seen in the electron microscope. The overall mechanism appears to be that of a modified Cairns mode. Intermediates resembling those in Figure 1B were observed at the 0.7% level in rat liver mitochondrial DNA preparations (14). The scheme derived for the replication of mtDNA in mouse cells (Figure 2) has several implications which have been experimentally verified. The first is that replication is unidirectional (15, 16). This was originally verified by analyzing expanded D-loop (Exp-D) molecules from a cell line with dimeric mitochondrial DNA, which often contains paired D loops 180° apart. The nonexpanding small D loop served as a marker for the expansion that occurs in replication. This result has been confirmed and extended with mouse monomeric mtDNA using the restriction enzyme R1 (16a). The specific cleavage served as references in the examination of fragmented replicating intermediates in the electron microscope (17). The pattern of fragments showed that replication occurred in the same direction in all molecules.

The second implication is that the asymmetry between displacement synthesis and duplex synthesis is specific with respect to strandedness (12). The displacement synthesis in A through C has been shown to produce a heavy (H) displaced strand,

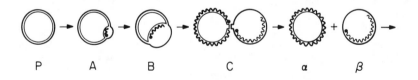

P   A   B   C   α   β

Figure 2   A modified Cairns mode for the replication of mitochondrial DNA. The sequence, $P \rightarrow \alpha$ and β, differs from that in Figure 1 in that a high degree of asymmetry is present and persists during separation to form β progeny molecules which have to be completed by a polymerase and a ligase. A, B, and C are referred to in the text as D-loop DNA (D-DNA), expanded D-loop DNA (Exp-D DNA), and expanded D-loop DNA after duplex synthesis [Exp-D (l)], respectively.

whereas the duplex synthesis in C involves the formation of a light (L) strand. The strand designations relate to their buoyant densities in alkaline CsCl gradients. Separation after displacement synthesis is largely completed results in two daughter molecules, one of which is a β-gapped circle which contains a complete parental circular H strand and an incomplete progeny L strand. Hybridization experiments evaluated with the aid of the electron microscope have verified the expected strandedness in the large-gapped circles.

As was first demonstrated in the replication of papova viral DNA (18, 19), it has been found that the two parental strands in the replicating forms B and C are for the most part covalently closed when the replicating forms are isolated by mild procedures (16).

Despite the asymmetry, the replication is semiconservative. This result has been demonstrated in density-labeling experiments in the mitochondria of *Neurospora crassa* and *Saccharomyces* (20–22) and later in those of HeLa cells (23). A unique feature of mtDNA replication is that displacement synthesis temporarily terminates after a small percentage of the genome has been replicated as in structure A (Figure 2) to form D-loop DNA (24), which accumulates awaiting some signal for the extension.

Within the above general framework, we now compare the results found in several different systems. We include also the results obtained in in situ experiments where isolated mitochondria are incubated with radioactive DNA precursors.

D-LOOP DNA   When isolated from cells in culture or from tissue rich in dividing cells, a large fraction of mtDNA consists of a homogeneous replicating intermediate referred to as D-loop DNA (24, Figure 2A). This replicating intermediate contains the first stretch of the H strand synthesized in the replication process. It may be released by partial denaturation with the formation of a circular DNA of increased degree of supercoiling and a homogeneous single-stranded DNA, 7S, of approximately 450 nucleotides in mouse L cells (24), HeLa cells, and chick liver cells (25). The size of the released pieces corresponds to the size of the D loop.

When the mouse L cells are pulse-labeled in a suspension culture and the closed DNA from the lower band of an EthBr-CsCl gradient is partially denatured and sedimented through sucrose gradients, the small single-stranded 7S DNA contained 4 to 7 times as much radioactivity as the closed circular DNA (26). Since the ratio of label was several times higher than the value expected, the result was taken as evidence that the 7S DNA turns over (26). The expected value was calculated with the assumption that suitably sized circles containing gaps were completed, closed, and D-loop synthesis initiated and continued without intervening delays. A delay in the closing step would have increased the radioactivity in the 7S DNA relative to that in closed circles and would correspondingly reduce the weight of the evidence for turnover of the 7S DNA.

Small progeny strands can be lost if branch migration (26a) occurs during isolation or specimen preparation.[2] This process, however, appears to be inhibited in closed

---

[2] The phenomenon of branch migration at a three-stranded junction can occur with the rupture of progeny-parental base pairs and simultaneous formation of parental-parental base pairs.

circular molecules because of the positive free energy associated with twisting when the bases in the parental strands pair. When D-loop DNA is nicked, branch migration is facilitated (26). The extent of nicking depends on the method of isolating and preparing the DNA. The reported frequencies which therefore should be regarded as minimum values vary between 25 and 65% in mouse L cells. In HeLa cells, D-loop DNA is approximately 10% of total mitochondrial DNA (P. John Flory, Jr., unpublished data). The D-loop frequency in mouse cells drops to 2% very late in the stationary state (27). Rat liver contained 42% D-loop DNA (28), whereas regenerating rat liver was found by other authors to contain 27% D-loop DNA (13, 29, 30). The Chang rat solid hepatoma and the Novikoff rat ascites hepatoma contained 9 and 25%, respectively (13, 29, 30). Chick liver contained approximately 25% (31). This D-loop DNA was found in mtDNA from human and beef thyroid (32) and also in *Tetrahymena* mtDNA (33). This latter DNA was isolated as a 15 μm linear species and contains up to six D loops per molecule (33). During oogenesis of the sea urchin, *Strongylocentrotus purpuratus*, the D-loop frequency in mtDNA is about 7% (34). No D loops could be detected in mature unfertilized sea urchin eggs released from the animal by injection of KCl (34). On the other hand, *Xenopus laevis* ovulated eggs obtained by hormone treatment contained 57% D-loop DNA in the mtDNA (35).

In view of the recently discovered role of the synthesis of short polyribonucleotides that serve as primers for DNA synthesis, the isolated progeny H strand, 7S DNA, from D-loop DNA has been examined for the presence of the terminal ribonucleotides after radioactive labeling with polynucleotide kinase (27). The terminal label remained attached after alkali treatment, indicating the absence of ribonucleotides close to the 5′ terminus. This result, of course, does not exclude the possibility that the ribonucleotides are removed while the D loop is held in the cell waiting for the extension. The 5′ terminal nucleotides of 7S DNA were analyzed and their bases consisted of 50% G, 20% A, 20% T, and 10% C (27). The base composition of the 7S DNA was not the same as the distribution of bases at the 5′ termini nor the distribution in the full H strand (27).

D-LOOP EXPANSION   Because of the high frequency of closed D-loop DNA relative to larger replicative intermediates, the former structure has been regarded as one that accumulates in the regulation of the replication cycle. Replication continues as indicated in Figure 2. A nicking-closing system is continuously operative in D-loop expansion (Figure 2, $A \rightarrow C$). A large asymmetry develops in mouse L cells during D-loop expansion (12). There appears to be some system that inhibits initiation of duplex synthesis on the first six tenths of the displaced strand. Initiation in the last 0.4 is not obligatory and in some molecules takes place after a larger expansion has occurred. The asymmetry is smaller in mtDNA replicating intermediates isolated from Novikoff rat ascites hepatoma, Chang rat solid hepatoma, regenerating rat liver, normal rat liver (13, 29), and sea urchin, *Strongylocentrotus purpuratus* (34). In the above systems, duplex synthesis can be initiated in smaller displacement loops. Very few Exp-D molecules free of visible single-stranded regions are seen in preparations from mouse L cells, but they are frequently seen in preparations of Novikoff rat ascites hepatoma, Chang rat solid hepatoma, regenerating rat liver, normal rat liver and *Strongylocentrotus purpuratus* in which approximately one third of the larger replicating intermediates are symmetric (Figure 1*B* and *C*) among expansions that vary in size from 3% to almost full size. Nevertheless, in all species so far studied, a substantial fraction of the replicating intermediates are asymmetric. It has been proposed that mtDNA (13, 29, 30) replication proceeds via two routes with intermediates that differ in the degree of asymmetry as indicated in the lower part of Figure 1 and in Figure 2. These two routes, however, may be regarded as a manifestation of the nonrigorous control for the initiation of duplex synthesis.

The accumulation of D-loop DNA has been explained as resulting from a physical barrier associated with positive superhelical free energy that developed during replication in closed circles in the absence of a nicking-closing cycle (24). In order to relieve the topological stress, a nick must be introduced in the unreplicated region. Other explanations for the accumulation are, however, possible, and the above has not yet been proven.

Because the Exp-D molecules (Figure 2*B* and *C*) were found in the upper band of an EthBr-CsCl gradient, it was originally assumed that these were nicked molecules (12). It was found, however, by Robberson & Clayton, who used thymidine kinase negative (Tk$^-$) cell lines which do not incorporate radioactive thymidine into the nuclear DNA, that the replicating intermediates prepared by procedures entirely at low temperature were largely closed (16) as had been previously found in the replication of papovaviruses (18, 19). Further examination of upper band molecules isolated after exposure to 37°C revealed that approximately 50% were covalently closed while the remainder were nicked close to a fork (27). The sensitivity of mtDNA replicating intermediates to nicking during incubation of the mitochondria at 37°C appears to vary with the cellular origin. HeLa and mouse LD cells appear to be less sensitive than mouse LA9 cells (unpublished result), and Tk$^-$ LD cells are less sensitive than Tk$^-$ L cells (26).

When DNA is isolated, there is always the possibility of inducing nicks in the molecule. This possibility confounds the problem of determining the position of biological nick(s) which occur during replication (27). Furthermore, the biological nicks appear to be transient and largely absent if the entire isolation procedure is carried out at 4°C (16). If, on the other hand, a 37°C incubation step is introduced in the isolation of mtDNA from LA9 cells, approximately 50% of the replicating intermediates are nicked in the nonreplicated region (27). From 80 to 90% of these nicked intermediates contained a nick within 50–100 nucleotides from a fork (27). At the present time it is not known which fork is involved. The nick appears to occur on L strands of parental molecules more frequently than on H strands. The above results suggest that the introduction of the nick near the fork occurred in a complex of DNA and enzymes in which replication takes place.

Electron microscope observations of clean nicked circles which had been treated with HCHO at alkaline pH to generate denaturation loops over 80% of the contour length showed that few molecules contained more than one nick (27).

STEPWISE CHAIN ELONGATION   Wolstenholme et al observed (29, 30) that both single- and double-stranded expansions increase in size in a stepwise fashion, the step corresponding to 4–7% of genome size in the first half of the replication. The result can be interpreted as indicating that displacement synthesis as well as duplex synthesis is discontinuous, as proposed by Okazaki et al (11). The displacement synthesis could have occurred in a closed molecule generating positive superhelical turns and a positive free energy that balances the free energy of nucleotide addition and causes the process to stop until a nicking-closing cycle occurs. The latter explanation was previously postulated for the accumulation of D loops in mouse L-cell mtDNA. Alternatively, some other regulatory system may operate to slow down net synthesis while the nucleotide addition rate remains rapid.

SEPARATION OF PROGENY MOLECULES FROM THE EXPANDED REPLICATING INTERMEDIATE
The mechanism involved in the separation of the largest replicating intermediate into two progeny molecules presents a problem which has been under consideration ever since Cairns' demonstration of the mode of replication of *Escherichia coli* DNA (8). A significant development occurred when the β circles containing large gaps were observed and attributed to the occurrence of the separation prior to the completion of duplex synthesis. It is apparent that the H strand was intact in the β progeny at the time of separation. In principle, only two possibilities remain for the state of the L parental strand in the α circle at separation. Either it is nicked during this process to form a linear intermediate, or is nicked and closed after the last parental turn is unwound. Separation could then occur if the remaining nonreplicated region contained nine or less base pairs in a Watson–Crick structure. Interwinding of less than 360° is inadequate to form a topological bond. Separation would produce two progeny molecules with the α progeny containing a very small gap in the H strand. The gap would be larger if an unwinding protein moved ahead of the growing point (36). The above postulate for the separation process does not require that synthesis occur over a nick in the template.

In Situ Studies

Partially purified mitochondria were incubated with DNA precursors originally to establish that mtDNA replicates autonomously (28, 31, 37–54). Incorporation of [^{3}H] thymidine into mitochondrial DNA depends partially on the addition of all four deoxyribonucleoside triphosphates and particularly on the maintenance of the ATP level in the mitochondria. Inhibition or uncoupling of electron transport inhibits incorporation. Inhibitors of DNA synthesis reduce incorporation. Linear rates of incorporation have been observed over 15 and 120 min periods.

Karol & Simpson demonstrated that DNA synthesis occurred in purified rat liver mitochondria and that a long strand of BUdR containing DNA was synthesized (44). Borst and collaborators incubated mitochondria from chick liver to obtain labeled mitochondrial DNA which they isolated by MAK column chromatography (28, 31, 48, 50–52). The radioactive DNA sedimented in sucrose gradients in two bands with sedimentation coefficients of 39 and 27S, the same as closed and open mtDNA. A low concentration of pancreatic DNase converted the fast material, I*, to a species that sedimented as 27S, indicating that I* was a covalently closed circular DNA. The BUdR labeling indicated that only a small stretch of newly synthesized DNA was added to pre-existing DNA (50). Upon denaturation of the I* component, the label appeared in a small 7S progeny strand, while the parental DNA, which remained closed, was found to be unlabeled, indicating that closed DNA was not formed from labeled replicating intermediates during the incubation, nor was the closed DNA labeled by a repair process. I* was identified as D-loop DNA (31). Since the D-loop size and frequency were not affected by the incubation (31, 51), it appears that only a small extension of the 7S DNA, estimated at 15 nucleotides (51), had occurred. The size of the D loops was about 3.5% of the genome, as previously found in mouse L cells. D-loop DNA was a substrate for the *E. coli* exonuclease III, which indicates that the displacing strand contains a 3′-end available for the enzyme (51). Additionally, DNA polymerase I promotes a short addition to the 3′-OH end (51). Heating I* to remove the displacing strand did not increase the sedimentation coefficient of the resulting clean closed circular DNA (51), as occurred in mtDNA isolated from mouse L cells. This suggests that the D-DNA from the in situ experiments had a higher degree of supercoiling than the D-DNA from mouse L cells. It has been established that the sedimentation coefficient rises steeply as the absolute degree of supercoiling increases from 0 to 2% but remains essentially unchanged as it increases from 2 to 4% (51a).

The slow DNA component, II*, formed a 7–8S H strand and a 25S L strand upon denaturation (52). The small H strands are presumably small displacing strands, while the longer L strands are the incomplete strands in circles containing large gaps. The latter forms were present at about 3% in frequency and the gap size decreased from a mean of 0.36 μm to 0.26 μm after incubation (28). Expanded D-loop molecules were not present, possibly because of a selective loss during isolation.

Sodium dodecylsulfate lysis and pronase digestion of incubated rat liver mitochondria yielded labeled mitochondrial DNA with properties similar to those observed with chick liver mitochondrial DNA described above, except that evidence was obtained for the presence of closed Exp-D molecules (53). These were found in an intermediate region between the lower and the upper band in EthBr-CsCl gradients. The DNA from this intermediate band, freed of EthBr, contained some single-stranded DNA as indicated by the buoyant density in the CsCl. Recently the D-loop DNA isolated from incubated rat liver mitochondria has been re-examined with the finding that the closed DNA freed of 7S strands by heating was labeled (54), a result which the authors claim is due to completion of replication to the closed circle stage. An extensive buoyant shift upon incorporation of BdUTP was observed and used to rule out a repair process for the incorporation of label into the closed circles (54). The variability of the results obtained in different laboratories presents a problem still unresolved.

THE REPLICATION OF PAPOVAVIRUS DNA

Introduction

The monkey virus, SV40, and the mouse virus, polyoma, have recently been the subject of intensive investigation because of their oncogenicity. These small DNA viruses replicate in the nucleus of susceptible cells. Injection of SV40 virus into hamsters gives rise to tumors at the site of injection and at multiple sites in the case of polyoma. The viruses also transform certain cells in culture via a lysogenic process in which the whole or part of the viral DNA becomes integrated into the nuclear DNA (55, 56). The transformed cells lose contact inhibition, contain viral antigens, and form tumors upon injection into rodents.

The wild-type viruses contain covalently closed circular DNA with a molecular weight of about 3×10^6. The DNA in papovaviruses rapidly becomes defective with continued passage at high multiplicity in susceptible host cells (57–61). The defective viruses appear to have exchanged genetic material with chromosomal DNA, as well as to have suffered deletions and internal recombinations (62, 63). The low molecular weight defective viral DNA may be isolated from the slower part of the main band sedimenting through sucrose gradients. The formation of defective molecules containing chromosomal DNA indicates that events resembling those that occur in transformation also occur during the lytic process. The defective molecules always retain the sequence of viral DNA in which the site for initiation of replication occurs (64).

When plaque-purified virus is isolated and the DNA extracted, a supercoiled molecule with a sedimentation coefficient of 21S is obtained (65–68). Polyoma DNA usually contains 14S linear DNA which derives from pseudovirions—host DNA encapsidated in viral coats (69–71). Small amounts of 16S nicked circular DNAs are often formed during the isolation process. The closed DNAs can be readily purified by buoyant centrifugation in EthBr-CsCl or PDI-CsCl gradients.

Replication

METHODS OF ISOLATION OF REPLICATING INTERMEDIATES   Infected confluent cells labeled for a few minutes approximately one day after infection are lysed with detergents while the cells are still attached to the plate (72, 73). Intracellular viral DNA or virion DNA is selectively isolated from the cellular DNA by the procedure described by Hirt (72). The supernatant fluid is then concentrated and sedimented in sucrose gradients. The replicating forms are collected as a broad band which sediments at 25S (74–76). Some replicating forms are present in the 21 and 16S bands. These DNAs may be further purified and segregated by degree of replication in an EthBr-CsCl density gradient (18, 19) or electrophoretically in gels (78). Replicating forms are also isolated on benzoylated, naphthoylated DEAE-cellulose columns (74, 75).

UNIQUE ORIGIN AND BIDIRECTIONALITY IN REPLICATION   In the absence of any coupling between replication and transcription, there is in principle no need for a specific site for the initiation of replication in either the unidirectional or bidirectional modes. Nevertheless, as in mitochondrial DNA (24), there is indeed a unique site for the initiation of replication in SV40 and polyoma DNA (79–82). It has been found that replication proceeds bidirectionally (19, 80, 82, 84, 85) in both of these animal viruses at approximately an equal rate and that separation of progeny molecules occurs approximately 180° from the initation site (80, 82, 84).

Four different methods have been used to establish the uniqueness of the initiation site and bidirectionality. The first method depends on the observation that the *Hemophilus influenzae* restriction endonuclease (83), *Hin,* cleaves SV40 DNA into

11 unique fragments of different size and electrophoretic mobility (86). Covalently closed circular DNA obtained after a 5 min pulse, a time less than that required for one replication round, was purified, cleaved, and the specific activity of the fragments determined. Since the fragments studied were prepared from closed 21S molecules, the specific activity should be highest near the termination site and lowest near the initiation site if these sites are unique. The results showed that the initiation site was indeed localized within *Hin* fragment C (81).

The *Hin* fragments were ordered from a study of the large fragments obtained upon limited digestion (87). The specific activity of the ordered fragments obtained from the above pulse-labeled covalently closed circular DNA decreased linearly with distance from the termination site to the origin. This demonstrated that the replication is bidirectional, terminates at a specific site in the *Hin* G fragment (84), and also that size distribution of replicating intermediates is flat. The maturation time of SV40 DNA was found to be 10–15 min, compared with earlier values of 5–10 min (74, 75, 78, 81). The maturation time measured above is the time from the onset of replication to the appearance of the progeny molecule in the closed form.

A second method is based on the finding that replicating intermediates have a buoyant density in EthBr-CsCl which increases inversely with the degree of replication (18). The pulse-labeled progeny strands are obtained by dissociating replicating intermediates into denatured double-stranded, closed parental molecules and slower moving progeny strands by sedimentation in an alkaline sucrose gradient. The kinetic complexity of the sized progeny strands, measured in self-annealing experiments, increased with progeny size as if initiation occurred at a single site (79). If initiated at randon, each sized sample would have been a pool of segments of the entire sequence, and the kinetic complexity would have been independent of progeny size.

The third method involves treatment of unfractionated replicating intermediates with the *E. coli* restriction enzyme R1 (16a), which linearizes SV40 DNA and polyoma DNA at a unique site (79a–80, 82). The ends of such linear molecules constitute markers which can be used to map the position of the initiation site and the direction of elongation from data obtained with the electron microscope. The results with SV40 and polyoma DNA showed unambiguously that replication begins at a unique site, proceeds bidirectionally at an approximately equal rate, and is completed at about 180° from the initiation site. The initiation and termination sites for replication in SV40 DNA, as well as the site of the R1 restriction enzyme cut, have been located on maps of the ordered *Hin* fragments and the ordered fragments obtained from the restriction enzyme *Hemophilus parainfluenzae, Hpa* (87).

The fourth method is based on the assumption that displacement synthesis is always somewhat ahead of duplex synthesis (Figure 3). In the case of bidirectional synthesis, single strandedness at the forks would be in *trans* configuration (Figure 1*B*). Single strandedness has been observed at two forks in 34% of SV40 DNA replicating intermediates and at one fork in 40% of the replicating intermediates (19). The existence of *trans* asymmetric forks indicates that some of the molecules replicate bidirectionally. The absence of asymmetry at a fork can be taken to indicate that electron microscopy procedures were inadequate. Single strandedness was found by

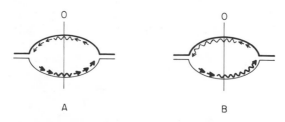

Figure 3 Discontinuous synthesis at the growing points in bidirectional replication. (*A*) Totally discontinuous model in which both strands at both forks are in small fragments. (*B*) Semidiscontinuous model in which one strand at each fork is synthesized in small fragments. Note the *trans* configuration of the fragments. *O* represents the origin for replication.

chromatographic means to be present in replicating intermediates of polyoma DNA (77). The enriched replicating intermediates were treated with a single-strand-specific endonuclease to form molecules which appeared first as circles containing tails (85) in the electron microscope. Further digestion converted these molecules to linear molecules (85).

CHAIN ELONGATION   While chain elongation appeared to be a continuous process as described above in SV40 DNA replicating intermediates labeled for 5 min, it became apparent that elongation is really discontinuous (88, 89) as in *E. coli* (11, 90), when the pulse time was shortened to as little as 15 sec. At short pulse times, 4S alkaline-resistant DNA fragments were obtained and accounted for 30% of the incorporated label (88). These fragments were 80% self annealable (89), indicating that both growing chains are synthesized discontinuously at both forks (Figure 3*A*, 89). The high degree of self annealing seems to be in conflict with the extent of single strandedness observed at the forks by electron microscopy (19). The asymmetric structures would not release complementary fragments. It is, of course, possible that multiple 4S fragments are contained in a few replicating intermediates and account for the observed high degree of self annealing.

Prior treatment of SV40-infected cells with FUdR followed by a short pulse of [³H] thymidine after removal of FUdR resulted in the formation of a fast sedimenting band of replicating intermediates, a 16S component II-like band and a band at 6–8S (91). The label was found largely in the 4S region when the replicating intermediates were sedimented through alkaline sucrose. This result indicated that the joining process was inhibited by the analog (91). Prior treatment of the infected cells with hydroxyurea for a short time also inhibited the joining of 4S fragments to the longer progeny strands present in the replicating intermediates (91a). Eighty percent of the radioactivity in the 4S DNA of purified large replicating intermediates is converted into longer progeny strands when the intermediates are incubated with T4 DNA polymerase, *E. coli* DNA ligase, and deoxyribonucleoside triphosphates. This result demonstrates that the 4S fragments are separated from the longer progeny strand by more than one nucleotide and that the gap termini bear 3'-OH and 5'-phosphate groups (91a, 91b).

The chain elongation of polyoma DNA was studied in two in vitro systems: isolated nuclei (92, 93) and hypotonically lysed cell suspensions (94). As was shown in an in vivo system (96), DNA synthesis was semiconservative in both systems where chain elongation occurs but chain initiation does not (94, 95). Maturation to the closed DNA stage occurs extensively in the lysate system but to a smaller extent in the isolated nuclei system.

The 4–5S fragments were observed in both systems after short pulses. These fragments were found in replicating intermediates of all sizes (98). The fragments were, in part, self annealable and could be chased into longer chains. The low extent of self annealing of the small fragments, 20%, obtained in the lysed cell system lead the authors to propose that the replication occurred by a semidiscontinuous mechanism (98, Figure 3B). This proposal was supported by the results from a study of the kinetics of incorporation of [α-^{32}P] dCTP into long and short (5S) progeny strands (98). The extent of incorporation of label into the long and the short chains was initially equal, continued linearly into the long chains, but reached saturation in the short chains after about 1 min (98). If the replication had been totally discontinuous (Figure 3A), incorporation would have been faster into short than into long chains. Incorporation into long chains would have increased exponentially (98). The time required for the formation of the 5S fragment was estimated to be about 1 min (98).

A higher extent of self annealing of small fragments, 70%, in the isolated nuclei system gave rise to the proposal that replication occurred discontinuously at both strands at both forks (101, Figure 3A), as was proposed for SV40 DNA replication based on in vivo results (89). Hydroxyurea treatment of infected cells prior to isolation of the nuclei led to an accumulation of 4S fragments (97, 99, 100). The 4S fragment from the Hirt supernatant of the treated nuclei was shown to be single-stranded DNA, completely digestible by E. coli exonuclease I and resistant to E. coli exonuclease III (99). The two in vitro systems have given contradictory and still unresolved results regarding short fragments at the growing point.

The 4S fragments isolated by thermal instead of alkaline denaturation of short pulse-labeled replicating intermediates contained polyribonucleotides covalently attached to DNA fragments (97, 102), as indicated by the methods described by Sugino et al (103). The nature of the ribonucleotides and deoxyribonucleotides at the switch point was determined by the nearest neighbor analysis using a single [α-^{32}P] deoxyribonucleoside triphosphate in each experiment (97, 102, 103a). There was no preference for the ribonucleotide at the switch; however, dCMP was the slightly preferred deoxyribonucleotide (99, 102). It was also found in further studies (103a) that each progeny strand contained on the average one RNA–DNA link at the 5' end of DNA, irrespective of the chain length, and that the progeny strands contained all four ribonucleotides and deoxyribonucleotides at the RNA–DNA switch point. All four 5'-deoxyribonucleotides were labeled with ^{32}P when alkali-treated progeny strands were incubated with [γ-^{32}P] ATP and polynucleotide kinase.

The direction of chain elongation was determined in experiments where the replicating intermediates were first labeled at 25°C in the lysate with [³H] TTP for 5 min and then with [α-³²P] dCTP for 30 sec to label the growing end (98). Short and long progeny strands were isolated and separately subjected to digestion with venom phosphodiesterase which degrades from the 3' end. The pulse label was degraded faster than the long-term label in both the short and long progeny strands. This result confirms in this animal system that DNA synthesis occurs only in a 5' to 3' direction (98).

The relative frequency of replicating intermediates in SV40 replication increases (104) with size in a steady fashion. This result appears to be inconsistent with the interpretation of findings in the study of the specific activity of ordered *H. influenzae* restriction enzyme fragments from purified pulse-labeled covalently closed circular SV40 DNA (81). In the latter experiments, described earlier, the specific activity was linearly related to the distance from the origin, which implies that, during the pulse, there is approximately an equal frequency of all sizes of replicating intermediates.

SEPARATION   From 1 to 3% of both the short- and long-term label in SV40 Hirt lysates appears at 16S open circles after purification through EthBr-CsCl and neutral sucrose gradients (105). All of the pulse-labeled DNA in these duplex circles was found in approximately full-length linear strands in alkaline sucrose gradients, while approximately half the long-term labeled parental strands were circular and half full-length linear. Treatment of the duplex circles with the *E. coli* R1 restriction enzyme gave rise to a single duplex linear species, which in alkali sedimentation forms two slower bands of progeny DNA and a band of full-length parental DNA. The molecular weights of the progeny fragments calculated from the sedimentation data demonstrated that the nick or gap in the progeny strand was at the replication terminus, 180° from the initiation site. It follows that such molecules must be an immediate product of the separation process with a discontinuity in the progeny strand near the termination site. The 16S post-separation replicative intermediate, when isolated from the cells treated with either FUdR or hydroxyurea, gives rise in alkaline sucrose gradients to approximately full-length single-stranded labeled linear DNA, and a broad labeled peak in the 5S region (91, 91b). The 16S intermediates isolated from FUdR-treated cells could not be closed by a crude ligase extracted from SV40-infected cells (91). However, in vitro incubation of the 16S DNA with *E. coli* DNA ligase, T4 DNA polymerase, and nonlabeled deoxyribonucleotides converted 60% of the radioactivity into the fast sedimenting covalently closed circular DNA species (91b). These results demonstrate that at least a major part of the 16S replicating intermediate contains one or more gaps in the newly synthesized strand.

Intracellular Nucleoprotein Complexes

When monolayers of polyoma and SV40-infected cells are lysed with a non-ionic detergent such as Triton X-100, both mature DNA and replicating intermediates can be isolated as nucleoprotein-DNA complexes in appropriate sucrose gradients (106–109). The replicating complexes sediment somewhat faster than the complexes of mature DNA and substantially faster than the free DNA. The protein contents determined from the buoyant density of formaldehyde or glutaraldehyde fixed complexes varied from 50 to 66% (108, 109). These proteins, prior to fixation, may be removed with sodium dodecylsulfate (SDS) and are, in part, digestible with proteases. It is anticipated that comparison of the proteins with those recovered from the corresponding virions will distinguish adventitiously or temporarily complexed proteins from those associated with viral DNA replication and encapsidation.

COVALENTLY CLOSED REPLICATING INTERMEDIATES

In the course of semiconservative replication, the parental strands are necessarily unwound from each other. If unwinding occurs in a closed duplex circle, a requirement arises for the winding of duplexes around each other in order to maintain the original winding number (8). The winding of duplexes around each other is referred to as supercoiling and is accompanied by the development of positive free energy which, in simple closed circles, increases quadratically as a function of the number of superhelical turns (110). For example, 80 kcal/mole of free energy are associated with the approximately 20 negative supercoils in viral SV40 DNA, which contain about 500 Watson–Crick turns.

The replication of 10 base pairs in a closed molecule with no superhelical turns would remove one Watson–Crick turn from the parental strand and introduce one positive superhelical turn. Continuation of replication in the closed circle could occur only until the buildup of superhelical free energy matched the free energy of nucleotide addition. Synthesis would then stop. It would be necessary, therefore, to release the twisting by a nicking process in which a swivel is developed (8).

All sizes of SV40, polyoma, and mitochondrial DNA replicating intermediates contain closed circular and still interwound parental strands (16, 18, 19, 111, 111a). From this observation, it is clear that the replicating intermediates were nicked in the nonreplicated region and closed periodically during replication with corresponding decreases in the topological winding number, α. The result, however, does not require that chain elongation occur preferentially in either the closed or the nicked form. There is no evidence at the present time that bears on this problem. In addition to endonucleases and ligases, two proteins, one in *E. coli* (112) and one in mouse cells (113), have been isolated which reduce the superhelix density of closed circles by a process involving nicking, relaxation, and reclosing.

The first evidence for closed parental strands in replicating intermediates arose in the study of sedimentation properties of the parental SV40 DNA in alkaline sucrose gradients in which it was found that the sedimentation velocity of the parental DNA was an inverse function of the degree of replication as determined by the sedimentation velocity of the released progeny strands in the same experiment (18, 104, 111, 111a). It was anticipated that the sedimentation coefficient of a closed DNA in alkali would decrease as the topological winding number, α, became smaller. This had been shown experimentally (113a) over a range of the normalized value of α from 1.0 to 0.9 and, more recently, over the same range with SV40 (18, 113b) and polyoma DNA (111a).

A further line of evidence was the observation that the replicating forms decrease in density in EthBr-CsCl gradients as the degree of replication increases (16, 18, 19, 104, 111, 111a). These molecules are positioned between closed viral DNA and nicked viral DNA in the gradients. This method provides a convenient procedure for isolating replicating forms of different degrees of replication. The decreased buoyant density is, in part, due to the fact that the replicated region consists of two DNA segments that are individually not a part of the two-stranded closed circular system and have the unrestricted dye binding properties of linear DNA (114).

When Cairns type closed replicating intermediates are spread for electron microscopy on an aqueous hypophase, they have a unique appearance with two open lobes and one highly twisted region (18, 19, 111, 111a), as illustrated diagrammatically in Figure 4C. Nicked replicating molecules assume the well-known appearance of Cairns forms (8, 14, 74–76, 115). When closed replicating intermediates are partially heat denatured so as to denature the progeny regions with the release of the progeny strands, the product molecules appear in formamide spreads as twisted duplex circles containing several single-stranded loops (16). The presence of loops is associated with the closed circularity because nicked mitochondrial DNA under these spreading conditions does not visibly denature. Presumably, these molecules renature with base pairing only until the gain in positive superhelical free energy becomes greater than the gain in negative free energy associated with base pairing.

The structure of the closed replicating intermediates can be represented by the three topologically equivalent forms shown in Figure 4. We take the molecule in Figure 4A, with a replicated region and an unwound single-stranded loop, as a model for the precursor of the isolated replicating intermediates. Duplex formation in the loop region, or, alternatively, an appropriate pitch change in the non-replicated duplex region would require that supercoiling occur as in Figure 4B and C. It is apparent in Figure 4B that interwinding in the replicated regions is only half as efficient as in the nonreplicated region. In the latter region, four parental strands cross at each node, whereas in the former, only two parental strands cross at a node. Therefore, twice the number of supercoils are required in order to accomplish the same unwinding effect. In solution, it is likely that an intermediate between the two

A   B   C

Figure 4   A model for the formation and structure of superhelical closed replicating intermediates. (*A*) Relaxed replicating intermediate with a single-stranded loop representing an unwound region topologically equivalent to a reduced duplex rotation angle in the entire unreplicated region. (*B* and *C*) Alternative superhelical molecules generated when the single-stranded loop in *A* winds up to form a duplex as described in the text.

forms occurs but that the spreading forces in electron microscope mounting procedures favor the structure in Figure 4C, which is actually seen in specimens prepared by the basic film aqueous technique.

If the supercoiling in the replicating intermediates arises because of a change in pitch of the unreplicated duplex upon isolation, as could occur if a nucleoprotein were dissociated, the number of superhelical turns would decrease linearly with the degree of replication (111, 111a, 116).

Puromycin-treated polyoma and SV40 DNA and cycloheximide-treated SV40 DNA give rise to the formation of finished closed DNAs having a superhelix density close to zero (117–119). These results suggest that action of some protein is necessary for the formation of superhelical molecules.

KINETOPLAST DNA AND CHLOROPLAST DNA

Unlike the circular DNAs discussed so far, which have been mainly in the form of circular molecules with only a fraction of molecules present as catenated and concatenated oligomers, kinetoplast DNA (kDNA) is a bewildering large network of interlocked small circular DNAs and some high molecular weight linear DNA (120–124). The kDNA also contains free small circular DNAs and simple catenated and concatenated circular oligomers (122, 125–129). This DNA is found in a condensed form in a single mitochondrion of protozoa (130). The DNA may be visualized by light microscopy after staining (131, 132). The organization of the network and its duplication and segregation is poorly understood at the present time.

kDNA from *Trypanosoma cruzi*, separated from nuclear DNA in a Hg^{2+}-Cs_2SO_4 buoyant density gradient was found to contain a few extremely rare Cairns replicating intermediates in the electron microscope (133). The frequency of this form was increased upon treating the culture for several days with the trypanocidal drug, Berenil, (4,4'-diazoamino dibenzamide diaceturate·4 H_2O) (133, 134). The Cairns mode appeared to be operative for the replication of the minicircle with a contour length of 0.5 μm. Based on the appearance of several maxima in the frequency distribution of the sizes of the replicated region, the authors suggest that replication proceeds stepwise in units of approximately 80 nucleotides (133). This conclusion must be regarded as tentative in view of the paucity of replicating intermediates measured and the problem of measurement accuracy. It is interesting that this drug inhibits chain elongation without a corresponding decrease in chain initiation. A density shift experiment in the presence of 5-BUdR showed that the overall DNA replication is semiconservative (135).

In another protozoan, *Leischmania tarentolae*, a similar network of circles as well as free circles occurs (122). The minicircle is 0.3 μm in contour length and contains 840 ±40 base pairs (122). The minicircle can be isolated free of nuclear DNA by treatment with alkali under conditions in which all the DNA becomes denatured (129). The material is then neutralized with renaturation of the closed circles. These were freed of single-stranded DNA and sedimented in a sucrose gradient to obtain monomeric minicircles and well-separated oligomers. EthBr-CsCl buoyant density gradient sedimentation revealed that approximately half the monomeric circular

DNA was nicked (129, 136). Extensive renaturation of the denatured DNA did not give rise to a multinucleate high molecular weight DNA. This result indicated that the strands were unbroken or nicked only once and that the nicked molecules might be uniquely nicked replicating intermediates (136). It is interesting to note that the small closed circles reannealed on neutralizing alkali denatured forms in contrast to larger closed circles which assumed a denatured form after such a pH cycle (136a).

It has been known for some time that chloroplasts contain DNA (137–146). The chloroplast DNA has been shown to replicate autonomously and semiconservatively in *Chlamydomonas reinhardi* (144). This DNA is of high molecular weight, 8.8×10^7 in the case of *Euglena gracilis* (146). In the latter organism, the DNA occurs extensively in a circular form (146) and appears to replicate by the Cairns mode (personal communication of J. E. Manning, D. R. Wolstenholme, and O. C. Richards).

RIBOSOMAL DNA AMPLIFICATION

The existence of the large amount of extrachromosomal but intranuclear DNA has been noted cytochemically in the oocyte germinal vesicle of many animal species (147–151). In some of these, it has been proven by DNA–RNA hybridization that the large relative amount of extrachromosomal DNA in the germinal vesicles is due to the synthesis of the rDNA in the early period of oogenesis prior to rRNA synthesis (150, 152–159). In *Xenopus laevis*, rDNA amplification occurs in the pachytene stage of oogenesis and the large amount of synthesized rDNA is contained in the many nucleoli spread over the inside of the germinal vesicle envelope (148–150, 152, 153, 157, 158, 160, 162, 163). The 1500-fold amplification occurs in the cell which contains 12 pg of DNA, of which approximately 0.1% is rDNA before the amplification process. Circular DNAs that still contain transcribed RNA and have different contour lengths capable of coding for up to 175 rRNAs tandemly were observed in preparations of dispersed germinal vesicles (161).

Twisted closed circular rDNAs isolated by CsCl buoyant centrifugation isolated from *Xenopus* oocytes have also been seen in the electron microscope (164–166). The amplified rDNA with a buoyant density (167) of 1.729 g/ml contained covalently closed circular DNA and open circular DNA in a frequency of 2–5%, while the remainder was linear (164). All of the DNA consisted of genes for rRNA as shown by the formation of the characteristic loop patterns upon partial denaturation (163).

About one sixth of the circles contained an attached duplex tail with a length that varied up to several times the contour length of the circle (164). These observations indicate that gene amplification occurs by a rolling circle mode of DNA replication (Figure 1). It is assumed that the replication begins in a circular DNA template formed by some unknown mechanism. Tails that consist of concatamers of rDNA apparently may be cyclized to form the circular concatamers that occur in moderate frequency among replicating and nonreplicating circles. These circular concatamers occur in all multiples including odd and even multiples of the gene length. This observation rules out the possibility that concatamers form by a doubling process which occurs at each round of replication. Circular concatamers could have arisen by excision of a DNA length containing tandemly arranged genes in the chromosome.

Similar forms were observed in the electron microscope in the fractionated rDNA isolated from ovaries labeled in organ culture for 6 hr early in oogenesis (165). Electron microscope autoradiographs showed directly that the rolling circle structures were replicating structures (165). The authors determined that the rate of rDNA replication was 10 μm/hr by light microscope autoradiography. This rate was adequate to label one strand in the tail of a 60 μm circle and most of both strands of smaller circles. The grain density of the small rolling circles appeared to be double that of the large ones.

REPLICATION OF CIRCULAR OLIGOMERS

Two kinds of circular oligomers were discovered in 1967 in the preparations of mitochondrial (168, 169) and bacterial plasmid DNA (170–172). Concatenated oligomers in which two or more genomes are arranged in a continuous circle, usually in a head-to-tail configuration (173), are often called circular oligomers. Catenated oligomers in which the circular genomes are interlocked with each other are also common. Higher catenated oligomers include those in which the genomes are interlocked in linear, branched, or circular arrays (168, 174).

Both catenated and concatenated oligomers are found among mitochondrial DNAs (33, 168, 169, 175–182), kinetoplast DNAs (183–188), and intracellular viral DNA (76, 77, 189–193). As previously stated, amplified ribosomal DNA contains circular concatamers. Concatenated dimers in animal cell tissue are infrequent and have been reported to be absent in several animal tissues (177), including immature human granulocytes (169). Their presence has been reported in peripheral leucocytes of patients having granulocytic leukemia and were observed to decrease in frequency during treatment with nucleic acid analogs and other drugs (180). Circular dimers have also been reported in nonmalignant human and beef thyroid tissue, indicating that the presence of circular dimers is not an unambiguous sign of malignancy (33).

Catenated dimers and higher oligomers occur in low frequency in animal tissue and high frequency in established cell lines (178, 179, 181). The oligomers present two problems relative to replication. Do they replicate? Are they formed by an aberration in replication or by recombination? The mitochondria of a mouse L cell line, the LD line, contain all the mitochondrial DNA in the form of circular dimers, some of which are catenated to each other. These dimers replicate by a modified Cairns mode as described previously for monomeric mtDNA. Displacement replication continues through both genomes (15, 16). Only one of the two D loops, which are often found in dimer molecules, appears to expand. The second D loop is displaced or incorporated in the elongation of the heavy strand.

The catenane appears to replicate by the normal mechanism for mitochondrial DNA replication (23, 76, 133, 179, 181, 190). The separation process here, however, is still unknown. It has been pointed out that in the absence of a linearization process, the two progenies of a submolecule in a catenane should remain catenated to the nonreplicated submolecule and form a linear catenated trimer (168). Replication of a submolecule in a trimer should lead to a branched tetramer with the exclusion of linear tetramers. These, however, have been found (168), indicating that linearization and recombination processes occur. In certain established cell lines, up to 40% of the mitochondrial genomes are present in the interlocked form. Early in replication in the presence of BUdR, a quarter dense species develops which consists of a single hybrid submolecule linked in a catenane to an unlabeled submolecule (23). This result also suggests that a linearization process occurs in the replication of mitochondrial catenated DNA. It is also clear that some mechanism must exist for maintaining the distribution of the several kinds of interlocked oligomers.

Both catenated and concatenated oligomers occur at a frequency of approximately 1–3% in the intracellular DNA after infection with SV40 and polyoma (117, 189, 192, 193). Cycloheximide and puromycin treatment during infection raises the oligomer level severalfold (117, 189, 192, 193). These results form the basis for the proposal that oligomers are due to an aberration in the replication process rather than the recombination process (76, 190, 192, 193).

The low frequency of trimeric SV40 molecules relative to dimer and tetramer forms after cycloheximide treatment was used to argue that the circular oligomer was formed from a replication aberration rather than a recombination process (190, 192). The catenated dimer possessed a higher specific activity than the circular dimer when pulsed for a short time in the presence or absence of cycloheximide. The radioactivity of catenanes decreased more rapidly during the chase, a result consistent with linearization of catenated molecules into simple molecules (117).

When ts-a polyoma-transformed mouse 3T3 cells were brought to permissive low temperature, a high frequency of concatenated dimer and trimer closed circles were formed (189), possibly, as suggested by the authors, by excision of tandemly lysogenized genomes in the chromosomes. The circular DNAs appear to replicate independently; the amount of label pulsed into each species remains constant during the cold chase. However, only the monomeric genomes are packaged into virions.

$E.$ $coli$ spheroplasts coinfected with wild-type ϕX DNA and deletion mutants contained catenanes with both kinds of submolecules and circular dimers with only one kind (194). This direct experimental result indicates that catenanes in this bacterial viral system are formed by a recombination mechanism, and concatamers are formed by a replication mechanism in which the two progeny genomes in the replicating complex combine to form circular dimers.

In the animal system cited above, the formation of both kinds of oligomers has been attributed to replication aberrations and, once formed, both kinds replicate. Coon & Dawid, however, have observed monomeric recombinant mitochondrial DNA in hybrid human-rat and human-mouse cell lines (195), a result which indicates that the recombination mode is feasible for the formation of oligomers.

ACKNOWLEDGMENTS
It is a pleasure to acknowledge the expert assistance provided by Patricia Ruda in the preparation of the manuscript. The authors are grateful to many colleagues who supplied preprints of manuscripts prior to publication and to Phillip A. Sharp for

critical comments on this article. This work was supported, in part, by USPHS Grant No. GM 15327 from the National Institute of General Medical Sciences and by CA 08014 from the National Cancer Institute.

This is contribution No. 4785 from the Division of Chemistry and Chemical Engineering.

Literature Cited

1a. Younghusband, H. B., Inman, R. 1974. *Ann. Rev. Biochem.* 43:605–19

1b. Bonhoeffer, F., Messer, W. 1969. *Ann. Rev. Genet.* 3:233–46

2. Helmstetter, C. E. 1969. *Ann. Rev. Microbiol.* 23:223–38

3. Lark, K. G. 1969. *Ann. Rev. Biochem.* 38:569–604

4. Gross, J. D. 1972. *Curr. Top. Microbiol. Immunol.* 57:39–74

5. Goulian, M. 1971. *Ann. Rev. Biochem.* 40:855–98

6. Richardson, C. C. 1969. *Ann. Rev. Biochem.* 38:795–840

7. Klein, A., Bonhoeffer, F. 1972. *Ann. Rev. Biochem.* 41:301–32

8. Cairns, J. 1963. *Cold Spring Harbor Symp. Quant. Biol.* 28:43–46

9. Gilbert, W., Dressler, D. 1968. *Cold Spring Harbor Symp. Quant. Biol.* 33:473–84

10. Eisen, H., da Silva, L. P., Jacob, F. 1968. *Cold Spring Harbor Symp. Quant. Biol.* 33:755–64

11. Okazaki, R. et al 1968. *Cold Spring Harbor Symp. Quant. Biol.* 33:129–43

12. Robberson, D. L., Kasamatsu, H., Vinograd, J. 1972. *Proc. Nat. Acad. Sci. USA* 69:737–41

13. Wolstenholme, D. R., Koike, K., Couchran-Fouts, P. 1973. *J. Cell Biol.* 56:230–45

14. Kirschner, R. H., Wolstenholme, D. R., Gross, N. J. 1968. *Proc. Nat. Acad. Sci. USA* 69:1466–72

15. Kasamatsu, H., Vinograd, J. 1973. *Nature New Biol.* 241:103–5

16. Robberson, D. L., Clayton, D. A. 1972. *Proc. Nat. Acad. Sci. USA* 69:3810–14

16a. Yoshimori, R. N. 1971. PhD thesis. University of California, San Francisco Medical Center

17. Robberson, D. L., Clayton, D. A., Davis, R. W., Morrow, J. F. In preparation

18. Sebring, E. D., Kelly, T. J. Jr., Thoren, M. M., Salzman, N. P. 1971. *J. Virol.* 8:478–90

19. Jaenisch, R., Mayer, A., Levine, A. 1971. *Nature New Biol.* 233:72–75

20. Reich, E., Luck, D. J. L. 1966. *Proc. Nat. Acad. Sci. USA* 55:1600–8

21. Gross, N. J., Rabinowitz, M. 1969. *J. Biol. Chem.* 244:1563–66

22. Corneo, G., Moore, C., Sanadi, D. R., Grossman, L. I., Marmur, J. 1966. *Science* 151:687–89

23. Flory, P. J. Jr., Vinograd, J. 1973. *J. Mol. Biol.* 74:81–94

24. Kasamatsu, H., Robberson, D. L., Vinograd, J. 1971. *Proc. Nat. Acad. Sci. USA* 68:2252–57

25. Ter Schegget, J., Flavell, R. A., Borst, P. 1971. *Biochim. Biophys. Acta* 254:1–14

26. Robberson, D. L., Clayton, D. A. 1973. *J. Biol. Chem.* 248:4512–14

26a. Lee, C. S., Davis, R. W., Davidson, N. 1970. *J. Mol. Biol.* 48:1–22

27. Kasamatsu, H., Grossman, L. I., Robberson, D. L., Watson, R., Vinograd, J. 1974. *Cold Spring Harbor Symp. Quant. Biol.* 38: In press

28. Arnberg, A. C., Van Bruggen, E. F. J., Flavell, R. A., Borst, P. 1973. *Biochim. Biophys. Acta* 308:276–84

29. Wolstenholme, D. R., Koike, K., Couchran-Fouts, P. 1974. *Cold Spring Harbor Symp. Quant. Biol.* 38: In press

30. Koike, K., Wolstenholme, D. R. *J. Cell Biol.* Submitted

31. Arnberg, A. C., Van Bruggen, E. F. J., Ter Schegget, J., Borst, P. 1971. *Biochim. Biophys. Acta* 246:353–57

32. Paoletti, C., Riou, G., Pairault, J. 1972. *Proc. Nat. Acad. Sci. USA* 69:847–50

33. Arnberg, A. C., Van Bruggen, E. F. J., Schutgens, R. B. H., Flavell, R. A., Borst, P. 1972. *Biochim. Biophys. Acta* 272:487–93

34. Matsumoto, L. H., Pikó, L., Kasamatsu, H., Vinograd, J. In preparation

35. Hallberg, R. In preparation

36. Alberts, B. M., Frey, L. 1970. *Nature* 227:1313–18

37. Schneider, M., Neubert, D. 1966. *Arch. Pharmakol. Exp. Pathol.* 255:68

38. Wintersberger, E. 1966. *Biochem. Biophys. Res. Commun.* 25:1–7

39. Parson, P., Simpson, M. V. 1967. *Science* 155:91–93

40. Brewer, E. N., DeVries, A., Rusch,

H. P. 1967. *Biochim. Biophys. Acta* 145: 686–92

41. Parson, P., Simpson, M. V. 1968. In *Round-Table Discussion on Biochemical Aspects on the Biogenesis of Mitochondria*, ed. E. C. Slater, J. M. Tager, S. Papa, E. Quagliariello, 171. Bari: Adriatica Editrice

42. Winterberger, E. Ibid

43. Neubert, D., Oberdisse, E., Bass, R. Ibid

44. Karol, M. H., Simpson, M. V. 1968. *Science* 162: 470–72

45. Neubert, D., Teske, S., Schneider, M., Köhler, E., Oberdisse, E. 1969. *Arch. Pharmakol. Exp. Pathol.* 262: 264

46. Nass, M. M. K. 1969. *Science* 165: 25–35

47. Mitra, R. S., Bernstein, I. A. 1970. *J. Biol. Chem.* 245: 1255–60

48. Ter Schegget, J., Borst, P. 1971. *Biochim. Biophys. Acta* 246: 239–48

49. Parson, P., Simpson, M. V. 1973. *J. Biol. Chem.* 248: 1912–19

50. Ter Schegget, J., Borst, P. 1971. *Biochim. Biophys. Acta* 246: 249–57

51. Ter Schegget, J., Flavell, R. A., Borst, P. 1971. *Biochim. Biophys. Acta* 254: 1–14

51a. Upholt, W. B., Gray, H. B. Jr., Vinograd, J. 1971. *J. Mol. Biol.* 62: 21–38

52. Flavell, R. A., Borst, P., Ter Schegget, J. 1972. *Biochim. Biophys. Acta* 272: 341–49

53. Gauze, G. G. Jr., Dolgilevich, S. M., Fatkullina, L. G., Mikhailov, V. S. 1973. *Biochim. Biophys. Acta* 312: 179–91

54. Hamilton, F., Vissering, F., Simpson, M. V. 1973. *Int. Congr. Biochem. Stockholm* (Abstr.)

55. Sambrook, J., Westphal, H., Srinivasan, P. R., Dulbecco, R. 1968. *Proc. Nat. Acad. Sci. USA* 60: 1288–95

56. Aloni, Y., Winocour, E., Sachs, L., Torten, J. 1969. *J. Mol. Biol.* 44: 333–45

57. Yoshiike, K. 1968. *Virology* 34: 391–401

58. Yoshiike, K. 1968. *Virology* 34: 402–9

59. Thorne, H. V. 1968. *J. Mol. Biol.* 35: 215–26

60. Thorne, H. V., Evans, J., Warden, D. 1968. *Nature* 219: 728–30

61. Blackstein, M. E., Stanners, C. P., Farmilo, A. J. 1969. *J. Mol. Biol.* 42: 301–13

62. Tai, H. T., Smith, C. A., Sharp, P. A., Vinograd, J. 1972. *J. Virol.* 9: 317–25

63. Lavi, S., Winocour, E. 1972. *J. Virol.* 9: 309–16

64. Brockman, W. W., Lee, T. N. H., Nathans, D. 1973. *Virology* 54: 384–97

65. Weil, R., Vinograd, J. 1963. *Proc. Nat. Acad. Sci. USA* 50: 730–38

66. Dulbecco, R., Vogt, M. 1963. *Proc. Nat. Acad. Sci. USA* 50: 236–43

67. Crawford, L. V., Black, P. H. 1964. *Virology* 24: 388–92

68. Vinograd, J., Lebowitz, J., Radloff, R., Watson, R., Laipis, P. 1965. *Proc. Nat. Acad. Sci. USA* 53: 1104–11

69. Michel, M. R., Hirt, B., Weil, R. 1967. *Proc. Nat. Acad. Sci. USA* 58: 1381–88

70. Winocour, E. 1967. *Virology* 31: 15–28

71. Winocour, E. 1968. *Virology* 34: 571–82

72. Hirt, B. 1967. *J. Mol. Biol.* 26: 365–69

73. Bourgaux, P., Bourgaux-Ramoisy, D., Dulbecco, R. 1969. *Proc. Nat. Acad. Sci. USA* 64: 701–8

74. Levine, A. J., Kang, H. S., Billheimer, F. E. 1970. *J. Mol. Biol.* 50: 549–68

75. Bourgaux, P., Bourgaux-Ramoisy, D., Seiler, P. 1971. *J. Mol. Biol.* 59: 195–206

76. Meinke, W., Goldstein, D. A. 1971. *J. Mol. Biol.* 61: 543–63

77. Bourgaux-Ramoisy, D. 1971. *Biochim. Biophys. Acta* 254: 412–14

78. Tegtmeyer, P., Macasaet, F. 1972. *J. Virol.* 10: 599–604

79. Thoren, M. M., Sebring, E. D., Salzman, N. P. 1972. *J. Virol.* 10: 462–68

79a. Mulder, C., Delius, H. 1972. *Proc. Nat. Acad. Sci. USA* 69: 3215–19

79b. Morrow, J., Berg, P. 1972. *Proc. Nat. Acad. Sci. USA* 69: 3365–69

79c. Griffin, B., Fried, M., Robberson, D. 1973. *Tumor Virus Meeting* (Abstr.) Cold Spring Harbor, NY: Cold Spring Harbor Lab.

80. Fareed, G. C., Garon, C. F., Salzman, N. P. 1972. *J. Virol.* 10: 484–91

81. Nathans, D., Danna, K. J. 1972. *Nature New Biol.* 236: 200–2

82. Crawford, L. V., Syrett, C., Wilde, A. 1973. *J. Gen. Virol.* 21: 515–21

83. Smith, H. O., Wilcox, K. W. 1970. *J. Mol. Biol.* 51: 379–91

84. Danna, K. J., Nathans, D. 1972. *Proc. Nat. Acad. Sci. USA* 69: 3097–3100

85. Bourgaux, P., Bourgaux-Ramoisy, D. 1971. *J. Mol. Biol.* 62: 513–24

86. Danna, K. J., Nathans, D. 1971. *Proc. Nat. Acad. Sci. USA* 68: 2913–17

87. Danna, K. J., Sack, G. H. Jr., Nathans, D. 1973. *J. Mol. Biol.* 78: 363–76

88. Fareed, G. C., Salzman, N. P. 1972. *Nature New Biol.* 238: 274–77

89. Fareed, G. C., Khoury, G., Salzman, N. P. 1973. *J. Mol. Biol.* 77: 457–62

90. Okazaki, R., Okazaki, T., Sakabe, K., Sugimoto, K., Sugino, A. 1968. *Proc. Nat. Acad. Sci. USA* 59: 598–605

91. Salzman, N. P., Thoren, M. M. 1973. *J. Virol.* 11: 721–29

91a. Laipis, P., Levine, A. J. 1973. *Virology.* 56:580–94

91b. Laipis, P., Levine, A. J. 1974. *NATO Advanced Study Institute, Tumor Virus-Host Cell Interaction.* New York: Plenum

92. Winnacker, E. L., Magnusson, G., Reichard, P. 1971. *Biochem. Biophys. Res. Commun.* 44:952–57

93. Winnacker, E. L., Magnusson, G., Reichard, P. 1972. *J. Mol. Biol.* 72: 523–37

94. Hunter, T., Francke, B. 1974. *J. Virol.* 13:125–39

95. Magnusson, G., Winnacker, E. L., Eliasson, R., Reichard, P. 1972. *J. Mol. Biol.* 72:539–52

96. Hirt, B. 1966. *Proc. Nat. Acad. Sci. USA* 55:997–1004

97. Magnusson, G., Pigiet, V., Winnacker, E. L., Abrams, R., Reichard, P. 1973. *Proc. Nat. Acad. Sci. USA* 70:412–15

98. Francke, B., Hunter, T. 1974. *J. Mol. Biol.* In press

99. Magnusson, G. 1973. *J. Virol.* 12:600–8

100. Magnusson, G. 1973. *J. Virol.* 12:609–15

101. Pigiet, V., Winnacker, E. L., Eliasson, R., Reichard, P. 1973. *Nature New Biol.* 245:203–5

102. Hunter, T., Francke, B. 1974. *J. Mol. Biol.* In press

103. Sugino, A., Hirose, S., Okazaki, R. 1972. *Proc. Nat. Acad. Sci. USA* 69: 1863–67

103a. Pigiet, V., Eliasson, R., Reichard, P. 1974. *J. Mol. Biol.* In press

104. Mayer, A., Levine, A. J. 1972. *Virology* 50:328–38

105. Fareed, G. C., McKerlie, M. L., Salzman, N. P. 1973. *J. Mol. Biol.* 74:95–111

106. White, M., Eason, R. 1971. *J. Virol.* 8:363–71

107. Green, M. H., Miller, H. I., Hendler, S. 1971. *Proc. Nat. Acad. Sci. USA* 68: 1032–36

108. Goldstein, D. A., Hall, M. R., Meinke, W. 1973. *J. Virol.* 12:887–900

109. Hall, M. R., Meinke, W., Goldstein, D. A. 1973. *J. Virol.* 12:901–8

110. Bauer, W., Vinograd, J. 1970. *J. Mol. Biol.* 47:419–35

111. Bourgaux, P., Bourgaux-Ramoisy, D. 1972. *J. Mol. Biol.* 70:399–413

111a. Roman, A., Champoux, J. J., Dulbecco, R. 1974. *Virology* 57:147–60

112. Wang, J. C. 1971. *J. Mol. Biol.* 55: 523–33

113. Champoux, J., Dulbecco, R. 1972. *Proc. Nat. Acad. Sci. USA* 69:143–46

113a. Smith, C. A., Jordan, J. M., Vinograd, J. 1971. *J. Mol. Biol.* 59:255–72

113b. Schmir, M., Révet, B. M. J., Vinograd, J. 1974. *J. Mol. Biol.* In press

114. Bauer, W., Vinograd, J. 1970. *J. Mol. Biol.* 54:281–98

115. Hirt, B. 1969. *J. Mol. Biol.* 40:141–44

116. Salzman, N. P., Sebring, E. D., Radonovich, M. 1973. *J. Virol.* 12: 669–76

117. Bourgaux, P., Bourgaux-Ramoisy, D. 1972. *Nature* 235:105–7

118. White, M., Eason, R. 1973. *Nature New Biol.* 241:46–49

119. Jaenisch, R., Levine, A. J. 1973. *J. Mol. Biol.* 73:199–212

120. Laurent, M., Steinert, M. 1970. *Proc. Nat. Acad. Sci. USA* 66:419–24

121. Renger, H. C., Wolstenholme, D. 1970. *J. Cell Biol.* 47:689–702

122. Simpson, L., da Silva, A. 1971. *J. Mol. Biol.* 56:443–73

123. Renger, H. C., Wolstenholme, D. 1972. *J. Cell Biol.* 54:346–64

124. Delain, E., Brack, C., Lacome, A., Riou, G. 1972. In *Symposium on the Comparative Biochemistry of Parasites,* ed. H. Van den Bossche, 167–84. New York: Academic

125. Riou, G., Delain, E. 1969. *Proc. Nat. Acad. Sci. USA* 62:210–17

126. Riou, G., Delain, E. 1969. *Proc. Nat. Acad. Sci. USA* 64:618–25

127. Renger, H. C., Wolstenholme, D. 1971. *J. Cell Biol.* 50:533–40

128. Renger, H. C., Wolstenholme, D. 1972. *J. Cell Biol.* 54:346–64

129. Wesley, R. D., Simpson, L. 1973. *Biochim. Biophys. Acta* 319:237–53

130. Simpson, L. 1972. *Int. Rev. Cytol.* 32: 165–71

131. Breslau, E., Scremin, L. 1924. *Arch. Protistenk.* 48:509–15

132. Baker, J. 1961. *Trans. Roy. Soc. Trop. Med. Hyg.* 55:518–24

133. Brack, C., Delain, E., Riou, G. 1972. *Proc. Nat. Acad. Sci. USA* 69:1642–46

134. Brack, C., Delain, E., Riou, G., Festz, B. 1972. *J. Ultrastruct. Res.* 39:568–79

135. Riou, G. 1973. *C. R. Acad. Sci. Paris* 277:745–47

136. Wesley, R. D., Simpson, L. 1973. *Biochim. Biophys. Acta* 319:254–66

136a. Pouwels, P. H., Knijnenburg, C. M., Van Rotterdam, J., Cohen, J. A., Jansz, H. S. 1968. *J. Mol. Biol.* 32: 169–82

137. Chun, E. H. L., Vaughan, M. H., Rich, A. 1963. *J. Mol. Biol.* 7:130–41

138. Sager, R., Ishida, M. R. 1963. *Proc.*

Nat. Acad. Sci. USA 50: 725–30

139. Ray, D. S., Hanawalt, P. C. 1964. *J. Mol. Biol.* 9: 812–24

140. Brawerman, G., Eisenstadt, J. M. 1964. *Biochem. Biophys. Acta* 91: 477–85

141. Edelman, M., Cowan, C. A., Epstein, H. T., Schiff, J. A. 1964. *Proc. Nat. Acad. Sci. USA* 52: 1214–19

142. Ray, D. S., Hanawalt, P. C. 1965. *J. Mol. Biol.* 11: 760–68

143. Edelman, M., Epstein, H. T., Schiff, J. A. 1966. *J. Mol. Biol.* 17: 463–69

144. Chiang, K. S., Sueoka, N. 1967. *Proc. Nat. Acad. Sci. USA* 57: 1506–13

145. Wells, R., Birnstiel, M. 1969. *Biochem. J.* 112: 777–86

146. Manning, J. E., Wolstenholme, D. R., Ryan, R. S., Hunter, J. A., Richards, O. C. 1971. *Proc. Nat. Acad. Sci. USA* 68: 1169–73

147. King, H. D. 1908. *J. Morphol.* 19: 369–438

148. Brachet, J. 1940. *Arch. Biol.* 51: 151–65

149. Painter, T. S., Taylor, A. N. 1942. *Proc. Nat. Acad. Sci. USA* 28: 311–17

150. Gall, J. G. 1968. *Proc. Nat. Acad. Sci. USA* 60: 553–60

151. Macgregor, H. C. 1968. *J. Cell Sci.* 3: 437–44

152. Brown, D. D., Dawid, I. B. 1968. *Science* 160: 272–80

153. Evans, D., Birnstiel, M. 1968. *Biochim. Biophys. Acta* 166: 274–76

154. Vincent, W. S., Halvorson, H. O., Chen, H. R., Shin, C. D. 1968. *Biol. Bull.* 135: 441

155. Gall, J. G., Macgregor, H. C., Kidston, M. E. 1969. *Chromosoma* 26: 169–87

156. Lima-de-Faria, A., Birnstiel, M., Jaworska, H. 1969. *Genetics* Suppl. 61: 146–159

157. Gall, J. G. 1969. *Genetics* Suppl. 61: 121–32

158. Gall, J. G., Pardue, M. L. 1969. *Proc. Nat. Acad. Sci. USA* 63: 378–83

159. John, H. A., Birnstiel, M. L., Jones, K. W. 1969. *Nature* 223: 582–87

160. Perkowska, E., Macgregor, H. C., Birnstiel, M. L. 1968. *Nature* 217: 649–50

161. Miller, O. L., Beatty, B. R. 1969. *Genetics* Suppl. 61: 133–43

162. Dawid, I. B., Brown, D. D., Reeder, R. 1970. *J. Mol. Biol.* 51: 341–60

163. Wensink, P., Brown, D. D. 1971. *J. Mol. Biol.* 60: 235–47

164. Hourcade, D., Dressler, D., Wolfson, J. 1973. *Proc. Nat. Acad. Sci. USA* 70: 2926–30

165. Bird, A. P., Rochaix, J. D., Bakken, A. H. 1973. *Molecular Cytogenetics Symposium.* New York: Plenum

166. Hourcade, D., Dressler, D., Wolfson, J. 1974. *Cold Spring Harbor Symp. Quant. Biol.* 38: In press

167. Wallace, H., Birnstiel, M. L. 1966. *Biochim. Biophys. Acta* 114: 296–310

168. Hudson, B., Vinograd, J. 1967. *Nature* 216: 647–52

169. Clayton, D., Vinograd, J. 1967. *Nature* 216: 652–57

170. Rush, M. G., Kleinschmidt, A. K., Hellman, W., Warner, R. C. 1967. *Proc. Nat. Acad. Sci. USA* 58: 1676–83

171. Rush, M. G., Warner, R. C. 1967. *Proc. Nat. Acad. Sci. USA* 58: 2372–76

172. Roth, T. F., Helinski, D. R. 1967. *Proc. Nat. Acad. Sci. USA* 58: 650–57

173. Clayton, D. A., Davis, R. W., Vinograd, J. 1970. *J. Mol. Biol.* 47: 137–53

174. Hudson, B., Clayton, D. A., Vinograd, J. 1968. *Cold Spring Harbor Symp. Quant. Biol.* 33: 435–42

175. Radloff, R., Bauer, W., Vinograd, J. 1967. *Proc. Nat. Acad. Sci. USA* 57: 1514–21

176. Pikó, L., Blair, D. G., Tyler, A., Vinograd, J. 1968. *Proc. Nat. Acad. Sci. USA* 59: 838–45

177. Clayton, D. A., Smith, C. A., Jordan, J., Teplitz, M., Vinograd, J. 1968. *Nature* 220: 976–79

178. Nass, M. M. K. 1969. *J. Mol. Biol.* 42: 521–28

179. Nass, M. M. K. 1969. *Nature* 223: 1124–29

180. Clayton, D. A., Vinograd, J. 1969. *Proc. Nat. Acad. Sci. USA* 62: 1077–84

181. Nass, M. M. K. 1970. *Proc. Nat. Acad. Sci. USA* 67: 1926–33

182. Wolstenholme, D. R., McLaren, J. D., Koike, K., Jacobson, E. L. 1973. *J. Cell Biol.* 56: 247–55

183. Riou, G., Delain, E. 1969. *Proc. Nat. Acad. Sci. USA* 62: 210–17

184. Riou, G., Delain, E. 1969. *Proc. Nat. Acad. Sci. USA* 64: 618–25

185. Renger, H. C., Wolstenholme, D. 1970. *J. Cell Biol.* 47: 609–702

186. Renger, H. C., Wolstenholme, D. 1971. *J. Cell Biol.* 50: 533–40

187. Simpson, L., da Silva, A. 1971. *J. Mol. Biol.* 56: 443–73

188. Renger, H. C., Wolstenholme, D. 1972. *J. Cell Biol.* 54: 346–64

189. Cuzin, F., Vogt, M., Dieckmann, M., Berg, P. 1970. *J. Mol. Biol.* 47: 317–33

190. Jaenisch, R., Levine, A. J. 1971. *Virology* 44: 480–93

191. Rush, M. G., Eason, R., Vinograd, J. 1971. *Biochim. Biophys. Acta* 228:585–94
192. Jaenisch, R., Levine, A. J. 1972. *Virology* 48:373–79
193. Bourgaux, P. 1973. *J. Mol. Biol.* 77:197–206
194. Benbow, R. M., Eisenberg, M., Sinsheimer, R. L. 1972. *Nature New Biol.* 237:141–43
195. Dawid, I. B., Horak, I., Coon, H. G. 1974. *International Symposium on the Biogenesis of Mitochondria, Genetics,* Suppl. In press

THE SELECTIVITY OF TRANSCRIPTION

× 864

Michael J. Chamberlin

Department of Biochemistry, University of California, Berkeley, California

CONTENTS

INTRODUCTION

Studies of the synthesis of macromolecules in the genetic pathway have revealed two levels of specificity. DNA, RNA, and proteins contain unique sequences of monomer units; the positioning of these units in the polymer chain during replication, transcription, and translation is under the control of the DNA or RNA template. The mechanism by which the template dictates such a unique sequence involves both base pairing interactions and specific interactions between proteins and nucleic acids (1–3). However, in all three synthetic pathways a second level of specificity is present; each kind of macromolecular chain is initiated at a specific site or sites on the template. In addition, the progress of chain growth during transcription and translation is subject to termination at unique sites on the template, defining units of transcription and translation. I refer to this level of specificity as selectivity. This review considers the selectivity involved in transcription of genetic sequences in certain bacterial and bacteriophage systems and discusses information about the mechanism by which selectivity may be imposed.

In our current model for selective transcription, initiation and termination of RNA transcripts is dictated by specific signals on the DNA template that are recognized and translated into biochemical events by the transcription apparatus. The site or region governing initiation of RNA synthesis has been designated the promoter (4–6), the region that governs termination as the terminator (6). The coupling of a promoter region with a terminator region at some distance downstream (i.e. in the direction of transcription) may be taken as defining a unit of transcription. In many instances, regulation of synthesis of a given transcript is effected at these same loci; the classical example of this is the operon, defined as a cluster of genes transcribed and regulated as a unit. Although the term operon might be extended to include all genetic units governed by a specific initiation and termination site, the term seems more appropriately restricted to units under the control of an operator sequence, and I simply use the term transcription unit in the general instance.

Fundamental questions can be posed about the selectivity of transcription in terms of this model: 1. What are the components of the transcription system responsible for selective transcription? 2. What are the essential structural features of promoter and terminator sites? 3. How are these sites read by RNA polymerase in the initiation and termination of RNA synthesis? 4. How is the synthesis of RNA regulated at these sites? 5. How is the specificity of transcription altered in differentiating systems? In addition, for each organism of interest one can ask: 6. What is the map of transcription units and regulatory sites on the chromosome(s)?

Although the selectivity of transcription has been reviewed extensively in recent years (1, 7–10), in general, decisive new information about the process has not emerged. In fact, many of the "exciting new breakthroughs" have proved to be dead ends or at best paths leading back into an unknown part of the maze. However, this period has seen a steady development of techniques and approaches and a reformulation of the basic questions at the molecular level. This consolidation should bear fruit over the next few years in the form of rather explicit answers to the basic questions for at least some transcription systems.

STRUCTURE OF THE TRANSCRIPTION APPARATUS

Structure of Escherichia coli RNA Polymerase

Transcription in bacterial cells is mediated by DNA-dependent RNA polymerase. The enzyme has been isolated from a large number of bacterial species (for review, see 3); however, the properties of the enzyme from *E. coli* have been most thoroughly explored and serve as the prototype for enzymes in this review.

A remarkable feature of the bacterial transcription apparatus is that purified bacterial RNA polymerase is able to carry out selective transcription of some helical DNAs in the absence of any other added components. This contrasts with the known components of the bacterial translation and replication systems. In the cell, both of the latter processes involve active complexes in which a number of protein components interact with DNA or the RNA message; however, the complexes appear to be quite labile to isolation and most knowledge of selective translation and replication has come from the properties of the single isolated components or from studies of complex reconstituted systems (11, 12). The ability of bacterial RNA polymerase to carry out selective transcription has focused a great deal of attention on the structure and physical properties of the purified enzyme.

The basic unit of selective transcription is thought to be the RNA polymerase holoenzyme (3, 8). The enzyme has a protomer of about 480,000–500,000 mol wt and aggregates reversibly to form a dimer of the protomer in a reaction favored at ionic strengths that are below 0.12 (13, 14). Each protomer (the term monomer is confusing since each enzyme contains several subunits) appears to be able to initiate (15) and elongate (16–18) an RNA chain, which suggests that this is the functional unit of the enzyme. However, the possibility that a dimer of the protomer form plays a role in site selection or in some regulatory event cannot be ruled out (18, 19). Because of the rapid reversibility of the aggregation process, references to purification procedures which yield the "monomer" or "dimer" forms of the enzyme are misleading.

RNA polymerase holoenzyme preparations contain four major kinds of subunits, designated β', β, σ, and α; these are present in the molar ratios $1:1:1:2$ (14, 20, 21). RNA polymerase preparations contain tightly bound Zn^{2+} ion, and RNA polymerase activity is inhibited by o-phenanthroline (22). Hence it is likely that RNA polymerase is a Zn-containing metalloenzyme. Direct evidence that the major polypeptide components of RNA polymerase holoenzyme are functional subunits is available only for β and σ subunits. All transcription in *E. coli* is sensitive to inhibitors of the purified enzyme, such as rifampicin and streptolydigin (23–25), which block RNA synthesis through interaction with the β subunit of the enzyme (26). Hence all RNA synthesis appears to depend on a unit in which the β subunit is functional. The β subunit is also altered in enzyme from mutants resistant to rifampicin (27) or streptolydigin (28, 29). The σ subunit is essential for promoter site selection and its function is reviewed below. In addition to β and σ, both β' and α subunits are required for reconstitution of the catalytically active holoenzyme from its isolated components. This provides indirect evidence that β' and α play a functional role in the holoenzyme molecule. In addition β' is synthesized coordinately with β in vivo and the structural gene for β' is located near that of β (30, 31). There are as yet no confirmed mutants of *E. coli* defective in β', α, or σ subunits.

Preparations of RNA polymerase contain both RNA polymerase holoenzyme and RNA polymerase lacking σ subunit (core polymerase) (32). This may reflect either loss of σ subunit during enzyme purification or a lack of equivalence of the components in vivo. RNA polymerase in ternary complexes with template DNA and nascent RNA chains lacks σ subunit (33–37), supporting the notion that the catalytic unit involved in chain elongation is the core polymerase (8, 9).

Dissociation and Reconstitution of RNA Polymerase

The complexity of RNA polymerase holoenzyme makes it difficult to assign functions to the individual components of the molecule. This is especially important when one comes to the study of selective transcription and its alteration, where RNA polymerase subunits may be replaced or altered. The ability to dissociate the enzyme into its components and to reconstitute active RNA polymerase provides an essential technique for the study of the function of RNA polymerase subunits.

The first example of the value of a dissociation-reconstitution procedure came in the study of the properties of σ subunit. The removal of σ subunit from RNA polymerase holoenzyme is accomplished under nondenaturing conditions simply by chromatography of the holoenzyme on phosphocellulose (21, 32). The resulting RNA polymerase core enzyme is enzymatically active but does not display a high degree of selectivity for initiation of RNA chains on most templates (8, 9). The reconstitution of holoenzyme from free σ subunit occurs rapidly when the two are mixed in solution (21, 32). The reconstituted holoenzyme regains the ability to initiate specific RNA chains. These results demonstrated the essential role of σ subunit in the process of site selection, and also showed that the core polymerase structure contains the basic catalytic unit for RNA chain elongation.

Chromatography on phosphocellulose does not remove σ subunit from some bacterial RNA polymerases (38a–38c). In addition, most preparations of the *E. coli* enzyme retain traces of σ subunit (32). This contamination can be of major significance when studying RNA synthesis by these preparations, since holoenzyme is up to 50 times more active than core polymerase and since σ can be used catalytically (34).

E. coli RNA polymerase can be dissociated into smaller components by treatment with a variety of agents including sodium dodecyl sulfate (SDS), urea, LiCl, and sulfhydryl reagents (20, 21, 39–41). Dissociation with low levels of urea, LiCl, and sulfhydryl reagents leads to a variety of complexes containing two or more kinds of subunits (41); in general, these lack RNA polymerase activity and do not carry out any of the partial reactions involved in the steps of RNA synthesis. However, many of the partially dissociated components as well as free β and β' bind to DNA and to polyanions in general (20, 26, 41, 42). It is not clear whether this binding is related to the binding of the enzyme to the DNA template and the RNA product that occurs during normal RNA synthesis. Dissociation of the enzyme with SDS or high concentrations of urea leads to complete separation of all the subunits of the enzyme; these can be resolved from each other by electrophoresis or acrylamide or cellogel or by sizing chromatography (21, 27, 43). Active σ subunits can be prepared after isolation of the subunit following denaturation with SDS (44). Under appropriate conditions, removal of the urea after complete dissociation of RNA polymerase by that agent leads to restoration of a major portion of the RNA polymerase activity (40). Reconstitution of enzymatic activity is also observed when the dissociated subunits are separated by electrophoresis and mixed in the proper proportions (27, 43). Efficient reconstitution of enzymatic activity requires the simultaneous presence of all four kinds of subunits (β', β, α, and σ) in the solution. Ishihama and his collaborators have been able to carry out stepwise assembly of the enzyme according to the general reaction (43)

$$2\alpha + \beta \rightarrow \alpha_2\beta$$
$$\alpha_2\beta + \beta' \rightarrow \alpha_2\beta\beta' \text{ (core polymerase)}$$
$$\alpha_2\beta\beta' + \sigma \rightarrow \alpha_2\beta\beta'\sigma$$

The second step is greatly enhanced in the presence of σ subunit.

Heterogeneity of Bacterial RNA Polymerase

While there is good evidence that the basic structure of RNA polymerase holoenzyme known from in vitro studies is part of the transcription apparatus in vivo, important questions can still be posed about its uniqueness and its functional structure in vivo. In particular, is there only a single RNA polymerase in the normal bacterial cell, or can multiple forms be distinguished? Such multiple species might differ in their subunit structures by virtue of the modification or the replacement of one of the known subunits of the RNA polymerase holoenzyme. A related question concerns the possibility that essential components of the transcription apparatus might have been separated from the RNA polymerase complex during fractionation or might not bind to RNA polymerase at all. Possible accessory components of the transcription system are discussed in a later section.

One can set several criteria for judging whether an altered subunit or a new component is a functional part of the transcription apparatus. Ideally, 1. The component should be purified and shown to be chemically distinct from the normal RNA polymerase subunits or accessory components. 2. The component should be required for some aspect of selective transcription in vitro. In the case of an altered subunit or a new subunit, for example, reconstitution of the enzyme in the presence or absence of the new component should give active enzyme that possesses or lacks the new specificity, respectively (27–29). 3. A corresponding function for the new component should be identifiable in vivo. Bacterial mutants or cells growing under conditions where the new polymerase component is not observed should lack the corresponding in vivo function. These are stringent criteria, since of the known subunits of E. coli RNA polymerase holoenzyme, only β fits all three tests. However, past experience suggests that any putative component that does not meet these criteria should be regarded with caution.

HETEROGENEITY OF BACTERIAL RNA POLYMERASE DURING NORMAL CELL GROWTH
Purified RNA polymerase holoenzyme is the only component required for the selective transcription of a variety of phage transcription units (7–9, 45). For other units where accessory factors are required, these are needed in addition to the holoenzyme. No bacterial mutants have yet been identified which alter a subunit of RNA polymerase and lead to a differential effect on the syntheses of different classes of

cellular RNA. Despite this evidence that RNA polymerase holoenzyme is generally required for transcription, there is no compelling evidence against the possible existence of multiple forms of the enzyme. Analytical studies of the subunit structure of the enzyme or on the chemical properties of the major subunits have not progressed to the point that permits one to exclude their heterogeneity. Current purification procedures seldom yield more than 30–50% of the enzyme in the cell in a homogeneous form (3, 8), and consequently the preferential loss of other forms of RNA polymerase, especially of minor species, cannot be ruled out. In fact, a variety of reports describe the detection or isolation of altered forms of bacterial RNA polymerase differing in their sedimentation properties, template specificity, or metal ion requirements (46–49). Since RNA polymerase tends to aggregate both with itself (13, 14) and with many other components of cell extracts, evidence of heterogeneity based on sedimentation properties or template specificity must be considered of doubtful significance until direct evidence of heterogeneity or alteration of subunits is provided. Even so, several specific kinds of artifacts must be considered. The σ subunit appears to be more sensitive to some kinds of inactivation than the components of the core polymerase (50), and enzyme has also been isolated which contains σ but behaves as core polymerase in its template specificity (51). The properties of RNA polymerase containing inactive σ subunits have not been thoroughly studied. The possible alteration of RNA polymerase in vitro after breakage of the cells needs careful attention. Extracts of E. coli (52), Bacillus brevis (53), and sporulating B. subtilis (54, 55) contain components which can inactivate or alter bacterial RNA polymerase. In the latter instance, selective cleavage of the β' subunit in vitro in the cell extracts may have led for some time to misleading conclusions concerning the role of the modified subunit in vivo (55).

There is some evidence that a substantial pool of nonfunctional RNA polymerase may exist in E. coli. Direct measurement of the concentration of $\beta' + \beta$ chains in the cell by gel electrophoresis gives an estimate of about 7000 enzyme equivalents per cell for rapidly growing E. coli (56, 57). In contrast, estimates of the number of enzyme molecules actively growing RNA chains range from 2000 to 5600 (55, 58). The procedures used to estimate the number of active RNA polymerase molecules in the cell are indirect. The result is affected by differences in the intrinsic chain growth rate for different RNA species (59) and might have a systematic error large enough to account for the apparent difference. Within the accuracy of the current determinations, both the RNA polymerase subunit concentration and the number of active enzyme molecules appear to increase somewhat with growth rate, and the fraction of active enzyme molecules appears to increase from about 20% at low growth rates to 50% at high growth rates. It may be that either some component of the transcription apparatus other than β and β' is limiting in the cell, or there is a sizable pool of nonfunctional enzyme present (56, 57). The presence of a large pool of inactive enzyme in the cell could serve as evidence of its heterogeneity if there were rapid activation of these molecules during metabolic shifts. It was previously believed that when slowly growing cells were shifted to rich medium, a rapid increase occurred in the absolute number of active RNA polymerase molecules. This is now disputed; there are substantial alterations in the rate at which exogenous precursors are taken up by cells in growth transitions and it appears that there may be little or no immediate increase in the absolute number of active enzyme molecules (58, 60).

ALTERATION OF THE STRUCTURE OF BACTERIAL RNA POLYMERASES   Under certain conditions there is good evidence for alteration of the structure and selectivity of bacterial RNA polymerase. Alterations follow infection by T4 and λ phages, and during the process of bacterial sporulation in B. subtilis; these have been studied extensively.

Alterations of the selectivity of RNA synthesis occur during lytic growth of T4 and λ phages in several stages, giving rise to different transcripts during the course of phage growth. During infection by T4 phage, there is a sequential alteration of each of the known subunits of RNA polymerase holoenzyme (61–63, for review see 64). The chemical nature of the alteration is known only for the α subunits and consists of the addition of an ADP ribose residue to the molecule (62, 63a). These modifications lead to a reduced affinity for σ subunit and its consequent loss from the bacterial enzyme during purification (39, 62, 65, 66). However, active σ subunit can be recovered from infected cells (67) and may still be functional in vivo. Several new polypeptides become tightly associated with RNA polymerase during T4 development (67). At least one of these is a protein (gene 33 protein) required in vivo for optimal synthesis of T4 late mRNA (68). While a rough correlation exists between the progression of enzyme modifications and the alteration of the T4 transcription program in the infected cell (69), there is no confirmed in vitro demonstration of an alteration (as opposed to loss) of the transcriptional selectivity due to any of the reported modifications. Reports of factors in infected cells which stimulate core polymerase to form delayed early (70) or late (71) T4 mRNA species have not been confirmed.

Two λ proteins, products of the phage N and Q genes, are required for optimal synthesis of all but the earliest species of λ mRNA. Neither protein is found complexed with RNA polymerase isolated from infected cells. However, there is indirect evidence that N protein must function through interaction with bacterial RNA polymerase (72, 73), possibly to prevent the normal reading of termination signals by RNA polymerase in concert with ρ factor (74, 75). An in vitro assay for N protein has been devised which depends on its ability to stimulate formation of mRNA for active enzyme synthesis in a coupled transcription-translation system (76, 77); hence the properties of purified N protein may soon be known.

The disappointing progress in elucidating the mechanisms of alteration of bacterial RNA polymerase selectivity in the E. coli phages has led several groups to explore these transitions in other bacteria, primarily Bacillus species. Encouraging progress has been forthcoming. B. subtilis phages SP01 and SP82 show a regulated program of transcriptional alteration during normal growth (78, 79). B. subtilis RNA polymerase transcribes predominantly early RNA from the SP01 DNA, while the enzyme purified from SP01-infected cells directs synthesis of middle RNA transcripts in vitro, as judged by asymmetry and competition hybridization. The infected cell enzyme contains two polypeptides not present in the normal RNA polymerase and, in addition, selectivity is not lost after phosphocellulose chromatography, which removes the σ subunit of the normal holoenzyme (78a). RNA polymerase extensively purified from cells infected with the related phage SP82 also lacks σ subunit, yet it has a higher specific activity with phage DNA templates than the normal holoenzyme (79). Three polypeptide components were detected in addition to the normal core polymerase subunits; the major additional polypeptide had a mol wt of 16,000 and was present in molar amounts equivalent to α subunit. The purified RNA polymerase from SP82-infected cells is able to transcribe asymmetrically classes of RNA not formed by the host cell RNA polymerase.

The two modified *B. subtilis* RNA polymerases represent the first well-documented examples of a modified bacterial enzyme able to display altered selectivity in vitro. Although the mechanism of this alteration is not yet known, it is clearly accessible to critical study and should yield important information about the selective process.

Regulation of sporulation in *B. subtilis* has been studied as a model differentiating system (9). New RNA and protein species appear during the process, suggesting that regulation may involve transcription. An early event in sporulation is the loss of sensitivity of the cells to φe phage (80); this event may be correlated with some alteration which leads to the loss of the σ subunit from the bacterial RNA polymerase when the enzyme is purified from sporulating cells (55). However, purified enzyme which has lost σ subunit is able to transcribe φe DNA if σ subunit is added to it. It is not clear whether an alteration in σ subunit, the core polymerase, or some other cellular component leads to loss of σ during purification. A new polypeptide component of unknown function becomes associated with the core polymerase during the third hour of sporulation (81).

When *B. subtilis* RNA polymerase is isolated from sporulating cells, the β' subunit of the enzyme is missing and a new 110,000 mol wt protein is present (82); the new subunit can be derived from β' by proteolytic cleavage of the subunit in the vegetative RNA polymerase with a protease present in sporulating cells (54). This alteration was thought to play a role in the regulation of transcription in the sporulation process, since some rifampicin-resistant mutants of *B. subtilis* neither sporulate nor exhibit the β' subunit cleavage (83). However, β' subunit in RNA polymerase isolated from sporulating cells in the presence of an excess of vegetative cells is not initially altered; during purification there is progressive cleavage to the 110,000 mol wt fragment (55). Thus the cleavage of β' subunit appears to be an in vitro process which leaves the functional role of such an alteration in vivo in great doubt. At this time, with the exception of the enhanced loss of σ by sporulating enzyme, there is no clear correlation between alterations in the bacterial RNA polymerase and alterations in transcriptional specificity during sporulation.

Stimulatory Factors of Unknown Function

The observation that σ subunit could be lost from RNA polymerase holoenzyme during purification suggested that other subunits might have been lost during purification. Alternatively, other factors homologous to σ were postulated to exist in cells undergoing growth transitions or differentiation during phage infections. Many workers have sought to explore these possibilities by looking for factors that could enhance or depress the activity of core polymerase or RNA polymerase holo-enzyme. Studies with core polymerase have proved technically difficult and un-rewarding to date. A variety of components in normal and phage-infected cells stimulate core polymerase (84), and in all cases that have been carefully studied these components appear to have nucleolytic activity or cannot be distinguished from the bacterial σ subunit. Core polymerase is dependent on the presence of single-strand interruptions for activity with most DNA templates (85–87).

A large variety of factors have been isolated which stimulate RNA synthesis by RNA polymerase holoenzyme (88–93a). Some of these components are highly purified. The mechanism by which activity is enhanced is not yet known for any of these components, nor has alteration in transcriptional specificity been shown to occur.

The most notable success with this approach has been the isolation of the termination factor ρ (74), which depresses the amount of RNA formed in the in vitro reaction. The ρ factor meets at least two of the criteria for the participation of a component in the transcription system; this evidence is discussed below.

Accessory Components of the Transcription Apparatus

In a variety of instances, accessory components have been identified which meet all criteria for participation in selective transcription of some transcription units but are not subunits of bacterial RNA polymerase and probably do not interact directly with the enzyme. These include: 1. Specific repressor proteins—a large variety of repressor genes are known from genetic and physiological studies (6). The *lac i* gene (94), *galR* gene (95), and λ phage C_1 (96) repressor proteins have been purified extensively and shown to repress specific *lac* (97), *gal* (95), and λ (98, 99) mRNA synthesis, respectively, in vitro. 2. Activator proteins—a good deal of genetic evidence exists which demonstrates that transcription of some operons is obligatorily linked to the presence of an activator protein. The first documented case was the *araC* gene (100) required for synthesis of mRNA for the *ara* operon (101). An analogous case is the gene for the catabolite activator protein (CAP) required for transcription of the *lac* and *gal* operons. Both *araC* and CAP proteins have been purified (102–104) and function to enhance specific transcription in vitro (101, 105–107). The CAP protein has been more extensively studied (103–108). 3. Phage-specific RNA polymerases—while most phage genes implicated in the alteration of transcriptional selectivity seem to function through alteration of the specificity of the bacterial RNA polymerase, T7 and T3 phages induce new phage-specific RNA polymerases (109–111).

The phage RNA polymerases have been purified to homogeneity and consist of single protein components of mol wt 100,000–110,000. The purified enzymes are highly selective for transcription of the homologous phage DNA template (109–112). The phage polymerases show a high degree of selectivity in vitro both in RNA chain initiation and termination (109, 113, 114).

In summary, the transcription apparatus in *E. coli* consists of RNA polymerase holoenzyme for nearly all known transcription units. For some transcription units, the holoenzyme appears to be completely sufficient to catalyze the initiation, growth, and termination of specific RNA chains. For other transcription units, accessory components are required for selective transcription or for regulation of transcription; in some instances these factors are well characterized. In the case of virulent phages T3 and T7, late transcription is mediated by an entirely new transcriptional apparatus.

THE TRANSCRIPTION SEQUENCE: OUTLINE OF THE STEPS IN THE IN VITRO REACTION

The DNA-directed synthesis of RNA by RNA polymerase may be broken down into a number of substeps (3, 8, 45). Each of these steps is itself complex. For the reading of transcription units requiring only RNA polymerase holoenzyme for selective transcription, the commonly accepted steps are: 1. Template binding, in which the RNA polymerase holoenzyme attaches to the DNA template, locates a specific site at which chain initiation can occur, and assumes an active conformation. 2. RNA chain initiation, in which the enzyme catalyzes the coupling of ATP or GTP with a second ribonucleoside triphosphate to eliminate inorganic pyrophosphate and generate a dinucleoside tetraphosphate of the structure pppPupX; this moiety remains tightly bound to the RNA polymerase DNA complex. 3. RNA chain elongation, in which successive nucleoside monophosphate residues are added from substrate nucleoside triphosphates to the initial dinucleoside tetraphosphate at its 3'-OH terminus to elongate the nascent RNA chain. 4. RNA chain termination and enzyme release, in which the newly formed RNA chain and RNA polymerase are released from the template DNA.

The in vitro reaction may consist of many cycles of this basic reaction sequence or only the first three steps, depending upon the nature of the transcription units on the DNA template, its conformation, the length of time synthesis proceeds, and reaction conditions such as ionic strength, temperatures, and so forth. The complexity of the overall reaction makes it essentially impossible to determine the effect of various factors (alteration of assay conditions, stimulatory factors, drugs, etc) on in vitro transcription simply by measuring the total amount of RNA formed in a given period of time, as in the standard RNA polymerase assay (3, 8, 45). To obtain an unambiguous result, one must not only follow each step of the reaction, but also know the sizes and properties of all the transcription units on the DNA template. This is a formidable undertaking, especially with very complex DNA templates.

In considering the selectivity of transcription, we are concerned primarily with steps 1 and 4, that is with template binding and RNA chain termination. It has been shown convincingly that the selection of a specific site for RNA chain initiation occurs during the initial step of the reaction in vitro and does not require the presence of the ribonucleoside triphosphates (26, 116, 117). Similarly, the correct termination of a growing RNA chain is probably mediated by the interaction of a specific site on the DNA template with RNA polymerase or with an accessory component of the transcription system such as the ρ factor.

Several points of nomenclature need to be clarified at this point. I refer to the region on a DNA template that governs the correct initiation of RNA chains as the promoter region. This region may contain a variety of discrete functions; in some cases these are physically separate from each other. Among other functions, one must include RNA polymerase recognition site(s), RNA chain initiation sites (initiators or start sites), sites for binding of positive control proteins (activators), and sites (operators) for the binding of negative control proteins (repressors). While the term promoter originally meant simply a "site at which transcription is initiated" (5), it seems appropriate to extend the term to cover the entire region at the beginning of a transcription unit which promotes and regulates the initiation of specific RNA transcripts. Similar considerations hold for terminator regions (5, 6); these too may be functionally and spatially complex units.

MEASUREMENT OF SELECTIVE TRANSCRIPTION

The Technical Problem

The study of selective transcription depends on methods to accurately measure the process. Ideally, for a collection of RNA species transcribed from a particular DNA template, one would like to know the number of discrete RNA species present, the relative amounts of each, and the location of sites for initiation and termination of the transcripts on the chromosome. The molar yield of RNA chains should be known relative to the number of molecules of RNA polymerase and the DNA template present in the reaction. This information would allow one to clearly define changes in specificity brought about by alterations in the reaction conditions, the transcription apparatus, or the DNA template. In addition, it would provide a direct measurement of the number of active promoter and terminator regions on the chromosome, thus defining a partial transcription map for that chromosome. In no instance has this ideal been achieved and very little hard information is available on the parameters important for selective initiation. While those involved in the study of translation can follow the appearance of specific proteins by measuring enzymatic activity, measurement of discrete RNA species is far more difficult. Most quantitative procedures for following the selectivity of transcription, such as DNA-RNA hybridization, measure only the mass of RNA transcribed from a particular genetic region, and alterations in this parameter usually reveal little about the steps at which selectivity is imposed or lost. More specific procedures, such as direct sequencing of transcripts, are far more difficult to quantitate, especially for routine measurements. Finally, the fraction of RNA polymerase molecules involved actively in the synthesis of a given transcript is seldom measured, yet without this information it is possible that the transcription events observed are relatively improbable and mask the true reactions of major importance.

These technical difficulties serve to emphasize that, to a considerable extent, progress in understanding the mechanism of selective transcription and its regulation has been limited by the sensitivity, specificity, and precision of techniques for measuring the process.

The second source of difficulty in the study of selective transcription lies in the nature of the DNA template itself. The Watson–Crick model for the secondary structure of DNA is pleasing and deceptively simple. However, at the level of transcription even the simplest phage DNA molecule is a highly complex apparatus and merits the appellation chromosome. Contained in each DNA nucleotide sequence are specific promoter and terminator regions representing a multitude of regulatory sites. The arrangement of these sites on the chromosome and the manner in which they respond to the transcription apparatus governs the transcription of the chromosome and hence the regulated growth of the organism. Thus transcription of DNA is an in vitro playback of the growth program for that particular organism. The difference between different organisms at this level is profound. For example, what is now known of the structures of the transcription maps and their regulation for the coliphages reveals a bewildering diversity of design. A consequence of this transcriptional complexity of chromosomes is that each DNA template, taken with its transcriptional apparatus, must be treated as a unique system. There is probably no such thing as a typical or standard DNA template. There is no common set of components or conditions which will allow complete, selective transcription of all DNA templates in vitro. Finally, the yet unknown general principles governing the arrangement of transcription units on the chromosome are likely to be as deeply rooted in the biology of the organism as in the physical or biochemical requirements of the transcription system.

In a practical sense, measurement of selectivity cannot be separated from measurement of the number of promoter and terminator regions or construction of a transcription map. For example, techniques which measure asymmetry in vitro depend on knowing that the transcription map is asymmetric. To review the limitations and developments of different approaches, I somewhat arbitrarily divide these problems as shown by the headings below. However, the reader should keep the circularity of the problem well in mind.

Measurement of Selective Transcription in vitro

In most instances transcription of even simple chromosomes in vitro leads to a mixture of several RNA species. Identification of individual species in such a mixture is difficult. In the past few years the development of gel electrophoresis in polyacrylamide gels has provided a procedure of great resolving power for separating RNA species on the basis of size and charge (118–122). The procedure not only allows the isolation of individual transcripts for study by hybridization or sequencing techniques, but in some instances individual transcripts can also be directly identified with a given genetic region by using a series of templates containing known genetic deletions (123–125) and simply looking for the corresponding alteration of individual transcripts. The simplicity and quantitative potential of the analytical gel procedure makes it the method of choice in the initial study of any new transcription system.

Two technical problems should be noted which affect the interpretation of gel electropherograms. The first affects nearly all procedures used in the study of RNA transcripts, namely, some transcription units may be quite long and others quite short (114, 125, 126). When RNA species are labeled with nucleoside monophosphates, the amount of label in a small transcript may be less than a few percent of that in larger species even though equal numbers of molecules are formed. This problem can be overcome by the use of several kinds of terminal labeling procedures (114, 123, 127, 128). A second problem is that because of the excellent resolution obtained on gels, it is easy to mistake a sharp band for a major component when a large background of random chains remains undetected. It is thus essential that the molar yields of individual bands be monitored relative to the amount of input RNA.

The earliest demonstrations of selective transcription in vivo and in vitro used DNA-RNA hybridization to demonstrate that RNA transcripts anneal preferably to one of the two DNA strands for a given genetic region (129–132). The technique of DNA-RNA hybridization has continued to be a major tool for studying the origins of RNA transcripts; a large variety of techniques based on hybridization have been devised. The general aspects of these procedures and their limitations have been reviewed by others (133–135a) and are not considered in detail here.

Since many phages use only one strand of the chromosome (129, 131, 132, 136) or restrict transcription of different strands to different temporal classes of RNA (137–139), selectivity is often monitored simply by following the asymmetry of transcript annealing to the separate DNA strands. Procedures have been described that allow separation of the DNA strands for a large variety of bacterial and phage species (140). Lack of asymmetry, which can reflect lack of selectivity, can also be measured directly by taking advantage of the ability of symmetric transcripts to form an RNase-resistant RNA duplex (141). However, not all symmetric transcription is nonselective (141a).

When specific genes (142, 143) or DNA fragments (144, 145) can be isolated, it is possible to follow transcription of these regions in a DNA template by direct hybridization (146). The isolation of sequence-specific restriction nucleases (147a–147d) has provided a source of unique DNA fragments, which should prove extremely useful in the study of selective transcription (144–146b). Isolated fragments will not only allow direct hybridization procedures to be employed, but also facilitate the location of promoter, operator, and terminator regions (144).

More generally, where a particular sequence is present on one DNA molecule but not on another, otherwise isogenic, DNA, transcription of that region can be followed by differential hybridization. This has been extensively used with phage deletions and with transducing phages (135). While the method has been very valuable, it has certain drawbacks. The two-stage hybridization procedure is laborious and there are always traces of RNA remaining from the phage genome, hence the final values are usually obtained by subtracting substantial blanks. Since only the genetic region of interest is normally followed, this technique does not usually distinguish between loss of selectivity in a system and an alteration of selectivity in which the phage genes are preferentially read.

Under certain conditions DNA-RNA hybrids can be visualized directly with the electron microscope. This technique can give direct information on the genetic origin of a transcript and may also allow one to position the promoter and terminator regions on the chromosome (148, 149). Direct mapping is potentially one of the strongest techniques for following selectivity and deserves a great deal more attention. In practice, it is difficult to determine accurately the points at which hybrid duplex begins and ends. In addition, the method has not been used enough for one to have confidence in its ability to detect minor transcripts and distinguish them from artifacts.

A common approach to the measurement of the selectivity of in vitro transcription depends on the ability of in vivo transcripts to compete with identical in vitro transcripts for hybridization to complementary DNA sequences (competition hybridization). In principle the method is excellent, since it involves a direct comparison between in vivo species and in vitro products, and until recently much of our information about the selectivity of the in vitro reaction has depended on this procedure. While the limitations of the method have been discussed in detail elsewhere (134, 135, 150, 151), several should be noted here. First, the sensitivity of the method is not high; it is quite difficult to detect a small amount of a selective transcript against a large background of random transcription. This can be ameliorated where the exact genetic region of interest can be isolated either physically (142–144) or by inclusion on a transducing phage (135), but this problem lies at the heart of a number of disputed issues (70, 71, 91, 143). A second limitation is that the RNA isolated in vivo may not contain significant amounts of all the transcripts actually read in vivo. Some transcripts or portions thereof may be broken down exceedingly rapidly in the cell (152, 153) or produced only in trace amounts.

Where genetic and physiological data indicate that transcription of a region depends on or is repressed by an accessory component of the transcription system, these proteins may be used in vitro to monitor selectivity (97–99).

The properties of the binary complex formed between RNA polymerase and a promoter region differ from those formed by core polymerase, for example, at some incorrect initiation sites. The former complexes are less sensitive to inactivation by certain inhibitors such as rifampicin (112, 154, 155), heparin (26), or poly [rI] (116), and these inhibitors are frequently used as rough indicators of selectivity in the in vitro system.

Where individual RNA transcripts can be obtained, the most definitive method of determining selectivity is by sequence determination. Procedures for determining rather large portions of the 5′ terminus of RNA replicates and transcripts are now well developed (156, 157). The application of these techniques to the study of transcription of λ phage has already led to precise information as to the number of promoter and terminator regions on the λ chromosome which are read by E. coli RNA polymerase (126) and provides a direct method of measuring the effect of different factors on the loss and alteration of selectivity for this template. Once the sequence of an RNA is known, it can be determined quantitatively with little effort, even in mixtures, by looking for characteristic oligonucleotides after RNase cleavage and fingerprinting (158).

The difficulty of resolving and sequencing all the transcripts for a template has led many workers to use techniques that depend on partial sequencing. It is less difficult to measure the formation of different triphosphate-containing 5′-terminal oligonucleotides after cleavage with T1 or pancreatic RNAse (159, 160). The number of such termini may reveal selectivity or its loss, although this is a difficult procedure to quantitate. Some information may also come simply from the determination of the ratios of incorporation of ATP to GTP into the 5′- terminal positions of transcripts. For example, core polymerase appears to initiate chains primarily with GTP (112, 161, 162).

The genetic content of an RNA transcript can be monitored sensitively by measuring its activity in an in vitro protein synthesizing system (76, 101, 163–168). This is most efficient when translation and transcription are coupled (76, 101, 163, 164); however, isolated transcripts are also active (165–168). The relationship between the amount of a specific mRNA present and the amount or activity of any given protein synthesized is usually not known; hence, the presence of a small amount of a rapidly translated contaminant or trace synthesis of a very active enzyme can produce misleading results (168). However, this procedure for measuring the presence of a specific transcript is not highly sensitive to interference from the presence of other transcripts.

The genetic content of an RNA can also be determined by using it to protect a complementary DNA sequence from nuclease attack. The DNA then can be analyzed by its ability to transform a competent cell (169).

Enumeration of Promoter and Terminator Regions

A variety of procedures have been employed to measure the number of promoter sites on procaryotic chromosomes. Characterization and positioning of both kinds of sites first came from genetic and physiological data. These kinds of approaches are valuable in defining in vivo transcription units and provide an independent approach to the problem of determining what components of the transcription system are required for selective transcription (5, 6). In vitro the most rigorous approach to the problem again involves resolution, sequencing, and mapping of all transcripts, and this approach has already borne fruit in some cases (126). In general, with less rigorous methods, it is essential to use several different techniques and compare the results.

Ternary complexes containing nascent RNA chains attached to RNA polymerase-DNA complexes have been visualized using the electron microscope (148, 170, 171), and this may allow a rough enumeration and positioning of the promoter and terminator regions. The method can also be applied to the study of binary RNA polymerase-DNA complexes (172). This can give a measurement of the number of sites on the DNA where a highly stable complex can form; promoter regions are one such class of sites (116), but discontinuities in DNA such as ends (172) and breaks (86, 173, 174) also bind the enzyme tightly.

The number of tight binding sites on a DNA template can also be measured directly by use of the filter binding assay (116). More recently another procedure has been devised in which the number of regions on a DNA where highly stable complexes can form is measured using restriction nuclease fragments of the chromosome (144). Finally, the number of such sites can be determined by directly isolating DNA regions protected from nuclease cleavage by RNA polymerase (19).

RNA polymerase holoenzyme bound at a promoter region in an activated form (see section on functional interaction of RNA polymerase with promoters) is able to initiate an RNA chain so rapidly that addition of nucleoside triphosphates with the initiation inhibitor rifampicin results in little inactivation of such holoenzyme molecules until high concentrations of the drug are used (175, 176). The exact relationship is complicated by the fact that 1. some holoenzyme is inactivated in such an experiment even at modest drug concentrations, and 2. at lower concentrations of the drug, several RNA polymerase holoenzyme molecules can use the same promoter region/start site before inactivation occurs. By measurement of the number of RNA chains initiated by $[\gamma^{32}P]$ ATP and $[\gamma^{32}P]$ GTP at a variety of drug and enzyme concentrations under different reaction conditions, the number of independent, rapidly initiated chains can be calculated (175). Initial applications of this technique were clouded by the incorrect assumption that binary complexes of RNA polymerase at the promoter were truly resistant to attack by rifampicin (7–9, 155). This method will give a minimum number of independent start sites for a given DNA molecule. Under any given set of conditions, some promoter regions may be poorly used (for example, if they have a high transition temperature). Alternatively, two discrete start sites may be close enough together to preclude independent function.

An analogous method employs heparin, a potent inhibitor of free RNA polymerase or RNA polymerase bound nonspecifically to DNA (26). Holoenzyme bound in activated complex at a promoter region is attacked by heparin at a greatly reduced rate (26). However, RNA polymerase molecules bound near the promoter but not at an initiator site may also be less sensitive to heparin, hence the number of RNA chains initiated in the presence of heparin may be some multiple of the true number of promoter regions or initiator sites.

Actinomycin D blocks RNA chain elongation and has also been used to preclude repeated initiations at start sites and allow their enumeration (159, 159a). Ideally the number of RNA chains initiated by an excess of enzyme in the presence of drug should equal the number of independent start sites. However the method can give either an underestimate or overestimate of the true number. Some start sites may contain an actinomycin binding site which will block initiation. Alternatively the first actinomycin binding site may be far enough from a start site to permit initiation of more than one chain. The number of start sites found for fd RF DNA by this method agrees with estimates obtained with other techniques (19, 159). However only one start site is found per T7 DNA molecule in the presence of actinomycin (159a), while other methods show three or four independent sites (123, 128, 175).

The rate of RNA chain initiation under optimal conditions is directly proportional to the concentration of the initial nucleoside triphosphate (ATP or GTP) (176). Consequently, reduction of the ATP or GTP concentration to a very low level can suppress de novo chain initiation while permitting RNA chain growth. The K_s values for RNA chain elongation are quite low compared to the nucleotide concentrations needed for a substantial value of RNA chain initiation (177, 178). The de novo initiation of RNA chains can also be blocked with rifampicin. Under these conditions introduction of a dinucleotide or oligonucleotide into the system provides a primer and bypasses the initiation step of synthesis (179–181). Different dinucleotides seem to serve as primers for RNA polymerase molecules bound at different promoter regions on DNA (123, 180, 182, 183). This presumably depends on the ability of the dinucleotide to form base pairs with the exact sequence of nucleotides at the start site opened in the binary complex (180, 183). By adding a particular primer to the reaction, only one or a limited number of transcription units can be read, which allows selection of the transcriptional specificity of the system. This procedure has been used to direct selective initiation with a variety of DNA templates and extensive studies have been carried out with *lac* DNA and T7 DNA (182, 183); the nature of the dinucleotides used and the sizes of the resulting RNAs can give information about the number of start sites and their nucleotide sequence. However, it has not been adequately shown that all start sites employed in the dinucleotide-primed reaction correspond to sites used with nucleoside triphosphates. There is some discrepancy between results obtained with dinucleotides and other methods for the early region of T7 DNA in particular (123, 128).

The problem of directly measuring terminator regions has seen little study. Several terminator sites are known on phage DNAs which are read by *E. coli* RNA polymerase in the absence of added factors (125, 126, 184, 185). The effectiveness of these sites normally has been monitored by following the size of the RNA formed in the reaction; however, direct biochemical assays for the termination reaction are badly needed. Termination by ρ factor has normally been studied by following inhibition of RNA synthesis with an appropriate DNA template, although the ability of ρ to block chain elongation also allows one to assay for loss of mRNA sequences distal to a ρ terminator region (186).

Phage Transcription Maps

For several of the smaller coliphages enough is known about selective transcription in vivo and in vitro to justify drawing preliminary transcription maps. A map for fd RF DNA has appeared elsewhere (9, 185) and data for ϕx174 should shortly permit a complete transcription map to be constructed (144). In some respects the picture for λ phage appears more complete than that of T7 phages since it is based on a combination of RNA sequencing techniques with a very large amount of genetic information about promoter and terminator regions on this phage. However T7 phage, despite the handicaps in its study imposed by its virulence, is probably a simpler and more primitive phage and the regulatory system it uses is currently more accessible to biochemical study.

T7 PHAGE   T7 is a virulent phage which contains a linear DNA molecule of unique, nonpermuted nucleotide sequence. The transcription map of T7 has been extensively studied (for reviews, see 187, 188). The transcription map shown here (Figure 1) is a composite drawn from work in several laboratories. Only a single set of promoter sites is shown at the leftmost end of the early region, although minor internal initiation sites are suggested by several groups (123, 171, 191). Distances on the map are given in terms of percentages of the r strand beginning at the left end of the molecule as the genetic map is written (124). The five early genes of T7 are clustered at the leftmost end of the genome (123, 124, 190). In vivo and at low enzyme to DNA ratios in vitro, a single transcription unit of 2.5×10^6 mol wt (128, 184, 191) is read

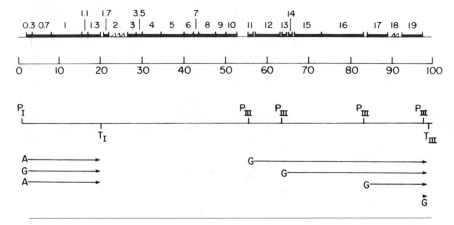

Figure 1   Partial Transcription Map of T7 Phage. T7 genes shown in the map are represented (heavy lines) by the lengths on the DNA molecule calculated from the polypeptide chain sizes (124, 189). Distances along the map are given as percent of the molecule from the left end. P_I and T_I are early promoter and terminator regions read by *E. coli* RNA polymerase (123, 124, 128, 148, 149, 184, 190); P_{III} and T_{III} are late promoter and terminator regions read by T7 RNA polymerase (114, 115). Arrows indicate the length of known transcripts read from these transcription units on the *r* strand; A or G is the initial 5′-terminal nucleotide.

from the *r* strand (136); it is controlled by three initiator sites in the promoter region at the left end of the DNA molecule (123, 128, 175). Termination requires no added components and is effected by a terminator at 20.2% (149). The T7 early mRNAs found in vivo are monocistronic (124, 128); identical RNAs can be generated by post-transcriptional cleavage in vitro on treatment with RNase III (128). The unprocessed in vitro transcript (2.5×10^6 mol wt) is also found in *E. coli* mutants defective in RNase III (192). However, it has also been reported that initiation at discrete sites in the interior of the early region can take place (123, 171). Smaller RNA transcripts are formed when elevated enzyme/DNA ratios are used (123, 191), with dinucleotides as initiators, or when ρ factor is included in the reaction (128, 148). However, at least one group finds no correspondence between the sizes of the in vivo RNAs and the transcripts formed in the presence of ρ (128).

There is at least one RNA chain initiated in vitro by *E. coli* RNA polymerase which begins with GTP and is complementary to the *l* strand (123, 175). No corresponding RNA transcript has yet been identified in vivo, nor has the map position of the transcript been located.

The late genes of T7 are transcribed by a phage-specific RNA polymerase; the RNA formed in vivo and in vitro is complementary solely to the *r* strand (109, 193). The T7 RNA polymerase is highly selective in vitro; seven major transcripts are formed, designated species I, II, IIIA, IIIB, and IV–VI; these range in apparent molecular weight (determined by gel electrophoresis) from 5.5×10^6 to 0.2×10^6, respectively (114). Four of these transcripts (I, II, IIIB, and VI, mol wt 5.5, 4.5, 2, and 0.2×10^6, respectively) appear to be read from sequential promoter regions on the T7 chromosome which share a common terminator just inside the right end of the molecule (115). The remaining major in vitro transcripts may well be derived from the region from 40–55% on the chromosome, leaving a substantial part of the late region (20–40%), which is transcribed very poorly in vitro. The latter region may be governed by intrinsically weak promoters or there may be some mechanism for activation of this region at the appropriate time in vivo. The late RNAs extracted from infected cells are much smaller than the large in vitro transcripts (114, 194), which suggests that these RNAs too are subject to processing.

LAMBDA PHAGE   The transcription map of λ phage (Figure 2) is considerably more complex than that of T7 yet is certainly the most thoroughly characterized chromosome now known. Maps containing genetic and physical information and the positions of the known regulatory sites have been prepared by several reviewers (195–198). In addition to the biochemical characterization of the λ transcripts formed in vitro (126) there is extensive genetic and physiological data about λ

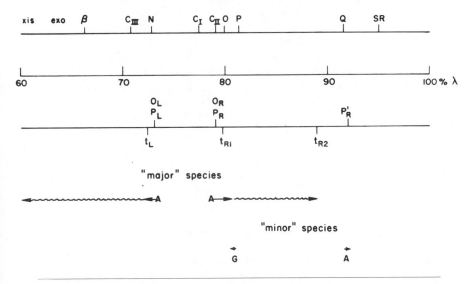

Figure 2   Partial Transcription Map of λ Phage. The genetic map shows the rightmost 40% of the λ phage chromosome. Gene positions and positions of promoters and terminators taken from the review by Davidson & Szybalski (196). Distances along the map (note shortening) are measured in percent from the left end of the molecule. P_L and t_L define the major leftward (N gene) transcript, P_R and t_{R1} the major rightward (*cro* or *tof* gene) transcript. Arrows indicate the approximate lengths of the major transcripts to scale, wavy lines are sequences transcribed in presence of N protein. Locations of the minor transcripts are approximate (126, 197, 199) and lengths are not drawn to scale. A or G at the end of the RNAs indicates the 5′-terminal nucleotide in vivo and in vitro.

transcription and its regulation (195, 198, 200). There are four large transcription units on the λ chromosome if one neglects transcription from the b_2 region (201). In vivo, λ genes are transcribed from both strands of λ DNA: from the l strand, genes c_I and rex (governed by two promoters between c_{II} and c_{III}) and early genes N through int, (governed by P_L). The other group of early genes, cro through Q, and the late genes (S, R, and A through J) are transcribed from the r strand (governed by promoters P_R and $P_{R'}$, respectively). Two short ("minor") transcription units first detected as in vitro transcripts are also found in vivo but are not yet identified with any known λ genes. Four transcription units are read in vitro by RNA polymerase holoenzyme (125, 126). The remaining units require accessory factors to be read in vivo (202, 203). Extensive nucleotide sequences are known for all four transcription units read in vitro (125, 126, 158). Two of the major transcription units (major leftward; governed by P_L, located at about 73.1% and major rightward, governed by P_R, located at about 79.1%) are under repressor-operator control (195, 198, 200). Neither operator nor promoter mutations in these units lead to alterations in the 5′ sequence of the transcripts (126). It is concluded that neither promoter nor operator regions are transcribed for these units. It was originally suggested that the sites of promoter mutations for leftward transcription were a considerable distance from the initiator site (197); more recent evidence (199) indicates that this distance is on the order of 50 nucleotides or less.

The so-called minor rightward and minor leftward RNAs (198 and 81 nucleotides, respectively) read by RNA polymerase in vitro (125, 126) are also synthesized in vivo (203a, b). At least the former species does not appear to be under control of the λ repressor (c_I gene product) (203b). The designation as minor species refers to their molecular weights and is confusing; the number of these chains formed is comparable to those of the major transcripts (158), hence these are governed by strong promoter regions. Neither transcript has been mapped precisely on the λ chromosome; the minor rightward RNA appears to map between Q and S but does not contain Q or S genetic sequences (199). Its position corresponds closely to that defined for $P_{R'}$ (198), and it may be an initiator RNA for transcription of the late λ genes. The minor leftward RNA originates near the ori site (about 80%) and may be involved in initiation of λ DNA replication (126, 203a, b).

Both minor leftward and minor rightward transcription units are governed by strong terminator regions read directly by RNA polymerase (125, 126, 158). In contrast, the major rightward and major leftward transcripts are continued in vitro past the in vivo termination sites at t_L (72.4%) and t_R (79.8%) unless ρ factor is added (186). In the presence of ρ discrete transcripts are formed. In vivo and in vitro in the presence of ρ factor, the λ N protein seems to be needed for transcription to be extended past the ρ termination regions (76, 77, 198, 200).

Conditions Affecting Selectivity of in vitro Transcription

Altering conditions used for in vitro transcription or the components of the reaction can affect the selectivity of transcription. Two kinds of alterations will be distinguished: 1. Those for which biologically incorrect transcription is enhanced or suppressed; that is, initiation or termination events not duplicated in the cell; and

2. Those for which the selectivity of transcription is altered by enhancing or suppressing some classes of transcripts relative to others, but where both classes are correct. The first of these situations is referred to as nonselective transcription or loss of selectivity, the second as alteration of selectivity.

With the exception of the direct sequencing procedures and to some extent, direct visualization of hybrid molecules, it is difficult to determine whether preferential synthesis of a given region involves selective initiation, selective termination, or both. Similarly, loss of selective synthesis may be due to alteration of either initiation or termination events. Even with direct sequencing techniques it is difficult to estimate the true frequency of correct initiation as compared to incorrect initiation. Consequently, this very basic information is not yet available for in vitro transcription with any template even under optimal conditions. Thus in discussing the effect of different factors on selectivity, it must be noted a priori that little concrete information is really known.

The most striking alterations in selectivity come from changes in RNA polymerase or the DNA template. Removal of σ subunit to give core polymerase has a decided effect on the ability of the enzyme to transcribe selectively, due to alteration of the initiation specificity. Core polymerase is greatly reduced in its ability to transcribe certain DNA templates (21, 32) and selective transcription measured by hybridization (86, 105, 106, 161), competition hybridization (65, 204), or end-group sequencing (160) procedures is reduced or lost. The primary initiation sites for core polymerase on helical DNA are single- or double-strand breaks (85, 86).

However, the possibility that core polymerase carries out a low level or rate of selective initiation is not ruled out. Few of the techniques used to follow selectivity would detect 10% selective initiation and in fact in many instances where asymmetric transcription is followed, core polymerase does give what appears to be partial selectivity (86, 161). In at least one instance, that of ϕX174 phage RFI, there is definite asymmetry of the RNA formed by core polymerase from *Azobacter vinelandii* (205). Selectivity by core polymerase could result from a partial recognition and utilization of promoter regions (26) or from a trace contamination of the core polymerase by σ subunit, which is not quantitatively removed even by repeated phosphocellulose chromatography (32). In the case of ϕX RFI cited above, it is possible that the supercoiling of this template may serve to bypass the strand separation function of σ subunit and provide an open region or regions in an otherwise intact helical template (206, 207). In view of the evidence that open regions in supercoiled DNAs can be fixed at unique sites on the molecule (208–209), the possibility of a high degree of site selectivity by core polymerase is not unreasonable.

Alteration of the DNA template structure can also lead to loss in the selectivity of RNA chain initiation. This is primarily true when the DNA is completely denatured. Denatured DNA can be transcribed almost completely to produce a DNA-RNA hybrid (210). However, here too some selectivity is suggested, since relatively little late T4 mRNA is made even with denatured T4 DNA (211). The molecular weights of the RNAs synthesized with denatured templates are much lower than those found with helical DNAs (127, 212); hence, termination too is altered with single-stranded templates (45).

Introduction of single-strand breaks (86, 204) or other lesions (213, 214) into DNA has relatively little effect on the selectivity of RNA synthesis by RNA polymerase holoenzyme, although these lesions serve as tight binding sites for the enzyme. This suggests that σ subunit may constrain the enzyme to carry out the initiation reaction only at a given nucleotide sequence.

While considering the specificity of synthesis with denatured templates, it may be important to note that selective synthesis can result from random initiation coupled with selective termination. If transcriptional termination regions occur frequently in sequences on the noncoding DNA strand, transcription of this strand would be reduced except where excess enzyme was present.

Altering the ionic composition of the reaction has a profound effect on the synthesis of RNA in that medium. Elevated ionic strengths alter the binding of RNA polymerase to DNA (116, 215, 216), the transition into an active complex at the promoter (176, 205, 217), the rate of RNA chain initiation (176), the rate of RNA chain elongation (178, 218), and the termination of RNA chains (217–220). The synthesis of RNA from different templates and with RNA polymerases from different bacteria (205, 221–224) shows great differences in sensitivity to ionic strength. This probably reflects differences among different DNAs and RNA polymerases in the transition into the open promoter complex (see below), although this has not been shown directly. These results almost certainly mean that altering the ionic strength can alter the transcriptional selectivity to favor one promoter region as opposed to another. λ Transcription is highly selective, yet greatly reduced at salt concentrations optimal for selective T4 transcription (224). A switch in promoter selection induced by altered ionic conditions on a single DNA has not yet been clearly shown.

It has been suggested that at low ionic strengths nonspecific binding of RNA polymerase to DNA is enhanced and hence nonspecific initiation is found. At very high enzyme to DNA ratios there may be some loss in enzyme activity at low ionic strengths (216). However, I know of no good evidence that under normal conditions there is preferential reduction of nonspecific binding by holoenzyme at elevated ionic strengths. The number of RNA polymerase molecules bound per DNA is reduced at elevated ionic strength when transport procedures are used and may approach the number of promoter regions on the DNA (215, 216). However, this reflects a general reduction in binding affinity for both specific and nonspecific sites and the total number of both kinds of sites is probably not altered (45).

In fact, it seems quite likely that elevated ionic strengths depress the true binding constants for specific promoter regions relative to nonspecific regions. Nonspecific complexes do not involve opening of DNA base pairs (116, 225). If this were true for both kinds of complexes, the relative change in electrostatic free energy with ionic strength (probably the major source of alterations in binding constant) would probably be similar for both since the geometry of the protein interactions with the phosphate residues would be similar. This result is found for binding of *lac* repressor to operator and nonspecific DNA regions (226). In fact, the specific RNA polymerase complexes involve opening of base pairs, which requires more energy at elevated salt concentrations and is disfavored at least in the case of T7 DNA

(176). In addition, the positioning of the enzyme between the two DNA strands is likely to increase the favorable electrostatic interactions in the complex (between basic enzyme groups and the DNA phosphates) which would be opposed at high counterion concentrations. Other lines of evidence also weaken the notion that elevated ionic strengths favor selective initiation by reducing nonspecific binding. Tight binding of RNA polymerase holoenzyme to DNA, for example at single-strand breaks, is not sufficient to lead to chain initiation (86, 204, 213, 214). Thus the ability of low salt concentrations to permit slightly enhanced binding at nonspecific sites on an intact template seems unlikely to favor random initiation. However, the activity of core polymerase relative to holoenzyme is enhanced at low ionic strengths (162) and most preparations of RNA polymerase contain some core polymerase (32).

Actual measurements of selectivity at a variety of ionic strengths have not yet been carried out by sequencing or gel analysis procedures. The initial ratio of ATP to GTP initiation events with T7 DNA is not varied between 0 and 0.2 M KCl (175), nor is asymmetry of hybridization under similar conditions (220, 227). However, several workers have reported a reduction in asymmetry of T7 transcription at lower ionic strengths which may be correlated with high molar ratios of enzyme to DNA (224, 228). Variation of ionic strength has little effect on the hybridization specificity of T4 RNA or on the amount of antimessenger measured by RNA duplex formation (204). The amount of $tryp$ RNA transcribed from $\phi80_{Ptryp}$ DNA is reduced at low ionic strength, but this is possibly due to an alteration of selectivity to favor reading of phage promoters (229).

Alterations in the specificity of termination at elevated ionic strengths are somewhat better documented. The chains of T4 RNA are larger at elevated ionic strengths than at low ionic strengths (218, 220, 230), although the length of T7 RNA chains is not changed. Termination is enhanced at elevated ionic strengths (218–220) even with dAT copolymer (231) where sequence-specific events cannot be involved. Thus high ionic strengths probably favor both spontaneous dissociation of the ternary complex (incorrect termination) and possibly also more rapid dissociation at specific termination sites. Rho factor-induced termination is suppressed in some instances at elevated ionic strengths but not in others (232–234).

Alterations in the selectivity of transcription at different temperatures have not been explored extensively. The cooperative nature of the transition involved suggests that temperature may be a critical variable in the transcription of any particular promoter (see section on functional interaction of RNA polymerase with promoters).

The presence of organic solvents in the reaction during in vitro transcription can have a profound effect on the selectivity of transcription. These solvents serve as DNA denaturing agents (235) although it has not yet been shown that their effects in altering transcriptional selectivity derive solely from this property. Transcription from the leftward promoter of λ_{sex} mutant DNA is activated by the presence of glycerol or other solvents in the reaction (236). The λ_{sex} mutant contains a promoter mutation at P_L that normally blocks leftward transcription from the site in vitro and in vivo. The presence of these same solvents also eliminates the requirement for CAP protein in the transcription of gal mRNA from that operon in vitro (236).

It is attractive to assume that these solvent effects are due to alterations in the stability of the DNA template in the promoter region. Fifteen percent glycerol (2.0 M) lowers the overall melting temperature of DNA 2° (235) and might have a much larger effect on the cooperative melting of a small, AT-rich region of DNA, for example in a promoter region.

Ethylene glycol alters the nature of RNA chains initiated with T4 DNA as template (237). Shorter T4 RNA chains are made which may signal an effect of the solvent on termination. In addition, the chains formed may be more frequently initiated incorrectly as monitored by asymmetry and competition hybridization.

THE STRUCTURE OF PROMOTER AND TERMINATOR REGIONS

Evidence from Study of Intact Chromosomes

Since the locations of at least some promoter and terminator regions are known for several phage chromosomes, it is possible to compare these transcription maps with other maps of unique DNA regions. For example, the separated strands of phage and bacterial DNAs contain sites that bind poly[rG] or poly[rUG] (140). These sites predominate on the DNA strand most extensively transcribed (136–139), and hence might be loci associated with transcriptional regulation sites (238). A careful analysis of the location of the poly G binding sites on λ DNA indicates that they are not correlated with promoter, operator, or normal terminator regions but may correspond to the sites on the genome where mRNAs are divided to smaller sizes (239).

This division might in principle occur during transcription at a site that could direct RNA chain release but where RNA polymerase could reinitiate or continue to elongate a fragment of the original chain (148, 239). However, it is more likely that these poly G binding sites are transcribed into mRNA and serve as recognition regions for cleavage by processing nucleases. *E. coli* and T4 tRNAs are transcribed as larger precursors and are subsequently processed by cleavage to yield the final tRNA (240–242). Very large mRNAs have been identified in λ infected cells (243). *E. coli* mutants defective in ribonuclease III accumulate ribosomal precursor RNAs (192, 242a). When mutants are infected by phage T7, a single mRNA derived from the entire early genetic region is found in place of the five monocistronic RNAs obtained in normal cells (192). A corresponding T7 transcript formed in vitro, is processed to RNA fragments identical in size and map position by ribonuclease III (128).

A second kind of physical mapping of phage chromosomes has been carried out by partially denaturing the DNA in the presence of formaldehyde. The opened loops, which correspond to regions rich in dAT residues (244), can be positioned on the chromosome to provide a physical map. In the case of phage T7 there is a remarkable correspondence between the known early and late promoter and terminator regions (Figure 1) and preferential melting regions (245); however, not all preferential melting regions correspond to known promoter or terminator regions. In the case of λ phage the correlation is not complete; there is a preferential melting region near the sites of P_L and P_R (which span the promoters prm and pre) and another region near the minor rightward promoter, but none near the minor leftward promoter (Figure 2, 244). Studies with supercoiled DNAs may also be relevant here; these molecules are cleaved by single-strand-specific nucleases; presumably because of transiently single-stranded regions (206–209). For SV40 supercoils these are unique sites on the genome (208, 209) and one of these sites is at or near the region at which *E. coli* RNA polymerase selectively initiates RNA chains (146). Again the converse correlation is not found; the major site of nucleolytic attack, which is also the binding site for the T4 gene 32 DNA melting protein, is apparently not a site where RNA polymerase initiates chains (208–209). Overall, these results suggest that promoter regions and possibly terminator regions are frequently associated with AT-rich regions of DNA templates which have relatively low melting temperature. However, not all AT-rich regions are promoter regions.

Isolation, Sequencing, and Synthesis of Promoter Regions

In the past few years several quite different approaches have been taken to elucidate the nucleotide sequences in promoter regions. One such approach has involved the isolation of these regions by extensive cleavage of the rest of the DNA molecule under conditions where RNA polymerase or a repressor is bound tightly to the region and can protect it from cleavage. There are several possible approaches to the sequencing of such a fragment (180, 183, 246–248). A second line of attack has come from the chemical synthesis of tRNA genes (249, 250). The strands of a fragment from the $tRNA_{tyr}$ gene will bind to the complementary strands of DNA from the transducing phage $\phi80_{pSuIII}$ which contains the gene for $tRNA_{tyr}$ (251). Elongation of the one complex with DNA polymerase I should give the sequence of nucleotides complementary to those in the promoter region; elongation of the second complex should provide the sequence of nucleotides in the terminator (252). The principle is, of course, not limited to fragments obtained from chemical synthesis (183). The combination of these two methods should give exact nucleotide sequences for several promoter and terminator regions in the near future.

The regions of a DNA template protected from nucleolytic attack by RNA polymerase might represent either the region of the promoter that dictates recognition by the polymerase or those at which RNA chain initiation takes place. The two sites may not be congruent; this possibility has been proposed and discussed by several groups (197, 253). However there is probably not as large a distance between promoter (recognition) and initiator elements on λ DNA as originally suggested (199). Evidence from direct binding studies with *E. coli* RNA polymerase and T7 DNA suggests that there are no more than two sites where RNA polymerase can bind tightly for each independent RNA chain initiation site, which is evidence against a large preinitiator region in this instance (166, 175). The protected regions isolated at 37° with RNA polymerase holoenzyme are probably the regions associated with RNA chain initiation (initiators). The major helical fragment obtained with fd RF can specify the initial nucleotide sequence $_{ppp}G_pU_pA$ (254) that is associated with one of the three major transcripts of the intact fd RF (159). In addition, the rate of RNA chain initiation under these conditions is exceedingly rapid and is unlikely to be preceded by translocation of the enzyme on the DNA (176).

DNA fragments protected from DNase cleavage by RNA polymerase have been isolated by a number of workers (18, 19, 144, 254–260). Earlier studies conflict somewhat with more recent results; possibly earlier enzyme preparations contained core polymerase which binds with reduced specificity. More recent results using RNA polymerase holoenzyme show many common features. Protection requires elevated temperatures in addition to holoenzyme. The sequences protected with T5, T7 (260), λ (259), and fd RF (19, 254) DNAs are AT rich (about 60–65%) and consist of helical fragments of about 40 base pairs. The fragments obtained with fd RF do not anneal to φX174 DNA (19). Hence these initiator regions may not possess a simple DNA nucleotide sequence in common with the initiator regions of other DNA templates. The pyrimidine tracts of the fd binding sites do not reveal evidence of an exceptional nucleotide sequence (19).

A priori it seemed that the nucleotide sequence at the 5′ end of RNA transcripts might incorporate the nucleotide sequence of the recognition or initiation portions of the promoter region. No common structural features are evident from studies of the 5′ termini of a number of transcripts. More important, promoter and operator mutations are not reflected in these nucleotide sequences in at least some instances (126), although in one instance the *lac* operator sequences are transcribed (183).

This approach has given more promising information in the case of the 3′ termini of RNA chains which are terminated at signals read only by RNA polymerase. These transcripts terminate with the sequence UUUUUUA in the two minor transcripts of λ (125, 126, 158). A similar 3′-terminal sequence ending with U may be found at the end of the major T7 early transcript formed in vitro (230). Thus it is possible that an extended sequence of uridine residues may be related to some portion of a terminator structure.

The 26 nucleotide sequence distal to the 3′-OH terminus of tRNA$_{tyr}$ has been determined by the synthesis-elongation technique mentioned earlier (Figure 3, 252). These sequences include the 3′-OH terminus of the tRNA$_{tyr}$ precursor (240), which is three nucleotides from the –CCA end of the tRNA and hence may well reflect the terminator sequence. However, the possibility that transcription in vivo continues well past this point cannot yet be eliminated. The possible terminator sequence does not include a poly A sequence longer than three (which would form a poly U terminus on the RNA), but includes a symmetrical sequence of six base pairs.

```
3'-G-G-T-A-G┌─T-G-A-A-A┐G-T┌─T-T-T-C-A┐G-G-G-A-C-T-T-G-A-5'
            │         ┊   ┊          │
5'-C-C-A-T-C┤A-C-T-T-T├C-A┤A-A-A-G-T├C-C-C-T-G-A-A-C-T-3'
            └─────────┘   └──────────┘
```

Figure 3   Nucleotide sequence adjoining the 3′-terminal CCA of *E. coli* tRNA$_{tyr}$. Sequences are taken from reference 252. The underlined sequence is the terminal CCA of the tRNA; dotted lines indicate a twofold symmetry region.

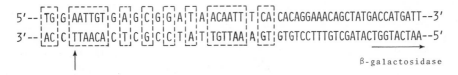

Figure 4   DNA nucleotide sequences in the *lac* promoter-operator region. Sequence taken from reference 183. Approximately the leftmost 24 nucleotides are those protected by *lac* repressor from DNase cleavage; remaining regions are determined from *lac* mRNA sequences. The underlined sequence indicates the codons corresponding to the N-terminal sequence of β-galactosidase. The vertical arrow indicates the initiation point for mRNA synthesis. Dotted regions indicate sequences with twofold symmetry.

This sequence is interesting in view of the possibility that this kind of locus may be associated with a recognition site for a protein containing identical subunits. Rho factor may be required for $tRNA_{tyr}$ termination (261) and is an oligomeric enzyme (186, 262, 263).

Binding of repressor proteins to DNA protects a variable length of duplex from cleavage. This region should encompass the operator and possibly adjacent loci. These regions have been isolated and studied for the λ (264, 265) and *lac* (183) repressors. The λ repressor protects a basic unit of 35 nucleotides at the two λ c_i-dependent operators (O_L and O_R); increases in repressor concentration lead to longer protected regions (265). It is thought that the initial interaction is with a repressor dimer and that several more dimers can then be added to the DNA at the original site. The protected regions are DNA duplexes and do not form unusual (hairpin) structures after denaturation. Partial sequencing of the protected regions O_R and O_L shows that they are not identical. The λ repressor is not a DNA melting protein but binds preferentially to helical DNA and may well stabilize the helical structure at the operator site.

The region protected by *lac* repressor has been completely sequenced along with adjacent regions and provides the first complete structural determination of a regulatory DNA sequence (183). The characterization was carried out with *lac* operator containing fragments derived by sonication from *lac* operon containing transducing phages. From such fragments, regions protected from DNase by *lac* repressor were obtained. These fragments are helical duplexes of about 27 base pairs and have been sequenced by a combination of DNA and RNA sequencing (after transcription) techniques (Figure 4). Remarkably, this entire region is transcribed in the *lac* mRNA. [These studies were done with a mutant (UV-5) in the promoter region which also does not require CAP protein (266); other evidence suggests that the wild-type operon is transcribed from a site or sites to the left of this region (183, 267)]. Several O^C mutants alter the initial sequence of the *lac* mRNA (183). The coincidence of *lac* operator and initiator regions provides an adequate explanation for *lac* repressor function; tightly bound repressor would physically exclude RNA polymerase from this site. However, the binding and action of *lac* repressor with the normal *lac* promoter region are not blocked by prior binding of RNA polymerase (97), and these sites are probably physically separate.

Twenty of 24 base pairs in the *lac* operator region are symmetrically arranged. Since the *lac* repressor is a tetramer, it is likely that this symmetry reflects the fact that binding occurs on two symmetrical sites of the protein; genetic evidence that the *lac* operator is symmetrical had previously been advanced by Sadler & Smith (268). This kind of recognition site probably represents a response to two opposing requirements: 1. The need for a large site—at least 12 nucleotide pairs are required for a unique recognition sequence in a DNA molecule the size of the *E. coli* genome (269); and 2. The need for a small site—the size of the protein required increases roughly as an exponent of the length of the linear recognition sequence. In addition there may be genetic selection against a common DNA sequence of over 12 base pairs which must be repeated at several places on the chromosome. Such a sequence might introduce a point at which intrachromosomal recombination could occur. These requirements can be met most easily by utilization of a dimer (or tetramer) with two identical binding sites and a recognition sequence that contains two symmetrical regions. In addition, if the recognition regions are spaced 10 base pairs apart, both sites will be present on the same side of the DNA helix, which is sterically pleasing.

The elucidation of the complete sequence of the *lac* operator region is an important milestone. There is no reason to suppose that a number of promoter, operator, and terminator sequences will not be accessible with careful application of similar techniques.

The actual nucleotide sequence of the region, taken with the properties of both *lac* and λ c_I repressor-protected fragments, make it likely that the recognition process involves direct identification by the repressor of an intact duplex. Recognition involving strand separation or formation of hairpin structures (270, 271) is probably ruled out. Similarly, it is unlikely that a unique helical DNA conformation (272) is responsible for recognition; there are many distinct protein repressors known and it is difficult to imagine a corresponding multiplicity of distinguishable DNA conformations. Thus repressors must recognize specific groups on the base pairs which can be read from without either in the major (273) or the minor grooves of the DNA helix. This recognition is highly specific and requires little time (274).

FUNCTIONAL INTERACTION OF RNA POLYMERASE WITH PROMOTERS

In the initial phase of RNA synthesis, RNA polymerase locates a promoter region on the DNA template and forms a productive binary complex which can initiate an RNA chain when presented with the ribonucleoside triphosphates. This step has been referred to above and elsewhere (3, 45) as the template binding step. Since several different kinds of interaction take place between RNA polymerase and the template during selective transcription, it would be more accurate to refer to this step as promoter site selection and activation. This selection process has also been referred to by the general term initiation (6), which is entirely appropriate for a discussion of the in vivo process, since productive complex formation then leads immediately to RNA chain initiation. However in vitro the steps in site selection are separable from the true initiation event in which the first phosphodiester bond is formed; hence, in this discussion the term initiation will be reserved for the latter process.

The properties of the binary complexes formed by RNA polymerase holoenzyme at promoter regions have been studied with a variety of templates. The interpretation of these experiments is made difficult by the same factors that complicate the measurement of selectivity, namely, the many steps in the in vitro reaction, the complexity of the DNA template structure, variations between different templates, and the technical difficulty of measuring selective transcripts. To facilitate the discussion, I begin by discussing the complexes formed with T7 DNA. However, many of the features of the holoenzyme-promoter interaction were first noted with other templates.

Binding of RNA Polymerase Holoenzyme to T7 DNA

By many of the criteria discussed above (see section on the transcription sequence) E. coli RNA polymerase holoenzyme initiates RNA chains selectively at promoter regions used in vivo for formation of T7 early mRNA. The properties of the binary complexes formed between low molar ratios of E. coli RNA polymerase and T7 DNA at 37° in solutions of low or moderate ionic strength are consistent with the hypothesis that these are productive complexes formed by the enzyme at the initiator sites of the promoter regions (116, 175). The interpretation of binding studies is also facilitated with T7 DNA, since under appropriate reaction conditions a great majority (70–90%) of the RNA polymerase molecules added to the reaction bind at T7 early promoter regions and are able to initiate an RNA chain selectively (175).

Three kinds of assay procedures have been employed to study binary complex formation. They take advantage of different properties of the final binary complex between enzyme and the promoter region. 1. Filter binding—binding of an RNA polymerase molecule to radioactively labeled T7 DNA leads to retention of the resulting complex on nitrocellulose filters; free RNA polymerase is retained in such a filter while free DNA is not (116). 2. Rapid RNA chain initiation—under certain conditions RNA polymerase bound at the promoter region is able to initiate an RNA chain rapidly and can then initiate chains even when challenged with the inhibitor rifampicin (176, 275). 3. Relative resistance to polyanionic inhibitors—inhibitors such as heparin (26, 253) and poly [rI] (116) attack free RNA polymerase far more rapidly than enzyme bound tightly to DNA.

The Kinetics of Promoter Site Selection on T7 DNA

When E. coli RNA polymerase holoenzyme and labeled T7 DNA are mixed together at 25°C, the kinetics with which a nonfilterable complex is formed are first order, that is, the rate of the reaction does not depend on the DNA concentration; the reaction has a half time of 15–20 sec (117). The kinetics are essentially unchanged when the reaction is terminated by adding an excess of unlabeled DNA (a procedure which measures the rate at which enzyme becomes permanently attached to a particular DNA molecule) or when the ability to initiate RNA chains rapidly (assay 2) or resistance to displacement by poly [rI] is observed (assay 3). Since the overall reaction between T7 DNA and RNA polymerase must be second order, a first-order reaction is rate limiting both for the initial selection of a DNA molecule by the RNA polymerase and for formation of a functional complex. This reaction is not a first-order sliding of RNA polymerase on a rapidly encountered DNA molecule since it is blocked by adding unlabeled DNA. Consequently, the rate-limiting reaction is probably the dissociation of RNA polymerase holoenzyme from randomly encountered, nonpromoter regions of the DNA molecule, to which it binds with a weak but detectible affinity. Such complexes are too transient to be nonfilterable.

The number of such Class B binding sites on a T7 DNA molecule is difficult to specify exactly (45, 116, 117). The number must be at least 1300, since the entire DNA molecule can be saturated with RNA polymerase under proper conditions (215, 216). However, if RNA polymerase can interact with essentially any 20–30 nucleotide region of a DNA helix, the majority of sites will overlap and hence not be revealed at enzyme saturation (275a). In this case the number of such Class B sites would be nearly equal to the number of total nucleotides in the DNA (7.7×10^4 sites per T7 DNA). The problem of determining the basic binding parameters for a protein which interacts nonspecifically with multiple units on a linear polymer is complex and is thoughtfully reviewed in (275a).

Since only 15–20 sec are required for one half the enzyme molecules to locate the promoter region (117), the rate of dissociation for Class B sites must be quite rapid. A half time for dissociation of between 1 and 100 msec can be estimated, depending on whether one chooses the larger or smaller number of Class B binding sites, respectively. Despite this rapid rate of dissociation the second-order rate with which RNA polymerase encounters DNA must be considerably more rapid, at least $2 \times 10^8 \, M^{-1} \, sec^{-1}$ (117). Thus, although the nonspecific binding of RNA polymerase to the DNA helix is quite weak, it is the rate-limiting step in locating a promoter region. Similar kinetics are found for the binding of *lac* repressor to operator sites on DNA when the repressor concentration exceeds a certain value (274); under these conditions the rate-limiting step in repressor binding is also the rate of dissociation from nonspecific sites on the DNA.

The nonspecific binding of RNA polymerase holoenzyme to DNA does not appear to open base pairs or require their prior opening. RNA polymerase is not a generalized melting protein; it does not appreciably lower the T_m of helical DNA (215). The limited opening of base pairs on λ DNA brought about by RNA polymerase holoenzyme is fewer than one base pair per enzyme molecule when a large excess of enzyme is present (225). In addition there is no appreciable variation in the rate of promoter site location, measured by the filter binding assay between $15°$ and $37°$ (227). Thus nonspecific binding of RNA polymerase to DNA occurs on the outside of the helix structure.

Since the rate of promoter site selection depends on the ratio of nonspecific binding sites to specific sites, the rate can only be increased by increasing the number, the reactivity, or the size of promoter regions. The reactivity of these regions, judged from the rate of site location and the number of nonspecific sites, is already quite high. T7 phage appears to have three independent promoter regions governing a common transcription unit read by *E. coli* RNA polymerase (128, 175) and this duplication may function primarily to increase the slow rate of site selection. In vivo, some *E. coli* promoters are read at rates comparable to the in vitro rate of site location found with T7 DNA (276); however, ribosomal transcription units are read an order of magnitude more rapidly, implying that RNA polymerase finds these promoter regions 10 times more rapidly (59, 277, 278). These regions may be served by multiple independent promoter sites or by some kind of physically large preinitiator (253) or collection region (197) at which more rapid initial binding of RNA polymerase can take place.

Properties of RNA Polymerase-Promoter Complexes; The Open Complex

The properties of the complexes formed at low molar ratios between RNA polymerase holoenzyme and T7 DNA in solutions of low ionic strength at temperatures over 20°C suggest that these are active promoter complexes in which the holoenzyme has separated the DNA strands at the initiator site and is poised for RNA chain initiation. In terms of the model presented below, these are open promoter complexes.

STABILITY AND STOICHIOMETRY OF OPEN COMPLEXES WITH T7 DNA   Direct binding studies show that the RNA polymerase holoenzyme-T7 DNA complexes formed at 25–37° and low ionic strength ($\mu = 0.07$) dissociate exceedingly slowly (116). A half time for dissociation of 30–60 hr is estimated; this corresponds to a binding constant of about 10^{14} M^{-1}. These highly stable binary complexes do not dissociate prior to selective chain initiation, hence they involve complexes in the promoter region. In fact, since these complexes initiate RNA chains rapidly in a reaction limited in rate only by the ATP or GTP concentrations (176) (see below), it is likely that they involve RNA polymerase bound at initiator sites.

Titration studies show that only about eight RNA polymerase holoenzyme molecules can interact with each T7 DNA molecule to form this highly stable binary complex (116). These binding sites have been referred to as Class A sites. The apparent stability of these complexes is dependent on the temperature and ionic strength of the solution; the rate of dissociation is increased about 100-fold at 15°C or in the presence of 0.2 *M* NaCl.

Similar highly stable binary complexes are formed with RNA polymerase holoenzyme and fd RF I DNA (19), T4 DNA, and λ DNA (279) templates.

THE APPARENT RESISTANCE OF OPEN COMPLEXES TO RIFAMPICIN; RAPID RNA CHAIN INITIATION   In an important series of experiments Sippel & Hartmann (154) showed that when binary complexes between RNA polymerase holoenzyme and DNA were challenged with a mixture of rifampicin and the four ribonucleoside triphosphates, enzyme in the complexes was able to initiate RNA chains without inactivation by the drug. Under comparable conditions if free RNA polymerase is added to a mixture of DNA, rifampicin, and triphosphates, all of the enzyme is inactivated. Rifampicin blocks RNA synthesis at the level of RNA chain initiation (154, 280).

The result led to the conclusion that the binary complex of RNA polymerase holoenzyme at the promoter was insensitive to attack by rifampicin (7–9, 154, 155). However, if the enzyme in such complexes is mixed with the drug it is rapidly inactivated (154, 155) in a second-order reaction (275). At the concentrations of rifampicin commonly employed (10 μg/ml), this inactivation is over 10^6 times more rapid than the rate of dissociation of RNA polymerase holoenzyme from the DNA. Hence inactivation cannot involve a free enzyme as intermediate (116, 275).

The paradox of the apparent resistance of the binary complex was resolved when it was realized that two competing reactions occur when binary complexes are presented with a mixture of rifampicin and the nucleoside triphosphates (176, 275). The first reaction, inactivation, leads to an inert RNA polymerase holoenzyme-DNA complex

$$\text{E} \sim \text{DNA} + \text{Rifampicin} \xrightarrow{k_2} \text{Rifampicin} - \text{E} \sim \text{DNA} \qquad 1.$$

The second, competing reaction, RNA chain initiation, generates a ternary complex completely resistant to rifampicin

$$\text{E} \sim \text{DNA} + 4\text{XTP} \xrightarrow{k^*} \frac{\text{E} \sim \text{DNA}}{\text{RNA}} \qquad 2.$$

If the intrinsic rate of RNA chain initiation is rapid, then a mjaor fraction of all binary complexes will be able to initiate an RNA chain before significant inactivation by the drug. In fact, the equations give an exact relationship between the fraction of binary complexes (C^*) able to initiate an RNA chain in such an experiment, the total complex concentration C_T, and the rifampicin concentration, namely

$$\frac{C_T - C^*}{C^*} = \frac{k_2[\text{Rifampicin}]}{k^*} \qquad 3.$$

This relationship has been measured for T7 DNA and holds under a variety of conditions (176). Thus the lack of sensitivity of RNA polymerase holoenzyme to rifampicin in binary complexes is not due to an intrinsic resistance to rifampicin, but to the rapid rate of RNA chain initiation by these complexes. This has led to the designation of these productive binary complexes as rapidly starting (RS) complexes.

Measurement of the sensitivity of binary holoenzyme-DNA complexes to rifampicin in challenge experiments leads to several useful results. The parameter k^*, the intrinsic rate of RNA chain initiation, can be measured in such experiments. For T7 DNA and dAT copolymer, k^* is directly proportional to nucleoside triphosphate concentration (176, 231) and in particular to the concentration of the substrate incorporated at the 5′ terminus of the RNA chain (231). Therefore it appears that the intrinsic rate of RNA chain initiation by an RS complex is determined by the second-order interactions

$$\text{Complex} + \text{ATP} \rightarrow \text{Complex} \sim \text{ATP} \qquad 4.$$

or

$$\text{Complex} + \text{GTP} \rightarrow \text{Complex} \sim \text{GTP} \qquad 5.$$

which are determined by the ATP or GTP concentrations. (The choice of ATP or

GTP is fixed by the exact DNA template base at the initiation site.) Evidence has accumulated from several studies that the binding of the 5′-terminal ATP or GTP to the enzyme is rather weak (177, 179, 281).

At 0.4 mM triphosphate, $k*$ determined with T7 DNA template is about 5 sec^{-1}, corresponding to a half time for chain initiation of 0.2 sec. This is far faster than expected from previous studies. This rapid rate implies that where the rate of chain initiation is still rate limiting, even at elevated ATP or GTP concentrations, the true rate-limiting step precedes the actual chain initiation reactions (Equations 4 and 5). The intrinsic rate of RNA chain initiation is not highly sensitive to alterations in the temperature or ionic strength of the medium (176).

A second important parameter measured by the rifampicin challenge experiment is the fraction of RNA polymerase molecules in binary complexes which can initiate an RNA chain rapidly. For RNA polymerase holoenzyme bound to T7 DNA at 25–37° in solutions of low ionic strength ($\mu = 0.07$), essentially all molecules of holoenzyme in binary complexes with T7 DNA are able to initiate an RNA chain rapidly (176).

RESISTANCE OF OPEN COMPLEXES TO POLYANIONIC INHIBITORS   Free bacterial RNA polymerase is inhibited by low concentrations of a number of polyanions, including RNA homopolymers (116, 179, 282–284) and the polyanions heparin (26, 285) and polyethanesulfonate (286). These polyanions bind tightly to the enzyme at a site or sites which are blocked when the enzyme is in a ternary complex growing an RNA chain (26, 285–287). Binary complexes formed under conditions of elevated temperature (25–37°) and low ionic strength are also relatively resistant to attack (26, 116). Since this resistance does not require that triphosphates be added together with the inhibitor, it is due to a true reduction in sensitivity to the inhibitors, not simply to rapid chain initiation as with rifampicin.

OPENING OF DNA BASE PAIRS IN THE OPEN COMPLEX   A good deal of indirect evidence has accumulated that transcription of the DNA helix structure must involve localized opening of the DNA strands (for review see 2). Evidence that the initiation process itself requires strand separation has come from several kinds of experiments, including the dependence of the rate and kinetics of RNA synthesis on temperature (285) and the effect of temperature on the sensitivity of binary complexes to poly[rI] (284), heparin (285), and rifampicin (176, 205). In addition, the stability of binary holoenzyme-T7 DNA complexes (116) is greatly reduced at temperatures below 20°C. Each of these experiments showed that the initiation process involved a temperature-sensitive step which might well be a localized denaturation event. However, interpretation of these results was limited because there was no direct demonstration of base pair opening. In addition, the effects of temperature on the rate of transcription might be due either to temperature sensitivity of the rate of a particular step (a high activation energy) or to the temperature sensitivity of an equilibrium process (a high enthalpy). Several new studies clarify this situation considerably.

Saucier & Wang (225) have shown directly that when excess RNA polymerase holoenzyme binds to λ DNA about 24 base pairs are opened per DNA molecule.

Binding of RNA polymerase to λ DNA prior to closing of the covalent duplex introduces negative supercoil twists into the molecule when the protein is removed; the extent of this supercoiling can be measured sensitively. Although the interpretation of their results was made difficult by the problem of measuring the exact number of active RNA polymerase molecules bound per mole of enzyme added, other evidence suggests that only four promoter regions are read by RNA polymerase under these conditions (126). Thus an average of six base pairs are opened in the binary complex for each promoter region. As discussed below, this is a minimum estimate since the exact transition temperature for these promoter regions is not known. No base pairs are opened when binding occurs at 0°C.

Less direct evidence that the rapidly starting binary complexes involve opening of base pairs comes from the study of RNA chain initiation by these complexes. While the intrinsic rate of RNA chain initiation by RS complexes is not strongly affected by temperature or elevated ionic strengths, the fraction of RNA polymerase holoenzyme molecules in RS complexes decreases dramatically at temperatures below 20°C or at salt concentrations above 0.1 M NaCl (176, 205). Under these same conditions enzyme in binary complexes becomes more sensitive to attack by polyribonucleotides (116, 284) or heparin (26, 253). These properties of holoenzyme-DNA mixtures could all be explained if binding failed to occur or was greatly reduced under these conditions, or if the conditions favored binding at nonspecific (Class B) sites on DNA. However in the case of holoenzyme-T7 DNA complexes formed at low temperatures or elevated salt concentrations, binding can be directly demonstrated (116) and relaxation of the enzyme to a rapidly starting conformation occurs from such complexes without dissociation from the template (176). Thus binding of RNA polymerase holoenzyme at or near the promoter region is indicated even at low temperatures where neither base pair opening nor rapid RNA chain initiation seems to occur. The properties of this complex are explored in the next section; however, for the purposes of identification it has been designated as I complex (initial and inert).

Properties of RNA Polymerase-Promoter Complexes; The Closed Complex

The properties of the binary complexes formed between RNA polymerase holo-enzyme and T7 DNA at elevated ionic strength ($\mu = 0.2$) or low temperatures suggest that these are complexes in which the enzyme has bound at the promoter region but has not opened the DNA strands and therefore cannot carry out RNA chain initiation. In terms of the model presented below, these are closed promoter complexes.

SENSITIVITY OF CLOSED COMPLEXES TO HEPARIN   Binary complexes formed at low temperatures or elevated ionic strength between RNA polymerase holoenzyme and a variety of templates including T7 DNA are much more sensitive to attack by polyanions such as heparin (26) and poly[rI] than are the corresponding complexes found at elevated temperature and low ionic strength (116, 284). This is probably a displacement reaction involving direct attack by the inhibitor on the binary complex rather than trapping of free enzyme after dissociation. Dissociation of binary holoenzyme-T7 DNA complexes even at 0°C requires many minutes (116), while attack by heparin or poly[rI] on these complexes is much more rapid. This reaction can be pictured (217) if one imagines that RNA polymerase possesses two DNA binding sites corresponding to the two DNA strands. In the open complex these would both be engaged, while in the closed complex one site is available for attack by an exogenous polyanion. It is interesting to note that core polymerase-DNA complexes are quite sensitive to displacement by polyanions (116). Core polymerase cannot normally open the strands of the helical DNA, hence it is always in a state analogous to closed complex on a helical DNA.

INABILITY OF CLOSED COMPLEXES TO INITIATE RNA CHAINS   When binary holoenzyme-T7 DNA complexes are formed in 0.2 M NaCl or at temperatures below 20°C, the fraction of enzyme which can initiate an RNA chain is very low; these are the conditions used to define I complexes as described above. This failure to initiate chains could be due to the inability of closed complexes to carry out the initiation reaction or to a far greater rate of attack by rifampicin on the complex. The latter possibility can be ruled out by direct measurement of the (second-order) rate of rifampicin attack on I complexes; when this rate is used in rifampicin challenge experiments to calculate the rate of T7 RNA chain initiation by I complexes, the latter rate is found to be approximately equal to that at which I complex can relax to RS complex (227). It is therefore likely that closed complexes are unable to carry out the RNA chain initiation reaction except by transition into the open complex.

THE ENTHALPY OF TRANSITION BETWEEN OPEN AND CLOSED COMPLEXES FOR THE HOLOENZYME T7 DNA INITIATION   If the open and closed promoter complexes represent two distinct states of the binary complex which are in equilibrium, then following the variation in their concentrations with temperature will give the enthalpy of the transition. The use of rifampicin challenge experiments allows the determination of the fraction of holoenzyme in RS and I complexes; the enthalpy for formation of the RS complex with T7 DNA is 57 kcal/mole with a transition temperature of 20°C (176). The enthalpy of DNA melting for complete unstacking of the base pairs in the helix is about 7–8 kcal/mole of base pairs (288; 288a); if it is assumed that binding of RNA polymerase to DNA in these complexes does not involve a large enthalpy change, the result is consistent with the opening of seven or eight base pairs in the transition from closed to open complex. For comparison, Saucier & Wang (225) found no detectible opening of base pairs at 0°C and about six base pairs opened for each initiator at 37° for λ DNA.

THE STABILITY OF BINARY COMPLEXES   If the binary complexes formed between holoenzyme and T7 DNA under different conditions are fundamentally different, the effect of these conditions on the stability of the complexes must be re-examined. For example, it is not possible to conclude (116) that the reaction temperature strongly affects the stability of the open T7 promoter complex when there is a dramatic reduction in the concentration of that complex as the temperature decreases. In fact, there is little change in the rate of dissociation of the highly stable holoenzyme T7 DNA complex between 37 and 20°C, and the dramatic change between 25 and 15°C is due almost entirely to a change in the equilibrium to favor the closed complex. Early estimates of the stability of the closed complexes formed with T7 DNA at 15°C gave a dissociation half-time of less than 1 hr (116). More recent determinations show that the stability of T7 DNA-holoenzyme complexes is essentially constant from 0 to 15° with a dissociation half-time of about 40 min (288b). This corresponds to a binding constant of about 10^{12} for RNA polymerase in the closed complex which is comparable to that found for binding of core polymerase (116) and about 50 times lower than that for the holoenzyme in the open complex. For comparison, the binding of *lac* and λ repressors to operator sites on DNA, which involves closed complexes, is not highly temperature-sensitive (289, 290).

A Model for the Interaction Between RNA Polymerase and a Promoter Region

THE MODEL   The properties of RNA polymerase-DNA complexes suggest a rather explicit model for the interactions between the enzyme and a promoter region (Figure 5). The model postulates that promoter regions are located during the initial phase of selection by repeated association-dissociation events at random sites on different DNA molecules. When the RNA polymerase encounters a promoter region, recognition occurs and a binary complex is formed between the enzyme and the region in its helical conformation (closed promoter complex). This complex is in equilibrium with a complex in which the DNA strands have opened to allow RNA polymerase direct access to the template bases (open promoter complex). It is the latter complex that is able to initiate an RNA chain; the rate of the initiation reaction is governed by the concentration of open complex and the concentration of the 5'-terminal nucleoside triphosphate. The major features of the model were originally suggested by Zillig and his collaborators (285) and have been used as a working hypothesis since that time by many investigators.

NOMENCLATURE   There is somewhat of a problem concerning the nomenclature to be used in describing the different forms of binary complexes. A diversity of terms has been used, including primary complex, secondary complex, initiation complex, preinitiation complex, highly stable complex, rapidly starting complex (RS complex) and I complex. In most cases these terms have been defined with respect to a particular assay procedure in which some property of the complexes was measured, for example, the rate or extent of RNA synthesis or the rate of enzyme dissociation from the complex.

Any particular procedure may or may not reveal intrinsic properties of the complex; for example, as discussed above it was believed that the rate of dissociation of the open RNA polymerase-T7 promoter complex was highly temperature dependent (116). It is more likely that the dissociation rate of this complex is not temperature dependent but that the complex is in a highly temperature-dependent equilibrium with a complex which has a far more rapid intrinsic rate of dissociation (176). Again, as an example, the term initiation complex was originally used to describe a complex formed between RNA polymerase and DNA in the presence of

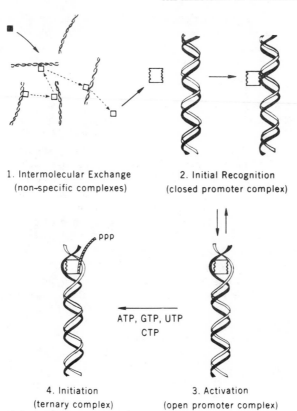

1. Intermolecular Exchange
(non-specific complexes)

2. Initial Recognition
(closed promoter complex)

ppp

ATP, GTP, UTP
CTP

4. Initiation
(ternary complex)

3. Activation
(open promoter complex)

Figure 5   Promoter site location and activation by RNA polymerase.

a purine nucleoside triphosphate. These complexes displayed enhanced stability to dissociation by salt and were believed to represent a very stable complex between enzyme, DNA, and the purine nucleoside triphosphate. In fact, these complexes are probably quite weak (177, 179, 281) and the initiation complexes were probably ternary complexes in which trace contamination by other ribonucleoside triphosphates had allowed chain initiation and limited elongation to occur (45, 290a).

Because of the limitation associated with designations based on direct measurements, I chose the terms closed and open to reflect what I hope are intrinsic properties of the respective complexes which are not tied to a particular experimental procedure. The terms are defined solely in conjunction with the model and refer to hypothetical states which may or may not exist. Where a particular experimental measurement is made, for example, the fraction of RNA polymerase bound in binary complexes which is not inactivated by heparin or which can initiate an RNA chain in a rifampicin challenge experiment, I try to use trivial or descriptive designations, i.e. heparin resistant complex and rapidly starting complex, until it is clear that the property measured is an intrinsic property of, for example, the hypothetical open promoter complex.

FUNCTIONAL ASPECTS OF THE MODEL   The properties of different binary complexes corresponding to closed and open promoter complexes have been discussed in previous sections. Consider now the specific interactions involved in the different complexes which govern the selection process. The model postulates a compound recognition process involving two distinguishable recognition steps: promoter region location and transition into the functional open complex. The steps involve different DNA structures and probably different informational interactions; in fact, the sites on the DNA might even be physically distinct. Several groups have discussed models in which recognition occurs at one site (entry) and initiation at yet another (start) (197, 253). The advantages of a compound selection process have been discussed by von Hippel & McGhee (2). The notion that initial recognition occurs with an intact helical promoter is supported by the properties of binary complexes formed at low temperatures and elevated ionic strengths. Kinetic arguments also make it unlikely that initial recognition can occur only if the DNA strands are open. If this were so, the rate of site location would be reduced by a large factor (at least 10^3) over the rate of site encounter, since only a small fraction of DNA base pairs are open at any given moment (2). For T7 DNA the second-order rate of promoter location (estimated at over 2×10^8 M^{-1} sec^{-1}) is already quite close to the maximum value for a diffusion-limited reaction (117). In addition, the rate of binary complex formation between T7 DNA and RNA polymerase holoenzyme, measured by the sensitivity of the complexes to dilution, does not vary over the temperature range from 37 to 15°C (227). This temperature alteration dramatically changes the equilibrium between the two binary complexes (176) and would be expected to affect any reaction in which DNA strand opening was a rate-limiting step.

Despite the evidence that RNA polymerase could recognize an intact helix, it can still be argued that this interaction involves a transient strand separation in which the enzyme determines the potential to be opened. Distinguishing between this kind of recognition and recognition of the intact helix will require detailed knowledge of exactly what reactive groups on the DNA are involved in the process.

If initial recognition does involve interactions between RNA polymerase and a helical promoter region, where does the specificity of the process lie? Repressors carry out an analogous recognition reaction with the operator region and the considerations offered to account for that process apply equally here. Both reactions appear to be extremely rapid and specific and may involve identical kinds of information. Three major mechanisms have been considered: 1. recognition of a unique tertiary structure [hairpin loop, etc (270, 271)]; 2. recognition of a unique DNA secondary structure [C form of DNA, etc (272)]; and 3. recognition of a unique nucleotide sequence utilizing reactive groups in either the major or minor grooves of the helix (2, 183, 273). Of these, the third now seems heavily favored in the repressor-operator interaction (see earlier section on the structure of promoter and terminator regions) and identical reasons can be advanced for the RNA polymerase-promoter interaction.

If the transition into the open form of the promoter complex involves opening DNA base pairs at the initiator, what provides the driving force for the reaction and where is the specificity of the process determined? The functional interaction of DNA melting proteins with DNA has been discussed by von Hippel & McGhee (2). The simplest mechanism involves preferential binding of the protein to the single DNA strands, which shifts the helix-coil equilibrium in the direction of strand separation. A melting protein such as RNA polymerase could carry out a sequence-specific melting process if it were to bind tightly to groups available on a single DNA strand (or both strands) and if those groups were base specific. This would be the case if RNA polymerase were to bind tightly to a specific, single-stranded DNA sequence. The reactive groups involved in this binding might or might not involve the hydrogen bonding groups of the bases but probably would not involve only the same groups on the bases that are involved in initial recognition. These latter groups are all available on the helical DNA structure as well as the separated strands, and would not account for the differential binding of RNA polymerase to the single strands which must drive the melting reaction.

An interesting problem is posed if the transition into the open complex opens only 6 to 8 base pairs and this is the recognition sequence for melting. A minimum of about 12 base pairs is required to form a unique sequence which is unlikely to be generated by random assortment of nucleotides on the E. coli chromosome (269). It may be that the initial recognition region for the closed complex is much larger and provides an important portion of the overall specificity in site selection, while the actual initiator or melting recognition sequence is rather short.

This is esthetically pleasing in several respects. Repetition of a common, linear, 12+ nucleotide sequence in a chromosome might provide nucleation sites at which stable base pairing could occur between adjacent regions of the chromosome (269). This could lead to intrachromosomal recombination, which would probably be quite undesirable. However a recognition sequence on the outside of a helix need not involve adjacent nucleotides. For example one could employ two unique hexanucleotide sequences spaced 10 nucleotides apart to generate a unique 12-nucleotide unit which would not contain a linear sequence long enough to serve as a nucleation point for base pairing. If the recognition sequence regions were 10 nucleotides apart they would occupy equivalent positions on the same side of the DNA helix and this might facilitate steric interactions between the protein and these sequences (183). The recognition unit in the above example would be about 70–80 Å in overall length, well within the estimates of the size of bacterial RNA polymerase.

The Transition Temperature of Promoter Regions

The foregoing discussion suggests that one can treat the transition between closed and open promoter complexes as a melting event involving a cooperative unit of about 7–8 DNA base pairs. This melting can be conveniently characterized by the temperature at which half the binary complexes are in the open state; this is termed the transition temperature of the particular promoter-enzyme system. The melting event occurs, as it were, in a solvent of RNA polymerase (2). Tight binding of RNA polymerase to the separated strands relative to the helix will lower the transition temperature, whereas weaker relative binding will raise the transition temperature. The parameters of the system that can be varied are: 1. the size of the cooperative unit; 2. the intrinsic T_m of the region (primarily AT content); and 3. the tightness of the relative binding by RNA polymerase. While little concrete information is known about these parameters it is clear that the transition temperature does depend, as predicted, on both the RNA polymerase and the promoter sequence. For example, *Bacillus stearothermophilus* RNA polymerase has an elevated transition temperature with T4 phage promoter regions, indicating a lower relative binding in the open complex (2, 291). The promoter regions for PM2 DNA (and probably other circular phage DNAs, see below) have a much higher transition temperature in the relaxed DNA conformation than those for T7 DNA, while those for the linear DNAs from T4 and T5 phages have much lower transition temperatures under similar conditions (292).

The opening of the DNA base pairs at the initiator site to form the open promoter complex depends on a balance of forces. Because of the cooperativity of the transition, it can be dramatically affected by modest changes in the ionic strength (176, 205), the ambient temperature (176, 205, 292), the presence of solvents (235), the presence of DNA-binding proteins that have melting or antimelting properties (2), the conformation of the DNA (see below), and of course by alterations of either the RNA polymerase or promoter structures. Transcription from a promoter region below its transition temperature is quite slow (39), hence any factor which shifts this equilibrium even slightly could potentially turn on or turn off transcription at that promoter in an essentially all or none manner. It seems very likely that many regulatory processes in which gene expression is turned on or off act directly in some way to shift this critical equilibrium.

One can picture the effect of a promoter-type mutation easily from this situation; replacement of an AT pair by a GC pair, for example, would raise the free energy change on melting of the region by about 1 kcal (293) and might simultaneously lower the binding of RNA polymerase to the open strands by changing the recognition sequence. This would also increase the free energy change for the transition, possibly by several kilocalories. Thus the overall effect of such a mutation would be to shift the transition temperature upward by 10–20°, which might well exceed the physiological temperature range and permanently close the promoter.

It has been shown recently that certain λ promoter mutations which block λ transcription for specific promoter regions in vitro can be phenotypically reversed in vitro by the addition of glycerol or other solvents to the reaction (236). These solvents are DNA-denaturing agents (235–236) and may well function by lowering the transition temperature of the altered promoter region.

Binary Complexes with Circular DNA Templates

Many prokaryotic chromosomes are found in vivo as covalently closed circular duplex structures. The isolated molecules contain negative supercoil twists (294,

295). The negative twists introduce a strain into the helical duplex structure which can be relieved by the opening of base pairs (294, 298), and a certain fraction of the base pairs in a twisted closed duplex is probably open at all times (206). While it is unlikely that a unique set of base pairs (for example, at a single site on the molecule) is continuously open (2), the fraction of time that different regions are open probably varies appreciably. Single-strand specific nucleases can introduce breaks in twisted, covalently closed DNAs (207–209). In the case of SV40 DNA the breaks are at unique sites (208, 209) and only one such break is made per molecule, hence this is not a sequence-specific cleavage but depends on the twisted state of the DNA duplex.

A consequence of the strain induced by twisting is that agents that bind preferentially to single strands, such as DNA melting proteins (2, 208a, 297), or agents that unwind the duplex structure, such as intercalating agents (294, 296), are much more strongly bound to negatively supercoiled DNA than to the equivalent circular relaxed duplexes (298). As might be expected from the picture of bacterial RNA polymerase as a specific DNA melting protein, the closure of a DNA template into a twisted duplex structure has a profound effect on its transcriptional properties.

Two groups of small DNA phages have been extensively studied with respect to their closed duplex structures. The one group contains the spherical phages ϕX174 and S13 (299), the second the filamentous phages including M13 and fd (300). Both groups contain a single-stranded circular DNA in the virion which is converted rapidly to a covalently closed duplex (RF I) after infection. Isolated RF I molecules are negatively supercoiled. The onset of phage DNA replication appears to be correlated with the opening of one of the strands of RF I to produce the relaxed circular duplex RF II (301).

Both RF I and RF II forms of ϕX174 are asymmetrically transcribed in vitro by bacterial RNA polymerase (302, 303); the transcripts are rather heterogeneous in size and include some species longer than the template strand (304). A tentative map of the binding sites on the ϕX174 chromosome is available (254). Transcription of both relaxed and supercoiled fd RF I is also asymmetric (19, 159). Transcription of fd RF I gives three major RNA species (185) and three corresponding 5'-terminal RNA sequences (159). A similar distribution of mRNA sizes is found in vivo for the closely related phage M13 (305). A preliminary transcription map of fd RF has been proposed (185). The binding sites for RNA polymerase on both fd RF and ϕX RF molecules have been isolated (19, 144, 254).

In an important experiment Hayashi & Hayashi (303) showed that ϕX RF I is much more rapidly transcribed than ϕX RF II. This kind of result has since been obtained for a variety of phage DNAs which exist in vivo in covalently closed duplex forms, including the fd RF species (305a) and analogous covalently closed and relaxed circular duplexes from phage PM-2 (306) and λ phage (307). In each instance the twisted covalently closed duplex is a far more active template than the relaxed form of the molecule. The low rate of ϕX RF II transcription is due to failure of RNA polymerase to form appreciable concentrations of the rapidly starting binary complex with this template, as shown by appropriate rifampicin challenge experiments (303). Since such complexes are almost quantitatively formed with

RNA polymerase and linear T7 DNA under identical conditions (175), the structure of ϕX174 promoters must be quite different.

If the rapidly starting binary complex is equated with the open promoter complex (Figure 5), ϕX174 promoters appear to be intrinsically more stable to denaturation and should have a higher transition temperature in the untwisted DNA molecule (linear or RF II). Twisting the molecule in the RF I form would stabilize the denatured form of the promoter region relative to the helical form and hence lower the transition temperature. This result is obtained for the supercoiled and relaxed forms of PM-2 DNA; when the transition between I and RS complexes is measured by rifampicin challenge experiments, transition temperatures of 13° and 28°C, respectively, are obtained (292). The effect of temperature on the alteration in the transcriptional properties of twisted duplexes is additional strong support for the hypothesis that productive complexes between RNA polymerase holoenzyme and promoter regions involve limited strand separation to form an open complex (Figure 5).

One difficulty in assessing the biological role of supercoiling comes from the possibility that the twisted molecules characterized in vitro are actually relaxed in vivo through a change in DNA conformation or binding of a denaturing protein (295). The number of twists in a duplex can be varied in vitro by alterations in salt and temperature (308), but there is no set of quasiphysiological conditions that can completely relax supercoiled molecules. Recent evidence indicates that the RF I and RF II forms of S13 phage DNA differ in vivo in their transcriptional properties in exactly the manner predicted from in vitro studies (309). This supports the notion that covalently closed duplexes are twisted in vivo. In S13 infected cells, the cleavage of RF I to RF II is accompanied by a sharp reduction in the intrinsic rate of transcription per DNA. This cleavage of RF I does not occur in A gene mutants which maintain a high rate of transcription per mole of RF I (309). The result suggests one reason why small DNA phages might possess promoters read only from supercoiled DNA templates. The RF I form of the DNA can be viewed as the transcriptive form of the phage DNA; it is formed immediately after infection and transcribed at a high rate to produce phage mRNA. By opening the DNA to give RF II, new transcription is terminated and the DNA replication process can be activated on this true replicative form without interference between the two processes.

Recent studies suggest that *E. coli* DNA may have a twisted conformation in the cell (310). Twisting need not necessarily depend on formation of a circular structure but could result from any situation in which two or more segments of a helix are fixed, for example, through attachment to a membrane structure. These considerations suggest that in the study of selective transcription from any particular promoter region, the effects of DNA conformation as well as of temperature and ionic conditions need to be carefully studied.

The Role of Sigma Subunit in Promoter Selection and Activation

Since σ subunit appears to play a vital role in nearly all selective transcription it is important to know its functional role in this process. At least three distinguishable processes are altered by binding of σ subunit to RNA polymerase, although all might have a common origin: 1. The nonspecific binding of RNA polymerase to the DNA helix is reduced by σ (116, 311); the rate of dissociation of these complexes increases by a factor of up to 10^6. 2. The σ subunit appears to be required for formation of highly stable (116), polyanion-resistant (26) binary complexes, hence it probably is required to form the open promoter complex. 3. The binding of σ subunit to core polymerase suppresses its ability to initiate chains at random single-strand breaks (86), hence imposes a more stringent initiation specificity on the RNA polymerase.

The enormous reduction brought about by σ in the affinity of core polymerase for nonspecific regions of DNA suggests that formation of holoenzyme leads to a conformational change in the RNA polymerase (116, 311). This may represent a response to the conflicting requirements placed on RNA polymerase which must carry out both generalized transcription and specific site selection (1). The conformation of core polymerase is probably optimal for the tight nonspecific binding to DNA which is needed for generalized transcription (elongation state), while that of RNA polymerase holoenzyme is optimal for specific site selection and initiation (selection initiation state). Quantitatively, the reduction in the nonspecific binding of core polymerase is essential to site selection. Since the probability of locating a promoter region on T7 DNA is between 1 in 100 and 1 in 10,000 (see section on the kinetics of promoter site selection above), core polymerase would require from 30–3000 hr to locate a promoter at 37° by random exchange (117). In practice, core polymerase undergoes a displacement reaction with single-stranded DNA (116) and possibly with single-stranded regions of helical DNA which can enhance the rate of equilibration.

The recognition properties of RNA polymerase are altered when σ is removed; however it is difficult to determine exactly which step in recognition is blocked. The core polymerase might actually be capable of forming closed and open complexes, for example Figure 3, but be kinetically unable to locate helical promoter regions. Even if core polymerase possesses an affinity for promoter regions identical to that of holoenzyme, the equilibrium distribution of core polymerase on a mixture of 10^4 nonspecific sites $(k = 10^{12})$ and 10 promoter sites $(k = 10^{14})$ would greatly favor nonspecific binding.

Despite these complexities, the fact that binding of σ subunit to core polymerase suppresses initiation at single-strand breaks does suggest that holoenzyme has a more rigid specificity for an initiator sequence. It is this kind of sequence-specific interaction that may well play a pivotal role in the transition from closed to open binary complex, where strand separation may depend on the binding between RNA polymerase and the initiator sequence. This transition could be bypassed by introducing an open region into the DNA duplex, and this kind of lesion is exactly that needed for synthesis by core polymerase (85–87).

These indirect arguments suggest that core polymerase is blocked in the promoter opening transition if not elsewhere in the selection process. The role of σ in the opening process could be direct (DNA melting protein) or indirect, (i.e. altering the conformation of the RNA polymerase to express an inherent specificity). The latter possibility is favored by studies of the stimulation of E. coli and B. subtilis core polymerases by the corresponding σ subunits (311a, 311b). Both template specificity and sensitivity to elevated ionic strength are determined by the core polymerase; either σ subunit will activate this potential.

In several instances where the structure of bacterial RNA polymerase is altered, σ subunit is not found in the isolated enzyme (78, 79, 81). Of course this does not necessarily reflect lack of a functional interaction with σ in vivo, since binding may be too weak to allow isolation of holoenzyme yet still adequate for σ function. However in at least two cases σ subunit appears unnecessary for selective transcription in vitro (78a, 79), and new polypeptide components are found associated with the enzyme. In this instance the associated components may serve one or more of the functional roles of σ subunit. However, only where a single component is homologous to σ subunit in function is the appellation σ-like (or σ factor) warranted.

The Effect of Repressor and Activator Proteins

Positive and negative control of transcription at operons is mediated by specific activators and repressors, or proteins that act to alter the functional interaction of RNA polymerase at the promoter region. In several instances the mechanism of this process has been explored in vitro. The lac transcription system has been most thoroughly studied in this regard. The normal template used is a ϕh80dlac or ϕ80dlac transducing phage (105, 106). Transcription of asymmetric lac RNA is dependent on 3'–5' cyclic AMP, CAP protein, and RNA polymerase holoenzyme; the former two regulators are expected from in vivo genetic and physiological studies and provide strong evidence for the selectivity of the system. The CAP protein is required to form a complex which can initiate when rifampicin is added together with triphosphates (by my previous nomenclature a rapidly initiating complex; however, several authors term this a preinitiation complex). A promoter mutation (p^r UV-5), which leads to catabolite insensitivity and lack of CAP protein requirement in vivo (266), bypasses the requirement for CAP in vitro as well (106). In this mutant a new transcription initiation site has been introduced to the downstream (z gene) side of the normal initiator (183).

Addition of lac repressor to the system before RNA polymerase prevents lac-specific transcription (97, 105). Repression is also observed if lac repressor is added to the RS complex 5 min before the initiation of transcription (97). In a lac promoter mutant [lac p^s (312)], however, the RS complex is not affected by repressor added 5 min prior to initiation (97).

In the case of the gal operon, identical requirements are found for synthesis of gal selective mRNA from λ_{pgal} DNA : RNA polymerase holoenzyme, CAP, and cyclic AMP (95). These components are also required for formation of a rapidly starting, poly [rI]-insensitive complex. The gal repressor has no effect on gal RNA synthesis by the RS complex after its formation (95).

Two matters concern us: the mechanism by which activator proteins function, and the mechanism of repression. With respect to the first problem, for both lac and gal systems the active form of the activator protein (CAP+cAMP) is needed for RS complex formation (95, 105, 106). According to the model in Figure 5, two steps are most likely to be affected: 1. initial recognition of the lac promoter (closed complex formation), or 2. activation (formation of the open complex). In the former instance, CAP-cAMP might function to enhance initial recognition and binding, for example, by interacting with the RNA polymerase to fix it to the promoter or by altering the conformation of the helix itself in some way. CAP protein binds to DNA and the binding is enhanced by cAMP (107, 108). However, conditions are not yet known under which the binding is site specific. In the second possible mode of action, CAP could facilitate open complex formation. This could occur through interaction with the RNA polymerase but is most easily imagined if CAP functions as a melting protein to effectively lower the transition temperature of the *lac* promoter [see (2) for general discussion]. By this view, RNA polymerase can bind to the *lac* promoter but cannot open it. The CAP protein provides the essential activating energy and RNA polymerase the specificity of binding.

Neither model can be rejected by current information; however, experiments to discriminate between a requirement for CAP for initial binding (insensitivity to dilution) as opposed to open complex formation (reduced sensitivity to poly [rI] or rifampicin + 4 triphosphates) are clearly feasible. The finding that glycerol and other solvents are able to suppress both the CAP-cAMP requirement and some promoter mutations (236) favors the second possibility, since these are potent DNA-denaturing agents (235). Since the equilibrium between open and closed promoter complexes is readily visualized as reflecting a balance of forces, it is easy to see how adjacent binding of a denaturing protein could facilitate the transition. The structure of P^S (312) and UV-5 (266) promoter mutations could be explained by this model as base pair changes from GC to AT which lower transition temperatures of the region.

The evidence concerning the action of repressors on the *lac, gal,* and λ systems seems initially conflicting. A resolution seems possible if one accepts the view that the geometry of the system is crucial (95). In the *lac* system the normal arrangement of activator, promoter (more precisely initiator), and operator probably involves nonoverlapping sites (97, 106, 267). Repression can be established before or after formation of an RS complex. In the UV-5 mutant the RS complex itself is not sensitive to repression; but repression can block its formation (97). Here the two sites, initiator and operator, are overlapping (183). The stability of RNA polymerase binding in the RS complex at the *lac* initiator is not known, but if it is comparable to the analogous T7 complex, dissociation would be very slow (116) and it is easily understood why RNA polymerase bound to an overlapping operator-initiator system would be insensitive to repressor. The converse is also true. The rate of dissociation of *lac* repressor from its operator [$t_{1/2} \sim$ 15–20 min (274)] is slow compared to the times employed in most experiments.

Studies of in vitro repression in the *gal* system give results comparable to the case of the *lac* UV-5 mutant and suggest that for *gal* too there are overlapping initiator and operator sites (95).

The λ repressor (c_I protein) can act either before or after formation of an RS complex (98, 99). In the latter instance repressor must be added to the RS complex for some period of time prior to addition of triphosphates. Simultaneous addition of repressor and triphosphates gives no repression (99). This is expected from the relative rates of repressor binding [$t_{1/2} \sim$ 1 min (290)] and RNA chain initiation by the complex [$t_{1/2} \sim$ 0.2 sec (176)]. Several repressor dimers can bind to the λ

operator region, protecting, progressively, a DNA duplex from 35 to 105 nucleotides in length (265). The sensitivity of binary complexes to λ repressor action suggests that at least some of these sites do not overlap with the initiator.

Direct steric interference is a sufficient explanation for the mechanism of repressor action for overlapping initiator-operator regions and might also hold for non-overlapping sites if the initiator were located upstream of the operator, and repressor could block the pathway of RNA polymerase (313). However, in the *ara* operon the operator site lies upstream of the promoter region (100) and may signal a different mode of repressor action.

The observations that *lac* and λ repressors bind specifically only to helical DNA (289, 314) and are not melting proteins (265) suggest that they could also act as antimelting proteins to block opening of the DNA strands in the promoter-operator region in the transition from closed to open promoter complex. A protein which binds tightly to helical DNA but not denatured DNA should raise the stability of the helix, both by virtue of its ability to shift the equilibrium between the two forms (2) and also because by fixing a certain DNA region in a helical conformation, free unwinding will be hindered in adjacent regions. This latter factor can be quite large; joining the end of dAT molecules to form a mini-circle of only a few base pairs raises their T_m substantially above the T_m for very long poly dAT helices (315, 316). Direct evidence that *lac* repressor can serve as a general antimelting protein has recently been obtained (316a).

FUNCTIONAL INTERACTION OF RNA POLYMERASE WITH TERMINATORS

Direct Termination

The selective termination of RNA chains is well documented and poorly understood. A variety of termination processes have been described for in vitro transcription (3, 8, 9, 45); only two are reviewed here. The first of these, direct termination of RNA chains, is mediated by the DNA template acting directly on RNA polymerase. Direct sequence termination was not detected for many years and in fact termination was considered to be completely lacking in vitro. This was due in part to the prevalent use of T4 DNA as template (217–219, 317), where direct sequence termination is not obtained at least at low ionic strengths, and in part to the fact that the transcripts obtained with other DNAs such as T7 appeared much larger than those isolated in vivo. This too suggested that some termination component was lacking. It is now known that the latter result is due in some instances to the fact that the smaller mRNAs found are derived in vivo by nucleolytic cleavage after transcription from large transcription units (128, 192, 240–243).

Termination must actually represent three processes; in principle, these are separable: 1. termination of RNA chain elongation, 2. release of the nascent RNA chain, and 3. release of RNA polymerase. All three of these reactions can occur at direct terminator sites in vitro. However, the sequence and efficiency of the different steps at different terminators have not been studied. Templates for which direct sequence termination with bacterial RNA polymerase is known include T7 DNA

(123, 230), fd RF I (234), λ DNA (125, 126), and the ternary *E. coli* DNA complexes which form ribosomal transcripts (36). In the cases of the major T7 early transcript and the two minor λ transcripts, specific 3'-terminal nucleotide sequences have been identified (125, 126, 230). The T7-specific RNA polymerase uses at least two different direct terminator sites in the selective transcription of the late genetic region of T7 (114, 115).

Early studies of RNA chain termination depended primarily on the measurement of repeated chain initiation by RNA polymerase. At elevated ionic strengths continual reinitiation occurs, whereas at lower ionic strengths the reaction terminates. This has led many groups to conclude that chain termination or RNA polymerase release (218–220, 317) does not occur at low ionic strengths. This result is partly due to the extensive use of T2 or T4 DNAs as templates; with other templates termination and enzyme release can occur at low ionic strengths (127, 175, 185, 230). Where the number of active RNA polymerase molecules added to the reaction is known, it has been shown with T7 DNA that even at low ionic strengths each enzyme can initiate, on the average, several T7 RNA chains (175). The enhancement of repeated chain initiation at elevated ionic strengths is probably due in large part to a depression by salt of the inhibition of RNA polymerase by product RNA after its release from the template (219, 287, 318). However, salt does have some effect on the termination process itself (45, 178) since it causes termination of growing chains of [rAU] copolymer by a ternary poly [dAT] complex. Schäfer & Zillig (220) have followed chain termination, transcript release, and enzyme release with T4, T5, and T7 DNAs and conclude that at low salt concentrations with T4 DNA there is little transcript or enzyme release after elongation ceases. With T5 and T7 DNAs, transcript release is observed but relatively little enzyme release. Their results suggest that salt can affect several steps in the termination process. The elucidation of the sequence of the three steps in chain termination and release and the clarification of the effects of salt on the process await the application of more direct assay procedures for the individual reactions.

A puzzling phenomenon which may be related to direct sequence termination in some way is the transient termination or hesitation of ternary complexes during growth of a transcript (158, 183, 319). The process may be enhanced at low triphosphate concentrations and gives rise to discrete species of partial transcripts during the reaction. These transcripts can be elongated further, hence do not reflect true termination events. Since no sequence similarities are apparent at these hesitation points (183), it may be that either the tertiary structure of the nascent RNA or its rate of displacement from the template DNA strand plays some role in the elongation process.

Rho-Induced Termination

Termination of certain transcripts in vitro appears not to occur at the sites employed in vivo but continues into adjacent regions. This is best documented in the case of λDNA where the major P_L and P_R transcripts are terminated at t_L and t_{R1} in vivo (Figure 4), but can continue past these points in vitro (74). Transcription of T3 early message in vitro by bacterial RNA polymerase also continues into genetic regions not read by that enzyme in vivo (320). Thus some component(s) of the transcription apparatus involved in selective termination appear to be missing from RNA polymerase holoenzyme. One such component may be ρ factor (74).

ρ Factor is an oligomeric protein [monomer mol wt 50,000 (74)] which depresses the amount of RNA formed in the in vitro reaction by causing RNA chain termination (74, 232–234, 321). The resulting RNA chains can be released from the DNA but the RNA polymerase remains bound to the DNA in an inactive complex. Thus ρ factor by itself does not allow continued recycling of RNA polymerase through the transcription sequence, and reconstruction of the physiological sequence either requires additional components of the termination-release step or some novel set of reaction conditions.

ρ Factor preparations aggregate reversibly to form complexes with sedimentation coefficients up to 9S (74, 262, 263), and ρ factor activity depends on its concentration in a manner which suggests that an oligomer is the active form (263). Highly purified ρ factor preparations contain components with a hexagonal subunit structure, as visualized with the electron microscope. A limited number of these structures can be observed to bind to fd RF I DNA (262). These complexes may be related to selective termination for ρ factor but direct evidence for their specific nature has not yet been presented. Formation of ρ-DNA complexes requires high protein concentrations in most (262, 263, 321) but not all (322) reports. This may reflect the necessity for an oligomeric form of the molecule (263) but may also simply reveal an inefficient nonspecific interaction (321). ρ Factor is active at much lower protein concentrations (74, 321) and can cause incorrect termination at high protein concentrations.

ρ Factor depresses RNA synthesis with a large variety of helical DNA templates (321) but not with synthetic templates such as dAT copolymer. There is little question that ρ leads to termination of RNA chains with all of these DNA templates, however the selectivity of the process in each instance is far less clear. Evidence that ρ can mediate termination at specific, biologically functional regions comes primarily from the study of λ transcription. ρ Factor diminishes transcription of λ DNA regions distal to the terminators t_L and t_{R1} (74). Early experiments in which these transcripts were resolved by sedimentation have been refined using the more sensitive techniques of gel electrophoresis to show that discrete λ RNA species are formed of the sizes expected for P_L and P_R RNAs (323).

Additional evidence that ρ can reproduce physiologically correct termination events comes from the study of polar insertion mutations. These mutations arise from insertion of a large DNA segment of unknown origin into an operon. In at least one instance such an insertion in the *gal* operon blocks transcription of distal *gal* genes in vivo and this can be mimicked in an in vitro transcription system by adding low concentrations of ρ factor (324).

In several instances ρ factor brings about termination events which appear to be nonphysiological. Transcription of T7 DNA in vitro in the presence of ρ yields discrete RNAs smaller than the normal 2.5×10^6 transcript (128). These RNAs do not, however, correspond to any of the early T7 mRNAs found in vivo.

ρ Factor can also bring about termination of transcription at sites in the interior on both *lac* and *gal* operons. It is possible that such sites might act partially in vivo to depress the amount of RNA for distal genes on the operon. However at elevated ρ concentrations all *gal* transcription distal to the initial (E) gene is blocked and *lac* mRNA is terminated in the interior of the z gene (324). This clearly represents a physiologically incorrect situation.

These results suggest that while ρ factor at low concentrations may cause termination at terminators used in vivo, higher concentrations of the factor can bring about termination at a variety of sites on DNA, some of which are not biologically active terminators.

ACKNOWLEDGMENTS

The author is indebted to a large number of scientists in the field who generously shared their ideas and their unpublished results during the preparation of this review.

Literature Cited

1. Yarus, M. 1969. *Ann. Rev. Biochem.* 38: 841
2. von Hippel, P., McGhee, J. 1972. *Ann. Rev. Biochem.* 41:231
3. Chamberlin, M. 1974. *Enzymes* 10: 333
4. Jacob, F., Ullman, A., Monod, J. 1964. *C.R. Acad. Sci. Paris* 258:3125
5. Epstein, W., Beckwith, J. 1968. *Ann. Rev. Biochem.* 37:411
6. Reznikoff, W. 1972. *Ann. Rev. Genet.* 6: 133
7. Travers, A. 1971. *Nature New Biol.* 229: 69
8. Burgess, R. 1971. *Ann. Rev. Biochem.* 40:711
9. Losick, R. 1972. *Ann. Rev. Biochem.* 41: 409
10. Bautz, E. 1972. *Progr. Nucl. Acid Res.* 12:129
11. Kurland, C. 1972. *Ann. Rev. Biochem.* 41:377
12. Klein, A., Bonhoeffer, F. 1972. *Ann. Rev. Biochem.* 41:301
13. Richardson, J. 1966. *Proc. Nat. Acad. Sci. USA* 55:1616
14. Berg, D., Chamberlin, M. 1970. *Biochemistry* 9:5055
15. Chamberlin, M., Ring, J. 1972. *J. Mol. Biol.* 70:221
16. Anthony, D., Goldthwait, D. 1970. *Biochim. Biophys. Acta* 204:156
17. Ruet, A., Sentenac, A., Fromageot, P. 1970. *FEBS Lett.* 3:169
18. Bibilashvili, R., Savotchkina, L. 1973. *Biochim. Biophys. Acta* 294:434
19. Heyden, B., Nüsslein, C., Schaller, H. 1972. *Nature New Biol.* 240:9
20. Burgess, R. 1969. *J. Biol. Chem.* 244: 6168
21. Burgess, R., Travers, A., Dunn, J., Bautz, E. 1969. *Nature* 221:43
22. Scrutton, M., Wu, C., Goldthwait, D. 1971. *Proc. Nat. Acad. Sci. USA* 68: 2497
23. Hartmann, G., Honikel, K., Knüsel, F., Nüesch, J. 1967. *Biochim. Biophys. Acta* 157:322
24. Zuno, S. M., Yamazaki, H., Nitta, K., Umezawa, H. 1968. *Biochim. Biophys. Acta* 157:322
25. Siddhikol, C., Erbstoeszer, J., Weisblum, B. 1969. *J. Bacteriol.* 99:151
26. Zillig, W. et al 1970. *Cold Spring Harbor Symp. Quant. Biol.* 35:47
27. Heil, A., Zillig, W. 1970. *FEBS Lett.* 11: 165
28. Rabussay, D., Zillig, W. 1970. *FEBS Lett.* 5:104
29. Iwakura, Y., Ishihama, A., Yura, T. 1973. *Mol. Gen. Genet.* 121:181
30. Matzura, H., Molin, S., Maaløe, O. 1971. *J. Mol. Biol.* 59:17
31. Nakamura, Y., Yura, T. 1973. *Biochem. Biophys. Res. Commun.* 53:645
32. Berg, D., Barrett, K., Chamberlin, M. 1971. *Methods Enzymol.* 21:506
33. Krakow, J., Daley, K., Karstadt, M. 1969. *Proc. Nat. Acad. Sci. USA* 62:432
34. Travers, A., Burgess, R. 1969. *Nature* 222:537
35. Gariglio, P., Roodman, S., Green, M. 1969. *Biochem. Biophys. Res. Commun.* 37:945
36. Pettijohn, D., Stonington, O., Kossman, C. 1970. *Nature* 228:235
37. Gerard, G., Johnson, J., Boezi, J. 1972. *Biochemistry* 11:989
38a. Johnson, J., DeBacker, M., Boezi, J.

1971. *J. Biol. Chem.* 246:1222
38b. Herzfeld, F., Zillig, W. 1971. *Eur. J. Biochem.* 24:242
38c. Bendis, I., Shapiro, L. 1973. *J. Bacteriol.* 115:848
39. Walter, G., Seifert, W., Zillig, W. 1968. *Biochem. Biophys. Res. Commun.* 30:240
40. Lill, U., Hartmann, G. 1970. *Biochem. Biophys. Res. Commun.* 39:930
41. Ishihama, A. 1972. *Biochemistry* 11:1250
42. Yarbrough, L. 1973. *Fed. Proc.* 32:509
43. Ishihama, A., Ito, K. 1972. *J. Mol. Biol.* 72:111
44. Weber, K., Kuter, D. 1971. *J. Biol. Chem.* 246:4504
45. Chamberlin, M. 1970. *Cold Spring Harbor Symp. Quant. Biol.* 35:851
46. Pene, J. 1969. *Nature* 223:705
47. Snyder, L. 1973. *Nature* 243:131
48. Chao, L., Speyer, J. 1973. *Biochem. Biophys. Res. Commun.* 51:399
49. Travers, A., Buckland, R. 1973. *Nature New Biol.* 243:257
50. Sümegi, J., Sanner, T., Pihl, A. 1972. *Biochim. Biophys. Acta* 262:145
51. Nüsslein, C., Heyden, B. 1972. *Biochem. Biophys. Res. Commun.* 47:282
52. Chelala, C., Hirschbein, L., Torres, H. 1971. *Proc. Nat. Acad. Sci. USA* 68:152
53. Sarkar, N., Paulus, H. 1972. *Proc. Nat. Acad. Sci. USA* 69:3570
54. Leighton, R. J., Freese, P. K., Doi, R., Warren, R., Kelln, R. 1972. *Spores* 5:238
55. Linn, T., Greenleaf, A., Shorenstein, R., Losick, R. 1973. *Proc. Nat. Acad. Sci. USA* 70:1865
56. Matzura, H., Hansen, B., Zenthen, J. 1973. *J. Mol. Biol.* 74:9
57. Dalbow, D. 1973. *J. Mol. Biol.* 75:181
58. Bremer, H., Berry, L., Dennis, P. 1973. *J. Mol. Biol.* 75:161
59. Dennis, P., Bremer, H. 1973. *J. Mol. Biol.* 75:145
60. Nierlich, D. 1972. *J. Mol. Biol.* 72:751, 765
61. Seifert, W. et al 1969. *Eur. J. Biochem.* 9:319
62. Rabussay, D., Mailhammer, R., Zillig, W. 1972. In *Metabolic Interconversions of Enzymes,* ed. O. Wieland, E. Helmreich, H. Holzer, 213. Berlin/Heidelberg/New York: Springer
63. Goff, C., Weber, K. 1970. *Cold Spring Harbor Symp. Quant. Biol.* 35:101
63a. Goff, C. *J. Biol. Chem.* In press
64. Wu, R., Geiduschek, E. P., Rabussay, D., Cascino, A. 1973. *Virus Research; 2nd ICN-UCLA Symp. Mol. Biol.,* ed. C. F. Fox, W. S. Robinson, 181. New York: Academic

65. Bautz, E., Bautz, F., Dunn, J. 1969. *Nature* 223:1022
66. Crouch, R., Hall, B., Hager, G. 1970. *Nature* 223:476
67. Stevens, A. 1972. *Proc. Nat. Acad. Sci. USA* 69:603
68. Horvitz, H. 1973. *Nature New Biol.* 244:137
69. Schachner, M., Seifert, W., Zillig, W. 1971. *Eur. J. Biochem.* 22:520
70. Travers, A. 1970. *Nature* 225:1009
71. Travers, A. 1970. *Cold Spring Harbor Symp. Quant. Biol* 35:241
72. Georgopoulos, C., Herskowitz, I. 1971. In *The Bacteriophage Lambda,* ed. A. D. Hershey. N.Y., Cold Spring Harbor, NY: Cold Spring Harbor Press
73. Ghysen, A., Pironio, M. 1972. *J. Mol. Biol.* 65:259
74. Roberts, J. 1969. *Nature* 224:1168
75. Portier, M., Marcaud, L., Cohen, A., Gros, F. 1972. *Mol. Gen. Genet.* 117:72
76. Greenblatt, J. 1972. *Proc. Nat. Acad. Sci. USA* 69:3606
77. Dottin, R., Pearson, M. 1973. *Proc. Nat. Acad. Sci. USA* 70:1078
78. Gage, L., Geiduschek, E. P. 1971. *J. Mol. Biol.* 57:279
78a. Duffy, J., Geiduschek, E. P. 1973. *FEBS Lett.* 34:172
79. Spiegelman, G., Whiteley, H. R. 1974. *J. Biol. Chem.* In press
80. Losick, R., Sonenshein, A. 1969. *Nature* 224:35
81. Greenleaf, A., Linn, T., Losick, R. 1973. *Proc. Nat. Acad. Sci. USA* 70:490
82. Losick, R., Shorenstein, R., Sonenshein, A. 1970. *Nature* 227:910
83. Sonenshein, A., Losick, R. 1970. *Nature* 227:906
84. Hager, G., Hall, B., Fields, K. 1970. *Cold Spring Harbor Symp. Quant. Biol.* 35:233
85. Vogt, V. 1969. *Nature* 223:854
86. Hinkle, D., Ring, J., Chamberlin, M. 1972. *J. Mol. Biol.* 70:197
87. Ishihama, A., Murakami, S., Fukuda, R., Matsukage, A., Kameyama, T. 1971. *Mol. Gen. Genet.* 111:66
88. Leavitt, J., Moldave, K., Nakada, D. 1972. *J. Mol. Biol.* 70:15
89. Davison, J., Pilarski, L., Echols, H. 1969. *Proc. Nat. Acad. Sci. USA* 63:168
90. Mahadik, S., Srinivasan, P. 1971. *Proc. Nat. Acad. Sci. USA* 68:1898
91. Travers, A., Kamen, R., Schleif, R. 1970. *Nature* 228:21
92. Murooka, Y., Lazzarini, R. 1972. *Proc. Nat. Acad. Sci. USA* 69:2336
93. Ghosh, S., Echols, H. 1972. *Proc. Nat. Acad. Sci. USA* 69:3660

93a. Cukier-Kahn, R., Jacquet, M., Gros, F. 1972. *Proc. Nat. Acad. Sci. USA* 69: 3643

94. Müller-Hill, B., Beyreuther, K., Gilbert, W. 1971. *Methods Enzymol.* 21:483

95. Nakanishi, S., Adhya, S., Gottesman, M., Pastan, I. 1973. *Proc. Nat. Acad. Sci. USA* 70:334; *J. Biol. Chem.* 248 : 5937

96. Ptashne, M., Pirrotta, V., Hopkins, N. 1971. *Methods Enzymol.* 21:487

97. Chen, B. et al 1971. *Nature New Biol.* 233:67

98. Wu, A., Ghosh, S., Echols, H. 1972. *J. Mol. Biol.* 67:423

99. Steinberg, R., Ptashne, M. 1971. *Nature New Biol.* 230:76

100. Englesberg, E. 1971. *Metab. Pathways* 5:257

101. Greenblatt, J., Schleif, R. 1971. *Nature* 233:166

102. Wilcox, G., Clemetson, K., Santi, D., Englesberg, E. 1971. *Proc. Nat. Acad. Sci. USA* 68:2145

103. Anderson, W., Schneider, A., Emmer, M., Perlman, R., Pastan, I. 1970. *J. Biol. Chem.* 246:5959

104. Zubay, G., Schwartz, D., Beckwith, J. 1970. *Proc. Nat. Acad. Sci. USA* 66:104

105. de Crombrugghe, B. et al 1971. *Nature New Biol.* 231:139

106. Eron, L., Block, R. 1971. *Proc. Nat. Acad. Sci. USA* 68:1828

107. Nissley, S., Anderson, W., Gottesman, M., Perlman, R., Pastan, I. 1971. *J. Biol. Chem.* 246:4671

108. Riggs, A., Reiness, G., Zubay, G. 1971. *Proc. Nat. Acad. Sci. USA* 68:1222

109. Chamberlin, M., McGrath, J., Waskell, J. 1970. *Nature* 228:227

110. Dunn, J., Bautz, F., Bautz, E. 1971. *Nature New Biol.* 230:94

111. Maitra, U. 1971. *Biochem. Biophys. Res. Commun.* 43:443

112. Chamberlin, M., Ring, J. 1973. *J. Biol. Chem.* 248:2235

113. Dunn, J., McAllister, W., Bautz, E. 1972. *Virology* 48:112

114. Golomb, M., Chamberlin, M. 1974. *J. Biol. Chem.* 249:2858

115. Golomb, M., Chamberlin, M. 1974. *Proc. Nat. Acad. Sci. USA* 71:760

116. Hinkle, D., Chamberlin, M. 1972. *J. Mol. Biol.* 70:157

117. Hinkle, D., Chamberlin, M. 1972. *J. Mol. Biol.* 70:187

118. Dingman, C., Peacock, A. 1968. *Biochemistry* 7:668

119. Dahlberg, A., Dingman, C., Peacock, A. 1969. *J. Mol. Biol.* 41:139

120. De Wachter, R., Fiers, W. 1971. *Methods Enzymol.* 21:167

121. Fisher, M., Dingman, C. 1970. *Biochemistry* 10:1896

122. Dingman, C., Fisher, M., Kakefuda, T. 1972. *Biochemistry* 11:1242

123. Minkley, E., Pribnow, D. 1973. *J. Mol. Biol.* 77:255

124. Simon, M., Studier, F. W. 1973. *J. Mol. Biol.* 79:249

125. Lebowitz, P., Weissman, S., Radding, C. 1971. *J. Biol. Chem.* 246:5120

126. Blattner, F., Dahlberg, J. 1972. *Nature New Biol.* 237:227

127. Maitra, U., Nakata, Y., Hurwitz, J. 1967. *J. Biol. Chem.* 242:4908

128. Dunn, J., Studier, F. 1973. *Proc. Nat. Acad. Sci. USA* 70:1559

129. Hayashi, M., Hayashi, M., Spiegelman, S. 1963. *Proc. Nat. Acad. Sci. USA* 50:664

130. Hayashi, M., Hayashi, M., Spiegelman, S. 1964. *Proc. Nat. Acad. Sci. USA* 51:354

131. Tocchini-Valentini, G. et al 1963. *Proc. Nat. Acad. Sci. USA* 50:935

132. Marmur, J., Greenspan, C. N. 1963. *Science* 142:387

133. Gillespie, D. 1968. *Methods Enzymol.* 12B:641

134. McCarthy, B., Church, R. 1970. *Ann. Rev. Biochem.* 39:131

135. Bøvre, K., Szybalski, W. 1971. *Methods Enzymol.* 21:350

135a. Kennel, D. E. 1971. *Progr. Nucl. Acid Res.* 11:259

136. Summers, W. C., Szybalski, W. 1968. *Virology* 34:9

137. Taylor, K., Hradecna, Z., Szybalski, W. 1967. *Proc. Nat. Acad. Sci. USA* 57:1618

138. Guha, A., Szybalski, W. 1968. *Virology* 34:608

139. Lozeron, H. A., Szybalski, W. 1969. *Virology* 39:373

140. Szybalski, W., Kubinski, H., Hradecna, Z., Summers, W. 1971. *Methods Enzymol.* 21:383

141. Colvill, A., Kanner, L., Tocchini-Valentini, G., Sarnat, M., Geiduschek, E. P. 1965. *Proc. Nat. Acad. Sci. USA* 53:1140

141a. Spiegelman, W. G. et al 1972. *Proc. Nat. Acad. Sci. USA* 69:3156

142. Shapiro, J. et al 1969. *Nature* 224:768

143. Haseltine, W. 1972. *Nature* 235:329

144. Chen, C., Hutchison, C., Edgell, M. 1973. *Nature New Biol.* 243:233

145. Danna, K., Sack, G., Nathans, D. 1973. *J. Mol. Biol.* 78:363

146a. Zain, B., Dhar, R., Weissman, S., Lebowitz, P., Lewis, A. 1973. *J. Virol.* 11:682

146b. Khoury, G., Martin, M., Lee, T., Danna, K., Nathans, D. 1973. *J. Mol. Biol.* 78:377

147a. Smith, H., Wilcox, K. 1970. *J. Mol. Biol.* 51:379

147b. Yoshimori, R. N. 1971. PhD thesis, University of California Medical Center, San Francisco

147c. Gromkova, R., Goodgal, S. 1972. *J. Bacteriol.* 109:987

147d. Sharp, P. A., Sugden, B., Sambrook, J. 1973. *Biochemistry* 12:3055

148. Davis, R., Hyman, R. 1970. *Cold Spring Harbor Symp. Quant. Biol.* 35:269

149. Hyman, R. 1971. *J. Mol. Biol.* 61:369

150. Darnell, J. 1968. *Bacteriol. Rev.* 32:262

151. Brody, E., Geiduschek, E. 1970. *Biochemistry* 9:1300

152. Altman, S. 1971. *Nature New Biol.* 229:19

153. Morse, D. 1971. *J. Mol. Biol.* 55:113

154. Sippel, A., Hartmann, G. 1970. *Eur. J. Biochem.* 16:152

155. Bautz, E., Bautz, F. 1970. *Cold Spring Harbor Symp. Quant. Biol.* 35:227

156. Sanger, F., Brownlee, G., Barrell, B. 1965. *J. Mol. Biol.* 13:373

157. Billeter, M., Dahlberg, J., Goodman, H., Hindley, J., Weissmann, C. 1969. *Cold Spring Harbor Symp. Quant. Biol.* 34:635

158. Dahlberg, J., Blattner, F. 1973. In *Virus Research,* ed. W. Robinson, C. Fox. New York: Academic

159. Sugiura, M., Okamoto, T. Takanami, M. 1969. *J. Mol. Biol.* 43:299

159a. Hyman, R., Davidson, N. 1970. *J. Mol. Biol.* 70:421

160. Sugiura, M., Okamoto, T., Takanami, M. 1970. *Nature* 225:598

161. Goff, C., Minkley, C. 1970. *Le Petit Colloquium on RNA Polymerase,* 1:124. Amsterdam: North-Holland

162. Hoffman, D., Niyogi, S. 1973. *Biochim. Biophys. Acta* 299:588

163. Zubay, G., Chambers, D., Cheong, L. 1970. In *The Lactose Operon,* ed. J. Beckwith, D. Zipser, 375. Cold Spring Harbor, NY: Cold Spring Harbor Press

164. Gold, L., Schweiger, M. 1971. *Methods Enzymol.* 20:535

165. Young, E. 1970. *J. Mol. Biol.* 51:591

166. Wilhelm, J., Haselkorn, R. 1971. *Methods Enzymol.* 20:531

167. Black, L., Gold, L. 1971. *J. Mol. Biol.* 60:365

168. O'Farrell, P., Gold, L. 1973. *J. Biol. Chem.* 248:5512

169. Jayaraman, R., Goldberg, E. 1970. *Cold Spring Harbor Symp. Quant. Biol.* 35:197

170. Harford, A., Beer, M. 1972. *J. Mol. Biol.* 69:179

171. Delius, H., Westphal, H., Axelrod, N. 1973. *J. Mol. Biol.* 74:677

172. Murti, K., Prescott, D., Pene, J. 1972. *J. Mol. Biol.* 68:413

173. Hagen, U., Ullrich, M., Petersen, E., Werner, E., Kroeger, H. 1970. *Biochim. Biophys. Acta* 199:115

174. Goddard, J., Weiss, J., Wheelen, C. 1970. *Biochim. Biophys. Acta* 199:125, 139

175. Chamberlin, M., Ring, J. 1972. *J. Mol. Biol.* 70:221

176. Mangel, W., Chamberlin, M. 1974. *J. Biol. Chem.* In press

177. Anthony, D., Wu, C., Goldthwait, D. 1969. *Biochemistry* 8:246

178. Rhodes, G., Chamberlin, M. 1974. *J. Biol. Chem.* In press

179. Niyogi, S., Stevens, A. 1965. *J. Biol. Chem.* 240:2593

180. Terao, T., Dahlberg, J., Khorana, H. G. 1972. *J. Biol. Chem.* 247:6157

181. Van Kreijl, C., Borst, P. 1973. *Biochem. Biophys. Res. Commun.* 54:17

182. Downey, K., Jurmark, B., So, A. 1971. *Biochemistry* 10:4970

183. Gilbert, W., Maxam, A. 1973. *Proc. Nat. Acad. Sci. USA* 7:3581; Maizels, N. Ibid. 7:3585

184. Schweiger, M., Herrlich, P., Millette, R. 1971. *J. Biol. Chem.* 246:6707

185. Okamoto, T., Sugiura, M., Takanami, M. 1969. *J. Mol. Biol.* 45:101

186. Roberts, J. 1969. *Cold Spring Harbor Symp. Quant. Biol.* 35:121

187. Studier, F. W. 1972. *Science* 176:367

188. Summers, W. 1972. *Ann. Rev. Genet.* 6:191

189. Studier, F. W., Maizel, J. R. 1969. *Virology* 39:575

190. Hyman, R., Summers, W. 1972. *J. Mol. Biol.* 71:573

191. Minkley, E. 1974. *J. Mol. Biol.* 83:289, 305

192. Dunn, J., Studier, F. W. 1973. *Proc. Nat. Acad. Sci. USA* 70:3296

193. Summers, W., Siegel, R. 1970. *Nature* 228:1160

194. Summers, W. 1969. *Virology* 39:175

195. Szybalski, W. et al 1970. *Cold Spring Harbor Symp. Quant. Biol.* 35:341

196. Davidson, N., Szybalski, W. 1971. See Ref. 72

197. Blattner, F., Dahlberg, J., Boettiger, J., Fiandt, M., Szybalski, W. 1972. *Nature New Biol.* 237:232

198. Herskowitz, I. 1973. *Ann. Rev. Genet.* 7:289

199. Blattner, F. 1973. Personal communication

200. Echols, H. 1971. *Ann. Rev. Biochem.* 40:827
201. Tonegawa, S., Hayashi, M. 1968. *Proc. Nat. Acad. Sci. USA* 61:1320
202. Eisen, H., Ptashne, M. 1971. *Proc. Nat. Acad. Sci. USA* 61:239
203. Echols, H., Green, L. 1971. *Proc. Nat. Acad. Sci. USA* 68:2190
203a. Champoux, J. 1970. *Cold Spring Harbor Symp. Quant. Biol.* 35:319
203b. Hayes, S., Szybalski, W. 1973. *Mol. Gen. Genet.* 126:275
204. Brody, E., Diggelmann, H., Geiduschek, E. P. 1970. *Biochemistry* 9:1289
205. Domingo, E., Escarmis, C., Warner, R. 1974. *J. Biol. Chem.* In press
206. Dean, W., Lebowitz, J. 1971. *Nature New Biol.* 231:5
207. Kato, A., Bartok, K., Fraser, M., Denhardt, D. 1973. *Biochim. Biophys. Acta* 308:68
208. Beard, P., Morrow, J., Berg, P. 1973. *J. Virol.* 12:1303
208a. Morrow, J., Berg, P. 1973. *J. Virol.* 12:1631
209. Danna, K., Nathans, D. 1971. *Proc. Nat. Acad. Sci. USA* 68:2913
210. Chamberlin, M., Berg, P. 1964. *J. Mol. Biol.* 8:297
211. Brody, E., Geiduschek, E. P. 1970. *Biochemistry* 9:1300
212. Wood, W. B., Berg, P. 1964. *J. Mol. Biol.* 9:452
213. Chessin, H., Summers, W. 1970. *Biochem. Biophys. Res. Commun.* 38:40
214. Boule-Charest, L., Mamet-Bratley, M. 1972. *Biochim. Biophys. Acta* 277:276
215. Richardson, J. 1966. *J. Mol. Biol.* 21:83
216. Pettijohn, D., Kamiya, T. 1967. *J. Mol. Biol.* 29:275
217. Fuchs, E., Millette, R., Zillig, W., Walter, G. 1967. *Eur. J. Biochem.* 3:183
218. Richardson, J. P. 1970. *Nature* 225:1109
219. Maitra, U., Barash, F. 1969. *Proc. Nat. Acad. Sci. USA* 64:779
220. Schäfer, R., Zillig, W. 1973. *Eur. J. Biochem.* 33:215
221. Whiteley, H., Hemphill, H. 1970. *Biochem. Biophys. Res. Commun.* 41:647
222. Avila, J., Hermoso, J., Vinuela, E., Salas, M. 1971. *Eur. J. Biochem.* 21:526
223. Pene, J., Barrow-Carraway, J. 1972. *J. Bacteriol.* 111:15
224. Matsukage, A. 1972. *Mol. Gen. Genet.* 118:11
225. Saucier, J., Wang, J. 1972. *Nature New Biol.* 239:167
226. Lin, S., Riggs, A. 1972. *J. Mol. Biol.* 72:671
227. Chamberlin, M., Ring, J. Unpublished observation
228. Dausse, J., Sentenac, A., Fromageot, P. 1972. *Eur. J. Biochem.* 26:43
229. Pannekoek, H., Pouwels, P. 1973. *Mol. Gen. Genet.* 123:159
230. Millette, R., Trotter, C., Herrlich, P., Schweiger, M. 1970. *Cold Spring Harbor Symp. Quant. Biol.* 35:135
231. Rhodes, G., Chamberlin, M. Unpublished studies
232. Goldberg, A. 1970. *Cold Spring Harbor Symp. Quant. Biol.* 35:157
233. Richardson, J. 1970. *Cold Spring Harbor Symp. Quant. Biol.* 35:127
234. Takanami, M., Okamoto, I., Sugiura, M. 1971. *J. Mol. Biol.* 62:81
235. Levine, L., Gordon, J., Jencks, W. 1963. *Biochemistry* 2:168
236. Nakanishi, S., Adhya, S., Gottesman, M., Pastan, I. 1974. *J. Biol. Chem.* In press
237. Brody, E., Leautey, J. 1973. *Eur. J. Biochem.* 36:347
238. Szybalski, W., Kubinski, H., Sheldrick, P. 1966. *Cold Spring Harbor Symp. Quant. Biol.* 31:123
239. Champoux, J., Hogness, D. 1972. *J. Mol. Biol.* 71:383
240. Altman, S., Smith, J. D. 1971. *Nature New Biol.* 233:35
241. Robertson, H., Altman, S., Smith, J. D. 1972. *J. Biol. Chem.* 247:5243
242. Nierlich, D., Lamfrom, H., Sarabhai, A., Abelson, J. 1973. *Proc. Nat. Acad. Sci. USA* 70:179
242a. Nikolaev, N., Silengo, L., Schlessinger, D. 1973. *Proc. Nat. Acad. Sci. USA* 70:3361
243. Chowdhury, D., Guha, A. 1973. *Nature New Biol.* 241:196
244. Inman, R., Schnös, M. 1970. *J. Mol. Biol.* 49:93
245. Gomez, B., Lang, D. 1972. *J. Mol. Biol.* 70:239
246. Sanger, F., Donelson, J., Coulson, A., Kössel, H., Fischer, D. 1973. *Proc. Nat. Acad. Sci. USA* 70:1209
247. Salser, W., Fry, K., Brunk, C., Poon, R. 1972. *Proc. Nat. Acad. Sci. USA* 69:238
248. Kleppe, R., Khorana, H. G. 1972. *J. Biol. Chem.* 247:6149
249. Khorana, H. et al 1972. *J. Mol. Biol.* 72:209
250. Besmer, P. et al 1971. *Fed. Proc.* 30:1314
251. Besmer, P. et al 1972. *J. Mol. Biol.* 72:503
252. Loewen, P., Sekiya, T., Khorana, H.

1974. *J. Biol. Chem.* 249:217

253. Schafer, R., Zillig, W., Zechel, K. 1973. *Eur. J. Biochem.* 33:207

254. Okamoto, T., Sugiura, M., Takanami, M. 1972. *Nature New Biol.* 237:108

255. Nakano, E., Sakaguchi, K. 1969. *J. Biochem.* 65:147

256. Matsukage, A., Murakami, S., Kameyama, J. 1969. *Biochim. Biophys. Acta* 179:145

257. Ruger, W. 1971. *Biochim. Biophys. Acta* 238:202

258. Shishido, K., Ikeda, Y. 1971. *Biochem. Biophys. Res. Commun.* 44:1420

259. Le Talaer, J. Y., Jeanteur, P. 1971. *Proc. Nat. Acad. Sci. USA* 68:3211

260. Le Talaer, J. Y., Kermici, M., Jeanteur, P. 1973. *Proc. Nat. Acad. Sci. USA* 70:2911

261. Ikeda, H. 1971. *Nature New Biol.* 234:198

262. Oda, T., Takanami, M. 1972. *J. Mol. Biol.* 71:799

263. Minkely, E. 1973. *J. Mol. Biol.* 78:577

264. Pirotta, V. 1973. *Nature New Biol.* 244:13

265. Maniatis, T., Ptashne, M. 1973. *Proc. Nat. Acad. Sci. USA* 70:1531

266. Silverstone, A., Arditti, R., Magasanik, B. 1970. *Proc. Nat. Acad. Sci. USA* 66:773

267. Beckwith, J., Grodzicker, T., Arditti, R. 1972. *J. Mol. Biol.* 69:155

268. Sadler, J. R., Smith, T. F. 1971. *J. Mol. Biol.* 62:139

269. Thomas, C. 1966. *Progr. Nucl. Acid Res.* 5:315

270. Gierer, A. 1966. *Nature* 212:1480

271. Sobel, H. 1972. *Proc. Nat. Acad. Sci. USA* 69:2483

272. Bram, S. 1971. *Nature New Biol.* 232:174

273. Adler, K. et al 1972. *Nature* 237:322

274. Riggs, A., Bourgeois, S., Cohn, M. 1970. *J. Mol. Biol.* 53:401

275. Hinkle, D., Mangel, W., Chamberlin, M. 1972. *J. Mol. Biol.* 70:209

275a. McGhee J., von Hippel, P. 1974. *J. Mol. Biol.* In press

276. Baker, R., Yanofsky, C. 1972. *J. Mol. Biol.* 69:89

277. Geiduschek, E. P., Haselkorn, R. 1969. *Ann. Rev. Biochem.* 38:647

278. Pettijohn, D., Clarkson, K., Kossman, C., Stonington, O. 1970. *J. Mol. Biol.* 52:281

279. Hinkle, D. 1971. PhD thesis, University of California, Berkeley

280. di Mauro, E. et al 1969. *Nature* 222:533

281. Wu, C., Goldthwait, D. 1969. *Biochemistry* 8:4458

282. Fox, C., Robinson, W., Haselkorn, R., Weiss, S. 1964. *J. Biol. Chem.* 239:186

283. Stevens, A. 1969. *J. Biol. Chem.* 244:425

284. Hirschbein, L., Dubert, J., Babinet, C. 1967. *Eur. J. Biochem.* 1:135

285. Walter, G., Zillig, W., Palm, P., Fuchs, E. 1967. *Eur. J. Biochem.* 3:194

286. Chambon, P., Ramuz, M., Mandel, P., Doly, J. 1967. *Biochim. Biophys. Acta* 149:584

287. Tissieres, A., Bourgeois, S., Gros, F. 1963. *J. Mol. Biol.* 7:100

288. Scheffler, I., Sturtevant, J. 1969. *J. Mol. Biol.* 42:577

288a. Shiao, D., Sturtevant, J. 1973. *Biopolymers* 12:1829

288b. Chamberlin, M. Unpublished studies

289. Riggs, A., Suzuki, H., Bourgeois, S. 1970. *J. Mol. Biol.* 48:67

290. Chadwick, P., Pirrotta, V., Steinberg, R., Hopkins, N., Ptashne, M. 1970. *Cold Spring Harbor Symp. Quant. Biol.* 35:283

290a. McConnell, D., Bonner, J. 1973. *Eur. J. Biochem.* 38:111

291. Remold-O'Donnell, E., Zillig, W. 1969. *Eur. J. Biochem.* 7:318

292. Richardson, J. 1973. Personal communication

293. Crothers, D., Zimm, B. 1964. *J. Mol. Biol.* 9:1

294. Bauer, W., Vinograd, J. 1968. *J. Mol. Biol.* 33:141, 173

295. Wang, J. 1969. *J. Mol. Biol.* 43:263

296. Waring, M. 1970. *J. Mol. Biol.* 54:247

297. Alberts, B., Frey, L. 1970. *Nature* 227:1313

298. Davidson, N. 1972. *J. Mol. Biol.* 66:307

299. Sinsheimer, R. 1969. *Progr. Nucl. Acid Res. Mol. Biol.* 8:115

300. Marvin, D., Hohn, B. 1969. *Bacteriol. Rev.* 33:172

301. Francke, B., Ray, D. 1971. *J. Mol. Biol.* 61:565

302. Warnaar, S., Mulder, G., Van der Sluys, I., van Kestern, L., Cohen, J. 1969. *Biochim. Biophys. Acta* 174:239

303. Hayashi, Y., Hayashi, M. 1971. *Biochemistry* 10:4212

304. Hayashi, Y., Hayashi, M. 1970. *Cold Spring Harbor Symp. Quant. Biol.* 35:17

305. Jacob, E., Hofschneider, P. 1969. *J. Mol. Biol.* 46:359

305a. Warner, R. Personal communication.

306. Richardson, J., Parker, S. 1973. *J. Mol. Biol.* 78:715

307. Botchan, P., Wang, J., Echols, H. 1973. *Proc. Nat. Acad. Sci. USA* 70:3077

308. Wang, J., Baumgarten, D., Olivera, B. 1967. *Proc. Nat. Acad. Sci. USA* 58: 1852
309. Puga, A., Tessman, I. 1973. *J. Mol. Biol.* 75:83
310. Worcel, A., Burgi, E. 1972. *J. Mol. Biol.* 71:127
311. Mueller, K. 1971. *Mol. Gen. Genet.* 111: 273
311a. Whiteley, H., Hemphill, H. 1971. *Biochem. Biophys. Res. Commun.* 41: 647
311b. Shorenstein, R., Losick, R. 1973. *J. Biol. Chem.* 248:6163
312. de Crombrugghe, B. et al 1971. *Nature* 230:37
313. Reznikoff, W., Miller, J., Scaife, J., Beckwith, J. 1969. *J. Mol. Biol.* 43: 201
314. Ptashne, M. 1967. *Nature* 214:232
315. Scheffler, I., Elson, E., Baldwin, R. 1970. *J. Mol. Biol.* 48:145

316. Baldwin, R. 1971. *Accounts Chem. Res.* 4:265
316a. Wang, A., Gross, C., Revzin, A., von Hippel, P. Personal communication
317. Bremer, H., Konrad, M. 1964. *Proc. Nat. Acad. Sci. USA* 51:801
318. So, A., Davie, E., Epstein, R., Tissieres, A. 1967. *Proc. Nat. Acad. Sci. USA* 58:1739
319. Darlix, J., Fromageot, P. 1972. *Biochimie* 54:47
320. Dunn, J., McAllister, W., Bautz, E. 1972. *Eur. J. Biochem.* 29:500
321. Goldberg, A., Hurwitz, J. 1972. *J. Biol. Chem.* 247:5637
322. Darlix, J., Sentenac, A., Fromageot, P. 1971. *FEBS Lett.* 13:165
323. Roberts, J. 1973. Personal communication
324. de Crombrugghe, B., Adhya, S., Gottesman, M., Pastan, I. 1973. *Nature New Biol.* 241:250

BIOCHEMISTRY OF MAMMALIAN FERTILIZATION[1]

Robert A. McRorie and William L. Williams

Reproduction Research Laboratories, Department of Biochemistry
University of Georgia, Athens, Georgia

CONTENTS

INTRODUCTION

Research on mammalian fertilization is extensive in most of the biological disciplines (1–6). The research in reproductive physiology, in particular, has revealed many exciting problems in fertilization that have been relatively neglected

[1] Abbreviations not explained in the text are: ANA, arginine-β-naphthylamide; APB, p-amidinophenacyl bromide; BAEE, N-benzoylarginine ethyl ester; BANA, N-benzoylarginine-β-naphthylamide; BAPA, N-benzoylarginine p-nitroanilide; DFP, diisopropylfluorophosphate; GPB, p-guanidinophenacyl bromide; LNA, lysyl β-naphthylamide; NPGB, p-nitrophenyl-p-guanidinobenzoate; SDS, sodium dodecylsulfate; TAME, N-tosylarginine methyl ester; TLCK, N-α tosyllysine chloromethylketone; TPCK, N-tosylphenylalanine chloromethylketone; ZANA, N-carbobenzoxyarginine-β-naphthylamide.

regarding the molecular events, i.e. the biochemistry. Considering the importance of the problems, they should be a fertile field for additional biochemical endeavor. We believe that for ultimate understanding and control, a biochemical approach is required and hope that this review might help in a small way to make the phrase "reproductive biochemistry" as common as "reproductive physiology."

Prior to fertilization sperm must undergo capacitation (7, 8), a process in the female reproductive tract whereby sperm achieve the ability to penetrate the ovum. The male sheds sperm incapable of fertilization and the female processes them for some hours until fertilizing ability is acquired. Since no changes visible in the phase or electron microscope take place in sperm as a result of capacitation (9–13), the process appears to be biochemical. As is developed later, we regard capacitation as activation of sperm acrosomal, ovum-penetrating enzymes as well as labilization of sperm membranes (12). From a biochemical viewpoint one of the more interesting and intriguing aspects of mammalian fertilization is that the sperm hydrolytic enzymes involved in penetrating the investments of the ovum are inactive (as they must be) when sperm are in the male and when first deposited in the female. Experimental evidence, although still incomplete, strongly suggests that the enzymes are either released, exposed, or activated as they are needed to attack each succeeding egg investment, i.e. each obstacle to fertilization.

Before considering the details of certain molecules and molecular events in mammalian fertilization, it is useful to look at the ultrastructure of sperm and egg and the changes in sperm membranes during ovum penetration. As diagrammed in Figures 1 to 3, fertilization takes place in most species, including the human and rabbit, with the egg investments, the cumulus oophorus, corona radiata, and zona pellucida, intact.

Figure 1 shows a capacitated spermatozoon sectioned through the thin flat head with the acrosome, a cap-like structure, and membranes intact. The plasma and acrosomal membranes are labilized and hyaluronidase is available to digest the

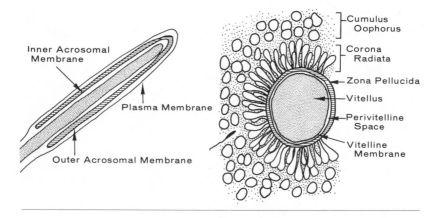

Figure 1   Status of capacitated sperm (*left*) as it penetrates cumulus (*right*).

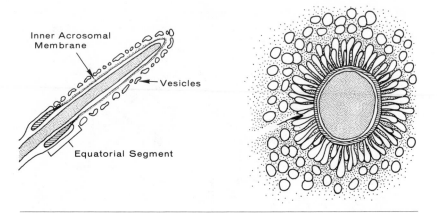

Figure 2   Acrosome reaction of sperm (*left*) as it penetrates corona (*right*).

hyaluronic acid between the cells of the cumulus oophorus as shown on the right. Penetration of the rabbit cumulus can be blocked by antibodies to rabbit hyaluronidase but not by antibodies to hyaluronidase of other species (14).

In Figure 2 the acrosome reaction is underway and vesicles are being formed by fusion of the plasma membrane and the outer acrosomal membrane. This has been observed in the rabbit (12) and the hamster (15). In the ovum the cementing substance between the corona cells is being digested by corona penetrating enzyme (CPE) (16–18). Penetration can be blocked prior to this stage by purified decapacitation factor (DF) (18). If the corona radiata is removed mechanically, DF preparations no longer inhibit fertilization with capacitated sperm (19). In Figure 3 the acrosome reaction has been completed by action of substances from the

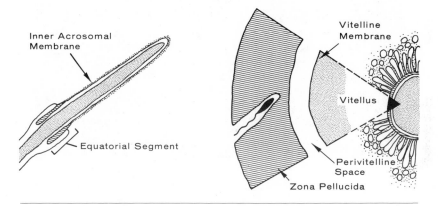

Figure 3   Status of reacted sperm (*left*) as it penetrates zona (*right*).

corona radiata and/or cumulus cells, and the remnants of the vesicles are left at the surface of the zona pellucida. Proteolytic enzymes are on the surface of the inner acrosomal membrane either being gradually released or acting in a bound

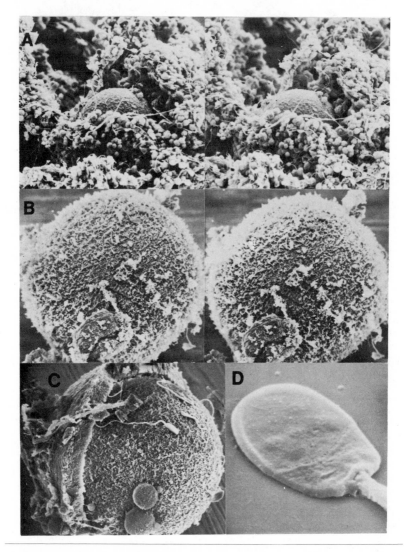

Figure 4 (3D image can be seen with stereo viewer used for protein molecular models or with 2–3 min of staring, the images can be superimposed by crossing the eyes.) *A* = Rabbit ovum with follicle cells. *B* = Vitellus without zona pellucida. *C* = Zona pellucida partially removed. *D* = Rabbit sperm with acrosome intact. All photos are scanning electron micrographs taken by Dr. K. G. Gould (19, 27).

state (20). In the ovum a limited fissure is being digested through the zona pellucida, very likely by an array of proteolytic enzymes. Penetration can be blocked at the surface of the zona by synthetic (21, 22) and natural proteinase inhibitors (23–25).

The stereo pair in Figure 4A shows an egg that was seminated when completely surrounded by cumulus and corona cells, the mass of cells being more than tenfold that of the egg. In vitro fertilization is taking place but many times more sperm than would approach the egg normally were added, hence the cumulus oophorus and corona radiata are falling away from the surface of the zona pellucida. Figure 4B shows a fertilized rabbit egg from which the zona pellucida has been removed by microsurgery. The surface of the vitellus has two sperm entangled with the upper middle surface. The structure in the lower center is unidentified. It may be a polar body devoid of outer membrane. Figure 4C shows the zona partially removed from a fertilized rabbit egg (26), exposing two polar bodies. Figure 4D shows a washed ejaculated rabbit sperm. During sample preparation (26) the acrosome contents concentrated on the anterior edge, although the acrosome can be seen to extend over the entire anterior portion of the head (27).

The vitelline membrane represents the last barrier to sperm penetration. Sperm penetration of this membrane has been the subject of many ultrastructural studies (e.g. 5, 13, 28), but little is known of the biochemistry of the process. It is significant that crude extracts of the sperm acrosome frequently digest this membrane (29; S. O. Newell and W. L. Williams, unpublished). How the sperm activates the egg, that is, how it initiates the biosynthesis of protein and other cell constituents, and how fusion of the male and female pronuclei takes place remain biochemical problems worthy of study.

Before considering the acrosomal enzymes in detail, an important point should be emphasized. In the study of the function, i.e. the biochemical effects of sperm acrosomal enzymes on the egg and its investments, it is necessary to treat a few ova with an extract of many million sperm, sometimes for several hours. It is fully appreciated that this type of assay is an imperfect approximation of a single sperm digesting a pathway through the various ovum layers, but so far, in spite of critical examination, this approach appears valid.

EXTRACTION OF ENZYMES FROM SPERM ACROSOMES

Since Yamane (30, 31) first obtained an extract of sperm that dispersed the egg investments (using Tyrode's solution saturated with toluene), many different methods have been reported for removal or extraction of one or more acrosomal enzymes. Other early methods included use of digitonin to remove acrosomes of rat, rabbit, and hamster sperm and release hyaluronidase (32), 0.1 N NaOH for guinea pig acrosomes (33), and 0.0125 N NaOH for bull and ram acrosome removal (34).

The use of the detergent, Hyamine 2389, to selectively remove acrosomes from ram sperm (35) and to release several acrosomal enzymes (29) from ram, bull, and rabbit sperm that dispersed all the egg investments sparked a new era of interest in these enzymes. Sonication plus Hyamine treatment yielded primarily a proteinase and hyaluronidase (36). The method consisted of sonication of washed rabbit sperm, separation of heads by gradient centrifugation, and extraction of heads with

Hyamine, according to Srivastava et al (29). Special merit was claimed for separation of heads to obtain more specific extraction of acrosomes to avoid extracting mito-chondrial enzymes (36, 37). Unfortunately intactness of the acrosomal and plasma membranes on such separated heads was not documented by electron microscopy. Even gentle sonication of boar sperm that does not break heads from tails extensively fragments plasma and outer acrosomal membranes (38). Such fragmentation may occur with sonication of rabbit sperm and may explain why many acrosomal enzymes are missing from extracts of sonicated heads (39), i.e. only hyaluronidase, a proteinase, and a trace of acid phosphatase are present. Two recent studies correlate ultrastructural changes with enzyme extraction. Pedersen (40) extracted human acrosin by thorough washing of sperm followed by rapid freezing in liquid nitrogen. A fairly selective removal of the plasma and acrosomal membranes was shown by electron microscopy, but unfortunately extraction of only the one proteinase was studied. Substituting barbituric acid for Hyamine gave more rapid and complete acrosome detachment and extraction of bull, human, and rabbit sperm, with little effect on midpiece or tail membranes (41). The barbituric acid was followed by a Triton X-100 treatment and the extracts contained several proteinases, acid phospha-tase, hyaluronidase, and β-glucuronidase.

Detergent extraction and sonication methods were compared regarding rupture of plasma and acrosomal membranes and release of selected mitochondrial enzymes (42). Detergent concentration, three times that recommended (43), gave an extract of human sperm in which 0.3% of the protein was lactate dehydrogenase (LDH) but controls without detergents were 0.1% LDH. Detergent extraction followed by standard alcohol precipitation (43) eliminated mitochondrial enzymes. Sonication insufficient to break heads from tails gave the greatest membrane damage and release of mitochondrial enzymes (42). A combination of Hyamine and Triton X-100 proved necessary to extract neuraminidase which, therefore, appeared to be the most tightly bound (44). Sequential extraction of ram and rabbit sperm first with 0.05 M $MgCl_2$ at 4° showed plasma and some outer acrosomal membrane rupture and released hyaluronidase, arylsulfatase, acrosin, other proteinases, and corona penetrating enzyme. A second extraction with Hyamine plus Triton X-100 yielded additional acrosin, other proteolytic activity, hyaluronidase, and neuraminidase concomitant with rupture of the inner acrosomal membrane, as shown by electron microscopy (20, 45).

SPERM HYALURONIDASE

Hyaluronidase is the most readily released enzyme from sperm. Seminal plasma from fresh bull semen frozen and thawed once is a rich source of hyaluronidase (46). Recent research reveals that about half the hyaluronidase in intact rabbit sperm can react with antihyaluronidase antibody (47), suggesting that a major portion of the enzyme is on the plasma membrane. Such antibody also agglutinates intact sperm (C. B. Metz, personal communication). The most exposed and loosely bound enzyme is needed first to start sperm penetration of the egg investments.

It is well established that sperm hyaluronidase is the enzyme responsible for penetration of the spermatozoon through the cumulus oophorus. The first hypothesis was that the hyaluronidase carried by sperm denuded the cumulus cell layer (48, 49), as a preliminary to further penetration of the eggs. It can be readily seen from Figure 4A why such a hypothesis was adopted, since the initial experiments involved large quantities of sperm or sperm extracts placed with ova in vitro. This hypothesis is no longer tenable, since Austin (50) has shown that the fertilizing sperm arrives at the vitellus before the cumulus cells are dispersed. In addition, only a few hundred sperm are present in the vicinity of the egg, far too few to disperse the cumulus layer (51). Retention around the egg of the cumulus and corona radiata has a favorable effect on fertilization. It has been suggested that the cumulus oophorus favors the directing of sperm toward the zona (52). A higher incidence of fertilization of rabbit ova in vitro is observed if the surrounding cells are retained (52a). More recently this has been confirmed in vivo with a report that 67% of eggs with cumuli were penetrated in the rabbit oviduct compared with 39% of those without cumuli (53). Austin (32) established the connection between capacitation and availability of hyaluronidase to act on the cumulus. Motile, intact, ejaculated sperm approached and adhered to the cumulus but failed to penetrate, whereas capacitated sperm could pass freely among the follicle cells of ova from rabbits, rats, and hamsters. In vitro release of hyaluronidase into the medium required disruption of the sperm membranes and acrosome (32). Research with mouse gametes has suggested that the cumulus cannot be penetrated until 2 hr or more after exposure to capacitated sperm (54–56). Instead of the postulated maturation of the cumulus, it is possible that the action is on sperm which may require contact with the cumulus before hyaluronidase release begins. In other species a 2 hr delay does not exist (13).

A recent claim has been made that postovulatory uterine fluid actively promotes the release of hyaluronidase from rabbit sperm, and that this release is related to capacitation in the uterus (56a). The data are not convincing, since the uterine fluid was obtained by ligating the uterus for 7–14 days, and under such conditions capacitation is markedly delayed (57). Hyaluronidase action takes place in oviduct fluid, therefore the fertilizing sperm would not normally come in contact with post-ovulatory uterine fluid. Although preovulatory fluid had an adverse effect on hyaluronidase release, best release was obtained with a mixture of post- and preovulatory fluid (58).

The best preparation of bull hyaluronidase has been obtained from seminal plasma (46). The enzyme requires Na^+ or K^+ and has a specific activity of 68,700 National Formulary (NF) units/mg. The seminal plasma enzyme is undoubtedly derived from the sperm acrosome. It was 50% more active than the enzyme obtained earlier (59). The optimum is pH 3.8 for the bull enzyme and pH 4.3 for the ram enzyme. This is unexpectedly low since the pH of oviduct fluid is 7.8 (60). Testicular and sperm acrosomal hyaluronidase are the same (61) but different from lysosomal hyaluronidase present in organs other than the testis.

Although all recent reports indicate a mol wt of about 60,000 or more (46, 61, 62), earlier work had indicated a mol wt of 11,000 (63), a figure still favored by some (37). The early studies were done on very crude preparations and enzyme activity was not established for the 11,000 mol wt unit. A recent report provides a basis for the earlier claims of low molecular weight by establishing that bull testicular hyaluronidase consists of four subunits, each with a mol wt of $\sim 14,000$ (62). An interesting recent report on sperm hyaluronidase is that hyaluronidase has properties immunologically unique for each species of sperm studied (14). Antibodies to rabbit hyaluronidase inhibit cumulus dispersion by the rabbit enzyme in vitro but do not react with hyaluronidase from other species. Treatment of ejaculated rabbit sperm with such antibodies prevents fertilization. Sera from some infertile women inhibit human sperm hyaluronidase, implying that isoimmunization of the female with human sperm hyaluronidase may be an approach to immunological control of fertility.

CORONA PENETRATING ENZYME (CPE)

The discovery of corona penetrating enzyme (16–18) resulted from attempts to repeat the observations of Srivastava et al (29) that 0.05% Hyamine 2389 extracts of ram sperm remove the investments of mammalian eggs. We were fortunate in selecting ejaculated rabbit sperm, since our crude extracts did not digest the zona pellucida or hydrolyze BAEE because the proteinases were inhibited, as discovered later (64). Had proteolytic activity been present, we might have concluded as have others (31) that a proteinase was removing the corona radiata. Using highly purified hyaluronidase to first remove the cumulus cells, it was demonstrated that a new enzyme that dispersed the cells of the corona radiata without effect on the zona was present in crude and purified fractions of acrosomal extracts (18, 65). Although the existence of CPE has not been completely accepted by other reviewers (13, 66), it has been independently demonstrated to exist by several investigators and to be inhibited by DF preparations (18, 65; J. C. Sadana, M. M. Bradford, S. O. Newell, and W. L. Williams, unpublished observations).

CPE acts on the corona radiata in a manner analogous to the action of hyaluronidase on the cumulus oophorus, i.e. dissolution of the intercellular material causing dispersal of intact corona cells. In normal in vivo fertilization in rabbits and humans, it is this unidentified material through which individual sperm digest a small passage.

Pincus & Enzmann (67) first demonstrated that heat-labile substances were present in sperm that could disperse both the cumulus and corona radiata surrounding the ovum. About the same time Yamane (30) also demonstrated that large numbers of live sperm in vitro dispersed the cumulus and corona cells and later reported (31) that lysing rabbit sperm membranes with toluene gave a proteinaceous solution which readily dispersed the two cellular layers of rabbit ova. The optimum was pH 7.7, a pH close to that of the fluid in the rabbit oviduct in which fertilization takes place (60). The cell dispersing activity was heat labile. For 30 years these observations were relatively neglected in so far as the corona radiata was concerned, until Srivastava et al (29) demonstrated that extracts of ram, bull, or rabbit sperm acrosomes dispersed the corona cells of rabbit ova.

In initial reports (16, 17) the designation corona removing enzyme was used. The designation has been changed to corona penetrating enzyme (CPE) to more properly describe the function of the enzyme in fertilization (18) and to distinguish the sperm enzyme from the corona removing system in the oviduct which acts hours after fertilization has occurred. The presence of CPE activity was confirmed in the rabbit, bull, and human and was also found in stallion and boar sperm acrosomes (65). CPE was demonstrated to be distinct from hyaluronidase and acrosin and present in both rabbit and human sperm. Of special significance was the observation that corona removal by CPE was inhibited by partially purified DF preparations from rabbit and bull seminal plasma and by whole seminal plasma. It has recently been shown (S. O. Newell, M. M. Bradford, and W. L. Williams, unpublished) that the inhibition of CPE parallels the potencies of the DF preparations in the in vivo anti-fertility assay (68, 69). None of the DF preparations inhibit hyaluronidase or acrosin. Perchloric acid destroyed both the CPE inhibiting activity and the antifertility activity of the DF pellet obtained from rabbit seminal plasma (70). There was no indication at this point that CPE might be a proteolytic enzyme. Soybean trypsin inhibitor (SBTI) prevented removal of the zona by crude acrosomal extracts but not removal of the corona radiata, i.e. SBTI did not inhibit CPE.

The factor (71) that removes the corona radiata in the oviduct has been reported to be bicarbonate (72). Bicarbonate is not involved in fertilization, since the corona cells are removed in the oviduct some time after sperm penetration (50).

The excellent work of Tillman (65) has added considerably to the knowledge of bull sperm CPE and led to the discovery of a new bull seminal plasma enzyme, designated corona dispersing enzyme (CDE), that also disperses the corona radiata yet differs in many properties from bull CPE.

Progress in the study of CPE has been slow because of the scarcity of the starting material, sperm acrosomes, and the semiquantitative nature of the assay, removal of coronae radiatae from rabbit ova. In spite of this, CPE was purified about 1000-fold based on whole rabbit sperm, giving a concentrate 20 μg of which in 0.2 ml completely removed the corona radiata of four ova (18) in 24 hr. In contrast to rabbit CPE (18), bull CPE was inhibited by naturally occurring proteinase inhibitors (65) (Table 1). Other differences among rabbit CPE, bull CPE, and bull CDE are listed in Table 1.

Table 1 Differences among enzymes that remove the rabbit corona radiata

Occurrence	Rabbit CPE Acrosome	Bull CPE Acrosome	Bull CDE Seminal plasma
Bull and rabbit DF preparations	Inhibited (16–18)	Inhibited (65)	No effect
Effect of SH groups	Not tested	Restored lost activity	None
Natural proteinase inhibitors	No effect of soybean trypsin inhibitor	Inhibited by lima bean, ovomucoid, pancreatic, and soybean trypsin inhibitors	No effect of these four
Potency of crude sources	See bull CPE	Acrosomal extracts are twice that of rabbit	Seminal plasma is twice that of stallion

As apparent from Table 1, bull CDE is quite a different enzyme from bull CPE. It has been purified fortyfold from bull seminal plasma, a rich source (65). Although the biological role of CDE has not been established, it is tempting to connect CDE with the fact that the bovine ovum is denuded of the corona radiata at the time of fertilization (73). Is enough CDE transported with the sperm to disperse the corona or is there another CDE in the cow oviduct fluid? Removal of the corona by bicarbonate (72) has not been studied using bovine ova.

The nature of the intercellular cementing substance of the corona radiata was sought by testing commercial pure enzymes on ova. Although hyaluronidase, neuraminidase, collagenase, lysozyme, β-glucuronidase, β-amylase, pancreatic trypsin, and aminopeptidase M were inactive, α-amylase (Calbiochem) was active. Heating the α-amylase preparation at 70° for 45 min destroyed the activity on the corona radiata, but the α-amylase activity was unchanged (65). The nature of the heat-labile contaminating enzyme in the α-amylase preparation has not yet been elucidated.

Bull CPE, rabbit CPE, and bull CDE have been purified to about the same extent, the minimum effective dose on four ova in 0.2 ml in 12 hr being 58, 40, and 20 μg, respectively.

In those few species in which the sperm do not traverse the corona radiata, the sperm would not require CPE. It would be interesting to determine whether such species remove the corona with CDE or bicarbonate. There is no correlation between manner of deposition of semen, presence of CDE, and corona removal, since myomorph rodents, the pig, horse, and dog deposit semen in the uterus, while the human, rabbit, cow, and sheep deposit semen solely in the vagina.

PROTEOLYTIC ENZYMES AND ZONA PELLUCIDA PENETRATION

Detailed study of the composition of the zona pellucida has not been made, however zonae are believed to consist primarily of neutral or acidic glycoproteins, as judged by proteolytic digestion (Table 2) and staining with protein (89) and carbohydrate stains (90). Major disulfide links must exist since zonae are readily removed by cysteine and mercaptoethanol (89, 91). Neuraminidase releases sialic acid from zona slices, but in intact ova only surface sialic acid residues are exposed to the enzyme (92). Four major proteins were detected on electrophoresis of sulfhydryl-solubilized rabbit zonae, and ninhydrin active material accounts for at least 22% of material in zona hydrolysates (89). Manually isolated mouse zonae also showed several classes of proteins of mol wt over 65,000 on SDS electrophoresis (93). Mouse zonae were soluble at low pH levels (4.5–5.5) without added enzymes (80), rat zonae were dissolved below pH 5 (94) and rabbit below pH 3 (83).

Hyaluronic acid has also been detected histochemically on the zona pellucida of rabbit (83) and cat (95) ova, but the partial and total removal of zonae by hyaluronidase suggest that the enzyme preparation contained proteinases (96) that could change interpretation of results. Other glycosidic components were also present.

Table 2 Removal of zona pellucida by proteolytic enzymes[a]

Species	Bromelin	Chymotrypsin	Ficin	Papain	Pronase	Trypsin	References
Hamster						+	(74–76)
		+	+	−		+	(77)
					+		(78)
	+		+	+	+	+	(79)
Mouse		−	+	+	+	−	(80)
		+	+	−		+[b]	(81)
					+		(82)
						+[b]	(83, 84)
		+					(85)
					+	+	(86)
Rabbit		−				+	(83)
		−	−	−		+[b]	(77)
				+		+	(87)
Rat	+					+	(83)
		−	−	−		+[b]	(77)
	+						(88)
					+		(82)

[a] Zona removed (+), not removed (−) or untested ( ).
[b] Trypsin was less effective in removing zona from fertilized eggs.

Proteolytic activity in sperm has been known since 1930, when Yamane (97) demonstrated that the trypsin activity of pancreatin mimicked the activity of rabbit, rat, and horse spermatozoa in causing second polar body extrusion by rabbit ova. This comparison to trypsin has been both a blessing and a curse. Established techniques for comparison to known proteolytic enzymes have made experimentation far too easy and delayed progress on the specificity and multiplicity of sperm proteolytic enzymes. In fact, virtually all biochemical experimentation to date has been mainly a comparison of sperm enzymes to pancreatic, liver, and lysosomal enzymes, without considering the individual mission-oriented biochemical functions of sperm enzymes. Subsequent experiments by Yamane (31) showed the presence of a heat-labile material in epididymal rabbit sperm extracts with high zona dissolution activity on rabbit ova. Over the next 25 years in vitro evaluation of the proteolytic nature of the zona lysin was made in a number of mammalian species (Table 2), using proteinases from many sources. Variability arose as a result of protein concentrations and purity in earlier studies. Reports of nonproteolytic enzyme digestion were apparently due to proteolytic contaminants (96). Current investigations began with the observation that sperm extracts hydrolyze synthetic proteinase substrates (87) and that inactive extracts contained enzyme-inhibitor complexes that could be dissociated at pH 3 (98).

Results from many laboratories (Table 3) have confirmed that crude acrosomal extracts from a number of mammalian species hydrolyze synthetic trypsin substrates and do not hydrolyze chymotrypsin substrates. Similarly, the trypsin inhibitor,

Table 3   References on presence of proteolytic, amidase, and esterase activities of sperm

Substrate	Rabbit	Boar	Human	Bull	Ram	Rat	Rooster	Guinea Pig	Mouse	Hamster	Horse	Monkey
TAME	36, 87, 99		40	100								
BAEE	99, 101–107	64, 98	40, 64, 101, 108	64, 100	43, 64	64	64	64		64	64	64, 101
BANA	105, 109		111	100, 110								
BAPA	103	114, 115	112, 113	100								
Gelatin	102, 106		106, 108		106	106		106	106			
Hemoglobin	36, 87			100, 116	116, 120							
Casein	41, 99	118	41	41, 100			76	117				

TLCK, is a potent inhibitor of acrosin, whereas the chymotrypsin analog, TPCK, is ineffective (99). Natural inhibitors of trypsin also exhibit antiacrosin activity and chymotrypsin inhibitors show no effect. A few of these investigators have also examined the proteolytic activity of extracts (Table 3), but the direct comparison to trypsin specificity was not possible until highly purified acrosin was available from rabbit (99, 107) and boar (118, 119) acrosomes.

Properties of proteolytic enzymes of known and potential activity in zonae digestion are detailed below.

Rabbit Acrosin[2] (99, 107)

Highly purified acrosin (trypsin-like enzyme, acrosomal proteinase) from rabbit acrosomes was found to have a mol wt of 55,000 by gel filtration and 27,300 by SDS-polyacrylamide gel electrophoresis, suggesting monomeric and dimeric forms. The optimum for BAEE hydrolysis was pH 8.0. The enzyme is apparently a serine proteinase with histidine in the active site, as ascertained by DFP and TLCK inhibition. In contrast to trypsin, BAEE is hydrolyzed twice as rapidly as TAME. Variations also exist in inhibitory effects of natural inhibitors on the two proteinases. The enzyme is uniquely inhibited by L-arginine and L-lysine has no effect (121, 122). Fingerprint analysis of digests of RNase by the purified enzyme showed that acrosin gave very limited digestion as compared to trypsin (99), and subsequent studies showed that boar acrosin hydrolyzed arginyl peptide bonds almost exclusively but lysyl bonds only when adjacent to arginyl or another lysyl bond (119, 122). Ca^{2+} enhances activity.

Two bands hydrolyzing BANA have been detected in acrylamide gels after electrophoresis at pH 4.3. A monomer-dimer relationship is indicated but not confirmed by the data presented (105).

[2] *Enzyme Nomenclature: Recommendations of the International Union of Biochemistry,* 1972, pp. 240–41. New York: Elsevier, indicates that acrosin is the recommended name for the proteinase, EC 3.4.21.10.

Boar Acrosin (115, 118, 119, 123)

Acrosin of high purity has been obtained from the acrosin-inhibitor complex of ejaculated boar sperm as judged by specific activity and protein components using chromatographic and electrophoretic criteria (118). The molecular weight estimated by gel filtration and SDS gel electrophoresis was 30,000. Small amounts of a presumed dimer were observed on gel filtration (118). Purification of the enzyme by affinity chromatography on cellulose-soybean trypsin inhibitor columns gave higher yields of the dimeric form, mol wt 56,000 (123). Benzamidine-cellulose affinity chromatography gave a threefold increase in potency (124). In addition to competitive inhibition of BAEE ($K_m = 5 \times 10^{-5}$ M) by L-arginine ($K_i = 3$ mM), D-arginine and L-homoarginine are also strong inhibitors of amidase and esterase activity. L-Lysine has no effect on enzyme activity, the enzyme is inhibited by DFP and TLCK, but TPCK is ineffective. The pH optimum for BAEE hydrolysis is 8.5 and for azocasein hydrolysis 8.7. Tests with a series of active site inhibitors supplied by Dr. E. Shaw indicate that the active site of acrosin is more closely related to that of thrombin than trypsin or plasmin. Although calcium and other divalent ions (Mg, Mn, Ni, and Co) enhance activity, addition of or dialysis against metal chelators does not markedly decrease activity. Hg is inhibitory (119).

Fingerprint analysis of lysozyme digests by the enzyme showed that 12 of the 13 digestion products detected with ninhydrin also gave a positive Sakaguchi test for arginine. Subsequent digestion with carboxypeptidase B and amino acid analysis for terminal residues showed nine terminal arginine residues of the ten cleaved, since the dipeptide Asn^{113}-Arg^{114} is not readily cleaved by carboxypeptidase B (119). Of the two lysine residues detected after the carboxypeptidase B digestion, one was shown to arise from cleavage of the Lys^{96}-Lys^{97} bond by isolation of the peptide Asn^{74}-Lys^{96}. The second lysine residue could arise from release of a residue from Lys^{13}-Arg^{14} by carboxypeptidase B, or combined cleavage of the Lys^{13} and Lys^{97} bonds. Only C-terminal leucine was observed after a second digestion with carboxypeptidase A.

N-terminal lysine is not released by acrosin digestion of lysozyme. No cleavage of the terminal Arg^{128}-Leu^{129}-OH sequence occurred as shown by the absence of leucine in acrosin digests. Arginine is not cleaved from Arg-Gly-Gly or from bradykinin that contains an N-terminal Arg residue, confirming the endoproteinase nature of acrosin. Polyarginine is readily cleaved by acrosin but lysyl peptides are only very slowly released from polylysine.

Human Acrosin

Acrosin from human sperm has been less extensively studied although comparison has been made to the enzyme from other species. The enzyme has a molecular weight of 30,000 as estimated by gel filtration, a pH optimum for BAEE hydrolysis of 8.0, and requires Ca^{2+} for maximum activity (104). DFP and TLCK inhibit BAEE hydrolysis and TPCK is ineffective. The effects of natural inhibitors on human acrosin parallel those on boar acrosin (113). Similar comparison can also be made with data from separate reports on inhibition of human acrosin (104, 113) and rabbit acrosin (99, 107). Human acrosin is also inhibited by L-arginine (L. J. D. Zaneveld, personal communication).

Bull Alkaline Proteinase

Although one of the earlier species in which sperm proteolytic activity was studied (116), detailed properties of bovine sperm extract activity have only recently been investigated (100, 110, 125). Acrosomal extracts hydrolyze BANA at pH 7.5 (110). Others have found an optimum for BANA, BAPA, and TAME hydrolysis around pH 8.4 (100). Calcium ion was not added (110). Unexpectedly, Ca^{2+} has been found inhibitory along with Cu^{2+} and Hg^{2+} (100) by other investigators although a complete Ca^{2+} concentration curve was not presented. Proteolytic activity on hemoglobin and casein had optima in the range reported for boar acrosin. Natural trypsin inhibitors have similar potencies (100) to those found for the rabbit (102), boar (119), and human (113) acrosin but no site-directed inhibitor work has been detailed. Inhibition by arginine was not tested in the purest preparations reported to date (125). Multiple molecular forms have been found by both gel filtration (100) and polyacrylamide gel electrophoresis techniques (110, 125). Monomer-dimer relationships are indicated for two bands and partial autodigestion for the third (110).

Avian Sperm Proteinases

Avian sperm proteinases are included since they appear to be similar to those of mammalian sperm. Proteolytic activity on casein was first reported for duck and rooster sperm in 1956 (117). Although the presence of proteolytic activity has been confirmed for both epididymal and ejaculated sperm (64), only one investigation of the enzyme has been conducted to date (111). Four bands, one at the origin, hydrolyzed BANA and ZANA after polyacrylamide gel electrophoresis of rooster sperm extracts and all were inhibited by soybean trypsin inhibitor. One band also hydrolyzed ANA and LNA but was not inhibited by trypsin inhibitor. BANA activity was not present in seminal plasma. Electrophoretic mobility of the three migrating zones is similar to that observed in the bull (110, 111).

Proacrosin (109)

Rabbit seminiferous tubules, homogenized with 0.25 N H_2SO_4 and fractionated with ammonium sulfate, conditions which should separate acrosin from inhibitors, yielded a fraction, IIIb (20–80% ammonium sulfate), that showed a marked increase in acrosin activity after incubation with trypsin, suggesting the possibility of a pro-acrosin form. Incubation of the fraction alone or with chymotrypsin did not increase activity on BANA. Thirty-six hour dialysis against 0.001 N HCl (a condition that removes inhibitor) gave a fivefold activity increase. High ratios of activated IIIb to unactivated IIIb (2.5:1) had activity equal to the activated fraction. At a ratio of 1:1, 40% inhibition was observed. Incubation with trypsin removed the inhibition and gave the anticipated activation. The results indicate that fraction IIIb contains either a zymogen form of acrosin or an acrosin inhibitor that can be either activated or removed by trypsin. Unfortunately, more definitive experiments to dissociate the inhibitor, such as preincubation at pH 3, and quantitation of the proteolytic assays and activations were not carried out.

Proteinase Inhibitors

Natural inhibitors of trypsin have been extensively studied especially in early studies where acrosin and trypsin were compared to demonstrate differences between the two enzymes (37, 104, 113, 126). Both immediate and progressive, i.e. increasing with time, inhibitions have been determined (102). The finding of progressive inhibition explains the reported lack of inhibition by α_1 antitrypsin (101).

The presence of proteinase inhibitors was first noted in human seminal plasma (100). Subsequent work showed that the inhibitor was present in seminal plasma of a number of mammalian species (64, 108). Although free in seminal plasma, the inhibitor binds either in or on membranes of the sperm acrosome (e.g. 87) probably as acrosin-inhibitor complexes (127, 128). The inhibitor complex is also present in small quantities in epididymal sperm of certain species (64).

Inhibitors were first isolated from human (113) and boar (129) seminal plasma. More detailed investigations have revealed that boar seminal plasma contains five trypsin-acrosin inhibitors and human seminal plasma contains eight or more (124). Since inhibitors are not species specific, studies of fertilization effects can be made on isolated ova. Pancreatic trypsin inhibitor and seminal plasma trypsin inhibitor inhibit in vivo fertilization of rabbit ova (23), and the seminal plasma acrosin inhibitor is normally removed or destroyed during capacitation of rabbit sperm (130).

Use of synthetic inhibitors of trypsin to block acrosin has led to interesting possibilities for contraception. Initial observations demonstrated that TLCK reduced the fertilizing capacity of capacitated rabbit sperm inseminated into oviducts of ovulated does (21, 131). Ejaculated sperm treated with TLCK or NPGB and inseminated into the rabbit failed to fertilize. TLCK was also effective in a vaginal contraceptive administered prior to breeding.

More extensive vaginal contraceptive experiments (132, S. O. Newell, R. T. Robertson, and W. L. Williams, unpublished) have demonstrated that other active site-directed inhibitors for trypsin, GPB and APB, also reduce fertility as do the competitive inhibitors arginine, benzamidine, and polyarginine. Mixtures of the inhibitors consistently reduce fertility by 80–97% in rabbits, a species in which human vaginal products are ineffective (133, 134). The current demonstration that other proteinases exist in acrosomal extracts of rabbit sperm (122, R. A. McRorie and P. N. Srivastava, unpublished) and the detection of an aminoacylarylamidase in bull sperm extracts (135) indicate that alternate enzymes for zona penetration probably exist.

Several facts must be considered in evaluating experiments on proteolytic sperm enzymes and proteinase inhibitors.

1. Crude acrosin preparations are relatively stable at pH 3–7 but the enzyme is unstable at alkaline pH. Activity of purified preparations with a specific activity of 5 or more for BAEE hydrolysis is lost in a matter of minutes in glass apparatus (118, 128).

2. In polyacrylamide gel electrophoresis the gel apparently protects the enzyme, the fluorometric BANA assay is the most sensitive available and the coupled dye reaction measures very small quantities of enzyme (110, 136), so major destruction might occur and the enzyme remain detectable.

3. The existence of a monomer-dimer relationship seems well established (118, 119, 124). The presence of additional forms is suggested by the fact that after a few minutes in the presence of substrate, acrosin is slowly released from natural inhibitor complexes. Any activity observed in incubations over 5 min must take into account the possibility that acrosin may combine with a number of inhibitors (124).

4. Acrosin-inhibitor complexes are readily dissociated at pH 3 or below and reassociation is negligible at optimum alkaline pH levels in the presence of synthetic or protein substrates. This test should be a routine procedure where the possibility of acrosin-inhibitor complexes exists.

5. Acrosin is the only known proteinase hydrolyzing basic aminoacyl residues that is inhibited by arginine. Investigators claiming the existence of other sperm proteinases should consider this property.

Other Proteolytic Enzymes of Sperm

An arylamidase that readily hydrolyzes L-methionyl-β-naphthylamide and to a lesser extent the β-naphthylamides of Ile > Leu, Val > Phe > α-Asp, but not of Ala, Arg, Lys, or Pro at pH 8.0 has been purified from Hyamine extracts of bull sperm (135). Amidase activity is greatly enhanced by dithiothreitol and inhibited by o-phenan-throline, TLCK, and DFP. These reagents produced no change in α-aspartamide activity, suggesting the existence of a second enzyme for aspartylamidase activity although the enzyme apparently differs from β-aspartylglucosamine amido hydrolase not detected in bull acrosomal extracts (137). No correlation between arylamidase and acrosin activity was observed in bull (135) or rabbit extracts (R. A. McRorie and P. N. Srivastava, unpublished) that slowly hydrolyze the methionyl amide.

Collagenase-Like Peptidase (CLP) (138)

A second proteolytic enzyme with an alkaline pH optimum (7.5) that hydrolyzes the chromophoric collagen substrate PZ-pentapeptide (4-phenyl-azo-benzyloxy-carbonyl-L-Pro-L-Leu-Gly-L-Pro-D-Arg) at the Leu-Gly bond has been localized as a sperm enzyme from epididymal human, rat, and bull sperm, although cytoplasmic droplets were not completely ruled out as a source. Rat testis extracts also hydrolyze casein but the CLP/caseinolytic ratio changes on purification, indicating the presence of other proteolytic enzymes in the crude extracts. The molecular weight of purified CLP was found to be 110,000 by gel filtration. Purified CLP does not hydrolyze native collagen, bovine serum albumin, or porcine insulin. A natural substrate for the enzyme has not been detected.

Sperm Acid Proteinases

Whole bull and ram semen and sperm from the two species were found to possess two pH optima for hemoglobin hydrolysis, 3.5 and 6.5. Cysteine enhanced activity in the more alkaline digests (116). Hemoglobin digestion by ram acrosomal extracts exhibit pH optima in the range of 3.8–4.5 and 5.4–5.6 as well as at the alkaline optimum of acrosin. Cysteine gave increased activity at all pHs (120).

BANA hydrolyzing activity at pH 5.5 has been found in bull sperm extracts for three electrophoretic bands showing major hydrolysis at pH 7.5 (110). Discounting minor variations, BAPA, TAME, and BANA were partially hydrolyzed at pH 4.4–5.3 by enzyme extracts of bull sperm. Hemoglobin was also hydrolyzed at pH 3.8–4.2.

Cysteine enhanced activity at alkaline pH; however, iodoacetate gave only weak inhibition. The authors suggest that the cysteine aids in cleavage of inhibitor-enzyme complexes since no effect is observed after removal of the inhibitor by dialysis (100). This suggestion is very reasonable since no SH activation or inhibition by SH inhibitors was observed with pure acrosin (119).

Azocasein is digested by acrosomal extracts of rabbit, boar, stallion, rooster, human, ram, and boar sperm at pH 2.8. The optimum for hemoglobin digestion is pH 3.5. The enzyme partially purified from boar sperm shows activation by cysteine and inhibition by iodoacetate. No metal requirement or inhibition by chelating agents was observed (139).

Complexity of Zona Pellucida Penetration

Although acrosin has been shown to remove the zona pellucida from rabbit ova in vitro, the physiological concentration necessary for sperm to digest a discrete pathway is unknown. More recent evidence indicates that, at least for the rabbit, zona penetration in normal fertilization is more complex. Acrosin inhibitors markedly reduce fertility in vaginal contraception experiments but total fertility blocks are seldom observed when inhibitors are used singly or in combinations (132, S. O. Newell and W. L. Williams, unpublished). Srivastava (45) has shown that zona lysis in vitro does not correlate with the acrosin content of sequential acrosomal extracts. Preliminary digestions of lysozyme by some of these extracts and amino acid analysis show the presence of hydrolytic activity for 20 aminoacyl residues (G. E. Chism and R. A. McRorie, unpublished). The broad activity of these extracts gives more rapid zona lysis in vitro. The requirement for more than one proteinase in normal fertilization or alternate zona digestion enzymes must be considered.

Other current investigations also point to a more complex process than simple proteolytic digestion for zona penetration in the hamster (140–143). Binding sites on zonae, exposed by hyaluronidase removal of cumulus cells, are apparently destroyed by proteolytic enzymes and differences observed between sperm binding to isolated zonae and cumulus-free ova indicate that a factor released from the vitellus is involved. Since the effects observed are more pronounced for uncapacitated sperm than for capacitated sperm, the applicability of the facts observed to normal fertilization is still in doubt.

Wheat germ agglutinin (WGA) binds as a layer to cumulus-free hamster ova and prevents fertilization (79). The effect is partially reversed by incubating WGA-treated ova with N-acetylglucosamine. The authors suggest that WGA binding occurs at N-acetylglucosamine sites on zonae. No data are presented that permit comparison of the binding sites studies in the two laboratories.

Acrosin-Hyaluronidase (36, 37)

The claim has been made that a single pure proteinase-hyaluronidase complex of mol wt 59,000 on sucrose density centrifugation is the sole protein component of rabbit acrosomes extracted with Hyamine. Justification for the proposed complex was based on the assumption that an acrosomal proteinase present had the same or lower molecular weight as pancreatic trypsin (23,800) and that sperm hyaluronidase would have the minimum molecular weight of testicular hyaluronidase (11,000), theoretically providing "space ... for at least one molecule and possibly more of each enzyme." With the isolation of highly purified hyaluronidase, mol wt 55,000 to 61,000 from testes (59, 62) and 62,000 from sperm (144), the proposed complex with acrosin monomer (mol wt 27,300) or dimer (mol wt 55,000) would be highly unlikely even using the minimum molecular weight for hyaluronidase monomer of 14,000 (62). No proteinase activity exists in testicular hyaluronidase of mol wt 55,000 (62), although the authors proposing the complex (37) have suggested that the hyaluronidase of 61,000 mol wt and high potency of 45,000 NF units/mg (59) may have been a proteinase complex since no assay for trypsin activity was performed. Highly purified sperm hyaluronidase had activity of 38,280 NF units/mg (144).

From activity considerations the claim of isolation of a pure complex is impossible. The purest preparation of rabbit acrosin has a specific activity of 15,000 BAEE milliunits/mg (107). Pure bull testicular hyaluronidase has a specific activity of 22,000 Tolkdorf units/mg (62), and the purest rabbit preparation to date has a specific activity of 3500 turbidity reducing units (TRU)/mg (14). The so-called pure hyaluronidase proteinase complex contains 48.5 Tolkdorf units and 1460 BAEE milliunits/mg (36), values clearly showing contamination with other materials.

It would appear that a mol wt of 59,000 for the complex was simply the mean of a mixture. Unfortunately the tenuous hypothesis that a pure proteinase-hyaluronidase complex has been isolated and identified has been widely and extensively quoted as fact in many scientific papers and reviews (37, 66, 145, 146). This complex matter should be laid to rest.

SPERM NEURAMINIDASE

The function in fertilization of acrosomal sperm neuraminidase (SN) is unknown. SN was originally investigated with the probability in mind that it was involved in vitelline membrane penetration in a fashion similar to that of viral neuraminidase. Unexpectedly, it was found that rabbit SN inhibited in vitro fertilization of rabbit ova and altered the chemical nature of the zona pellucida (89). SN is the most tightly membrane-bound enzyme of those discussed. Whereas Hyamine readily extracts hyaluronidase, acrosin, and CPE, Triton X-100 is required to extract SN, suggesting that it is bound to the inner acrosomal membrane (44). As first extracted from sperm acrosomes, full neuraminidase activity is not expressed. A substance in the extracts designated neuraminidase-like factor (NLF) reacts with bound neuraminic acid (44, 147), rendering it reactive in the colorimetric test (148) that ordinarily measures only free neuraminic acid (149). Upon fractionation of extracts of acrosomes, SN activity develops and neuraminic acid bound in the $2 \rightarrow 6$ linkage in mucoproteins is liberated. Either NLF is converted to SN or an inhibitor of SN is being removed. At present the first possibility is favored (44).

Using in vitro fertilization (150) of rabbit ova, 0.015 units/ml of purified SN completely inhibited fertilization (151). Boiling the SN preparation eliminated the inhibition of fertilization as did the addition of the neuraminidase inhibitors, β-(p-nitrophenyl) α-mercaptoacrylic acid and 1-(o-aminophenyl) 3,4-dihydroisoquinoline. The above two inhibitors in the absence of SN had no effect on fertilization, suggesting that SN does not play a positive role in fertilization (19). Exposure of zonae pellucidae of rabbit ova to SN preparations renders them relatively resistant to dissolution by mercaptoethanol or trypsin (89). It is tempting to speculate that SN might be involved in the block to polyspermy but, unfortunately, in normal fertilization the rabbit zona is penetrated by 20 to 60 sperm and the block is at the vitelline membrane (84). It has been shown that neuraminic acid is a component of the zona pellucida (92) and that treatment of ova with bacterial neuraminidase will inhibit sperm penetration in vivo (152).

OTHER SPERM ENZYMES

Activities analogous to arylsulfatase A and B (pH optima 4.8 and 5.6, respectively) have been highly purified from detergent extracts of rabbit sperm acrosomes. Properties of the enzymes that migrate as a single band on acrylamide gel electrophoresis are intermediate between enzyme types isolated from other sources (153, 154). Arylsulfatase is also present in rat sperm acrosomes (155). The role of the enzyme in normal fertilization has not been clearly established although it may play an adjunct role in penetration of the cumulus cell layer (120). Sulfated polysaccharides were detectable histochemically between corona cells and in the zona pellucida (53, 95, 156). An amidase, β-aspartylglucosylamine amido hydrolase, has been detected in acrosomal extracts of the boar, squirrel monkey, human, and ram (137, 157) and highly purified from boar acrosomal extracts (158). Activity was not detected in rabbit or bull acrosomal extracts. No distinct role for the enzyme has been established. Acid phosphatase, phospholipase A, and β-N-acetylglucosaminidase as well as arylsulfatase have been detected in extracts of ram spermatozoa (120). Rabbit sperm extracts contain detectable quantities of catalase, carbonic anhydrase, and lactic dehydrogenase isozymes (36). The acid phosphatase activity of rabbit, bull (159), and mouse (160) spermatozoa has been localized in the subacrosomal space and postacrosomal regions with concentration at the equatorial segment of the rabbit (159) and hamster (160), areas that involve initial contact with the vitelline membrane. Nonspecific esterases hydrolyzing α-naphthol esters have been observed in extracts of mouse (160), human (161), and bull (162) spermatozoa. As many as eleven esterases exist in bull sperm (162) and no correlation with known aminoacyl esterases was observed (135). Although bull esterase has been localized in the acrosome (163), no role for these enzymes in normal fertilization has been proposed.

CAPACITATION AND DECAPACITATION FACTOR

As mentioned earlier, capacitation is a process that normally takes place in the female reproductive tract whereby the sperm achieve the ability to penetrate the ovum. Six years after this process was described and defined (7, 8, 164), Chang (165) showed that capacitation could be functionally reversed by immersing uterine sperm in seminal plasma. Bedford & Chang named the naturally occurring antifertility substance present in rabbit and several other species' seminal plasma, decapacitation factor (DF) (166, 167). Capacitation was first shown to be required in the rabbit (7, 8) and in the rat (7). Other species are the mouse (52), golden hamster (168), cat (169), ferret (170), sheep (171), cow (172), and pig (173). Bedford (13) has recently reviewed the time required and the evidence for capacitation in the above species.

Although the evidence is indirect, human sperm appear to require capacitation. Convincing evidence is in a recent report of Brackett (174) that human sperm require 4–5 hr incubation in the Rhesus monkey uterus before in vitro fertilization of human ova is successful. Austin et al (175) have observed a marked delay of 7 hr between addition of ejaculated sperm to an in vitro fertilization system using human ova and penetration of the ova by sperm, and they regard this delay as the time required for capacitation, acrosome reaction, and penetration.

A second biochemical line of reasoning has been used to assign capacitation to a species, that is, the presence of DF in the species' seminal plasma as determined by the rabbit assay (176, 177). This reasoning confirmed that capacitation is required for sperm of the bull, boar, rabbit, and monkey and added stallion sperm to the list (178). It was first concluded that human and dog seminal plasma did not contain DF activity (178), but testing at higher levels established the presence of DF in human (179) and dog (65) seminal plasma.

DF was considered for many years to be a single substance (167, 180, 181) but at present it is established that the antifertility action of seminal plasma can be attributed to different substances. Polypeptide trypsin inhibitors block fertilization in the in vivo DF assay, but the DF entity being purified does not inhibit trypsin (18). Seminal plasma contains both a CPE inhibitor (18) and a membrane stabilizer (12) that have antifertility action and are not likely to be the same substance.

It is important to bear in mind the original definition of DF (165). Besides its antifertility action on capacitated sperm, it does not adversely affect motility and its action is reversible, i.e. after decapacitation sperm can be recapacitated by a second sojourn in the uterus. These properties have been maintained by the DF preparations throughout our purification studies (182, 183). The most useful first step for DF purification is ultracentrifugation which sediments all the DF activity from rabbit seminal plasma (167, 180) and part of the activity from bull (182) and human (179) seminal plasma. Although DF activity in the rabbit ultracentrifuged pellet is tightly bound and only released by proteolytic digestion, DF is readily extracted with water from the bull pellet as a mixture of soluble proteins (182). Further column fractionation and electrophoresis of the proteolytic digest of bull soluble proteins yielded three active peptides of mol wts about 1000 (126, 183–185).

The most active peptide had a minimum effective dose of 2 μg/10^5 capacitated sperm (V. K. Bhalla and W. L. Williams, unpublished) compared to 5000 μg for whole bull seminal plasma (178, 186). The rabbit ultracentrifuged pellet after proteolytic digestion and several purification steps yielded a peptide active at 50 μg/10^5 sperm (131, 184, 187). The very low yield of peptides obtained from both bull and rabbit sources was insufficient for amino acid sequence determination. The present approach is to first purify the DF protein extractable from the bull pellet with water or that from the rabbit pellet with detergents (W. F. Lehnhardt, V. K. Bhalla, and W. L. Williams, unpublished) and then the active peptide of minimum molecular weight, hopefully synthesizable. Davis (188) has attributed DF activity solely to a large (86S) glycoprotein, a vesicle surrounded by a membrane, purified from rabbit seminal plasma. Although it is possible that such a particle is active (it would occur in the DF pellet), motility after decapacitation was not reported and recapacitation was not achieved. The report that rabbit DF has a mol wt of about 170,000 (189) is in conflict with the above (188) as are the observations that extensive proteolysis or solubilization of the rabbit DF pellet with detergent does not destroy activity (181, 190).

In addition to the DF antifertility activity measured in the involved in vivo assay, DF preparations inhibit in vitro fertilization of both rabbit (151) and hamster ova (191).

The molecular mode of action of the forms of DF isolated by the Georgia group appears to be inhibition of CPE. This is supported by two lines of evidence. The relative potency of various DF preparations in the in vivo assay (68, 69) parallels the potency of the DF preparations as in vitro inhibitors of CPE (S. O. Newell, M. M. Bradford, and W. L. Williams, unpublished). In rabbit in vitro fertilization DF no longer blocks fertilization by capacitated sperm if the coronae radiatae are previously removed (19).

In experiments designed to explore the possible contraceptive use of DF, it was found that the rabbit uterus has a high capacity to remove or destroy DF (192). Rabbit DF pellet purified from 40 ejaculates was ineffective in preventing capacitation if instilled with ejaculated sperm in the rabbit uterus. DF from 80 ejaculates was effective but sperm migrating from the high DF concentration in the uterus to the oviduct were capacitated and fertilized ova. If a small synthetic DF peptide were available, it possibly could prove useful as a postcoital contraceptive (126).

FERTILE AREAS FOR FURTHER BIOCHEMICAL RESEARCH

Besides the problems mentioned as being unsolved in various parts of this review, there are several physiological phenomena which appear to us to be important and languishing because of neglect by biochemists.

Eggs are normally penetrated by only one sperm after which an event occurs called "block to polyspermy." Upon fertilization of sea urchin eggs (193) the cortical granules of the vitellus release a heat-labile substance and sperm no longer attach to or penetrate the modified fertilization membrane of the vitellus. Similarly, in the golden hamster (194) the block to polyspermy occurs at the surface of the zona pellucida which upon fertilization is modified by a heat-labile substance released again by cortical granules. This substance evidently prevents sperm attachment and zona penetration. Pancreatic trypsin mimics the action of the substance, possibly by proteolytic digestion of binding sites on the surface of the zona. Wheat germ agglutinin forms a layer over the zona of the golden hamster and prevents fertilization (79) presumably by covering binding sites for sperm attachment. Thus, there are two possibilities as to the mode of action of the unidentified polyspermy-preventing substance, proteolytic destruction or protein masking of binding sites. The existence of such a proteinase or masking protein should be demonstrated in other species, particularly in the rabbit where in contrast to the hamster the block to polyspermy does not occur at the zona surface but like the sea urchin occurs at the vitelline membrane. Identification of the substance involved in the block to polyspermy may yield means of enhancing its action, thereby preventing fertilization or inhibiting its action, thus allowing polyspermy and consequently death of the egg. In addition, demonstration and identification of the sperm component responsible for initiating the block would be of great interest and possible utility.

Although considerable evidence indicates that acrosin is essential for sperm penetration of the zona pellucida, the role of the other acrosomal proteinases is not known. It is quite possible that the acid proteinase takes over digestion of the zona as the sperm creates an acidic microenvironment deep in the zona. The role in fertilization of other enzymes such as the sulfatases, amidases, esterases, and phosphatases remains a crucial question that should be attacked by isolating the enzymes and seeking the substrates in the egg or its surrounding cells. Inhibitors for the other proteinases should be sought in seminal plasma or epididymal fluid.

The acrosomal enzymes involved in fertilization so far obtained in pure or partially purified form have unique properties compared to similar enzymes elsewhere in the mammalian organism. Specific synthetic inhibitors, individually or in combination, may be specific and innocuous agents useful in a new contraceptive approach. There appears to be no technological reason why they could not be used postcoitally in the female or systemically in the male. Delivery of such inhibitors to the site of fertilization or to the sperm enzymes in the male has not yet been attacked as a major biochemical effort and awaits investigation. In this connection further research is needed on the transport of sperm through the cervix (which appears to involve sperm or seminal plasma substances), uterus, and oviduct to the site of fertilization. If sperm can be so transported by the female, it seems likely that particles containing loosely bound inhibitors of fertilization could be transported with the sperm.

Although salt and detergent extraction as well as subcellular fractionation have given strong indications as to the location of acrosomal enzymes on the various sperm membranes, the results are still far from conclusive. During such drastic processing it is quite feasible that enzymes could migrate from one membrane layer to another or, for example, from the mitochondrial sheath to the sperm head. At this point sophisticated cytobiochemical techniques are needed to demonstrate the exact location of the various enzymes on sperm. We believe it is possible that the sperm enzymes are on, within, or under membranes and there is no basis at present for making a choice. Such demonstration on sperm before and after capacitation as well as before and after the acrosome reaction would be very revealing as to the biochemical events taking place during these essential physiological events.

In no species has in vitro capacitation, in vitro fertilization, embryo culture and transfer with the production of live young been achieved, although each of these processes has been achieved with one or more species. Many important problems remain. For example, although hamster ova can be readily fertilized in vitro with in vitro capacitated sperm, cleavage of the egg past the two-cell stage has not been observed. Is a cleavage factor missing or have improper physical conditions been used for culture of the fertilized ovum? Unpublished work at Georgia has revealed that fertilized rabbit ova that failed to cleave will cleave in minutes to the four-cell stage by treatment with commercial β-amylase.

From work with the sea urchin (3), it appears that egg activation, that is, the initiation of protein synthesis in the fertilized egg, results from a contribution of proteolytic enzymes from the sperm which unmask and activate the polyribosomes. This has yet to be investigated in the mammalian egg.

The cumulus and corona cells around the ovum apparently secrete substances that cause capacitation and the acrosome reaction of sperm of certain species. Are these substances enzyme activators or substances that remove inhibitors from the essential acrosomal enzymes? Present indications are that there are at least four different types of natural antifertility substances in seminal plasma. Is the membrane stabilizing factor of seminal plasma the same as decapacitation factor? Do acrosomal proteolytic enzymes destroy DF?

These and many other problems await the involvement of additional biochemists in mammalian reproduction research. Finally, we wish to emphasize again that each of the complex biochemical steps involved in sperm penetration of ova offer a target for control of fertilization.

ACKNOWLEDGMENTS

The research from our laboratories referred to in this review has been supported by research grants from the Population Council, a Career Development Award to Dr. Williams, and a Training Grant to the Reproduction Research Laboratories. More recently crucial support was supplied by the Ford Foundation (680-0805A), without which we would have been forced to drastically curtail our research, and by the outstanding contract program of the Center for Population Studies at NIH (NIH-69-2103 and NIH-70-2147).

We wish to acknowledge the work of our colleagues on DF, which over a period of a decade ultimately led to the pure DF peptide. They are Drs. D. E. Weinman, H. N. Chernoff, W. R. Dukelow, M. C. Pinsker, R. T. Robertson, and V. K. Bhalla and, finally, W. F. Lehnhardt, whose research is yielding pure DF protein. Dr. L. J. D. Zaneveld discovered CPE and Dr. K. L. Polakoski isolated acrosin and acrosin inhibitors. In vitro fertilization techniques were developed and utilized by Drs. B. G. Brackett, E. M. Cline, and K. G. Gould. W. L. Tillman extended the CPE work and discovered CDE. C. H. Yang isolated bull acrosomal hyaluronidase.

Sally O. Newell, DVM, developed proteinase inhibitors as the best known rabbit vaginal contraceptive. Finally, we must acknowledge the present enthusiasm, cooperation, and industry of Dr. P. N. Srivastava, whose early studies at Cambridge were so stimulating to us.

Literature Cited

1. Rothschild, L. 1956. *Fertilization.* London: Methuen. 170 pp.
2. Mann, T. 1964. *The Biochemistry of Semen and of the Male Reproductive Tract.* London: Methuen. 493 pp.
3. Monroy, A. 1965. *Chemistry and Physiology of Fertilization.* New York: Holt. 150 pp.
4. Metz, C. B. 1967. *Fertilization I, II,* ed. A. Monroy. New York: Academic. 489 pp., 553 pp.
5. Austin, C. R. 1968. *Ultrastructure of Fertilization.* New York: Holt. 183 pp.
6. Balin, H., Glasser, S. 1972. *Reproductive Biology.* Amsterdam: Excerpta Med. 534 pp.
7. Austin, C. R. 1951. *Aust. J. Sci. Res. B* 4:581
8. Chang, M. C. 1951. *Nature* 168:697
9. Adams, C. E., Chang, M. C. 1962. *J. Exp. Zool.* 151:159
10. Bedford, J. M. 1963. *J. Reprod. Fert.* 6:245
11. Bedford, J. M. 1964. *J. Reprod. Fert.* 7:221
12. Bedford, J. M. 1969. *Advan. Biosci.* 4:35–50
13. Bedford, J. M. 1972. *Reproductive Biology,* ed. H. Balin, S. Glasser, 338–92. Amsterdam: Excerpta Med. 734 pp.
14. Metz, C. B. 1973. *Fed. Proc.* 32:2057
15. Franklin, L. E., Austin, C. R. 1967. *J. Cell Biol.* 34:C1
16. Zaneveld, L. J. D., McRorie, R. A., Williams, W. L. 1968. *Fed. Proc.* 27:567
17. Zaneveld, L. J. D., McRorie, R. A., Williams, W. L. 1969. *Fed. Proc.* 28:705
18. Zaneveld, L. J. D., Williams, W. L. 1970. *Biol. Reprod.* 2:363
19. Gould, K. G. 1973. *Studies on the interaction of mammalian gametes in vitro.* PhD thesis. Univ. London. 208 pp.
20. Srivastava, P. N., Munnell, J. F., Yang, C. H., Foley, C. W. 1974. *J. Reprod. Fert.* 36:363
21. Zaneveld, L. J. D., Robertson, R. T., Williams, W. L. 1970. *FEBS Lett.* 11:345
22. Newell, S. O., Polakoski, K. L., Williams, W. L. 1972. *Proc. Int. Congr. Anim. Reprod. Artif. Insem., 7th, 1972.*
23. Zaneveld, L. J. D., Robertson, R. T., Kessler, M., Williams, W. L. 1971. *J. Reprod. Fert.* 25:387
24. Zaneveld, L. J. D., Polakoski, K. L., Robertson, R. T., Williams, W. L. 1971. *Proc. Int. Res. Conf. Proteinase Inhibitors, 1st, 1970.*
25. Stambaugh, R., Brackett, B. G., Mastroianni, L. 1969. *Biol. Reprod.* 1:223
26. Gould, K. G. 1973. *Fert. Steril.* 24:448
27. Gould, K. G., Williams, W. L. 1972. *Proc. Int. Congr. Anim. Reprod. Artif. Insem., 7th, 1972.*
28. Zamboni, L. 1971. *Fine Morphology of Mammalian Fertilization.* New York: Harper. 222 pp.
29. Srivastava, P. N., Adams, C. E., Hartree, E. F. 1965. *J. Reprod. Fert.* 10:61
30. Yamane, J. 1935. *Cytologia* 6:233
31. Yamane, J. 1935. *Cytologia* 6:474
32. Austin, C. R. 1960. *J. Reprod. Fert.* 1:310
33. Clermont, Y., Clegg, R. E., Leblond, C. P. 1955. *Exp. Cell Res.* 8:453
34. Hathaway, R. R., Hartree, E. F. 1963. *J. Reprod. Fert.* 5:225
35. Hartree, E. F., Srivastava, P. N. 1965. *J. Reprod. Fert.* 9:47
36. Stambaugh, R., Buckley, J. 1969. *J. Reprod. Fert.* 19:423
37. Stambaugh, R. L. 1972. *Biology of Mammalian Fertilization and Implantation,* ed. K. S. Moghissi, E. S. E. Hafez, 185–212. Springfield: Thomas. 509 pp.
38. Lunstra, D. D., Clegg, E. D., Morré, D. J., Williamson, F. A., Malven, P. V. 1973. *Fed. Proc.* 32:284
39. Stambaugh, R., Smith, M. 1972. *Biol. Reprod.* 7:100
40. Pedersen, H. 1972. *J. Reprod. Fert.* 31:99
41. Bernstein, M. H., Teichman, R. J. 1973. *J. Reprod. Fert.* 33:239
42. Churg, A., Zaneveld, L. J. D., Schumacher, G. F. B. 1974. *Biol. Reprod.* In press
43. Srivastava, P. N. 1973. *J. Reprod. Fert.* 33:323
44. Srivastava, P. N., Zaneveld, L. J. D., Williams, W. L. 1970. *Biochem. Biophys. Res. Commun.* 39:575

45. Srivastava, P. N. 1973. *Biol. Reprod.* 9: 84
46. Yang, C. H. 1972. *Sperm acrosomal enzymes involved in fertilization.* MS thesis. Univ. Georgia, Athens. 116 pp.
47. Metz, C. B., Seiguer, A. C., Castro, A. E. 1972. *Proc. Soc. Exp. Biol. Med.* 140: 776
48. Swyer, G. I. M. 1947. *Nature* 159: 873
49. Swyer, G. I. M. 1947. *Biochem. J.* 41: 413
50. Austin, C. R. 1948. *Nature* 162: 63
51. Chang, M. C. 1951. *Ann. Ostet. Ginecol.* 73: 918
52. Austin, C. R., Braden, A. W. H. 1954. *Aust. J. Biol. Sci.* 7: 195
52a. Chang, M. C., Bedford, J. M. 1962. *Fertil. Steril.* 13: 421
53. Harper, M. J. K. 1970. *J. Exp. Zool.* 173: 47
54. Braden, A. W. H., Austin, C. R. 1954. *Aust. J. Biol. Sci.* 7: 543
55. Braden, A. W. H. 1959. *Symp. Genet. Biol. Ital.* 9: 1
56. Zamboni, L. 1970. *Biol. Reprod. Suppl.* 2: 44
56a. Lewis, B. K., Ketchel, M. M. 1972. *Proc. Soc. Exp. Biol. Med.* 141: 719
57. Soupart, P., Orgebin-Crist, M. C. 1966. *J. Exp. Zool.* 163: 311
58. Lewis, B. K., Ketchel, M. M. 1972. *Proc. Soc. Exp. Biol. Med.* 141: 712
59. Borders, C. L. Jr., Rafferty, M. A. 1968. *J. Biol. Chem.* 243: 3756
60. Hamner, C. E., Williams, W. L. 1965. *Fert. Steril.* 16: 170
61. Zaneveld, L. J. D., Polakoski, K. L., Schumacher, G. F. B. 1973. *J. Biol. Chem.* 248: 564
62. Khorlin, A. Y., Vikha, I. V., Milishnikov, A. N. 1973. *FEBS Lett.* 31: 107
63. Malmgren, H. 1953. *Biochim. Biophys. Acta* 11: 524
64. Zaneveld, L. J. D., Polakoski, K. L., Foley, C. W., Williams, W. L. 1971. *Fed. Proc.* 30: 595
65. Tillman, W. L. 1972. *Enzymology of spermatozoan penetration of mammalian ova.* MS thesis. Univ. Georgia, Athens. 105 pp.
66. Hartree, E. F. 1971. *Of Microbes and Life,* ed. J. Monod, E. Borek, 271–303. New York: Columbia Univ. Press. 538 pp.
67. Pincus, G., Enzmann, E. V. 1936. *J. Exp. Zool.* 72: 195
68. Dukelow, W. R., Chernoff, H. N., Williams, W. L. 1966. *Am. J. Physiol.* 211: 826
69. Bhalla, V. K., Dohanian, A., Newell, S. O., Williams, W. L. 1972. *Fed. Proc.* 31: 278

70. Pinsker, M. C., Williams, W. L. 1967. *Arch. Biochem. Biophys.* 122: 111
71. Mastroianni, L., Ehteshamzadeh, J. 1964. *J. Reprod. Fert.* 8: 145
72. Stambaugh, R., Noriega, C., Mastroianni, L. 1969. *J. Reprod. Fert.* 18: 51
73. Bedford, J. M. Personal communication
74. Austin, C. R. 1956. *J. Roy. Microsc. Soc.* 75: 141
75. Hanada, A., Chang, M. C. 1972. *Biol. Reprod.* 6: 300
76. Yanagimachi, R., Nicolson, G. L., Noda, Y. D., Fujimoto, M. 1973. *J. Ultrastruct. Res.* 43: 344
77. Chang, M. C., Hunt, D. M. 1956. *Exp. Cell Res.* 11: 497
78. Yanagimachi, R. 1972. *J. Reprod. Fert.* 28: 477
79. Oikawa, T., Yanagimachi, R., Nicolson, G. L. 1973. *Nature* 241: 256
80. Gwatkin, R. B. L. 1964. *J. Reprod. Fert.* 7: 99
81. Smithberg, M. 1953. *Anat. Rec.* 117: 554
82. Mintz, B. 1962. *Science* 138: 594
83. Braden, A. W. H. 1952. *Aust. J. Sci. Res. Ser. B* 5: 460
84. Braden, A. W. H., Austin, C. R., David, H. A. 1954. *Aust. J. Biol. Sci.* 7: 391
85. Pavlok, A., McLaren, A. 1972. *J. Reprod. Fert.* 29: 91
86. Bowman, P., McLaren, A. 1970. *J. Embryol. Exp. Morphol.* 24: 331
87. Stambaugh, R., Buckley, J. 1968. *Science* 161: 585
88. Toyoda, Y., Chang, M. C. 1968. *Nature* 220: 589
89. Gould, K., Zaneveld, L. J. D., Srivastava, P. N., Williams, W. L. 1971. *Proc. Soc. Exp. Biol. Med.* 136: 6
90. Da Silva Sasso, W. 1959. *Acta Anat.* 36: 352
91. Hauschka, S. D. 1963. *Biol. Bull.* 125: 363
92. Soupart, P., Noyes, R. W. 1964. *J. Reprod. Fert.* 8: 251
93. Inoue, M. 1973. *Biol. Reprod.* 9: 80
94. Hall, B. V. 1935. *Proc. Soc. Exp. Biol. Med.* 32: 747
95. Konecny, M. 1959. *C.R. Soc. Biol.* 153: 893
96. Bedford, J. M. 1968. *Am. J. Anat.* 123: 329
97. Yamane, J. 1930. *Cytologia* 1: 394
98. Polakoski, K. L., Zaneveld, L. J. D., Williams, W. L. 1971. *Biochem. Biophys. Res. Commun.* 45: 381
99. Polakoski, K. L., Zaneveld, L. J. D., Williams, W. L. 1972. *Biol. Reprod.* 6: 30
100. Multmäki, S., Niemi, M. 1972. *Int. J. Fert.* 17: 43

101. Stambaugh, R., Buckley, J. 1970. *Biol. Reprod.* 3:275
102. Schumacher, G. F. B. 1971. *Contraception* 4:67
103. Stambaugh, R., Buckley, J. 1972. *Biochim. Biophys. Acta* 284:473
104. Zaneveld, L. J. D., Dragoje, B. M., Schumacher, G. F. B. 1972. *Science* 177:702
105. Garner, D. L., Graves, C. N. 1972. *J. Anim. Sci.* 35:241
106. Gaddum, P., Blandau, R. J. 1970. *Science* 170:749
107. Polakoski, K. L., Zaneveld, L. J. D., Williams, W. L. 1972. *Biol. Reprod.* 6:23
108. Zaneveld, L. J. D., Dragoje, B. M., Schumacher, G. F. B. 1972. *Biol. Reprod.* 7:101
109. Meizel, S. 1972. *J. Reprod. Fert.* 31:459
110. Garner, D. L., Salisbury, G. W., Graves, C. N. 1971. *Biol. Reprod.* 4:93
111. Ho, J. J. L., Meizel, S. 1970. *J. Reprod. Fert.* 23:177
112. Ingrisch, H., Haendle, H., Werle, E. 1970. *Andrologie* 2:103
113. Fritz, H. et al 1972. *Z. Physiol. Chem.* 353:1953
114. Schleuning, W. D., Schiessler, H., Fritz, H. 1973. *Z. Physiol. Chem.* 354:550
115. Schiessler, H., Fritz, H., Arnhold, M., Fink, E., Tschesche, H. 1972. *Z. Physiol. Chem.* 353:1633
116. Dott, H. M., Dingle, J. T. 1968. *Exp. Cell Res.* 52:523
117. Buruiana, L. M. 1956. *Naturwissenschaften* 43:523
118. Polakoski, K. L., McRorie, R. A., Williams, W. L. 1973. *J. Biol. Chem.* 248:8178
119. Polakoski, K. L., McRorie, R. A. 1973. *J. Biol. Chem.* 248:8183
120. Allison, A. C., Hartree, E. F. 1970. *J. Reprod. Fert.* 21:501
121. Polakoski, K. L., Williams, W. L., McRorie, R. A. 1972. *Fed. Proc.* 31:278
122. Polakoski, K. L., McRorie, R. A. 1972. *Biol. Reprod.* 7:100
123. Fink, E., Schiessler, H., Arnhold, M., Fritz, H. 1972. *Z. Physiol. Chem.* 253:1633
124. Fritz, H. et al 1973. *Biol. Reprod.* 9:64
125. Garner, D. L. 1973. *Biol. Reprod.* 9:71
126. Williams, W. L. 1972. *Biology of Mammalian Fertilization and Implantation,* ed. K. S. Moghissi, E. S. E. Hafez, 19–53. Springfield: Thomas. 509 pp.
127. Zaneveld, L. J. D., Polakoski, K. L.,
128. Fritz, H. et al 1974. *Advan. Biosci.* In press
129. Polakoski, K. L. 1972. *Properties and function of enzymes and inhibitors involved in mammalian fertilization.* PhD thesis. Univ. Georgia, Athens. 182 pp.
130. Zaneveld, L. J. D., Srivastava, P. N., Williams, W. L. 1969. *J. Reprod. Fert.* 20:337
131. Robertson, R. T. 1971. *Naturally occurring antifertility substances from bovine seminal plasma and antifertility effects of active site directed reagents.* PhD thesis. Univ. Georgia, Athens. 123 pp.
132. Newell, S. O., Williams, W. L. 1972. *Fed. Proc.* 31:278
133. Chang, M. C. 1960. *Fertil. Steril.* 11:109
134. Hartman, C. G. 1961. *Fertil. Steril.* 12:170
135. Meizel, S., Cotham, J. 1972. *J. Reprod. Fert.* 28:303
136. Greenberg, L. J. 1962. *Biochem. Biophys. Res. Commun.* 9:430
137. Bhalla, V. K., Tillman, W. L., Williams, W. L. 1973. *J. Reprod. Fert.* 34:137
138. Koren, E., Milković, S. 1973. *J. Reprod. Fert.* 32:349
139. Polakoski, K. L., Williams, W. L., McRorie, R. A. 1973. *Fed. Proc.* 32:310
140. Hartmann, J. F., Gwatkin, R. B. L. 1971. *Nature* 234:479
141. Hartmann, J. F., Gwatkin, R. B. L., Hutchison, C. F. 1972. *Proc. Nat. Acad. Sci. USA* 69:2767
142. Hartmann, J. F., Hutchison, C. F. 1972. *J. Cell Biol.* 55:107a
143. Hartmann, J. F., Hutchison, C. F. 1973. *Biol. Reprod.* 9:85
144. Yang, C. H. 1973. *Biol. Reprod.* 9:71
145. Stambaugh, R., Buckley, J. 1971. *Fed. Proc.* 30:1184
146. Yanagimachi, R., Teichman, R. J. 1972. *Biol. Reprod.* 6:87
147. Srivastava, P. N., Goetsch, D. D. 1972. *Fed. Proc.* 31:813
148. Warren, L. 1959. *J. Biol. Chem.* 234:1971
149. Brown, C. R., Srivastava, P. N., Hartree, E. F. 1970. *Biochem. J.* 118:123
150. Brackett, B. G., Williams, W. L. 1965. *J. Exp. Zool.* 160:271
151. Gould, K. G., Srivastava, P. N., Cline, E. M., Williams, W. L. 1971. *Contraception* 3:261
152. Soupart, P., Clewe, T. H. 1965. *Fertil. Steril.* 16:677
153. Yang, C. H., Srivastava, P. N.,

Williams, W. L. 1973. *Biol. Reprod.* 9:219

Williams, W. L. 1973. *Fed. Proc.* 32: 310

154. Yang, C. H., Srivastava, P. N. 1974. *Proc. Soc. Exp. Biol. Med.* In press

155. Seiguer, A. C., Castro, A. E. 1972. *Biol. Reprod.* 7:31

156. Wislocki, G. B., Bunting, H., Dempsey, E. W. 1947. *Am. J. Anat.* 81:1

157. Bhalla, V. K. 1972. *Biol. Reprod.* 7:101

158. Bhalla, V. K., Tillman, W. L., Williams, W. L. 1973. *J. Reprod. Fert.* 34:137

159. Teichman, R. J., Bernstein, M. H. 1971. *J. Reprod. Fert.* 27:243

160. Bryan, J. H. D., Unnithan, R. R. 1973. *Histochemie* 33:169

161. Beckman, L., Jessler, B. K. 1968. *Hum. Hered.* 18:55

162. Meizel, S., Boggs, D., Cotham, J. 1971. *J. Histochem. Cytochem.* 19:226

163. Bryan, J. H. D., Unnithan, R. R. 1972. *Histochem. J.* 4:413

164. Austin, C. R. 1952. *Nature* 170:326

165. Chang, M. C. 1957. *Nature* 179:258

166. Bedford, J. M., Chang, M. C. 1961. *Fed. Proc.* 20:418

167. Bedford, J. M., Chang, M. C. 1962. *Am. J. Physiol.* 202:179

168. Yanagimachi, R., Chang, M. C. 1964. *J. Exp. Zool.* 156:361

169. Hamner, C. E., Jennings, L. L., Sojka, N. J. 1970. *J. Reprod. Fertil.* 23:477

170. Chang, M. C., Yanagimachi, R. 1963. *J. Exp. Zool.* 154:175

171. Mattner, P. E. 1963. *Nature* 199:772

172. Edwards, R. G., Unpublished. Quoted by C. R. Austin, 1969. *Advan. Biosci.* 4:5–12

173. Hunter, R. H. F., Dziuk, P. J. 1968. *J. Reprod. Fertil.* 15:199

174. Brackett. B. G., Seitz, H. M. Jr., Rocha, G., Mastroianni, L. Jr. 1972. *Biology of Mammalian Fertilization and Implantation*, ed. K. S. Moghissi, E. S. E. Hafez, 165–184. Springfield: Thomas. 509 pp.

175. Austin, C. R., Bavister, B. D., Edwards, R. G. 1973. *The Regulation of Mammalian Reproduction*, ed. S. J. Segal, R. Crozier, P. A. Corfman, P. G. Condliffe, 247–254. Springfield: Thomas. 586 pp.

176. Dukelow, W. R., Chernoff, H. N.,

Williams, W. L. 1967. *Am. J. Physiol.* 213:1397

177. Williams, W. L., Dohanian, A., Newell, S. O. 1972. *Biol. Reprod.* 7:102

178. Dukelow, W. R., Chernoff, H. N., Williams, W. L. 1967. *J. Reprod. Fert.* 14:393

179. Pinsker, M. C., Williams, W. L. 1968. *Proc. Soc. Exp. Biol. Med.* 129:446

180. Weinman, D. E., Williams, W. L. 1964. *Nature* 203:423

181. Williams, W. L., Abney, T. O., Chernoff, H. N., Dukelow, W. R., Pinsker, M. C. 1967. *J. Reprod. Fert. Suppl. Ser.* 2:11

182. Pinsker, M. C., Williams, W. L. 1967. *Arch. Biochem. Biophys.* 122:111

183. Robertson, R. T., Bhalla, V. K., Williams, W. L. 1971. *Biochem. Biophys. Res. Commun.* 45:1331

184. Williams, W. L., Robertson, R. T., Dukelow, W. R. 1970. *Advan. Biosci.* 4:61–72

185. Srivastava, P. N., Williams, W. L. 1971. *Nobel Symposium 15: Control of Human Fertility*, ed. E. Diczfalusy, U. Borell, 73–87. Stockholm: Almqvist & Wiksell. 354 pp.

186. Dukelow, W. R., Chernoff, H. N., Williams, W. L. 1966. *Proc. Soc. Exp. Biol. Med.* 121:396

187. Robertson, R. T. 1969. *Purification and fractionation of decapacitation factor.* MS thesis. Univ. Georgia, Athens. 58 pp.

188. Davis, B. K. 1971. *Proc. Nat. Acad. Sci. USA* 68:951

189. Hunter, A. G., Nornes, H. O. 1969. *J. Reprod. Fert.* 20:419

190. Chernoff, H. N., Pinsker, M. C., Dukelow, W. R., Williams, W. L. 1966. *Fed. Proc.* 25:284

191. Gwatkin, R. B. L., Williams, D. T. 1970. *Nature* 227:182

192. Abney, T. O., Williams, W. L. 1970. *Biol. Reprod.* 2:14

193. Vacquier, V. D., Tegner, M. J., Epel, D. 1972. *Nature* 240:352

194. Gwatkin, R. B. L., Williams, D. T., Hartmann, J. F., Kniazuk, M. 1972. *J. Cell Biol.* 55:101A

THE MOLECULAR ORGANIZATION OF MEMBRANES

S. J. Singer

Department of Biology, University of California, San Diego, La Jolla, California

INTRODUCTION

The molecular biology of membranes is a remarkably vital and active area at the present time. This is in part reflected in the books (1–6) and many reviews that have appeared recently covering various aspects or applications of membrane science:[1] membrane proteins and membrane structure (7–14); physical studies of membrane systems (11,15); membrane lipids (16, 17); bacterial membranes (18, 19); membrane transport (20, 21); membranes and cell motility (22–24); and membranes and immunology (25, 26). Since January 1973, the journal *Hospital Practice* has been publishing monthly articles on many aspects of membrane structure and function. It is not the intention of this article to review the most recent progress in an exhaustive manner, nor to stress the compilation of a large body of facts, figures, and references. Instead, I have tried to examine a few selected hypotheses and proposals about the structures and arrangements of the molecules, particularly the proteins, of membranes, and to review some recent investigations concerning these proposals. No claim is made, therefore, for comprehensiveness in this article. Some useful

[1] A very useful guide to the current literature on membrane studies is the monthly collection of references entitled "Cell Membranes" issued by the Biomedical Information Project, University of Sheffield, Sheffield, SI0 2TN, England.

generalizations about membrane structure and function are beginning to emerge from recent and current studies, and it is to these that I direct the discussion in this article.

THE PROTEINS OF MEMBRANES

Proteins constitute the largest fraction by weight of most functional cell membranes (8). Despite the great heterogeneity of proteins that exists in most individual membranes, a good beginning has recently been made in understanding the structures and arrangements of membrane proteins. Some aspects of this problem are considered below.

Peripheral and Integral Proteins

It has been suggested (7, 27) that the proteins associated with membranes can be classified into two broad categories termed peripheral and integral. The experimental criteria for this classification are given in Table 1. In short, peripheral proteins are those that appear to be only weakly bound to their respective membranes and do not appear to interact with the membrane lipids, whereas the integral proteins are ordinarily more strongly bound to the membrane and exhibit functionally important interactions with the membrane lipids. It is presumed that there is a structural basis for this classification; that peripheral and integral proteins are attached to the membrane in distinctly different ways.

Integral Proteins

By one or more of the criteria listed in Table 1, about 70–80% of membrane proteins are easily recognized as integral. These include most membrane-associated enzymes, antigenic proteins, transport proteins, drug and hormone receptors, and receptors for lectins. Enzymes that are integral proteins usually require lipids (sometimes specific lipids) for their activities to be expressed (13, 28, 29). Certain protein antigens, such as the Rh antigen of human erythrocyte membranes (30), show a similar require-

Table 1   Criteria for distinguishing peripheral and integral membrane proteins

Property	Peripheral protein	Integral protein
Requirements for dissociation from membrane	Mild treatments sufficient: high ionic strength, metal ion chelating agents	Hydrophobic bond-breaking agents required: detergents, organic solvents, chaotropic agents
Association with lipids when solubilized	Usually soluble free of lipids	Usually associated with lipids when solubilized
Solubility after dissociation from membrane	Soluble and molecularly dispersed in neutral aqueous buffers	Usually insoluble or aggregated in neutral aqueous buffers

ment. Rather than to compile a list of integral proteins, it is easier and more useful to list those known proteins that appear to be peripheral (see Table 2), most of the remainder being integral. As yet, because of their insolubility in ordinary aqueous media, not many integral proteins have been adequately purified and structurally characterized. However, the finding that a number of nonionic detergents can molecularly disperse the protein components of membranes, and often permit them to retain their specific enzymatic or binding functions (see below), has provided the major advance needed to overcome these problems. Some examples of the recent rapid progress in the isolation of integral membrane proteins include enzymes (31–41), hormone receptors (42, 43), lectin receptors (44–46), and transport components (32, 33, 47, 48).

In certain cases, the criteria of Table 1 have to be applied with some care. For example, the acetylcholinesterase (AChE) activity of bovine erythrocyte membranes is extensively solubilized by treatment of the membranes with 1.2 M NaCl in the absence of divalent cations (49), which might suggest that the enzyme is a peripheral protein in this case. However, substantial amounts of lipid are also solubilized in this process; apparently membrane fragments are released which can be enriched in the AChE component. Other cases exist where membrane components, which by several criteria are integral proteins, may be released spontaneously by a process of shedding membrane fragments (50). The mechanisms involved are not clear; perhaps the release occurs from specialized regions of the membrane such as microvilli. In any event, such apparent ready solubilization of a membrane protein should not be used to infer the peripheral nature of the protein involved.

In other cases, enzymes found in soluble form may have been derived by proteolysis from an originally intact membrane-bound form. The UDP-galactose:N-acetyl-glucosamine galactosyltransferase in milk consists of several soluble species of different molecular weights (50), probably all derived from an integral form of the enzyme found in mammary gland membranes (51, 51a, 52).

It is assumed from their properties that the integral proteins interact directly with at least some of the membrane lipids and are involved with the lipids in making up the structural matrix of the membrane. What are the special structural features of integral proteins that make this possible? Very rarely, such a protein may be covalently bound to large nonpolar structures that might anchor the protein to the lipid bilayer (53); generally, however, the structures of the folded polypeptide chains themselves must provide the answer. It was proposed in 1966 (54, 55) that as a general rule, the molecules of integral proteins were more or less globular and amphipathic; that is, their folded three-dimensional structures were segmented into hydrophilic and hydrophobic ends, with the hydrophilic ends (containing essentially all the ionic and highly polar residues of the protein) protruding from the membrane into the aqueous phase, and the hydrophobic ends embedded within the nonpolar interior of the lipid bilayer. The possibility was also recognized (54) that a molecule of an integral protein might span the membrane protruding from both surfaces if it had two hydrophilic ends separated by an appropriate hydrophobic middle segment.

Evidence has begun to appear supporting these structural proposals. Electron microscopy, utilizing the technique of freeze-fracture-etching, has revealed the fact

Table 2 Some peripheral proteins and complexes

Peripheral Protein or Complex	Membrane Localization	Receptor or Membrane Binding Site	Remarks	References
Cytochrome c	Outer surface of inner mitochondrial membrane	Cytochrome c reductase and cytochrome oxidase	Conformation depends on oxidation state (85)	83, 84
Spectrin	Cytoplasmic surface of erythrocyte	Intramembranous particle? (95)	Actomyosin-like complex	8, 22, 88, 93, 94
α-Lactalbumin	Mammary gland	Protein A galactosyl-transferase	Part of lactose synthetase enzyme complex	52
HPr protein	Cytoplasmic surface of bacterial membranes	Enzyme II complex	Part of PEP-sugar phospho-transferase system in bacteria. May bind preferentially when phosphorylated?	207
D-Glyceraldehyde-3-phosphate dehydrogenase	Cytoplasmic surface of erythrocyte membranes	Unknown	A glycolytic enzyme extracted from human ghosts by 0.1 M EDTA at pH 7.9, or 1 mM ATP	208–210
Aldolase	Same	Unknown	Another glycolytic enzyme of erythrocyes. All others seem to be cytoplasmic proteins?	209, 210
Ribosomes	Cytoplasmic side of rough endoplasmic reticulum (RER)	Not known precisely, but numbers and binding affinity measured (211); may be the sites which permit the transfer of newly synthesized polypeptide across RER membrane? (213)	Ribosomes are stripped from RER by puromycin and high salt. Reassociation occurs most avidly to stripped membranes; is saturable; and is inhibited by pretreatment of stripped membranes with trypsin	211–214
Nectin	Plasma membranes of *Streptococcus faecalis*	Unknown	Apparently required to bind ATPase to membrane. May act as linkage protein, binding to integral protein in membrane and to ATPase. mol wt $\sim 37,000$	215

Oligomycin-sensitivity conferring protein (OSCP)	Matrix side of inner mitochondrial membrane	Integral protein(s) of the rutamycin-sensitive ATPase system? Synthesis of these required for attachment of F_1-ATPase (217)	OSCP may be comparable to bacterial nectin. It may link the F_1-ATPase complex in the membrane. Stalk and headpiece arrangement controversial (218)	216–218
Monoamine oxidase	Outer membrane of mitochondria	Unknown	Released from membrane by 1.5% digitonin, or by sonication at pH 9.6. Purified enzyme reassociated with stripped outer membranes, but not with inner membranes. Showed saturable binding to outer membranes. The enzyme is not inhibited by treatment with phospholipase A.	219
Phytochrome	Unknown	Unknown	Phytochrome undergoes reversible conformational change with light between two states, $P_R \rightleftarrows P_{FR}$. Some seems to be membrane-bound, the rest soluble. Membrane-bound molecules are oriented differently in P_R and P_{FR} states.	220–222
Periplasmic binding proteins	Plasma membrane of gram-negative bacteria	Unknown	Proteins appear upon shock treatment of gram-negative bacteria. Specific ligand binding to these proteins is involved in transport of ligands through plasma membrane. See text for details.	General (20, 105, 223); sulfate (106, 107); phosphate (113); leucine (LIV protein) (20, 111) and (L protein) (224); histidine (112); arginine (225–227); glutamine (228, 229); cystine (230); galactose (110, 115, 231–233); arabinose (234, 235)

that substantial numbers of particles deeply penetrate most functional membranes, and in at least one case evidence has been obtained that these particles contain protein (56). Furthermore, by chemical labeling methods (57, 58) and proteolytic cleavage experiments (59), it has been shown that two major proteins of the human erythrocyte membrane span the entire thickness of the membrane. In their present state, these experiments show that the gross morphological features of certain integral proteins are as predicted: they are globular and deeply embedded in the membrane and, in certain instances, span the membrane. It is in regard to the amphipathy of integral proteins, however, that other evidence has recently appeared.

Cytochrome b_5 of microsomal membranes was first obtained in soluble form by enzyme treatments of the intact membranes (60, 61). In this form, it consists of a polypeptide chain of about 100 amino acid residues. The amino acid sequences of this form of the enzyme from several different species have been determined (62–66), and the X-ray crystallographic structure of the enzyme at 2.0 Å resolution has been obtained (67). However, the experiments of Ito & Sato (68) and of Strittmatter and his co-workers (34, 69) showed that the enzyme extracted by detergent treatment of the membranes was larger than that obtained by trypsin treatment. It contained about 40 additional amino acid residues, predominantly hydrophobic, attached at the carboxyl end of the tryptic fragment (34). The entire cytochrome b_5 molecule could bind spontaneously at 37°C to microsomal membranes and to mitochondria, but if the hydrophobic 40 amino acid fragment was first removed, the amino-terminal hydrophilic fragment would not bind to the membranes under the same conditions. Closely similar results have been obtained with the NAD-cytochrome b_5 reductase of microsomal membranes (35). It is a considerably larger protein than cytochrome b_5. When solubilized by proteolysis, it has a molecular weight of 33,000. Upon extraction with Triton X-100, however, its molecular weight is 43,000. The difference is due to an apparently membrane-bound fragment that has a disproportionately large fraction (65%) of nonpolar amino acid residues. This hydrophobic fragment must be present on the intact protein molecule for it to bind to microsomal membranes.

The major glycoprotein of human erythrocyte membranes, following earlier studies by Winzler and his colleagues (70) and others (71), has been isolated and characterized by Marchesi and co-workers (45, 56, 58). It consists of 60% carbohydrate and 40% peptide with a total molecular weight of about 55,000. The peptide is apparently a single polypeptide chain, but perhaps with some microheterogeneity. By CNBr and tryptic cleavage experiments and partial amino acid sequence determinations, it has been shown that the carbohydrate is present in multiple short oligosaccharide chains confined approximately to the NH_2-terminal half of the polypeptide chain. The amino-terminal CNBr fragment is labeled by the ^{125}I-lactoperoxidase technique on the intact cell and on resealed erythrocyte ghosts, and hence is presumably exposed at the exterior surface of the cell membrane. This is consistent with electron microscopic experiments using ferritin-labeled lectins (72) and with other chemical evidence (59, 70, 73), which show that essentially all the bound saccharide moieties of the membrane are on the exterior surface. On the other hand, the COOH-terminal CNBr fragment is also labeled by the lactoperoxidase technique if leaky erythrocyte ghosts are used, which would suggest that this portion of the glycophorin molecule is exposed at the cytoplasmic surface of the membrane. In addition, there is an intervening sequence of about 23 amino acid residues which are predominantly hydrophobic. It is suggested that if this stretch of residues was arranged in a continuous α helix, it would be long enough to span the membrane. There is also electron microscopic evidence that glycophorin is associated with the intramembranous particles in the erythrocyte membrane (56).

With these three integral proteins, the evidence is strong that they achieve an amphipathic structure in the intact membrane, with one (or two, in the case of glycophorin) exposed hydrophilic segment and one membrane-embedded hydrophobic segment. Furthermore, this three-dimensional amphipathy is correlated with the linear amphipathic structure of the molecule. It seems very likely, therefore, that in these cases the characteristically hydrophilic and hydrophobic linear sequences of the chain can fold independently of one another to form more or less autonomous domains, a possibility which might be important in the mechanisms whereby integral proteins are inserted into membranes after synthesis.

Although so far only these three proteins have been sufficiently well characterized to be informative about the proposed amphipathic structure of integral proteins, some other evidence has been obtained which lends further support to the proposal. The lipoprotein membranes of certain animal viruses contain relatively few proteins; in the case of the Semliki Forest Virus (SFV), only one, or perhaps two (74–76). By differentially labeling several of the hydrophobic amino acids of the virus during growth, Gahmberg et al (77) showed that proteolysis of the intact virus left behind in the viral membrane a fragment of the protein that was markedly enriched in these hydrophobic residues, consistent with an amphipathic structure for this protein.

Experiments have been reported with rhodopsin (78), an integral protein of the retinal rod-disk membrane, concerning energy transfer between certain covalently bound fluorescent molecules and the 11-*cis* retinal moiety of the protein. Three different fluorochromes were attached at three different sites on the exposed portion (presumably hydrophilic) of the rhodopsin molecule, while the retinal is presumably bound in the hydrophobic interior of the molecule. The experiments indicated that the fluorochromes were all about 73–77 Å away from the retinal, suggesting that the fluorochrome attachment sites were clustered at one end of the molecule, and the hydrophobic region at the other, as expected for an amphipathic structure.

That the amphipathy of integral membrane proteins may be a very general feature of their structure is indirectly suggested by many studies that have recently been carried out using nonionic detergents, which have shown that a particular integral protein may bind to, and be dissolved in, such detergents with essentially complete retention of its enzymatic or antigenic activity. Deoxycholate shows similar properties. Furthermore, it has been shown that ordinary soluble proteins bind very little, if any, nonionic detergent (79). These results taken together suggest that 1. integral proteins may be generally amphipathic; 2. their exposed hydrophilic ends, whose native conformations determine their respective enzymatic or antigenic activity, do not bind the nonionic detergent and hence are unaffected by it; and 3. their membrane-embedded hydrophobic ends bind the nonionic detergent, but since these hydrophobic ends may have no direct influence on the structure of the hydrophilic ends, the detergent binding does not affect the enzymatic or antigenic activity.

By contrast, ionic detergents such as sodium dodecylsulfate not only disperse membrane proteins but often inactivate them. This could be attributed to the interaction of such detergents with both the hydrophobic and hydrophilic (active) ends of the amphipathic proteins, since it is well known that ionic detergents bind to, and alter the conformations of, ordinary soluble proteins.

It has been suggested (26) that there may be several distinct ways that integral proteins may be associated with membranes, two of which are shown in Figure 1. If the protein exists as a monomer in the intact membrane, it may have the simple amphipathic structure represented in Figure 1, which accords, for example, with the information about cytochrome b_5 and cytochrome b_5 reductase discussed above. On the other hand, by analogy with multisubunit soluble proteins, specific subunit aggregates may exist in membranes, which, if they span the membrane, may generate water-filled pores through the membrane (27, Figure 1). On the lining of such pores, ionic groups of the subunits might well be present, since they would be in contact with water. The resultant structure of the subunit would therefore be more complex than simply amphipathic. Subunit aggregates such as those schematically represented in Figure 1 may be characteristic of those proteins that are the specific transport components of membranes (see below).

In recent pictorial representations of membrane models, protein molecules have sometimes been drawn as completely embedded within the hydrophobic interior of the membrane. However, according to the thermodynamic arguments that originally led to the concept of amphipathic integral proteins (27), for completely buried proteins to exist they would have to contain very few ionic amino acid residues. Although there is a general trend for integral membrane proteins to contain a larger fraction of hydrophobic amino acid residues than do soluble proteins (80), those so far analyzed also contain substantial numbers of ionic residues. For example, one of the most hydrophobic integral proteins known is the C_{55}-isoprenoid alcohol phosphokinase of *Staphylococcus aureus* (37); nonetheless, it has between 16 and 32 ionic residues (depending upon the unknown state of amidation of its aspartic and glutamic acid moieties) per molecule of 17,000 daltons. The total free energy

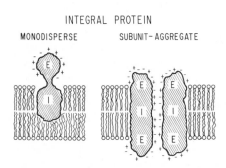

INTEGRAL PROTEIN

MONODISPERSE   SUBUNIT-AGGREGATE

Figure 1 Two of the ways that integral proteins might be associated with membranes, depicted schematically. Depending on their amino acid sequences and structures, they might exist as single molecules (monodisperse) or as subunit aggregates of identical or nonidentical subunits. The subunit aggregates, if they span the membrane, can generate water-filled pores through the membrane. E and I refer to exterior (exposed) and interior regions of the protein, respectively. See text and (26) for further details.

required to embed all or most of these residues within the hydrophobic interior could be as much as several hundred kilocalories per mole (27).

Therefore, although the volume fraction of an integral protein embedded in the membrane interior may vary widely for different proteins, there is no theoretical or experimental reason to expect any known proteins to be completely embedded.

In a related connection, it is entirely possible for an integral protein to have an overall amino acid composition which is highly hydrophilic, depending on the volume fraction and the composition of the exposed portion. Thus, calsequestrin (81), the calcium binding protein of sarcoplasmic reticulum, which is an integral protein because it requires deoxycholate for its release from the membrane, has only 46% of its amino acids hydrophobic (compared to 69%, for example, for the C_{55}-isoprenoid alcohol phosphokinase) and is unusually rich in aspartic and glutamic acid residues (which are probably associated with the large binding capacity of this protein for calcium).

Peripheral Proteins

Perhaps one should first ask: do peripheral proteins really exist, and why make the distinction? The answer is best given with some clear examples, each of which will be analyzed extensively before turning to a general discussion of peripheral proteins.

Cytochrome c, as a component of the mitochondrial electron transport chain, is associated with the inner mitochondrial membrane where the other components of the chain have also been localized (82). Unlike the other protein components of the chain, however, it is a peripheral protein since it can be rapidly released from the membrane, molecularly intact and free of lipids, by increasing the ionic strength to 3 M KCl. From studies with the labeling reagent ^{35}S-diazobenzenesulfonate, which crosses the outer mitochondrial membrane but cannot penetrate the inner, Schneider et al (83) have inferred that cytochrome c is attached to the surface of the inner membrane that faces the outer membrane, while the F_1-ATPase (see below) is on the opposite surface. Four different specific-site-directed antibodies have been prepared to human cytochrome c, and the Fab fragments of all four have been prepared (84). All four Fab preparations together inhibit both the reduction of cytochrome c by the succinate cytochrome c reductase and the oxidation of cytochrome c by cytochrome oxidase. One of the four types of Fab fragments, directed to a determinant controlled by the isoleucine residue at position 58 in the amino acid sequence of cytochrome c, has by itself no effect on either the reduction or oxidation of the protein. Another purified Fab blocks the oxidase reaction stoichiometrically but has no effect on the reductase reaction. These investigators suggest, therefore, that there seem to be two distinct binding regions on the cytochrome c molecule, which are quite far apart, by which the protein is attached on the one hand to the reductase and on the other to the oxidase components of the chain. Another factor to be taken into account is the significant conformational difference between the reduced and oxidized forms of the cytochrome c molecule (85), and the possible role of such differences in the binding to membrane sites.

Another reasonably well-studied example of a set of peripheral proteins is the complex called spectrin (86) or tektin (87, 88), associated with erythrocyte membranes and accounting for about 30% of the total membrane protein. Some properties of this complex have been discussed in Guidotti's review (8). It appears to consist primarily of two large polypeptide chains, of molecular weights about 230,000–250,000, and a smaller chain of about 43,000, corresponding to components 1, 2, and 5, respectively, on SDS-polyacrylamide electropherograms (89). Other minor components may be involved in this noncovalently bound complex. Components 1 and 2 have some of the physical properties of striated muscle myosin, such as the unusually high molecular weight and rod-like characteristics, but no detectable ATPase activity has been associated with spectrin. Component 5 resembles muscle actin in that it can be polymerized into microfilaments that can be decorated with heavy meromyosin (M. Sheetz, R. G. Painter, and S. J. Singer, unpublished results), a powerful diagnostic test for actin-like properties (22). The entire complex is peripheral because it is rapidly released free of lipids from the erythrocyte ghost membranes by the addition of 1 mM EDTA in water, although in time the entire membrane becomes fragmented under these conditions. Furthermore, ferritin-antibody staining of components 1 and 2 on intact erythrocyte ghosts (90) shows these components to be attached as long ($\sim$600–1000 Å) filamentous structures to the inner cytoplasmic surface of the membrane.

There is much interest in the possibility that the spectrin complex is responsible for certain mechanochemical activities of erythrocyte membranes. It is known that metabolic depletion of intact normal human erythrocytes, such as is produced by storage of heparinized blood at 4°C for 24 hr, results in the conversion of the cells from their usual biconcave disk shape to a highly crenated one (91, 92). These mechanical shape changes are accompanied by a drastic loss of intracellular ATP and a large increase in intracellular Ca^{2+} (91), and they can be rapidly reversed by the addition of adenine to the intact cells. These mechanochemical effects appear to be localized in the erythrocyte membranes; addition of ATP to ghosts from metabolically depleted cells converts their mechanical properties rapidly to those of ghosts prepared from normal, or metabolically repleted, cells. In other studies (93) it has been shown that the addition of γ ^{32}P-labeled ATP to erythrocyte ghosts labels component 2 of the spectrin complex. The inference to be drawn from these studies is that a reversible metabolically regulated phosphorylation of one or more components of the spectrin complex (perhaps mediating some actomyosin-like interaction) is involved in mechanical shape changes of the intact erythrocyte.

A related role proposed for the spectrin complex in erythrocytes is in controlling the lateral mobility of components in the erythrocyte membrane (26, 94, 95). The adult human erythrocyte is highly unusual (see below) among mammalian cells in the inability of many of its antigenic components or lectin receptors in the membrane to become clustered in the plane of the membrane upon binding of their respective antibodies or lectins (26, 96). On the other hand, such clustering can be induced in well-washed ghosts prepared from adult human erythrocytes (97), although the fluidity of the bulk lipid of both ghosts and intact cells appears to be essentially the same (98). Furthermore, such clustering appears to be inducible with the intact mature erythrocytes of newborn humans (99). The capacity to be clustered must reflect the translational mobility of certain components in the plane of the membrane, and the results quoted show that this mobility is rather finely regulated in the erythrocyte membrane. It appears that this mobility can be markedly altered without a change in lipid composition or fluidity. How might this occur?

Evidence has been obtained by electron microscopic experiments with erythrocyte ghosts (95) that the binding of antibodies specific to spectrin components 1 and 2 (on the cytoplasmic surface of the membrane) causes a redistribution of membrane-bound sialyl residues (presumably on glycophorin molecules) on the exterior surface of the ghosts. There is evidence that glycophorin is a component of the intra-membranous particles of the erythrocyte membrane (56). This suggests, therefore, that one or more components of the spectrin complex may be directly attached to one or more of the integral proteins in the intramembranous particles on the cyto-plasmic surface of the erythrocyte membrane. Some aggregation-disaggregation equilibrium of the spectrin complex itself might thereby control the lateral mobility of the intramembranous particle, but other integral proteins also might not diffuse as readily through this superstructure.

It has been suggested that components resembling the spectrin complex may be associated with membranes other than those of erythrocytes, but in much smaller amounts, and may be involved in such diverse mechanochemical activities as cell motility and cell ruffling, cytokinesis, and pinocytosis (22–24). The recent observa-tions—that the number of intramembranous particles of smooth muscle cell plasma membranes markedly increases in regions of the membrane undergoing pinocytosis (100); that actin-like filaments are associated with the contractile ring in cytokinesis (101); that there is a change in cell surface glycoproteins during mitosis (coincident with cytokinesis) (102); and that the intramembranous particles of lymphocyte mem-branes are collected in the cap produced by phytohemagglutinin or concanavalin A (103)—are all consistent with the hypothesis that in eukaryotic cells generally there may be a physical interaction between glycoprotein-containing intramembranous particles exposed at the exterior cell surface and an actomyosin-like system peri-pheral to the cytoplasmic surface, as is postulated for the erythrocyte membrane. This is certainly an intriguing possibility for further investigation.

Another protein which exhibits the properties of a peripheral protein is α-lactal-bumin of milk. This is a highly soluble and well-characterized protein for which no function was known until a few years ago. It was then shown that it is a component of the lactose synthetase-enzyme complex in the mammary gland (52). There is a membrane-bound galactosyltransferase (protein A) which by itself cata-lyzes the reaction

UDP-galactose + N-acetylglucosamine → N-acetyllactosamine + UDP

However, in the presence of α-lactalbumin, this enzyme carries out the lactose synthetic reaction

UDP-galactose + glucose → lactose + UDP

α-Lactalbumin shows considerable amino acid sequence homologies with egg white lysozyme (52), an oligosaccharide degrading enzyme, and presumably in combina-tion with protein A exhibits an active site capable of binding the specific saccharide moieties involved in the lactose synthetic reaction. α-Lactalbumin therefore can be characterized as a peripheral protein of membranes within the mammary gland.

Before discussing other proteins which may be membrane peripheral proteins, it is useful to examine those features that the three examples discussed have in common. In each case, the highly soluble peripheral protein appears to become attached to the membrane by specific binding to a particular integral protein of that membrane and, furthermore, achieves its function by virtue of that binding. This is not to be confused with a situation in which two integral proteins interact while both are embedded in the membrane matrix. Instead, it is proposed that the peripheral protein, when bound, remains completely outside the lipid matrix of the membrane and binds to a region of an integral protein that protrudes into the aqueous phase. Such specific binding, occurring in the aqueous phase, presumably involves the same kinds of interactions that hold together the subunits of soluble proteins. Such interactions are often mainly hydrophobic, occurring between specific regions of two homologous polypeptide chains, as between the α and β chains of hemoglobin (104).

I suggest that peripheral proteins will generally be attached to membranes by binding to the exposed hydrophilic ends of specific amphipathic integral proteins of the membrane. To account for this specific binding, it may turn out in some cases that a peripheral protein is structurally homologous (i.e. evolutionarily related) to the hydrophilic portion of the integral protein to which it specifically binds. (In particular, it is predicted that α-lactalbumin is homologous to the hydrophilic portion of protein A in the example discussed above.) In some cases, the peripheral protein may exist in two conformational states (perhaps influenced by binding of some small ligand) and be preferentially bound to the integral protein when it is in one of these states. Since each integral protein is often localized to certain intracellular or organelle membranes and is likely to be asymmetrically disposed on the surfaces of that membrane (see below), it follows that each peripheral protein attached by the proposed mechanism would therefore be localized to specific membranes and membrane surfaces. An alternative possibility, that peripheral proteins are attached not to integral proteins but to the polar heads of lipids, is not very attractive because it would not readily account for a high degree of specificity of peripheral protein attachment to membranes nor for functional specificity of the protein.

It is this suggested mode of attachment and function that, in fact, makes it meaningful to distinguish peripheral proteins as a separate class of membrane proteins. It is not their exposure to the aqueous phase that makes them special, since it is argued above and elsewhere (27) that all integral proteins have portions of their structures (their hydrophilic ends) also exposed. Nor is it only their weakness of binding to the membrane that distinguishes them. Rather, as is the case with the three examples discussed above, their relatively weak binding to specific integral proteins in the membrane is their special feature, which serves to modulate and regulate specific membrane functions.

If this structural and functional role is taken as the true definition of a peripheral protein, it is evident that by the operational criteria of Table 1 alone, and without additional structural or functional information, it may sometimes be difficult to distinguish whether a protein is truly peripheral or is either: 1. a contaminant attached during the isolation of the membrane but not actually associated with it in situ; or 2. an integral protein only slightly embedded into the lipid matrix of the membrane and hence easily dissociated from it. Two additional criteria that may prove useful in this connection are the specificity and stoichiometry with which the isolated peripheral protein reassociates with the stripped membrane. With either of the two alternative possibilities just mentioned, the protein should bind equally well to any membrane of appropriate lipid content and fluidity, whereas only that membrane stripped of the peripheral protein would have specific integral binding sites for it. Furthermore, such specific binding sites should show saturation of binding at stoichiometric concentrations of added peripheral protein, whereas either spurious or integral attachment to a membrane should not be as readily saturable.

The proteins and protein complexes listed in Table 2 are proposed to be linked peripherally to their respective membranes. They represent only those systems known to the author, and no doubt others should be added. They satisfy all or most of the criteria of Table 1. On the other hand, certain other proteins which appear to be only weakly bound to their membranes are deliberately omitted from Table 2 because they do not satisfy some of the criteria of Table 1. Two examples will suffice: succinic dehydrogenase of mitochondrial membranes can be fairly easily extracted, but a chaotropic agent, 0.4 M NaClO$_4$, is required (38). Similarly, the receptor immunoglobulin on B lymphocytes is spontaneously released from the cell surface, but apparently by a process of shedding membrane fragments; otherwise 0.5% Nonidet P-40 is required to solubilize this protein (50). Both proteins are therefore likely to be relatively weakly bound integral proteins of their respective membranes. Further discussion of the binding of the receptor immunoglobulin to B lymphocytes is given in (26).

In the case of protein (or nucleoprotein) complexes such as the F$_1$-ATPase or the ribosome, presumably only one or a few subunits of the complex is directly bound to some integral component(s) in the membrane. Possibly, however, further information will indicate that some of these systems are actually integral to the membrane but weakly bound (that is, only slightly embedded in the lipid matrix). The main reasons to propose such a category as peripheral membrane proteins and to compile the list in Table 2 are 1. to suggest new mechanisms by which these proteins may function and 2. to encourage the experimental search for the particular integral proteins presumed to function as the specific binding sites for the peripheral protein in question. To illustrate these points, only the periplasmic proteins are discussed in detail.

The periplasmic binding proteins of gram-negative bacteria are among an important set of proteins and enzymes selectively released when the bacteria are subjected to treatment with EDTA and lysozyme (which removes much of the cell wall) or upon osmotic shock, neither of which treatments, if properly carried out, ruptures the plasma membrane of the organism (105). There is a considerable number of different proteins released, but together they normally amount to only about 5% of the total cell protein, and they appear largely to fall into two functional categories: 1. degradative enzymes such as alkaline phosphatase, 5'-nucleotidase, ribonuclease I, DNA endonuclease I, etc; and 2. binding proteins, each of which is capable of specifically binding some small ligand such as sulfate ion, individual amino acids, or certain sugars. None of these binding proteins have been shown to have any enzymatic activity, but many of them have been implicated in the transport of their specific ligands across the plasma membrane. However, neither their mode of attachment to the membrane nor the mechanism of their transport functions is known. A review containing most of the salient facts known about the periplasmic binding proteins has recently appeared (20).

It is suggested that the periplasmic binding proteins are peripheral proteins on the outer surface of the plasma membrane of the bacterium; that their peripheral attachment is to certain integral proteins that form water-filled channels through the membrane; and that this association is responsible for the transport of the specific ligand of the binding protein. Some or all of the other periplasmic proteins (e.g. the degradative enzymes) may also be peripheral proteins of the plasma membrane or they may be weakly bound to other structures of the cell envelope, such as the murein or peptidoglycan layers (105). Little information is presently available to allow a choice among these possibilities. Attention in this section is therefore focused only on the binding proteins.

The periplasmic binding proteins appear to be bound to the outer surface of the plasma membrane of the bacterium. For example, the sulfate binding protein of *Salmonella typhimurium* (106) is labeled upon treatment of intact cells with ^{35}S-diazobenzenesulfonate, which penetrates the outer lipopolysaccharide-containing membrane but not the inner plasma membrane (107). The binding proteins are released from shocked cells under mild aqueous conditions requiring the presence of EDTA (105); under these conditions, neither known integral proteins of the plasma membrane [such as the M protein (108)] nor cytoplasmic proteins are released. The binding proteins are lipid-free, soluble, and unaggregated in aqueous buffers. The sulfate binding protein of *S. typhimurium* (109) and one of the leucine binding proteins of *Escherichia coli* (20, 110) have been crystallized. Several others have now been isolated in a pure state and partially characterized (20). Their molecular weights are remarkably similar, between 25,000 and 40,000, with many close to 35,000.

Transport of small molecules across bacterial membranes is remarkably complex. There are often several different transport systems for a particular molecule and sometimes these can be differentially inducible. Some of these systems, such as that for proline in *E. coli* (111), remain firmly bound to the membranes after osmotic shock treatment and show no evidence for a periplasmic binding protein involved in the transport process. For those systems where periplasmic binding proteins have been found, however, the following biochemical and genetic evidence strongly implicates them in the transport process: 1. osmotic shock treatment causes a loss in the transport activity corresponding to the binding protein released; 2. apparent K_m values for cellular transport and equilibrium constants for binding activity by the binding proteins are closely similar, and vary in a similar manner for the several ligands that may be transported by the same system; and 3. in one of the histidine transport systems of *S. typhimurium*, Ames & Lever (112) have shown that a particular gene, the *hisJ* gene, codes for the periplasmic histidine binding protein or *J* protein. Mutants defective in transport can be isolated which are defective in the *J* protein; revertants that have restored transport activity have recovered the ability to make normal *J* protein. Related genetic studies have been carried out in several other systems involving periplasmic binding proteins (20), but it has not been established as clearly that a particular gene codes for the binding protein in those cases.

On the other hand, there is also good evidence that these periplasmic binding proteins are not the only components involved in their respective transport processes. In the histidine transport system studied by Ames & Lever (112), for example, there are two other genes besides *hisJ*, termed *hisP* and *hisK*, that are implicated. The nature of the *hisK* gene product is not yet clear, but the *hisP* locus codes for a protein *P* which is apparently necessary for the *J* protein to be functional in transport. I suggest that the *P* protein is the integral protein of the plasma membrane to which the peripheral *J* protein binds, and that the combination is the functional transport unit which operates by a mechanism discussed in a later section of the article (see Figure 3). Another interesting case involves phosphate transport in *E. coli*. Medveczky & Rosenberg (113) have shown that mild shock treatment allowed the partial restoration of transport activity by incubating the treated cells with a pure phosphate binding protein isolated from shock fluids. Two different mutants were isolated which were defective in phosphate transport: one lacking the binding protein, the other having normal amounts of it. Upon shocking both types of cells, restoration of transport activity by the addition of phosphate binding protein occurred only with the cells originally lacking it. This suggests that the other mutant lacked another component essential to the transport activity; this is proposed to be the specific integral protein to which the binding protein must be attached to make an effective transport unit. I suggest that this will turn out to be a general property of such binding protein transport systems.

If this hypothesis is correct, it is somewhat surprising that it is only a relatively rare finding (113, 114) that exogenously supplied binding proteins can restore the specific transport activity lost from a shocked cell. Perhaps this is due to some modification of the integral proteins that occurs upon shock treatment (such as the proteolysis of their exposed hydrophilic regions by possible periplasmic proteases that may be released during shocking). Or, perhaps the binding protein attaches effectively to the integral protein only when in some particular conformational state (115).

In a related connection, Hazelbauer & Adler (116) have provided evidence that the galactose binding protein of *E. coli* is involved not only in β-methylgalactoside transport but also in galactose chemotaxis. On the other hand, there are mutants in which this transport activity is lost without any effect on galactose chemotaxis, and vice versa (117). This suggests that the same galactose binding protein may become attached to two different integral protein sites in the membrane: one mediating transport and the other chemotaxis.

It may not be far-fetched to realize that in well-recognized situations, a number of soluble proteins function as peripheral proteins in the same sense as is described in this article. These include polypeptide hormones, such as ACTH, insulin, glucagon, etc, which combine with receptor proteins that are integral proteins of the appropriate cell plasma membranes (118, 119); and protein antigens, which combine with their specific receptor IgM molecules on B lymphocyte membranes (120, 121). In each case this binding reaction, by some as yet poorly understood mechanisms, is the trigger for a complex cascade of events leading to the activation, differentiation, and proliferation of the cell.

MEMBRANE FLUIDITY

The concept that functional membranes are a kind of two-dimensional solution of globular integral proteins dispersed in a fluid lipid matrix, as embodied in the fluid mosaic model of membrane structure (7, 122), has received much attention recently. Part of the evidence in support of this model is discussed above in the section on integral proteins. Other supporting evidence comes from studies on the redistributions that lipids and proteins can be made to undergo in the plane of the membrane, and on the influence of lipid fluidity and phase changes on membrane functions. A complete list of appropriate references for the last two years would be too long to be practical. A partial list of topics follows: general references (7, 9, 11); redistributions of membrane lipids (123–129) and membrane proteins (97, 130–133); redistribution of membrane receptors on lymphocyte surfaces (96, 120, 121, 134–138) and other cells (139–141); influence of redistributions on the lectin-agglutinability of normal and malignantly transformed cells (142–146); influence of lipid fluidity and phase changes on membrane transport (147–152), enzyme activities (153–155), colicin-*E. coli* interactions (156), and insertion of integral proteins into membranes (34, 157); and influence of lipid fluidity on the mechanical properties of cell membranes (158, 159). There seems little doubt that fluidity of membranes and lateral mobility of lipids and proteins in the plane of the membrane are factors of prime importance in membrane biology.

McConnell and his colleagues (126, 127, 152) have emphasized the difference between first order phase transitions that occur at sharply defined temperatures with pure phospholipids, and phase separations that occur over a broad temperature range with binary mixtures of phospholipids. In phase separations, solidus and liquidus phases of different characteristic lipid compositions coexist at equilibrium at a given temperature, and the lateral mobility of the lipid molecules in the bilayer is required to achieve these separations. With bacterial mutants defective in the synthesis of unsaturated fatty acids (160), membranes can be generated with restricted fatty acid compositions in their phospholipids, which exhibit such lipid phase separations over temperature ranges characteristic of the particular lipid mixtures. Arrhenius plots of the transport rates in several independent transport systems each showed a sharp inflection at the temperature corresponding to the onset of the lipid phase separation (152), i.e. when very little *solidus* phase had yet formed and the composition of the *liquidus* phase was hardly altered. On the other hand, several enzyme activities of these membranes showed little or no effect of lipid phase separations on their rates (153, 154). These lipid effects no doubt reflect the different mechanisms of transport and enzymatic activities in membranes. If the enzymes involved are amphipathic integral proteins with their active sites within the hydrophilic exposed segments, then little effect might be expected of lipid phase separation on the activity. On the other hand, if the transport components spanned the membrane, lipid phase separations might significantly affect transport rates (see below). However, the possibility that in real membranes the lipids are asymmetrically distributed in the two halves of the bilayer (see below) may also play an as yet unanticipated role in lipid phase separation phenomena.

Although membrane fluidity and lateral mobility of membrane components appear to be general and functionally important phenomena, there is clear evidence that fluidity or mobility is restricted in certain membranes, or in regions of membranes, under particular conditions. Attention is therefore being directed to understanding the nature and mechanisms of such restrictions.

Long Range Protein-Protein Aggregation

There are certain regions of specialized and ordered structure within eukaryotic cell membranes that are well recognized, such as synapses (161) and gap junctions (162, 163). In certain prokaryotic membranes, such as with the halobacterium, similarly large regions of ordered structure (plaques) occur as part of an otherwise randomly ordered membrane matrix (164). It seems likely now that these ordered structures are special in that they consist of a single protein, or very few species of protein molecules, which form a large two-dimensional regular lattice by virtue of specific oriented protein-protein interactions in the plane of the membrane. Such a lattice could have pockets of lipid in the interstices between the protein subunits. In other words, it is a mosaic structure with the protein functioning as the rigid matrix, by contrast to a mosaic with a fluid lipid as the matrix; these could exist side by side in the same membrane. This picture is consistent with the findings that single protein species predominate in the gap junctions of liver membranes (165) and in the plaques of halobacterium membranes (166), whereas the proteins of the rest of these membranes are extremely heterogeneous. There have also been findings that the rhodopsin-like molecules in the halobacterium plaques have a rotational diffusion constant at least 1000 times smaller than the rhodopsin molecules in the fluid membranes of retinal rod outer disks (167). None of this information is therefore irreconcilable with a fluid mosaic model for the usual functional membranes, although this is sometimes disputed (168).

Peripheral Interactions with Membrane Components

In a previous section, the peripheral complex spectrin in erythrocyte membranes and its possible role in inhibiting the translational mobility of integral membrane components have been discussed, and it was suggested that similar actomyosin-like systems may operate with other cells and their membranes, perhaps at only certain times in the cell cycle (102) and on specialized regions of the membrane (101). The conversion of receptor immunoglobulin molecules from a translationally mobile to an immobile state in lymphocyte plasma membranes by the addition to the intact cells of a less than saturating amount of the lectin concanavalin A (169) may be due to the attachment or activation of such an actomyosin-like system on the cytoplasmic surface of the membrane. Furthermore, the apparently contradictory findings that concanavalin A binding to normal 3T3 cells does (144–145a) and does not (146) cause a clustering of their membrane receptors may be associated with the attachment of an actomyosin-like system to the plasma membrane in the latter case (in which cells attached to a substrate were examined), but not in the former (in which free cells in suspension were used).

In other cases, other types of peripheral attachments may inhibit the mobility of integral components in the membrane. The remarkable differences in the distribution of intramembranous particles in contiguous portions of the same chloroplast membrane, depending on whether the membrane is isolated or part of a membrane stack (170), may be due to external interactions between two closely opposed membranes. Connections between the plasma membrane and other segments of the cell envelope of bacteria may also serve such organizing functions, although there appears to be a translational mobility and random distribution for the intra-membranous particles in bacterial membranes (18).

MEMBRANE ASYMMETRY

There is a large body of evidence that the components of biological membranes are asymmetrically disposed across the membrane. Although membrane asymmetry has been discussed for many years, it is only recently that direct evidence has been obtained confirming asymmetry at the molecular level, and its fuller implications in membrane structure, function, and biosynthesis have been appreciated.

Three methods have been employed in these more recent studies. 1. Chemical and enzymatic methods have been used to differentially label or modify the components of the plasma membranes of intact cells compared to broken cells or organelles. The chemical methods used include covalent labeling with reagents impermeable to the plasma membrane of the intact cell, such as stilbene-4-acetamido-4′-thiocyano disulfonate (SITS) (171), ^{35}-S-diazoniumbenzenesulfonate (172), ^{35}-S-formylmethionylsulfone methylphosphate (173), and pyridoxal phosphate (174). The enzymatic methods involve restricted proteolytic digestion (59, 175); selective glycolytic degradation [e.g. with neuraminidase for terminal sialyl residues (73) or β-galactosidase for terminal β-galactosyl residues (176)]; specific lypolytic digestions of different phospholipids and their various covalent linkages (177); iodination with ^{125}I using the lactoperoxidase-H_2O_2 method (178) and its more gentle modification (179); enzymatic oxidation of specific saccharide moieties followed by reduction with ^{3}H-NaBH$_4$ (59); and transamidation of glutaminyl residues of membrane proteins using a transglutaminase enzyme and any of a variety of primary amines (180). Most of these studies have been carried out by comparing the patterns of labeling or modification of intact erythrocyte membranes to those of erythrocyte ghosts.

2. Right-side-out and inside-out vesicles can be separately prepared from isolated membranes under slightly different conditions, and such that they are largely impermeable to various reagents (59). They can then be compared, for example, in their ability to utilize an exogenously added and impenetrable substrate or

effector of a membrane-bound enzyme. They can also be advantageously used with the labeling and modifying reagents described above to circumvent the problem of whether any molecular reorganization has occurred upon isolating the membrane from the intact cell. (This problem is not necessarily trivial for the erythrocyte membrane because it is known, as discussed earlier in this article, that at least enough change occurs in the membrane upon ghosting to allow components that are originally translationally immobile to become mobile.) A disadvantage with this procedure, however, is that there is some loss of protein in preparing the inside-out vesicles.

3. The third method involves use of the electron microscope. Histochemical methods specific for particular enzymes can be used to determine the sidedness of their active sites on membranes in situ. An excellent example is the staining of the active site of the Na^+,K^+-dependent ATPase in erythrocytes (181). Another powerful electron microscopic method, where it can be applied, is the use of electron-dense-labeled agents, specific for particular membrane components, such as antibodies, lectins, toxins, hormones, etc. For example, ferritin-labeled antibodies can be used to determine the orientation of specific antigens or antigenic determinants in membranes (90, 182), and ferritin-labeled lectins have been used to demonstrate the exclusive localization of bound saccharide moieties to single surfaces of plasma and intracellular membranes (72, 183).

The three methods, where they have so far overlapped in application, have happily yielded the same conclusions. One result has been a fairly detailed description of the orientations of the major proteins and glycoproteins of the human erythrocyte membrane. These conclusions have been reviewed, (8, 12, 59) and are not extensively discussed here. Briefly, only two of the major proteins of the human erythrocyte membrane appear to be exposed at the outer surface of the membrane: component 3 (molecular weight about 100,000) and PAS-1, in the terminology of Steck (89). [These correspond to component a of Bretscher (12) and glycophorin (56), respectively.] Both components are glycoproteins and span the entire membrane; evidence suggests that glycophorin is a component of the intramembranous particle seen in freeze-fracture experiments (56), and it is possible that component 3 is also, but there is no direct evidence for this. Recent affinity labeling studies of intact erythrocytes with p-isothiocyano benzene sulfonic acid (184) and 4,4'-di-isothiocyano-2,2'-dihydrostilbene disulfonate (48) have provided strong evidence that component 3 is a major anion transport component of the membrane. The other seven or so major noncarbohydrate-containing proteins are accessible only at the cytoplasmic surface of the erythrocyte membrane. Of these, component 7 definitely seems to be an integral protein. The saccharide moieties of the glycoproteins and glycolipids are exclusively on the outer surface of the erythrocyte membrane, as well as those of the plasma membranes of many other types of mammalian cells (183).

Although the two major proteins exposed at the exterior surface of the erythrocyte membrane span the entire membrane, it is unwarranted to assume that all integral proteins exposed at the outer surface span the erythrocyte membrane. Many other known erythrocyte membrane proteins are present at 10^4 molecules or fewer per cell, amounts too small to be detected by standard staining procedures on SDS-polyacrylamide gels of whole membrane protein mixtures. Some of these, such as the enzyme acetylcholinesterase (49), the Rh antigen (185), and the major histocompatibility antigen (186, 187), have the properties of integral proteins and are known to be accessible on the outer surface of the membrane. It is not known whether any of these spans the entire membrane, but it seems likely that at least some integral proteins exposed at the outer surface will have their polypeptide chains embedded only part way through the erythrocyte membrane. In other membranes this is almost certainly the case. The receptor IgM molecule on the outer surface of B lymphocyte membranes is an example. Contrary to the manner in which it is sometimes drawn as deeply embedded in the membrane by its Fc fragment, the molecule appears to be integral but only slightly embedded in the surface of the membrane, according to its ^{125}I-lactoperoxidase labeling characteristics and its relatively easy release from the membrane by 0.5% Nonidet P-40 (25). It is clearly of great importance to an eventual understanding of the mechanisms by which integral proteins are incorporated into membranes to determine whether proteins only partly embedded in a membrane are found exposed at both or only one surface of the membrane. Another related question is whether integral proteins exposed at the outer surface of the plasma membrane of a eukaryotic cell are all necessarily glycoproteins.

The erythrocyte membrane is so far the most thoroughly investigated membrane with respect to the composition and asymmetrical distribution of its protein components. A beginning has been made in the distribution analysis of the electron transport chain components and of oxidative phosphorylation on the inner membrane of mitochondria (83), but so far the only firm conclusions are that the peripheral protein cytochrome c is exposed on the inner membrane surface that faces the outer membrane, while the F_1-ATPase complex, which is probably peripheral (Table 2), is located on the opposite surface. The functional asymmetry of the components of the electron transport chain across the mitochondrial inner membrane is a critical element in the chemiosmotic hypothesis of Mitchell (188).

It has been suggested that the major phospholipids of the erythrocyte membrane are asymmetrically distributed across the two halves of the bilayer (12) and by inference that this is true of other biological membranes. Although this may well turn out to be correct, the evidence first presented for it (189) was not compelling. It was shown that the amino-group-containing phospholipids (phosphatidylethanolamine and phosphatidylserine) were essentially unreactive toward a membrane-impermeable reagent in intact erythrocytes, whereas they reacted readily in ghost preparations. Similar results were obtained with another reagent (189a). This was interpreted to mean that these amino phospholipids were concentrated exclusively in the cytoplasmic half of the lipid bilayer, and by inference that the phosphatidylcholine and sphingomyelin were concentrated in the outer half. However, many instances have been reported with model systems of the chemical and enzymatic reactivities of lipids markedly affected by their local environment in the bilayers containing them (cf 190); the different physical properties of the membranes of intact and lysed erythrocytes have been discussed earlier. These chemical labeling results could therefore be ascribed to such physical differences of the membranes rather than to membrane asymmetry (cf 190a). Similar chemical experiments with right-side-out and inside-out erythrocyte membranes (mentioned above) could resolve the ambiguities involved.

More recently Zwaal et al (190b) have found that the phospholipase A_2 of *Naja naja* venom converts 70% of the phosphatidylcholine of intact human erythrocytes to lysolecithin without much effect on sphingomyelin, phosphatidylserine, or phosphatidylethanolamine, and without lysing the cell. Furthermore, treatment with a sphingomyelinase from *S. aureus* degrades 80% of the sphingomyelin without lysing the cell. Neither of these treatments degrades more than 10% of the phosphatidylethanolamine. On the other hand, if the lysed cell membranes are subjected to the phospholipase A_2 treatment, essentially all classes of phospholipids are extensively degraded. This evidence, a good deal more compelling than the chemical data just mentioned, is therefore consistent with the asymmetrical distribution of phospholipids in the erythrocyte membrane, as formulated by Bretscher (12).

There is no evidence at present for an asymmetrical distribution of phospholipids in cell membranes other than in erythrocytes. Ferritin antibody experiments, using antibodies directed to specific lipid polar head groups, may be useful in this regard. If phospholipid asymmetry is the result of an originally asymmetric synthesis of phospholipids on the two surfaces of a membrane, then it should of course be a general phenomenon. On the other hand, there are many enzymatic modifications and interconversions known to occur with the lipids in intact membranes by lipases as well as by other enzymes (191). In addition, exchanges of intact phospholipids between serum and erythrocyte membranes (192, 193) and between different intracellular membranes (194, 195) are known to occur. To what extent such enzymatic and exchange processes remodel the membrane lipids and affect the symmetry of their distribution in particular membranes is not known. If exchange processes are specific, that is, if only intact phospholipids with the same structure are transferred in a single exchange, they would have no effect on existing asymmetry. [In this regard and in connection with the proposed phospholipid asymmetry of the erythrocyte membrane discussed above, it is interesting that in man and dog the phosphatidylserine and phosphatidylethanolamine of erythrocytes do not exchange with serum, while phosphatidylcholine and sphingomyelin do (196).]

If an asymmetrical distribution of phospholipids is a general phenomenon of cell membranes, then its implications for the structure and function of membranes are considerable. For example, lipid asymmetry has not previously been considered a factor in lipid phase separation experiments (150, 152). In *E. coli* membranes, with which most of these experiments were carried out, the major phospholipid (70–80%) is phosphatidylethanolamine and the only other important phospholipids are phosphatidylglycerol (5–15%) and cardiolipin (5–15%). If the two latter anionic lipids were concentrated in one layer of the bilayer, with the phosphatidylethanolamine distributed in both layers, then the two layers and surfaces could have distinctly different properties. Lipid asymmetry would also pay a role in the kinds of direct short range lipid-protein interactions that some integral enzymes of membranes may require to express their activities (28, 29). Obviously, lipid asymmetry would also have to be taken into account in reconstitution experiments between isolated proteins and lipids (14).

TRANSMEMBRANE ROTATIONS AND THE MECHANISMS OF TRANSPORT

The existence of a stable asymmetrical distribution of proteins and lipids across a membrane bilayer implies that these components do not rotate from one surface of the membrane to the other at significant rates. Such rates have been examined directly for phospholipids in synthetic bilayers (197) and have indeed been shown to be very slow.[2] Where such transmembrane rotations of integral proteins have been sought (59, 198a), they have also been found to be undetectable. This is especially remarkable in view of the rapid translational and in-plane rotational mobilities of proteins (199–201) and lipids (202) in many membranes under physiological conditions, and the conclusive evidence that the bulk of the lipid in a membrane is physically independent of the membrane protein (cf 7). It is therefore clear that the negligibly slow rates of transmembrane rotations cannot be attributed to the presence of a rigid structure preventing such rotations. Rather, it has been suggested (7) that the passage through the hydrophobic membrane interior of the polar head groups and ionic residues of the lipids and proteins is thermodynamically a highly unfavorable process. The free energies of activation required are so large that the rates of such rotations are negligibly slow. The thermodynamic arguments are those which led originally to the suggestion that the molecules of integral membrane proteins are generally amphipathic (27, 54).

These considerations bear on the molecular mechanisms involved in the facilitated diffusion or active transport of small hydrophilic molecules through membranes. It is virtually certain that these processes are mediated by a set of proteins associated with the membrane, each with a characteristic binding site capable of specifically binding a small molecule. The problem is, how is the small molecule translocated; that is, how does it get from one side of the membrane to the other? The many specific mechanisms proposed for the translocation process fall into two broad categories: 1. carrier mechanisms, in which the binding site gets from one side of the membrane to the other by a rotation of the entire transport protein across the membrane; and 2. fixed pore mechanisms, in which a conformational change within the protein-lined pore translocates the binding site across the membrane. The experimental facts and thermodynamic arguments discussed above would argue against carrier mechanisms that require transmembrane rotations of integral protein molecules, but there has been no firm evidence to discriminate between these two

[2] In similar recent experiments with a spin-labeled phospholipid incorporated into membrane vesicles from the electroplax of the electric eel (198), it was concluded that the transmembrane rotation was relatively rapid, with a half-time on the order of 5 min. There may be artifacts in these experiments, however; whether the spin-labeled phospholipid actually penetrated into the inner half of the vesicle membrane bilayer was not directly demonstrated.

categories. Recently, however, Kyte (182) has prepared antibodies to the purified Na^+,K^+-dependent ATPase of canine kidney (32), the enzyme involved in the active coupled transport of Na^+ and K^+ through the membrane. He has shown conclusively that some of these antibodies can bind firmly to the enzyme (as measured by complement fixation and by ferritin-antibody staining of ATPase-containing membrane vesicles) and yet have no effect on the kinetic parameters of the ATPase enzymatic activity. Since there is a considerable body of evidence that the ATPase enzyme activity and the coupled Na^+,K^+ transport are expressions of the same event, these results strongly argue against a carrier mechanism for the transport process in this case. It is highly likely that the attachment of one or more large antibody molecules to the enzyme would so markedly inhibit its transmembrane rotation rate as to make it rate limiting in the transport process, and hence affect the kinetics of the ATPase activity. [There are also antibodies known which can inactivate this enzyme (203–205), but because there are many possible explanations for such inactivation, this tells nothing definitive about the transport mechanism.]

These results suggest that further investigation of fixed pore mechanisms in transport would be justified. Jardetzky (206) and Singer (27, 94) have suggested essentially similar models in which an integral protein, which is a specific aggregate of subunits, spans the membrane and generates a continuous pore through the membrane down the central axis of the aggregate, much as in the case of the central channel through the hemoglobin molecule (104). A specific binding site within this pore faces one side of the membrane when the protein is in one conformational state (Figure 2), but a quaternary rearrangement of the subunits with respect to one another (triggered perhaps by some enzymatic reaction in the case of active transport) translocates the binding site to the other side of the membrane. Completion of the enzymatic reaction returns the protein to its initial state.

A simple extension of this model affords a reasonable explanation of the role of periplasmic binding proteins (see earlier section in this article) in some transport

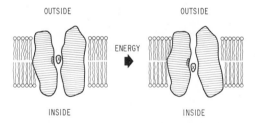

OUTSIDE   OUTSIDE

ENERGY

INSIDE   INSIDE

Figure 2 A schematic mechanism for the translocation event in active transport (26, 27, 206). A specific site for a hydrophilic ligand X exists on the surface of a pore formed by a particular subunit aggregate (see Figure 1); the aggregate might actually be tetrameric, like hemoglobin. Some energy-yielding process is then converted into a quaternary rearrangement of the subunits, which translocates the binding site and X from one side of the membrane to the other. Reversal of the protein change restores the subunit aggregate to its initial state.

processes through bacterial plasma membranes, a role presently not understood (20, 21). Although these binding proteins appear to carry the specific binding sites involved in the transport process, they are highly soluble and hydrophilic proteins which are, for thermodynamic reasons (7, 27), not likely to diffuse through or rotate across the membrane at a significant rate. Elsewhere in this article, I have suggested that these proteins are peripheral proteins attached to specific integral proteins exposed at the outer surface of the bacterial membrane. It is further proposed that these integral proteins are subunit aggregates which span the membrane (Figure 3). This system could function in transport in the following way: Without the peripheral binding protein attached to it, this aggregate forms a pore which is closed to, and has no specific affinity for, any small hydrophilic molecules other than water. With the peripheral protein attached, the complex can exist in one of two conformational states. In one, the pore remains closed, and the binding sites on the binding protein face the outer surface of the membrane. In the other (triggered by some enzymatic reaction in the case of active transport), the pore formed by the integral protein is open, and the binding site on the binding protein is altered to face the other side of the membrane. Many variations of this scheme are conceivable, but the basic notions are clear. The binding protein is responsible for the specificity of transport, but does not itself cross the membrane during transport. It is an essential component of the transport system, but the integral protein to which it binds, and which forms the transmembrane channel, is equally essential. The energy requirement for the active transport serves to produce a quaternary rearrangement of the subunits forming the channel, and translocates the ligand.

These proposals have the virtue of providing a unified scheme for the critical

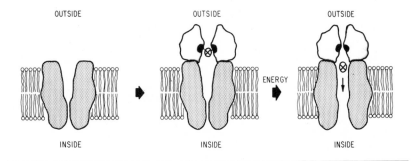

Figure 3   A schematic mechanism for the translocation event in active transport, where a periplasmic binding protein is obligatorily involved. It is proposed that there is present in the membrane subunit aggregates of specific integral protein(s) (stippled) which span the membrane, forming a water-filled pore (Figures 1 and 2). This pore (left) is initially closed to the diffusion of molecules other than water. A binding protein (shaded), with an active binding site for ligand X, attaches specifically to the exposed surface of the integral protein (center), the pore remaining closed. Some energy-yielding step results in a quaternary rearrangement of the subunits of this structure (right), opening the pore and releasing X to the other side of the membrane.

translocation events in transport phenomena that has some precedent in ordinary solution protein chemistry and is consistent with experimental and theoretical analyses of membrane structure. There is at present, however, no direct evidence to support these proposals. An experimental test of this proposal, analogous to that of Kyte (182), would be to prepare an antibody to a binding protein, such as the phosphate binding protein, that did not affect the capacity of the binding protein either to bind its specific ligand or to bind to the bacterial membrane. Addition of this antibody to a reconstituted system that had been mildly shocked and had excess binding protein added back (113) might not markedly change the transport rate. If such a result were obtained, it would strongly indicate that the binding protein did not traverse the membrane during transport.

Literature Cited

1. Rothfield, L. I., Ed. 1971. *Structure and Function of Biological Membranes.* New York: Academic
2. Fox, C. F., Ed. 1972. *Membrane Research.* New York: Academic
2a. Fox, C. F., Keith, A. D., Eds. 1972. *Molecular Biology of Membranes.* Stanford, Conn.: Sinauer
3. Bolis, L., Keynes, R. D., Eds. 1972. *Role of Membranes in Secretory Processes.* New York: Elsevier
4. Kreuger, F., Slegers, J. F. G., Eds. 1972. *Biomembranes,* Vol. 3: *Passive Permeability of Cell Membranes.* New York: Plenum
5. Wallach, D. F. H., Ed. 1973. *Plasma Membranes: Dynamic Perspectives, Genetics, and Pathology.* New York: Springer
6. Nystrom, R. A. 1973. *Membrane Physiology.* New Jersey: Prentice-Hall
7. Singer, S. J., Nicolson, G. L. 1972. *Science* 175:720–31
8. Guidotti, G. 1972. *Ann. Rev. Biochem.* 41:731–52
9. Gitler, C. 1972. *Ann. Rev. Biophys. Bioeng.* 1:51–92
10. Wallach, D. F. H. 1972. *Biochim. Biophys. Acta* 265:61–83
11. Oseroff, A. R., Robbins, P. W., Burger, M. M. 1973. *Ann. Rev. Biochem.* 42:647–82
12. Bretscher, M. S. 1973. *Science* 181:622–29
13. Coleman, R. 1973. *Biochim. Biophys. Acta* 300:1–30
14. Razin, S. 1972. *Biochim. Biophys. Acta* 265:241–96
15. Levine, Y. K. 1972. *Progr. Biophys. Mol. Biol.* 24:1–74
16. Bangham, A. D. 1972. *Ann. Rev. Biochem.* 41:753–76
17. Cronan, J. E. Jr, Vagelos, P. R. 1972. *Biochim. Biophys. Acta* 265:25–60
18. Remsen, C. C., Watson, S. W. 1972. *Int. Rev. Cytol.* 33:253–96
19. Machtiger, N. A., Fox, C. F. 1973. *Ann. Rev. Biochem.* 42:575–600
20. Oxender, D. L. 1972. *Ann. Rev. Biochem.* 41:777–814
21. Lin, E. C. C. See Ref. 1, 285–341
22. Huxley, H. E. 1973. *Nature* 243:445–49
23. Wessells, N. K., Spooner, B. S., Luduena, M. A. 1973. *Ciba Found. Symp.* 14:53–82
24. Allison, A. C. See Ref. 23, 109–48
25. Vitetta, E. S., Uhr, J. W. 1973. *Transplant. Rev.* 14:50–75
26. Singer, S. J. 1974. *Advan. Immunol.* In press
27. Singer, S. J. See Ref. 1, 145–222
28. Rothfield, L. I., Romeo, D. See Ref. 1, 251–84
29. Triggle, D. J. 1970. *Recent Progr. Surface Sci.* 3:273–90
30. Green, F. A. 1968. *J. Biol. Chem.* 243:5519–21
31. MacLennan, D. H., Seeman, P., Iles, G. H., Yip, C. C. 1971. *J. Biol. Chem.* 246:2702–10
32. Kyte, J. 1971. *J. Biol. Chem.* 246:4157–65
33. Hokin, L. E. et al *J. Biol. Chem.* 248:2593–2605
34. Strittmatter, P., Rogers, M. J., Spatz, L. 1972. *J. Biol. Chem.* 247:7188–94
35. Spatz, L., Strittmatter, P. 1973. *J. Biol. Chem.* 248:793–99
36. Blumberg, P. M., Strominger, J. L. 1972. *Proc. Nat. Acad. Sci. USA* 69:3751–55
37. Sandermann, H., Strominger, J. L. 1971. *Proc. Nat. Acad. Sci. USA* 68:2441–43
38. Davis, K. A., Hatefi, Y. 1971. *Biochemistry* 10:2509–16
39. Scandella, C. J., Kornberg, A. 1971. *Biochemistry* 10:4447–56
40. Wickner, W. T., Kennedy, E. P. 1971. *Fed. Proc.* 30:1119

41. Müller, E., Hinckley, A., Rothfield, L. 1972. *J. Biol. Chem.* 247:2614–22
42. Cuatrecasas, P. 1972. *Proc. Nat. Acad. Sci. USA* 69:1277–81
43. Lindstrom, J., Patrick, J. 1973. In *Synaptic Transmission and Neuronal Interaction,* ed. M. V. L. Bennett. New York: Raven
44. Schmidt, J., Raftery, M. A. 1973. *Biochemistry* 12:852–56
45. Marchesi, V. T., Andrews, E. P. 1971. *Science* 174:1247–48
46. Allan, D., Anger, J., Crumpton, M. S. 1972. *Nature New Biol.* 236:23–25
47. Akedo, H., Mori, Y., Tanigaki, Y., Shinkai, K., Morita, K. 1972. *Biochim. Biophys. Acta* 271:378–87
48. Cabantchik, Z. I., Rothstein, A. 1974. *J. Membrane Biol.* In press
49. Burger, S. P., Fujii, T., Hanahan, D. J. 1968. *Biochemistry* 7:3682–3700
50. Vitetta, E. S., Uhr, J. W. 1972. *J. Immunol.* 108:577–79
51. Barker, R., Olsen, K. W., Shaper, J. H., Hill, R. L. 1972. *J. Biol. Chem.* 247:7135–47
51a. Bergeron, J. J. M., Ehrenreich, J. H., Siekevitz, P., Palade, G. E. 1973. *J. Cell Biol.* 59:73–88
52. Brew, K., Vanaman, T. C., Hill, R. L. 1968. *Proc. Nat. Acad. Sci. USA* 59:491–97
53. Braun, V., Rehn, K. 1969. *Eur. J. Biochem.* 10:426–38
54. Lenard, J., Singer, S. J. 1966. *Proc. Nat. Acad. Sci. USA* 56:1828–35
55. Wallach, D. F. H., Zahler, P. H. 1966. *Proc. Nat. Acad. Sci. USA* 56:1552–59
56. Marchesi, V. T., Tillack, T. W., Jackson, R. L., Segrest, J. P., Scott, R. E. 1972. *Proc. Nat. Acad. Sci. USA* 69:1445–49
57. Bretscher, M. S. 1971. *J. Mol. Biol.* 59:351–57
58. Segrest, J. P., Kahane, I., Jackson, R. L., Marchesi, V. T. 1973. *Arch. Biochem. Biophys.* 155:167–83
59. Steck, T. L. See Ref. 2, 71–93
60. Strittmatter, P., Velick, S. F. 1956. *J. Biol. Chem.* 221:253–64
61. Kajihara, T., Hagihara, B. 1968. *J. Biochem.* 63:453–61
62. Ozols, J., Strittmatter, P. 1969. *J. Biol. Chem.* 244:6617–18
63. Tsugita, A., Kobayashi, M., Kajihara, T., Hagihara, B. 1968. *J. Biochem.* 64:727–30
64. Ozols, J. 1970. *J. Biol. Chem.* 245:4863–74
65. Nobrega, F. G., Ozols, J. 1971. *J. Biol. Chem.* 246:1706–17
66. Ozols, J. 1972. *J. Biol. Chem.* 247:2242–45
67. Mathews, F. S., Argos, P., Levine, M. 1971. *Cold Spring Harbor Symp. Quant. Biol.* 36:387–95
68. Ito, A., Sato, R. 1968. *J. Biol. Chem.* 243:4922–23
69. Spatz, L., Strittmatter, P. 1971. *Proc. Nat. Acad. Sci. USA* 68:1042–46
70. Winzler, R. J. 1969. In *Red Cell Membrane,* ed. G. A. Jamieson, T. J. Greenwalt, 157–71. Philadelphia: Lippincott
71. Morawiecki, A. 1964. *Biochim. Biophys. Acta* 83:339–47
72. Nicolson, G. L., Singer, S. J. 1971. *Proc. Nat. Acad. Sci. USA* 68:942–45
73. Eylar, E. H., Madoff, M. A., Brody, O. V., Oncley, J. L. 1962. *J. Biol. Chem.* 237:1992–2000
74. Simons, K., Kääriäinen, L. 1970. *Biochem. Biophys. Res. Commun.* 38:981–88
75. Strauss, J. H., Burge, B. W., Darnell, J. E. 1970. *J. Mol. Biol.* 47:437–48
76. Schlesinger, M. J., Schlesinger, S., Burge, B. W. 1972. *Virology* 47:539–41
77. Gahmberg, C. G., Utermann, G., Simons, K. 1972. *Fed. Eur. Biochem. Soc. Lett.* 28:179–182
78. Wu, C. W., Stryer, L. 1972. *Proc. Nat. Acad. Sci. USA* 69:1104–8
79. Helenius, A., Simons, K. 1972. *J. Biol. Chem.* 247:3656–61
80. Capaldi, R. A., Vanderkooi, G. 1972. *Proc. Nat. Acad. Sci. USA* 69:930–32
81. MacLennan, D. H., Wong, P. T. S. 1971. *Proc. Nat. Acad. Sci. USA* 68:1231–35
82. Ernster, L., Kuylenstierna, B. 1970. In *Membranes of Mitochondria and Chloroplasts,* ed. E. Racker, 172. New York: Van Nostrand
83. Schneider, D. L., Kagawa, Y., Racker, E. 1972. *J. Biol. Chem.* 247:4074–79
84. Smith, L., Davies, H. C., Reichlin, M., Margoliash, E. 1973. *J. Biol. Chem.* 248:237–43
85. Takano, T., Swanson, R., Kallai, O. B., Dickerson, R. E. 1971. *Cold Spring Harbor Symp. Quant. Biol.* 36:397–404
86. Marchesi, V. T., Steers, E. Jr. 1968. *Science* 159:203–4
87. Mazia, D., Ruby, A. 1968. *Proc. Nat. Acad. Sci. USA* 61:1005–12
88. Clarke, M. 1971. *Biochem. Biophys. Res. Commun.* 45:1063–70
89. Steck, T. L. 1972. *J. Mol. Biol.* 66:295–305
90. Nicolson, G. L., Marchesi, V. T., Singer, S. J. 1971. *J. Cell Biol.* 51:265–72
91. Weed, R. I., La Celle, P. L., Merrill, E. W. 1969. *J. Clin. Invest.* 48:795–809
92. Hochmuth, R. M., Mohandas, N. 1972. *Microvasc. Res.* 4:295–97

93. Williams, R. O. 1972. *Biochem. Biophys. Res. Commun.* 47:671–78
94. Singer, S. J. 1973. *Hospital Practice* 8 (5):81–90
95. Nicolson, G. L., Painter, R. G. 1973. *J. Cell Biol.* 59:395–406
96. Loor, F., Forni, L., Pernis, B. 1972. *Eur. J. Immunol.* 2:203–12
97. Pinto da Silva, P. 1972. *J. Cell Biol.* 53: 777–87
98. Landsberger, F. R., Paxton, J., Lenard, J. 1972. *Biochim. Biophys. Acta* 266:1–6
99. Blanton, P. L., Martin, J., Haberman, S. 1968. *J. Cell Biol.* 37:716–28
100. Orci, L., Perrelet, A. 1973. *Science* 181:868–69
101. Schroeder, T. 1973. *Proc. Nat. Acad. Sci. USA* 70:1688–92
102. Fox, T. O., Sheppard, J. R., Burger, M. M. 1971. *Proc. Nat. Acad. Sci. USA* 68:244–47
103. Loor, F. 1973. *Eur. J. Immunol.* 3:112–16
104. Perutz, M. F. 1969. *Proc. Roy. Soc. Edinburgh B* 173:113–40
105. Heppel, L. A. See Ref. 1, 223–47
106. Pardee, A. B., Prestidge, L. S., Whipple, M. B., Dreyfus, M. B. 1966. *J. Biol. Chem.* 241:3962–69
107. Pardee, A. B., Watanabe, K. 1968. *J. Bacteriol.* 96:1049–54
108. Jones, T. H. D., Kennedy, E. P. 1969. *J. Biol. Chem.* 244:5981–87
109. Langridge, R., Shinagawa, H., Pardee, A. B. 1970. *Science* 169:59–61
110. Anraku, Y. 1968. *J. Biol. Chem.* 243: 3116–22
111. Piperno, J. R., Oxender, D. L. 1966. *J. Biol. Chem.* 241:5732–34
112. Ferro-Luzzi-Ames, G., Lever, J. 1970. *Proc. Nat. Acad. Sci. USA* 66:1096–1103
113. Medveczky, N., Rosenberg, H. 1970. *Biochim. Biophys. Acta* 211:158–68
114. Anraku, Y. 1967. *J. Biol. Chem.* 242: 793–800
115. Boos, W., Gordon, A. S. 1971. *J. Biol. Chem.* 246:621–28
116. Hazelbauer, G. L., Adler, J. 1971. *Nature New Biol.* 230:101–4
117. Adler, J. 1969. *Science* 166:1588–97
118. Karlin, A. 1973. *Fed. Proc.* 32:1847–53
119. Rodbell, M. 1973. *Fed. Proc.* 32:1854–58
120. Diener, E., Paetkau, V. H. 1972. *Proc. Nat. Acad. Sci. USA* 69:2364–68
121. Dunham, E. K., Unanue, E. R., Benacerraf, B. 1972. *J. Exp. Med.* 136: 403–8
122. Singer, S. J. 1972. *Ann. NY Acad. Sci.* 195:16–23
123. Overath, P., Hill, F. F., Lamnek-Hirsch, I. 1971. *Nature New Biol.* 234: 264–67
124. Scandella, C. J., Devaux, P., McConnell, H. M. 1972. *Proc. Nat. Acad. Sci. USA* 69:2056–60
125. Träuble, H., Sackmann, E. 1972. *J. Am. Chem. Soc.* 94:4499–4510
126. Shimshick, E. J., McConnell, H. M. 1973. *Biochem. Biophys. Res. Commun.* 53:446–51
127. Shimshick, E. J., McConnell, H. M. 1973. *Biochemistry* 12:2351–60
128. Ohnishi, S., Ito, T. 1973. *Biochem. Biophys. Res. Commun.* 51:132–38
129. Lee, A. G., Birdsall, N. J. M., Metcalfe, J. C. 1973. *Biochemistry* 12: 1650–59
130. Junge, W. 1972. *Fed. Eur. Biochem. Soc. Lett.* 25:109–12
131. Poo, M. M., Cone, R. A. 1973. *J. Supramolecular Struct.* 1:354
132. Nicolson, G. L. See Ref. 2, 53–70
133. Edidin, M., Fambrough, D. 1973. *J. Cell Biol.* 57:27–37
134. Taylor, R. B., Duffus, W. P. H., Raff, M. C., de Petris, S. 1971. *Nature New Biol.* 233:225–29
135. Unanue, E. R., Perkins, W. D., Karnovsky, M. J. 1972. *J. Exp. Med.* 136:885–906
136. Neauport-Sautes, C., Lilly, F., Silvestre, D., Kourilsky, F. M. 1973. *J. Exp. Med.* 137:511–26
137. de Petris, S., Raff, M. C. 1973. *Nature New Biol.* 241:257–59
138. Roelants, G., Forni, L., Pernis, B. 1973. *J. Exp. Med.* 137:1060–77
139. Sundqvist, K. G. 1972. *Nature New Biol.* 239:147–49
140. Edidin, M., Weiss, A., 1972. *Proc. Nat. Acad. Sci. USA* 69:2456–59
141. Becker, K. E., Ishizaka, T., Metzger, H., Ishizaka, K., Grimley, P. M. 1973. *J. Exp. Med.* 138:394–409
142. Nicolson, G. L. 1971. *Nature New Biol.* 233:244–46
143. Bretton, R., Wicker, R., Bernhard, W. 1972. *Int. J. Cancer* 10:397–410
144. Nicolson, G. L. 1973. *Nature New Biol.* 243:218–20
145. Rosenblith, J. Z., Ukena, T. E., Yin, H. H., Berlin, R. D., Karnovsky, M. J. 1973. *Proc. Nat. Acad. Sci. USA* 70: 1625–29
145a. Noonan, K. D., Burger, M. M. 1973. *J. Cell Biol.* 59:134–42
146. de Petris, S., Raff, M. C., Mallucci, L. 1973. *Nature New Biol.* 244:275–78
147. Schairer, H. U., Overath, P. 1969. *J. Mol. Biol.* 44:209–14
148. Wilson, G., Fox, C. F. 1971. *J. Mol. Biol.* 55:49–60

149. Krasne, S., Eisenman, G., Szabo, G. 1971. *Science* 174:412–15

150. Träuble, H., Overath, P. 1973. *Biochim. Biophys. Acta* 307:491–512

151. Overath, P., Träuble, H. 1973. *Biochemistry* 12:2625–34

152. Linden, C. D., Wright, K. L., McConnell, H. M., Fox, C. F. 1973. *Proc. Nat. Acad. Sci. USA* 70:2271–75

153. Esfahani, M., Limbrick, A. R., Knutton, S., Oka, T., Wakil, S. J. 1971. *Proc. Nat. Acad. Sci. USA* 68:3180–84

154. Mavis, R. D., Vagelos, P. R. 1972. *J. Biol. Chem.* 247:652–59

155. Kimelberg, H. K., Papahadjopoulos, D. 1972. *Biochim. Biophys. Acta* 282:277–92

156. Cramer, W. A., Phillips, S. K., Keenan, T. W. 1973. *Biochemistry* 12:1177–81

157. Tsukagoshi, N., Fox, C. F. 1973. *Biochemistry* 12:2816–22

158. Adams, K. H. 1973. *Biophys. J.* 13:209–17

159. Skalak, R., Tozeren, A., Zarda, R. P., Chien, S. 1973. *Biophys. J.* 13:245–64

160. Silbert, D. F., Vagelos, P. R. 1967. *Proc. Nat. Acad. Sci. USA* 58:1579–86

161. Whittaker, V. P. 1969. In *The Structure and Function of Nervous Tissue,* ed. G. H. Bourne, 3:1–24. New York: Academic

162. Revel, J. P., Karnovsky, M. J. 1967. *J. Cell Biol.* 33:C7

163. Goodenough, D. A., Revel, J. P. 1970. *J. Cell Biol.* 45:272–90

164. Blaurock, A. E., Stoeckenius, W. 1971. *Nature New Biol.* 233:152–55

165. Goodenough, D. A., Stoeckenius, W. 1972. *J. Cell Biol.* 54:646–56

166. Oesterhelt, D., Stoeckenius, W. 1971. *Nature New Biol.* 233:149–52

167. Cone, R. A., Stoeckenius, W. Personal communication

168. Esser, A. F., Lanyi, J. K. 1973. *Biochemistry* 12:1933–39

169. Yahara, I., Edelman, G. M. 1972. *Proc. Nat. Acad. Sci. USA* 69:608–12

170. Goodenough, U. W., Staehelin, L. A. 1971. *J. Cell Biol.* 48:594–619

171. Maddy, A. H. 1964. *Biochim. Biophys. Acta* 88:390–99

172. Berg, H. C. 1969. *Biochim. Biophys. Acta* 183:65–78

173. Bretscher, M. S. 1971. *J. Mol. Biol.* 58:775–81

174. Rifkin, D. B., Compans, R. W., Reich, E., 1972. *J. Biol. Chem.* 247:6432–37

175. Bender, W. W., Garan, H., Berg, H. C. 1971. *J. Mol. Biol.* 58:783–97

176. Spiro, R. G. 1962. *J. Biol. Chem.* 237:646–52

177. Gatt, S., Barenholz, Y. 1973. *Ann. Rev. Biochem.* 42:61–90

178. Phillips, D. R., Morrison, M. 1971. *Biochemistry* 10:1766–71

179. Hubbard, A. L., Cohn, Z. A. 1972. *J. Cell Biol.* 55:390–405

180. Dutton, A., Singer, S. J. In preparation

181. Marchesi, V. T., Palade, G. E. 1967. *J. Cell Biol.* 35:385–404

182. Kyte, J. 1974. *J. Biol. Chem.* In press

183. Nicolson, G. L., Singer, S. J. 1974. *J. Cell Biol.* 60:236–48

184. Guidotti, G. Personal communication

185. Nicolson, G. L., Masouredis, S. P., Singer, S. J. 1971. *Proc. Nat. Acad. Sci. USA* 68:1416–20

186. Aoki, T., Hämmerling, U., de Harven, E., Boyse, E. A., Old, L. J. 1969. *J. Exp. Med.* 130:979–1001

187. Nicolson, G. L., Hyman, R., Singer, S. J. 1971. *J. Cell Biol.* 50:905–10

188. Mitchell, P. 1966. *Biol. Rev.* 41:445–502

189. Bretscher, M. S. 1972. *J. Mol. Biol.* 71:523–28

189a. Gordesky, S. E., Marinetti, G. V. 1973. *Biochem. Biophys. Res. Commun.* 50:1027–31

190. Papahadjopoulos, D., Weiss, L. 1969. *Biochim. Biophys. Acta* 183:417–26

190a. Schmidt-Ullrich, R., Knüfermann, H., Wallach, D. F. H. 1973. *Biochim. Biophys. Acta* 307:353–65

190b. Zwaal, R. F. A., Roelofsen, B., Colley, C. M. 1973. *Biochim. Biophys. Acta* 300:159–82

191. McMurray, W. C., Magee, W. L. 1972. *Ann. Rev. Biochem.* 41:129–60

192. Sakagami, T., Minaro, O., Orü, T. 1965. *Biochim. Biophys. Acta* 98:111–16

193. Shohet, S. B., Nathan, D. G. 1970. *Biochim. Biophys. Acta* 202:202–5

194. Tarlov, A. 1968. *Fed. Proc.* 27:458

195. Zilversmit, D. B. 1971. *J. Lipid Res.* 12:36–42

196. Reed, C. R. 1968. *J. Clin. Invest.* 47:749–60

197. Kornberg, R. D., McConnell, H. M. 1971. *Biochemistry* 10:1111–20

198. McNamee, M. G., McConnell, H. M. 1973. *Biochemistry* 12:2951–58

198a. Bennett, F., Cuatrecasas, P. 1973. *Biochim. Biophys. Acta* 311:362–80

199. Frye, C. D., Edidin, M. 1970. *J. Cell Sci.* 7:319–33

200. Brown, P. K. 1972. *Nature New Biol.* 236:35–38

201. Cone, R. A. 1972. *Nature New Biol.* 236:39–43

202. Kornberg, R. D., McConnell, H. M. 1971. *Proc. Nat. Acad. Sci. USA* 68: 2564–68
203. Askari, A., Rao, S. N. 1972. *Biochem. Biophys. Res. Commun.* 49: 1323–28
204. Jørgensen, P. L., Hansen, O., Glynn, I. M., Cavieres, J. O. 1973. *Biochim. Biophys. Acta* 291: 795–800
205. Smith, T. S., Wagner, H., Young, M., Kyte, J. 1973. *J. Clin. Invest.* 52: 78a
206. Jardetzky, O. 1966. *Nature* 211: 969–70
207. Roseman, S. 1969. *J. Gen. Physiol.* 54: 138s–84s
208. Tanner, J. J. A., Gray, W. R. 1971. *Biochem. J.* 125: 1109–17
209. Shin, B. C., Carraway, K. L. 1973. *J. Biol. Chem.* 248: 1436–44
210. Duchon, G., Collier, H. B. 1971. *J. Membrane Biol.* 6: 138–57
211. Borgese, N., Kreibich, G., Sabatini, D. 1972. *J. Cell Biol.* 55: 24a
212. Shires, T. K., Pitot, H. C. 1972. *J. Cell Biol.* 55: 238a
212a. Shires, T. K., Ekren, T., Narurkar, L. M., Pitot, H. C. 1973. *Nature New Biol.* 242: 198–201
213. Kreibich, G., Sabatini, D. 1973. *Fed. Proc.* 32: 2133–38
214. Robash, M., Penman, S. 1971. *J. Mol. Biol.* 59: 227
215. Baron, C., Abrams, A. 1971. *J. Biol. Chem.* 246: 1542–44
216. MacLennan, D. H., Tzagaloff, A. 1968. *Biochemistry* 7: 1603–10
217. Tzagoloff, A., Meagher, P. 1972. *J. Biol. Chem.* 247: 594–603
218. Telford, J. N., Racker, E. 1973. *J. Cell Biol.* 57: 580–86
219. Erwin, V. G., Hellerman, L. 1967. *J. Biol. Chem.* 242: 4230–38
220. Smith, H. 1970. *Nature* 227: 665–68
221. Haupt, W., Mörtel, G., Winkelnkemper, I. 1969. *Planta* 88: 183–86
222. Quail, P. H., Marmé, D., Schäfer, E. 1973. *Nature New Biol.* 245: 189–91
223. Pardee, A. B. 1968. *Science* 162: 632–37
224. Furlong, C. E., Weiner, J. H. 1970. *Biochem. Biophys. Res. Commun.* 38: 1076–83
225. Wilson, O. H., Holden, J. T. 1969. *J. Biol. Chem.* 244: 2737–42
226. Wilson, O. H., Holden, J. T. 1969. *J. Biol. Chem.* 244: 2743–49
227. Rosen, B. P. 1971. *J. Biol. Chem.* 246: 3653–62
228. Weiner, J. H., Furlong, C. E., Heppel, L. A. 1971. *Arch. Biochem. Biophys.* 124: 715–17
229. Weiner, J. H., Heppel, L. A. 1971. *J. Biol. Chem.* 246: 6933–41
230. Berger, E. A., Heppel, L. A. 1972. *J. Biol. Chem.* 247: 7684–94
231. Kalckar, H. M. 1971. *Science* 174: 557–65
232. Anraku, Y. 1968. *J. Biol. Chem.* 243: 3123–27
233. Boos, W., Gordon, A. S., Hall, R. E., Price, H. D. 1972. *J. Biol. Chem.* 247: 917–24
234. Hogg, R. W., Englesberg, E. 1969. *J. Bacteriol.* 100: 423–32
235. Schleif, R. 1969. *J. Mol. Biol.* 46: 185–96

INTRACELLULAR PROTEIN DEGRADATION IN MAMMALIAN AND BACTERIAL CELLS [1]

A. L. Goldberg and J. F. Dice

Department of Physiology, Harvard Medical School, Boston, Massachusetts

CONTENTS

INTRODUCTION

Living systems are in a continual state of turnover at all levels of organization from populations of whole organisms to populations of molecules within cells. More than thirty years ago Schoenheimer (1) coined the phrase "dynamic state of body constituents" to indicate that cellular constituents are continually degraded and replaced by new synthesis. Prior to that time, the cell was generally considered to be analogous to a chemical engine which burned exogenous nutrients to provide

[1] This article is the first half of a two-part review of intracellular protein breakdown. In a subsequent publication, we will discuss topics not covered here, including recent findings on the regulation of average rates of protein breakdown and the biochemical mechanisms of degradation in animal and bacterial cells.

energy but was itself not subject to metabolic transformations. The pioneering work of Schoenheimer (1) and Borsook (2) indicated instead that the cell's enzymatic machinery was also continually turning over. There is now convincing evidence that all proteins within mammalian cells are degraded to their constituent amino acids at characteristic rates (3–10). Although this process has been less thoroughly studied in bacteria (11–18) and even less in fungi (19, 20) or plants (21–24a), catabolism in such cells also appears to be extensive and to share many common features with the degradative process in mammalian tissues.

In this review "protein degradation" and "protein turnover" refer strictly to hydrolysis of intracellular proteins to their component amino acids. We do not discuss several other physiological processes which have frequently been called protein turnover, including turnover of whole cells, loss of proteins by secretion, protein flux from one cell compartment to another, degradation of extracellular proteins, or inactivation of enzymes by covalent modifications. In addition, we do not review the interesting literature on proteolytic cleavage of large precursors to form biologically active polypeptides [e.g. in phage morphogenesis (25) or synthesis of viral proteins (26–28), certain extracellular proteins (29–32), polypeptide hormones (28, 33), and digestive enzymes (28)]. Such processes serve a biosynthetic function that appears distinct from the degradation of proteins to individual amino acids. Recent studies of intracellular protein degradation have served to emphasize three important features of this process which are discussed in the following paragraphs.

Where it has been studied carefully, the breakdown of enzymes obeys first order kinetics, which means that newly synthesized proteins are as likely to be degraded as old ones (3–5). In other words, degradation of polypeptides is a random event unlike aging of organisms or red cells. The suggestion that myofibrillar proteins have a finite lifespan (34), rather than turning over randomly, has not been substantiated in more recent studies (35, 52, 58), but the possibility that certain proteins may turn over in other fashions cannot yet be eliminated (35a).

The degradation rates of different enzymes within animal and bacterial cells can vary over a wide range. For instance, proteins in rat liver have an average half-life of 3.5 days (3), but the half-lives of specific enzymes may vary from 11 min for ornithine decarboxylase (36) to 19 days for isozyme 5 of lactate dehydrogenase (37). Different subunits within a multimeric protein may also have distinct half-lives as in the case of hepatic isozymes of lactate dehydrogenase (37). Furthermore, the different proteins composing a single organelle are degraded heterogeneously. There is strong evidence in a variety of tissues that proteins of the plasma membranes (38–40), endoplasmic reticuli (38, 41–44), ribosomes (45), mitochondria (46, 47), and chromosomes (48) turn over at distinct rates. For example, in mitochondria of cardiac muscle, δ-aminolevulinate synthetase and ornithine transcarbamylase turn over much more rapidly than cytochromes and other proteins of the inner mitochondrial membrane (47). Several reports have indicated that myofibrillar proteins also turn over at heterogeneous rates (49–51), but contradictory evidence on this point also exists (52). Exceptions to this general conclusion may be the peroxisome (53), which in liver has been suggested to turn over as a unit, and microtubules, whose two main components are degraded with similar half-lives (54).

In addition, these rates of degradation vary under different physiological conditions. For example, in both bacterial and many mammalian tissues, the average rates of protein catabolism increase reversibly during starvation (6–8, 55–58). Similarly, the average rates of protein degradation in muscle can be influenced by changes in nutrient supply (56), hormones (e.g. insulin) (57, 58), denervation (58–60), and contractile activity (61). The rates of catabolism of individual proteins also vary in characteristic fashions. For example, starvation retards degradation of rat liver arginase (62), but accelerates the catabolism of acetyl coenzyme A carboxylase in the same tissue (63). Similarly, the degradative rate of ferritin changes with the age of the animal (64) and the supply of dietary iron (65).

The last six years have witnessed a marked growth of interest in protein catabolism and a general recognition that it represents a fundamentally important cellular process. A number of excellent reviews of protein catabolism with different orientations than the present one have appeared in the past few years (3–11, 13, 66, 67); those of Schimke (3, 4, 66, 67), Rechcigl (5), and Pine (13) are highly recommended. Despite our growing knowledge about intracellular protein catabolism, we still have only vague ideas about (a) the physiological significance of this process, (b) the reasons why different proteins have different half-lives, (c) factors regulating protein breakdown, and (d) biochemical mechanisms of the degradative process. In this review,[1] we focus on several topics: the possible physiological functions served by protein breakdown, problems in the measurement of degradative rates, and the structural factors which may determine a protein's catabolic rate.

PHYSIOLOGICAL SIGNIFICANCE OF PROTEIN DEGRADATION

Although the continuous degradation of cell proteins at first glance may seem highly wasteful, this process must be of some selective advantage to the organism. Recent studies, in fact, have indicated several important functions which may be served by protein turnover.

1. One important function of protein catabolism is probably the removal of abnormal proteins which might arise by mutations, errors in gene expression, denaturation, or chemical modification. As discussed in a subsequent section, there is now strong evidence that bacterial and animal cells can selectively hydrolyze abnormal proteins. Such a mechanism for eliminating proteinaceous waste or the potentially harmful consequences of imprecise synthesis would be favored by the same selective pressures that caused the evolution of highly accurate systems for gene expression (68, 68a). Degradation of abnormal cell constituents would appear especially important in slowly growing cells (e.g. most mammalian tissues) which, unlike bacteria, cannot dilute out such proteins by rapid growth.

2. The continuous turnover of proteins must significantly increase the organism's ability to adapt readily to changes in its environment, just as the turnover of individual organisms within a population allows the characteristics of the population to evolve. Schimke and co-workers (3, 69) have pointed out that the more rapidly

an enzyme is degraded, the faster its intracellular concentration can change in response to hormonal stimuli, nutritional influences, etc. Thus, short half-lives may have evolved in order that certain crucial enzymes can fluctuate rapidly with changing physiological situations. In rat liver, where half-lives for nearly 40 different enzymes have been determined, degradative rates appear to correlate in an interesting fashion with the enzyme's role in intermediary metabolism. The 12 liver proteins with the shortest reported half-lives catalyze either the first or the rate-limiting step in metabolic pathways (70), and thus their concentrations would be expected to be precisely regulated. In contrast, the dozen proteins most slowly degraded in liver do not include any enzymes known to be carefully controlled. Similar principles may apply to the selection of overall rates of protein turnover in different tissues. Thus rapid degradation and synthesis of proteins in liver compared to other organs may be crucial in this organ's ability to adapt rapidly in composition and size to changing dietary intake.

Protein catabolism must be especially important in an organism's adaptation to poor environments. Rates of degradation increase in bacteria (11–18), yeast (19, 20), and certain mammalian tissues (55, 56) during starvation in part to provide essential amino acids for the synthesis of enzymes appropriate to such conditions. Thus *Escherichia coli* deprived of nitrogen can still synthesize β-galactosidase in response to an inducer because of the increased catabolism (11–15), but when degradation is blocked, synthesis of this enzyme is prevented (71, 179). Protein catabolism probably plays similar adaptive roles in other organisms. For example, mass degradation of normal cell protein occurs in many bacterial species during sporulation to provide amino acids for spore formation (72). Other cases are known where more complex cells undergo cytodifferentiation or synthesize new structure proteins although starved. For example, *Euglena,* if exposed to light, form elaborate photosynthetic organelles even when completely starved, presumably at the expense of pre-existent cell proteins (73). Finally, in mammalian liver, enzymes appropriate for catabolic reactions are synthesized during fasting (e.g. gluconeogenic enzymes) (74) while many proteins involved in anabolic processes are degraded more rapidly (e.g. ribosomal proteins) (75).

3. Protein degradation may in another important way help the organism to withstand hard times. In all organisms cell protein constitutes a major energy reservoir that can be mobilized in times of decreased caloric intake. Thus in mammals, protein reservoirs in muscle and liver provide amino acids that are either oxidized directly or can be converted into glucose (76). There is now convincing evidence that the supply of such amino acids is hormonally regulated through effects on rates of protein degradation as well as synthesis. For example, insulin, which is probably the most important hormone influencing protein balance in most tissues (77), has been found to retard protein catabolism in liver (78), cardiac muscle (79, 80), skeletal muscle (57), adipose tissue (81), and fibroblasts (82). Regulation of both synthesis and degradation allows the organism to accumulate or mobilize protein rapidly.

Further progress in understanding protein degradation may clarify the relative importance of the physiological functions suggested above and may suggest other important reasons why continuous protein catabolism is advantageous to the organism.

TECHNIQUES FOR MEASURING PROTEIN DEGRADATION

One factor responsible in part for our slow progress in understanding protein breakdown has been various methodological difficulties involved in measuring this process. Many of these difficulties are not simply problems in biochemical technique, but actually result from our limited knowledge concerning the degradative process and the function of intracellular amino acid pools. It should be pointed out that the phrase "rate of protein degradation" in this review and in the entire literature of protein catabolism has distinct connotations in different contexts. In discussing turnover of specific enzymes, degradative rate generally refers to the rate-limiting cleavage of a peptide bond that leads to loss of a recognizable protein (see below). In determining the degradation of average cell proteins or of the activity of proteolytic enzymes, rate of catabolism is used to indicate the release rate of free amino acids. These definitions are used interchangeably even though they refer to biochemical events that may or may not be closely coupled within the cell. Further progress in understanding intracellular proteolysis hopefully will clarify this important issue and may require use of more precise terminology to distinguish these processes.

Schimke (3, 67) has elegantly reviewed in detail the various techniques for determining the degradative rates of specific proteins. The present discussion will therefore emphasize new developments and certain difficulties in using these approaches.

Nonisotopic Techniques

One commonly used method for estimating degradative rates involves analysis of the rate at which the concentration of an enzyme changes from one steady state to another. Such changes in enzyme content can be described by the following kinetics (3, 4, 69): $dE/dt = k_s - k_d E$, where E is the number of enzyme molecules, k_s is the rate of synthesis of the protein (a zero order reaction), and k_d is the constant for the degradative process (a first order reaction). By measuring the enzyme concentration at different times (e.g. during induction by a hormone), the rate of protein degradation can be calculated. One great advantage of this approach is that it can be applied to enzymes that have not been purified.

Unfortunately, in most studies the crucial assumptions for this approach have not been critically evaluated, and it may be invalid for many experimental situations. For example, this approach assumes that synthetic rates and degradative rates are constant during enzyme adaptation. Yet in growing, hormone-treated, or atrophying tissues, synthetic rates may vary continually with the changes in tissue mass. Even though it may be possible to calculate such degradative constants, they might not accurately reflect the physiological process. Another frequent misuse of this approach has been the unjustified assumption by many workers that the rate of enzyme loss after removing the inducing agent is equivalent to the rate of protein degradation. As Haining has clearly emphasized (83), this simple interpretation is invalid since the decrease in activity reflects both protein catabolism and synthesis of some new enzyme molecules.

Perhaps the simplest method for analyzing enzyme loss is by assaying protein concentration as a function of time after protein synthesis has been inhibited. This technique, while not requiring purification of proteins, does require appreciable caution in interpretation. One serious complication is that inhibition of synthesis with puromycin, cycloheximide, or chloramphenicol can also reduce protein degradation in both mammalian and bacterial cells (84–88). Thus the values obtained for degradative rate may be somewhat prolonged. The magnitude of this effect varies with different enzymes for reasons that are unclear, and the effect may not be general to all cellular proteins (88, 89). Despite these problems, many agents or physiological conditions that alter degradative rates can do so even when synthesis is prevented by cycloheximide (56, 57, 61, 89, 90). The effects of inhibitors of protein and RNA synthesis on degradation depend on the nutritional status of the cell (84, 85, 91). It appears likely that such inhibitors affect degradative rate through some indirect mechanism (84, 85).

In addition, it is important to note that the rates calculated by this approach are for enzyme inactivation, which are not necessarily the same as the rates of polypeptide degradation. In a variety of cell types, there is now strong evidence for specific mechanisms for inactivating enzymes (e.g. phosphorylation) (92, 92a), and as discussed above, an important unsolved problem is whether enzyme inactivation and protein hydrolysis occur at identical or even similar rates.

As an alternative to measuring enzymatic activity, changes in protein levels frequently have been estimated with immunological techniques. Certain immunological methods [e.g. use of antibodies as inhibitors of catalytic activity (93) or complement fixation] can be applied even when the antigen has not been purified completely. Like enzyme activity, antigenic activity can be lost through processes other than degradation, including conformational changes in the protein such as denaturation (94). Obviously the simultaneous use of several techniques for determination of catabolic rate will give more reliable values than any single approach.

For certain proteins, it has been possible to follow specifically the loss of a protein by acrylamide gel electrophoresis without measuring either antigenic or enzymatic activity. These electrophoretic methods have frequently been used in combination with isotopic techniques to identify the protein under study. They can also be used when the protein is present in the cell in large amounts (95, 96) or when antibody precipitation can be used to purify the protein prior to electrophoresis (97). Such methods have been used successfully to follow the degradation of altered *lac* repressor (97) or nonsense fragments of β-galactosidase (95, 96). In the latter experimental system, it has also been possible to measure degradation of different peptide fragments of β-galactosidase, since the amino terminal fragments of the molecule can be selectively assayed by complementation in vitro with fragments containing the carboxyl-terminal region (98–102). This method thus permits comparison of the disappearance rates of different portions in the same molecule.

One potential approach for measuring degradation of certain proteins (e.g. collagen, histones, myosin) takes advantage of the fact that they contain unique amino acids formed by post-translational modifications such as methylated arginine, methylated histidine, and hydroxylysine. Measuring the production of such amino acids permits an estimate of the extent of protein degradation since these residues cannot be reincorporated. For example, hydroxyproline in the urine has long been useful clinically as an assay of collagen degradation. Analogous methods for measuring methylhistidine in urine (103) are presently under study in several laboratories for estimating the degradation of the contractile proteins. Unfortunately, the relative concentration of this amino acid in muscle proteins is very low (104–106), and the feasibility of this approach is still unclear.

Isotopic Techniques

The first definitive demonstration of protein turnover in animals was made by Schoenheimer & Rittenberg (1), who measured the incorporation and subsequent loss from tissue proteins of precursors labeled with heavy isotopes. Isotopic techniques remain the most popular methods for measuring protein catabolism. Three variations on this approach have been used and are discussed separately.

CONTINUOUS LABELING   One approach for estimating degradative rates involves continuous exposure of a cell or organism to labeled precursors. With intact animals, maintaining a constant intracellular specific activity of an amino acid (3) is difficult to achieve. Rats or mice have been maintained on diets containing radioactive precursors (62, 107, 108) or alternatively the animals received a continuous infusion of radioactive amino acids usually by the jugular or tail veins (109–112). From measurements of the rate at which the specific activity of cell protein approaches its maximal value, the degradative rate can be calculated. A clear exposition of this method can be found in the review by Schimke (3). One problem in using this method has been that an appreciable time lag (several hours) is often required for some labeled precursors to reach their maximal intracellular specific activity. Thus the method's usefulness has been limited for investigating components that turn over rapidly or in tissues where amino acid transport is especially slow. Proper choice of precursors may avoid many such difficulties. For example, even though many amino acids equilibrate only slowly between blood and brain, glucose rapidly enters this tissue and has been used to label intracellular pools of glutamic acid in studies of protein degradation (113).

PULSE-LABELING AND AMINO ACID REUTILIZATION   The most common isotopic technique used is the pulse-labeling approach in which a cell or organism is administered a radioactive precursor for a short time to label cell protein. Then the radioactivity within a specific protein is assayed as a function of time, usually after purification with a specific antibody, by electrophoretic separation, or by other methods. Decrease in the specific activity of the protein or total amount of radioactive protein is taken as a measure of the protein degradation rate, on the assumption that the new proteins being synthesized are nonradioactive. At steady state where the total level of protein is constant, the specific activity and total radioactivity in proteins decrease at equivalent rates. However, when the protein studied is either increasing or decreasing, then the fall in specific activity should reflect both rates of synthesis and degradation, and it is essential to consider the fall in specific activity resulting from dilution of the radioactive proteins with new unlabeled ones.

The fundamental assumption of these approaches is that after the initial incorporation of labeled amino acids, protein synthesis utilizes only nonradioactive precursors. This ideal situation has not been easy to achieve experimentally, since labeled amino acids released by the degradation of protein in the organism can be reutilized in another round of protein synthesis. For the biochemist, such reincorporation of radioactive amino acids represents a major nuisance, but for the organism this recycling process must be highly advantageous. In the face of continuous protein turnover, failure of the cell to reutilize the liberated amino acids would be highly wasteful; it would necessitate much larger dietary intake of protein and much larger energy utilization in the uphill transport of amino acids. It is not surprising therefore that decreased protein intake in mammals leads to enhanced reutilization of several amino acids (114–116).

As a consequence of extensive reutilization of radioactive precursors, many studies probably have significantly overestimated protein half-lives. As Koch has pointed out, this artifact is most serious in measuring half-lives of the more stable cell proteins (117), and this point has been re-emphasized recently by Poole (118). One reason this problem is particularly serious is that physiological conditions that influence degradative rates may also influence rates of protein synthesis and amino acid transport as well as sizes of precursor pools and the pattern of amino acid metabolism. Thus, the extent of reutilization may change under conditions where degradative rates are being measured.

A variety of experimental approaches have been used to minimize the reutilization of radioactive residues. In bacterial or mammalian cells in culture, it appears feasible to add large excesses of nonradioactive amino acid to compete with the labeled residues for incorporation and to chase them out of the cell. Large amounts of nonradioactive amino acids in the medium are essential for measuring protein breakdown in E. coli (119, 120), which in the absence of such a chase can completely reutilize the labeled residues for further protein synthesis (11, 120). Presumably the concentration of nonradioactive amino acid which gives the greatest apparent rate of protein breakdown successfully prevents the reutilization of labeled amino acids. Evidence for the efficiency of this approach has been published for leucine in E. coli (119), although it should be noted that the required concentration of the non-radioactive chase probably varies with cell concentration, growth rate, and specific precursor used. For example, the concentration of extracellular leucine effective in chasing radioactive residues from nongrowing cultures of E. coli is insufficient for cultures growing at high density (149).

For cultures of mammalian cells, the extent to which exogenous nonradioactive amino acid can prevent reutilization has not been carefully studied. In cultured cells (121–123) as in mammalian tissues in vivo (124, 125), at least 50% of the amino acids in the intracellular pools originate from protein breakdown. Consequently after a short exposure to a labeled precursor, the specific activity of the total intra-cellular pool remains significantly higher than that in the medium, despite the presence of unlabeled amino acids in most culture media (126). Even greater con-centrations of nonradioactive amino acids in the medium should promote the uptake of these compounds and increase competition for reutilization (127). Nevertheless, pulse-chase measurements of protein breakdown with cultured human fibroblasts have given slower values for the observed rate of degradation than measurements based on continuous infusion (128). This discrepancy decreased when the medium containing nonradioactive amino acids was changed frequently, and such pro-cedures may be necessary to prevent significant reutilization of label by the cultured cells. In this context, the choice of amino acids appears important since transport rates of different amino acids across the cell membranes vary over a wide range and appear fastest for neutral, hydrophobic residues (129, 130).

One successful strategem used to reduce reincorporation in mammalian tissues has been to prelabel the tissues in vivo (e.g. the liver with ^{14}C-valine) and then to perfuse the organ in vitro in a medium containing much larger concentrations of unlabeled amino acid than are found in plasma (78, 90a, 126).

In addition, several groups have had some success in chasing radioactive amino acids out of the intact animals. Thus the apparent half-life of proteins in muscle is decreased if the rat is maintained on a high protein diet or when large amounts of the corresponding unlabeled amino acid are given daily following the initial radioactive injection (114, 115, 131). In such regimens, the large amounts of cold amino acids may also be reducing reutilization by inducing enzymes that degrade the labeled residues. Unfortunately, such dietary regimens are not applicable for many experimental studies (e.g. those concerned with the effects of diet). One additional problem with these approaches is that large amounts of one amino acid may indirectly affect intracellular protein turnover. For example, high concentrations of leucine (132) or tryptophan (132a) can reduce degradation of tyrosine amino trans-ferase, and in intact organisms amino acids can cause release of insulin (77), an inhibitor of protein degradation in many tissues (57, 78–82).

The extent of reutilization of amino acids after incorporation into protein and release into the intracellular pools is difficult to estimate. About 50–80% of the amino acids within the intracellular pools of rat liver are derived from protein degradation (109, 124). This fraction varies in different tissues and with changes in endocrine and nutritional status (124). Righetti, Little & Wolf (122) have suggested that in HeLa cells amino acids from pre-existent proteins might be reincorporated into new proteins without first mixing with the intracellular pool. Selective re-incorporation may be explained by the suggestion of Walter (133) that intracellular protein hydrolysis might involve transfer of amino acids from protein directly to tRNA. Although such a mechanism would allow the cells to conserve the ATP necessary for amino acid activation, it fails to explain how labeled amino acids in proteins can be recovered as free residues. Furthermore, recent measurements of the specific activity of the amino acids bound to tRNA do not support this suggestion (134, 135).

These observations nevertheless raise the fundamental issue of whether the free pool of amino acids within the cell is in fact the precursor for proteins. This question has been investigated by a number of groups who have used a variety of approaches and have come to very different conclusions. Certain studies on incubated skeletal muscle (136, 137), perfused liver (138), perfused heart (134, 135), and intestine (139) have suggested that amino acids may enter protein from a compartment with a higher specific activity than the average intracellular pool. On the other hand, recent studies on incubated skeletal muscles (140, 141), intestine (142), and the perfused heart (79) have concluded that the free intracellular amino acids resemble the precursor pool for synthesis. This fundamental question clearly has not been resolved. Greater information about the precursor pool for protein synthesis, the physiological function of the intracellular amino acids, the factors that influence the exchange of intracellular and extracellular amino acids, and the fate of amino acids released by the protein degradation should help improve methods for measuring degradative rates.

One additional form of amino acid reutilization occurs in intact multicellular organisms and has not previously received wide attention. Labeled amino acids released from one tissue might subsequently be reincorporated into protein of another tissue. For example, upon intraperitoneal injection many labeled amino acids (e.g. lysine) are accumulated and incorporated into protein by the liver. Subsequently, the liver slowly releases the labeled residues into the blood for use by peripheral tissues. Thus, while specific activity of liver proteins falls, that in muscle may continue to rise (143). During fasting, on the other hand, peripheral tissues may release the labeled amino acids back into the blood. The magnitude of such fluxes varies with the specific amino acid and the tissue.

Proper choice of amino acid precursor may also significantly reduce the extent of reutilization and thus give a shorter apparent half-life. For example, Penn et al (144) studied the turnover of plasma albumin by measuring the amounts of seven different labeled amino acids. These values varied over a threefold range; glycine gave the shortest value while lysine gave the longest. For extracellular proteins, it is possible to measure degradation rates definitively by labeling tyrosine moieties with radioactive iodine, since iodinated tyrosine cannot be used in protein synthesis. For intracellular proteins, this approach is obviously impossible, and the ideal radioactive precursor for pulse-labeling studies would be one that is immediately destroyed by the cell after its release from proteins. In general, amino acids that are rapidly metabolized give shorter values for protein half-lives than more stable precursors, presumably because of different rates of reincorporation. Unfortunately, the same properties that minimize reutilization also decrease the initial incorporation of label into protein and thus make such compounds expensive to use. Few systematic investigations comparing different labeled precursors have been carried out, and insufficient information is presently available to predict the amino acid most appropriate for studies in specific tissues.

Swick (145) first introduced the use of ^{14}C-guanido arginine for the study of degradation in the liver because this organ contains large amounts of arginase, which cleaves the guanido group to produce urea. Several groups have reported shorter mean half-lives for liver proteins with this compound than with uniformly labeled arginine or ^{3}H-leucine (41, 146). Guanido-labeled arginine is, however, subject to some degree of reutilization that varies with the nutritional state of the animal (114, 116, 147, 148) and dietary intake of arginine (114). The usefulness of guanido-labeled arginine is limited to tissues rich in arginase. Thus in skeletal muscle, which has little of this enzyme but oxidizes leucine and glutamate at a rapid rate, leucine-1-^{14}C (149) or ^{14}C-glutamate (35, 114) gives shorter estimates of half-lives than ^{14}C-guanido arginine.

Virtually all body tissues rapidly catabolize glutamate and aspartate by transamination and entrance into the Krebs cycle. Several groups have used these amino acids labeled in the C-1 position for measuring degradative rates in various tissues. Since the total intracellular concentration of bicarbonate (and carbonate) is 30- to 100-fold greater than those of the amino acids, the chances of reutilization of ^{14}CO$_2$ released by amino acid degradation should be greatly reduced. ^{14}CO$_2$ appears to be a highly useful precursor for measurements of turnover in liver (147) as well as in peripheral tissues such as muscle (35) or adrenal gland (148). The liver can incorporate ^{14}C-carbonate into a large number of amino acids, while muscle, probably like most tissues, incorporates the ^{14}C into only glutamate or aspartate (150). By measuring the specific activity of glutamate in muscle proteins, Millward (35) obtained an average half-life of six days, which is much shorter than the value obtained with guanido-labeled arginine. Alternatively, ^{14}C-l-aspartate or ^{14}C-l-glutamate (114, 131) can be administered directly to the animal; this approach probably eliminates the necessity of isolating the specific labeled residues from the proteins, but the carboxyl-labeled precursors tend to be much more expensive than ^{14}CO$_2$. Of related interest is the observation of Nettleton (114) that ^{3}H-3-glutamate gave good estimates of protein half-lives in muscle although ^{14}C-U-glutamate was extensively reutilized.

It is also possible to use labeled precursors other than amino acids for following degradation of certain proteins. ^{14}C-δ-aminolevulinate, a precursor of the heme moiety, has been widely used to study the turnover of heme-containing proteins such as the cytochromes (47, 52) or catalase (146) on account of the rapid degradation of heme to bilirubin in liver. However, the heme group and polypeptide may not turn over at the same rate. For example, the finding that the heme moiety can give a shorter half-life than a protein labeled with an amino acid may indicate that the amino acids are reutilized more than the heme moiety, or alternatively, that the heme group turns over more rapidly than the polypeptide.

Another new method for minimizing reutilization problems involves labeling with D$_2$O. Williams & Neidhardt (151) have studied the degradation of aminoacyl tRNA synthetases in E. coli after growth on heavy water and transfer to normal medium (or vice versa). The proportion of heavy enzyme was estimated by density gradient centrifugation as a function of time. Because of the enormously high concentration of water, reutilization of deuterium can be discounted. This elegant approach is probably limited to studies with cells in culture. In addition, it is presently unclear whether deuterium by itself may affect the degradative process.

One other promising new approach utilizes H$_2$^{18}O to estimate amino acid reincorporation (152). This approach was introduced for studies in sporulating

Bacillus lichenformis where conventional methods cannot be used because amino acids do not exchange rapidly between the cells and the medium. The great advantage of this method is that water exchanges rapidly across all membranes. In cells growing in $H_2{}^{18}O$, the transfer of ^{18}O from water onto the carboxyl group of amino acids occurs during protein hydrolysis. Thus the fraction of ^{18}O amino acids in protein at different times can be used to calculate the rates of protein breakdown and amino acid reutilization. The main disadvantage of this approach is the expense and difficulty in assay of ^{18}O and its probable restriction to studies of cultured cells.

Pine (152a) has recently introduced an analogous approach in which cells are suspended in medium containing ^3H_2O of high specific activity. Because of the continuous transamination of amino acids within cells, tritium rapidly becomes bound to the α-carbons of free amino acids and is then incorporated into protein. However, once peptide linkages are formed, the tritium on the α-carbons can no longer exchange freely with the surrounding medium. After the cells are transferred to normal medium containing H_2O, protein breakdown can be measured from the loss of radioactivity from protein. Although this clever approach appears to offer many advantages, it is presently unclear whether it will prove free from practical problems and applicable to all cell types.

DOUBLE-LABEL TECHNIQUES   An elegant application of the pulse-chase method which has been used to compare relative degradation rates of various proteins within the same cell is the double-isotope technique described by Arias, Doyle, & Schimke (41). In this procedure, two isotopic forms of an amino acid, usually ^3H- and ^{14}C-leucine, are used to establish two time points on the curve describing degradation of the protein. ^{14}C-Leucine is first administered to an animal and then some days later ^3H-leucine is given to the same animal just before it is killed. Proteins with rapid turnover rates [i.e. those which are synthesized rapidly (high ^3H) and degraded rapidly (low ^{14}C)] will have high $^3H/^{14}C$ ratios.

The assumptions inherent in the use of the double-isotope technique have been described in detail (41, 118, 153). Briefly, they are: 1. The labeled precursors are not metabolized to other compounds which are also incorporated into the proteins; 2. The rates of protein synthesis and breakdown are the same when the first and second injections of isotope are given; 3. The proteins under study follow exponential decay kinetics; and 4. At the time the animal is killed, the amounts of ^3H and ^{14}C in tissue proteins are decreasing at identical rates. The first three assumptions are valid for rat liver proteins using leucine as the precursor. The fourth assumption is not strictly valid for reasons discussed by Poole (118). Since leucine is a reutilized amino acid in liver (see above), the time at which a protein reaches its maximum radioactivity is a function of the degradative rate. Proteins with slow turnovers will take longer to reach maximal labeling than will proteins that turn over rapidly. Therefore, as shown experimentally by Glass & Doyle (153), the interval separating the two isotope injections must be varied depending upon the range of half-lives being examined.

Compared to the pulse-decay approach, the double-isotope technique may yield more reproducible results since animal to animal variations are minimized. Also, the technique has provided information about general features of the degradative process, since degradative rates of many proteins can be compared in a single experiment. For instance, the approach has been used to determine whether the components of an organelle are degraded at uniform rates and to relate physical properties of the proteins with their catabolic rates (see below).

Recently, Glass & Doyle (153) have described procedures which make it possible to convert $^{3}H/^{14}C$ ratios into actual protein half-lives in rat liver. They correct for reutilization of leucine by showing that a linear relationship exists between $^{3}H/^{14}C$ ratios of several proteins and the proteins' half-lives determined with ^{14}C-guanido arginine, which is only slightly reutilized. By constructing such standard curves, it is possible to use the double-isotope method to determine both relative and absolute rates of degradation.

Techniques for Measuring Average Rates of Protein Degradation

In a subsequent article,[1] we will review recent findings on the regulation of average rates of protein catabolism in mammalian tissues and bacteria. Both the constant infusion technique and pulse-labeling method discussed above have been used frequently to determine such average rates of degradation. However, the definition of the mean rate of degradation presents certain methodological problems, since the individual cell proteins turn over at very different rates. Thus, in a pulse-labeling experiment, the loss of label from cell proteins need not follow exponential decay kinetics, even if each component protein does. In addition, the observed decay rate should depend upon the length of the initial exposure to the radioactive precursor, and such measurements are often biased to reflect the loss of more labile cell components. For instance, exposure of E. coli to labeled precursors for 5 min or for 2 hr gives vastly different values for rates of protein catabolism during subsequent growth or starvation (13, 149, 244, 257).

Recently other approaches have been introduced for measuring degradative rates in tissues incubated or perfused in vitro. Their great advantage is that they are inexpensive (requiring little, if any, isotope) and allow measurements of breakdown over short periods of time in response to various nutritional or hormonal factors (14, 126). The net balance between protein synthesis and breakdown in a cell can be estimated from the net uptake or release of amino acids. The validity of this approach depends upon the proper choice of amino acid. For example, measurements of total ninhydrin-reactive material can be misleading, since a variety of intracellular materials (e.g. taurine, ammonia) react with this reagent but are probably unrelated to protein turnover (154). This input-output approach requires that the amino acids 1. should exchange rapidly across the cell membrane, and 2. should be neither synthesized nor degraded in the tissue. This latter requirement is essential in order that changes in amino acid content in the cell and medium necessarily reflect protein synthesis and degradation. Valine has been shown suitable for such experiments in liver (78), phenylalanine in hearts (79), and tyrosine (57) and phenylalanine (155) in skeletal muscle.

The rate of protein breakdown can then be determined by simultaneously measuring the rate of protein synthesis with labeled amino acids. By subtracting the measured molar rate of amino acid incorporation from the net release of that residue from tissue proteins, one obtains the actual rate of protein catabolism. These methods assume that the specific activity of intracellular amino acids represents that of the precursor pool for protein synthesis under the conditions used (78, 79, 138).

A major simplification of this approach is in measuring the net production of the amino acid under conditions where protein synthesis is blocked. For example, in rat muscles incubated with cycloheximide, tyrosine is released into the medium at a linear rate corresponding to the rate of proteolysis (57). The major limitation of this simple approach is that cycloheximide, as discussed above, may reduce the rates of degradation to some extent (56, 57, 84–91). However, the rates of degradation obtained in this fashion agree well with those found in vivo by other techniques (57). Furthermore, the physiological factors that influence catabolism in the presence of cycloheximide appear to have qualitatively similar effects in its absence (56, 57, 61, 89, 90, 90a). Data obtained by this simple approach tend to be more reproducible than results dependent upon the simultaneous measurement of specific activity of pools and incorporation rates.

RELATIONSHIP BETWEEN PROTEIN STRUCTURE AND DEGRADATIVE RATE

Proteins within animal and bacterial cells vary widely in their rates of degradation. The simplest explanation for these marked differences in stability is that the half-life of a protein, like its catalytic and regulatory properties, is an inherent feature of the protein structure. In fact, appreciable evidence now exists for the conclusion that the selectivity of the degradative process resides primarily in the conformations of the individual proteins. This conclusion is supported by four types of evidence: 1. alterations in normal protein conformation can markedly affect rates of protein degradation, 2. the degradative rate of a protein correlates with its size, 3. interactions of a protein with small molecules can change the protein's conformation and its rate of degradation, and 4. the rates of protein catabolism appear to correlate with their sensitivity to well-characterized proteases in vitro. These related topics are discussed below.

Degradation of Abnormal Proteins

As mentioned earlier, there is now appreciable evidence that one useful function served by intracellular protein turnover is the elimination from the cell of abnormal and potentially deleterious proteins. The clearest evidence for such selective removal of aberrant proteins has been obtained from studies in *E. coli* of mutant proteins present in lower concentrations than normal gene products. Platt et al (97) investigated the altered *lac* repressor produced by strains carrying the deletion *L-1*. This repressor can bind the inducer but is inactive in repressing the genome (156). Unlike the wild-type protein which is stable in growing cells, the *L-1* repressor was found to be degraded with a half-life of about 20 min. The normal and mutant repressors are both synthesized at the same rate; the approximately tenfold lower concentration of the *L-1* protein results from its rapid degradation.

Similarly, Goldschmidt (95, 96) and Lin & Zabin (100) have shown that various nonsense fragments produced by either *amber* or *ochre* mutations in β-galactosidase are rapidly degraded in growing *E. coli*. Decay of the incomplete proteins was demonstrated either by the disappearance of the specific peaks on gel electrophoresis or the loss of material that could complement with polypeptides containing the carboxyl-terminal portion of the molecule (99–102). This behavior clearly contrasts with wild-type β-galactosidase, which has long been recognized to be stable under these conditions. In fact, ironically, the marked stability of this enzyme originally led early workers (157, 158) to conclude prematurely that protein degradation did not occur in bacteria.

Altogether nearly a dozen nonsense mutations in β-galactosidase were examined; the half-lives of the resulting proteins ranged from a few minutes to several hours and did not correlate with the length of the fragment. For example, the mutation *X-90* produces a protein resembling normal β-galactosidase in weight, which is nevertheless degraded with a half-life of 7 min at 37°C (95, 96). Upon growth at 30 or 24° the *X-90* fragment was much more stable (100). These dramatic effects of temperature may result from the unfolding of the incomplete polypeptides at the higher temperature, since overall protein degradation does not show such a marked temperature dependence (159). Under these conditions, the various fragments of β-galactosidase, like the *L-1* repressor, are all synthesized at normal rates, and their low intracellular levels correlated with their rapid rates of catabolism (100).

Similar effects probably occur in a variety of other mutations in *E. coli* where investigators have failed to find anticipated gene products (e.g. temperature-sensitive proteins at nonpermissive temperatures). Many examples are known where missense mutations also reduce the concentrations of the gene product. Such observations have generally been explained by hypothetical effects on protein synthetic rates (160), although they may simply reflect more rapid hydrolysis of the abnormal protein. Thus far, clear evidence for rapid catabolism has not been obtained for products of simple missense mutations in *E. coli*, perhaps because the latter have less dramatic effects on protein structure than nonsense mutations or deletions.

Bukhari & Zipser (102) have taken advantage of the low concentrations of these nonsense fragments to select strains defective in their ability to degrade abnormal proteins. This selection depended upon the ability of such incomplete proteins to form active β-galactosidase upon in vivo complementation with distal fragments of enzyme produced by episomes within the same cell (99, 101, 102). In the mutant where degradation was reduced two- to fivefold, sufficient active enzyme was formed for the strain to grow on lactose. The *deg⁻* mutants are defective in their ability to degrade abnormal proteins arising in different fashions (102, 149) (see below). Interestingly, all the *deg⁻* mutants isolated also appear to be *lon⁻*. The *lon* locus is a pleiotropic gene affecting UV sensitivity, mucoid morphology, wall glycoproteins, cell division, and plasmid incorporation (160a); furthermore, *lon* mutants show some reduction in their ability to catabolize the nonsense fragments (161). These studies may indicate an important role for protein degradation in determining the *lon⁻* phenotype; alternatively, the degradative process may share common features with these various *lon* characteristics (e.g. all may involve some aspect of membrane function). The characterization of these interesting mutants will hopefully be a powerful tool in understanding the mechanisms of degradation.

It appears noteworthy that all *deg* mutants isolated thus far can still degrade the β-galactosidase fragments albeit at a reduced rate. The failure to isolate nonleaky mutants may indicate that some protein catabolism is essential for cell viability or may indicate the existence of multiple degradative systems. It is of interest that intracellular complementation (i.e. formation of active enzyme) decreased the rate of degradation (101, 162). In such complemented proteins, the overlapping regions of the two polypeptides were selectively lost (101, 162). Such redundant regions of the complementing polypeptides also have been found to be more sensitive to proteases in vitro, and thus may represent unfolded regions of the molecule (see subsequent section) (163). An unexpected discovery of particular interest has recently been reported by Apte & Zipser (162). These workers found that *E. coli* were capable of forming intact single polypeptides resembling normal β-galactosidase out of certain complementing polypeptides. The mechanism of this polypeptide ligation is presently unclear, although it may be related to the proteolytic removal of the overlapping regions of the two polypeptides.

Pine (164) and Goldberg (165) have presented evidence that *E. coli* selectively hydrolyze abnormal proteins produced by a variety of mechanisms. To study the fate of prematurely terminated polypeptides, these workers treated growing cells with low concentrations of puromycin. This antibiotic is incorporated into the growing polypeptide and causes its premature release from the ribosome. The polypeptides made in the presence of puromycin were subsequently degraded to acid-soluble form with the release of incorporated puromycin (165). Incomplete polypeptides may also fall off the ribosome under normal conditions (166). Presumably such unfinished chains containing tRNA would be rapidly catabolized like those containing puromycin. In fact, Chapeville et al (166) have described a soluble enzyme in *E. coli* capable of cleaving the tRNA from such polypeptides, and this enzyme may play a similar role in the release of the puromycin.

Complete proteins with aberrant structures have been induced by exposing growing auxotrophs to analogs of the required amino acid (164, 165). Proteins containing the analogs were degraded two- to tenfold more rapidly than normal protein conformations. A large variety of amino acid analogs have been found to have such consequences, including canavanine (an arginine analog), 7-azatryptophan or 5- or 6-fluorotryptophan, P-fluorotyrosine, azetidine carboxylic acid (a proline analog), O-methylthreonine (an isoleucine analog), D-*threo*-α-amino-β-chloro-butyric acid (a valine analog), S-(β-aminoethyl)cysteine (a lysine analog), and fluorophenylalanine or thienylalanine (phenylalanine analogs) (164, 165, 167). These compounds appear to affect the catabolism of only the proteins that had incorporated them. Thus, polypeptides containing ^{3}H-canavanine or ^{3}H-fluorophenylalanine were catabolized far more rapidly than proteins in the same cells that had incorporated ^{14}C-arginine and ^{14}C-phenylalanine (167). Interestingly, the incorporation of certain analogs (e.g. canavanine or azetidine carboxylic acid) promoted degradation to a greater extent than incorporation of others (e.g. fluorotryptophan or fluorotyrosine). Presumably these differences depend upon the extent to which incorporation of these analogs perturbs protein conformations, although direct evidence for such a conclusion is lacking. It is interesting that those analogs promoting degradation to the greatest extent also do not permit the formation of enzymatically active β-galactosidase, while those analogs with least effect on degradation cause the synthesis of temperature-sensitive enzyme.

In the course of this work, Prouty & Goldberg (168, 169) unexpectedly found that a large portion (20–50%) of the analog-containing proteins accumulate prior to hydrolysis in rapidly sedimenting fractions of the cell. These structures, sedimenting at 10,000 g to 20,000 g, correspond to the dense, amorphous intracellular inclusions that have been previously reported (170, 171) in bacteria growing on certain amino acid analogs. The abnormal proteins initially associated with these granules can account for most, if not all, of the proteins subsequently hydrolyzed (168, 169). The isolated granules resemble dense aggregates of denatured proteins bound together by hydrophobic linkages and not proteins within a membrane-bound organelle. These structures are present only in E. coli that are forced to make large amounts of abnormal proteins. The formation of such granules does not require metabolic energy (unlike the subsequent hydrolysis of the analog-containing proteins) and probably results from spontaneous aggregation (169) of the denatured proteins. Similar inclusions have been reported in many human diseases, where abnormal proteins accumulate intracellularly [e.g. Heinz-Bodies in red cells (172)]. Possibly the proteins within such inclusions are also rapidly turning over in mammalian cells (169).

E. coli also degrades proteins resulting from mistakes in gene translation or transcription. Thus, Goldberg (165) demonstrated increased protein breakdown in strains carrying ribosomal mutations that reduce fidelity of translation (ram^-). Similar results were obtained with a missense suppressor mutation, which should cause miscoding by a specific tRNA. Pine (164) has reported increased degradation rates of proteins made in the presence of streptomycin or fluorouracil, both of which can induce errors in protein or RNA synthesis, although they also have other known effects (e.g. causing release of unfinished polypeptides). In addition, Pinkett & Brownstein (173) demonstrated rapid catabolism of proteins synthesized in the presence of streptomycin in a strain where this drug induces high rates of miscoding.

One physiological condition where miscoding may occur at increased rates is during starvation for a required amino acid. When an essential residue is lacking, an available amino acid may be incorporated in its place or the incomplete polypeptide may be released from the ribosome. In fact, Bremer et al (174) observed that the proteins synthesized by starving E. coli are on the average smaller and turn over more rapidly than those synthesized by normal cells. In addition, Hall & Gallant (175) have shown that rel^- but not rel^+ strains synthesize a temperature-sensitive β-galactosidase during amino acid starvation. The rel^- strains, unlike wild-type cells, do not increase protein degradation upon deprivation of an essential amino acid (14, 176); therefore, in rel^- the required amino acids must be present in much lower concentrations and the chances of translational errors are likely to increase.

Presumably abnormal proteins also arise by the spontaneous denaturation of cell enzymes. One physiological condition where this process may occur at accelerated rates would be during growth of thermophilic bacteria at high temperatures. Possibly the thermophiles survive at such temperatures in part because of the rapid turnover and replacement of heat-sensitive cell components. Bubela & Holdsworth (177) have reported high rates of protein breakdown under such conditions. But another group (178) failed to support this observation, and instead concluded that the survival of thermophiles depends on the temperature resistance of their proteins. Unfortunately, the two studies involved different thermophilic strains and studied different classes of cell proteins.

The systems in *E. coli* responsible for degrading these various types of aberrant proteins share several common features: 1. The *deg* mutants (102) have a decreased ability to hydrolyze analog- or puromycin-containing peptides as well as nonsense fragments (149); 2. In all cases examined, the catabolism of abnormal proteins requires metabolic energy (102, 165); and 3. Unlike normal *E. coli* proteins, whose catabolism increases severalfold on carbon or nitrogen deprivation, the abnormal components are degraded at similar or even reduced rates upon starvation (149, 165). In addition, protease inhibitors that reduce protein degradation in starving cells failed to affect the degradation of analog-containing proteins in growing cells (179). For these latter reasons, it was suggested (165, 179) that *E. coli* contain two distinct degradative systems, one activated in poor environments and one present in all cells for catabolizing abnormal protein. However, the deg^- mutants also show slightly decreased rates of protein catabolism upon starvation (149). Thus, these two proteolytic responses are probably not completely independent.

A variety of observations indicate that mammalian cells have a similar capacity for degradation of aberrant proteins. Rabinowitz & Fisher (180, 181) observed that the incorporation of a valine or lysine analog into hemoglobin by reticulocytes leads to rapid catabolism of this protein. In addition, incorporation of canavanine by cultured human fibroblasts (128) and baby hamster kidney cells (149) promotes protein degradation significantly. On the other hand, Johnson & Kenney (182) have studied the incorporation of several tryptophan analogs on the half-life of tyrosine aminotransferase in rat hepatoma cells; although the enzyme became more sensitive to high temperatures, its half-life in vivo was not shortened. The significance of this negative observation on a single enzyme is unclear. These experiments would not have detected analog-containing proteins that were so abnormal that they failed to react with antibody; such proteins may well be the ones selectively degraded by the cell. Unfortunately, the effects of the tryptophan analogs on the average degradation rate in the cells were not examined. In addition, the tryptophan analogs studied have only mild effects on the average rate of degradation in *E. coli* where they also do not promote the catabolism of all cellular proteins (167).

Mammalian cells can also selectively degrade aberrant proteins resulting from mutational events. One interesting example comes from studies of the maturation of attenuated polio virus in HeLa cells (27). Wild-type viral proteins are synthesized and undergo maturational cleavages at similar rates in cells infected at 35 and 39°. The attenuated virus, however, produces only 20% as much viral protein at the higher temperature because the mutant polypeptides, unlike those of virulent strains, are rapidly degraded by the cell. Similarly human hemoglobin (Hb) Ann Arbor is a variant resulting from a single amino acid shift. This protein is found in low concentrations primarily as a consequence of more rapid catabolism (183). In addition, a large number of human mutations are known, in which the gene product has not been demonstrated or is present in decreased amounts, possibly because of rapid degradation. For example, several forms of glucose 6-phosphate dehydrogenase deficiency appear to result from production of an unstable enzyme (184). In addition, it has long been known that in heterozygotes for sickle cell hemoglobin, HbS is present at 70% or less of the amount of the normal Hb (172). This discrepancy is much larger under conditions that promote deoxygenation (185, 186) where the conformation and solubility of HbS should be altered (172).

IMPLICATIONS   Together these observations indicate that one important role of intracellular proteolysis is the elimination of abnormal proteins. It is unclear, however, to what extent the degradation of such proteins may contribute to the observed basal rates of catabolism. Reliable estimates of the frequency of translational or transcriptional errors, premature terminations, or silent mutations are not possible at present. Previous attempts to determine such rates (187) have probably given false estimates due to the rapid elimination of the proteins containing serious errors. Denaturation, spontaneous unfolding, or chemically induced alterations of proteins [e.g. acylation, deamidation (187a, 187b), cleavage, phosphorylation (92, 92a)] possibly also occur at significant rates in vivo. More importantly, the degradation of aberrant proteins may involve the same mechanisms responsible for catabolism of normal cell proteins for which the initial degradative step may well be denaturation or chemical modification (188). Recently, evidence has been obtained for such a mechanism for the degradation of tryptophan oxygenase in cell-free extracts (94). Virtually nothing is known about the rates at which proteins may unfold irreversibly in vivo. In vivo degradative rates may correlate with a protein's susceptibility to thermal inactivations (188a) although contradictory evidence also exists (189).

The ability of cells to hydrolyze proteins containing translational and transcriptional errors is of special interest in light of Orgel's suggestion (190, 191) that aging may involve an increased frequency of protein synthetic errors, leading eventually to a lethal error catastrophe. As originally formulated (190), however, this mathematical theory failed to consider continuous protein degradation. A mechanism for removing abnormal proteins could prevent the inexorable accumulation of aberrant proteins and their catastrophic consequences. Two groups have presented evidence for inactive or temperature-sensitive enzymes in aging organisms (192, 193). Other investigators, however, find no evidence for increased frequency of protein synthetic errors in old cells (194). Theoretically, accumulation of abnormal proteins may also result from a decreased degradative ability in aged cells.

The finding that deviations from normal structures markedly shorten protein half-lives in vivo also implies that normal proteins share certain conformational features that protect them from degradation (165, 195). However, the features that differentiate the abnormal components and lead to their rapid hydrolysis remain to be identified. These features may be similar to those distinguishing normal proteins that turn over rapidly from more stable ones.

Protein Size and Degradative Rate

One unanticipated characteristic of protein degradation which has been discovered with the double-isotope technique described earlier is that proteins of larger molecular size are generally degraded more rapidly than smaller ones. This correlation was first noted by Dehlinger & Schimke in rat liver (197), and has since been reported in a number of other experimental systems (24, 38–40, 45, 48, 199, 201). In a typical experiment, a rat was first administered ^{14}C-leucine and 4 days later ^{3}H-leucine. Then the animal was killed 4 hr later and soluble liver polypeptides fractionated according to their molecular weight by polyacrylamide gel electrophoresis in the presence of sodium dodecylsulfate (SDS) (196). Gel fractions containing polypeptides of larger molecular weight had up to twofold higher ^{3}H/^{14}C ratios than those containing smaller proteins (24). This result suggests that larger proteins were degraded on the average more rapidly than smaller ones.

A correlation between molecular size and degradation rate was also observed when the soluble proteins of liver were separated by Sephadex chromatography (197). In fact, the ^{3}H/^{14}C ratios varied over a similar range when the proteins were fractionated according to the sizes of their protein subunits (in the presence of SDS) or according to the sizes of the protein multimers (in the absence of SDS). Dice et al (24) subsequently reported that multimers of a specific size had variable degradation rates depending upon the size of their subunits. Conversely, subunits of a certain size had similar rates of degradation regardless of the size of the multimer from which they were derived. Therefore, the degradation rate of a protein appears to be related to its subunit size rather than its multimer size. The earlier finding of a correlation between size and degradation rate of undissociated multimers may be explained by the observation that larger multimeric proteins tend to be composed of larger subunits (198, 210).

The relative degradation rates of soluble proteins from other tissues were studied with similar techniques and also appeared to be related to their subunit molecular weights. Soluble proteins from rat muscle, kidney, brain, and testes showed a correlation similar to that found with liver proteins (24). In addition, more rapid degradation of large proteins was observed with soluble proteins of HeLa cells (24), pea stem cells (24), and baby hamster kidney cells (149). Thus, a correlation between molecular size and degradative rate may be a general characteristic of protein turnover in eukaryotic cells. However, it is unknown whether the accelerated degradation of cell proteins seen under starvation (6–8, 55–58) or other physiological conditions (72–82) is related to molecular size. Whether or not such a correlation holds for protein catabolism in prokaryotes also remains to be shown. Prouty & Goldberg (169) have found, however, that after incorporation of the amino acid analogs, the larger analog-containing proteins are degraded more rapidly than smaller ones in E. coli.

Larger protein components of certain organelles also appear to be degraded more rapidly than the smaller proteins. Correlations between size and degradative rates have been demonstrated for the protein subunits of the endoplasmic reticulum and plasma membrane from rat (38) and mouse (39) liver, for plasma membrane proteins from cultured baby hamster kidney cells (40), and for membrane proteins in brush borders of rat intestine (199). One factor that may contribute to the correlation for membrane proteins is that glycoproteins artifactually appear to be of large size when separated by SDS gel electrophoresis (200, 200a). Such glycoproteins turn over more rapidly than other membrane components (39). A similar correlation between size and degradation rate was observed among the protein components of rat liver ribosomes (45), chromosomes (48), and the multienzyme complex, fatty acid synthetase (201). It is presently unknown whether this correlation holds for the proteins in other organelles of rat liver or in organelles of other tissues or even other species. Myofibrillar proteins from muscle may be degraded at rates unrelated to molecular weight (49, 51), although the limited data available are contradictory (50, 52).

All reports using the double-isotope technique have taken the $^3H/^{14}C$ ratios as a measure of relative rates of protein degradation. A number of alternative explanations for the correlation between molecular size and $^3H/^{14}C$ ratio have been considered (24) but do not appear to explain this relationship.

1. The larger soluble proteins are possibly secreted from the cell preferentially. In most of these experiments, however, the animal was killed at a time when most of the labeled secretory protein had already been secreted. In addition, the correlation was found among soluble proteins from tissues in which little protein secretion occurs.

2. Possibly, intracellular proteins are cleaved into smaller fragments that exist for a finite period before further degradation. Thus, newly synthesized proteins (labeled with ^3H) would be larger than older proteins (labeled with ^{14}C). If such cleaved proteins exist within a cell, they may contribute to the observed correlation between $^3H/^{14}C$ ratio and molecular size. A number of proteins are synthesized as larger precursor molecules (26–33), but it seems unlikely that such proteins and their cleaved product both exist within ribosomes, chromosomes, and membranes in sufficient amounts to entirely explain the correlation. Tabor (202) has recently proposed that many soluble proteins from HeLa cells are formed by proteolytic cleavage of larger precursors within minutes of synthesis. In the long labeling periods used in the double-isotope studies, however, such large precursors should not be detectable.

3. Alternatively, proteins may accumulate breaks along their polypeptide chain which do not alter the protein's structure or biological activity (203–205). Such proteins may remain associated with organelles, but in the presence of SDS the newly synthesized proteins (^3H) would once again be larger on the average than older, nicked ones (^{14}C). However, Glass & Doyle (153) have failed to detect nicks within three rat liver enzymes whose half-lives correlate with their subunit sizes. In addition, this general correlation holds whether or not conditions are used which would disrupt nicked proteins into smaller fragments (197).

The above double isotope experiments all compared degradative rates of a great many cell proteins. A correlation between protein size and degradative rate may (153) or may not (209) be evident when only a few enzymes are compared.

If a sufficient number of enzymes within one tissue were examined, however, such a correlation should become evident. Rates of degradation have been determined for a large number of enzymes from rat liver, and for 34 of these, data are available on the molecular weight of the enzyme and its subunits. Although many of these values may not be exact, half-lives of these enzymes do correlate with their molecular size (210). The correlation was significantly better with log subunit size of the protein ($r = 0.554$, $p = 0.01$) than with log multimer size ($r = 0.315$, $p = 0.05$), in agreement with previous findings for total soluble proteins from rat liver (24). In fact, when the dependence of multimer size on subunit size was taken into account by calculation of partial correlation coefficients, the correlation between multimer size and half-lives was abolished, while that between subunit size and degradative rate remained highly significant. In other words, the apparent correlation between multimer size and degradative rate (24, 197) may be entirely explained on the basis of large multimers being composed of large, rapidly degraded subunits.

The best correlation between subunit size and half-lives of soluble proteins (determined by SDS gel electrophoresis) in rat liver was significantly better ($r = 0.840$) (153) than that obtained by surveying the 34 specific enzymes. However, when all soluble proteins are examined by the usual labeling techniques, degradative rates for all proteins of the same size are averaged. When the half-lives of known proteins of similar size (± 1000 daltons) are also averaged, the correlation with size improves ($r = 0.629$). This survey also demonstrated several examples of small proteins having short half-lives; such examples further indicate that size is only one factor influencing catabolic rates. It is interesting that no exceptions have been found so far where proteins with large subunits have long half-lives. Although this correlation appears to be general, factors other than size clearly can also influence rates of protein degradation. For instance, the degradation of a protein may be altered by binding various ligands (discussed below) or possibly by association with other macromolecules within an organelle (206). Also, proteases specific for individual proteins (207) or groups of proteins (208) may determine degradation of certain molecules selectively.

A possible basis for the correlation between size and degradative rate is suggested by the work of Dice et al (24), who have shown that larger proteins on the average are more susceptible to proteolytic degradation in vitro than smaller ones (see below).

It is possible that the more rapid degradation of large proteins evolved because of some special function that large proteins may serve in the cell and that may require rapid turnover. For example, as discussed above, the rat liver enzymes known to have rapid rates of degradation tend to be rate-limiting or first enzymes in pathways (70). Their rapid turnover presumably evolved in order that their concentrations might rise and fall quickly in response to changing physiological conditions. Such proteins may also be of larger average size as a consequence of their more complex catalytic and regulatory responsibilities. According to such a model, rapid rates of catabolism and greater size would have evolved in parallel.

Alternatively, the more rapid degradation of large polypeptides may be a necessary consequence of their greater size. By this model, evolutionary pressures favoring rapid degradation of an enzyme may even select for proteins of larger size. There are, in fact, several possible chemical reasons why larger proteins may be degraded more rapidly than smaller ones:

1. Larger proteins may contain more protease-sensitive sites than smaller ones, thereby acting as larger targets for an initial, rate-limiting "hit" by a protease.

2. Certain conformations of proteins may render a protein more susceptible to proteolysis, and larger proteins may tend to have larger amounts of their polypeptide chain in this sensitive conformation, whatever it may be.

3. The native conformations of larger proteins may tend to be less stable than those of smaller proteins so that large proteins are more likely to denature within the cell. Fisher (211, 212) has pointed out that large proteins will be more unstable in an aqueous environment unless they have a decreased ratio of polar to nonpolar residues or alternatively, very long, thin tertiary structures. It is also possible that once denatured, larger proteins may be less likely to refold into their native conformations. In this respect it is interesting that in vitro renaturation of unfolded molecules has been most successful for small polypeptides (216). As discussed below, denatured proteins are generally more susceptible to proteolysis than proteins in their native conformations in vitro and probably in vivo.

4. Larger proteins may be more likely to contain errors in amino acid sequence or amino acid modifications which alter the protein's conformation and thus lead to rapid degradation. Random transcriptional or translational errors in amino acid placement along the peptide chain or random post-translational modifications of amino acid residues should be more common per molecule in larger polypeptides.

Whatever its basis, this correlation between size and degradative rate represents additional evidence that the half-life of a protein is determined at least in part by its structure. Further knowledge about the unique properties of large proteins should provide important insights into the mechanisms of intracellular degradation and the selectivity of the degradative process.

Influence of Ligands upon Rates of Degradation

A number of examples are known in mammalian cells where the rate of protein catabolism changes in response to supply of substrates, coenzymes, or other factors that bind to the polypeptide. There is strong evidence to indicate that these effects on degradation are a consequence of ligand-induced alterations in the protein's conformation.

It is now generally accepted that protein structures are highly flexible (213) and can change dramatically upon binding of small molecules (214–216). For example, studies of deuterium or tritium exchange rates (217) indicate that proteins are continually "breathing" (i.e. partially unfolding and refolding), and the frequency of such macromolecular breathing is influenced by the presence of ligands. Interactions with small molecules can alter a number of other chemical and physical properties of a polypeptide, including its sensitivity to proteolytic enzymes (218). For instance, glucose decreases the sensitivity of hexokinase to trypsin (219), substrate analogs make aldolase more resistant to carboxypeptidase (220), while the binding of various ligands reduces the protease sensitivity of serum albumin (221), aspartate transcarbamylase (222), staphylococcal nuclease (223), and ribonuclease (224).

Such effects appear to result from changes in the protein's overall conformation rather than from changes in highly localized regions of the molecule. Thus, competitive inhibitors of RNase can alter the protease sensitivity of the S peptide (224) located on the surface of the enzyme at some distance from the active site (216). Furthermore, binding of a single ligand to a protein can reduce its sensitivity to proteolytic enzymes with very different chemical specificities (218, 221, 223).

Such effects may explain the ability of substrates or cofactors to influence degradative rates in vivo. It has long been recognized that administration of tryptophan (225, 226) or α-methyltryptophan to a rat leads to increased levels of tryptophan oxygenase as a result of decreased rates of enzyme degradation (225, 226) as well as increased rates of enzyme synthesis (227). The classical studies of Schimke, Sweeney & Berlin (225, 226) showed that in cell-free extracts, tryptophan or its analog makes tryptophan oxygenase more resistant to digestion by trypsin, chymotrypsin, or protease from *Streptomyces*. Recently, Li & Knox (94) reported that tryptophan in cell-free preparations also reduces the spontaneous loss of tryptophan oxygenase assayed immunologically. This loss of antigen in the absence of substrate does not reflect proteolytic degradation of the enzyme but instead results from some sort of spontaneous denaturation. These findings suggest that the enzyme's conformation is altered by binding of tryptophan in a manner that is probably responsible for the decreased degradation in vivo.

Similar findings have been obtained for a number of other proteins. Drysdale & Munro (65) showed that administration of iron to rats increases the concentration of ferritin in liver in part by decreasing the degradation of this protein. Those ferritin molecules that bound the largest number of iron atoms were stabilized to the greatest extent. In vitro the supply of iron also protected ferritin from digestion by trypsin or chymotrypsin, presumably by altering the conformation of the protein (227a). The degradation of pyruvate carboxylase in rat liver decreases upon administration of tryptophan (228). This amino acid or more likely a product of its metabolism, quinolinate, can bind to pyruvate carboxylase and thereby alter its conformation. In this case, the stabilizing ligands are structurally unrelated to the substrate or cofactor, and probably therefore do not bind to the active site. The intracellular activity of dihydrofolate reductase increases in a variety of cultured cells in response to methotrexate, a competitive inhibitor of the enzyme. It is likely that the increased enzyme activity results in part from a decreased rate of degradation (229). In vitro methotrexate and other folate analogs bind to dihydrofolate reductase and stabilize it against digestion by trypsin (230), chymotrypsin (230), elastase (231), and subtilisin (231). Finally, in rat liver and kidney, administration of thymidine or thymidylate (232) increases the levels of thymidylate kinase. The activity of this enzyme is also stabilized in homogenates by thymidine (233). It thus appears likely that thymidine decreased intracellular degradation of thymidilate kinase by some effect on the enzyme's conformation.

In these instances, the catabolism of the protein appears to change as a consequence of ligand-induced stabilization. Thus far there have been no reports of physiological examples where a substrate or other ligand promotes degradation of a protein in vivo. Presumably such effects may occur; in vitro, for instance, oxygen binding increases hemoglobin's susceptibility to carboxypeptidase (234) and glucose binding increases the sensitivity of glycogen phosphorylase to cleavage by trypsin (234a).

In addition, coenzymes have been found to stabilize protein conformations in vitro in a fashion that may influence intracellular degradative rates. Bond, for example (209), has reported that pyridoxal phosphate reduces the sensitivity of serine dehydratase and tyrosine aminotransferase to trypsin and chymotrypsin. Perhaps of greater physiological relevance is the recent finding by Katunuma and co-workers (208, 235, 236) of proteases that can inactivate selectively pyridoxal-requiring enzymes and proteases for NAD-requiring enzymes. The sensitivity of several such enzymes to these group-specific proteases increases upon removal of their cofactor. The greater sensitivity of the apoenzymes to such proteases can explain the low levels of vitamin-requiring enzymes in Vitamin B_6 and niacin deficiency (208, 235, 236a) or the decreased degradation of serine dehydratase upon administration of pyridoxine (236b).

Recently, Litwack and co-workers proposed that the tightness of cofactor binding could be a major determinant of degradative rates in vivo. They point out that those liver enzymes which are inducible by glucocorticoids have short half-lives and dissociable cofactors, while enzymes which are poorly inducible by glucocorticoids have long half-lives and nondissociating cofactors (237). In addition, these workers showed that the relative rate of dissociation of the cofactor roughly correlated with the half-life of five liver enzymes (238). These data suggest that the formation of the apoenzymes might be the rate-limiting step in the degradative process.

Thus, ligand binding may be of general import in retarding protein degradation. It is conceivable that changes in the average rates of protein degradation in a cell could also occur in this fashion. For instance, in both animal and bacterial cells, the average rates of protein catabolism increase during starvation (11–15, 55, 56). Such effects could possibly result from changes in the intracellular concentrations of specific ligands or simply from the greater susceptibility of cell enzymes to proteolysis in the absence of their normal substrates or cofactors. A similar model may explain the increased degradation of proteins in unused or denervated muscle (58–61), where the average frequency with which cell proteins interact with substrates is likely to be reduced.

It is presently unclear whether these effects of ligands are in any way related to the other parameters that influence degradative rates in vivo. For example, one feature of abnormal proteins that might contribute to their rapid catabolism may be their reduced ability to bind stabilizing ligands. Similarly, the more rapid catabolism of large proteins may reflect some general differences in their affinities for cofactors. Possibly the various ligands that reduce protease sensitivity in vitro promote certain conformations especially resistant to proteolytic attack. Studies of these ligand-induced changes in protein structure should help clarify those conformational features that determine protease sensitivity. Such information appears of special interest since, as suggested below, the enzymes responsible for intracellular degradation probably act in similar fashion to the well-characterized proteases used in these in vitro studies.

Correlation between Degradative Rates and Susceptibility to Proteases in vitro

The findings reviewed in the preceding sections indicate that the structure of a protein is a major determinant of its degradation rate in vivo. Perhaps the simplest explanation of the diversity of protein half-lives would be that degradative rates are determined by inherent differences in the susceptibility of cell proteins to a general intracellular proteolytic system (195). During the past two years, appreciable evidence has accumulated to indicate that differences in the degradative rates can be explained to a large extent by this simple model. This evidence has consisted of the demonstration in both bacterial and mammalian cells that the catabolic rates of different proteins correlate well with their relative sensitivities to well-characterized proteolytic enzymes.

As discussed above, bacterial and animal cells selectively degrade abnormal proteins. Goldberg (195) has shown that the various conditions that induce the formation of abnormal proteins and thus promote protein degradation also increase the sensitivity of average cell proteins to a variety of endoproteases. For instance, incorporation of amino acid analogs into E. coli proteins not only leads to their rapid intracellular hydrolysis but also makes the proteins of cell-free extracts more sensitive to trypsin, pronase, chymotrypsin, or subtilisin. The abnormal proteins specifically degraded by the cell appear responsible for this enhanced protease sensitivity. As the analog-containing proteins were degraded within the cell, cell-free extracts decreased in sensitivity to trypsin and chymotrypsin (195). Furthermore, those analogs that stimulated degradation to the greatest extent in vivo also had the greatest effects on susceptibility to these proteases (167). In similar experiments the incorporation of amino acid analogs into cultured hamster kidney cells (149) or human fibroblasts (128) was found to increase intracellular degradation as well as the sensitivity of cell lysates to proteolytic digestion in vitro.

In addition, the incorporation of puromycin into E. coli polypeptides and production of frequent errors in protein synthesis due to presence of the *ram* mutation increased both intracellular degradation and protease sensitivity of average cell proteins (195). Finally, the abnormal *lac* repressor (*L-1*) which turns over rapidly in vivo has also been found more susceptible to a variety of proteolytic enzymes (167). Thus, under all conditions, increased protease sensitivity correlated with and may explain the rapid hydrolysis of the abnormal components within the cell.

Since these various abnormal proteins were more sensitive to endopeptidases having quite different specificities, it appears likely that major conformational features differentiate the normal cell constituents from those rapidly degraded by the cell. The difference in protease sensitivity between normal and canavanine-containing proteins in E. coli disappeared when the cell-free extracts were heated to 55°C or treated with low concentrations of SDS prior to incubation with proteases (167). These treatments increased protease susceptibility of the normal extracts without significantly affecting that of extracts of cells growing on canavanine. Thus, the analog-containing proteins appear to resemble denatured polypeptides.

It has long been recognized that denatured polypeptides are far more rapidly hydrolyzed by proteolytic enzymes than proteins in their native conformations (239–242). The unfolding of certain polypeptides can increase proteolytic sensitivity by 2–3 orders of magnitude (240–242). This well-documented effect may also explain the selective degradation of abnormal proteins within the cell (195). According to this simple model, normal proteins must share certain conformational features which prevent their rapid hydrolysis. Deviations from such conformations would make them more sensitive to the cell's degradative machinery. This simple mechanism would be entirely analogous to the elegant use of proteases by Anfinsen and co-workers (243) to purify an enzyme away from molecules that had failed to fold into their normal conformations.

In addition to these findings with abnormal proteins, strong correlations between intracellular half-lives and protease sensitivity have been demonstrated for normal proteins in *E. coli* (195) and in several eukaryotic organisms. Using pulse-chase protocols, it is possible to label differentially proteins of different biological stabilities. Those proteins in *E. coli* that turn over rapidly (119, 244) were found to be more sensitive to proteases in vitro than cell components with slower degradative rates (195). Similar findings were obtained with the soluble proteins from rat liver (14, 24). Dice et al (24) used the double-isotope approach, so that proteins which were degraded more rapidly in the liver had high $^3H/^{14}C$ ratios. When these proteins were incubated with pronase or trypsin, the radioactivity that initially appeared in acid-soluble form had relatively high $^3H/^{14}C$ ratios. Therefore, the soluble proteins degraded more rapidly in the liver must have been the ones more susceptible to proteolysis in vitro. However, when the native conformations of the proteins were first altered by addition of SDS or by blocking sulfhydryl groups, the correlation between degradative rate and protease sensitivity was less strong (44).

Similar experiments have also been carried out on proteins from rat kidney (14), HeLa cells (24), and pea stem cells (24). Once again those proteins subject to rapid intracellular catabolism were also more susceptible to digestion by trypsin or pronase. As discussed above, larger molecular weight polypeptides in a number of eukaryotic cells turn over on the average more rapidly than smaller ones. This correlation between molecular size and degradative rate may result from a greater protease sensitivity of the large polypeptides (24). Dice et al (24) separated soluble proteins from rat liver into five different molecular size fractions using Sephadex chromatography and then incubated each fraction with pronase. Both the initial rate and total extent of digestion of the larger proteins exceeded those of the smaller ones (24). These differences in protease susceptibility were not evident when the experiments were performed in the presence of SDS.

Further support for this general conclusion has been obtained through studies of specific mammalian enzymes. Bond (209) has reported that the half-lives of five rat liver enzymes correlated with their relative susceptibilities to inactivation by trypsin and chymotrypsin. However, such a precise correlation was not obtained with pronase or subtilisin, which for unknown reasons inactivated lactate dehydrogenase especially rapidly. Because only five enzymes were studied, the anomalous findings with pronase and subtilisin are of uncertain significance. Finally, as discussed earlier, in all cases where a ligand is known to influence half-lives of an enzyme in vivo, it also affects the protein's susceptibility to model proteases. Thus, the decreased rates of degradation of tryptophan oxygenase (225, 226), ferritin (65), dihydrofolate reductase (229), thymidylate kinase (232, 233), and pyruvate carboxylase (228) induced by small molecules may result from the established ability of the ligand to reduce protease sensitivity in vitro.

Double-label experiments were performed to determine whether the degradative rates of proteins in isolated endoplasmic reticuli (44), ribosomes (45), or chromosomes (48) also correlated with their relative sensitivities to proteases, as was found with soluble proteins (24). However, when the isolated organelles were incubated with protease, those components rapidly degraded in vivo were not those most susceptible to digestion. These findings may simply indicate that such proteins are not degraded within the cell while they are bound to organelles. In fact, when proteins found in chromatin were extracted, their relative sensitivities to pronase or trypsin correlated with their intracellular degradative rates. These observations thus suggest that chromosomal proteins and perhaps the components of other organelles are degraded in vivo only if they are dissociated from such structures. In fact, when rat liver ribosomes (251), membranes (44), and chromatin (48, 244a) are incubated with proteases, the proteins tend to be relatively resistant to digestion. Disruption of the organelles significantly increases their protease sensitivity. It is possible that when the proteins are within the intact organelles, many proteins or especially sensitive peptide bonds are shielded from added proteases. Alternatively, the proteins may assume distinct conformations when in the dissociated state and when organelle bound. Evidence does exist for cytoplasmic pools of proteins of ribosomes (45, 245–247), chromosomes (248), and possibly membranes (249). Presumably these cytoplasmic pools exchange continually with proteins within the organelles; such exchange has been demonstrated for the ribosome-associated proteins (45, 250) and for subunits of multimeric proteins such as lactate dehydrogenase (253, 256) and glyceraldehyde 3-phosphate dehydrogenase (254). Thus, organelles and multimeric proteins must be viewed as highly dynamic structures which continually exchange their protein constituents. This model is in accord with the finding of heterogeneous degradation rates for different organelle-associated proteins and for different subunits of multimeric enzymes. If, as appears likely, the degradation of such proteins occurs when they are dissociated from the organelle, it is not surprising that the half-lives of such polypeptides correlate with their molecular size in similar fashion to the half-lives of soluble proteins.

POSSIBLE LIMITATIONS   These extensive correlations between degradative rates in vivo and protease susceptibility in vitro can clearly account for many of the known features of intracellular proteolysis. Since much of the selectivity of the cellular process has been reproduced with proteases of differing specificities, these studies have provided little information about the enzymes actually responsible for degradation in mammalian or bacterial cells. It should be noted that the results discussed above were all obtained with neutral proteases (most of which were serine proteases). It is unknown whether similar correlations may be obtained with lysosomal proteases under highly acidic conditions found in this organelle (255). In any case, the finding that a rather general proteolytic system may account for most, if not all, of the selectivity of intracellular degradation simplifies this problem and makes it unnecessary to hypothesize the existence of a large number of highly specific proteases that degrade a single or even several cell proteins (3).

A variety of observations, however, suggest that inherent differences in protease sensitivity by itself cannot explain all the specificity or regulation of intracellular proteolysis. One interesting observation that may be difficult to explain in this fashion is the finding that a single enzyme may have very different half-lives in different tissues. For example, alanine aminotransferase (189) appears to have a much longer half-life in rat skeletal muscle (20 days) than in rat liver (3 days). Similarly, the half-lives of lactic dehydrogenase isozyme 5 (LDH5) are very different in rat liver (16 days), heart (1.6 days), and skeletal muscle (31 days) (252). It is possible that the levels of ligands or activity of modifying enzymes differ in these several tissues and cause differences in susceptibility to a general proteolytic system. Alternatively, these different half-lives may reflect differing specificities or amounts of the rate-limiting protease systems in the several tissues.

It also appears unlikely that changes in the average rate of intracellular proteolysis under different conditions can be explained solely by variations in overall protease sensitivity. In *E. coli* deprived of a carbon source the degradation of proteins (11–15), especially of more stable cell components (244, 257), increases, but thus far attempts to demonstrate changes in protease sensitivity of cell lysates have not been successful (149). In these instances (71, 179), as in the variations in proteolysis induced by hormones in mammalian tissues (258, 259), the regulation and specificity may reside to a large extent within changes in the degradative system (258, 259). In sporulating bacteria, there is strong evidence that the selective degradation of cell enzymes results from the appearance of new proteolytic machinery (72).

Finally, strong evidence for highly selective intracellular proteases has been obtained by Katunuma and co-workers (208, 235, 236), who have demonstrated serine proteases specific for pyridoxal- or NAD-requiring enzymes in several mammalian tissues. Since these enzyme levels change upon decreased dietary intake of Vitamin B_6 or niacin, these proteases may play a crucial role in removal of apoenzymes lacking their cofactors. The generality of such group-specific proteases is an important unsolved problem, although a protease specific for pyridoxal enzymes has also been recently described in yeast (208). It is unclear how such enzymes might be related to other proteolytic activities of the cell. Such proteases appear to cleave substrate enzymes at several places, after which the polypeptides are especially sensitive to trypsin or other proteases to which they were previously resistant (208). The group-specific proteases thus may represent a mechanism superimposed upon the cell's more general proteolytic mechanisms. According to this model (208), the selective inactivation of enzymes by the group-specific proteins (like selective covalent modifications) would under certain physiological conditions make otherwise stable enzymes especially susceptible to digestion by the cell's less specific proteolytic machinery.

POSTSCRIPT

The investigations summarized here emphasize the important influence of protein conformation on degradative rates. Traditionally biochemists and geneticists have clearly distinguished between point mutations that affect an enzyme's function and control-gene mutations that determine a protein's concentrations. The demonstration that conformation and thus primary sequence can also influence a protein's intracellular level weakens this distinction and emphasizes a physiologically important aspect of protein structure that has generally been overlooked. A major goal for future studies of protein structure should be to elucidate those conformational features that determine a protein's susceptibility to proteolytic digestion and thus its degradative rate in vivo.

The recognition that nearly all cell proteins are continually being degraded and replaced and that protein degradation is a major factor determining intracellular protein concentrations has led to a dramatic growth of interest in this process. The recent improvements in techniques for measuring degradation, the clarification of the physiological functions served by protein breakdown, and the demonstration that a protein's conformation influences its catabolic rate have all helped to define the degradative process more precisely. But knowledge about the biochemical mechanisms of intracellular protein catabolism and about the regulation of this process within the cell is still rudimentary. The subsequent part of this review will attempt to summarize our present understanding of these unsolved problems.

ACKNOWLEDGMENTS

This article has been prepared with the aid of grants from the National Institute of Neurological Disease and Stroke (NINDS), the Muscular Dystrophy Associations of America (MDAA), and the Air Force Office of Scientific Research. A. L. Goldberg holds a Research Career Development Award from NINDS; J. F. Dice is a Postdoctoral Fellow of the MDAA. We are grateful to Drs. Robert Jungas, Robert Perlman, Joel Kowit, and John Burr for their helpful criticism and to Mrs. Elsa Fox, Dr. Joan Goldberg and Mrs. Elizabeth Howell for their invaluable assistance in the preparation of this manuscript. We are also deeply indebted to Dr. Jeanne B. Li for her assistance and valuable suggestions in the planning of this review.

Literature Cited

1. Schoenheimer, R. 1942. *Dynamic State of Body Constituents.* Cambridge: Harvard Univ. Press
2. Borsook, H., Keighley, G. L. 1936. *Proc. Roy. Soc. B* 118:488
3. Schimke, R. T. 1970. *Mammalian Protein Metab.* 4:177
4. Schimke, R. T. 1973. *Advan. Enzymol.* 37:135
5. Rechcigl, M. Jr. 1971. In *Enzyme Synthesis and Degradation in Mammalian Systems,* ed. M. Rechcigl, 237. Baltimore: University Park Press
6. Rechcigl, M. Jr. 1968. *Enzymologia* 34:23–39
7. Bohley, P. et al 1972. *Tissue Proteinases,* ed. A. J. Barrett, J. T. Dingle. Amsterdam: North-Holland

8. Biochemical Society of the German Democratic Republic. 1974. *Intracellular Protein Catabolism.* In press
9. Marks, N., Lajtha, A. 1971. *Handb. Neurochem.* 5: 49–139
10. Bohley, P. 1968. *Naturwissenschaften* 55: 211–17
11. Mandelstam, J. 1960. *Bacteriol. Rev.* 24: 289
12. Willetts, N. S. 1965. *Biochem. Biophys. Res. Commun.* 20: 692–96
13. Pine, M. J. 1972. *Ann. Rev. Microbiol.* 26: 103
14. Goldberg, A. L., Howell, E. M., Martel, S. B., Li, J. B., Prouty, W. F., 1974. *Fed. Proc,* 33. In press
15. Goldberg, A. L. 1974. In *Neurosciences, Third Study Program,* ed. F. O. Schmitt, F. G. Worden, 3: 827–34. Cambridge: MIT Press
16. Willetts, N. S. 1967. *Biochem. J.* 103: 453–61
17. Schlessinger, D., Ben-Hamida, F. 1966. *Biochim. Biophys. Acta* 119: 171–82
18. Pine, M. J. 1966. *J. Bacteriol.* 92: 847–50
19. Halvorson, H. 1958. *Biochim. Biophys. Acta* 27: 267–76
20. Halvorson, H. 1958. *Biochim. Biophys. Acta* 27: 255–66
21. Steward, F. C., Bidwell, K. G. S., Yemm, E. W. 1956. *Nature* 178: 734
22. Chibnall, A. C., Wiltshire, G. H. 1954. *New Phytol.* 53: 38
23. Kemp, J. D., Sutton, D. W. 1971. *Biochemistry* 10: 81–88
24. Dice, J. F., Dehlinger, P. J., Schimke, R. T. 1973. *J. Biol. Chem.* 248: 4220
24a. Sitz, T. O., Molloy, G. R., Schmidt, R. R. 1973. *Biochim. Biophys. Acta* 319: 103
25. Laemmli, U. K. 1970. *Nature* 227: 680–85
26. Jacobson, M. F., Baltimore, D. 1968. *J. Mol. Biol.* 33: 369
27. Garfinkle, B. D., Tershak, D. R. 1972. *Nature New Biol.* 233: 206–8
28. Steiner, D. F. et al 1969. *Recent Progr. Horm. Res.* 25: 207
29. Drescher, D., DeLuca, H. F. 1971. *Biochemistry* 10: 2308
30. Russell, J. H., Geller, D. M. 1973. *Biochem. Biophys. Res. Commun.* 55: 239
31. Judah, J. D., Gamble, M., Steadman, J. H. 1973. *Biochem. J.* 134: 1083
32. Layman, D. L., McGoodwin, E. B., Martin, G. R. 1971. *Proc. Nat. Acad. Sci. USA* 68: 454
33. Tager, H. S., Steiner, D. F. 1973. *Proc. Nat. Acad. Sci. USA* 70: 2321
34. Dreyfus, J. C., Kruh, J., Schapira, G. 1960. *Biochem. J.* 75: 574
35. Millward, D. J. 1970. *Clin. Sci.* 39: 577–90
35a. Young, R. W., Bok, D. 1969. *J. Cell. Biol.* 42: 392
36. Russell, D. H., Snyder, S. H. 1971. *Mol. Pharmacol.* 5: 253
37. Fritz, P. J., White, E. L., Pruitt, K. M., Vesell, E. S. 1973. *Biochemistry* 12: 4034
38. Dehlinger, P. J., Schimke, R. T. 1972. *J. Biol. Chem.* 246: 2574
39. Gurd, J. W., Evans, W. H. 1973. *Eur. J. Biochem.* 36: 273
40. Kaplan, J. Personal communication
41. Arias, I. M., Doyle, D., Schimke, R. T. 1969. *J. Biol. Chem.* 244: 3303
42. Omura, T., Siekevitz, P., Palade, G. E. 1967. *J. Biol. Chem.* 242: 2389
43. Kuriyama, Y., Omura, T., Siekevitz, P., Palade, G. E. 1969. *J. Biol. Chem.* 244: 2017
44. Taylor, J. M., Dehlinger, P. J., Dice, J. F., Schimke, R. T. 1973. *Drug Metab. Disposition* 1: 84
45. Dice, J. F., Schimke, R. T. 1972. *J. Biol. Chem.* 247: 98
46. Swick, R., Rexroth, A. K., Strange, J. L. 1968. *J. Biol. Chem.* 243: 3581
47. Druyan, R., DeBernard, B., Rabinowitz, M. 1968. *J. Biol. Chem.* 244: 5874
48. Dice, J. F., Schimke, R. T. 1973. *Arch. Biochem. Biophys.* 158: 97
49. Low, R. B., Goldberg, A. L. 1973. *J. Cell Biol.* 56: 590–95
50. Wikman-Coffelt, J., Zelis, R., Fenner, C., Mason, D. T. 1973. *J. Biol. Chem.* 248: 5206
51. Funabiki, R., Cassens, R. G. 1972. *Nature New Biol.* 236: 249
52. Rabinowitz, M. 1973. *Am. J. Cardiol.* 31: 202
53. Poole, B. 1971. See Ref. 5, pp. 375–402
54. Hemminki, K. 1973. *Biochim. Biophys. Acta* 310: 285
55. Millward, D. J. 1970. *Clin. Sci.* 39: 591–603
56. Li, J. B., Goldberg, A. L. 1974. Submitted for publication
57. Fulks, R. M., Li, J. B., Goldberg, A. L. 1974. *J. Biol. Chem.* In press
58. Goldberg, A. L. 1969. *J. Biol. Chem.* 244: 3223–29
59. Pearlstein, R. A., Kohn, R. R. 1964. *Am. J. Pathol.* 48: 823
60. Schapira, S., Dreyfus, J. C., Couszet, J., Schapira, F. 1953. *Bull. Soc. Chim. Biol.* 35: 1309
61. Goldberg, A. L., Jablecki, C., Li, J. B. 1974. *Ann. NY Acad. Sci.* In press
62. Schimke, R. T. 1964. *J. Biol. Chem.*

239:3808–17
63. Majerus, P. W., Kilburn, E. 1970. *J. Biol. Chem.* 244:6254
64. Ove, P., Obenrader, M., Lansing, A. 1972. *Biochim. Biophys. Acta* 277:311
65. Drysdale, J. W., Munro, H. N. 1966. *J. Biol. Chem.* 241:3630
66. Schimke, R. T. 1969. *Curr. Top. Cell. Regul.* 1:77
67. Schimke, R. T., Doyle, D. 1970. *Ann. Rev. Biochem.* 39:929
68. Goldberg, A. L., Wittes, R. B. 1966. *Science* 153:420
68a. Woese, C. R. 1970. *Biol. Sci.* 20:471
68b. Crick, F. H. C. 1968. *J. Mol. Biol.* 38:367
69. Berlin, C. M., Schimke, R. T. 1965. *Mol. Pharmacol.* 1:149
70. Goldberg, A. L., Li, J. B. 1974. In preparation
71. Goldberg, A. L. 1971. *Nature New Biol.* 234:51
72. Kornberg, A., Spudich, J. A., Nelson, D. L., Deutscher, M. P. 1968. *Ann. Rev. Biochem.* 37:51
73. Zeldin, M. H., Skea, W., Matteson, D. 1973. *Biochem. Biophys. Res. Commun.* 52:544
74. Knox, W. E. 1972. *Enzyme Patterns in Fetal, Adult and Neoplastic Rat Tissues.* Basel: Karger
75. Hirsch, C. A., Hiatt, H. H. 1966. *J. Biol. Chem.* 241:5936
76. Cahill, G. F. 1970. *N. Engl. J. Med.* 282:668–75
77. Steiner, D. F., Freinkel, N. 1972. *Handb. Physiol.* 7 Vol. 1
78. Mortimore, G. E., Mondon, C. E. 1970. *J. Biol. Chem.* 245:2375–83
79. Morgan, H. E. et al 1971. *J. Biol. Chem.* 246:2152–62
80. Morgan, H. E. et al 1972. *Insulin Action,* 437–59. New York: Academic
81. Minemura, T., Lacy, W. W., Crofford, O. B. 1970. *J. Biol. Chem.* 245:3872–81
82. Hershko, A., Mamont, P., Shields, R., Tomkins, G. M. 1971. *Nature New Biol.* 232:206–11
83. Haining, J. L. 1971. *Arch. Biochem. Biophys.* 144:204–8
84. Goldberg, A. L. 1971. *Proc. Nat. Acad. Sci. USA* 68:362
85. Hershko, A., Tomkins, G. M. 1971. *J. Biol. Chem.* 246:710–14
86. Barker, K. L., Lee, J. L., Kenney, F. T. 1971. *Biochem. Biophys. Res. Commun.* 43:1132–38
87. Kenney, F. T. 1967. *Science* 156:525
88. Feldman, M., Yagil, G. 1969. *Biochem. Biophys. Res. Commun.* 37:198–203
89. Seglen, P. O. 1971. *Biochim. Biophys. Acta* 230:319–26
90. Levitan, I. B., Webb, T. E. 1969. *J. Biol. Chem.* 244:4684–88
90a. Woodside, K. H., Ward, W. F., Mortimore, G. E. 1974. *J. Biol. Chem.* In press
91. Auricchio, F., Martin, D., Tomkins, G. 1969. *Nature* 224:806–8
92. Holzer, H., Duntze, W. 1971. *Ann. Rev. Biochem.* 40:345
92a. Huijing, F., Lee, E. Y. C., Eds. 1973. *Protein Phosphorylation in Control Mechanisms.* New York: Academic
93. Li, J. B., Knox, W. E. 1972. *J. Biol. Chem.* 247:7550
94. Li, J. B., Knox, W. E. 1972. *J. Biol. Chem.* 247:7546
95. Goldschmidt, R. 1970. *Nature* 228:1151–56
96. Goldschmidt, R. 1970. *Protein Breakdown in E. coli.* PhD thesis. Columbia Univ., New York
97. Platt, T., Miller, J., Weber, K. 1970. *Nature* 228:1154
98. Ullman, A., Perrin, D. 1970. *The Lactose Operon.,* ed. J. R. Beckwith, D. Zipser, 143. New York: Cold Spring Harbor Lab.
99. Morrison, S. L., Zipser, D., Goldschmidt, R. 1971. *J. Mol. Biol.* 60:485
100. Lin, S., Zabin, I. 1972. *J. Biol. Chem.* 247:2205–11
101. Villarejo, M., Zamenhof, P. J., Zabin, I. 1972. *J. Biol. Chem.* 247:2212
102. Bukhari, A. I., Zipser, D. 1973. *Nature New Biol.* 243:238–41
103. Young, V. R., Alexis, S. D., Baliga, B. S., Munro, H. N. 1972. *J. Biol. Chem.* 247:3592–3600
104. Johnson, P., Harris, C. I., Perry, S. V. 1967. *Biochem. J.* 105:361–70
105. Huszar, G., Elzinga, M. 1971. *Biochemistry* 10:229
106. Reporter, M., Corbin, J. L. 1971. *Biochem. Biophys. Res. Commun.* 43:644–56
107. Swick, R. W. 1958. *J. Biol. Chem.* 231:751–64
108. Buchanan, D. L. 1961. *Arch. Biochem. Biophys.* 94:500–11
109. Gan, J. C., Jeffay, H. 1971. *Biochim. Biophys. Acta* 252:125–35
110. Garlick, P. J. 1969. *Nature* 223:61–62
111. Garlick, P. J., Marshall, I. 1972. *J. Neurochem.* 19:577–83
112. Waterlow, J. C., Stephen, J. M. L. 1968. *Clin. Sci.* 35:287–305
113. Austin, L., Lowry, O. H., Brown, J. G., Carter, J. G. 1972. *Biochem. J.* 126:351–59

114. Nettleton, J. A. 1973. *Protein Catabolism and Amino Acid Recycling During Protein and Calorie Restriction.* PhD thesis. Harvard Univ., Cambridge
115. Fashakian, J. B., Hegsted, D. M. 1970. *Nature* 228:1313–14
116. Dallman, P. R., Manies, E. C. 1973. *J. Nutr.* 103:257–66
117. Koch, A. L. 1962. *J. Theor. Biol.* 3:283
118. Poole, B. 1971. *J. Biol. Chem.* 246:6587
119. Nath, K., Koch, A. L. 1970. *J. Biol. Chem.* 245:2889–2900
120. Mandelstam, J. 1958. *Biochem. J.* 69:110–19
121. Eagle, H., Piez, K. A., Fleichman, R., Oyama, V. I. 1959. *J. Biol. Chem.* 234:592–97
122. Righetti, P., Little, E. P., Wolf, G. 1971. *J. Biol. Chem.* 246:5724–32
123. Klevecz, R. R. 1971. *Biochem. Biophys. Res. Commun.* 43:76–81
124. Gan, J. C., Jeffay, H. 1967. *Biochim. Biophys. Acta* 148:448–59
125. Loftfield, R. B., Harris, A. 1956. *J. Biol. Chem.* 219:151–59
125a. Pine, M. J. Personal communication
126. Rannels, D. E., Li, J. B., Morgan, H. E., Jefferson, L. S. 1974. *Methods Enzymol.* In press
127. Poole, B., Wibo, M. 1973. *J. Biol. Chem.* 248:6221
128. Bradley, M. O., Dice, J. F., Hayflick, L., Schimke, R. T. In preparation
129. Christensen, H. N., Handlogten, M. E. 1968. *J. Biol. Chem.* 243:5428
130. Eavenson, E., Christensen, H. N. 1967. *J. Biol. Chem.* 242:5386
131. Goldberg, A. L. 1969. *J. Biol. Chem.* 244:3217–22
132. Lee, K., Kenney, F. T. 1971. *J. Biol. Chem.* 246:7595–7601
132a. Pitot, H. C., Kaplan, J., Čihàk, A. 1971. See Ref. 5, p. 216
133. Walter, H. 1960. *Nature* 188:643–45
134. Martin, A. F., Prior, G., Zak, R. 1973. *Fed. Proc.* 32:532
135. Rannels, D. E., Morgan, H. E. 1973. *Fed. Proc.* 32:532
136. Hider, R. C., Fern, E. B., London, D. R. 1969. *Biochem. J.* 114:171–78
137. Kipnis, D. M., Reiss, E., Helmreich, E. 1961. *Biochim. Biophys. Acta* 51:519–24
138. Mortimore, G. E., Woodside, K. H., Henry, J. E. 1972. *J. Biol. Chem.* 247:2776–84
139. Fern, E. B., Hider, R. C., London, D. R. 1971. *Eur. J. Clin. Invest.* 1:211
140. Fern, E. B., Garlick, P. J. 1973. *Biochem. J.* 134:1127–30
141. Li, J. B., Fulks, R. M., Goldberg, A. L. 1973. *J. Biol. Chem.* 248:7272–75
142. Alpers, D. H., Thier, S. O. 1972. *Biochim. Biophys. Acta* 262:535–45
143. Waterlow, J. C., Stephen, J. M. L. 1966. *J. Nutr.* 20:461–84
144. Penn, N. W., Mandeles, S., Anker, H. S. 1957. *Biochim. Biophys. Acta* 26:349
145. Swick, R., Handa, D. T. 1956. *J. Biol. Chem.* 218:557
146. Poole, B., Leighton, F., DeDuve, C. 1969. *J. Cell Biol.* 41:536
147. McFarlane, A. S. 1963. *Biochem. J.* 89:277–90
148. Canick, J. A., Villee, D. B. 1974. Submitted for publication
149. Goldberg, A. L. Unpublished observations
150. Manchester, K. L., Young, F. G. 1959. *Biochem. J.* 72:136–41
151. Williams, L. S., Neidhardt, F. C. 1969. *J. Mol. Biol.* 43:529–50
152. Bernlohr, R. W. 1972. *J. Biol. Chem.* 247:4893–99
152a. Pine, M. J. Personal communication
153. Glass, R. D., Doyle, D. 1972. *J. Biol. Chem.* 247:5234
154. Goldspink, D. F., Goldberg, A. L. 1973. *Biochem. J.* 134:829
155. Li, J. B. Personal communication
156. Miller, J. H., Platt, T., Weber, K. 1970. See Ref. 98, 343
157. Hogness, D. S., Cohn, M., Monod, J. 1955. *Biochim. Biophys. Acta* 16:99
158. Rotman, B., Spiegelman, S. 1954. *J. Bacteriol.* 68:419
159. Pine, M. J. 1973. *J. Bacteriol.* 115:107–16
160. Ames, B. N., Hartmann, P. E. 1963. *Cold Spring Harbor Symp. Quant. Biol.* 28:349
160a. Walker, J. R., Ussery, C. L., Allen, J. S. 1973. *J. Bacteriol.* 113:1326–32
161. Steinberg, A., Zipser, D. 1973. *J. Bacteriol.* 116:1469
162. Apte, B. N., Zipser, D. 1973. *Proc. Nat. Acad. Sci. USA* 70:2969–73
163. Goldberg, M. E. 1970. See Ref. 98, 273
164. Pine, M. J. 1967. *J. Bacteriol.* 93:1527
165. Goldberg, A. L. 1972. *Proc. Nat. Acad. Sci. USA* 69:422
166. Chapeville, F., Yota, P., Paulin, D. 1969. *Cold Spring Harbor Symp. Quant. Biol.* 34:493–98
167. Goldberg, A. L. 1974. In preparation
168. Prouty, W. F., Goldberg, A. L. 1972. *Nature New Biol.* 240:147
169. Prouty, W. F., Karnovsky, M., Goldberg, A. L. 1974. Submitted for publication

170. Schachtele, C. F., Anderson, D. L., Rogers, P. 1970. *J. Mol. Biol.* 49:255
171. Rabinowitz, M., Finkelman, A., Regan, R. L., Breitman, T. R. 1969. *J. Bacteriol.* 99:336
172. Harris, J. W. 1972. *The Red Cell.* Cambridge: Harvard Univ. Press
173. Pinkett, M. O., Brownstein, B. L. 1974. *J. Bacteriol.* In press
174. Bremer, H., Brunshede, H. 1971. *J. Mol. Biol.* 57:35–57
175. Hall, B., Gallant, J. 1972. *Nature New Biol.* 237:131–35
176. Sussman, A. J., Gilvarg, C. 1969. *J. Biol. Chem.* 244:6304–6
177. Bubela, B., Holdsworth, E. S. 1966. *Biochim. Biophys. Acta* 123:364–75
178. Epstein, I., Grossowicz, N. 1969. *J. Bacteriol.* 99:418–21
179. Prouty, W. F., Goldberg, A. L. 1972. *J. Biol. Chem.* 247:3341
180. Rabinowitz, M., Fisher, J. M. 1961. *Biochem. Biophys. Res. Commun.* 6:449
181. Rabinowitz, M., Fisher, J. M. 1964. *Biochim. Biophys. Acta* 91:313
182. Johnson, R. W., Kenney, F. T. 1973. *J. Biol. Chem.* 248:4528
183. Adams, J. G., Winter, W. P., Ricknagel, D. L. Spencer, H. H. 1972. *Science* 176:1427
184. Yoshida, A., Stamatoyannopoulos, G., Motulsky, A. G. 1967. *Science* 155:97
185. Levere, R. D., Lichtman, H. C., Levine, J. 1964. *Nature* 202:499
186. Zuelzer, W. W., Neel, J. V., Robinson, A. R. 1956. In *Progress in Hematology,* ed. L. M. Tocantius, 1:91. New York: Grune and Stratton
187. Loftfield, R. B., Vanderjagt, D. 1972. *Biochem. J.* 128:1353
187a. Robinson, A. B., McKerrow, J. H., Legaz, M. 1974. *Int. J. Peptide Protein Res.* 6:31
187b. McKerrow, J. H., Robinson, A. B. 1974. *Science* 183:85
188. Robinson A. B., McKerrow, J. H., Cary, P. 1970. *Proc. Nat. Acad. Sci. USA* 66:753–57
188a. Bond, J. S. Personal communication
189. Segal, H. L., Matsuzawa, T., Haider, M., Abraham, G. J. 1969. *Biochem. Biophys. Res. Commun.* 36:764–70
190. Orgel, L. E. 1963. *Proc. Nat. Acad. Sci. USA* 49:517–21
191. Orgel, L. E. 1973. *Nature* 243:441–45
192. Holliday, R., Tarrant, G. M. 1972. *Nature* 238:26
193. Gershon, H., Gershon, D. 1973. *Proc. Nat. Acad. Sci. USA* 70:909
194. Holland, J. J., Kohne, D., Doyle, M. V. 1973. *Nature* 245:316
195. Goldberg, A. L. 1972. *Proc. Nat. Acad. Sci. USA* 69:2640
196. Weber, K., Osborn, M. 1969. *J. Biol. Chem.* 244:3406
197. Dehlinger, P. J., Schimke, R. T. 1971. *Biochem. Biophys. Res. Commun.* 40:1473
198. Darnell, D. W., Klotz, I. M. 1972. *Arch. Biochem. Biophys.* 149:1
199. Alpers, D. H. 1972. *J. Clin. Invest.* 51:2621
200. Bretscher, M. S. 1971. *Nature New Biol.* 231:229
200a. Segrest, J. P., Jackson, R. L., Andrews, E. P., Marchesi, V. T. 1971. *Biochem. Biophys. Res. Commun.* 44:390
201. Tweto, J., Dehlinger, P. J., Larrabee, A. R. 1972. *Biochem. Biophys. Res. Commun.* 48:1371
202. Taber, R., Wertheimer, R., Golrick, J. 1973. *J. Mol. Biol.* 80:367
203. Richards, F. M., Vithayathil, P. J. 1960. *Brookhaven Symp. Biol.* 13:115
204. Chaiken, I. M., Anfinsen, C. B. 1970. *J. Biol. Chem.* 245:2337
205. Uchida, T., Pappenheimer, A. M., Harper, A. A. 1973. *J. Biol. Chem.* 248:3851
206. Siekevitz, P. 1972. *J. Theor. Biol.* 37:321
207. Brush, J. S. 1971. *Diabetes* 20:140
208. Katunuma, N. 1973. *Curr. Top. Cell. Regul.* 7:37–51
209. Bond, J. S. 1971. *Biochem. Biophys. Res. Commun.* 43:333–39
210. Dice, J. F., Goldberg, A. L. In preparation
211. Fisher, H. F. 1964. *Proc. Nat. Acad. Sci. USA* 51:1285
212. Fisher, H. F. 1965. *Biochim. Biophys. Acta* 109:544
213. Koshland, D. E. Jr., Neet, K. E. 1968. *Ann. Rev. Biochem.* 37:359
214. Paulus, H., Alpers, J. B. 1971. *Enzyme* 12:385–401
215. Grisolia, S. 1964. *Physiol. Rev.* 44:657
216. Anfinsen, C. B. 1972. *Science* 181:223
217. Englander, S. W., Downer, N. W., Teitelbaum, H. 1972. *Ann. Rev. Biochem.* 41:903
218. Citri, N. 1973. *Advan. Enzymol.* 37:397
219. Trayser, K. A., Colowick S. P. 1961. *Biochim. Biophys. Acta* 94:169
220. Adelman, R. C., Morse, D. E., Chan, W., Horecker, B. L. 1968. *Biochim. Biophys. Acta* 126:343
221. Markus, G. 1965. *Proc. Nat. Acad. Sci. USA* 54:255

222. McClintock, D. K., Markus, G. 1968. *J. Biol. Chem.* 243:2855
223. Taniuchi, H., Moravek, L., Anfinsen, C. B. 1969. *J. Biol. Chem.* 244:4600
224. Markus, G., Barnard, E. A., Castellani, B. A., Saunders, D. 1968. *J. Biol. Chem.* 243:4070
225. Schimke, R. T., Sweeney, E. W., Berlin, C. M. 1965. *J. Biol. Chem.* 240:322
226. Schimke, R. T., Sweeney, E. W., Berlin, C. M. 1965. *J. Biol. Chem.* 240:4609
227. Pirias, M., Knox, W. E. 1967. *J. Biol. Chem.* 242:2959
227a. Azari, P. R., Feeney, E. 1958. *J. Biol. Chem.* 232:293
228. Ballard, F. J., Hopgood, M. F. 1973. *Biochem. J.* 136:259
229. Hillcoat, B. L., Swett, V., Bertino, J. R. 1967. *Proc. Nat. Acad. Sci. USA* 58:1632–37
230. Hillcoat, B. L., Marshall, L., Gauldie, J., Hiebert, M. 1971. *Ann. NY Acad. Sci.* 186:187
231. Hakala, M. T., Soulinna, E. M. 1966. *Mol. Pharmacol.* 2:465
232. Hiatt, H. H., Bojarski, T. B. 1960. *Biochem. Biophys. Res. Commun.* 2:35
233. Bojarski, T. B., Hiatt, H. H. 1960. *Nature* 188:1112
234. Zito, R., Antonini, E., Wyman, J. 1964. *J. Biol. Chem.* 239:1804
234a. Graves, D. J., Mann, S., Ann, S., Philip, G., Oliveira, R. J. 1968. *J. Biol. Chem.* 243:6090
235. Katunuma, N., Kito, K., Kominami, E. 1971. *Biochem. Biophys. Res. Commun.* 45:76
236. Katunuma, N., Kominami, E., Kominami, S. 1971. *Biochem. Biophys. Res. Commun.* 45:71
236a. Greengard, O. 1964. *Advan. Enzyme Regul.* 2:277
236b. Khairallah, E. A., Pitot, H. C. 1968. In *Symposium on Pyridoxal Enzymes,* ed. Yamada, Katsunama, Wada, 159. Tokyo: Maruzui
237. Litwack, G., Singer, S. 1972. *Biochem. Actions Horm.* 2:113
238. Litwack, G., Rosenfeld, S. 1973. *Biochem. Biophys. Res. Commun.* 52:181

239. Linderstrom-Lang, K. 1950. *Cold Spring Harbor Symp. Quant. Biol.* 14:117
240. Green, N. H., Neurath, H. 1954. *Proteins B.* 2:1059
241. Bennett, J. C. 1967. *Methods Enzymol.* 11:211
242. Ripley, J. A. 1967. *Methods Enzymol.* 11:905–17
243. Anfinsen, C. B. 1972. *Biochem. J.* 128:737
244. Pine, M. J. 1965. *Biochim. Biophys. Acta* 104:439–56
244a. Bartley, J., Chalkley, R. 1970. *J. Biol. Chem.* 245:4286
245. Warner, J. R. 1966. *J. Mol. Biol.* 19:383
246. Warner, J. R. 1971. *J. Biol. Chem.* 246:447
247. Gupta, R. S., Singh, U. N. 1972. *J. Mol. Biol.* 69:279
248. Seale, R. L., Aronson, A. I. 1973. *J. Mol. Biol.* 75:633
249. Ray, T. K., Lieberman, I., Lansing, A. I. 1968. *Biochem. Biophys. Res. Commun.* 31:54
250. Garrison, N. E., Bosselman, R. A., Kaulenas, M. S. 1972. *Biochem. Biophys. Res. Commun.* 49:171
251. Ostner, U., Hultin, T. 1968. *Biochim. Biophys. Acta* 154:376
252. Fritz, P. J., Vesell, E. S., White, E. L., Pruitt, K. M. 1969. *Proc. Nat. Acad. Sci. USA* 62:558–65
253. Millar, D. B., Summers, M. R., Niziolek, J. A. 1971. *Nature New Biol.* 230:117
254. Lebherz, H. G., Savage, B., Abacherli, E. 1973. *Nature New Biol.* 245:271
255. DeDuve, C., Wattiaux, R. 1966. *Ann. Rev. Physiol.* 28:435
256. Fritz, P. J., White, E. L., Vesell, E. S., Pruitt, K. M. 1971. *Nature New Biol.* 230:119
257. Nath, K., Koch, K. L. 1971. *J. Biol. Chem.* 246:6956–67
258. Mortimore, G. E., Neely, A. N., Cox, J. R., Guinivan, R. A. 1973. *Biochem. Biophys. Res. Commun.* 54:189
259. Morgan, H. E. Personal communication

ELECTRON TRANSPORT PHOSPHORYLATION

✷ 868

Herrick Baltscheffsky[1] and Margareta Baltscheffsky[1]

Bioenergetics Group, Departments of Biochemistry and Plant Physiology
University of Stockholm, Stockholm, Sweden

CONTENTS

INTRODUCTION

Recent developments within the complex field of biological electron transport phosphorylation seem to epitomize certain concepts linked to the general course of scientific progress. According to one such concept, the steady accumulation of experimental data tends to lift successively various areas of knowledge to new and increasingly high levels of perception, with intervals of chaos occurring after the breakdown of an old pattern and before the establishment of a new. We will not try to suggest the appropriate place for electron transport phosphorylation in this

---

[1] Present address: Department of Biochemistry, Arrhenius Laboratory, University of Stockholm, Fack, S-104 05 Stockholm, Sweden.

scheme of things but rather leave it to the choice of the reader (a scientist, after all, is an optimist). Another well-known concept is the Newtonian "Natura enim simplex est" (1), which was applied more than two decades ago to energy conversion reactions of the living cell by Krebs (2). Current progress in electron transport and energy conservation appears to show both an increasing degree of complexity in experimentation as well as in interpretation, and an emergence, still only dimly seen, of new understanding at the fundamental level of molecular structure.

It is the aim of this article to describe selected basic properties and reactions of electron transport phosphorylation systems from bacteria, plants, and animals. Most new information comes from the electron transport phosphorylation systems involved in oxidative phosphorylation by animal mitochondria and in photophosphorylation by chromatophores from photosynthetic bacteria or chloroplasts from green plants. Thus, data is presented from any of these or related systems, which are all embedded in their respective polyphasic subcellular structures. In line with this approach, particular attention is focused on new structural information about some of the proteins involved. As the picture currently is developing, the following question may well be of more than passing interest: will increasing knowledge about the evolution of biological electron transport phosphorylation from simpler to more complex systems contribute to a better understanding of the detailed mechanisms involved in this energy transduction process?

Numerous useful review articles, symposia volumes, and books covering the whole field or selected parts have appeared in recent years. There are some very pertinent ones from recent issues of the *Annual Review of Biochemistry* (3–8) as well as some particularly important ones covering the area of interest in a broad and fundamental manner (9–27). It is hoped that these references provide an acceptable excuse for the fact that we have seen it necessary to treat some areas in detail at the expense of other equally significant parts of the field.

IRON-SULFUR CENTERS (COUPLING SITE I)

An impressive proliferation of iron-sulfur centers in schemes of mitochondrial respiration is surely one of the new features in biological electron transport. Thus, with submitochondrial particles from beef heart and with NADH-ubiquinone reductase [complex I in the generally accepted terminology of Hatefi (28)] Orme-Johnson et al (29) were able to demonstrate by electron paramagnetic resonance (EPR) spectroscopy the existence of four iron-sulfur centers, all associated with complex I. Ohnishi et al (30) had earlier obtained similar data with submitochondrial particles from yeast, indicating at least two iron-sulfur centers in the NADH dehydrogenase region of the respiratory chain. This approach paved the ground for closer examination of electron transport at coupling site I, where increasingly strong evidence indicated that iron-sulfur protein was involved (31–33).

The four iron-sulfur centers of the NADH-ubiquinone reductase were termed centers 1, 2, 3, and 4, based on their EPR-derivative peak positions (29). In both beef heart and yeast particles, center 1 has the lowest oxidation-reduction potential and center 2 the highest (29, 34). In yeast mitochondria and submitochondrial particles a potential gap of 190 mV existed between center 1 and center 2 (34). Gutman et al (35), using beef heart inner mitochondrial membrane preparations, found that center 1 was reoxidized by the respiratory chain when NADH was exhausted but that center 2 remained reduced. As addition of ATP reoxidized center 2, they suggested that coupling site I is located on the substrate side of center 2 and on the O_2 side of center 1. Ohnishi and her co-workers (36, 37) demonstrated that phosphorylation at coupling site I, energy-dependent reduction of NAD, piericidin A sensitivity, and EPR-detectable iron-sulfur proteins are simultaneously induced by aeration of *Candida utilis* cells grown under iron stress, and they concluded that EPR-detectable iron-sulfur proteins may play a role in energy conservation at coupling site I. Of particular interest was the finding that the half-reduction (midpoint) potential of what they termed iron-sulfur center 1a in intact pigeon heart mitochondria appeared more negative upon addition of ATP, whereas centers 1b, 2, 3, and 4 were not significantly affected (38). Oligomycin and uncoupler prevented (or reversed, if added subsequently) the effect of ATP, which influenced only about half the center 1 signal; thus the designation center 1a. The remaining 50% of center 1 not affected by ATP was designated center 1b. On the basis of indications that center 1a had a half-reduction potential dependent on the phosphate potential, Ohnishi et al (38) suggested that iron-sulfur center 1a is involved in energy transduction of coupling site I.

Further detailed resolution of iron-sulfur center in the coupling site I region of the respiratory chain by Ohnishi & Pring (39) has revealed the apparent presence of at least two additional centers in pigeon heart mitochondria, as evaluated from new EPR signals detected when the temperature of liquid helium (5°K) was approached. Designating iron-sulfur centers associated with the NADH dehydrogenase by the prefix N-, there were now the following seven iron-sulfur centers: N-1a, N-1b, N-2, N-3 and 4, and N-5 and 6 (N-5 and 6 being the two newly detected centers, which like N-3 and 4 had identical $E_{m7.2}$ or half-reduction potentials at pH 7.2). By redox titration according to the potentiometric procedure of Wilson et al (40) and Dutton (41) and by using a computer program originally devised to resolve titrations of cytochromes b_T and b_K (42), Ohnishi & Pring were able to obtain half-reduction potential values of the seven suggested iron-sulfur centers as given in Table 1 (39). Furthermore, they were able to show that center N-2 appeared to shift the half-reduction potential upon addition of ATP in appropriately

Table 1 Half-reduction potential ($E_{m7.2}$) values of iron-sulfur centers in the coupling site I region of pigeon heart submitochondrial particles. From Ohnishi & Pring (39).

Iron-Sulfur Centers	$E_{m7.2}$	Effect of ATP on $E_{m7.2}$
N-1a (−380 mV)	−380 ± 10	more negative
N-1b (−250 mV)	−250 ± 10	none
N-2	−30 ± 15	more positive
N-3.4	−245 ± 10	none
N-5.6	−270 ± 10	none

equilibrated pigeon heart mitochondrial suspensions. Whereas the $E_{m7.2}$ of N-1a appeared to become more negative when ATP was added, that of N-2 became more positive. These two iron-sulfur centers may be involved in the energy-transducing reaction at coupling site I.

The iron-sulfur centers N-1a and N-2 at coupling site I in the mitochondrial respiratory chain thus appear to react in close analogy to certain cytochromes at coupling sites II and III. A discussion of electron transport at these latter sites is a necessary prerequisite to a consideration of apparent molecular responses during electron transport to pressure from energy conversion as well as the possible significance of such responses for the coupling mechanism in electron transport phosphorylation.

B-TYPE CYTOCHROMES (COUPLING SITE II)

In the last few years much attention has been devoted to elucidating the role of b-type cytochromes in electron transport and energy transduction at coupling site II. A wealth of information, not always unambiguous, has emerged and given new insight into the functions of these cytochromes. Excellent and extensive review articles on the subject of mitochondrial cytochrome b have recently been published by Wilson et al (43), Erecińska et al (44), and Wikström (45).

Nomenclature

The recent discovery of more than one species of cytochrome b in animal mito-chondria has caused some confusion with regard to the identities of the b-type cytochromes described in the literature. It is established that at least two, and possibly three, main different cytochromes b exist. The terms b_K and b_T have been designated by Chance et al (46) for the classical cytochrome b first discovered by Keilin with a difference absorbance maximum at 562 nm (b_K), and the energy transducing cytochrome b with a difference absorbance maximum at 565 nm (b_T) and a 558 nm shoulder, respectively. A third species of cytochrome b with a difference absorbance maximum at 558 nm has been proposed but it is still far from clear whether this is a separate cytochrome (47, 48) and not the 558 shoulder of cyto-chrome b_{565} (b_T). With the development of the potentiometric titration technique (41, 49) it has been possible to differentiate cytochromes b with respect to their potential. It has been found by Wilson et al (50) that in uncoupled preparations of rat liver and pigeon heart mitochondria, cytochrome b_{562} (b_K) has an $E_m = 40$ mV and cytochrome b_{565} (b_T) an $E_m = -30$ mV measured at pH 7.2. Two or more species of b-type cytochromes have been found in a large variety of mitochondrial sources, such as yeast mitochondria (50, 51), mitochondria from Ehrlich ascites tumor cells (52), and mitochondria from *Neurospora crassa* (53). Usually these b-type cyto-chromes have been termed b_T and b_K respectively, on the basis of their reactivity and absorption maxima.

In the following we assume that two, not three, different b-type cytochromes participate in animal mitochondrial electron transport and phosphorylation. We also use the terms b_{565} and b_{562} for b_T and b_K, respectively.

In plant mitochondria (mung bean and skunk cabbage) there are three different cytochromes b with different midpoint potentials. Their difference absorbance maxima at 77°K are 553, 557, and 562 nm (54). At room temperature the maxima are 556, 560, and 565 nm, respectively. Their midpoint potentials have been determined by Dutton & Storey (55) and at pH 7.2 are as follows: b_{556}, 75 mV; b_{560}, 42 mV; and b_{565}, -75 mV.

Reactions

MITOCHONDRIAL CYTOCHROMES b A major breakthrough to elucidating the possible role of cytochromes b in the energy transduction process came in 1970 with the finding of Wilson & Dutton (56) that in pigeon heart mitochondria there are two distinct species of cytochromes b with different midpoint potentials, and one of these

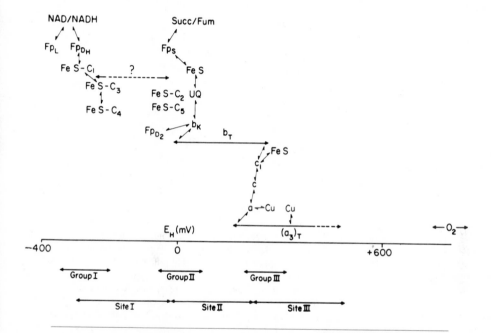

Figure 1   Scheme proposed by Chance for the mitochondrial electron transport. The electron carriers of the respiratory chain arranged as groups of fixed potential (Groups I, II, and III) and individual components of variable midpotentials: cytochromes b_T and $(a_3)_T$ for Sites II and III and (?) for Site I. The components on the substrate side of Site I are F_{P_L}, the highly fluorescent lipoate dehydrogenase flavoprotein; F_{PDH}, the NADH dehydrogenase flavoprotein, and iron-sulfur proteins here given as C-1, and C-4. On the oxygen side of Site I are F_{P_S}, the succinate dehydrogenase flavoprotein, with associated iron-sulfur proteins (Fe-S); UQ, ubiquinone; F_{PD2}, the fluorescent flavoprotein; C-2 iron-sulfur protein; cytochromes b_K and b_T. On the oxygen side of Site II are the four cytochromes, c_1, c, a, and a_3, with associated copper. [By courtesy of Dr. B. Chance (24).]

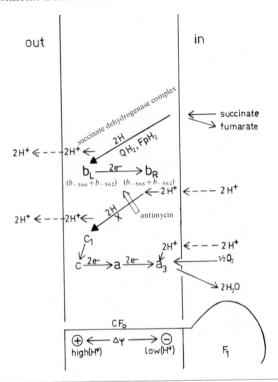

Figure 2   Scheme proposed by Wikström for the arrangement of the *b* cytochromes in the mitochondrial membrane. The model proposes that the two groups of cytochrome *b* are localized near each interphase of the mitochondrial membrane (b_L and b_R; L = left, R = right) in such a way that forward electron transfer between the groups (from left to right) is opposed by the build-up of an electrostatical potential difference between the two regions of the membrane occupied by b_L and b_R, respectively.

Arrows with solid pointers represent hydrogen transfer reactions. Arrows with thin pointers indicate flow of electrons, protons or substrates. The oligomycin-sensitive ATPase complex (CF_0-F_1) is indicated in the lower part of the figure with symbols showing the electrical and chemical components of an electrochemical potential difference of protons between the outer and inner portions of the membrane. [By courtesy of Dr. M. K. F. Wikström (45).]

species changes its apparent midpoint potential from -30 mV to $+240$ mV in the presence of ATP. This dependence on the phosphate potential $[\text{ATP}]/[\text{ADP}][\text{P}_i]$ of the half-reduction potential of cytochrome b_{566} shows a linear relationship between the logarithm of the phosphate potential and the experimentally measured half-reduction potential (43). Essentially equilibrium conditions are claimed to prevail (43).

Slater et al (57) found that addition of ATP or antimycin to anaerobic mitochondria further reduced cytochrome b with an absorption maximum at 565 nm.
In the mid-1960s Chance and co-workers (58–60) found a high-energy form of
cytochrome b with an absorbance maximum at 555 nm (77°K), and this could now
be correlated with the high potential form of cytochrome b_{565}, which has a shoulder
at 555 nm at low temperature (558 nm at room temperature). Earlier work by Tyler
et al (61) had also indicated the existence of multiple forms of cytochrome b in
submitochondrial particles, with a possible role in energy transduction for the
somehow energy-linked form.

Cytochrome b_{565} was subsequently found not to be completely reducible by
substrate (succinate) or on anaerobiosis, but to be rapidly reduced in anaerobic
mitochondria or submitochondrial particles upon addition of ATP (47). Chance et
al (46) named it cytochrome b_T and proposed the direct participation of this cytochrome in the energy transduction at site II via formation of $b_T \sim$ I, the high-energy
high-potential form of b_{565}. In the same year, Slater et al (57, 62) proposed a theory
for energy transduction at site II which involved the high-energy form of both high
and low potential cytochromes b as a dimer.

Wikström (47) has suggested, on the basis of the relative contribution of the two
different cytochromes b to the absorption spectrum, that both b_{562} and b_{565} show
rises in their apparent midpoint potentials in the presence of ATP. He proposed
that the term b_T thus involved both cytochromes b_{562} and b_{565}. This suggestion has
been further supported by Berden et al (63), who also reported that in freshly
prepared mitochondria (beef heart) cytochrome c shows lowered apparent midpoint
potential upon addition of ATP. This effect is lost in aged mitochondria. However,
Dutton et al (64) found no effect of ATP on the potential of cytochrome c in the same
material.

In contrast to animal mitochondria, plant mitochondria show no increase in the
midpoint potential of cytochromes b under the influence of ATP, as shown by
Dutton & Storey (55) and emphasized by Bonner et al (65). Likewise, Sato et al (51)
were unable to demonstrate an ATP-induced change in the midpoint potential of
any of the b-type cytochromes in mitochondria from $C.$ $utilis.$

CYTOCHROMES b IN CHLOROPLASTS AND CHROMATOPHORES   In plant chloroplasts
two cytochromes b have long been identified. They are named after their room
temperature spectral properties as cytochrome b_{559} and cytochrome b_{563}. Cytochrome b_{563} was originally found and named cytochrome b_6 by Hill (66).
Cytochrome b_{559} has been shown to participate exclusively in noncyclic electron
transport, whereas cytochrome b_{563} has been assigned to the cyclic system. Both
cytochromes have recently been shown to have potentials dependent on the
energetic state of the chloroplast membrane, so that when decoupled both by aging
and by uncouplers, the apparent midpoint potential is substantially lowered.
Böhme & Cramer (67), using the potentiometric titration technique, have found that
the cyclic cytochrome b_{563} in uncoupled chloroplasts has an apparent midpoint
potential 100–150 mV lower than in well-coupled chloroplasts. Likewise, the noncyclic cytochrome b_{559} (68) shows a lowered midpoint potential under uncoupled

conditions compared to that in coupled chloroplasts. Antimycin and hydroxylamine cause a lowering of the midpoint potential of cytochrome b_{559}. In neither case has it been possible to find any connection between energy transduction and the change in redox potential, apart from the fact that uncoupling causes the transition from high to low potential.

The role of cytochrome b_{559} in coupled ATP formation is undecided. Ben-Hayyim & Avron (69) found that addition of $ADP + P_i$ as well as certain uncouplers caused reduction of this cytochrome, placing a rate-limiting step in electron transfer between photosystem II and cytochrome b_{559}. On the contrary, Böhme & Cramer (70), when localizing a coupling site for ATP synthesis between plastoquinone and cytochrome f using the crossover approach, found no redox changes associated with cytochrome b_{559} under conditions where the two abovementioned electron carriers readily responded with such redox change. Thus evidence seems to exist for two sites of phosphorylation between system II and cytochrome f. If this is so, the failure of Böhme & Cramer to observe any steady-state redox change of cytochrome b_{559} could possibly be explained by concomitant release of electron transport on both sides of this cytochrome at both rate-limiting steps.

At 77°K, where both water splitting and electron transfer between the two photosystems are inhibited, the high potential form of cytochrome b_{559} is photo-oxidized by system II (71, 72). At room temperature the cytochrome is reduced by system II and oxidized by system I (69, 73). Present ideas about the function of cytochrome b_{559} (68, 74) tend to place it in its high potential form in a separate pathway at system II, giving it a regulatory role with a pool function. Only the low potential form would participate in the electron transport between the two photosystems. The high potential form of cytochrome b_{559} has also been suggested as directly involved in the electron transport between system II and oxygen evolution (71).

In coupled *Rhodospirillum rubrum* chromatophores, the cytochrome b that participates in cyclic electron transport on the reducing side of a coupling site does not change its midpoint potential to any appreciable extent in the presence of ATP (75), nor does addition of an uncoupler alter its potential (M. Baltscheffsky, un-published observation).

MIDPOINT POTENTIAL CHANGE OR REVERSED FLOW   The lack of ATP-induced change in the midpoint potential of cytochromes b which otherwise show reactivities assumed to be associated with energy transduction brings up the recurring question of whether the increase in the apparent midpoint potential observed in animal mitochondria is due to ATP-induced reversed electron flow, as proposed by Caswell (76), by Wikström (77, 78), and by Bonner et al (65). This could be the case if the redox-mediating system does not directly equilibrate with all forms of cyto-chrome b, but rather through the whole coupling reaction. It is difficult with the present experimental evidence to rule out this possibility. However, the potential changes found in cytochromes b_{559} and b_{563} in plant chloroplasts occur when the system becomes uncoupled and cannot be easily explained by the reversed electron flow alternative.

THE AEROBIC REDUCTION OF CYTOCHROME b A rather anomalous behavior of cytochrome b has been reported repeatedly in the literature over the last twenty years (79–82), but only with recent development and insight in the reactions of cytochrome b has attention been sharply focused on this problem. Introduction of O_2 (78, 79, 83) or ferricyanide (81–83) to an antimycin-inhibited electron transport system results in a rapid reduction of cytochrome b. This reduction appears simultaneously with oxidation of cytochrome c_1 (84) and also concomitant with the oxidation of ubiquinone, which very likely serves as the reductant for cytochrome b, as first proposed by Baum et al (85). With a functioning cytochrome oxidase, O_2 can serve as the oxidant. If the oxidase system is inhibited or impaired, O_2 cannot induce the reduction of cytochrome b (83), but instead it can be induced by ferricyanide, as is also the case in complex III from mitochondria (50, 82, 83, 85). The cytochrome b reduced was identified as b_{566} (42).

Baum & Rieske (81) as well as Wilson et al (83, 86) have shown that oxidation of cytochrome c_1 in complex III appears simultaneously with reduction of cytochrome b. Baum et al (85) suggested that transient oxidation of complex III caused a conformation change which affected the reactivity of cytochrome b so that it became reducible by ubiquinone. Rieske (82) reported that antimycin induced a positive change in the apparent midpoint potential of cytochrome b, using varied mixtures of Q_1H_2 (ubiquinol-5) and Q_1 as the redox couple. He also suggested that the oxidation state of an unknown compound X controls the midpoint potential of cytochrome b. The high potential form of cytochrome b would require this component in the oxidized state. Dutton et al (87), using the anaerobic potentiometric titration technique, found no increase in the midpoint potential of cytochromes b in the presence of antimycin.

The requirement for antimycin lies apparently in a necessary restriction of electron flow. Erecińska et al (84) have shown that lowering the temperature to 0°C has the same effect as antimycin, and at the same low temperature they have been able to resolve kinetically the sequence of events leading to the cytochrome b reduction. The first reaction that occurs is the oxidation of cytochrome c_1; only following this does the reduction of cytochrome b_{566} occur.

Conflicting with these results are those of Wikström & Berden (88) who have found that the O_2 + antimycin-enhanced reduction of cytochrome b_{566} is stable for some time in the presence of reduced cytochrome c_1 in beef heart mitochondria. They have concluded that the oxidation of cytochrome c_1 is not a prerequisite for the complete reduction of cytochrome b_{566}, and have strongly emphasized that cytochrome b_{562} also undergoes an increased reduction in the presence of antimycin + oxidant. They suggest that the component that controls the redox state of cytochromes b may be ubiquinone (UQ) as the semiquinone/quinone redox couple.

Control exerted by ubiquinone of the b-c_1 complex has been suggested earlier by Ernster et al (89), who compared the effect of antimycin on the interaction of succinate and NADH dehydrogenases with cytochrome b in normal and UQ-depleted submitochondrial particles.

Results with plant mitochondria are very similar to those in beef heart mitochondria insofar as the reactions of cytochrome b_{566} are concerned. Storey (90) and Lambowitz & Bonner (91) have shown that in antimycin-treated mung bean mitochondria, cytochrome b_{566} becomes oxidized upon attainment of anaerobiosis. According to Storey this oxidation is concomitant with reduction of cytochrome c_{551}. Lambowitz & Bonner, on the other hand, find that cytochrome b_{566} undergoes a biphasic oxidation with cytochrome c_1 reduction. Furthermore, they ascribe this fast oxidation at least partly to spectral interference from cytochrome c_1. In view of the finding of Wikström & Berden (88) that both cytochromes b in beef heart mitochondria participate in these types of reactions, it also seems possible that the biphasicity can be caused by different kinetics of two different cytochromes. An oxidant-induced reduction of cytochrome b in R. rubrum chromatophores has also been observed (M. Baltscheffsky, unpublished observations). Here the reaction appears to be strictly dependent on the redox potential of the system and occurs only within the span of 80–160 mV.

A recent investigation by Chance (92) concerns the kinetic and stoichiometric relationships between cytochromes b (565 and 562), cytochrome c_1, and ubiquinone.

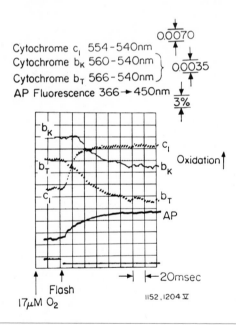

Figure 3 Comparison of the kinetics of cytochromes c_1, b_{565} (b_T) and b_{562} (b_K), and ubiquinone oxidation (here measured as fluorescence quenching of the added probe, anthracene palmitic acid), following an oxygen pulse and flash photolysis of a_3:CO in a suspension of pigeon heart mitochondria. 0.3 mM succinate, 13 mM malonate, 3 μM rotenone, 17 μM pentachlorophenol, 0.7 μM antimycin/1 mg protein, 400 μM CO, and in the lower trace 13 μM anthracene palmitic acid. [By courtesy of Dr. B. Chance (92).]

The results indicate that although both cytochromes b participate in the reduction coupled to oxidation of cytochrome c_1, the close time correlation between c_1 oxidation and b_{565} reduction is not found for b_{562} whose reduction lags behind the reduction of b_{565}. The oxidation of c_1 kinetically precedes the reduction of both cytochromes b. The extent of reduction of the b-type cytochromes after an oxygen pulse depends on their oxidation level before the pulse.

A close kinetic as well as stoichiometric correlation is found between UQ oxidation and the reduction of both cytochromes b.

Chance (93) ascribes the effect of action of cytochrome c_1 to a heme-heme interaction between this cytochrome and cytochrome b_{565}, whereby the redox state of cytochrome c_1 would control the midpoint potential of cytochrome b_{565}. Wilson et al (86) propose that electron flow through coupling site II, activated by the oxidation of cytochrome c, raises the half-reduction potential of b_{565}.

It is evident that the cytochromes b and their reactions in energy-transducing membranes share the fate of most other reactions in these structures of being highly controversial. Possibly the approach of comparative biochemistry, looking for a system as simple as possible, may be rewarding in future attempts to obtain final answers to questions concerning the function of b-type cytochromes in cellular electron transport and energy coupling.

CYTOCHROME OXIDASE (COUPLING SITE III)

The terminal reaction in respiratory electron transport is the reduction of molecular O_2 to water. This reaction is carried out by the cytochrome oxidase complex. In its isolated form this protein complex contains two hemes, a and a_3, two atoms of Cu, and varying amounts of phospholipid (94–96), part of which appears to be essential for the full enzymic activity (97).

Wikström & Saris (98) found that addition of ATP to rat liver mitochondria caused an energy-dependent spectral shift in the absorption spectrum of cytochrome a_3. Subsequently, Wilson and co-workers (99, 100) discovered that the phosphate potential, when regulated by addition of ATP, influences the half-reduction potential of cytochrome a_3, lowering it from 375 to 155 mV, which led to the suggestion that energy conversion takes place either between cytochromes a and a_3 or between cytochrome a_3 and oxygen, in both cases involving the heme of this cytochrome directly in the energy-coupling mechanism.

Thermodynamic considerations as well as experimental evidence had earlier led Slater and co-workers to favor the placement of a phosphorylation site between cytochrome a_3 and oxygen (101, 102).

The influence of the phosphate potential on cytochrome oxidase was further substantiated by the finding that addition of ATP to pigeon heart mitochondria induced a transition from high to low spin state in a heme of oxidized cytochrome oxidase (103, 104), as judged from the analogy with cyanide-induced spectral changes. Addition of cyanide to oxidized cytochrome oxidase induces spectral changes compatible with a high to low spin transition in a heme both in the isolated enzyme (105) and in intact pigeon heart mitochondria (106). The effect of ATP on

cytochrome a_3 also results in a slight rise in the half-reduction potential of cytochrome a as well as an increase in the extinction coefficient and a shift in the
absorption spectrum of this cytochrome (100).

Wikström (107) found that in intact mitochondria the transition from aerobic to
anaerobic conditions resulted in the reduction of cytochrome a_3 before cytochrome
a. In the presence of uncoupler or oligomycin (98), the time separation between the
reduction of the two hemes could not be seen. He also found that the addition of
ATP to anaerobic mitochondria results in an oxidation as well as what was interpreted as a red shift of the cytochrome a absorption, possibly due to a conformational
change involving the heme iron. Lindsay & Wilson (100) reported a red shift in the
cytochrome a_3 absorption and a blue shift in cytochrome a absorption upon ATP
addition to anaerobic mitochondria. They suggested that these changes were due to
an apparent ligand binding of ATP to the a_3 heme, which in turn modified the
spectrum of the a heme through heme-heme interaction. Further investigations are
needed to resolve these differences.

Wilson & Brocklehurst (108) recently found that the effect of ATP on the spectral
change of cytochrome a_3 was stoichiometric, so that one molecule of ATP was
hydrolyzed for each cytochrome a_3 modified. They proposed a ligand exchange
mechanism for the energy transduction in the oxidase moiety in situ.

A lowering of the midpoint potential of cytochrome a_3 has been suggested by
Chance (109) to occur as an early event in the reaction between cytochrome oxidase
and oxygen. Low-temperature studies (110) with intact mitochondria suggest that
the first product of the reaction between the oxidase and oxygen has a 428 nm
absorption band and is $a_3^{2+} \cdot O_2$, an unstable intermediate in the oxidase reaction.

Two recent reports claim that reconstituted cytochrome c-cytochrome oxidase
complexes perform electron transport-linked formation of ATP. Yong & King (111)
reported that the reconstituted complex as such exhibited both respiratory control
and ATP formation as well as oligomycin sensitivity with ascorbate as electron
donor. Total phosphorylation activity was very low, however, compared to total
potential cytochrome oxidase activity. Racker & Kandrach (112) showed that
reconstitution of cytochrome oxidase with cytochrome c in phospholipid vesicles
combined with a crude hydrophobic coupling factor protein fraction from mitochondria gave a preparation capable of ATP formation. The P/O values reported
were as high as 0.4–0.6 in both cases. The most important difference appears to be
the requirement for vesicles and added coupling factor in the latter system. Results
of this nature show great promise for closer insight into the mechanism of electron
transport phosphorylation coupled to a specific site in the respiratory chain.

FUNCTIONAL COUPLING FACTORS

Structures and Subunit Structures

Mitochondrial coupling factors have been thoroughly discussed in a number of
recent review articles (3, 113–115). Their purification and properties have been
described in great detail by Beechey & Cattell (115), who compiled various data
on mitochondrial coupling factors with ATPase activity, or "functional coupling

factors," and listed suggestions concerning their subunit structure, showing from three to ten subunits proposed in reports from various laboratories (115–117). In the last few years a wealth of new information has been forthcoming, also from studies with functional coupling factors from higher plant chloroplasts and membrane ATPases from bacterial sources. For the sake of simplicity we use the rather generally accepted term "coupling factors" for membrane ATPases isolated from animals, plants, and bacteria, where they appear to function in electron transport-coupled energy transduction reactions.

Recent work with the coupling factor F_1 from rat liver (117, 118) and bovine heart (119–122) mitochondria and the corresponding coupling factor CF_1 from spinach chloroplasts (123, 124) has demonstrated a remarkable similarity with respect to molecular weight and subunit size between the proteins from these different sources (Table 2). Available information about bacterial coupling factors and their subunits indicates that fundamental structural properties of coupling factors with ATPase activity existing already at the prokaryotic level of cellular organization appear to have been retained during the evolution of higher plants and animals (125–128).

Thus, both in *Micrococcus lysodeikticus* (125) and *Bacillus megaterium* (126), membrane ATPases solubilized and analyzed for subunit structures showed two major polypeptides with molecular weights from 60,000 to 68,000. With a *Micrococcus* preparation (125), one to three additional, minor bands were seen in the polyacrylamide SDS-gel when a "shock wash" method rather than an "*n*-butanol type" method was used for the solubilization of the coupling factor. In addition, the earlier work of Schnebli et al (127) with *Streptococcus faecalis* ATPase also indicated that this protein consisted of different subunits, and Hanson & Kennedy (128) recently found that the membrane ATPase from *Escherichia coli,* which appears to function in both oxidative phosphorylation and active transport, can dissociate into four subunits of different size on electrophoresis of the purified protein in gels containing SDS. The molecular weight of the *E. coli* coupling factor protein was about 360,000. For the subunits the values obtained were 60,000, 56,000,

Table 2 Subunit sizes of F_1 and CF_1

Source	Coupling Factor	Molecular Weight	Method for Subunit Isolation	Subunits (molecular weights)				
				(1)(A)(α)	(2)(B)(β)	(3)(C)(γ)	(4)(D)(δ)	(5)(E)(ε)
				1	2	3	4	5
Beef heart mitochondria (122)	F_1	347,000	SDS gels, gel filtration, amino acid analysis	59,400 53,200 53,300	54,400 50,800 49,000	33,000 33,000 33,160	13,600 17,300 16,100	5570 5850
Beef heart mitochondria (119)	F_1	360,000	SDS gels	53,000	50,000	25,000	12,500	7500
Rat liver mitochondria (118)	F_1	384,000	SDS gels	A 62,500	B 57,000	C 36,000	D 12,000	E 7500
Spinach chloroplasts (124)	CF_1	325,000	SDS gels	α 59,000	β 56,000	γ 37,000	δ 17,500	ε 13,000

35,000, and 13,000. Bragg et al (129) very recently confirmed preliminary data (130) showing five subunits in the E. coli ATPase, with mol wt 56,800, 51,800, 30,500, 21,000, and 11,500. This shows some resemblance to the molecular weights of the five subunits that Tzagoloff & Meagher (131) obtained from Saccharomyces cerevisiae coupling factor F_1: 58,500, 54,000, 38,500, 31,000, and 12,000. From this eukaryotic organism, at an intermediate evolutionary level between bacteria and higher organisms, a rutamycin-sensitive complex was also isolated (131), which in addition to the five subunits given above contained subunits of the molecular weights 29,000, 22,000, and 18,500. The smallest of these polypeptides corresponds to the oligomycin-sensitivity-conferring protein (OSCP), first isolated and purified from beef heart submitochondrial particles (132, 133) and subsequently purified from the yeast complex (134). OSCP is required for reconstitution of oligomycin- and rutamycin-sensitive ATPase and oxidative phosphorylation activities to depleted membranes.

Another protein of low molecular weight that has been isolated from animal mitochondria and extensively purified is the ATPase inhibitor discovered by Pullman & Monroy (135). It is a trypsin-sensitive protein with mol wt 10,500 (119, 136) extensively purified by Brooks & Senior (137), who showed that its presence in or absence from mitochondrial coupling factors was entirely dependent on the technique of purification. The protein may be considered a sixth component of some oligomycin-insensitive coupling factor F_1 preparations (compare Table 2). The ATPase inhibitor has been assumed to modify F_1 such that only the forward reaction with ADP can proceed while the reverse reactions with ATP are inhibited (135, 136). According to Asami et al (138) the inhibitor may have an important physiological role as a unidirectional regulator of electron transport-linked energy transfer, a suggestion supported by recent evidence that such a regulation may take the form of a control of F_1 turnover capacity (139).

This ATPase inhibitor protein from mitochondria has recently found its functional counterpart from the spinach chloroplast coupling factor CF_1, from which Nelson et al (123), using 7 M urea in Tris buffer, were able to detach a corresponding inhibitor. It was found to be the smallest of the five CF_1 subunits (the ε chain of Table 2, mol wt = 13,000) after purification by extraction with pyridine, precipitation with ethanol, and elution from DEAE-cellulose column. The purified CF_1 inhibitor was insoluble in water but soluble in the Tris-urea buffer, in which it could be stored at $-70°C$ for several weeks without loss of activity.

The molecular properties of CF_1 inhibitor, compared to those of mitochondrial F_1 inhibitor, show differences and similarities. A pronounced difference concerns solubility properties. Only the mitochondrial F_1 inhibitor is water soluble (135, 136), whereas the CF_1 inhibitor requires the presence of urea or detergents for solubility. Its great hydrophobicity may be responsible for its comparatively strong affinity to coupling factor CF_1 and also for the fact that it was not discovered previously. Activation of the latent ATPase by heat treatment in the presence of 1% digitonin suggests that the CF_1 inhibitor had either been destroyed or removed from the active site of the enzyme (123). Dithiothreitol (DTT) does not appear to activate the ATPase by removal of the CF_1 inhibitor, since removal of DTT brings back the latency of the ATPase activity. Rather, DTT may prevent the interaction between the inhibitor and the active site involved in the ATPase activity of CF_1. This interpretation is supported by the observation that inhibition of heat-activated CF_1 by added inhibitor is counteracted by DTT (123). Of the antisera against the individual subunits of CF_1, anti-α, anti-β, anti-γ, anti-δ, and anti-ε, only the antisera against the γ and ε subunits effectively interfered with the inhibitory effect of the CF_1 inhibitor on the ATPase activity (123, 124).

As is seen in Table 2, the apparent evolutionary conservation of coupling factor structure is particularly pronounced in the three largest subunits (A, B, and C) (1, 2, and 3) (α, β, and γ). The evolutionary argument is strengthened further by the great similary that exists between the sizes and substructure sizes of these eukaryotic, animal and plant, coupling factors and the abovementioned E. coli coupling factor. In addition, a coupling factor of molecular weight 350,000 (140) and containing five subunits (B. C. Johansson, unpublished) is obtained from chromatophore membranes of the photosynthetic bacterium R. rubrum.

When Catterall & Pedersen (117) in 1971 reported the amino acid composition of coupling factor F_1 from rat liver mitochondria, they drew attention to the similarities with respect to this parameter between membrane ATPases from phylogenetically different sources and proposed the following hypothesis: a central, still unknown mechanism of energy coupling and transduction is operating in mitochondrial oxidative phosphorylation, photophosphorylation in chloroplasts, and ATP-driven translocation of cations, phosphate, and some amino acids in bacterial membranes.

The subsequent rapid development in this area supports and extends this view. Today the amino acid compositions of several purified subunits from various eukaryotic sources have been determined (120, 122–124, 137). However, evaluation of the final significance of apparent similarities between different substructures from a coupling factor, and corresponding substructures from different sources will have to await determinations of primary, secondary, and tertiary structures of the proteins involved.

Nucleotide Binding Properties

In both coupling factor F_1 from mitochondria and coupling factor CF_1 from chloroplasts the binding of adenine nucleotides and related compounds has been investigated with increasing intensity since Zalkin et al (141) studied the formation and properties of a complex between F_1 and ADP. This nucleotide had earlier been found to be a strong specific inhibitor of ATP and ITP hydrolysis catalyzed by purified F_1 (142). To this date, nucleotide binding has not been reported below the structural level of intact coupling factors, but the availability of isolated and highly purified subunits will no doubt stimulate the search for a specific subunit with nucleotide binding capacity.

F_1 FROM MITOCHONDRIA   In mitochondrial coupling factors originating from beef heart or rat liver, the existence of one (143) or two (144, 145) binding sites for ADP (and none for ATP or AMP) has been reported from several laboratories. Very recently the coupling factor F_1 isolated from beef heart was found by Harris et al

(146) to contain as many as 5 moles of tightly bound adenine nucleotide per mole of enzyme. In addition to 2 moles of ADP there remained 3 moles of ATP bound to the coupling factor enzyme even after repeated precipitation with $(NH_4)_2SO_4$.

The characteristics of nucleotide binding to F_1 are important, since they may reflect the molecular mechanism involved in the final, ATP-forming reaction of electron transport phosphorylation. The observed distinction between a tight and a loose binding site (145) in solubilized beef heart mitochondrial F_1 for ADP, in the presence of Mg^{2+} ion, with dissociation constants at pH 8.0 of 0.28 μM and 47 μM, respectively, is of great interest. The only site observed in F_1 from rat liver mitochondria (143) gave a dissociation constant of 0.94 μM and was thus more of the tight type. This raised the question whether a corresponding loose type of binding might be absent from the rat liver F_1 due to an inherent difference between the coupling factors from rat liver and beef heart, to the method of preparation, or to a combined effect of these two alternatives. A second, loose binding site for ADP has now been found also in rat liver mitochondria (P. L. Pedersen, personal communication).

Hilborn & Hammes (145) found that the tight coupling site of isolated beef heart F_1 may be missing or have a reduced binding affinity when the ATPase is attached to its native mitochondrial membrane. In earlier studies they had observed that the competitive inhibition of the ATPase reaction exerted by the product ADP is much stronger in the soluble than in the membrane-bound enzyme (147), with a K_i at pH 8.0 of 30 μM as determined with the soluble and 80 μM with the membrane-bound ATPase. The similarity between the K_i and the dissociation constant for ADP at the loose binding site suggests that this is the catalytic site for the ATPase reaction (145). This conclusion is supported also by the fact that the hydrolysis reaction is not markedly inhibited when the tight site is occupied by ADP or IDP (145, 147). The apparent existence of only one active site and two nucleotide binding sites per molecule would not, however, necessarily suggest an asymmetric subunit structure for F_1. Studies of the binding of the ADP fluorescent analog 3-β-D-ribofurano-sylimidazo(2,1-i)-purine 5'-diphosphate (ε-ADP) to the tight binding site of F_1 revealed an unusually low value for the rate constant (148). Tondre & Hammes (148) suggested that a conformational change may thus limit the rate of the binding process. They pointed out that the binding of ADP to CF_1 isolated from spinach chloroplasts had earlier been found by Nelson et al (149) to have a large effect on the rate of the ATPase reaction of CF_1 from these photosynthetic organelles.

CF$_1$ FROM CHLOROPLASTS   Roy & Moudrianakis (150, 151) showed in 1971 that CF_1 or the Ca^{2+}-dependent ATPase, which Vambutas & Racker (152) had isolated from spinach chloroplasts, has two sites for ADP binding. With two molecules of ADP it can carry out a myokinase-like transphosphorylation reaction by which the β-phosphate of one bound ADP molecule is transferred to that of the other to yield ATP and AMP. The products formed show such a high affinity for the solubilized enzyme that it must be destroyed by 10 M urea for their release and detection after the reaction. In contrast to added ATP, added AMP does not bind to CF_1, it is only as a product of the transphosphorylation reaction within the enzyme that

AMP is firmly bound. From labeling experiments with illuminated chloroplast membranes and subsequent isolation of the CF_1-nucleotide complex, it was concluded that enzyme-bound, labeled ADP is generated in light from AMP and P_i in the photosynthetic energy conversion apparatus (150). The constraints on the release of ATP and AMP by isolated CF_1 appeared to be relaxed in the membrane-bound state of the coupling factor (151). The results indicated that CF_1 can bind "two potentially unequal classes of ADP" (151). One was the enzyme-bound ADP in equilibrium with AMP, P_i, and the energy supply driving this photophosphorylation reaction. The second was the ADP acting as the terminal phosphate acceptor in the subsequent transphosphorylation step. It may be bound in a quite different location on the CF_1 placed in an appropriate position for the subsequent transphosphorylation reaction, which was assumed on the basis of the data given above to be the final step of light-induced formation of ATP in chloroplasts.

In a study of photophosphorylation and related hydrolytic and exchange reactions, Shahak et al (153) found great differences in the capability of certain fluorescent analogs of ADP and ATP to substitute as substrates for their nucleotide. It was shown that the fluorescent ATP derivative was a poor substitute for ATP in various ATPase reactions and the ATP-P_i exchange reaction, in spite of the fact that it apparently could bind to the proteins involved. On the other hand, the fluorescent ADP derivative replaced ADP quite well in photophosphorylation. It was concluded that at least two types of active sites may be involved and that the high specificity site may not directly participate in the process of ATP formation.

CF_1, when compared with F_1, has manifested unique properties with respect to the type of nucleotide binding leading to transphosphorylation, and thus to this pattern of electron transport-coupled formation of ATP, but has shown similarity with F_1 from beef heart in the apparent existence of two nonequivalent binding sites for ADP.

Metal Activation and Other Functional Properties

Whereas Mg^{2+} is known to be the divalent cation required for activation of the ATPase reaction in F_1 from mitochondria, Ca^{2+} serves for activation of CF_1 from chloroplasts. When the first coupling factor was isolated from chromatophores of photosynthetic bacteria by Baccarini-Melandri et al (154), Mg^{2+} ion was shown to be the activating metal ion. In contrast to this situation in the *Rhodopseudomonas capsulata* coupling factor, the corresponding solubilized coupling factor from *Rhodospirillum rubrum* chromatophores found by Johansson (155) required Ca^{2+} ion for ATPase activity (156).

Since Mg^{2+} is known to be the divalent cation in both oxidative phosphorylation and photophosphorylation, the activation of ATPase activity by Ca^{2+} in certain isolated coupling factors seems somewhat surprising. The fact that rebinding of coupling factor to chloroplasts and chromatophores results in both masking of the Ca ATPase activities and re-emergence of Mg ATPase activity (140, 149) indicates that the Ca ATPase, with respect to metal specificity, is either a distorted expression of the physiological activities of the coupling factors or a reflection of a regulatory capability of this cation.

This conclusion would seem to be supported by the demonstration by Nelson et al (149) that heat treatment of purified CF_1 (4 min at 64°C) in a suitable medium containing among other components 0.5–1% digitonin caused the appearance of a Mg^{2+}-dependent ATPase activity which, in the presence of sodium bicarbonate, exceeded 15 μmoles of ATP cleaved per mg of CF_1 protein per min. Several organic acids also strongly stimulated the activity. The demonstration of Mg ATPase activity in CF_1 has removed an apparent difference between mitochondrial F_1 and chloroplast CF_1.

As shown recently by Adolfsen & Moudrianakis (157) the ATPase activity of F_1 from beef heart mitochondria and from the similar coupling factor of oxidative phosphorylation from *Alcaligenes faecalis* was enhanced by both monovalent and divalent cations. With a low concentration of added ATP (5×10^{-5}–10^{-4} M), K^+ activated the ATPase activity much more strongly than Mg^{2+}. Ca^{2+} activated just as well as Mg^{2+} in the absence of K^+ but had no effect in its presence. These interesting observations indicate that much is yet to be learned about metal activation of coupling factor ATPase activity. Incisive studies in this area may contribute much to our knowledge about coupling factor function. As was pointed out by Adolfsen & Moudrianakis (157), the activation of ATPase activity by K^+ (and Na^+) poses the question of whether the enzyme is involved in active transport of monovalent cations.

The first direct evidence for a conformational change in a coupling factor protein was obtained by Ryrie & Jagendorf (158) in 1971. They found that in the light a small amount of tritium was incorporated from labeled water medium into the coupling factor CF_1 in spinach chloroplasts. This tritiation did not occur in darkness and was markedly inhibited by uncouplers. In an extended study (159), it was shown that uptake of tritium by the membrane-bound CF_1 also occurred during acid-base transition and light-induced ATPase activity. Apparently, induction of the energized state in chloroplasts leads to conformational change in the protein, parts of which become exposed and tritiated, and subsequently refolded into regions of the molecules inaccessible to solvent hydrogens. Several pieces of evidence support the concept that tritium is retained because of shielding effects: 1. disordering of the protein structure with urea causes rapid release of the radioactivity; 2. failure of tritium released by urea to absorb to Norite and its complete evaporation during lyophilization shows that the released tritium was in the form of ^3H_2O; 3. the exchange into CF_1 bound to illuminated chloroplast membranes is more than 1300-fold faster than the back-exchange out of solubilized CF_1; and 4. there is a rapid discharge of label during reillumination of chloroplasts in a nonradioactive medium (158, 159).

ADP and P_i limited the extent of the light-induced conformational change in CF_1 (159). The data obtained indicated that a common region of the protein must be exposed in both the presence and the absence of these agents and that a further region of the protein must be exposed only in the absence of ADP and P_i.

It was suggested that the conformational change of CF_1 may be involved in the phosphorylation mechanism (158, 159).

Two recent publications show that some significant progress may have been made toward elucidating the detailed molecular mechanisms involved in coupling factor function. One concerns the beef heart F_1 (160) and the other the γ subunit of spinach CF_1 (161).

Senior (160) found that, out of a total of eight sulfhydryl groups and two disulfide bonds per molecule of beef heart F_1, two of the sulfhydryl groups are freely accessible to iodoacetate and Ellman's reagent, or 3,3-dithiobis(6-nitrobenzoic acid). Upon heating of the enzyme, a conformational change occurs which involves burial of the two sulfhydryl groups inside the molecule and an increase in ATPase activity. The ATPase activity was very resistant to sulfhydryl reagents such as N-ethylmaleimide, mercurials, and Ellman's reagent. On the other hand, with tetranitromethane and iodoacetate, rapid inactivation occurred which apparently was not due to modification of sulfhydryl groups. However, when inactivation by tetranitromethane was almost complete and if ATP was absent, a total of 9–10 tyrosine residues were nitrated. ATP reduced the amount of inactivation by half and reduced the number of tyrosines nitrated to 6–7. These results led to the suggestion that tyrosine residues, but not sulfhydryl groups, may be involved in the binding of ATP and its hydrolysis. It is still an open question whether the active or catalytic site for the ATPase reaction should be distinguished from the ATP binding site, as a large conformational change may occur after the binding of the ATP and change its location (160). If this would be the case, then sulfhydryl groups might play a more or less direct role in the hydrolysis but not in the primary binding of ATP.

It is tempting to speculate that a relationship may exist between indications that three tyrosine groups are involved in binding of ATP to F_1 (160), three molecules of ATP may be bound to F_1 (146), and three molecules of each of the two largest subunits of F_1 exist in one intact coupling factor molecule (118). One may very tentatively imagine that each of the three polypeptides in one of the two largest F_1 subunits may bind one ATP, with a tyrosine residue participating in this binding.

McCarty & Fagan (161) found that light markedly enhanced the incorporation of tritiated N-ethylmaleimide (MalNEt) into one of the five subunits of CF_1 (the γ subunit) in spinach chloroplasts previously incubated with labeled MalNEt in the dark. The conditions for this effect were similar to those required for the earlier discovered light-dependent inhibition by MalNEt of photophosphorylation (162). Uncouplers that abolished the light-dependent inhibition of photophosphorylation by MalNEt strongly inhibited the light-dependent incorporation of MalNEt into the γ subunit of CF_1, as did ADP in the presence of P_i. The results indicate that a conformational change is induced by illumination of chloroplasts which exposes a group (or groups) in the γ subunit to react with MalNEt. It was suggested that the reaction of MalNEt with an SH group in the γ subunit causes the inhibition of photophosphorylation by this reagent. The approximate molar ratio between bound MalNEt and CF_1 was assumed to be 0.6. The question arises whether these data indicate a very special role for the γ subunit in photophosphorylation, as is indicated also by the strong inhibition of photophosphorylation by antiserum against this subunit (124).

The data obtained by Senior (160) and McCarty & Fagan (161) may perhaps, although they emanated from as different physiological sources as beef and spinach, be compared with respect to the question of SH-group requirements and participation in coupling factor function. Indeed, with respect to content of cysteine plus half-cysteine residues remarkable similarities exist between F_1 and CF_1. According to Knowles & Penefsky (122) and Senior (160) there is a total of 12 such residues in F_1. Farron & Racker earlier obtained the same number in CF_1 (163). In each factor there appears to be two free sulfhydryl groups (reacting relatively easily with SH reagents) as well as two disulfide bonds, giving a total of eight sulfhydryl groups per molecule. The data of McCarty & Fagan (161) indicated a connection between light-induced conformational change, resulting in an apparent SH-group exposure, and the light-induced activation of Mg ATPase (164) as well as $(^{32}P)P_i$-ATP exchange (165) occurring in the presence of sulfhydryl compounds. This brings into focus earlier data on the role of thiols in energy conservation, for example those of Gautheron et al (166). However, as discussed above, although thiols may have a position of great importance for the coupling factor functions, they do not participate in ATP binding. Interestingly, in lactate dehydrogenase Adams et al (167) recently found the essential thiol group to be not directly involved in binding of the substrate or nucleotide coenzyme, however, modification of this SH group inactivates LDH of most species.

Studies with mutants of *Escherichia coli* may open up a fascinating line of investigations on coupling factor function. In one mutant strain phosphorylation is uncoupled from electron transport with D-lactate as substrate and also MgCa ATPase activity is lost (168). Butlin et al (169) recently described another mutant strain of *E. coli* in which phosphorylation is also uncoupled from D-lactate oxidation but the MgCa ATPase activity retained. Both mutants have an impaired energy-linked transhydrogenase activity. If, as remains to be determined, these results mean that two different coupling factor substructures are impaired in the two mutants, this would be another possible indication that ATPase may involve proper functioning of one particular subunit as compared to two subunits for ATP formation. The earlier indication was found in chloroplasts, where the α subunit might possibly have a special relationship to ATP binding and ATPase function, whereas both the α and γ subunits may be specially related to ATP formation (124, 161). Although caution should always prevail when coupling factor properties from different energy-transducing systems are compared, the interchangeability of coupling factors in photosynthetic and respiratory energy conversion (170) constitutes an additional source of inspiration for the evolutionary approach.

CONFORMATIONAL CHANGE AND COUPLING MECHANISM

Among the three well-known competing hypotheses for the electron transport phosphorylation mechanism, the classical chemical and Mitchell's chemiosmotic hypotheses are older and more often discussed at length than the conformational change hypothesis proposed in 1965 by Boyer (171). He suggested that energy transformation or coupling processes may essentially involve protein conformational changes. Very recently, the conformational change hypothesis has been extended to include a radically new concept for energy coupling in oxidative phosphorylation — ATP formed at the coupling site may be released to the medium by an energy-driven conformational change (172, 173). Attractive features of this concept, also introduced by Boyer, are the focusing of attention on both retained hydrophobicity in energy coupling and the H_2O molecule in the coupling mechanism.

Our choice to treat only the conformational change hypothesis in some detail does not reflect any lack of appreciation of the two older alternatives or other important variations on the coupling mechanism theme. However, as pointed out below, several kinds of new experimental evidence seem to justify serious consideration that protein conformational changes may be involved in the coupling mechanism at the levels of both electron transport and phosphorylation.

Electron Transport Level

On the electron transport level few facts are known about protein conformational changes, but some current hypotheses pointing out analogies with other systems known to involve changes in protein conformation make it seem possible that they can be an obligatory part of energy coupling.

The only carrier known with respect to its three-dimensional structure in both the oxidized and reduced form, cytochrome c, does indeed show a dramatic change in protein conformation associated with its oxidation-reduction function (174). It is of course an open question whether this is a general phenomenon for cytochromes.

The change in apparent midpoint potential at all three coupling sites of the respiratory chain involves potential change in two carriers. Ohnishi & Pring (39) find changes in the iron-sulfur centers N-1b and N-2. Dutton & Lindsay (175) have shown that a slight change in the apparent midpoint potential of cytochrome c_1 upon addition of ATP accompanies the large one in cytochrome b_{565}. The postulated heme-heme interactions between cytochromes b_{565} and c_1 (93) and between the two hemes of cytochrome oxidase (100) would conceivably be associated with conformational changes in the protein structures associated with these hemes (24). Interesting in this connection is the finding of Kassner (176) that transfer from a polar to a nonpolar environment drastically changes the redox potential of heme groups.

Rydström et al (177) have suggested conformational change coupling for the energy-linked transhydrogenase reaction. Lee & Slater (178) have reported results indicating that there exists a restriction in accessibility for electrons from substrates to cytochrome b_{565} in the nonenergized state, and Slater (179) has suggested the possibility of different subunit interaction in the energized and nonenergized states.

Lumry (180) has reviewed many of the possibilities for conformationally transmitted energy in proteins and suggested a mechanism for the generation of free energy by conformational changes related to oxidation-reduction of cytochromes fixed in a rigid matrix. It would be premature to speculate about the requirement and the detailed molecular nature of constraints on protein conformational change for involvement in energy conservation.

Coupling Factor Level

The apparently quite drastic conformational change occurring in CF_1 when the photosynthetic apparatus of spinach chloroplasts is energized, notably by light (159, 161), may well be related to the mechanism of electron transport-coupled formation of ATP. This conclusion is supported by the great sensitivity of the change to uncouplers and by the findings concerning the protective effect of added $ADP + P_i$.

A first step from a more general acknowledgement of the existence of such a pronounced conformational change toward a more specific view of its molecular nature and possible role in electron transport phosphorylation may be the demonstration of light-induced exposure of a binding site for the uncoupler MalNEt in the γ subunit of CF_1 (161) under conditions similar to those required for the inhibition of photophosphorylation by this agent.

The significance of tyrosine residues in the binding of ATP to coupling factor F_1 (160) should also be closely scrutinized in the protein conformation context. Tyrosine residues are of interest in connection with both the electron transport and the phosphorylation levels. Attention has been drawn by Takano et al (174) to the involvement of a tyrosine residue in the conformational change cytochrome c undergoes during electron transport, and the location of tyrosine residues close to the iron-sulfur clusters of bacterial ferredoxin has been noted by Adman et al (181). An exciting connection with energy-rich phosphate compounds has been demonstrated by Mantel & Holzer (182), who found that the adenylyl-O-tyrosine bond in adenylylated glutamine synthetase was energy rich. Attention was drawn to the possibility that such a protein-bound energy-rich structure may participate in a conformational change accompanying electron transport phosphorylation (183, 184).

Thus there are now at least indications that conformational changes in electron carrier proteins may induce midpoint potential changes at all three coupling sites in the respiratory electron transport chain and that ATP formation involves conformational change at the coupling factor level. Whether the cooperative capacities of biological membranes for electron transport phosphorylation allow a direct conformational coupling between the levels of electron transport and ATP formation will probably remain an open question for some time to come. At present we divide the problem about the nature of the coupling mechanism into three consecutive levels: 1. electron transport, 2. intermediary, and 3. ATP formation.

Even if, as current information may indicate, the conformational change hypothesis could account for the coupling mechanism at levels 1 and 3, there might well remain a level 2 for any alternative hypothesis, such as the chemical and chemiosmotic ones. On the other hand, as suggested, the existence of direct coupling between levels 1 and 3 would eliminate the necessity for a level 2. The lack of decisive information in this regard will undoubtedly stimulate further experimentation.

EVOLUTIONARY EPILOGUE

There are many variations on the theme of continuity, often expressed as "where there is history, there is future." The current situation in the field of electron transport phosphorylation appears to indicate that research at both electron transport and phosphorylation levels will increasingly include the evolutionary approach as a new focal point in attempts to unravel the actual structures and mechanisms involved. This approach has as a logical background the discovery of the universal genetic code for living organisms on this earth and has gained momentum in recent years also in connection with studies on electron transport phosphorylation.

Ferredoxins and c-type cytochromes are the two electron carriers extensively studied from an evolutionary point of view. For cytochrome c an evolutionary tree was constructed in 1967 (185). The ferredoxins have been investigated in great detail with respect to evolution within the group and connection to prebiological amino acid formation (186) as well as other iron-sulfur proteins (187, 188). Recently, evolutionary relationships between electron carriers along and across electron transport chains have been suggested (189). According to a new hypothesis for the origin and evolution of biological electron transport, it may have originated close to the potential of the hydrogen electrode and evolved in various advantageous directions including, when molecular oxygen became available, that of the oxygen electrode (189, 190). Evidence supporting such evolutionary relationships has been given for both iron-containing (190) and nucleotide-containing (191) electron carriers. Amazing three-dimensional similarities between flavodoxin and NAD-linked dehydrogenases have been discovered (192).

At the level of electron transport phosphorylation, inorganic pyrophosphate (193, 194) and high molecular weight polyphosphates (195) have been implicated as evolutionary predecessors of ATP.

Finally, an example may be given to elucidate the possibility that information about the structure and evolution of proteins involved in oxidation-reduction may be of potential interest also in considerations of protein structure and function at the phosphorylation level. Studies of the conformation of the coenzyme fragments AMP and ADP when bound to lactate dehydrogenase have revealed that two sites exist on the protein for the second phosphate of ADP (196). The free energies of binding AMP and ADP to LDH are nearly equal, as found by McPherson (197). The recent demonstration of common "super-secondary structures" in nucleotide binding proteins (192), including both dehydrogenases (191, 198) and kinases (199, 200), direct attention to the unknown structures of the nucleotide binding coupling factor proteins. When taken together, these findings could be hinting at a similar nucleotide binding region in coupling factors, where adenine nucleotide + P_i might be able to occupy such a region. The possibility of exploring this question experimentally in the not too distant future may well depend on whether the catalytic nucleotide binding center of a coupling factor complex contains amino acid residues from a single subunit or from two different subunits, as in D-glyceraldehyde-3-phosphate dehydrogenase (191).

Literature Cited

1. Newton, I. 1687. *Philosophiae Naturalis Principia Mathematica,* 402. London: S. Pepys

2. Krebs, H. A. 1953. *Brit. Med. Bull.* 9: 97–104

3. Lardy, H. A., Ferguson, S. M. 1969. *Ann.*

Rev. Biochem. 38:991–1034
4. van Dam, K., Meyer, A. J. 1971. Ann. Rev. Biochem. 40:115–60
5. Walker, D. A., Crofts, A. R. 1970. Ann. Rev. Biochem. 39:389–428
6. Bishop, N. I. 1971. Ann. Rev. Biochem. 40:197–226
7. Dickerson, R. E. 1972. Ann. Rev. Biochem. 41:815–42
8. Orme-Johnson, W. H. 1973. Ann. Rev. Biochem. 42:159–204
9. Quagliariello, E., Papa, S., Rossi, C. S. 1971. Energy Transduction in Respiration and Photosynthesis. Bari: Adriatica Editrice. 1029 pp.
10. Azzone, G. F., Carafoli, E., Lehninger, A. L., Quagliariello, E., Siliprandi, N. 1972. Biochemistry and Biophysics of Mitochondrial Membranes. New York: Academic. 714 pp.
11. Azzone, G. F., Ernster, L., Papa, S., Quagliariello, E., Siliprandi, N. 1973. Mechanisms in Bioenergetics. New York: Academic. 609 pp.
12. Sanadi, D. R. 1971. Curr. Top. Bioenerg. Vol. 4
13. Sanadi, D. R., Packer, L. 1973. Curr. Top. Bioenerg. Vol. 5
14. Chance, B., Lee, C. P., Blasie, J. K. 1971. Probes Struct. Funct. Macromol. Membranes. Vol. 1
15. Chance, B., Yonetani, T., Mildvan, A. S. 1971. Probes Struct. Funct. Macromol. Membranes. Vol. 2
16. Ernster, L., Estabrook, R. W., Slater, E. C. 1974. Dynamics of Energy Transducing Membranes. Amsterdam: Elsevier
17. Åkesson, Å., Ehrenberg, A. 1972. Structure and Function of Oxidation-Reduction Enzymes. Oxford: Pergamon. 777 pp.
18. Slater, E. C. 1973. Biochim. Biophys. Acta 301:129–54
19. Dayhoff, M. O. 1972. Atlas of Protein Sequence and Structure, Vol. 5. Washington, D.C.: Nat. Biomed. Res. Found. 124 pp.
20. Lemberg, R., Barrett, J. 1973. Cytochromes. New York: Academic. 580 pp.
21. Kamin, H. 1971. Flavins and Flavoproteins. Baltimore: University Park. 712 pp.
22. Slater, E. C. 1971. Quart. Rev. Biophys. 4:35–71
23. Witt, H. T. 1971. Quart. Rev. Biophys. 4:365–477
24. Chance, B. C. 1972. FEBS Lett. 23:3–20
25. Mitchell, P. 1972. Bioenergetics 3:5–24
26. Green, D. E., Ji, S. 1972. J. Bioenerg. 3:159–202

27. Williams, R. J. P. 1973. Biochem. Soc. Trans. 1:1–26
28. Hatefi, Y. 1966. Compr. Biochem. 14:199–231
29. Orme-Johnson, N. R., Orme-Johnson, W. H., Hansen, R. E., Beinert, H., Hatefi, Y. 1971. Biochem. Biophys. Res. Commun. 44:446–52
30. Ohnishi, T., Asakura, T., Wohlrab, H., Yonetani, T., Chance, B. 1970. J. Biol. Chem. 245:901–2
31. Butow, R., Racker, E. 1965. J. Gen. Physiol. Suppl. 49:149–62
32. Light, P. A., Ragan, C. J., Clegg, R. A., Garland, P. B. 1968. FEBS Lett. 1:4–8
33. Gutman, M., Mayr, M., Oltzik, R., Singer, T. P. 1970. Biochem. Biophys. Res. Commun. 41:40–44
34. Ohnishi, T., Asakura, T., Wilson, D. F., Chance, B. 1972. FEBS Lett. 21:59–62
35. Gutman, M., Singer, T. P., Beinert, H. 1971. Biochem. Biophys. Res. Commun. 44:1572–78
36. Ohnishi, T., Chance, B. 1971. In Flavins and Flavoproteins, ed. H. Kanin, 681–89. Durham, N.C.: University Park Press.
37. Ohnishi, T., Panebianco, P., Chance, B. 1972. Biochem. Biophys. Res. Commun. 49:99–106
38. Ohnishi, T., Wilson, D. F., Chance, B. 1972. Biochem. Biophys. Res. Commun. 49:1087–92
39. Ohnishi, T., Pring, M. See Ref. 16
40. Wilson, D. F., Erecińska, M., Dutton, P. L., Tsudzuki, T. 1970. Biochem. Biophys. Res. Commun. 41:1273–78
41. Dutton, P. L. 1971. Biochim. Biophys. Acta 226:63–80
42. Dutton, P. L. et al 1972. Biochim. Biophys. Acta 267:15–24
43. Wilson, D. F., Dutton, P. L., Wagner, M. 1973. Curr. Top. Bioenerg. 5:233–65
44. Erecińska, M., Wagner, M., Chance, B. 1973. Curr. Top. Bioenerg. 5:261–303
45. Wikström, M. K. F. 1973. Biochim. Biophys. Acta 301:155–93
46. Chance, B., Wilson, D. F., Dutton, P. L., Erecińska, M. 1970. Proc. Nat. Acad. Sci. USA 66:1175–82
47. Wikström, M. K. F. 1971. Biochim. Biophys. Acta 253:332–45
48. Yu, C. A., Yu, L., King, T. E. 1971. Biochim. Biophys. Acta 267:300–8
49. Caswell, A. H. 1968. J. Biol. Chem. 243:5827–36
50. Wilson, D. F., Dutton, P. L., Erecińska, M., Lindsay, J. G., Sato, N. 1972. Accounts Chem. Res. 5:234–41
51. Sato, N., Ohnishi, T., Chance, B. 1972.

Biochim. Biophys. Acta 275: 288–97
52. Cittadini, A., Galeotti, T., Chance, B., Terranova, T. 1971. FEBS Lett. 15: 133–36
53. von Jagow, G., Klingenberg, M. 1972. FEBS Lett. 24: 278–82
54. Bonner, W. D. 1965. Plant Biochemistry, ed. J. Bonner, J. R. Warner, 89–123. New York: Academic
55. Dutton, P. L., Storey, B. T. 1971. Plant Physiol. 47: 282–88
56. Wilson, D. F., Dutton, P. L. 1970. Biochem. Biophys. Res. Commun. 39: 59–64
57. Slater, E. C., Lee, C. P., Berden, J. A., Wegdam, H. J. 1970. Nature 226: 1248–49
58. Chance, B., Schoener, B. 1966. J. Biol. Chem. 241: 4567–73
59. Chance, B., Lee, C. P., Schoener, B. 1966. J. Biol. Chem. 241: 4574–76
60. Chance, B., Schoener, B. 1966. J. Biol. Chem. 241: 4577–87
61. Tyler, D. D., Estabrook, R. W., Sanadi, D. R. 1966. Arch. Biochem. Biophys. 114: 239–51
62. Slater, E. C., Lee, C. P., Berden, J. A., Wegdam, H. J. 1970. Biochim. Biophys. Acta 223: 354–64
63. Berden, J. A., Opperdoes, F. R., Slater, E. C. 1972. Biochim. Biophys. Acta 256: 594–99
64. Dutton, P. L., Wilson, D. F., Lee, C. P. 1971. Biochem. Biophys. Res. Commun. 43: 1186–91
65. Bonner, W. D., Lambowitz, A. M., Wikström, M. K. F. 1974. See Ref. 16
66. Hill, R. 1954. Nature 174: 501–3
67. Böhme, H., Cramer, W. A. Submitted for publication
68. Cramer, W. A., Böhme, H. 1972. Biochim. Biophys. Acta 256: 358–69
69. Ben-Hayyim, G., Avron, M. 1970. Eur. J. Biochem. 14: 205–13
70. Böhme, H., Cramer, W. A. 1972. Biochemistry 7: 1155–60
71. Bendall, D. S., Sofrová, D. 1971. Biochim. Biophys. Acta 234: 371–80
72. Erixon, K., Butler, W. L. 1971. Biochim. Biophys. Acta 234: 381–89
73. Cramer, W. A., Butler, W. L. 1967. Biochim. Biophys. Acta 143: 332–39
74. Cramer, W. A., Fan, H. N., Böhme, H. 1971. J. Bioenerg. 2: 289–303
75. Dutton, P. L., Baltscheffsky, M. 1972. Biochim. Biophys. Acta 267: 172–78
76. Caswell, A. H. 1971. Arch. Biochem. Biophys. 144: 445–41
77. Wikström, M. K. F., Berden, J. A. 1973. Mechanisms in Bioenergetics, ed. L. F. Azzone, L. Ernster, S. Papa, E. Quag-

liariello, M. Siliprandi, 545–60. New York: Academic
78. Wikström, M. K. F., Lambowitz, A. M. 1974. FEBS Lett. In press
79. Chance, B. 1952. 2nd Int. Congr. Biochem. Paris Abstr. 32
80. Pumphrey, A. M. 1962. J. Biol. Chem. 237: 2384–90
81. Baum, H., Rieske, J. S. 1966. Biochem. Biophys. Res. Commun. 24: 1–9
82. Rieske, J. S. 1971. Arch. Biochem. Biophys. 145: 179–93
83. Wilson, D. F., Koppelman, M., Erecińska, M., Dutton, P. L. 1971. Biochem. Biophys. Res. Commun. 44: 759–66
84. Erecińska, M., Chance, B., Wilson, D. F., Dutton, P. L. 1972. Proc. Nat. Acad. Sci. USA 69: 50–54
85. Baum, H., Rieske, J. S., Silman, H. J., Lipton, S. H. 1967. Proc. Nat. Acad. Sci. USA 57: 798–805
86. Wilson, D. F., Erecińska, M., Leigh, J. S., Koppelman, M. 1972. Arch. Biochem. Biophys. 151: 112–21
87. Dutton, P. L. et al 1972. Biochim. Biophys. Acta 267: 15–24
88. Wikström, M. K. F., Berden, J. A. 1972. Biochim. Biophys. Acta 283: 403–20
89. Ernster, L. et al 1971. Probes of Structure and Function of Macromolecules and Membranes, ed. B. Chance, C. P. Lee, J. K. Blasie, 377–89. New York: Academic
90. Storey, B. T. 1972. Biochim. Biophys. Acta 267: 48–64
91. Lambowitz, A. M., Bonner, W. D. 1973. Biochem. Biophys. Res. Commun. 52: 703–11
92. Chance, B. 1974. See Ref. 16
93. Chance, B. 1972. FEBS Lett. 23: 3–20
94. Okunuki, K., Sekuzu, J., Yonetani, T., Takemori, S. 1958. J. Biochem. Japan 45: 847–54
95. Lemberg, M. R. 1969. Physiol. Rev. 49: 48–121
96. Takemori, S. 1960. J. Biochem. Japan 47: 382–90
97. Awashti, Y. C., Chuang, F. F., Kleenan, T. W., Crane, F. L. 1970. Biochim. Biophys. Acta 226: 42–52
98. Wikström, M. K. F., Saris, N. E. L. 1970. Electron Transport and Energy Conservation, ed. E. Quagliariello, E. C. Slater, 77–88. Bari: Adriatica Editrice
99. Wilson, D. F., Dutton, P. L. 1970. Arch. Biochem. Biophys. 136: 583–84
100. Lindsay, J. G., Wilson, D. F. 1972. Biochemistry 11: 4613–21
101. Slater, E. C. 1969. The Energy Level and Metabolic Control in Mitochondria, ed.

896   BALTSCHEFFSKY & BALTSCHEFFSKY

S. Papa, J. M. Tager, E. Quagliariello, E. C. Slater, 255–59. Bari: Adriatica Editrice
102. Muraoka, A., Slater, E. C. 1969. *Biochim. Biophys. Acta* 180:227–36
103. Wilson, D. F., Erecińska, M., Nicholls, P. 1972. *FEBS Lett.* 20:61–65
104. Erecińska, M., Wilson, D. F., Sato, N., Nicholls, P. 1972. *Arch. Biochem. Biophys.* 151:188–92
105. Van Buuren, K. J. H., Nicholls, P., van Gelder, B. F. 1972. *Biochim. Biophys. Acta* 256:258–76
106. Wilson, D. F., Erecińska, M., Brocklehurst, E. S. 1972. *Biochim. Biophys. Acta* 151:180–87
107. Wikström, M. K. F. 1972. *Biochim. Biophys. Acta* 283:385–90
108. Wilson, D. F., Brocklehurst, E. S. 1973. *Arch. Biochem. Biophys.* 158:200–12
109. Chance, B. 1972. *Molecular Basis of Electron Transport,* ed. J. Schultz, B. F. Cameron, 65–89. New York: Academic
110. Erecińska, M., Chance, B. 1972. *Arch. Biochem. Biophys.* 151:304–15
111. Yong, F. C., King, T. E. 1972. *Biochem. Biophys. Res. Commun.* 47:380–86
112. Racker, E., Kandrach, A. 1973. *J. Biol. Chem.* 248:5841–47
113. MacLennan, D. H. 1970. *Curr. Top. Membranes Transp.* 1:177–232
114. Tzagoloff, A. 1971. *Curr. Top. Membranes Transp.* 2:157–205
115. Beechey, R. B., Cattell, K. J. 1973. *Curr. Top. Bioenerg.* 5:305–57
116. Lambeth, D. O., Lardy, H. A. 1971. *Eur. J. Biochem.* 22:355–63
117. Catterall, W. A., Pedersen, P. L. 1971. *J. Biol. Chem.* 246:4987–94
118. Catterall, W. A., Coty, W. A., Pedersen, P. L. 1973. *J. Biol. Chem.* 248:7427–31
119. Senior, A. E., Brooks, J. C. 1971. *FEBS Lett.* 17:327–29
120. Brooks, J. C., Senior, A. E. 1972. *Biochemistry* 11:4675–78
121. Knowles, A. F., Penefsky, H. S. 1972. *J. Biol. Chem.* 247:6617–23
122. Knowles, A. F., Penefsky, H. S. 1972. *J. Biol. Chem.* 247:6624–30
123. Nelson, N., Nelson, H., Racker, E. 1972. *J. Biol. Chem.* 247:7657–62
124. Nelson, N., Deters, D. W., Nelson, H., Racker, E. 1973. *J. Biol. Chem.* 248:2049–55
125. Salton, M. R. J., Schor, M. T. 1972. *Biochem. Biophys. Res. Commun.* 49:350–57
126. Mirsky, R., Barlow, V. 1973. *Biochim. Biophys. Acta* 291:480–88
127. Schnebli, H. P., Vatter, A. E., Abrams, A. 1970. *J. Biol. Chem.* 245:1122–27
128. Hanson, R. L., Kennedy, E. P. 1973. *J. Bacteriol.* 114:772–81
129. Bragg, P. D., Davies, P. L., Hou, C. 1973. *Arch. Biochem. Biophys.* 159:664–70
130. Bragg, P. D., Hou, C. 1972. *FEBS Lett.* 28:309–12
131. Tzagoloff, A., Meagher, P. 1971. *J. Biol. Chem.* 246:7328–36
132. MacLennan, D. H., Tzagoloff, A. 1968. *Biochemistry* 7:1603–10
133. Bulos, B., Racker, E. 1968. *J. Biol. Chem.* 243:3891–3900
134. Tzagoloff, A. 1970. *J. Biol. Chem.* 245:1545–51
135. Pullman, M. E., Monroy, G. C. 1963. *J. Biol. Chem.* 238:3762–69
136. Horstman, L. L., Racker, E. 1970. *J. Biol. Chem.* 245:1336–44
137. Brooks, J. C., Senior, A. E. 1971. *Arch. Biochem. Biophys.* 147:467–70
138. Asami, K., Juntti, K., Ernster, L. 1970. *Biochim. Biophys. Acta* 205:307–11
139. van der Stadt, R. J., de Boer, B. L., van Dam, K. 1973. *Biochim. Biophys. Acta* 292:338–49
140. Johansson, B. C., Baltscheffsky, M., Baltscheffsky, H., Baccarini-Melandri, A., Melandri, B. A. 1973. *Eur. J. Biochem.* 40:109–17
141. Zalkin, H., Pullman, M. E., Racker, E. 1965. *J. Biol. Chem.* 240:4011–16
142. Pullman, M. E., Penefsky, H. S., Datta, A., Racker, E. 1960. *J. Biol. Chem.* 235:3322–29
143. Catterall, W. A., Pedersen, P. L. 1972. *J. Biol. Chem.* 247:7969–76
144. Sanadi, D. R., Sani, B. P., Fisher, R. J., Li, O., Taggart, W. V. 1971. See Ref. 9, 89–107
145. Hilborn, D. A., Hammes, G. G. 1973. *Biochemistry* 12:983–90
146. Harris, D. A., Rosing, J., van der Stadt, R. J., Slater, E. C. 1973. *Biochim. Biophys. Acta* 314:149–53
147. Hammes, G. G., Hilborn, D. A. 1971. *Biochim. Biophys. Acta* 233:580–90
148. Tondre, C., Hammes, G. G. 1973. *Biochim. Biophys. Acta* 314:245–49
149. Nelson, N., Nelson, H., Racker, E. 1972. *J. Biol. Chem.* 247:6506–10
150. Roy, H., Moudrianakis, E. N. 1971. *Proc. Nat. Acad. Sci. USA* 68:464–68
151. Roy, H., Moudrianakis, E. N. 1971. *Proc. Nat. Acad. Sci. USA* 68:2720–24
152. Vambutas, V. K., Racker, E. 1965. *J. Biol. Chem.* 240:2660–67
153. Shahak, Y., Chipman, D. M., Shavit, N. 1973. *FEBS Lett.* 33:293–96
154. Baccarini-Melandri, A., Gest, H., San

Pietro, A. 1970. *J. Biol. Chem.* 245: 1224–26
155. Johansson, B. C. 1972. *FEBS Lett.* 20: 339–40
156. Johansson, B. C., Baltscheffsky, M., Baltscheffsky, H. 1972. *Proc. 2nd Int. Congr. Photosyn. Res.,* ed. G. Forti, M. Avron, A. Melandri, 1203–9. The Hague: Dr. W. Junk
157. Adolfsen, R., Moudrianakis, E. N. 1973. *Biochemistry* 12: 2926–33
158. Ryrie, I. J., Jagendorf, A. T. 1972. *J. Biol. Chem.* 246: 3371–74
159. Ryrie, I. J., Jagendorf, A. T. 1972. *J. Biol. Chem.* 247: 4453–59
160. Senior, A. E. 1973. *Biochemistry* 12: 3622–27
161. McCarty, R. E., Fagan, J. 1973. *Biochemistry* 12: 1503–7
162. McCarty, R. E., Pittman, P. R., Tsuchiya, Y. 1972. *J. Biol. Chem.* 247: 3048–51
163. Farron, F., Racker, E. 1970. *Biochemistry* 9: 3829–36
164. Petrack, B., Lipmann, F. 1961. *Light and Life,* ed. W. D. McElroy, B. Glass, 621–30. Baltimore: Johns Hopkins
165. Carmeli, G., Avron, M. 1966. *Biochem. Biophys. Res. Commun.* 24: 923–28
166. Gautheron, D. C. et al 1971. See Ref. 9, 807–16
167. Adams, M. J. et al 1973. *Proc. Nat. Acad. Sci. USA* 70: 1968–72
168. Butlin, J. D., Cox, G. B., Gibson, F. 1971. *Biochem. J.* 124: 75–81
169. Butlin, J. D., Cox, G. B., Gibson, F. 1973. *Biochim. Biophys. Acta* 292: 366–75
170. Melandri, B. A., Baccarini-Melandri, A., San Pietro, A., Gest, H. 1971. *Science* 174: 514–16
171. Boyer, P. D. 1965. *Oxidases and Related Redox Systems,* ed. T. E. King, H. S. Mason, M. Morrison, 2: 994–1008. New York: Wiley
172. Boyer, P. D., Cross, R. L., Momsen, W. 1973. *Proc. Nat. Acad. Sci. USA* 70: 2837–39
173. Boyer, P. D. 1974. *Biochim. Biophys. Acta.* In press
174. Takano, T., Swanson, R., Kallai, O. B., Dickerson, R. E. 1972. *Cold Spring Harbor Symp. Quant. Biol.* 36: 397–404
175. Dutton, P. L., Lindsay, J. G. See Ref. 77, 535–44
176. Kassner, R. J. 1972. *Proc. Nat. Acad.*

Sci. USA 69: 2263–67
177. Rydström, J., Teixeira da Cruz, A., Ernster, L. 1970. *Eur. J. Biochem.* 17: 56–62
178. Lee, I. Y., Slater, E. C. 1972. *Biochim. Biophys. Acta* 283: 223–33
179. Slater, E. C. See Ref. 77, 405–31
180. Lumry, R. 1972. *Electron and Coupled Energy Transfer in Biological Systems,* ed. T. E. King, M. Klingenberg, 1–116. New York: Dekker
181. Adman, E. T., Sieker, L. C., Jensen, L. H. 1973. *J. Biol. Chem.* 248: 3987–96
182. Mantel, M., Holzer, H. 1970. *Proc. Nat. Acad. Sci. USA* 65: 660–67
183. Holzer, H. 1971. *Freiburg. Universitätsblätter* 33: 19–27
184. Holzer, H., Wohlhueter, R. 1972. *Advan. Enzyme Regul.* 10: 121–32
185. Fitch, W. M., Margoliash, E. 1967. *Science* 155: 279–84
186. Hall, D. O., Cammack, R., Rao, K. K. 1971. *Nature* 233: 136–38
187. Dayhoff, M. O. 1971. *Chemical Evolution and the Origin of Life,* ed. R. Buvet, C. Ponnamperuma, 1: 392–419. Amsterdam: North-Holland
188. Hall, D. O., Cammack, R., Rao, K. K. 1973. *Theory and Experiment in Exobiology,* 2: 67–85. Broningen: Wolters-Noordhoff
189. Baltscheffsky, H. 1974. *Origins of Life,* Vol. 4
190. Baltscheffsky, H. See Ref. 16
191. Buehner, M., Ford, G. C., Moras, D., Olsen, K. W., Rossman, M. G. 1973. *Proc. Nat. Acad. Sci. USA* 70: 3052–54
192. Rao, S. T., Rossman, M. G. 1973. *J. Mol. Biol.* 76: 241–56
193. Baltscheffsky, H. 1967. *Acta Chem. Scand.* 21: 1973–74
194. Baltscheffsky, H. 1971. See Ref. 187, 466–74
195. Kulaev, I. S. 1971. See Ref. 187, 458–65
196. Chandrasekhar, K., McPherson, A. Jr., Adams, M. J., Rossman, M. G. 1973. *J. Mol. Biol.* 76: 503–18
197. McPherson, A. Jr. 1970. *J. Mol. Biol.* 51: 39–46
198. Brändén, C. I. et al 1973. *Proc. Nat. Acad. Sci. USA* 70: 2439–42
199. Schulz, G. E., Elzinga, M., Marx, F., Schirmer, R. H. 1974. In press
200. Liljas, A., Rossmann, M. G. 1974. *Ann. Rev. Biochem.* 43: 475–507

BIOSYNTHESIS OF WATER-SOLUBLE VITAMINS

Gerhard W. E. Plaut[1] *and Colleen M. Smith*
Department of Biochemistry, Temple University School of Medicine
Philadelphia, Pennsylvania

William L. Alworth[2]
Department of Chemistry, Tulane University, New Orleans, Louisiana

CONTENTS

Discussions in this review are confined to recent work on the biosynthesis of Coenzyme A, riboflavin, Vitamin B_{12}, and Vitamin B_6. Literature in these areas has been covered up to August 1973.

FORMATION OF COENZYME A

Coenzyme A (CoA) and 4'-phosphopantetheine, the cofactor forms of pantothenic acid, participate in the enzyme-catalyzed reactions of several metabolic pathways by formation of thiolester bonds with the substrates. The tissues of higher animals lack the capacity for synthesis of pantothenic acid from pantoic acid and β-alanine,

[1] The experimental work from this laboratory and the preparation of this article have been assisted by grants from the National Institute of Arthritis, Metabolic and Digestive Diseases (AM 15404).

[2] WLA wishes to acknowledge the helpful comments of B. Guirard and H. C. Friedmann who read early drafts of the material on B_6 and B_{12}, respectively. WLA also wishes to thank H. A. Barker and other members of the Biochemistry Department at Berkeley for providing the space and facilities that enabled him to work on the review while on leave from Tulane. The preparation of this article was assisted by a grant from the National Institute of Arthritis, Metabolic and Digestive Diseases (5 K04 AM70280).

the precursors established in microorganisms. This section is restricted primarily to recent experiments pertaining to the portion of the biosynthetic pathway from pantothenic acid to CoA. Earlier work on the biosynthesis of pantothenic acid and CoA has been reviewed in detail by Brown & Reynolds (1), Winestock & Plaut (2), and more recently by Brown (3). Summaries of methods for the synthesis of CoA analogs (4, 5) and procedures for preparation of labeled CoA (6, 7) have also appeared recently.

The pathway proposed by Hoagland & Novelli (8) for the synthesis of CoA from pantothenic acid followed the sequence

pantothenate → pantothenylcysteine → pantetheine →
4'-phosphopantetheine → dephospho-CoA → CoA

Their proposal was based on the demonstration that the sulfhydryl-containing fragment was derived from cysteine (8–10) and on the reformation of CoA in crude extracts from components obtained by degradation with phosphatases and a nucleotide pyrophosphatase (11). Evidence in support of the order for this series of reactions included experiments which showed conversion of pantetheine and ATP to CoA by crude extracts in low yield (11–13), separation of a fraction from pigeon liver which could convert pantetheine plus ATP to 4'-phosphopantetheine (14), separation of enzyme fractions that could convert 4'-phosphopantetheine or dephospho-CoA to CoA (8), and the demonstration that a rat liver enzyme fraction could produce CoA from ATP, pantothenate, and cysteine, or ATP and panto-thenylcysteine (8). Ward, Brown & Snell (15) found, however, that in the absence of cysteine, extracts of *Proteus morganii* would also phosphorylate pantothenate. They suggested an additional or alternate route for the series of reactions, which is shown in Figure 1.

An examination of the specificity of the synthetase and decarboxylation enzymes led Brown to conclude that this is the only route operative in either microbial or animal systems (16). He obtained an enzyme fraction from *P. morganii* and rat liver that could condense cysteine with 4-phosphopantothenate but not pantothenate, and an enzyme fraction from *P. morganii,* rat liver, and rat kidney that could de-carboxylate phosphopantothenylcysteine but not pantothenylcysteine.

These results were later confirmed by Abiko using phosphopantothenylcysteine-condensing enzyme purified 26-fold from crude liver extract and free of kinase activity (17), and phosphopantothenylcysteine decarboxylase purified 112-fold relative to the crude extract (18, 19). Although the synthetase would not act on pantothenylcysteine (18), it could remove the carboxyl moiety from α-carboxyde-phospho-CoA (20), the hypothetical intermediate if phosphopantothenylcysteine instead of phosphopantetheine were linked to the 5'-phosphoadenosine fragment. Partially purified dephospho-CoA pyrophosphorylase could not, however, catalyze formation of α-carboxydephospho-CoA when incubated with phosphopantothenyl-cysteine and ATP (20).

Using rat liver phosphopantothenylcysteine-condensing enzyme purified free of pyrophosphatases and adenylate kinase (21, 22), Abiko was able to demonstrate that the ATP required for the condensation of phosphopantothenate and cysteine is

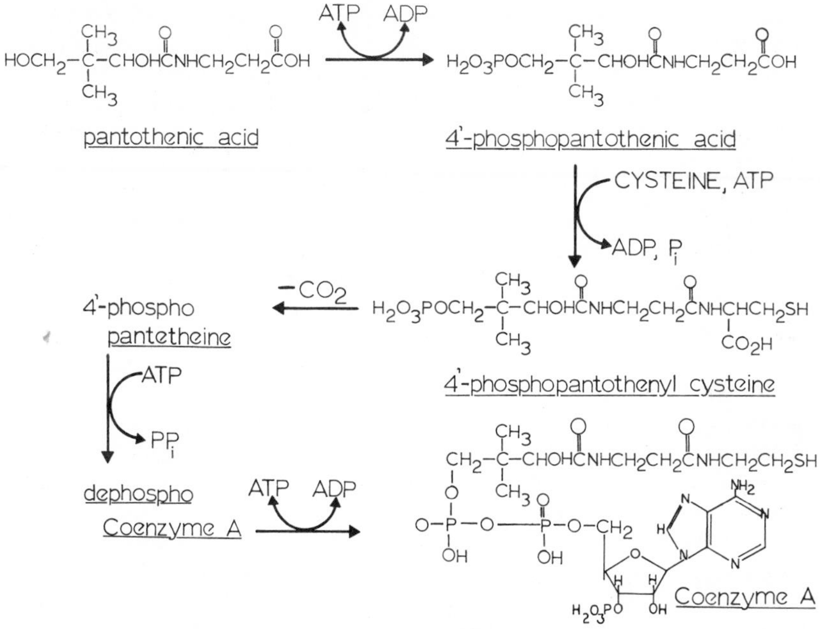

Figure 1   Biosynthesis of Coenzyme A from pantothenic acid.

stoichiometrically cleaved to ADP and inorganic phosphate (21). The mechanism of the decarboxylase reaction remains an open question. Scandurra et al (23) postulated that the decarboxylase contained pyridoxal phosphate which would react with the free amino group of an enzymatically formed thiazoline derivative of phosphopantothenylcysteine. But 12 N HCl and high temperatures were required for nonenzymatic rearrangement of phosphopantothenylcysteine (23) as well as pantetheine (24). Furthermore, Abiko (18) was unable to demonstrate a requirement for pyridoxal phosphate by dialyzing the enzyme or by adding semicarbazide to the assay mixture to inhibit the enzyme, although these experiments were by no means conclusive.

Dephospho-CoA pyrophosphorylase and dephospho-CoA kinase have also been obtained in purer form (25). Suzuki, Abiko & Shimizu found that the kinase and pyrophosphorylase activities cochromatograph on CM-cellulose, DEAE-cellulose, and Sephadex G-200 (25), and suggested that the two enzymes exist as a bifunctional enzyme complex (25, 26). Dephospho-CoA kinase activity has been reported in guinea pig liver mitochondria (27) and in rat liver mitochondria (28). The relationship of the mitochondrial activities to the enzymes purified by Abiko and co-workers (25, 26) from a rat liver 59,000 × g supernatant fraction has not been established.

Abiko (17) also confirmed the lack of specificity of the pantothenate kinase discovered by Ward et al (15). Enzyme purified 27-fold over the crude extract could phosphorylate D-pantothenate, pantetheine, and pantothenylcysteine (17), but not L-pantothenate or ketopantothenate (29). The activity of pantothenate kinase with several of the CoA precursors explains the conversions of pantetheine and pantothenylcysteine to CoA observed by Hoagland & Novelli (8). As in the experiments of Brown (16), the rate of phosphorylation of one precursor was inhibited by the addition of one of the other precursors (17), suggesting that the kinase activities are the property of a single enzyme. Although pantothenylalcohol (panthenol), the source of pantothenate activity commonly used by pharmaceutical companies (30), can also be phosphorylated by pantothenate kinase (31), it is probably first converted to pantothenate by liver alcohol dehydrogenase (32). The 112-fold purified pantothenate kinase reached maximal activity when Mg^{2+} and ATP were present in a 1:1 ratio, suggesting that $Mg:ATP$ is the substrate (29). Stimulation of the rate of CoA formation by the addition of Mg^{2+} to crude extracts of liver has also been observed (33).

The theory was advanced by Karasawa et al that the rate of CoA biosynthesis is regulated by feedback inhibition of pantothenate kinase by CoA (34). For example, 10 μM CoA inhibited the partially purified pantothenate kinase from rat kidney by 30% in the presence of 0.45 mM pantothenate and 10 mM ADP. This was much stronger than the inhibition by either dephospho-CoA, 4'-phosphopantetheine, or 4'-phosphopantothenate. Abiko, Ashido & Shimizu, using pantothenate kinase purified 700-fold from rat liver, also found an inhibition by several of the intermediates; however, 50 μM CoA inhibited the reaction by only 10% (40 μM pantothenate, 2 mM ATP); 4'-phosphopantothenate, 4'-phosphopanto-thenylcysteine, and 4'-phosphopantetheine were stronger inhibitors (29). Several investigators have measured the distribution of CoA precursors in rat liver (35–40) and found that 4'-phosphopantetheine is present in much higher concentrations than the others [i.e. 220 nmole 4'-phosphopantetheine and 360 nmol CoA per g wet weight (38)]. Thus, either or both CoA and 4'-phosphopantetheine might inhibit pantothenate kinase in vivo.

In support of their theory, Karasawa and co-workers reported that neither pantothenate deficiency nor excess pantothenate alter the concentration of CoA in rat livers or kidneys (34). However, other investigators have found a graded response of the CoA levels in liver, kidney, heart, and other tissues of young rats to the amount of pantothenate given the rat (35, 37, 41, 42, 43). Nakamura et al measured CoA and the sum of 4'-phosphopantetheine plus dephospho-CoA in the livers of young pantothenate-deficient rats given various amounts of pantothenate or pantetheine in their diet (35). CoA and the sum of 4'-phosphopantetheine plus dephospho-CoA increased to a maximum, whereas the ratio of CoA to 4'-phospho-pantetheine plus dephospho-CoA decreased to a minimum with the progressive increases in the supply of pantothenate. At excessive concentrations of pantothenate the trend was reversed (35). With large concentrations of pantothenate, there was an increase of phosphopantothenate and phosphopantothenylcysteine (37). The time course for the response of CoA levels to intraperitoneal injection of calcium pantothenate or S-benzoylpantetheine in young pantothenate-deficient rats was studied by another group of workers (41). The CoA concentrations in the liver reached a maximum 2 hr after injection; there was a decrease to control values at 6 hr followed by a second maximum at 18 hr. Several other variables have also been found to alter the concentration of CoA and 4'-phosphopantetheine (42, 43). These observations of the relationship between precursor and CoA levels have not yet been interpreted in terms of a comprehensive theory for the regulation of CoA biosynthesis. Neither has an explanation been advanced for changes in total CoA levels known to occur during starvation, with diet, or from acute alcohol stress (44, 45).

C^{14}- and H^3-pantothenate have been used to establish precursor relationships for the pantothenate derivatives in vivo (39, 40, 46, 47). The specific activity of the 4'-phosphopantetheine pool rises more rapidly than that of CoA and more slowly than that of pantothenate (39, 40). However, Powell et al reported for E. coli (46) and Tweto et al for pigeon liver (47, 48) that although the 4'-phosphopantetheine bound to acyl carrier protein is exchangeable, its specific activity rises more slowly than that of CoA. Vagelos suggested that this cofactor form of 4'-phosphopantetheine is therefore derived from enzymatic cleavage of CoA (46).

Little work has been done on the pathway for degradation of CoA, since phosphatases, a "peptidase," and a pyrophosphatase were used to establish the structure of the CoA molecule (11). However, an interesting report by Bremer, Wojtczak & Skrede has appeared recently (28). These investigators have identified CoA phosphatase activity in the lysosomal fraction which is inhibited by phosphate and a P_i-insensitive dephospho-CoA degrading activity in nuclear and microsomal fractions. There are also nuclear and microsomal CoA-degrading activities not inhibited by phosphate. The specificity of these enzymes was not determined.

BIOSYNTHESIS OF RIBOFLAVIN

The biosynthesis of riboflavin has been reviewed in 1971 and 1972 (49, 50). The preparation, assay, and properties of substrates and enzymes of riboflavin metabolism have been described in a series of articles in *Methods in Enzymology* (51). Demain (50) has presented a comprehensive discussion of the status of riboflavin oversynthesis. The present summary focuses mainly on recent developments in the field.

Bacher, Lingens, Oltmanns, and their colleagues have continued their systematic studies of riboflavin-deficient mutants of *Saccharomyces cerevisiae*. From investigations of accumulation products detected with individual riboflavinless mutants they (52, 53) have proposed the formation of riboflavin from a guanine precursor. The nomenclature used in Figure 2 to designate these mutants (rib_1 to rib_7) is that adopted by these investigators. The sequence of steps depicted in Figure 2 has received support by further genetic analyses (53) in which haploid strains with two different rib genes were formed. Tetrad analysis of these strains and studies of the accumulation products showed that genes rib_1, rib_7, and rib_2 are not linked to each other. Strains rib_1-rib_7 and rib_1-rib_2 had the phenotypic properties of rib_1

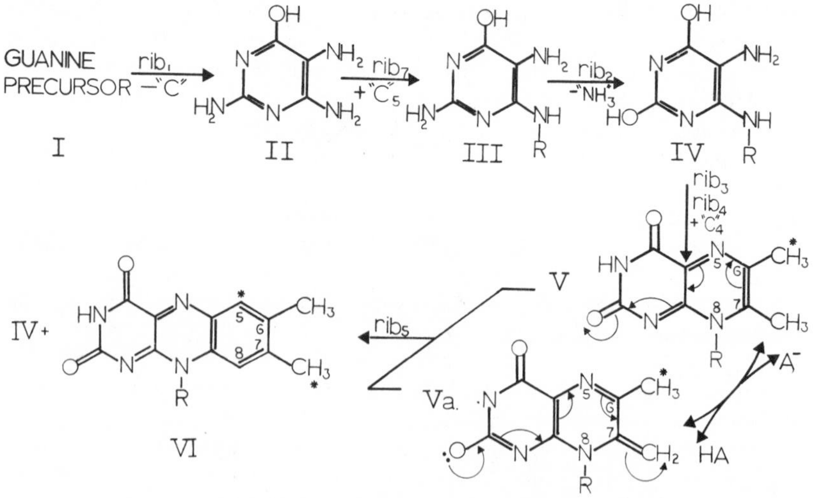

Figure 2 Proposed scheme of riboflavin biosynthesis from a guanine precursor. Genes involved in *S. cerevisiae* (53) are designated by *rib*. R is the ribityl group. The distribution of deuterium in riboflavin formed enzymatically from 6-deuteriomethyl-7-methyl-8-ribityllumazine (54) is shown by *. I, Guanine precursor; II, 6-hydroxy-2,4,5-triaminopyrimidine; III, 2,5-diamino-6-hydroxy-4-ribitylaminopyrimidine; IV, 4-ribitylamino-5-amino-2,6-di-hydroxypyrimidine; V, 6,7-dimethyl-8-ribityllumazine; Va, 6-methyl-7-methylene-8-ribityl-lumazine; VI, riboflavin.

strains, and rib_7-rib_2 strains accumulated 6-hydroxy-2,4,5-triaminopyrimidine (II) (rib_7, Figure 2). This is in accord with a sequence in which genes rib_1 and rib_7 code for the first and second enzymes, respectively, of riboflavin synthesis (Figure 2).

Earlier experiments of Lingens et al (55) had shown accumulation of 6,7-dimethyl-8-ribityllumazine by rib_5 mutants. These results were interpreted to mean that the riboflavin synthetase reaction (56–58), which catalyzes the conversion of 6,7-dimethyl-8-ribityllumazine (V) to riboflavin (VI) and 4-ribitylamino-5-amino-2,6-dihydroxypyrimidine (IV) (Figure 2), was missing. This proposal has now received direct support from the demonstration by Baur, Bacher & Lingens (59) that haploid strains of the yeast defective in genes rib_1, rib_2, rib_3, rib_4, or rib_7 contained levels of riboflavin synthetase comparable to that of the wild strain, whereas this activity could not be detected in dialyzed extracts or ammonium sulfate fractions of rib_5 mutants. The genetic studies of Baur et al (59) indicate, therefore, that a single structural gene, rib_5, codes for riboflavin synthetase; an earlier proposal (60, 61) that an additional gene (rib_6) may be involved at this step was disproved by subsequent investigations.

Baur et al (59) also reported that two diploid strains formed from a combination of haploid strains deficient in gene rib_5 with genotypes rib_{5-3}/rib_{5-7} and

rib_{5-3}/rib_{5-4}, respectively, did not require riboflavin for growth. However, the level of riboflavin synthetase was markedly lower than that of the wild type in one strain and undetectable in the other. The authors suggest that these results could be explained by interaction of identical subunits of riboflavin synthetase with different structural defects (intragenic complementation). This hypothesis will need direct experimental support by studies of the subunit structures of the normal enzyme as compared to defective enzymes obtained from different rib_5 mutants. A molecular weight of 70,000–80,000 has been estimated from sedimentation data for riboflavin synthetase from baker's yeast (58), and it is quite likely that the enzyme is composed of two or more subunits.

Evidence has been presented that the level of riboflavin synthetase in *B. subtilis* is controlled by repression and that with flaviogenic mutants of the microorganism, the level of the enzyme is derepressed when incubated in riboflavin-free medium (61a, 61b). Riboflavinless mutants of *B. subtilis* that accumulated 6,7-dimethyl-8-ribityllumazine were found to lack riboflavin synthetase (61b).

The enzymic and nonenzymic conversion of 6,7-dimethyl-8-ribityllumazine (V) to riboflavin (VI) has been studied in considerable detail. An improved method of purification of riboflavin synthetase from baker's yeast has appeared; the final preparation shows a single band of protein in disc electrophoresis containing the activity (62). The stoichiometry of the reaction catalyzed by riboflavin synthetase

2   6,7-dimethyl-8-ribityllumazine → riboflavin +
4-ribitylamino-5-amino-2,6-dihydroxypyrimidine

indicates that there should be a second order dependency of reaction rate on substrate concentration. However, measurement of initial velocity over a wide range of lumazine concentrations shows zero to first order kinetics with the enzymes from yeast (58), spinach (63), and *A. gossypii* (56). This implies that the dissociation constants of the binary and ternary enzyme-lumazine complexes are dissimilar. Thus, with the yeast enzyme, the lumazine molecule donating a four-carbon unit appears to be bound more tightly than the lumazine molecule which accepts it (58). However, in the enzyme from *E. ashbyii,* binding of one molecule of lumazine increases the affinity for the second molecule (64).

The formation of riboflavin from 6,7-dimethyl-8-ribityllumazine involves the removal of four atoms of hydrogen from the methyl groups of the molecule. Chemical experiments have shown that the hydrogens of the methyl group at carbon 7 (but not at carbon 6) exchange with protons of the medium (65–68) and that the exchange occurs more rapidly under acidic and basic than under neutral conditions (66). Stewart & McAndless (69) have studied systematically this unusual example of general acid–general base catalysis with the model compound 6,7,8-trimethyllumazine and have proposed that the 7-exomethylene group containing intermediate (Va) is formed by interaction of a base (A^-) with the neutral molecule (V) and with a protonated form of the compound (not shown) in the base-catalyzed and acid-catalyzed routes, respectively. The base-catalyzed type of reaction is shown in Figure 1 (V and Va). The presence of nonenzymically formed 7-exomethylene compounds is indicated by the loss of ultraviolet absorption in the range of 400 nm and the appearance of a maximum in the 375 nm range (66, 70–72). It is confirmed by the appearance of a doublet in the 3.6 to 4.2 ppm range in NMR spectra taken in basic media (65–68). Pfleiderer, Mengel & Hemmerich (72) were able to isolate in crystalline form 3,6-dimethyl-8-phenyl-7-methylene pteridine showing these spectral character-istics. The 7-exomethylene form predominates in alkaline solution in 6,7-dimethyl-lumazines substituted at position 8 with any group lacking a 2′-hydroxyl substituent. However, in the presence of a 2′-hydroxyl group, an equilibrium mixture is found containing the 7-exomethylene derivative and, predominantly, a form of the compound in which an intramolecular ether is formed between the 2′-hydroxyl group of the side chain and carbon 7 of the pyrazine ring of the lumazines with the characteristic 7-methyl absorption of 1.37 ppm in the NMR spectra (73).

The exchange reaction is also catalyzed by purified riboflavin synthetase but only at the 7-methyl group of those lumazine derivatives that are substrates or can be bound by the enzyme (62). Furthermore, the enzymatic (54) and nonenzymatic (67) formation of the aromatic ring of riboflavin proceeds by condensation of the 6- and 7-methyl groups of one molecule of 6,7-dimethyl-8-ribityllumazine with carbons 7 and 6, respectively, of a second molecule of the substrate. Thus, 6-deuteromethyl-7-methyl-8-ribityllumazine (V) has been converted enzymatically to riboflavin labeled with deuterium in the 7-methyl group and at carbon 5 (Figure 2). On the basis of the chemical properties of the lumazine and the enzyme-catalyzed reaction, a mechanism of riboflavin synthesis has been proposed. The initial step involves reaction of a proton acceptor group on the enzyme (A$^-$) with the 7-methyl group of the lumazine (V) to form the corresponding 7-methylene derivative (Va) (Figure 2), which could condense with carbon 6 of a second lumazine molecule. The condensation of the 6-methyl group with carbon 7 and aromatization to riboflavin would then follow these steps (54).

No definitive product of accumulation has been identified so far in mutants of *S. cerevisiae* with a block at gene rib$_1$ (Figure 2). However, there is considerable circumstantial evidence that the first committed step of riboflavin biosynthesis involves elimination of carbon 8 of a guanine-containing compound. Numerous studies have shown that purine precursors and purine derivatives are precursors of riboflavin (for review see 74), and Howells & Plaut (75) demonstrated with a purineless mutant of *E. coli* that a compound at or subsequent to the level of IMP is an intermediate. Support for the participation of a guanine derivative comes from the following observations: (*a*) Bacher & Lingens (76) showed that a mutant of *A. aerogenes* with a genetic block between XMP and GMP was able to incorporate label from [2-^{14}C]guanine, but not [^{14}C]xanthine, into riboflavin. Similar results were obtained by Baugh & Krumdieck (77) with a *Corynebacterium* species inhibited by decoyinin. (*b*) The pyrimidine compounds identified at the initial stages of ribo-flavin synthesis (Figure 2), 6-hydroxy-2,4,5-triaminopyrimidine (II) and 2,5-diamino-6-hydroxy-4(1′-D-ribitylamino)pyrimidine (III), are related structurally more closely to guanine than to the subsequent deaminated product, 4-ribitylamino-5-amino-2,6-dihydroxypyrimidine (IV). (*c*) A ring opening reaction leading to elimination of carbon 8 from GTP has been shown to be the first committed step in the synthesis of a number of pteridines, azapteridines, and pyrrolopyrimidine nucleosides (78–84).

Although a definitive involvement of GTP has not been reported so far, it is tempting to believe that GTP is also the starting point of riboflavin biosynthesis. Initial elimination of carbon 8 from the ring of GTP could lead to formation of a product unique for the riboflavin pathway. This would be consistent with other GTP 8-formylhydrolase-type reactions where the nature of products formed varied with different enzymes. For example, diaminouracil has been reported as the major product of the toxoflavin-producing organism *Pseudomonas cocovenenans* (83). The triphospho ester of 2-amino-4-hydroxy-6-(*erythro*-1',2',3'-trihydroxypropyl) dihydropteridine formed by the enzyme from *E. coli* (78) and *S. rimosus* (84) appears to participate in folate and pyrrolopyrimidine nucleoside synthesis, respectively, and a cyclic phosphate of neopterin was obtained with *Comamonas* sp. (80), which is presumably an intermediate in the formation of L-threoneopterin.

The isolation of a radioactive blue fluorescent guanine-containing compound from cells of *E. ashbyii* incubated with [U-^{14}C]GMP has been reported by Mehta et al (85). They have reported conversion of this material to riboflavin by cell-free extracts of *E. ashbyii*. Inorganic phosphate, radioactive guanine, and ribitol were identified by paper chromatography as products of hydrolysis of this compound, and these groups might contribute to formation of both the heterocyclic ring and ribityl portions of flavin. If the blue fluorescent material is an intermediate, this may suggest a difference in the pathways of riboflavin formation in *E. ashbyii* and in *S. cerevisiae*, since the first ribityl-containing compound (2,5-diamino-6-hydroxy-4-ribitylaminopyrimidine, III) recovered in mutants of *S. cerevisiae* is at least one step beyond the ring opening stage where loss of carbon 8 from the purine is presumed to occur (Figure 2). Mehta et al suggest (85) that the blue fluorescent compound may be a nucleotide derivative resembling 9-phosphoribitylpurine. It should be possible to test whether 9-(5'-phosphoribityl)guanine per se is identical with the blue fluorescent compound, since the dephosphorylated substance, 9-ribitylguanine, has been synthesized by Davoll & Evans (86).

The mechanism and stage of formation of the ribityl group of riboflavin is still uncertain. Studies of Ali & Al-Khalidi (87) with cell suspensions of *E. ashbyii* have shown that 20–30% of the label from [1-^{14}C]glucose, [6-^{14}C]glucose, or [1-^{14}C]-ribose recovered in riboflavin is located in the ribityl group. These results and those obtained with *A. gossypii* (88) are consistent with the possibility that the ribityl group can be derived from a pentosyl precursor. However, it is questionable whether the ribosyl group of a guanosine derivative is used directly for formation of the ribityl group. Forrest & McNutt (89) had found that intact cells of *E. ashbyii* incorporated labels from [U-^{14}C]guanosine into the ring system of riboflavin more efficiently than into the ribityl substituent. This could be due to exchange of the labeled ribosyl group of added guanosine with the unlabeled pentose pool of the organism, and the ribosyl group of a purine nucleoside or nucleotide could still be the direct precursor of the ribityl group, perhaps by way of reduction of an intermediary Amadori compound.

So far, studies with *S. cerevisiae* do not permit a decision on the source of the ribityl group, since no accumulation products have been isolated from mutant rib$_1$ cultures (Figure 2). It is also uncertain whether 6-hydroxy-2,4,5-triaminopyrimidine, identified as the corresponding pteridine derivatives by trapping with added glyoxal or 2,3-butanedione (52), accumulates as such in mutants of type rib_7 (Figure 2) or if it is formed as an artifact of isolation of a labile precursor compound containing a pentose moiety (52). If 6-hydroxy-2,4,5-triaminopyrimidine is an intermediate, formation of a ribityl group must take place in its conversion to 2,5-diamino-6-hydroxy-4-(1'-D-ribitylamino)pyrimidine (90) (Figure 2). This could occur by condensation of the 4-amino group of the pyrimidine with a ribosyl compound, rearrangement of the ribosylamino derivative to an Amadori compound followed by reduction at the ribityl group. Alternatively, direct introduction of the group from a ribityl precursor compound could occur at the 4-amino group of the pyrimidine. The possibility that a nucleoside diphosphate-ribitol derivative may be such a precursor of the ribityl group has been suggested by a number of reviewers of this field (50, 91). Indications that ribitol is a better precursor than D-ribose of the ribityl group of riboflavin have been obtained in studies with intact organisms by Miersch (92) and Mehta et al (85).

Using cultures of *Candida guilliermondii*, Miersch (92) found more extensive incorporation of radioactivity into riboflavin from [U-^{14}C]ribitol than from [U-^{14}C]ribose. This occurred particularly when the iron level of the media was relatively high, a condition which depressed the overproduction of riboflavin (93). The cells from low iron media contained rather large amounts of free ribitol (about 1.5% of the dry weight), whereas practically no ribitol was found in the high iron cells. Less dilution by the ribitol pool may thus explain the larger incorporation of radioactivity from [^{14}C]ribitol into riboflavin by cells grown on high iron media. Radioactivity from [^{14}C]ribose or [^{14}C]ribitol was recovered almost exclusively in the ribityl portion of riboflavin, favoring formation of the ribityl group through an intermediate at the level of oxidation of ribitol. The almost complete absence of incorporation of label from [^{14}C]ribose into the heterocyclic ring portion of riboflavin, however, suggests a fundamental difference in ribose utilization or in riboflavin biosynthesis between *C. guillermondii* and *E. ashbyii*, since Ali & Al-Khalidi (87) found that resting cells of *E. ashbyii* incorporated radioactivity from [1-^{14}C]-ribose effectively into the ribityl and the *o*-xylene ring of riboflavin.

Preferential utilization of ribitol rather than ribose for riboflavin formation is also suggested by Mehta et al (85). They found that unlabeled ribitol added to cultures of *E. ashbyii* decreased incorporation of radioactivity from [1-^{14}C]ribose into riboflavin, while incorporation into other cellular components (nucleotides and nucleic acids) appeared unchanged. It is not certain whether this can be attributed solely to a preferential role of ribitol in the formation of the ribityl group, since the distribution of label in riboflavin was not determined. However, such a role for ribitol is favored by the fact that a number of enzymes which could be involved in formation of a ribityl group precursor from a ribosyl compound (GMP nucleosidase, TPN-specific ribose reductase, ribose-5-phosphatase and ribitol kinase) increase in activity during growth before riboflavin formation became maximal. Moreover, addition of ribitol induced the formation of both ribitol kinase and riboflavin. The flaviogenic purines, guanine and xanthine, also activated ribitol kinase in vitro.

These observations may indicate utilization of the ribosyl group of nucleotides with ribitol compounds as intermediates for formation of the ribityl moiety of riboflavin. However, the activities were measured in extracts of the organisms, and the individual enzymes remain to be characterized. Thus, the TPN-linked aldose reductase catalyzed the reduction of ribose 5-phosphate and ribulose 5-phosphate in addition to the preferred substrate ribose, and the ribitol kinase activity also catalyzed phosphorylation of xylitol. The nucleosidase appears highly unspecific or represents a mixture of enzymes, since a number of nucleoside mono- and poly-phosphates served as substrates yielding free ribose as the product. The role of this nucleosidase in riboflavin synthesis is uncertain, since studies by Mehta et al (85) on incorporation of [U-^{14}C]GMP by intact cells of E. ashbyii did not show label in the expected product guanine, although radioactivity was recovered in GTP, GDP, GMP, riboflavin, and the blue fluorescent compound discussed previously.

The four carbons required for formation of 6,7-dimethyl-8-ribityllumazine (V) from 4-ribitylamino-5-amino-2,6-dihydroxypyrimidine (IV) (Figure 2) may be derived from a carbohydrate precursor (87, 94). However, it may be pertinent to the problem of formation of the lumazine compounds that a number of 6-substituted-7-hydroxy-8-ribityllumazine derivatives have been isolated from Pseudomonas ovalis, in which the substituent at position 6 is a methyl, hydrogen, 2-carboxyethyl, 3-indolyl, or p-hydroxyphenyl group (95–98). The methyl derivative has been detected previously in a number of riboflavin overproducers (for review see 50). The indolyl and p-hydroxyphenyl substituents in particular could suggest an origin of carbons 6 and 7 with the attached side chains of these lumazines from an amino acid precursor.

FORMATION OF VITAMIN B$_{12}$[3]

The role of Vitamin B$_{12}$ and related compounds in enzymatic reactions has been recently reviewed by Barker (99); biosynthesis of these compounds has been reviewed by Friedmann & Cagen in 1970 (100) and by Friedmann in a forthcoming book (101). Recent developments in the area of B$_{12}$ biosynthesis are discussed in this review in terms of the mode of formation of the individual structural units illustrated in Figure 3.

Shemin and co-workers established that the corrinoid macrocycle, like the structurally related porphyrin macrocycle, is derived from 5-aminolevulinate (103–105). In contrast to porphyrins, the corrinoids lack a one-carbon bridge between rings A and D (Figure 3) and contain seven methyl groups (attached at C-1, 2, 5, 7, 12, 15, and 17) which cannot be derived from decarboxylation of the acetate side chains of a macrocyclic precursor. In 1972 Brown, Katz & Shemin (107) and Scott et al (108) investigated the ^{13}C NMR spectra of B$_{12}$ isolated from Propionibacterium shermanii after exposure to ^{13}C-enriched precursors. Contrary to expectation based upon the earlier investigations of Bray & Shemin (105, 106) with ^{14}C-labeled precursors, label from [5-^{13}C]5-aminolevulinate was incorporated into only seven positions of the isolated B$_{12}$, and not into any of the methyl groups (107, 108).

[3] Abbreviations used in this review are: α-ribazole [(1-α-D-ribofuranosyl)5,6-dimethyl-benzimidazole]; DBI (5,6-dimethylbenzimidazole).

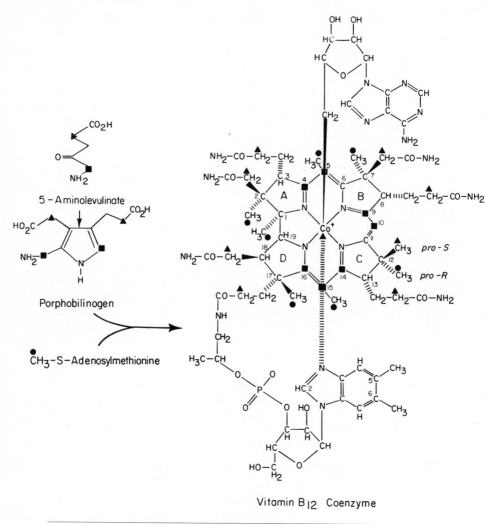

Vitamin B₁₂   Coenzyme

Figure 3   5'-Deoxyadenosylcobalamin (Vitamin B₁₂ Coenzyme). For nomenclature rules for Vitamin B₁₂ and related compounds, see (102). The biosynthetic origins of selected carbon atoms from 5-aminolevulinate, porphobilinogen, and S-adenosylmethionine are indicated.

Furthermore, all seven of the extra methyl groups, including that at C-1, were derived from [CH₃-¹³C]methionine (108). This observation has been confirmed by Brown, Katz & Schemin (109) and by Battersby et al (110). Battersby et al (110) and Scott et al (111) have also established that the *pro-R* methyl at C-12 of ring C is derived from methionine.[4] This requires that the methylation process for ring C proceed by formal *anti* addition of CH_3 and H. Based upon the structure of B_{12} established by X-ray crystallography (112), formal *anti* addition of CH_3 and H also must occur in rings A, B, and D during the formation of the corrinoid (Figure 3).

Scott et al (113) confirmed an earlier report (114) that porphobilinogen is converted into B_{12}. In addition, they found that a mixture of chemically formed ^{14}C-labeled uroporphyrinogens I, II, III, and IV, and a mixture of enzymatically produced uroporphyrinogens I and III yielded ^{14}C-labeled B_{12} with *P. shermanii* under carefully controlled culture conditions (113). Neither uroporphyrinogen I alone nor a mixture of uroporphyrins I, II, III, IV was incorporated. They attributed the failure of other investigators (115, 116) to detect the incorporation of radioactivity from uroporphyrinogen into B_{12} to differences in the culture incubation conditions. Scott et al also established that [8-^{13}C]porphobilinogen specifically labeled four methylene carbons of B_{12}, and that a ^{13}C-enriched uroporphyrinogen I–IV mixture (prepared chemically from [8-^{13}C]porphobilinogen) specifically labeled the same methylene carbons (113). They concluded that biosynthesis of the corrinoid ring of B_{12} proceeds via the sequence

5-aminolevulinate → porphobilinogen → uroporphyrinogen III → B_{12}

and involves a reductive contraction of the uroporphyrinogen III ring (113).

These experiments substantiate the proposals of Burnham (117, 118) and Porra (119) that uroporphyrinogen III lies at the branch point of heme and corrinoid biosynthesis. These workers found that cell suspensions and cell-free extracts prepared from *Clostridia,* organisms not known to make heme compounds, convert 5-aminolevulinate to uroporphyrinogen.

As shown in Figure 3, these recent results (107–111, 113) establish that: 1. All seven extra methyl groups of the corrinoid ring are derived from methionine. 2. Carbon 5 of one of the eight 5-aminolevulinate molecules utilized in the formation of the macrocycle is lost. 3. Uroporphyrinogen III serves as a precursor in *P. shermanii.* 4. The acetate and propionate side chains of porphobilinogen are incorporated; decarboxylation of one of the incorporated acetate groups yields the *pro-S* methyl group at C-12.[5]

The earlier mechanistic proposals for corrinoid ring formation (see review 120) are not compatible with these recent findings. Scott and co-workers have proposed several alternate pathways from uroporphyrinogen III to the corrinoids which are consistent with the results summarized in Figure 3 (113, 121); two involve hydrolytic ring contractions (113) and one a 16 π electron conrotatory electrocyclic ring closure (121).

Details which remain obscure about the biosynthetic transformation of uroporphyrinogen III into cobyrinic acid include the sequence of the seven methylations, the reactions resulting in the loss of the carbon bridging rings A and D from the uroporphyrinogen precursor, including the oxidation state of the

---

[4] Brown, Shemin & Katz have recently proposed the opposite interpretation (192).

[5] See, however, 192.

eliminated carbon atom, and the precise role of cobalt in the biosynthesis of the corrinoids. The report by Scott et al (122) of a stable cell-free *P. shermanii* extract that forms cobyrinic acid when incubated with S-adenosylmethionine and either 5-aminolevulinate or uroporphyrinogens is a major advance in the study of corrinoid ring biosynthesis and will permit these details to be investigated.

No definite role for the cobalt-free corrinoids first isolated by Toohey from *Chromatium* (123, 124) has been established (for review see 100). The biosynthetic paths proposed by Scott et al (113, 121) involve the insertion of cobalt into a preformed corrinoid ring to yield cobyrinic acid. Consistent with this view, Burnham has reported that a ^{14}C-labeled cobalt-free corrinoid isolated from *Chromatium* is converted into ^{14}C-labeled B_{12} when added to a *Clostridium perfringens* culture (120, p. 469). In contrast, Dolphin has postulated that cobalt plays a special role in the cyclization processes that yield the corrinoid macrocycle (125), and Koppen-hagen & Pfiffner have reported that cobalt-free corrinoids are not B_{12} precursors in *P. shermanii* (126).

In the next phase of the B_{12} biosynthetic pathway, cobyrinic acid is converted to 5′-deoxyadenosylcobinamide. Six of the carboxylic acid residues of cobyrinic acid are converted to primary amides, and one to a secondary amide with (*R*)-1-amino-2-propanol (also designated D_g-isopropanolamine). The 5′-deoxyadenosyl group also is attached to the corrinoid during this stage of the biosynthesis. These conversions may involve concurrent pathways of assembly rather than a linear sequence (see 100, 120). Rapp's (127) recent studies dealing with this aspect of B_{12} biosynthesis in *P. shermanii* agree with the previously postulated sequence of amidations (100, 120) leading to cobinamide in *P. shermanii*.

The origin of the (*R*)-1-amino-2-propanol moiety that becomes attached to the propionate side chain at C-17 has been the subject of several recent papers. This molecule corresponds in chemical and stereochemical structure to the decarboxy-lation product of L-threonine (Figure 4), and label from L-[^{15}N]threonine and L-[U-^{14}C]threonine is incorporated into the aminopropanol unit of B_{12} (128–130). However, the direct decarboxylation of L-threonine has not been detected (131, 132), and experiments indicate that neither threonine nor O-phosphothreonine are de-carboxylated after attachment to the corrinoid (133).

Neuberger & Tait (134) proposed that L-threonine could be converted into (*R*)-1-amino-2-propanol via 2-amino-3-ketobutyrate and aminoacetone. This scheme, shown in Figure 4, is supported by the discovery of L-threonine dehydrogenase activity (134–136), the fact that 2-amino-3-ketobutyrate decarboxylates spon-taneously to yield aminoacetone (137), and the presence of 1-amino-2-propanol dehydrogenase activity in *E. coli* extracts (138–140). Campbell & Dekker (141) have obtained L-threonine dehydrogenase and (*R*)-1-amino-2-propanol dehydrogenase in purified form, each free of the other, and they demonstrated the conversion of L-threonine into (*R*)-1-amino-2-propanol in a coupled system. However, Lowe & Turner (142) found that the (*R*)-1-amino-2-propanol dehydrogenase purified from *E. coli* was more active with alcohols such as (*RS*)-propan-1,2-diol and glycerol. They also could not detect L-threonine dehydrogenase or (*R*)-1-amino-2-propanol dehydrogenase activities in extracts of anaerobically grown *P. shermanii* or *P.*

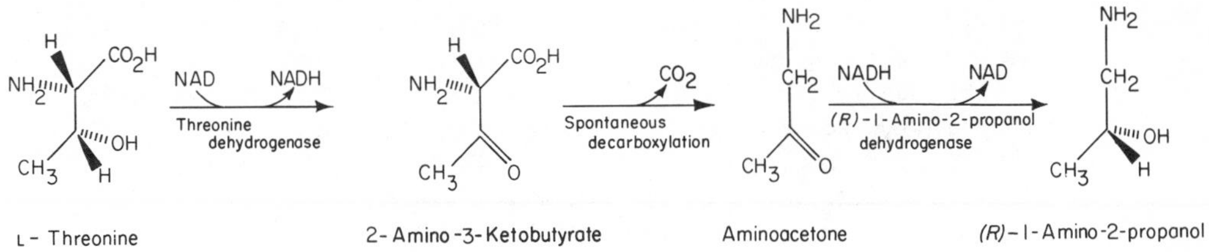

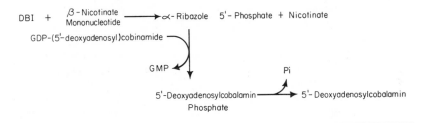

L - Threonine   2- Amino -3- Ketobutyrate   Aminoacetone   (R) - 1 - Amino-2-propanol

Figure 4   Proposed biosynthetic route to (R)-1-amino-2-propanol involving L-threonine dehydrogenase and (R)-1-amino-2-propanol dehydrogenase.

freudenreichii (142), although under such conditions these organisms require only the presence of DBI to form the complete cobalamin (see discussion in 100). Furthermore these dehydrogenases are present in *E. coli,* which cannot form B_{12} de novo (143, 144) or attach the 1-amino-2-propanol moiety to cobyric acid to form the vitamin (145, 146). On the basis of these observations, Lowe & Turner (142) concluded that it is unlikely that L-threonine dehydrogenase and (R)-1-amino-2-propanol dehydrogenase are involved in B_{12} biosynthesis.

Two groups of workers who attempted to detect incorporation of radioactivity from (R)-[U-^{14}C]-1-amino-2-propanol into the aminopropanol unit of B_{12} reached different conclusions. Lowe & Turner (129) did not find significant incorporation with *Streptomyces olivaceus* and concluded that 1-amino-2-propanol was not a direct precursor of B_{12}; Müller, Gross & Siebke (130) detected slight incorporation with *P. shermanii* and concluded that it was. In summary, a role for L-threonine dehydrogenase, (R)-1-amino-2-propanol dehydrogenase, and free (R)-1-amino-2-propanol in the biosynthesis of B_{12} remains to be established.

In the final stages of B_{12} biosynthesis, 5′-deoxyadenosylcobinamide is phosphorylated and the product condenses with GTP to form GDP-(5′-deoxyadenosyl)-cobinamide (147, 148). This activated cobinamide then combines with α-ribazole 5′-phosphate to form 5′-deoxyadenosylcobalamin phosphate (149–151) (Figure 5). This latter compound was shown to be an obligatory intermediate by Friedmann and co-workers (150, 151), and the structure of the cyano derivative was established

DBI + β - Nicotinate Mononucleotide ⟶ α - Ribazole 5′ - Phosphate + Nicotinate

GDP-(5′-deoxyadenosyl)cobinamide

GMP

Pi

5′-Deoxyadenosylcobalamin ⟶ 5′- Deoxyadenosylcobalamin Phosphate

Figure 5   Biosynthetic sequence from 5′-deoxyadenosylcobinamide to the completed B_{12} coenzyme, 5′-deoxyadenosylcobalamin.

by X-ray diffraction (152). Although a specific phosphatase for the conversion of 5'-deoxyadenosylcobalamin phosphate to 5'-deoxyadenosylcobalamin (B_{12} coenzyme) remains to be characterized, Schneider & Friedmann (153) have described the enzymatic dephosphorylation of cobalamin phosphates with purified *E. coli* alkaline phosphatase and have observed the slow hydrolysis of 5'-deoxyadenosylcobalamin phosphate by a heat-labile activity in *P. shermanii* extracts. The α-ribazole 5'-phosphate required in this sequence is formed by a displacement reaction involving free DBI and β-nicotinate mononucleotide (154–156). The *trans*-N-glycosidase catalyzing this reaction in *Clostridium sticklandii* has been extensively purified (156); a similar enzyme in *P. shermanii* has also been described and partially purified (154, 155).

Biosynthesis of the DBI involved in formation of B_{12} (Figure 5) occurs by way of riboflavin or its immediate precursor. In 1967 Renz & Reinhold (157) studied the distribution of ^{14}C from lactate into DBI and compared it to the distribution of radioactivity in the structurally related 4,5-dimethyl-1,2-phenylene portion of riboflavin (ring A) previously obtained by Plaut with labeled acetate (94). They concluded that formation of DBI and riboflavin occurred via related pathways (157). This conclusion was supported by the observation of Alworth et al (158) that precursors of the dimethylphenylene unit of DBI had also been found to be efficient precursors of ring A of riboflavin (87). A direct link between DBI biosynthesis and riboflavin was then established by Renz, who demonstrated the incorporation of radioactivity from [U-^{14}C]riboflavin into the DBI moiety of B_{12} by disrupted *P. shermanii* cells (159).

Subsequent investigations carried out in the laboratories of Renz (160, 161) and Alworth (162–164) established interesting details of the relationship between the biosyntheses of riboflavin and DBI. [CH$_3$-^{14}C]6,7-Dimethyl-8-ribityllumazine was an efficient precursor of DBI (160, 162, 163); the resulting labeling pattern in the dimethylphenylene structure corresponded to that previously determined by Plaut (56) in ring A of riboflavin. Lu & Alworth (164) observed that ^{14}C label from [1'-^{14}C,5-^{15}N]6,7-dimethyl-8-ribityllumazine was recovered exclusively in the C-2 position (see Figure 3) of the DBI. Furthermore, the $^{15}N/^{14}C$ enrichment ratios of the lumazine precursor and the DBI product indicated that the C-1' and the N-5 atoms were incorporated as a unit. These observations establish that all of the atoms of DBI may be derived from 6,7-dimethyl-8-ribityllumazine, the known precursor of riboflavin.

As in riboflavin biosynthesis (56), the 4,5-dimethyl-1,2-phenylene structure is formed by the combination of the four-carbon units of two lumazine molecules (Figure 2). The C-2 carbon of DBI is derived from C-1' of the ribityl side chain, one of the nitrogen atoms of DBI comes from N-5 of the lumazine precursor, and the other nitrogen atom presumably from N-8.

The labeling patterns in DBI obtained with labeled 6,7-dimethyl-8-ribityllumazines are consistent with formation of DBI via riboflavin. Renz & Weyhenmeyer have found incorporation of radioactivity from [1'-^{14}C]riboflavin into C-2 of DBI (161). The efficient conversion of riboflavin labeled in various positions into DBI by *P. shermanii* (159, 161) strongly indicates that riboflavin is an obligatory intermediate, since the conversion of riboflavin to 6,7-dimethyl-8-ribityllumazine has never been observed. Alworth & Lu (164) had argued that incorporation of 51% of the specific activity of the added $[1'-^{14}C,5-^{15}N]$-6,7-dimethyl-8-ribityllumazine precursor into DBI by *P. shermanii* cultures was too high to be consistent with incorporation through the riboflavin pool. However, considering the total amount of riboflavin present in *P. shermanii* cultures [1.89 mg/liter (165)] and the amount of labeled 6,7-dimethyl-8-ribityllumazine added [152.1 mg/12 liters (164)], the riboflavin pool could have been sufficiently labeled to account for the high incorporation. The biosynthetic connection between riboflavin and DBI has been reviewed recently by Schlee (166).

ASPECTS OF VITAMIN B_6 BIOSYNTHESIS[6]

The biosynthesis of Vitamin B_6 was reviewed briefly in *Nutritional Reviews* in 1972 (168). The compounds that comprise the Vitamin B_6 group are metabolically interconvertible (169–171), and it has been difficult to establish the first cyclic compound formed on the biosynthetic pathway. Wild-type *E. coli* extracts rapidly phosphorylate pyridoxol to pyridoxol 5'-phosphate and rapidly oxidize pyridoxol 5'-phosphate to pyridoxal 5'-phosphate, but cannot oxidize pyridoxol to pyridoxal (172). A mutant of *E. coli* (B-WG2) that lacks pyridoxol 5'-phosphate oxidase activity and will not grow on pyridoxol can grow on pyridoxal or pyridoxamine and excretes pyridoxol 5'-phosphate when shifted to a pyridoxal-free medium (172, 173). Dempsey therefore proposed that the terminal steps are

$$\text{pyridoxol} \xrightarrow{\text{kinase}} \text{pyridoxol 5'-phosphate} \xrightarrow{\text{oxidase}} \text{pyridoxal 5'-phosphate}$$

This proposal is also supported by the relative K_m values for pyridoxol, pyridoxal, and pyridoxamine of several kinases (169). However, it has also been suggested that pyridoxol 5'-phosphate is directly formed as the initial cyclic product and is the precursor of the unphosphorylated pyridoxol (174).

The incorporation of label from many compounds into Vitamin B_6 by a variety of organisms has been tested (174–186). The results[7] have led to varying and at times contradictory conclusions. For example, $[3-^{14}C]$serine was found to be the most effective precursor of those tested in *Candida utilis* (180), but was a relatively poor precursor in *Saccharomyces fragilis* (181). Dempsey reported that the specific radioactivity of pyridoxol isolated from *E. coli* was 12–53% that of $[3-^{14}C]$serine administered (182, 184), incorporations that were two orders of magnitude greater than that found for the $[2-^{14}C]$pyruvate precursor (184). In contrast, Hill et al concluded that in *E. coli*, $[3-^{14}C]$serine was a poorer precursor of pyridoxol than $[2-^{14}C]$pyruvate (174). They found that the label from $[3-^{14}C]$serine was specifically incorporated into pyridoxol, with 92% being localized in C-2 + C-2' (174), whereas

---

[6] Vitamin B_6 is the recommended generic name for all 2-methylpyridine derivatives with biological activity; pyridoxine or pyridoxol specifically refers to 3-hydroxy-4,5-bis(hydroxymethyl)-2-methylpyridine (167).

[7] These compilations are available upon request from Dr. W. L. Alworth.

Schroer & Frieden concluded that in *S. fragilis* label from [3-^{14}C]serine was randomly incorporated into pyridoxol (181). The investigations involving labeled aspartate precursors (174, 176–178, 180, 181, 184) also illustrate the range of variation found in published results.

A most comprehensive study of B$_6$ biosynthesis has been carried out by Hill, Spenser, and their associates (174, 179, 185, 186) using washed cells of *E. coli* B-WG2 (172, 173) suspended in a medium lacking pyridoxal. In these experiments the introduction of label from potential precursors should have coincided with the maximal rate of pyridoxol 5'-phosphate formation (172). Hill et al found that the best incorporation of isotope into B$_6$ was obtained with glycerol, pyruvate, and glucose. However, recalculation of the data obtained by Hill, Spenser, and associates[7] indicates that total incorporations of radioactivity into pyridoxol was slight. Furthermore, the relative extent of labeling of pyridoxol phosphate by the presumed distant (leucine, aspartate) and close (glycerol, pyruvate) precursors differed maximally by only one order of magnitude. The yields of pyridoxol were not determined and, therefore, it is impossible to determine the actual incorporations into the newly synthesized pyridoxol.

On the basis of specific labeling patterns obtained with glycerol, pyruvate, and glucose, Hill et al (174, 186) proposed that dihydroxyacetone phosphate condenses with acetaldehyde derived from pyruvate or with active glycolaldehyde derived from a hexose to yield pentulose 1-phosphate intermediates. The intermediate would be D-5-deoxyxylulose 1-phosphate or D-xylulose 1-phosphate depending on the condensation of dihydroxyacetone phosphate occurring with acetaldehyde or glycolaldehyde, respectively. The pentulose 1-phosphate intermediate serves as the biosynthetic precursor of C-2, 3, 4, 2', 4' of pyridoxol. The pentulose intermediate undergoes a condensation with glyceraldehyde 3-phosphate (the proposed precursor of C-5, 5', 6) to yield an eight-carbon branched chain saccharide. This intermediate is then converted into pyridoxol 5'-phosphate.

The experimental observations of Hill, Spenser, and associates (174, 179, 185, 186) were as follows: Label from [1,3-^{14}C]glycerol is recovered in pyridoxol at C-2' (18%), C-4' (22%), and C-5' (22%), and from [2-^{14}C]glycerol at C-2 (33%) and C-4 (33%).[8] The remainder of the label from the [1,3-^{14}C]glycerol is probably located equally (20%) at C-3 and C-6 of pyridoxol, while that from [2-^{14}C]glycerol (33%) is probably at C-5. The partial labeling patterns resulting from [1,3-^{14}C]glycerol and [2-^{14}C]glycerol incorporations into pyridoxal in *Flavobacterium* reported by Suzue & Haruna (177) are consistent with these observations. Label from [2-^{14}C]-pyruvate was localized in the C-2+C-2' fragment (94%) and that from [3-^{14}C]-pyruvate was in C-2' (88%), suggesting that the C-2+C-2' portion of the pyridoxol molecule was formed from a two-carbon unit such as acetaldehyde derived from

---

[8] The amount of label localized at C-2'+C-2 has been used as a test for random incorporation (181). As illustrated by the more complete labeling pattern that was determined by Hill et al in the case of [1,3-^{14}C]glycerol incorporation (174), the fact that about one fourth of the incorporated label is in C-2+C-2' is not adequate proof that the incorporation is random.

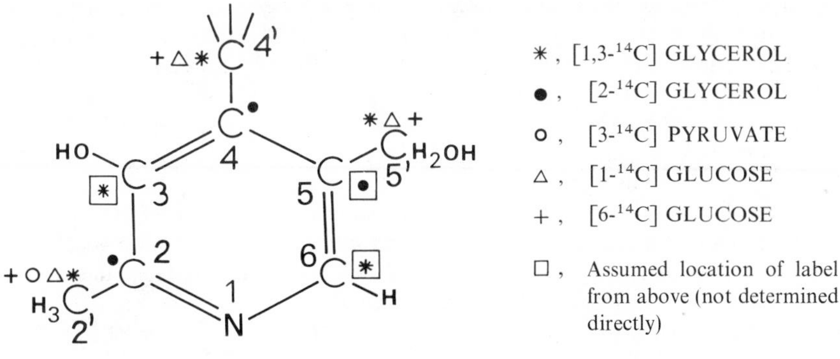

*, [1,3-^{14}C] GLYCEROL

•, [2-^{14}C] GLYCEROL

○, [3-^{14}C] PYRUVATE

△, [1-^{14}C] GLUCOSE

+, [6-^{14}C] GLUCOSE

□, Assumed location of label from above (not determined directly)

Figure 6 Locations of label in Vitamin B$_6$ incorporated from various precursors by *E. coli* B-WG2, according to Hill, Spenser and associates (174, 179, 185, 186).

the C-2 + C-3 atoms of pyruvate. [1-^{14}C]Glucose was found to specifically label the C-2' (36%), C-4' (38%), and C-5' (26%) positions of pyridoxol, and [6-^{14}C]glucose was found to label the C-2' (26%), C-4' (27%), and C-5' (48%) positions. These labeling results are summarized in Figure 6.

Based upon these quantitative labeling results, it was postulated that the three-carbon unit incorporated to form C-5 + C-5' + C-6 of pyridoxol was closely related to glyceraldehyde 3-phosphate and that the three-carbon unit incorporated to form C-3 + C-4 + C-4' of pyridoxol was closely related to dihydroxyacetone phosphate. Since the labeling at C-2' produced by the incorporation of either [1-^{14}C]glucose or [6-^{14}C]glucose is equivalent to the labeling at C-4' but differs significantly from the labeling at C-5', Hill et al made the additional proposal that the C-2 + C-2' fragment could also be derived from the C-1 + C-2 position of glucose in a process that did not proceed via pyruvate and acetaldehyde. It was postulated that the action of a transketolase on a hexose such as fructose 6-phosphate could yield an active glycolaldehyde unit that could function as the precursor of C-2 + C-2' of pyridoxol (186).

To the limited extent that the proposals have been tested, the pathways of Hill, Spenser, and associates have been difficult to corroborate. Schroer & Frieden (181) found that the presence of unlabeled glycolaldehyde significantly decreased the amount of [U-^{14}C]glucose incorporated into pyridoxal by *S. fragilis,* and Tani & Dempsey have found that both [2-^{14}C]glycolaldehyde and [U-^{14}C]glycolaldehyde served as highly efficient precursors of labeled pyridoxol in *E. coli* WG3[9] (183). According to the pathway proposed by Hill et al, label from glycolaldehyde should only be incorporated into the C-2 + C-2' positions, yet less than 3% of the radioactivity in the pyridoxol samples isolated by Tani & Dempsey is located in these

---

[9] This mutant is a pyridoxal auxotroph in which glycolaldehyde can replace the pyridoxal requirement (187).

positions (186). Furthermore, although a number of pentoses were tested, none were able to replace the pyridoxal requirement of several *E. coli* B_6 auxotrophs (183). The pentoses tested included L-xylulose, D-xylulose, D-5-deoxyxylose, D-xylulose 5-phosphate, D-ribulose 5-phosphate, L-fucose plus D-arabinose, L-arabinose, D-xylose, and D-ribose.

In a series of papers, Dempsey and his colleagues have described the isolation and characterization of a number of *E. coli* mutants which are B_6 auxotrophs (172, 173, 182, 187, 191). As a result of transduction studies with the phage Plbt, Dempsey determined that B_6 auxotrophs of *E. coli* B could be divided into five genetically unlinked groups (189). From the transduction results he estimated there were between six and ten enzymatically catalyzed transformations between the first committed step and pyridoxal phosphate. Among the different classes of *E. coli* B_6 auxotrophs are some in which glycolaldehyde will restore B_6 biosynthesis without restoring normal growth. Others will grow normally in the presence of either 0.4 mM glycolaldehyde or 0.15×10^{-3} mM pyridoxal (187). The possibility was raised that 3-phosphoserine could be a biosynthetic intermediate. Mutants lacking 3-phospho-hydroxypyruvate:glutamate transaminase require both pyridoxal and serine for growth, suggesting that these two substances have a common precursor following the transaminase step (182, 188, 190, 191). A serine auxotroph that lacked 3-phosphoserine phosphatase produced pyridoxal at twice the normal rate during serine starvation (191). This implies that serine per se is not a precursor of Vitamin B_6. Nevertheless, in a preliminary report Dempsey (182) indicated that the specific radioactivity of pyridoxol formed by *E. coli* mutant WG1100 was about 50% that of added DL-[3-^{14}C]serine or 15% that of L-[U-^{14}C]serine. This result was interpreted as an incorporation of C-3 of serine into C-5' of B_6 as a one-carbon unit. Dempsey later reported that the specific radioactivity of the pyridoxol was only 12% that of added L-[3-^{14}C]serine (184), closer to the dilution expected. Dempsey (187) has also reported that thiamine is required for B_6 biosynthesis in some *E. coli* auxotrophs, while in others glutamate or α-ketoglutarate is required.

At the present time more precursors have been proposed for B_6 than can be used in formulating a single biosynthetic pathway. The studies carried out by Hill et al have established that the three-carbon metabolic units related to glycolytic intermediates can serve as the ultimate source of cyclic precursors for the pyridoxol molecule. However, specific intermediates of the biosynthetic pathway have yet to be identified.

Literature Cited

1. Brown, G. M., Reynolds, J. J. 1963. *Ann. Rev. Biochem.* 32:419–62
2. Winestock, C. H., Plaut, G. W. E. 1965. *Plant Biochemistry*, ed. J. Bonner, J. E. Varner, 391–437. New York, London: Academic. 1054 pp.
3. Brown, G. M. 1971. *Compr. Biochem.* 21:73–80
4. Shimizu, M. 1970. *Methods Enzymol.* 18A:322–37
5. Mautner, H. G. 1970. *Methods Enzymol.* 18A:338–49
6. Chester, C. J., Butterworth, P. H. W., Porter, J. W. 1970. *Methods Enzymol.* 18A:371–78
7. Schweizer, E., Lerch, I., Kroeplin-Rueff, L., Lynen, F. 1970. *Eur. J. Biochem.* 15:472–82
8. Hoagland, M. B., Novelli, G. D. 1954. *J. Biol. Chem.* 207:767–73

9. Pierpoint, W. S., Hughes, D. E. 1954. *Biochem. J.* 56:130–35
10. Brown, G. M., Snell, E. E. 1953. *J. Am. Chem. Soc.* 75:2782–83
11. Novelli, G. D., Schmetz, F. J., Kaplan, N. O. 1954. *J. Biol. Chem.* 206:533–45
12. Govier, W. M., Gibbons, A. J. 1951. *Arch. Biochem. Biophys.* 32:347–48
13. King, T. E., Strong, F. H. 1951. *J. Biol. Chem.* 189:325–33
14. Levintow, L., Novelli, G. D. 1954. *J. Biol. Chem.* 213:761–65
15. Ward, G. B., Brown, G. M., Snell, E. E. 1955. *J. Biol. Chem.* 213:869–76
16. Brown, G. M. 1959. *J. Biol. Chem.* 234:370–78
17. Abiko, Y. 1967. *J. Biochem.* 61:290–99
18. Abiko, Y. 1967. *J. Biochem.* 61:300–8
19. Abiko, Y. 1970. *Methods Enzymol.* 18A:354–57
20. Abiko, Y., Suzuki, T., Shimizu, M. 1967. *J. Biochem.* 61:309–12
21. Abiko, Y., Tomikawa, M., Shimizu, M. 1968. *J. Biochem.* 64:115–17
22. Abiko, Y. 1970. *Methods Enzymol.* 18A:350–53
23. Scandurra, R., Marcucci, M., Federici, G. 1970. *Acta Vitaminol. Enzymol.* 24:118–22
24. Jones, D. H., Nelson, W. L. 1969. *Biochemistry* 8:2622–26
25. Suzuki, T., Abiko, Y., Shimizu, M. 1967. *J. Biochem.* 62:642–49
26. Abiko, Y. 1970. *Methods Enzymol.* 18A:358–63
27. Masiarz, F. R., Hajra, A. K., Agranoff, B. W. 1972. *Biochem. Biophys. Res. Commun.* 46:992–98
28. Bremer, J., Wojtczak, A., Skrede, S. 1972. *Eur. J. Biochem.* 25:190–97
29. Abiko, Y., Ashida, S., Shimizu, M. 1972. *Biochim. Biophys. Acta* 268:364–72
30. Bird, O. D., Thompson, R. Q. 1967. *Vitamins* 7:209–41
31. Abiko, Y., Tomikawa, M., Shimizu, M. 1969. *J. Vitaminol.* 15:59–69
32. Abiko, Y., Tomikawa, M., Hosokawa, Y., Shimizu, M. 1969. *Chem. Pharm. Bull.* 17:200–1
33. Andrieux-Domont, C., Van Hung, L. 1971. *Arch. Sci. Physiol.* 25:47–57
34. Karasawa, T., Yoshida, K., Furukawa, K., Hosoki, K. 1972. *J. Biochem.* 71:1065–67
35. Nakamura, T., Kusunoki, T., Soyama, K. 1967. *J. Vitaminol.* 13:289–97
36. Nakamura, T., Kusunoki, T., Soyama, K., Tsujita, K., Tanaka, K. 1968. *Bitamin* 38:45–49
37. Nakamura, T., Kusunoki, T., Soyama, K., Tsujita, K., Tanaka, K. 1969. *Bitamin* 40:354–57
38. Nakamura, T., Kusunoki, T., Soyama, K., Kuwagata, M. 1969. *Bitamin* 40:412–15
39. Kuwagata, M. 1971. *Bitamin* 43:78–86
40. Nakamura, T., Kusunoki, T., Soyama, K., Kuwagata, M. 1972. *J. Vitaminol.* 18:34–40
41. Agafonova, N., Kopilevich, V., Rozanov, A., Yakovlev, V. 1971. *Ukr. Biokhim. Zh.* 43:725–29
42. Rozanov, A., Savluchinskaia, L., Zhdanovich, E., Kopelevich, V. 1970. *Biokhimiya* 35:58–63
43. Nakamura, T. 1969. *Bitamin* 40:1–16
44. Tubbs, P. K., Garland, P. B. 1964. *Biochem. J.* 93:550–56
45. Kondrup, J., Grunnet, N. 1973. *Biochem. J.* 132:373–79
46. Powell, G. L., Elovson, J., Vagelos, P. R. 1969. *J. Biol. Chem.* 244:5616–24
47. Tweto, J., Liberati, M., Larrabee, A. R. 1971. *J. Biol. Chem.* 246:2468–71
48. Tweto, J., Larrabee, A. R. 1972. *J. Biol. Chem.* 247:4900–4
49. Plaut, G. W. E. 1971. *Compr. Biochem.* 21:11–45
50. Demain, A. L. 1972. *Ann. Rev. Microbiol.* 11:369–88
51. McCormick, D. B., Wright, L. D., Eds. 1971. *Methods Enzymol.* 18B:253–98
52. Bacher, A., Lingens, F. 1971. *J. Biol. Chem.* 246:7018–22
53. Oltmanns, O., Bacher, A. 1972. *J. Bacteriol.* 110:818–22
54. Beach, R. L., Plaut, G. W. E. 1970. *J. Am. Chem. Soc.* 92:2913–16
55. Lingens, F., Oltmanns, O., Bacher, A. 1967. *Z. Naturforsch. B* 22:755–58
56. Plaut, G. W. E. 1963. *J. Biol. Chem.* 238:2225–43
57. Wacker, H., Harvey, R. A., Winestock, C. H., Plaut, G. W. E. 1964. *J. Biol. Chem.* 239:3493–97
58. Harvey, R. A., Plaut, G. W. E. 1966. *J. Biol. Chem.* 241:2120–36
59. Baur, R., Bacher, A., Lingens, F. 1972. *FEBS Lett.* 23:215–16
60. Oltmanns, O., Bacher, A., Lingens, F., Zimmermann, F. K. 1969. *Mol. Gen. Genet.* 105:306–13
61. Oltmanns, O. 1971. *Mol. Gen. Genet.* 111:301–2
61a. Bresler, S. E., Glazunov, E. A., Perumov, D. A. 1972. *Genetika* 8:109–18
61b. Bacher, A., Eggers, U., Lingens, F. 1973. *Arch. Mikrobiol.* 89:73–77
62. Plaut, G. W. E., Beach, R. L., Aogaichi, T. 1970. *Biochemistry* 9:771–85
63. Mitsuda, H., Kawai, F., Suzuki, Y., Yoshimoto, S. 1970. *J. Vitaminol.* 16:285–92

64. Suzuki, Y., Mitsuda, H. 1971. *Biochim. Biophys. Acta* 242:500–3
65. Beach, R. L., Plaut, G. W. E. 1969. *Tetrahedron Lett.* 40:3489–92
66. Beach, R. L., Plaut, G. W. E. 1970. *Biochemistry* 9:760–70
67. Paterson, T., Wood, H. C. S. 1969. *Chem. Commun.* 6:290–91
68. McAndless, J. M., Stewart, R. 1970. *Can. J. Chem.* 48:263–70
69. Stewart, R., McAndless, J. M. 1973. *J. Chem. Soc.* 376–79
70. Winestock, C. H., Plaut, G. W. E. 1961. *J. Org. Chem.* 26:4456–62
71. Pfleiderer, W., Bunting, J. W., Perrin, D. D., Nübel, G. 1966. *Chem. Ber.* 99:3503–23
72. Pfleiderer, W., Mengel, R., Hemmerich, P. 1971. *Chem. Ber.* 104:2273–92
73. Beach, R. L., Plaut, G. W. E. 1971. *J. Org. Chem.* 36:3937–43
74. Plaut, G. W. E. 1961. *Ann. Rev. Biochem.* 30:409–46
75. Howells, D. J., Plaut, G. W. E. 1965. *Biochem. J.* 94:755–59
76. Bacher, A., Lingens, F. 1969. *Angew. Chem. Int. Ed. Engl.* 8:371–72
77. Baugh, C. M., Krumdieck, C. L. 1969. *J. Bacteriol.* 98:1114–19
78. Burg, A. W., Brown, G. M. 1968. *J. Biol. Chem.* 243:2349–58
79. Baugh, C. M., Shaw, E. 1963. *Biochem. Biophys. Res. Commun.* 10:28–33
80. Cone, J., Guroff, G. 1971. *J. Biol. Chem.* 246:979–85
81. Shiota, T., Palumbo, M. P. 1965. *J. Biol. Chem.* 240:4449–53
82. Dalal, F. R., Gots, J. S. 1965. *Biochem. Biophys. Res. Commun.* 20:509–14
83. Levenberg, B., Kaczmarek, D. K. 1966. *Biochim. Biophys. Acta* 117:272–75
84. Elstner, E. F., Šuhadolnik, R. J. 1971. *J. Biol. Chem.* 246:6973–81
85. Mehta, S. U., Mattoo, A. K., Modi, V. V. 1972. *Biochem. J.* 130:159–66
86. Davoll, J., Evans, D. D. 1960. *J. Chem. Soc.* 4:5041–49
87. Ali, S. N., Al-Khalidi, U. A. S. 1966. *Biochem. J.* 98:182–88
88. Plaut, G. W. E., Broberg, P. L. 1956. *J. Biol. Chem.* 219:131–38
89. Forrest, H. S., McNutt, W. S. 1958. *J. Am. Chem. Soc.* 80:739–43
90. Bacher, A., Lingens, F. 1970. *J. Biol. Chem.* 245:4647–52
91. Plaut, G. W. E. 1961. *Metab. Pathways* 2:673–712
92. Miersch, J. 1973. *Phytochemistry* 12:1595–96
93. Van Lanen, J. M., Tanner, F. W. Jr. 1948. *Vitam. Horm.* 6:163–224
94. Plaut, G. W. E. 1954. *J. Biol. Chem.* 211:111–16
95. Takeda, I., Hayakawa, S. 1968. *Agr. Biol. Chem.* 32:873–78
96. Suzuki, A., Goto, M. 1970. *Nippon Kagaku Zasshi* 91:404–5
97. Suzuki, A., Goto, M. 1971. *Bull. Chem. Soc. Japan* 44:1869–72
98. Suzuki, A., Miyagawa, T., Goto, M. 1972. *Bull. Chem. Soc. Japan* 45:2198–99
99. Barker, H. A. 1972. *Ann. Rev. Biochem.* 41:55–90
100. Friedmann, H. C., Cagen, L. M. 1970. *Ann. Rev. Microbiol.* 24:159–208
101. Friedmann, H. C. 1974. *Vitamin B_{12}: Biochemistry and Biochemical Pathology,* ed. B. M. Babior. New York: Wiley. In press
102. IUPAC-IUB Tentative Rules on Nomenclature of Corrinoids. 1966. *J. Biol. Chem.* 241:2992–94
103. Shemin, D., Corcoran, J. W., Rosenblum, C., Miller, I. M. 1956. *Science* 124:272
104. Corcoran, J. W., Shemin, D. 1957. *Biochim. Biophys. Acta* 25:661–62
105. Bray, R. C., Shemin, D. 1958. *Biochim. Biophys. Acta* 30:647–48
106. Bray, R., Shemin, D. 1963. *J. Biol. Chem.* 238:1501–8
107. Brown, C. E., Katz, J. J., Shemin, D. 1972. *Proc. Nat. Acad. Sci. USA* 69:2585–88
108. Scott, A. I. et al 1972. *J. Am. Chem. Soc.* 94:8267–69
109. Brown, C. E., Katz, J. J., Shemin, D. 1973. *Fed. Proc.* 32:661 (Abstr.)
110. Battersby, A. R., Ihara, M., McDonald, E., Stephenson, J. R., Golding, B. T. 1973. *Chem. Commun.* 404–5
111. Scott, A. I., Townsend, C. A., Cushley, R. J. 1973. *J. Am. Chem. Soc.* 95:5759–61
112. Lenhert, P. G., Hodgkin, D. C. 1961. *Nature* 192:937–38
113. Scott, A. I., Townsend, C. A., Okada, K., Kajiwara, M., Cushley, R. J. 1972. *J. Am. Chem. Soc.* 94:8269–71
114. Schwartz, S., Ikeda, K., Miller, I. M., Watson, C. J. 1959. *Science* 129:40–41
115. Müller, G., Dieterle, W. 1971. *Z. Physiol. Chem.* 352:143–50
116. Franck, B., Gantz, D., Montforts, F. P., Schmidtchen, F. 1972. *Angew. Chem. Int. Ed. Engl.* 11:421–22
117. Burnham, B. F. 1965. *Fed. Proc.* 24:223
118. Burnham, B. F., Plane, R. A. 1966. *Biochem. J.* 98:13C–15C
119. Porra, R. J. 1965. *Biochim. Biophys. Acta* 107:176–79

120. Burnham, B. F. 1969. *Metab. Pathways* 3:403–537
121. Scott, A. I., Townsend, C. A., Okada, K., Kajiwara, M. 1973. *Trans. NY Acad. Sci.* 35:72–79
122. Scott, A. I., Yagen, B., Lee, E. 1973. *J. Am. Chem. Soc.* 95:5761–62
123. Toohey, J. I. 1966. *Fed. Proc.* 25:1628–32
124. Toohey, J. I. 1965. *Proc. Nat. Acad. Sci. USA* 54:934–42
125. Dolphin, D. 1973. *Bioorg. Chem.* 2:155–62
126. Koppenhagen, V. B., Pfiffner, J. J. 1971. *Fed. Proc.* 30:1088 (Abstr.) and personal communication from V. B. Koppenhagen
127. Rapp, P. 1973. *Z. Physiol. Chem.* 354: 136–40
128. Krasna, A. I., Rosenblum, C., Sprinson, D. B. 1957. *J. Biol. Chem.* 225:745–50
129. Lowe, D. A., Turner, J. M. 1970. *J. Gen. Microbiol.* 64:119–22
130. Müller, G., Gross, R., Siebke, G. 1971. *Z. Physiol. Chem.* 352:1720–22
131. Gale, E. F. 1946. *Advan. Enzymol.* 6:1–32
132. Guirard, B. M., Snell, E. E. 1964. *Compr. Biochem.* 15:138–99
133. Bernhauer, K., Wagner, F. 1962. *Biochem. Z.* 335:325–39
134. Neuberger, A., Tait, G. H. 1960. *Biochim. Biophys. Acta* 41:164–65
135. Elliott, W. H. 1960. *Biochem. J.* 74:478–85
136. Neuberger, A., Tait, G. H. 1962. *Biochem. J.* 84:317–28
137. Laver, W. G., Neuberger, A., Scott, J. J. 1959. *J. Chem. Soc.* 1483–91
138. Turner, J. M. 1966. *Biochem. J.* 99:427–33
139. Turner, J. M. 1967. *Biochem. J.* 104:112–21
140. Dekker, E. E., Swain, R. R. 1968. *Biochim. Biophys. Acta* 158:306–7
141. Campbell, R. L., Dekker, E. E. 1973. *Biochem. Biophys. Res. Commun.* 53:432–38
142. Lowe, D. A., Turner, J. M. 1970. *J. Gen. Microbiol.* 63:49–61
143. Foster, M. A., Tejerina, G., Guest, J. R., Woods, D. D. 1964. *Biochem. J.* 92:476–88
144. Volcani, B. E., Toohey, J. I., Barker, H. A. 1961. *Arch. Biochem. Biophys.* 92:381–91
145. Bernhauer, K., Becher, E., Gross, G., Wilharm, G. 1960. *Biochem. Z.* 332:562–72
146. Bernhauer, K., Wagner, F. 1960. *Z. Physiol. Chem.* 322:184–89
147. Boretti, G. et al 1960. *Biochim. Biophys. Acta* 37:379–80
148. Ronzio, R. A., Barker, H. A. 1967. *Biochemistry* 6:2344–54
149. Renz, P. 1968. *Biochem. Biophys. Res. Commun.* 30:373–78
150. Friedmann, H. C. 1968. *J. Biol. Chem.* 243:2065–75
151. Ohlenroth, K., Friedmann, H. C. 1968. *Biochim. Biophys. Acta* 170:465–67
152. Coulter, C. L., Hawkinson, S. W., Friedmann, H. C. 1969. *Biochim. Biophys. Acta* 177:293–302
153. Schneider, Z., Friedmann, H. C. 1972. *Arch. Biochem. Biophys.* 152:488–95
154. Friedmann, H. C., Harris, D. L. 1965. *J. Biol. Chem.* 240:406–12
155. Friedmann, H. C. 1965. *J. Biol. Chem.* 240:413–18
156. Fyfe, J. A., Friedmann, H. C. 1969. *J. Biol. Chem.* 244:1659–66
157. Renz, P., Reinhold, K. 1967. *Angew. Chem. Int. Ed. Engl.* 6:1083
158. Alworth, W. L., Baker, H. N., Lee, D. A., Martin, B. A. 1969. *J. Am. Chem. Soc.* 91:5662–63
159. Renz, P. 1970. *FEBS Lett.* 6:187–89
160. Kühnle, H. F., Renz, P. 1971. *Z. Naturforsch. B* 26:1017–20
161. Renz, P., Weyhenmeyer, R. 1972. *FEBS Lett.* 22:124–26
162. Lu, S. H., Winkler, M. F., Alworth, W. L. 1971. *Chem. Commun.* 191–92
163. Alworth, W. L., Lu, S. H., Winkler, M. F. 1971. *Biochemistry* 10:1421–24
164. Lu, S. H., Alworth, W. L. 1972. *Biochemistry* 11:608–11
165. Janicki, J., Chelkowski, J., Nowakowska, K. 1966. *Acta Microbiol. Pol.* 15:249–54
166. Schlee, D. 1973. *Pharmazie* 28:284–87
167. IUPAC-IUB Tentative Rules for the Nomenclature of Vitamins B_6 and Related Compounds. 1970. *J. Biol. Chem.* 245:4229–31
168. *Nutr. Rev.* 1972. 30:238–40
169. Snell, E. E., Haskell, B. E. 1970. *Compr. Biochem.* 21:47–71
170. Goodwin, T. W. 1963. *The Biosynthesis of Vitamins and Related Compounds,* 159–66. New York: Academic. 366 pp.
171. Brown, G. M., Reynolds, J. J. 1963. *Ann. Rev. Biochem.* 32:447–50
172. Dempsey, W. B. 1966. *J. Bacteriol.* 92:333–37
173. Dempsey, W. B., Pachler, P. F. 1966. *J. Bacteriol.* 91:642–45
174. Hill, R. E., Rowell, F. J., Gupta, R. N., Spenser, I. D. 1972. *J. Biol. Chem.* 247:1869–82
175. Sato, K., Suzuki, T., Sahashi, Y. 1966. *Vitamins Kyoto* 33:357–60

176. Suzue, R., Haruna, Y. 1970. *J. Vitaminol.* 16:154–59
177. Suzue, R., Haruna, Y. 1970. *J. Vitaminol.* 16:161–63
178. Stanley, S. B. 1969. *Studies on the Biosynthesis of Vitamin B$_6$*, PhD thesis. Smith College, Univ. Mass., Northampton. 175 pp.
179. Hill, R. E., Spenser, I. D. 1970. *Science* 169:773–75
180. Lunan, K. D., West, C. A. 1963. *Arch. Biochem. Biophys.* 101:261–68
181. Schroer, R. A., Frieden, E. H. 1973. *Proc. Soc. Exp. Biol. Med.* 142:369–73
182. Dempsey, W. B. 1970. *Biochim. Biophys. Acta* 222:686–87
183. Tani, Y., Dempsey, W. B. 1973. *J. Bacteriol.* 116:341–45
184. Dempsey, W. B. 1972. *Biochim. Biophys. Acta* 264:344–53
185. Hill, R. E., Gupta, R. N., Rowell, F. J., Spenser, I. D. 1971. *J. Am. Chem. Soc.* 93:518–20
186. Hill, R. E., Spenser, I. D. 1973. *Can. J. Biochem.* 51:1412–16
187. Dempsey, W. B. 1971. *J. Bacteriol.* 108:1001–7
188. Dempsey, W. B., Itoh, H. 1970. *J. Bacteriol.* 104:658–67
189. Dempsey, W. B. 1969. *J. Bacteriol.* 97:1403–10
190. Dempsey, W. B. 1969. *J. Bacteriol.* 100:1114–15
191. Dempsey, W. B. 1969. *Biochem. Biophys. Res. Commun.* 37:89–93
192. Brown, C. E., Shemin, D., Katz, J. J. 1973. *J. Biol. Chem.* 248:8015–21

DNA SEQUENCING TECHNIQUES

Winston A. Salser

Department of Biology and Molecular Biology Institute
University of California at Los Angeles, Los Angeles, California

CONTENTS

INTRODUCTION

In the past few years a great variety of approaches have become available for nucleotide sequencing. RNA sequencing techniques have been used most extensively and have been reviewed recently by Barrell (1), Brownlee (2), Gilham (49), and Mandeles (3). Considerable progress in sequencing DNA molecules has also been made by developing new techniques. Although most of these techniques are still very new, a number of biologically interesting sequences have been determined. Those determined by "pure" DNA sequencing techniques include short sequences at the ends of some bacteriophage genomes (4–10), sequences of restriction enzyme cleavage sites (11–15a), purine and pyrimidine tracts from phages such as ϕX and fd (16), repeating sequences from several highly repetitive nuclear satellite DNAs (17–19), a ribosomal binding site from phage fd (20), and a nucleotide sequence from another region of the phage fd genome (21). A number of other long sequences have been determined by "hybrid" techniques, in which in vitro transcription with RNA polymerase is used to solve a DNA sequencing problem by RNA sequencing techniques (22–27).

923

Use of DNA sequencing techniques is essential when one investigates a sequence not transcribed in vivo. But in many other cases as well, use of an appropriate DNA sequencing approach enables the investigator to overcome some of the most serious limitations of conventional RNA sequencing techniques. In this review, we emphasize especially the following topics:

(a) The use of restriction enzymes and other sequence-specific deoxyribonucleases (DNases). This class of enzymes makes it possible for a molecule whose sequence is much too large for analysis by conventional RNA sequencing methods to be dissected in an orderly fashion into smaller ordered fragments of sizes convenient for sequencing studies.

(b) The use of oligonucleotide primers of defined sequence. Such primers permit the investigator to focus his attention on a particular region by obtaining synchronous in vitro synthesis starting at specific locations.

(c) The use of reverse transcriptase. The ability to make complementary DNA copies of mammalian mRNAs permits study of these interesting molecules (which are extremely difficult to effectively label with ^{32}P in vivo) by a variety of in vitro labeling techniques.

(d) The use of ribosubstitution techniques. The ability to incorporate ribonucleotides into DNA synthesized in vitro permits the experimenter to obtain specific cleavages at either G, C, or A residues.

These are four especially promising areas where DNA sequencing techniques may offer the most striking advantages over conventional RNA sequencing methods in determining long nucleotide sequences. We mention only briefly the more specialized methods for sequencing short runs of nucleotides at the ends of phage genomes and the technical details of sequencing the cleavage sites of the restriction enzymes, since these methods have recently been reviewed extensively (28a–31, 153).

What Biological Questions may be Answered with DNA Sequencing Techniques?

It may be helpful to review some of the biologically significant questions for which the newly available DNA sequencing techniques might be especially useful. In prokaryotic cells we need to know the sequences of binding sites for the various repressors and positive control elements in order to understand the molecular details of their action. Preliminary work suggests that these molecular details may be substantially different even for molecules with analogous functions, such as the *lac* and *λ* repressors. For instance, the *lac* repressor binds to a single site of about 21–27 nucleotides in length (24, 25), while each of the two phage *λ* operators appears to be much larger, consisting of a series of six similar repressor binding sites. The repressor first binds to a unique site within this array and then there is cooperative binding to the other five sites (32). It is not yet clear whether this large array of repressor binding sites provides an explanation for the transcriptionally "silent" regions reported to exist between operator mutant sites and the start of transcription in *λ* (22, 23), or whether there are still other silent regions of unknown function.

Within the transcribed regions of bacterial genomes more data is available from conventional RNA sequencing; however, we still do not know enough ribosomal attachment site sequences to confidently formulate any general rules about why one such signal may be several hundred times more effective than another in mobilizing ribosomes for protein synthesis. Nor do we adequately understand the untranslated intercistronic regions (e.g. see 33) or the role of highly base-paired regions found within mRNA molecules (e.g. 34, 35). In many cases it may soon be easier to obtain such answers by DNA sequencing techniques than by conventional RNA sequencing.

We are beginning to learn more about the complex and distinctive organization of eukaryotic genomes, and it is clear that they have many features not found in bacteria. For instance, we need more information about the variants of the repeated sequences in highly repetitive nuclear satellite DNAs so that we can test the different models which attempt to explain their biological role and the manner in which millions of such closely related sequences can be maintained in the cell (17–19).

The discovery that mammalian RNAs are derived from much larger heterogeneous nuclear RNA (HnRNA) molecules, and that most of them have extensive tracts of poly(A) at their 3'-ends, suggests a variety of questions (36). We would like to determine the sequences that specify addition of poly(A), the sequences that specify how the mRNA is cleaved from the large HnRNA precursor and, of course, those that specify the sites of ribosome attachment and termination of protein synthesis. Further, it appears that mRNAs themselves may include extensive sequences not translated into protein. Whether these untranslated sequences are located at the 5'-end of the mRNA, between the structural gene and the poly(A) sequence, or both, is not known. Nor do we know what role, if any, these sequences may play in translational control. It has been suggested that the sequences within the structural gene itself may play a part in translational control by virtue of their participation in highly base-paired "hairpin" and other folded secondary structures. Direct sequence analysis will allow us to study all of these questions in detail.

One might logically suppose that questions concerning the sequences of RNAs from mammalian cells could best be answered by sequencing RNA labeled in vivo. In actual practice, however, it has proven very difficult to obtain mRNA preparations with the requisite high ^{32}P specific activity and purity needed for rapid sequencing; thus, none of these questions have yet been answered using standard RNA sequencing techniques. As we discuss below, it may be easier to start with unlabeled mRNA preparations and use reverse transcriptase to permit an RNA sequencing problem to be solved by DNA sequencing techniques. Alternatively, transcription of DNA fragments into RNA as well as pure DNA sequencing approaches are being applied to the genomes of small DNA viruses such as Simian Virus 40 (SV40), where it may be possible to determine the complete nucleotide sequence in a relatively few years.

Three General Approaches

As will be seen, the in vitro DNA sequencing techniques discussed have great advantages for sequencing the many interesting eukaryotic nucleic acids which are difficult or impossible to label in vivo at the very high specific activities needed for rapid sequence determination. But the main advantage of DNA sequencing methods is probably the ease with which a very large sequencing problem can be simplified into a number of smaller problems, each of which can be solved much more efficiently. Attempts to sequence large RNA molecules such as phage MS2 or 16S ribosomal RNA by conventional RNA sequencing techniques have been critically dependent upon the success of attempts to subdivide the molecule into fragments of reasonable size by partial digestions with RNase T1 (34, 37, 38). A great deal has been accomplished, but such an approach is far from ideal; very complex mixtures of products are obtained, many products usually are present in very low yield, and especially sensitive portions of the RNA molecule may be lost entirely. By contrast, restriction enzymes permit even very large DNA molecules to be dissected into fragments of a size appropriate for rapid sequence determination while being subject to none of the above disadvantages.

In their investigation of the $Q\beta$ phage genome, RNA sequencers have used an alternative approach involving the synchronous replication of the phage genome by $Q\beta$ replicase (39a). Although this is a very elegant method, the $Q\beta$ replicase is highly specific for initiation at the termini of $Q\beta$ RNA and it has not yet been possible to use this approach with other RNA molecules. The analogous DNA sequencing technique involves the use of oligonucleotides of defined sequence to prime in vitro synthesis starting at a specific location on a DNA template. The advantage of this approach over use of the $Q\beta$ replicase is that it should be possible to obtain synchronous synthesis from defined positions on any DNA molecule. By using different primer sequences, it should be possible to obtain synthesis from different locations within the DNA template almost at will.

In another section of this review we consider in detail these as well as several other approaches for simplifying the attack on large nucleotide sequences. If sequencing techniques are so well adapted to the determination of long sequences, one might ask why long sequences have not been determined already using these approaches. One important factor delaying such efforts has been the lack of base-specific DNases with which to determine the actual primary sequence of DNA fragments in the 50–200 nucleotide size range. As remarked in an earlier review, although many DNases show some base specificity "Unfortunately none of the known deoxyribonucleases possesses the absolute base specificity found in some RNases", such as RNase T1 which cleaves specifically at G residues (28). This basic limitation has now been overcome in three independent ways, which are mentioned below and considered in detail in the following section.

(*a*) Ribosubstitution techniques permit the synthesis in vitro of heteropolymers which can be cleaved specifically at either G, C, or A residues.

(*b*) Partial digestion with exonucleases followed by analysis of the partial digests by electrophoresis and homochromatography works best on fragments with less than 15 nucleotides and must therefore be used in combination with procedures such as T4 endonuclease IV digestion or ribosubstitution cleavage which yield fragments in this size range.

(*c*) The problem can be circumvented entirely by transcribing the DNA sequence into RNA in vitro, thus permitting all the standard RNA sequencing techniques to be used.

THREE VARIATIONS ON THE OVERLAP METHOD AS APPLIED TO DNA SEQUENCING

Ribosubstitution Sequencing

Berg, Fancher & Chamberlin (39b) first reported that, in the presence of Mn^{2+} and appropriate mixtures of ribo- and deoxyribonucleoside 5′-triphosphates, *Escherichia coli* DNA polymerase could synthesize a product in which one of the deoxyribonucleotides was replaced by the corresponding ribonucleotide. They suggested this as an aid to DNA sequence analysis, but it was many years before this hope was realized. Recognizing that this could provide an extremely powerful DNA sequencing approach when teamed with specific cleavages by restriction enzymes or with initiation of DNA synthesis with specific primers, Salser et al (40) analyzed ribosubstitution techniques in greater detail. They were interested especially in testing the overall fidelity of the in vitro reaction carried out under ribosubstitution conditions. With in vitro synthesis at 37°C, it was found that a bacteriophage M13 DNA template could be extensively copied in ribosubstitution reactions with rGTP, rCTP, or rATP replacing the corresponding deoxyribonucleotides. From a number of tests it was concluded that the fidelity of the ribosubstitution synthesis was adequate for sequencing studies.

Two papers appearing subsequently questioned both the fidelity and extent of the reaction. Wu and his collaborators (28a) found it necessary for technical reasons to carry out the synthesis at 5°C and showed that the synthesis stopped whenever the template called for incorporation of two or three ribonucleotides in series. Van de Sande, Lowewen & Khorana (41) observed poor fidelity at 37°C with both rCMP and rGMP incorporation. At 10°C they obtained good fidelity with the rCTP substitution reaction, but misincorporation with rGTP. They also observed that the presence of two or more successive ribonucleotides in the product tended to block the synthesis at the low temperatures.

Where it is necessary to use low temperatures to preserve the fidelity of the ribosubstitution reaction, the tendency of the reaction to stop when two or three adjacent ribonucleotides are encountered would appear to severely limit the application of this method to long sequences. The results of Van de Sande et al (41) also disagreed with our own in that they had difficulty with rA incorporation and concluded that "as a tool for DNA sequencing, the method (ribosubstitution) would appear to give complete fidelity only for rC incorporation at 10°C." This would further limit the usefulness of the method since one of the greatest potential values of the ribosubstitution technique is that one can carry out specific cleavages at the sites of rGMP, rCMP, and rAMP incorporation in different experiments, thus obtaining enough overlap data to reassemble even quite large sequences.

Because of these problems, my own laboratory decided to carry out an extensive further examination of the fidelity of the ribosubstitution reaction at 37°C. For these experiments a highly repetitive satellite DNA was chosen for independent sequence analysis by DNA ribosubstitution sequencing and by in vitro transcription into RNA. The two methods independently yielded the same sequence data, and there was evidence of good fidelity at 37°C with both rCMP and rGMP substitution (18). In addition to the agreement between RNA sequencing and ribosubstitution sequencing results, the sets of ribo G and ribo C cleavage data were themselves internally consistent, and both were consistent with the available ribo A cleavage data (ribo A data were from an experiment with a single label rather than a complete set of experiments). Such complete agreement would seem very unlikely if appreciable levels of infidelity were occurring (18, 182).

These experiments also clearly showed that neither long runs of G or C residues seriously block the reaction at 37°C. The satellite sequence analyzed contained runs of three, four, and five dGMP residues in one strand (and, of course, corresponding runs of dCMP in the complementary strand).

We have therefore concluded that, as long as the DNA polymerase is working on a long DNA template, the ribosubstitution fidelity is adequate for sequencing studies. It has occurred to us that, in both cases where a lack of fidelity has been noted at 37°C (28a, 41), the DNA polymerase was polymerizing nucleotides very close to the end of its template. It is possible that the strength of the interaction between DNA polymerase I and its template decreases as the enzyme approaches the end of its template (a situation which presumably seldom occurs in vivo) with a resultant decrease in the fidelity of ribosubstitution synthesis.

We are further encouraged in our assessment of the fidelity of the ribosubstitution reaction by our observation of very clean unique fingerprints from DNA molecules of higher complexity (about 1600 nucleotides) (42, 43). Had low fidelity been a problem, these fingerprints would have contained not only the major legitimate spots but also many faint spots, whose presence could have been detected with high sensitivity.

Partial Digestions with Exonucleases

Partial digestion of RNA with exonucleases provides one of the most useful approaches for analyzing RNA fragments from RNase T1 cleavages and can be used in a very similar way for DNA sequencing. In theory the partial digestion is carried out to obtain all the possible intermediates, each differing from the preceding one by loss of one nucleotide. By analyzing the base compositions of each of these partial products, the sequence can be established. From partial spleen enzyme digests of RNase T1 fragments, it is possible to read off the sequence from the mobilities of successive products except for ambiguities introduced by the fact that this enzyme removes 5'-terminal C residues much more slowly than A or (especially) U residues so that some of the expected products may be overlooked (2). In the case of partial digests with either the spleen or venom diesterases, additional "illegal" products may also confuse the issue. For instance, both venom and spleen phosphodiesterases are frequently contaminated with endoribonuclease activity cleaving predominantly at Py–A bonds (1).

In the process of adapting these approaches for DNA sequencing, many of the problems associated with earlier use of the techniques have been solved. For example, the exonuclease preparations seem to be less contaminated with DNA-specific endonucleases than with RNA-specific endonucleases, so fewer illegal products are obtained when DNA fragments are digested. Also, the use of homochromatography permits the analysis of much larger fragments. Ling (16) was able to determine the sequences of pyrimidine tracts up to 20 nucleotides in length from bacteriophage fd DNA using this approach. The sequence could be read directly from inspection of the fingerprint patterns since loss of C or T residues had quite different effects on the fragment mobilities.

Similar techniques have been used for determining the sequences of DNA fragments produced by the sequence-specific DNase T4 endonuclease IV (45, 47) or by ribosubstitution cleavages. In these cases, interpretation of the data is more complex. Such fragments can contain any of the four DNA bases, and it is much more difficult to read off the sequence simply by inspecting the homochromatography fingerprints (for instance, removal of a G residue may be confused with removal of a T residue). Consequently, in the analysis of the T4 endo IV fragments, the final sequences had to be checked by determining the base compositions or depurination products for each of the partial spleen or snake venom phosphodiesterase digestion products (21, 46). In the case of digestions with spleen phosphodiesterase, the sequences were also checked by determination of the 5′-terminal dinucleoside sequence of each of the partial digestion products (46).[1] In the case of fragments resulting from RNase A cleavages of ribo C-substituted DNA synthesized in vitro, nearest neighbor analysis using different (^{32}P) deoxyribotriphosphates and analysis of pyrimidine tracts were performed in addition to direct inspection of the fingerprints of partial exonucleolytic digestions (21).

Use of enzymes such as T4 endonuclease IV with partial digestion techniques has the advantage that it can be applied directly to DNA labeled in vivo, thus obviating any problems of in vitro fidelity. Of course in using this approach it is necessary to have available some preliminary fragmentation technique, such as use of T4 endonuclease IV, which is able to give specific fragments in the desired size range (about 5–20 nucleotides).

Sequencing DNA by Using it as Template for in Vitro RNA Synthesis

If the DNA sequence can be converted to a highly labeled RNA sequence by in vitro transcription, all of the standard RNA sequencing techniques can be used. Of course, there is the additional advantage of being able to obtain nearest neighbor labeling data in vitro, which greatly facilitates the sequence analysis (39a). Available evidence suggests that the fidelity of in vitro transcription is excellent, so sequences obtained in this way should accurately reflect the sequences of the DNA templates. In the first place, unique sequences have been obtained for all in vitro transcripts studied so far, which would appear to rule out random mistakes as a serious problem. The fact that active proteins can be synthesized in vitro in coupled systems strongly argues that systematic base substitutions are not a problem. As a more

---

[1] This was accomplished by treating each of the partial products with exonuclease I which digests all but the 5′-terminal dinucleoside monophosphate. This compound is isolated and its structure determined by complete digestions with spleen and venom phosphodiesterases.

direct proof, Lozeron, Szybalsky & Dahlberg (48) have compared sequences of bacteriophage λ transcripts made in vitro and in vivo and found excellent agreement.

Since standard RNA sequencing techniques have been reviewed recently (1–3, 49), this review concentrates on special problems peculiar to DNA sequencing: the forcing of RNA synthesis from small DNA fragments which lack promoters, special template properties of DNA made using reverse transcriptase, and methods for obtaining synchronous synthesis from specific primers or from natural promoter sequences in larger DNA molecules.

STRATEGIES FOR SIMPLIFYING SEQUENCING FROM LARGE DNA MOLECULES

In this section we consider a variety of ways in which one may convert a long molecule into fragments in the range of 40–400 nucleotides, thus simplifying the sequencing problem. This can be accomplished by using ribosubstitution techniques, in vitro transcription, or cleavage by enzymes such as T4 endonuclease IV as discussed in the previous section. In the detailed discussions of the approaches considered below, the reader should keep in mind that different sequencing projects may have very different goals. In some cases, such as the SV40 genome, the starting molecule is sufficiently small (about 5000 base pairs) and of sufficiently great biological interest that one may wish to ultimately determine the entire nucleotide sequence. In this case, it might be adequate to use restriction enzymes to cleave the molecule into sets of ordered fragments of convenient size and methodically sequence all of these. Even if the whole molecule is ultimately to be sequenced, the experimenter would like to at least start with those regions he considers most biologically interesting. In many other cases, the sequencer is only interested in determining a small sequence of particular biological interest, current examples being ribosomal attachment sites, the operator sequences in phage λ and the *lac* operon, and the nucleotide sequences preceding and following the structural genes for *E. coli* tyrosine tRNA. Extremely useful here are techniques in which the desired nucleotide sequence may be isolated by its specific binding to a ribosome, repressor, or other specific binding protein. Alternatively, by using primer sequences complementary to a specific template region, one may restrict in vitro synthesis of labeled nucleic acid to the region immediately following the primer binding site. With techniques such as this, one may focus attention on the sequence of a small specific region to the exclusion of all others in a very large molecule.

It should be emphasized that not only have many different methods for simplifying large sequencing problems recently become available, but the number of different approaches and combinations of approaches is likely to continue to increase rapidly. Thus, while there appears to be no lack of methods, choice of the best one for any particular problem deserves careful thought and may save a great deal of subsequent work.

Specific Priming

The specific priming approach takes advantage of the fact that synthesis of DNA by DNA polymerase I is absolutely dependent upon the presence of a primer complementary to the template and bearing a 3′-OH group. By providing a specific primer sequence which binds to the DNA template at a unique position, it is possible to insure that in vitro DNA synthesis starts at a single location chosen by the experimenter. Some advantages of this approach are similar to those achieved by RNA sequencers using Qβ replicase (39a). This has permitted the rapid determination of two long sequences, but the approach has so far been limited to the two ends of the Qβ genome because of the great specificity of the replicase. The use of specific primers would be more versatile since it should be successful with any DNA molecule, and the experimenter should be able to use a variety of different primers to obtain synchronous initiation at practically any point desired.

SOME GENERAL EXAMPLES   Specific primers may be used in a variety of ways as exemplified in recent articles from the laboratories of Khorana and Sanger. Loewen & Khorana (50) and Loewen, Sekiya & Khorana (51) are attempting to determine the nucleotide sequence which follows the *E. coli* tyrosine tRNA structural gene and presumably signals termination of transcription. This is done by annealing a synthetic deoxyribopolynucleotide having a 22-nucleotide sequence corresponding to the 3′-end of the tRNA to the strand of phage φ80 psuIII DNA. DNA polymerase I is then used to elongate the primer, synthesizing a ^{32}P-labeled copy of the unknown sequence which follows the tRNA structural gene.

The first 12 nucleotides adjoining the structural gene were determined (50) using synthesis with a restricted number of nucleoside triphosphates. These techniques were pioneered by Wu and his collaborators for sequencing the sticky ends of the lambdoid phages (see 28a for a review of these techniques). One of the difficulties of this approach is illustrated by the fact that in a two step synthesis, contaminating nucleotides from the first step led to an incorrect sequence (50) corrected in the subsequent study (51). Correction of this initial sequence and determination of an additional 11 nucleotides were accomplished by a combination of approaches. These included ribosubstitution sequencing with rCTP substituted for dCTP, partial step-wise degradations with spleen or venom phosphodiesterases, and separation of products of various length from a timed synthesis. These products were then analyzed by ribosubstitution cleavage and fingerprinting or by nearest neighbor analysis.

Sanger et al, in an earlier paper (52), had shown that an octadeoxyribonucleotide sequence (ACCATCCA) was active in priming DNA synthesis starting at a unique position on the bacteriophage f1 genome. (These authors had intended to initiate synthesis at a Trp–Met–Val sequence reported to exist within the major coat protein of the phage. The initiation actually observed, although unique, was not at the expected location.) The sequence determination depended heavily on the use of ribosubstitution sequencing methods and yielded a 50-nucleotide sequence believed to have come from an intercistronic region in the f1 genome.

When synthesis of ribo C-substituted DNA was carried out for 10 min at 25°C and the product was digested with pancreatic RNase A and fractionated by electrophoresis and homochromotography, seven major spots contained most of the radioactivity. These were subsequently ordered and shown to account for the first 50 nucleotides added to the primer. A very elegant way of ordering the ribo C cleavage products was introduced, involving a brief in vitro synthesis (4 min at 0°C) of ribo

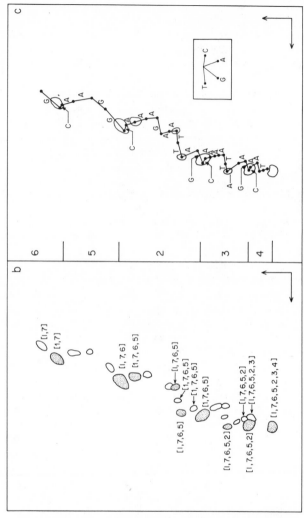

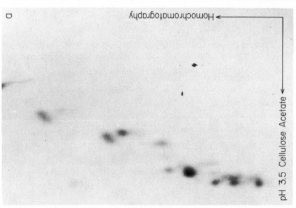

Figure 1

(a) Ribo C-substituted DNA was synthesized from an fl-DNA template by *E. coli* DNA polymerase I primed by the octanucleotide ACCATCCA. DNA resulting from a brief synthesis at 0°C was fingerprinted (by electrophoresis and homochromatography) to separate the products of various lengths.

(b) Composite diagram summarizing a number of experiments as in *a* with varying incubation conditions and varying labeled triphosphates. Strippling indicates the 9 products routinely obtained in high yield. Each of the products was eluted, digested with pancreatic RNase A to cleave at the ribo C residues, and the resulting fragments characterized. The numbers in brackets beside each spot in *b* indicate the pancreatic digestion fragments obtained. The order of these numbered fragments was established to be 1-7-6-5-2-3-4 where the sequence of each fragment is as shown below:

(primer   )
(A-C-C-A-T-C-C-A)-A-T-A-A-A-T-rC/A-T-A-rC/A-G-G-rC/A-A-G-G-rC

fragment number: 1   7   6   5

/A-A-A-G-A-A-T-T-A-G-rC/A-A-A-A-T-T-A-A-G-rC/A-A-T-A-A-A-G-rC/rC
           2   3   4

(c) Diagram showing the relationship between the positions and sequences for the spots on the fingerprints. The inset shows the expected effect of the addition of a single residue on the position of an oligonucleotide. The line connecting the spots on the diagram shows the expected positions of those products not detected because of low yield and the letters adjacent to the line show the base residues by which successive products differ. The difference in position caused by the addition of a single residue is greater for the smaller oligonucleotides (top of fingerprint) and smaller for larger ones.

Redrawn with permission from Figure 13 of Sanger, F. 1973. *Virus Research, Second ICN-UCLA Symposium on Molecular Biology.* © New York: Academic

C-substituted DNA so that products of varying length were obtained. These products were separated by electrophoresis and homochromatography. Each product was then analyzed by RNase A digestion and separation of the cleavage products.

One might have anticipated that the radioactivity would be rather widely distributed among the 50 possible products, each product being one nucleotide longer than the preceding one, and that the low individual yields and difficulty of separating so many compounds would complicate the further analysis. Fortunately, the radioactivity was mostly concentrated in about nine strong spots well separated from each other, as shown in Figure 1. Further analysis suggests that this is because the in vitro synthesis seems to pause at certain points, predominantly at positions where rCMP is to be inserted. (Synthesis also appeared to be held up, in some experiments, at the positions of the radioactively labeled residues, presumably because these were supplied at very low concentration.)

The smallest of the major products (top of Figure 1) yielded oligonucleotides 1 and 7 when cleaved by pancreatic RNase A. Sequence analysis established that fragment 1 contained the primer oligonucleotide, and so was at the 5′ end of the product. The next larger of the major products contained oligonucleotides 1, 7, and 6, establishing that these appeared in the order 1–7–6 as indicated by the numbers in brackets in the figure. By further analysis of larger and larger products it was possible to establish that the seven oligonucleotides resulting from the ribo C cleavage were arranged in the order 1–7–6–5–2–3–4.

HOW LARGE MUST A PRIMER BE TO GIVE SPECIFICITY? The specificity and stability with which a particular primer forms an association with the DNA template depends, among other things, upon the size of the primer, its base composition (number of GC pairs formed), and the temperature at which the primer is hybridized to the template. Such considerations have effects which may be predicted at least in part. Unfortunately, the outcome must also be heavily influenced by the local secondary structure of the template DNA molecule, and this is difficult or impossible to predict.

Some of the most systematic studies of the effects of size and base composition upon the stability and specificity of complex formation between oligonucleotides and their complementary DNA sequences have been carried out with RNA fragments from phages T2, T5, and T7 RNA (53–55). RNA fragments were prepared by partial digestion with RNase T1 and fractionated according to chain length. The various size classes were then tested for their ability to form RNase-resistant complexes at different temperatures with denatured DNAs immobilized on membrane filters. It was noted that chain lengths of about 10 or more ribonucleotides are required to form ribonuclease-resistant complexes in the absence of Mg^{2+}, and that even the complexes formed with the shortest fragments were highly species specific (i.e. oligoribonucleotides from phage T5 did not anneal to either T2 or T7 DNA).

The temperature for optimum complex formation and the melting temperatures of the complexes increase as expected with increasing chain lengths. The efficiency of annealing to homologous DNA when the optimum temperature is used increases dramatically as the size of the RNA fragments is increased from 9 to 14 nucleotides

(T5 RNA, 55), suggesting that at the lower temperatures many incorrectly base-paired complexes are stable but sensitive to RNase.

The effects of different G + C content on the melting temperature were also studied, and it was shown that for any given fragment size there is a linear relationship between the percentage of G + C and the melting temperature. The slope of the curve is strongly dependent on fragment size.

The fact that the studies above were carried out with RNA does not lessen their interest since it has been shown that RNA fragments are excellent primers for DNA polymerase I (see below). But, the relevance of this work is lessened by the fact that the conditions used (no divalent cation, RNase treatment of the complexes) are inappropriate for most in vitro priming situations. As pointed out by Nyogi (55), in 0.01 M MgCl$_2$ octamers can form complexes at 30°C, while oligonucleotides as short as pentamers can form complexes if the temperature is lowered to 0°C, but the background of nonspecific complex formation is very high.

Besmer et al (56) have investigated the specificity of binding between poly-deoxynucleotide sequences corresponding to the tyrosine suppressor tRNA gene and the separated strands of phage ϕ80 psuIII$^+$ DNA. Radioactively labeled segments ranging in length from 8–30 nucleotides were hybridized at various temperatures ranging from 15 to 60°C to the r or l strands, with or without competing tRNA. With the octanucleotide sequence CCCCACCA, it was possible to obtain largely specific hybridization at 15 or 25°C followed by isolation at 5°C (as indicated by the detection of five times as much hybridization to the phage ϕ80 psuIII$^+$ r strand as to the phage 80 r strand). No hybridization was observed at 35°C followed by isolation at 20°C. With the addition of three more nucleotides to this fragment (CCCCACCACCA), hybridization could be shown at 45°C with isolation at 20°C. By contrast, a different undecadeoxynucleotide (TCGAATCCTTC) from positions 12 to 22 of the tRNA sequence showed no detectable hybridization under a variety of conditions, including hybridization at 25°C and isolation at 5°C.

This latter difference may partly reflect the fact that the undecadeoxynucleotide forming a stable hybrid has a much higher G + C content. Unfortunately, different stabilities for different primer binding could also be strongly influenced by whether the complementary sequence in the phage ϕ80 psuIII$^+$ DNA strand was normally single stranded or tightly base-paired with an adjacent sequence. Even though this binding appeared to be quite specific (as indicated by the fact that very much less binding occurred to the opposite strand or to phage ϕ80 DNA), the authors observed that for several of the fragments, 2 or 3 moles were bound per mole of template; this has not been adequately explained. In these cases, binding to several sites was shown not only by the stoichiometry of binding, but also by the fact that part of this binding could not be completed by adding excess tyrosine tRNA. Since Loewen, Sekiya & Khorana (51) were interested in determining a particular sequence embedded in a rather large template (roughly 50,000 nucleotides), it was especially important to obtain high specificity, and they used a primer 22 nucleotides long. In earlier work (56) this primer had annealed at two sites, but apparently only one sequence was obtained in the in vitro synthesis (50, 51).

Another difficulty expected to cause trouble is priming by the 3′-OH ends of the template. Englund (57) has shown that single-stranded DNA molecules with free 3'-OH groups are often self-priming, presumably because the sequence at the 3'-OH end primes by finding a complementary or nearly complementary sequence. Englund has pointed out that the fit need not be exact since DNA polymerase I is well suited for an exonucleolytic attack on an unpaired 3'-OH end which can be continued until an accurately base-paired sequence (which can act as primer) is reached. Fortunately, the newly synthesized material attached to the oligonucleotide primer can be separated from that synthesized by elongation of the 3' end of the template by virtue of their large difference in size under denaturing conditions.

Sanger and his collaborators (52) used a circular single-stranded template, thus avoiding the problem of free 3'-OH ends. Their conditions also differ from those used by Loewen et al in that the molar concentrations of template and primer were much higher (about eightfold and 280-fold, respectively). It seems possible that the primer used, ACCATCCA, may have been only imperfectly paired to the template; there is only about one chance in ten that a particular octanucleotide would find an exact fit anywhere in a sequence the size of the f1 genome. It is interesting to speculate that perhaps any randomly chosen octanucleotide might serve as a useful template for sequencing part of a small genome like f1. It may be that by proper choice of primer concentration one may frequently find conditions such that any particular primer gives predominantly a single start. Even if a particular oligonucleotide primed in roughly equal amounts at two different locations, sequencing might still be relatively easy using the approach of Sanger et al (52). If priming occurred at two sites, an experiment analagous to that shown in Figure 1 should show two distinct trails of spots, each giving its own characteristic products after complete digestion to cleave at all ribose linkages.

THE SYNTHESIS OF PRIMERS OF DEFINED SEQUENCE   The polynucleotides of defined sequence synthesized to date have almost all been made using the methods developed by Khorana and his collaborators (58–60). We do not have space to describe these methods in detail here nor is this necessary, they are very well known because of their use in the first organic synthesis of the DNA sequence coding for valine tRNA (59, 60). Since considerable effort is required for the preparation and purification of such sequences due to the necessity for chemical protection of the bases and the occurrence of side reactions and losses which decrease the yields, it may be useful to mention some alternative methods of synthesizing primers. Gilham & Smith (61) have reported a technique for stepwise enzymatic synthesis of deoxyoligonucleotides of defined sequence using an enzyme (62) capable of the unprimed polymerization of deoxyribonucleoside 5'-diphosphates. This enzyme permits the predominant addition of a single deoxyribonucleotide to the 3' terminus of a deoxyribo-oligonucleotide. Sequences ranging from 4 to 8 nucleotides were tested and found to be equally efficient acceptors.

Stepwise synthesis of ribo-oligonucleotides provides another approach (63, 64, 67, 68). A number of laboratories (e.g. 65, 66) have shown that oligoribonucleotides will serve as very efficient primers for DNA polymerase I. Mackey & Gilham (67) and Bennett et al (68) have investigated the synthesis of oligonucleotides of defined sequence, using polynucleotide phosphorylase in combination with nucleoside 5'-diphosphate substrates having their 2'- and 3'-hydroxyl groups chemically blocked. In trial reactions it was possible to obtain the quantitative conversion of the trinucleotide pApApA to the tetranucleotide pApApApU–ME (where ME denotes the –methoxyethyl blocking group which limits the additions to one nucleotide at each step of the reaction). The product pApApApU–ME was then deblocked by mild acid treatment and used as acceptor for reaction with ppA–ME to yield pApApApUpA–ME, also quantitatively. If the reaction works properly for acceptor molecules of various chain lengths and nucleotide sequences, and if a solid support system can be devised, this reaction could form the basis for rapid automated synthesis of ribo-oligonucleotides of defined sequence.

OTHER SOURCES OF SPECIFIC PRIMERS   Maps of restriction enzyme fragments for many phage and virus genomes are rapidly becoming available so that such fragments should provide a convenient source of large and highly specific primers with well-characterized points of initiation. A complicating factor is that such fragments are double stranded, and if used with template containing both strands, will give an independent initiation from each strand. This is not a serious problem in the case of many DNAs whose strands may be conveniently separated.

SPECIFIC PRIMING WHEN THE SEQUENCE IS UNKNOWN   As outlined above, if part of the biologically interesting target sequence is known, either from some other nucleic acid sequencing approach or imperfectly from amino acid sequencing, it may be possible to construct primers specifically made to give the desired initiations. Such information is only infrequently available so that one normally does not know which oligomeric primer sequence to synthesize to start at a particular point. Short nucleotide sequences sometimes may be deduced from other data as when runs of tryptophan and methionine residues (each of which has a single codon) occur in a protein sequence, or by the analysis of amino acid changes resulting from frame-shift mutations. Wu (69) has suggested that it may be possible to deduce acceptable primer sequences from amino acid sequence data in a more general way by taking advantage of the fact that GT base pairs are thought not to seriously destabilize base-paired nucleic acid structures. Thus, where the amino acid sequence predicts that either C or T may be present, one places a G in the primer sequence. If it is known that either G or A are present in the template, one places a T at the corresponding primer position. Since a GT base pair is considerably less stable than GC or AT pairs, primers synthesized according to these rules may be considerably less stable and less specific than those with an exact fit. This would be a particularly serious problem if the template itself contains highly base-paired structures, which may be a common occurrence (34, 35, 71, 72).

Restriction Enzymes

It now appears that bacterial restriction enzymes may provide methods permitting the rapid and systematic dissection of DNA molecules many thousands of nucleotides long. Smith & Wilcox (73) showed that a restriction enzyme preparation from

Haemophilus influenzae cleaves double-stranded bacteriophage T7 DNA into about 40 different fragments. It was subsequently found that the cleavage sites had the sequence

$$5'\ldots G_pT_pPy\downarrow_p Pu_pA_pG\ldots3'$$

$$3'\ldots C_pA_pPu_p\uparrow Py_pT_pG\ldots5'$$

where Pu and Py stand for purine and pyrimidine, respectively, and the arrows indicate the sites of cleavage.

Danna & Nathans (74) found that the *H. influenzae* restriction enzymes introduce 11 specific cleavages into the SV40 genome. The sizes of these fragments were accurately determined by electrophoresis on acrylamide gels, and it was possible to establish the order of some of these fragments in the genome by analysis of partial digests (75). Partial digests of the labeled SV40 DNA were electrophoresed in parallel with complete digests. Bands present in the partial digests but not in a complete digestion mixture were eluted, digested to completion, and electrophoresed again to determine which fragments were adjacent.

Some of the desired partial digestion products could not be obtained in adequate yield, apparently because certain cleavage sites were more sensitive than others. This resulted in a few ambiguities in the fragment order which have been completely resolved by analysis of products resulting from cleavage with a restriction enzyme from *H. parainfluenzae* which cleaves SV40 DNA in three places (76). The resulting physical map of the SV40 genome is shown in Figure 2.

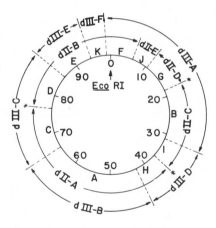

Figure 2   A cleavage map of the SV40 genome. Map units are given as a percentage of the distance from the *Eco*$_{RI}$ site. Fragments A to K result from cleavage of SV40 DNA by Endo R·*Hin*$_d$ (74, 75). The fragments dII-A through dII-E result from cleavage with purified Endo R·*Hin*$_d$II, whereas the fragments dIII-A through dIII-F are the products of purified Endo R·*Hin*$_d$III. Also indicated (asterisks) are the sites of cleavage of SV40 DNA by Endo R · *Hpa*I (74–75a). Cleavage map courtesy of D. Nathans.

Several new restriction enzymes have now been characterized and the sequences of the cleavage sites of six of them have been determined. Together they provide an extremely impressive arsenal for the systematic dissection, mapping, and (in combination with other techniques) sequence determination of even quite large DNA molecules. The most important characteristics of some of these enzymes are summarized in Table 1.

It is clear that analysis of a DNA molecule by these techniques will be considerably more difficult if too many fragments are produced. For instance, if 25 fragments are produced it might be possible to resolve them by gel electrophoresis, but it would be extremely difficult to resolve the much larger number of fragments produced by partial digestions. Thus it is probably desirable to carry out the analysis of large DNA molecules in several stages. For the first stage one might wish to use the Eco_{RI} or *Serratia* endonuclease R enzymes which make 5 and 11 breaks, respectively, in the adenovirus 2 genome, and 5 and 12 breaks, respectively, in phage λ DNA (see Table 1). After electrophoresis on acrylamide gels to purify the fragments obtained with one of these enzymes (see 95 for an example of fractionation of fragments in this size range), one might want to treat each of the purified fragments, or only those from an especially interesting part of the molecule, with one of the other restriction enzymes which produces a larger number of fragments. For example, the *H. aegyptius* enzyme, endonuclease Z, and the enzyme Eco_{RII} cleave the roughly 5000-base pair SV40 genome into 10 and 17 fragments, respectively.

It is clear from an inspection of Table 1 that such choices will have to be largely empirical since the frequency of a particular cleavage may be very different in different DNAs. For instance, the *H. influenzae* enzymes (Hin_dII and Hin_dIII) make similar numbers of cleavages in the SV40 genome (5 and 6 cleavages, respectively), whereas Hin_dII makes 40 or more cleavages in phage T7 DNA but Hin_dIII makes few or no cleavages in this DNA.

In addition to constructing physical maps showing the position of each fragment, it is of great interest to make whatever correlations may be possible between this physical map and the gene products produced from each region. So far it has been possible to make such correlations by a variety of techniques. Three particularly valuable approaches are the use of genetic deletion strains, RNA-DNA hybridization techniques, and genetic transduction assays.

In bacteriophage λ, a very rich supply of well-characterized deletion mutants is available, and one can determine the location of fragments by comparing the electrophoretic patterns of fragments obtained by digesting phage λ genomes that differ by appropriate single-deletion mutations; bands missing from the digest of the deletion strain can be assumed to represent fragments that fall at least partly within the deletion. This technique was useful in ordering the six fragments produced by digestion of phage λ with Eco_{RI} (95) as well as the more difficult task of ordering many of the more than fifty fragments produced by the $HpaII$ enzyme (86).

It has also been shown that the $HpaII$ enzyme cuts the phage λ immunity region, permitting separation of the leftward promoter and operator from the rightward promoter and operator. In similar studies it has been shown (79, 96) that digestion of phage λ DNA with the enzymes Hin_dII and Hin_dIII cleaves the λ operator

Table 1   Restriction enzyme cleavage specificities

Bacterial Source of Enzyme and Common Name	Formal Designation of Enzyme	Sequence of Cleavage Site
H. influenzae D (note that the common name endonuclease R refers to a mixture of Hin_dII and Hin_dIII) (See note *d*)	Hin_dII Hin_dIII	5′....G_pT_pPy ↓$_p$ Pu_pA_pC...3′ 3′....$C_pA_pPu_p$↑Py_pT_pG....5′ (11) 5′....$N_pA_pG_pC_p$ T_pN....3′ 3′....N_pT $_pC_pG_pA_pN$....5′ (78)
H. influenzae B (Rb)		Unknown
H. influenzae C	Hin_cII	
H. influenzae F		Unknown
H. suis		Unknown
H. aphirophilus (Endo AP)		See note *e*
H. aegypticus (endonuclease Z)		5′....$G_pG_pC_pC$....3′ 3′....$C_pC_pG_pG$....5′ (81)
H. parainfluenzae (endonuclease Hp)	*Hpa*I *Hpa*II	See notes *b* and *d* See note *c*
H. parahemolyticus		Unknown
E. coli carrying RTF-1 (RI restriction endonuclease)	*Eco*RI	5′....A/ T_pG ↓$A_pA_pT_pT_pC_pT$/ A....3′ 3′....T/ $A_pC_pT_pT_pA_pA_p$↑G_pA/ T....5′ (13)
E. coli carrying R factor RII (RII restriction endonuclease)	*Eco*RII	5′.... ↓ $C_pC_pA_pG_pG$ ....3′ 3′.... $G_pG_pT_pC_pC$↑....5′ (14)
B. subtilus		5′....$G_pG_pC_pC$....3′ 3′....$C_pC_pG_pG$....5′ (94)
Serratia		Unknown

^{a} Hin_dIII "is not active (or very low in activity) on T7 DNA" (Hamilton Smith, personal communication).

^{b} The *Hpa*I enzyme cleaves the sequence: pApApCp . . . (K. Murray, personal communication).

^{c} The *Hpa*II enzyme cleaves the sequence: pCpGpGp...(K. Murray, personal communication).

Number of Cleavage Sites in Selected Genomes

Phage λ (about 50,000 base pairs)	Phage φX174 (about 5,000 base pairs)	SV40 (about 5,000 base pairs)	Adenovirus 2 (about 40,000 base pairs)	Phage T7 (about 38,000 base pairs)
Hin_dII and Hin_dIII together give about 39 breaks (79)	both enzymes together give 12 fragments (80a)	5 (75)		≥ 40 (77, 15)
		6 (75)		0^a (77)
7 (80)	0 (80)			

Hin_cII has the same specificity as Hin_dII—*H. influenzae* C appears to lack an Hin_dIII-type activity (79)

| | ≥ 9 (80) | | | |
| 7 (80) | 0 (80) | | | |

13 fragments from phage fd RF-I DNA and about 50 fragments from T3 DNA (15a)

	11 (82)	10 (83)		
	both enzymes together give	4 (76) 3 (75a)		
≥ 50 (86)	8 fragments (84)	1 (85, 75a)		
	4 (84)			
5 (90, 13, 91) 6 (91)	0 (80, 93)	1 (87, 88, 89)	5 (92)	0 (93)
20 in λh 80 (14)		~ 17 (12)		
1 or 2 (93)	0 (93)	0 (93)	11 (93)	0 (93)

[d] The results of Sharp et al (75a) and of Murray (see note *b*) suggest that the *Hpa*I enzyme cleaves at the GpTpT/pApApC site of Hin_dII. It is not clear whether the Hin_dII enzyme is a mixture of two or more enzymes (75a) or a single activity which can also cleave at GpTpC/pGpApC sites.

[e] *H. aphirophilus* has been shown to cleave at the site: (5′) pNpC/pCpGpGpN (3′). The site is thus symmetrical and the fragments have overlapping single-strand ends (15a).

sequences themselves, a fortunate accident which may be very useful in the detailed sequence analysis of this very interesting region.

Khoury et al (97) have determined which of the 11 SV40 DNA fragments resulting from digestion with the Hin_dII plus Hin_dIII enzymes are involved in early and late RNA synthesis in SV40-infected cells. When transcripts labeled before or after the onset of virus DNA synthesis were hybridized to purified DNA fragments, it was found that fragments A, H, I, and B (see Figure 2) were active in early RNA synthesis, whereas fragments A, C, D, E, K, F, J, G, and B were active in late RNA synthesis.

Edgell, Hutchison and their collaborators (98–100) have used a genetic approach (101) to directly correlate particular restriction enzyme fragments with the known genetic map of bacteriophage ϕX174. In this procedure, restriction enzyme fragments from wild-type phage DNA are isolated as purified bands after acrylamide gel electrophoresis. Each band is eluted, heat denatured, and annealed with single-stranded circular phage DNA mutant for a particular gene locus. These complexes are then infected into *E. coli* spheroplasts and the phage progeny obtained are plated on a restrictive indicator bacteria (which will not support the growth of the mutant phage). When the complex (single-stranded circle to which a complementary fragment is hydrogen bonded) enters the spheroplast, it will be converted to a fully double-stranded replicative form (RF) with two possible outcomes. If the fragment does not include the region of the mutant gene, then the formation of the RF molecule will involve direct duplication of the mutant sequence so that both strands are mutant and no wild-type progeny will be produced. If, however, the fragment does include the region of the mutant gene, then the RF molecule will contain one mutant strand and one wild-type strand. Further replication will then yield some mutant and some wild-type progeny as detected by their growth on the restrictive indicator. In principle, very similar approaches could be used with mammalian viruses such as SV40 or polyoma, where one cannot carry out conventional genetic mapping of the sort possible in bacterial systems.

Other Nucleases

The restriction enzymes provide very powerful tools, but they do not, at present, allow DNA sequences to be cleaved into fragments of average size much smaller than 500 nucleotides. Fortunately, a number of other enzymes are available which may be useful in cleaving DNA into smaller specific fragments. Bacteriophage T4 endonuclease IV is the best known of these enzymes (45, 102, 103) and has been shown to cleave denatured DNA at most (but not all) TpC sequences in DNA (20, 46). For many sequencing problems it would be extremely helpful to be able to use T4 endonuclease IV digestions in combination with ribosubstitution sequencing. Fedoroff & Salser (manuscript in preparation) have shown that this is possible since the enzyme will digest ribo C-substituted hemoglobin DNA to yield the characteristic large specific fragments.

Now that suitable techniques of analysis are available, other DNases producing fragments in this size range may be discovered. For instance, Robertson et al (20) found that streptococcal nuclease ("streptodornase") cleaved a 23-nucleotide T4 endonuclease IV fragment at three principal sites.

There have also been tantalizing results in the search for base-specific DNases. Kato, Ando & Ikeda (104, 105) reported the isolation of deoxyribonucleases K1 and K2 which preferentially split the GpG and GpA bonds in DNA (106). Unfortunately, the purified enzymes are rather unstable, and neither the crude nor the purified preparations are commercially available at the present time (106).

Specific Binding Techniques

RIBOSOME BINDING SITES   In the specific binding approach, one takes advantage of the ability of a particular protein to bind specifically to the region whose sequence is desired. This has been used extensively by RNA sequencers to isolate ribosome binding sites from bacteriophages R17 (44, 107), $Q\beta$ (44, 108, 109), f2 (110), and T7 (111).

From recent work by Bretscher (112), it is clear that these methods can be applied in isolating the DNA sequences that code for the ribosomal binding sites. Robertson and his collaborators (20) used this approach to isolate a DNA sequence from the ϕX174 phage genome protected from nuclease attack by ribosomal binding. Subsequent sequence analysis showed that the protected DNA sequence corresponded to the initiation site for the gene G spike protein (113).

RNA POLYMERASE BINDING SITES   Other laboratories have attempted to study the structure of promoter regions by isolating RNA polymerase binding sites (114–121). In the most recent of these papers (114), the material isolated from the phage fd genome was studied by fingerprint analysis of its pyrimidine tracts. It consisted of a homogeneous piece of DNA of 40 nucleotide pairs isolated from a single site. In these experiments viral single-stranded DNA was used as template for the in vitro synthesis of radioactive replicative form (RF) DNA by E. coli DNA polymerase I and polynucleotide ligase. Complexes of this DNA with RNA polymerase were digested with DNase and the protected DNA was separated from digested fragments by passage over a Sephadex G-100 column, after which it was depurinated and fingerprinted. Use of higher levels of RNA polymerase resulted in protection of more DNA, suggesting that there is also a second binding site with weaker affinity for the enzyme. This was confirmed by the appearance of a second set of pyrimidine tracts when the material isolated in this way was fingerprinted. A variety of experiments show that E. coli RNA polymerase maintains its in vivo specificity in vitro under the proper conditions, and from this it is hoped that the sequences isolated in this way will in fact prove to correspond to in vivo promoters.

REPRESSOR BINDING SITES   Several groups have taken advantage of the very tight complexes formed between repressors and their corresponding operators to isolate operator regions for sequence studies. The complex between a phage λ repressor dimer and its operator has a half-life of about 170 min at $0°C$ (122), while the lactose operator-repressor complex has a half-life of about 15 min under the conditions used by Gilbert, Maizels & Maxam (24). Since the operator sequence represents such a very small fraction of the total genome from which it is to be isolated, it is useful to carry out the isolation in stages to reduce the amount of background due to material binding nonspecifically to the filters. Similar methods are used for isolating both the λ and *lac* operator fragments. Pirrotta (122) sonically treats DNA to give fragments of molecular weight $3–6 \times 10^5$, binds them to repressor, and filters through a nitrocellulose membrane so that only those fragments containing an operator sequence are retained on the filter by virtue of attached repressor molecules. These DNA fragments are then released from the filter by treatment with sodium dodecylsulfate (SDS). After removal of the SDS they are again bound to repressor, treated with pancreatic DNase I for 3 min at 0°C, and then rapidly filtered through a membrane filter. The filter is washed and rapidly extracted with buffer containing SDS to remove the purified operator fragments. Controls in which repressor is omitted or DNA missing one or both operator sites is used, indicate that the material isolated in this manner is pure. Acrylamide gel electrophoresis of fragments isolated in this way show a sharp peak with a size of roughly 70 base pairs. The binding of these fragments to the repressor is less stable (half-life about 10 min at 0°C rather than 170 min) than that observed with the whole DNA molecule. This destabilization has obvious practical implications in making the isolation more difficult, as well as theoretical implications in suggesting that the DNA sequences near the operator may also be important for repressor binding.

The *lac* operator can be purified in a manner similar to that discussed above. It is much smaller than the λ operator fragments and the sequence has been determined (24, 25).

```
           ┌─────────┬─┬─┬─┬─┬─┬─┬─┬─┬─┬─────────┐
(Promotor) │ A A T T G T │G│A│G│C│G│G│A│T│A│ A C A A T T │ (Structural Gene)
           │ T T A A C A │C│T│C│G│C│C│T│A│T│ T G T T A A │
           └─────────┴─┴─┴─┴─┴─┴─┴─┴─┴─┴─────────┘
```

The boxes indicate the striking amount of symmetry in the structure, as predicted from the analysis of operator mutants (123, 124). The lactose messenger sequence from the UV-5 CAP independent promoter mutant was shown to start at the extreme promoter end of this operator sequence, running through the operator to the right where a 22 nucleotide sequence corresponding to the N-terminal amino acid sequence of β-galactosidase was found to follow 20 nucleotides after the end of the operator (70).

Although these sequencing studies are not yet complete, several very striking contrasts are already apparent between the λ and the *lac* operator-promoter combinations. As noted above, the operator fragment protected by λ repressor binding is several times larger than the *lac* operator fragment, and evidence is accumulating that, unlike the *lac* operator, it consists of a repeated series of closely related sequences which bind a series of repressor molecules (32, 96, 125). Further, it has been suggested by Blattner & Dahlberg (22) and Blattner et al (23) that the λ promoter and operator sequences are unlike the situation in *lac* in that they are located well upstream (roughly 200 base pairs) from the starting points of the transcription they control. This finding has been questioned recently by Maniatis,

Ptashne & Maurer (96) who estimate that the separation is no more than 30–50 base pairs. Thus there will be considerable interest in further investigation of the molecular details of control in both the λ and *lac* systems as well as comparison of those with other systems where repressors have been purified and offer the possibility of selective isolation of operator fragments.

Use of Reverse Transcriptase

Most rapid sequencing techniques in use at present rely heavily on the use of material very highly labeled with ^{32}P, and progress in sequencing mammalian mRNA species has been very slow until recently because of the difficulty in obtaining such highly labeled material. One approach used for hemoglobin mRNA is to label in vivo by injecting ^{32}P orthophosphate into anemic rabbits (126). The fraction of the total ^{32}P that can be recovered in hemoglobin mRNA after such a procedure is quite small, however, and the specific activities obtained are limited by the large pools of unlabeled phosphate and the toxicity of the ^{32}P. The use of polynucleotide kinase to introduce ^{32}P at the 5′ termini of fragments from unlabeled RNA pieces was developed by Szekely & Sanger (127) for sequence analysis of molecules otherwise difficult to label. Neither of these approaches has yet proven sufficiently powerful to permit rapid sequence analysis. It now appears that these problems can be circumvented and rapid progress achieved by converting mRNA sequencing to a DNA sequencing problem. RNA-dependent DNA polymerase from avian myeloblastosis virus has been used to synthesize complementary DNA from hemoglobin mRNA (128–130), immunoglobin mRNA (131–133) and other mRNAs (134–136a) using oligo(dT) as primer.

This approach takes advantage of the fact that RNA-dependent DNA polymerase will not copy a template except by elongation of a primer sequence complementary to the template. Since most eukaryotic mRNAs contain poly(A) at their 3′ terminus [see 136b for a recent enumeration of metazoan mRNAs shown to contain poly(A)], an oligo(dT) primer can be used to initiate synthesis at the 3′ end of the template so that in principle the entire mRNA sequence could be copied.

Some of the most important general features of the reaction when rabbit globin mRNA is used as the template are (Kacian, personal communication, 128–130, 137): (*a*) The synthesis of complementary DNA is primer dependent; (*b*) The DNA synthesized in the presence of actinomycin D is perhaps as long as the mRNA template; (*c*) If actinomycin D is omitted from the reaction mixture, double-stranded DNA is obtained and the product is shorter; (*d*) The DNA synthesized appears to be a faithful copy of the mRNA as estimated by RNA-DNA hybridization experiments; and (*e*) Both the double- and single-stranded DNA products can be transcribed into RNA by *E. coli* RNA polymerase in vitro.

Spiegelman, Watson & Kacian (138) showed that the enzyme could be used with a variety of RNA templates. For a recent review of the extent of synthesis obtained with various RNA templates, with or without added primers, see (137). Because this approach appears to provide widely useful means of obtaining the highly radioactive material necessary for rapid sequence determinations, it is essential to discuss in more detail what is known about the fidelity of the RNA-dependent DNA polymerase as well as the novel features of its products, including (a) the extent to which various RNAs are copied, (b) effects of the run of dT residues at the 5′ terminus of the product, (c) differences between the products synthesized with or without actinomycin, and (d) possibilities of using primers other than oligo(dT).

FIDELITY   RNA-DNA hybridization studies do not provide an adequate test of the fidelity of complementary DNA synthesis by the avian myeloblastosis enzyme. One might expect excellent hybridization even if many random mistakes were made in the synthesis of hemoglobin complementary DNA; however, these mistakes should be easily detectable if such material is cleaved at a specific base and fingerprinted. If the error frequency is high, one would expect to see many faint spots in the fingerprint in addition to the major cleavage products (since there are many different ways of making a mistake in each fragment). We have looked for such evidence of low fidelity by fingerprinting RNase T1 digests of RNA synthesized from the hemoglobin complementary DNA template and by fingerprinting ribo G- or ribo C-substituted DNA synthesized from the same template. In all cases, the results suggest that the fidelity of the avian myeloblastosis virus enzyme is adequate for sequencing studies.

Detailed sequencing studies are now in progress in our laboratory (139, unpublished work of R. Poon, G. Paddock, and H. Heindell carried out in collaboration with D. Kacian and A. Bank), and we have found the expected amount of agreement between individual fragment sequences and the known amino acid sequences.[2] Similar detailed analysis of the nucleotide sequences of human globin mRNA is being carried out by Marotta et al (27). In our own laboratory, the sequences of almost all of the fragments resulting from cleavages at G are now complete. In many cases, these sequences are rather long and their correspondence with the known amino acid sequences provides strong preliminary evidence that

---

[2] It is expected that as much as 30% of the sequences in hemoglobin mRNA is not translated into protein. The most recent estimates of the size of the rabbit globin mRNA indicate that it is 270–280 nucleotides longer than the structural gene itself (Gould & Hamlyn, 140), and the data of Lim & Canellakis (141) suggest that only about 50–70 nucleotides of this extra material is poly(A). Thus one of our main goals is to determine the structure, location, and hopefully the function of these large untranslated regions. If one suggested that there were no untranslated sequences except poly(A) in the globin mRNA, it would be necessary to choose the smallest estimate of the mRNA size (190,000 daltons, corresponding to about 195 nucleotides in addition to the structural gene, 58) and the largest estimate of the poly(A) tract size in any eukaryotic mRNA [about 200 nucleotides of poly(A) in newly synthesized mRNA from HeLa cells, 142]. In fact the larger of the estimates of globin mRNA size obtained by Gould & Hamlyn (140) might be expected to be more accurate since these authors are the first to have carried out size measurements on denatured mRNA so that the size estimated would not be affected by mRNA secondary structure. The smaller of the estimates of the length of the poly(A) tract may be more likely since even the 200-nucleotide poly(A) sequences in HeLa cell mRNA decrease during the life of the mRNA (143). Note added in proof: Recent determinations by Hunt (141a) indicate that the poly(A) in rabbit globin mRNA is only 30–40 nucleotides long.

reverse transcriptase is making few, if any, systematic errors (e.g. always inserting C in place of T at a particular position in the sequence).

THE EXTENT OF COPYING   Several laboratories have observed that synthesis by RNA-dependent DNA polymerase in the absence of actinomycin D results in double-stranded DNA smaller in size than the single-stranded DNA synthesized in the presence of the antibiotic (137, D. Kacian, personal communication). This double-stranded DNA renatures rapidly as if it has a hairpin structure. Such a structure could arise if the growing DNA chain folded back upon itself to provide the primer for synthesis of the second strand. According to this model, actinomycin, by its inhibition of double-stranded DNA synthesis, forces the enzyme to copy a larger portion of the template. The single-stranded DNA synthesized in the presence of actinomycin appears to be roughly comparable in size to the mRNA sequence. Verma et al (129) give one of the smaller size estimates, about 450 nucleotides, roughly 100–200 nucleotides shorter than the intact mRNA itself. Other laboratories have estimated that the product obtained using rabbit globin mRNA is 500–600 nucleotides in length (130).

With other mRNA templates, products have been obtained which are often un-ambiguously shorter than the template. The product obtained with duck globin mRNA was estimated to be about 250 nucleotides long and that from a myeloma light-chain mRNA about 325 nucleotides long (roughly 40% as large as the mRNA) (131, 132). At present it is not known whether failure to synthesize a full-length complementary DNA results from a particular block in the RNA sequence, inactivation of the enzyme (131), or some other factor.

DOUBLE-STRANDED PRODUCTS   As mentioned above, short double-stranded hairpin products are synthesized in the absence of actinomycin. Bishop and his colleagues (144–146) have shown that the double-stranded DNAs synthesized from the Rous Sarcoma Virus (RSV) RNA using RSV RNA-dependent DNA polymerase are not only quite short, but they are also copied from very limited portions of the RSV genome. Depending on whether natural or exogenous primers were used, the products were estimated to have complexities of 820 or 130 base pairs, roughly 3 or 0.5% of the total genome size! Such reactions may be extremely useful in allowing the sequencer to concentrate his attention on a small region adjacent to the poly(A) sequences used for the priming reaction.

SYNTHESIS OF POLY(rA) FROM THE OLIGO(dT) PRIMER   The presence of the oligo(dT) primer sequence at the 5'-end of a DNA template made with RNA-dependent DNA polymerase leads to some minor annoyances by catalyzing poly(rA) synthesis by *E. coli* RNA polymerase. Chamberlin & Berg (147) noted that *E. coli* RNA polymerase could synthesize poly(rA) from oligo(dT) tracts in a DNA template. Such poly(rA) synthesis is strongly inhibited by the presence of the other three ribonucleoside triphosphates, but this is not true with the hemoglobin complementary DNA template.

If the oligo(dT) tracts priming the poly(rA) synthesis are in the interior of the template, then addition of the other three ribonucleoside triphosphates strongly inhibits poly(rA) synthesis, presumably by allowing the enzyme to proceed into other regions of the template (147). This interpretation predicts that when oligo(dT) is at the extreme 5′-end, as with hemoglobin complementary DNA, the presence of a full complement of nucleoside triphosphates should not inhibit poly(rA) synthesis, and we find this to be the case (unpublished results of R. Poon). The poly(rA) synthesis observed is template specific and not inhibited by high levels of inorganic phosphate, so it cannot be due to the poly(rA) synthesizing activity noted by Ohasa & Tsugita (148).

Poly(rA) is known to be a specific and efficient inhibitor of *E. coli* RNA polymerase (149), so synthesis of poly(rA) might be expected to cause a premature cessation of RNA synthesis. This seems to be the case since we find that RNA synthesis is almost complete in about 30–60 min at 37°C. In order to minimize the number of RNA polymerase molecules initiating synthesis directly in the oligo(dT) region, we first incubate with GTP for 10 min, and then add the remaining three nucleoside triphosphates to start the in vitro synthesis. This increases the yields of RNA relative to poly(rA), and routinely gives 1–3 μg product per microgram of template. (Note that these results are for synthesis in low ionic strength buffer from single-stranded hemoglobin complementary DNA templates.)

The enzyme preparations used in RNA sequence analysis may frequently contain traces of activity which cleave poly(A) sequences so that, when the product has been labeled with ^{32}P ATP, partial digestion of large amounts of radioactive poly(rA) may produce compounds of the form A–(A–)$_n$AOH and A–(A–)$_n$Ap. Such fragments overlap and interfere with the analysis of many of the legitimate cleavage products. Passage of the digest over an oligo(dT) cellulose column (150) provides a convenient way of removing interfering oligo(rA) fragments. It must be kept in mind, however, that absorption and elution from such a column is also an excellent way to preferentially isolate any heteropolymeric sequences attached to poly(A).

SYNTHESIS OF POLY(dA) AND POLY(dT) FROM HEMOGLOBIN COMPLEMENTARY DNA   An earlier paper (139) discussed the problem of "escape synthesis" of homopolymers by DNA polymerase I. With a hemoglobin complementary DNA (C-DNA) template we have observed that when we synthesize a ribosubstituted DNA product a substantial but variable fraction of the synthesis by DNA polymerase I is devoted to poly(A) (unpublished results of N. Fedoroff). This is an expected consequence of slippage of the product on the oligo(dT) portion of the template. Slightly more surprising was the finding that quantities of poly(dT) are also synthesized, suggesting that the poly(dA) synthesized from oligo(dT) stretches is itself serving as a template.

HEMOGLOBIN SINGLE-STRANDED COMPLEMENTARY DNA IS A SELF-PRIMING TEMPLATE   It has been shown that many of the short double-stranded DNA molecules synthesized in the absence of actinomycin are rapidly renaturing (137, 144), suggesting that in in vitro reactions RNA-dependent DNA polymerase is prone to copy its DNA product rather than the RNA template so that hairpin structures are formed. Further evidence suggests that even the single-stranded DNA products formed in the presence of actinomycin D may have short hairpin structures at the 3' end. For instance, when single-stranded DNA synthesized in the presence of actinomycin D is purified and used as template for further DNA synthesis by the avian myeloblastosis enzyme, no primer is needed. The DNA synthesized in this way is covalently linked to the template DNA and has the rapidly renaturing characteristics of hairpin structures (137).

If there is a short hairpin structure at the 3' end of the single-stranded complementary DNA, it would help explain some of the surprising features of DNA synthesis by *E. coli* DNA polymerase I from this template (unpublished results of N. Fedoroff). This synthesis starts very rapidly with no dependence on added primer and is quickly completed. Again, virtually all of the product is covalently attached to the template as shown by rapid renaturation after heating.

The formation of rapidly renaturing structures in itself is not surprising since a similar result is observed when single-stranded linear fragments of phage M13 DNA are used as template. In this case, it is presumed that the 3' OH end of the molecule forms chance base pairs with some other part of the template sequence. The 3'–5' exonuclease activity of DNA polymerase I would then be expected to remove any mispaired bases at the 3'-terminus until it reaches a stretch of base pairing sufficiently accurate to cause initiation of a round of DNA synthesis. But in the case of the M13 DNA fragments, the addition of a mixture of oligodeoxynucleotide primers of random sequence stimulates DNA synthesis and suppresses synthesis of the rapidly renaturing product resulting from self priming. The same primer mixture has little if any effect on synthesis from the hemoglobin DNA template. Even when a large amount of exogenous primer (either random oligodeoxynucleotides, as mentioned above, or fragments of hemoglobin mRNA) is added, the *E. coli* DNA polymerase exhibits a strong preference for the endogenous priming activity of the hemoglobin C-DNA molecule.

PRIMERS OTHER THAN OLIGO(dT)   Because most eukaryotic mRNA possess poly(rA) sequences, and because these are located at the 3' termini (exactly where one wants to prime in order to attempt synthesis of a complete copy of such an mRNA), oligo(dT) primers will obviously play a role of unique importance. On the other hand, RNA-dependent DNA polymerase would appear to accept a wide variety of primers, and the use of primers other than oligo(dT) may be very useful in permitting copying of specific regions of mRNA molecules. Of the other homopolymers, little priming activity has been obtained with oligo(dC) or oligo(dA), but oligo(dG) gives synthesis from rabbit globin mRNA which ranges from 10–50% of that obtained with oligo(dT) primers (129, 130, 137). Both oligo(dG) and oligo(dT) products were rendered completely resistant to single-stranded specific S1 nuclease by hybridization to 10S globin RNA (137). The oligo(dG)-primed product was only slightly smaller in size than the oligo(dT)-primed product, suggesting that the oligo(dG) primer must bind to a C-rich region rather near the 3'-end of the mRNA. Competition hybridization experiments suggest that approximately 30% of the sequences in oligo(dG)-primed DNA do not compete with oligo(dT)-primed DNA.

We have confirmed that there are important differences between the oligo(dT)- and oligo(dG)-primed hemoglobin C-DNAs by direct sequencing studies. These differences are shown in the fingerprints of RNAs transcribed in vitro from the (dT)- and (dG)-primed C-DNAs (unpublished observations of R. Poon). Equally striking differences are obtained when ribosubstituted DNA synthesized from the two C-DNAs is fingerprinted (unpublished results of N. Fedoroff). It is possible that the new fragments seen in these fingerprints indicate that the oligo(dG)-primed synthesis by reverse transcriptase continues further toward the 5′ end of the mRNA. In this case the additional data obtained will be very helpful in determining the sequence of hemoglobin mRNA. Alternatively, we must not forget that those sequences not found in the oligo(dT)-primed material might result from copying some unknown species of RNA present in the 10S globin mRNA preparation.

Synthesis primed by oligo(dG) has also been observed with immunoglobin light-chain mRNA from MOPC41 cells (131, 132). In one case it was shown that this oligo(dG)-primed product is comparable in size to oligo(dT)-primed material from the same template.

GENERAL METHODS FOR PURIFYING mRNAs   In the preceding section we indicated our optimism that the availability of complementary DNAs made with reverse transcriptase would permit rapid progress in sequencing eukaryotic mRNAs. Other approaches, such as use of the E. coli or Micrococcus lysodeikticus RNA polymerases to synthesize RNA in vitro from an RNA template (151, 152) or labeling of mRNA itself by reaction with radioactive iodine (154) or mercury (155a), may also prove to be extremely helpful.

In all of these approaches it is necessary to have pure mRNA preparations. In most cases the methodology for isolation of a specific mRNA has been heavily dependent on the choice of situations where a particular cell type is devoted almost entirely to the synthesis of one or two individual proteins (for example, hemoglobin, myeloma proteins, lens crystallines, and silk fibroin). It would obviously be highly advantageous to be able to obtain a much wider range of mRNAs available for sequence analysis, and this may now be possible through two new approaches.

The first of these takes advantage of the fact that the polysomes synthesizing a particular protein carry nascent polypeptide chains which can be specifically reacted with antibody prepared against the native protein. Several groups have reported success with variations of this method. Delovitch and others (155b, 156) reported that precipitation of intact polyribosomes with whole antisera lacks specificity due to the binding of the Fc portion of IgG to ribosomes. They reported good specificity in isolating MOPC149 mouse plasmacytoma immunoglobulin light-chain mRNA by reacting the polysomes with F(ab′)2 fragments obtained by trypsin treatment of a rabbit anti-MOPC149 L-chain antiserum. Optimum amounts of the L-chain antigen were added to obtain precipitation of the polysomes. Other groups used purified rabbit γ-globulins against chicken ovalbumin (157) or glutamine synthetase (158) to precipitate polysomes carrying specific mRNAs. The work with glutamine synthetase (158) is especially interesting since it appears that the mRNA was obtained in reasonable purity (although probably not sufficiently free from other mRNAs for sequence studies) from cells in which less than 5% of total protein synthesis is devoted to this protein.

In other work, different methods have been used to isolate the polysome-antibody complexes. Uenoyama & Ono (159) used chicken *anti*-rabbit γ-globulin to precipitate specific complexes between rabbit γ-globulin and rat liver polysomes synthesizing either catalase or albumin. When the RNA extracted from the polysomes isolated with *anti*-catalase γ-globulin was placed in an in vitro system, as much as 46% of the protein synthesized was catalase, compared with 2.4% with RNA extracted from whole polysomes. Palacios and his collaborators (160) have used immuno-absorption rather than immunoprecipitation. Hen oviduct polysomes were reacted with goat *anti*-ovalbumin γ-globulin and the complexes were absorbed on a matrix of glutaraldehyde cross-linked ovalbumin. Elution with EDTA dissociates the ribosomes, releasing the mRNA. Of course all of these methods yield mRNA contaminated with ribosomal RNAs, but such contamination can be readily removed by present techniques. Some of the authors mentioned above have used velocity sedimentation (156) or absorption of the poly(A)-containing mRNA on Millipore filters (160) to remove ribosomal RNA, but chromatography on oligo(dT) cellulose columns (150) may be the method of choice.

One of the limitations of the approach discussed above is that only mRNAs attached to polysomes can be isolated. The second approach considered allows the experimenter to isolate not only a specific mRNA, but also its HnRNA presursors. It takes advantage of the discovery of a mechanism of translational control of protein synthesis by the immunoglobulin H-chain mRNA (161, 162). Complete myeloma protein (H2L2) binds specifically to the heavy-chain (H-chain) mRNA, and this mRNA may be isolated by precipitating the complex with antiserum directed against the myeloma protein. The RNA thus purified from total cellular RNA can be resolved into three bands by acrylamide gel electrophoresis. The two bands corresponding to lower molecular weights are found in the cytoplasm and code for complete H-chain protein when injected into oocytes (163a). The third band contains much larger molecules confined to the nucleus (163a). Thus it might appear to be the HnRNA precursor for the myeloma H-chain mRNA, a view supported by the finding that this RNA also codes for H-chain synthesis in the frog oocyte system (163b).

It is reasonable to believe that a variety of very interesting eukaryotic mRNAs and mRNA precursors will soon be available in purity adequate for sequencing studies; therefore it seems that the use of reverse transcriptase for sequence determinations will be widely applicable to the whole range of problems associated with eukaryotic mRNA synthesis and expression.

Obtaining Specificity by in Vitro Transcription of the DNA

In this section we consider ways in which in vitro RNA transcription can be used alone or in combination with a variety of other methods to focus on a particularly interesting small DNA sequence or to enable a large DNA sequencing task to be simplified more efficiently.

NATURAL PROMOTER SITES   One approach is to use conditions in which RNA polymerase will initiate at natural promoter site(s). In this way, the sequence specificity of the holoenzyme can be used to restrict the incorporation of radioactive label to a very small and interesting part (or parts) of a large genome. Using this approach, it has been possible to determine long sequences starting from the 5′ end of several natural mRNAs synthesized from phage λ DNA (22, 23, 26). Because of the possibility of obtaining synchronous initiation of synthesis, this approach allows the experimenter to specifically label different parts of the sequence through the use of appropriate pulse-labeling schedules. In this respect, it shares some of the power of the specific primer approach, but the specific primer approach is better insofar as one can use several different primers in different experiments to obtain as many different start points as desired for sequencing. Of course, the natural RNA polymerase start points are extremely interesting since they constitute the 5′-termini of natural mRNA sequences.

A similar approach has been used by Zain and his collaborators (164), who found that *E. coli* RNA polymerase initiates in vitro RNA transcription at predominantly a unique site in the SV40 genome (164, 165). In this case there is no reason to believe that the sequence obtained with the bacterial polymerase corresponds to the 5′ end of a transcript normally produced in SV40-infected cells.

If RNA polymerase is found to initiate at several sites on a particular genome, the usefulness of this approach will depend upon techniques for separating the various transcripts. In the case of RNA synthesized in vitro from phage λ, this has been done by RNA-DNA hybridization using the separated strands of the phage genome or DNA from suitable deletion strains. In some cases, the transcripts can be separated on the basis of size (some promoters are closely followed by ρ-independent termination signals so that in vitro transcription gives small molecules of discrete size). Such approaches are more difficult with the many other genomes for which the desired deletions are not available or the strands cannot easily be separated. In addition, some genomes may have a much larger number of initiation signals; for instance, bacteriophage T4 appears to have 50 polymerase initiation sites (166, 167), and even the very much smaller ϕX174 genome appears to have three sites (166). The lambdoid phage genomes are not unique in having such a small ratio of initiation sites to total genome size; the phage T7 genome appears to have only six initiation sites recognized by the *E. coli* RNA polymerase, with five of them appearing to serve the same transcription unit (168)!

DNA FRAGMENTS PRODUCED BY RESTRICTION ENDONUCLEASES A more general approach to separating particular in vitro RNA transcripts from a mixture might involve RNA-DNA hybridization to specific DNA fragments obtained by digestion with restriction enzymes. There has been very rapid improvement in the technology for producing such fragments in quantity. A more important use of these fragments may be to serve as actual templates for in vitro RNA synthesis by RNA polymerase (169), as shown by the work of several laboratories (169, 170; Fiers, personal communication).

One might wonder whether a small double-stranded DNA fragment lacking a natural promoter sequence would be a very poor template for RNA synthesis, and whether or not RNA synthesis might be initiated at random locations along the fragment. Kleppe & Khorana (171) and Terao, Dahlberg & Khorana (172) used short DNA fragments of known sequence as templates and showed that (*a*) adequate amounts of synthesis can be obtained even from short fragments, and (*b*) synthesis does not start at random positions. In this particular case (172) synthesis predominantly started at the 5th, 7th, and 9th nucleotides of a 29-nucleotide template.

If there are three initiation points in such a small fragment, it might seem that transcription of one of the much larger restriction enzyme fragments from SV40 would yield a large number of different products; however, in all but one of several SV40 fragments tested (169), the transcripts obtained contained as major components a very limited number of electrophoretic bands (as expected if there are large numbers of different initiation sites, but these sites have a very broad range of affinities for RNA polymerase so that a few of the strongest sites always dominate the synthesis). When several bands from a particular DNA template are eluted, digested with RNase, and fingerprinted, the patterns from the different bands should show differences which will be extremely useful in establishing the relative positions of each of the RNase fragments in the larger sequence.

SINGLE-STRANDED DNA TEMPLATES   At present most of the available information about sequencing transcripts from single-stranded DNA comes from two types of projects: 1. in vitro transcripts of hemoglobin complementary DNA (Hgbn C-DNA) synthesized by RNA-dependent DNA polymerase on a hemoglobin mRNA template in the presence of oligo(dT) primer and actinomycin D (55, 139), and 2. in vitro transcription of separated strands of highly repetitive satellite DNAs (18, unpublished data of R. Poon and J. Idriss). When the synthesis from a Hgbn C-DNA template is carried out in high ionic strength, a heterogeneous product of low molecular weight is obtained (with a size distribution peaking at roughly 60–70 nucleotides). Synthesis in low-ionic strength buffer yields much larger products (ranging up to the size of the Hgbn C-DNA template, 139 and unpublished data of R. Poon and N. Fedoroff). This observation seems to be generally applicable to single-stranded DNA templates rather than just to Hgbn C-DNA per se, since very similar results were obtained with single-stranded bacteriophage M13 DNA (unpublished observations of N. Fedoroff). This effect of high ionic strength probably explains why Verma et al observed synthesis of only low molecular weight RNA from Hgbn C-DNA (137). Interestingly, the shorter RNA transcripts produced by synthesis in high ionic strength do not seem to come from a more restricted part of the genome than the products of a low ionic strength incubation, since both types of preparation give very similar fingerprints (164, unpublished results of R. Poon).

We have also observed the synthesis of poly(U) from the Hgbn C-DNA template (unpublished observation of R. Poon). It is possible that the poly(A) synthesized iteratively, as discussed above, may serve as a template for poly(U) synthesis since a number of authors (151, 152, 173–175) have observed that RNA polymerase can synthesize RNA from an RNA template (albeit under somewhat different conditions than we used).

In experiments where the ^{32}P label is introduced on A residues, the synthesis of large quantities of poly(A) can interfere with sequence analysis. The RNase A, *E. coli* alkaline phosphatase or other enzyme preparations used in digesting the product may have contaminating activities which produce highly radioactive oligo(A) runs of various lengths, possibly appearing in unfortunate proximity to some of the legitimate products on the fingerprint. This may be avoided by special attention to the purity of the enzymes used; for instance, treatment of commercial *E. coli* alkaline phosphatase with diethyl pyrocarbonate (176). Alternatively, we have preferred to pass the enzyme digests over a column of oligo(dT) cellulose to remove poly(A) fragments (unpublished observations of R. Poon). Conversely, the oligo(dT) columns provide a convenient way of isolating the mRNA sequences located next to the 3′-terminal poly(A) sequence in the mRNA. (For obvious reasons, this is best done with material labeled in some residue other than A.) With controlled partial digestions prior to the use of the oligo(dT) column it has been possible to isolate regions of greater or lesser size adjacent to the poly(A) region (unpublished results of R. Poon).

SINGLE STRANDS FROM SATELLITE DNAs AS TEMPLATES   Frequently the strands of the highly repetitive satellite DNAs can be separated by equilibrium sedimentation in alkaline gradients due to the extreme bias in the base compositions of the complementary strands. Knowledge of the strand assignments of each of the RNase fragments obtained is very useful in deriving the basic repeat sequence and deciding what "mutational" variants of this sequence are present. We have found that the separated strands of some satellite DNAs provide good templates for RNA polymerase, but others do not. In the case of the HSβ satellite from *Dipodomys ordii*, the double-stranded DNA provided an excellent template for either RNA or DNA polymerases, whereas the separated strands were templates for DNA polymerase but not RNA polymerase. Consequently, the strand assignments for the various fragments could only be made by using ribosubstitution-DNA sequencing methods (18, 18a).

USE OF DINUCLEOSIDE MONOPHOSPHATES TO PRIME RNA SYNTHESIS FROM DUPLEX DNA The K_m for initiation by RNA polymerase is apparently much greater than the K_m for RNA chain elongation. Thus, when the nucleoside triphosphate concentration is lowered to 1–5 μM, the enzyme can elongate chains but cannot initiate normally. It will initiate if it is presented with a dinucleotide (in practice, these are usually supplied at concentrations of about 100 μM or greater) or oligonucleotide which it can incorporate at the beginning of an RNA chain (177). This has been used in several cases to restrict initiation by *E. coli* RNA polymerase to particular sites on duplex DNA templates (24, 70, 168, 178).

The principles involved here differ in some important respects from using specific oligonucleotide primers to direct the initiation of DNA synthesis by DNA polymerase I to a specific location. In the first place, since dinucleotides are active, it is clear that the oligonucleotide used to prime RNA synthesis does not have to form a stable association with its complementary DNA template. Secondly, with duplex DNA most of the initiation specificity still appears to derive from the ability of the RNA polymerase to recognize specific regions of DNA rather than from recognition of complementary sequences by the primers. For instance, when the 16 ribodinucleoside monophosphates were tested individually for their ability to stimulate initiation of RNA synthesis on intact bacteriophage T7 DNA, only six distinct RNA initiation sites were observed (168). As suggested earlier (178), it appears that such initiations can occur only at normal promoter sites where the RNA polymerase binds and locally denatures the DNA. In the absence of satisfactory threshold levels of the ribonucleoside triphosphate, initiation at such sites depends upon whether one supplies a dinucleoside monophosphate (or longer oligonucleotide) which can base pair with the denatured sequence. Consequently it may be possible to restrict in vitro RNA synthesis to one out of several possible promoter sites on a given template by using the properly chosen dinucleoside monophosphate as primer.

The work of Gilbert, Maizels & Maxam (24, 70), who have used initiation with GpA to sequence the 5'-end of the *lac* mRNA, illustrates some other advantages of this approach. In their case, the use of specific binding techniques enabled them to isolate DNA fragments which contained only the *lac* promoter. The use of GpA increased the specificity of initiation, helping to eliminate nonspecific background, and it enabled them to use very low levels of nucleoside triphosphates. This not only increased the efficiency of utilization of the radioactive precursors, but also slowed down the reaction and caused the RNA polymerase to pause at specific sites along the DNA so that the products could be separated as a series of discrete bands after acrylamide gel electrophoresis. As with the analogous ribosubstituted DNA synthesis (see 21), the analysis of different products of increasing length is extremely valuable in providing a partial ordering of the oligonucleotides produced by cleavage at specific base residues.

USE OF RIBOPOLYNUCLEOTIDES TO PRIME RNA SYNTHESIS ON DENATURED DNA   When a single-stranded DNA template is used, a dinucleoside monophosphate or larger ribopolynucleotide will prime synthesis by RNA polymerase wherever there is an available complementary sequence. Thus dinucleoside monophosphates will be expected to prime synthesis at many points (about once every 16 nucleotides on the average), and it is necessary to use larger fragments if priming at specific locations is desired. Gilbert and his collaborators used two pentanucleotide primers to obtain specific initiations within the roughly 50 nucleotides of template obtained by denaturing their *lac* operator fragments (24, 25).

The Use of Electron Microscopy for DNA Sequencing

The use of high resolution electron microscopy for the determination of nucleotide sequences was proposed in 1962, several years before the first RNA sequence was actually determined by Holley and his collaborators using other methods (179). The average distance between nucleotides in extended single-stranded nucleic acid is about 5.5 Å (180), a distance within the resolving power of available instruments. It is clear that electron microscopy techniques, if they can be made to work at all, might work so very well that large amounts of sequence information are obtained very rapidly. The serious problems which have so far prevented this are reviewed by Beer et al (181a). In the first place, each nucleotide is an extended object several angstroms in size, so that the 5.5 Å spacing mentioned above does not mean that individual nucleotides could be resolved with available instruments even if contrast were not a problem. Consequently, the approach must be to somehow stain one of the fouf nucleotides with a heavy atom so that its distribution along the chain can be determined precisely. If this could be repeated with other individual nucleotides and carried out independently with the separated strands of a DNA molecule, one might hope to reconstruct the complete sequence.

A number of laboratories have published data showing that single heavy atoms can be visualized using the conventional transmission electron microscope (CTEM) (see, for instance, 181b–d). Beer & Moudrianakis (182a) and their later co-workers have developed heavy atom marker stains specific for guanine (182b) and for uracil and thymine (182c). But, so far, none of these efforts have aided in any actual sequence determinations. One problem is the low ratio of signal from the heavy atom market to background noise from the supporting film. In some of the many papers on this subject (only a few of which are cited here), it takes a truly "loving eye" to distinguish the characteristic configurations of heavy atoms in the model compounds used.

The paper of Whiting & Ottensmeyer (181c) is one of the best in this regard but they, like others, find it impossible to visualize the sugar-phosphate backbone of the nucleic acid. This makes it very difficult to follow the DNA chain (181b). In some cases where such visualization has been claimed, it may in fact be due to the great natural affinity of many metal ions for the DNA chain.

Finally, microscopy with the CTEM unavoidably involves problems of beam damage which make it impossible to take two meaningful successive photographs of the same area with single-atom resolution. Thus, none of the published work has included convincing proof that such damage is not seriously distorting the first as well as successive photographs.

New encouragement that electron microscopy will ultimately make a contribution to nucleotide sequencing comes from recent technical developments which make possible the design of scanning electron microscopes able to operate at high resolution in the transmission mode (182d). As pointed out in (183b), because of the reciprocity theorem, the scanning transmission electron microscope (STEM) has the same theoretical resolution as the CTEM but can achieve several other important advantages:

1. Because the efficiency of detecting elastically scattered electrons is higher for the STEM than the CTEM the amount of radiation damage is reduced roughly five-fold or greater for instruments operating with resolutions of 2 to 3 Å (see 183a and especially Figure 5 of 184f).

2. At the extremely high vacuum (about 10^{-10} torr) used in the STEM the problem of contamination buildup on the sample is very much less than in existing CTEM equipment (183b).

3. The inelastically scattered electron signal has both lower resolution and lower contrast (for heavy metal marker atoms) than the signal from elastically scattered electrons. One of the greatest superiorities of the STEM over the CTEM is that the elastically and inelastically scattered electrons in the signal can be efficiently resolved. Simple removal of the inelastically scattered electrons from the signal results in increased resolution and roughly twofold better contrast than in the CTEM (as can be derived from Table 1 of 183c).

4. Since the inelastically scattered electrons may also be collected as a separate signal in the STEM, and since the ratio of elastically to inelastically scattered electrons is proportional to Z (thus for the carbon film and a heavy metal stain such as mercury, this ratio differs by about 13-fold), the inelastically scattered electron signal may be used to accurately subtract the background introduced by the carbon film. This results in very striking increases in contrast in addition to the roughly twofold improvement mentioned previously (183d).

Langmore, Isaacson & Crewe (183b) have recently shown that STEM techniques give adequate contrast to permit visualization of marker atoms as light as silver ($Z = 47$) when 43 kev electrons are used. The increased resolutions which can be achieved with 100 kev electrons should further increase the signal-to-noise ratio. Therefore it would seem that the usefulness of the STEM to provide sequence data is assured if dependable ways can be found to reproducably label specific nucleotides. As mentioned above, Beer and his collaborators have developed stains permitting specific nucleotides to be labeled with uranyl atoms or with osmium tetroxide. The reaction with osmium tetroxide is only about 80% complete and the products are hydrolytically unstable. The latter problem has now been solved by using pyridine rather than thiocyanate ligands to facilitate the reaction (184a). More recently, techniques have been developed for selectively labeling nucleotides with mercury (155a, 184b).

It remains to be shown, however, that the heavy metal labels now available will remain stably associated with the desired nucleotide when placed in the electron beam. For instance, even with the low beam dosages used in the STEM, uranium and mercury atoms are frequently moved by the beam. Thorium and silver are superior in this respect and it has been shown that single silver atoms will remain in the same location during 10 successive scans of the STEM (183b). [This might suggest that in selectively labeling guanine residues with diazotized 2-amino-p-benzenedisulphonic acid (182b, 184c) it might be preferable to form the thorium rather than the uranyl salt for visualization in the STEM.]

Preliminary evidence suggests that at least in some cases when mercury-labeled DNA is examined in the STEM most of the mercury atoms are seen to be dissociated from the DNA (184d). This appears to result from contact with the reactive carbon film (for a discussion of the catalytic properties of such surfaces, see 184e) rather than from beam damage (184d). In some previous experiments the appearance of rows of heavy metal atoms has been taken as evidence that the heavy metal ions were still specifically linked to particular nucleotides. This is not an adequate criterion for many heavy metals which, like uranium, have a strong tendency to bind nonspecifically to DNA.

Perhaps one of the most convincing ways in which the validity of a specific labeling technique could be verified is by examining a labeled nucleic acid molecule whose sequence has already been established by other means. For this purpose, the highly repetitive satellite DNAs are especially suitable since one could examine several repeats of the same sequence in each electron micrograph. The HSβ satellite sequences from *Dipodomys ordii* (18), the satellite sequences from *Drosophila virilus* (19), and perhaps the satellite of the guinea pig (17) are known with sufficient accuracy to provide good test objects. In the case of the HSβ satellite the major repeat sequence is

```
5'   ACACAGCGGG   3'
3'   TGTGTCGCCC   5'
```

so that when observing single strands of denatured DNA labeled at either G or C residues, one would expect to see repeated patterns of heavy atoms with 10, 5, 5, and 30 Å spacings in one strand and 15, 25, and 10 Å spacings in the other strand (assuming about 5 Å internucleotide spacings in denatured DNA).

It does not seem likely that it will soon be possible to label more than one of the four bases so that they can be distinguished in the same micrograph. Consequently the determination of complex nucleotide sequences completely by electron microscopy will require the experimenter to be able to accurately place in the proper register several different sets of data obtained with different heavy atom labels.

Perhaps a more immediate and simple application of electron microscopy sequence data will be to provide the overlap information needed so that oligonucleotide fragments sequenced by conventional techniques can be assembled into a complete sequence. Obtaining such overlap data is the most difficult part of normal RNA and DNA sequencing techniques, so this would constitute a very important advance. Since such use of the electron microscopy data would be largely self-checking, it would provide a good test of the dependability of more ambitious sequencing by electron microscopy. Finally, it may be important that electron microscopy could make important contributions of this sort as soon as even a single dependable heavy atom label is available.

SEQUENCE ANALYSIS OF SMALL FRAGMENTS

This review emphasizes methods of simplifying large sequencing problems since it is here that conventional RNA sequencing techniques are most limited and the various DNA sequencing approaches have the most to offer. It must not be forgotten, however, that in any sequencing operation, no matter how elegant the original dissection of the molecule, the sequencer is ultimately confronted with the problem of determining the actual primary sequences of large numbers of fragments from 2 to roughly 10 or 20 nucleotides in length. There is no need to describe in detail how this is done for the operations involved have been adequately discussed in a number of other reviews cited below for the reader's convenience.

One of the advantages of using in vitro transcription from DNA templates is that the short fragments resulting from digestion of the labeled transcript can be analyzed with the familiar techniques developed for conventional RNA sequencing (1–3, 49). The techniques developed for analysis of DNA fragments themselves are also rapid and effective. There are two major approaches for DNA fragments, one depending upon partial digestions with exonucleases, the other depending upon overlapping di- and tri-nucleotide sequences obtained from nearest neighbor analysis of micrococcal nuclease digests. Partial digestions with venom or spleen phosphodiesterases (21, 44, 46) have been discussed in an earlier section and, as mentioned, this approach is the method of choice for large fragments if relatively large amounts of radioactivity are available. Most of the fragments resulting from a ribosubstitution cleavage will be less than 10 nucleotides in length, however, and these can be conveniently sequenced using micrococcal nuclease digestions as described in detail by Whitcome, Fry & Salser (42).

PERSPECTIVES FOR THE FUTURE

Previous sequencing efforts have concentrated mainly on bacterial systems. It is clear that the new techniques described here will encourage even more rapid progress in this area. Perhaps more important is the fact that the new techniques give us powerful approaches for studying some eukaryotic nucleic acid sequences which have been almost entirely inaccessible. Here we attempt to briefly sum up our present capabilities in this area and speculate about ways in which it may be possible to attack even more difficult problems in the future.

There seems to be no reason why it should not be possible to determine complete sequences for a few eukaryotic mRNAs in the next few years, even with no further improvements in methodology. Hemoglobin and immunoglobin mRNAs are currently being investigated (42, 139, 169, 185), and improved methods for purifying individual mRNAs may soon make it possible for investigators to systematically sequence any one of a large number of mRNAs (156–160, 163). Thus we may expect a rapid increase in our knowledge of eukaryotic mRNA sequences and hope that this will lead to a better understanding of the various recognition signals discussed earlier and of the regulatory role (if any) of the large untranslated regions known to be present.

We also anticipate that large portions of the small DNA viruses can now be sequenced in a fairly straightforward manner. Indeed, if several laboratories concentrate on SV40, as now seems likely, it is possible that the complete sequence of this virus (roughly 5000 base pairs) could become known over the next few years. Such a concentrated effort may well be worthwhile since a complete sequence would give us for the first time an opportunity to study and try to understand, at the sequence level, the workings of an entire interrelated set of genes.

As impressive as such achievements could be, they might shed very little light on other fundamental questions about genomes of higher organisms. We would like to know the sequence and the importance of the DNA located between structural genes. There appears to be much more of this "extra DNA" of unknown function than there is of DNA that actually codes for proteins. Some of it appears to never even be transcribed into RNA. Other regions are transcribed to give HnRNA of which very roughly 90% is in sequences which appear to never participate in protein synthesis since they are degraded in the nucleus. It has been variously proposed that such sequences have important control functions (186) or that such sequences are only remnants of genes that lost their function through mutation (187).

It might appear impossible to approach such questions by nucleotide sequence analysis; however, several laboratories are developing techniques which may make this possible. The basic difficulty is that eukaryotic genomes are so large; a human diploid cell contains about 5 pg of DNA, or roughly a million times as much DNA as the SV40 genome. The size of the genome makes it impractical to purify a single region for sequence analysis, and even if we could purify such a sequence with 100% yield, the amount of starting DNA required to yield a few micrograms of a 1000 base-pair fragment is enormous.

These problems could be circumvented if it were possible to insert a fragment of eukaryotic DNA into the genome of bacteriophage λ or one of the plasmids of *E. coli.* Once carried on a self-replicating genome of this sort the fragments could be grown in large quantities in bacterial hosts and purified by standard techniques. Simple cloning of the phage or plasmid genomes carrying eukaryotic DNA sequences would permit the investigator to select individual sequences for intensive study.

Insertion of fragments of eukaryotic DNA into carrier bacterial phage or plasmid genomes may in theory be accomplished by a variety of methods (188–190). Lobban & Kaiser (189) and Jackson, Symons & Berg (190) have shown that two linear duplex DNA molecules can be linked to form circular dimers by the combined action of exonuclease, terminal transferase, DNA polymerase I, and polynucleotide ligase. The terminal transferase is used to add blocks of poly(dA) or poly(dT) to 3'-OH termini of the two groups of molecules to be joined. When the two groups of molecules are mixed and annealed the blocks of complementary poly(dA) and poly(dT) hybridize to form circles. DNA polymerase is used to fill in any gaps, and polynucleotide ligase can then covalently close the circular hybrid molecules. This approach has been used to insert λ phage genes and the galactose operon of *E. coli* into the circular SV40 genome (190).

Sgaramella (188) has shown that the polynucleotide ligase coded by bacteriophage T4 is capable of joining together pairs of completely duplex DNA molecules (with no single-stranded "cohesive" ends). Finally, it has been shown by a number of laboratories that the restriction enzymes Eco_{RI} and Eco_{RII} make staggered cleavages to yield DNA fragments with cohesive ends. The sequences of these cohesive ends are

$$5'....A/\,T_pG\overset{\downarrow}{\cdot}A_pA_pT_pT_pC_pT/\,A....3' \quad \text{and} \quad 5'....\overset{\downarrow}{\cdot}C_pC_pA_pG_pG\,....3'$$
$$3'....T/\,A_pC_pT_pT_pA_pA_p\overset{\uparrow}{\cdot}G_pA/\,T....5' \quad\quad\quad 3'....\ G_pG_pT_pC_pC_{\overset{\uparrow}{\,}}....5'$$

respectively (12–14). As shown by Mertz & Davis (92), any two DNA molecules with R1 (or R2) sites can be recombined at their restriction sites by the sequential action of R1 endonuclease and DNA ligase to generate hybrid DNA molecules. Cohen et al (188a) have recently demonstrated the construction of functional bacterial plasmids, carrying added DNA sequences, using this approach.

We think it likely that the application of these techniques will soon lead to very elegant detailed analysis and dissections of individual parts of eukaryotic genomes. At present there is no way that a particular genetic locus can be selectively isolated in this way, but in some organisms, such as *Drosophila,* it should be possible to determine which genetic region has been isolated by in situ hybridization (Wensink, personal communication; 191). It should also be possible to assemble an extensive collection of strains carrying many different fragments from a small region. This would be done by choosing a strain carrying a large eukaryotic DNA fragment from an interesting region and using it as the source of DNA fragments for the construction of other strains. Mapping a series of such strains by heteroduplex mapping and detailed sequence analysis might enable us to greatly extend our very limited present understanding of the organization and function of eukaryotic genomes.

ACKNOWLEDGMENTS

I am especially indebted to all of those who made their manuscripts available and allowed me to cite their results prior to publication. I also owe a special debt of gratitude to John Langmore and Albert Crewe for their patient help and advice on the electron microscopy section. The review could not have been completed without long hours spent by Stephanie Day, Terryl Chandler, Shirley Chu, and Hannia Kempfer coaxing the UCLA Campus Computing Network's computer to edit the manuscript (an idea whose time has still not come). The author wishes to acknowledge the support of USPHS Career Development Award GM-70045. Research that originated in the author's laboratory was supported by research Grants GM-18586 and CA-15940 from the USPHS and by NSF Grants GB-27512 and GB-40312X.

A more complete bibliography, containing many papers not cited here because of limitations of space, may be obtained from the author.

Literature Cited

1. Barrell, B. G. 1971. *Proc. Nucl. Acid Res.* 2:171
2. Brownlee, G. G. 1972. *Laboratory Techniques in Biochemistry and Molecular Biology,* ed. T. S. Work, E. Work. Vol. 3, Part 1. Amsterdam: North-Holland
3. Mandeles, S. 1972. *Nucleic Acid Sequence Analysis.* New York: Columbia Univ. Press
4. Padmanabhan, R., Padmanabhan, R., Wu, R. 1972. *Biochem. Biophys. Res. Commun.* 48:1295
5. Padmanabhan, R., Wu, R. 1972. *J. Mol. Biol.* 65:447
6. Wu, R., Taylor, E. 1971. *J. Mol. Biol.* 57:491
7. Englund, P. T. 1971. *J. Biol. Chem.* 246:3269
8. Price, S. S., Schwing, J. M., Englund, P. T. 1973. *J. Biol. Chem.* 248:7001
9. Weigel, P. H., Englund, P. T., Murray, K., Old, R. W. 1973. *Proc. Nat. Acad. Sci. USA* 70:1151
10. Englund, P. T. 1972. *J. Mol. Biol.* 66:209
11. Kelly, T. J., Smith, H. O. 1970. *J. Mol. Biol.* 51:393
12. Boyer, H. W., Chow, L., Dugaiczyk, A., Hedgpeth, J., Goodman, H. M. 1973. *Nature New Biol.* 244:40
13. Hedgpeth, J., Goodman, H. M., Boyer, H. 1972. *Proc. Nat. Acad. Sci. USA* 69:3448
14. Bigger, C. H., Murray, K., Murray, N. E. 1973. *Nature New Biol.* 244:7
15. Roy, P. H., Smith, H. O. 1973. *J. Mol. Biol.* 81:445
15a. Sugisaki, M., Takanami, M. 1973. *Nature New Biol.* 246:138
16. Ling, V. 1972. *J. Mol. Biol.* 64:87
17. Southern, E. M. 1970. *Nature* 227:794
18. Fry, K. et al 1973. *Proc. Nat. Acad. Sci. USA* 70:2642
18a. Salser, W., Poon, R., Whitcome, P., Fry, K. 1973. In *Virus Research, Second ICN-UCLA Symp. Mol. Biol.,* ed. C. F. Fox, W. S. Robinson, 573. New York: Academic
19. Gall, J. G., Atherton, D. D. 1974. *J. Mol. Biol.* In press
20. Robertson, H. D., Barrell, B. G., Weith, H. L., Donelson, J. E. 1973. *Nature New Biol.* 241:38
21. Sanger, F., Donelson, J. E., Coulson, A. R., Kossel, H., Fischer, D. 1973. *Proc. Nat. Acad. Sci. USA* 70:1209
22. Blattner, F. R., Dahlberg, J. E. 1972. *Nature New Biol.* 237:227
23. Blattner, F. R., Dahlberg, J. E., Boettiger, J. K., Fiandt, M., Szybalski,

W. 1972. *Nature New Biol.* 237:232
24. Gilbert, W., Maizels, N., Maxam, A. 1974. *Cold Spring Harbor Symp. Quant. Biol.* 38: In press
25. Gilbert, W., Maxam, A. 1973. *Proc. Nat. Acad. Sci. USA* 70: 3581
26. Lebowitz, P., Weissman, S. M., Radding, C. M. 1971. *J. Biol. Chem.* 246:5120
27. Marotta, C., Verma, I., McCaffrey, R. P., Forget, B. G. 1973. *Fed. Proc. Abstr.* 32:455
28a. Wu, R., Donelson, J., Padmanabhan, R., Hamilton, R. 1972. *Bull. Inst. Pasteur Paris* 70:203
28b. Murray, K., Old, R. W. 1974. *Progr. Nucl. Acid Res. Mol. Biol.* 14: In press
29. Murray, K. 1973. *Biochem. J.* 131:569
30. Brezinski, D. P., Wang, J. C. 1973. *Biochem. Biophys. Res. Commun.* 50:398
31. Donelson, J. E., Wu, R. 1972. *J. Biol. Chem.* 247:4654
32. Maniatis, T., Ptashne, M. 1973. *Proc. Nat. Acad. Sci. USA* 70:1531
33. Bronson, M. J., Squires, C., Yanofsky, C. 1973. *Proc. Nat. Acad. Sci. USA* 70:2335
34. Min Jou, W., Haegeman, G., Ysebaert, M., Fiers, W. 1972. *Nature* 237:82
35. Ricard, B., Salser, W. 1973. *Nature.* In press
36. Jelinek, W. et al 1973. *J. Mol. Biol.* 75:515
37. Fellner, P., Ehresman, C., Ebel, J. P. 1970. *Cold Spring Harbor Symp. Quant. Biol.* 35:29
38. Ehresman, C., Stiegler, P., Fellner, P., Ebel, J. P. 1972. *Biochimie* 54:901
39a. Billeter, M. A., Dahlberg, J. E., Goodman, H. M., Hindley, J., Weissman, C. 1969. *Nature* 224:1083
39b. Berg, P., Fancher, H., Chamberlin, M. 1963. In *Symposium on Informational Macromolecules,* ed. H. Vogel, V. Bryson, J. O. Lampen, 467. New York: Academic
40. Salser, W., Fry, K., Brunk, C., Poon, R. 1972. *Proc. Nat. Acad. Sci. USA* 69:238
41. van de Sande, J. H., Loewen, P. C., Khorana, H. G. 1972. *J. Biol. Chem.* 247:6140
42. Whitcome, P., Fry, K., Salser, W. *Methods Enzymol.* 29:295
43. Salser, W., Fry, K., Wesley, R., Simpson, L. 1973. *Biochim. Biophys. Acta* 319:277
44. Steitz, J. A. 1973. *J. Mol. Biol.* 73:1
45. Ling, V. 1971. *FEBS Lett.* 19:50
46. Ziff, E. B., Sedat, J. W., Galibert, F.

1973. *Nature New Biol.* 241:34
47. Sanger, F. 1973. See Ref. 18a, p. 573
48. Lozeron, W., Szybalsky, W., Dahlberg, J. E. In preparation
49. Gilham, P. T. 1970. *Ann. Rev. Biochem.* 39:224
50. Loewen, P. C., Khorana, H. G. 1973. *J. Biol. Chem.* 248:3489
51. Loewen, P. C., Sekiya, T., Khorana, H. G. 1974. *J. Biol. Chem.* 249:217
52. Sanger, F., Donelson, J. E., Coulson, A. R., Kossel, H., Fischer, D. 1973. *Proc. Nat. Acad. Sci. USA* 70:1209
53. Niyogi, S. K., Thomas, C. A. Jr. 1967. *Biochem. Biophys. Res. Commun.* 26:51
54. Niyogi, S. K. 1969. *J. Biol. Chem.* 244:1576.
55. Niyogi, S. 1973. *J. Biol. Chem.* 248:2323
56. Besmer, P. et al 1972. *J. Mol. Biol.* 72:503
57. Englund, P. T. 1971. *J. Biol. Chem.* 246:5684
58. Khorana, H. G. 1968. *Pure Appl. Chem.* 17:349
59. Khorana, H. G. et al 1972. *J. Mol. Biol.* 72:209
60. *J. Mol. Biol.* 1972. Vol. 72, No. 2
61. Gilham, S., Smith, M. 1972. *Nature New Biol.* 238:233
62. Hsieh, W. T. 1971. *J. Biol. Chem.* 246:1780
63. Thach, R. E. 1966. *Proc. Nucl. Acid Res.* 1:520
64. Kaufmann, G., Fridkin, M., Zutra, A., Littauer, U. Z. 1971. *Eur. J. Biochem.* 24:4
65. Roychoudhury, R., Kössel, H. 1973. *Biochem. Biophys. Res. Commun.* 50:259
66. Keller, W. 1972. *Proc. Nat. Acad. Sci. USA* 69:1560
67. Mackey, J. K., Gilham, P. T. 1971. *Nature* 233:551
68. Bennett, G. N., Mackey, J. K., Wiebers, J. L., Gilham, P. T. 1973. *Biochemistry* 12:3956
69. Wu, R. 1972. *Nature* 236:198
70. Maizels, N. M. 1973. *Proc. Nat. Acad. Sci. USA* 70:3585
71. Ricard, B., Salser, W. In preparation
72. Salser, W., Ricard, B. In preparation
73. Smith, H. O., Wilcox, K. W. 1970. *J. Mol. Biol.* 51:379
74. Danna, K., Nathans, D. 1971. *Proc. Nat. Acad. Sci. USA* 68:2913
75. Danna, K., Sack, G., Nathans, D. 1973. *J. Mol. Biol.* 78:363
75a. Sharp, P., Sugden, B., Sambrook, J. 1973. *Biochemistry* 12:3055
76. Sack, G. H., Nathans, D. 1973. *J. Virol.* 51:517
77. Smith, H. O. Personal communication

78. Old, R. W., Roizes, G., Murray, K. In preparation
79. Landy, A. Personal communication
80. Middleton, J. H., Edgell, M. H., Hutchison, C. A. Personal communication
80a. Edgell, M. H., Hutchison, C. A., Sclair, M. 1972. *J. Virol.* 9:574
81. Murray, K., Morrison, A. 1974. In preparation
82. Middleton, J. H., Edgell, M. H., Hutchison, C. A. 1972. *J. Virol.* 10:42
83. Huang, E., Newbold, J. E., Pagano, J. S. 1973. *J. Virol.* 11:508
84. Johnson, P. H., Lee, A. S., Sinsheimer, R. L. 1973. *J. Virol.* 11:596
85. Sharp, P. A., Sugden, B., Sambrook, J. 1973. *Biochemistry* 12:3055
86. Allet, B. 1973. *Biochemistry* 12:3972
87. Morrow, J. F., Berg, P. 1972. *Proc. Nat. Acad. Sci. USA* 69:3365
88. Mulder, C., Delius, H. 1972. *Proc. Nat. Acad. Sci. USA* 69:3215
89. Fareed, G. C., Garon, C. F., Salzman, N. P. 1972. *J. Virol.* 10:484
90. Greene, P. J., Betlach, M. C., Goodman, H. M., Boyer, H. W. *Methods Mol. Biol.* In press
91. Marx, J. L. 1973. *Science* 180:482
92. Mertz, J. E., Davis, R. W. 1972. *Proc. Nat. Acad. Sci. USA* 69:3370
93. Mulder, C. Personal communication
94. Boon, S., Trautner, T., Murray, K. In preparation
95. Allet, B., Jeppesen, P. G. N., Katagiri, K. J., Delius, H. 1973. *Nature* 241:120
96. Maniatis, T., Ptashne, M., Maurer, R. 1974. *Cold Spring Harbor Symp. Quant. Biol.* 38: In press
97. Khoury, G., Martin, M. A., Lee, T., Danna, K. J., Nathans, D. 1973. *J. Mol. Biol.* 78:377
98. Edgell, M. H., Hutchison, C. A. III, Sclair, M. 1972. *J. Virol.* 9:574
99. Middleton, J. H., Edgell, M. H., Hutchison, C. A. III 1972. *J. Virol.* 10:42
100. Hutchison, C. A. III, Edgell, M. H. 1971. *J. Virol.* 8:181
101. Weisbeck, P. J., Van de Pol, J. H. 1970. *Biochim. Biophys. Acta* 224:328
102. Sadowski, P. D., Hurwitz, J. 1969. *J. Biol. Chem.* 244:6192
103. Sadowski, P. D., Bakyta, I. 1972. *J. Biol. Chem.* 247:405
104. Kato, M., Ikeda, Y. 1968. *J. Biochem.* 64:321
105. Kato, M., Ando, T., Ikeda, Y. 1968. *J. Biochem.* 64:329
106. Ikeda, Y. Personal communication
107. Steitz, J. A. 1969. *Nature* 224:957
108. Staples, D. H., Hindley, J. 1971. *Nature New Biol.* 234:211
109. Staples, D. H., Hindley, J., Billeter, M. A., Weissman, C. 1971. *Nature New Biol.* 234:202
110. Gupta, S. L., Chen, J., Schaefer, L., Lengyel, P., Weissman, S. M. 1970. *Biochem. Biophys. Res. Commun.* 39:883
111. Arrand, J. R., Hindley, J. 1973. *Nature New Biol.* 244:10
112. Bretscher, M. S. 1969. *Cold Spring Harbor Symp. Quant. Biol.* 34:651
113. Air, G. M., Bridgen, J. 1973. *Nature* 241:40
114. Heyden, B., Nusslein, C., Schaller, H. 1972. *Nature New Biol.* 240:9
115. Novak, R. L. 1967. *Biochim. Biophys. Acta* 149:593
116. Okamoto, T., Sugiura, M., Takanami, M. 1972. *Nature New Biol.* 237:108
117. Matsukage, A., Murakami, S., Kameyama, T. 1969. *Biochim. Biophys. Acta* 179:145
118. Nakano, E., Sakaguchi, K. 1969. *J. Biochem.* 65:147
119. Ruger, W. 1971. *Biochim. Biophys. Acta* 238:202
120. LeTalaer, J., Jeanteur, P. 1971. *Proc. Nat. Acad. Sci. USA* 68:3211
121. Sentenac, A., Ruet, A., Fromageot, P. 1968. *FEBS Lett.* 2:53
122. Pirrotta, V. 1973. *Nature New Biol.* 244:13
123. Smith, T. F., Sadler, J. R. 1971. *J. Mol. Biol.* 59:273
124. Sadler, J. R., Smith, T. F. 1971. *J. Mol. Biol.* 62:139
125. Maniatis, T., Ptashne, M. Submitted
126. Labrie, F. 1969. *Nature* 221:1217
127. Szekely, M., Sanger, F. 1969. *J. Mol. Biol.* 43:607
128. Kacian, D. L. et al 1972. *Nature New Biol.* 235:167
129. Verma, I. M., Temple, G. F., Fan, H., Baltimore, D. 1972. *Nature New Biol.* 235:163
130. Ross, J., Aviv, H., Scolnick, E., Leder, P. 1972. *Proc. Nat. Acad. Sci. USA* 69:264
131. Diggelmann, H., Faust, C. H., Mach, B. 1973. *Proc. Nat. Acad. Sci. USA* 70:693
132. Aviv, H., Packman, S., Swan, D., Ross, J., Leder, P. 1973. *Nature New Biol.* 241:174
133. Faust, C. H., Diggelmann, H., Mach, B. 1973. *Biochemistry* 12:925
134. Harrison, P. R., Hell, A., Birnie, G. D., Paul, J. 1972. *Nature* 239:219
135. Bishop, J. O., Rosbash, M. 1973. *Nature New Biol.* 241:204
136a. Zassenhaus, P., Kates, J. 1972.

964 SALSER

Nature New Biol. 238:139

136b. Slater D. W., Slater, I., Gillespie, D. 1972. *Nature* 240:333

137. Verma, I. M., Temple, G. F., Fan, H., Baltimore, D. 1973. In *Viral Replication and Cancer, Proc. 2nd Duran-Reynals Int. Symp., Barcelona,* ed. J. L. Melnick, S. Ochoa, J. Oro.

138. Spiegelman, S., Watson, K. F., Kacian, D. L. 1971. *Proc. Nat. Acad. Sci. USA* 68:2843

139. Salser, W., Poon, R., Whitcome, P., Fry, K. 1973. See Ref. 18a, p. 45

140. Gould, H. J., Hamlyn, P. H. 1973. *FEBS Lett.* 30:301

141. Lim, L., Canellakis, E. S. 1970. *Nature* 227:710

141a. Hunt, J. A. 1973. *J. Biochem.* 131:327

142. Molloy, G. R., Darnell, J. E. 1973. *Biochemistry* 12:2324

143. Sheiness, D., Darnell, J. E. 1973. *Nature New Biol.* 241:265

144. Taylor, J. M. et al 1973. *Biochemistry* 12:460

145. Taylor, J. M., Faras, A. J., Varmus, H. E., Levinson, W. E., Bishop, J. M. 1972. *Biochemistry* 11:2343

146. Varmus, H. E., Levinson, W. E., Bishop, J. M. 1971. *Nature New Biol.* 233:19

147. Chamberlin, M., Berg, P. 1964. *J. Mol. Biol.* 8:708

148. Ohasa, S., Tsugita, A. 1972. *Nature New Biol.* 240:35

149. deRobertis, E., Ezcurra, P. M., Judewicz, N. D., Pucci, P. R., Torres, H. N. 1972. *FEBS Lett.* 25:175

150. Aviv, H., Leder, P., 1972. *Proc. Nat. Acad. Sci. USA* 69:1408

151. Melli, M., Pemberton, R. 1972. *Nature New Biol.* 236:172

152. Robertson, H. D. 1971. *Nature New Biol.* 229:169

153. Englund, P., Price, S., Schwing, J., Weigel, P. 1973. In *DNA Synthesis in Vitro,* ed. R. D. Wells, R. B. Inman, 35. Baltimore: University Park Press

154. Robertson, H. D., Dickson, E., Model, P., Prensky, W. 1973. *Proc. Nat. Acad. Sci. USA* 70:3260

155a. Dale, R. M. K., Livingston, D. C., Ward, D. C. 1973. *Proc. Nat. Acad. Sci. USA* 70:2238

155b. Holme, G., Delovitch, T. L., Boyd, S. L., Sehon, A. H. 1971. *Biochim. Biophys. Acta* 247:104

156. Delovitch, T. L., Davis, B. K., Holme, G., Sehon, A. H. 1972. *J. Mol. Biol.* 69:373

157. Palmiter, R. D., Palacios, R., Schimke, R. T. 1972. *J. Biol. Chem.* 247:3296

158. Sarkar, P. K., Moscona, A. A. 1973. *Proc. Nat. Acad. Sci. USA* 70:1667

159. Uenoyama, K., Ono, T. 1972. *J. Mol. Biol.* 65:75

160. Palacios, R., Sullivan, D., Summers, N. M., Kiely, M. L., Schimke, R. T. 1973. *J. Biol. Chem.* 248:540

161. Stevens, R. H., Williamson, A. R. 1973. *J. Mol. Biol.* 78:505

162. Stevens, R. H., Williamson, A. R. 1973. *J. Mol. Biol.* 78:517

163a. Stevens, R. H., Williamson, A. R. 1973. *Proc. Nat. Acad. Sci. USA* 70:1127

163b. Stevens, R. H., Williamson, A. R. 1973. *Nature New Biol.* 245:101

164. Zain, B. S., Dhar, R., Weissman, S. M., Lebowitz, P., Lewis, A. M. 1973. *J. Virol.* 11:682

165. Westphal, H. 1970. *J. Mol. Biol.* 50:407

166. Takanami, M., Okamoto, T., Sugiura, M. 1970. *Cold Spring Harbor Symp. Quant. Biol.* 35:179

167. Ruger, W. 1971. *Biochim. Biophys. Acta* 238:202

168. Minkley, E. G., Pribnow, D. 1973. *J. Mol. Biol.* 77:255

169. Marotta, C. A., Lebowitz, P., Dhar, R., Zain, S., Weissman, S. M. 1974. *Methods Enzymol.* In press

170. Defilippes, F. M. 1972. *Biochim. Biophys. Acta* 272:125

171. Kleppe, R., Khorana, H. G. 1972. *J. Biol. Chem.* 247:6149

172. Terao, T., Dahlberg, J. E., Khorana, H. G. 1972. *J. Biol. Chem.* 247:6157

173. Straat, P., Tso, P. O. P., Bollum, F. J. 1968. *J. Biol. Chem.* 243:5000

174. Niyogi, S., Stevens, A. 1965. *J. Biol. Chem.* 240:2587

175. Niyogi, S., Stevens, A. 1965. *J. Biol. Chem.* 240:2593

176. Wimmer, E. 1972. *J. Mol. Biol.* 68:537

177. Downey, K. M., So, A. G. 1970. *Biochemistry* 9:2520

178. Downey, K. M., Jurmark, B. S., So, A. G. 1971. *Biochemistry* 10:4970

179. Holley, R. W. et al 1965. *Science* 147:1462

180. Highton, P. J., Beer, M. 1963. *J. Mol. Biol.* 7:70

181a. Beer, M., Bartl, P., Koller, T., Erickson, H. P. 1971. *Methods Cancer Res.* 6:283

181b. Henkelman, R. M., Ottensmeyer, F. P. 1971. *Proc. Nat. Acad. Sci. USA* 68:3000

181c. Whiting, R. F., Ottensmeyer, F. P. 1972. *J. Mol. Biol.* 67:173

181d. Formanek, H., Muller, M., Hahn,

M. H., Koller, T. 1971. *Naturwissenschaften* 58:339

182a. Beer, M., Moudrianakis, E. N. 1962. *Proc. Nat. Acad. Sci. USA* 48:409

182b. Moudrianakis, E. N., Beer, M. 1965. *Biochim. Biophys. Acta* 95:23

182c. di Giamberardino, L., Koller, T., Beer, M. 1969. *Biochim. Biophys. Acta* 182:523

182d. Crewe, A. V., Wall, J., Langmore, J. P. 1970. *Science* 168:1338

183a. Crewe, A. V. 1973. *J. Mol. Biol.* 80:315

183b. Wall, J., Langmore, J. P., Isaacson, M., Crewe, A. V. 1974. *Proc. Nat. Acad. Sci. USA* In press

183c. Wall, J., Isaacson, M., Langmore, J. P. 1973. *Optik* 38:In press

183d. Crewe, A. V., Isaacson, M., Langmore, J. P. 1974. *Physical Techniques in Electron Microscopy,* ed. B. Siegel. New York: Wiley

184a. Subbaraman, L. R., Subbaraman, J., Behrman, E. J. 1971. *Bioinorg. Chem.* 1:35

184b. Langmore, J. P., Cozzarelli, N. R., Crewe, A. V. *Proc. 30th Ann. EMSA Conf.,* 184

184c. Moudrianakis, E. N., Beer, M. 1965. *Proc. Nat. Acad. Sci. USA* 53:564

184d. Langmore, J. P. Personal communication

184e. Boehm, H. P., Hofmann, U., Clauss, A. 1957. *Proc. Conf. Carbon, 3rd, Buffalo, NY,* 241. New York: Pergamon

184f. Langmore, J. P., Wall, J., Isaacson, M. S. 1973. *Optik* 38:335

185. Brownlee, G. G., Cartwright, E. M., Cowan, N. J., Jarvis, J. M., Milstein, C. 1973. *Nature New Biol.* 244:236

186. Crick, F. H. C. 1971. *Nature* 234:25

187. Ohno, S. 1972. *J. Med. Genet.* 9:254

188. Sgaramella, V. 1972. *Proc. Nat. Acad. Sci. USA* 69:3389

188a. Cohen, S., Chang, A., Boyer, H., Helling, R. 1973. *Proc. Nat. Acad. Sci. USA* 70:3240

189. Lobban, P. E., Kaiser, A. D. 1973. *J. Mol. Biol.* 78:453

190. Jackson, D. A., Symons, R. H., Berg, P. 1972. *Proc. Nat. Acad. Sci. USA* 69:2904

191. Pardue, M. L., Gall, J. G. 1970. *Science* 168:1356

REGULATION OF STEROID BIOSYNTHESIS[1]

Mary E. Dempsey[2]

Department of Biochemistry, University of Minnesota, Minneapolis, Minnesota

CONTENTS

INTRODUCTION

To set the stage for this review of current knowledge in the area of steroid bio-synthesis regulation, it is essential to summarize certain facts regarding the most ubiquitous and abundant steroid in mammalian tissues, cholesterol. The structure of cholesterol has been known for decades. It is commercially available in kilogram or larger quantities. The biological sources of each of its 27 carbon atoms (i.e. from C_1 and C_2 of acetate) and its oxygen atom (i.e. from O_2) were established 15 or more years ago (1–3). Recently the absolute stereochemistry and biological sources of each of its hydrogen atoms were elucidated through the contributions of numerous investigators (4). In spite of extensive chemical knowledge of cholesterol, we know little of its biological function on a molecular basis. The principal pathway of its biosynthesis also is unclear; apparently, enzymic synthesis may occur by various pathways following squalene cyclization (5, 6). All the intermediates in cholesterol

---

[1] Abbreviations: HMG-CoA, β-hydroxy-β-methylglutaryl-Coenzyme A; VLDL, very low density lipoproteins; LDL, low density lipoproteins; HDL, high density lipoproteins; SCP, squalene and sterol carrier protein.

[2] Studies from the author's laboratory described here were supported by NIH Grant HL-8634 and The NHLI Lipid Research Clinics Program.

967

synthesis are presently unknown (7). A number of enzymes catalyzing individual steps in cholesterol biosynthesis have been purified and characterized; many more remain unpurified and their reactions are studied only with crude enzyme preparations, e.g. tissue homogenates (6, 8). Great interest and research activity are centered in the areas of cholesterol biosynthesis regulation and the design of hypocholesterolemic agents (9), primarily due to close correlations of hypercholesterolemia with development of arteriosclerosis and coronary artery disease (10). Again, notwithstanding innumerable studies, we do not fully understand the molecular mechanism(s) regulating the biosynthesis of either cholesterol or its steroid products, bile acids and steroid hormones. It should not be surmised that the studies on steroid biosynthesis regulation performed in the 1950s and early 1960s were of poor quality. Many of these are classic works in a difficult area; they laid the necessary foundation for current attacks on this problem. Recently, there have been several important preliminary breakthroughs which hopefully will lead to rapid elucidation of the regulatory mechanisms. Investigations discussed here are chosen to reflect the current state of the art in this area. Because of space limitations, it is impossible to cover all pertinent topics, and chosen topics are often not discussed in depth. Unfortunately, many excellent reports are not included. Stress is placed on recent studies; comprehensive reviews of earlier investigations are mentioned in the discussion of each topic.

The reader unfamiliar with this overall subject is referred to several excellent reviews (11–13)—Dietschy & Wilson (11) is a comprehensive discussion of cholesterol metabolism regulation with emphasis on human studies; Siperstein (12) deals with previous studies on regulation of cholesterol synthesis by cholesterol and bile acids; and Rodwell, McNamara & Shapiro (13) summarize work on one major regulatory enzyme of cholesterol synthesis, HMG-CoA reductase. Briefly, cholesterol synthesis by liver enzymes is suppressed by cholesterol feeding; measures that remove cholesterol from the body increase enzymic cholesterol synthesis. Fasting and bile acid administration also decrease cholesterol synthesis. Cholesterol biosynthesis is subject to diurnal changes in rate. These synthetic rate modifications are reflected by the activity of HMG-CoA reductase.

ENZYMIC CONVERSION OF ACETATE TO CHOLESTEROL

The first part of this discussion deals with recent studies on tissue preparations—homogenates, slices, and perfused organ and isolated cell systems. It might be assumed that since a major regulatory enzyme, HMG-CoA reductase, has been identified, there is little need for further studies of other enzymic steps in sterol biosynthesis. However, evidence is accumulating for the occurrence of regulatory steps in addition to HMG-CoA reductase. Furthermore, actions of factors affecting sterol biosynthesis (e.g. hormonal and potential hypolipidemic agents) may be better characterized in the presence of all enzymes required for overall synthesis.

It has been over twenty years since the ingenious discovery by Bucher and her colleagues (14–16) that using a loose fitting homogenizer to prepare liver homogenates yielded an enzyme preparation capable of converting acetate to cholesterol. This preparation is still used for studies with factors affecting cholesterol synthesis (cf Table 1, 9), as a source of individual enzymes of cholesterol biosynthesis, and to

Table 1   Selected list of recent studies using liver homogenates to study factors affecting cholesterol synthesis in vitro (9)[a]

Compound or Factor Tested	Substrate	Reference
Sex hormones, fatty acids	Mevalonate	Carroll & Pritham (17)
Bile salts	Lanosterol	Miller & Gaylor (18)
Phenylalanine	Mevalonate	Shah et al (19)
Vitamin E, selenite, lipoic acid	Acetate, mevalonate	Eskelson & Jacobi (20)
Phenformin and related biguanides	Cholest-5,7-dienol, cholest-7-enol	Dempsey (21)
Chlorophenoxyacetic acids	Acetate	Witiak et al (22)
Cholestane derivatives	Acetate, mevalonate, cholest-7-enol	Dempsey et al (23) Witiak et al (24, 25)
Ethanol	Acetate, mevalonate	Eskelson et al (26)
Fat soluble, vitamins	Acetate, mevalonate	Eskelson et al (27)
Plasma lipoproteins	Mevalonate	Onajobi & Boyd (28)
Renal factors, kidney microsomes	Acetate, mevalonate	Haven & Jacobi (29), Thuy et al (30)
Steroid hormones	Mevalonate	Ono & Imai (31)
3,5-Dihydroxy-3,4,4-trimethylvaleric acid	Mevalonate	Hulcher (32)
Cyclic AMP	Acetate, squalene, lanosterol	Bloxham & Akhtar (33), Bloxham, Wilton & Akhtar (34)
(−) Hydroxycitrate	Acetate	Barth et al (35)
Liver factors	Acetate, mevalonate	Goodwin & Margolis (36, 37)

[a] This list does not include the numerous studies in which animals are fed or administered an agent and the effects of the agent in vivo examined by preparing homogenates of the animal livers to determine the possible occurrence of decreased or increased synthesis rates (e.g. conversion of acetate or mevalonate to cholesterol; effects on HMG-CoA reductase activity).

demonstrate that a steroid compound is a possible intermediate in cholesterol synthesis (7). In this regard, it is important to mention that at elevated levels (greater than millimolar) a variety of compounds interfere with cholesterol synthesis in vitro by blocking more than one enzymic step in a nonspecific manner, e.g. by detergent effects, chelation, etc. Recently, Cornforth (38) also indicated that caution should be used in interpretating results obtained with liver homogenates, regarding pathways and intermediates of cholesterol synthesis. Of particular interest in the area of sterol biosynthesis regulation are recent studies by Goodwin & Margolis (36, 37) using rat liver homogenates. These investigators found that preincubation of a liver homogenate increased the subsequent conversion of labeled acetate to cholesterol eight- to twentyfold. Similar activation did not occur with mevalonate as substrate; however, conversion of acetate to fatty acids and CO_2 was somewhat increased. The evidence indicated that a decrease in the size of the acetyl-CoA pool during preincubation could not account for these observations. These results suggest that the activation of cholesterol synthesis during preincubation is specific for the enzymes that catalyze conversion of acetyl-CoA to mevalonate.

Goodwin & Margolis (37) propose that their observations reflect reversal in vitro of a physiological control over cholesterol biosynthesis involving cyclic AMP. For example, one of the enzymes catalyzing mevalonate synthesis (see following sections) could be regulated by phosphorylation due to the action of a cyclic AMP-dependent protein kinase and dephosphorylation by a phosphoprotein phosphatase. The phosphatase would be activated during preincubation of a homogenate, resulting in dephosphorylation of the control enzyme and increased cholesterol synthesis. This mechanism, by analogy to other cyclic AMP-dependent enzyme systems, could account for known hormonal effects on cholesterol biosynthesis in vivo (11). Preliminary evidence by other workers indicates that a mechanism similar to this may occur with acetyl-CoA carboxylase, an important regulator of fatty acid synthesis (39, 40). If such a phosphorylation-dephosphorylation mechanism were to regulate one of the early enzymes in cholesterol synthesis (HMG-CoA synthetase or reductase), it would be a more rapid regulatory process than that of enzyme turnover [e.g. HMG-CoA reductase is known to have a half-life of about 4 hr (41)]. Indeed, Beg et al (165) found that HMG-CoA reductase activity of liver homogenates, slices, and cells is inhibited by preincubation with cyclic AMP. They also showed that HMG-CoA reductase activity of the washed liver microsomal fraction is decreased by preincubation in the presence of ATP, Mg ions, and a protein component isolated from the liver soluble fraction. This reduction in enzymic activity could be reversed by treatment of the microsomal fraction with a second protein component also obtained from the liver soluble fraction.

Before leaving the topic of liver homogenate experiments, some preliminary results by Frantz & Ener[3] should be noted. They found that when the liver high-speed soluble fraction obtained from rats fed a high-cholesterol diet was combined with the microsomal enzyme fraction prepared from liver homogenates of rats fed a low cholesterol diet, the observed rate of cholesterol synthesis was markedly less than that seen when the soluble fraction from rats fed a low-cholesterol diet was used. These intriguing results will be recalled during the subsequent discussion of effects by lipoprotein fractions on the activity of HMG-CoA reductase (see section on Enzymic Reduction of HMG-CoA to Mevalonate).

The tissue slice technique is also an older procedure still favored by a number of investigators. It has the advantage of maintaining a close structural similarity (though not necessarily functional similarity) to the intact organ, while permitting experiments in vitro. Usually, slices are prepared from liver or intestinal tissues after dietary or operative manipulation of an animal, and the incorporation of labeled

---

[3] Frantz, I. D. Jr., Ener, M. Personal communication.

acetate into cholesterol, fatty acids, and CO_2 determined. It should be noted that if an experimental manipulation changes the size of the acetate pool in the tissue under study, altered synthetic rates may be observed and attributed erroneously to enzymes acting beyond acetate (e.g. HMG-CoA reductase). A similar comment is applicable to using liver homogenates for studying effects of dietary or operative manipulations on overall lipid synthesis in the intact animal (cf also 79). It is now becoming common to use ^3H_2O for quantitation of overall cholesterol synthesis, thus avoiding possible errors due to changes in the size of substrate pools. Another general comment on the slice technique is that it is not usually possible to know the level of viable cells present in the slice, and problems with transport of substrates across cell membranes could occur.

In spite of these possible drawbacks, the literature on regulation of steroid synthesis is replete with results obtained using the tissue slice; these studies have yielded much useful information. Recently, Weis & Dietschy (42) and Harry, Dini & McIntyre (43) used the liver slice technique to demonstrate that liver cells retain their ability to respond to cholesterol in biliary stasis. Liver and intestinal slices were studied by Cayen (44) and Fishler-Mates, Budowski & Pinsky (45) to show that compounds that interfere with intestinal absorption of cholesterol cause increased liver cholesterol synthesis. White (46, 47) examined the enzymic action sites of two commonly prescribed hypolipidemic agents by using liver homogenates and slices. Bricker & Levey (48, 49) reported inhibition of cholesterol synthesis by cyclic AMP in liver slices from normal rats, but not in slices from rat hepatoma [cf Goodman & Margolis (37)].

While the homogenate and tissue slice techniques have certain advantages in studying regulation of steroid biosynthesis, it may be desirable or essential to relate synthetic rates in vitro with those observed in vivo. The latter correlation is usually not possible with tissue homogenates and slice preparations; it is possible with the perfused liver technique. Brunengraber, Boutry & Lowenstein (50) described detailed conditions for perfusing rat liver, resulting in rates of fatty acid and sterol synthesis that equal those observed in vivo. Sterol synthesis is determined by adding ^3H_2O to the perfusate at an appropriate time during perfusion. By this technique, Liersch et al (51), showed that infusion of physiological levels of bile salts at a constant rate through the portal vein did not inhibit cholesterol synthesis. They concluded that inhibition of cholesterol synthesis by bile salts in vivo is not mediated by a direct (e.g. allosteric) effect on a regulatory enzyme. The perfused liver system was also utilized by Bricker, Kozlovskis, & Goodman (52) to obtain evidence that rats fed a high-cholesterol diet incorporated less labeled acetate into sterols than low cholesterol-fed rats. Again, the disadvantage of not knowing the initial pool size of acetate in these preparations is apparent. More recently, using ^3H_2O as a tracer, hypercholesterolemic serum was shown to inhibit sterol synthesis in a perfused rat liver system (53). Effects of the serum occurred rapidly, in 1 hr, and perhaps were due to direct inhibition of a controlling enzyme. The liver perfusion system should prove valuable in further attempts to elucidate mechanisms of steroid synthesis regulation. This technique is mainly useful for measuring effects of possible regulators over a rather long time period (hours); more rapidly acting effectors need to be studied with isolated enzyme systems (see following sections).

Finally, an exciting experimental tool or model being exploited for studies on the regulation of steroid biosynthesis is the isolated cell or cell culture system. Several comprehensive reviews have appeared describing lipid uptake, synthesis, metabolism, and excretory properties of cells in culture from normal or malignant tissue sources (54–61). Selected examples of current findings obtained with isolated or cultured cell systems are described here. Many types of cells grown in culture require cholesterol for maximal growth [e.g. human skin fibroblasts (62, 63), virus-transformed fibroblasts (64), human kidney cells (65), and minimal deviation hepatoma cells (66)]. In the presence of delipidized growth medium (serum), these cells synthesize cholesterol at a rate considerably faster than when cholesterol is added to the medium. The rate of cholesterol synthesis is most probably regulated by cholesterol uptake. In a recent carefully designed study, Sokoloff & Rothblat (67) showed that the patterns of flux, content, and synthesis of sterol in mouse fibroblasts (L cells) were readily modified by variations in the lipid (cholesterol and phospholipid) composition of the growth medium. These cells do not synthesize cholesterol; rather $\Delta^{5,24}$-cholestadienol, an immediate precursor of cholesterol in normal cells, is the endproduct of their sterol synthesis. L cells incorporate cholesterol from the growth medium, followed by reduced synthesis of $\Delta^{5,24}$-cholestadienol. These cells, therefore, offer an important tool for basic studies on the control steroid synthesis and uptake, since these processes can be readily differentiated.

With regard to regulation of sterol synthesis in cultured cells by lipids in the growth medium, several investigators found that specific plasma lipoprotein fractions cause decreased sterol synthesis [VLDL and LDL (61, 68)]. As discussed later, the presence of specific lipoproteins in the growth medium of cultured cells can be shown to influence the activity of this regulatory enzyme (e.g. 69). In addition to cultured cell systems, other isolated cell systems are potentially valuable tools for probing regulatory mechanisms. Fogelman et al (70), working with human leukocytes, discovered that sterol synthesis from acetate but not mevalonate was enhanced by carrying out incubations in lipid-free serum. Cycloheximide blocked the latter effect, suggesting that an enzyme acting prior to mevalonate was induced. Their data further indicated that leukocytes collected from hypercholesterolemic (Type II, lipoprotein pattern) patients gave an exaggerated response to the presence of lipid-free serum, i.e. a much greater enhancement of sterol synthesis than exhibited by cells from normal individuals. Isolated liver cells appear to be another promising tool for sterol synthesis studies. For example, Capuzzi et al (71) reported reduced levels of cholesterol synthesis from acetate by isolated rat liver cells in the presence of dibutyryl cyclic AMP. Other cell systems yielding important information on regulation of sterol biosynthesis are from lower forms of life, e.g. protozoa (see section on Enzymic Conversion of Squalene to Cholesterol; Squalene and Sterol Carrier Protein) and yeast (72).

ENZYMIC SYNTHESIS OF HMG-CoA

There are two enzymic steps required for synthesis of HMG-CoA from acetyl-CoA, acetoacetyl-CoA thiolase and HMG-CoA synthase (cf Figure 1). McNamara & Rodwell (73) have reviewed thoroughly earlier studies on the properties and cellular

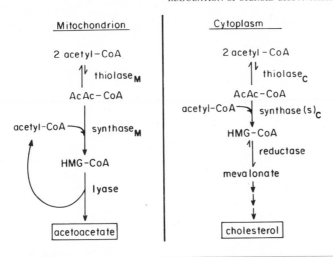

Figure 1   Pathways of HMG-CoA synthesis. Reprinted from Clinkenbeard et al (75) with permission of the author and the *Journal of Biological Chemistry*.

location of these enzymes. It was previously thought that the major supply of HMG-CoA was synthesized in the mitochondria and then transported to the cytoplasm for cholesterol synthesis. Now evidence is accumulating that there are two sources of HMG-CoA: one process in the mitochondria involved in ketogenesis and another independent process in the cytoplasm involved in cholesterol synthesis (Figure 1). Lane et al (74), Clinkenbeard et al (75), and Sugiyama et al (76) demonstrated that the thiolase and synthase are present in both the mitochondrial and cytoplasmic fractions of avian liver (Table 2). These enzymes were isolated and

Table 2   Intracellular distribution of thiolase, HMG-CoA synthase, and HMG-CoA lyase in avian liver (74)

		Distribution	
	Total Activity	Particulate (mitochondrial)	Cytoplasmic
	Units per Gram Wet Weight	Percentage	Percentage
I. Enzymes of Cholestero-Ketogenesis			
AcAc-CoA thiolase	88	53 (matrix)	47 (cytosol)
HMG-CoA synthase	1.9	82 (matrix)	18 (cytosol)
HMG-CoA lyase	6.9	96 (matrix)	4 —
II. Enzyme markers			
Citrate synthase	5.6	98 (matrix)	2 —
Lactate dehydrogenase	234	10 —	90 (cytosol)

Table 3 Properties of cytosolic and mitochondrial HMG-CoA synthases from chicken liver (74)

	Synthase	
	Cytosolic	Mitochondrial
Intracellular distribution	18%	82%
Multiple forms	yes	no
Molecular weight (M_{SD}, M_{GF})	~ 100,000	104,000
Subunit weight	55,000	52,000
Isoionic point (pI)	~ 5.3	7.3
Relative mobility (R_m) (gel electrophoresis, pH 8.5)	0.29, 0.32	0.1
Proposed physiological function	Cholesterogenesis	Ketogenesis

purified from both cellular fractions and shown to be molecularly distinct; see Table 3 for properties of the synthases. Definitive proof for distinct cytosolic and mitochondrial thiolases in avian liver was the presence of one cytosolic and two mitochondrial enzymes, each having a different isoionic point by isoelectric focusing (75). It is probable that one of the mitochondrial thiolases is involved in β oxidation of fatty acids and the other in ketogenesis (75). In addition, Lane et al (74) showed that HMG-CoA lyase is almost exclusively located in the mitochondrial fraction (cf Table 2).

The studies of Barth et al (35, 77) and Sullivan et al (78), using the same mammalian liver enzymes, are compatible with conclusions drawn from experiments with avian liver enzymes (74–76), i.e. that cholesterol synthesis from acetyl-CoA is exclusively an extramitochondrial process. As a consequence it appears that HMG-

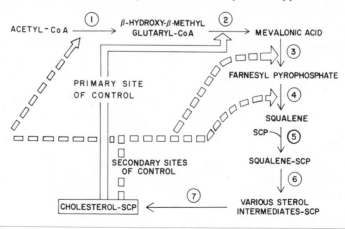

Figure 2 Possible regulatory roles of SCP in cholesterol synthesis

CoA reductase is not a branch point between ketogenesis and cholesterolgenesis and also not the first unique step in sterol synthesis. As suggested by Lane et al (74), the thiolase-synthase enzyme system commits the sterol synthesis pathway and probably is the primary site of regulation. Indeed, cholesterol feeding does decrease the activity of this enzyme system, in particular the synthase (74–76, 79). It is apparent that further studies are needed to elucidate whether this enzyme system is regulated, for example, by synthesis and degradation of enzyme protein, an allosteric type mechanism, or both types of mechanisms. The relative roles of the thiolase-synthase system and HMG-CoA reductase in regulating overall cholesterol synthesis also must be assessed. Possible effects of a hormonally influenced cyclic AMP participation in the thiolase-synthase system is suggested by work showing inhibition of overall cholesterol synthesis by cyclic AMP or its derivatives (e.g. 48, 49, 71). Finally, a cholesterol-SCP complex could influence the activity of the thiolase-synthase system as indicated in Figure 2 and discussed in a later section.

ENZYMIC REDUCTION OF HMG-CoA TO MEVALONATE

Earlier studies on the properties, assay, cyclic rhythm, and developmental pattern of HMG-CoA reductase were reviewed in detail by Rodwell, McNamara & Shapiro (13). Hamprecht (80) has also discussed regulation of cholesterol synthesis with emphasis on the role of the reductase. Only more recent studies on purification and properties of the reductase, as well as modes of regulation, are summarized here.

One of the most important recent developments in this area is the solubilization of liver microsomal HMG-CoA reductase by several techniques, thus permitting purification of the enzyme and application of immunochemical techniques. Currently, four methods of solubilization are available: deoxycholate treatment (81); acetone powder preparation (82); high salt, glycerol, or snake venom treatment (83); and freeze-thaw treatment (84). The first method (81) results in low yields of enzyme and the final specific activity is not as high as with other methods (83, 84). However, the preparation of Kawachi & Rudney (81) is homogeneous—mol wt 200,000 with subunits of 65,000 to 70,000 (85). Antibodies were prepared against the purified HMG-CoA reductase; studies on the synthesis and degradation of the enzyme using these antibodies are discussed in the following paragraph (86). The acetone powder preparation (82) permits solubilization of the enzyme; reports of purification following solubilization have not appeared. The remaining methods (83, 84) [with the exclusion of the snake venom digestion (83)] involve mild treatments, suggesting that HMG-CoA reductase is bound to the surface of the microsomal membrane and can be released without covalent bond cleavage. Once the enzyme is solubilized, it dissolves in aqueous buffers and does not require detergents. The fact that HMG-CoA reductase readily dissociates from microsomal membranes indicates that it could be degraded faster than the bulk of membrane protein, in keeping with the short half-life of the enzyme (hours vs days for total microsomal protein) (83). The preparation of Brown et al (83) has a higher specific activity than that of Kawachi & Rudney (81); however, it is inactivated by chilling to 4° unless high salt (4 M) is present and fairly heat stable. The molecular weight of the Heller

& Gould (84) preparation is similar to that of HMG-CoA isolated by the other techniques (81, 83); it is also fairly heat stable. Recent evidence indicates that this preparation is also reversibly inactivated by cold.[4]

As stated in the introduction, the molecular mechanism(s) regulating HMG-CoA reductase activity are not well understood. The occurrence of a circadian rhythm in both cholesterol synthesis and the reductase are now widely accepted (e.g. 13, 80, 87). This rhythm is thought to be due to synthesis and degradation of enzyme protein; for example, injection of protein synthesis inhibitors into animals prevents the rhythm. Factors that affect the rate of cholesterol synthesis (cholesterol feeding, fasting, and cholesterol removal by biliary diversion or treatment) also affect the diurnal rhythm. For the past several years, most investigators held that regulation of sterol synthesis was controlled solely by turnover of HMG-CoA reductase; now reports are appearing that additional more rapid mechanisms of regulation may be operative. Recent studies by Higgins & Rudney (86) support the latter proposal. These authors followed the synthesis, decay, and activity of HMG-CoA reductase in rats accustomed to a light-dark cycle. A specific antiserum to purified HMG-CoA reductase (see preceding paragraph) was used to demonstrate the presence of enzyme protein. Rats were followed before and after cholesterol feeding. The results in Figure 3 demonstrate that reductase activity and the antigen-antibody reaction

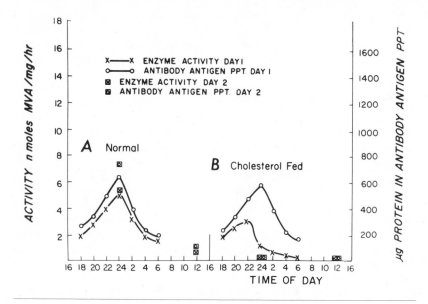

Figure 3 HMG-CoA reductase activity and immunochemical detection of HMG-CoA reductase before and after cholesterol feeding. Reprinted from Higgins & Rudney (86) with permission of the authors and *Nature New Biology.*

[4] Heller, R., Gould, R. G. Personal communication.

follow the same time pattern in normal animals; after feeding cholesterol, enzymic activity falls off rapidly and is significantly depressed within 4 hr. In contrast, the antigenic protein is not decreased initially and decays at the same rate as for normal animals. After 24 hr the rise in HMG-CoA reductase activity and antigenic protein are markedly depressed in cholesterol-fed animals. These data indicate that HMG-CoA reductase is under more than one control—an immediate effect independent of protein synthesis and a slower effect causing inhibition of protein synthesis. Similar findings were described by Tanabe et al (88): cycloheximide acted more slowly than injected cholesterol in blocking cholesterol synthesis by intact rats.

Several authors have reported the occurrence of tissue factors capable of inhibiting HMG-CoA reductase in vitro (e.g. 13, 29, 30). Most recently, Carlson et al (89) and Carlson & Dempsey (90) presented evidence that SCP is capable of modifying HMG-CoA reductase activity in vitro (cf Figure 2). Direct effects on enzyme activity could, for example, result from dissociation of the subunit structure of HMG-CoA reductase. Clearly, there is sufficient evidence to warrant further careful examination of possible direct effectors, using the highly purified HMG-CoA reductase preparations now available. It is possible that effects of modifiers will be observed only with unpurified microsomal HMG-CoA reductase, indicating an essential role for membrane structures in the regulatory process [cf also the work of Beg et al (165) discussed previously].

In addition to studies on sterol synthesis regulation by hepatic tissue, several recent reports describe the properties of HMG-CoA reductase from intestinal tissue. Shefer et al (91–93) noted that reductase activity was located in both the mitochondrial and microsomal cells of rat intestinal crypt cells but not of villi cells. Intestinal HMG-CoA reductase had kinetic characteristics similar to the rat liver enzyme; however, the liver enzyme activity varied with the age and sex of the animals. Cholestyramine treatment enhanced reductase activities of both tissues (91). Intestinal HMG-CoA reductase also exhibited a diurnal rhythm coinciding with the rhythm of the liver enzyme but with a lower amplitude (92). Similar results were obtained by Edwards et al (94) when they examined the circadian rhythm of overall cholesterol biosynthesis in liver and intestine. Bile acid feeding did not reduce the basal activity of the intestinal reductase, but did reduce the normal diurnal rise in activity. Feeding both cholesterol and bile acids resulted in marked suppression of intestinal HMG-CoA reductase activity, probably due to an increase in sterol concentration within crypt cells (93). Similar findings were reported by Hamprecht et al (95). In related studies Shefer et al (96) concluded that the composition of the bile acid pool influences liver concentrations of both HMG-CoA reductase and cholesterol 7α-hydroxylase (considered to be an important regulatory enzyme in bile acid synthesis, see later sections).

With regard to recent work on the regulation of the diurnal rhythm of HMG-CoA reductase and sterol synthesis, Huber, Latzin & Hamprecht (97) concluded that a light-dark cycle or other environmental change is not the cause of the rhythm; the rhythms are thought to be caused by other rhythms endogenous in the animal. McNamara et al (13, 98) showed that HMG-CoA reductase activity and rhythm vary markedly prior to birth, after the first postnatal week, and following weaning.

Craig et al (99), Dugan et al (100), and Slakey et al (101) examined effects of nutritional states on activities and diurnal changes of HMG-CoA reductase as well as other enzymes catalyzing cholesterol and fatty acid synthesis. They concluded that the fat and cholesterol composition of the diet influences activities of the enzymes catalyzing cholesterol synthesis in a manner different from its influence on the enzymes catalyzing fatty acid synthesis (99). Careful kinetic analysis showed that the diurnal variation of HMG-CoA reductase activity is due to an increased rate of enzyme formation, not increased enzyme breakdown (100). The data of Dugan et al (100) also suggested that the diurnal variation in cholesterol synthesis from acetate in fed animals may be controlled by an enzyme acting prior to HMG-CoA reductase (cf discussion of the preceding section, Enzymic Synthesis of HMG-CoA). Related studies showed that HMG-CoA reductase is the only enzyme in the cholesterol-genesis pathway beyond HMG-CoA that undergoes diurnal variations. However, as discussed in the following section, additional regulatory sites most probably occur after mevalonate formation.

In other recent investigations, possible hormonal influences on HMG-CoA reductase activity and its rhythm are indicated. Edwards (102), for example, showed that adrenalectomy abolished the circadian rhythm of HMG-CoA reductase and hypophysectomy resulted in a further decrease of enzyme activity. Microsomal HMG-CoA reductase isolated from normal and adrenalectomized rats exhibited the same kinetic properties; loss of activity after adrenalectomy was probably not due to altered enzyme structure. Reductase activity could not be restored by pharmacological doses of adrenal hormones. Administration of epinephrine to intact rats did increase reductase activity severalfold and appeared to depend on protein synthesis, i.e. it was blocked by cycloheximide (102).

Lakshmanan et al (103, 104) demonstrated that HMG-CoA reductase activity is increased two- to sevenfold after administration of insulin to normal or diabetic animals and the increased activity is blocked by simultaneous administration of glucagon (103). Administration of glucagon, hydrocortisone, and cyclic AMP also abolished the diurnal rise in HMG-CoA reductase of normal rats (104, 105). Dugan et al (106) found that a single dose of L-triiodothyronine, T_3, administered to hypophysectomized animals increased markedly the barely measurable reductase activity of these animals. Similar studies with hypophysectomized diabetic animals showed that both T_3 and insulin were required for restoration of reductase activity (106). Thus, insulin and thyroid hormone increase reductase activity; glucagon and glucocorticoids reduce reductase activity. These hormonal effects should be considered in conjunction with the suggested effects of cyclic AMP on cholesterol-genesis discussed in previous sections.

In an attempt to further define the molecular events giving rise to the diurnal rhythm of HMG-CoA reductase, Edwards & Gould (107) studied the effects of high doses of actinomycin D on the normal rhythm and the increased reductase activity after epinephrine administration (102). Actinomycin D did not affect the basal enzyme activity or the rate of decline after peak activity was reached. The antibiotic did partially suppress the normal rhythmic increase in activity and prevented the rise following epinephrine administration, thus suggesting a role for mRNA in these processes. Related studies indicated that effects of cholesterol feeding may stimulate degradation or inactivation of the reductase as well as blocking synthesis [cf results of Higgins & Rudney described above (85, 86)]. Edwards & Gould (107) also correlated the decrease in microsomal HMG-CoA reductase activity 4–7 hr after cholesterol feeding with an increase in the level of microsomal cholesterol esters. As mentioned, Harry et al (43) also suggested that cholesterol esters may block cholesterol synthesis.

In another intriguing study, Davison & Gould (108) offered preliminary evidence for the occurrence of chromatin-bound cholesterol in rat liver nuclei following cholesterol feeding, suggesting a mechanism for regulation of HMG-CoA reductase synthesis at the level of DNA. Whether the occurrence of a diurnal rhythm in reductase activity is the essential characteristic of cells able to regulate cholesterol synthesis is not entirely clear. Sabine et al (109) showed that two minimal deviation hepatomas did not exhibit a diurnal rhythm in cholesterol synthesis, whereas Goldfarb & Pitot (110) were able to show a diurnal rhythm in HMG-CoA reductase with two other hepatomas. However, the response to dietary cholesterol (a reduction in sterol synthesis) by these tumors was not normal. Similar results were obtained by Kandutsch & Hancock (111) with mice hepatomas. These authors also demonstrated that the kinetic and structural properties of HMG-CoA reductase isolated from hepatomas or normal livers were the same. It is apparent that the basic cause leading to lack of response to dietary cholesterol may be a defect in cholesterol transport across the cell membranes of intact tumors.

As alluded to previously, cultured hepatoma cells (in contrast to intact tumors) are able to respond to lipoprotein fractions in the growth medium. Watson (61), for example, has shown that HMG-CoA reductase activity of cultured hepatoma cells is increased severalfold when the cells are grown in media relatively free of lipids. Brown et al (69), working with cultured human fibroblasts, presented evidence that HMG-CoA reductase of these cells is suppressed by serum and enhanced by delipidized serum. The lipoprotein fraction of human serum or plasma most effective in suppressing reductase activity is the LDL fraction (69). With regard to hormonal effects in cultured cells, Bhathena et al (112) reported a marked increase in HMG-CoA reductase activity by human skin fibroblasts in response to insulin (cf results describing hormonal effects on HMG reductase in intact animals). Finally, the potential importance of membrane binding and transport of cholesterol into a cell has been mentioned. Evidence that the ability of cells to accumulate cholesterol may be the essential element in regulating sterol synthesis is derived from the studies of Brown & Goldstein (113) and Goldstein & Brown (114), using fibroblasts cultured from patients with familial hypercholesterolemia. Their work indicates that cells of these patients (in particular homozygous Type II patients) do not respond in the same manner as normal cells to the presence of LDL in their growth media: there is little or no decrease in HMG-CoA reductase activity. Further studies with fibroblasts from familial hypercholesterolemic patients suggest that these cells do not bind LDL molecules as well as cells derived from normal patients (114). Goldstein & Brown (114) suggest that cells derived from hypercholesterolemic patients require an increased level of LDL to turn off their HMG-CoA reductase and cholesterol synthesis. Similar results were obtained with leukocytes from hypercholesterolemic patients by Fogelman et al (70), as discussed previously. The possible roles of membrane receptors and carrier proteins in the regulation of sterol synthesis and metabolism are presented in the review by Bailey (60) and are discussed in the following sections.

Some caution is warranted in the interpretation of results in the previous paragraphs. First, care should be taken in ascribing causes of findings obtained by administration of protein and polynucleotide synthesis inhibitors, especially if the levels of these inhibitory compounds are toxic, or, in the case of intact animals, cause changes in the dietary habits of animals under study. Second, cultured cells from normal, hypercholesterolemic, or malignant tissue sources after several passages in culture do not necessarily reflect the genetic expression of the original cells in situ. Third, in order to ascertain effects of agents added to growth media or fed to experimental animals, rather lengthy time periods of study are required; rapidly acting effectors cannot be assayed—a similar comment was made previously with regard to organ perfusion experiments (see section on Enzymic Conversion of Acetate to Cholesterol).

ENZYMIC CONVERSION OF MEVALONATE TO SQUALENE; TRANS-METHYLGLUTACONATE SHUNT

The possibility was expressed in previous sections that sites of sterol synthesis regulation exist in addition to HMG-CoA reductase. Indeed, a number of years ago Gould & Swyryd (115) offered evidence for secondary sites of regulation following mevalonic acid. Recently, Slakey et al (101) measured the activity levels of six enzymes in the pathway from mevalonate to squalene. Effects of fasting, refeeding a fat-free diet, and time of day were noted. As mentioned before, none of these enzymes exhibited a significant diurnal variation in activity; however, all enzymes tested, except mevalonate and phosphomevalonate kinase, declined in activity on fasting and returned to normal or supernormal levels on refeeding a fat-free diet. Slakey et al (101) also determined rate constants for the formation and degradation of a number of the enzymes—pyrophosphomevalonate decarboxylase, isopentenyl-pyrophosphate isomerase, dimethylallyltransferase, and squalene synthetase. These authors concluded that the level of liver HMG-CoA reductase activity is sufficient to account for observed rates of overall conversion of acetate to cholesterol. However, their data also demonstrated that there are at least two secondary regulatory sites, not yet defined, controlling the flux of compounds to squalene.

Recently, Popjak & Edmond (116) and Edmond & Popjak (117) presented evidence for the existence of a pathway leading intermediates arising from mevalonate back to HMG-CoA, ketone bodies, and acetyl CoA, rather than to squalene and cholesterol (Figure 4). This pathway is termed the *trans*-methylglutaconate shunt. Results supporting the occurrence of this pathway are the detection of label from mevalonate-2-^{14}C in fatty acids. Furthermore, $[3',4-^{13}C_2]$-, $[5-^{14}C]$-, and $[4R-4-^3H]$ mevalonate, but not $[4S-4-^3H]$ mevalonate, yielded labeled fatty acids in accordance with predictions (117). Popjak & Edmond (116) estimate, based on their experimental observations in young and adult rats, that at least 20% of mevalonate could be shunted through the new pathway. The discovery of the *trans*-methylgluta-conate shunt revealed another process capable of regulating steroid biosynthesis. The quantitative role of the shunt under varying physiological conditions needs to be assessed; cf the recent studies by Edmond (166) in brain tissue.

Finally, Rilling (118) recently noted a possible requirement for SCP in the conversion of farnesylpyrophosphate to presqualene pyrophosphate and squalene (cf

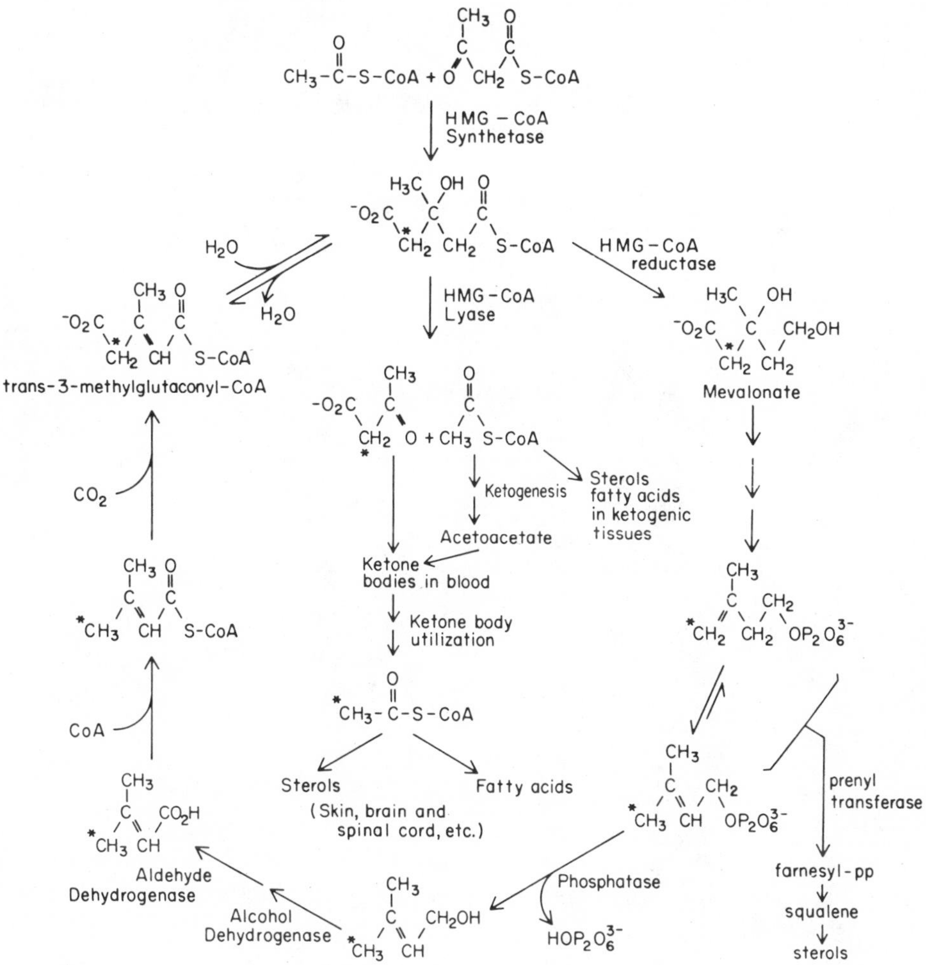

Figure 4   *Trans*-methylglutaconate shunt. Reprinted from Edmond & Popjak (117) with permission of the authors and the *Journal of Biological Chemistry*.

Figure 5). This finding again suggests that SCP could have regulatory roles in sterol synthesis prior to its recognized participation in the pathways converting squalene to cholesterol and in the conversion of cholesterol to bile acids and steroid hormones (cf following sections and Figure 2).

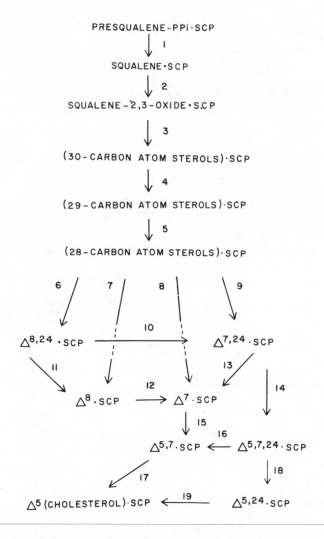

Figure 5 Outline of pathways and intermediate compounds in the later stages of cholesterol biosynthesis; role of SCP. Intermediates are shown as SCP complexes and positions of unsaturation are indicated by the delta (Δ) symbol.

ENZYMIC CONVERSION OF SQUALENE TO CHOLESTEROL; SQUALENE AND STEROL CARRIER PROTEIN

Several comprehensive reviews are available covering the subject of this section and its related topics (1–9, 119–123).

This discussion stresses, in particular, recent developments with regard to the structural and functional properties of SCP. As mentioned, SCP is a strong candidate for a regulatory role in steroid biosynthesis (Figure 2). SCP, first discovered in the soluble fraction of liver homogenates, was shown to be a ubiquitous, fairly heat-stable protein which binds sterols and other water-insoluble lipids; it is essential for conversion of water-insoluble precursors to cholesterol by microsomal enzymes (Figure 5) (124–127). Scallen et al (128, 129) and Gavey (130) confirmed these observations, except for the temperature stability property, the ubiquitous occurrence in mammalian tissues and molecular weight of the protomer form of SCP (i.e. sterol-free form). Shah (131) and Johnson & Shah (132) also interpreted their data with nervous and other mammalian tissue to indicate that SCP is not present in extrahepatic tissue. In contrast, Tabacik, Descomp & Crastes de Paulet (133) demonstrated a requirement for SCP in human term placenta for maximal conversion of squalene to sterols. SCP was also purified from human liver and shown to be functionally and structurally similar to rat liver SCP (134).

Rat and human liver SCP recently were purified to homogeneity, e.g. the purified protein migrated as a single band during urea-polyacrylamide gel electrophoresis at various pH values (135, 136). A protein functionally and structurally similar to liver SCP is present in various mammalian tissues (135, 136), protozoa (137–139), and yeast (118). The apparent differences in observed properties of SCP reported by various laboratories are readily accounted for by variations in assay conditions for SCP functional activity [e.g. by conversion of squalene to cholesterol (129) vs effects on individual enzymes (124–127)] and methods used to determine molecular weight [e.g. ultracentrifugation (129) vs gel filtration and sodium dodecylsulfate gel electrophoresis (127)]. In addition, SCP has a marked tendency to aggregate; it is therefore common to observe multiple molecular forms. The ubiquitous occurrence of SCP in mammalian tissues (135, 136), as well as in lower forms of life (118, 137–139), suggests that a requirement for SCP or an SCP-like protein may be a general biological requirement for water-insoluble steroid synthesis, metabolism, and the transport and regulation of these processes (123).

The form of SCP that interacts with microsomal enzymes is considered to be a high molecular weight sterol-SCP complex formed from the protomer form of SCP (mol wt 16,000) (Figure 6). Formation of this complex is facilitated by the presence of phospholipids and sterols (127). Carlson et al (89, 90) recently reported that a spectrum of lipid-SCP complexes was formed during incubations of pro-SCP and lipids. The molecular weight range of these complexes was $2.5 \times 10^5 – 2 \times 10^6$, as determined by density gradient centrifugation. It was proposed (123–127, Figure 6) that the high molecular weight complex of water-insoluble cholesterol precursor

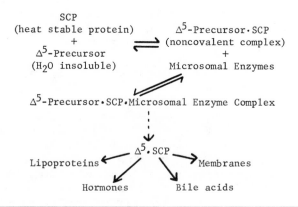

Figure 6   Proposed biological roles of SCP in cholesterol (Δ^5) biosynthesis and metabolism

and SCP combine with a specific microsomal enzyme. The precursor SCP is converted to its product, the following precursor SCP, which then combines with the microsomal enzyme next in the sequence of cholesterol synthesis. Finally, cholesterol SCP results (cf also Figure 5). Cholesterol SCP could then participate in the regulation of its own synthesis by direct effects on the activities of individual regulatory enzymes or indirect effects on the synthesis or degradation of regulatory enzymes (Figure 2). Preliminary findings on effects of SCP on HMG-CoA reductase activity have been mentioned previously (89, 90); Ritter et al (140) also noted that the level of cellular SCP could regulate the pathways by which cholesterol is synthesized; at high levels of SCP the pathway involving saturated side-chain sterol intermediates would predominate (cf Figure 5).

Recent evidence further suggests that SCP may participate in formation of lipoproteins, initial stages of cholesterol metabolism to steroid hormones and bile acids (see following section), and membrane synthesis [protozoan studies (137–139); discussed below]. With regard to lipoprotein formation, the HDL fraction of plasma was shown to contain SCP-like activity (126, 134) associated with one of the major HDL apopeptides, apo-Gln-II (also designated apo-A-II) (134). Apo-Gln-II (16,000 daltons) was completely sequenced by Brewer et al (141). The Gln-II molecule contains two identical polypeptide chains joined by a single disulfide bond. The other major HDL apopeptide, apo-Gln-I (Apo-A-I), is a single polypeptide chain with a molecular weight of approximately 28,000 (142). The possible structural similarity or identity of SCP and Apo-Gln-II deserves further careful study. In this regard Frnka & Reiser (143) recently presented evidence that the synthesis rate of HDL peptides is reduced by cholesterol feeding and this process may participate in the regulation of cholesterolgenesis. Furthermore, Onajobi & Boyd (28) found that high levels of serum or HDL could inhibit cholesterol synthesis in vitro by binding squalene and removing it from the synthetic pathway.

As mentioned, SCP has been isolated from protozoa (137–139). The strain of protozoa used was *Tetrahymena pyriformis*. In extensive studies Caspi (144) and Conner, Mallory and associates (145–146) showed that when cholesterol is supplied to the growth media of a *Tetrahymena* culture, synthesis of tetrahymanol (the normal pentacyclic triterpenoid alcohol of this strain) is inhibited and cholesterol is converted to $\Delta^{5,7,22}$-cholestatrienol. The induced Δ^7- and Δ^{22}-dehydrogenases are microsomal and require protozoan SCP (137–139). Interestingly, protozoan SCP has the same molecular weight and similar functional properties as liver SCP; e.g. protozoan SCP will substitute for liver SCP in reactions catalyzed by liver enzymes, and liver SCP will substitute in the same manner for protozoan SCP (137–139). This protozoan system appears to be an excellent tool for future studies on mechanisms of steroid biosynthesis regulation.

Before leaving the subject of SCP, it is important to differentiate this protein from other recently discovered lipid-carrying proteins. Ehnholm & Zilversmit (147) have purified a phospholipid exchange protein from the soluble fraction of beef heart homogenates. The molecular weight of this protein is 21,000; it is thought to participate in the exchange of phospholipid between mitochondria and microsomes and is assayed using synthetic liposomes (147). Another protein (mol wt 12,000), isolated from the soluble fraction of intestine and other tissue homogenates, binds long-chain fatty acids preferentially and also other anions, e.g. sulfobromophthalein (BSP) (148). This protein may participate in fatty acid absorption and transport. It appears to be the same as the "Z" protein studied by Reyes et al (149). SCP does bind fatty acids and phospholipids (127), but the molecular weight of the protomer form is 16,000. These proteins, differing somewhat in molecular weight from SCP, appear to be distinct from SCP; however possible similarities should be assessed further. In addition, Bloch and his colleagues (167, 168) reported that a soluble protein is required for microsomal squalene epoxidase. The latter protein is apparently heat-labile and of higher molecular weight (44,000) than protomer SCP. As it is possible to show conversion of squalene to cholesterol and other sterols by microsomal enzymes in the presence of SCP and appropriate cofactors (126, 169), a relationship of SCP to the protein required by the epoxidase is plausible.

INITIAL ENZYMIC STEPS IN CONVERSION OF CHOLESTEROL TO BILE ACIDS AND STEROID HORMONES

Regarding the initial stages of cholesterol metabolism to steroid hormones, Boyd & Trzeciak (152) summarized recent findings and proposed a mechanism of regulation involving ACTH stimulation of adrenal gland enzymes. These authors showed that ACTH stimulation of the adrenal leads to an increase in cyclic AMP; in stressed rats protein kinase activity increases as well as cholesterol esterase activity, followed by metabolism of cholesterol by mitochondrial side-chain cleavage enzymes. Boyd & Trzeciak (152), Ungar et al (153), and Garren et al (154) propose that a protein required for transport of free cholesterol from lipid stores to the mitochondrial enzymes could regulate steroid hormone biosynthesis. Beall & Sayers (155) used isolated adrenal cortex cells to demonstrate that the response to ACTH is rapid; cyclic AMP and steroid hormone synthesis occur within a few minutes. Thus, if regulation is to involve activation or synthesis of a cholesterol carrier protein, this process must also occur rapidly. It is possible that SCP could fulfill such a regulatory role. For example, Kan et al (156) and Ungar et al (153) showed that a cholesterol-liver SCP complex is metabolized by adrenal mitochondrial enzymes to pregneno-lone. Steroids formed from pregnenolone are progressively more water soluble and

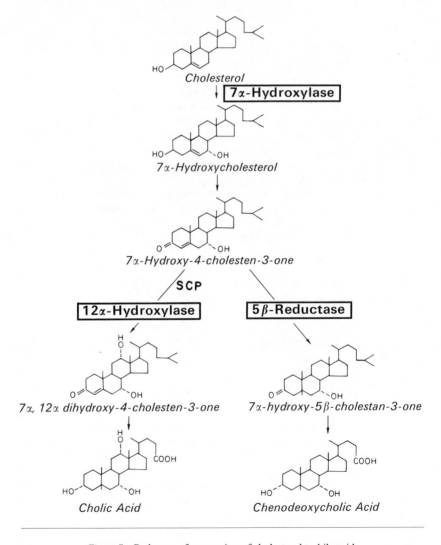

Figure 7   Pathways of conversion of cholesterol to bile acids

not bound by SCP (126). In addition, Kan & Ungar (157) isolated from adrenal tissue a heat-stable protein structurally similar, and perhaps identical, to liver SCP. The adrenal SCP functions with adrenal enzymes catalyzing cholesterol side-chain cleavage-yielding pregnenolone.

Two recent comprehensive reviews on the regulation of bile acid synthesis were prepared by Elliott & Hyde (158) and Mosbach (159). Pathways of bile acid synthesis are outlined in Figure 7. Recent evidence implicates the enzymic step from cholesterol to 7α-hydroxycholesterol (7α-hydroxylase) as the major regulatory site. Also, there are additional possible regulatory steps at the branch points leading to either cholic acid or chenodeoxycholic acid (cf Figure 7). As mentioned, Shefer et al (96) found that dietary manipulations causing changes in the composition of the bile acid pool led to changes in the activities of both HMG-CoA reductase and 7α-hydroxylase. Furthermore, it is now recognized that feeding chenodeoxycholate to patients with cholesterol gallstone disease leads to a dissolution of gallstones (160), perhaps due to decreased cholesterol synthesis. In this regard, Salen et al (161) found a decrease in liver HMG-CoA reductase and cholesterol level in gallstone patients fed chenodeoxycholate. Extensive further studies are needed to obtain a more complete elucidation of the processes regulating bile acid synthesis and interrelationships to the regulation of cholesterol synthesis. Properties of liver microsomal 7α-hydroxylase (which has not been solubilized or purified) were examined by Boyd et al (162) and Mitropoulos & Balasubramaniam (163). The enzyme has been difficult to assay because of the unknown level of substrate (endogenous cholesterol) in microsomal preparations. A double-isotope derivative method was developed to circumvent this problem (163). A possible regulatory role for SCP in bile acid synthesis (Figure 6) is indicated by recent data showing that 7α-hydroxycholesterol and other early water-insoluble bile acid intermediates (Figure 7) are bound by liver SCP (150, 151). Indeed, microsomal 12α-hydroxylase (Figure 7) was shown to specifically require SCP for activity (150, 151, 164). Other early enzymic steps in bile acid synthesis may also require SCP (e.g. 7α-hydroxylase).

Literature Cited

1. Bloch, K. 1965. *Science* 150: 19–38
2. Popjak, G., Cornforth, J. W. 1960. *Advan. Enzymol.* 22: 281–335
3. Clayton, R. B. 1965. *Quart. Rev. Chem. Soc.* 19: 168–200
4. Fiecchi, A. et al 1972. *Proc. Roy. Soc. London B* 180: 147–66
5. Dempsey, M. E. 1965. *J. Biol. Chem.* 240: 4176–88
6. Gaylor, J. L. 1973. *Int. Rev. Sci., Biochem. Sect.—Lipids,* ed. T. W. Goodwin, 1–42. Oxford: Med. Tech. Publ.
7. Schroepfer, G. J. Jr. et al 1972. *Proc. Roy. Soc. London B* 180: 125–46
8. Gaylor, J. L. 1972. *Advan. Lipid. Res.* 10: 89–141
9. Dempsey, M. E. 1974. *Pharmacology of Hypolipidemic Agents,* ed. D. Kritchev-

sky. Berlin: Springer. In press
10. Frantz, I. D. Jr., Moore, R. B. 1969. *Am. J. Med.* 46: 684–90
11. Dietschy, J. M., Wilson, J. D. 1970. *N. Engl. J. Med.* 282: 1128–38, 1179–83, 1241–49
12. Siperstein, M. D. 1970. *Curr. Top. Cell. Regul.* 2: 65–100
13. Rodwell, V. W., McNamara, D. J., Shapiro, D. J. 1973. *Advan. Enzymol.* 38: 373–412
14. Bucher, N. L. 1953. *J. Am. Chem. Soc.* 75: 498
15. Frantz, I. D. Jr., Bucher, N. L. 1954. *J. Biol. Chem.* 206: 471–81
16. Bucher, N. L., McGarrahan, K. 1956. *J. Biol. Chem.* 222: 1–15
17. Carrol, J. J., Pritham, G. H. 1966.

Biochim. Biophys. Acta 115:320–28
18. Miller, W. L., Gaylor, J. L. 1967. *Biochim. Biophys. Acta* 137:400–2
19. Shah, S. N., Peterson, N. A., McKean, C. M. 1968. *Biochim. Biophys. Acta* 164:604–6
20. Eskelson, C. D., Jacobi, H. P. 1969. *Physiol. Chem. Phys.* 1:487–94
21. Dempsey, M. E. 1969. *Drugs Affecting Lipid Metabolism*, 511–20. New York: Plenum
22. Witiak, D. T., Hackney, R. E., Whitehouse, M. W. 1969. *J. Med. Chem.* 12:697–99
23. Dempsey, M. E., Ritter, M. C., Witiak, D. T., Parker, R. A. 1970. *Atherosclerosis Proceedings 2nd International Symposium,* ed. R. Jones, 290–95. Berlin: Springer
24. Witiak, D. T. et al 1971. *J. Med. Chem.* 14:216–22
25. Witiak, D. T., Parker, R. A., Dempsey, M. E., Ritter, M. C. 1971. *J. Med. Chem.* 14:684–93
26. Eskelson, C. D., Cazee, C., Towne, J. C., Walske, B. R. 1970. *Biochem. Pharmacol.* 19:1419–27
27. Eskelson, C. D., Jacobi, H., Cazee, C. R. 1970. *Physiol. Chem. Phys.* 2:135–50
28. Onajobi, F. D., Boyd, G. S. 1970. *Eur. J. Biochem.* 13:203–22
29. Haven, G. T., Jacobi, H. P. 1971. *Lipids* 6:751–57
30. Thuy, L. P., Haven, G. T., Jacobi, H. P., Ruegamer, W. R. 1973. *Res. Commun. Chem. Pathol. Pharmacol.* 5:5–18
31. Ono, T., Imai, Y. 1971. *J. Biochem.* 70:45–54
32. Hulcher, F. H. 1971. *Arch. Biochem. Biophys.* 146:422–27
33. Bloxham, D. P., Akhtar, M. 1971. *Biochem. J.* 123:275–78
34. Bloxham, D. P., Wilton, D. C., Akhtar, M. 1971. *Biochem. J.* 125:625–34
35. Barth, C., Hackenschmidt, J., Ullmann, H., Decker, K. 1972. *FEBS Lett.* 22:343–46
36. Goodwin, C. D., Margolis, S. 1973. *Fed. Proc.* 32:671
37. Goodwin, C. D., Margolis, S. 1973. *J. Biol. Chem.* 248:7610–13
38. Cornforth, J. W. 1973. *Chem. Soc. Rev.* 2:1–20
39. Allred, J. B., Roehrig, K. L. 1973. *J. Biol. Chem.* 248:4131–33
40. Carlson, C. A., Kim, K. H. 1973. *J. Biol. Chem.* 248:378–80
41. Edwards, P. A., Gould, R. G. 1972. *J. Biol. Chem.* 247:1520–24
42. Weis, H. J., Dietschy, J. M. 1971. *Gastroenterology* 61:77–84
43. Harry, D. S., Dini, M., McIntyre, N.

1973. *Biochim. Biophys. Acta* 296:209–20
44. Cayen, M. N. 1971. *J. Lipid Res.* 12:482–90
45. Fishler-Mates, Z., Budowski, P., Pinsky, A. 1973. *Lipids* 8:40–42
46. White, L. W. 1971. *J. Pharmacol. Exp. Ther.* 178:361–70
47. White, L. W. 1972. *Circ. Res.* 31:899–907
48. Bricker, L. A., Levey, G. S. 1972. *Biochem. Biophys. Res. Commun.* 48:362–66
49. Bricker, L. A., Levey, G. S. 1972. *J. Biol. Chem.* 247:4914–15
50. Brunengraber, H., Boutry, M., Lowenstein, J. M. 1973. *J. Biol. Chem.* 248:2656–69
51. Liersch, M. E., Barth, C. A., Hackenschmidt, H. J., Ullmann, H. L., Decker, K. F. 1973. *Eur. J. Biochem.* 32:365–71
52. Bricker, L. A., Kozlovskis, P. L., Goodman, M. G. 1973. *Proc. Soc. Exp. Biol. Med.* 173:375–78
53. Zavoral, J. H., Ener, M. A., Frantz, I. D. Jr. 1973. *Circulation* 48:IV-233
54. Robertson, A. L. 1967. *Lipid Metabolism in Tissue Culture Cells,* ed. G. H. Rothblat, D. Kritchevsky, 115–25. Philadelphia: Wistar
55. Rothblat, G. H., Kritchevsky, D. 1968. *Exp. Mol. Pathol.* 8:314–29
56. Rothblat, G. H. 1969. *Advan. Lipid Res.* 7:135–65
57. Cristofalo, V. J., Howard, B. V., Kritchevsky, D. 1970. *Res. Prog. Org. Biol. Med. Chem.* 2:95–140
58. Rothblat, G. H. 1972. *Growth, Nutrition and Metabolism of Cells in Culture,* ed G. H. Rothblat, V. Cristofalo, 297–315. New York: Academic
59. Howard, B. V., Butler, J. D., Bailey, J. M. 1973. *Tumor Lipids: Biochemistry and Metabolism,* ed. R. Wood. Champaign: Am. Oil Chem. Soc. In press
60. Bailey, J. M. 1973. *Atherogenesis Initiating Factors,* 63–92. New York: Plenum
61. Watson, J. A. 1973. See Ref. 59
62. Holmes, R., Helms, J., Mercer, G. 1969. *J. Cell Biol.* 42:262–71
63. Williams, C. D., Avigan, J. 1972. *Biochim. Biophys. Acta* 260:413–23
64. Rothblat, G. H., Boyd, R., Deal, C. 1971. *Exp. Cell Res.* 67:436–40
65. Gonzalez, R., Dempsey, M. E., Elliott, R. Y., Fraley, E. E. 1974. *Exp. Cell Res.* In press
66. Watson, J. W. 1972. *Lipids* 7:146–55
67. Sokoloff, L., Rothblat, G. H. 1972. *Biochim. Biophys. Acta* 280:172–81
68. Watson, J. A. 1973. *Fed. Proc.* 32:480

69. Brown, M. S., Dana, S. E., Goldstein, J. L. 1973. *Proc. Nat. Acad. Sci. USA* 70:2162–66
70. Fogelman, A. M., Edmond, J., Polito, A., Popjak, G. 1973. *J. Biol. Chem.* 248: 6928–29
71. Capuzzi, D. M., Rothman, V., Margolis, S. 1971. *Biochem. Biophys. Res. Commun.* 45:421–29
72. Berndt, J., Boll, M., Lowel, M., Gaumert, R. 1973. *Biochem. Biophys. Res. Commun.* 51:843–48
73. McNamara, D. J., Rodwell, V. W. 1972. *Biochemical Regulatory Mechanisms in Eukaryotic Cells,* ed E. Kun, S. Grisolia, 206–43. New York: Wiley
74. Lane, M. D., Clinkenbeard, K. D., Reed, W. D., Sugiyama, T., Moss, J. 1973. *Int. Congr. Biochem. 9th* Abstr. 412
75. Clinkenbeard, K. D., Sugiyama, T., Moss, J., Reed, W. D., Lane, M. D. 1973. *J. Biol. Chem.* 248:2275–84
76. Sugiyama, T., Clinkenbeard, K., Moss, J., Lane, M. D. 1972. *Biochem. Biophys. Res. Commun.* 48:255–61
77. Barth, C. A., Hackenschmidt, J., Weis, E. E., Decker, K. F. 1973. *J. Biol. Chem.* 248:738–39
78. Sullivan, A. C., Hamilton, J. G., Miller, O. N., Wheatley, V. R. 1972. *Arch. Biochem. Biophys.* 150:183–90
79. White, L. W., Rudney, H. 1970. *Biochemistry* 9:2725–31
80. Hamprecht, B. 1969. *Naturwissenschaften* 56:398–405
81. Kawachi, T., Rudney, H. 1970. *Biochemistry* 9:1700–5
82. Linn, T. C. 1967. *J. Biol. Chem.* 242: 984–89
83. Brown, M. S., Dana, S. E., Dietschy, J. M., Siperstein, M. D. 1973. *J. Biol. Chem.* 248:4731–38
84. Heller, R. A., Gould, R. G. 1973. *Biochem. Biophys. Res. Commun.* 50: 859–63
85. Higgins, M., Rudney, H. 1973. *Int. Congr. Biochem. 9th* Abstr. 412
86. Higgins, M., Rudney, H. 1973. *Nature New Biol.* 246:60–61
87. Kandutsch, A. A., Saucier, S. E. 1969. *J. Biol. Chem.* 244:2299–2305
88. Tanabe, D. S., Ener, M. A., Frantz, I. D. Jr. 1972. *Circulation* 46:II–252
89. Carlson, J. P., McCoy, K. E., Dempsey, M. E. 1973. *Circulation.* 48:IV–70
90. Carlson, J. P., Dempsey, M. E. 1974. *J. Biol. Chem.* In press
91. Shefer, S., Hauser, S., Lapar, V., Mosbach, E. H. 1972. *J. Lipid Res.* 13: 402–12
92. Shefer, S., Hauser, S., Lapar, V., Mosbach, E. H. 1972. *J. Lipid Res.* 13: 571–73
93. Shefer, S., Hauser, S., Lapar, V., Mosbach, E. H. 1973. *J. Lipid Res.* 14: 400–5
94. Edwards, P. A., Muroya, H., Gould, R. G. 1972. *J. Lipid Res.* 13:396–401
95. Hamprecht, B., Nussler, C., Waltinger, G., Lynen, F. 1971. *Eur. J. Biochem.* 18:10–14
96. Shefer, S., Hauser, S., Lapar, V., Mosbach, E. H. 1973. *J. Lipid Res.* 14: 573–80
97. Huber, J., Latzin, S., Hamprecht, B. 1973. *Int. Congr. Biochem. 9th* Abstr. 395
98. McNamara, D. J., Quackenbush, F. W., Rodwell, V. W. 1972. *J. Biol. Chem.* 247:5805–10
99. Craig, M. C., Dugan, R. E., Muesing, R. A., Slakey, L. L., Porter, J. W. 1972. *Arch. Biochem. Biophys.* 151:128–36
100. Dugan, R. E., Slakey, L. L., Briedis, A. V., Porter, J. W. 1972. *Arch. Biochem. Biophys.* 152:21–27
101. Slakey, L. L. et al 1972. *J. Biol. Chem.* 247:3014–22
102. Edwards, P. A. 1973. *J. Biol. Chem.* 248:2912–17
103. Lakshmanan, M. R., Nepokroeff, C. M., Ness, G. C., Dugan, R. E., Porter, J. W. 1973. *Biochem. Biophys. Res. Commun.* 50:704–10
104. Lakshmanan, M. R., Nepokroeff, C. M., Ness, G. C., Dugan, R. E., Porter, J. W. 1973. *Fed. Proc.* 32:1379
105. Porter, J. W., Ness, G. C., Dugan, R. E., Nepokroeff, C. M., Lakshmanan, M. R. 1973. *Int. Congr. Biochem. 9th* Abstr. 412
106. Dugan, R. E., Ness, G. C., Lakshmanan, M. R., Nepokroeff, C. M., Porter, J. W. 1973. *Abstr. Am. Chem. Soc.* 46
107. Edwards, P. A., Gould, R. G. 1974. *J. Biol. Chem.* In press
108. Davison, A. M., Gould, R. G. 1973. *Circulation* 48:IV–68
109. Sabine, J. R., Horton, B. J., Hickman, P. E. 1972. *Eur. J. Cancer* 8:29–32
110. Goldfarb, S., Pitot, H. C. 1971. *Proc. Am. Assoc. Cancer Res.* 12:31
111. Kandutsch, A. A., Hancock, R. L. 1971. *Cancer Res.* 31:1396–1401
112. Bhathena, S. J., Avigan, J., Schreiner, M. 1973. *Circulation* 48:IV–68
113. Brown, M. S., Goldstein, J. L. 1973. *Circulation* 48:IV–69
114. Goldstein, J. L., Brown, M. S. 1973. *Proc. Nat. Acad. Sci. USA* 70:2804–8
115. Gould, R. G., Swyryd, E. A. 1966. *J. Lipid Res.* 7:698–707
116. Popjak, G., Edmond, J. 1973. *Circulation* 48:IV–69

117. Edmond, J., Popjak, G. 1974. *J. Biol. Chem.* 249: 66–71
118. Rilling, H. C. 1972. *Biochem. Biophys. Res. Commun.* 46: 470–75
119. Olson, J. A. 1965. *Ergeb. Physiol. Biol. Chem. Exp. Pharmakol.* 56: 173–215
120. Frantz, I. D. Jr., Schroepfer, G. J. Jr. 1967. *Ann. Rev. Biochem.* 36: 691–726
121. Gaylor, J. L., Delwiche, C. V. 1973. *Ann. NY Acad. Sci.* 212: 122–38
122. Gaylor, J. L., Hsu, S. T., Delwiche, C. V., Comai, K., Seifried, H. E. 1973. *Oxidases and Related Redox Systems,* ed. T. E. King, H. S. Mason, M. Morrison, 431–44. Baltimore: University Park
123. Dempsey, M. E. 1974. *Sub-Unit Enzymes: Biochemistry and Function,* ed. K. E. Ebner, New York: Dekker. In press
124. Dempsey, M. E. 1971. *Chemistry of Brain Development,* 31–39. New York: Plenum
125. Ritter, M. C., Dempsey, M. E. 1970. *Biochem. Biophys. Res. Commun.* 38: 921–29
126. Ritter, M. C., Dempsey, M. E. 1971. *J. Biol. Chem.* 246: 1536–47
127. Ritter, M. C., Dempsey, M. E. 1973. *Proc. Nat. Acad. Sci. USA* 70: 265–69
128. Scallen, T. J., Schuster, M. W., Dhar, A. K. 1971. *J. Biol. Chem.* 246: 224–30
129. Scallen, T. J., Srikantaiah, M. V., Skralant, H. B., Hansbury, E. 1972. *FEBS Lett.* 25: 227–33
130. Gavey, K. L. 1973. *Fed. Proc.* 32: 519
131. Shah, S. N. 1972. *FEBS Lett.* 20: 75–78
132. Johnson, R. C., Shah, S. N. 1973. *Fed. Proc.* 32: 519
133. Tabacik, C., Descomps, B., Crastes de Paulet, A. 1973. *Int. Congr. Biochem, 9th Abstr.* 333
134. Dempsey, M. E., Ritter, M. C., Lux, S. E. 1972. *Fed. Proc.* 31: 430
135. McCoy, K. E., Koehler, D. F., Carlson, J. P. 1973. *Fed. Proc.* 32: 519
136. McCoy, K. E., Dempsey, M. E. 1974. *J. Biol. Chem.* In press
137. Calimbas, T. 1972. *Fed. Proc.* 31: 430
138. Calimbas, T. 1973. *Fed. Proc.* 32: 519
139. Calimbas, T., Dempsey, M. E. 1974. *J. Biol. Chem.* In press
140. Ritter, M. C., Dempsey, M. E., Frantz, I. D. Jr. 1972. *Fed. Proc.* 31: 430
141. Brewer, H. B. Jr., Lux, S. E., Ronan, R., John, K. M. 1972. *Proc. Nat. Acad. Sci. USA* 69: 1304–8
142. Frederickson, D. S., Lux, S. E., Herbert, P. N. 1972. *Advan. Exp. Med. Biol.* 26: 25–56
143. Frnka, J., Reiser, R. 1974. *Biochim. Biophys. Acta* In press
144. Caspi, E. et al 1968. *J. Am. Chem. Soc.* 90: 3563–64
145. Conner, R. L., Mallory, F. B., Landrey, J. R., Iyengar, C. W. 1969. *J. Biol. Chem.* 244: 2325–33
146. Mallory, F. B., Conner, R. L. 1971. *Lipids* 6: 149–53
147. Ehnholm, C., Zilversmit, D. B. 1973. *J. Biol. Chem.* 248: 1719–24
148. Ockner, R. K., Manning, J. A., Poppenhausen, R. B., Ho, W. K. 1972. *Science* 177: 56–58
149. Reyes, H., Levi, A. J., Gotmaitan, Z., Arias, I. M. 1971. *J. Clin. Invest.* 50: 2242–52
150. Grawbowski, G. A., Dempsey, M. E., Hanson, R. F. 1973. *Fed. Proc.* 32: 520
151. Grawbowski, G. A., McCoy, K. E., Dempsey, M. E., Hanson, R. F. 1974. *J. Biol. Chem.* In press
152. Boyd, G. S., Trzeciak, W. H. 1973. *Ann. NY Acad. Sci.* 212: 361–77
153. Ungar, F., Kan, K. W., McCoy, K. E. 1973. *Ann. NY Acad. Sci.* 212: 276–89
154. Garren, L. D., Gill, G. N., Masui, H., Walton, G. M. 1971. *Progr. Hormone Res.* 27: 433–43
155. Beall, R. J., Sayers, G. 1972. *Arch. Biochem. Biophys.* 148: 70–76
156. Kan, K. W., Ritter, M. C., Ungar, F., Dempsey, M. E. 1972. *Biochem. Biophys. Res. Commun.* 48: 423–29
157. Kan, K. W., Ungar, F. 1973. *J. Biol. Chem.* 248: 2868–75
158. Elliott, W. H., Hyde, P. M. 1971. *Am. J. Med.* 51: 568–79
159. Mosbach, E. H. 1972. *Arch. Intern. Med.* 130: 478–88
160. Thistle, J. L., Hofmann, A. F. 1973. *N. Engl. J. Med.* 289: 655–59
161. Salen, G., Nicolau, G., Shefer, S. 1973. *Clin. Res.* 21: 523
162. Boyd, G. S., Scholan, N. A., Mitton, J. R. 1969. *Advan. Exp. Med. Biol.* 4: 443–56
163. Mitropoulos, K. A., Balasubramaniam, S. 1972. *Biochem. J.* 128: 1–9
164. Dempsey, M. E., McCoy, K. E., Calimbas, T. D., Carlson, J. P. 1973. *Int. Congr. Biochem. 9th* Abstr. 411
165. Beg, Z. H., Allmann, D. W., Gibson, D. M. 1973. *Biochem. Biophys. Res. Commun.* 54: 1362–69
166. Edmond, J. 1974. *J. Biol. Chem.* 249: 72–80
167. Yamamoto, S., Bloch, K. 1970. *J. Biol. Chem.* 245: 1670–74
168. Tai, H., Bloch, K. 1972. *J. Biol. Chem.* 247: 3767–73
169. Dempsey, M. E., McCoy, K. E., Carlson, J. P., Calimbas, T. D. 1974. *Fed. Proc.* In press

REPRINTS

The conspicuous number aligned in the margin with the title of each article in this volume is a key for use in ordering reprints.

Available reprints are priced at the uniform rate of $1 each postpaid. Payment must accompany orders less than $10. A discount of 20% will be given on orders of 20 or more. For orders of 200 or more, any Annual Reviews article will be specially printed.

The sale of reprints of articles published in the Reviews has been expanded in the belief that reprints as individual copies, as sets covering stated topics, and in quantity for classroom use will have a special appeal to students and teachers.

AUTHOR INDEX

SUBJECT INDEX

CUMULATIVE INDEXES

CONTRIBUTING AUTHORS   VOLUMES 39-43

CHAPTER TITLES VOLUMES 39-43